MAMMALOGY

MAMMALOGY

Adaptation, Diversity, Ecology

Third Edition

George A. Feldhamer
Southern Illinois University at Carbondale

Lee C. Drickamer
Northern Arizona University

Stephen H. Vessey
Bowling Green State University

Joseph F. Merritt
Illinois Natural History Survey

Carey Krajewski
Southern Illinois University at Carbondale

The Johns Hopkins University Press
Baltimore

The Johns Hopkins University Press
2715 North Charles Street
Baltimore, Maryland 21218-4363
www.press.jhu.edu

Library of Congress Cataloging-in-Publication Data

Mammalogy : adaptation, diversity, and ecology /
George A. Feldhamer... [et al]. — 3rd ed.
        p. cm.
 Includes bibliographical references and index.
 ISBN-13: 978-0-8018-8695-9 (hardcover : alk. paper)
 ISBN-10: 0-8018-8695-3 (hardcover : alk. paper)
 1. Mammalogy. I. Feldhamer, George A.
 QL703.M36 2004
 599—dc22   2007018297

A catalog record for this book is available from the British Library.

For charts on weights & measures and conversions,
see pages 646–47.

Special discounts are available for bulk purchases of this book. For more information,
please contact Special Sales at 410-516-6936 or specialsales@press.jhu.edu.

Brief Contents

Contents

Preface

As we noted in the preface to the second edition of this textbook, research on all aspects of mammals continues at a rapid pace. Given the accelerating amount of new material on mammals, balancing breadth and depth of coverage in a textbook for a one-semester upper-level undergraduate or graduate mammalogy course continues to be a challenge. In this third edition, we have retained the general format of 29 chapters arranged in 5 parts. Part 1 (Chapters 1 through 5) introduces the subject of mammalogy, history of the discipline, current methods and molecular techniques important in systematics and population analyses, and the evolution of mammals. We have moved the chapter on zoogeography into Part 1 as it provides a natural extension of the evolution chapter. Part 2 (Chapters 6 through 10) covers biological functions and the physical structure of mammals. Adaptive radiation in form and structure among the currently recognized mammalian orders is covered in Part 3 (Chapters 11 through 20). Morphology, fossil records, conservation and economics, and a brief synopsis of all extant families are included for each order. Part 4 focuses on behavior and ecology (Chapters 21 through 26). Finally, in Part 5 (Chapters 27 through 29), we explore mammalian parasites and diseases, including zoonoses; domestication of mammals; and conservation issues. As in previous editions, all literature citations are collected at the end of the text to avoid redundancy. Technical terms throughout each chapter are in boldfaced type when they are first introduced, and those terms are defined in both the text and the glossary. Although there is continuity between sections and chapters of the text, instructors can select certain chapters based on individual interest, emphasis, or time constraints without sacrificing clarity and understanding.

Changes to the Third Edition

The major impetus for this edition of *Mammalogy: Adaptation, Diversity, Ecology* was the significant changes in higher-level taxonomy of mammals. As noted in the introduction to Part 3 (Adaptive Radiation and Diversity), the number of recognized mammalian species has increased from about 4600 to over 5400. Additionally, this edition includes 5 new orders and 20 new families resulting from the considerable molecular and morphological work of the past decade. We have updated recent advances in mammalian anatomy and physiology, behavioral ecology, conservation, zoogeography, paleontology, and other areas of mammalogy. The third edition includes hundreds of new citations to recent literature, many new photos and figures, study questions designed to help generate critical thinking and discussion, and suggested readings. More specifically, Part 1 includes a completely new chapter on molecular techniques, as well as added coverage and current thinking concerning mammalian evolution. Part 2 has additional new information on locomotion. Taxonomic revisions with new orders and families are discussed in Part 3. New fossil evidence and updated phylogenies are given for many orders, as are recent advances in natural history for numerous families along with the current conservation status of threatened species. Part 4 has been updated and in Part 5 current trends in Lyme disease, SARS, and other zoonoses are discussed along with updated material on domesticated species and conservation issues of various mammalian groups.

In terms of form and function, feeding and locomotion, the approximately 5400 species of mammals represent the most diverse class of vertebrates. Mammals are terrestrial, arboreal, or marine; they burrow, run, or fly; and they feed on meat, nectar, blood, pollen, leaves, or a variety of other things. They range in size from 2-gram white-toothed pygmy shrews and hog-nosed bats to 160 million-gram blue whales. We continue to explore the diversity and complexity of all aspects of mammals in this third edition of *Mammalogy: Adaptation, Diversity, Ecology*. We hope this book does justice to past and present mammalogists on whose research and teaching efforts it is largely based. We also trust that it will continue to prove useful to students, the mammalogists of the future, as they better appreciate and explore the mysteries of mammals—those "fabulous furballs."

Acknowledgments

In this edition, we welcome Dr. Carey Krajewski as an additional author. His expertise and insight in molecular systematics and other areas of mammalogy have greatly strengthened the book. The five authors bring a combined total of nearly 170 years of field and laboratory experience with mammals in a variety of settings to the collaborative endeavor of this book. The benefits of this collaboration, including the experience and insights gained from writing the first and second editions, are reflected in this revised edition. Also, we have each gained through the years from the suggestions, ideas, discussions, and constructive criticism of many teachers, colleagues, students, and friends. Many individuals reviewed parts of the manuscript or the entire text of the first or second edition. Their input continues to be reflected in this third edition.

First Edition Reviewers

David M. Armstrong, *University of Colorado at Boulder*
Richard Buchholz, *Northeast Louisiana University*
Jack A. Cranford, *Virginia Polytechnic Institute and State University*
Jim R. Goetze, *Laredo Community College*
Dalton R. Gossett, *Louisiana State University*
Kay E. Holekamp, *Michigan State University*
Carey Krajewski, *Southern Illinois University*
Thomas H. Kunz, *Boston University*
Peter L. Meserve, *Northern Illinois University*
Christopher J. Norment, *SUNY College at Brockport*
Larry S. Roberts, *University of Miami*
Robert K. Rose, *Old Dominion University*
Michael D. Stuart, *University of North Carolina at Asheville*
John A. Vucetich, *Michigan Technological University*
Wm. David Webster, *University of North Carolina at Wilmington*
John O. Whitaker, Jr., *Indiana State University*
Bruce A. Wunder, *Colorado State University*

Second Edition Reviewers

Anthony D. Barnosky, *University of California, Berkeley*
Bruce E. Coblentz, *Oregon State University*
John D. Harder, *Ohio State University*
Lynda A. Randa, College of DuPage
Thomas McK. Sproat, *Northern Kentucky University*
Michael D. Stuart, *University of North Carolina at Asheville*
Christopher J. Yahnke, *University of Wisconsin–Stevens Point*
John A. Yunger, *Governors State University*

In addition, several colleagues reviewed revised chapters for this edition.

Third Edition Reviewers

William H. Brown, *Shawnee Community College*
Rainer Hutterer, *Zoologisches Forschungsinstitut und Museum Alexander Koenig*
Zhe-Xi Luo, *Carnegie Museum of Natural History*
Bruce D. Patterson, *Field Museum of Natural History*
Brett Riddle, *University of Nevada, Las Vegas*
Eric Schauber, *Southern Illinois University Carbondale*
Hugh Tyndale-Biscoe, *Fellow of the Australian Academy of Sciences*
George Waring, *Southern Illinois University Carbondale*
Michael Westerman, *LaTrobe University*
Patricia A. Woolley, *LaTrobe University*

We are most grateful for the assistance of Vince Burke and the staff at the Johns Hopkins University Press for all their efforts toward production of this edition. Kelly Paralis Keenan, Penumbra Design, Inc., and Anne Gibby were crucial in all phases of production. We also thank Science Librarian Kathy Fahey and her staff at Southern Illinois University at Carbondale for their help. The invaluable assistance of Lisa Russell of the Environmental Studies Program at SIUC in all phases of the third edition—including providing many original drawings—is gratefully acknowledged. We extend a special thanks to Rexford Lord, Jr., for the use of his photos that grace many of the chapters. Hal S. Korber, Galen B. Rathbun, Patricia Woolley, and Michael Westerman also provided use of their photos.

PART 1

Introduction

CHAPTER 1

The Study of Mammals

What Is Mammalogy?

Mammalogy is the study of the animals that constitute the Class Mammalia, a taxonomic group of vertebrates (Phylum Chordata, Subphylum Vertebrata) within the Kingdom Animalia. Humans (*Homo sapiens*) are mammals, as are many domesticated species of pets and livestock as well as wildlife, such as deer and squirrels, with whom we share our natural surroundings (figure 1.1). Many of the species of animals that have aroused public concern for their survival, such as elephants, whales, large cats, gorillas, and the giant panda, are mammals. Mammals share a number of common features, including (1) the capacity for internal temperature control, often aided by a coat of fur; (2) the possession of mammary glands, which, in females, provide nourishment for the young during early development; and (3) with a few exceptions, the ability to give birth to live young. These and many other features of mammals are discussed in detail in chapter 4 and in parts 2 and 3.

Animal biology can be studied from a taxonomic perspective, that is, by concentrating on groups of organisms, such as mammals (mammalogy) or birds (ornithology). Or, the functional perspective can be used, concentrating on processes, as in physiology and ecology. In this book, we combine both approaches. The disciplines of biochemistry, physiology, animal behavior, and ecology, among many others, all contribute to mammalogy. Our goal is to explore and integrate discoveries from all these disciplines to provide the most comprehensive and productive approach to the study of mammals.

Throughout the book, we weave together at least four major themes: evolution, methods for investigating mammals, diversity, and the interrelationships of form and function. A basic underlying theme for all of biology is evolution by natural selection. Beginning with chapter 4, we take up the thread of evolutionary thought, giving particular emphasis to both speciation and adaptations of mammals. Chapter 3 begins the second thread, scientific methods, which deals with how mammalogists formulate questions (hypotheses) for investigation and

A

B

C

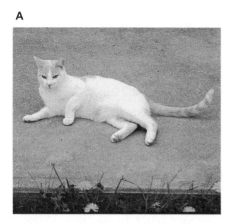  

Figure 1.1 Mammals with whom we share the world. In addition to our own species, mammals with whom we share our world can be grouped roughly into (A) domestic pets and livestock, such as a house cat (*Felis catus*), (B) wildlife in our familiar environment, which we may see often or in other cases rarely, such as an Abert's squirrel (*Sciurus aberti*), and (C) wildlife from other lands, particularly endangered or threatened species, such as an African elephant (*Loxodonta africana*).

what methods they use to answer these questions. The third thread, which is covered in parts 2 and 4, involves how form, function, and behavior are tightly interwoven and shaped by natural selection to provide solutions to the key problems of survival and reproduction in mammals. Our fourth thread, mammalian diversity, is emphasized in part 3, but examples offered throughout the text further underscore this theme.

Why Study Mammals?

Most of us have at least a passing interest in mammals, but we seldom stop to think why the formal study of mammalogy is important. Mammalogy can be approached from a variety of directions and for diverse reasons (Wilson and Eisenberg 1990). Mammals were a resource for early humans. Knowledge about them was important if humans were to hunt or trap them successfully. Some mammals, such as the saber-toothed cats that coexisted with early humans, were potential predators on humans. Knowledge of their habits was important for survival. Indeed, there are still locations throughout the world where wild animals, including grizzly bears (*Ursus arctos*) in western North America and tigers (*Panthera tigris*) in India, may attack and kill humans. Mammals, both wild and domesticated, continue to be important to humans as food. People with a subsistence way of life may depend on capturing or killing free-ranging mammals. More industrialized cultures depend on domesticated livestock for food. In addition, humans have a long tradition of using mammals in numerous ways, including use of their hides, bones, fur, or blubber from whales and seals.

Mammals serve the needs of humans as pets, for transport, and for recreational hunting. Humans keep many types of mammals as pets, ranging from cats, dogs, and

mice to more exotic species, such as large cats, primates, and even skunks. Much of the practice of veterinary medicine, which developed originally to serve the needs of agriculture, is now devoted to the diagnosis and treatment of illnesses and injuries affecting our mammalian pets. Many species are hunted for sport in North America, including the cottontail rabbit (*Sylvilagus floridanus*), the white-tailed deer (*Odocoileus virginianus*), and elk (*Cervus elaphus*). Exotic forms of mammalian wildlife, including free-ranging populations of fallow deer (*Dama dama*), sika deer (*Cervus nippon*), and feral hogs (*Sus scrofa*), have been introduced in several states, most notably Texas, to provide additional game species. Some exotics have become major pests after introduction for sport or trade because their interactions with native species were unforeseen. An example is the release of the Indian mongoose (*Herpestes javanicus*) on many islands in the Caribbean Sea and on the Hawaiian Islands. Mongooses were introduced to control rodents that had, in turn, been brought to the islands by humans. Mongooses consume the eggs and young of many native bird species, however, as well as competing with other native animals.

Some mammals pose risks for humans and other animals because they serve as reservoirs or vectors for a variety of diseases and parasites (e.g., the black rat [*Rattus rattus*] or black-tailed prairie dog [*Cynomys ludovicianus*] are vectors for plague). Knowledge of the life cycles of parasites and the symptoms of various mammal-borne diseases is necessary for humans to avoid and treat these health hazards.

Some mammals can damage portions of our environment or negatively affect other mammals. Rats, mice, and occasionally other small mammals with which we share our living areas do great harm to both our property and food stores. Some rodents exhibit explosive population growth and overrun large areas of planted cropland. A better understanding of the reproductive and population biology of such agricultural pests can lead to means for controlling them. Moles or gophers may damage our lawns, and beavers can cause flooding of forests and croplands. It is sometimes difficult to realize that these mammals are just carrying out their normal activities, which, unfortunately, often lead them into conflict with humans. Our anthropocentric (human-centered) perspective of life leads us to view many "normal" activities of nonhuman mammals as being in conflict with our goals.

Another currently important reason for studying mammals is conservation. After driving many species into or close to extinction, some effort is being made to reverse the trend. Toward that goal, some people work to understand and protect the habitats of endangered or threatened species. Others study social and reproductive biology under natural conditions or to establish captive breeding programs designed to eventually reintroduce species into their natural habitats. Good examples are current efforts involving the black-footed ferret (*Mustela nigripes*) and red wolf (*Canis lupus*).

In addition to examining the loss of species through extinction, a further result of conservation efforts in recent years is a broad-based attempt to account for all living species of animals and plants. This effort has resulted in the discovery of previously unrecorded species. For example, a new species of rodent, *Tapecomys primus* from Bolivia, new species of lemurs (*Microcebus lehilahytsara* and *Mirxa zaza*), and a new genus of African monkey (Genus *Rungwecebus*) all were described in the past several years (Anderson and Yates 2000; Kappelar et al. 2005; Davenport et al. 2006). The distinctions needed to differentiate these new taxa from those that were previously described involve traditional measures of morphology and modern molecular techniques based on DNA.

Because we are mammals, we can learn much about ourselves by studying similar processes that occur in other mammals. Some animals serve as models for various diseases or as subjects for developing or testing vaccines for eventual use on humans. We also maintain large colonies of some mammals in captivity to better study a whole variety of physiological, behavioral, and related medical phenomena. Work on particular species broadens and enhances our knowledge about such basic processes as developmental biology, immunology, endocrinology, and reproduction.

Resources for Mammalogists

A variety of resources is available to help us learn about mammals. Those who study mammals over many years develop personal libraries of pertinent materials, including general reference works and guidebooks containing keys for identifying mammals and providing basic information on the habits of particular species.

A great deal of literature is available on all aspects of mammals. Some volumes encompass worldwide coverage, such as *Walker's Mammals of the World* (Nowak 1991); *Encyclopedia of Marine Mammals* (Perrin, Wursig, and Thewissen 2002); and Wilson and Reeder's (2005) *Mammal Species of the World*. Other books provide coverage of a particular continent or faunal region, such as Hall and Kelson's *The Mammals of North America* (1959), Hall's second edition of the same work (1981, 2001), Burt and Grossenheider's *A Field Guide to the Mammals of North America North of Mexico* (1980), and *Wild Mammals of North America* (Feldhamer et al. 2003), covering North America; *Mammals of South America* (Lord 2007); *A Field Guide to the Mammals of Australia* (Menkhorst and Knight 2005); *The Mammals of the Palearctic Region* (Corbet 1988); and *Wild Cats of the World* (Sunquist and Sunquist 2002). In the United States, many books cover the mammals of various regions; for example, *Bats of the Rocky Mountain West* (Adams 2003), *Mammals of the Intermountain West* (Zeveloff 1988), *Mammals of the Great Lakes Region* (Kurta 1995), and *Guide to the Mammals of the Plains States* (Jones et al. 1985), as well as of practically every state, for example, *Mammals of California* (Eder 2005), *Mammals of Michigan* (Tekiala 2005), *Mammals of Illinois* (Hoffmeister 1989), *Guide to the Mammals of Pennsylvania* (Merritt 1987), and *The Mammals of Texas* (Schmidly 2004). Other works are specialized treatises on particular taxonomic groups (e.g., *The Natural History of Badgers* by Neal [1986]) or even monographs on particular species (e.g., *White-Tailed Deer Ecology and Management*, edited by Halls [1984]). There are books on practically every mammalian order, on particular families, and on many individual species.

A variety of journals are devoted strictly to mammals, such as *Journal of Mammalogy*, *Acta Theriologica*, *Mammalia*, and *Mammal Review*. In addition, a number of professional organizations promote the study of mammals through national and international meetings and publications, as well as by mentoring younger scientists interested in the discipline. Most notable among these organizations is the American Society of Mammalogists, founded in 1919 (see Birney and Choate 1994). Mammalian biology is part of other disciplines ranging from ecology and behavior to neurobiology and studies of biological rhythms. Results from these sorts of studies are published in numerous journals worldwide.

Mammalogists rely on good university or college libraries for extensive collections of books and journals containing information on mammals. The *Zoological Record*, which began in 1848 and is issued annually, is the best overall source for literature on mammals. It provides information by species, subject, author, and geographic area. Computerized bibliographic databases are very useful for locating literature on a particular subject or a specific mammalian species or works by a particular author. A note of caution is in order; most computerized bibliographic databases are limited to information going back only 20–25 years. Works published earlier than that are not recorded. Because mammalogy has a history of significant work dating back more than a century, additional sources beyond databases should be consulted. One method uses reference lists in books or papers to compile a retrospective list of articles that encompasses a broad time period. As time passes, more information, recent and older, is now accessible via the World Wide Web.

Mammals are studied as both living organisms and preserved specimens. Living mammals are studied in various situations, including in the wild in their natural habitats. Observing mammals directly through such means as trapping or radiotelemetry can provide particular insights and may involve travel to exotic places or simply being "out in the field." Other types of research are conducted in a captive setting, usually a laboratory facility, zoological park, or aquarium. Some species are either rare enough in the wild or are small, nocturnal, or secretive enough to necessitate the use of a zoological park or laboratory setting to conduct research. Many people who visit zoos are unaware that they also function as places of scientific study. In some exhibit buildings or enclosed areas not open to the public, species conservation and related investigations are taking place (figure 1.2). Today, a number of zoos have separate facilities for the study of species that may be endangered or threatened. An excellent example is the Smithsonian Institution (National Zoological Park) facility at Front Royal, Virginia, which was formerly a major horse-breeding station for the U.S. Army cavalry. Investigations of domestic animals, both livestock and pets, also are carried out under more controlled conditions in a variety of laboratory research settings and on farms.

The field of mammalogy is fortunate that, beginning several centuries ago, museum collections of preserved mammals were started, and public and private menageries, the forerunners of the modern zoological parks, came into existence. For a short but thorough summary of this topic, with particular emphasis on North America, consult

Wilson and Eisenberg (1990). Today, a large network of museums and related collections of mammals provides study skins, whole mounts, skeletal remains, and, in some instances, preserved soft anatomy and tissues for genetic studies (Hafner et al. 1997). These collections also include large numbers of fossils discovered by paleontologists. Without such materials, we would not be able to discern very much about the evolutionary history of mammals. Although we generally see such museums in their role as educational institutions, they also contain vast storage and work areas where professionals curate and study the preserved materials (figure 1.3). In several countries (e.g., Canada), standardized, computerized databases exist of all of the preserved and fossil materials from most museums, permitting investigators throughout the country to access the location of and information about particular specimens. A global network of this sort may be developed in the coming decades.

Organization of the Book

This book is divided into five parts. In the first part, after these introductory remarks, we examine the history of mammalogy (chapter 2). Chapter 3 explores how mammalogists proceed with their investigations; what questions they ask, and how they answer them. Chapter 4 deals with the evolution of mammals and the diagnostic characteristics that define them. Chapter 5 covers the biogeography of mammals—where they are found around the world and in what sorts of habitats.

Part 2 integrates the morphological features of mammals with their physiological functions and behavior. This

Figure 1.2 Research at zoos. Exhibit areas of many zoological parks that are not open to the public are important in terms of the studies taking place there on breeding and social biology. Often these areas involve breeding programs of endangered species, such as that for the golden lion tamarin (*Leontopithecus rosalia*), which has now been successfully reintroduced to its native habitat in South America.

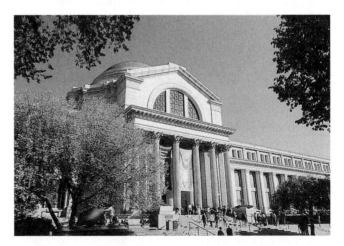

Figure 1.3 National Museum of Natural History, Washington, D.C. Museums are important repositories for large collections of preserved and fossilized specimens of mammals and other animals. The more familiar function of such facilities, serving to educate and entertain the public, may also be viewed as an important reason for the study of mammals.

material and the examples provide a foundation for structure and function and illustrate variation among mammals, including specialized adaptations that serve as solutions to particular problems.

Part 3 is a taxonomic examination of biodiversity among the 30 orders of mammals, emphasizing traits that characterize each mammalian family. Methods for discerning relationships among mammals at various levels of taxonomic classification are covered, including such recent developments as the use of protein allozymes and DNA.

Part 4 examines the interactions of mammals from behavioral and ecological perspectives. Vaughan (1986) commented that behavior and related topics were slow in developing. In earlier textbooks by Cockrum (1962) and Gunderson (1976), as well as earlier editions of Vaughan's text, we find scant coverage of behavior and only modest coverage of ecology. As is true for ornithology and herpetology, a shift in emphasis has taken place in the last quarter century; behavior and ecology are now receiving considerably more attention. An examination of recent

issues of the *Journal of Mammalogy* further illustrates our point. Fifty percent or more of the papers deal with these topics. Some earlier books on mammals, including the text by Davis and Golley (1963) and the volume on natural history by Bourliére (1970), provided substantial coverage of ecology and behavior. In part 4, we build on the foundations of classical mammalogy and combine description, systematics, form, and function to examine the ecology and behavior of mammals.

The final section (part 5) covers several specialized topics. Animal diseases and parasites (chapter 27) are important from both biological and practical perspectives. Domestication of mammals (chapter 28) has long been a part of human life. Recently, emphasis on management of animals in zoological parks and on game farms and ranches worldwide has been renewed. Conservation biology (chapter 29) takes on renewed importance each time we read about the threat to an endangered species. This last chapter attempts to apply what we have learned throughout the book to the issues of habitat conservation, reproductive biology, and related species preservation efforts.

SUMMARY

Mammals are one of the classes of vertebrates or animals with backbones. All mammals share a series of common characteristics, including internal control of body temperature (often aided by an insulating layer of fur), mammary glands, and (with a few exceptions) live birth of young. Mammals are studied for a variety of reasons, including their use for food and other products; as subjects of recreational hunting; as pets; as pests that cause damage; their conservation; their role in disease and related health considerations; because we are mammals ourselves; and for aesthetic interest.

Our study of mammals
1. Examines the history of the discipline, the methods used by mammalogists, and the evolution and characteristics of mammals
2. Explores the details of relationships between structure and function in the morphological and physiological systems of mammals
3. Reviews the taxonomic subdivisions of mammals

4. Explores the behavior and ecology of mammals
5. Provides special chapters on diseases, parasites, domesticated mammals and those kept on game ranches, and conservation.

Our approach weaves four major themes together: the process of evolution by natural selection as it has shaped mammals, in terms of both adaptation and speciation; how mammalogists ask and go about answering questions; the interrelationships of morphology, physiology, and behavior; and the diversity of mammals.

A variety of helpful resources is available for students and professionals interested in mammalogy. These include places to study mammals, such as zoological parks, aquaria, laboratories, and the natural setting. Excellent collections of preserved material from mammals are located in museums. Finally, modern libraries, with their collections of books, journals, and computerized databases, contain vast quantities of readily accessible information.

SUGGESTED READINGS

Bourliére, F. 1970. The natural history of mammals. Alfred A. Knopf, New York.

Macdonald, D. (ed.). 1984. The encyclopedia of mammals. Facts on File Publications, New York.

Wilson, D. E., and J. F. Eisenberg. 1990. Origin and applications of mammalogy in North America. Pp. 1–35 in Current mammalogy, vol. 2 (H. H. Genoways, ed.). Plenum, New York.

DISCUSSION QUESTIONS

1. Make a list of all of the mammals (use common names for now) that you have encountered in the past month. Note also where you saw them and any key features you used to distinguish them first as mammals and then as individual species.

2. After reading the section on reasons for studying mammals, compile your own list of reasons for investigating this group of animals. You can use these reasons as a starting point and provide some more detailed purposes, or you can start from scratch to come up with some reasons for the study of mammals and see how your list compares with the one provided. For each reason on your list, provide a brief specific example of how that rationale for the study of mammals has already affected you.

3. At your library, locate the computer terminals that access bibliographic databases. Select four topics related to mammalogy. Search these topics and the several permutations of those topics that occur to you as you peruse the search output. This exercise should familiarize you with the use of such databases and can provide references for topics that may become part of a required paper in the course you are taking.

CHAPTER 2

History of Mammalogy

Early in human evolution, our ancestors became aware of other mammals with whom they shared various habitats. In some cases, this knowledge about mammals served to determine possible prey that could be food. In other instances, information about potential predators was necessary for humans to avoid becoming food for larger carnivores. Other mammals have always been competitors with humans for food and shelter. Over time, some mammals were domesticated, serving humans in various capacities.

Early recorded indications of human knowledge about mammals come from cave paintings and petroglyphs (figure 2.1). Other evidence comes from sites where humans either drove herds of mammals over cliffs to their death or forced them through a narrow passage to be trapped and clubbed or otherwise killed. Prior knowledge of the behavior and movement patterns of mammals was crucial to the development of an effective hunting strategy. Some relationships between early humans and mammals were even closer. Dogs were domesticated from wolves (*Canis lupus*), and sheep (*Ovis aries*), goats (*Capra hircus*), and cattle (*Bos taurus*) were domesticated from wild ungulates (chapter 28). Religious icons, such as small statues (figure 2.2), indicate that humans ascribed certain mystical powers to the animals with whom they shared their world. Throughout much of human history, many mammalian species have also been used to convey people and goods. It has only been within the last several hundred years that other major modes of transportation have replaced mammals. Even today in many locations around the world, mammals are a major means of transport.

In the sections that follow, we trace the development of mammalogy from the days of classical Greece and Rome through the period of exploration and the dawn of natural history, and we end with a brief examination of the discipline of mammalogy today.

Mammalogy and the study of mammals are important to humans for at least three reasons. First, as humans, we are mammals, and thus we share similar physiological, behavioral, social, and ecological traits. Gaining knowledge about nonhuman mammals aids in understanding ourselves and human evolution. Second, some species of mammals, perhaps a disproportionately high number relative to other vertebrates,

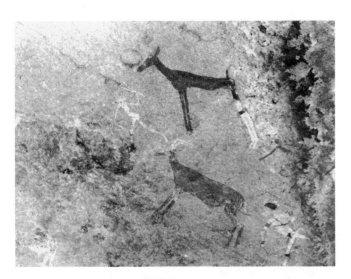

Figure 2.1 Cave painting of a mammal. Early humans daily dealt with other living mammals in their environment. The large ungulate mammal represented here in a cave painting from Alpera, Spain, was likely hunted by the people who painted it. The animal was eaten for food, and its bones, fur, and other products were also put to good use as tools and clothing.

are endangered in our world today. If we are to conduct effective programs to preserve these species, we need an enhanced understanding of mammals. Third, mammals have been most useful to scientists in the search for general principles of evolution, ecology, and behavior. This last point is covered in greater detail in part 4.

First Interest in Mammals

From the days when most humans were hunters and gatherers through the beginnings of agriculture about 9,000–10,000 years ago, a body of knowledge about mammals developed and was passed from generation to generation. With the advent of written language, some of this knowledge was recorded in glyphs, replacing or augmenting earlier depictions in art and oral stories. Particularly important in these early times was knowledge about mammals as food sources and work animals. Mammals were first domesticated in the Middle East and Asia. These early rudiments of formal interest in mammals, including scholarly writings, were later fostered in Egypt, Greece, and Rome as part of a growing body of information about the natural world.

Interest in mammals involved both curiosity about living forms and attention to various fossils that were discovered, collected, and passed among what were then called natural philosophers, a group that included Hippocrates (460–377 B.C.) and Aristotle (384–322 B.C.). Although Aristotle did not actually generate a classification scheme for living organisms, he did group animal

Figure 2.2 Early mammalian carving. The mammal shown in this statue from a Meso-American archeological site is a cat and is represented because it shared the environment with the people who carved it. The fact that it is depicted in this small carving suggests that it also may have been the subject of some form of religious importance or served as a clan totem.

forms as he saw them (Singer 1959). The category he labeled as having red blood and viviparous reproduction (mammals) included three major groups: (1) viviparous quadrupeds, subdivided into the ruminants with cutting teeth in the lower jaw only and having cloven hoofs (e.g., sheep, oxen), the solid-hoofed animals (e.g., horses), and other viviparous quadrupeds; (2) cetaceans (whales and their relatives); and (3) humans.

Fossilized bones and teeth were discovered from time to time in the ancient world. These remains often raised questions about the origin of these animals and their relationships to existing animal life. Even early in the history of mammalogy as a science, fossils were important in attempts to understand mammals, their history, and their distribution around the globe (Miller and Gidley 1934).

Fossil remains of mammals continue to be an important topic for study, including determining relationships between living and past forms. A fossil discovery in Wyoming, an insectivore, may be the smallest mammal ever known (Bloch et al. 1998). Even though body size does not fossilize, it is possible to use relationships between tooth size (or leg bone size) and body mass to predict the fossil organism's mass. Such a calculation relies on a range of living organisms on which we can measure both the teeth and the body mass. Only by continued work in the study of fossils, a field called paleontology, can we eventually piece together the evolutionary histories of mammals, both extinct and extant.

Aristotle and Pliny the Elder (A.D. 23–79) made extensive records of what they observed and heard about various mammals. Although these writings are primarily anecdotal, they are interesting, as the following quote from Pliny illustrates:

In the mountains of Mauretania it is said that the herds of elephants move at the new moon down to a river by the name of Amilo, ceremoniously cleanse themselves there by spraying one another with water, and after having thus paid their respect to the heavenly light return to the forests bearing their weary calves with them. It is also said that when they are to be transported overseas, they refuse to go onboard until the master of the ship has given them a promise under oath to convey them home again. Further, they are so modest that they never mate except in secluded spots, while adultery never occurs amongst them. Towards weaker animals they show compassion, so that an elephant when passing through a flock of sheep will with his trunk lift out of the way those he meets, for fear of trampling on them. (Nordenskiöld 1928:55)

Other early natural historians, such as the Roman anatomist Galen (A.D. 130–201), performed dissections, thereby generating new knowledge about the structure and function of different organ systems of animals, including mammals. Over the next several centuries, various expeditions, undertaken primarily for trade and warfare, brought the Europeans into greater contact with areas of North Africa and the Middle East and introduced them to new varieties of animals. Between the 5th century and the 16th century, very little new knowledge was gained. Natural philosophy and the various scientific disciplines waited more than 1,000 years before reemerging.

17th- and 18th-Century Natural History

NATURALISTS

By the 1600s, interest in natural history had reawakened. European explorers were traveling to many parts of the world during this time. The materials they discovered, described, and brought back from various countries for further study served to stimulate interest in the life sciences. Although none of these early natural historians was strictly a mammalogist, their broad interests included mammals. They undertook the early study of mammals around the

Figure 2.3   Illustration from Mark Catesby. During his two visits to the southeastern portion of North America, Catesby painted and described a number of mammals, including the chipmunk (*Tamias striatus*) shown here.

world and the development of a number of concepts that are still applied broadly to all living organisms.

Notable among the early naturalists to visit North America was Mark Catesby (1683–1749). He made two lengthy trips from England to the colonies, one from 1712 to 1719 and a second from 1722 to 1726. He was fortunate to have the financial backing of several wealthy English patrons, enabling him to travel and explore freely, particularly in what is now the southeastern United States. In 1748, Catesby completed a three-volume treatise entitled *The Natural History of Carolina, Florida, and the Bahama Islands*, in which he provided original descriptions and illustrations of a number of North American mammals (figure 2.3).

Among the Europeans who contributed to the growing body of information on mammals was Georges Buffon (1707–1788), a Frenchman who compiled and wrote a 44-volume *Histoire Naturelle*. The following passages dealing with mammals illustrate both the state of knowledge at the time and the style of writing that characterized science during this period. The first concerns the lion (*Panthera leo*):

Both the ancients and the moderns allow that the Lion, when newly born, is in size hardly superior to a weasel; in other words, that he is not more than six or seven inches long; and if so, some years at least must necessarily elapse before he can increase to eight or nine feet. They likewise mention that he is not in condition to walk till two months after he is brought forth; but, without giving entire credit to these assertions, we may, with great appearance of truth, conclude that the Lion, from the largeness of his size, is at least three or four years in growing, and that, consequently, he must live seven times three or four years, that is, about twenty-five years. (Buffon 1858, vol. II:28; a later translation from the French).

This second passage from Buffon concerns the European badger (*Meles meles*), and although the exact terms are not used, he describes aspects of imprinting and omnivory:

The young badgers are easily tamed; they will play with young dogs, and, like them, will follow any person whom they know, and from whom they receive their food; but the old ones, in spite of every effort, still remain wild. They are neither mischievous nor voracious, as the fox and wolf are, yet they are carnivorous; and though raw meat is their favourite food, yet they will eat anything that comes in their way, as flesh, eggs, cheese, butter, bread, fish, fruit, nuts, roots, &xc. They sleep the greatest part of their time, without, however, being subject, like the mountain rat or the dormouse, to a torpor during the winter; and thus it is that, though they feed moderately, they yet are always fat. (vol. I:255)

Several key developments in Europe contributed to the renewed interest in biology (Gunderson 1976). John Ray (1627–1705) first attempted to define a species as a group of organisms that can interbreed. As more and more distinct types of organisms were discovered and described, the need became apparent for some system of **taxonomy**; a way of classifying and organizing information on all plants and animals. From the 1500s onward, a number of individuals proposed schemes for classification, although none proved entirely satisfactory. A Swedish botanist, Carl von Linné (known generally by his Latinized name, Carolus Linnaeus; 1707–1778; figure 2.4) published a series of editions of a classification scheme. His tenth edition of *Systema Naturae*, published in 1758, is generally cited as the basis for modern taxonomy of living animals. The Linnaean system considered species to be fixed, discrete, individually created entities. A key feature of this system was that it involved a hierarchical arrangement of the various levels of classification. Thus, species were grouped into genera, genera into families, and families into orders. Together, the genus and species constituted the **binomial nomenclature** (two names) that scientists use today. After classification systems like that proposed by Linnaeus were adopted, patterns became evident with regard to the grouping together of organisms that shared similar traits. These groupings reflected possible relationships between organisms. Linnaeus and some of his contemporaries were the first true systematists.

In addition to those who traveled and those who worked in Europe, there were some notable naturalists in North America. Thomas Jefferson (1743–1826) had a keen interest in all of the sciences, including natural history. As vice president of the United States, he published a paper in the *Transactions of the American Philosophical Society* on fossil ground sloths (Genus *Megalonyx*) (Jefferson was president of the society at the time). The story is often told that when he was president of the United States, Jefferson spread out the bones of several fossil mammals on the floor of the East Room of the White House. Included in this array were mastodons, bison, deer, and many smaller mammals. These materials were part of the trove of materials recovered from Big Bone Lick (now in Kentucky), excavated by William Clark (of Lewis and Clark fame). Besides his scientific contributions, Jefferson sponsored expeditions of exploration. He was responsible, as president, for sending Meriwether Lewis and William Clark on their journey westward to the Pacific from 1804 to 1806 and arranged for Clark to explore Big Bone Lick in 1807.

CONCEPTS AND IDEAS

The numerous discoveries of the 17th and 18th centuries led to the formulation of key concepts and ideas that con-

Figure 2.4 Carolus Linnaeus. Linnaeus is considered by many as a founder of modern systematics. He formalized the system of binomial nomenclature by which all distinct species are given two names. He also developed a hierarchical scheme for classifying living and fossil organisms.

tinue to influence mammalogy today. The Linnaean system of classification eventually made apparent the relationships between various groups of animals. New attention was turned to the origins of these animals and to the idea of evolution. Three individuals who contributed to the development of the theory of evolution deserve special mention: Erasmus Darwin, the grandfather of Charles Darwin; Thomas Malthus; and Charles Lyell. Their work helped provide the bases for development of the theory of evolution by natural selection. Erasmus Darwin (1731–1802), an Englishman and a physician by profession, published a number of papers on scientific topics. His comprehensive treatise, *Zoonomia*, explored all the laws of organic life. Although he did not attempt to explain the origin of species, Erasmus Darwin proposed that the diversity of living organisms resulted from influences of the various environments in which they lived.

Thomas Malthus (1766–1834), also an Englishman, was a university professor in London. His primary contribution, *The Principle of Population*, argued that the human population had the potential to grow beyond its limits. He reasoned that self-control and restraint were necessary if humans were to avoid the problems stemming from overpopulation. The notion of overpopulation was part of the rationale that Charles Darwin used to formulate the theory of evolution by natural selection.

Charles Lyell (1797–1875; figure 2.5), often considered to be the founder of modern geology, proposed that processes that influenced the physical world in the past are still active in the present. He called this process "uniformitarianism." Lyell also contended that these changes occurred gradually rather than catastrophically, meaning that time was necessary for such changes. These thoughts had a major influence on Charles Darwin.

19th-Century Mammalogy

EXPLORATIONS AND EXPEDITIONS

Although this section concentrates primarily on events that occurred in North America, expeditions to other continents and to islands of the Pacific Ocean contributed greatly to existing knowledge. Discoveries from expeditions that were important to mammalogy in North America include those by (1) Lewis and Clark and, soon thereafter, Zebulon Pike; (2) trappers and fur traders who moved into the West in the early 1800s, as well as whalers and other seafaring explorers; and (3) naturalists who accompanied U.S. Army troops in the West or who were members of survey parties working to find routes for railroads. Together, these various expeditions provided an opportunity to discover, collect, and describe new species of mammals, their habitats, and their habits.

Figure 2.5 Charles Lyell. Lyell proposed that to understand the present state of things, particularly with respect to geology, we needed to examine the progression of changes that lead to the present state. This idea of change through time was relatively novel and served to further the thinking that lead to the theory of evolution by natural selection.

Lewis and Clark, and Pike

Shortly after the Louisiana Purchase in 1803, President Jefferson authorized an expedition to explore this new territory. The party, led by Meriwether Lewis (1774–1809) and William Clark (1770–1838), traveled from St. Louis to the Pacific Ocean and back between May 1804 and September 1806. Their route took them through much of the northern Great Plains and the northwestern United States. No other single exploration of North America added as much information about natural history and ethnology. It is noteworthy that this and many subsequent expeditions of discovery were funded, at least in part, by the federal government. In the case of the Lewis and Clark expedition, the total cost to the U.S. government was reported to be $2,500.

Lewis, Clark, and the members of their party were the first to report on and describe a number of small and large mammals. Among these were pronghorn (*Antilocapra americana*), grizzly bear (*Ursus arctos*, formerly *Ursus horribilis*), eastern woodrat (*Neotoma floridana*), and black-tailed prairie dog (*Cynomys ludovicianus*). The following passage from the diary of Meriwether Lewis for 14 May 1806, when the party was in Idaho on its return journey eastward, illustrates the types of observations made of mammals (figure 2.6):

*The hunters killed some pheasants, two squirrels, and a male and female bear, the first of which was large, fat, and of a bay color; the second meager, grizzly, and of smaller size. They were of the species [*Ursus horribilis*] common to the upper part of the Missouri, and might well be termed the variegated bear for they were found occasionally of a black, grizzly, brown, or red color. There is every reason to believe them to be of precisely the same species. Those of different colors are killed together, as in the case of these two, and as we found the white and bay associated together on the Missouri; and some nearly white were seen in this neighborhood by the hunters. Indeed, it is not common to find any two bears of the same color; and if the difference in color were to constitute a distinction of species, the number would increase to almost twenty. Soon afterward the hunters killed a female bear with two cubs. The mother was black, with a considerable intermixture of white hairs and a white spot on the breast. One of the cubs was jet black and the other of a light reddish-brown color.* (Lewis and Clark 1979, vol. II:1010–1011)

Sadly, few of the mammal specimens gathered during the Lewis and Clark expedition survive. Materials the expedition brought back were deposited in Peale's Museum at Philadelphia. When that museum was dissolved in 1846, the contents were sold at public auction. Half the collection, containing most of the mammals, went to the showman P. T. Barnum and subsequently was destroyed in a fire in 1865. The Library of the American Philosophical Society in Philadelphia still has the diaries of Lewis and Clark. These have been published in several editions and are a rich source of information on the mammals of the American West.

Shortly after Lewis and Clark had begun their travels, President Jefferson sent a 26-year-old army officer, Lieutenant Zebulon Pike (1779–1813), to find the source of the Mississippi River. That expedition lasted approximately 6 months from the fall of 1805 into spring of 1806, during which Pike and his associates thought that they had found the source. They had not, although this mistake was not known for some years. Pike, after whom Pike's Peak in Colorado is named, was sent almost immediately on a second journey to explore the region south of the Missouri River and west of the Mississippi. This area included parts of what are now the Great Plains, the American Southwest, and northern Mexico. These travels added to the growing body of knowledge on mammals and other natural history of the western United States.

Trappers, Fur Traders, and Whalers

Beginning in the late 1700s and continuing until the mid-1800s, the beaver (*Castor canadensis*) became the center of an entire industry (Gunderson 1976; Chittenden 1986). Initially, the fur trade in North America involved bartering with the native peoples. As animal numbers declined in the eastern United States, the fur trade shifted westward. Eventually, individual trappers braved the rigors of life alone or in small groups in the mountains and valleys of the West. Their life was uncluttered by "modern" conveniences, and in good years, the payoff was excellent. During the early 19th century, John Jacob Astor's Pacific Fur Company, the Hudson Bay Company, the Missouri Fur Company, and a number of smaller firms turned the beaver pelt operations into a well-organized economic boom (Chittenden 1954). The records from the fur-trading companies for beaver and other mammals (e.g., Canadian lynx [*Lynx canadensis*] and snowshoe hare [*Lepus americanus*]) brought in by the mountain men who trapped and hunted them, are a valuable data set for historical population estimates of these mammals. By the 1850s, the fur trade was almost nonexistent because the supply of animals had been so depleted by overtrapping. At about this same time, garments made of beaver fur went out of style in Europe, eliminating the demand for the pelts.

Figure 2.6 Grizzly bears. These two photos illustrate the sort of color differences that exist within the North American species of grizzly bear. It is easy to see that the passage quoted in the text has a real basis in terms of variation in color pattern of these bears.

Whales, particularly sperm whales (*Physeter catodon*), were the basis for a 19th-century industry on the high seas (figure 2.7). Again the primary focus was economic, but significant data were gathered on populations of sperm whales and other related species. The wide-ranging voyages of the whaling ships resulted in considerable knowledge about the oceans and the movements and habits of sea mammals. Sadly, with the advent of more efficient hunting methods in the 20th century, populations of many species of whales were badly overexploited. This led to the current bans on whaling, observed by all but a very few countries in the world (see chapter 29).

Army and Railroad Survey Expeditions

In the 19th century, a number of U.S. Army expeditions traveled the western states on reconnaissance or in search of suitable locations for forts. Many of these expeditions benefited from the presence of medical personnel who often were also natural historians. These medical officer/naturalists numbered more than 100 and included names familiar to students of mammalogy, such as Say, Baird, and Mearns. Thomas Say (1787–1843) was one of the first of these surgeon/naturalists. He accompanied Major Stephen Long on expeditions to the Rocky Mountains and up the Mississippi and Minnesota Rivers. Although he is rightly known more for his contributions to entomology, he supplied descriptions of and data on living mammals and participated in fossil finds important in describing the phylogenies of several mammalian groups.

Spencer Fullerton Baird (1823–1887) helped to found the U.S. National Museum (now the National Museum of Natural History) within the Smithsonian Institution. He published a monograph entitled *General Report on*

Figure 2.7 Sperm whale. Sperm whales, among the largest of all mammals, were hunted heavily during the 19th-century, and with modern technology, their populations were decimated during the first half of the 20th century. Due to current restrictions on whaling, adhered to by almost all countries, populations of this species appear to be making a comeback.

North American Mammals in 1859, with descriptions of more than 730 species of mammals. Many of the mammals described in the monograph were discovered during railroad surveys searching for the best route to the Pacific Ocean.

Edgar Alexander Mearns (1856–1916) served as the medical officer and naturalist for the Mexico–United States International Boundary Commission. Mearns published *Mammals of the Mexican Boundary of the United States* (1907) as a result of this service and collected over 7,000 mammal specimens. A substantial portion of the early collection of mammals at the American Museum of Natural History in New York City is the result of his work.

Another young physician, who was influenced by Baird and deserves special mention for his contributions to mammalogy, was C. Hart Merriam (1855–1942; Brown and Wilson 1994). In 1889, Merriam initiated a new publication series, *North American Fauna*, which continues today, covering a wide range of topics related to aspects of systematics, taxonomy, and natural history of North American mammals. He is probably best known for an 1890 paper on the effects of changes in elevation and latitude on the presence of certain plants and animals, which served as the basis for the concept of life zones. He developed and refined a number of the techniques used in systematic mammalogy, including an emphasis on cranial traits and dentition. It was under Merriam's auspices that the Division of Economic Ornithology and Mammalogy of the U.S. Department of Agriculture became the Bureau of Biological Survey (in 1905). His efforts did much to foster the rapid development of mammalogy as a distinct science, and he served as the first president of the American Society of Mammalogists.

Other explorers also made contributions, including those who spent decades attempting to find the elusive Northwest Passage through the Arctic to connect the Atlantic and Pacific Oceans. They encountered animals such as the polar bear (*Ursus maritimus*), arctic fox (*Alopex lagopus*), arctic ground squirrel (*Spermophilus parryii*), and numerous other small mammals, some of which had been known to science, but others that were recorded for the first time. Other explorers penetrated interior regions of Africa, Asia, Australia, and South America, finding and describing numerous mammals that were unknown to scientists in western Europe and North America.

There were also enterprising, private, individual collectors. One of whom, Martha A. Maxwell of Colorado, may be considered a pioneer female in mammalogy (Schantz 1943). She spent many years constructing what we now would call a diorama (in this instance, a very large one with 100 mammals and 400 birds). At the request of the Colorado Legislature, her work was featured at the Centennial Exhibition in Washington, D.C., in 1876.

MUSEUMS

The great expeditions of discovery and exploration of the 19th century resulted in the collection of literally thousands of mammalian specimens, many of which were deposited in museums. Most of these repositories initially were in the eastern United States, but over time, major museums were established west of the Mississippi, such as those at Lawrence, Kansas, and Berkeley, California. Financed by individuals, governments, or universities, museums served as repositories for specimens and for developing collections of written materials on mammals and other fauna and flora. Books and papers were important sources of information for those who worked to describe and document new species; avoiding renaming a type of mammal already described was a common problem. The collections and accompanying libraries were modeled after those developed in Europe in the 18th century. Resources for such endeavors were usually only available in or near larger population centers. Hence, Boston, New York, Philadelphia, Pittsburgh, Washington, D.C., and several smaller centers became the initial focal points for museum activity in the United States.

Additional information on these museums and how they began can be found in the edited volume on the history of the American Society of Mammalogists (Birney and Choate 1994). One example of this process should be instructive. The founding of a major center for the study of mammalogy and training of mammalogists at the University of California at Berkeley was due, in large measure, to the efforts of Annie M. Alexander (1867–1950). An early interest in travel and natural science resulted in her support for and leadership of three expeditions to Alaska in 1906, 1907, and 1908. She became friends with C. Hart Merriam and discussed with him her idea of a museum for the study and preservation of the rapidly disappearing wildlife of the western states. Her proposal to the University of California at Berkeley was accepted, and in 1908, the museum was established with Joseph Grinnell as its first director. Its early growth was fostered by Alexander's substantial financial contributions and the hundreds of specimens she helped to collect (Stein 1996).

Museums have traditionally had two major functions. One has been to educate the public by exhibiting their collections. Recently, this role has been expanded, and conservation education is becoming a key focus for many museums. A second major function has been to provide a base for research on mammals and for training new generations of mammalogists. These two functions are still priorities for many major museums. Our ability to study various mammalian systems or to work on the systematics and phylogeny of a particular group of mammals depends on the continued existence and maintenance of these museum collections. Mammal collections at universities range from small to modest holdings used primarily for teaching in such courses as mammalogy and vertebrate natural history to large, specialized research collections of mammals from an entire region of the world.

ZOOS

Around the world are several known instances of people, usually royalty, who captured and housed collections of animals, many of them mammals, in their gardens or as small menageries. The earliest of these collections date from more than 3,000 years ago in places such as Egypt and China. Aristotle's work on natural history was facilitated in part by various collections of living animals. His pupil, Alexander the Great, was likely the first person to use zoo collections as an educational tool. In Roman times, the use of wild animals developed into staged hunts of animals in captive settings, into combats between animals such as a rhinoceros and a bear in amphitheaters, and into the well-known gladiatorial encounters between humans and animals such as lions.

After the decline of the Roman Empire, menageries continued to be in fashion with rulers and other members of the nobility. A prime example is the Tower of London Zoo maintained by members of the British royalty during the 13th and 14th centuries. At about the same time, the Grand Khan in the Mongol Empire in Asia maintained a vast collection of birds, mammals, and reptiles. When explorers from Europe reached the New World, they found an enormous zoo, the royal menagerie of Montezuma, at the site of what is now Mexico City. By the 16th and 17th centuries, during the age of exploration, some adventurers from Europe brought back specimens of heretofore-unknown creatures from faraway places. Under Louis the XIV in France, the zoo established at Versailles became the first truly public zoological park in the western world.

From those beginnings have sprung a large number of zoos, safari parks, and wild animal farms all around the world. Major zoos, which started in Europe, such as those at London, Paris, and Berlin, were joined by those in major U.S. cities such as Washington, D.C., New York, Chicago, New Orleans, San Diego and Philadelphia. During the 20th century, the functions of zoos changed dramatically. For much of the period, animals were displayed in relatively small cages and the primary purpose was for the public to be able to see these wild creatures from around the world. Since 1970, as biodiversity and conservation became major issues, zoos gradually changed the manner in which the animals were displayed, and the functions of zoos shifted. Animals were now housed in more naturalistic settings and in larger cages, often with outdoor enclosures attached to indoor facilities. Education became a paramount function for zoos, particularly for school-age children but also for adults. Understanding the biology and needs of the many captive animals enabled zookeepers to do a better job of housing and feeding, and veterinary medicine for zoo animals grew as a new specialty. Finally, the need for work on threatened and endangered species resulted in a greater understanding of the social and reproductive biology of zoo animals. Many zoos started captive breeding programs (see figure 1.2) for sustaining certain species and for providing sufficient animals to reintroduce them into the wild; many zoos also work on specialized topics such as *in-vitro* fertilization and molecular genetics.

NEW THEORIES AND APPROACHES

Biology was revolutionized during the second half of the 19th century. Charles Darwin (1809–1882), Alfred Russell Wallace (1823–1913), and other naturalists accompanied ships from European countries on journeys to many regions of the world. In the process, these ship's scientists were able to observe and collect specimens of plants and animals. From those collections, from extensive observations made of native fauna and flora in many distant lands, from observations of domestic animals and plants, and from knowledge of earlier work (particularly the ideas of Malthus, Lyell, and others), Darwin and Wallace independently arrived at the theory of evolution by natural selection. Their joint paper, presented to the Linnean Society in London in 1858, began a process of debate and acceptance concerning the ways in which living organisms adapt to changing conditions and the manner by which new species arise. The resulting theory of evolution by natural selection has become the unifying principle for all of life science.

The plant-breeding trials of the Augustinian monk Gregor Mendel (1822–1884) on the inheritance of traits in peas helped set the stage for modern genetics. He described dominant and recessive characters in plants and formulated the laws of segregation and independent assortment. The subfield of population genetics has become very important in modern mammalogy, both for understanding the interplay of ecology and evolution and as a foundation for conservation biology (chapter 29).

The process of conducting science was also undergoing dramatic changes. Experimental manipulations to test specific hypotheses became standard procedure. This augmented, but did not replace, reliance on observations as the basis for understanding natural phenomena. In addition, more refined equipment and technology, such as better microscopes, provided new insights that changed the way experimental science was conducted, paving the way for modern mammalogy. In chapter 3, we will examine in more detail the variety of research methods and techniques that are used by mammalogists. Because the older techniques remain a valuable part of today's research tools in mammalogy, we also review much of the history of methods in mammalogy.

EARLY WRITTEN WORKS ON MAMMALS

During the 19th century, a number of monographs were written on the mammalian fauna of North America. Some of these were portions of multivolume sets covering all of the plant and animal life on the continent. Sir John Richardson's (1787–1865) series entitled *Fauna Boreali Americana* (1829) contained a volume on mammals. Examples of more specialized books dealing only with quadrupeds are Thomas Bewick's (1753–1828) *A General History of Quadrupeds* (1804), the first truly American book on mammalogy, and *The Viviparous Quadrupeds of North America*, produced by John James Audubon (1785–1851) and John Bachman (1790–1874) between 1846 and 1854.

The first books dealing specifically with mammalogy appeared during the second half of the 19th century. Most of these were compendia, some several volumes in length. They provided descriptions of the mammals of North America, including those known for several centuries from the eastern portion of the country, to which were added all of the new forms discovered during western explorations. Many of these books follow a typical pattern, true even of some, but not all, subsequent textbooks of mammalogy. For instance, the first 75 pages of the 750-page book by William Henry Flower (1831–1899) and Richard Lydekker (1849–1915), *An Introduction to the Study of Mammals Living and Extinct* (1891), provide an introduction to the structure and function of mammals, with an emphasis on skeletal and dental traits. The remainder of this text is devoted to accounts of the various orders of mammals. With the vast increase in knowledge during the 20th century, some authors (e.g., D. E. Davis and F. B. Golley [1963], *Principles in Mammalogy*), as well as our own treatment of the subject, have taken a different approach, focusing on comparisons of physiology, anatomy, ecology, and behavior across the orders of mammals.

In addition to more general works, some treatises dealt with smaller groups of mammals or even single species. An example of the first type is the work of Elliott Coues (1842–1899) on a rodent family, the Muridae (1877). Coues later became the first curator of mammals at the U.S. National Museum in Washington, D.C. A prime example of the single-species treatment is *The American Beaver and His Works* (1868) by Lewis H. Morgan (1818–1881). Morgan summarized what was known about the beaver (*Castor canadensis*), exemplified in this short passage about beaver dams and lodges (figure 2.8):

The dam is the principal structure of the beaver. It is also the most important of his erections as it is the most extensive and because its production and preservation could only be accomplished by patient and long-continued labor. In point of time, also, it precedes the lodge since the floor of the latter and the entrances to its chamber are constructed with reference to the level of water in the pond. The object of the dam is the

Figure 2.8 Beaver. The beaver was the center of a major industry for much of the first half of the 19th century. Trappers and fur traders, both individuals and companies, constituted a key segment of the economy of the Rocky Mountains and northwestern United States, declining quickly as the number of beaver dwindled to isolated populations.

formation of an artificial pond, the principal use of which is the refuge it affords to them when assailed, and the water connection it gives to their lodges and to their burrows in the banks. Hence, as the level of the pond must, in all cases, rise from one to two feet above these entrances for the protection of the animal from pursuit and capture, the surface level of the pond must, to a greater or lesser extent, be subject to their immediate control. (Morgan, 1868:82–83)

Volumes like those noted are the forerunners of specialized monographs that abound today in mammalogy. Any mammalogy student conducting literature research will find entire books on particular topics, for example on reproductive physiology or foraging behavior. As noted in chapter 1, books are available for almost every taxonomic group, and a mammal guide exists for just about every state or region of North America as well as for most geographical and faunal regions of the world.

Emergence of Mammalogy as a Science

The extensive survey work and exploration that occurred in the 1800s and just after the turn of the century resulted in a critical mass of specimens, records, and other information on mammals. That information, combined with increasing numbers of professional scientists working almost exclusively on mammals, led to the emergence of mammalogy as a distinct discipline. These developments helped foster the formation of the American Society of Mammalogists in 1919. The history of the society and its role in furthering the study of mammals has been thoroughly covered in a volume edited by Birney and Choate (1994) to commemorate the seventy-fifth anniversary of its founding. Other histories of North American mammalogy have been written by Storer (1969), Hoffmeister (1969), and Hamilton (1955). The American Society of Mammalogists has become the largest professional group in the world whose main focus is the biology of mammals.

Collections of mammals in museums, particularly those at universities, played a pivotal role in establishing mammalogy as an academic discipline within the life sciences. One of the most distinguished of those early mammalogists was Joseph Grinnell (1877–1939). The son of a physician, Grinnell grew up on and around Indian reservations in Oklahoma, Nebraska, South Dakota, and North Dakota (figure 2.9). He spent most of his adult life in an academic career at the University of California at Berkeley. One of his key scientific contributions was the concept of the niche: the idea that organisms have functional roles within the framework of the community.

A   B   C   D   E

Figure 2.9 Distinguished mammalogists. The five men shown here were among a number of professionals who played key roles in establishing mammalogy as a distinct subdiscipline within vertebrate zoology. Each of them was associated with a particular museum, and each trained a number of graduate students; their academic descendants are numerous among today's mammalogists. (A) Joseph B. Grinnell; (B) William H. Burt; (C) Lee R. Dice; (D) E. Raymond Hall; (E) William J. Hamilton, Jr.

While at Berkeley, Grinnell began the museum collection, introduced courses in vertebrate zoology to the curriculum, and trained graduate students. Among Grinnell's graduate students were William H. Burt (1903–1987), who played a vital role in developing the notions of home range and territory; Lee R. Dice (1887–1977), whose efforts contributed to our knowledge about interspecific competition and its effects on the structure of communities; and E. Raymond Hall (1902–1986), who conducted extensive research on the taxonomy and distribution of mammals (figure 2.9). Together, the academic descendants of the Grinnell group constitute the single largest branch of the genealogical tree within mammalogy (Whitaker 1994a). Another major lineage was formed from the graduate students of W. J. Hamilton, Jr. (1902–1990) at Cornell University (figure 2.9). Hamilton's major emphasis was on life history traits and ecology.

The presence and role of women in mammalogy has changed dramatically in the past several decades. As noted earlier, several women played instrumental roles in the early days of the discipline; however, few women were trained in the major academic and professional centers for mammalogy before the late 1960s. The history of women in mammalogy and the shift toward greater involvement of women has been well documented by many articles in the *Journal of Mammalogy* (Horner et al. 1996; Kaufman et al. 1996; Smith and Kaufman 1996; Stein 1996). These changes are reflected in the membership composition of the American Society of Mammalogists as well as in authorship of papers published in the *Journal of Mammalogy* and presented at annual meetings.

Excellent museums at Harvard, Yale, Michigan, Cornell, Kansas, Texas Tech, and many other universities contributed to the growth of mammalogy, both through their collections and through the training of students by faculty associated with the museums. At various times during the first half of the 20th century, courses in mammalogy were started, often growing out of courses in vertebrate zoology. After World War II, the number of mammalogy courses grew rapidly until the 1980s, after which the number declined slightly.

The emergence and growth of mammalogy as a distinct subdiscipline within zoology can be characterized by a progression of trends in the accumulation of knowledge. Initially, most information came from observation and description. With time, approaches involving experimental manipulations contributed greatly to our knowledge about mammals. Ecology and life history research were major foci for mammalogists before and for several decades after World War II. This was followed by an upsurge of interest in physiological processes during the 1960s and 1970s. At about the same time, and continuing into the 1980s, research began on models of various types, aided by the introduction of computers. Processes that use computer modeling ranged from population biology to energetics. Most recently, considerable attention has been devoted to molecular genetics. Fostered by the development of a variety of DNA-based techniques, molecular genetics has aided investigations in areas ranging from population genetics to reproductive success. No approach has been dropped from the mammalogist's repertoire in this research progression. Because of the cumulative nature of the process, mammalogists need to be trained in a wide variety of techniques and approaches. Mammalogy today involves a broad range of scientists who study systematics, paleontology, behavior, physiology, ecology, anatomy, biochemistry, and other biological topics, although the degree of individual specialization varies. Modern mammalogy integrates knowledge across all of these disciplines.

Throughout the 20th century and into the 21st, the importance of mammalogy has grown as scientists have found new areas in which they can apply their knowledge about mammals (Wilson and Eisenberg 1990). Wild mammals have been food for humans for millennia. Domesticated livestock and their products have been critical to human cultural and economic development for 10,000 years (chapter 28). Other mammals, such as the beaver and the mink (*Mustela vison*), provided the basis for fur trapping and fur farming, although the demand for these products has declined in recent years. Mammals also

Figure 2.10   White-footed mouse. The white-footed mouse shown here and the white-tailed deer are distributed throughout much of the eastern United States. Both species are vectors of Lyme disease, carried to humans by ticks.

provide us with oils (whales, seals) and ivory (elephants, walruses), although suitable substitutes are now available. Interest in mammals extends to their actions as pests in terms of livestock predation, crop depredation, and destruction of the environment. Rodents of various species cause billions of dollars in damage annually to cereal crops and grain stores in many parts of the world. Many home owners have had to deal with burrows made by animals such as gophers and moles. On the other hand, many rodents have vital ecological roles in terms of insect and seed control.

Mammals are also of interest to us in several ways in connection with medical science. A number of species serve as useful laboratory models for various human diseases or physiological functions. Several species of rodents, most notably house mice (*Mus musculus*) and Norway rats (*Rattus norvegicus*), have been the primary subjects for many laboratory investigations of general biological principles and more specifically of studies in human medicine. Many procedures used to treat coronary ailments were first tested on domestic dogs (*Canis* [*familiaris*] *lupus*) or pigs (*Sus scrofa*). Some mammals are known to be reservoirs for or vectors of diseases. A good example of this is the rat (*Rattus* spp.), which carries the fleas responsible for transmitting bubonic plague (chapter 27). Rabies, which is carried by mammals, also poses a serious threat to humans. The recent outbreak of a hantavirus, involving several species of Genus *Peromyscus*, is another example of a disease carried by a mammal (figure 2.10). White-tailed deer (*Odocoileus virginianus*) and white-footed mice (*Peromyscus leucopus*) are primary reservoirs of the larval and adult ticks that carry Lyme disease, which has become a major health concern as it spreads in the eastern United States. The virus that leads to AIDS in humans likely had its origin as a similar virus in African primates.

Lastly, mammals have become valuable to humans for recreation. Sport hunting is a major industry in many regions of the United States and throughout the world. In the past 20 years, particularly in such places as Africa and South America, mammals have become a valuable resource in ecotourism (figure 2.11). Mammals are also prized for their aesthetic value; most people enjoy seeing a herd of deer feeding, a female bear and her cubs ambling across a mountain meadow, or a cat silently stalking its prey. Television offers considerable numbers of programs dealing with the nonhuman mammals with whom we share the world.

Mammalogists work in a variety of settings today. Many are at universities and colleges, and a large number are employed by state and federal agencies. On the state or provincial level, agencies that deal with wildlife and conservation have mammalogists on their staffs, as do those that deal with parks and recreation. Positions for natural heritage biologists have expanded in recent decades. At the federal level, the talents of mammalogists are utilized in programs in agriculture and forestry, in which mammals are studied with regard to pest manage-

Figure 2.11 Herd of zebras and wildebeests.
The mixed herd of zebras (*Equus burchelli*) and wildebeests (*Connochaetes taurinus*) are a major feature of the Masai Mara-Serengeti ecosystem. They are among the many East African mammals that became a valuable resource as tourists have flooded the region in recent decades.

ment, and in programs of wildlife management of nature preserves, forests, and other federally owned lands. With the advent of laws to protect the environment and the need to rehabilitate human-degraded lands, a variety of private environmental and consulting firms have hired mammalogists. Finally, many nongovernmental organizations, such as the Nature Conservancy and the National Audubon Society, employ scientists trained in mammalogy. These individuals work in local, national, or international operations primarily surveying mammalian fauna and striving to protect and conserve mammalian species.

Among recent technological advances used by mammalogists to push the frontiers of the discipline forward are night-vision scopes and radiotelemetry to watch and monitor, respectively, the movements and behavior of animals. Beginning in the 1950s, molecular techniques, such as protein electrophoresis, were applied to the study of mammals. Current DNA-based techniques are used to assign parentage for individual animals, and both nuclear and mitochondrial DNA sequencing techniques aid in studying phylogenies of various mammalian groups. These and related topics pertaining to the methods we use to study mammals are the subjects of the next chapter.

In closing, we take note of the fact that although the discovery and recording of new species of mammals reached its peak during the 19th century and early years of the 20th century, new mammals are still being discovered. More than 100 previously undescribed mammals have been recorded since 1980 (Morell 1996). Among these are rodents from Madagascar and the Philippines as well as several primates from South America and Madagascar. Work still remains to be done to complete our inventory of mammalian diversity.

SUMMARY

The earliest humans were aware of and interested in other mammals that shared their environment. They needed knowledge about mammals to obtain meat and avoid becoming prey. The ancient Greeks and Romans (e.g., Aristotle and Pliny) recorded their observations of the known mammals of their time. After a hiatus in discovery during the Dark Ages, extensive interest in natural history was revived in the 1600s. Expeditions by Europeans resulted in the collection of plant and animal specimens from previously unknown places. The diversity of mammals and other animals represented in these new collections stimulated interest in general zoological principles. Among those whose contributions are singled out in this chapter were Mark Catesby, who visited America in the first half of the 18th century; Georges Buffon, who wrote an extensive series of volumes summarizing what was known of living organisms, including mammals; and Carolus Linnaeus, who developed a hierarchical system of classification based on binomial nomenclature. Several ideas developed in the late 18th and early 19th centuries by individuals such as Malthus and Lyell, who served to stimulate thinking about the origins of biological diversity.

The 19th century was an age of exploration in North America. The century began with the travels of Lewis and Clark, supported by President Jefferson. Others who contributed to the progress of natural history during this century included trappers and fur traders, whalers, surveyors for the army, and those who searched for the Northwest Passage in the Arctic. Preserved materials and fossil specimens from these expeditions were often deposited in museums. The museums and their associated libraries became key centers of learning about mammals.

The cornerstone for all of modern biology, the theory of evolution by natural selection, was first presented in 1858 by Darwin and Wallace. Other scientific developments, including the beginnings of modern genetics and the use of experimental manipulations in the course of conducting research, characterized natural science during the last half of the 19th century. It was during this period that the first written works devoted solely to mammals were published; some of these were summary volumes, covering all known mammals from North America or the world, whereas others dealt with specific groups of mammals or even individual species.

The 20th century saw mammalogy emerge as a distinct discipline within the life sciences. The American Society of Mammalogists, founded in 1919, has served as the primary scholarly organization for mammalogists in North America. Current mammalogists are a diverse array of scientists working on research ranging from molecular genetics to ecology and systematics, but sharing the common subject of mammals. University museums in particular have become important foci for mammalogy, both for their education and conservation functions and as fertile grounds for training new mammalogists and conducting research.

Mammals are of interest to humans for a variety of reasons. They are used for food and other products. In medical science, nonhuman mammals are models for some human diseases, and they are reservoirs for or vectors of disease. Mammals are also pests, destroying grain and other crops, and they have an aesthetic value in terms of recreation and sport hunting. New techniques have been developed in recent decades for monitoring mammal movements and activities and for studying molecular phylogeny to understand the evolutionary relationships among mammals, both living and extinct.

SUGGESTED READINGS

Birney, E. C., and J. R. Choate (eds.). 1994. Seventy-five years of mammalogy (1919–1994). Special Pub. No. 11, Am. Soc. of Mammal.

Gunderson, H. I. 1976. The evolution of mammalogy, a history of the science. Pp. 3–38 in Mammalogy. McGraw-Hill, New York.

Journal of Mammalogy. 1996. Special Section on Women in Mammalogy. 77(4).

Wilson, D. E., and J. F. Eisenberg. 1990. Origin and applications of mammalogy in North America. Pp. 1–35 in Current mammalogy, vol. 2 (H. H. Genoways, ed.). Plenum, New York.

DISCUSSION QUESTIONS

1. Suppose that you were living in the 19th century and were about to travel across the country from St. Louis to the area that is now Los Angeles via a southern route through Texas, New Mexico, and Arizona. Using any of several mammal guides that you may have available, generate a list of the mammals that you might expect to encounter.

2. Select any three of the 19th-century naturalists mentioned in this chapter and, using available reference books and the Internet, write one-page synopses of their contributions to biology and to mammalogy in particular.

3. In the manner of the early naturalists, spend some time watching members of one or more mammal species, preferably in a field setting. Write down your observations in a journal format. Using your observations, what sort of questions can you generate about various aspects of the biology of these mammals?

4. The most important theory for biology is evolution by natural selection, first postulated by Darwin and Wallace. From your previous biology background, write down what you believe are the major components of the thinking that resulted in this theory. Check your answer by referring to an introductory biology textbook or other reference that contains a section on the theory of evolution and how it developed.

5. Using a good historical atlas, guides to living mammals, and the Internet, construct a list of the mammals that Lewis and Clark might have encountered on their western expedition. For each mammal, list the state or states where it might have been found.

CHAPTER 3

Methods for Studying Mammals

Like most fields of biology, mammalogy relies on a diverse array of research methods, most of which cross disciplinary boundaries. No single chapter could describe them all, so we present a selection of techniques that are widely used by researchers in different branches of mammalogy, especially methods that have had a major impact on our knowledge of mammalian biology. For convenience, we divide this chapter into three major sections: field methods, laboratory and museum methods, and systematic methods. These categories are not disciplinary—a single research project might employ methods from each. For example, a phylogenetic study of a particular group of mammals might begin with field work in which specimens are collected, proceed to museum work in which those specimens are prepared and examined, and culminate with a reconstructed phylogeny, proposed classification, and estimated divergence times for major groups.

An overview such as we provide here will inevitably omit many important topics and lack detail on those that are discussed. Indeed, some of the techniques we describe are quite complex and scarcely amenable to cursory treatment. Our goal is to present these methods in the context of the research questions they were developed to address. We have included fairly extensive references to more thorough and advanced descriptions of each. Of course, no textbook treatment can replace hands-on experience and we strongly encourage mammalogy students to take full advantage of any field, laboratory, or data-analysis opportunities associated with their course.

We cannot overemphasize the importance of mathematics and statistics in modern biology. Any student aspiring to a career in mammalogy *must* obtain sufficient training in modeling, probability theory, statistical inference, and experimental design. These topics are too extensive for us to cover here, but readers may consult any number of helpful texts at introductory or advanced levels (e.g., Motulsky 1995; Sokal and Rohlf 1995; Morrison et al. 2001; Quinn and Keough 2002; Gotelli and Ellison 2004; Zar 2006).

Field Methods

Mammalogists often ask such questions as: How many squirrels live in a particular forest or woodlot? What are the sizes and shapes of their home ranges? Does their socio-spatial system include dominance hierarchies or territories? When are the squirrels most active? Answering questions like these usually entails identifying and monitoring individual animals in the field. But most mammals are difficult to observe directly in the wild because they spend at least part of their time in inaccessible places, are active at night, or are simply too small to see easily in their natural habitat. For these reasons, researchers have developed a wide range of indirect methods for studying mammals in the field.

TRAPPING AND MARKING

Trapping

Methods for capturing wild mammals include a variety of trapping and netting techniques (Bush 1996; Lehner 1996; Schemnitz 1996). Whether mammals are live-trapped or killed depends on the nature of the study and the reason for their capture. Although many studies now employ live-trapping, kill-trapping is often necessary and justified. For instance, museum collections maintain reference specimens for taxonomic studies and biotic surveys, but the skin, skeleton, or tissue materials required for such analyses cannot be obtained from live animals. If trapped specimens are to be used for a museum collection, it is necessary to ensure that the specimen is not damaged by the capture procedure. Trapping to remove animals, such as pest species, from an area can best be accomplished with kill traps, guns, or poison. Mammal trapping of any sort usually requires permits issued by government conservation authorities. Most mammalogists follow standard practices for humane handling of trapped mammals, such as those published by the American Society of Mammalogists (Animal Care and Use Committee 1998).

Live traps come in a variety of types and sizes. For small rodents, the most widely used are Longworth, Sherman, and Tomahawk traps (Chambers et al 2000; Lambert et al. 2005). For species of intermediate size, such as raccoons (*Procyon lotor*), traps with wire mesh sides are available, such as the Havahart or National traps (Baldwin et al. 2004). Box traps (figure 3.1) are constructed for capturing larger mammals such as ungulates or carnivorans (Grassman et al. 2005). The dimensions and operation of such traps are adapted to the subject species. Box traps can also be used to capture groups of smaller animals. Rood (1975) used large box traps to capture banded mongoose (*Mungos mungo*) groups in East Africa, and enclosures have been used to trap groups of ungulates (Taber and Cowan 1969) and primates (Rawlins et al. 1984). Pitfall traps, consisting of a can or bucket buried in

Figure 3.1 Stephenson box trap. Large box traps of the type shown here are used to capture larger mammals, such as white-tailed deer. Animals captured in this manner can be measured, tagged, or dyed for individual identification, or fitted with collars for tracking by radiotelemetry.

the ground, are used to capture very small mammals, such as mice and shrews (Brannon 2000; Umetsu et al. 2006). Pitfall traps are most frequently employed as kill traps, but may function as live traps if checked frequently. Padded leghold traps are used for canids and felids (Thornton et al. 2004). Mist nets are often used to catch bats, particularly when the bats follow a regular flight path to and from their roost (Kunz 1988). Larger nets, fired by guns, have been used to capture ungulates such as bighorn sheep (*Ovis canadensis*; DeCesare and Pletscher 2006). The various types of kill ("snap") traps used with small mammals include Museum Special, Victor (Stancampiano and Schnell 2004), and McGill types. Several kinds of traps are available for capturing burrowing mammals, including the harpoon mole and the Macabee-type gopher trap. Descriptions of these and other special-purpose traps and trapping methods are provided by Martin et al. (2001).

Guns that fire tranquilizer darts are often used to immobilize larger mammals (Bush 1996). This technique is helpful when animals need to be held for only a brief period, after which they may be given an antagonist drug to reverse the anesthetic. For example, Zedrosser et al. (2006) anesthetized female brown bears (*Ursus arctos*) with darts shot from a helicopter, then obtained tissue samples and body measurements while the bears were incapacitated. To study water flux in Arabian oryxes (*Oryx leucoryx*), Ostrowski et al. (2002) used darts to administer doses of radioactive hydrogen as well as anesthetic. Dart guns can also be used to catch animals for translocation or captive study. Specially modified guns have been used to shoot marking devices into whales (Brown 1978).

The care of wild animals that have been captured and held in captivity has been important since the advent of zoological parks and the use of mammals as laboratory subjects. In a landmark volume, Crandall (1964) spelled

.out many of the procedures to be followed in caring for captive mammals. This topic has become increasingly significant in the past 25 years as zoos and wildlife parks have expanded their mission to include conservation as well as exhibition. Working with captive mammals has enabled us to get a better understanding of their physiology and behavior, information that is critical for (among other things) developing captive breeding programs (Elias et al. 2006). Kleiman et al. (1996) provide a more up-to-date summary of procedures use to care for mammals.

Marking

Appropriate marking techniques vary with the species of mammal being studied, and whether individuals are free-ranging, held in zoos, or part of laboratory stocks (Stonehouse 1978). For techniques used with zoo and laboratory animals, see Rice and Kalk (1996) or Lane-Petter (1978), respectively. Here we are concerned only with free-ranging mammals and those maintained in seminatural conditions.

In some instances, physical features of individual mammals can be used for identification. Individuals of large mammal species (e.g., ungulates, primates) can be identified by a profile of observable characteristics such as size, coloration, scars or other marks of injury, and behavior patterns. Vibrissae spot patterns have been used to identify individual lions (Rudnai 1973). This approach is well developed in studies of cetaceans, many of which show natural color or shape variations on their tail flukes (Hammond et al. 1990) and dorsal fins (Balcomb and Bigg 1986; Gubbins 2002). Such markings can be photographed and the images archived to form a permanently accessible record for identification of individuals (Mizroch et al. 2004). Photographic identification has been enhanced recently by the development of computerized matching techniques. Kelli (2001) used a three-dimensional computer matching system to assess similarities among 10,000 photographs of cheetahs (*Acinonyx jubatus*).

In species with more cryptic habits and appearances, artificial marking devices are used to identify individuals after they have been captured and released. The most common devices are coded metal or plastic tags attached to the ears (Bradshaw et al. 2003), webbing of feet, neck collars (figure 3.2), or leg bands (Kunz and Robson 1996). Animals tagged in this way can only be identified if they are recaptured. For field identification at a distance, investigators use larger tags, dyes, and flags (Bookhout 1996; Lehner 1996; Stonehouse 1978). Michener (2004) used both ear tags and dye to mark North American badgers (*Taxidea taxis*). In some studies, individuals have been identified with radioisotopes placed in tags or subcutaneous implants (Linn 1978). Cantoni (1993) used ear tags containing radioactive filaments of different energy levels to monitor the movements of shrews. Individuals were identified by the energy output of their tag, detected with a portable scintillation counter.

Figure 3.2 **Animal marking.** A cougar (*Puma concolor*) shown wearing a neck collar.

Researchers engaged in long-term field studies may give animals permanent brands or tattoos. Raum-Suryan et al. (2002) branded over 8,500 pups in a study of lifetime dispersal by Steller sea lions (*Eumetopias jubatus*) in Alaska. Freeze-branding with liquid nitrogen results in permanent white hairs growing where the liquid was applied. Some investigators remove tissue as a means of marking mammals (Twigg 1978). Clipping fur patterns and using depilatories to remove patches of hair have been used as marking techniques in several species (Glennon et al. 2002), as have toe clipping (Granjon et al. 2005) and ear notching (Berry 1970). The latter marks may be visible from a distance in larger mammals, particularly with the use of binoculars.

MONITORING

Methods for monitoring the movements of wild mammals may be divided into two broad categories: those that involve tracking physical signs of animals and those that involve radio tagging. Both approaches are widely used, and the choice of method is dictated by the nature of mammal species studied, the kind of research question posed, and the cost of conducting the research. Researchers must also ensure that the monitoring procedure does not alter the behavior or survivorship of the animals being studied.

Powdertracking (Lemen and Freeman 1985) involves coating a small mammal with a fluorescent dust; its movements after release can then be traced at night with an ultraviolet light. This technique has been used to study home ranges (Mikesic and Drickamer 1992b), foraging (Hovland and Andreassen 1995), seed caching (Longland and Clements 1995), and dispersal (Jacquot and Solomon 2004). Bait-marking can also be used to assess spatial relations. When small plastic pellets are ingested with bait food, the locations of feces containing pellets mark areas visited by individual mammals (Delahay et al. 2000). Fecal pellet surveys and tracks have been used to monitor wild populations of ungulates (Neff 1968; Mayle et al. 2000; McShea et al. 2001), while tracking tubes (or tunnels) have been used with smaller mammals (Diaz 1998; Nams and Gillis 2003). Aquatic species pose special problems in terms of field-monitoring. Churchfield et al. (2000) used baited tubes to survey water shrews (*Neomys fodiens*), and Giraudoux et al. (1995) estimated numbers of water voles (*Arvicola terrestris*) by using a surface index (numbers of earth mounds).

The use of **passive integrated transponder (PIT)** tags began as a means of identifying individuals in the field (Prentice et al. 1990; Neubaum et al. 2005), but it has developed into an effective monitoring technique for small mammals. PIT tags are small (1 cm), glass-encased, electronic devices implanted beneath the skin. Each PIT tag contains an integrated circuit with a digital identification code and an antenna that transmits the code when it is activated by the electric field of a transceiver. If an animal is close enough to a transceiver for its PIT to be activated, the tag transmits its unique code to a data-logging system that identifies the animal. Harper and Batzli (1996) put PIT tags into voles (*Microtus pennsylvanicus* and *M. ochrogaster*) and placed transceivers along runways. They found that several individuals used the runway system each day and that voles were most active near sunrise or sunset. Rehmeier et al. (2006) described an activity-monitoring system based on PIT technology that allowed them to study movements of individual deer mice (*Peromyscus maniculatus*) in burrows.

A **radiotelemetry** system includes a battery-powered radio transmitter attached to an individual mammal, the signal of which is detected by an antenna connected to a receiver (White and Garrott 1990; Kenward 2000). Transmitters can be placed in collars (figure 3.3) or implanted. There are four major types of receivers for monitoring radiotransmitter signals: handheld antennae, antennae mounted on ground vehicles or aircraft, antennae mounted on towers, and satellites (Samuel and Fuller 1996). The receiver translates the signal into audible sound or stores it digitally for analysis. The position of an

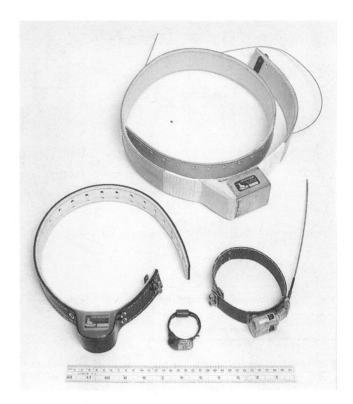

Figure 3.3 **Radio collars.** A variety of radio collars have been developed to use with mammals of different sizes. The collar, battery, and transmitter should not exceed about 10% of the weight of the animal. Thus, with constraints on collar size in smaller mammals, there are often limitations on the size of the battery and, consequently, on the length of time the transmitter will be functional and the distance over which the signal can be detected.

animal is determined by triangulation using two or more bearings taken with one antenna moved to different locations, or multiple fixed antennae at different locations.

In addition to pinpointing the locations of specific animals, radiotelemetry can be used to assess behavioral traits in the field. Information on rates of movement can be obtained by calculating time intervals and distances between location fixes, or by using an accelerometer (Dyhrepoulsen et al. 1994; Bradshaw et al. 1996). Radiotelemetry can be used for studying dispersal, home-range area (figure 3.4), habitat use, homing behavior, and survival rates (Wilson et al. 1996; Krebs 1999). For example, Ilse and Hellgren (1995) used radiotransmitters to study space use by, and interactions between, feral swine (*Sus scrofa*) and collared peccaries (*Pecari tajacu*) in south Texas. In a study of two *Peromyscus* species, Ribble et al. (2002) found that estimates of home-range size from radiotelemetry were significantly larger than those obtained by trapping, provided that the mice populations were at low densities; at high densities, radiotelemetry estimates were similar to those from trapping. Physiological measurements such as body-core temperature and heart rate can also be relayed by implanted transmitters (Folk and Folk 1980; Lefcourt and Adams 1996) and used to assess activity levels and energy expenditures.

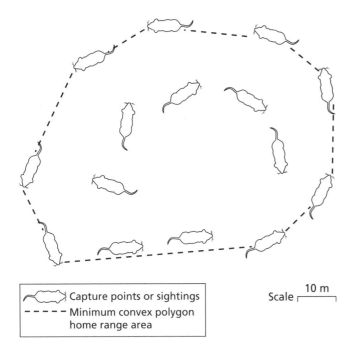

Capture points or sightings
- - - - Minimum convex polygon home range area

Scale |— 10 m —|

Figure 3.4 Home range. The home range of a mammal can be plotted from a series of radiotransmitter locations. Home-range area can vary with the age, sex, season, population density, and reproductive condition of the animal.

Geographic information systems (GISs) are computer software programs used to store and manipulate geographic data (Koeln et al. 1996; Demers 1997; Goodchild 2003). First developed in the 1970s (Tomlinson et al. 1976), the GIS implementation procedure involves taking data from maps, aerial photographs, satellite images, ground surveys, and other sources and converting it to a digital format. Geographic information includes spatial coordinates, topography, environmental variables, locations of discrete objects, or virtually any data that can be displayed on a map. The data are integrated, analyzed, and visualized with computer software. GIS databases are now available for much of the world. The **global positioning system (GPS)**, derived from a satellite navigation system developed by the U.S. Department of Defense in the 1970s, uses satellite-based radio signals to pinpoint the latitude, longitude, and elevation of a receiver on the ground (El-Rabbany 2006). GPS information can be stored for subsequent analysis, frequently in conjunction with GIS data.

GIS and GPS technologies have developed rapidly during the past 30 years, and their application in mammalogy has expanded (Rogers et al. 1996; Kenward 2000). They are used to study home range (Ostro et al. 1999), population density (Radeloff et al. 1999), movement patterns (Pepin et al. 2004), predation (Sand et al. 2005), and habitat use (Clark et al. 1993). Much of this work relies on collars, each fitted with a GPS receiver and battery power supply, that are attached to animals. D'Eon and Serrouya (2005) followed the movements of 12 mule deer (*Odocoileus hemionis*) fitted with GPS collars in a British

Columbia population over a 4-year period. GPS data allowed these researchers to track seasonal movements of the deer, documenting the locations and characteristics of their summer and winter ranges. Although radiotelemetry and GPS technology are very powerful, the equipment is costly. As a result, GPS researchers are often forced to monitor fewer individuals than are typical of physical tracking studies.

Methodological and statistical problems can complicate the collection and analysis of radiotelemetry data. As noted previously, the transmitter device should not interfere with an animal's normal activity, nor should the animal's behavior be affected by the capture, handling, and attachment procedures (Fuller 1987). Mikesic and Drickamer (1992a) noted a reduction in wheel-running activity for at least 96 hours after transmitter collars were placed on house mice (*Mus musculus*), and radio collars produced acute infestations of ticks in yellow-necked mice (*Apodemus flavicollis*; DeMendonca 1999). Electrical lines or obstacles (e.g., hills, trees) between the antenna and the transmitter may cause inaccurate triangulation. Autocorrelation occurs when locations ("fixes") are obtained too close together in time, such that each location is strongly dependent on the previous one. Such lack of independence among data points causes home-range size to be underestimated. The appropriate interval between locations depends on the behavior pattern of the species being studied, which in turn may be affected by environmental variables. Ribble et al. (2002) used an interval of 30 minutes between locations in their study of *Peromyscus* home ranges.

Data obtained from trapping, marking, and monitoring are used to address many questions in mammalogy, but they are especially important for studying habitat use and biotic diversity (Lancia et al. 1996; Rhodes et al. 1996; Wilson et al. 1996; Buckland et al. 2001). One of the most common approaches is a **mark-recapture study** in which individuals from a population are captured, marked, released, and then recaptured at later times (Sutherland 1996). Recapture rates can be used to estimate population size, survivorship, seasonal reproductive patterns, and home-range size (Slade and Russel 1998). Population size is usually estimated from such data with one of several standard methods: Lincoln-Petersen (Krebs 1999), Schnabel (1938), or Jolly-Seber (Seber 1982). Analysis of mark-recapture data may be complicated by the assumption that all members of a population are equally catchable. This assumption is violated if, for instance, animals of different sexes or ages have different capture probabilities. In addition, some individuals become more catchable after repeated captures ("trap happy") or avoid traps after initial capture ("trap shy"; Nichols and Hines 1984). The impact of "catchability" assumptions on estimates of population parameters has been mitigated by incorporating models of the trapping process into the estimation procedure (Hammond and Anthony 2006), an approach implemented by the computer program CAPTURE (Otis et al. 1978; Rexstad and Burnham 1991).

OBSERVATIONAL METHODS

Many mammals can be observed directly in the wild, particularly those that are large and live in open habitats. Diurnal species are easiest to observe, but nocturnal mammals can be viewed with night-vision scopes (Arletazz 1996; Rancourt et al. 2005) or monitored with infrared imaging (Sabol and Hudson 1995). Watching mammals provides data on life history characteristics such as movements, habitat use, foraging, territoriality, reproduction, and interspecific interactions. Behavior, both individual and social, is most commonly recorded using observational methods.

Basic methods for making observational records have been summarized by Altmann (1974), Crockett (1996), and Lehner (1996). The choice of which observational sampling technique to use should be based on what information is needed and what behavioral activities can be adequately seen and recorded. Behavioral traits can be variable, with differences within the same individual over time, among populations, and over seasons. Thus, a well-formulated hypothesis and an appropriate experimental design are important considerations in selecting an observational method. Observations can be made with the unaided eye, through binoculars or telescopes, with still photographs, or with video recording. The latter has proven especially valuable because it allows multiple observers to study one observational sequence repeatedly and perhaps in slow motion.

Some observational sampling techniques record only behavioral states (general categories of behavior) such as "grooming" or "feeding," other methods note events (specific brief actions) such as a "yawn" or "tail flick," and some methods record both. The two most common methods of observational sampling are focal animal and scan sampling. **Focal animal sampling** is the recording of specified behavior states or events by a given individual (or group) in bouts of prescribed length (figure 3.5). Often only the behaviors related to a hypothesis being tested are recorded. Focal animal sampling has been used extensively to study ungulate behavior. Weckerly et al. (2001) observed feeding and aggressive behaviors in male and female Roosevelt elk (*Cervus elaphus roosevelti*) to test hypotheses of sexual segregation, the spatial separation of males and females in mixed-sex groups. They found that closer proximity of males and females was associated with reduced feeding rates in males and increased aggression between females. Keeley and Keeley (2004) observed mating of Brazilian free-tailed bats (*Tadarida brasiliensis*) on a highway bridge in Texas and discovered two distinct copulatory behavior patterns (passive and aggressive).

Scan sampling involves recording the behavioral state (e.g., resting, grooming, moving) of each animal in a small group at predetermined intervals (such as every 15 seconds) or over a predetermined block of time (such as 30 minutes). Bearzi (2006) used scan sampling to study feeding associations between groups of dolphins and California sea lions (*Zalophus californianus*) in Santa

Figure 3.5 Social interactions among female baboons. Focal animal sampling can be used to examine interactions among members of a group. Watching a single individual for a prescribed time period provides an accurate picture of its activities. In the case shown here, two females are engaged in grooming behavior.

Monica Bay, California. Data were recorded on the behavior of all individuals in dolphin groups during observation periods longer than 25 minutes. Together with observations on the proximity and behavior of nearby sea lions, Bearzi (2006) concluded that sea lions follow and exploit the superior food-finding ability of dolphin groups. This study also employed video recording, GPS monitoring, and GIS analysis. Scan sampling and focal animal sampling are often used together. In a study of Iberian red deer (*Cervus elaphus hispanicus*), Sánchez-Prieto et al. (2004) used scan sampling to record the sex and age-class of deer in an observation area and then used focal animal sampling to determine the incidence of specific aggressive and reproductive behaviors.

Other observational methods include instantaneous sampling (sequentially recording the behavioral state of an animal at the end of each observation interval), all-occurrences sampling (recording all occurrences of a particular event in a group of animals), and ad libitum sampling (recording all the states or events of all organisms in a group). Such methods are described in more detail by Altmann (1974) and Lehner (1996). In addition to passive observation, behavior researchers may use sound playback to test hypotheses that involve vocalizations. In this method, calls that have been recorded are played to animals in particular situations and the animals' reactions monitored. For example, Weary and Kramer (1995) demonstrated that eastern chipmunks (*Tamias striatus*) respond to playback of trill vocalizations by becoming alert.

One key issue that arises in observational studies, regardless of sampling method, is the **observability** of animals under study. "Observability" has two meanings in behavioral research. In the first sense, it refers to whether the habitat permits regular, direct observation of an animal. For example, foliage can interfere with an investigator's ability to make behavioral observations of arboreal mammals or those that live in dense vegetation. This is especially problematic when the sampling method requires observations at specified intervals. In such cases, it may be necessary to correct for missing observations. A second meaning of "observability" refers to the fact that when animals are watched, some of them may be more conspicuous than others, perhaps depending on their age, sex, or dominance status. For example, male rhesus macaques (*Macaca mulatta*) of high social status are seen more frequently than those of low social status in the central area of a group (Drickamer 1974). If the relationship between dominance status and mating activity is being studied, differential observability could lead to erroneous conclusions. If high-ranking males were seen more often engaging in mating activities with females, we might decide that social rank confers an advantage in mating, but such a conclusion would be suspect unless we controlled for the differential observability of high-ranking males. In fact, when observability is taken into account, there is no relationship between social rank and mating success in this species (Drickamer 1974).

Laboratory and Museum Methods

PHYSIOLOGICAL MEASURES

Physiology is the study of how tissues and organs function. The scope of this discipline, even if restricted to mammals, is extremely broad, as is the range of experimental methods that physiologists employ. Modern, mainstream physiology relies heavily on molecular, cellular, and biochemical techniques in addition to the more traditional, whole-organism approaches of previous decades (Randall et al. 2002). Mammalogists have focused on a smaller, but still diverse, set of topics including nutrition, energy metabolism, reproduction, behavior, sensory mechanisms, growth, locomotion, disease, stress, and water balance. An overview of methods employed in the first three of these areas illustrates how mammalogists approach the study of physiology.

Nutrition

Nutrients are substances that an animal must obtain from its diet and that serve as raw materials for energy production, growth, and other metabolic functions. Nutrients include water, proteins, lipids, carbohydrates, nucleic acids, salts, vitamins, and trace elements. **Nutritional requirements**, both the quantity and kinds of nutrients, may vary significantly among species and among (or within) individuals in different physiological states. For this reason, it is difficult to determine the precise nutritional

requirements of a species, though it is usually possible to identify major dietary components. Thus, carnivores have diets higher in protein and vitamins, but lower in carbohydrates, compared to herbivores (Hume 2003).

Dietary preferences of wild mammals are usually determined by direct observation or by examination of scats and stomach contents. Stomach content analysis provides a more complete assessment of diet if a sufficient number of individuals can be examined, but field mammalogists more often analyze scats to estimate dietary preferences (Wiens et al. 2006). Fecal sampling is nondestructive and noninvasive, though only partially digested food items can be identified by visual inspection (Dickman and Huang 1988). As with any sampling strategy, scat analysis will provide a more accurate picture of dietary components if data are available from a large number of individuals of known physiological states. For captive mammals, researchers may infer nutritional requirements through food-preference experiments. Astúa de Moraes et al. (2003) found that 12 species of Neotropical didelphids display a gradient of dietary specialization from frugivory to carnivory based on the preferences of captive animals for different proportions of fruit, roots, leaves, tubers, arthropods, and meat (figure 3.6).

Researchers are frequently interested in assessing the **nutritional condition** of wild mammals (Harder and

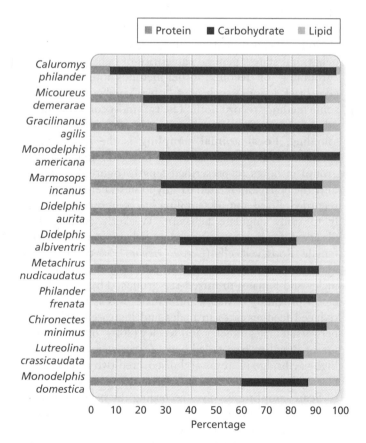

Figure 3.6 Dietary components of opossums.
Proportions of protein, carbohydrate, and lipids in the diets of 12 didelphid marsupials. *Adapted from Astua de Moraes et al. (2005).*

Kirkpatrick 1994) as a means of understanding how food limitation influences population dynamics. Nutritional condition is a composite of several variables that reflect how well an animal has assimilated the raw materials needed for metabolism and can be indexed by morphometric measures, urine chemistry, assays of electrical conductivity, or direct chemical analyses. Cook et al. (2004) assessed the nutritional condition of female elk in Yellowstone National Park using three morphometric variables: body fat, body mass, and thickness of the longissimus dorsi muscle. The body fat index was a function of a "rump body condition score" and "subcutaneous rump fat thickness," the latter measured with ultrasound imaging. Ultrasonography was also used to measure thickness of the longissimus dorsi, an index of protein catabolism. Body mass was a function of chest girth circumference, body fat percentage, age, and pregnancy status. Cook et al. (2004) found that nutritional condition was independent of age in Yellowstone elk but strongly related to lactational status and pregnancy. They suggested that nutritional limitation during severe winters is a major density-dependent factor limiting elk population size. DelGiudice et al. (2001) found that levels of nutritional restriction in Yellowstone elk and bison during winter were reflected by changes in urinary potassium and nitrogen relative to creatinine. Researchers have also developed methods that use electrical conductivity to estimate fat reserves in live animals. These include bioelectric impedance analysis (Pitt et al. 2006) and total body electrical conductivity (Wirsing et al. 2002). Lipid and protein contents can be assayed by direct chemical analysis of dried carcasses if destructive sampling is feasible.

Metabolism

Metabolism is the sum total of all chemical reactions taking place in an organism. The rate of these reactions is affected by body temperature, body mass, muscular activity, and the energetic demands of reproduction. Thus, the metabolic rate—the amount of heat energy released per unit time, usually expressed as kilojoules per day (kJ/d)—of a mammal is a fundamental consideration in studies of thermoregulation, hibernation, torpor, locomotor performance, foraging, and reproductive behavior. Mammalogists are concerned with two measures of metabolic rate: **basal metabolic rate (BMR)** is the rate of energy conversion in a resting animal, with no food in its intestine, at an ambient temperature that causes no thermal stress; **field metabolic rate (FMR)** is the rate of energy use in an animal engaging in normal activities under natural conditions (Randall et al. 2002). FMR is typically higher, more variable, and more difficult to measure than BMR.

BMR is commonly determined in the laboratory with a respirometer, a device that measures the rate of oxygen and carbon dioxide exchange during normal breathing (McNab 1995). This approach exploits the direct relationship between oxygen consumption and energy production

by aerobic metabolism. A test animal is confined to an air-filled chamber for a specified period during which the amount of oxygen in the chamber is monitored. BMR is a function of oxygen flux in units of ml O_2/h. Because of the correlation between body mass and BMR (McNab 1989), oxygen flux is typically standardized by the organism's weight; mass-specific BMR is expressed as ml O_2/(g•h). Geiser (2003) used respirometer measurements to show that regressions of BMR on body mass are indistinguishable for carnivorous versus omnivorous/herbivorous marsupials but significantly different than those for eutherians (figure 3.7). Indeed, BMRs of marsupials are generally lower than similarly sized eutherians (Haysen and Lacy 1985). In contrast, Harrington et al. (2003) showed that the mass-specific BMR in two mustelids was comparable to that predicted for other placental mammals of similar size. They suggested that previous results showing an elevated BMR in mustelids were due to induced stress during respirometer experiments.

The most popular method for measuring mammal FMRs involves the use of "doubly labeled water" (DLW; Nagy 1989), often in combination with studies of physiological water flux. Water molecules with radioisotopes of oxygen (^{18}O) and hydrogen (deuterium or tritium) are injected into an animal, and repeated blood samples are taken at intervals of hours or days. The loss of isotopic oxygen in blood over time is related to the rate of CO_2 loss (L/d), which in turn is proportional to metabolic rate (kJ/d). Williams et al. (1997) used this technique to measure seasonal variation in metabolic rates of South African

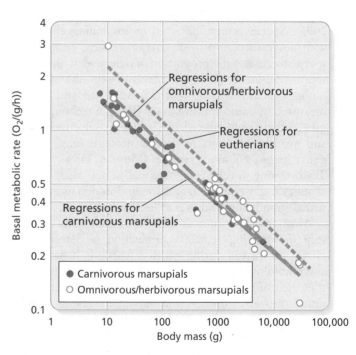

Figure 3.7 Basal metabolic rate (BMR) as a function of body mass. Solid circles represent species of carnivorous marsupials, open circles represent omnivorous/herbivorous marsupials. Solid, broken, and dotted lines are regressions for carnivorous marsupials, omnivorous/herbivorous marsupials, and eutherians, respectively. *Adapted from Geiser (2003).*

aardwolves. During summer, the average aardwolf produced 112.5 L CO_2/d, or 2891.2 kJ/d of energy; in winter the corresponding values were 71.8 L/d and 1,844.8 kJ/d. These are significantly lower than values predicted for other eutherians of comparable size and mirror the reduced BMRs characteristic of most myrmecophagus (ant- or termite-eating) mammals (McNab 1984), including the marsupial numbat (Hume 2003). Nagy and Gruchacz (1994) used DLW to show that kangaroo rats (*Dipodomys merriami*) experienced little energy or water stress despite the extreme aridity of their desert habitat.

Reproduction

Because production of progeny is a basic measure of individual fitness and is important to understanding population dynamics, much attention has been given to mammalian reproductive biology. Data on the number of females producing young, litter size, and juvenile survivorship provide the basic information for studies of population growth. Numerous techniques have been developed for assessing the reproductive condition of males and females.

For many male mammals, the testes are only scrotal just prior to and during the breeding season; thus, determining whether the testes of an individual are scrotal ("descended") or abdominal provides an important clue to its reproductive status. In male marsupials, reproductive maturity is marked by spermatorrhoea, the discharge of sperm into the urine (McAllan 2003). The weights of testes and accessory sex glands are also useful indices of reproductive activity (Krutzsch et al. 2002; Murphy et al. 2005). Sperm counts, obtained by examining ejaculates with a microscope, provide data on relative numbers of spermatozoa and the proportion of defective or abnormal sperm (Zenuto et al. 2003). Levels of reproductive hormones also reveal male reproductive status and can be measured by **radioimmunoassay (RIA)**, a technique that uses radioactively labeled antibodies to selectively bind and quantify amounts of specific proteins (e.g., androgens) from a blood sample (Muteka et al. 2006). Testosterone levels vary seasonally but also exhibit changes due to stress, aggression and dominance status, sexual stimulation, and overall health.

Most female mammals also breed seasonally. A first step in assessing female reproduction is detection of estrus, the period during which a female is behaviorally and physiologically receptive to mating attempts. Estrus is the result of hormonal changes associated with the maturation of ovarian follicles and subsequent release of ova. During estrus, females exhibit greater tolerance to mating solicitation by males—they may adopt a lordosis posture with the head and hindquarters elevated and back depressed to permit intromission. One consequence of the increase in circulating estrogens during estrus is the cornification of cells that line the vaginal walls as ovulation approaches. Thus, vaginal epithelial cells obtained with a moist swab can be examined under a microscope,

with appropriate histological staining, to provide a quick indication of whether the animal is, or has recently been, in estrus (Kennelly and Johns 1976; Valdespino et al. 2002).

Ovulation rate, or the number of ova released at the onset of estrus, can be determined in live animals by flushing the reproductive tract immediately after ovulation. If the animal is sacrificed, ovaries can be examined histologically to study follicular development and check for the presence of corpora lutea ("yellow bodies"), the remnants of ruptured follicles (Cowles et al. 1977). Counting placental scars (figure 3.8) in the uterus indicates the number of fetuses that implanted (Oleyar and McGinnes 1974; Sacks 2005).

In living female mammals that must be examined repeatedly, several other procedures can be employed to assess reproduction. **Laparotomy** is a surgical examination, performed by trained personnel, of the reproductive tract to determine the condition of the ovaries and uterus (Adams et al. 1989; Green et al. 2002). With the invention of fiber-optic techniques, laparotomy can be replaced by

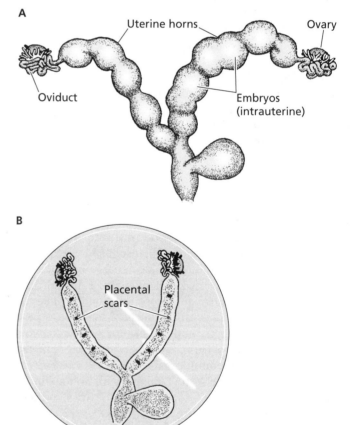

Figure 3.8 Indications of pregnancy. (A) The swellings in the two uterine horns from a pregnant white-footed mouse; each swelling represents a fetus and associated placenta. (B) The uterine horns, dissected from the abdominal cavity, are compressed between the bottom and inverted top of a Petri dish, revealing the placental scars along both horns. This information can be used to determine the number of embryos that implanted in the uterus.

laparascopy. For larger mammals, ultrasonography is now used to determine the morphology of the reproductive tract, as well as the presence and condition of developing fetuses (Adams et al. 1991; McNay et al. 2006).

We can use RIA or enzyme immunology assay (EIA; Nelson 2000) on blood samples to determine circulating levels of reproductive hormones such as estrogens, progesterone, and prolactin (Austin and Short 1984; Muteka et al. 2006). More recently, it has become possible to make these hormone determinations using urine and feces (Wasser et al. 1991; Scheibe et al. 1999; DeMatteo et al. 2006). This noninvasive technique has been applied to both males and females and allows investigators to obtain hormone measurements on free-ranging mammals.

GENETICS AND MOLECULAR TECHNIQUES

For most of its history, mammalian biology focused on studying whole animals or their preserved remains. As understanding of the biochemical basis of heredity developed throughout the 20th century, however, progressively more sophisticated laboratory methods were developed to assay variation in chromosomes, proteins, and nucleic acids. Mammals have been a central group in such research since its inception (Wahrman and Zahavi 1955; Harris 1966; Avise et al. 1979), and the use of "molecular markers" has allowed mammalogists to develop new and powerful approaches to the study of population genetics, speciation, and phylogeny.

Chromosomes

The modern study of mammalian cytology involves visualization of chromosomes obtained from cells in the metaphase portion of mitotic division (Hsu 1979). At this stage, the chromosomes are condensed (i.e., the DNA is tightly wound around nucleosome proteins within sister chromatids) and appear as rod-shaped structures when viewed under a light microscope. Typically, photomicrographs of chromosome spreads are manually or digitally edited such that images of each chromosome from an individual cell can be arranged in order of size. Such an image is termed the **karyotype** of a cell. The karyotype is usually the same for all cells of an individual, but sometimes it shows variation (polymorphism) among individuals from the same species. Karyotypes may vary considerably among species, and such variation may provide insights on evolutionary relationships (Andrades-Miranda et al. 2001).

Karyotypes contain two major types of information—the diploid (2N) number of chromosomes within a cell and the morphology of each chromosome. A chromosome's morphology is described by its size, its pattern of secondary constrictions, and the position of its centromere: metacentric if the centromere is at the center of the chromosome or acrocentric if the centromere is closer

to one end (Hartwell et al. 2008). More detailed information about the chromosomes can be revealed by differential staining prior to visualization. One procedure, named after the chemical dye Giemsa, produces "G-bands" on chromosomes; the dark-staining bands correspond to DNA regions that are low in guanine (G) and cytosine (C) nucleotides and that usually contain few genes. The G-banding pattern of each pair of homologous chromosomes in a cell is unique (figure 3.9). Similarly, "C-banding" reveals areas of repetitive DNA sequences associated with centromeres (Lewin 2006).

The diploid numbers of mammals range from 2N = 6 in the Indian muntjac deer (*Muntiacus muntjak*) to 2N = 92 in the Ecuador fish-eating rat (*Anotomys leander*). Within groups, karyotypes may be highly conserved (e.g., dasyurid marsupials, felids) or highly variable (e.g., diprotodontian marsupials, canids, gibbons). Where karyotypes of closely related species differ, they can usually be related to one another by simple rearrangements such as fusions, fissions, inversions, and translocations. Indeed, careful comparison of morphology and banding patterns may reveal the evolutionary history of rearrangements within such a group (Qumsiyeh 1994; Faria and Morielle-Versute 2006).

During the past three decades, **molecular cytogenetics** has exploited the **in situ hybridization** technique to map the location of specific DNA sequences on chromosomes (Sessions 1996). In this method, a single-stranded DNA fragment of known sequence (the probe) is "labeled" with a radioisotope or biotin molecule that will allow it to be visualized. A solution containing the probe is then applied to a chromosome preparation that has been denatured (i.e., the DNA has been made single-stranded). During subsequent incubation, the probe binds specifically to its homologue(s) on the chromosome(s) by virtue of their complementary sequences. Upon visualization, the probe appears as a dark-staining or fluorescent region on the chromosome spread, revealing the physical location of its homologous sequence. Although early studies employed repetitive DNA sequences, such as ribosomal RNA genes, as probes, a wide variety of probes is currently in use, including many single-copy genes. Most

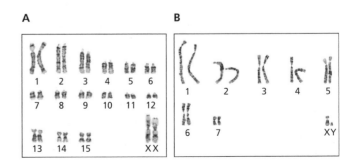

A   B

Figure 3.9 G-banded karyotypes of marsupials.
(A) Rufous bettong (*Aepyprymnus rufescens*); (B) tammar wallaby (*Macropus eugenii*). The chromosomes are arranged in homologous pairs and numbered sequentially from largest to smallest.

recent studies use biotin-labeled probes that are visualized with fluorescently labeled antibodies (*fluorescence* in situ hybridization, or FISH). Digital imaging of FISH preparations involving multiple probes is referred to as chromosome painting and can yield precise physical maps for several genes on a single chromosome spread (Sessions 1996). Such studies have been extremely useful for establishing patterns of **synteny**, the localization of homologous genes on the same chromosome in different species, often among very distantly related taxa (Yang et al. 2003).

Protein Electrophoresis

Evolutionary geneticists have employed several methods for studying variation at the protein level, including comparative immunology (Maxson and Maxson 1990) and direct amino acid sequencing (Goodman 1978). By far the most frequently used approach, however, has been **protein** (allozyme) **electrophoresis** (Murphy et al. 1996). Allozymes are alleles of enzymes, and electrophoresis is a method used to separate macromolecules based on their charge and mobility through a porous medium (e.g., a starch gel). Distinct protein alleles have, by definition, different amino acid sequences, and these in turn may result in different chemical properties (e.g., charge, size, shape) for the alleles. When placed in a slab of starch or cellulose acetate that is immersed in an aqueous buffer and subjected to an electric current, allozyme molecules will move through the medium at rates determined by their physical properties. After the gel is run for a specified period, different allozymes will be physically separated, occurring at different locations within the gel. Enzymes catalyze specific biochemical reactions, and these reactions can be coupled to specific staining reactions to reveal the precise location of allozymes on a gel. Such histochemical staining allows different alleles at a particular enzyme locus to be identified by their mobility (figure 3.10), and the organism from which those alleles came to be characterized as a homozygote (two copies of the same allele) or a heterozygote (two alleles).

By staining thin slices of a starch gel on which all the proteins from a tissue extract have been separated, or by using manufactured cellulose acetate strips, a large number of loci can be visualized from a single tissue sample (Harris and Hopkinson 1976). Most gels are wide enough to accommodate extracts from 20 or more individuals, and replicate staining of multiple gels can accommodate even larger sample sizes. These sampling considerations—scoring many loci from many individuals—make allozyme electrophoresis an excellent and cost-effective method for estimating allele frequencies from populations of organisms. Allele frequencies, in turn, are the starting point for analyses of genetic variation within and among populations (see Intraspecific Variation below).

Prior to the development of DNA assays, allozyme electrophoresis was the method of choice for a wide range of genetic problems in mammalogy—from parentage-dispersal studies (Patton and Feder 1981) to phylogenetic

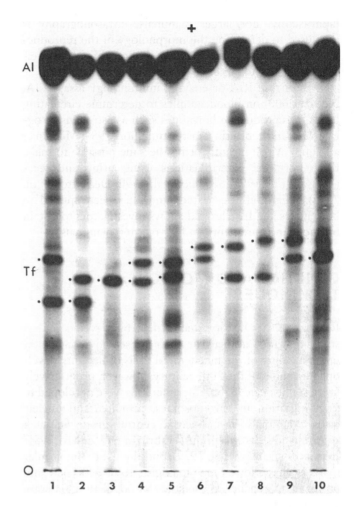

Figure 3.10 Protein electrophoresis. Evidence from protein electrophoresis of albumin (Al) and transferrin (Tf) in white-footed mice. These are two of the most common types of proteins found in mammalian blood. Note the variation in Tf alleles among individuals 1–10. Individuals 3 and 10 are homozygotes; the others are heterozygotes.

reconstruction (Baverstock et al. 1982). In most of these areas, allozymes have now been abandoned in favor of DNA markers. This derives in large part from the increased information-content of DNA data, especially the ability to quantify differences among alleles and the generally higher level of variation among organisms at the DNA level. However, allozyme electrophoresis continues to be a valuable tool in conservation genetics (Tiedemann et al. 1996) and systematic analyses of closely related species (Macholán et al. 2001).

DNA Sequences

Two indirect measures of DNA sequence variation led to major advances in mammalian systematics and population biology during the 1980s and 1990s, but they, like allozymes, have lately given way to direct amplification and sequencing methods. Whole-genome DNA hybridization (Springer and Krajewski 1989; Werman et al. 1996) was used primarily for phylogeny reconstruction in marsupials

(Kirsch et al. 1997), rodents (Catzeflis et al. 1993), bats (Kirsch et al. 1995), and primates (Caccone and Powell 1989). Restriction-site analysis (Dowling et al. 1996) led to the development of DNA fingerprinting and intraspecific phylogeography and has been applied widely in studies of parentage, kinship, speciation, hybridization, conservation genetics, and phylogeny (Avise 2003, 2004). Both methods helped lay the foundation for many hypotheses about mammalian evolution that are currently being investigated with DNA sequence and microsatellite data. Some of these are described in later chapters.

Development of the **polymerase chain reaction** (**PCR**; Mullis and Faloona 1987) brought DNA sequence and fragment analysis within the technological reach of most evolutionary geneticists. Prior to this, in vivo cloning procedures were required to isolate particular DNA fragments and manufacture sufficient copies of them for visualization or sequencing. In PCR, target sequences are amplified in vitro with a thermostable DNA polymerase that extends short, oligonucleotide primers that have annealed to their homologues at sites flanking the target region (hence, primers are used in pairs, with each member corresponding to one of the two complementary DNA strands of the target sequence). Initial PCR mixtures contain a small amount of genomic DNA, many copies of each primer, free nucleotides, and DNA polymerase in an aqueous solution with appropriate pH and Mg^{+2} concentration. Typically, PCR targets a specific gene or region for amplification; thus, primer design requires that the sequence of short regions ("primer sites") flanking the target be known in advance. Primers are usually 15–25 nucleotides in length, just long enough to ensure specific and stable annealing to the primer sites. After an initial denaturation of the genomic DNA, PCR proceeds through multiple cycles of primer annealing, extension, and denaturation carried out in an automated thermocycler. During each cycle, the polymerase adds nucleotides to the free sugar (3′) ends of the primers using the complementary strand of target DNA as a template (figure 3.11). At the completion of a cycle, the region of DNA between primers has been copied and is available to serve as a template in the next cycle. By repeating this process, the number of copies of the target sequence increases geometrically—30 cycles is sufficient to produce a billion copies of a single template molecule. This vast excess of amplified DNA fragments can be separated from the original genomic DNA, visualized by gel electrophoresis, and (if necessary) purified for subsequent sequencing (Palumbi 1996). The sensitivity of PCR has made it possible to recover and analyze DNA sequences from the remains of long-dead, or even extinct, mammals (Noonan et al. 2005; Poinar et al. 2006).

Most DNA studies in mammalogy now focus on either **microsatellite markers** or DNA sequences. The former are derived from genomic loci showing tandem (back-to-back) repeats of 2–4 base pair (bp) motifs spanning less than about 150 bp (Krane and Raymer 2003). Because these loci are highly prone to slip-strand mispairing dur-

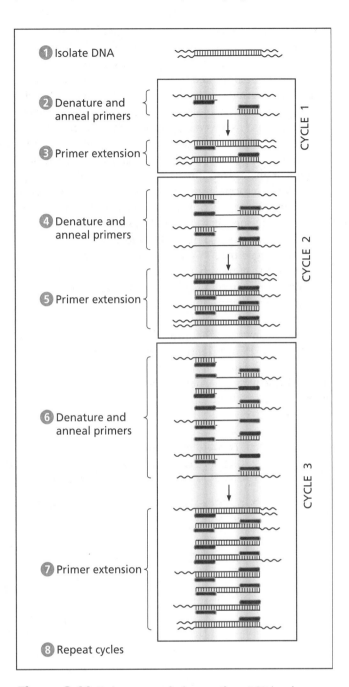

Figure 3.11 Polymerase chain reaction. PCR involves multiple cycles of denaturation, primer annealing, and primer extension, resulting in a geometric increase in the number of copies of a target sequence. *Adapted from Oste (1988).*

ing DNA replication (Graur and Li 2000), they have a high mutation rate that generates considerable variation in allele size (i.e., number of repeat units) among individuals. Microsatellite markers tend to be most useful for population-level problems such as estimating gene flow (Goldstein and Pollock 1997). DNA sequences, on the other hand, have been used at all levels of the taxonomic hierarchy, though they have been most widely applied to phylogeographic and phylogenetic questions.

In principle, assays of microsatellite variation are straightforward. Given primers that amplify a locus of interest, the assay consists simply of high-resolution gel electrophoresis to determine the allele size(s) found in a particular individual. This is accomplished by comparing the mobility of amplified fragments with DNA sequences of known length ("size ladders"), on the principle that the rate at which DNA molecules migrate through an electrophoretic gel is inversely proportional to their length (figure 3.12). Once the genotypes of all individuals in a sample have been scored, the size- and frequency-distributions of alleles lead directly into analyses of parentage, gene flow, or population history. In practice, however, there are some technical hurdles. For one thing, microsatellite loci that are highly variable in one species may be much less so in other, closely related species ("ascertainment bias"). This means that useful loci, and the primers that amplify them, often need to be identified de novo for different species, a process that requires preparation and screening of genomic libraries (Dowling et al. 1996). PCR amplification of microsatellite loci also entails idiosyncrasies that can affect interpretation of data. For example, PCR is inefficient for alleles with very large or very small repeat numbers; when such alleles fail to appear on scoring gels, heterozygotes may be mistaken for homozygotes ("allelic dropout"). Alternatively, the PCR

itself may be subject to strand slippage and produce amplification products that contain more or fewer repeat units than the template, which appear as multiple "stutter" bands on a scoring gel. Investigators now regularly monitor their data for such phenomena, and the value of microsatellites as highly variable markers has far outweighed these complications.

Currently, the most commonly used DNA sequencing method is derived from that developed by Sanger et al. (1977). In this approach, an oligonucleotide primer is annealed to purified, denatured template DNA (e.g., that obtained from a PCR reaction) and incubated in a solution with DNA polymerase, free nucleotides (dNTPs), and fluorescently labeled dideoxy nucleotides (ddNTPs). As in PCR, the polymerase extends the primer at its 3′ end by incorporating free nucleotides and forming a sequence complementary to that of the template (Hillis et al. 1996a). However, when a ddNTP is added to a growing chain, elongation is terminated because the ddNTP lacks the 3′ hydroxyl necessary for addition of another nucleotide. By adjusting the relative amounts of dNTPs and ddNTPs, sequencing reactions produce an array of DNA fragments such that every nucleotide position in the template sequence is represented by a fragment that

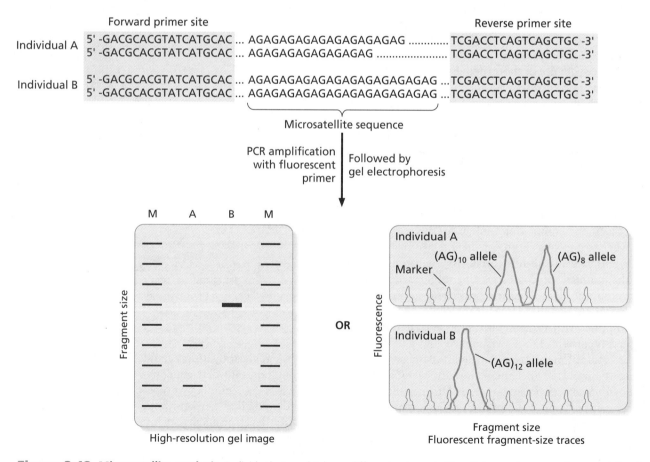

Figure 3:12 Microsatellite analysis. Individuals A and B have different microsatellite alleles as shown at the top: A is a heterozygote with one $(AG)_{10}$ and one $(AG)_8$ allele; B is a homozygote with two $(AG)_{12}$ alleles. Dots indicate DNA sequence between PCR primer sites and the microsatellite locus. After amplification, the genotypes of A and B can be visualized as lanes in an electropherogram or as separate fluorescent traces from an automated fragment analyzer. "M" indicates a series of DNA size standards in 1-base increments.

terminates at that position. Thus, for example, sequencing a 100-base-pair template would produce fragments of length 1, 2, . . . , 100 (ignoring the primer length). When electrophoresed through a high-resolution gel, these fragments separate in 1-base increments and the sequence of terminal ddNTPs in the fragments corresponds to the sequence of the template. Because each ddNTP carries a different fluorescent label, the position of each fragment in the gel and the identity of its terminal base can be detected by fluorescence imaging.

Technological advances have greatly improved the efficiency of Sanger sequencing. For example, **cycle sequencing** combines PCR with dideoxy chain-termination chemistry to allow sequencing templates in very small amounts. In this approach, a thermostable DNA polymerase is used to incorporate dNTPs and ddNTPs into growing chains, but the reaction is repeated many times in a thermocycler to produce a linear amplification of DNA fragments, which can then be separated by electrophoresis. Most laboratories now employ *automated* cycle sequencing with *capillary* electrophoresis, in which separation, visualization, and electronic documentation of sequencing products takes place within a single, compact instrument (figure 3.13). Tubes containing extension fragments from cycle-sequencing reactions are placed in a robotic device that successively loads each sample into one end of a capillary tube filled with a separation polymer. Electrophoresis takes place in the capillary; as fragments pass a fixed point, their fluorescent labels are excited by a laser, and the distinct color emitted by each ddNTP is detected by a charge-coupled camera and stored as a digital record. Computer software integrates these records and produces a "trace" of each reaction, from which a base-calling algorithm reconstructs the sequence of nucleotides in the original template. It is now common to process 96 or more samples in a single "run" of a capillary sequencer, greatly increasing the throughput of samples over early methods, and research into even more cost-effective methods for DNA sequencing continues at a rapid pace (Service 2006).

The most widely studied sequence in mammalogy is that of the **mitochondrial DNA (mtDNA)** molecule. Mitochondria, organelles in which the reactions of cellular respiration take place, are the result of an ancient symbiosis between an aerobic prokaryote and the evolving eukaryotic cell (Lewin 2006). Each mitochondrion includes one or more copies of a circular DNA molecule, a remnant of the ancestral symbiont's chromosome, consisting of approximately 16,000 bases. The gene-content of this molecule is highly conserved among mammals (figure 3.14), consisting of 22 transfer RNA, 2 ribosomal RNA, and 13 protein-coding genes, as well as a noncoding "control" region. With minor exceptions, the arrangement of these loci on the mtDNA molecule is also conserved. Several properties of mtDNA make it attractive to mammalian evolutionary geneticists. First, it is a haploid, homoplasmic, nonrecombining marker—that is, the vast majority of mtDNA molecules within an organism have identical sequences (Birky 1991). Second, it has a higher average rate of sequence

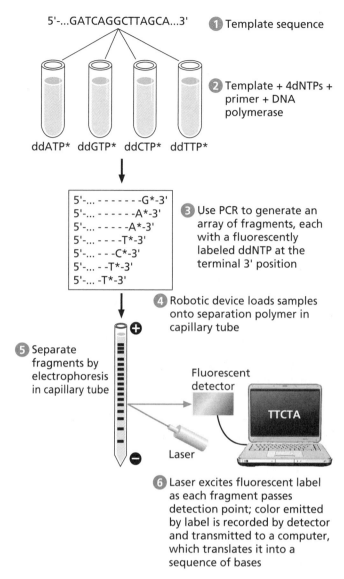

Figure 3.13 Automated DNA sequencing. An amplified template sequence is subjected to cycle-sequencing reactions that include free nucleotides (dNTPs), one primer, DNA polymerase, and one of four chain-terminating dideoxy nucleotides (ddNTP). Each ddNTP has a different fluorescent label (asterisks). Resulting DNA fragments are separated by size with capillary electrophoresis. The sequence of fluorescent signals emitted by each fragment as it passes the detector is assembled into a DNA sequence by the computer. *Adapted from Atherly et al. (1999) and Hartwell et al. (2004).*

evolution than nuclear DNA, probably due to lack of a DNA repair mechanism. Even so, mtDNA shows a considerable range of rate variation among its loci, from very rapidly evolving portions of the control region to some of the more conserved protein-coding genes (e.g., cytochrome oxidase subunit II). Third, it is transmitted to offspring through the egg cytoplasm (maternal inheritance), making it a marker that tracks the history of maternal lineages. Collectively, these properties make mtDNA relatively easy to isolate experimentally and ensure that variation among mtDNA sequences (**haplotypes**) can be used to address genetic problems at levels from intraspecific phylogeography to intraordinal phylogeny (Avise 2004).

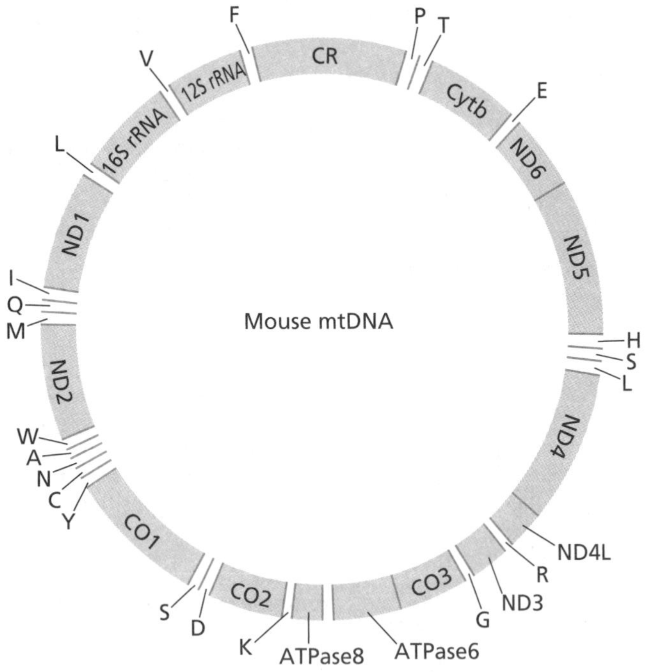

Figure 3.14 The mouse mitochondrial genome.
The mouse mtDNA molecule is just over 16,000 bases long. Protein-coding, ribosomal RNA (rRNA), and noncoding (CR) regions are labeled inside the circle; transfer RNA (tRNA) genes are indicated outside the circle by the one-letter abbreviations of their corresponding amino acids. *Adapted from Bibb et al. (1981).*

Although mtDNA sequences have had a major impact in studies of mammalian evolution, they are not suitable for all research questions. In particular, molecular systematists concerned with relationships among major mammalian groups (e.g., orders), for which common ancestors existed in the Cretaceous or early Tertiary, have turned to more slowly evolving genes in the nucleus. In part as a result of their lower average rates of change, nuclear coding sequences are less prone than mtDNA to experiencing multiple substitutions at the same sites over long periods of evolutionary time (Springer et al. 2001). As such, they preserve a phylogenetic "signal" that can be used to relate very divergent groups of mammals. In addition, the nuclear genome (about 3 billion bases in a haploid human cell) is vastly more diverse in structure and properties than mtDNA. It contains sequences that evolve very rapidly (e.g., introns, pseudogenes, intergenic regions), at a range of intermediate rates (e.g., most exons), or very slowly (e.g., conserved exons, small subunit ribosomal RNAs); each group includes markers with levels of variation appropriate for specific evolutionary questions. For example, multiple nuclear sequences have been used to test phylogenetic hypotheses derived from mtDNA, an application that is especially important given that mtDNA loci are permanently linked and so do not provide independent estimates of relationships (Krajewski et al. 2004). Perhaps the most exciting recent development in the study of nuclear DNA is the availability of complete genome sequences from several species of mammals. As of 2006, genomic sequences

have been completed for humans (Lander et al. 2001; Venter et al. 2001) and the house mouse (*Mus musculus*; Mouse Genome Sequencing Consortium 2002); those of another 22 species are in the assembly stage (i.e., the sequences of fragments are being connected to reconstruct chromosome—length, contiguous sequences), and those of 23 other species are in progress (National Center for Biotechnology Information, *www.ncbi.nlm.nih.gov/genomes*). Although most studies involving mammalian genomic data sets have focused on fundamental questions in genetics, a few researchers have begun to assess the genomic changes that underlie the evolutionary divergence of species (Prabhakar et al. 2006).

Analyses of Museum Specimens

Systematic Collections

In 1994, a group of systematists organized by three professional societies produced a report entitled *Systematics Agenda 2000: Charting the Biosphere*, a declaration of priorities for dealing with the worldwide loss of biodiversity. Among the major components of their action plan was the goal of enhancing systematic collections, which was justified as follows (p. 16): "Collections contain the primary evidence for the existence of species, document their presence at particular sites, and serve as the ultimate standards for comparisons and identifications of species." The central role of carefully prepared specimens, representing the preserved remains of organisms in well-organized collections, has been characteristic of biological taxonomy for centuries and remains a critical part of systematic mammalogy.

The phrase "museum specimen" refers not only to specimens housed in institutions called museums (see chapters 1 and 2), but also to those in university departments, government agencies, and private holdings. In mammalogy, such specimens are of several major types (Hall 1962). A **study skin** is the preserved integument (pelage, epidermis, and superficial dermis) of a mammal, from which muscles, internal organs, and most bones have been removed (figure 3.15). For small mammals, study skins are usually stuffed with cotton after chemical treatment, then dried in a flattened, linear pose. Larger skins may be dessicated or tanned, then stored as hides. Skulls and skeletons are dried bones recovered from an animal that has been skinned and defleshed. **Spirit** (fluid) **specimens** are carcasses, or portions thereof, preserved in alcohol. Many spirit specimens are initially fixed in formalin, a chemical that induces cross-linking of proteins and thereby immobilizes the cell contents of tissues. Fossils are the remains of dead organisms, usually preserved in rock. Tissue samples are pieces of flesh, aliquots of blood, or hairs (Milius 2002) taken from live or freshly killed animals, then frozen or immersed in a solution that prevents degradation of protein and DNA molecules. In recent years, mammalogists have begun to extract, purify, and preserve (usually frozen) DNA samples

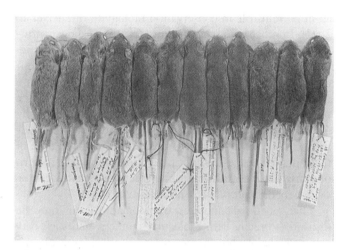

Figure 3.15 Study skins. This series of study skins from white-footed mice could be used to examine differences in pelage characteristics, which might vary with habitat or locality.

from animal tissues (Dessauer et al. 1996). However, a tissue or DNA sample is of limited utility unless it is accompanied by a **voucher specimen**—a skin, skull, or other whole-organism remnant that serves to identify the source of the sample.

Curation is the task of maintaining a collection of museum specimens, including their initial *preparation*, safe storage, organization, and accessibility to researchers. The value of museum specimens comes not only from the anatomical structures and taxonomic history that they preserve, but also from the documentation they provide that a particular species occurred in a particular place at a particular time. Thus, one of the most important jobs of a curator is to maintain an accurate, complete, and accessible specimen database. Typically, each specimen is labeled with unique alphanumeric code that ties it to a record that includes the name of the collector, date of collection, collection locality (often with habitat information), and perhaps other data such as field measurements or behavioral observations. Modern collections make efficient use of computer database technology, in some cases making complete lists of specimens available on the World Wide Web.

Specimen-Based Studies

Museum collections represent the archived results of field collections. Mammals taken from the wild and preserved as museum specimens are the primary documentation of mammalian biodiversity in specific areas of the world. Regional surveys of mammal diversity are based on information derived primarily from these collections, either through an author's original observations or previously published records. Flannery (1995), for example, documented the occurrence of nearly 190 living mammal species indigenous to New Guinea. The record of museum specimens allowed Flannery to describe key aspects of the biology of each species, including its distribution, alti-

tudinal range, habitat association, taxonomic history, and measurements indicative of body size. Faunal surveys of this sort comprise another major goal set forth in the *Systematics Agenda 2000* document.

Museum specimens are a primary source of information on the morphology of mammals, and they have been particularly valuable for systematic studies involving comparative anatomy. Descriptions of new mammal species (see below) typically rely on painstaking comparison of all skins and skulls from a particular taxon available in museum collections, comparisons that may include the postcranial skeleton and soft tissues as well. Craniodental characters scored from skulls have been used extensively for phylogenetic analysis of living and fossil mammals, the latter frequently represented *only* by their teeth (see chapter 4). Skins and spirit specimens can be examined for parasites, pathogens, environmental pollutants, or other trace chemicals.

Skins, skulls, and skeletons are also the basis of **morphometric** studies—studies that analyze the *shape* of anatomical structures (Lestrel 2000). Morphometrics has been an invaluable tool in research on mammalian developmental anatomy (Lucio et al. 2003), functional morphology (Couette et al. 2005; Shapiro et al. 2005), shape evolution (Macholán 2006), and sexual dimorphism (Motokawa et al. 2003). One of the most widespread applications of morphometrics in mammalogy is the study of intraspecific variation (see below), particularly that associated with subspecies designations. For example, Wehausen and Ramey (2000) used multivariate statistical analysis of 17 skull measurements taken from 694 bighorn sheep (*Ovis canadensis*) housed in 36 different collections to evaluate the distinctness of four recognized subspecies. Their findings led them to recommend that one of the subspecies (*O. c. auduboni*) be merged with another (*O. c. canadensis*), and clarified the geographic ranges of the remaining three.

Museum collections have become the largest repository of tissue and DNA samples for modern molecular analyses of mammals, including molecular systematics, molecular ecology, conservation genetics, and forensics (see Systematic Methods below). These samples, tied to voucher specimens and preserved over long periods of time, will become increasingly significant as more species of wild mammals decline and access to them becomes restricted by conservation management. Moreover, it is now a common practice to recover DNA from bits of study skins (Thomas et al. 1989), dried flesh adhering to bones (Krajewski et al. 1997), or even the bones themselves (Hadly et al. 2003). In some cases, DNA has been extracted from spirit specimens fixed in formalin (Westerman et al. 1999). Thus, specimens prepared long before the development of modern biotechnology represent an immense archive of genetic data. Because recovering DNA from such specimens entails the permanent loss of some preserved tissue, however, most museums have formulated explicit policies for **destructive sampling** that weigh this loss against the potential gain of information from genetic analysis.

Systematic Methods

INTRASPECIFIC VARIATION

Taxonomists and evolutionary biologists share a preoccupation with the variation (**polymorphism**) found in natural populations. Indeed, a major component of the 20th century "Modern Synthesis" in evolutionary biology was documentation that patterns of phenotypic variation in wild organisms are consistent with the predictions of evolutionary theory (Dobzhansky 1937; Mayr 1942, 1963). Evolutionary interpretation of such variation in turn provided a conceptual basis for modern taxonomy (Simpson 1961; Mayr 1969). Methods for measuring genetic variation became available in the 1960s (Harris 1966), and the data they provided immediately reinvigorated discussion about the roles of mutation, natural selection, and genetic drift in determining levels of polymorphism (Lewontin 1974; Kimura 1983). Moreover, the development of molecular population genetics provided new technological and analytical tools with which to study phenomena like inbreeding, migration, population subdivision, and effective population size. Thus, interpreting patterns of intraspecific variation is a central theme in modern mammalogy.

The study of phenotypic variation in mammals emphasizes interindividual, temporal, and spatial dimensions. An especially significant form of interindividual variation is **sexual dimorphism**, the possession of different phenotypic characteristics by males and females of the same species. This phenomenon is most common in large mammals and the most frequently studied morphological features include body size (males are often larger than females) and ornamental, secondary sex characteristics (males tend to have them and females do not). Differentiation between the sexes is thought to result from sexual selection, adaptation to different parental roles, or intersexual competition for food (Derocher et al. 2005). Although most studies have focused on dimorphic traits of the phenotype, some researchers have begun to explore the underlying genetic bases for anatomical, physiological, and behavioral differences between the sexes (Rinn and Snyder 2005). Research on temporal variation is dominated by studies of growth, particularly as it relates to postnatal development and adaptation to variable food resources (Knott et al. 2005).

Analyses of spatial variation in mammals focus on the extent to which geographically separated populations are taxonomically, genetically, or adaptationally distinct. It is quite common for mammals of the same species to show different characteristics in different portions of their range. When geographically disjunct populations display diagnosably different traits, they may be recognized taxonomically as **subspecies**. Subspecies definitions are based on discrete morphological (Sutton and Patterson 2000), morphometric (Wan et al. 2005), behavioral (Pluháček et al. 2006), or genetic (Elrod et al. 2000) differences, and formal recognition of subspecies as units of biodiversity continues to play an important role in conservation biology (González et al. 2002). In other cases, one or more traits may show continuous variation along some geographic axis, forming a **cline**. Clines may be the result of secondary contact between populations that diverged in isolation, primary contact with a gradient of selection pressures among populations along a transect, selection against hybrid individuals from partially isolated populations, or mixing ("diffusion") of individuals from two currently isolated populations. Although each of these models makes distinct predictions about the pattern of clinal variation, distinguishing them in practice can be extremely difficult (Owen and Baker 2001).

The genetic distinctness of populations can now be measured directly with molecular markers, and such data may provide insights on current or historical levels of gene flow among populations. Allozymes, mtDNA, and microsatellite loci have been extensively applied to such problems. Because it is often difficult to measure migration directly (e.g., by tagging individuals), geneticists use the frequencies and divergence of alleles to quantify levels of population subdivision and from these make inferences about gene flow. The most commonly used measures of population subdivision are Wright's (1951) F statistics, correlation coefficients that partition allele-frequency variation among three levels (total population, subdivisions, and individuals within subdivisions) in a hierarchically structured population. One of these, F_{ST}, is interpreted as the genetic differentiation of subpopulations, with $F_{ST} = 0$ indicating panmixia and $F_{ST} = 1$ indicating complete isolation (Hedrick 2000). In practice, values of F_{ST} are usually intermediate. Estimates of effective population size or effective number of migrants are obtained by relating an observed F_{ST} to its expectation under specific models of population structure, such as Wright's (1940) "island model." Conceptually similar approaches have been formulated for DNA sequence (Lynch and Crease 1990) and microsatellite (Feldman et al. 1999) variation. In addition to allele frequencies, these data types include measures of sequence or repeat-length differences between pairs of alleles. Partitioning sequence and microsatellite variation among levels of a hierarchically sampled geographic region has been termed "analysis of molecular variance" or AMOVA (Excoffier et al. 1992), and this method is now widely used in mammalogy to assess whether populations are genetically distinct (Ducroz et al. 2005). Moreover, a plot of the frequency of haplotype pairs showing increasing numbers of sequence differences—a mismatch distribution (Rogers and Harpending 1992)—can be used to infer whether individual populations have undergone expansions or bottlenecks in the recent past (Westlake and O'Corry-Crowe 2002).

INTRASPECIFIC PHYLOGEOGRAPHY

Avise et al. (1987) coined the phrase "intraspecific **phylogeography**" to denote study of the geographical distribution of genealogical lineages within species. Most animal phylogeography studies employ mtDNA because the unique properties of this marker make it ideal for identifying genealogical lineages. Specifically, animal mtDNA is maternally inherited, nonrecombining, and rapidly evolving; thus, the mutation history of mtDNA haplotypes corresponds to the genealogical relationship of the maternal lineages within a species through which those haplotypes are transmitted. Methods of phylogenetic reconstruction (discussed below) can be applied to haplotypes associated with animals from specific localities, and the "gene tree" obtained then related back to the geographic origin of the animals. For example, Firestone et al. (1999) obtained control region sequences from 20 tiger quolls (*Dasyurus maculatus*, a dasyurid marsupial), sampled across the species range in eastern Australia from northern Queensland to Tasmania (figure 3.16). Phylogenetic analysis of these sequences revealed that haplotypes from the mainland and Tasmania formed separate genealogical groups corresponding to the separation of these areas by Bass Strait in the Pleistocene (12,000–15,000 years ago). Interestingly, the phylogeographic groups do not correspond to morphologically defined subspecies, which distinguish populations in northern Queensland from those in southeastern Australia.

Although mtDNA is still the most widely used marker for studies of mammalian phylogeography, many recent investigations have also utilized nuclear DNA. Firestone et al. (1999) supplemented mtDNA sequences for tiger quolls with six microsatellite loci. The latter showed the major divergence of Tasmanian and mainland population found with mtDNA but also detected significant differentiation between Queensland and southeastern mainland populations. Sipe and Browne (2004) combined mtDNA sequences with amplified fragment length polymorphism (AFLP) data in a study of shrew phylogeography in the southern Appalachian Mountains. AFLPs are variably sized DNA fragments amplified by PCR from genomic DNA that has been digested with restriction endonucleases. PCR primers corresponding to recognition sequences of the endonucleases amplify anonymous (mostly nuclear) sequences that show size variation due to the evolutionary gain or loss of restriction sites. Sipe and Browne (2004) demonstrated that Appalachian masked shrew (*Sorex cinereus*) populations exhibited little phylogeographic structure, whereas smoky shrews (*S. fumeus*) were highly differentiated, perhaps as a result of their lower population densities in the region. Brändli et al. (2005) examined sequences from mtDNA, X-chromosome, and Y-chromosome loci to study the history of greater white-toothed shrews (*Crocidura russula*) in northern Africa and Europe. Although most of the sex-chromosome sequences were noncoding, they showed considerably less variation than those from mtDNA. Nevertheless, the combination of loci clearly separated Moroccan and western European shrews from those in Tunisia and Sardinia, lending support to the hypothesis that these shrews colonized mainland Europe from northwestern Africa. Blacket et al. (2006) complemented mtDNA sequences with those from an intron of the ω-globin gene to show that common dunnarts (*Sminthopsis murina*) are phylogeographically distinct on the east and west sides of the Great Dividing Range in Australia. These examples illustrate that molecular analyses of population structure and history are becoming increasingly multifaceted, incorporating both phylogeographic and population-genetic approaches, and producing hypotheses that are much more detailed than those from single genetic markers.

SPECIES BOUNDARIES

The recognition and naming of species has been a cornerstone of taxonomy for centuries. Species are the units of

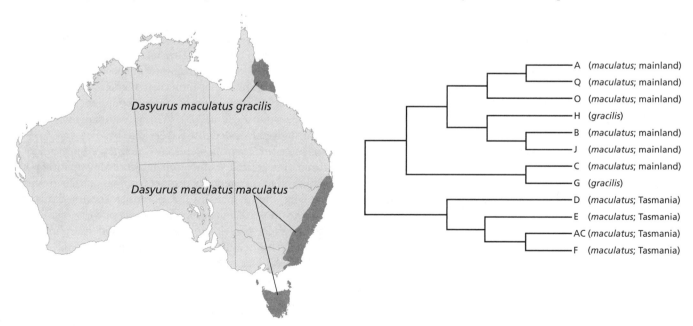

Figure 3.16 Intraspecific phylogeography. At left, the range map of the spotted quoll (*Dasyurus maculatus*) shows that subspecies *D. m. gracilis* and *D. m. maculata* occur in isolated areas of northeastern and southeastern Australia, respectively. At right, the relationships of mtDNA haplotypes (letters at ends of branches) suggest that the deepest genealogical division within *D. maculatus* is between Tasmanian and mainland quolls, not between the recognized subspecies. *Adapted from Firestone et al. (1999).*

biodiversity, the outer limits of population genetics, the constituents of higher taxa, and the entities that are related by a phylogeny. The processes by which new species arise from preexisting species (**speciation**) has been the subject of intense theoretical and empirical research since the time of Darwin (Coyne and Orr 2004). Ironically, even as our understanding of speciation mechanisms increases, biologists remain sharply divided on the question of what constitutes a species (Baker and Bradley 2006). The dispute centers on the value of different criteria for defining the biological term *species* versus those used for recognizing species in nature. The extremes of this argument were apparent in two early, alternative species concepts. Simpson (1951) defined an *evolutionary species* as "a lineage (an ancestral-descendant sequence of populations) evolving separately from others and with its own unitary evolutionary role and tendencies." This definition is theoretically well motivated, but entirely nonoperational. In contrast, the traditional *morphospecies* as described by Cain (1954) is operational, in that species are delimited by diagnostic morphological traits, but has no connection with evolutionary theory. Species concepts currently in use have more common ground with theory and practice, but they often conflict when applied to populations at intermediate stages of speciation. The impossibility of predicting the outcome of incipient speciation led O'Hara (1993) to suggest that no solution to "the species problem" would be forthcoming. Nevertheless, the science of recognizing species continues and is an integral part of systematic mammalogy.

The **biological species concept (BSC)** of Mayr (1942) has been very popular in mammalogy. In this concept, populations are considered distinct species when they are *reproductively isolated* from one another. The BSC is easiest to apply when two or more populations occur in the same area but do not hybridize. It is more problematic when limited hybridization takes place between populations, but it is entirely nonoperational when populations are geographically isolated. In applying the BSC, few studies directly assess the extent of reproductive isolation. Rather, the maintenance of diagnostic phenotypes in different populations is taken as evidence of a barrier to gene flow (i.e., a species boundary). Given the difficulties inherent in applying the BSC, many taxonomists have recently advocated the **phylogenetic species concept (PSC)**: "a species is the smallest diagnosable cluster of individual organisms within which there is a parental pattern of ancestry and descent" (Cracraft 1983:170). The PSC emphasizes diagnosability and common ancestry as criteria for establishing a species boundary; the former has been standard practice in systematics for centuries, but the latter relies on a phylogenetic interpretation of character variation. Whichever concept is employed—and there are many others (Coyne and Orr 2004)—the standard for describing new species of mammals has been, and continues to be, diagnosis. Can we identify characteristics of individual organisms that will allow us, with a high degree of confidence, to assign those organisms to one group or another?

Mammalogists adhere to the practice of designating **type specimens** as described by the International Commission of Zoological Nomenclature (1999). Most type specimens are preserved carcasses, skins, skulls, skeletons, or other anatomical remains in a museum collection. Under the rules of nomenclature, the type is strictly the "name-bearing" specimen and functions as a historical reference for the Latin binomial used to denote a particular species. The type concept, however, has engendered a distinct format for species description that is widely followed by mammalogists. Descriptions of new species are frequently embedded in a **taxonomic revision** of some larger group, such as a genus. In conducting a traditional revision, taxonomists examine as many specimens as possible from the group under review and identify features that diagnose each putative species. In addition, descriptive information is provided that will assist in the identification of future specimens. Thus, for example, when Woolley (2005) identified a new species of three-striped dasyure (*Myoictis leucura*), the formal description included a summary of the type material (i.e., relevant museum voucher-specimens), geographic distribution of the new species, diagnostic traits, description of other anatomical features, and a comparison of the new species to the three other recognized species of *Myoictis* (box 3.1) Such thorough and careful description helps ensure that species-level classification is as objective and stable as possible.

The widespread application of molecular methods in mammalian systematics has had a profound effect on the study of species boundaries during the past few decades. Molecular markers can be used to assess directly the level of gene flow among populations, information that is crucial to informed application of the BSC. Phylogeographic analysis of haplotypes from specific loci may reveal the extent to which members of different populations are genealogically distinct, providing critical insight for use of the PSC. Indeed, it is now quite common to find morphological species descriptions accompanied by phylogeographic analyses confirming the **reciprocal monophyly** of haplotypes from all species in the study group (i.e., all haplotypes recovered from one species are more closely related to one another than to those from any other species). Some taxonomists have argued that molecular markers are critical for thorough documentation of species boundaries (Avise and Ball 1990; Tautz et al. 2003). Within mammalogy, this view is strongly advocated by Baker and Bradley (2006:643), whose "genetic species concept" defines a species as "a group of genetically compatible interbreeding natural populations that is genetically distinct from other such groups." Applying this concept to a potential species boundary requires molecular markers, and its popularity testifies to the potential that integrated genetic and morphological data have for helping mammalogists "get over" the species problem (O'Hara 1993).

PHYLOGENY RECONSTRUCTION

A **phylogeny** is a speciation history for some group of species. Phylogenies are usually depicted as trees, on which current species are represented as terminal branches,

Description of a new species of dasyurid marsupial, *Myoictis leucura*, by Woolley (2005).

Myoictis leucura **n. sp.**

Type material. HOLOTYPE AM 17122. Skin and skull of adult male. Collected in 1985 by K. Aplin at Agofia, Mt Sisa (Haliago), Papua New Guinea, 06°17'S 142°45'E, 650 m. The tip of tail has been damaged in preparation of the skin (white portion reduced in length from 9 mm, when the specimen was first examined in spirit, to 5 mm on the prepared skin). PARATYPE AM 18091, adult female in alcohol, skull extracted, collected in 1985 by K. Aplin at Namosado, Mt Sisa (Haliago), 06°142'S 142°47' E, 750–1000 m.

Distribution. Southern side of the central mountain ranges in Papua New Guinea from Mt. Bosavi in the west to Mt. Victoria/Vanapa R. in the east. Altitude records range from 650 to 1600 m.

Diagnosis. *Myoictis leucura* differs from other species of *Myoictis* in having a white-tipped tail with long hairs on the top and sides of the tail, the hairs decreasing in length towards the tip.

Description. The external appearance of the holotype is similar to the specimen shown in Fig. 3c [of original] except that the portion of the tail that is white is shorter. The general coat colour is dark reddish brown above, with brighter, reddish hairs between the black dorsal stripes, and lighter below. The dorsal stripes extend from behind the ears to the rump, and the median stripe extends forward on the head. Red auricular patches are absent. The ears and feet are dark. The first interdigital and thenar footpads are generally not fused. Body dimensions can be found in Table 2 [of original]. The posterior palatal foramina are large, and P_3 is single rooted. Females have four nipples.

Comparison with other species. Differences between *Myoictis leucura* and other species are summarized in Table 2 [of original]. *Myoictis leucura* can be distinguished from *M. wallacei*, *M. wavicus*, and *M. melas* by the form of the tail. *Myoictis leucura* is larger than *M. wavicus* but similar in size to *M. wallacei* and *M. melas* with respect to mass, head-body length, foot length, basicranial length and length of the lower molar tooth row. The females of *M. leucura* (and *M. wavicus*) differ from *M. wallacei* and *M. melas* in having four rather than six nipples. *Myoictis leucura* can be distinguished from *M. melas* by the larger size of the posterior palatal foramina and by the presence of the third premolar tooth. *Myoictis leucura* can be distinguished from both *M. wallacei* and *M. melas* by the upper premolar tooth row gradient, and from *M. wallacei* in having a single rooted, as opposed to a double rooted, lower third premolar tooth. Differences in coat color between the species are described earlier.

Box 3.1 This text follows similar descriptions of three recognized species in the same genus and is accompanied by photographs of study skins, skulls, and teeth from each species. A table summarizes 11 key characters that serve to differentiate the four species.

ancestral species as internal branches, speciation events as nodes, and the common ancestor of all species under study as the root (figure 3.17). Every phylogeny implies a time dimension in which the tips of terminal branches are closest to the present and the root is furthest in the past. Understood in this way, a phylogenetic tree illustrates that the measure of evolutionary relationship for any two species is the time elapsed since their most recent common ancestor: If species A and B are more closely related to each other than either is to C, this means precisely that A and B diverged from a common ancestor more recently than did A and C, or B and C. A group of species that includes a common ancestor, all of its descendants, and nothing else is a **monophyletic group** or **clade**. Two monophyletic groups that are each other's closest relatives are called **sister groups**. Phylogenetics is the research program that seeks to reconstruct the "Tree of Life" (i.e., the phylogeny of all species), though individual phylogenetic studies are much smaller in scope. Mammalian phylogenetics has been contentious for most of its history (see chapter 4).

Phylogenetics became widely acknowledged as a distinct branch of systematics following the publication of Hennig (1966), though inferential methods of phylogeny reconstruction had begun to be developed even before Hennig's work (e.g., Edwards and Cavalli-Sforza 1963). To estimate a phylogeny, systematists rely on evidence

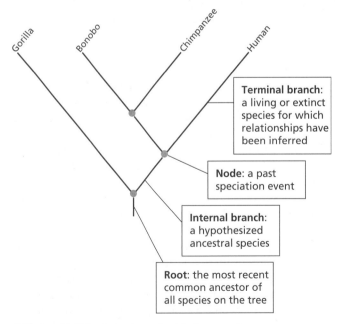

Figure 3.17 The parts of a phylogenetic tree.
Relationships among four species of hominoid primates are depicted as a tree, with the time dimension moving from bottom to top. The degree of phylogenetic relationship among any set of species is given by age of their most recent common ancestor. Monophyletic groups include an ancestral species, all of its descendants, and nothing else. On this tree, (bonobo + chimpanzee) and (bonobo + chimpanzee + human) are examples of monophyletic groups.

presented by the species whose relationships are in question. This evidence consists of **characters**—heritable features of organisms that vary among species and that, if interpreted in light of reasonable assumptions about evolution, reveal patterns of common ancestry. Most methodological progress in phylogenetics has involved the development of analytical tools to estimate relationships from patterns of character variation. This is now a vast and complex literature, of which we will discuss only the most fundamental aspects.

Characters come in many forms, but now the most frequently studied are anatomical and molecular. An anatomical character may be any feature of the body of an organism that is heritable, and descriptions of such characters are correspondingly diverse. Molecular characters are ultimately protein or DNA sequences, but variation in those sequences has been assayed in many ways: presence of particular allozymes or restriction sites, relative strength of antibody affinities for albumin proteins, thermal stability of artificially hybridized DNA strands from different species, or (most commonly now) direct comparison of amino acid or nucleotide sequences. It is significant that characters are *selected* for phylogenetic analysis by investigators, usually because the characters have desirable properties (e.g., heritability, variation) for the phylogenetic problem at hand; not all characters are equally useful. Variation is captured by descriptions of the *states* of a character in different species. For example, Wroe et al. (2000) found two states for the character "shape of upper incisors" in dasyuromorphian marsupials: peg-shaped and spatulate. Typically, each species possesses a single state of each character, but sometimes researchers include *polymorphic* characters (i.e., those that show more than one state in some species).

Perhaps the most critical consideration in selecting a character for phylogenetic analysis is that the character be **homologous** among the species studied. A homologous character is one that species share by inheritance from a common ancestor that also possessed it. For example, wings are homologous among bats because bats are descended from a common ancestor that had wings. Tusks are not homologous between elephants and walruses—elephant tusks are incisors, walrus tusks are canines, and the most recent common ancestor of elephants and walruses had neither type. Variation in the form of wings among bats might reveal something about their phylogeny, but the form of tusks in elephants and walruses has no bearing on their relationship to one another. Homology of molecular characters must be considered at two levels, the locus and the site. For example, mammalogists have made extensive use of cytochrome *b* gene sequences, but always on the assumption that species share cytochrome *b* genes because their ancestor had one. This is reasonable for loci that have not been *duplicated* in the genome. Thus, sequences of myoglobin (an oxygen-binding protein in muscle cells) should not be compared with sequences of hemoglobin (an oxygen-binding protein in blood cells) for reconstructing phylogeny: myoglobin and hemoglobin are related by a gene duplication that predates the diversification of mammals (Graur and Li 2000). Even when comparing homologous loci, the homology of individual sites is a crucial consideration. Is position 1143 in the cytochrome *b* gene of the Virginia opossum (*Didelphis virginiana*) homologous to position 1143 in the Norway rat (*Rattus norvegicus*)? When sequences differ in length due to insertions or deletions in their evolutionary history, site homology may be difficult to assess. Making this assessment is the goal of sequence **alignment** (Mindell 1991).

Unfortunately, it is often impossible to know that a particular character is homologous across all species in a group unless one already knows something about phylogeny. For example, Pettigrew (1986) argued that megachiropteran bats are more closely related to primates than to microchiropterans. If this hypothesis were true, wings would not be homologous across the two bat clades, instead having arisen by **convergent evolution** from separate ancestors. Thus, assumptions about character homology are really *hypotheses* that may be, to some extent, tested by phylogenetic analysis. Hennig (1966) insightfully referred to the relationship between hypotheses of homology and phylogeny as one of "reciprocal illumination." A mistaken hypothesis of homology, in which similar derived states are shared by species due to convergent evolution, is referred to as **homoplasy**.

A second critical feature of characters selected for phylogenetic analysis is that they be *independent* of one another, such that patterns of covariation among species can legitimately be attributed to coancestry rather than other causes. Characters may be correlated for several reasons. For example, "3rd premolar = present or absent" and "3rd premolar size = large or small" are *logically* correlated characters: premolars are necessarily present if they are also small. Some characters may be *functionally* correlated: gliding locomotion occurs in mammals, such as colugos and flying squirrels, that also have a patagium (a flap of skin that stretches from wrists to their ankles). Characters may be *structurally* correlated. Ribosomal RNA sequences maintain their secondary structures by complementary base pairing of nucleotides in stem regions: an "A" at one stem position necessitates a "T" at the complementary position elsewhere in the sequence (Wheeler and Honeycutt 1988). Characters may be *developmentally* correlated: levels of the cell signaling protein ectodysplasin during embryonic development affect the number and shape of cusps on mouse molars (Kangas et al. 2004). Although character correlation is seldom an all-or-nothing relationship, systematists try to select characters that are not tightly correlated.

In addition to the problem of selecting characters, systematists usually must make choices about which species to include in a phylogenetic study, a consideration called **taxon sampling**. Suppose we wish to study relationships among five species in a genus that is known to be monophyletic. In this situation, exhaustive sampling is feasible—we can include all species in our study. Exhaustive sampling may be infeasible for more diverse groups, particularly in molecular studies for which fresh tissue is required, and in these cases investigators employ **exemplars**—species that represent known or suspected clades within a larger taxon of interest. For example, a study of interordinal relationships among marsupials might use one or more species from each of the seven living, monophyletic orders. As with any sampling scheme, phylogenetic accuracy increases with denser taxon-sampling; the use of exemplars represents a trade-off between feasibility and thoroughness. A further consideration in taxon sampling is choice of **outgroups**, species that are clearly not part of the group under study (the "ingroup"), but are closely related enough that their position on an estimated phylogeny serves to locate the ingroup root or helps to infer the primitive states of ingroup characters (Maddison et al. 1984; Smith 1994).

Once appropriate sets of taxa and characters have been identified, and all states determined, the results are assembled into a **data matrix** in which species define rows, characters define columns, and states are the cell values. For molecular sequences, the data matrix is an alignment. From this point, phylogenetic analysis can take several different directions. The most fundamental distinction among methods currently in use is *model-based* versus *model-free* approaches. Model-based procedures are explicitly statistical and include additive-distance (Nei and Kumar 2000), maximum-likelihood (Swofford et al. 1996), and Bayesian methods (Holder and Lewis 2003). Each relies on a probabilistic model of evolution that specifies relative rates of change among the states of characters, and perhaps sets other constraints on character evolution. Model-based methods are most often applied to molecular sequences because alignments represent relatively large numbers of characters with regularities in their substitution patterns. As just one example, transition substitutions (A-G and C-T) occur at measurably higher rates than transversion substitutions (all other pairs) in animal mtDNA (Moritz et al. 1987); this pattern can be incorporated into a model that informs our choice of a phylogeny. Molecular systematists expend considerable effort identifying models that accurately and economically apply to particular data sets (Posada and Crandall 1998). The computational requirements of model-based analyses are considerable, but they are facilitated by software packages such as PHYLIP (Felsenstein 2006), PAUP* (Swofford 2002), and MRBAYES (Ronquist and Huelsenbeck 2003).

The most widely used model-free approach to phylogenetic inference is **parsimony**, the principle that the best estimate of phylogeny from a particular data matrix is that which requires the fewest character-state changes. Although its methodological underpinnings have been intensely debated (Farris 1983; Felsenstein 1983, 2004; Sober 1988), parsimony is the dominant analytical framework for morphological characters and is frequently employed with molecular data as well. Parsimony analysis of anatomical traits is often combined with *character analysis* of the sort championed by Hennig (1966). Typically, *uninformative* characters—those that require the same

number of state changes on all possible trees—are excluded from a data matrix. Investigators may use fossil, outgroup, or developmental data to decide which state of a character is phylogenetically primitive (ancestral) for the ingroup (i.e., they may *polarize* the states). In some cases, a stepwise series of evolutionary transformations between states can be inferred and the states *ordered* for phylogenetic analysis. Researchers may decide that some characters are less prone to homoplasy than others, and therefore apply *differential weights* across columns of the data matrix. Polarization, ordering, and weighting use a priori hypotheses about character evolution to constrain the search for a minimum-length tree. However, character hypotheses are often difficult to defend, and many investigators perform unpolarized, unordered, and uniformly weighted parsimony analyses. This is especially true, if not universally appropriate, for parsimony analyses of molecular sequence data (Cracraft and Helm-Bychowski 1991).

Obtaining an optimal phylogeny for a particular data matrix under any of the preceding criteria is not the end of phylogenetic inference. Investigators want to know how much better the best tree is than other trees, or how much *support* the data provide for specific groups on the optimal tree. There are several ways to make a statistical assessment of whether the difference in optimality scores between alternative trees is significant (Swofford et al. 1996). Such tree-comparison tests allow researchers to decide whether an a priori hypothesis of phylogeny is consistent with a particular data matrix, even if it does not match the optimal tree for that matrix. The most widely used method of assessing support for individual branches on an optimal tree is **bootstrapping** (Felsenstein 1985). Characters are resampled with replacement from the original data matrix to create many *pseudoreplicate* matrices, and an optimal tree is obtained for each. This process mimics the sampling of new characters that have the same properties as the ones actually sampled, and thus approximates the level of character-sampling error inherent in the original data. The frequency with which a particular clade on the optimal tree appears among pseudoreplicates in the bootstrap analysis is a measure of that clade's *resolution* by the original data. Clades that appear on an estimated tree, but are poorly resolved, must be viewed with caution.

CLASSIFICATION

Since the time of Linneaus, biologists have grouped species into progressively more inclusive categories to produce hierarchical classifications. Thus, Mammalia is one of several groups within Chordata, Artiodactyla is one of several groups within Mammalia, Bovidae is one of several groups within Artiodactyla, *Bos* is one of several groups within Bovidae, and *Bos taurus* (the cow) is one of several species within *Bos*. Although the concept of hierarchical classification is almost universally accepted by systematists, the criteria used to place groups within groups have been extremely controversial. Perhaps the strongest argument in Hennig's (1966) foundational work is that biological classifications should accurately reflect phylogeny. In other words, *only monophyletic groups should be recognized* (i.e., named). For most of the history of biological taxonomy, such a tenet was impractical because detailed phylogenies were unknown. "Evolutionary" classifications (Simpson 1961), developed after Darwin, emphasized groups delineated by shared anatomical adaptations, whether those adaptations involved primitive or derived characters (but to the exclusion of convergent characters). However, after robust methods of phylogenetic inference began to be developed in the 1960s, Hennig's call for a **phylogenetic classification** gained many adherents and is currently advocated by many, perhaps most, practicing taxonomists.

Unfortunately, most major taxonomic groups (including those of mammals) were named long before Hennig, and many such names have attained the status of tradition. Moreover, the difficulty in applying monophyly as a criterion for classification is that our understanding of phylogeny, though far more advanced than it was a century ago, is still undergoing revision. Sometimes traditional and phylogenetic groups coincide, as in the case of bats (Order Chiroptera), primates (Order Primates), or rabbits (Order Lagomorpha). In other cases, relationships have become sufficiently established that phylogenetic groups have supplanted or restricted traditional ones. Thus, mammalogists no longer use "Insectivora" to denote a group that includes tree shrews. The latter are recognized as a major clade (Order Scandentia) more closely related to primates than to ordinary shrews. When phylogenetic results conflict, classificatory issues become contentious. Such is the case for whales and dolphins versus even-toed ungulates: morphological evidence (Novacek 1992) favors treating these as sister groups (Orders Cetacea and Artiodactyla, respectively), whereas molecular data (Gatesy and O'Leary 2001) support inclusion of the former within the latter (giving a single order Cetartiodactyla). In any event, it appears that the stability of classification is now tied to progress in phylogenetic reconstruction: As the tree of life begins to take on a more definite shape, so too will the names applied to its branches.

ESTIMATING DIVERGENCE TIMES

Systematists have long been interested in knowing the absolute ages of phylogenetic events. For example, how long ago did monotremes branch off from the lineage leading to therian mammals? For much of the history of mammalogy, answering such questions has been the province of paleontology. Fossils provide direct evidence that members of specific groups existed at specific times in the past. Thus, fossil triconodonts, multituberculates, and holotherians from the Jurassic show that theriimorphan mammals were distinct from monotremes by approximately 175 million years ago (mya) (Pough et al. 2005). Two considerations are critical for interpreting fossil evidence

for divergence times (figure 3.18). First, the cladogenic event must be defined by phylogenetic analysis. The age of triconodont and multituberculate fossils would not help us date the origin of monotremes if we did not think that the former groups were theriimorphans, and that Theriimorpha is the sister group of the monotreme clade. Second, the age of the oldest fossils represent a *minimum* estimate of divergence time—the lineages may have diverged earlier, but they left no fossils or fossils that have yet to be discovered. Perhaps theriimorphans or monotremes from the early Jurassic will be found that push the minimum divergence of these groups further back in time.

A phylogeny gives the framework for interpreting first-occurrences of fossils. Given that monotremes and holotherians are sister groups, both lineages existed (by definition) after their separation from a common ancestor. Thus, the oldest fossils representing *either* lineage provide a minimum divergence date. In fact, fossil monotremes do not appear until the mid-Cretaceous some 100 mya, but the monotreme lineage must have been present at least since the mid-Jurassic; from the mid-Jurassic to the mid-Cretaceous, monotremes are a **ghost lineage** (Norrell 1992).

Fossil first-occurrences provide *point estimates* of divergence times that are likely to be underestimates due to the existence of lineages prior to the age of their oldest fossils. Paleontologists have attempted to obtain more meaningful *interval estimates* by modeling the fossil record as the outcome of a stochastic process such that the times (or stratigraphic levels) at which fossils occur are randomly distributed (Marshall 1990). This approach allows one to calculate a confidence interval for the stratigraphic range of a lineage. Depending on the stochastic model employed and the distribution of actual fossil horizons, a 95 or 99% confidence interval on a lineage's duration may indicate that the lineage could be much older than its oldest fossils. Prior to 1985, the monotreme fossil record consisted of three localities from the Pleistocene and mid-Miocene, but Marshall (1990) noted that an estimated 99% confidence interval on this range extended back to the late

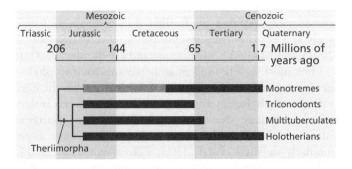

Figure 3.18 Phylogeny and fossil record of Crown Group Mammalia. Periods of time bracketed by fossils of each group are shaded. The gold portion of the monotreme branch indicates the period during which this group is a ghost lineage. Thin lines indicate phylogenetic relationships but not divergence times. *Adapted from Pough et al. (2005).*

Jurassic (figure 3.19). *Steropodon*, a fossil platypus from the Cretaceous of Australia discovered by Archer et al. (1985), fell squarely within this interval and extended the first-occurrence age of monotremes by over 80 million years. This example illustrates the pitfalls of a too literal reading of the fossil record in estimating divergence times.

The advent of molecular data in phylogenetics brought with it the possibility of estimating temporal divergence from sequence divergence. Zuckerkandl and Pauling (1965) first proposed the existence of a **molecular clock** (i.e., that the rate of sequence evolution is roughly the same for any given protein in all lineages). Many studies have evaluated the molecular clock for different kinds of molecular data, different loci, different groups of organisms, and different models of evolution (Avise 2004). The almost universal realization has been that there is considerable variation in the rate of sequence evolution among loci and among lineages, although there are examples of *local clocks*—loci that show roughly uniform rates of change within a particular group of closely related species. In these cases, researchers have applied molecular-clock dating to obtain estimates of divergence times.

If a local clock can be documented, we can calculate divergence times if we know the divergence rate. For example, if sequences diverge at 2% of their sites per million years (my), sequences that are 10% different must have separated from a common ancestor $10\%/(2\% \cdot \mathrm{my}^{-1}) = 5$ mya. Of course, the divergence rate is never known and must be estimated, usually by *calibration* of observed sequence differences against some independently estimated divergence date. Calibration points are most often obtained from the fossil record or past geographic events (e.g., continental separations) that can be stratigraphically or radiometrically dated. This simple logic has been justly criticized on numerous methodological grounds (e.g., Hillis et al. 1996b), but it has nonetheless produced many reasonable (if very approximate) timescales for phylogenies of closely related species (e.g., Krajewski et al. 2000; Mercer and Roth 2003).

At deeper phylogenetic levels, the compound sources of error in molecular-clock estimates become much more problematic. For a given locus, rates of change usually are not uniform *among* branches of the phylogenetic tree, and sometimes they are not constant over time *within* branches of the tree. Moreover, the fossil record may be so poor that few reliable calibration points can be obtained. As hopeless as this might seem, new analytical methods for sequence data allow divergence times to be estimated, along with confidence intervals, in the face of such complications (Sanderson 1997). The most promising *relaxed* clock method is the Bayesian approach of Kishino et al. (2001), which employs a probabilistic model of changing evolutionary rates over a phylogeny (rather than assuming a uniform constant rate) and uses fossils to place *constraints* on specific divergence times (rather than as fixed calibration points). Springer et al. (2003) used this method to argue that the living orders of placental mammals diverged in the Cretaceous, considerably earlier than had been assumed based on fossil first-occurrences.

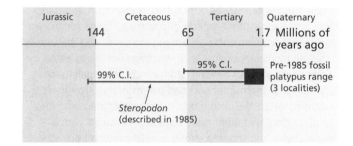

Figure 3.19 The fossil record of platypuses. Confidence intervals are shown for the pre-1985 stratigraphic range of platypus fossils. The arrow shows the stratigraphic level of Steropodon, an Australian platypus fossil described in 1985. *Adapted from Marshall (1990).*

SUMMARY

Research in mammalogy relies on many specific techniques for studying mammals in the field or laboratory, assembling and studying preserved specimens, and analyzing data obtained to address hypotheses about evolution and ecology. A background in statistics and experimental design is essential for modern mammalogists.

Mammals are captured in the field using techniques adapted for the size of the mammal and the purpose of collection. Small and intermediate-sized terrestrial species are taken in live traps or kill traps placed on the ground and checked regularly by investigators. Large terrestrial mammals may be anesthetized with dart guns, trapped with large nets, or herded into enclosures. Bats are most frequently captured with mist nets. Mammalogists now follow standard procedures for the safe and humane treatment of wild animals in the field or captive settings.

Individuals of large, active species may be identified in the field by characteristics such as size, coloration, or behavior patterns. The development of digital photographic archives and computerized image-matching analysis has greatly enhanced this approach for some groups (e.g., cetaceans, felids). Smaller, more cryptic mammals are identified with tags, collars, dyes, fur shavings, toe clips, ear notches, and even radioisotopes. For long-term projects, individuals may be permanently marked with brands or tattoos.

The movements of mammals in the field can be monitored by physical tracking procedures or with radio signals. Examples of the former include powdertracking, bait-marking, fecal pellet surveys, tracking tubes, and location of surface indices. Passive integrated transponder (PIT) tags are small devices that are usually implanted beneath the skin and transmit a unique code signal when excited by the electronic field of a reader. The reader decodes the signal, thereby identifying the animal and its location. Radiotelemetry uses radio transmitters attached to animals, the signals of which are detected with antennae connected to receivers. The animal's location is determined by triangulation after taking fixes from two or more locations. The global positioning system allows animals fitted with GPS collars to be located by satellite. Analyses of animal movement and habitat use have been greatly facilitated by the development of computer geographic information systems. Geographic data can be integrated with mark-recapture studies to estimate population size, home range size, and other life history traits. Such analyses are enhanced by statistical modeling of the trapping process.

The behavior of wild mammals is most often studied by direct observation and video recording. Given a hypothesis about behavior, researchers perform observational sampling by noting the occurrence of behavioral states or events in the animals under study. The most general observational sampling techniques are focal animal and scan sampling, though several others (including sound playback) are used to address specific kinds of questions. Researchers must take account of animals' observability in analyses of behavioral data.

Within mammalogy, physiology research focuses on whole-organism analyses of nutrition, metabolism, reproduction, and several other aspects of organ and tissue function. Nutritional requirements vary among mammals and are usually assessed by dietary preferences, which in turn are assessed by direct observation of feeding, stomach content analysis, or scat analysis. The nutritional condition of mammals reflects the extent to which their nutritional requirements are being met. Basal (resting) metabolic rate is typically measured with a respirometer and shows a strong correlation with body size. Field metabolic rates, those characteristic of mammals engaging in normal activities, are measured with doubly labeled water and show considerably more variation than BMRs. Depending on the species, the reproductive condition of males can be determined by descended testes, spermatorrhea, or sperm counts. Female estrus is indicated by behavioral changes, cornified vaginal epithelium cells, or surgical examination. Levels of reproductive hormones in both sexes can be monitored with immunological assays.

Molecular markers have become important tools for studying mammalian evolution. The number and morphology of chromosomes within a mammalian cell is its karyotype. Details of karyotypes can be studied with staining or probing techniques; the latter identifies the location of specific genes or other DNA sequences on the chromosomes. Protein electrophoresis allows researchers to determine the genotype of many individuals at many enzyme loci with relative ease, providing valuable insight on levels of genetic variation in populations. Development of the polymerase chain reaction enabled researchers to assay DNA sequence variation without in vivo cloning, and DNA markers are currently the most widely employed tool in mammalian evolutionary genetics.

Microsatellites are regions of DNA in which short sequences (1–4 bases) occur as tandem repeats. The high mutation rate of microsatellite loci generates variation in repeat number among individuals, making these markers extremely useful for population genetics. Technological advances, such as automated cycle sequencing with capillary electrophoresis, have made it possible to obtain DNA sequences from PCR-amplified targets quickly and easily. Mitochondrial DNA has been the most frequently studied sequence in mammalogy, principally due to its haploid, nonrecombining, and maternally transmitted mode of inheritance. However, mtDNA is most informative for closely related taxa, and mammalogists have begun to explore the nuclear genome to address a wider range of evolutionary questions.

Systematic collections of museum specimens are central to the study of mammalian diversity. Museum specimens include study skins, skulls, skeletons, spirit specimens, fossils, tissue specimens, and DNA samples. Curation of such collections includes maintaining an accurate and accessible specimen database as well as preserving the specimens themselves. Museum specimens form the basis of regional biotic surveys and research in comparative anatomy, systematics, morphometrics, and evolutionary genetics.

The study of intraspecific variation in mammals includes sexual dimorphism, growth, and population subdivision. In the latter, molecular markers have come to play important roles in assessing gene flow, effective population size, and geographic structuring of genetic variation. Intraspecific phylogeography is the study of genealogical and geographical relationships among lineages within a species, often with a view to understanding the history of population subdivision. Although initially dominated by mtDNA, mammalian phylogeography now exploits a variety of nuclear markers. Like other biologists, mammalogists have struggled to formulate a coherent and operational definition of species. Several definitions are current, but all of them emphasize the ability to diagnose species by heritable characteristics. Mammalogists adhere to the practice of designating type specimens and follow the rules of the International Code of Zoological Nomenclature in giving formal names to species and higher taxa. Species descriptions and analyses of species boundaries are specimen-based undertakings that, in mammalogy, have come to rely heavily on the use of molecular markers. The goal of phylogenetics is to reconstruct precise evolutionary relationships among species using the evidence provided by characters. Characters are selected for phylogenetic analysis on the basis of homology, levels of variation, and independence. Modern phylogenetic inference methods are model-based (e.g., additive-distance, maximum likelihood, Bayesian analysis) or model-free (e.g., parsimony) and include some measure of support for individual clades. Phylogeny is gradually becoming the basis for supraspecific classification, though uncertainty about many mammalian relationships has produced conflict about the composition of specific groups. Statistical analyses of the fossil record and molecular sequence divergence can provide reasonable estimates of divergence times on a phylogeny, though confidence intervals on these estimates may be quite large.

SUGGESTED READINGS

Avise, J. C. 2004. Molecular markers, natural history, and evolution, 2nd ed. Sinauer Associates, Sunderland, MA.

Bookhout, T. A. (ed.). 1996. Research and management techniques for wildlife and habitats, 5th ed. rev. Wildlife Society, Bethesda, MD.

El-Rabbany, A. 2006. Introduction to GPS: the global positioning system, 2nd ed. Artech House Publishers, Norwood, MA.

Hillis, D. M., C. Moritz, and B. K. Mable (eds.). 1996. Molecular systematics, 2nd ed. Sinauer Associates, Sunderland, MA.

Holder, M., and P. O. Lewis. 2003. Phylogeny estimation: Traditional and Bayesian approaches. Nature Rev. Genetics 4:275–284.

Jones, M., C. Dickman, and M. Archer (eds.). 2003. Predators with pouches: The biology of carnivorous marsupials. CSIRO Publishing, Collingwood, Australia.

Kenward, R. E. 2000. A manual for wildlife radio tagging, 2nd ed. Academic Press, San Diego.

Lehner, P. N. 1996. Handbook of ethological methods, 2nd ed. Cambridge Univ. Press, New York.

Martin, R. E., R. Pine, and A. F. DeBlase. 2001. A manual of mammalogy with keys to the families of the world, 3rd ed. McGraw-Hill, Dubuque, IA.

DISCUSSION QUESTIONS

1. Suppose you must determine which species of mammals occur in a 10-acre plot, along with estimates of the relative abundance of each species. The plot includes a variety of habitat types such as forest, brush, and open grassy areas. Outline a proposal for this research project, noting which methods described in this chapter would be most helpful. How would published works or reference collections facilitate your study?

2. As a result of the survey you performed in the previous question, you discover some mice that you cannot identify, and you suspect they represent a new species (i.e., one that has not yet been formally described). What steps would you take to determine whether this is true?

3. Suppose the mice you discovered in question 2 do represent a new species, which you describe and include in a genus that contains nine other species, all of which are relatively common but have disjunct ranges scattered across your continent. No phylogeny of this genus has ever been suggested, and your university invites you to prepare a proposal to obtain such a phylogeny. What are the major activities that this project will entail? What logistic and financial considerations are relevant?

4. Many species of mammals are currently endangered, facing the very real possibility of extinction, often due to destruction of their habitat by human activities. What, if any, special considerations should be given to studying endangered species with the methods described in this chapter? Make a series of general policy recommendations.

5. You are asked to develop a general activity time budget for pronghorn antelope (*Antilocapra americana*) living at several sites in northern Colorado. Describe the methods you might use to conduct such a study. What problems might be encountered? What measures could you adopt to resolve those difficulties?

CHAPTER 4

Evolution and Dental Characteristics

Mammals evolved from a lineage of tetrapod during the 100-million-year period from the late Paleozoic to the early Mesozoic era. The distribution, adaptive radiation, and resulting diversity seen in today's mammalian fauna are products of the evolutionary process operating throughout hundreds of millions of years—a process that continues today. In this chapter, we describe evolutionary morphological changes from amniotes to mammalian structural organization, including the development of mammalian traits and the emergence of early mammals. We continue tracking changes in mammals throughout the Mesozoic era and the explosive adaptive radiation of mammals beginning in the early Cenozoic era. We also describe dental characteristics of early mammals and how these teeth developed into the dentition seen in modern mammals.

Synapsid Lineage

The Amniota are a monophyletic group that arose from amphibian—like tetrapods in the early Carboniferous period of the Paleozoic (table 4.1). This group showed a key morphological change adaptive for reproduction on land, namely, development of the cleidoic, or shelled, egg. The amniotes represent the common ancestor of all reptiles and mammals. By the late Carboniferous period, the amniotes had diverged into three lineages: the synapsids, anapsids, and diapsids (figure 4.1). These groups were distinguished by the number, size, and position of lateral temporal openings (fossa) in the skull used to facilitate attachment of jaw muscles. Mammals arose from the phylogenetic lineage or clade called the **Synapsida** (Pough et al. 2005; Prothero 1998; Hickman et al. 2004; Benton 2005; Kemp 2005). The synapsids were the first group of amniotes to radiate widely in terrestrial habitats. They first appeared in the late Paleozoic era, about 320 million years ago (mya) in North America (see table 4.1). They were the dominant land animals for 70 million years, but had passed their evolutionary peak by the time of emergence of the dinosaurs. Early synapsids diverged into diverse herbivorous and carnivorous forms: the pelycosaurs and therapsids. The **Pelycosauria** was the most primitive of the two groups known from fossil remains in North

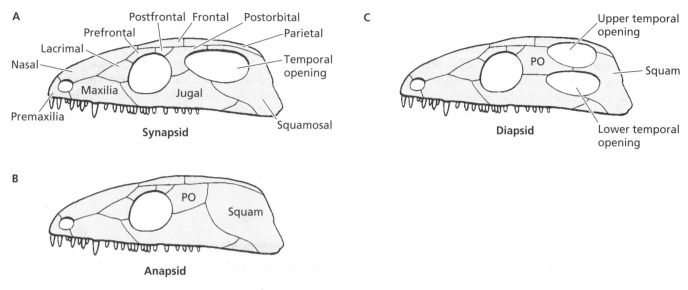

Figure 4.1 Temporal openings of skulls in three groups of amniotes. (A) Synapsida, the group from which mammals arose. Note the articulation of the postorbital (PO) and squamosal (Squam) bones above the single temporal opening. (B) Anapsida, with no temporal opening. This group led to turtles. (C) Diapsida, named for the two temporal openings on each side of the skull. Lizards and snakes arose from the diapsids. *Adapted from D. Kermack and K. Kermack, 1984,* Evolution of Mammalian Characters, *Croom Helm (Kapitan Szabo).*

Table 4.1   Geologic time divisions

Era	Period	Epoch	Myr BP (approx.)	Biological Events
Cenozoic	Quaternary	Recent	0.01	
		Pleistocene	1.8	
	Neogene	Pliocene	5	
		Miocene	24	Most mammalian families in evidence
	Tertiary	Oligocene	37	
	Paleogene	Eocene	54	Origin of most mammalian orders
		Paleocene	65	Mammalian radiation, Dinosaurs extinct
Mesozoic	Cretaceous		144	Dinosaurs abundant
	Jurassic		213	
	Triassic		248	First mammals
Paleozoic	Permian			
	Carboniferous	Pennsylvanian	320	Synapsids
		Mississippian	360	
	Devonian		408	Devonian tetrapods
	Silurian		438	First jawed fishes
	Ordovician		505	First vertebrates and land plants
	Cambrian		590	Invertebrates

America and South Africa (Dilkes and Reisz 1996). The **Therapsida**, the more advanced group, were the top carnivores in the food web. They are known from fossil remains in Russia, South Africa (Carroll 1988), and China (Rubidge 1994; Jinling et al. 1996).

Therapsids have traditionally been referred to as "mammal-like reptiles" (Romer 1966; Kemp 1982). In recent years, however, new phylogenetic techniques have overturned much of what we thought about mammalian evolution. Current research in cladistic zoology focuses on the heirarchical arrangement of monophyletic groups. Reptiles are not a monophyletic group but rather paraphyletic (not all members are descendents of a single common ancestor). As a result, the Class Reptilia is no longer recognized as a valid taxon by cladists; reptiles are referred to as amniotes—neither birds nor mammals. As customarily thought, mammals and our transitional "mammal-like reptiles" did not evolve from reptiles; rather, reptiles and mammals shared a common ancestor (the Amniota) from which each group evolved in divergent ways (Hickman et al. 2004; Pough et al. 2005).

Pelycosaurs and Therapsids

Pelycosaurs, like all synapsids, were distinguished by a single lateral temporal opening, with the postorbital and squamosal bones meeting above. They were common by the end of the Pennsylvanian epoch (figure 4.2). By that time, they had radiated into three suborders. The Ophiacodontia, with several known families, were the most primitive. They were semiaquatic and ate fish. The Edaphosauria were terrestrial herbivores and probably

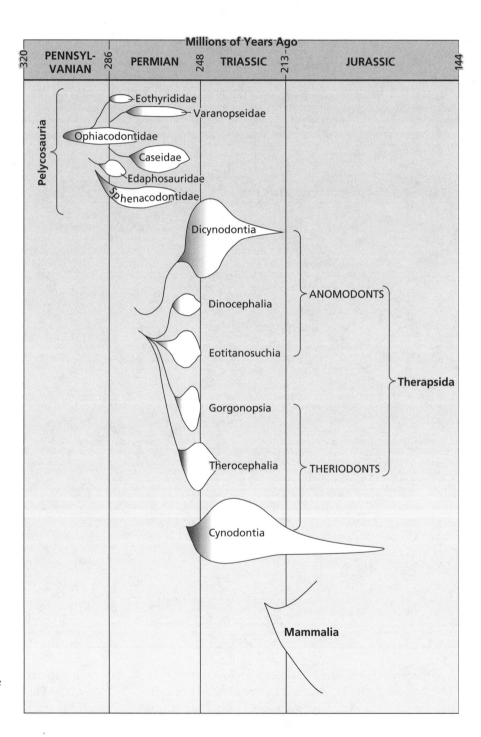

Figure 4.2 The major groups of synapsids. The Pelycosauria were early synapsids. The more advanced Therapsida arose from the carnivorous group of pelycosaurs called sphenacodonts. Therapsids included lineages of both herbivores (the anomodonts) and carnivores (the theriodonts). It was from a branch of the theriodonts, specifically the cynodonts, that mammals arose over 200 mya. Widths of each lineage suggest relative abundance. *Adapted from R. L. Carroll, 1988,* Vertebrate Paleontology and Evolution, *W. H. Freeman.*

were preyed on by the third group—the carnivorous Sphenacodontia (Kermack and Kermack 1984). The sphenacodonts, such as Genus *Dimetrodon* (figures 4.3A and 4.4A), were the dominant carnivores throughout the early Permian period. All sphenacodonts shared a unique feature—a reflected lamina of the angular bone in the lower jaw. This feature was to become part of the development of the middle ear in later synapsids and mammals. The sphenacodonts eventually gave rise to the Therapsida. The therapsids are thought to be monophyletic (Benton 2005). By the middle to late Permian period, all pelycosaurs were replaced by the more advanced therapsids (figure 4.3B) (Rowe 1993; Wible et al. 1995; Benton 1997).

The oldest known therapsids date from the late Permian period (see table 4.1). Therapsids may be divided into two suborders: **Anomodontia** and **Theriodontia** (see figure 4.2). The anomodonts included two lineages, the Dinocephalia and the Eotitanosuchia, neither of

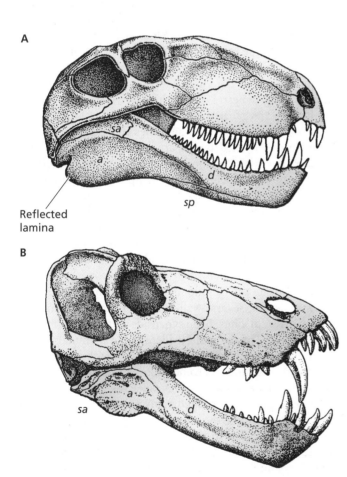

A

Reflected lamina

B

Figure 4.3 Contrast between pelycosaurs and therapsids. (A) Skull of *Dimetrodon*, a common carnivorous sphenacodont Pelycosaur. Relatively primitive features include the small temporal opening and large angular bone (a). Other postdentary (d) bones include the surangular (sa) and the splenial (sp). These bones are larger than those in (B) *Titanophoneus*, a more advanced therapsid, with a large temporal opening and smaller postdentary bones. *Note:* Not to the same scale. *Adapted from A.S. Romer, 1966,* Vertebrate Paleontology, *3rd ed., Univ. Chicago Press.*

which left Triassic descendents. The largest and most successful group of anomodonts was the Dicynodontia. They enjoyed a worldwide distribution (although the continents were not in their current positions) and were the dominant terrestrial herbivores for 60 million years from the mid-Permian until the late Triassic period, when the last of the various lines of anomodonts became extinct.

The other suborder of therapsids was the Theriodontia. They were primarily carnivorous and much more diverse and successful than the herbivorous anomodonts. Several different theriodont lines are recognized in the fossil record (see figure 4.2). The gorgonopsians were the prevalent theriodonts throughout the late Permian period, but they did not survive into the Triassic period. The therocephalians were a much more advanced and diverse group. They paralleled the other advanced theriodont group, the **Cynodontia**, in some of their mammal-like characteristics, including a secondary palate and complex **cheekteeth** (postcanine teeth; the premolars and molars). Therocephalians did not show changes in position of the jaw muscles, however, as did cynodonts. The therocephalians were extinct by the early Triassic period, and only the cynodonts possessed the specialized cranial and skeletal features that eventually led to the evolution of the mammals.

CYNODONTIA

Cynodonts existed for 70 million years, throughout the Triassic to the middle Jurassic period. During this time, the several recognized families of cynodonts developed many of the transitional anatomical features leading from synapsids to the earliest mammals (Hotton et al. 1986). Later cynodonts included diverse herbivores (gomphodonts and tritylodonts) as well as carnivores (cynognathids and tritheledonts). Several cynodont characteristics approached mammalian grade, that is, a level of organization similar to mammals. These characteristics included changes in dentition to tricuspid (**cusps** are the projections, or "bumps," on the **occlusal** [chewing] surface of a tooth) and double-rooted cheekteeth; jaw structure and masseter muscles (increased dentary size with reduction in postdentary bones and development of a glenoid fossa on the squamosal bone); hearing; postcranial skeleton (differentiated vertebrae, including modification of the first two vertebrae—the atlas/axis complex, modified pectoral and pelvic girdles, and thoracic ribs); and their phalangeal formula (Crompton and Jenkins 1979; Dawson and Krishtalka 1984). Several of these changes are discussed in more detail later in this chapter. Changes in cynodonts and early Mesozoic mammals also included more erect posture and efficient movement as well as increased adaptability in feeding. For example, development of the masseter muscles was associated with changes in the jaw, especially enlargement of the dentary bone and reduction in the number and size of postdentary bones (see figure 4.4), and development of the

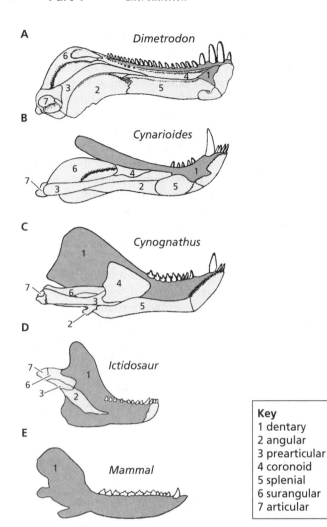

Figure 4.4 Enlargement of the dentary bone. The progressive enlargement of the dentary bone (shaded) and reduction in postdentary bones is evident when comparing jaws of primitive mammal-like reptiles: (A) *Dimetrodon*, an early Permian pelycosaur; (B) *Cynarioides*, a late Permian therapsid; (C) *Cynognathus*, an early Triassic cynodont; and (D) *Ictidosaur*, a late Triassic-early Jurassic cynodont. The dentary is the sole bone in the jaw of mammals (E). *Adapted from R. Savage and M. Long, 1986*, Mammal Evolution: An Illustrated Guide, *Facts on File.*

Key
1 dentary
2 angular
3 prearticular
4 coronoid
5 splenial
6 surangular
7 articular

zygomatic arch, temporal openings of the skull, and the eventual development of the lateral wall of the braincase. These developments reduced stress on the jaw joint, increased the force of the bite, and protected the brain.

The evolution of synapsids to mammals was a continuum, with a *mammal* necessarily (if somewhat arbitrarily) defined by the articulation of the squamosal and dentary bones. As noted previously, synapsids had several bones in the lower jaw in addition to the dentary (see figures 4.4 and 4.5A). The joint between the lower jaw and the cranium was formed by the quadrate and articular bones. In transitional forms, as cynodonts became more mammal-like, postdentary bones continued to decrease in size. The primitive quadrate-articular joint remained, and an additional joint formed between the cranial squamosal and the surangular bones of the jaw. This occurred because of the

progressive enlargement of the dentary bone and concurrent reduction in size of the postdentary bones. Certain lineages of cynodonts had a double-hinged jaw joint; the dentary bone articulated with the squamosal laterally, and the quadrate and articular bones also formed a medial jaw articulation (Crompton 1972).

This later joint served not only as a hinge but also to transmit sound to the tympanic membrane. This membrane was supported by the reflected lamina of the angular bone (figure 4.6). Only one relatively large middle ear bone, the **stapes**, conducted sound from the tympanic membrane to the inner ear in stem mammals. Eventually, sound transmission became the only function of the quadrate and articular bones as the dentary became the only bone in the lower jaw. The articulation between the dentary and squamosal bones is the characteristic used to define a mammal. The position and reduction in size of the quadrate-articular joint was associated with the transformation of these bones into **ossicles** in the mammalian middle ear (see figure 4.5B). The long attachment arm of the **malleus** to the tympanic membrane in modern mammals is called the **manubrium** and is derived from the former retroarticular process of the articular bone. In conjunction with the large tympanic membrane and the much smaller fenestra ovalis, this lever system (enhanced by the long "lever" arm of the manubrium) not only transmits sound waves but also amplifies them. Low-pressure sound waves carried by air are increased to the higher pressure necessary for conduction through the fluid of the inner ear—the cochlear endolymph. This results in increased auditory acuity. As with many other changes from reptilian to mammalian organization, these two features are interrelated. That is, the reduction in size of the postdentary jaw bones increased not only auditory acuity, especially of high-frequency sound, but also the efficiency of chewing.

By the time the cynodonts became extinct in the mid-Jurassic period, several well-defined groups of mammals already existed. Thus, from synapsids that existed 320 mya to emergence of the earliest identifiable mammals about 70 million years later, evidence in the fossil record indicates several anatomical trends in organization that resulted in a mammalian grade (see the next section). Paleontologists have isolated many important osteological characters defining the mammalian grade of evolution. Some studies have focused on early amniotes (Hopson 1991; Laurin and Reisz 1995), whereas others delineate characters and relationships during the nonmammal-to-mammal transition period (Wible 1991; Crompton and Luo 1993; Luo and Crompton 1994; Wible et al. 1995; Kielan-Jaworowska 1997). Associated changes must also have occurred in the soft anatomy, physiology, metabolism, and related features of synapsids that are not evident from fossil remains. These features were interrelated in terms of increased efficiency of metabolism needed for endothermy, better food gathering and processing methods, increased auditory acuity, and other adaptations to maintain internal homeostasis and enhance survival.

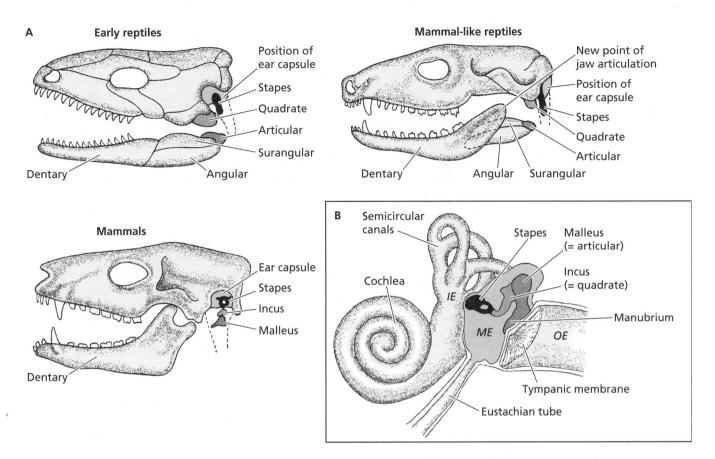

Figure 4.5 Transition of the jaw and history of the ear ossicles. Please note that we use traditional terminology in the following caption as pertains to Early Reptiles (= amniotes), and Mammal-like reptiles (= synapsids). (A) Simplified transition of the jaw structure from reptiles through mammal-like reptiles to mammals, showing the increase in size of the dentary bone and decrease in postdentary bones. The quadrate and articular bones of mammal-like reptiles eventually changed from their dual role of jaw joint and sound transmission to solely sound transmission in mammals. (B) Outer ear (*OE*), middle ear (*ME*), and inner ear (*IE*) of modern mammals. The tympanic membrane is now supported by the tympanic bone, derived from the former reflected lamina of the angular bone (see figure 4.6). The articular bone has become the first of the three small bones (ossicles) in the middle ear, specifically, the malleus. The second ossicle, the **incus**, is derived from the quadrate bone. The mammalian stapes, much reduced in size from the reptilian stapes, connects the incus to the inner ear through the fenestra ovalis (the "oval window"). Thus, in mammals, the stapes is not connected directly to the tympanic membrane as in reptiles, but instead is connected through a lever system of two small bones—the malleus (former articular bone) and the incus (former quadrate bone)—the familiar "hammer, anvil, and stirrup." *Adapted from G. Simpson and W. Beck, 1965,* Life, *2nd ed., Harcourt, Brace, World.*

MONOPHYLETIC OR POLYPHYLETIC ORIGIN OF MAMMALS?

During the early Jurassic period, one of two events happened. Either a single lineage of therapsids gave rise to early mammals, or two or more therapsid lines independently achieved the mammalian grade of organization (figure 4.7) (Rowe and Gauthier 1992). In the first case, early mammals subsequently split into two divergent lines, the prototherians (which today include the monotremes—see chapter 11), and another line, which after many adaptive radiations, gave rise to therians (the metatherians and eutherians, or "placental," mammals). This scenario posits a monophyletic origin for mammals (Rowe and Gauthier 1992). In the second case, the characteristics of the three major mammalian groups (infraclasses) are seen as strictly

convergent, mammals having a polyphyletic origin (Cifelli 2001). The question of monophyly persists because different interpretations can be drawn from a fragmentary and incomplete fossil record. If the mammal node is supported by characteristics that evolved at different times in different lineages, such as three bones in the middle ear that were derived from the jaw joint, then by definition, mammals are polyphyletic. Most monophyletic groups are supported by multiple characters, and they are presumed to be independent. It is their separate derivation in independent groups that makes these characters indicate polyphyly. As noted, however, most authorities consider mammals to be the taxa sharing the single dentary bone with a squamosal-dentary articulation (Rowe 1996) and their common ancestor. This definition, based on common ancestry, makes mammals monophyletic (Rowe and Gauthier 1992).

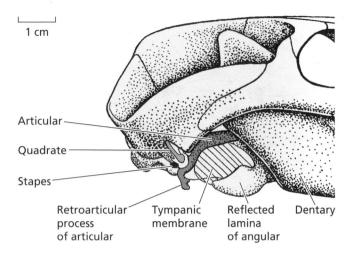

1 cm

Articular

Quadrate

Stapes

Retroarticular   Tympanic   Reflected   Dentary
process   membrane   lamina
of articular   of angular

Figure 4.6 Tympanic membrane in a mammal-like reptile (= synapsids). Posterior part of the cranium in an advanced mammal-like reptile, *Thrinaxodon liorhinus*. Note the jaw joint, the location of the tympanic membrane, and its relationship to the postdentary bones. The quadrate and articular bones form the jaw joint as well as transmit sound from the tympanic membrane to the stapes. *Adapted from J. Lillegraven et al., 1979, Mesozoic Mammals, Univ. California Press.*

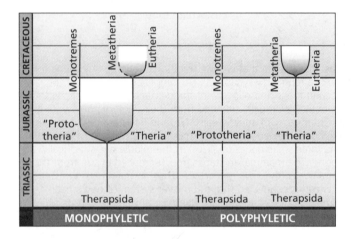

Figure 4.7 Diagram of alternative hypotheses of the origin of mammals. In the monophyletic scheme, the major groups of mammals alive today diverged from a common ancestral line of amniotes. Conversely, the three major groups may have converged from at least two distinct, separate lines of amniotes. Most authorities suggest mammals had a monophyletic origin.

The First Mammals

EARLY MESOZOIC MAMMALS

As we have seen, therapsids achieved many mammalian features; however, only mammals had the primary jaw joint composed of the dentary and squamosal bones. Also, they had **diphyodont** dentition (only two sets of teeth during an individual's lifetime) with complex occlusion. The

mammalian fauna in the Mesozoic era and for the first 100 million years of the existence of mammals, however, was strikingly different from what we see today (Prothero 1998). Until the early Cenozoic era, mammals were a relatively rare and insignificant part of the fauna compared with the larger, widespread, specialized, and well-adapted lineages of reptiles. Mammals were small (mouse-sized), relatively uncommon, and probably nocturnal. Because the majority of the terrestrial reptiles were diurnal, mammals could better avoid predation by being nocturnal. Throughout this era, mammals appeared to be a narrowly restricted offshoot, confined to a few phyletic lines. It is interesting that the fossil record shows little overlap in size between the smallest dinosaurs and the largest mammals for the 140 million years these two groups shared the terrestrial environment. In general, the smallest dinosaurs were many. times larger than the largest mammals. Mammals did not attain large body sizes until after the extinction of the dinosaurs (Lillegraven 1979).

Mammalian phylogeny may be depicted as in figure 4.8. The great evolutionary "bush" of diverse Mesozoic mammalian clades is the dominating feature in their taxic evolutionary pattern (Cifelli 2001; Luo et al. 2001; Ji et al. 2002). The almost fully resolved cladogram (figure 4.8) of all Mesozoic mammal groups, together with their improved records of temporal distribution of fossils, suggest that mammalian diversification occurred episodically during the entire span of Mesozoic mammalian history. Five episodes of diversifications occurred (see figure 4.8). The earliest diversifications of stem eutherians and stem metatherians, as documented by the currently available fossil record, predate the likely time window estimated by molecular studies, which indicates that some superordinal clades of placental mammals may have extended back into the Cretaceous (e.g., split of the earliest placental superordinal clades around 108 ± 6 mya; Murphy et al. 2001a,b).

EARLY PROTOTHERIANS

Members of the Family Morganucodontidae are among the most primitive known mammals, with the Genus *Morganucodon* abundant in the fossil history (see figure 4.8). Members of this family from the late Triassic period of Europe represent the earliest known mammals (Mammaliaformes). Morganucodontids may have been the ancestors of later major groups, including the unknown ancestors of **monotremes**. Relationships among lines of Mesozoic mammals are unclear, however, and affinities of the morganucodontids to later groups remain uncertain. They may be considered "prototherian" only in that they did not form the ancestry of the marsupials and placental mammals. Although monotremes today are numerically an insignificant part of the modern mammalian fauna, prototherian mammals during the Mesozoic era were numerous and diverse.

Several distinct mammalian structural features are present in morganucodontidae and evolved among several dif-

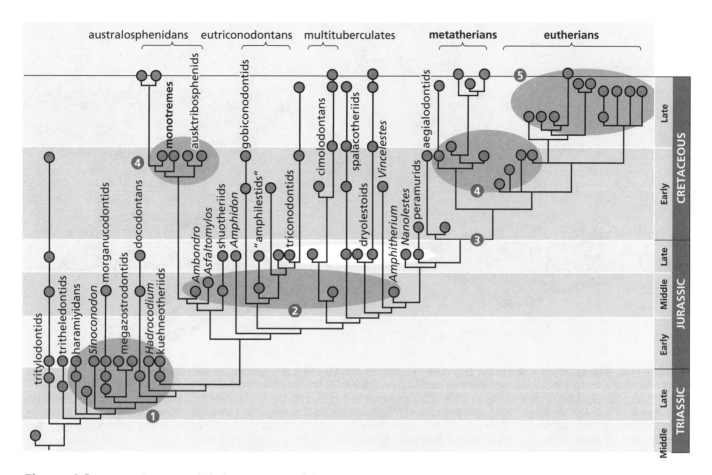

Figure 4.8 Mesozoic mammal clades. Overview of the temporal distribution and relationships among Mesozoic mammal clades. The five episodes of diversification are as follows:

1. The earliest-known episode of diversification occurred in the **Late Triassic-Early Jurassic** on a global scale, when haramiyidans, morganucodontans, kuehneotheriids, and docodontans (including *Woutersia*) appeared.
2. The next episode occurred globally in the **Middle Jurassic**, characterized by the appearance of shuotheriids, the earliest australosphenidans, eutriconodontans, putative multituberculates, and amphitheriid "eupantotherians."
3. The **Late Jurassic** diversification occurred primarily in Laurasia among eutriconodontans, spalacotheriids, dryolestoids, and peramurans.
4. The **Early Cretaceous** episode saw diversification within australosphenidans and toothed monotremes on the Gondwanan continents, and the basal splits of eutherians and metatherians and diversification of tricondontids on the Laurasian continents.
5. The **Late Cretaceous** episode witnessed diversification within metatherians, within eutherians, and within cimolodontan multituberculates on the northern continents and of gondwanatherians on the southern continents. *Adapted from Z.-X Luo et al. (2002) and Q. Ji et al. (2002).*

ferent lineages (Crompton and Jenkins 1979; Carroll 1988). These features included dentary-squamosal articulation, although involvement of the quadrate and articular bones remained. Besides incisors and canines, the cheek-teeth were differentiated into premolars and molars. The occlusal surfaces (the portion of the crowns that contact each other when an animal chews) of the upper and lower molars were clearly mammalian. The cochlear region was large relative to skull size. The first two vertebrae were similar to those seen in later mammals, and two occipital condyles were present. Thoracic and lumbar vertebrae and the pelvic region were distinct from the reptilian pattern. Morganucodontids had a mammalian posture, with the legs beneath the body, not splayed out as in reptiles. Also, the vertebrae allowed flexion and extension of the spine during locomotion. Thus, many interrelated features of the skull and postcranial skeleton that define mammals

were evident in morganucodontids (as well as in late cynodonts). These features continued to be refined in later groups of late Jurassic and early Cretaceous lineages of mammals. The most prominent of these lines were the triconodonts, amphilestids, docodonts, and multituberculates groups defined on the basis of tooth structure and associated adaptive feeding types.

The **Triconodonta** were a successful lineage that extended from the late Triassic to the late Cretaceous period—120 million years (Cifelli and Madsen 1998). They included the early Morganucodontidae. The triconodonts were small, carnivorous mammals named for their molars, which had three cusps arranged in a row (figure 4.9A). The **amphilestids** occurred from the mid-Jurassic to the early Cretaceous periods. They had a linear row of cusps much like morganucodontids. The amphilestid Genus *Gobiconodon*, from the early Cretaceous period, is

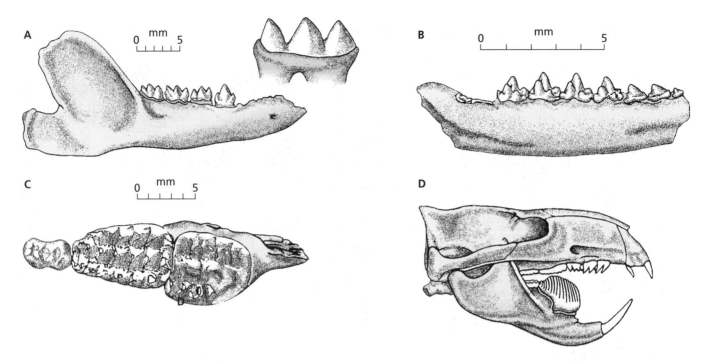

Figure 4.9 **Representative dentition from the "prototherian" line of early mammals.** (A) Lateral view of the right jaw and medial view of a lower molar from Genus *Triconodon*, a triconodont; (B) lingual view of the left mandible from the docodont *Borealestes serendipitus*; (C) occlusal view of the upper premolar and molars of the multituberculate *Meniscoessus robustus*; and (D) lateral view of the skull of Genus *Ptilodus*, showing the enlarged, shearing posterior lower premolar. Note the completely formed braincase for muscle attachment and the large dentary with coronoid process. The actual length of the skull is approximately 8 cm. *Adapted from J. Lillegraven et al., 1979,* Mesozoic Mammals, *Univ. California Press.*

noteworthy because it had deciduous molars (Carroll 1988). The **Docodonta** are known only from the late Jurassic period and may have arisen from the triconodonts. Based only on teeth and jaws from which they have been described, they appear to have been omnivores. The lower molars were rectangular with prominent cusps. These are the most complex teeth seen in fossils from the Jurassic period and achieve the same complexity as later Cretaceous therians (figure 4.9B). The advanced teeth of docodonts were in contrast to the retention of reptilian jaw articulation in this group. This demonstrates an important point. The suite of characteristics seen in mammals evolved at different rates within and among a number of lineages, and often in association with retained primitive reptilian characters.

A third order, the **Multituberculata**, was another successful mammalian line that extended about 120 million years from the late Jurassic period to the late Eocene epoch, concurrent with the emergence and radiation of flowering plants. Multituberculates were primarily herbivorous and had a single pair of large, procumbent lower incisors, as do modern rodents (Wall and Krause 1992; Kielan-Jaworowska 1997). They also had as many as three pairs of upper incisors, however. The order is named for the molariform teeth, which had up to eight large, conical cusps (figure 4.9C). These cusps were arranged in triangles in anterior molars but in longitudinal rows in the posterior teeth. The posterior lower premolar often was very large and was used for shearing (figure 4.9D). Multituberculates

and eutherian mammals coexisted for more than 70 million years. The decline of multituberculates began in the late Paleocene epoch. The last multituberculates appeared in the early Oligocene epoch of North America (Van Valen and Sloan 1966). Multituberculates probably were replaced by true rodents, primates, and other eutherian herbivores. The Triconodonta, Haramiyoidea, Docodonta, and Multituberculata have unclear phylogenetic placement (Jenkins et al. 1997).

EARLY THERIANS

Remains of the earliest therians (formerly known as pantotheres) occur in rock strata that also contain the prototherian morganucodontids. Two orders are known only from teeth and jaw fragments: the **Symmetrodonta** and the **Eupantotheria**. The earliest known symmetrodonts, within the Family Kuehnoetheriidae, are the Genera *Kuehneotherium* and *Kuhneon*—very small carnivores or insectivores from the late Triassic period (Hu et al. 1997). Later Jurassic pantotheres radiated into numerous different lines and adaptive feeding niches during the Cretaceous period. A significant feature of pantotheres was their molars, which had three principal cusps in triangular arrangement. This tribosphenic tooth pattern (the basic pattern for later mammals; see the next section) allowed for both shearing and grinding food. The most diverse family of eupantotheres, the **Dryolestidae**, may have been omnivorous and survived into the early Cretaceous period. Based on derived dental characteristics, advanced therians, that is, distinguishable **metatherians** and **eutherians**, probably originated within the eupantothere Family **Peramuridae** by the middle to late Cretaceous period, if not before (see figure 4.8). Peramurids are known only from the late Jurassic Genus *Peramus*. Genetic evidence (timed by the molecular clock) indicates that the marsupial lineage split from a therian ancestor about 173 mya, much earlier than estimated by fossil evidence (Kumar and Hedges 1998).

TRIBOSPHENIC MOLARS

As noted previously, early mammals had tooth cusps arranged longitudinally (see figure 4.9A and C). Metatherians and eutherians, and their immediate ancestors in the Cretaceous period (referred to generally as "theria of metatherian-eutherian grade"), had more advanced **tribosphenic** (or tritubercular) molars (Butler 1992; Smith and Tchernov 1992; Muizon and Lange-Badre 1997; Luo et al 2001). These are named for the three large cusps arranged in a triangular pattern. A tribosphenic upper molar (figure 4.10) consists of a **trigon** with three cusps, a **protocone** that is lingual (the apex of the triangle points inward toward the tongue), and an anterior, labial (outward toward the cheek) **paracone** and posterior **metacone**. A lower molar, or **trigonid** (an *-id* suffix always denotes mandibular dentition), consists of these three cusps and a "heel," or talonid basin. In lower molars, the protoconid is labial (not lingual, as in the upper molars), whereas the paraconid and metaconid are lingual. The **talonid** of the lower molars also has smaller accessory cusps. These often include a labial **hypoconid**, a posterior **hypoconulid**, and a lingual **entoconid** (see figure 4.10). Thus, the occlusal view of the trigon(id) of tribosphenic molars is a somewhat asymmetrical, three-cusped triangle (see figure 4.10A,C). The apex of the triangle points lingually (inward toward the tongue) in upper molars and labially (outward toward the cheek) in lower molars. During occlusion, a crushing or grinding action occurs as the protocone of an upper molar contacts the talonid basin of the opposite lower molar. Food is not only crushed but also sheared. Shearing results from several facets of the upper and lower molars coming together (see Crompton and Hiiemae 1969; Bown and Kraus 1979), for example, the anterior face of a paracone and the posterior face of a protoconid and metaconid.

The basic pattern of the tribosphenic molar in early mammals was very important because it is believed to be the ancestor of modern therian mammals (Hopson 1994). It is seen today in lineages such as marsupials and insectivores and has been modified in other modern mammals. For example, molars have become square (**euthemorphic**) with the addition of another main cusp (the **hypocone**) posterior to the protocone. Such four-cusped (**quadritubercular**) molars occur in many species of modern mammals, including humans. Cusps are often connected by a series

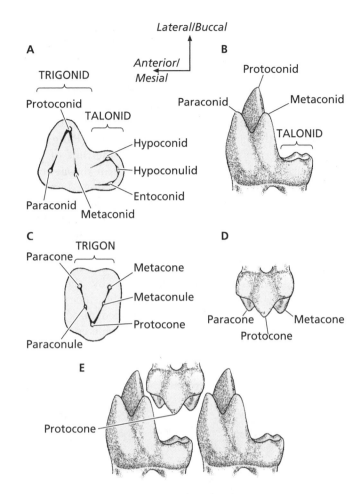

Figure 4.10 Nomenclature of the cusps of tribosphenic dentition. A lower molar in (A) occlusal view and (B) lingual view, and an upper molar in (C) occlusal and (D) lingual views. The upper and lower dentition is shown in occlusion in (E). The addition of a fourth cusp, the hypocone, forms a quadrituberculate upper molar. *Adapted from D. Kermack and K. Kermack, 1984,* Evolution of Mammalian Characters, *Croom Helm (Kapitan Szabo).*

of crests or ridges, as in many "insectivores" (see figure 1.1). Hershkovitz (1971) provided an exhaustive treatment of cusp patterns, homologies, and associated terminology.

Cenozoic Mammals and Mammalian Radiation

The different mammalian lineages seen today began with diversification of mammals during the early Cenozoic era. This radiation resulted from two major events that occurred worldwide. The first was the extinction of the dominant terrestrial vertebrate fauna, the dinosaurs, at the end of the Cretaceous period. There are several hypotheses as to why dinosaurs died out so quickly. Nobody knows for sure, although evidence exists for the hypothesis that extinction of the dinosaurs was caused by a large asteroid that struck the earth, resulting in major climate and vegetation changes. Whatever the reason, disappearance of the dominant Mesozoic reptiles opened new adaptive opportunities and resulted in a worldwide mammalian radiation. Rapid expansion and divergence of the mammals was also facilitated by the breakup of the large continental land mass (**Pangaea**) that had been in place during much of the time of the dinosaurs (Fooden 1972). Continental drift throughout the early Cenozoic era (figure 4.11) allowed major genetic differentiation of the various phyletic lines to proceed in relative isolation. These two factors, in addition to ever-expanding faunal and floral diversity worldwide, allowed mammals to occupy increasingly specialized ecological roles. As a result, for most modern mammals, ordinal differentiation was underway by the early Cenozoic era, and for many groups probably since the late Mesozoic era. Most extant orders are recognized in the fossil record by the beginning of the Eocene epoch, and most families date from before the Miocene. Mammals have been the dominant terrestrial vertebrates ever since—for the last 65 million years.

INTERRELATIONSHIP OF CHARACTERISTICS AND INCREASED METABOLISM

The changes in skeletal features noted in the following section occurred in association with metabolism, physiology, and reproduction—all of which were related to and developed concurrently with maintenance of endothermy (Grigg et al. 2004; Kemp 2006). Features of the soft anatomy related to endothermy are not visible in the fossil record because these organ systems do not fossilize; however, many of these features can be inferred. Evolutionary changes from reptiles to mammals can be related to increased metabolic demands of mammals. Mammals need approximately 10 times the amount of food and oxygen that reptiles of similar size need to

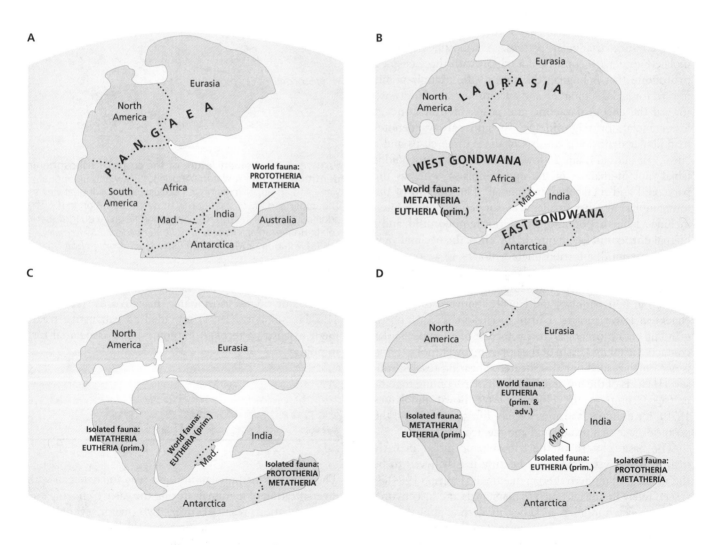

Figure 4.11 Early continental land masses. The breakup of the single large land mass (Pangaea) beginning about 200 mya and eventual isolation of the continents that promoted differentiation of the various mammalian phyletic lines following the Cretaceous period. (A) $2.00 \cdot 10^8$ years ago; (B) $1.8 \cdot 10^8$ years ago; (C) $1.35 \cdot 10^8$ years ago; (D) $6.5 \cdot 10^7$ years ago. *Sources: Data from J. Fooden, 1972, "Breakup of Pangea and Isolation of Relict Mammals in Australia, South America, and Madagascar" in Science 175:894–898.*

maintain their high body temperature. Endothermy demands an efficient supply of oxygen to the lungs for aerobic metabolism, a widespread and constant food supply, and the ability to obtain and process that food quickly and efficiently (McNab 2002; Kemp 2005, 2006). Thus, from reptilian to mammalian organization, most of the trends summarized in the following section relate directly to efficient homeostasis. All these trends are no doubt interrelated in a much more complicated and sophisticated manner than can be appreciated from a simple reconstruction from fossil history. The adaptive significance and interrelationship in postdentary bones and increase in the size of the dentary bone offer an excellent example of this. These interrelated changes in anatomy not only increased efficiency of chewing and digestion but also directly enhanced auditory acuity through greater efficiency of vibrations from the tympanic membrane. Enhanced hearing can help an individual avoid predators or capture prey more efficiently.

Summary of Anatomical Trends in Organization from Mammal-Like Amniotes to Mammals

Several morphological trends were evident in the evolution of mammals from their mammal-like reptilian ancestors (figure 4.12). Many of these trends have been noted, as have some of the interrelationships among them. Also, remember that different characters appeared at different times and in different phyletic lines. The process of change from reptile to mammal certainly did not proceed in orderly, progressive, or easily defined steps. The following trends are evident, however, in the evolution of mammals from reptiles:

1. The temporal opening of the skull of therapsids is enlarged (see figure 4.3). This was associated with eventual movement of the origin of the jaw muscles from the inner surface of the temporal region in

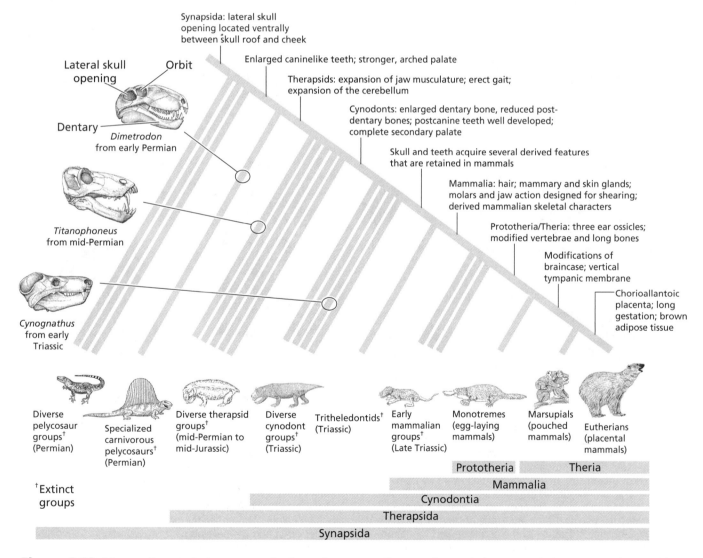

Figure 4.12 Mammalian evolutionary trends. General summary of mammalian evolutionary trends from ancestral synapsids to modern mammals.

mammal-like reptiles to the outer surface of the brain case and newly developed zygomatic arch in mammals (figure 4.13).

2. These changes parallel development of a larger, heavier dentary bone for processing the food necessary for higher metabolic activity and maintenance of homeostasis. Thus, the dentary bone became progressively larger as the postdentary bones were reduced in size (see figure 4.4). As noted earlier, the articular and quadrate bones diminished in size and became part of the middle ear. Remember, however, that several other bones in the lower jaw were retained for a long time, even after the emergence of the dentary-squamosal articulation.

3. The maxillary and palatine bones extended posteriorly and medially, forming a secondary palate (figure 4.14). This resulted in more efficient airflow, allowing a constant supply of oxygen to the lungs while permitting chewing and thus enhancing metabolism. It also may have affected suckling in neonates (Maier et al. 1996).

4. Dentition changed from **homodont** (uniform, peglike tooth structure with little occlusion) to strongly **heterodont** (teeth differentiated on the basis of form and function) in association with obtaining and processing foods more efficiently. Chewing efficiency also was enhanced by the development of tribosphenic molars (Luo et al. 2001).

5. A change occurred from one occipital condyle in amniotes to two in advanced synapsids and mammals (see figure 4.14). This reduced tension on the spinal cord when the head is moved up and down and allowed finer control of head movements, but decreased lateral movement.

6. Limbs rotated 90° from the "splayed" reptilian stance (i.e., horizontal from the body and parallel to the ground) to directly beneath the body (perpendicular to the ground; figure 4.15). Additional changes resulted in the pectoral and pelvic girdles, including loss of the coracoid, precoracoid, and interclavicle bones in the pectoral girdle, although monotremes still retain them. In the pelvic girdle, the separate bones found in reptiles fused in mammals and moved to a more anteriodorsal orientation. Mammals can therefore move with less energy expenditure than reptiles.

7. Cervical and lumbar ribs were lost completely, and the number and size of thoracic ribs were reduced. In association with changes in the vertebrae and scapula (figure 4.16), as well as others, this again allowed for more flexibility in movement, especially dorsoventral flexion of the spine.

8. The number of carpal and tarsal bones was reduced, and the phalangeal formula was reduced from the reptilian 2-3-4-5-3 (forefeet) and 2-3-4-5-4 (hind feet) to the typical 2-3-3-3-3 found in most mammals (Hopson 1995).

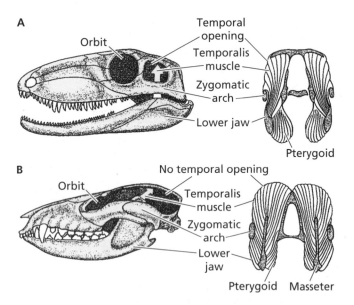

Figure 4.13 Muscle attachment and the temporal opening. Lateral view and cross section of the skulls of (A) a mammal-like reptile and (B) a mammal showing movement of the origin of the muscle attachment to the lower jaw from inside the cranium to outside the cranium. Muscle attachment was around the edge of the temporal openings in mammal-like reptiles. Muscle attachment moved to the outside of the cranium with complete ossification of the braincase and formation of the zygomatic arch in mammals. *Adapted from L. B. Radinsky, 1987,* Evolution of Vertebrate Design, *Univ. Chicago Press.*

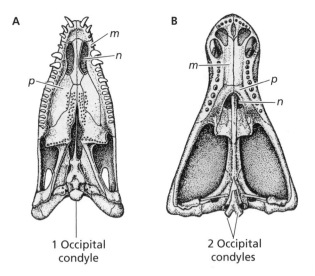

Figure 4.14 Formation of the secondary palate. Ventral views of the cranium of (A) Genus *Dimetrodon*, a pelycosaur, and (B) Genus *Cynognathus*, a more advanced cynodont. Note that the internal nares (*n*) open immediately to the front of the mouth in the primitive form, but in the cynodont, the air enters the back of the mouth because of the medial extension of the maxillary (*m*) and palatine (*p*) bones that form a secondary palate. Also notice the one occipital condyle in *Dimetrodon* and the development of two occipital condyles in *Cynognathus*. *Adapted from A. S. Romer, 1966,* Vertebrate Paleontology, *3rd ed., Univ. Chicago Press.*

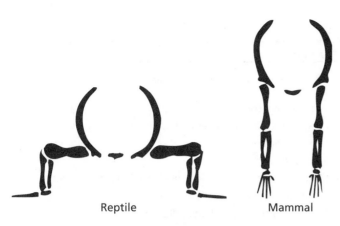

Figure 4.15 Conformation of limbs. A reptile adopts a sprawling gait with the limbs emerging horizontally from the body. A mammal adopts an upright stance with the limbs placed directly beneath the body; this is mechanically much more efficient. *Adapted from R. Savage and M. Long, 1986,* Mammal Evolution: An Illustrated Guide, *Facts on File.*

Many of these skeletal transformations, as well as differences in the soft anatomy, are evident between reptiles and mammals today (see table 4.2). The generally accepted feature used in the recognition of early mammals is a jaw joint with squamosal-dentary articulation. This is first seen in the Mesozoic era about 220 mya and was the result of a 100 million years process of change. Any single criterion separating mammal-like reptiles and early mammals becomes more arbitrary, however, as fossil history becomes more complete.

Characteristics of Modern Mammals

We noted several characteristic skeletal trends associated with the evolution of mammals (Hickman et al. 2004; Benton, 2005; Kemp 2005). The single dentary bone and three ossicles of the middle ear are unique to mammals. Two occipital condyles, epiphyses on many long bones (which result in determinant growth, unlike reptiles), and a tympanic bone are other mammalian skeletal characteristics. Among the vertebrates, several aspects of the soft anatomy of mammals are also unique.

Probably the most obvious mammalian feature is hair, or fur. These terms are synonymous—structurally, no difference exists between hair and fur. There are several types of hair, and one or more types make up the **pelage** (coat) of mammals. Mammals have a four-chambered heart, with a functional left aortic arch. Birds also have a four-chambered heart but with a functional right aortic arch. The biconcave **erythrocytes** (red blood cells) of mammals are **enucleate** (without a nucleus). Not having a nucleus enhances the oxygen-carrying capacity of these cells. Female mammals have milk-producing **mammary glands (mammae)**. This character is, of course, the basis for the name of the Class Mammalia. Finally, mammals have a muscular **diaphragm** separating the thoracic and abdominal cavities. Other aspects of the soft anatomy can be used to characterize mammals. These either are not unique to mammals or are not found in all mammals. For example,

Figure 4.16 Evolutionary trends in therapsids toward the development of mammals. (A) Genus *Lycaenops*, an early theriodont from the late Permian period, and (B) an early Triassic cynodont, Genus *Thrinaxodon*, showing formation of distinct mammalian characteristics, including enlargement of the dentary bone, with the coronoid process extending above the zygomatic arch (*a*); the second cervical vertebra (axis) with a spine (*b*); enlargement of the scapula (*c*); formation of distinct lumbar vertebrae and associated reduction in the number of ribs (*d*); enlargement of the pelvic bones (*e*); and formation of a heel bone (*tuber calcanei*) and distinct plantigrade feet (*f*). (C) Lifelike reconstruction (hair is hypothetical) of *Thrinaxodon*, which was about the size of a weasel. *Adapted from N. Hotton et al., 1986,* Ecology and Biology of Mammal-like Reptiles, *Smithsonian Institution Press.*

Table 4.2 Different characteristics of reptiles and mammals

Reptiles	Mammals
More than one bone in mandible; with quadrate-articular articulation of jaw joint	Single bone in mandible; with squamosal-dentary articulation
One occipital condyle	Two occipital condyles
Long bones without epiphyses (indeterminant growth)	Long bones with epiphyses (determinant growth)
Unfused pelvic bones	Fused pelvic bones
Secondary palate usually absent	Secondary palate present
Middle ear with one ossicle (stapes-columella)	Middle ear with three ossicles (malleus, incus, and stapes)
Phalangeal formula 2-3-4-5-3 (4)	Phalangeal formula usually 2-3-3-3-3
Dentition homodont and polyphyodont	Dentition often heterodont and diphyodont
Epidermis with scales	Epidermis with hair
Oviparous or ovoviviparous	Viviparous (except for the monotremes)
Three-chambered heart in most	Four-chambered heart with left aortic arch
Ectothermic with low metabolic rate	Endothermic with high metabolic rate
Nonmuscular diaphragm	Muscular diaphragm
No mammary glands	Mammary glands present
Relatively small, simple brain	Relatively large, complex brain

the **corpus callosum**, a bundle of nerve fibers that integrates the two cerebral hemispheres of the brain in eutherians, does not occur in monotremes and marsupials. Likewise, a true vascular chorioallantoic **placenta** occurs only in eutherians (except for marsupial bandicoots—see chapter 10). Aspects of mammalian dentition are discussed in the following sections. Other general mammalian characteristics, including **locomotion** (movement), feeding, hair, and reproduction, are examined in greater detail in part 2.

Dentition

Teeth are one of the most important aspects of living mammals. Also, many fossil lineages are described only on the basis of their teeth. Although all mammals begin life on a diet of milk, they eventually enter into one of a variety of adaptive feeding modes. An individual's teeth reflect its trophic level and feeding specialization. A number of different feeding niches are available, and as a result, mammalian dentition shows a number of different modifications. These modifications are derived in large part from the basic tribosphenic pattern, which allowed much more efficient processing of food necessary for endothermy and is retained in more primitive groups, such as "insectivores", tree shrews, elephant shrews (chapter 12), and some marsupials (chapter 11). Besides their role in feeding, teeth also may function secondarily in burrowing, grooming, and defending. Whereas mammals show little skeletal variation, except in their limbs, a great deal of variation occurs in their dental patterns.

Teeth may occur in three bones in mammals: the premaxilla and maxilla of the cranium and the mandible

(dentary bones). Most species have teeth in all three of these bones; others have a much reduced dentition in only one or two of these bones. Still other species are **edentate**, that is, they have no permanent teeth at all.

TOOTH STRUCTURE

The portion of the tooth above the gum line is the **crown**, and the **roots** are below the gum line (figure 4.17). In most species, **enamel** overlays **dentine** in the crown of

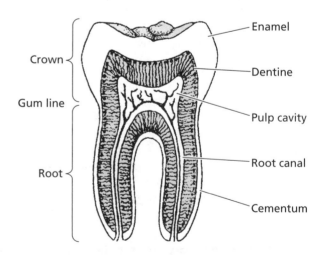

Figure 4.17 Longitudinal section of a mammalian molar. The tooth is seated in an alveolus (socket). With increased age, the enamel wears away, and progressively more dentine is exposed in most species. Also, an additional layer of cementum often is deposited each year. *Adapted from DeBlase and Martin, 1981, A Manual of Mammalogy, 2nd ed., Wm. C. Brown Publishers.*

the tooth. Enamel is harder, heavier, and more resistant to friction than any other vertebrate tissue. It is acellular, cannot regenerate, and is made up of crystallized calcium phosphate (hydroxyapatite). Enamel is ectodermal in origin, whereas dentine is of dermal origin and makes up most of the tooth. In some species, such as the aardvark and Order Cingulata (chapter 15), the teeth have no enamel. Rodent incisors have enamel only on the anterior surface, causing differential wear and continuous sharpening of the incisors for gnawing. Within the dentine is the **pulp cavity**, in which blood vessels and nerves maintain the dentine. In **open-rooted** teeth, growth is continual, and such teeth are termed "ever-growing." Incisors of rodents are a prime example of ever-growing, open-rooted teeth (see figure 18.1). Alternatively, teeth stop growing and begin to wear down with age when the opening to the pulp cavity closes. Wear on teeth that are **closed-rooted** may be used to estimate an individual's age. Just as the crown of a tooth usually is covered with enamel, the root is covered with **cementum**. This is a modified bony material deposited throughout an individual's life. Cementum annuli often are deposited much like the rings of a tree, and as with tree rings, the annuli can sometimes be used to determine an individual's age. A socket in a bone containing the tooth roots is an **alveolus**. Connective tissue between the cementum and the alveolar bone holds the tooth in place.

One of the anatomical trends noted previously was the change from homodont dentition of synapsids to heterodont dentition in mammals. Certain modern mammalian lines have homodont dentition, for example, the toothed whales (odontocetes) and armadillos (Order Cingulata: Dasypodidae). Still other groups, such as the platypus and spiny echidnas (monotremes), are edentate. So are several unrelated mammalian lineages that feed on ants and termites, for example, true anteaters (Order Pilosa: Myrmecophagidae) and pangolins, or scaly anteaters (Order Pholidota: Manidae). A special case of edentate mammals is the mysticete whales, in which teeth have been replaced with baleen in the upper jaw (see chapter 17). The edentate condition is secondarily derived; that is, teeth develop and sometimes emerge in embryos but are resorbed or lost prior to parturition.

Most mammals have heterodont dentition, with well-defined incisors, canines, premolars, and molars. **Incisors** are the anteriormost teeth, with the upper incisors rooted in the premaxilla. All lower teeth are rooted in the dentary bones. The incisors often function to cut or gnaw, as in rodents and lagomorphs. Incisors usually are structurally simple with a single root. Sometimes, though, they are highly modified and serve a variety of purposes. In shrews (Order Soricomorpha: Soricidae), the first pair of incisors are long and curved (Hutterer 2005c). They appear to function as forceps in seizing insect prey. Vampire bats have blade-like upper incisors for making incisions. Incisors are sometimes modified as tusks, as in elephants and male narwhals (*Monodon monoceros*). Some groups, such as deer, have lost the upper incisors but have retained the lower ones.

They clip vegetation by cutting against a tough, padlike tissue in place of the upper incisors.

Canines are posterior to the incisors. There is never more than one pair of upper and lower canines in modern mammals. These teeth generally are **unicuspid** (i.e., they have one cusp) with a single root. In carnivores and some other groups, canines are often enlarged and elongated for piercing and tearing prey. They may even form tusks in certain species, such as the walrus (*Odobenus rosmarus*) and pigs (Order Artiodactyla: Suidae). In species of deer without antlers, musk deer (Genus *Moschus*) and the Chinese water deer (*Hydropotes inermis*), males have elongated, tusk-like upper canines.

Premolars are posterior to the canines and anterior to the molars. Generally, teeth in the posterior part of a dental arcade are structurally more complex than anterior teeth. Premolars are generally smaller than molars and have two roots, whereas molars usually have three. Premolars may be unicuspid, or they may look the same as molars, but premolars have deciduous counterparts ("milk teeth"). **Molars** have multicusps and no deciduous counterparts; that is, they are not replaced. The premolars and molars are often considered together as "cheekteeth," "postcanine," or "molariform" teeth, especially in species in which they are difficult to differentiate. Molariform dentition is used for grinding food. As such, these teeth usually have the greatest degree of specialization in cusp patterns and ridges associated with a particular feeding niche. The height of the crown varies among species. Teeth with low crowns are termed **brachyodont** and often are found among omnivores. Herbivores consume forage that is often highly abrasive and contains large amounts of silica. This wears teeth down more rapidly than does a carnivorous diet, and it is therefore adaptive for an herbivore to have high-crowned or **hypsodont** cheekteeth (Martin et al. 2001).

In addition to crown height, occlusal surfaces are quite variable. Specifically, the cusp patterns are often highly modified (figure 4.18A–C). Brachyodont cheekteeth are often **bunodont**, with rounded cusps for crushing and grinding, as in most monkeys and pigs. Alternatively, the cusps may form continuous ridges, or **lophs**, such as occur in elephants, a pattern termed **lophodont**. Sometimes the lophs are isolated and crescent-shaped, as in deer, in which case they are called **selenodont**. Loph patterns may become so complex that it is difficult to discern the original cusp pattern. A great deal of diversity in cusp patterns is evident among individual families or within an order such as the rodents (e.g., see figure 18.2). Another example of specialization is found in many modern carnivores, which have **carnassial** or **sectorial** teeth for shearing. Carnassial teeth in modern carnivores are always the last upper premolar and the first lower molar. These teeth are particularly well developed in the cat (Family Felidae) and the dog (Family Canidae). They still occur but are less evident in more omnivorous groups, including many of the mustelids (see figure 16.4). When carnassials are used on one side of the mouth, they are not aligned on the

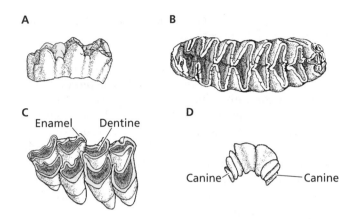

Figure 4.18 Occlusal surfaces. Teeth showing general types of occlusal surfaces: (A) a pig, with a bunodont surface; (B) the lophodont dentition of an African elephant, with cusps in the form of transverse ridges; and (C) a deer, with selenodont teeth forming crescent-shaped ridges. The enamel appears lighter than the dentine. (D) Dorsal view of the lower incisors of a white-tailed deer, showing the lateral "incisiform" canines. *Note:* The teeth are not to the same scale.

other side and cannot be used. Likewise, in most species, when the anterior dentition (incisors or canines) is used, cheekteeth do not occlude, and when the animal chews with the cheekteeth, incisors do not come together.

Many species have lost teeth through evolutionary time so that a gap, or **diastema**, occurs in the toothrow. All rodents and lagomorphs have lost their canines and have a diastema between their incisors and the anteriormost cheekteeth (see figure 18.1). Deer (Family Cervidae) also have a diastema between the lower incisiform teeth and the cheekteeth. Deer also can be used to illustrate that teeth in a given position may resemble the teeth next to them. What appears to be the last lower "incisor" in a deer jaw is actually a canine. Because it functions as an incisor, however, its form has changed through time to accommodate its function. It has become "incisiform," that is, indistinguishable from the three true incisors on each side of the midline (see figure 4.18D).

The structure of the lower jaw and primary use of different muscle groups differ between herbivores and carnivores. In herbivores, the mandibular condyle and its articulation with the fossa of the cranium is elevated above the mandibular dentition. This gives maximum advantage to the masseter muscles in closing the jaw. In carnivores, the temporal muscles are the primary muscle group closing the jaw, and the mandibular articulation is level with the dentition (figure 4.19). Also, carnivore jaws close in a shearing manner like scissors. Conversely, when the jaw of an herbivore closes, all the opposing teeth occlude together.

TOOTH REPLACEMENT

Generally, mammals have two sets of teeth during their lifetime; that is, they are **diphyodont**. The deciduous,

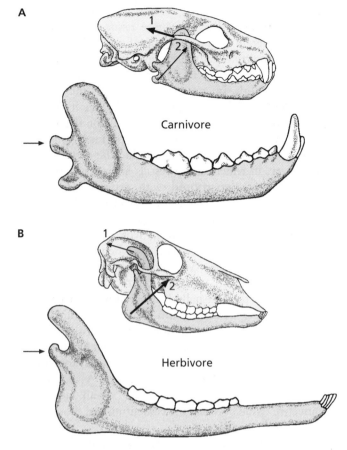

Figure 4.19 Mandibular condyle and muscle groups. The position of the mandibular condyle (*arrow*) relative to the plane of the teeth differs between (A) carnivores and (B) herbivores. Thus, the temporalis muscles (1) are the primary group of chewing muscles in carnivores, whereas the masseters (2) are the primary group in herbivores. *Adapted from L. B. Radinsky, 1987,* Evolution of Vertebrate Design, *Univ. Chicago Press.*

or "milk," teeth, are replaced by permanent dentition later in life. In many species, the pattern of tooth replacement is used to estimate the age of individuals. In eutherian mammals, all the teeth except the molars have deciduous counterparts. It is unclear, however, whether molars are permanent teeth without preceding milk teeth or "late" milk teeth without succeeding permanent teeth. In metatherians, only the last premolar is deciduous; the remaining postcanine teeth are not replaced. Deciduous incisors and canines do not erupt, and only the permanent anterior dentition is seen (Luckett and Woolley 1996). Certain eutherians, such as shrews, appear to have only permanent teeth because the deciduous teeth are resorbed during fetal development. With few exceptions, replacement of deciduous teeth by their permanent counterparts is vertical, the permanent tooth erupting from below and pushing out the worn-down deciduous tooth. In elephants (Order Proboscidea) and manatees (Order Sirenia: Trichechidae), however, tooth replacement is horizontal. Teeth in the posterior part of the mandible slowly move forward and replace anterior cheekteeth as they wear down and fall out (see figure 19.4).

DENTAL FORMULAE

A dental formula provides a useful shorthand description of the total number and position of teeth in a given species. Dental formulae are always given in the following order: incisors, canines, premolars, and molars. Thus, the dental formula for the brown hyena (*Parahyaena brunnea*) is 3/3, 1/1, 4/3, 1/1. This means there are 3 upper (numbers above the line) and 3 lower incisors (numbers below the line), 1 upper and lower canine, 4 upper and 3 lower premolars, and 1 upper and lower molar. Because dentition is bilaterally symmetrical, dental formulae are given for 1 side of the mouth only and may be multiplied by 2 to arrive at the total number of teeth in the mouth. The hyena has 17 teeth on each side of the mouth (9 upper and 8 lower), for a total of 34. Many species have lost teeth in a particular position through evolutionary time. For example, white-tailed deer (*Odocoileus virginianus*) have lost the upper incisors and canines. Thus, their dental formula is 0/3, 0/1, 3/3, 3/3 = 32 (figure 4.20). Again, because the tooth order given in a dental formula is always the same, words or abbreviations for incisors, canines, premo-lars, and molars are not necessary. If there are no teeth in a position, as in the white-tailed deer, a zero is shown, but the position is not deleted. Thus rodents, which have no upper or lower canines, have a canine formula of 0/0.

Abbreviations often are used when describing particular teeth. Superscripts and subscripts may be used with abbreviations for tooth type. For example, P^2 refers to the second upper premolar, whereas P_2 means the second lower premolar. Alternatively, you may see capital letters used to refer to upper teeth (e.g., P2), and lowercase letters for lower teeth (p2). Care must be taken to avoid confusion when referring to particular teeth.

PRIMITIVE DENTAL FORMULAE

For most species of extant mammals, there is a maximum number of each type of tooth. Presumably, these maxima represent the ancestral condition. Thus, although species tend to lose teeth (see the next section), very few exceed the primitive number of incisors, canines, premolars, and molars. In eutherians, the primitive dental formula is 3/3, 1/1, 4/4, 3/3 = 44. This means ancestral eutherian mammals had 44 teeth. Thus, most modern eutherians will not have more than 3 upper or lower incisors per quadrant, more than 1 upper and lower canine, and so forth. For metatherians, the primitive dental formula is 5/4, 1/1, 3/3, 4/4 = 50. Although very few species exceed the primitive number, the toothed whales often have more than 44 teeth; some species actually have over 200. Among terrestrial species, only the giant armadillo (*Priodontes giganteus*), bat-eared fox (*Otocyon megalotis*), and marsupial numbat (*Myrmecobius fasciatus*) exceed the primitive numbers of teeth.

The evolutionary trend is toward reduction from the primitive dental formula. In the white-tailed deer noted earlier, three upper and lower premolars occur in each quadrant (side of the mouth). One premolar (the first position) has been lost over evolutionary time. The most anterior upper and lower premolars actually are P^2 and P_2, although they may be described as the "first" premolars in the arcade (row). Similar examples could be cited for most other species.

DENTAL ANOMALIES

Occasionally, an individual's dental complement is different from that normally seen in the species. Such congenital anomalies or abnormalities may involve **supernumerary** dentition (extra teeth in a position) or, conversely, **agenesis** (reduced number of teeth in a position). Anomalies may be unilateral, occurring on one side of the jaw, or bilateral, occurring on both sides. Although they are rare, dental anomalies have been reported in representative species from most orders of mammals (Miles and Grigson 1990; Koyasu et al. 2005).

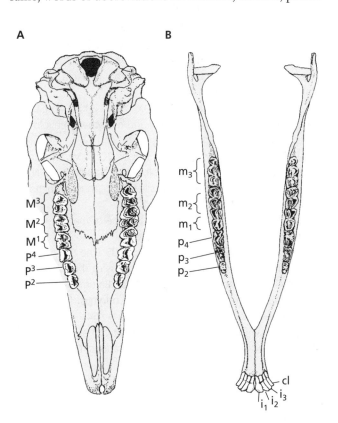

Figure 4.20 Tooth position and structure. (A) Ventral view of the upper dentition and (B) dorsal view of the lower dentition of the white-tailed deer, showing the number and structure of teeth in each position. For further discussion of the abbreviations, see text.

SUMMARY

The evolution of mammals from therapsids occurred during a 70-million year period from the late Paleozoic to the early Mesozoic era, with mammals appearing about 220 mya. During this time, numerous changes occurred in the skull, dentition, and skeleton from the synapsid to the mammalian form. These skeletal changes, and concurrent changes in soft anatomy, adapted mammals for improved ability to maintain homeostasis. Mammals became more efficient at gathering and processing food than synapsids and developed a much higher metabolic rate (although some dinosaurs may also have had high metabolic rates). This set the stage for the explosive adaptive radiation of mammals during the 70-million year period after their initial appearance. The radiation of numerous phylogenetic lines of mammals, from shrews to elephants and rodents to whales, occurred after the extinction of the dominant terrestrial vertebrates—the dinosaurs.

Mammalian radiation was further enhanced by genetic isolation of phylogenetic lines resulting from continental drift and the separation of the continental land masses as well as increased diversity of flowering plants throughout the world. Most mammalian orders today are recognized from the Eocene epoch, whereas most families are evident by the Miocene.

Diversity of form and function is manifested in the highly interrelated characteristics of modern mammals. The broad diversity of modern mammals in terms of their dentition, locomotion, pelage, feeding, and reproduction enables them to adapt to the wide range of biomes and habitats. Many of the general characteristics noted in this chapter are discussed in detail in part 2. The extent of mammalian diversity is seen in part 3, where the orders and families of extant mammals are examined.

SUGGESTED READINGS

Benton, M. J. 2005. Vertebrate palaeontology, 3rd ed. Blackwell Science Publishers, New York.

Cifelli, R. L. 2001. Early mammalian radiations. Journal of Paleontology 75:1214–1226.

Eisenberg, J. F. 1981. The mammalian radiations: An analysis of trends in evolution, adaptation, and behavior. Univ. Chicago Press, Chicago.

Hickman, C. P., Jr., L. S. Roberts, A. Larson, and H. I'Anson. 2004. Integrated principles of zoology. 12th ed., McGraw Hill Higher Education, Boston.

Kemp, T. S. 2005. The origin and evolution of mammals. Oxford Univ. Press, New York.

Kielan-Jaworowska, Z., R. L. Cifelli, and Z. Luo. 2004. Mammals from the age of dinosaurs: Origins, evolution, and structure. Columbia Univ. Press, New York.

Lillegraven, J. A., Z. Kielan-Jaworowska, and W. A. Clemens (eds.). 1979. Mesozoic mammals: the first two-thirds of mammalian history. Univ. California Press, Berkeley.

Lucas, S. G., and Z. Luo. 1993. *Adelobasileus* from the Upper Triassic of West Texas: The oldest mammal. J. Vert. Paleontol. 13:309–334.

Luo, Z.-X., R. L. Cifelli, and Z. Kielan-Jaworowska. 2001. Dual origin of tribosphenic mammals. Nature 409:53–57.

Luo, Z.-X., R. L. Cifelli, and Z. Kielan-Jaworowska. 2002. In quest for a phylogeny of Mesozoic mammals. Acta Palaeontologia Polonica 47:1–78.

Luo, Z.-X., Q. Ji, J. R. Wible, and C.-X. Yuan. 2003. An early Cretaceous tribosphenic mammal and metatherian evolution. Science 302:1934–1940.

Pough, F. H., C. M. Janis, and J. B. Heiser. 2005. Vertebrate Life, 2nd ed. Pearson/Prentice Hall, Upper Saddle River, NJ.

Prothero, D. R. 1998. Bringing fossils to life: An introduction to paleobiology. McGraw-Hill, Boston.

Rose, K. D. 2006. The beginning of the age of mammals. Johns Hopkins Univ. Press, Baltimore.

Rose, K. D., and J. D. Archibald (eds.). 2005. The rise of placental mammals: Origins and relationships of the major extant clades. Johns Hopkins Univ. Press, Baltimore.

Rowe, T. 1988. Definition, diagnosis and origin of Mammalia. Journal of Vertebrate Paleontology 8:241–264.

Savage, R. J. G., and M. R. Long. 1986. Mammal evolution: An illustrated guide. Facts on File Publications, New York.

DISCUSSION QUESTIONS

1. What reasons can you give for mammals being so much smaller than even the smallest dinosaurs for the 140 million years they were on earth together? What might have been the adaptive advantages to mammals of having been so small?

2. Because they were so small, what morphological and physiological characteristics were necessary for early mammals to survive?

3. Why does heterodont dentition of a mammal allow an individual a much broader range of feeding possibilities than the homodont dentition of synapsids?

4. Before you read chapter 7, list as many different mammalian feeding adaptations as you can think of.

5. How did the concurrent rise in the diversity of other fauna and flora early in the Cenozoic era affect the potential for early mammals to radiate into different lineages?

CHAPTER 5

Biogeography

How do scientists explain the abundance of marsupials in Australia and South America, along with their scarcity on northern continents? Why are there members of the Camelidae in central Asia, North Africa, and South America? What factors led to the present distribution of primates, extending from Japan to Africa and including South, but not North, America? **Biogeography** is the study of the distribution of organisms, both living and extinct, on the Earth (Lomolino et al. 2006).

We have already encountered geographic considerations in our discussion of home ranges (the "distributions" of individual organisms), faunal surveys, and phylogeographic analyses of species boundaries (chapter 3), as well as the influence of continental drift on mammalian radiations (chapter 4). Perhaps the most basic datum in biogeography is the **species range**—the complete area of the Earth over which individuals of a particular species occur. Species ranges are inferred primarily from museum-specimen records, but observational data are also important for larger mammals. Ranges are dynamic, changing over time as a consequence of abiotic and biotic factors. For example, lions were once widespread throughout Africa and southwestern Asia; today, they are restricted to several isolated populations scattered throughout Africa and one small population in northwest India (figure 5.1). A fundamental question posed by biogeography is: What factors determine the range of a species? The same question, but from a slightly different perspective, is also important: Why does a particular region harbor the particular set of species that we observe there? The answers invariably have to do with two kinds of causal factors, history and ecology, that define major research traditions within biogeography.

Historical biogeography emphasizes the study of changes in species ranges that have taken place over evolutionary time. It encompasses organismal history (phylogeny and phylogeography) and Earth history and brings information from both to bear on biogeographic problems. One of the distributional patterns most intriguing to historical biogeographers is **endemism,** the restriction of a species' range to a circumscribed area. Why, for example, are long-beaked echidnas (*Zaglossus bruijni*) found only in New Guinea? Even more striking are patterns of endemism that characterize areas—why are so *many* mam-

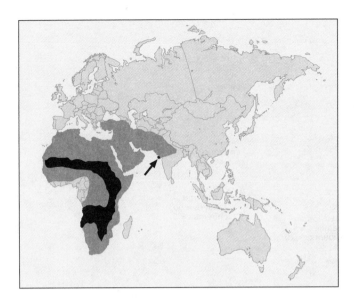

Figure 5.1 Changes in species range. The lion (*Panthera leo*) was once distributed throughout much of Africa and southwestern Asia, including the Arabian Peninsula (light shading). Today, lions still inhabit many areas of Africa (dark shading), but their range in Asia has been reduced to a small remnant population in the Gir Forest of India (black dot with arrow). *Data from Burton and Pearson (1987).*

mal species found only in New Guinea (Flannery 1995)? A second pattern of interest is the **disjunct distribution**—a gap in the range of related species or clades. Marsupials are now found in Australasia and South America. How does this distribution relate to the evolutionary history of marsupials? Do the species on each continent represent separate monophyletic groups that are each other's closest relatives? How did these groups, which clearly have a single common ancestor, become separated by two oceans? It is often the case that several groups show the same disjunctions: Monotremes, though currently endemic to Australasia, also have fossil representatives in South America (Pascual et al. 1992a).

Ecological biogeography focuses on the current distributions of species and seeks to explain those distributions in terms of community-level interactions among organisms and their environment. One common line of inquiry has to do with **species richness:** Why do some regions of the Earth (e.g., the tropics) harbor vastly more species than other regions (e.g., the Antarctic)? What determines the number and identity of species on an island? Because answers to such questions involve evolutionary adaptations, ecological biogeography frequently entails studying the patterns of morphological, physiological, or life-history variation among organisms in different places. Until recently, historical and ecological biogeography were largely separate disciplines (Posadas et al. 2006), but most practitioners now realize that both perspectives are necessary to arrive at complete explanations of geographic patterns (Ricklefs and Schluter 1993; Wiens and Donoghue 2004).

Global Provincialism of Mammal Distributions

BIOGEOGRAPHIC REGIONALIZATION

If one were to tabulate the numbers of species in major clades of virtually any group of animals or plants that occur in different continental regions of the Earth, two patterns would be readily apparent. First, different regions harbor distinct taxonomic assemblages (i.e., there is endemism on a worldwide scale). Second, there are dramatic differences in species richness among continental regions—some regions constitute **centers of diversity** and others do not. These observations, together with knowledge of phylogenetic relationships, demonstrate the **provincialism** of life on Earth, a pattern that has been evident in the vertebrate fossil record since the Mesozoic (Pough et al. 2005). Provincialism in terrestrial animal distributions led Wallace (1876) to divide the world into six faunal regions, each with a distinct assemblage of species: Palearctic, Nearctic, Neotropical, Ethiopian, Oriental, and Australian (figure 5.2). This was one of the first attempts at biogeographic **regionalization,** the estimation of boundaries between areas of endemism/centers of diversity. Darlington (1957) and Simpson (1965) provided important syntheses of descriptive information on vertebrate distributions, generally endorsing the regions recognized intuitively by Wallace. Procheş (2005) applied a more quantitative approach (cluster analysis) to regionalization and concluded that the global distribution of bats conforms fairly well to the traditional faunal regions.

Biogeographers have also been intrigued by transitional areas between regions, and what the species compositions of these areas can tell us about the historical-ecological determinants of biodiversity. Some transition areas are so broad that it is difficult to place the boundary between adjacent regions. This is true of Nearctic and Neotropical faunas, for which the transition area constitutes much of Central America and Mexico (Darlington 1957). Perhaps the most famous transition area is that between Oriental and Australian regions, where the position of Wallace's Line (figure 5.2) has stimulated over a century of studies on the mixture of faunal elements in the Malay Archipelago (Archer, 1984b). Biogeographers have also undertaken regionalizations on smaller scales. The Interim Biogeographic Regionalisation for Australia (IRBA; Thackway and Cresswell 1995), developed by Australia's Department of Environment and Heritage, identifies 85 bioregions across the continent based on fauna, flora, geomorphology, and climate. The IRBA serves as a framework for understanding species distributions and as a basis for conservation planning.

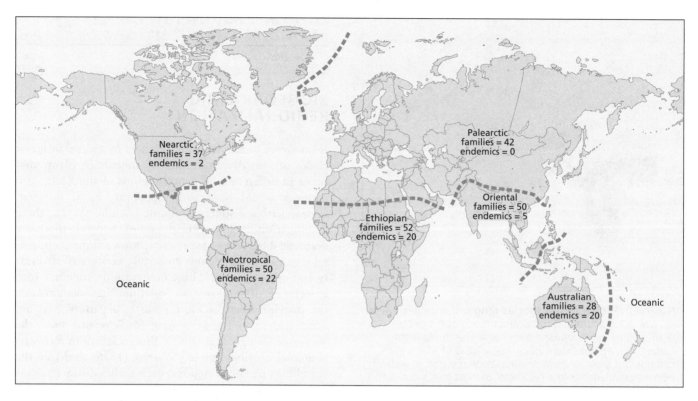

Figure 5.2 Faunal regions. The land surfaces of the world can be divided into six major faunal regions based on geographic barriers, geological history, and the distributions of vertebrate species. Oceanic islands and the open ocean can be considered as an additional region. Considerable differences are evident in mammalian family diversity and the number of endemic families among these regions. The boundary between Oriental and Australian regions is Wallace's Line.

FAUNAL REGIONS

Here we summarize the extent, ecosystem characteristics, and assemblage of mammals in each of Wallace's (1876) faunal regions. Continental ecosystems are conveniently classified as **biomes,** each with a specific type of plant community determined by climate and soil characteristics. The major terrestrial biomes are tropical rainforest, tropical deciduous forest, savanna, desert, chaparral, grasslands, temperate deciduous forests, temperate rainforests, taiga, and tundra. Of course, some mammals occupy freshwater aquatic and marine ecosystems. Aquatic species are usually considered part of the terrestrial biomes in which their lakes, rivers, etc. occur. Marine species may be associated with coastal waters or the open ocean; for convenience, we will discuss the ocean as if it were a single faunal region.

Palearctic

The Palearctic, largest of the faunal regions, consists of the northern Old World, including Europe, Russia, central Asia, and northern China. It is separated from the Ethiopian region by deserts, from the Oriental region by mountains, and from the Nearctic by the Bering Strait. An east-west band of taiga centered on 60°N latitude dominates the Palearctic, with tundra to the north. The southwestern Palearctic includes the Mongolian steppes, while temperate deciduous forest and chaparral occur in

Europe. There are no endemic mammal families in the Palearctic, and only one endemic subfamily (Spalacinae, the blind mole-rats). Species diversity is concentrated in the warm, wet areas of the southeast and southwest. The Palearctic mammal fauna represents a mixture of Nearctic, Ethiopian, and Oriental elements. About 50% of Palearctic families also occur in the Nearctic, primarily because these regions were connected for extended periods by the Bering land bridge between Siberia and Alaska. Species of Cervidae, Bovidae, and Ursidae occur in all three regions; Suidae, Hyaenidae, and Viverridae are shared with Oriental and Ethiopian regions; Gliridae, Dipodidae, and Procaviidae also occur in the Ethiopian region. The Palearctic includes widespread, continuously distributed species from the rodent families Muridae and Sciuridae, as well as from the carnivoran families Mustelidae, Canidae, and Felidae.

Nearctic

This region extends from the Arctic in northern Canada to the central Mexican plateau, and includes Greenland. It is separated from the Palearctic by the Bering Strait and from the Neotropics by the Mexican-Central American transition area. Like the Palearctic, the northern Nearctic consists of tundra and taiga, with temperate deciduous forest in the southeast, grassland in the south-central, and desert-chaparral in the southwest. Relatively few mammalian families occur in the Nearctic, and only two (Antilocapridae and Aplodontidae) are endemic. Species diversity is highest in the west and south due to topographic variation and mild climate, respectively. The biotic similarity of Nearctic and Palearctic regions is such that they are often grouped together as the Holarctic, and two mammal families (Ochotonidae and Castoridae) have Holarctic distributions. The Nearctic-Neotropical land connection was formed relatively recently and became the dispersal route for the Great American Interchange (see discussion later in this chapter). As a result of this event, several families (e.g., Didelphidae, Erethizontidae, Tayassuidae, Heteromyidae, Dasypodidae) now occur in both regions.

Neotropical

The Neotropics extend from central Mexico to South America. Except at its northern limit, this region is isolated entirely by oceans. The Neotropics have a large number of mammal families and nearly half of them are endemic. Most of the species diversity is concentrated at low latitudes in Amazonian rainforests and flanking savanna or tropical scrub forests. To the south, the Neotropics are dominated by grasslands and desert and along their western margin by alpine habitats associated with the Andes. Eleven families of hystricognath rodents are endemic to the Neotropics: Abrocomidae, Cuniculidae, Capromyidae, Caviidae, Chinchillidae, Ctenomyidae, Dasyproctidae, Dinomyidae, Echimyidae, Myocastoridae, and Octodontidae. Other endemics are pilosans (Bradypodidae, Megalonychidae, Cyclopedidae, Myrmecophagidae), primates (Cebidae, Atelidae, Aotidae, Pithecidae), marsupials (Caenolestidae, Microbiotheriidae), and soricomorphans (Solenodontidae). The Caribbean Islands and Patagonia are normally considered parts of the Neotropics, but the biotas of these areas are quite distinct. Indeed, the Caribbean bat fauna is more similar to that of North and Central America than that of South America (Proches 2005).

Ethiopian

This region includes Madagascar and Africa south of the Sahara Desert. The Arabian Peninsula is a transition area between Ethiopian and southern Palearctic regions. Sub-Saharan Africa includes a band of tropical rainforest centered on the equator and extending along the Gulf of Guinea north to the Senegal River. The rainforest is flanked by broad savannas covering midlatitude regions, and southwestern Africa is mostly desert. Madagascar has rainforest along its eastern coast and savanna in the west, both habitats running in north-south bands parallel to the central highlands. The Ethiopian region shares over 70% of its mammal fauna with the Orient and nearly the same percentage with the Palearctic. Its great familial diversity of mammals may be attributed in part to its warm climate. Among endemics, lemuriform primates (Cheirogaleidae, Lemuridae, Lepilemuridae, Indriidae), aye-ayes (Daubentoniidae), and sucker-footed bats (Myzopodidae) are restricted to Madagascar. The remaining Ethiopian endemics are rodents (Anomaluridae, Bathyergidae, Ctenodactylidae, Pedetidae, Petromuridae, Thryonomyidae), lorisiform primates (Galagidae, Lorisidae), artiodactyls (Giraffidae, Hippopotamidae), tenrecs (Tenrecidae), golden moles (Chrysochloridae), elephant shrews (Macroscelididae), and the aardvark (Orycteropodidae).

Oriental

The Oriental region consists of the Indian subcontinent, southeast Asia, the Malay Archipelago, the Philippine Islands, Sumatra, Java, Borneo, and islands south to Wallace's Line between Bali and Lombok. It is isolated from the Palaearctic by deserts and mountains, and from the Australian region by a transition area known as "Wallacea" (extending east from Wallace's Line to encompass the Lesser Sunda Islands, Sulawesi, and the Moluccas). Much of the Oriental region is tropical forest. Nearly 75% of its mammal families are shared with the Ethiopian region, and over 50% are shared with the Palearctic. Mammal diversity is high in the Orient, but only five families are endemic: colugos (Cynocephalidae), tree shrews (Tupaiidae), hog-nosed bats (Craseonycteridae), gibbons (Hylobatidae), and tarsiers (Tarsiidae). The Orient's position as a tropical crossroads between Palearctic, Ethiopian, and Australian regions may explain its high diversity, but low endemicity, of mammal families.

Australian

This region includes Australia and New Guinea. It is bounded to the northwest by Wallacea and on all other fronts by ocean. Most of Australia is desert, but coastal areas show a diverse set of biomes–tropical forests in the north (and also in New Guinea), temperate deciduous forest in the southeast, chaparral in the south, and temperate rainforest in eastern Tasmania. Australia is the most isolated faunal region, sharing less than 20% of its mammal families with other areas. The endemics are monotremes (Ornithorhynchidae, Tachyglossidae) and marsupials from the orders Dasyuromorphia (3 families), Notoryctemorphia (1 family), Peramelemorpia (3 families), and Diprotodontia (11 families). The only recent route of exchange for Australian mammals is via Wallacea, and two eutherian groups—bats and murid rodents—have invaded by this route. Indeed, based on bat distributions, Proches (2005) found that New Guinea is more closely allied with the Oriental than the Australian region. Areas adjacent to Australia are also distinct. New Zealand has only three native nonhuman mammals, all bats, and two of these represent the endemic family Mystacinidae (Flannery 1984). Melanesia also harbors a distinctive bat fauna (Proches 2005).

Oceanic

Oceanic mammals include those that live on islands remote from continents and those that are fully marine. The mammal faunas of Micronesia and Polynesia illustrate several general patterns for oceanic islands: There are few native mammals, those that do occur are mostly bats or small rodents, and human movements across the ocean have facilitated dispersal (Darlington 1957). Among marine groups, sirenians (manatees and dugongs) occur in tropical coastal waters and associated deep river areas. Pinnipeds (seals and sea lions) breed on pack ice, nearshore rocks, or coastal areas, but some forage at considerable distances from land. Most cetaceans (whales and dolphins) are denizens of the open ocean. Larger whale species show extensive migrations between high-latitude feeding and low-latitude breeding areas, whereas smaller toothed whales occur within latitudinal zones. The distributions of marine mammals seem to be tied to their food sources, and regionalization of oceans, in turn, is strongly influenced by seawater temperatures (Briggs 1995). Thus, large cetaceans move across multiple oceanic provinces whereas smaller species have more circumscribed ranges.

Historical Biogeography

ABIOTIC PROCESSES

Plate Tectonics and Continental Drift

The suggestion that land masses of the Earth move over geological time (**continental drift**) was formalized by

Wegener (1915) but not accepted until the 1960s, because no geological mechanism was known that could account for such movement. **Plate tectonics** (Dietz 1961, Hess 1962) provided the mechanism. The Earth's crust, including continents and ocean floors, is made of rocky plates that float on denser, partially melted mantle rock. There are some ten major plates and many minor ones, separated from one another by ridges, trenches, or faults (figure 5.3). As heat from the Earth's core radiates outward, it creates convection cells in viscous mantle rock. Midocean ridges are sites where molten basalt from the mantle spews onto the surface and pushes the crust laterally ("seafloor spreading"), driving the movement of plates. When plates collide, they may form mountains (e.g., the Himalayas) or subduction zones (trenches), the latter being sites where dense oceanic rock plunges below lighter continental plates, producing earthquakes and volcanic eruptions. The mountainous, volcanic islands of Japan mark a zone of subduction between Pacific and Eurasian plates.

As a result of tectonic processes, continents have collided, merged, and fragmented during Earth history. In doing so, they have carried their biota along with them and profoundly influenced the distribution of organisms. Continental fragmentation is a principal mechanism of **vicariance,** the geographic isolation of populations of a once-widespread species by the development of a physical barrier (e.g., an ocean) within the ancestral species range (Rosen 1978). Geographic isolation initiates speciation, and subsequent evolution within descendant lineages may produce

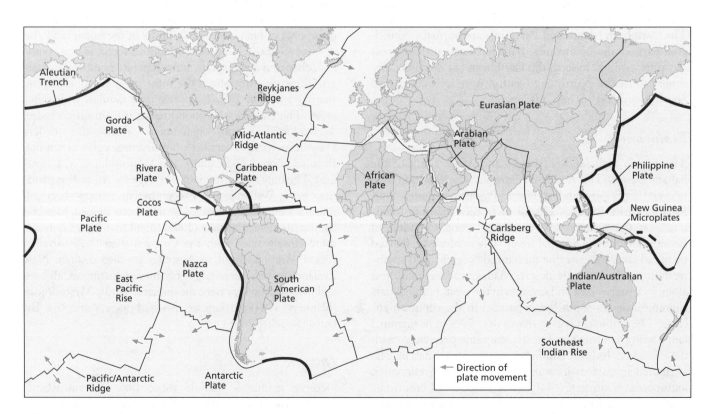

Figure 5.3 Tectonic plates. This map shows the boundaries of the major tectonic plates of the Earth's crust and the direction of their current movement (indicated by arrows). Oceanic trenches are shown as thickened lines.

diverse clades on different continents. Continental drift is only one of many ways to erect or destroy such a barrier, but it is the greatest historical determinant of provincialism among faunal regions. Vicariance on a smaller scale has been a major factor influencing biogeographic patterns within faunal regions.

The past movements and positions of continents can be inferred from paleomagnetic, petrological, stratigraphic, and structural information. When the first synapsids appeared in the Carboniferous, the land masses of North America, Greenland, Western Europe, and Siberia had moved close to the equator and eventually merged to form a supercontinent, Laurasia. South America, Antarctica, Africa, India, and Australia were similarly merged near the south pole to form the supercontinent Gondwana. By the Permian, Laurasia and Gondwana had coalesced to a single land mass, Pangaea, that would persist for some 160 million years (figure 5.4). In the Jurassic, when the world's synapsid fauna contained therapsids, cynodonts, and early mammals, Pangaea began to split apart. Separation of Laurasia and Gondwana opened a circumtropical waterway, the Tethys Sea, that moderated climate throughout the later Mesozoic, a period when elevated sea levels resulted in extensive epicontinental seas. In the Cenozoic, the continents fragmented, rotated, and drifted to their present locations; at the same time, mammals evolved, diversified, and spread over the globe.

Climate Change

The climate, or long-term weather pattern, of a particular area on Earth results from interactions among sunlight, air temperature, and water. A portion of the sun's infrared radiation that reaches the Earth is reflected, trapped by CO_2 and other greenhouse gases, and warms the atmosphere.

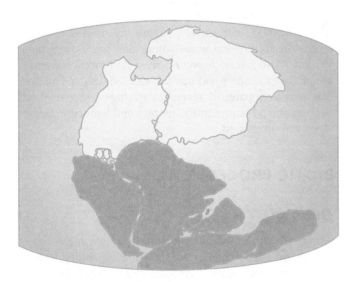

Figure 5.4 The supercontinent of Pangaea. A reconstruction of Pangaea in the Permian. Northern regions constituting Laurasia are shown in white; southern regions constituting Gondwana are shaded.

Because the Earth tilts on its axis, the amount of sunlight reaching the surface varies with latitude. At the equator, sunlight meets the Earth at nearly a right angle and the amount of solar energy per unit area is high, making the tropics warm. At higher latitudes, the incident angle is more oblique and temperatures are lower. Moreover, as the Earth's tilt changes throughout the year, warm and cold temperatures alternate seasonally between northern and southern hemispheres. As equatorial air rises, it cools and loses moisture, which falls as tropical rain. The same air masses sink at approximately 30° north and south latitude, but they are dry, and Earth's great deserts occur at these latitudes. Earth's rotation interacts with moving air masses to produce easterly or westerly winds, depending on latitude. Prevailing winds blowing over mountains drop moisture on the windward side, but create dry, rainshadow conditions on the leeward. These same factors move sea water in clockwise or counterclockwise gyres around the ocean basins and, by distributing warm water from the equator to coastal areas, provide a moderating influence on the climate of continental margins. To the extent that this interplay of variables creates a stable climate in a particular area, it also determines the abiotic selection pressures to which resident organisms must adapt.

However, climates are not stable over evolutionary time. Continental positions affect air and ocean circulation, profoundly influencing regional climates. Moreover, variations in Earth's orbit affect how much sunlight reaches different parts of the world and trigger climate changes by "orbital forcing"; variation in the energy output of the sun ("solar forcing") can have the same effect. The influence of climate forcing is mediated by conditions in the atmosphere (e.g., levels of greenhouse gases, tropospheric aerosols that influence cloud formation) and acts through the global carbon cycle (Pälike et al. 2006). As a result of variation in, and interaction among, these factors, Earth history is characterized by alternating periods of "icehouse" and "hothouse" conditions. Extreme icehouse conditions produced **ice ages,** extended periods during which global mean temperatures were low and glaciers expanded across continents. It was in the context of such climate variation that synapsids and their mammalian descendants originated and diversified.

Ice Ages

At several times in Earth history, levels of atmospheric greenhouse gases fell and the continents occupied positions blocking the flow of equatorial ocean water to the poles. In the early Carboniferous, much of Gondwana was near the south pole, while closure of the Tethys Sea blocked circumtropical currents and disrupted warm-water flow to the Antarctic. This caused cooler summers, the accumulation of winter snow in highland areas, and growth of alpine glaciers. At the same time, atmospheric CO_2 levels dropped, probably due to elevated oxygen produced by land plants in the world's extensive equatorial swamps (Berner 1997). As Earth became colder, glaciers spread from the highlands and covered much of southern Gondwana, initiating the Karoo Ice Age. In tropical Laurasia, far from the Karoo ice fields, we find the oldest fossil synapsids–pelycosaurs such as *Archaeothryis* and *Eothryis*. By the time the Karoo glaciers receded at the end of the Permian, all major lineages of pelycosaurs and noncynodont therapsids had appeared, many with representatives in Gondwana (Benton 2005). Ironically, few of them would survive into the Triassic (see discussion later in this chapter).

After a period of cooling in the late Cretaceous, hot-house conditions prevailed again until the Eocene. In the Oligocene, atmospheric CO_2 levels began to drop and fragmenting continents allowed cold polar water to move toward the equator. Glacial ice occurred in Antarctica and, by the Pliocene, covered the north pole as well. From the mid-Miocene on, the world became cooler and drier until, roughly 2 million years ago (mya), it plunged into the Pleistocene ice age. Glaciation during this period was most dramatic in the Nearctic, where ice covered most of modern Canada and the northern United States, shifting tundra and taiga habitats southward. In the Palearctic, ice covered northern Europe, but a unique steppe-tundra habitat developed in Siberia and Beringia. Between about 1.7 million and 10,000 years ago, continental glaciers advanced and retreated four times, giving rise to a cycle of glacial and interglacial periods that culminated in the Recent. The distributions of many modern mammals are still responding to the last glacial retreat.

Refugia

Refugia are circumscribed areas within a larger biome that preserve biodiversity during periods of environmental change (Lynch 1988). **Nunataks,** refugia within the continental ice sheets of glacial periods, were ice-free pockets of variable size. In them, remnants of the preglacial biota survived until the ice retreated. Nunataks such as the "driftless area" in southwestern Wisconsin, northwestern Illinois, and eastern Iowa, served as sources of new populations for surrounding areas once the glaciers receded (figure 5.5). The least weasel (*Mustela nivalis*) and Franklin's ground squirrel (*Spermophilus franklinii*) likely survived in the driftless area during the last glaciation. Mountaintops may also act as refugia for taxa with narrow altitudinal distributions—as communities shift their distribution to higher or lower elevations during periods of climate change, their constituent taxa are alternately isolated from and merged with those from nearby mountains (Lomolino et al. 2006).

Tropical rainforests around the world contain enormous species diversity. When Pleistocene glaciers covered much of the landscape in northern latitudes, rainforests became fragmented—in effect, islands of forest in large areas of open grassland—due to the cooler, drier conditions at tropical latitudes. During interglacial periods, forest refugia expanded and became connected. Haffer (1997) suggested that cycles of rainforest fragmentation were a major cause of vertebrate diversity in the Amazon.

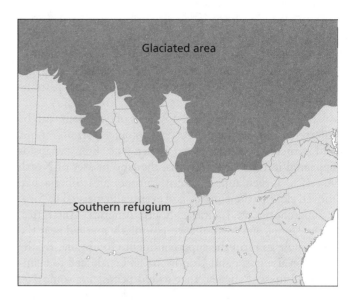

Figure 5.5 Driftless area. An area in northwestern Illinois, east central Iowa, and southwestern Wisconsin, known as the driftless area, remained free of ice sheets during the Pleistocene glaciations and served as a refugium. Several species of plants and animals, including mammals, spread outward from the driftless area as the glaciers receded.

According to this hypothesis, vicariant isolation led to differentiation among populations of the same species in different refugia. When the forests became contiguous again during interglacials, isolated populations had become reproductively isolated. This scenario has been challenged, however, by the apparent divergence of many Amazon species prior to the Pleistocene (Bush 1994).

Oceanic islands have been refugia, particularly islands close to larger land masses that experienced environmental change, as was the case for Madagascar (Eisenberg 1981). The Malagasy endemics Tenrecidae and Lemuriformes have fossil representatives from the Miocene of Africa (Savage and Long 1986) and Oligocene of Pakistan (Benton 2005), respectively, suggesting origins for both groups in continental areas where they are now extinct. Within Madagascar, however, they underwent extensive radiations and now occupy a wide variety of niches. Madagascar apparently served as a refuge where, isolated from mainland competitors, tenrecs and lemuriforms survived and diversified.

BIOTIC PROCESSES

Dispersal

The term "dispersal" has two closely related meanings in biology. In a later chapter, we describe how individuals or small groups may leave their natal area to breed elsewhere. Such movement occurs within the lifetime of an organism and constitutes **ecological dispersal.** "Dispersal" is also used in a biogeographic sense, referring to the extension of a species range (i.e., **species dispersal**). These processes are related in that individual movements are the basis for changing species ranges. Two kinds of species dispersal are usually recognized. **Passive dispersal,** such as rafting or human transport, involves movements in which the dispersing organisms have no active role. **Active dispersal** involves an accumulation of ecological dispersal events in which individuals move by their own locomotion.

Active dispersal occurs via several kinds of pathways (Simpson 1940). A **corridor route** provides minimal resistance to the passage of animals between two areas. The present connection between Europe and Asia is a corridor that allows extensive interchange of terrestrial animals. Many mammalian taxa, down to the level of genera and species, are distributed throughout Eurasia as a result of this corridor. A **filter route** allows only certain species to pass from one area to another. A good example is the Bering land bridge that connected Siberia and Alaska at times of lowered sea levels in the later Cenozoic. Only mammals such as voles (*Microtus, Clethrionomys*) that were adapted to the cold climate of Beringia could successfully cross between North America and Asia. Another example is the Panamanian land bridge that formed between North and South America 2.5 mya. Extreme habitats such as deserts or mountains may also act as filter routes that separate faunal regions and subregions. Waif dispersal takes place via **sweepstakes routes,** in which some unusual occurrence carries an organism or group of organisms beyond the limits of normal ecological dispersal and lands them in a habitat where they can survive and reproduce. The most common examples involve mammals moving across water barriers by rafting (terrestrial species) or wind-assisted flight (bats). The mystacinid bats of New Zealand have Miocene fossil representatives in Australia, suggesting that they crossed the Tasman Sea by an aerial sweepstakes route sometime in the later Tertiary (Hand et al. 1998).

Extinction and Diversification

Paleobiologists generally recognize two kinds of extinction. **Background extinction** refers to the incidental loss of species due to local factors such as habitat change, interspecific competition, predation, and so on. **Mass extinction** involves the simultaneous, catastrophic, and worldwide loss of species from many taxonomic groups. There have been several mass extinctions in the history of life, and synapsids have experienced at least four of them. The Permo-Triassic extinction, perhaps the result of a runaway greenhouse effect initiated by massive volcanism, eliminated the last pelycosaurs, most therapsid lineages, and a few early cynodont groups (Pough et al. 2005). The mid-Triassic extinction of herbivorous synapsids (e.g., dicynodonts, diademodontids) may have facilitated the early diversification of dinosaurs (Benton 2005). Several therapsid and cynodont lineages survived the end-Triassic extinction that wiped out numerous tetrapod families. The Cretaceous-Tertiary extinction (likely triggered by an asteroid impact) took a heavy toll on dinosaurs but cleared the way for subsequent diversification of therian mammals.

In the context of historical biogeography, extinction (especially background extinction) is often invoked to explain the *absence* of species from areas where, based on some biogeographic hypothesis, they are expected to occur. For example, monotremes and marsupials are two of several mammalian groups currently restricted to southern continents, and it is likely that this reflects a Gondwanan distribution of their ancestors (Long et al. 2002). A Gondwanan ancestry of marsupials is supported by the presence of living marsupial groups in South America and Australasia. Although living monotremes are restricted to Australasia, the Paleocene fossil platypus *Monotrematum* from Argentina documents a broader Gondwanan distribution for the group (Pascual et al. 1992a). It is often the case, however, that extinctions postulated by biogeographic hypotheses have not yet been confirmed by fossils.

Just as extinction reduces the species richness of a clade, evolutionary diversification increases it. Diversification in this sense is nothing more than speciation followed by genotypic and phenotypic divergence, but the most noteworthy cases took place rapidly, produced many descendants, and were geographically restricted. We often refer to them as *adaptive radiations*, and several of the most spectacular have already been mentioned—e.g., the lemuriform primates of Madagascar (60 species in four families), the New World monkeys (128 species in four families), and the Australasian marsupials (237 species in 18 families). In these and other cases, the ancestors of a lineage dispersed into a new region or became vicariantly isolated in a remote portion of their ancestral range. These progenitors may have encountered little competition in their new ranges if ecologically similar species were lacking, may have been competitively superior to resident species, or may have been preadapted to subsequent environmental changes that drove their competitors extinct. Interestingly, the result of diversification is not always divergence. A striking regularity in mammal phylogeny is the number of times ecologically similar species have arisen in different areas and from different ancestors as a result of **convergent evolution.** For example, myrmecophagy (ant- or termite-eating) and its specialized cranial morphology have evolved in six orders (figure 5.6): anteaters (Pilosa), pangolins (Pholidota), aardwolves (Carnivora), aardvarks (Tubulidentata), numbats (Dasyuromorphia), and echidnas (Monotremata).

BIOGEOGRAPHIC INFERENCE

Distributional Patterns and Historical Hypotheses

Research in historical biogeography seeks to explain current species distributions in light of phylogeny and Earth history. The former emphasizes the evolutionary past of

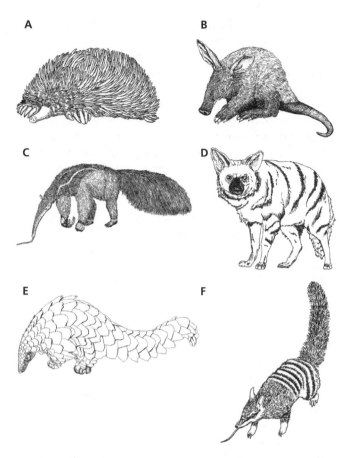

Figure 5.6 **Convergence.** Convergence is evident in these unrelated lineages of anteating mammals. All have a long rostrum and a sticky, extensible tongue: (A) echidna (Monotremata), (B) aardvark (Tubulidentata), (C) giant anteater (Pilosa), (D) aardwolf (Carnivora), (E) pangolin (Pholidota), (E) numbat (Dasyuromorphia). Not to same scale.

the species under study, and the latter stresses the evolutionary past of the regions those species occupy. Thus, like most areas of comparative biology, historical biogeography requires a phylogenetic framework. The distributional patterns we have discussed so far, provincialism and endemism, are predicated on phylogenetic hypotheses for major mammalian groups (e.g., monophyly of Monotremata, Marsupialia, Lemuriformes). Not all biogeographic problems are so straightforward. The most frequently studied distribution pattern is **disjunction,** the geographic isolation of sister groups, which can be observed at the level of species or larger clades. For example, the western quoll (*Dasyurus geoffroii*) and the bronze quoll (*D. spartacus*) are sister species of dasyurid marsupials (Krajewski et al. 2004); the western quoll is restricted to southwestern Australia and the bronze quoll is endemic to New Guinea. At the other extreme, Australasian marsupials are a monophyletic group whose sister is the South American Microbiotheria (Amrine-Madsen et al. 2003).

Prior to the acceptance of continental drift and awareness of past climate change, biogeographers assumed that dispersal and extinction were the major causes of disjunction. In a tectonically and climatically stat-

ic world, the simplest way to establish a disjunction is for an ancestral species that occurs in one area to send propagules into a second area, isolated from the first, that give rise to a new species. Alternatively, an ancestral species may expand its range into a new area, but later experience range contraction or extinction of intermediate populations, leaving an isolated colony that undergoes speciation (Darlington 1957). This dispersalist framework led early biogeographers to focus on "centers of origin" for particular taxa (i.e., the ranges of ancestral species). Unfortunately, the criteria used to infer such centers were not robust, and many dispersalist scenarios were overly speculative by modern standards. For example, Mathew (1915) argued that major mammalian groups arose in northern continents, with the more recently evolved and competitively superior forms displacing older, less well-adapted groups southward. Such scenarios relied on the types of dispersal routes described earlier or, where these did not suffice, posited ephemeral land bridges and lost continents (Archer 1984a). Analytical methods for inferring ancestral areas have been developed (Bremer 1992; Ronquist 1994; Hausdorf 1998), but most modern studies of historical biogeography focus on taxon-area relationships rather than centers of origin.

Vicariance and Dispersal

The nearly simultaneous development of phylogenetic systematics and plate tectonics theory in the 1960s gave a new conceptual framework to historical biogeography. It became reasonable to postulate that current species distributions reflect past fragmentation or connection of the continents on which ancestral species lived. Indeed, changing connections among Caribbean land masses was the driving force in Rosen's (1975) classic study of vicariance. Biogeographers realized that the causal links between geographic isolation, speciation, diversification, and extinction have the potential to explain many current distribution patterns. Out of this realization emerged *vicariance biogeography*, a research program committed to explaining distributions as a result of the interplay between phylogeny and area relationships. The logic of vicariance biogeography is straightforward (figure 5.7). Suppose species *ABC* occurs throughout an ancestral area that is split in two by a dispersal barrier. In one part of the range (area I), species *A* or its descendants persist; in the other part, a new species (*BC*) is derived by allopatric speciation. Now suppose that the area occupied by *BC* or its descendants is subdivided into areas II and III, resulting in a second isolation that produces species *B* in area II and species *C* in area III. In this scenario, the sequence of speciation events results in a phylogeny on which *B* and *C* are sisters apart from *A*; the sequence of range divisions results in area relationships in which II and III were more recently connected than either was with I. Because range fragmentation and speciation events occurred in parallel, area and phylogenetic relationships coincide.

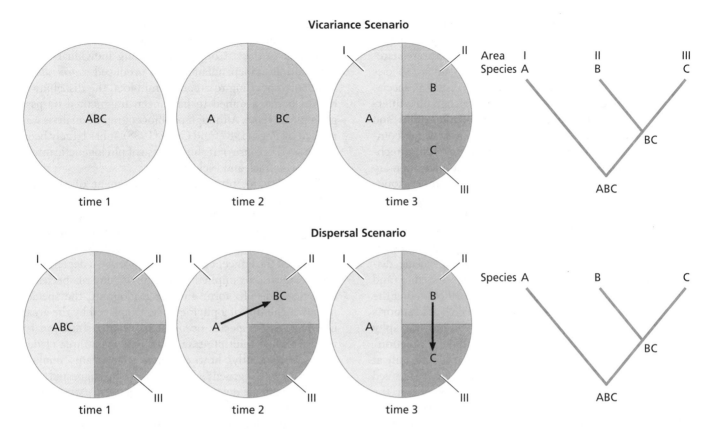

Figure 5.7 Vicariance and dispersal. Species *A*, *B*, and *C* evolve in areas I, II, and III. In the vicariance model, the ancestral species (*ABC*) is widespread and the areas are contiguous at time 1. At time 2, a barrier isolates species *A* and the ancestor of *B* and *C* (*BC*). At time 3, a second barrier results in species *B* and *C*. The phylogeny of species and the area cladogram are identical. In the dispersal model, the three areas are distinct at time 1. At time 2, a dispersal event (arrow) isolates ancestor *BC* in area II. At time 3, a second dispersal event isolates species *C* in area III. The phylogeny of species is consistent with both scenarios. *Redrawn from Brown and Gibson (1983).*

In the 1970s and 1980s, vicariance and dispersal hypotheses were often viewed as mutually exclusive. However, they differ only in the *timing* of dispersal: Vicariant models posit dispersal of an ancestral species throughout an ancestral area prior to the development of barriers; dispersalist models assume that the barriers came first (Kirsch 1984). Given the phylogeny and current distribution of species in the preceding example, a dispersalist scenario might posit that the three already distinct areas were simply colonized in the order I→II→III or I→III→II, with a speciation event corresponding to each colonization (figure 5.7). If both vicariance and dispersal scenarios are plausible, how do we choose between them? Vicariance biogeographers argued that, because a sequence of area fragmentations affects *all* species that occur in those areas, a vicariance hypothesis predicts that the phylogenies of multiple, independent groups will be *congruent* (i.e., that taxa from areas II and III will be sisters apart from their relatives in area I, and this will be true for many groups that inhabit the areas).

Several problems arose as biogeographers tried to implement vicariance reasoning. First, area relationships are usually difficult to work out, especially when the areas do not correspond to discrete land masses. Such areas may even be difficult to *define*. Continental biogeographers

attempt to do so from species distributions themselves, identifying *areas of endemism* within faunal regions (Hausdorf 2002). However, fine-scale regionalization often yields areas for some groups that are not congruent with those for others. Even if such areas can be identified, their history is often obscure. Therefore, vicariance biogeographers have relied on congruent phylogenies to reveal area relationships (Nelson and Platnick 1981). Of course, estimated phylogenies also have degrees of uncertainty that complicate the assessment of congruence among them. Moreover, it has become obvious that few species distributions were formed by vicariance alone—dispersal, change in community compositions over time, and evolving climatic conditions make inferring the geographic history of organisms a complicated endeavor. In response to these challenges, researchers have developed a variety of analytical techniques to extract historical information from current distributions and phylogenies.

Analytical Biogeography

Techniques of biogeographic and phylogenetic analysis have developed in parallel, and the two disciplines share many methodological approaches. Chief among these is the application of parsimony, the principle that the best estimate of relationships is the one that minimizes the number of events required to explain the observed data. In phylogeny reconstruction, the data are characters and the events are state-changes (chapter 3); in biogeography, the data are species distributions and the events are vicariance, dispersal, speciation, and extinction. In recent years, biogeographic inference has also come to rely on more mathematical and statistical techniques in attempting to cope with the complexity of evolutionary history. The literature on these techniques is quite large, and we provide only a brief summary of the major methods. Our discussion follows the "taxonomy of historical biogeographic approaches" proposed by Crisci et al. (2003) and Posadas et al. (2006).

Phylogenetic biogeography Brundin (1966, 1988) emphasized the predictive value of phylogeny for reconstructing biogeographic history: Within a single clade, disjunct sister groups provide evidence of vicariance and sympatric sisters suggest dispersal. Congruence among the phylogenies of independent, codistributed clades supports a common vicariant history. Phylogenetic biogeography was tightly linked, however, to the "peripheral isolation" model of speciation, in which new species originate only at the margins of their ancestor's range and possess derived characters (Hennig 1966). This model implies that the spatial distribution of species runs parallel to the primitive-derived sequence of their characters, and its practitioners emphasized dispersal and center-of-origin aspects of biogeographic history (Morone and Crisci 1995). In contrast, Croizat's (1964) **panbiogeography** assumed a strong causal link between Earth history and vicariance. Croizat

plotted the disjunct distributions of species within a clade on a map and connected occupied areas by lines to depict an *individual track*. Congruence among individual tracks for multiple, independent groups produced a *generalized track* corresponding to an ancestral biota, the distribution of which was assumed to have been fragmented by past geological events. Although panbiogeographic analysis was refined by Page (1987) and Craw (1988), it has largely been abandoned because of its limited use of phylogenetic information (Platnick and Nelson 1988).

Rosen (1978) initiated the development of **cladistic biogeography**, the primary goal of which was to infer *area relationships*. In this approach, the distribution and phylogeny of taxa are taken as evidence from which the vicariant history of areas is reconstructed. Rosen (1978) introduced the concept of an *area cladogram*, a depiction of area relationships implied by the phylogeny of species in a single clade. To form an area cladogram, the species names at the tips of a phylogeny are replaced by the areas in which those species occur. Area relationships can be inferred when multiple taxa show congruent area cladograms. Frequently, however, area cladograms contain unique or incongruent elements. Rosen suggested that such elements be deleted to produce a *reduced area* cladogram, the "residual congruence" of which shows vicariance relationships for some of the original areas (figure 5.8). Instances of incongruence may be explained by dispersal in one or more of the taxa.

Nelson and Platnick (1981) developed the logic of this approach into a procedure for constructing *general* area

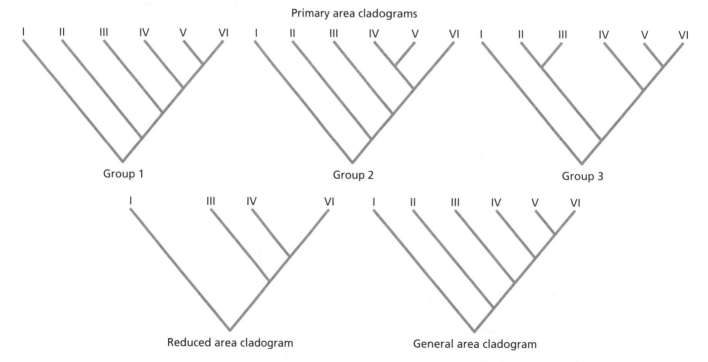

Figure 5.8 Area cladograms. Primary area cladograms (top) are the phylogenies of codistributed groups (Groups 1, 2, and 3) on which species names have been replaced by the areas (I–VI) in which the species occur. The primary area cladograms disagree on the relationships of areas II and VI. The reduced area cladogram (bottom left) is formed by deleting incongruent taxa. One possible general area cladogram, constructed using the cladistic biogeography approach, is shown at bottom left. This tree minimizes the number of dispersal events required to explain the incongruent relationships of areas II and VI in the primary area cladograms.

cladograms (figure 5.8), amenable to the inclusion of many independent taxa and to implementation by computer algorithms. Cladistic biogeography regularly encounters three major complications: widespread species (those present in multiple areas); missing areas (those absent in one or more area cladograms); and redundant distributions (areas with more than one species of a particular clade). These problems are addressed in the construction of general area cladograms by one of three approaches (dubbed assumptions 0, 1, and 2) that differ in how they interpret distribution patterns as evidence for vicariance, dispersal, or extinction. Area cladograms can be constructed by several methods, including component analysis (Nelson and Platnick 1981), Brooks parsimony analysis (Wiley 1987), three-area statements (Nelson and Ladiges 1991), and paralogy-free subtrees (Nelson and Ladiges 1996). Page (1990) noted that area relationships inferred from the branching orders of phylogenies also assume *temporal congruence* of the branch points, an idea that was often ignored in early cladistic biogeography studies (Donoghue and Moore 2003).

Parsimony analysis of endemicity (PAE; Craw 1988; Rosen 1988) uses the presence or absence of individual species in different areas as binary characters in a species × area data matrix from which a minimum-length tree of areas is constructed. This tree is taken as an estimate of the historical relatedness of areas. In its original formulation, PAE did not consider phylogenetic relationships, but Cracraft (1991) adapted it to incorporate hierarchical groupings of species into genera.

Ronquist and Nylin (1990) were the first to apply an explicitly statistical approach to reconstructing histories of cospeciation, ushering in a suite of **event-based methods** in historical biogeography (Crisci et al. 2003). Ronquist (1997) argued that areas seldom have a unique history that can be recovered from species phylogenies because different groups respond differently to dispersal barriers. His *dispersal-vicariance analysis* (DIVA) refocused attention on the distributional history of species within a single clade rather than area relationships. DIVA relies on a model in which speciation takes place by vicariance and biogeographic events are assigned different "costs" (e.g., vicariance = 0, dispersal = extinction = 1); a numerical algorithm identifies the minimum-cost area cladogram on which each event is mapped. The similarity between vicariance biogeography and host-parasite coevolution (both cospeciation processes), led Page (1994) to develop the *reconciled tree* method, in which one tree (e.g., the species phylogeny) is mapped onto another (e.g., area relationships). Instances of congruence are interpreted as vicariance, incongruence as dispersal, redundant areas as speciation without vicariance, and missing species as extinctions. Huelsenbeck et al. (2000) suggested that their Bayesian analysis of host-parasite cospeciation could similarly be extended to biogeography.

Comparative Phylogeography

In chapter 3 we described intraspecific phylogeography, the study of how lineages of conspecific organisms, iden-

tified by molecular markers, are spatially distributed and the processes that led to those distributions. Phylogeography is thus a way to study historical biogeography, but its emphasis on intraspecific patterns has made it most illuminating for recent divergences (e.g., within the last 25 million years). Phylogeographic inferences, like those drawn from analytical biogeography, are most robust when based on patterns observed in multiple, independently evolving, codistributed species (Cracraft 1989). Zink (1996) coined the phrase **comparative phylogeography** to describe this research program, but its roots go back to the concept of "genealogical concordance" articulated by Avise and Ball (1990). Avise (1996, 2000) defined four aspects of genealogical concordance that are the foundations of comparative phylogeography (figure 5.9): (1) concordance across sequence characters within a gene (i.e., lineages defined by individual gene trees are well-supported); (2) concordance in significant genealogical partitions across multiple genes within a species (i.e., the same well-supported lineages are recovered by several molecular markers); (3) concordance in the geography of gene-tree partitions across multiple codistributed species (i.e., well-supported lineages within several species have congruent distributions); and (4) concordance of gene-tree partitions with spatial boundaries between traditionally recognized biogeographic provinces (i.e., the distribution

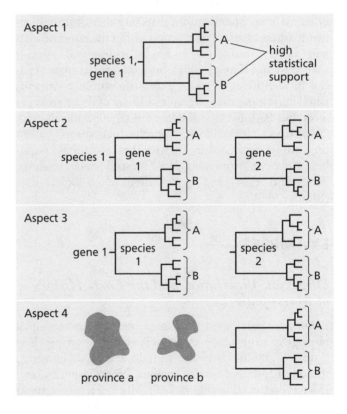

Figure 5.9 Genealogical concordance. These four aspects of genealogical concordance form the basis of comparative phylogeography. Together, they provide strong evidence that a common set of historical factors, such as vicariance, have shaped the biogeographic histories of codistributed species (here, species 1 and species 2). *Redrawn from Avise (2000).*

of well-supported lineages within several species corresponds to areas of endemism). When such concordance applies, phylogeographers infer that a common set of historical factors (e.g., vicariant events) have shaped the distributions of species in a region.

Cracraft (1988) suggested that the study of biogeography *within* faunal regions (e.g., continents) would be complicated by their long history of successive vicariance and dispersal episodes, ranging from the very old ("deep history") to the more recent ("shallow history"). Such layering of events produces incongruent distribution patterns when groups with deep and shallow histories are compared, thus confounding evaluation of genealogical concordance. For phylogeographers, the transition between deep and shallow history is roughly the Tertiary-Quaternary boundary, some 2 mya (Riddle 1996). Lineage divergences may be dated with respect to this boundary by molecular clock or population-genetic techniques (see chapter 3), providing an assessment of whether phylogeographic patterns are temporally, as well as spatially, congruent (Edwards and Beerli 2000).

Comparative phylogeography has been most successful in reconstructing events in the shallow part of deep history. Within the last 5–10 million years, many instances of *cryptic divergence* have taken place in mammals and other taxa. These are characterized by deeply divergent genealogical lineages within species that are not marked by phenotypic differences. When cryptic lineages occupy different areas of endemism, detecting them with molecular markers reveals a disjunction within the range of what had been considered a single widespread species (Arbogast and Kenagy 2001). Such studies have been crucial in identifying areas of endemism within continents and determining the geographic history of those areas (da Silva and Patton 1998; Riddle et al. 2000; Sullivan et al. 2000; Costa 2006). Phylogeographic analyses of shallow time are strengthened by analyses that incorporate population-genetic principles (e.g., nested clade analysis; Templeton 1998) and by inclusion of ecological data (Riddle 1996).

EXAMPLES

Dispersal, Vicariance, and the Early History of Marsupials

Patterson (1981) used traditional ideas about marsupial phylogeny to produce an area cladogram showing close historical affinities between Australia and New Guinea on the one hand, and South America, North America, and Eurasia on the other (figure 5.10). More or less congruent results were obtained for osteoglossine fishes, chelid turtles, galliform birds, ratite birds, and hylid frogs, implying a general area cladogram with the same structure. This cladogram reflects the then-current notion of a fundamental evolutionary split between Australasian and New World marsupials resulting from the late-Mesozoic

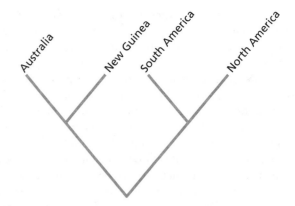

Figure 5.10 Marsupial area cladogram. Patterson's (1981) area cladogram for recent marsupial groups suggests that South America has stronger historical affinities with North America than with Australasia. These area relationships are predicated on the presumed reciprocal monophyly of New World and Australasian taxa.

breakup of Gondwana. However, knowledge of marsupial evolution has expanded considerably since 1981 and challenges this simple picture.

The oldest fossil marsupials are from the Cretaceous of North America and Asia, including such forms as *Kokopelia*, *Asiatherium*, pediomyids, glasbiids, peradectids, and stagodontids. The affinities of *Holoclemensia* and deltatheroideans are uncertain, but Rougier et al. (1998) placed these Asian-American fossils as primitive metatherians. The oldest South American marsupials are from the Paleocene Tiupampa Local Fauna of Patagonia (62 mya), which includes the orders Peradectida, Didelphimorphia, Sparassodonta, and Microbiotheria. The oldest Australian fossils are from the Eocene Tingamarra Local Fauna (55 mya) of Queensland, including *Thylacotinga* and *Djarthia* of uncertain ordinal affiliations. Is the history of these groups consistent with Patterson's (1981) area cladogram? To find out, we must consider marsupial phylogeny and Gondwanan geology.

Rougier et al. (1998) performed an extensive phylogenetic analysis of marsupial fossil taxa, while Amrine-Madsen et al. (2003) supplied the most complete molecular phylogeny of living marsupial orders to date. Integrating these results gives a modern, if not thoroughly validated, picture of marsupial phylogeny (figure 5.11). Fossil data suggest that predominantly North American and Asian taxa are the sister group of a South American-Australasian clade. Within the latter, Microbiotheria is sister to a monophyletic group of Australasian orders. These results reject part of Patterson's (1981) area cladogram—South American taxa are more closely related to Australasian than to North American ones. However, a close relationship between Australia and New Guinea is supported in that three of the four living australidelphian orders have representatives in both areas.

On the basis of this evidence, researchers generally agree that marsupials arose in what is now the Holarctic, then

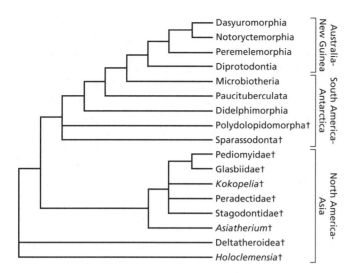

Figure 5.11 Marsupial phylogeny. This tree combines the morphological analysis of fossil groups (indicated with †) by Rougier et al. (1998) and the molecular analysis of living orders by Amrine-Madsen et al. (2003). Continental areas in which taxa occur are shown at right.

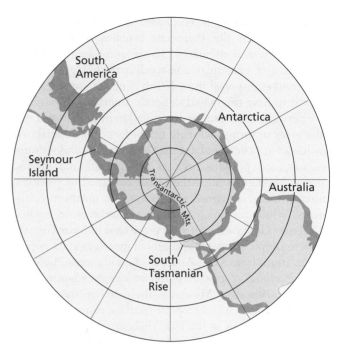

Figure 5.12 Gondwana in the early Eocene. Movement of marsupials from South America to Antarctica, and from Antarctica to Australia, likely took place just prior to this date. Dark shading indicates continental shelves less than 2000 m deep. *Redrawn from Woodburne and Case (1996).*

dispersed from North to South America (Goin 2003). The dates for this dispersal are bracketed by the Cretaceous Los Alamitos Local Fauna (73–83 mya) of Patagonia and the Paleocene Tiupampa fossils (62 mya): the former includes many mammals but no marsupials, whereas the latter includes the oldest South American marsupials (Wroe and Archer 2006). The nature of the north-to-south dispersal path is unclear; by the late Cretaceous, Laurasia and Gondwana were well-separated, and the existence of a corridor route is unlikely. Patterson and Pascual (1968) speculated that marsupials (and placental condylarths) moved via filter or sweepstakes routes over the water barrier between North and South America at this time, and that further separation of the continents produced vicariant isolation of the two faunas. Marsupials underwent a spectacular radiation in South America during the Tertiary, producing at least five major clades and numerous species that filled insectivore, omnivore, and carnivore niches. But how and when did they get to Australia?

Even when Pangaea was at its most extensive, South America and Australia were not contiguous—between them lay Antarctica, and it was via this landmass that marsupials must have dispersed to Australia (figure 5.12). Geologic evidence suggests that southern South America and the Antarctic Peninsula were contiguous from the Cretaceous until the late Eocene (36 mya) opening of the Drake Passage. Fossil polydolopoids, didelphimorphians, and perhaps microbiotheres have been recovered from the late Eocene (37–41 mya) of Seymour Island, adjacent to the Antarctic Peninsula. The anatomy of these animals, particularly the microbiotheres, suggests that they are too specialized to be direct ancestors of Australian marsupials. Their closest relatives are South American forms from 51–54 mya deposits, though the two faunas share no genera. Based on this evidence, Woodburne and Case (1996)

concluded that the ancestors of Australian marsupials dispersed across Antarctica no later than the early Eocene, and perhaps much earlier.

To reach Australia, marsupials must have moved from the Antarctic Peninsula across East Antarctica and into southeastern Australia via the South Tasman Rise. Geologic and climatic data suggest that the latter connection was submerged by the early Paleocene (64 mya) and that glacial ice was present in Antarctica by the late Eocene (35 mya). During this interval, overland dispersal became increasingly unlikely as the corridor evolved into a filter route. Woodburne and Case (1996) and Goin et al. (1999) argued for dispersal and subsequent vicariant isolation of Australian marsupials prior to the early Paleocene (64 mya). In contrast, Wroe and Archer (2006) cited potentially close relationships between the oldest Australian marsupials (*Thylacotinga* and *Djarthia*, 55 mya) and South American forms as evidence for a later, perhaps Eocene, vicariance. Modern Australian marsupial orders and many families first appear in late Oligocene (26 mya), though Woodburne and Case (1996) suggested that the ordinal radiation had taken place by the early Eocene (53 mya).

The Great American Interchange

After North and South America separated in the late Mesozoic, they remained isolated by a marine barrier until the Pliocene (2.5 mya). During this interval, mammal evolution on the two continents proceeded more or less independently. The Tertiary fossil record of South

America documents three phases of mammal evolution, beginning with the Paleocene remains of marsupials, condylarths, and edentates (Savage and Long 1986). As noted earlier, marsupials and condylarths likely arrived by sweepstakes dispersal from North America, but edentates appear to have originated in South America. Throughout the Tertiary, these groups diversified to produce a spectacular array of forms including five orders of marsupials, six orders of archaic ungulates such as litopterns and notoungulates, and pilosans such as glyptodonts and ground sloths. Marsupials filled the niches occupied by placental insectivores, carnivores, and rodents on northern continents, while placentals dominated herbivorous niches. The second phase began with the arrival of hystricognath ("caviomorph") rodents and ceboid monkeys via a sweepstakes route from West Africa in the Oligocene. Like the older mammal groups, these too underwent adaptive radiations during the Miocene—their living descendants comprise 11 and 4 Neotropical families, respectively. The third phase began in the late Miocene, when North and South America had drifted close enough to one another for limited mammalian dispersal across island chains. At this time, procyonids (raccoons) moved from north to south, and megalonychid ground sloths reached Florida. In the North American Tertiary, marsupials and condylarths went extinct while carnivorans, perissodactyls, artiodactyls, proboscideans, soricimorphs, lagomorphs, and sciuromorph rodents diversified.

In the Pliocene, some 2.5 mya, the land connection between North and South America was reestablished by emergence of the Isthmus of Panama, initiating extensive dispersal of mammals between the two continents, an event widely referred to as the "Great American Interchange" (Marshall 1988a). The isthmus was initially dominated by savannas similar to those in areas to the north and south, and thus constituted a corridor for southward-dispersing mammals like horses and deer, as well as northbound glyptodonts, ground sloths, and notoungulates (Pough et al. 2005). Development of tropical forests in Panama during the early Pleistocene, however, converted the isthmus into a filter route through which weasels, bears, cats, dogs, tapirs, llamas, peccaries, gomphotheres, shrews, rabbits, and voles invaded the south, while opossums, anteaters, armadillos, capybaras, and porcupines moved north (figure 5.13). In the Pliocene, these mixed faunas seemed to coexist on both continents, but Pleistocene extinctions took a heavier toll on South American forms: glyptodonts, ground sloths, and notoungulates disappeared, as did many of the Pliocene invaders from the north (e.g., horses, gomphotheres). Today, opossums (*Didelphis virginiana*), porcupines (*Erethizon dorsatum*), and armadillos (*Dasypus novemcinctus*) are the only South American mammals that have established ranges north of the Rio Grande (figure 5.14). In contrast, roughly 50% of modern South American genera are derived from North American immigrants.

Dispersal-Vicariance Analysis of Gibbons

Gibbons (Hylobatidae) are the living sister group of hominid primates (gorillas, chimpanzees, and humans). The 14 currently recognized species comprise four genera: *Bunopithecus* (1 species), *Hylobates* (7 species), *Nomascus* (5 species), and *Symphalangus* (1 species). These occur exclusively in southeast Asia, from easternmost India to southernmost China, south through Bangladesh, Myanmar, and Indochina to the Malay Peninsula, Sumatra, western Java, and Borneo. The fossil record of gibbons is extremely limited, and no comprehensive, paleogeographic scenario for their evolutionary history had been suggested until

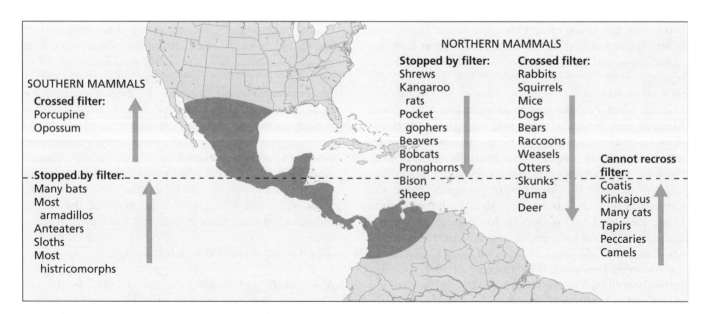

Figure 5.13 The Great American Interchange. During the Pleistocene, the tropical forests of Central America and the Panamanian isthmus were a filter route that permitted the dispersal of some mammals but a dispersal barrier for others. *From Vaughn (1986).*

Chatterjee (2006) used dispersal-vicariance analysis (DIVA) to obtain such a hypothesis. Chatterjee estimated phylogenetic relationships for 12 of the 14 species using

mitochondrial cytochrome *b* DNA sequences, employing the human sequence as an outgroup. Calibrating branch lengths to the presumed separation of hominids and gibbons some 15 mya provided molecular clock estimates of divergence dates on the gibbon tree. To implement DIVA, the phylogeny was converted to a taxon-area cladogram by replacing species with the areas in which they occur. Lacking any clear identification of areas of endemism in southeast Asia, Chatterjee employed geopolitical regions that circumscribe current gibbon distributions. DIVA provided parsimonious assignments of ancestral areas to each node of the tree from which vicariance, dispersal, and extinction events could be inferred (figure 5.15).

On the cytochrome *b* phylogeny, *Nomascus*, *Symphalangus*, and *Bunopithecus* form successive sister groups to a monophyletic *Hylobates*, with the entire gibbon radiation taking place during the past 10 million years. DIVA found many optimizations of ancestral areas, each requiring a minimum of 23 dispersal events to explain current distributions. These optimizations were constrained, for the sake of interpretability, to allow only two current areas of endemism to constitute any ancestral area. At the base of the tree, Vietnam and Laos are implicated as ancestral areas for gibbons. However, the oldest gibbon fossil (*Laccopithecus*, late Miocene; Wu and Pan 1985) is from Yunnan. When the algorithm was modified to allow three

Figure 5.14 Virginia opossum. The only living marsupial found in the United States, *Didelphis virginiana*, crossed the Panamanian isthmus from South America during the Great American Interchange.

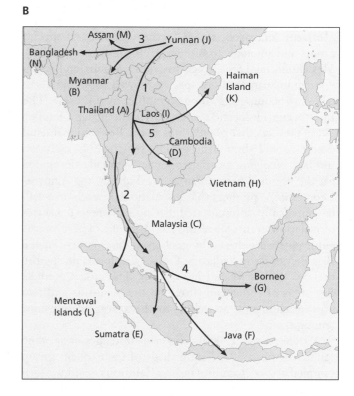

Figure 5.15 Biogeography of gibbons. (A) Potential ancestral areas reconstructed by dispersal-vicariance analysis are shown on the internal branches of gibbon phylogeny derived from cytochrome-*b* DNA sequences. Area abbreviations are given in the map at right. Two letters together indicate that each area is equally likely to constitute the ancestral area. (B) The geography of gibbon evolution suggested by DIVA analysis at left and the fossil record. Events: 1 = origin of gibbons in eastern Indochina (10.5 mya); 2 = dispersal of *Symphalangus* and *Hylobates* ancestors into Sumatra (8.6–10.5 mya); 3 = dispersal of *Bunopithecus* ancestor to Myanmar, Assam, and Bangladesh (7–8 mya); 4 = radiation and dispersal of *Hylobates* into Borneo (3–5 mya); 5 = radiation and dispersal of *Nomascus* into Cambodia and Hainan Island. *Redrawn from Chatterjee (2006).*

current areas to constitute an ancestral area, Yunnan appeared at the basal node. Thus, Chatterjee (2006) suggested that the ancestral area for gibbons included Vietnam, Laos, and Yunnan, with species of *Nomascus* diversifying in the former two areas during the Pleistocene (figure 5.15). *Symphalangus* originated by a late Miocene dispersal from southeast Asia into the Malay Peninsula, and *Bunopithecus* by a slightly later westward dispersal into Myanmar. Diversification of *Hylobates* species involved Pliocene dispersal from the Malay Peninsula southward and vicariant isolation in Sumatra, Java, and Borneo. This scenario is consistent with limited data on emergent land masses (Hall 1998) and rainforest distributions (Heaney 1991) in the late Cenozoic of southeast Asia.

Comparative Phylogeography of Baja California Rodents

The Sonoran Desert of Baja California, northwestern Mexico, and the southwestern United States is occupied by arid-adapted groups such as pocket mice (Heteromyidae). Indeed, several small-mammal species and species groups have ranges that cover much of this area and extend into the adjacent Mojave and Chihuahua Deserts. Riddle et al. (2000) conducted a comparative phylogeography study that included the cactus mouse (*Peromyscus eremicus*) species group, Bailey's pocket mouse (*Chaetodipus baileyi*) species group, little desert pocket mouse (*C. arenarius*), Merriam kangaroo rat (*Dipodomys merriami*) species group, and antelope ground squirrels (*Ammospermophilus*). Mitochondrial DNA trees for these groups and seven nonmammalian amniote species allowed Riddle et al. to test two competing biogeographic hypotheses. The Pleistocene-Holocene dispersal hypothesis (e.g., Findley 1969) predicts that populations in the Baja and mainland deserts would show little genetic structure. In contrast, the late Neogene vicariance model—initially developed for the regional herpetofauna (see Murphy and Aguirre-Léon 2002)—predicts that populations in each area will be genetically differentiated, having experienced a series of isolation events since the late Miocene. These vicariant events were driven by geological development of the Baja Peninsula and adjacent Gulf of California during the past 5.5 million years (figure 5.16). The model predicts five areas of endemism (three on the peninsula and two on the mainland), as well as the historical relationships among them.

Riddle et al. (2000) obtained mitochondrial gene sequences from 30–73 individuals of each rodent group sampled at 7–33 localities in Baja California, southwestern United States, and northwestern Mexico. Gene trees for these species and the nonmammalian taxa yielded area cladograms and sequence-divergence levels that are complex but broadly congruent with the Neogene vicariance model (figure 5.16). The results suggest that species in the Baja region have responded to a common set of historical factors, but they also exhibit species-specific instances of extinction, dispersal, or maintenance of gene flow between areas (Riddle and Hafner 2006).

Ecological Biogeography

Major topics that are usually discussed in the context of ecological biogeography include community assembly, island biogeography, and macroecology. Because these subjects are so deeply rooted in community ecology, we defer discussion of them to chapter 26. Here we summarize several key geographic phenomena that have been observed in mammals and are explained by ecological determinants.

ECOGEOGRAPHIC PATTERNS OBSERVED IN MAMMALS

Since the 19th century, a number of regularities have been noted in the ways that morphological characteristics of mammals vary with geography. Many of these patterns have been codified as *ecogeographic rules* and, for the sake of convenience, we use this terminology here. However, none of the patterns are invariant, some are of questionable generality, and all are the result of complex interactions among multiple historical and environmental factors. Thus, they are "rules" only in the loosest sense. Lomolino et al. (2006) provide a critical discussion of ecogeographic rules, including their applicability to nonmammalian groups, and conclude that there is as much to be learned about the causes of geographic variation from studying the exceptions as from the rules themselves.

The Island Rule

Small mammals on islands tend to have larger body sizes than their close relatives or ancestors on the mainland (insular gigantism). Large mammals show the opposite trend, with island species usually smaller than their mainland counterparts (insular dwarfism). Foster (1964) drew attention to the generality of this pattern, and Van Valen (1973) reified it as the **island rule** (figure 5.17). Gigantism has most frequently been described in rodents (Delaney and Healey 1967; Angerbjörn 1986; Michaux et al. 2002), though the pattern is not universal (Millien 2004). Dwarfism has been documented in proboscideans (Vartanyan et al. 1993), artiodactyls (Lister 1989; Simmons 1988), primates (Brown et al. 2004), pilosans (Anderson and Handley 2002), and carnivorans (Foster 1964). Both patterns can be found in bats (Krzanowski 1967). As just one example of the island rule, Croft et al. (2006) described variation among mainland and insular forms of *Bubalus* water buffaloes. Males of the mainland

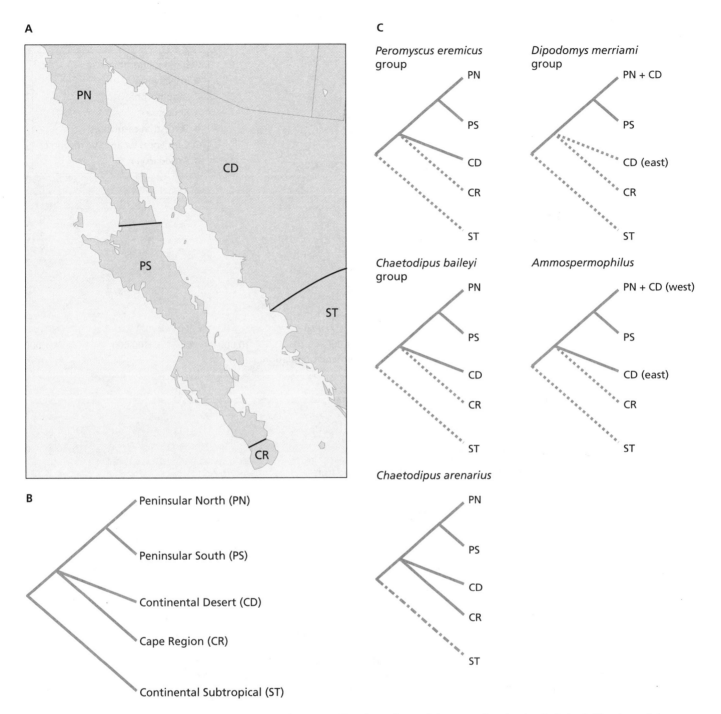

Figure 5.16 Comparative phylogeography of Baja California rodents. (A) Areas of endemism in Baja California and the adjacent mainland. (B) Area relationships reconstructed from geological history. (C) Area cladograms based on mitochondrial genealogies for five of the rodent groups studied by Riddle et al. (2000). Taxonomic names have been replaced by the areas in which species were sampled. Note the congruence among gene trees and area relationships. Dotted lines indicate missing taxa; dot-dash lines indicate missing taxa thought to occur in specific areas. *Redrawn from Riddle et al. (2000).*

Asiatic water buffalo (*B. bubalus*) weigh over 1000 kg, whereas other extant *Bubalus* species occupy islands and are much smaller: *B. mindorensis* from the Philippine island of Mindoro weighs about 210 kg; *B. quarlesi* and *B. depressicornis* from Sulawesi are less than 150 kg. Croft et al. (2006) also described a new fossil species of dwarf water buffalo (*B. cebuensis*) from the Pleistocene or

Holocene of Cebu Island (Philippines), with an estimated body mass of 150–165 kg.

A considerable literature exists on the evolutionary causes of insular gigantism and dwarfism (Lomolino et al. 2006). Heaney (1978) found a correlation between body and island sizes in insular tri-colored squirrels (*Callosciurus prevostii*) in southeast Asia. Because larger islands tend to

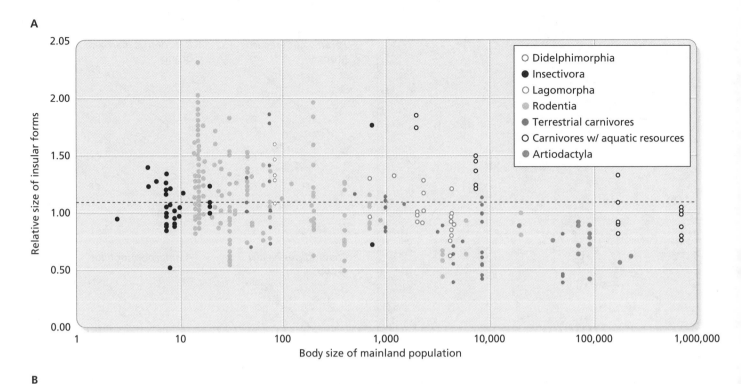

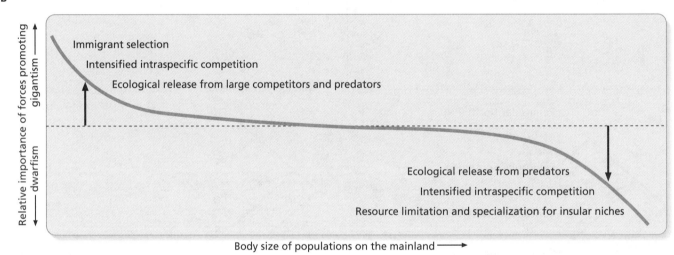

Figure 5.17 The island rule. (A) Relative body size of island mammals (proportion of body mass of mainland relatives) plotted as a function of the body mass of mainland relatives. (B) The effects of selective forces on the size of island mammals vary as a function of the ancestral (mainland) body size. *Redrawn from Lomolino (2005).*

have more resident species than smaller islands, Heaney argued that insular mammals experience variable levels of food limitation, predation, and competition depending on the size of the island they occupy. For large mammals, resource limitation on small islands should be the most intense selection pressure and favor reduced body size; for smaller mammals, reduced interspecific competition ("ecological release") should favor larger size. Lomolino (1985) noted that the latter trend could be reinforced by immigrant selection (i.e., larger individuals, especially in smaller species, have a better chance of colonizing islands in the first place). Moreover, the species composition of

islands will be affected by their distance from mainland source populations. Lomolino (2005) showed that the island rule is a general pattern emerging from this array of selective forces (figure 5.17) and suggested that insular species may evolve toward an optimal size determined by their ecological constraints.

Rapoport's Rule

Rapoport (1982) noted that the latitudinal breadth of species ranges in mammals tends to increase from the equator to the poles. Stevens (1989) called this pattern

Rapoport's rule, documented it in other groups of organisms, and argued that it results from species' responses to a gradient of increasing climatic variability at higher latitudes. Species living at high latitudes are adapted to a relatively broad seasonal range of environmental conditions and are able to expand their geographic ranges accordingly. In contrast, tropical species are adapted to a more stable climate, have evolved toward ecological specialization, and have narrower distributions. Rapoport's rule has been supported most consistently in quantitative studies of North American mammals (Arita et al. 2005; Pagel et al. 1991), as well as New World bats and marsupials (Lyons and Willig 1997). However, Gaston et al. (1998) argued that the pattern is limited to high-latitude taxa in the Holarctic and that alleged documentations of it are undermined by statistical flaws. Indeed, Smith et al. (1994) found that Rapoport's rule did not apply to Australian mammals. Colwell and Lees (2000) even suggested that this and other latitudinal diversity gradients are due to a statistical artifact, the "mid-domain effect," in which chance alone results in more and narrower ranges nearer to the equator than to the poles. Lomolino et al. (2006) acknowledge the limited generality of Rapoport's rule but argue that its application is sufficiently common to warrant continued investigation.

Bergmann's Rule

One of the earliest ecogeographic rules was that of Bergmann (1847), who observed that the body sizes of mammals and birds increase with increasing latitude. Although originally framed to describe trends among species, **Bergmann's rule** is now usually interpreted as a pattern of *intraspecific* variation (Mayr 1956; Lomolino et al. 2006). The trend has been shown to hold statistically for mammals in general, though there is much variation in its applicability among individual species. Blackburn and Hawkins (2004) demonstrated that Bergmann's rule was strongly supported by analyses of size variation in North American mammals. However, Meiri and Dayan (2003) found that only 65% of the 149 mammal species they tested conformed to the rule, and Meiri et al. (2004) found that it applied to only 50% of 44 carnivoran species. Moreover, variation in some small-bodied mammals—notably shrews (Ochocinska and Taylor 2003; Yom-Tov and Yom-Tov 2005) and heteromyid rodents (Ashton et al. 2000)—appears to contradict Bergmann's rule in that body size decreases with increasing latitude.

Bergmann's (1847) original explanation for a latitudinal size gradient was based on the superior heat-conserving capacity of large-bodied endotherms. A typical large mammal has a lower surface-to-volume ratio than a small mammal, and hence a smaller surface area across which to lose body heat at cold temperatures. Indeed, Blackburn and Hawkins (2004) found that average annual temperature is the strongest predictor, among six variables evaluated, of average body mass in North American mammals

(figure 5.18). Thermoregulatory constraints also appear to be causally related to latitudinal variation in the size of gray mouse lemurs (*Microcebus murinus*) in Madagascar (Lahann et al. 2006). However, McNab (1971) rejected Bergmann's hypothesis as a general explanation, noting that large mammals living in extreme environments at high latitudes face a greater challenge in meeting their energy requirements than do small mammals in the same environment. McNab suggested that large body size may be the evolutionary result of reduced interspecific competition in species-poor communities at high latitudes. More recent hypotheses emphasize the greater ability of large mammals to survive prolonged periods of energy deprivation or thermal stress by virtue of their lower critical temperatures and increased capacity to store food and water (McNab 2002). Very likely, Bergmann's rule and its exceptions are due to different combinations of causal mechanisms in different species.

Allen's Rule, Gloger's Rule, and Other Patterns

Extending Bergmann's reasoning about thermoregulatory adaptations in endothermic vertebrates, Allen (1877) observed that mammals and birds living in cold climates have shorter appendages than do their close relatives in warm climates. Long limbs, tails, ears, or other appendages increase the surface area for heat dissipation in mammals, which might be adaptive as a cooling mechanism in hot, dry environments (Schmidt-Nielsen 1979). The traditional

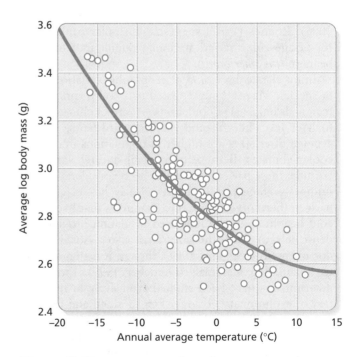

Figure 5.18 Bergmann's rule and temperature. The mean values of log(body mass in grams) as a function of average annual temperature experienced by species of northern North American mammals. The curve is a best-fit polynomial regression. *Redrawn from Blackburn and Hawkins (2004).*

example of **Allen's rule** is correlation of ear length with air temperature in hares of the genus *Lepus* and in foxes (Griffling 1974; Hesse 1937), but few other mammalian examples have been discovered. Stevenson (1986) confirmed the correlation for size-specific ear and tail lengths in *Lepus* but rejected it for cottontail rabbits (*Sylvilagus*), also showing that hind foot length in both groups was *negatively* correlated with temperature. Lindsay (1987) interpreted the results of a multivariate analysis of craniometric characters in Rocky Mountain red squirrels (*Tamiasciurus hudsonicus*) as consistent with Allen's rule. Lomolino et al. (2006) suggest several reasons why the original thermoregulatory explanation for Allen's rule may be overly simplistic.

Gloger (1883) noticed an apparent correlation between the plumage colors of closely related birds and the humidity levels of their habitats, with darker forms more frequently found in humid environments and lighter forms in dry areas. In mammals, the rule would apply to pelage color—for example, white polar bears (*Ursus maritinus*) inhabiting the dry Arctic contrast with brown grizzly bears (*U. arctos*) in the somewhat more humid tundra and boreal habitats of North America and Eurasia. However, the generality of **Gloger's rule** has not been assessed, and it may be a manifestation of selection for cryptic coloration in mammals (Lomolino et al. 2006). Several studies have documented an adaptive (as predator-avoidance) correlation between dorsal pelage color and soil color in rodents, including deer mice (*Peromyscus maniculatus*; Dice 1947), pocket gophers (*Geomys bursarius*; Krupa and Geluso 2000), and pocket mice (*Chaetodipus intermedius*; Hoekstra et al. 2005). Rounds (1987) suggested a similar explanation for pelage-color variation among populations of black bears (*Ursus americanus*).

In his classic discussion of geographic variation within species, Mayr (1942:92) listed two "rules applying to mammals only": (1) Populations in warmer climates tend to have less underfur and shorter guard hairs, and (2) average litter sizes are higher in populations inhabiting cooler climates. Because mean temperatures vary with latitude, these patterns have also been considered as latitudinal gradients. The former clearly has an adaptive basis in thermoregulation (i.e., mammals that live in the cold benefit from the increased insulation provided by dense pelage). The litter-size rule is also reasonably well documented (Cockburn 1991), though it may not apply to hibernating mammals (Derocher 1999). Litter-size determinants have been studied extensively in the context of evolutionary ecology (Pianka 2000) and life-history theory (Stearns 1992). High latitudes experience seasonal extremes of productivity, reducing life expectancy for small mammals and favoring large litters with minimal parental investment per offspring (*r*-selection). A similar effect has been shown in comparisons of mainland and island voles (e.g., Tamarin 1977): Island populations, which experience a more stable climate

than mainland populations, have smaller litters with higher per-offspring energy investment (*K*-selection).

GRADIENTS IN SPECIES DIVERSITY

The Latitudinal Gradient

Perhaps the first global ecological pattern described by naturalists was that species diversity (i.e., the number of species per unit area) decreases from the equator toward the poles, a trend that holds for most groups of organisms (Hawkins 2001). It is apparent in Darlington's (1957) counts of tropical and temperate mammal species in eastern Asia (180 and 100, respectively). A classic demonstration of this pattern for mammals was given by Simpson (1964), who tabulated the number of mammal species occurring in 150 square-mile quadrats throughout North America (figure 5.19). Simpson found that over 160 mammal species may occur within a single quadrat of Costa Rican tropical forest, with diversity dropping in an irregular but unmistakable trend to a low of about 20 species per quadrat in northern Canada and Alaska. Since Simpson's classic work, the latitudinal gradient of species diversity has been confirmed for North American mammals in general (Wilson 1974), and for specific New World groups such as forest-dwelling species (Fleming 1973), bats (Willig and Selcer 1989; Stevens and Willig 2002), and marsupials (Willig and Lyons 1998). Cowlishaw and Hacker (1997) showed the gradient in African primates, and, most recently, W. Sechrest and colleagues (in Lomolino et al. 2006) illustrated it for all terrestrial mammal species. It is important to note that the trend is not a monotonic decline along a transect from the equator to the poles, but one that displays considerable local variation. For many groups, it is dominated by an abrupt drop in species diversity as one moves from tropical to subtropical latitudes (Lomolino et al. 2006).

Despite the ubiquity of the latitudinal trend, it does not apply to all mammal groups or all regions. Marine mammals (pinnipeds and cetaceans), in particular, seem to show the opposite gradient, with species diversity peaking at high latitudes (Rohde 1992). Similarly, the diversity of soricomorphs (shrews and moles) is higher in temperate than in tropical areas (Cotgreave and Stockley 1994). In South America, drylands south of the tropics support a greater diversity of endemics than the lowland Amazon rainforest (Mares 1992). Andrews and O'Brien (2000) suggested that different latitudinal gradients apply to large versus small African mammals and to species with different feeding strategies.

Numerous explanations have been suggested for the latitudinal gradient. Lomolino et al. (2006) list 30 such hypotheses in four categories that were published through 2004: null models such as the mid-domain effect (see "Rapoport's Rule" above); community ecology models that emphasize species interactions; abiotic models based

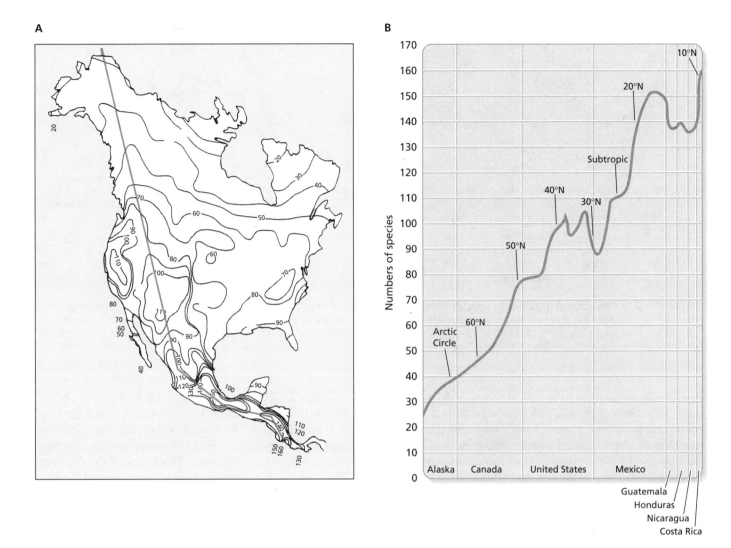

Figure 5.19 Latitudinal gradient in species diversity. (A) Contour map showing the number of species of mammals per 150 square-mile quadrats in North and Central America. (B) Plot of species diversity along the transect (solid line) shown in (A). *Redrawn from Simpson (1964).*

on environmental conditions, land area, and age of land masses; and integrative models involving combinations of factors in the other three categories. Some of these models have been cited frequently by mammalogists (Willig et al. 2003). In tropical areas, high primary productivity, climatic stability, and habitat heterogeneity may contribute to high species diversity (Currie 1991). Badgely and Fox (2000) found that combinations of abiotic factors had the strongest statistical associations with species diversity of mammals in North America, but individual factors varied among mammals of different sizes and feeding guilds. High species diversity may be a self-reinforcing pattern in that the presence of more species in an area (e.g., the tropics) results in complex interspecific interactions (competition, predation, mutualism, parasitism), which tend to promote species coexistence, perhaps by narrowing niche widths (Pianka 1966). Others have argued that the high diversity of tropical habitats reflects the greater area of tropical landmasses (Connor and McCoy 1975) or their

longer residence time in tropical latitudes (Pianka 1966). Lomolino et al. (2006) suggest that the integrative approaches of Rosenzweig (1992), Wiens and Donoghue (2004), and others have the greatest potential for explaining the latitudinal diversity gradient. These models attempt to evaluate the influence of latitudinally varying ecological factors on the evolutionary processes (speciation, extinction, and dispersal) that determine regional species richness.

Elevational and Peninsular Gradients

Ecologists have long been aware that elevation exerts a powerful influence on the composition of biological communities and the diversity of species within those communities (Whittaker 1975). Initial studies emphasized the species-poor nature of mountaintops compared with the more diverse habitats in adjacent lowlands (e.g., Patterson et al. 1996). In some cases, the elevational gradient of

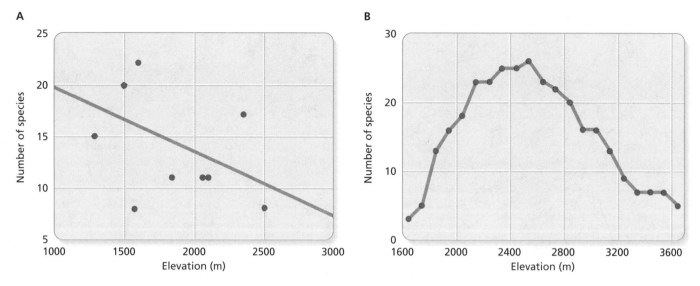

Figure 5.20 Elevational gradient in species diversity. (A) In Bwindi Impenetrable National Park, Uganda, the diversity of small rodents and shrews shows a steady decrease with altitude. Dots represent collecting localities and the line is a linear regression. (B) In the Uinta Mountains of the southwestern United States, the diversity of small nonflying mammals peaks at intermediate altitudes. *Redrawn from Kasangaki (2003) and Rowe (2005).*

species diversity seems to be a steady decrease from low to high elevations (figure 5.20A; Kasangaki et al. 2003). Explanations for this pattern have emphasized the changing climatic characteristics along a mountainside (Happold and Happold 1989). At higher elevations, temperature and oxygen levels drop while moisture becomes tied up in frosts. These abiotic conditions result in lower primary productivity on mountaintops and a correspondingly less diverse community. However, many studies with thorough geographic sampling (e.g., Rickart 2001) have suggested that species diversity peaks at middle, rather than low, elevations (figure 5.20B). Lomolino (2001) argued that this pattern is expected if one considers how the area and relative isolation of elevational zones interact with changing climate conditions. The conical shape of most mountains ensures that high-elevation zones are smaller and more isolated from one another than lower zones, resulting in lower immigration rates and higher extinction rates. Isolation, however, also promotes speciation. The net effect of these countervailing factors is that species diversity should be relatively high at the base of mountains, highest at middle elevations, and lowest on mountaintops. Elevational effects are also responsible for the increased species diversity in areas with topographic variation (Simpson 1964), such as the western United States (figure 5.19).

Simpson (1964) noted that the number of North American mammal species declines toward the terminal ends of peninsulas. This **peninsular effect** is evident in Simpson's map (figure 5.19) if one examines diversity contours in Baja California, Florida, and the Yucatan. The effect in Baja California was demonstrated for heteromyid rodents by Taylor and Regal (1978), for small mammals on the Iberian Peninsula by Barbosa and Benzal (1996), for rodents in Italy by Contoli (2000), and for nonvolant mammals in Europe by Baquero and Telleria (2001). On the other hand, Lawlor (1983) argued that there is no peninsular effect in Baja for soricomorphs, lagomorphs, geomyid and murid rodents, carnivorans, or bats. Simpson (1964) attributed the pattern to lower immigration rates and higher extinction rates as one moves closer to the terminus of a peninsula, both effects due to isolation from the mainland, reduced land area, and limited habitat diversity (Lomolino et al. 2006). Some authors (Taylor and Pfannmuller 1981; Lawlor 1983) have found this kind of explanation insufficient to account for the variation in peninsular diversity patterns among different groups, arguing instead that factors such as topographic complexity, ecological barriers, and the dispersal abilities of resident species determine whether a particular group will show the peninsular effect.

SUMMARY

Biogeography is the study of how organisms are distributed on the Earth and the causal processes that determine those distributions. The most fundamental datum for biogeography is the species range, the entire area over which individuals of a species occur. Historical biogeography is concerned with changes in species ranges over time and how such changes result from evolutionary processes (e.g., speciation, extinction, dispersal) and geological events (e.g., continental drift, climate change). Ecological biogeography focuses on how species ranges are influenced by factors such as migration, community interactions, and the physiological or morphological adaptations of organisms.

Mammal distributions, like those of most organisms, show varying degrees of endemism or restriction to specific areas. On a worldwide scale, mammal distributions show provincialism—the existence of distinct faunal regions or provinces that harbor diverse groups of closely related species. Biogeographic regionalization is the scientific process of identifying areas of endemism at different spatial scales. One of the most common regionalizations applied to mammals on different continents recognizes six faunal regions—the Palearctic, Nearctic, Neotropical, Ethiopian, Oriental, and Australian regions. These regions are most easily characterized in terms of their family-level diversity and endemicity. The Palearctic consists of North Africa and Eurasia north of the Indian subcontinent and southern China; it has a moderate diversity of mammals (42 families) but no endemic families. The Nearctic comprises North America and northern Mexico, is also moderately diverse (37 families), and endemic-poor (2 families). The Neotropical region is Central and South America; it harbors 50 families, 22 of them endemic. The Ethiopian region includes sub-Saharan Africa, Madagascar, and the southern Arabian Peninsula; it is the most diverse faunal region with 52 mammal families, including 20 endemics. The Oriental region consists of the Indian subcontinent, southeast Asia, and the western East Indies; it is home to 50 families, but only 4 are endemic. The Australia region, including the eastern East Indies, New Guinea, and Australia, has the fewest mammal families (28) but the highest proportion of endemics (20), largely as a result of resident marsupials and monotremes. Mammals are also found on oceanic islands (mostly rodents and bats), in coastal waters (e.g., pinnipeds and sirenians), or in the open ocean (cetaceans).

Scientific acceptance of continental drift and plate tectonics theory in the 1960s revolutionized our understanding of historical biogeography. Throughout the Phanerozoic eon, continents have moved, coalesced, and fragmented, carrying their biotas with them and producing the greatest historical influence on current provincialism. As a result of continental drift and other factors, climate conditions have also varied dramatically throughout Earth history, displaying alternating periods of "hothouse" and "icehouse" conditions. The most extreme icehouse conditions produced ice ages, during which glaciers advanced across portions of continents and many

species distributions were contracted into small refugia. Since their origin in the late Mesozoic, synapsids have experienced both continental movements and climate changes, and have responded to them evolutionarily in a variety of ways. One response is dispersal. Species dispersal involves the extension of a species range into previously unoccupied areas via one of several pathways. Corridor routes are pathways through which many species may move with little ecological resistance. Filter routes are accessible to some species but not others, depending on their ecological tolerances, and sweepstakes routes are the result of rare events that move individuals across a barrier into an area where they can successfully reproduce. Another possible response is vicariance, in which the ancestral range of a species is fragmented by a dispersal barrier, inducing speciation and perhaps subsequent diversification of lineages in isolated geographic areas. Vicariance followed by diversification is thought to be the primary mechanism by which patterns of endemism arise. Species may also respond to environmental changes by going extinct. Background extinction is the result of ecological processes that lead to the demise of individual species, whereas mass extinctions are worldwide events that result in the simultaneous loss of many species from different major groups.

Biogeographic hypotheses are usually framed to explain patterns of disjunction, the geographic separation of sister groups. Early biogeographers emphasized dispersal and extinction as the primary processes that determine species distributions, but awareness of continental drift and past climate change, coupled with the development of rigorous methods of phylogeny reconstruction, has led to a stronger emphasis on the role of vicariance. Cladistic biogeographers focus on reconstructing the history of areas from the phylogenies and current distributions of species in those areas. "Area cladograms" may be inferred from the congruent phylogenies of multiple, codistributed species. However, cladistic biogeography is complicated by widespread and missing species, species with multiple representatives in the same area, and—perhaps most significantly—the temporal incongruence of biogeographic events that affected different groups in a single region. Nevertheless, there are several powerful algorithmic approaches to biogeographic inference, including model-based methods that apply statistical principles. Molecular techniques have also been brought to bear on questions of historical biogeography. Comparative phylogeography uses principles of genealogical concordance to formulate and test hypotheses of recent vicariance and dispersal by examining the genetic structure of populations throughout the ranges of individual species.

The early evolutionary history of marsupials illustrates the value of combining paleontological, geological, and phylogenetic information to understand the biogeographic history of a major mammalian group. Marsupials probably originated in North America during the Cretaceous and dispersed by a sweepstakes route to South America, where they diversified during the Cenozoic. Some of these taxa, including microbiotheres, made their way across Antarctica and into Australia by the Paleocene or Eocene. Breakup of the Gondwanan continents isolated the common ancestor of Australian marsupials, which underwent a spectacular radiation throughout the later Tertiary Period as Australia drifted north to its current position. Along with marsupials, the Paleocene mammal fauna of South America contained archaic ungulates, diverse edentates, and condylarths. Hystricomorph rodents and ceboid primates arrived during the Oligocene via waif dispersal from Africa, and procyonids from North America invaded via island chains in the Miocene. In the Pliocene, the Panamanian land bridge emerged and formed a dispersal route for the Great American Interchange of mammals between the two continents. Pleistocene extinctions took a heavier toll on South American forms, however, resulting in a large proportion of current South American species being derived from northern ancestors. In the case of gibbons (Hylobatidae), a molecular phylogeny and dispersal-vicariance analysis provided a framework from which to estimate an ancestral area for the family, along with the timing and direction of dispersal for ancestors of the current genera in southeast Asia. A comparative phylogeography study of rodents and other tetrapods from the southwestern United States and Mexico revealed a pattern of cryptic vicariance resulting from geological events that shaped the Baja Peninsula and Gulf of California during the past five million years.

Since the 19th century, ecological biogeographers have noted geographical patterns in the variation of morphological and life history characters of mammals. These patterns have been codified as ecogeographic rules, although their generality is often quite limited. The island rule is that insular representatives of large-bodied species are often smaller than their mainland relatives (insular dwarfism), whereas small-bodied species on islands tend to be larger than their counterparts on the mainland (insular gigantism). Rapoport's rule is that the latitudinal breadth of species ranges tends to increase with increasing latitude. The same trend in body size—larger individuals tend to occur at higher latitudes—constitutes Bergmann's rule. According to Allen's rule, individuals that live in cold climates have shorter appendages than do conspecifics that live in warmer areas. Gloger's rule suggests that species living in dry areas have darker pelage than their relatives in humid areas. Mammals in cold climates also tend to have thicker underfur and larger litter sizes than those in warm climates. Ecological explanations of these trends emphasize the different community interactions in, and organismal adaptations to, island versus mainland habitats, high versus low latitudes, and so on.

Ecological biogeographers are also concerned with explaining large-scale patterns in species diversity. The most striking pattern is that the number of species per unit area tends to decrease with latitude, with tropical habitats harboring the greatest diversity. Similarly, species diversity changes with elevation in montane areas, usually peaking at midelevations and declining toward the summits. Species numbers also decline as one moves toward the terminus of a peninsula. Again, these trends are not universal, and the factors that produce them are probably complex. Models that integrate the effects of various ecological processes on rates of dispersal, extinction, and speciation hold the greatest promise for explaining gradients of species diversity.

SUGGESTED READINGS

Avise, J. C. 2000. Phylogeography: The history and formation of species. Harvard Univ. Press, Cambridge, MA.

Benton, M. J. 2005. Vertebrate paleontology, 3rd ed. Blackwell Publishing, Malden, MA.

Lomolino, M. V., B. R. Riddle, and J. H. Brown. 2005. Biogeography, 3rd ed. Sinauer Associates, Sunderland, MA.

Merrick, J. R., M. Archer, G. M. Hickey, and M. S. Y. Lee (eds.). 2006. Evolution and biogeography of Australasian vertebrates. Auscipub, Oatlands, N.S.W., Australia.

Wiens, J. J., and M. J. Donoghue. 2004. Historical biogeography, ecology, and species richness. *Trends Ecol. Evol.* 19:639–644.

DISCUSSION QUESTIONS

1. Using a field guide for mammals, select an order with numerous species in North America (or any other continent). Count the numbers of species in the order that occur at 5° or 10° increments of latitude. In light of the information in this chapter, what factors might explain differences in species richness at the different latitudes?

2. Some species of mammals, such as house mice (*Mus musculus*), have nearly cosmopolitan distributions. Others, such as the Kangaroo Island dunnart (*Sminthopsis aitkeni*), occur only in very small areas (e.g., Kangaroo Island off the coast of South Australia). Most mammal species have range sizes between these two extremes. What factors do you think help to explain the variation in range size among species?

3. Select your favorite mammalian family and determine the distribution of each species within it using guidebooks or the internet. Based on your knowledge of geography and climate, along with information about the niches and phylogeny of these species, what can you say about the factors that produced the current distribution of the family?

4. How is biogeography related to conservation biology? What kinds of information could biogeographers supply that might assist in developing a conservation management strategy for an individual species or a particular region?

PART 2

Structure and Function

In part 2, we examine the relationships between the anatomical parts of a mammal (its structure) and the way those parts work together to accomplish fundamental life processes (their function). Our goal is to understand how anatomy and physiology have been shaped by natural selection in response to varied environmental circumstances. We give special attention to the morphological adaptations of mammals in general, and to some of the specialized body structures associated with particular lifestyles. Later, in parts 3 and 4, we describe the anatomical features characteristic of major mammalian groups and explore how morphology and physiology are related to mammal behavior.

Chapter 6 describes the outer covering of a mammal's body (the skin, or integument), the support system provided by the axial and appendicular skeletons, and the muscles that power movement. Chapter 6 concludes with an examination of locomotor adaptations among mammals. Chapter 7 covers the ways in which mammals obtain and process food, including a discussion of how variation in tooth structure relates to different modes of chewing. This chapter also explores how the digestive system extracts nutrients from food and how mammals store energy. Chapter 8 is devoted to control and regulatory systems. The nervous system receives information from sensory receptors, processes and stores that information, and mediates responses. Endocrine organs produce hormones, chemical messages within the body that mediate physiological responses to external and internal stimuli. In chapter 9, we discuss the concept of homeostasis and how it is maintained. This includes a description of the circulatory, respiratory, and urinary systems and how these systems interact to accomplish respiration, osmoregulation, thermoregulation, and energy conservation. In chapter 10, we explore the morphological structures and physiological processes by which mammals reproduce.

CHAPTER 6

Integument, Support, and Movement

Imagine a cheetah (*Acinonyx jubatus*) slowly stalking a Thomson's gazelle (*Gazella thomsonii*) on the African savanna. The cheetah blends into the dappled shade under acacia trees and warily follows the gazelle at a distance. Suddenly, it sprints after the gazelle in an all-out dash to capture its prey (figure 6.1). We are amazed by the cheetah's burst of speed and by the graceful, powerful movements of its legs as it races toward the gazelle. What body structures enable the cheetah to make such a swift attack? What body processes are at work to produce the smooth, forceful run of this predator?

The cheetah's body, like that of other mammals, is adapted for its way of life. Its fur protects and camouflages it. As it runs or walks, each stride involves intricately coordinated movements of muscles and bones throughout its body. To understand the dynamics of locomotion, we must first examine the arrangement of skin, bones, and muscles in a mammal's body. This chapter starts with the outside of a mammal—the integument—and then moves under the skin to explore the skeleton and the muscles that power movement. In the final section, we show how the integument and musculoskeletal system relate to modes of locomotion.

Integument

The skin, or **integument,** is the interface between a mammal and the external environment. Its primary role is to separate the internal, homeostatically regulated milieu of tissues and cells from the vagaries of outside conditions. The integument of mammals also has two more specific functions: water conservation and insulation. An impervious outer layer of skin is a trait shared by all amniotes and is an adaptation for preventing evaporative loss of body water in the dry air of terrestrial habitats. As endotherms, mammals must also conserve body heat that is energetically expensive to produce. The outermost layer of their integument has evolved a relatively simple structure—hair— that traps a layer of air next to the skin and prevents convective heat loss. Feathers perform the

Figure 6.1 Cheetah and gazelle. This scene of a cheetah attempting to run down a gazelle shows the general body form of each animal, as well as their integuments and color patterns. Both species are cursorial, or running, mammals but the structure of their limb bones are extremely different.

same function in birds, the other group of endothermic vertebrates. The integument serves other critical functions as well, including the ability to dissipate excess heat by evaporative cooling, communication via pelage color or patterns, sensory capabilities for obtaining information about the environment, and production of diverse secretions from specialized glands. Moreover, the skin has produced some uniquely mammalian structures such as the surfaces of horns, antlers, nails, and hoofs.

STRUCTURE AND FUNCTION

Skin

Vertebrate skin consists of an outer epidermis and inner dermis, below which is a hypodermis or subcutaneous region that overlies muscle (figure 6.2; Kent and Carr 2001). The **epidermis** of mammals consists of three layers: an inner stratum basale (or germinativum), an intermediate stratum granulosum, and an outer stratum corneum. Cells in the *stratum basale* divide continually and their daughter cells migrate into the overlying *stratum granulosum* where they manufacture large amounts of the protein keratin (i.e., they become keratinized or cornified), a process that causes them to die by the time they reach the stratum corneum. Keratin is water-insoluble and thus the keratinized stratum corneum prevents dessication of the underlying skin. Cells of the stratum corneum are shed and replaced throughout the life of a mammal. The thickness of the *stratum corneum* varies among regions of the body and among species. It is very thick in the foot pads of most species. These surfaces experience regular abrasion against rough substrates during locomotion, and thickening of the stratum corneum provides protection against excess wear. Epidermal thickening may even result in a fourth cell layer—the *stratum lucidum*— above the

stratum granulosum. Cells in the stratum lucidum are translucent due to large amounts of keratohyaline, a precursor of keratin.

At the boundary between the epidermis and dermis are melanophores, cells that contain **melanin.** This brown pigment absorbs ultraviolet radiation from the sun that might otherwise damage the underlying dermal tissue. Melanin in the dark tongues of giraffes (*Giraffa camelopardalis*) protects the tissue from sunburn as the animals lift their heads into the top branches of trees to forage for leaves. In humans, synthesis of melanin increases upon exposure to sunlight and produces tanning. Extensions of melanophores reach into the epidermis where they inject their pigments into developing hair cells (see next section). In a few cases, such as the bright ischial callosities of baboons, pigmented patches of skin are used as visual signals. The blue and red coloration on the scrotum and perineal region of male vervets (*Chlorocebus aethiops*) is used in dominance displays.

The thick mammalian **dermis** contains connective tissue, blood vessels, nerves, and slips of integumentary muscle. The epidermis has no blood supply of its own, so metabolic needs of cells in the stratum basale are met by diffusive exchange of nutrients and waste products with the highly vascularized dermis. Constriction and dilation of dermal arterioles helps regulate heat loss by directing blood toward, or away from, the surface of the skin. Encapsulated sensory nerve endings in the dermis terminate at tactile receptors such as Pacinian (Vater-Pacini) corpuscles or the Meissner's corpuscles of primates (figure 6.2). Receptor stimulation results in firing of the associated nerve and transmission of an impulse to the central nervous system. A tiny smooth muscle, the arrector pili, inserts at the base of each hair within the dermis. When they contract, arrectores pilorum cause hairs to stand erect, a response that may conserve heat by thickening the dead-air layer above the skin, or act as a visual signal (e.g., when "hairs stand on end" on the neck of a snarling dog). Although hairs and integumentary glands are derived from epidermal cells, their bases grow down into the dermal layer as they develop (figure 6.2).

Below the skin lies a **hypodermis** consisting of loose connective and adipose (fat) tissues. Connective tissue causes the skin to adhere to underlying muscle, whereas subcutaneous fat serves as insulation and an energy reserve. Some sensory receptors (e.g., Pacinian corpuscles) also occur in the hypodermis.

Hair

Hair is one of the unique characteristics of mammals, but its evolutionary origin is obscure. Derived from the stratum corneum, the hairs of mammals occupy the same structural position, and perform some of the same functions, as the epidermal scales of nonavian reptiles and the feathers of birds. As noted previously, hair provides insulation for mammals, so perhaps its evolution was tied to the origin of endothermy. However, there is an inconsistency

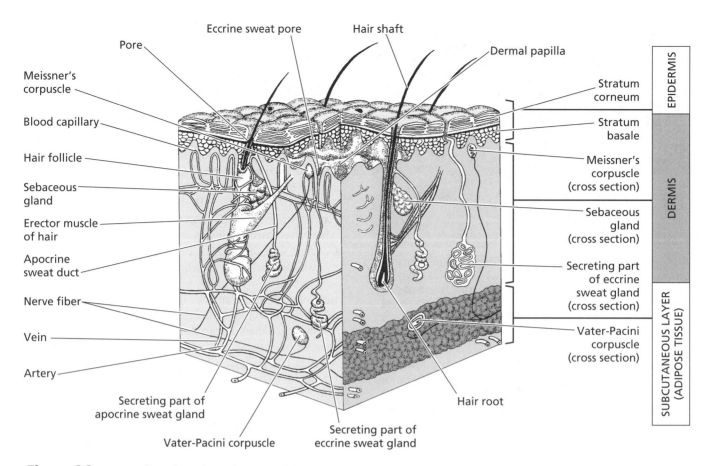

Figure 6.2 Mammalian skin. Three-dimensional section through the skin and subcutaneous region of a mammal. Glands and hairs are epidermal structures that grow into the dermis during development.

in this scenario: Insulatory fur would be maladaptive for an ancestral ectotherm that relied on efficient heat exchange with the environment for thermoregulation, but some form of insulation appears to be necessary for endothermy to be energetically cost-effective (Pough et al. 2005). So which came first—hair or endothermy? Maderson (1972) suggested that hairs originated as small tactile receptors between the scales of early synapsids, and only later were coopted as insulation along with the evolution of endothermy. The fossil record has been of little help on this question because hairs rarely fossilize. Interestingly, the Permian therapsid fossil *Estemmenosuchus* has an exceptionally well-preserved skin impression that lacks any trace of hair (Kardong 2006). Thus, it is unclear exactly when hair appeared in the evolutionary history of synapsids.

Although they exhibit positional and functional similarities, the hairs of mammals and the feathers of birds are not homologous. Feathers are homologous to the scales of reptiles, both structures having similar developmental origins that involve a contribution from the dermis. In contrast, hairs are entirely epidermal. A hair follicle begins its development in the stratum basale (Butcher 1951) and grows down into the dermis, inducing the formation of a dermal papilla (figure 6.3). The papilla becomes vascularized and serves as a conduit for nutrients and waste prod-

ucts with the developing hair. Where it reaches the base of the dermis, the follicle swells to form a *bulb* around the dermal papilla. Continual mitosis occurs within the bulb, where *root* cells synthesize keratin and grow outward to form a **shaft** of dead cells that eventually emerges from surface of the skin. As the hair is differentiating, so too are dermal cells that will form the arrector pili muscle and follicle cells that will form a sebaceous gland. The root of each hair becomes surrounded by sensory nerve endings that transmit tactile signals to the brain whenever the shaft is displaced (figure 6.2).

A typical hair has three structural layers. The **medulla** occupies the center of the shaft and consists of sparse, irregular cells connected by keratin strands and surrounded by air space. The medullas of some species, including deer, lack cells altogether and their hairs are hollow. The **cortex,** composed of tightly packed cells, surrounds the medulla and comprises most of the shaft. A thin, transparent **cuticle** forms a scalelike pattern on the surface of the hair. Cuticle patterns vary greatly among species (figure 6.4) and have been used to identify closely related forms (Wooley and Valente 1992).

Hair can be classified in a variety of ways (Martin et al. 2001). One distinction recognizes different growth patterns. **Angora** hair grows continuously to produce long, flowing shafts that may or may not be shed (e.g., a horse's

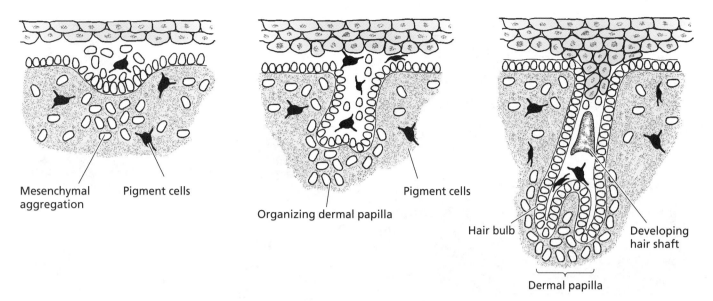

Figure 6.3 Hair development. At left, epidermis invades the dermis (shaded) at the site where an individual hair will form. At center, a dermal papilla organizes beneath the epidermal invagination, which is beginning to form a follicle. At right, a hair shaft begins to differentiate within the follicle and a bulb of epidermal cells surrounds the dermal papilla. *Adapted from Searle (1968).*

mane). **Definitive** hairs attain a fixed length before being shed and replaced (e.g., body hairs of a dog). The two main functional types of hairs are vibrissae and pelage (fur). Vibrissae are long, stiff hairs that function as tactile receptors and exhibit definitive growth. The most common are "whiskers" in the facial region of nocturnal species, but they are also found on the tails of some fossorial species or on the legs of others. *Active* vibrissae can be erected voluntarily, whereas *passive* ones cannot. **Pelage** consists of long, coarse **guard hairs** underlain by **underfur** that is short, dense, and fine. The most common type of guard hair is an **awn,** characterized by an expanded distal end, a weak base, and definitive growth. Awns usually lie in one direction, giving the pelage a distinct nap. **Spines** are the stiff, enlarged guard hairs that exhibit definitive growth and form the protective quills of porcupines, echidnas, and hedgehogs. **Bristles** are firm, generally long hairs that exhibit angora growth to form manes; they function as visual signals that augment facial expressions (e.g., lions) or body postures (e.g., horses). **Wool** is long, soft, and usually curly. **Velli** are the very short, fine hairs found on newborns (often referred to as "down" or "fuzz"). The "scales" of pangolins are agglutinated keratin fibers developmentally similar to hairs.

Hair color is determined by the distribution and density of melanin granules within the shaft, as well as the proportions of two forms of melanin. **Pheomelanin** (xanthophyll) produces shades of red and yellow, whereas **eumelanin** is black or brown. "Agouti" hair results from a mixture or banding of these pigments. Gray or white hair results from the absence of melanin and the increase in the volume of air vacuoles within the medulla. White hair (along with pale skin and pink eyes) may be due to a genetic condition, **albinism,** in which melanin production is blocked. Although rare overall, albinism has reached

high frequencies in some populations, such as the albino gray squirrels (*Sciurus carolinensis*) of Olney, Illinois. The opposite of albinism is **melanism,** in which the pelage is very dark due to high levels of melanin. Melanic forms of several mammal species have been described, including fox squirrels (*Sciurus niger*) and red foxes (*Vulpes vulpes*).

The pelage color of a species, or populations within a species, is an evolutionary response to several environmental pressures. One common factor seems to be the adapative value of camouflage. As noted in chapter 5, Gloger's rule predicts that mammals from arid regions will tend to be paler than those from more humid areas. This is probably an example of **crypsis,** in which the color of the animal matches that of the substrate (e.g., the desert floor is more lightly colored than that of forests or grasslands; Benson 1933). Mammals that display **countershading** have lighter-colored pelage on the ventral portion of the body than on the dorsal portion. This causes them to look white (like the sky) when viewed from below, but dark (like the ground) when viewed from above, and therefore they appear less conspicuous from either angle. Many species show **cryptic coloration** in which the pelage blends in with the horizontal background against which an animal is viewed. White-tailed deer fauns (*Odocoileus virginianus*) have light-colored spots on their brown fur that conceal them in the dappled shade characteristic of temperate forests. Other species have **disruptive coloration** patterns that distort the outline of their bodies. For example, the black-and-white banding pattern of zebras (*Equus* spp.) causes them to appear larger than their actual size in the eyes of predators. The hypothesis that pelage color is an antipredator adaptation was generally supported by Belk and Smith's (1996) study of oldfield mice (*Peromyscus polionotus*) in the southeastern United States, but their

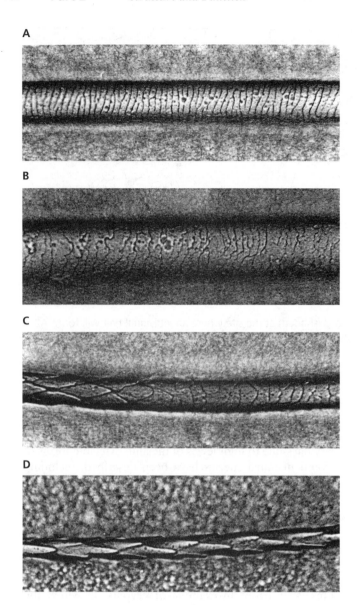

Figure 6.4 **Hair cuticles.** Scale patterns on the cuticles of primary guard hairs vary among species and even within regions of the same hair. These light micrographs show guard hairs of two dasyurid marsupials: (A) *Murexia melanurus*; (B) *Phascolosorex doriae* (distal portion); (C) *P. doriae* (transition region); (D) *P. doriae* (proximal region).

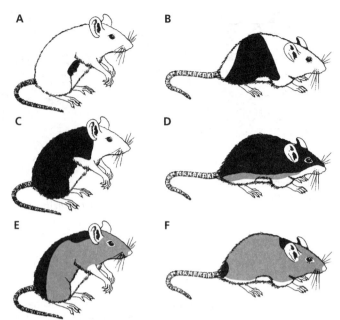

Figure 6.5 **Molting.** The postjuvenile molt of the golden mouse (*Ochrotomys nuttalli*) proceeds in a regular manner from the underside of the flanks dorsally, and finally to the back of the neck and base of the tail. Letters A–F refer to the chronological sequence of the molt. *Adapted from Linzey and Linzey (1967).*

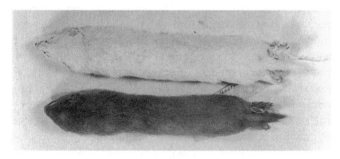

Figure 6.6 **Seasonal changes of coat color.** Weasels of the genus *Mustela* may change their coat color seasonally. They molt to a mostly white pelage (top) for the winter, which camouflages them against the snowy background. The summer pelage (bottom) is brown on the dorsal surface and flanks, and white on the ventral surface; this shading more closely resembles summer background colors.

results also suggest that other factors determine small-scale (local) variations in color.

Many mammals periodically replace their fur by **molting,** which occurs in three principal patterns. Young mammals, particularly rodents, undergo a **postjuvenile molt** that starts after weaning (figure 6.5). Such molts have been described for many species, including meadow voles (*Microtus pennsylvanicus*; Ecke and Kinney 1956), deer mice (*Peromyscus*; Sinclair et al. 1998), and pocket gophers (*Cratogeomys, Thomomys*; Davidow-Henry and Jones 1988). Mammals living in temperate latitudes undergo a fairly rapid **annual molt** during which most hairs are replaced. This process is important for animals in cooler environments because daily activities wear down

their hairs, resulting in less effective insulation during winter unless the hairs are replaced. Annual molts require considerable energy for the rapid growth of new hair and may be timed to coincide with periods of maximum food availability, as in Columbian ground squirrels (*Spermophilus columbianus*; Neuhaus 2000), or to coordinate with migrations as in hoary bats (*Lasiurus cinereus*; Cryan et al. 2004). Some mammals, such as Cape hares (*Lepus capensis*; Lu 2003), have a **seasonal molt** that replaces their pelage more than once a year. Some species have camouflage coloration that changes with season: Ermine (*Mustela vison*) and arctic foxes (*Alopex lagopus*) have a white coat during the winter, matching the snow-covered landscape they inhabit but molt into a darker coat for the summer (figure 6.6). Animals in molt can be recognized by short or partially elongated hairs, which are dark at the base, growing among mature hairs that are light to the base. Fully developed molt lines can be seen in short-haired species such as moles, shrews, and pocket gophers. The skin of a molting mammal is dark or speckled due to presence of pigment-injecting melanophores just below the surface. In the fur industry, a pelt is "prime" when the animal is not molting.

Integumentary Glands

Two types of glands are derived from the epidermis of mammalian skin—sebaceous and sweat glands. More specialized structures, such as scent and mammary glands, are derived from these basic types. Like hairs, epidermal glands have their base in the dermis but are connected to the surface of the skin or hair by a duct. **Sebaceous glands** are generally associated with hair follicles (figure 6.2) and secrete an oily product (sebum) that keeps the hair shaft moist and waterproof. These glands are situated such that contraction of integumentary muscles causes sebum to be squeezed onto the shaft. Sebaceous glands not associated with hairs occur on the lips, penis, labia minora, and nipples. Examples of specialized sebaceous glands include those that produce lanolin in sheep, ceruminuous glands that secrete a waxy protective lubricant (cerumen) into the outer ear canal, and Meibomian glands in the eyelid that moisten the conjunctiva.

Sweat glands produce a watery secretion (sweat), lie deep in the dermis, and are connected to the skin surface by coiled ducts (figure 6.2). **Eccrine** sweat glands produce thin sweat, are not associated with hairs, and function immediately after birth. Evaporation of eccrine secretions from the skin results in evaporative cooling, an important thermoregulatory mechanism in some mammals. Eccrine glands are most common on surfaces (e.g., soles and palms) that come in regular contact with a substrate, where their secretions increase adhesion and tactile sensitivity. **Apocrine** (sudoriferous) sweat glands produce viscous sweat, are located near hair follicles, and begin functioning at puberty. The distribution of sweat glands over the body varies, but they are most common in areas where fur is the least dense (e.g., the feet of cats, dogs, rodents, and primates; the lips of rabbits). In humans and chimpanzees, eccrine glands are broadly distributed over the body, but apocrine glands occur primarily in the axillae (armpits), anogenital region, naval areas, and nipples. Sweat glands occur on the ears of hippopotami, the heads of bats, and the snouts of platypuses. Ciliary glands that drain onto the eyelashes of many mammals are apocrine glands. Sweat glands are absent in pangolins, cetaceans, sirenians, and echidnas.

Scent glands may be either modified sebaceous glands or modified sweat glands. The composition and function of the secretions they produce vary widely, but many act as **pheromones** (chemical signals that convey information between members of the same species). Scents may be used to deter predators as in skunks, to mark territo-ries as in deer, to mark sites for spatial orientation as in weasels and badgers, or to mediate sexual and social interactions (Brown and Macdonald 1985; Ralls 1971). Scent glands can be found in almost any location on the body, including the anal area (rodents and mustelids), back (kangaroo rats), head (elephants, peccaries), fore-limbs (lemurs, carnivorans), and hind limbs (ungulates). Some mammals spread scented glandular secretions onto their fur, possibly to enhance the effectiveness or persistence of the chemical signal. The yellowish color often seen on the fur of opossums is the result of such behavior.

The structures for which Mammalia is named, **mammary glands,** are specialized epidermal glands. They develop at one or more points along parallel ridges ("milk lines") of embryonic skin in the ventrolateral body wall, with epidermal cells growing deep into the hypodermis to form an extensive branching system of ducts that terminate in alveoli and are surrounded by milk-producing cells. At the surface, the ducts extend through an elevated nipple or teat to open at its apex (figure 6.7). In many species, adipose tissue forms beneath the mammary gland to produce breasts. The number and location of glands varies among species: one axillary pair in dermopterans and marmosets; one thoracic pair in anthropoid primates; one inguinal pair in perissodactyls and cetaceans; one thoracic and one inguinal pair in soricomorphs and some lemurs; two pairs on the back in nutrias (*Myocastor coypus*); up to five pairs from the axillae to the inguinal regions in rodents, cats, dogs, pigs, and edentates; up to ten pairs in the pouches of some marsupials. The Virginia opossum (*Didelphis virginiana*) has 13 mammary glands—12 in a circle and one in the center. Monotreme mammary glands lack nipples, but the surface openings of milk ducts are marked by tufts of hair.

With few exceptions, mammary glands are functional only in females. Late in pregnancy, hormones such as prolactin stimulate the proliferation of alveoli, adipose tissue, and milk-producing cells. Milk is an aqueous mixture of protein, lipid, and carbohydrate that serves as nourishment for young. The relative proportions of different nutrients in milk vary among species (table 6.1). In general, young grow more quickly in species with higher proportions of protein in their milk (Boulieré 1964). The nutritional composition of milk also changes over the course of the suckling period (Lang et al. 2005; Sharp et al. 2005). When a female gives birth, mammary glands have produced a significant amount of milk that is stored in the alveoli. When suckling commences, nerve impulses travel from the nipple to the brain causing the hormone oxytocin to be released by the hypothalamus. Oxytocin stimulates contraction of muscle slips around the alveoli and forces milk into the ducts, a process known as "milk letdown." The flow of milk from mother to suckling is **lactation.**

The evolutionary origin of mammary glands is unclear. Although they are undoubtedly epidermal, their precise homology with sebaceous or sweat glands is not

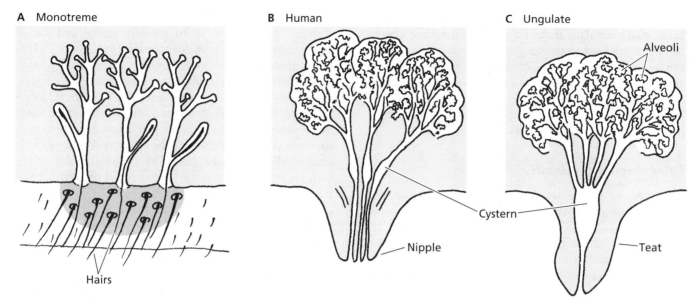

Figure 6.7 Mammary glands. Milk-producing cells surround the alveoli of mammary glands, which project into the hypodermal portion of the skin and may be surrounded by adipose tissue forming breasts. The arrangement of ducts varies among species, and monotremes lack nipples or teats. *Modified from Kent and Carr (2001).*

obvious, nor is it certain that the mammary glands of monotremes are homologous with those of therians. Even more problematic is inferring the timing and initial adaptive value of lactation. Pond (1977) argued that suckling must have preceded the appearance of diphyodont dentition (see chapter 4); the latter is a prerequisite for precise tooth occlusion and could not have evolved in neonates that used their teeth for chewing. If correct, this places the origin of lactation just prior to the emergence of diphyodont teeth with complex occlusal surfaces (i.e., near the evolutionary transition from cynodonts to true mammals; Pough et al. 2005). The origin of milk itself may relate to the early adaptive value of glandular secretions for eggs or hatchlings that were incubated by parents (Blackburn 1991; Blackburn et al. 1989). Modern milk contains proteins that are similar to lysozymes and act as antimicrobial agents (Vorbach et al. 2006). Such secretions may have protected the surface of eggs or young from infection in early mammals; if ingested by hatchlings, the same protection might have extended to the digestive tract. It is also possible that maternal antibodies could be transferred to young via an integumentary secretion, as they are in modern milk. Once ingestion of the secretion became established, evolution of the nutritional aspect of lactation and the associated anatomical features of mammary glands could proceed (Kardong 2006). Unfortunately, glands (like hair) rarely fossilize, so it is difficult to evaluate these hypotheses. They are at least consistent with the available data.

CLAWS, NAILS, AND HOOVES

Claws, nails, and hooves are keratinized structures derived from the stratum corneum at the ends of digits.

Claws are the most primitive form, appearing early in amniote evolution and still common in nonavian reptiles, birds, and mammals. A claw consists of two parts: a heavily keratinized, convex, dorsal **unguis** and a softer, ventral **subunguis** that is continuous with the pad at the end of a digit (figure 6.8A). Both layers are wrapped around the dorsolateral surface of the last phalanx. Mammals use their claws for climbing, digging, fighting, and defending. In some groups, such as felids, the sharp claws are retractable, an arrangement that may be advantageous during peaceful interactions with conspecifics and young (figure 6.9). Felids are much more adept at manipulating objects with their forepaws than are canids, which lack retractable claws. Russell and Bryant (2001) suggested that reversion to less retractable claws in cheetahs (*Acinonyx jubatus*) is associated with this species' unique hunting behavior, including a diminished use of the forepaws in feeding. Claws, like nails and hooves, are evergrowing in mammals and must be worn down by abrasion.

Nails cover only the dorsal surface of a phalanx, the unguis being broad and flat, and the subunguis reduced (figure 6.8B). They are modified claws that evolved in primates and facilitate gripping and object manipulation by the hands and feet. Along with the development of nails, the number of sensory nerve endings in the dermis at the ends of primate digits has greatly increased.

Hooves are characteristic of ungulates and are also derived from claws. They consist of a much-thickened, U- or V-shaped unguis that completely surrounds the subunguis, which in turn forms the sole of the foot (figure 6.8C). Hooves occur in conjunction with a reduction in the number of digits in ungulates, both features being adaptations for cursorial locomotion. A more primitive hoof structure appears in hyraxes (Hyracoidea).

Table 6.1 Milk composition*

	Water	Protein	Fat	Sugar	Ash
Marsupials					
Kangaroo (wallaroo)	73.5	9.7	8.1	3.1	1.5
Primates					
Rhesus monkey	88.4	2.2	2.7	6.4	0.2
Orangutan	88.5	1.4	3.5	6.0	0.2
Human	88.0	1.2	3.8	7.0	0.2
Xenarthrans					
Giant anteater	63.0	11.0	20.0	0.3	0.8
Lagomorphs					
Rabbit	71.3	12.3	13.1	1.9	2.3
Rodents					
Guinea pig	81.9	7.4	7.2	2.7	0.8
Rat	72.9	9.2	12.6	3.3	1.4
Carnivores					
Cat	81.6	10.1	6.3	4.4	0.7
Dog	76.3	9.3	9.5	3.0	1.2
European red fox	81.6	6.6	5.9	4.9	0.9
Pinnipeds					
California sea lion	47.3	13.5	35	0	0.6
Harp seal	43.8	11.9	42.8	0	0.9
Hooded seal	49.9	6.7	40.4	0	0.9
Cetaceans					
Bottle-nosed dolphin	44.9	10.6	34.9	0.9	0.5
Blue whale	47.2	12.8	38.1	?	1.4
Fin whale	54.1	13.1	30.6	?	1.4
Ungulates					
Indian elephant	70.7	3.6	17.6	5.6	0.6
Zebra	86.2	3.0	4.8	5.3	0.7
Black rhinoceros		1.5	0.3	6.5	0.3
Collared peccary		5.8	3.5	6.5	0.6
Hippopotamus	90.4	0.6	4.5	4.4	0.1
Camel	87.7	3.5	3.4	4.8	0.7
White-tailed deer	65.9	10.4	19.7	2.6	1.4
Reindeer	64.8	10.7	20.3	2.5	1.4
Giraffe	77.1	5.8	12.5	3.4	0.9
American bison	86.9	4.8	1.7	5.7	0.9
Cow	87.0	3.3	3.7	4.8	0.7

From F. Boulieré, The Natural History of Mammals, 3rd ed. Revised, 1964, Alfred A. Knopf. Reprinted by permission.

*The composition of milk in different mammal species varies in the proportions of protein, fat, and sugar. Notice high fat content for pinnipeds, cetaceans, and lagomorphs. In lagomorphs, the mother usually feeds her young once per day.

HORNS AND ANTLERS

Among extant mammals, head ornamentation occurs only in artiodactyls and perissodactyls. "Horn" refers to a surface made of keratin. True **horns,** found only in bovid artiodactyls (cattle), have an inner core of dermal bone (derived from the frontal bone of the skull) covered by a keratinized sheath (derived from epidermis; see figure 20.24). Bovid horns usually grow continuously throughout the life of the animal, remain unbranched, are never

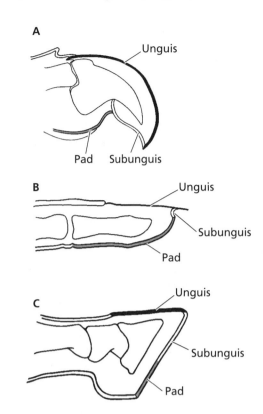

Figure 6.8 Claws, nails, and hooves. Longitudinal sections through the ends of digits. Unguis is solid black, subunguis is *white*, and digital pad is *dark gray*. (A) Claws are the most common cornified epidermal structure at the ends of digits in mammals. The tough outer unguis wraps around the surface and sides of the terminal phalanx; the softer subunguis is continuous with the pad on the underside of the digit. (B) Nails occur in primates. The unguis is flattened, the subunguis reduced, and the digital pad is highly innervated. (C) Hooves occur in ungulates. A thick unguis surrounds the terminal phalanx and the subunguis overlies a thickened foot pad.

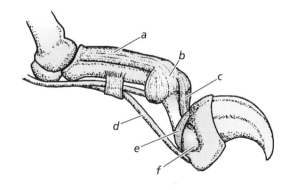

Figure 6.9 Retractable claws. The retractile claw mechanism of the mountain lion (*Puma concolor*) involves (A) an extensor muscle, (B) middle interphalangeal joint, (C) tendon of extensor muscle, (D) tendon of flexor muscle, (E) lateral dorsal elastic ligament, and (F) distal interphalangeal joint. *Adapted from Gonyea and Ashworth (1975).*

shed, and occur in both sexes. Growth rings at the base of the horns may be used to determine the age of an individual. In mountain goats (*Oreamnos americanus*), horn growth is significant only for the first five years of life, and growth rate is significantly influenced by sex and lactation status (Cote et al. 1998). Pronghorn antelope (*Antilocapra americana*), endemic to the American west, are unique in having two-branched horns, the keratinized sheath of which is shed and regrown each year (O'Gara and Matson 1975). Antilocaprids also display sexual dimorphism in horn size, with female horns being smaller and less forked than those of males. The **hair horns** of rhinoceroses (Rhinocerotidae) are masses of agglutinated, keratinized epidermal cells that develop above the dorsal surface of the nasal bones in both sexes and are not shed.

Antlers are branched head ornaments made entirely of bone and found in cervid artiodactyls (deer). As they grow, antlers are covered by a layer of "velvet," highly vascularized skin that supplies the growing bone tissue with nutrients. When bone growth is complete, the blood supply to velvet is cut off and the dead skin is worn away. After the annual rut, or mating season, hormones induce a weakening of the bone just above the base of each antler, such that the distal portion of the antlers are shed. New antlers develop each year, and their growth is one of the fastest examples of organogenesis known among animals (Price et al. 2005). Antlers develop only in male cervids, except for caribou (*Rangifer tarandus*), in which they occur in both sexes. Giraffes and okapi (Giraffidae) have two bony projections that develop from separate ossification centers and then fuse to the skull near the suture of the frontal and parietal bones. They are permanently covered by skin and hair and occur in both sexes. Although commonly called "horns," these structures are not homologous to bovid horns or cervid antlers.

Head ornaments can have a variety of functions, but they are all adapted for intraspecific competition among males (Kitchener 2000). This is especially evident in Rocky Mountain bighorns (*Ovis canadensis*), in which males compete during the rut by running toward each other and butting heads. Horns and antlers may be used defensively to ward off predators, as markers of social rank (Walther 1984), or as advertisements of male fertility (Malo et al. 2005). Some of these functions are discussed more thoroughly in part 4.

Basic Skeletal Patterns

A detailed description of the mammalian skeleton would require several volumes because an understanding of skeletal anatomy has been central to mammalian systematics, paleontology, and biomechanics for many decades. Our goal here is to provide an overview of the skeleton, with an emphasis on the relationship between structure and function. More detailed information on the skeletal morpholo-

gy of specific groups can be found in part 3. The skeleton is the body's framework: it provides structural support against the force of gravity, a system of levers that function in locomotion, attachment points for the muscles that drive movement, and a protective casing for vital organs. Vertebrate skeletons consist of two subdivisions: the **axial skeleton** corresponds to the skull, vertebral column, ribs, and sternum; the **appendicular skeleton** comprises bones of the pectoral and pelvic girdles and their associated limbs.

SKULL

Comparative anatomists divide the skull into three portions, each with a distinct developmental origin: the neurocranium (or primary braincase), the dermatocranium (membrane bones that surround the neurocranium), and the visceral skeleton (jaws and other derivatives of the embryonic pharyngeal arches). The neurocranium consists of bones, few of which are visible on the surface of the skull, that ossify in the shape of a bowl to hold the brain. Many of the bones are perforated by openings (**foramina**) that allow passage of nerves and blood vessels (figure 6.10). At the back of the skull, a ring of *occipital* bones (basioccipital, exoccipitals, supraoccipital) forms a **foramen magnum,** through which the spinal cord passes, and a pair of *occipital condyles* on either side that articulate with the vertebral column. Just anterior to these on both sides of the skull, a series of *otic* centers ossify around the inner ear and coalesce to form the petrosal bone of adults. *Sphenoid* bones form the anteromedial floor of the braincase (basisphenoid, presphenoids) and contribute to the wall of the orbit (orbitosphenoid). *Ethmoid* elements surround the nasal area, giving rise to scroll-like turbinal bones that support the olfactory and nasal epithelia, a perforated cribiform plate through which pass fibers of the olfactory nerve, and a mesethmoid bone forming the nasal septum.

Overlying the neurocranium, and in intimate association with it, are dermatocranial bones that form within the hypodermis of developing embryos. Paired *roofing bones* (nasals, frontals, parietals) occur on either side of the dorsal midline, forming a medial **sagittal crest** that marks the dorsalmost origin of the temporalis jaw muscle. At the rear angle of the skull (the *temporal* region), a squamosal bone contributes to the jaw joint and the posterior portion of the **zygomatic arch.** The arch, a point of origin for the masseter muscle, is completed anteriorly by a jugal bone. Lacrimal bones form in the anteromedial corners of each orbit. Remnants of embryonic cartilage in the *upper jaw* are invested by dermal bones to form paired, tooth-bearing premaxillae and maxillae; portions of the same embryonic cartilages ossify in the posterior walls of the orbits as alisphenoid bones. On the ventral portion of the cranium, bones of the *primary palate* (vomer, palatine, pterygoid) lie alongside neurocranial bones. The premaxillae, maxillae, and palatines develop winglike processes that grow ventrally and medially to meet at the midline and form a complete

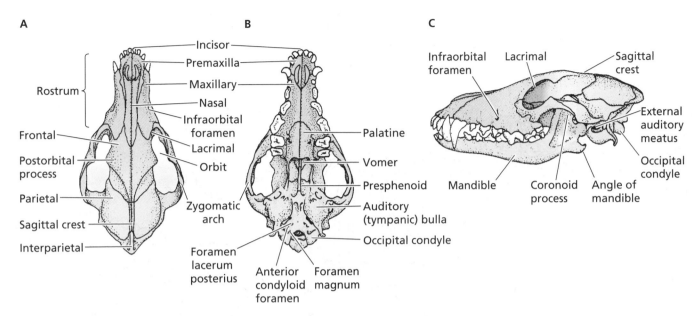

Figure 6.10 **Anatomy of the skull.** The skull of a coyote (*Canis latrans*). (A) Dorsal view of cranium. (B) Ventral view of cranium. (C) Lateral view of cranium and mandible. *Adapted from Gunderson and Beer (1953).*

secondary palate. The hollow tube between the primary and secondary palates is the nasal passageway for respiratory air between the external and internal nares.

The lower jaw, or **mandible,** consists of right and left tooth-bearing bones (dentaries) that meet anteriorly at the *mandibular symphysis.* Each dentary articulates with a squamosal bone of the cranium to form the characteristic "dentary-squamosal jaw joint" of mammals. A depression, the masseteric fossa, on the lateral surface of the dentary marks the insertion site of the masseter muscle; the temporalis muscle inserts dorsally on the coronoid process. The dentary bone develops by ensheathing the anterior portion of the embryonic lower-jaw cartilage. At its posterior end, however, remnants of this cartilage and that of the upper jaw ossify as small *ossicles* (malleus and incus,

respectively) within the middle ear cavity. These sound-transmitting ossicles are homologous to bones (articular and quadrate, respectively) that formed the jaw-joint of nonmammalian synapsids. A third ossicle, the stapes, is common to all tetrapods and is derived from the dorsalmost element of the second pharyngeal arch in fishes. A fourth lower-jaw bone of mammalian ancestors, the angular, is homologous to the tympanic bone that frames the eardrum (tympanum) of mammals.

The protrusion of the jaw and snout varies considerably among mammals, ranging from the nearly flattened face of primates to the elongated rostrum of carnivores (figure 6.11). These variations relate to differences in diet, the importance of olfaction, the angle of the skull relative to the vertebral column, and the location of muscles that move

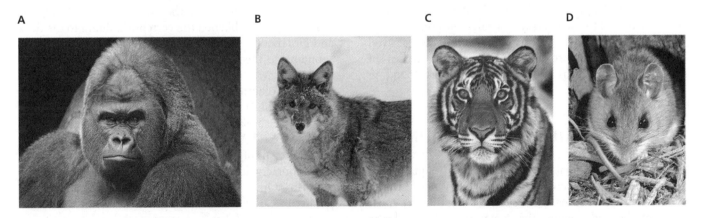

Figure 6.11 **Facial regions.** Shapes of facial regions vary widely among mammals, depending on such factors as diet, the positioning and importance of special sense receptors (i.e., eyes, ears, taste buds, olfactory organs), and the angle of the head relative to the vertebral column. (A) The gorilla (*Gorilla gorilla*) has a blunt rostrum and forward-facing eyes. (B) Coyotes (*Canis latrans*) and (C) Indian tigers (*Panthera tigris*) have long snouts with expanded areas for olfactory receptors and teeth. (D) A white-footed mouse (*Peromyscus leucopus*) has a pointed snout and lateral-facing eyes.

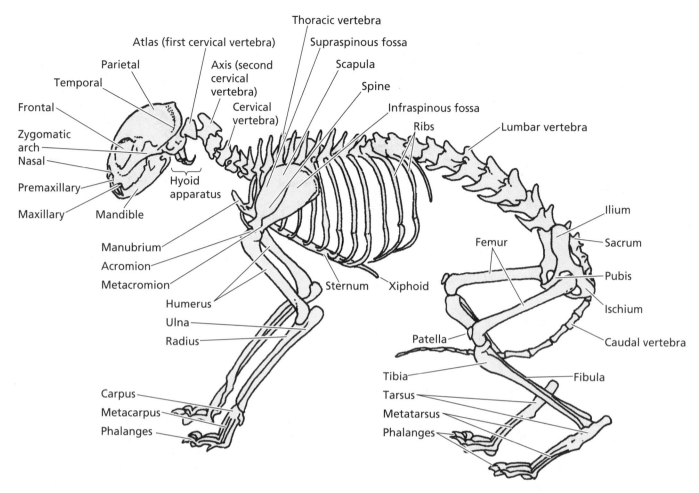

Figure 6.12 Cat skeleton. Lateral view of the skeleton of a domestic cat (*Felis catus*). Note the digitigrade foot posture and reduced clavicle (the small bone just anterior to the head of the humerus) that does not articulate with the rest of the skeleton. These are cursorial adaptations. *Adapted from Kistler et al. (1975).*

the skull. The flattened face of primates is an adaptation for binocular vision; a short snout allows the eyes to face forward and their visual fields to overlap, a prerequisite for stereoscopic perception.

The **hyoid apparatus,** located in the throat region, consists of an H-shaped array of small bones that support the tongue and larynx (figure 6.12). These are derived from the second through fifth pharyngeal arches of embryos, homologous to the skeleton of corresponding gill arches in fish.

VERTEBRAE, RIBS, AND STERNUM

The bodies of mammals that are permanently aquatic, like those of fish, can be supported by their buoyancy in water. For terrestrial mammals, the skeleton and muscles must support the body against the downward pull of gravity. The vertebral column ("backbone"), suspended above the ground by limbs, is the central element in this architecture. The column consists of a series of bony elements (**vertebrae**) separated from one another by cartilaginous

intervertebral discs, extending from the base of the skull to the tail. Within each intervertebral disc is a *pulpy nucleus,* a gelatinous remnant of the embryonic notochord. The entire column is ensheathed in longitudinal bands of ligaments and axial muscles that determine its flexibility. Thomson (1942) likened the vertebral column to the deck and girders of a suspension bridge, with the limbs corresponding to the bridge's pillars. Indeed, many organs in the body cavity are attached to, or suspended from, the vertebral column.

An individual vertebra consists of a circular *centrum* on top of which are a pair of *neural arches* that form a canal for the spinal cord. Various articular surfaces (*apophyses*) occur at the margins of centra, marking sites where the vertebrae make contact with one another or (in the thoracic region) with ribs. Other vertebral processes mark the attachment points of axial muscles and ligaments. Various foramina occur within and between centra, the most significant being the *intervertebral foramina* through which spinal nerves emerge.

Tetrapod vertebral columns display regional specialization (i.e., vertebrae in different parts of the column have distinct morphologies reflecting distinct functions). In mammals there are five regions, each with a corresponding type of vertebra (figure 6.12). All mammals except sloths and manatees have seven **cervical** vertebrae in the neck The first cervical vertebra, the **axis**, articulates anteriorly with the occipital condyles of the skull and posteriorly with the second cervical, the **atlas.** The structure of this joint allows mammals to move their skulls both vertically and horizontally, independent of the body trunk, improving their ability to position sensory receptors on the head for maximum sensitivity to external stimuli. Fish, amphibians, and most reptiles have much less flexibility. Posterior to the cervicals, 12–15 **thoracic** vertebrae occur in the chest region and articulate with ribs; the long, dorsocaudally oriented *spinous processes* of mammalian thoracics make these vertebrae readily identifiable. In the lower back region are 4–7 **lumbar** vertebrae, which may be partially or entirely fused to one another. In most mammals, the **sacral** vertebrae (usually 3–5, but as many as 13 in edentates) are fused to form a *sacrum* that articulates with the pelvic girdle. A variable number (3–50) of **caudal** vertebrae occur in the tail, usually diminishing in size and structural complexity toward the distal end. The 4–5 vestigial caudals of hominoid primates fuse to form a rigid *coccyx* ("tailbone") posterior to the sacrum.

The remainder of the axial skeleton consists of **ribs** attached to thoracic vertebrae at their dorsal ends and, usually, a **sternum** at their ventral ends. Most mammals have 12 pairs of ribs, but the number ranges from nine (in whales) to 24 (in sloths). The posterior ribs usually fail to reach the sternum, terminating instead as "floating" *costal cartilages*. Gorillas and chimpanzees have ribs on their first two lumbar vertebrae. The rib cage, or thoracic basket, surrounds and protects the heart and lungs. Ribs also provide attachment surfaces for one axial (iliocostalis) and several deep thoracic muscles that expand or compress the rib cage.

APPENDICULAR SKELETON

The **pectoral girdle** is the skeleton of the shoulder, forming a complex surface for muscle attachment and articulation of the forelimb. Primitively, the pectoral girdle consisted of an inner layer of replacement bones (procoracoid, coracoid, scapula) and an outer layer of dermal bones (clavicle, interclavicle). This arrangement occurred in therapsids and is still found in living monotremes. In therian mammals, however, the pectoral bones are typically reduced to just two—the clavicle ("collar bone") and the scapula ("shoulder blade"). In cursorial mammals (e.g., cats, ungulates) and cetaceans, the clavicle is reduced or lost, leaving the scapula as the only substantial bone of the girdle (figure 6.12). Mammalian pectoral girdles do not make contact with the axial skeleton; they are suspended in a sling of appendicular muscles that allows considerable flexibility in one or more planes of motion.

In contrast, the **pelvic girdle** is firmly braced against the sacrum at a broad *sacroiliac joint*. The pelvis consists of three fused bones on each side—an anteroventral *pubis*, a posteroventral *ischium*, and a dorsally oriented *ilium* that articulates with the sacrum. These elements coalesce during development into a pair of *innominate bones* ("coxae") that meet ventrally at the *ischiopubic symphysis*, forming a bony ring around the lower abdomen. The pelvic girdle is vestigial in cetaceans.

Forelimbs and hind limbs have a common architecture, consisting of a proximal *propodium*, an intermediate *epipodium*, and a distal *autopodium*. In the forelimb, the propodium consists of a single long bone, the humerus, which articulates proximally at the glenoid fossa of the scapula (figure 6.12). Distally the humerus articulates with the paired epipodial bones of the forearm, the medial radius and lateral ulna, forming the elbow joint. The ulna is sometimes reduced, as in bats. The proximal part of the autopodium consists primitively of three rows of small carpal (wrist) bones, but these are often reduced or fused in extant mammals. Beyond the wrist, there are primitively five metacarpal bones forming the palm, followed by two or more phalanges comprising each digit. Metacarpals and phalanges may be reduced or lost in some cursorial species that have fewer than five digits. In the hind limb, the propodial bone is the femur, articulating proximally at the *acetabulum* of the pelvis to form the ball-and-socket hip joint. Distally, the femur meets the paired epipodial bones (the tibia and fibula) to form the knee joint. Tarsal bones form the ankle joint and were primitively arranged much like the carpals. In modern terrestrial mammals, the number of tarsals is usually reduced, but one of the proximal bones is expanded posteriorly to form a heel, the insertion site of the powerful shank muscles (via Achilles' tendon). Metatarsals (one per digit) form the sole, articulating distally with one or more phalanges in the digits. Hind limbs are absent in cetaceans and sirenians.

Muscles

As with the skeleton, our presentation of the basic musculature of mammals is much briefer than what might be justified by the enormous body of information available (Hildebrand et al. 2001; Kent and Carr 2001; Liem et al. 2001; Romer and Parsons 1986). Muscles may be classified in several ways, but we are here concerned with **somatic muscles,** those that orient the body in the external environment. Somatic muscles are *striated* (i.e., histological preparations show the presence of sarcomeres), *skeletal* (i.e., attached to bones by tendons), and *voluntary* (i.e., can be contracted at will). As such, they provide the force to move the skeleton's levers, resulting in general body movements, locomotion, and other actions. Nonsomatic (i.e., visceral) muscles are smooth or cardiac types, nonskeletal, and involuntary.

The distribution of somatic muscles in mammals follows the architecture of the skeleton (figure 6.13). *Axial muscles* are those having their origins and insertions on the axial skeleton, or on connective tissues associated with it. Dorsally, these muscles are disposed as longitudinal bands lying above the transverse processes, and on either side of the neural spines, of vertebrae. In general, they serve to extend (straighten) the spine. Ventrally, axial muscles below the transverse processes of vertebrae flex (arch) the spine; others form the multilayered body wall. Anteriorly, the diaphragm and thoracic muscles power ventilation of the lungs. *Appendicular muscles* are those that insert on the girdles or limbs; their contraction drives limb movements and locomotion. In most mammals, the appendicular musculature is extremely well developed, resulting in the unparalleled locomotor diversity of mammals compared to other vertebrates.

Modes of Locomotion

Many mammals use walking as their primary means of locomotion, and this appears to be the primitive pattern. Others are specialized for running. Kangaroos, some rodents, rabbits, and hares employ hopping, jumping, or leaping to get around. Climbing is a primary activity of arboreal species such as many primates and sloths, and a secondary locomotor pattern for many other species that are mainly terrestrial. Some species are specialized for burrowing, including eutherian moles, marsupial moles, and many subterranean rodents. Dermopterans, some rodents, and some marsupials have the ability to glide, and one group (bats) has evolved powered flight. Other mammals spend all or most of their time in water, and for them swimming is the dominant form of locomotion. This group includes marine mammals (cetaceans, sirenians, and pinnipeds) as well as freshwater species such as

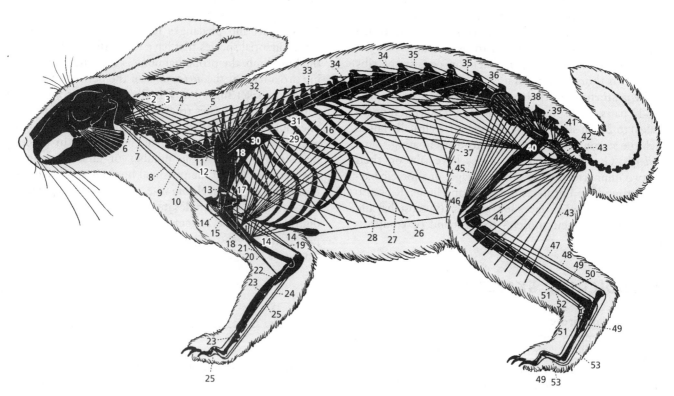

Figure 6.13 Muscle groups. Schematic representation of muscle groups in the European rabbit (*Oryctolagus cuniculus*). Notice the groups of muscles that operate the limbs and the manner in which they are positioned to work in opposition to one another. (1) masseter; (2) obliquus capitis; (3) splenius capitis; (4) semispinalis capitis; (5) longissimus cervicis; (6) longissimus capitis; (7) obliquus capitis inferior; (8) basioclavicularis; (9) levator scapulae; (10) sternomastoid; (11) scalenus; (12) supraspinatus; (13) infraspinatus; (14) pectoralis; (15) cleidohumeralis; (16) latissimus dorsi; (17) subscapularis (displaced caudally); (18) deltoid; (19) triceps; (20) biceps brachii; (21) brachialis; (22) extensor carpi ulnaris; (23) extensor digitorum communis; (24) flexor digitorum sublimis; (25) flexor digitorum profundus; (26) rectus abdominis; (27) transversus abdominis; (28) external oblique; (29) serratus anterior; (30) trapezius; (31) iliocostalis; (32) longissiumusi; (33) semispinalis dorsi; (34) longissimus dorsi; (35) multifidous; (36) sacrospinalis; (37) psoas major; (38) gluteus medius; (39) pyriformis; (40) gluteus maximus; (41) abductor caudae; (42) gemellus inferior; (43) biceps femoris; (44) adductors; (45) rectus femoris; (46) vastus intermedius; (47) gastrocnemius and plantaris; (48) soleus; (49) flexor digitorum longus; (50) peroneus muscles; (51) extensor digitorum; (52) tibialis anterior; (53) Achilles' tendon. *Reprinted from Young and Hobbs (1975b); reprinted by permission of Oxford Univ. Press.*

beavers and muskrats. We will examine each of these in turn. It is important to note, however, that most mammals use more than one means of movement.

WALKING AND RUNNING

Most mammals are quadrupedal, though a few (including humans) have evolved some form of bipedal locomotion. Species that move predominantly by walking are called **ambulatory,** and those with adaptations for running are **cursorial.** The primitive, walking gait is associated with a **plantigrade** foot posture in which all or most of the palms and soles are in contact with the substrate. This means that the metatarsals and phalanges of the hind foot, and the metacarpals and phalanges of the front foot, are oriented parallel to the ground. Cursorial mammals have one of two foot postures. **Digitigrade** species (e.g., cats) have elevated the metacarpals and metatarsals to an acute angle, leaving only the phalanges in contact with the substrate (see figure 6.12). Most digitigrade species have reduced one of their digits, leaving only four functional toes for locomotion. **Unguligrade** mammals (i.e., ungulates) have elevated the phalanges, as well as the metacarpals and metatarsals, such that only the tips of the phalanges are in contact with the ground. Along with their foot posture, ungulates have further reduced the number of digits to three (e.g., tapirs), two (i.e., artiodactyls), or one (i.e., horses); they have also developed hooves and increased the length of the functional metacarpals, metatarsals, and phalanges. Speeds achieved by mammals vary considerably with the different forms of locomotion (table 6.2).

Hildebrand (1985b) identified four functional requirements for animals that walk or run: (1) support and stability even though the feet make only intermittent contact with the substrate; (2) propulsion to move the body forward; (3) maneuverability; and (4) endurance (see also Hildebrand et al. 2001 and Liem et al. 2001). The first challenge is partly postural. Large, heavy species such as elephants and hippopotamuses are **graviportal:** Their legs are directly under the bodies, their propodial and epipodial bones are columnar, and their ankle and knee joints are nearly vertical. This arrangement allows the skeleton to bear most of the large body weight, taking the burden off postural muscles that would otherwise require large amounts of energy. Lighter mammals such as deer have their limbs positioned slightly outside the trunk axis and rely more on muscles for postural support and stability during locomotion.

Different approaches to propulsion result in distinct **gaits** (the pattern of regular oscillation of the limbs during forward movement) for walking and running mammals. In ambulatory species, each foot is on the ground for at least half the duration of a single stride cycle; in cursorial species each foot is on the ground less than half the time. Walking, pacing, and trotting all involve equal spacing of the feet making contact with the substrate, with the footfalls evenly spaced in time (Hildebrand et al. 2001). These are *symmetrical* gaits (figure 6.14). Walking is the most stable gait because of the prolonged contact of the feet with the ground (figure 6.14D). This pattern reduces the possibility of the front and hind limbs interfering with one another. Pacing, often seen in long-legged carnivorans and camelids, involves the simultaneous movement of both legs on the same side of the body (figure 6.14C). Trotting involves synchronously moving the two legs that are diagonally opposite one another (figure 6.14B) and provides somewhat more stability than pacing. Fast trotting (figure 6.14A) is how many short-legged cursorial species run. Gaits such as galloping and bounding, in which the footfalls are unevenly spaced in time, are *asymmetrical*. At moderate to high velocities, these gaits entail having all four feet off the ground simultaneously for a portion of the stride cycle.

Models of animal movement (Hildebrand 1980; Hildebrand et al. 2001; Liem et al. 2001) are based on the idea that the legs swing below the body like a pendulum. The jointed limbs and their associated muscles, tendons, and ligaments work like levers (Maynard Smith and Savage 1956). Several predictions from such models are supported by the comparative anatomy of mammals. Faster species have relatively longer limbs than slower species. Cursorial mammals tend to have limb joints that restrict movement to a single plane, forward and backward, parallel to the body. The insertion points of many muscles used for running have shifted so that they are nearer to and in the same plane as the lever joint they operate. Reducing weight in the distal portion of the limbs also increases running speed. This is accomplished by bone reductions and fusions in the lower legs (e.g., the cannon bone of horses) and by moving the appendicular muscles nearer to the body's center of mass, with long

Table 6.2 Running speeds of land mammals*	
Species	**Speed (km/h)**
Cheetah	110
Pronghorn antelope	98
Thompson's gazelle	80
Wildebeest	80
Elk (wapiti)	72
Coyote	70
European hare	65
Spotted hyena	64
Cape buffalo	55
Giraffe	50
Grizzly bear	50
Human	45
African elephant	40
Tree squirrel	20
Three-toed sloth	1

*Some of these speeds were measured over varying distances. For some species, the value represents a maximum speed that can be maintained only for short distances.

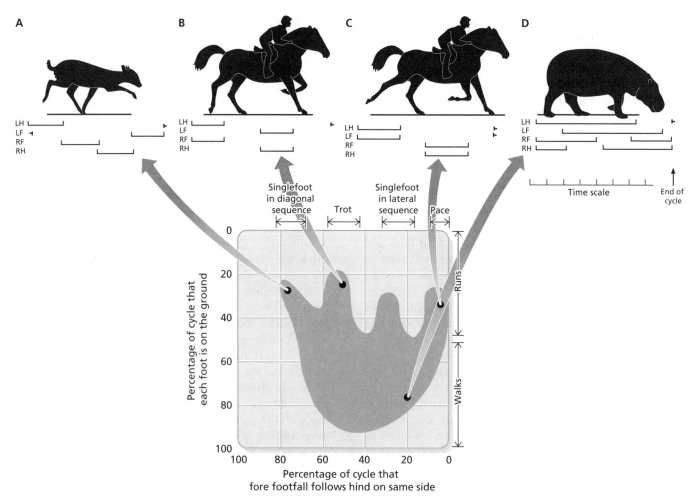

Figure 6.14 Symmetrical gaits for walking and running. The graph shows the percentage of time during a single stride cycle that each foot is in contact with the substrate (vertical axis) versus the percentage of the cycle during which feet on the same side of the animal touch the ground consecutively. (A) Fast running in a small antelope; (B) fast running trot in a horse; (C) moderate running pace in a horse; (D) moderate walking in a hippopotamus. *Adapted from Hildebrand et al. (1985).*

tendons stretching to the distal limb joints. Tendons and ligaments that cross joints can function like springs when the joints are flexed, storing energy that is released during subsequent extension (Alexander and Bennet-Clark 1977). Ungulates, for example, have a long "springing" ligament connecting their metacarpals or metatarsals with more distal phalanges.

Maneuverability, the capacity to change direction during locomotion, is required by both predators and prey. Most cursorial mammals can alter their gait momentarily, such that both legs on the same side (or both hind legs) strike the ground simultaneously, resulting in an angular shift in the direction of movement. Some mammals can turn their bodies (spinal flexion) while in the air, thereby altering their direction. Just as when we round a curve on a bicycle, mammals lean their weight in the direction of the turn to maintain balance and control. Running mammals that are large and heavy tend to be less maneuverable and agile than smaller, lighter ones.

Endurance results from integration of an animal's musculoskeletal and physiological adaptations. The musculoskeletal traits described here contribute to efficient motion of the limbs. Moreover, the appendicular muscles of cursorial mammals tend to be rich in "fast twitch" fibers capable of rapid, powerful contractions (Goldspink 1981; Kent and Carr 2001). The study of joint mechanics, models of ambulatory and cursorial movement, and research in bioenergetics have become quite sophisticated in recent years, incorporating principles from mechanical engineering (Hildebrand et al. 2001) and physics (Bejan and Marden 2006).

Consider again the example of the cheetah and gazelle from the beginning of this chapter. Both the predator and the prey are cursorial species, with digitigrade and unguligrade foot postures, respectively. The top speeds of these animals (table 6.2) might predict that the cheetah will invariably win the race, but the distance and length of time over which each species can maintain its top speed (i.e., its endurance) must also be considered. Cheetahs can move swiftly for relatively short distances, a few hundred meters at most, whereas the endurance of the gazelle is considerably greater. Thus, the cheetah needs to be close enough when it begins its dash so that it can overtake the gazelle quickly; if it fails, it will soon be outdistanced by its prey. Oddly enough, the fastest land mammals in North America—pronghorns—seem to lack a predator that would explain their great speed. Perhaps that predator existed among the diverse collection of Pleistocene carnivores that did not survive to the Recent.

JUMPING AND RICOCHETING

Jumping and ricocheting are forms of **saltatorial** locomotion. Jumping involves the use of all four feet, as in the case of lagomorphs. Ricocheting involves propulsion with only the hind limbs, such as in kangaroos, kangaroo rats, and jumping mice (figure 6.15). Mammals that employ ricochetal locomotion spend much of their time in a bipedal position, using the forepaws only occasionally for slow, short-distance movements. The forelimbs of such animals are shorter than the hind limbs and are often employed for manipulating objects such as food. Most mammals (indeed, most vertebrates) that use saltatorial locomotion are relatively small, with kangaroos (up to 90 kg body mass) being exceptions.

Because saltatorial movement has evolved in several different groups, the anatomical similarities shared by jumping or ricocheting species are the result of convergent evolution (Emerson 1985). The principal adaptation is lengthening one or more segments of the hind limbs, usually the tibia, and development of long, elastic tendons that stretch across the knee and ankle joints. Energy stored in these tendons as the limbs recover from one jump (i.e., during landing) is released during the next propulsive leap forward (Biewener et al. 1981). Other adaptations for saltatorial locomotion include: (1) a posterior shift in the center of body mass to avoid tumbling forward (or backward) when thrusting with the hind limbs; (2) an enlargement of muscles in the hip region, with lengthy tendons; (3) changes in the size and arrangement of bones in the pelvic girdle to accompany shifts in musculature and the center of body mass; (4) larger hind feet for takeoff and landing; and (5) a longer tail for balance.

Arboreal clinging and leaping become ricochetal locomotion in some lemurs (figure 6.16), tarsiers, and the white-faced saki (*Pithecia pithecia*). These animals often cling to a tree trunk with all four limbs, drop down to the ground, land on their hind feet, then bound back up into another tree. Primates that leap and ricochet have longer femurs than do other primates, allowing them to make longer leaps by increasing the lever action of their hip muscles (Connour et al. 2000). Tarsiers, which also leap great distances from branch to branch, also have elongated shank and tarsal bones such that their thighs, lower legs, and feet are of roughly equal length (Macdonald 2006).

CLIMBING

Climbing mammals use their limbs to move about in trees (i.e., they employ **arboreal** locomotion and display a suite of corresponding adaptations; Cartmill 1985). In species that also spend a substantial portion of their time on the ground, climbing is accomplished primarily by use of claws. Small arboreal mammals, such as squirrels, gain a holdfast in tree bark with their sharp claws and are able to maneuver on trunks and branches with considerable agility. Larger species, such as bears (figure 6.17), use their claws in a similar manner but are much less agile. These species possess footpads to provide friction that aids in gripping tree limbs securely, as well as increased numbers of sensory receptors on their palms, soles, and ventral surfaces of digits. Sloths employ their claws to hang underneath tree branches for extended periods. Arboreality in mammals with a plantigrade foot posture is facilitated by the flexible joint between metacarpals or metatarsals and phalanges, allowing the hands and feet to be used for grasping. Primate digits are even more flexible—mobile joints between phalanges in the hand allow them to wrap their fingers around branches or other objects. Such hands, and occasionally feet, are said to be **prehensile.** Old World monkeys and humans have evolved a fully opposable thumb that can be extended to touch each of the other digits. Their thumb has a saddle joint at the base of the proximal phalanx, is oriented at nearly a right angle to the first finger, and is attached to a powerful muscle (the adductor pollicis). Partial opposability is found in the thumbs of New World monkeys and nonhuman anthropoids, and in the great toes of many primates (Kent and Carr 2001). **Brachiation,** swinging from branch to branch using the forelimbs, is best developed in gibbons (Hylobatidae) but is also found in other primates. Brachiators have large clavicles anchored to the sternum, relatively long forelimbs, grasping hands, and opposable great toes. Their stout

Figure 6.15 Saltatorial locomotion. Kangaroos, such as the eastern gray kangaroo (*Macropus giganteus*) shown here, move by ricochetal (bipedal) hopping. Notice the robust hind legs and feet that are used to propel the animal forward with each bound.

A

B

Figure 6.16 Ricochetal locomotion. Lemurs such as the ring-tailed lemur (*Lemur catta*) use a specialized form of saltatorial locomotion involving ricochetal movement from one tree (A) to the ground (B), and then back to another tree.

pectoral girdle stabilizes the shoulder joint, allowing the forelimb to bear the weight of the animal (Kardong 2006).

Figure 6.17 Tree climbing. When a black bear (*Ursus americanus*) climbs a tree, it grasps the trunk with its forepaws and hind feet, aided by lateral pressure from the arms and legs, and by claws on the ends of its digits.

Humans have this kind of pectoral girdle, a reflection of our arboreal ancestry.

Many arboreal mammals have long tails, which they use for balance (e.g., harvest mice of the genus *Reithrodontomys*). Some lemurs and Old World monkeys use their tails as a brace against tree trunks during climbing. In a few species of South American monkeys, the tail has become a prehensile appendage used to grasp branches. The distal portion of the tail in these species has developed friction pads and an increased number of tactile receptors similar to the gripping hands and feet of other climbing mammals.

DIGGING AND BURROWING

Mammals that dig in the soil to find food or create shelter are called **fossorial**. It is useful to distinguish the terms "fossorial" and **subterranean**—the former refers to animals with adaptations for digging, whereas the latter refers to species that live virtually their entire lives underground (Lacey et al. 2000). Among mammals, there are more fossorial than subterranean species. The limbs of digging species are short and powered by strong appendicular muscles. Digging is usually accomplished by scratching at the soil with forepaws to excavate a hole or tunnel large enough to accommodate the animal's body. A series of rapid alternate strokes of the forelimbs loosens soil in front of the digger, pushing excavated material posteriorly under the belly, after which the hind limbs kick it further back or compact it. Marsupial moles (Notoryctemorphia) and golden moles (Chrysochloridae) use fore and hind limbs to dig through sand but do not leave a permanent burrow. Golden moles use their heads to push up the surface layer of soil and to compact excavated debris (Puttick and Jarvis 1977). Bateman (1959) described a golden mole (*Amblysomus hottentotus*) weighing less than 60 g that moved a 9 kg iron plate with its head in order to escape from a fishbowl filled with soil. In true moles (Talpidae), the short humerus is highly sculpted as a means of increasing surface area for muscle attachment, the distal bones of the forelimb are enlarged, and the wrist joint is rotated outward such that abduction and extension of the limb results in a lateral digging stroke; recovery is accomplished by rotating the limb downward (see figure 12.4). Some tunnel-diggers will periodically clear the passage behind them by turning around and pushing the excavated soil back out to the surface.

Many fossorial rodents also use their teeth as digging tools, including members of Spalacidae (e.g., root rats, bamboo rats, blind mole-rats), Geomyidae (pocket gophers), and Bathyergidae (African mole rats). These animals have large incisors external to the lips, allowing them to dig with their teeth while their mouths remain closed. Their skulls have a broad rostrum and stout zygomatic arches for attachment of powerful masseter (jaw-closing) muscles, as well as well-developed neck musculature for moving the head (Orcutt 1940; Macdonald 2006).

Subterranean species show additional morphological specializations (figure 6.18). Their eyes are small and probably sightless as in bathyergids and talpids, or vestigial as in marsupial moles and blind mole-rats. Tactile

receptors in the snout are well developed, and vibrissae may occur on the tail, body wall, or legs. Senses of hearing and smell are usually well developed, with most species lacking external pinnae and some possessing valvular nostrils that can be closed during digging.

Many other mammals—including monotremes (platypus and echidnas), marsupials (e.g., wombats), and numerous placental species—excavate soil for nesting, food storage, hibernation, refuge from predators, or other functions (Reichman and Smith 1990). Burrows may be used continuously over months or years, or they may be used briefly and abandoned. Some species construct different kinds of burrows for different purposes. For example, hamsters (Muridae) make separate chambers for sleeping, food storage, and defecation. Woodchucks (*Marmota*) usually have a summer burrow in an open area and a winter burrow in more forested habitat.

GLIDING AND FLYING

Gliding has evolved independently in several groups of mammals—gliding possums (Petauridae, Pseudocheiridae, Acrobatidae), colugos (Dermoptera), and members of Rodentia including "flying" squirrels. In each of these groups, the gliding species are arboreal and use their gliding ability as a means of moving among tree branches. The principal morphological adaptation is a **patagium,** an extension of skin that stretches from the lateral neck and body wall to the wrists and ankles, as well as to the tips of the fingers, toes, and tail in colugos (see figures 18.8 and 18.13). The animal leaps from a perch extending its limbs and tail such that the patagium acts as an airfoil. Aerodynamic control during gliding and landing is accomplished by adjusting the position of the limbs (Jackson 1999). The patagium may also be used for protection or insulation when wrapped around the body like a cloak; colugos also use it as a pouch in which to hold neonates.

Among mammals, only bats (Chiroptera) have evolved true powered flight, or **volant** locomotion. With the exception of swimming, flying is the most energetically efficient means of moving a body of a given mass between two points (Norberg 1985, 1990). A bat's wing is also a patagium, but its skeletal support—principally the autopodial bones—is more highly modified than in gliding mammals (figure 6.19). The broad and slightly keeled sternum serves as the point of origin for flight muscles. The shoulder includes a stout clavicle and a locking mechanism to keep the joint at an appropriate angle (locking devices may also occur in the elbow, wrist, and digits). The radius is thin and elongated, but the ulna is reduced distal to the elbow, allowing no rotational movement of the forearm. At the wrist, central and distal carpals are lost, and the three proximal carpals fuse to form a single bone. The first digit is unmodified and bears a claw, but digits 2–5 have greatly elongated metacarpals followed by

Figure 6.18 Burrowing. Moles, such as this eastern American mole (*Scalopus aquaticus*), live their entire lives underground. A key adaptation for this subterranean existence is the modification of the hands and feet for digging. Moles build extensive tunnel systems just below the surface, primarily as a means of finding food.

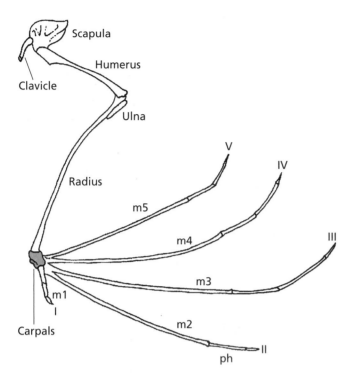

Figure 6.19 Bat wing skeleton. The pectoral girdle and right wing of a bat. Note the elongated radius and metacarpals (m2–m5) on digits II–V, the reduced ulna, and the fusion of carpals into a single wrist bone. *Redrawn from Kent and Carr (2001).*

two or three phalanges that are oval in cross section (Hildebrand et al. 2001). The radius and bones of digit 2 support the leading edge of the wing, while digits 3–5 form struts extending to the trailing edge. Digit 1 (the thumb) may function in the awkward terrestrial locomotion of bats, or aerodynamically as a leading-edge flap for the wing. The patagium stretches from the body wall anterior to the forelimb to the short hind limbs and a portion of the tail. Bats, like birds, have a body trunk that is compact and stiff (including several fused vertebrae and proximally fused ribs in some species) for aerodynamic efficiency. We discuss the aerodynamics of bat flight in chapter 13.

How did flight evolve in bats? Current authors agree that the common ancestor of all bats was a small, quadrupedal animal with a generalized forelimb anatomy (Speakman 2001). Unfortunately, the bat fossil record shows no transitional forms—the oldest fossils (e.g., *Icaronycteris* from the Eocene) have winglike forelimbs very similar to those of modern species. Sears et al. (2006) confirmed that the relative lengths of bones in digits 3–5 are the same in living and fossil bats. They also showed that embryonic digits in the hands of modern bats are initially similar to those of mice, but subsequently become elongated due to locally increased expression of a bone morphogenetic protein gene. Causal hypotheses for the evolution of flight in bats have linked this trait

to echolocation, a mechanism used by many bats for navigation (see chapter 13). Speakman (2001) reviewed five scenarios: (1) Echolocation may have evolved before flight as an adaptation for efficient hunting of insects from an arboreal perch. There are no data to support this idea and energetic considerations suggest that it is unlikely. (2) Flight may have evolved before echolocation, as the end point of a trend that began with arboreal leaping and gliding. Although this scenario has some support, it seems unlikely that bat ancestors could have successfully leaped between branches in the dark. Moreover, there is no morphological evidence that ancestral gliders could have slowed themselves for a stable landing. (3) It has been suggested that bats (and hence powered flight) evolved from two distinct ancestors (a "diphyletic" origin; Pettigrew 1986), one leading to large fruit bats (most of which do not echolocate) and the other to small, insectivorous forms (all of which echolocate). However, most systematists reject bat diphyly (e.g., Teeling et al. 2005); consequently, dual origins of flight appear untenable. (4) Echolocation and flight may have evolved together. Ancestral bats may have leaped and glided while relying on enhanced visual perception, thereby establishing a selection pressure favoring both echolocation and powered flight. Given that echolocation and flight seem to have originated only once among fossil and living bats (Springer et al. 2001), this scenario is plausible. (5) Speakman (2001) suggested that bat ancestors were arboreal, diurnal, and frugivorous. Once flight evolved, however, bats were forced into a nocturnal lifestyle because of competition with raptorial birds. From that point, some bats developed echolocation and others specialized vision. This hypothesis also appears consistent with the available data.

SWIMMING

All mammals that spend a significant portion of their time in water have evolved from terrestrial ancestors. Because water is generally cooler than average body temperatures and more thermally conductive than air, aquatic and marine mammals must conserve heat when they are in water. Most species have thick coats of fur or body fat that serve as insulation and provide buoyancy. **Amphibious** (semiaquatic) species occur in many groups (monotremes, marsupials, tenrecs, rodents, soricomorphs, carnivorans, and artiodactyls); most of these animals are equally at home in water or on land (figure 6.20). Webbing between the toes in many species increases the surface area in contact with the water for propulsion, especially significant in that semiaquatic mammals swim by alternating strokes of the limbs (paraxial swimming), much as they walk on land (Webb and Blake 1985). Water shrews (*Sorex palustris*) have stiff hairs between the toes (fimbriation) that serve the same function as webbing. Muskrat tails are laterally flattened, and those of beavers are dorsoventrally flattened;

Figure 6.20 Swimming. Muskrats (*Ondatra zibethicus*) are amphibious mammals, spending a portion of their time in the water. They have a tail that is laterally flattened, which can be used as a propeller and as a rudder (e.g., when the animal swims with its limbs).

both can be used for propulsion or like a rudder for directional control.

Pinnipeds (seals, sea lions, walruses) are fully aquatic mammals that move onto land only to breed and give birth. The autopodia of pinnipeds have evolved into flippers, with all five digits encased in a single sheath of integument; tails are absent or rudimentary. In the water, wriggling (earless, or true) seals (Phocidae) generate propulsion with their hind limbs, which are caudally directed. This orientation is permanent, the tibial portion of both hind limbs being enclosed with the proximal tail in a common neck of integument. Phocids swim by lateral undulations of the posterior trunk and hind limbs, facilitated by flexion at the knee and ankle joints that produces a sculling motion. On land, however, the hind limbs are of no use and phocids are forced to pull themselves along the substrate with their front flippers and wriggle with their trunk. Walruses (Odobenidae) also swim by sculling, though much more weakly than phocids. Eared (or fur) seals and sea lions (Otariidae) employ a modified form of paraxial swimming, generating thrust by moving their forelimbs in synchrony while their back and caudally directed hind limbs undulate dorsoventrally. Unlike phocids, otariids are able to rotate their rear flippers forward when they are on land, allowing for an awkward but distinctly quadrupedal gait. This ability is especially well developed in pups, giving them much more agility on land than is possible for adults.

Anatomical specializations for a fully aquatic life reach their pinnacle among marine mammals—those that never come onto land. There are two such groups, Cetacea (whales) and Sirenia (manatees and dugongs). The axial skeleton of these animals is simplified, with the cervical vertebrae partially fused and interlocking facets lost between trunk and tail vertebrae. Hind limbs and sacrum are absent, and the pelvic girdle is vestigial. The tails are modified into horizontal flukes that provide propulsion by dorsoventral undulation. Forelimbs are modified into flippers and frequently show an increase in the number of phalanges; they are used as stabilizers and have no role in propulsion. In cetaceans, the body trunk is fusiform, a friction-reducing adapation for fast swimming. The buoyancy provided by water has removed some of the restrictions on body size that would accompany a terrestrial tetrapod, allowing some whales to reach gigantic body sizes. Other noteworthy adaptations of marine mammals are discussed in chapters 17 and 19.

SUMMARY

Study of the integument and musculoskeletal system tells us a great deal about the integration of form and function in the evolution of mammalian body forms. This is especially evident in the context of locomotion, where mammals are unparalleled among vertebrates in their morphological and functional diversity.

The integument consists of an outer epidermis and an inner dermis. The epidermis contains a basal layer of dividing cells that become impregnated with keratin as they move superficially, ultimately forming a layer of dead tissue on the surface of the body that prevents dessication and damage to the underlying skin. Epidermal derivatives include many uniquely mammalian characteristics such as hair, sebaceous glands, sweat glands, mammary glands, nails, hooves, horns, and antlers. The dermis contains integumentary muscles that insert on hairs, general somatic receptors and associated sensory nerves, blood vessels, and connective tissue. Beneath the dermis, a layer of subcutaneous fat serves as insulation and an energy reservoir.

The mammalian skeleton has axial and appendicular divisions. The skull, the anteriormost portion of the axial skeleton, consists of a cranium and mandible. The cranium includes bones of two different embryonic origins: the neurocranium (occipital, otic, sphenoid, and ethmoid bones) forms the primary braincase; the dermatocranium consists of roofing, upper jaw, temporal, and secondary palate bones. The mandible is formed by dentary bones that articulate with the squamosal bone of the cranium to form the characteristic mammalian jaw joint. The remainder of the axial skeleton consists of the hyoid apparatus, vertebral column, ribs, and sternum. Bones of the pectoral and pelvic girdles, along with their associated limbs, constitute the appendicular skeleton. Fore- and hind limbs have a common structural design, with a proximal propodium, intermediate epipodium, and distal autopodium.

Like the skeleton, mammalian musculature has axial and appendicular divisions. Axial muscles are disposed as parallel bands along the vertebral column that flex or extend the spine, as a multilayered body wall in the trunk region, and as thoracic muscles associated with ventilation. Appendicular muscles insert on bones of the appendicular skeleton and power the limbs. Skeletal muscles are somatic (derived from mesodermal somites), striated (sarcomeres are visible in histological preparations), and voluntary (under the control of cranial or spinal nerves).

There are several major categories of locomotion in mammals. Walking mammals have an ambulatory gait and plantigrade foot posture. Running mammals are cursorial, and they display gaits and foot postures that vary with running speed and species. Some cursorial species, such as cats, have a digitigrade foot; others, such as ungulates, are unguligrade.

Cursorial locomotion is frequently associated with a reduction in the number of digits and lengthening of those digits that remain. Jumping and ricocheting mammals use saltatorial locomotion; they frequently have enlarged hind limbs and reduced forelimbs. Climbing mammals are arboreal, variously making use of claws, flexible autopodial joints, or prehensile tails for grasping branches. Brachiators have stout, weight-bearing shoulder joints and long forelimbs. Digging and burrowing species are fossorial, and those that spend nearly all of their time underground are subterranean. Most digging mammals scratch at the soil with their forelimbs, but several groups of fossorial rodents dig with their incisor teeth. Gliding mammals are arboreal species with a patagium that serves as an airfoil when extended during leaps between trees. Powered flight occurs in bats, the only volant mammals. Bat wings consist of a patagium, with skeletal support provided primarily by modified autopodial bones. Semiaquatic mammals are adapted for swimming and terrestrial locomotion; most swim by alternating strokes of their limbs. Pinnipeds are marine carnivores that leave the ocean only to reproduce. Their limbs are modified as flippers. The hind limbs of true seals are permanently reoriented caudally to function in sculling. Sea lions swim by paddling with the forelimbs and can rotate their hind limbs forward for terrestrial locomotion. Cetaceans and sirenians are fully marine. They swim by dorsoventral undulation of their tail, which is modified to form a horizontal fluke. Their hind limbs are absent and their forelimbs are flippers.

SUGGESTED READINGS

Hildebrand, M., D. M. Bramble, K. F. Liem, and D. B. Wake (eds.). 1985. Functional vertebrate morphology. Harvard Univ. Press, Cambridge, MA.

Hildebrand, M., G. E. Goslow, Jr., and V. Hildebrand. 2001. Analysis of vertebrate structure, 5th ed. John Wiley, New York.

Kardong, K. V. 2006. Vertebrates: Comparative anatomy, function, evolution, 4th ed. McGraw-Hill, New York.

Kent, G. C., and R. K. Carr. 2001. Comparative anatomy of the vertebrates, 9th ed. McGraw-Hill, New York.

Lacey, E. A., J. L. Patton, and G. N. Cameron (eds.). 2000. Life underground—The biology of subterranean rodents. Univ. Chicago Press, Chicago.

Liem, K. F., W. E. Bemis, W. F. Walker, Jr., and L. Grande. 2001. Functional anatomy of the vertebrates: An evolutionary perspective. Harcourt College Publ., Fort Worth, TX.

Macdonald, D. (ed.). 2006. Encyclopedia of mammals, 2nd ed. Facts on File, Oxford Univ. Press, Oxford, U.K.

DISCUSSION QUESTIONS

1. Using the internet or other reference material, compare the skeleton of a generalized mammal such as the Virginia opossum (*Didelphis virginiana*) to that of a generalized reptile such as a tuatara (*Sphenodon* spp.). What differences are apparent in bones of the cranium and mandible? How does the arrangement of bones in the pectoral and pelvic girdles differ, and how does this relate to locomotor differences in the two species?

2. Mammalian skin is unique in that it contains all of the tissue types recognized by histologists (i.e., epithelium, connective tissue, muscle, and nerves). Consult a reference text and identify the tissue type of each integumentary structure described in this chapter. How are the functions of these structures related to their histological properties?

3. Bodybuilding programs for humans attempt to increase the strength and mass of specific muscle groups in both the axial and appendicular divisions. Consult a trainer's handbook or similar reference and relate specific exercises to the muscles they are designed to enhance. Which of these exercises would be possible for a nonprimate mammal to perform?

4. Bats and birds are both capable of powered flight, but this capacity evolved independently in the two groups. Find illustrations of the skeleton of a bird's wing and compare it to the bat wing shown in this chapter. What are the anatomical similarities and differences? How do the two types of wing function as airfoils and propellers?

5. Marine mammals (e.g., pinnipeds) and marine birds (e.g., penguins) share many adaptations for swimming and survival in cold ocean water. Compare the integument and skeletal anatomy of these groups and note the similarities and differences in their locomotor and thermoregulatory adaptations.

CHAPTER 7

Modes of Feeding

Mammals, like all organisms, require energy and nutrients for maintenance, growth, activity, and reproduction, that is, for survival. Maintaining a high body temperature, however, which is a key feature of Class Mammalia, requires regular acquisition of food. The food of mammals ranges from microscopic forms such as diatoms and crustaceans—a staple in the diet of the largest mammals, the baleen whales—to sedentary forms such as plants used by the most abundant mammals, the rodents. Mammals consume food of high-energy content (blood of vertebrates and insects) as well as of low-energy value (grasses and stems). The food of mammals may be highly specialized and restricted (nectar of localized plants) or rather general and readily available (grasses and herbs). To meet their high-energy needs, mammals have evolved a diverse array of trophic, or nutritional, specializations. The adaptive radiation in food-gathering morphologies is diverse and reflects the diversity of available food.

In this chapter, we detail the feeding apparatus of mammals, focusing on the capturing (teeth, tongue, and jaw musculature) and processing (alimentary canal) of food. Feeding integrates the sense organs and locomotor adaptations (see chapters 6 and 8). Although different orders of mammals are sometimes grouped according to their modes of feeding (i.e., Carnivora), food habits cannot be employed as a systematic criterion because many members of an order may depart from these feeding generalizations. Thus, to enhance understanding of nutritional adaptations, we suggest consulting specific chapters to unite anatomical specializations of different groups with their dietary habits. At the end of the chapter, we will examine briefly some general principles regarding mammalian foraging strategies.

Foods and Feeding

We understand the life-history traits and food habits of extant mammals by examining their teeth. As you learned in chapter 4, all mammals, except certain whales, monotremes, and anteaters, have teeth, and these structures are inextricably linked with food habits. As mammals evolved in the Mesozoic era, major changes occurred in their dentition and jaw musculature; teeth became differentiated to perform specialized functions. Within extant species, several trophic groups can be recognized, namely insectivorous, carnivorous, herbivorous, and omnivorous mammals. Other specialized modes of feeding have evolved from these four basic plans (figure 7.1).

INSECTIVOROUS

Insectivory

Mammals that consume insects, other small arthropods, or worms are referred to as **insectivorous** (meaning "insect-eating"). We know from examination of Triassic mammals that the insectivorous feeding niche represented the primitive, or basal, condition of eutherian mammals. Today, this feeding niche is exploited by members of 11 orders of mammals: echidnas and platypuses (Order Monotremata); hedgehogs, shrews, and moles (Orders Notoryctemorphia, Afrosoricidae, Erinaceomorpha, and

Soricomorpha); most bats (Order Chiroptera); anteaters and armadillos (Orders Cingulata and Pilosa); pangolins (Order Pholidota); aardvarks (Order Tubulidentata); and the aardwolf (Order Carnivora; see figure 7.1). Many other orders of mammals also contain members that exhibit insectivorous habits. The dentition of hedgehogs, shrews, moles, and most bats is typified by numerous sharp teeth with sharp cones and blades for piercing, shearing, and ultimately crushing the tough chitinous exoskeletons of insects. In many forms, the lower incisors are slightly **procumbent** (pointing forward and upward) to aid in grasping prey (see figure 12.12). Because insectivorous mammals consume minimal amounts of fibrous vegetative material, prolonged fermentation is not required; their alimentary canals are short, and most insectivores and chiropterans lack a cecum (figure 7.2).

Aerial Insectivores

The most abundant foods are plants and insects; it is therefore not surprising that the most abundant mammals are rodents and bats. Chiropterans occupy ecological niches in almost all habitats of the world; the diversity of

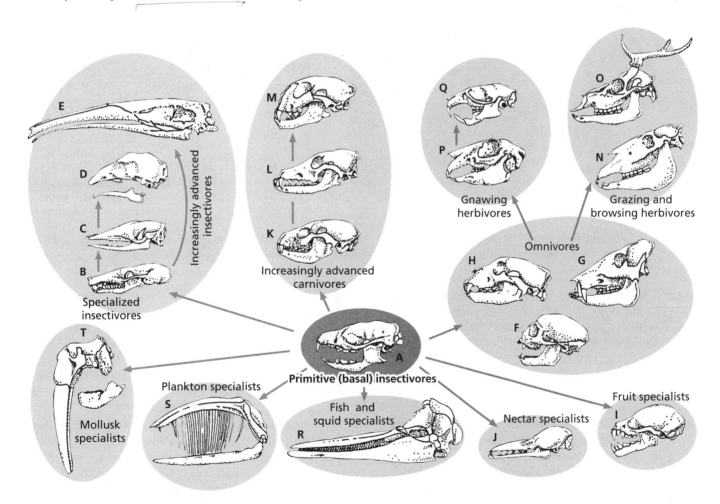

Figure 7.1 Skull and dentition specialization. Feeding specializations in the dentition and skulls of mammals relate to their dietary habits: (A) hedgehog, (B) mole, (C) armadillo, (D) anteater, (E) giant anteater, (F) marmoset, (G) peccary, (H) bear, (I) fruit-eating bat, (J) nectar-eating bat, (K) raccoon, (L) coyote, (M) mountain lion, (N) horse, (O) deer, (P) jackrabbit, (Q) woodrat, (R) porpoise, (S) right whale, and (T) walrus. *Adapted from E. Rogers, 1986,* Looking at Vertebrates, *Longman Group Ltd.*

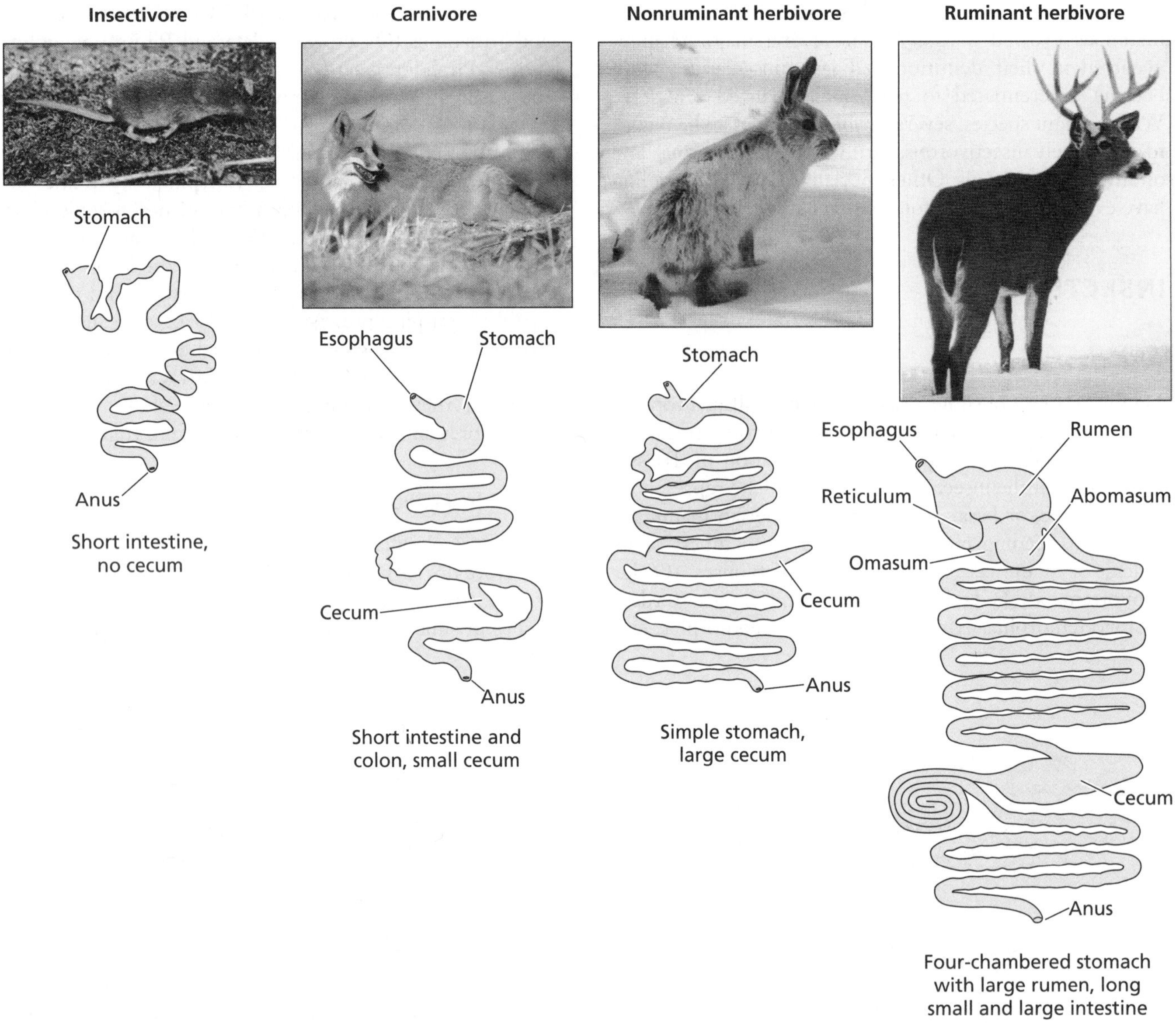

| **Insectivore** | **Carnivore** | **Nonruminant herbivore** | **Ruminant herbivore** |

Figure 7.2 Digestive system. The digestive systems of mammals, illustrating the differences in morphology that correspond to different diets. *Adapted from Hickman et al., 1995,* Integrated Principles of Zoology, *10th ed., Wm. C. Brown Publishers.*

their diets is unparalleled among extant mammals (chapter 13). The majority (70%) of "microchiropterans" are insectivorous (Black 1974; Whitaker 1988; Whitaker et al. 1996; Neuweiler 2000; Patterson et al. 2003). All bats residing north of 38°N and south of 40°S latitude are insectivorous. Throughout their range, insectivorous bats feed on a diverse array of arthropods, ranging from scorpions, spiders, and crustaceans to soft-bodied and hard-bodied insects (Whitaker 1994b; Neuweiler 2000; Schulz 2000). Insects are either captured in the mouth or trapped by a wing tip or the uropatagium (see figure 13.1). Foraging styles vary depending on the species. Insectivorous bats are voracious eaters: Mexican free-tailed bats (*Tadarida brasiliensis*) in central Texas, totaling some 20 million individuals, may consume up to a quarter of a million pounds of insects nightly and fly as high as 10,000 feet (3,000 m) in search of its prey (Kunz et al. 1995; Whitaker et al. 1996;

McWilliams 2005). Lactating female big brown bats (*Eptesicus fuscus*) nightly consume a quantity of insects equivalent to more than their body mass (Kurta et al. 1990). Within a given habitat, feeding assemblages of bats can be quite diverse and may be divided into different guilds, such as species that glean insects or those that forage within forest openings, over water, and in open-air zones above the forest canopy (Findley 1993; Nowak 1999).

The trophic niche of insectivorous bats may be assessed by determining echolocation calls plus morphological attributes of a species, including its wing and jaw and tooth morphology, brain size, and external dimensions (Findley and Wilson 1982; Entwistle et al. 1996; Bogdanowicz et al. 1999). The size of prey varies in relation to the predator's jaw morphology, from very small midges and mosquitoes to large beetles. For example, Freeman (1979, 1981, 1988) predicted food habits of molossids by assessing jaw structure and mechanics; beetle-eaters were characterized by more robust skulls and fewer but larger teeth, whereas moth-eaters had delicate skulls and numerous yet smaller teeth. Most insectivorous bats are generalists and opportunistic feeders, but remarkable specialists do occur (Whitaker 1994b). Pallid bats (*Antrozous pallidus*) of the southwestern United States, for example, feed on beetles, Jerusalem crickets, sphinx moths, scorpions, and small vertebrates gleaned from the ground. Golden-tipped bats (*Kerivoula papuensis*) of southeastern Australia feed by gleaning, flying slowly in dense vegetation and hovering and plucking orb spiders from their webs (Richards 1990; Strahan 1995).

Terrestrial Insectivores

The platypus is a semiaquatic insectivore that feeds on benthic worms, insects, mollusks, and small invertebrates—those creatures that live at the bottom of a body of water (chapter 11). Food obtained during a dive is stored in large cheek pouches that open to the rear of the bill. When the cheek pouches are full, the platypus rests on the surface of the water, and the food is transferred to the rear of the mouth and masticated by horny pads. As in other insectivores, the alimentary canal of platypuses is simple and lacks gastric glands. Cheek pouches are thought to replace the stomach as a food storage area (Harrop and Hume 1980).

Four species of mammal produce a venomous saliva: the northern short-tailed shrew (*Blarina brevicauda*) of North America, the European water shrew (*Neomys fodiens*), Mediterranean shrew (*Neomys anomalous*), and the Haitian solenodon (*Solenodon paradoxus*). The toxin of *Blarina brevicauda* was purified and characterized as a lethal mammalian venom possessing a tissue kallikrein-like protease activity derived from the submaxillary and sublingual glands (Kita et al. 2004, 2005). Research by Tomasi (1978) and Martin (1981) demonstrated the importance of venom in the hoarding behavior of *Blarina*. In both *Blarina* and *Neomys*, the toxin is stored in submaxillary glands and is administered to the prey through a concave medial surface in the first lower incisors. Extracts of this toxin administered to mice affect the nervous, respiratory, and vascular systems, causing irregular respiration, paralysis, convulsions followed by death (Lawrence 1945; Kita et al. 2004, 2005). Blarina bites its prey, immobilizing it, and caches it below ground in a comatose state. Caching sites are marked by defecation and urination and provide shrews with a source of fresh food for some time. The ability to cache unused prey ensures that a predictable quick energy source is accessible and readily available if prey is scarce (Churchfield 1990).

Several groups of insectivorous mammals are **myrmecophagous** (meaning "ant-eaters"). Representatives include the armadillo (*Dasypus*), silky anteater (*Cyclopes*), giant anteater (*Myrmecophaga*), pangolin (*Manis*), aardvark (*Orycteropus afer*), and numbat (*Myrmecobius fasciatus*), which feed on colonial insects, such as ants and termites. Reduction of teeth is common among myrmecophagous mammals, and their dentition departs from the "insectivorous" design of the hedgehogs, shrews, and moles. They possess numerous peglike teeth (armadillos) or no teeth at all (echidnas, anteaters, and pangolins). The marsupial numbat, the sole member of the Family Myrmecobiidae, possesses numerous, small, delicate teeth—the total number may be as high as 52. The aardvark (Order Tubulidentata) is a special case, characterized by columnar cheekteeth composed of vertical tubes of dentine within a matrix of pulp (see figure 15.11). Mammals that consume colonial insects such as termites and ants possess long, extendible, wormlike tongues (Chan 1995; Reiss 1997). Elongated snouts and strong front feet used as digging tools enable anteaters and aardvarks to burrow rapidly into and tear apart termite hills. Their highly maneuverable, sticky tongues are effective in reaching the inner recesses of ant and termite nests. The tongue may be three times the length of the head; in several groups of anteaters, the tongue is anchored at the posterior end of the sternum rather than the throat (Hildebrand 1995; figure 7.3). Greatly enlarged salivary glands situated in the neck produce a viscous, sticky secretion that coats the tongue and is important in the breakdown of chitin.

Echidnas consume ants, termites, and earthworms, but they do not have teeth or even horny grinding plates on the rear of their jaw, as do platypuses. Rather, a pad of horny spines on the back of the tongue grinds against similar spines on the palate to crush the exoskeletons of arthropods. The mouth of the echidna is positioned at the very tip of its elongated snout and can be opened only enough to permit passage of the long, sticky, protrusile tongue. Because there are no glands in the stomach of echidnas, digestive enzymes are not present. The amylase present in the saliva therefore assists in the breakdown of insect chitin within the stomach. Like the insectivores, the monotremes have a simple alimentary canal with a tiny, nonfunctional cecum (Harrop and Hume 1980).

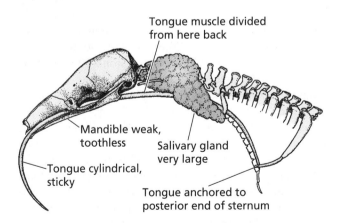

Figure 7.3 Anteater tongue specialization. The long, wormlike tongue of the collared anteater (Tamandua) is anchored to the posterior end of the sternum and can be protruded extensively to assist in capturing ants and termites. *Adapted from M. Hildebrand, 1995,* Analysis of Vertebrate Structure, *John Wiley & Sons.*

Insects represent a staple in the diets of many other mammals; for example, several species of nasute harvester termites (Genus *Trinervitermes*) form the chief food of the aardwolf (*Proteles cristatus*) of southern Africa (Koehler and Richardson 1990). A single aardwolf was estimated to consume about 105 million termites a year (Kruuk and Sands 1972). Slender-tailed meerkats (*Suricata suricatta*) are small, greparious mongooses residing in thorn and grassland savannah of the Kalahari Desert of southern Africa. Here, 78% of their diet comprised insects, namely larvae and adult Coleoptera supplemented by small reptiles (Doolan and Macdonald 1996). The bat-eared fox (*Otocyon megalotis*) of eastern and southern Africa consumes primarily termites and beetles; close to 70% of its diet consists of harvester termites (Genus *Hodotermes*) and dung beetles (Family Scarabaeidae). A combination of extremely long ears and small, numerous teeth enhances the bat-eared fox's ability to detect, capture, and consume its prey.

Members of Order Rodentia are notably omnivorous, but one member, the grasshopper mouse (Genus *Onychomys*) of North America, is unique in having a diet composed almost entirely of grasshoppers, crickets, and ground-dwelling beetles (McCarty 1975, 1978). Grasshopper mice have evolved specialized attack strategies to avoid the defensive secretions of insect prey such as beetles (Genera *Elodes* and *Chlaenius*). When attacking a whip-scorpion, *Onychomys* first immobilizes the tail and then attacks the head. A marsupial "equivalent" of the grasshopper mouse, the mulgara (*Dasycercus cristicauda*), resides in the arid sandy regions of central Australia and specializes in consuming large insects, spiders, and rodents (Chen et al. 1998; Dickman et al. 2001; Haythornthwaite and Dickman 2006).

Insectivorous mammals are broadly distributed throughout the class and exhibit remarkable adaptations for locating food. Like platypuses, which rely on tactile receptors on their bill to locate food under water, certain moles employ a similar system underground (Manger and Pettigrew 1996; Grand et al. 1998). Talpids have poor vision but acute hearing and touch. The snouts of moles and desmans are equipped with several thousand sensitive tactile organs, known as **Eimer's organs**, located on the nose (Quilliam 1966; Gorman and Stone 1990; Catania 1995, 2000; Catania and Kaas 1996). In the star-nosed mole (*Condylura cristata*) of North America, touch receptors are distributed among 22 fleshy, tentacle-like appendages around the tip of the nose. Eimer's organs appear as "… a mass of bulbous protuberances, reminiscent of a miniature cobbled street" (Gorman and Stone 1990:47). Each organ is surrounded at its base by a blood-filled sinus sitting on a network of sensory nerves. Nerve endings pass up from this network into a thick epidermal cap. When the mole touches an object, Eimer organs rock on their fluid foundation, transmitting the stimulus to the underlying nerve endings via sensory nerves to the central nervous system. At the CNS stimuli are received and integrated from other organs that have been "altered" thus providing information about the characteristics of the

A

B

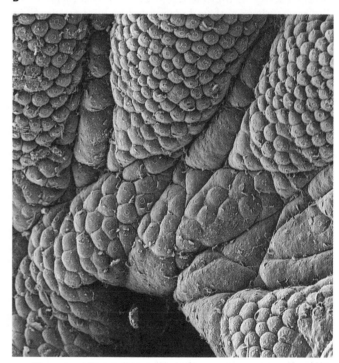

Figure 7.4 Mole specializations. (A) The star-nosed mole of North America is unique among mammals in possessing 22 fleshy, tentacle-like appendages surrounding the tip of its nose. (B) Eimer's organs on the nose of a European mole (*Talpa europaea*), as shown in a scanning electron micrograph.

stimulus (Gorman and Stone 1990) (figure 7.4). See chapter 21 for a discussion of electroreception in mammals.

Like the long protrusile tongues of anteaters, some arboreal primates and marsupials employ elongated digits to secure well-hidden insect prey. The third finger of the aye-aye (*Daubentonia madagascariensis*) of Madagascar and the fourth finger of two species of striped possums

A

B

Figure 7.5 **Insect-eaters.** (A) The aye-aye (*Daubentoria madagascariensis*) feeds primarily on the tree-burrowing larvae of beetles. (b) It bites into the bark with its powerful incisors and crushes and extracts insect larvae with its elongated third finger.

(*Dactylopsila*) from Australasia are uniquely adapted as probes for removing insects from the crevices of trees (figure 7.5). For example, using their keen hearing, aye-ayes detect larval insects hidden under the bark of dead branches. An aye-aye exposes prey by first gnawing off the overlying bark with its incisors, then inserting its third finger to crush and extract larvae, which it transfers to its mouth (Oxnard 1981; Erickson 1991, 1994; Krakauer et al. 2002). Aye-ayes may have filled this insect-eating niche on Madagascar, which elsewhere is occupied by woodpeckers (Macdonald 1984). Tarsiers (Genus *Tarsius*) from islands of southeast Asia appear to be exclusively insectivorous and carnivorous. Equipped with long legs (the name *tarsier* refers to the elongated tarsal, or ankle), exceedingly large eyes that face forward to permit stereoscopic vision, and a keen sense of hearing, tarsiers are exquisite predators. They capture arthropods, such as ants, beetles, and cockroaches, in trees or on the ground by leaping and pinning the prey down with both hands and quickly dispatching the victim with several bites.

CARNIVOROUS

Carnivory

Carnivorous (meaning "meat-eating") mammals feed primarily on animal material. Members of this group comprise the flesh- or meat-eating members of the Order Carnivora (chapter 16) and the marsupial dasyurids (chapter 11). The Carnivora include the canids, mustelids, felids, and allies. The Order Carnivora is represented by a diverse array of feeding types and dental morphologies, ranging from obligatory meat-eaters with large carnassial teeth, such as felids and hyaenids, to members such as the giant pandas (*Ailuropoda melanoleuca*) with crushing molars that feed exclusively on bamboo shoots (see figure 7.1). During their evolution, carnivores retained a versatile dentition, with different teeth adapted for cutting meat, crushing bone, and grinding insects and fruits (Van Valkenburgh 1989). Animal material is mostly protein and is converted to energy more efficiently than plant material is. Thus, like the insectivorous mammals, the alimentary canal of carnivorous mammals is short, and the cecum is small or absent (see figure 7.2).

Most carnivores are predators typified by strong skulls, jaws, and teeth—namely sharp incisors and canines—designed to kill and dismember prey. Killing techniques of predators differ; felids and mustelids kill with a single, penetrating bite, whereas hyaenids and canids may kill with several shallower bites. Once the prey is subdued, carnivores rely on their large, strong, pointed canine teeth to tear and shear flesh into hunks, which are then swallowed without being finely divided in the mouth. Also, most carnivores have a pair of carnassial teeth—a combination in which the last upper premolar and the first lower molar teeth form a powerful shearing mechanism when the mouth is closed (figure 7.6). The carnassials are most highly developed in the felids and canids and least developed in the more omnivorous families of ursids and procyonids (figure 16.4). Members of the Order Carnivora bite by employing a chopping motion. Because they have large, crushing molars in addition to carnassials, dogs can crush bones, whereas cats cannot.

Terrestrial Carnivores

With the exception of the otters (Subfamily Lutrinae), which are efficient marine and aquatic carnivores, most mustelids hunt on land. Mustelids are active, fierce hunters, many with specialized methods of killing prey (King 1990; Powell 1993). For example, the long-tailed weasel (*Mustela frenata*) of North America kills its victim by inflicting a rapid bite to the base of the skull or by severing the jugular vein with its sharp teeth. It first consumes the brain, then the heart, lungs, and ultimately the entire body, including bones and fur.

Felids are highly adapted for capturing and consuming vertebrate prey. Their senses of smell and hearing are acute. Their eyes, larger than those of most carnivores,

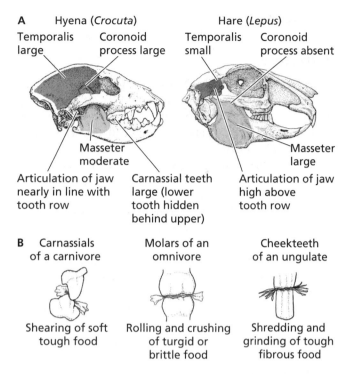

A

Hyena (*Crocuta*)

Temporalis large

Coronoid process large

Masseter moderate

Articulation of jaw nearly in line with tooth row

Carnassial teeth large (lower tooth hidden behind upper)

Hare (*Lepus*)

Temporalis small

Coronoid process absent

Masseter large

Articulation of jaw high above tooth row

B   Carnassials of a carnivore

Shearing of soft tough food

Molars of an omnivore

Rolling and crushing of turgid or brittle food

Cheekteeth of an ungulate

Shredding and grinding of tough fibrous food

Figure 7.6 Kinds of teeth. (A) Comparison of the jaw mechanics of a carnivore (*left*) and a herbivore (*right*). Note that the carnivore possesses a large temporalis muscle and a moderate masseter muscle attached to a large coronoid process. Herbivores possess a large masseter muscle and a small temporalis, and the coronoid process is absent. (B) The occlusal surface of teeth are adapted for processing three principal kinds of food: (*left to right*) shearing of soft, tough food (e.g., the carnassials of a carnivore); rolling and crushing of brittle food (e.g., molars of an omnivore); and shredding and grinding of tough fibrous food (e.g., cheekteeth of an ungulate). *Adapted from M. Hildebrand, 1995,* Analysis of Vertebrate Structure, *John Wiley & Sons.*

face forward, thus providing the binocular vision and depth perception vital to locating prey. Their long, stiff, highly sensitive vibrissae are especially useful for foraging at night. Long, sharp, usually retractile claws serve as effective meat hooks for capturing, slashing, and manipulating prey. Cats use their long, sharp canines for grasping prey and their well-developed carnassials for shearing food. The tongue of felids is covered with many sharply pointed papillae, which are well suited for holding prey and scraping meat from a carcass (Van Valkenburgh 1989; Dunstone and Gorman 1993; Pierce et al. 2000).

Canids are opportunistic hunters that rely on high intelligence, social organization, and superb behavioral adaptability (see chapter 23). Small canids hunt singly or in pairs, whereas larger canids, such as gray wolves (*Canis lupus*), hunt in packs of up to 30 members seeking prey that are far larger than themselves (Paradiso and Nowak 1982; J. D. Gittleman 1989, 1996).

The jaw muscles of carnivorous mammals differ from those of herbivores (see figure 7.6) with regard to the relative importance of the three major adductor muscles of the mandible: the **temporalis, masseter,** and **pterygoideus.** Carnivores must first seize and hold their prey

with the canines, which requires a large force at the front of the jaws. The very large temporalis muscles function in holding the jaws closed and aid in the vertical chewing action. In carnivores, the masseter muscle is comparatively small and serves to stabilize the articulation of the jaw, and the pterygoideus muscle helps position the carnassials. In contrast, herbivores rely heavily on a large masseter muscle to maintain a horizontal movement of molars for grinding fibrous food.

Aerial Carnivores

Most chiropterans feed on insects taken on the wing (Whitaker 1988). Certain species of bats are quite specialized, however, feeding on small vertebrates such as rodents, birds, frogs, lizards, small fish, and even other bats. The "carnivorous" bats (or animalivorous sensu; Patterson et al. 2003) comprise six families: Megadermatidae, Hipposideridae, Nycteridae, Phyllostomidae, Noctilionidae, and Vespertilionidae (Norberg and Fenton 1988; Pavey and Burwell 1997; Dondini and Vergari 2000; Bonato et al. 2004; Thabah et al. 2007). As noted earlier, the shape of the skull and teeth in bats is a good indicator of diet. For example, the carnivorous false vampire bat (*Vampyrum spectrum*), the largest bat in the New World, has a massive skull equipped with strong, sharp canine teeth and shearing molars adapted for crushing bones and cutting flesh. *Vampyrum* was once thought to be a true vampire bat, but it does not consume blood. Its diet consists of birds, bats, rodents, and some insects and fruit (Gardner 1977). Asian false vampire bats (*Megaderma lyra*) prey on small vertebrates, such as mice, baby birds, and frogs, which are carried to the roost to be eaten. The frog-eating bat (*Trachops cirrhosus*), a phyllostomid, primarily consumes insects and small vertebrates such as frogs and lizards (Tuttle and Ryan 1981; Bruns et al. 1989). Able to locate and distinguish between different species of frogs by listening for and analyzing their unique calls, *Trachops* can discriminate between poisonous and palatable species (Ryan and Tuttle 1983). Bird eating has been reported primarily from gleaning tropical bats that occasionally capture resting birds (Norberg and Fenton 1988). Recently, researchers have found birds to be an important item in the diet of giant noctule bats (*Nyctalus lasiopterus*) from temperate regions of Spain and Italy (Dondini and Vergari 2000) and great evening bats (*Ia io*) from India and China (Thabah et al. 2007). It is noteworthy that giant noctule bats may capture migrating birds while flying at high elevations (Ibáñez et al. 2001; Popa-Lisseanu et al. 2007).

The morphology of teeth, alimentary canal, and limbs is strongly correlated with food habits, a fact that is well illustrated by the only **sanguinivorous** (meaning "blood-eating") mammals—the vampire bats. Three species consume blood, and all are phyllostomids confined to the New World from Mexico to northern Argentina. The vampire bat (*Desmodus rotundus*) preys exclusively on mammals, but the white-winged vampire bat (*Diaemus youngi*) and the hairy-legged vampire bat (*Diphylla ecauda-*

ta) prefer avian prey (Altenbach 1979; Greenhall et al. 1983, 1984; Hermanson et al. 1993; Greenhall and Schutt 1996; Schutt and Altenbach 1997). *Desmodus rotundus* are medium-sized bats weighing between 25 and 40 g (figure 7.7). *D. rotundus* prey on primarily medium-sized and large terrestrial mammals (Mayen 2003). When found near human settlements, they will ingest blood from cattle, horses, mules, goats, pigs, sheep, and also humans and transmit pathogens such as the rabies virus (Mayen 2003; Messenger et al. 2003). Today, vampire bats occur in close proximity to livestock farming and thus show a high degree of preference for cattle (Voigt and Kelm 2006). Morphologically, *D. rotundus* are adapted for a diet of blood in the following ways:

1. The rostrum is reduced, supporting upper incisors and canines that are unusually large and knifelike, with the sharp points of the incisors fitting into pits in the lower jaw.
2. Cheekteeth are tiny.
3. The tongue possesses a pair of grooves at each border that function like drinking straws.
4. The stomach is long and tubular, highly distensible, and well vascularized to enhance the storage of blood and absorption of water.
5. The small intestine is thin-walled and twice as long as the stomach.
6. The kidneys have a unique excretory ability linked with feeding and roosting behavior (McFarland and Wimsatt 1969).

Also, the humerus is strong and well developed. It supports a thumb that is unusually long and equipped with three pads that function like a sole. The forelimbs are unique and greatly aid terrestrial locomotion (Altenbach 1979). *Desmodus rotundus* feeds chiefly on domestic livestock. They detect vascular areas of their victims by use of specialized heat-sensitive pits surrounding their noses (Kurten and Schmidt 1982). Bats typically land on the ground near the leg of the host and climb or jump to a feeding site, usually the legs, shoulders, or neck. The upper incisors and canines are used to remove a small piece of skin from the victim or make an incision several millimeters deep. Movement of blood is facilitated by several anticoagulants in the bat's saliva. The flow of blood is maintained by peristaltic waves of the tongue as the bat rapidly licks and continuously abrades the wound. Up to 13 *Desmodus* at a time have been observed feeding on the neck of a cow, with a feeding time of 9–40 minutes. Within a 3-hour period, seven bats may feed from the same wound, one after another.

The long and highly vascularized stomach of *Desmodus* does not function for protein digestion as in most mammals. Rather, it is important in the storage of blood and the absorption of water to concentrate the blood. Average consumption of blood in the wild is about 20 mL/day (Wimsatt and Guerriere 1962). Bats may consume up to 50% of their body mass in blood per night, which might impose serious constraints on their ability to fly. Vampire bats cope with this potential problem by employing a

A

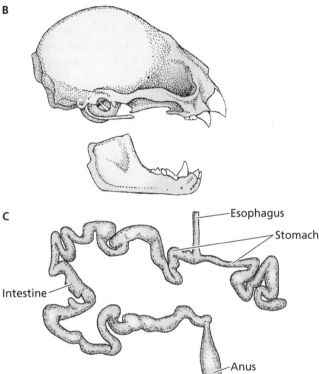

B

C

Esophagus

Stomach

Intestine

Anus

Figure 7.7 **Vampire bats.** (A) The vampire bat (Desmodus rotundus) occurs only in the New World, ranging from northern Mexico south to southern South America. (B) The skull of the vampire bat showing the bladelike upper incisors and canines. The sharp points of the upper incisors fit into distinct pits in the lower jaw behind the incisors. (C) The alimentary canal of the vampire bat. The stomach serves to store large amounts of blood and absorb water rather than to digest protein, as with most mammals. *Adapted from H. Gunderson, 1976,* Mammalogy, *McGraw-Hill.*

unique "two-phase" renal function: in the first phase, which occurs at the feeding site, water is excreted; during the second phase, which takes place at the roost, urine is concentrated. About an hour after feeding, bats rapidly lose much of the water taken in with the blood meal—about 25% of the ingested blood is excreted as urine. This weight loss is essential to enabling the bats to fly back to the roost. At the roost, digestion of the partially dehydrated blood continues. Bats concentrate wastes and thus excrete a highly concentrated urine. The kidney of vampire bats may surpass that of many desert mammals in its ability to concentrate urine and thus conserve water. In addition to relying on superb kidney function for water conservation, vampire bats practice blood sharing at the roosting site by regurgitating blood into the mouth of another bat (Wilkinson 1985, 1987; Altringham 1996).

Aquatic Carnivores

Cetaceans that feed on small marine organisms show a second type of specialized dentition called baleen. Baleen whales (the mysticetes) are filter, or suspension, feeders— that is, they strain small organisms, known collectively as plankton, from the water by use of the baleen sieves in the front of their mouths (Pivorunds 1979; Voelker 1986; Lambertsen et al. 1995). The structure and function of baleen and feeding habits of the mysticetes are detailed in chapter 17.

Although baleen whales are touted as the filter-feeders par excellence, they are equaled by one of the pinnipeds. The crabeater seal (*Lobodon carcinophagus*) is distributed along the leading edge of the Antarctic pack ice and is a major consumer of **krill**. The population consumes up to 160 million tons per year (Øritsland 1977; Bonner 1990; Croll and Tershy 2002; Heithaus and Dill 2002). The teeth of crabeater seals are well adapted for straining krill

from the sea. Their cheekteeth possess elaborate cusps (figure 7.8; see also figure 16.8) such that when the jaws close, the cusps intermesh to form an effective sieve for separating krill from the water being forced out of the mouth. Unlike baleen whales, crabeater seals are selective feeders, and foraging is directed at individual prey rather than large masses. This eliminates the amount of water taken in, and thus a less extensive filtering system is required. Seals locate a swarm of krill and direct their snouts to the prey. By depressing the floor of their mouth, they effectively suck the prey into their mouth. Once inside, the jaws are closed, and the tongue is raised, expelling excess water and filtering the krill through the matrix formed by the interlocking lobulate cheekteeth. When a sufficient bolus of food is collected in the mouth, it is swallowed. The average meal of krill for a crabeater seal is about 8 kg (Øritsland 1977).

Predators that eat fish are **piscivorous** (meaning "fish-eating"). At least three species of bats are known to capture and eat fish: *Noctilio leporinus*, *Myotis vivesi*, and *Myotis adversus* (Brooke 1994; Nowak 1994). Bulldog, or fisherman, bats (*N. leporinus*) are remarkable for their structural and behavioral modifications for capturing and consuming fish. Bulldog bats (the name stems from their large, jowl-like upper lip resembling that of a bulldog— see figure 13.18) are characterized by unusually long hind limbs and large feet equipped with sharp, recurved, and laterally flattened claws. Bulldog bats employ echolocation to detect ripples caused by fish swimming near the surface of the water (Brown et al. 1983; Wenstrup and Suthers 1984; Altringham 1996). They skim low, dragging their feet through the water with limbs and hooklike claws rotated forward, thus acting as a gaff (fishing spear). Once gaffed, fish are quickly transferred to the mouth, where long, thin canines combine with large upper lips and elastic cheeks to form a sort of internal pouch to secure the slippery fish. Fish up to 8 cm long may be captured, and from 30 to 40 fish may be taken per night. *Noctilio* usually forages over small pools, slow-moving rivers, or sheltered lagoons (Bloedel 1955; Wenstrup and Suthers 1984; Altenbach 1989; Findley 1993; Brook 1994; Nowak 1994, 1999; Schnitzler et al. 1994).

Among mammals, piscivory is common in seals, sea lions, and dolphins. Toothed whales, porpoises, and dolphins (Odontocetes) are fish and squid specialists (see figures 7.1 and 17.7). They have numerous, small, simple, teeth that are all alike (homodont). To optimize prey capture, the mouth of a porpoise or dolphin forms a fish trap similar to that used by other fish-eating vertebrates, such as gars, crocodiles, and mergansers. The adaptive value of this morphology is clear—fish are active and slippery and must be trapped and swallowed quickly to prevent their escape. Killer whales (*Orcinus orca*) are the largest predator among "warm-blooded" animals. These opportunistic feeders consume fish, squid, baleen whales and smaller cetaceans, pinnipeds, penguins and other aquatic birds, and marine invertebrates. Foraging success is optimized by cooperation. Pods vary from 4 to 40 individuals, with

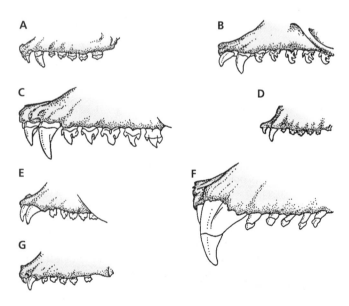

Figure 7.8 Pinniped dentition. Comparison of the dentition of seals: (A) harbor seal (*Phoca vitulina*), (B) crabeater seal (*Lobodon carcinophagus*), (C) leopard seal (*Hydrurga leptonyx*), (D) ringed seal (*P. hispida*), (E) Ross seal (*Ommatophoca rossii*), (F) elephant seal (Genus *Mirounga*), and (G) bearded seal (*Erignathus barbatus*). The teeth of the upper and lower jaws of the crabeater seal (B) intermesh to form sieves adaptive in straining krill from the sea. *Adapted from W. N. Bonner, 1990,* Natural History of Seals, *Facts on File.*

adults and older juveniles, aided by underwater vocalizations, cooperatively herding the fish. Leopard seals (*Hydrurga leptonyx*) represent the top-level predators in the antarctic ecosystem (Hiruki et al. 1999; Hall-Aspland and Rogers 2004) where they feed on krill, cephalopods, crustaceans, fish, seals, and penguins. The diet tends to vary with age of the seals, with juveniles consuming primarily krill and adults shifting to penguins and seals. From January to March, penguins form the staple in the diet of leopard seals in the antarctic (Siniff and Stone 1985). Sperm whales (Family Physeteridae) feed primarily on squid. When consumed, the beak of a squid may act as an irritant to stimulate the production of ambergris from the stomach and intestines of the whale (Macdonald 1984; Voelker 1986). **Ambergris,** once used as a fixative in the cosmetic industry, is a form of excrement from sperm whales. One lump taken from a sperm whale in the Antarctic weighed 421 kg (926 lb; Slijper 1979).

Like the odontocetes, the jaws and teeth of pinnipeds are adapted for grasping prey, not chewing it, and most prey are swallowed whole. Because meat requires only a short period for digestion compared with vegetable matter, one would expect pinnipeds to have a short alimentary canal (see figure 7.2). This is not the case with some seals, however; an adult male southern elephant seal (*Mirounga leonina*) possessed a small intestine 202 m (660 ft) in length—42 times its body length! The reason for this long gut is unknown. Like other carnivorous mammals, the cecum, colon, and rectum of seals are relatively short.

HERBIVOROUS

Herbivory

Herbivorous (meaning "plant-eating") mammals consume green plants and thus constitute the base of the consumer food web. Plant food is far more abundant than animal food, but its energy content is lower. Gaining access to the protein within leaves and stems is difficult due to the tough fibrous cell walls of plants. We can divide herbivores into two main groups: (1) browsers and grazers, such as the hooved mammals—the Perissodactyla and Artiodactyla (chapter 20)—and (2) the gnawers—the Rodentia and Lagomorpha (chapter 18). Other important herbivores are the kangaroos, wallabies, wombats, langurs, sloths, elephants and hyraxes, and the aquatic grazers such as manatees and dugongs. It is noteworthy that over one half of the extant species of bats use plants as food or for shelter (Fleming 1993; Kunz 1996; Kunz and Fenton 2003; Patterson et al. 2003). Herbivores feed on a great diversity of foods, including grasses, leaves, fruit, seeds, nectar, pollen, and even the sap, resins, or gums of plants. Herbivores share unifying characteristics in the design of the skull, teeth, and alimentary canal, which are adapted for feeding on cellulose-rich herbs and grasses for which mammals lack digestive enzymes. In general, herbivorous mammals are typified by skulls in which canines are reduced or absent (see figure 7.1), and broad molars are adapted for crushing, shredding, and grinding fibrous plant tissue (see figure 7.6). Rodents are characterized by the presence of a single pair of evergrowing, chisel-like incisors on both the upper and lower jaws. Lagomorphs have an additional pair of "secondary" upper incisors that are located immediately behind the first pair. Because canine teeth are absent, a wide gap (diastema) occurs between the incisors and cheekteeth. Plant-eaters typically possess a long intestine with either a simple stomach (nonruminant herbivores) or one that has internal folds and is divided into several functionally different chambers (ruminant herbivores; see figure 7.9).

Omnivorous mammals, such as primates, bears, pigs, and some rodents, possess teeth with low crowns (brachyodont), well-developed roots and root canals, and rounded cusps (bunodont) adaptive for a generalized diet (see figure 7.6). These teeth are not very effective in shredding and grinding tough fibrous plant tissue, and several modifications of this basic design have evolved among herbivores. In response to dietary needs, elephants and some rodents have cheekteeth characterized by transverse ridges, or lophs, on their grinding surface (lophodont). In the ungulates, wear on the surfaces of the teeth from chewing on course plant tissue produces and maintains the transverse ridges. Because enamel, dentine, and cement have different hardnesses, continuous grinding maintains the rough surface. In horses, these teeth are hypsodont (high-crowned) with complex, folded ridges on the surface. In **ruminant artiodactyls,** such as goats, cows, and deer, the grinding surfaces of the teeth have longitudinal crescents, or "half-moons" (selenodont), adaptive for breaking down tough plant material.

The jaw muscles of herbivores differ from those of carnivores. The movement in mastication is from side to side (not up and down, as in carnivores), and the upper cheekteeth slide across the complementary surfaces of the lower teeth in a sweeping motion. For herbivores, the major muscles involved in mastication are the masseter and pterygoideus, whereas the temporalis muscle is smaller than that found in carnivores (see figures 7.6 and 16.5). Nonruminant grazers such as horses use their large incisors to snip and cut tough, fibrous stems. They consume large quantities of fibrous food and have robust lower jaws supporting a large masseter muscle used primarily for closing the jaw. Because leaves require less mastication than grasses due to their lower fiber content, the lower jaw and masseter complex of deer and other browsers are not as pronounced as in horses. Unlike horses and other perissodactyls, ruminant artiodactyls have lost their upper incisors, and they crop foliage by use of the lower incisors biting against a callous pad on their upper gum, which acts as a sort of cutting board. Cows are well adapted for eating grasses, which they pull free by twisting them around their mobile tongue held against their lower incisors. After biting off the foliage, the cow holds it within the diastema before moving it back to the cheekteeth for grinding. A distance between the lower incisors and the cheekteeth offers the advantage of allowing for a narrow snout that can penetrate into small spaces to crop food, as seen in smaller deer, antelopes, and rodents (see figure 7.1).

Although the specialized teeth of herbivores effectively shred and grind the cell walls of plant tissue and release their contents, only certain enzymes can digest cellulose. Mammals, however, do not produce these **cellulolytic** (cellulose-splitting) enzymes, so they rely on symbiotic microorganisms residing in their alimentary canal. These microorganisms break down and metabolize the cellulose of plants and release fatty acids and sugars that can be absorbed and used by the mammal host. Rodents and lagomorphs become inoculated with the appropriate anaerobic protozoans and bacteria by eating maternal feces, whereas young ungulates commonly consume soil to acquire their microorganisms. Ungulates have evolved two different systems for breaking down cellulose: foregut fermentation (rumination) and hindgut fermentation (Putman 1988; Robbins 1993).

Foregut Fermentation

Rumination (digastric digestive system), also called **foregut fermentation,** is typified by artiodactyls such as camelids, giraffids, hippopotamuses antilocaprids, cervids, and bovids, as well as by kangaroos, sloths, and colobus monkeys (Freudenberger et al. 1989; Alexander, 1993b; Stevens and Hume 1996). Foregut fermenters possess a complex, multichambered stomach with cellulose-digesting microorganisms. After food is procured by cropping or grazing, it immediately passes to the first and largest chamber of the network, the **rumen** (see figures 7.2 and 7.9). Here, the food is moistened and kneaded, which mixes it thoroughly with microorganisms that ferment the food. Large particles of food float on top of the rumen fluid and pass to the second chamber, the **reticulum**—a blind-end sac with honeycomb partitions in its walls. The reticulum is where a softened mass called the "cud" is formed. Fermentation occurs in both the rumen and reticulum, and both absorb the main products of fermen-

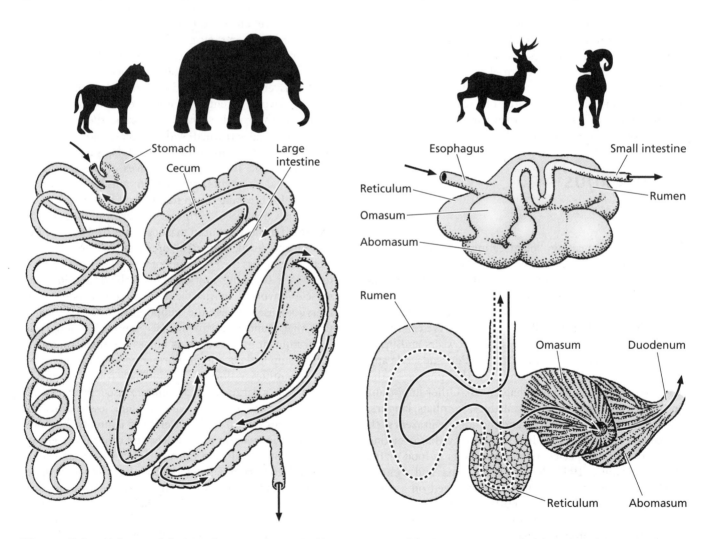

Figure 7.9 Hindgut and foregut fermentation. Two digestive systems of herbivorous mammals: (A) Hindgut fermentation (monogastric digestive system) is characteristic of perissodactyls, such as horses, zebras, asses, tapirs, and rhinoceroses, in addition to other herbivores such as elephants, lagomorphs, and rodents; (B) Foregut fermentation (digastric digestive system) is typified by artiodactyls, such as cervids and bovids, and by kangaroos and colobus monkeys. *Adapted from Pough et al., 2005,* Vertebrate Life, *7th ed.,* Prentice Hall.

tation, short-chain fatty acids. When the animal is at rest, this softened mass is regurgitated, allowing the animal to "chew its cud," or "ruminate." At this time, the mass is further broken down by a potent enzyme, **salivary amylase.** The food is then swallowed a second time and enters the third chamber, the **omasum,** where muscular walls knead it further. The fourth, and final, chamber, the **abomasum,** is the true stomach. Here, digestive enzymes that kill any escaping microorganisms are secreted, and protein digestion is completed. Digested material then passes into the small intestine, where the products of microbial digestion and acid digestion are absorbed. Some additional fermentation and absorption occur in the cecum.

Hindgut Fermentation

The **monogastric system,** also called **hindgut fermentation,** is characteristic of horses, tapirs, rhinoceroses, howler monkeys, elephants, lagomorphs, hyraxes, rodents, and some arboreal marsupials (Prins and Kreulen 1990; Alexander 1993b) (see figures 7.2 and 7.9). Hindgut fermenters masticate food as they eat, initiating digestion with salivary enzymes. Digestion continues by enzymatic activity within the simple stomach, and food then moves rapidly into the small intestine as new food is eaten. Unlike ruminant artiodactyls, hindgut fermenters do not regurgitate their food. Nutrients are absorbed in the small intestine. Finely ground particles of food pass from the small intestine into the cecum, and larger food particles move through the large intestine and are passed as feces. Among the hindgut fermenters, the colon acts as the principal fermentation chamber for the larger species, while the cecum fulfills this function in the smaller species (Hume 1989).

The two kinds of fermentation processes that take place in herbivores have clear advantages and disadvantages (Montgomery 1978; Dawson 1995; see figure 7.9). Foregut fermentation tends to be very efficient because microorganisms begin to break down the plant material before it reaches the small intestine, where it is absorbed. Furthermore, in foregut fermenters, microorganisms from the rumen are themselves broken down by acids in the true stomach (abomasum). The resulting material, which contains the carbohydrates and protein synthesized by the microorganisms as well as the products of fermentation, moves into the small intestine and colon. Lastly, the microorganisms in the rumen can detoxify many harmful alkaloids in the plants that foregut fermenters consume.

In contrast, food passes rapidly into the small intestine in hindgut fermenters and is then mixed with microorganisms in the cecum. These animals do not digest the microorganisms present in the cecum and thus cannot exploit this potential source of nutrients. In addition, hindgut fermenters must absorb toxic plant chemicals into the bloodstream and transport them to the liver for detoxification or sequestration.

Efficiency may indeed be the trademark of foregut fermenters; however, hindgut fermenters are able to process material much more rapidly. For example, food moves through the gut of a horse in about 30–45 hours, whereas it may take a cow from 70–100 hours to process food. Hindgut fermenters efficiently digest food high in protein because large volumes of food can be processed rapidly. Furthermore, hindgut fermentation is effective when forage is dominated by indigestible materials, such as silica and resins, because these compounds move quickly through the alimentary canal by bypassing the cecum. In sum, due to their lowered efficiency, hindgut fermenters must eat large volumes of food in a short time. The foregut system is comparatively slow because food cannot pass out of the rumen until it has been ground into very fine particles. Thus, ruminants do poorly on forage containing high levels of resins and tannins because these compounds inhibit the function of microorganisms in the rumen. Furthermore, plants with high silica content break down slowly and thus impede movement of food out of the rumen.

The digestive physiology of herbivores influences their ecology and distribution in several ways. Ruminants benefit most from foods that require an optimally efficient digestive system, whereas the best forage for hindgut fermenters is that which facilitates speed of digestion. Each strategy has advantages for survival in particular ecological niches. Speed of digestion may not be important to ruminant artiodactyls. When food is available in the form of tender, short herbage with high protein content, ruminant digestive efficiency pays off. The process of rumination also permits animals to feed quickly and then move to safe cover to chew the cud at leisure. For environments such as the arctic tundra, where food is limited but of high quality, ruminants such as the musk ox (*Ovibos moschatus*) and caribou (*Rangifer tarandus*) have an advantage. As we will learn in chapter 9, some ruminants, such as elands and oryx, also survive well in the deserts of Africa. Although absorption of the products of protein digestion is similar in ruminants and hindgut fermenters, ruminants have the advantage of being able to recycle urea. This allows these mammals to survive with very little water, whereas perissodactyls residing in xeric areas must drink daily to balance the urea in their urine.

When food is of low quality, with high fiber content, yet is not limited in quantity, a premium is placed on the ability to process large quantities quickly. Perissodactyls can survive in regions typified by seasonal drought and poor-quality food—places where ruminants could not process food fast enough to survive.

On the Serengeti Plains of East Africa, dense migrating herds of ungulates influence plant succession and finely partition available resources. They respond to growth of grasses in a predictable sequence (figure 7.10; Gwynne and Bell 1968; Bell 1971). First, perissodactyls such as zebras (*Equus burchellii*) enter the long-grass communities of the Plains and consume many of the longer stems of grasses. Next come large herds of wildebeests (*Connochaetes taurinus*), trampling and grazing the grasses to short heights. The last invasion of ungulates is Thomson's

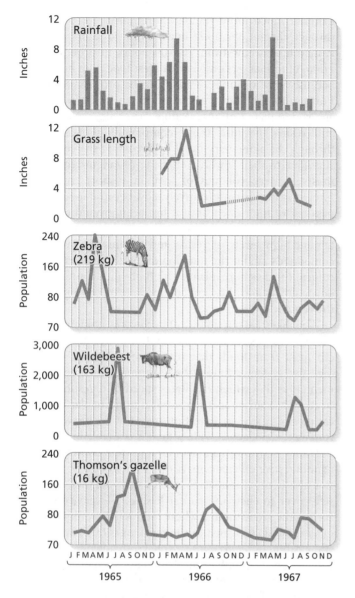

Figure 7.10 Ungulates of the Serengeti. Populations of migrating zebras, wildebeests, and Thomson's gazelles in relation to rainfall and length of grass on the Serengeti Plains of Africa. *Data from R. H. V. Bell, "A Grazing Ecosystem in the Serengeti" in* Scientific American, *225:86–93, 1971.*

gazelles (*Gazella thomsonii*), which feed on short grass during the dry season. In addition to different spatial and temporal division of resources, these ungulates sort out available food according to different parts of plants. Zebras consume mostly stems and sheaths of grasses and almost no leaves. Wildebeests eat great numbers of sheaths and leaves, and gazelles eat grass sheaths and herbs not consumed by the other two species. Because grass stems and sheaths are low in protein and high in lignin, and leaves are high in protein and low in lignin, it appears that zebras fare poorly and gazelles do quite well. As we learned earlier, however, perissodactyls, such as zebras, process twice the volume of plant material that ruminant artiodactyls can and therefore compensate with quantity (volume) for what their forage lacks in quality. In addition, because

zebras are larger than wildebeests and gazelles, they require less energy per unit mass than the smaller mammals.

There are more species of ruminant artiodactyls than perissodactyl hindgut fermenters among the ungulates. This may be due in part to the ability of ruminants to recycle nitrogen and digest protein-rich bacteria. As a result of this ability, they do not have to gain all of their amino acids in forage and can focus feeding on specific, preferred species of plants. This independence of forage selection permits ruminants the freedom to partition resources in finer fashion than most perissodactyls can.

Gnawing Mammals

One of the most successful groups of herbivores are the gnawing mammals—namely the rodents and lagomorphs. Like the ungulates, rodents and lagomorphs cannot produce the enzyme cellulase, so they facilitate the fermentation of fibrous forage with the aid of bacteria and protozoa. As with the perissodactyls, rodents and lagomorphs do not ruminate, and hindgut fermentation occurs in the colon and cecum (see figure 7.2). The only rodents that lack a cecum are dormice (Family Gliridae), indicating that their diet possesses little cellulose. Compared with ruminants, the stomach of rodents is simple but possesses from one to three chambers (Carleton 1973, 1985; Hume 1994). The small intestine is comparatively short, and the hindgut (colon and cecum) is complex, with the cecum having many spiral folds, recesses, and saclike expansions (Bjornhag 1994). Within the Order Rodentia, variation in morphology of digestive systems is correlated with diet. For example, the sciurids (squirrels, chipmunks, and marmots), which feed on a variety of seeds, nuts, fruits, and herbs, have a much simpler digestive system than grass-eating arvicolines do (voles and lemmings; Batzli 1985; Batzli and Hume 1994). Variation in the morphology and function of the gastrointestinal tracts of voles correlates with diet and thermal stress. For example, high-fiber diets coupled with cold acclimation of arvicoline rodents such as prairie voles (*Microtus ochrogaster*) may result in higher rates of food intake and increases in the size of the hindgut (Hammond and Wunder 1991).

Rodent skulls are characterized by their large gnawing incisors (see figure 18.1), but they also show features common to all herbivores. The incisors (see figure 7.1) are used to gnaw through hard plant coverings to reach the tender material inside as well as for nibbling grasses and shrubs. The lips can be folded in behind the incisors to prevent chips of bark or soil from entering the mouth during gnawing. As with other herbivores, a diastema posterior to the incisors results from the absence of canine and premolar teeth. Most mouselike rodents lack premolars, but jumping mice (Genus *Zapus*) have one on both sides of the upper jaw and squirrel- and cavy-like rodents have one or two premolars on either side of the jaw. Food can also be held in the diastema before it is passed back to the cheekteeth for processing—this is most obvious in grass-eating herbivores. Among rodents, members of at least four families (hamsters, pocket gophers, pocket mice, and squirrels) have either internal or external cheek pouches that open near the angle of the mouth. External cheek pouches can be everted for cleaning. Cheek pouches are well adapted for carrying food; an early biologist discovered a total of 32 beechnuts in the cheek pouches of an eastern chipmunk (*Tamias striatus*; Allen 1938). In herbivores, three main masticatory muscles—the masseter, pterygoideus, and digastricus—regulate how these animals shred and grind tough, fibrous food (see figure 7.6). The arrangement and function of these muscles are responsible for the forward and backward jaw movements of rodents, in contrast to the lateral chewing movements of lagomorphs.

Coprophagy

Digestion of cellulose in hindgut fermenters, such as rodents, lagomorphs, and the marsupial common ringtail, occurs in the cecum. Because there is no regurgitation and the rate of passage of forage is rapid, these mammals process a minimal amount of the fiber when they first ingest plants. As a result, **coprophagy** (refection), the feeding on feces, has evolved in lagomorphs, rodents, shrews, and some marsupials (McBee 1971; Kenagy and Hoyt 1980; Macdonald 1984; Proctor-Grey 1984; Alexander 1993a; Stevens and Hume 1995; Hirakawa 2001; Langer 2002). The cecum, located between the small and large intestines, houses bacteria that aid in digestion of cellulose (figure 7.11). Most products of digestion, except for certain nutrients such as essential B vitamins produced by microbial fermentation, pass through the gut into the bloodstream. Such minerals and vitamins would be lost if lagomorphs did not eat some of their feces and so pass them through the gut twice. To optimize the uptake of essential vitamins and minerals and enhance assimilation of energy, lagomorphs produce two kinds of feces. The first are moist, mucus-coated, black cecal pellets excreted and promptly eaten directly from the anus. These are stored in the stomach and mixed with food derived from the alimentary mass. Second, lagomorphs produce hard, round feces that are passed normally. The frequency of coprophagy in rabbits is usually twice daily. Prevention of coprophagy in laboratory rats resulted in a 15% to 25% reduction in growth. Another coprophagous mammal is the mountain beaver (*Aplodontia rufa*), which extract fecal pellets from their anus individually with their incisors. They cache these in underground fecal chambers, reingesting them at a later time. Coprophagy has been documented in about nine species of shrews, including the northern short-tailed shrew (*Blarina brevicauda*; figure 7.12; Merritt and Vessey 2000). The adaptive significance of coprophagy for shrews is unknown, but it may represent a technique of reducing daily food intake and extracting certain essential nutrients and vitamins from available food.

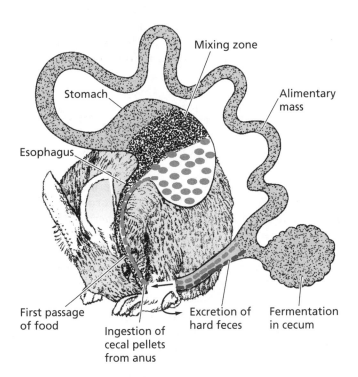

Figure 7.11 Coprophagy. Coprophagy occurs in shrews, rodents, and lagomorphs. The digestive tract of lagomorphs is highly modified for coping with large quantities of vegetation. The alimentary canal has a large cecum, which contains bacterial flora to aid in the digestion of cellulose. *Adapted from D. MacDonald, 1984*, Encyclopedia of Mammals, *Facts on File.*

Figure 7.12 Short-tailed shrew. The northern short-tailed shrew (*Blarina brevicauda*), one of the rare venomous mammals, is a common inhabitant of moist habitats of eastern North America. Its poison, produced by submaxillary glands, is administered to its victim though a medial groove in the lower incisors. The poison quickly immobilizes small prey, which may be cached in a comatose state to be available as a source of fresh food for some time after capture.

Herbivory is not confined to ungulates, rodents, and lagomorphs. Orders such as Carnivora and Chiroptera have species with strong herbivorous habits. The Orders Rodentia and Chiroptera comprise the greatest number of species of mammals; their success is due in part to adaptive radiation in feeding techniques, including insectivory, granivory, folivory, frugivory, nectarivory, and gummivory (Eisenberg 1981).

SPECIALIZATIONS IN HERBIVORY

Granivory

Herbivorous mammals that consume primarily fruits, nuts, and seeds are referred to as **granivorous** (meaning "seed-eating"). Equipped with large, external, fur-lined cheek pouches and a keen sense of smell, heteromyid rodents represent the most specialized seed-eaters. Kangaroo rats (Genus *Dipodomys*), kangaroo mice (Genus *Microdipodops*), and pocket mice (Genus *Perognathus*) of North American deserts are primarily granivorous (Vander Wall 1995; Schooley et al. 2000; Duval et al. 2005; Murray et al. 2006). Seeds are also the mainstay for tropical and subtropical species of heteromyids (Genera *Heteromys* and *Liomys*) that harvest fruits, nuts, and seeds from shrubs and trees and cache these propagules in underground burrows (Fleming 1970, 1974; Sanchez-Cordero and Fleming 1993). The diversity and availability of seeds in desert ecosystems is a key to the evolutionary success of the heteromyids (Price and Mittler 2006). In terms of the biomass of seeds harvested, heteromyids are rivaled only by ants as important granivores inhabiting North American deserts (Brown and Davidson 1977). Rodents are reported to use over 75% of all seeds produced at certain Mohave and Chihuahuan Desert sites (Brown et al. 1979). In the Mohave Desert of California, *Dipodomys merriami* consumed over 95% of the seeds produced by the annual *Erodium cicutarium* (Soholt 1973). Maximum numbers of seeds produced in desert habitats of North America range from 80 to 1480 kg/ha (Tevis 1958; French et al. 1974; M'Closkey 1978). Minimum densities of seeds remaining in the soil years after the last seed crop are rarely below 1,000 seeds/m² (Nelson and Chew 1977; Reichman and Oberstein 1977).

As a result of the abundant seed resources and competition with ants, birds, and other rodents, heteromyids have evolved fascinating morphological and behavioral adaptations to optimize their foraging success. All heteromyids cache seeds. They collect large quantities of seeds and store them either in larders within their burrows or scatter hoard them in small buried caches outside the burrow. They employ their large cheek pouches to collect many seeds in single foraging bouts (Brown et al. 1979). Seeds used by heteromyids are derived primarily from grasses and forbs and are quite small, usually less than 3 mm long and weighing less than 25 mg. Kangaroo rats collect most of their seeds directly from plants by clipping fruiting stalks and removing seeds from felled seed heads or by plucking seeds from fruit located close to the ground. Heteromyids may also collect seeds, primarily located in clumps, from the surface of the soil or strain

them from the soil (Eisenberg 1963; Randall 1993; Reichman and Price 1993). Seeds are relocated by olfactory cues coupled with memory (Jacobs 1992; Rebar 1995; Vander Wall 1995). *Dipodomys merriami* and *Perognathus amplus* may locate seeds below the surface of the soil by detecting concentrated odor characteristics of buried seeds (Reichman 1981; Smith and Reichman 1984; Reichman and Rebar 1985). Although quite opportunistic in their quest for food, squirrels and chipmunks (Family Sciuridae) are principally granivorous (Smallwood et al. 2001; Steele and Koprowski 2001; Vander Wall et al. 2006). Several species of the Families Cricetidae and Muridae readily cache seeds; for example, *Peromyscus* (Vander Wall et al. 2006), *Apodemus* (Abe et al. 2006; Xiao and Zhang 2006), *Leopoldamys* and *Rattus* (Cheng et al. 2005 Xiao and Zhang 2006).

Folivory

Animals that exhibit adaptations for consuming leaves, stems, buds, and other green portions of plants are referred to as **folivorous** (meaning "leaf-eater"). Within the Class Mammalia, about 42 genera, or 4%, of mammals specialize in the consumption of leaves and stems (Eisenberg 1978). Like grazing and browsing mentioned earlier, consumption of leaves and stems requires considerable morphological adjustment in dentition, jaw musculature, and gut morphology (Eisenberg 1978). In response to predation of leaves by herbivores, however, plants have evolved diverse chemical defenses (Freeland and Janzen 1974; Belovsky and Schmitz 1994; Foley and McArthur 1994). A detailed discussion of plant defenses to herbivory is not within the scope of this chapter. Instead, we choose to highlight specific mammalian practitioners that exemplify mechanisms adaptive in folivory. We encourage you to examine recent reviews of the ecological and evolutionary consequences of plant-herbivore interactions (Batzli 1985, 1994; Belsky 1986; Palo and Robins 1991; Coley and Barone 1996).

Leaves are difficult to digest and have poor nutritional value. In addition, many leaves contain potentially toxic phenolics and terpines. In spite of these obstacles, three species of marsupials subsist on seemingly unpalatable leaves of *Eucalyptus*: koalas (*Phascolarctos cinereus*), greater gliders (*Petauroides volans*), and common ring-tailed possums (*Pseudocheirus peregrinus*) of eastern Australia. These species have evolved a remarkable suite of anatomical, physiological, and behavioral adaptations to consume *Eucalyptus*. Of special interest among these is the koala (figure 7.13). Only about five species of *Eucalyptus* comprise the bulk of the diet of koalas. The koala's jaw is very powerful and is equipped with sharply ridged, high-cusped molars that finely grind *Eucalyptus* leaves. Koalas consume about 500 g of leaves each day. The stomach is small and the small intestine is of intermediate length; the colon and cecum, however, are extremely long and wide (figure 7.14). The cecum, which is the site of microbial fermentation, is the most capacious of any mammal, meas-

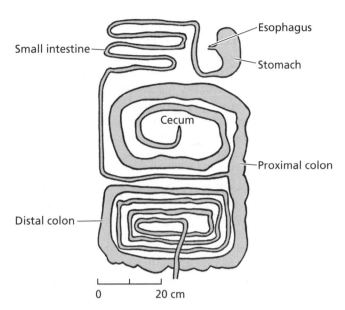

Figure 7.14 **Koala digestion.** The digestive tract of the koala is well adapted for digesting fibrous leaves of *Eucalyptus*, the staple in its diet. The cecum, measuring up to 2.5 m (8 ft), is the site of microbial fermentation. *Adapted from Harrop and Hume, 1980,* Comparative Physiology: Primitive Mammals, *Cambridge Univ. Press.*

Figure 7.13 **Koala.** The koala (*Phascolarctos cinereus*) occurs in *Eucalyptus* forests and woodlands of Australia. It lives alone or in small groups and is well adapted for arboreal life. Climbing is enhanced by its strong limbs, sharp claws, and use of two opposable digits on its forepaw and the first toe of the foot.

uring four times the koala's body length! Hindgut microflora are thought to detoxify certain essential oils of *Eucalyptus*—about 15% of which pass through the alimentary canal without transformation or absorption (Eberhard et al. 1975). Toxic compounds are inactivated in the liver through the action of glucuronic acid and then are excreted (Lee and Martin 1988). The cecum also plays a vital role in absorption of water from *Eucalyptus* leaves: The water content of fecal pellets of koalas is low (about 48%), similar to that of camels and kangaroo rats (Schmidt-Nielsen 1964). The frugal water economy of koalas is essential because they do not drink free water and must rely on water derived from their folivorous diet.

Folivory is represented in about 12% of the genera of primates. With this group, notable folivores include the indris, howlers, langurs, gorillas, and colobus and leaf monkeys of Africa and Asia. The diet of gorillas (*Gorilla gorilla*), the largest of all primates, consists of about 86% leaves, shoots, and stems (Fossey and Harcourt 1977), yet their close relatives, the chimpanzees, feed mainly on fruit. Leaves are the staple in the diet of all species of colobus monkeys except Brelich's snub-nosed monkey (*Pygathrix brelichi*), which feeds on fruits such as wild cherries, pears, and cucumbers. Colobus monkeys have flexible

food preferences. They consume fruits, flowers, buds, seeds, shoots, sap, and arthropods in addition to leaves. These monkeys are unusual because, like ruminant artiodactyls and kangaroos, they possess a greatly modified forestomach and thus are able to ruminate. Colobus monkeys have large stomachs with a large upper, sacculated region where fermentation by microorganisms occurs and a lower region typified by high concentrations of digestive enzymes. Digestion is further enhanced by enzymes produced by large salivary glands augmented by mastication with high, pointed cusps on their cheekteeth.

Two-toed and three-toed sloths of South America (Genera *Choloepus* and *Bradypus*, respectively) feed almost exclusively on leaves, stems, and fruit. Like other leaf-eating mammals, sloths possess an extremely large, compartmentalized stomach containing cellulose-digesting bacteria. As in colobus monkeys, the sloth's stomach may be one third its body mass. The colon and cecum of sloths are relatively simple. Feces and urine are passed only once per week, and thus the rectum is quite expanded, an adaptation for storing feces during prolonged periods between defecations.

Contrary to our stereotyped view of Order Carnivora, at least two species of "carnivores" practice folivory. The red panda (*Ailurus fulgens*) consumes mostly bamboo sprouts, roots, and fruit. The giant panda is well known for its consumption of bamboo shoots but feeds on only about 5 species out of the 20 available. Pandas are unique in possessing an extra "digit" on their forepaws (Schaller et al. 1989). This enlargement of one of the wrist bones acts as sort of a "thumb" to oppose the rest of the digits, enabling pandas to grip and manipulate slender pieces of bamboo with great dexterity. The carnassial teeth of *Ailuropoda* are well adapted for crushing and slicing fibrous plants. Because bamboo is low in nutritional value, giant pandas spend about 12 hours a day consuming up to 40 kg of bamboo, yet they digest less than 20% of what they eat. Much of the stem is passed through the gut relatively unchanged, resulting in many large feces.

Other notable arboreal folivores include the dermopterans (colugos) of southeast Asia, the prehensile and South American porcupines, tree hyraxes of Africa, bamboo lemurs of Madagascar, and about 18 species of rodents (Meier et al. 1987; Santini-Palka, 1994; Nowak 1999). Folivory in plant-visiting bats was once thought to be rare. However, leaf eating may be quite common and more widespread than we realize. Within the Chiropterans, leaves are consumed by 17 species of Old World "flying foxes" and 4 species of neotropical bats (Kunz and Ingalls 1994). Bats are known to masticate leaves of some 44 species of plants in 23 different families (Kunz and Diaz 1995).

Frugivory

Animals that exhibit adaptations to consume a diet of fruit, the reproductive part of plants, are referred to as **frugivorous** (meaning "fruit-eating"). Mammals from several orders are known to specialize in the consumption of fruit: pteropodid and phyllostomid bats; phalangerids; tupaiids; and primates such as indrids, lorisids, cercopithecids, colobine; and the pongids (Emmons 1991). Because fruit may have a hard outer covering, the teeth of some frugivores are adapted for piercing and crushing. Mammals that subsist on softer fruits typically possess a reduced number of cheekteeth with a bunodont occlusal pattern.

As mentioned earlier, many species of bats depend on plants as food. The frugivorous bats are distributed within two families: the Pteropodidae of the Old World tropics and Phyllostomidae of the New World tropics (Fleming 1982, 1988, 1993; Fleming and Sosa 1994; Racey and Swift 1995; Dumont 2003; Patterson et al. 2003). Frugivorous bats are reported to dominate assemblages in lowland forests of the Neotropics (Kalko 1998). No fruit- or nectar-feeding "microchiropterans" occur in the Old World (Nowak 1999). Bats play an essential role in the pollination of flowers (chiropterogamy) and dispersal of seeds. Because some seeds exhibit higher rates of germination after passing through the gut of a mammal, this method of dispersal is termed chiropterochory. Frugivorous bats comprise either principal or partial pollinators and dispersers of close to 130 genera of tropical and subtropical plants. Bats are particularly important to those species of plants that blossom only at night (e.g., avocados [Genus *Persea*], balsa [*Ochroma lagopus*], durian [*Durio zibethinus*], *Eucalyptus*, figs [Genus *Ficus*], guava [*Psidium guajava*], kopok [*Ceiba pentandra*], mangoes [*Mangifera indica*], papaya [*Carica papaya*], and wild bananas [*Musa paradisiaca*]).

"Megachiropterans" (Old World flying foxes and fruit bats) are restricted to the tropical forests of Africa, Asia, and the Australian region, where succulent fruits are plentiful. Marshall (1985) reported that "megachiropterans" feed on fruits of at least 145 plant genera in 50 families. Most species locate fruit by smell. Old World fruit bats possess comparatively few teeth, and the lower molars are reduced in number and possess large, flat grinding surfaces. The canines are the principal piercing teeth (see figure 7.1). Rather than biting off and swallowing mouthfuls of fruit, "megachiropterans" crush the pulp of ripe fruit in their mouth, swallow the juice, and spit out most of the pulp and seeds. They can bite into fruit while hovering, or they may hang onto a branch with one foot and press the fruit to their chest with the other foot and bite into it. If the fruit is small, they may carry it with them to a branch and hang their head downward while they consume it (Nowak 1999; Dumont 2003). Most "megachiropterans" crush pulp from soft fruits and may extract only the juice and reject the pulp and seeds or digest the pulp and excrete intact seeds. Because this activity often occurs at some distance from the harvesting site, seeds are dispersed great distances. Several species of Old World fruit bats (*Rousettus aegyptiacus*, *Epomophorus wahlbergi*, and *Eidolon helvum*) are principal agents of dispersal of the baobab (*Adansonia digitata*), an important tree in the African savanna (Start 1972; Nowak 1999).

Although phyllostomids may be viewed as generalists (chapter 13), over one half of the species consume some fruit. Most fruit-eating phyllostomids are in the Subfamilies Carolliinae or Stenodermatinae of the neotropics or the Brachyphyllinae of the Antilles (Gardner 1977; Findley 1993). Members of the Subfamily Carolliinae possess reduced molars and consume ripe, soft fruits such as bananas and figs. In contrast, members of the Subfamilies Stenodermatinae and Brachyphyllinae have more robust molars adapted for crushing fruit (see figure 7.1). *Artibeus*, a neotropical fruit bat, possesses a short rostrum and a high coronoid process on the dentary, which supports a strong temporalis muscle and enhances a vertical chewing motion. Like many of the "megachiropterans," *Artibeus* bites chunks from fruit and crushes the pulp for its juices with its broad, flat posterior teeth. They eat their own weight in fruit each night, and food passes through their alimentary tract rapidly, in about 15–20 minutes. *Artibeus* is important in the dispersal of seeds of tropical fruits. *Artibeus jamaicenis* is a generalist frugivore, consuming up to 92 taxa of plants, with figs forming a staple in its diet throughout the range (Gardner 1977; Fleming 1982, 1988; Handley et al. 1991; Kalko et al. 1996).

Nectarivory

Insects and hummingbirds are not the only animals that feed on flowers. Some mammals are also exquisitely adapted to feed on nectar and in the process transfer pollen. **Nectarivorous** (meaning "nectar-eating") mammals are represented by about six genera of bats (both mega- and "microchiropterans") and the marsupial honey possums. Their skulls are characterized by elongated snouts; small, weak teeth (bats); reduced numbers of teeth (honey possums); and poorly developed jaw musculature. The tongue is long, slender, and protrusile and typically has a brush tip (see figures 7.15, 11.27, and 11.29) consisting of many rows of hairlike papillae pointing toward the throat (Greenbaum and Phillips 1974; Hildebrand 1995). Bats of Subfamily Glossophaginae (Phyllostomidae) are the most specialized mammalian nectarivores. They are well adapted to feed on fruits, pollen, nectar, and insects (Fleming 1995; Freeman 1995). Flowers of the *Agave* and the saguaro cactus (*Carnegia gigantea*) are a staple in the diet of the big long-nosed bat (*Leptonycteris nivalis*; Hensley and Wilkins 1988), a bat (20–25 g) with high energy demands. During feeding, *L. nivalis* crawls down the stalk of an *Agave*, thrusting its long snout into the corolla of the flower. It licks nectar with its tongue, which can be extended up to 76 mm. Big long-nosed bats emerge from feeding covered with pollen. They are agile flyers and form foraging flocks containing at least 25 bats that feed at successive plants. While feeding, bats circle the plants and take turns feeding on the flowers. Flocks show a cohesiveness typified by minimal antagonistic behavior. Flocking seems to confer the adaptive advantage of an increased foraging efficiency, which is critical for minimizing energy expenditure. Another phyllostomid regarded as an obligate pollen feeder is the Mexican long-tongued bat (*Choeronycteris mexicana*). Analysis of stomach contents of bats from Central America showed a majority of pollen grains from pitahaya (Genus *Lemaireocereus*), cazahuate (Genus *Ipomoea*), *Ceiba*, *Agave*, and garambulla (Genus *Myrtillocactus*; Arroyo-Cabrales et al. 1987).

Possibly the most specialized mammalian nectar-feeding bat is the recently described glossophagine *Anoura fistulata*, an inhabitant of the cloud forests of the Andes Mountains of Ecuador (Muchhala et al. 2005). The tube-lipped nectar bat (*A. fistulata*) coinhabits the cloud forests with two other glossophagines, *A. caudifer* and *A. geoffroyi*. This unique bat possesses a tongue that extends 84.9 mm—150% of its body length and twice as long as its congeners (figure 7.15). Its protrusile tongue is longer, relative

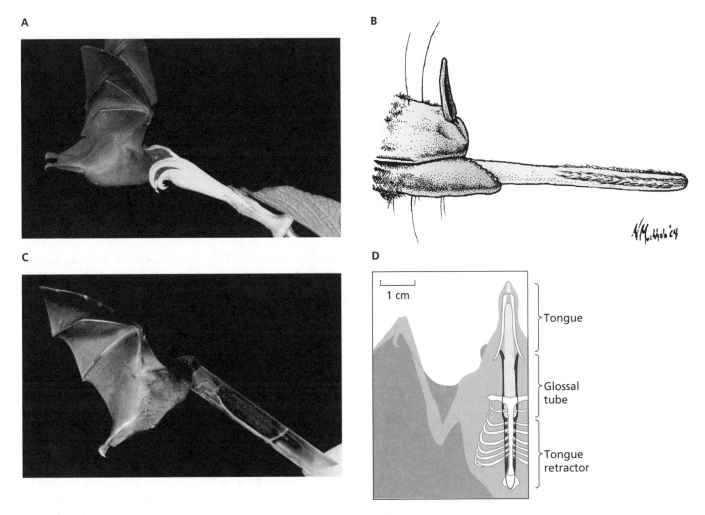

Figure 7.15 Bat tongues. Protrusible tongue of the tube-lipped nectar bat, *Anoura fistulata*. (A). *A. fistulata* pollinating the specialized flower of *Centropogon nigricans*; because of the long corolla, only *A. fistulata* can reach its nectar. (B). Lateral view of the noseleaf, lip, and partially extended tongue showing proximally facing papiillae (C). *A. fistulata* feeding from a test tube filled with sugar water; note its tongue (tan) can extend to 150% of body length. (D). Ventral view of *A. fistulata*, showing tongue (tan), glossal tube and tongue retractor muscle (black), and skeletal elements (white). *Modified from Muchhala (2006) and Muchhala et al. (2005).*

to body length, than any other mammal—surpassed only by the chameleon among vertebrates. The morphology of their tongue is indeed unique. In other nectarivarous bats, the base of the tongue attaches to the base of the oral cavity. However, the tongue of the nectar bat passes back through the neck attaching in the thoracic cavity where its distal portion is surrounded by a sleeve of tissue (the glossal tube) that parallels the ventral surface, with the base inserting between the heart and sternum. *A. fisulata* is the exclusive pollinator of the plant (*Centropogon nigricans*), which possesses corollas of matching length—80–90 mm. Both *A. fistulata* and pangolins (Pholidota:Manidae) possess glossal tubes to accommodate attachment of their long, protrusile tongues (Chan 1995). Tube-lipped nectar bats and pangolins represent excellent examples of convergent evolution—the evolution of similar morphologies (i.e., long protrusile tongues) by distantly related lineages as adaptive solutions to similar ecological pressures, namely availability of ant- and pollen-feeding niches (see chapter 5 for additional examples of convergent evolution).

Parallel adaptations for nectarivory are well known in "megachiropterans" such as the long-tongued fruit bats (Genus *Macroglossus*) and blossom bats (Genus *Syconycteris*). Common blossom bats (*S. australis*) of eastern Australia feed on nectar and pollen of a variety of rain forest plants, such as cauliflorous (Genus *Syzigium*), *Banksia*, *Melaleuca*, *Callistemon*, and certain *Eucalyptus*. Blossom bats locate nectar and pollen with their large eyes and keen sense of smell. *Syconycteris* typically lands on an inflorescence and gathers pollen and nectar by use of its long snout and brush-tipped tongue—it does not hover while feeding, as do many other nectarivores. These bats are unique in that they do not consume pollen directly from the flower. Their body hairs are covered with small, scalelike projections in which pollen lodges. This pollen is "consumed" while the animal grooms its fur and wings after a foraging episode. As with other nectarivores and frugivores, pollen rapidly passes through the gut and appears in the feces 45 minutes after ingestion. The majority of protein in the diet of these bats is provided by the pollen, whereas sugars in the nectar help to meet energy demands (Law 1992, 1993).

A species of marsupial is the premier terrestrial nectarivore. The honey possum (*Tarsipes rostratus*), the only species within the marsupial Family Tarsipedidae, is found only on the sand plain heaths of southwestern Australia (chapter 11). It weighs only 7–12 g (figure 7.16). Unlike most mammals, this mouse-sized marsupial does not climb by the aid of claws but has digits that are expanded at the tip with short, nail-like structures adaptive in gripping branches. The honey possum locates food by smell, inserting its long snout into a flower and using its protrusile tongue (reaching some 25 mm beyond the nose), which has bristles at the tip (see figure 11.29), to lick pollen from protruding anthers. The stomach is small and may act as a temporary storage compartment, the intestine is short, and there is no cecum. Pollen passes though the gut in about 6 hours.

Figure 7.16 Nectarivore. The honey possums, or noolbengers, inhabit the coastal sand plain heaths of southwestern Australia. With their thin tongue, bearing bristles and a tuft at the tip, honey possums lick nectar and pollen from the anthers of native plants, such as bottlebrush and *Banksia*. Honey possums are a significant pollinator of plants of southwestern Australia.

In addition to chiropterans and other previously mentioned mammals, the following groups are important seed dispersers: primates, tapirs (Genus *Tapirus*), African elephants (*Loxodonta africana*), one-horned rhinoceroses (*Rhinoceros unicornis*), European badgers (*Meles meles*), and foxes (*Genus Dusicyon*).

Gummivory

Animals that primarily consume plant exudates, such as resins, sap, or gums, are termed **gumivorous** (meaning "gum-eating") (Bearder and Martin 1980; Nash 1986). This peculiar dietary specialization occurs in eight species of marmosets (Genus *Cebuella*), bush-babies (*Galago senegalensis;* Bearder and Martin 1980), pottos (Genus *Perodicticus*), slow lorises (*Nycticebus coucang;* Wiens et al. 2006), four species of petaurid gliders (Genera *Petaurus*), and Leadbeater's possum (*Gymnobelideus leadbeateri;* Oates 1984). All members of the family Cheirogaleidae (dwarf and mouse lemurs) feed on tree exudates; however, consumption of gum is only occasional for *Micocebus*, *Cheirogaleus*, and *Mirza*. Gums are a staple in the diet of *Phaner* and *Allocebus* (Viguier 2004). For example, the diet of the fork-marked mouse lemur (*Phaner furcifer*) of Madagascar consists of close to 90% gum exudates from the trunks and branches of trees. Lemurs use a "tooth comb" (formed by their procumbent lower incisors and canines) to scrape off gum released from the surface of a plant. During feeding, lemurs easily cling to the surface of the trunk by use of needle-sharp claws. They digest gums within the enlarged cecum, which houses symbiotic bacteria. Although lemurs are able to scrape off saps and gums exuded as a result of damage from wood-boring insects, marmosets are the only primates that actually gouge holes to liberate plant juices. The incisors of marmosets (family Callitrichidae) are composed of thickened enamel on the outer surface and lack enamel on the inner surface, thus producing chisel-like instruments. By anchoring their upper incisors in the bark, marmosets use their lower incisors to gouge oval holes in the trunks of trees. Th. holes may measure 2–3 cm across, and certain trees .dled with channels of holes 10–15 cm in length 982). Like the lemurs, the clawlike nails of n   ential adaptations for clinging to vertical t   ding on sap, gums, and resins. Plant an   .tree make up the bulk of the diet of petaurid .tralia. Yellow-bellied gliders (*Petaurus australis*) obtain sap from *Eucalyptus* by biting out small patches of bark of the trunk or main branches. After the flow of sap dries up, they move on to a new area. As a result, some trees become heavily scarred after several years of feeding (Goldingay and Kavangah 1991).

Mycophagy

Animals that consume fungi are referred to as **mycophagous** (meaning "fungus-eating"). Fungi of various types are an important component in the diet of a diverse array of mammals, representing "insectivores", herbivores, carnivores, and omnivores. Fogel and Trappe (1978) provide a thorough review of mycophagy in mammals, detailing both the specific taxa of fungi consumed by mammals and the mammalian groups known to exhibit mycophagy. Fungi preferred by mammals include the higher Basidiomycetes, Ascomycetes, and Phycomycetes (Endogonaceae) and lichens (Maser et al. 1978). These groups of fungi are especially well represented in the diets of sciurids, murids, and members of the marsupial Families Potoroidae and Phalangeridae (Maser et al. 1978; Ure and Maser 1982; Maser et al. 1985; Maser and Maser 1987; Taylor 1992; Claridge and May 1994; Johnson 1994; Waters and Zabel 1995; Pastor et al. 1996; McIlwee and Johnson 1998; Mangan and Adler 1999; Orrock et al. 2003; Lehmkuhl et al. 2004; Izzo et al. 2005).

It is noteworthy that 22 species of primates consume fungi, for example, gorillas, bonobos, macaques, vervets, mangabeys, snub-nosed monkeys, marmosets, and lemurs (Hanson et al. 2006). For most, less than 5% of their feeding time is allocated to mycophagy. However, some primates spend from 12 to 95% of feeding time consuming fungi, namely buffy tufted-eared marmosets (12%), Japanese macaques (14%), Goeldi's monkeys (29%), and two species of snub-nosed monkeys (95%).

Fungi constitute a principal component of the diet of mammals on a year-round basis (Maser et al. 1986; Currah et al. 2000), with seasonal peaks in consumption reflecting availability (Merritt and Merritt 1978). When ingested by mammals, sporocarps pass through the digestive tract and are excreted without morphological change or loss of viability (Trappe and Maser 1976), while the other tissues are digested. The rate of passage of spores varies from 12 to 24 hours in the Cascade golden-mantled

ground squirrel (*Spermophilus saturatus*) and the deer mouse (*Peromyscus maniculatus*), respectively (Cork and Kenagy 1989b). Both subterranean fungi and fungi growing under the bark of trees are important source of food for red squirrels (*Sciurus vulgaris*) (Moller 1983; Gurnell 1987). Caching of such fungi is common, and *S. vulgaris* may hang fungi on branches of trees next to the trunk at heights of up to 8 meters—fungal stores will typically be short-lived with most fruiting bodies gone after 2 weeks (Lurz and South 1998).

Fleshy fungi are 70–90% water and provide protein and phosphorous to the consumer. Sporocarps of hypogeous (below ground) fungi are reported to contain high concentrations of nitrogen, vitamins, and minerals (Cork and Kenagy 1989a; Mcilwee and Johnson 1998; Claridge et al. 1999; Azcon et al. 2001). Fungi often contain complex carbohydrates associated with cell walls, however, and thus many small mammals are unable to efficiently digest and access the nutritious material of fungi. Some mammals possess modifications of the digestive tract that enable them to use fungi as a primary food. Potoroids, however, possess an enlarged foregut supporting fermentation of food by microbial symbionts. Long-nosed potoroos (*Potorous tridactylus*) are able to digest much of the cell wall and sporocarps of certain hypogeous fungi, thus accessing more of the fungi's available energy (Claridge and Cork 1994).

Certain coniferous and deciduous trees possess a relationship between their root systems and the mycelia of certain fungi. The combination of host and fungus is called a **mycorrhiza** (a mutualistic relationship), trees rely on filaments of the fungi to extract water and nutrients from the soil, and fungi derive nutrition from the sugars of the trees (Trappe and Maser 1976). Mycorrhizal fungi do not produce aboveground fruiting bodies, however, and thus rely on small mammals to disperse their spores. Many forest-dwelling small mammals, such as shrews, mice, voles, and squirrels, consume large quantities of spores of mycorrhizal fungi that pass unchanged through their digestive tracts. Detection of truffles by small mammals is primarily by olfaction (Pyare and Longland 2001, 2002). The dispersal of sporocarps by small mammals serves the essential function of reinoculating habitats for reestablishment of forests following natural catastrophes or deforestation (Green et al. 1999; Loeb et al. 2000; Pyare and Longland 2001, 2002).

OMNIVOROUS

Most mammals are **omnivorous** (meaning "everything-eating") and notably opportunistic. Each order of mammals contains omnivorous species; however, omnivory is best illustrated in opossums, primates such as humans and many monkeys, pigs, bears, and raccoons (see figure 7.1). The dentition is versatile, adapted to process a variety of foods. Omnivorous mammals retain piercing and ripping cusps in the anterior teeth but typically have flat, broad cheekteeth with low cusps (bunodont) adapted for crushing food (see figure 7.6). The stomachs of omnivores such as pigs (*Sus scrofa*) are comparatively simple. The small intestine is elongated, and the colon is large with many folds and bands of longitudinal muscle. The cecum of most omnivores is poorly developed due to the lack of fibrous plant material in the diet. The success of raccoons (*Procyon lotor*) in North America is a good example of a mammal that combines omnivory with opportunism to enhance survival. Although they exhibit distinct food preferences, availability largely dictates selection. During spring and summer in eastern North America, *P. lotor* feeds mainly on animal matter, including insects, earthworms, snails, bird eggs, and small mammals. They also feed on carrion and commonly visit creek edges to search for crayfish, frogs, fish, and other aquatic prey. During late summer, autumn, and winter, fleshy fruits and seeds, such as wild grapes, acorns, beechnuts, and berries, constitute the bulk of the raccoon's diet.

Foraging Strategies

During the preceding discussion of dietary types we learned many aspects of how mammals obtain their food. In that discussion, we developed some general rules to describe the processes that occur when an animal, such as a mammal, forages for food. These rules apply to all foraging behavior, regardless of the type of diet—to nectarivorous bats searching for flowers and to coyotes looking for small mammals or fruit. In addition, when some mammals discover a food source, they hoard the food they find. We conclude the chapter with an examination of patterns of hoarding behavior. Predictions derived from many of the general rules concerning foraging behavior have been tested on a variety of birds, on some other vertebrates, and on insects such as bees. To date, although mammalogists maintain a consistent interest in foraging theory, fewer aspects of this theory have been thoroughly tested on mammals. Part of the problem may be that when the various foraging models are applied to some mammals (ungulates in particular), the models have proved to be relatively poor predictors (Belovsky 1984; Owen-Smith 1993).

OPTIMAL FORAGING

Models of foraging behavior are created to generate testable predictions about such behavior that can be evaluated by empirical studies. One general theoretical framework for analyzing the foraging behavior of mammals is the notion of optimal foraging; these models attempt to predict the combination of costs and benefits that will ultimately maximize the animal's fitness (Charnov 1976; Pyke et al. 1977). Optimal foraging models involve three types of issues (Stephens and Krebs

1986): (1) The decisions an animal makes while it forages. Should it eat a particular type of food or prey? Should it remain in one place to capture its prey (sit-and-wait behavior) or actively seek out prey (stalking)? (2) The currency involved. What is being maximized? (3) How the constraints or limits of the animal affect its foraging pattern. How large a prey item can it handle? Does the fruit have a hard or soft covering?

For mammals, foraging decisions depend a great deal on past evolutionary history: What particular morphological, physiological, and behavioral traits adapt the mammal for a particular type of diet? Much of this chapter has dealt with the adaptations of the dentition, skulls, and alimentary canals of mammals with respect to various types of diets ranging from seeds utilized by heteromyids, gerbils, and chipmunks (Price and Correll 2001; Veech 2001; Ovadia and Dohna 2003; McAleer and Biraldeau 2006) to krill consumed by fur seals (Mori and Boyd 2004). Some mammals, such as those we labeled omnivores, are food generalists, consuming a wide variety of food types. Others, such as koalas, are food specialists, being highly adapted to exploit a particular type of food.

Two types of currencies are generally considered when examining foraging behavior: energy (calories) and time. Should the animal minimize its time hunting to gain a fixed quantity of food, or should it maximize its energy intake in a fixed amount of time? A lion spots a potential prey. What "decision rules" does it use to determine its actions? How close is the prey, and how much energy will be needed to pursue and capture it? How big of a meal does it represent? What are the dangers in pursuing this particular prey? Given the time that it will take to pursue and subdue this prey, would the lion be better off if, instead, it went off in search of some other prey? The lion does not actually compute a cost-benefit analysis before acting, but we can use the economic model as a means of generating testable predictions about its behavior.

Several key constraints influence a mammal's foraging strategy. The first are constraints related to the time spent traveling to find food-search time. Consider the differences between an insectivorous bat that must find swarms or patches with insects versus a nectarivorous bat searching for groups of flowers (Pleasants 1989; Jones 1990; Barclay and Brigham 1994; von Helversen and Winter 2003). For smaller mammals, the danger of starvation may be greater than for large mammals, and thus the constraint imposed by time and the need to regularly find food sources affects foraging decisions. Second, foods vary in how easily their energy "packets" can be extracted by a mammal-handling time. Some foods, such as grasses, require almost no manipulation; they are snipped or chewed off and swallowed. On the other hand, some fruits and seeds may require the animal to spend considerable time and energy to extract the nutrition. In an experiment that combined field and laboratory testing, *Peromyscus polionotus* made their food

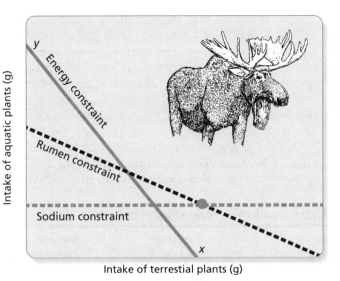

Figure 7.17 Moose feeding. The foraging constraints affecting daily feeding by moose involve (1) the need for sufficient energy (solid line), (2) the need for sufficient sodium intake (dashed line), and (3) the size of its rumen (broken line). Aquatic plants are bulkier but contain more sodium than terrestrial plants. The moose's average intake lies within the triangle defined by the three constraints, near the circle. *Data from G. E. Belovsky, "Diet Optimization in a Generalist Herbivore: The Moose" in* Theoret. Pop. Biol., *14:105–134, 1978.*

selections among millet, sunflower seeds, and peanuts based, in part, on handling time (Phelan and Baker 1992).

Third, mammals that are themselves potential prey must always balance the need for finding and consuming food with some degree of vigilance. Individual mammals may do this by keeping a wary eye or ear attuned to the surroundings as they feed. Group-living mammals may have various systems for detecting predators and alerting other members of the group. Meerkats (*Suricata suricatta*) feed in groups, and individual members take turns being the sentinel, watching for predators such as snakes or hawks. The sentinel, of course, is sacrificing valuable feeding time, and thus the meerkats rotate this duty to ensure that each individual has sufficient time for energy intake (Ewer 1968; Doolan and Macdonald 1996). Another way to reduce predation risk is to hoard food; by having food stores in a protected location, the mammal does not have to venture forth, exposing itself to potential predators. Hoarding behavior is discussed in more detail in the last section of this chapter. Fourth, some mammals have nutrient constraints. The contrasts between foregut and hindgut fermenters particularly exemplify this point. Moose (*Alces alces*) must consume both sufficient energy and sufficient sodium to meet their daily needs (Belovsky 1978), but the capacity of the rumen also must be considered (figure 7.17). A central challenge for classical foraging theory pertains to patch use decisions of large herbivores such as roe deer, and free-ranging bison (Illius et al. 2002; Fortin 2003).

MARGINAL VALUE THEOREM

A second foraging behavior model examines how animals make decisions about their feeding sites; these are often called patch models (Wiens 1976; Bell 1991; Nonacs 2001; Searle et al. 2005). Many food items consumed by mammals are found in patches, or clumps. Some questions are associated with locating patches, devising search patterns within patches, and deciding when to leave a patch and search for another and when to return to a patch (if the resource is renewable, such as nectar). For example, if a wild pig locates some acorns, it initially moves from one acorn to the next rather rapidly. Soon, however, the patch becomes depleted and the acorns are more difficult to find. Mapping the path used to find acorns can provide insights concerning the pig's foraging decisions. Finally, at what point should the pig give up on that patch and attempt to locate another, perhaps more profitable, patch?

The marginal value theorem deals with the decisions about when to leave patches (Charnov 1976; Searle et al. 2005). The model relates the energy an animal spends moving between resource patches to the energy that can be obtained in that patch. At some point, the gain in energy relative to the costs of continued foraging in that patch becomes marginal; this is often referred to as the giving-up time (McNair 1982). Leaving a patch that is becoming depleted of food means the animal is confronted with the costs of searching for another patch and of traveling to another patch. The basic question is, based on diminishing returns within a patch, when should the animal leave?

Our discussion of foraging behavior can best be summarized by the following formula:

$$\text{Net rate of energy intake} = \frac{\text{Energy from food} - \left(\begin{array}{c}\text{Search} \\ \text{energy}\end{array} + \begin{array}{c}\text{Pursuit} \\ \text{energy}\end{array} + \begin{array}{c}\text{Handling} \\ \text{energy}\end{array} + \begin{array}{c}\text{Eating} \\ \text{energy}\end{array}\right)}{\begin{array}{c}\text{Search} \\ \text{time}\end{array} + \begin{array}{c}\text{Pursuit} \\ \text{time}\end{array} + \begin{array}{c}\text{Handling} \\ \text{time}\end{array} + \begin{array}{c}\text{Eating} \\ \text{time}\end{array}}$$

Exploring and testing the various components of this formula have been and will continue to be significant research problems for mammalogists. Feeding is, after all, of primary importance for fitness, measured as survival and reproductive success.

FOOD HOARDING

During autumn in seasonal environments, certain mammals, such as squirrels, are commonly observed harvesting fallen nuts and establishing food stores in preparation for winter. This habit, called **food hoarding,** or **caching,** has been reported for members of 6 orders and 30 families of mammals (Vander Wall 1990). Caching confers the advantage of providing a reserve food supply during lean periods. Further, the larder is isolated from competitors and offers protection from predators during feeding. Caching techniques, location, composition, season, and how food is used vary at both the individual and species levels. Caching is performed by employing either larder hoarding (concentration of all food at one site) or scatter hoarding (one food item stored at each cache site). Vander Wall (1990) documents the mammalian taxa that store food and summarizes their food-storing behavior (table 7.1).

Most mammals that cache food are members of the Orders Rodentia or Carnivora. Only one marsupial is reported to store food, the pygmy possum (*Burramys parvus*). The pygmy possum is the only mammal restricted to the alpine and subalpine zones of Australia. It is a true hibernator, and in autumn, it caches seeds and fruits of heathland shrubs as a food reserve during its winter dormancy.

Of the 312 species of shrews, only 7 cache food. This behavior has been documented primarily in the laboratory with food provided ad libitum (Merritt and Vessey

Table 7.1 Mammalian taxa known to store food and a summary of their food-storing behavior

Order Family	Dispersion*	Food Type†	Substrate/ Location‡	Storage Duration§
Marsupials				
Pygmy opossums (Burramydae)	L	S	N	L
Insectivores				
Shrews (Soricidae)	L	I,SM,A,Fi,S	N,B	L
Moles (Talpidae)	L	I	B,N	L
Primates				
Squirrel monkey (Cebidae)	S?	Mi	?	S
Green monkey (Cercopithecidae)	L?	Fr	C	S
Chimpanzee (Hominidae)	S?	Me	?	S?
Carnivores				
Foxes, wolves (Canidae)	S,L	SM,MM,Bi,E,Re	S,Sn,C	S,L
Bears (Ursidae)	S,L	LM	S,L	S
Weasels, mink (Mustelidae)	L	SM	B,N,S,C	S,L
Hyenas (Hyaenidae)	S	Ca	W	S
Tigers, bobcats (Felidae)	S	MM,LM	S,L,Sn,T	S
Rodents				
Mountain beaver (Aplodontidae)	L	V	B	S,L
Squirrels and chipmunks (Sciuridae)				
Chipmunks	L,S	S,Nu,Bu	B,N,S	L
Red squirrels	L	Co,Fu	L,F,T	L
Tree squirrels	S	Nu,Fr,Fu	S,T	L
Ground squirrels	L,S	S,Nu,V	B,S	L
Flying squirrels	S,L	Nu	S,L,T	L
Pocket gophers (Geomyidae)	L	Bu,Ro	B,Sn	L
Kangaroo rats (Heteromyidae)	L,S	S	B,S	L
Beavers (Castoridae)	L	WV	W	L
Mice, hamsters (Cricetidae)				
New World mice	L,S	S,Nu,I	B,S	L
Woodrats	L	V,S,Nu,Fr	N	L
Hamsters	L	S	B	L
Gerbils	L	S,Nu,Ro,V	B	L
Mole-rats	L	Bu,Ro,V	B	L
Voles and muskrats	L	V,Bu,Ro,WV	B,G,Sn,F	L
Old World rats and mice	L,S	S,Nu	B,S	L
Dormice (Gliridae)	L	Nu,Fr,Mi	C	L
Jerboas (Dipodidae)	L?	?	B	L?
Old World porcupines (Hystricidae)	S	?	S	?
Agoutis and acouchis (Dasyproctidae)	S	Nu,Fr	S	L
Octodonts (Octodontidae)	L	Bu	B	L
Tuco-tucos (Ctenomyidae)	L	Bu	B	L?
African mole-rats (Bathyergidae)	L	Ro,Bu	B	L
Lagomorpha				
Pikas (Ochotonidae)	L	V	G,B	L

From S. B. Vander Wall, Food Hoarding in Animals, 1990. Copyright ©1990 Univ. of Chicago Press. Reprinted by permission.
*Dispersion patterns: L = larder; S = scattered.
†Food types: A = amphibians; Bi = birds; Bu = bulbs; Ca = carrion; Co = cones; E = eggs; Fi = fishes; Fr = fruit; Fu = fungi; I = invertebrates; LM = large mammals; Me = meat; Mi = miscellaneous; MM = medium mammals; Nu = nuts; Re = reptiles; Ro = roots; S = seeds; SM = small mammals; V = green vegetation; WV = woody vegetation.
‡Substrates and locations: B = burrow chambers; C = cavity or chamber (not in burrow); F = foliage; G = ground surface; L = litter; N = nest; Sn = snow; S = soil; T = tree trunk and branches; W = water.
§Storage duration: S = short term (generally < 10 days); L = long term (generally > 10 days).

2000). Among shrews, the premier scatter hoarder is the northern short-tailed shrew. European moles (*Talpa europaea*) collect and cache earthworms and insect larvae. After mutilating the head segments of earthworms, moles cache the worms in chambers and walls of galleries near the nest as food for the winter. When soil temperature increases in the spring, some of the remaining worms, trapped in the galleries, may regenerate a head and burrow to escape (Gorman and Stone 1990).

Members of five families of carnivores cache food; the canids, ursids, mustelids, hyaenids, and felids (Vander Wall 1990; Skelpkovych and Montevecchi 1996). Groups of predators follow several distinctive caching patterns: (1) canids bury prey in shallow surface depressions; (2) felids and ursids do not excavate a hole but rather rake soil and leaf litter over the prey; (3) mustelids either cache the prey in their dens or, like canids, bury it in shallow surface pits; and (4) hyenas may submerge prey in water (Macdonald 1976; Elgmork 1982). Because the food items captured by these predators are usually single carcasses, these predators practice larder hoarding.

The most common hoarders are rodents. They store food in many different locations, usually for periods of more than 10 days. Larder and scatter hoarding techniques are employed to cache foods ranging from seeds and nuts to woody vegetation, roots, and invertebrates (Steele and Koprowski 2001). As winter approaches, eastern chipmunks carry large amounts of food in their cheek pouches and cache the food in their burrows for winter use. Preferred items in their winter diet include hickory nuts (Genus *Carya*), beechnuts (Genus *Betula*), maple seeds (Genus *Acer*), acorns (Genus *Quercus*), and a long list of seeds of woody and herbaceous plants.

Gray squirrels (*Sciurus carolinensis*) are principal consumers of acorns of red and white oaks in eastern North America. Steele et al. (2001) have demonstrated the importance of embryo excision as a means of long-term cache management by gray squirrels. Red oaks exhibit delayed germination of acorns and can be stored up to 6 months before they begin to germinate. Acorns of white oaks show no dormancy and germinate in autumn soon after the seeds fall; however, if the embryo is excised, these acorns will remain intact for up to 6 months. *S. carolinensis* cached significantly more acorns of red oak species than those of white oak species, caching white oak acorns after excision of the embryos. Squirrels excised embryos of red oaks only when the acorns began to germinate following winter. Naive captive gray squirrels, without previous experience with acorns, also cache red oak acorns over those of white oaks and attempted embryo excision on white oaks, suggesting a strong innate tendency for the behavior. Such attempts, however, are often unsuccessful, indicating that the behavior is likely perfected through learning (Steele et al. 2006).

Flying squirrels (*Glaucomys volans*) are more selective than eastern chipmunks; hickory nuts may comprise up to 90% of the nuts they store for the winter. Unlike

chipmunks (Genus *Tamias*) and tree squirrels (Genus *Sciurus*), red (*Tamiasciurus hudsonicus*) and Douglas (*T. douglasii*) squirrels establish large surface middens to cache conifer cones and mushrooms. Heteromyids primarily cache seeds, but some species also store fruit, dried vegetation, and even fungi (Rebar and Reichman 1983; Reichman et al. 1985; Reichman and Rebar 1985). Foods are cached in burrows or shallow pits on the surface of the ground near entrances to the burrow, or possibly grouped in small haystacks on the ground.

Some small mammals store food in scattered surface caches and underground chambers, whereas others concentrate their cache in a single or just a few larder sites. These foods are consumed during the winter when food supplies are scarce. Eastern woodrats (*Neotoma floridana*) gather and store large quantities of fruit, seeds, leaves, and twigs in their large surface dens constructed of sticks (Wiley 1980). Perishability and nutrient content of food influence caching decisions of woodrats (Family Cricetidae; Post and Reichman 1991; Post et al. 1993, 2006) and kangaroo rats (Heteromyidae; Reichman and Rebar 1985; Reichman 1991; Brown and Harney 1993; Jenkins and Breck 1998; Price et al. 2000). Mole-rats (*Spalax microphthalmus*) of the steppes of Eurasia are noteworthy for establishing many storerooms up to 3.5 m in length, which they pack with rhizomes, roots, and bulbs. These large caches are essential for meeting the energy demands of mole-rats during the long winter on the steppes, when the ground surface is frozen and foraging is restricted.

Among lagomorphs, only the pikas (Genus *Ochotona*) establish food caches, called hay piles. Hay-gathering behavior is common to most species of pikas. *Ochotona princeps* of North America collect green vegetation during late summer and autumn and establish hay piles under overhanging rocks within talus (rubble or scree) deposits (chapter 18). Hay piles function as a source of food during winter as well as serve as a safeguard against an unusually harsh or prolonged winter (Conner 1983; Dearing 1997). Size and placement of hay piles vary greatly among different species of pikas. *Ochotona princeps* establishes comparatively large hay piles, weighing up to 6 kg (Millar and Zwickel 1972), and up to 30 species of plants may be found in one hay pile (Beidleman and Weber 1958). Daurian pikas (*O. dauurica*) inhabiting the steppes of northern Manchuria establish hay piles weighing 1–2.5 kg (Vander Wall 1990). Pallas's pikas (*O. pallasi*) make hay piles measuring up to 100 cm in height placed on the ground over burrow entrances. They carry pebbles (up to 5 cm in diameter) in their mouth, placing them near the burrow entrance to prevent the hay from being scattered by the wind.

Beavers (Genus *Castor*) stockpile in submerged caches below the ice near their dens woody vegetation, which remains a fresh and handy food supply through the winter. The phrase "busy as a beaver" refers to the energetic activity of beavers generating these food stockpiles.

SUMMARY

Just as mammals range in size from shrews to elephants, their food ranges from microscopic organisms to prey that is larger than the predator. Types of food range from those low in calorie content, such as grasses and nectar, to proteinaceous foods, such as insects, blood, and pollen. Food may be highly restricted, such as underground fungi available only during summer, or ubiquitous and available year-round, such as insects in tropical environments.

Mammals have evolved many fascinating adaptations to procure food. Evolutionarily, the precursor of all mammalian feeding groups was represented by insectivores of the Triassic. Insectivores today include aerial forms (bats) and terrestrial forms (platypuses, shrews, moles, anteaters, echidnas, aardwolves, etc.). These mammals have one thing in common—teeth equipped with sharp cones and blades for crushing insect exoskeletons. Insectivorous mammals also possess additional adaptations that optimize their ability to capture and process prey, such as the toxic bites of shrews for immobilizing prey and the sensitive tactile organs (Eimer's organs) of some talpids for efficiently locating insects below ground.

Mammals that feed primarily on flesh are called carnivores and include the canids, mustelids, felids, hyaenids, and dasyurid marsupials. Carnivorous mammals have sharp incisors and canines as well as a pair of carnassial teeth—a characteristic of meat-eaters. Carnivores possess a keen sense of smell and acute vision and hearing. Some carnivores such as canids rely on intelligence, social organization, and behavioral adaptability to optimize their hunting success. Carnivores hunt prey on land, in the air, and in the water. Three species of "carnivorous" mammal specialize in the consumption of blood—the vampire bats. The largest of all mammals, the baleen whales, consume the smallest prey—diatoms and krill.

Herbivorous mammals consume plants and include the ungulates and gnawing mammals. The artiodactyls and perissodactyls possess high-crowned cheekteeth with transverse ridges, which are well adapted for shredding and grinding coarse plant material. Ungulates, rodents, and lagomorphs have microorganisms in their alimentary canals that break down and metabolize cellulose. The ruminant artiodactyls (foregut fermenters) have a very efficient but slow digestive process. The perissodactyls (hindgut fermenters) have a faster but less efficient digestion. Gnawing mammals (lagomorphs and rodents) and the ungulates are unable to produce the enzyme cellulase. Thus, they ferment fibrous plant material using bacteria and protozoa in the alimentary canal. Rodents and lagomorphs possess large gnawing incisors followed by a conspicuous gap (diastema) and a battery of cheekteeth characterized by many peculiar patterns adapted to consuming a diverse array of foods. Herbivorous mammals that consume seeds are called granivores and include mostly the desert mammals in the Family Heteromyidae. Leaf-eaters, or folivores, must work very hard for minimal nutritional reward. The koala of Australia, with its elongated cecum, and many species of primates, sloths, and the pandas are noteworthy folivores. Fruit eating (frugivory) is characteristic of pteropodid and phyllostomid bats, phalangerids, and certain primates. Megachiropterans feed on fruits of some 145 genera of plants and are important pollinators of flowers and dispersers of plant seeds. This frugivorous niche is filled in the New World by the phyllostomids, notably the Genus *Artibeus*. Nectarivorous mammals are represented in about six genera—the flying foxes, the "microchiropterans", and the diminutive honey possums of Australia. Adaptations for consuming nectar and pollen include greatly elongated snouts and long, slender, protrusile tongues equipped with a brush tip for gleaning pollen and nectar from flowers. The champion nectarivore is arguably the tube-lipped nectar bat, *Anoura fistulata* of the Andes Mountains of Ecuador. Mammalian gumivores consume the exudates of plants, such as resins, sap, or gums. This unusual dietary specialization is represented by various marmosets, bush-babies, mouse lemurs, pottos, petaurid gliders, and Leadbeater's possum.

Fungi are important in the diet of many insectivores, herbivores, carnivores, and omnivores and the marsupial Families Potoroidae and Phalangeridae. Many species of rodents, called mycophagists, act as important consumers and dispersers of mycorrhizal fungal spores and thus contribute to forest succession. Further, some eight groups of primates are reported to consume fungi; however, for most, less than 5% of their feeding time is allocated to mycophagy.

Various models developed to explain foraging behavior have been tested for mammals. These include the optimal foraging theory, which models dietary decisions, the currency that is maximized (energy or time), and the constraints that affect feeding behavior. Patch models are used where food is distributed in clumps. The marginal value theorem makes predictions about when a mammal should give up feeding in one location and find another patch. These models have been applied to mammalian foraging with varying degrees of success.

Food hoarding, or caching, occurs in 6 orders and 30 families of mammals. The techniques, location, and composition of the cache and how the food is used vary greatly depending on the individual, species, and season.

SUGGESTED READINGS

Chivers, D. J., and P. Langer. 1994. The digestive system in mammals: Food, form, and function. Cambridge Univ. Press, New York.

Clutton-Brock, T. H. (ed.) 1977. Primate ecology: Studies of feeding and ranging behaviour in lemurs, monkeys, and apes. Academic Press, London.

Dawson, T. J. 1995. Kangaroos: Biology of the largest marsupials. Comstock Publ. Assoc., Ithaca, NY.

Gittleman, J. L. (ed.). 1989. Carnivore behavior, ecology, and evolution. Comstock Publ. Assoc., Ithaca, NY.

Gittleman, J. L. (ed.). 1996. Carnivore behavior, ecology, and evolution, vol. 2. Comstock Publ. Assoc., Ithaca, NY.

Greenhall, A. M., and U. Schmidt (eds.). 1988. Natural history of vampire bats. CRC Press, Boca Raton, FL.

Kunz, T. H., and M. B. Fenton. 2003. Bat ecology. University of Chicago Press, Chicago.

Owen, J. 1980. Feeding strategy. Univ. Chicago Press, Chicago.

Popa-Lisseanu, A.G., A. Delgado-Huertas, M. G. Forero, A. Rodríguez, R. Arlettaz, and C. Ibáñez. 2007. Bats' conquest of a formidable foraging niche: The myriads of nocturnally migrating songbirds. PLoS ONE :2 (2): e205.

Rogers, E. 1986. Looking at vertebrates: A practical guide to vertebrate adaptations. Essex, England.

Stevens, C. E., and I. D. Hume. 1996. Comparative physiology of the vertebrate digestive system, 2nd ed. Cambridge Univ. Press, New York.

Vander Wall, S. B. 1995. Influence of substrate water on the ability of rodents to find buried seeds. J. Mammal. 76:851–856.

DISCUSSION QUESTIONS

1. What would the diet of primitive mammals likely have been? What modern mammals possess teeth like those of primitive mammals?

2. Compare the digestive systems of Artiodactyla and Perissodactyla. Comment on their digestive efficiency and speed of processing food. How do the differences in processing food of foregut and hindgut fermenters relate to the ecology of ungulates?

3. How can you distinguish between the skull of a herbivore and that of a carnivore?

4. Of a group of mammals consisting of a bovid, mustelid, primate, and pinniped, which do you predict would have the shortest gut in proportion to its body length? Why?

5. Compare the feeding behavior of right whales to that of rorquals.

6. Explain the mechanism and adaptive significance of the toxic bite of the northern short-tailed shrew (*Blarina brevicauda*).

CHAPTER 8

Control Systems and Biological Rhythms

Three major control processes regulate internal events in mammals: the nervous system, the endocrine system, and the immune system. Part 3, on mammalian diversity, covers a number of specialized aspects of these systems, as do several chapters within this part. This chapter examines some general aspects of these three body control systems that are important to mammals. We begin by exploring the sensory mechanisms of mammals and the features of their central nervous system. We then treat the key aspects of the mammalian endocrine system and some of the functional processes regulated by hormones. We next make a brief examination of the mammalian immune system. The activity patterns of mammals are governed by a series of biological rhythms. These rhythms are, in turn, regulated by activities in the nervous and endocrine systems. We conclude the chapter with an exploration of both the mechanisms that underlie these rhythms in mammals and their functional consequences. For background in the general biology of the nervous and endocrine systems, consult Hickman and colleagues (1997).

Nervous System

SENSORY SYSTEMS

Each of the major sensory systems in mammals has undergone adaptations that help to define the group. The importance of a particular sensory modality varies among mammals; vision, olfaction, and hearing predominate to varying degrees in most groups.

Vision

Detection and use of radiant energy in the form of light of particular wavelengths is important to most but not all mammals. The mammalian eye is generally similar to the eye of other amniote vertebrates. Because humans use vision perhaps more than any other sense, we may tend to

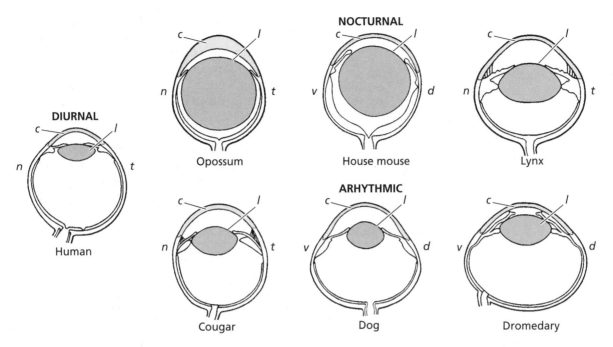

Figure 8.1 **Eye proportions.** The eyes of vertebrates vary in the intraocular proportions depending on normal light intensity at the time of activity. *d* = dorsal (top) side of eyeball; *n* = nasal (nose) side of eyeball; *t* = temporal (outer) side of eyeball; *v* = ventral (bottom) side of eyeball; *c* = cornea; *l* = lens. *Adapted from Walls, 1942*, The Vertebrate Eye, *Cranbrook Institute.*

overemphasize its importance. Among some insectivores that are primarily fossorial and cetaceans living in the aquatic environment, the eyes are greatly reduced and can really only differentiate light from dark. At the other extreme, primates, felids, and some other mammals have binocular, stereoscopic vision. Many nocturnal mammals possess a tapetum lucidum, usually in the choroid; this reflective layer lies outside the receptor layer of the retina. The **tapetum lucidum** provides the typical eye shine when light, for example from car headlights, strikes the retina at night. Hence, the often used expression, "like a deer in the headlights," used for someone who is caught in a situation where they are uncertain how to act. This structure aids in night vision by reflecting light that has passed through the receptor layer back toward the retina.

The proportions of the eye vary among mammals depending on environmental light intensity (figure 8.1). The cornea and lens are positioned farther from the retina in diurnal mammals than in nocturnal mammals and are also less sharply curved to provide the appropriate focal length. Together, these adaptations provide for a larger visual image in diurnal mammals. Mammals possess both **rods** for black-and-white vision and **cones** for color vision. Diurnal mammals have a greater proportion of cones, and nocturnal mammals have a greater proportion of rods. Color vision is not very useful at low light levels, and the predominance of rods provides a sharper black-and-white image for nocturnal mammals.

Hearing

Reception and use of auditory cues, along with olfaction, are the two sensory systems most prominent in the majority of mammals. Auditory signals are used for a variety of functions, including predator detection and communication. Mammals are unique in possessing pinnae (external ears), which function to focus and direct sound and thus aid in their acute sense of hearing. Pinnae have been lost secondarily in some insectivores, fossorial mammals, cetaceans, and phocid seals. The positioning of the ears, with one on each side of the head allows for directional hearing; this ability to localize sound in terms of the direction from which it came is a distinct advantage for detecting prey as well as potential predators. For some mammals, as in many ungulates and particularly bats, the ears can be rotated to facilitate reception of sound waves in order to determine the direction of a sound source. Mammals can detect a wide range of wavelengths and intensities of sound. Hearing ranges extend from **infrasound**, frequencies of less than 20 Hz, in elephants (Payne et al. 1986; Langbauer et al. 1991) to **ultrasound**, frequencies of greater than 20,000 Hz, in bats and cetaceans (Gould 1983). Echolocation in bats and cetaceans is discussed in chapters 13 and 17.

The internal ear in mammals has several specialized functions (figure 8.2). The semicircular canals are the body's balancing mechanism. Three ossicles, or bones, are present in the ears of mammals: the malleus, incus, and stapes. The incus and malleus are unique to mammals. The stapes is homologous with the hyomandibular bone of fish and the columella of reptiles and amphibians. The ossicles transmit sound pressure information impinging on the **tympanum** (eardrum) at the interior end of the external auditory canal through the middle ear to the inner ear. The coiled cochlea of the inner ear contains the organ of Corti, which transduces (changes) pressure information into neural impulses.

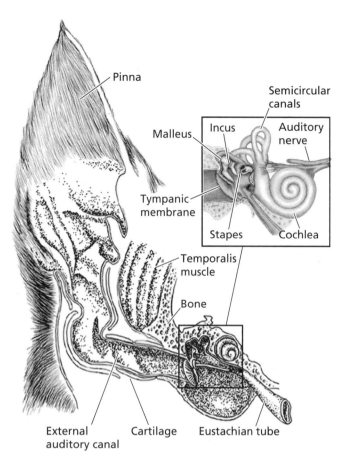

Figure 8.2 Mammalian ear. The features of the ear include an external pinna for collecting sound, the ossicles of the middle ear for transmitting the sound pressure, and the organ of Corti in the inner ear for transducing the pressure signal to neural impulses. The semicircular canals, important for balance, are located in the middle ear. *Adapted from W. C. Stebbins, 1983,* The Acoustic Sense of Mammals, *Harvard Univ. Press.*

Taste

Although the sense of taste is quite important for some mammals in dietary selection, it is secondary in most groups. Varying types of sensory receptors are located on different portions of the tongue, usually in discrete patches. For some mammals, including humans, these receptors detect five basic classes of substances (flavors): sweet, sour, salty, bitter, and unami (we know this as monosodium glutamate or MSG). Recent research indicates that individual taste receptors on the tongues of mammals may be capable of detecting multiple flavors (Adler et al. 2000).

Olfaction

Chemoreception via olfactory receptors located in the nasal cavity is well developed in most mammals. Smell is important for a variety of activities, including locating food and detecting danger, as well as in social relations. Mammals have evolved a variety of glands that produce odorous substances, such as sebaceous glands in hamsters and male goats; preorbital glands in deer; interdigital

glands in sheep; preputial glands in deer, beaver, and mice; anal glands in mustelids; and perineal glands in guinea pigs and rabbits (figure 8.3). Many mammals also obtain chemosensory information via the vomeronasal organ (Jacobson's organ), which is located near the floor of the nasal cavity on each side of the nasal septum and attached to the vomer. In whales and some higher primates, the sense of smell is reduced, and it is absent in dolphins and porpoises. Although some aspects of the olfactory epithelium, where chemical information is transduced to neural impulses, are modified or specialized in mammals, the general morphology is similar to that in reptiles.

Touch

In addition to pressure receptors associated with hairs, most mammals possess vibrissae. Vibrissae (whiskers) are elongated hairs located on the muzzle and around the eyes and ears. Some fossorial mammals also have vibrissae on their tails. These tactile organs enable the animal to detect nearby objects and may also function to protect the face and eyes. Many mammals, ranging from moles to elephants, have specialized touch sensors in the snout and lips. Tactile sensory systems are most highly developed, however, in smaller mammals that burrow and tunnel.

Other mammals have highly evolved tactile receptors in their digits that can detect heat, cold, pressure and other stimuli that contact the skin. The skin in these regions has lost its hairy cover, and receptors are concentrated near the surface. This is particularly true for primates, but it also characterizes other groups, including carnivores. This enhanced sense of touch is useful in locomotion, food processing, and social relations.

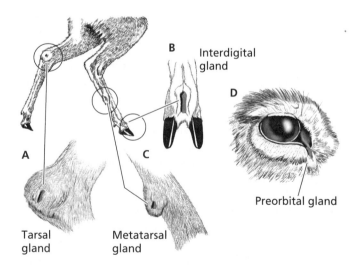

Figure 8.3 Scent glands of white-tailed deer. (A) Tarsal glands are located on the inner surfaces of the hind legs. (B) Interdigital glands are located between the hooves of each foot. (C) Metatarsal glands are found on the outer surface of the hind legs. (D) Preorbital glands are located at the front of each eye. *Adapted from Lowell K. Halls (ed.), 1984,* White-tailed Deer: Ecology and Management, *Stackpole Books.*

CENTRAL NERVOUS SYSTEM

The central nervous system consists of the brain and spinal cord. Although some generally minor aspects of the mammalian spinal cord are specialized and help to differentiate mammals from other vertebrates, major changes have occurred in brain morphology and organization. The mammalian brain can be subdivided into five major regions (figure 8.4). A central, unique feature of mammalian evolution has been the expansion of the cerebral hemispheres (neocortex) of the telencephalon. All the other brain regions have been virtually surrounded by this new gray matter. Cells in the cerebral hemispheres are arranged in layers, with all the cells in each particular region of the cortex devoted to processing the same type of information.

The amount of cerebral tissue in each major region varies considerably among different mammals (figure 8.5). The general structures depicted in figure 8.4 are all underneath the cortex shown in figure 8.5, as the cortex wraps around these more ancestral brain regions. In the Norway rat (*Rattus norvegicus*) and even tarsiers (*Tarsius* spp.), large proportions of the cerebral cells are in regions pertaining to the different sensory modalities. Very little **association cortex**—regions of the cortex that are not specifically identifiable as sensory or motor cortex—occurs where information is processed and integrated across the various sensory modalities. The rat has considerably more olfactory cortex than the tarsier, but the tarsier has more visual cortex. These patterns reflect the relative use of these sensory modalities by rats (rodents) and tarsiers (primates). In the chimpanzee (*Pan troglodytes*) and human (*Homo sapiens*), the area devoted to association cortex has increased enormously (Camhi 1984). In addition, of course, the total size of the brain has increased from a volume of about 20 cm³ in the rat to about 1,200

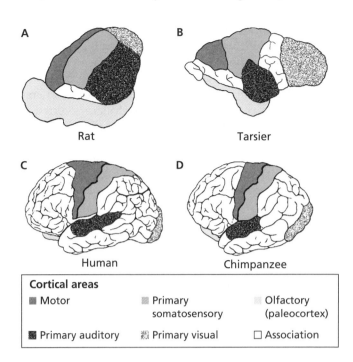

Cortical areas

■ Motor	▨ Primary somatosensory	▧ Olfactory (paleocortex)
▨ Primary auditory	▨ Primary visual	□ Association

Figure 8.5 **Brain areas of neocortex.** Considerable differences occur among the four mammals shown here with respect to both (a) the relative areas devoted to different sensory systems in terms of storage areas in the cerebrum and (b) the amount of association cortex. *Adapted from J. M. Camhi, 1984,* Neuroethology, *Sinauer.*

cm³ in the human. Both absolute and relative changes must be considered. There is a significant trend toward increase in the ratio of brain size to body mass among mammals over evolutionary time (Jerison 1973; Eisenberg 1981). Although only partially visible in figure 8.5, brain surfaces in rats and even in tarsiers are rather smooth, which contrasts with the multiple infoldings and overall rough surface appearance of the brains of chimpanzees and humans. These differences in texture reflect a vast increase in the overall surface area of the brain, much like the difference between traveling along a straight coastline and one with many bays and coves. The increased surface area reflects the presence of many more neurons and more complex interconnections between various brain regions.

Because of the increased association areas and modest but significant changes in the underlying neural structures in mammalian brains, mammalian sensory pathways are more complex than those in most other vertebrates (Camhi 1984). An impulse arriving from the visual or auditory nerves splits and goes to a number of areas of the brain, rather than to one or a few locations, as happens in lower vertebrates. These multiple pathways provide for at least two major characteristics of mammalian central nervous systems. First, considerably more storage area is available for retaining stimulus information, permitting easier recall and later use. Second, incoming information is better integrated, both across modalities and with respect to relating current and past stimuli.

The various brain structures of mammals have not all evolved at one time; rather, they follow a pattern called

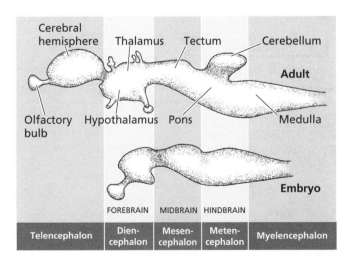

Figure 8.4 **Mammalian brain.** The five major regions of the mammalian brain. A rather primitive mammal is depicted here; considerable additional material is found in the cerebral hemispheres of higher mammals. *Adapted from A. S. Romer, 1962,* The Vertebrate Body, *W. B. Saunders.*

mosaic evolution, in which different brain structures and features developed at different points in the evolutionary sequence (Barton and Harvey 2000). So, for example, in insectivores there are clear connections between the evolution of the medulla, cerebellum, and diencephalon, whereas in primates correlated evolutionary changes occurred in the medulla, cerebellum, and neocortex.

In mammals, both the mode of impulse transmission and the manner in which the effector neurons convey information to the muscle at the motor end plates are generally similar to these mechanisms in other vertebrates. Results from research in the past several decades indicate that more types of neurotransmitters are involved in nerve impulse transmission across synapses in mammals than in other vertebrate groups. To date, more than 20 different neurotransmitters have been identified in mammalian central nervous systems. Nerve regeneration occurs in the peripheral nervous system of mammals, much as it does in other vertebrates, but it does not occur readily in the central nervous systems of mammals. Because of the difficulties associated with spinal cord injuries in humans, a great deal of research is currently devoted to studying this problem, using nonhuman mammals and other animals as models.

Endocrine System

HYPOTHALAMIC-PITUITARY SYSTEM

The master control mechanisms for the endocrine system in mammals are the hypothalamus and the pituitary gland (figure 8.6). The **hypothalamus**, part of the forebrain, is located below the thalamus and contains collections of neuron cell bodies (nuclei). Sensory input to the hypothalamus comes from other brain regions and from cells within the hypothalamus that monitor conditions in the blood that passes through the region. The **pituitary gland** is located directly below the hypothalamus and is connected to it by both blood vessels and neurons. Thus, the hypothalamus serves as the connector between the brain (nervous system) and the pituitary (endocrine system). The vascular connection is known as the **hypothalamic-pituitary portal system**. The pituitary gland is divided into anterior and posterior regions. Endocrine functions of the anterior pituitary are regulated by releasing neurohormones that are carried in the blood from the

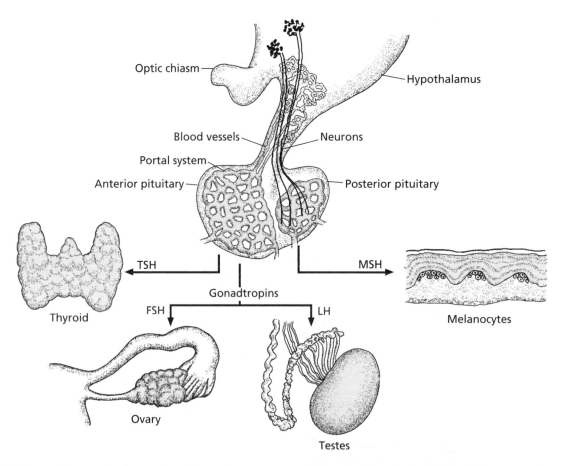

Figure 8.6 Mammalian endocrine system. Shown here are some of the endocrine glands and the influences on them from the two major regions of the pituitary (anterior and posterior). In particular, the systems shown are either unique to mammals or specialized in mammals, for example, the thyroid gland and its control of homeothermic metabolism. *Abbreviations:* TSH = thyroid-stimulating hormone; FSH = follicle-stimulating hormone; LH = luteinizing hormones; and MSH = melanocyte-stimulating hormone. *Adapted from Enger and Ross, 2000,* Concepts in Biology, *7th ed., McGraw-Hill.*

hypothalamus (see figure 8.6). The cells of the anterior pituitary are neurons that have been modified through evolution to serve a slightly different function, integrating the nervous and endocrine systems. Neurons that pass from the hypothalamus to the posterior pituitary release hormones (neurosecretions) directly into that region. These neurosecretions, in turn, influence the production and release of hormones into the bloodstream passing through this anterior portion of the pituitary. Thus, the hypothalamus controls the pituitary gland, and in turn, the pituitary gland and the hormones it releases play critical roles in regulating the endocrine processes of many tissues and organs of the body (see figure 8.6).

GLANDS AND HORMONES

Endocrine glands are specialized groups of cells that produce chemical substances that are released into the bloodstream. These substances, called **hormones**, act as messengers as they travel via the blood to cells throughout the body. Mammals possess most of the same types of peripheral endocrine glands as other vertebrates, and they have both steroid and protein hormones (see figure 8.6). The mammalian endocrine system has some unique aspects, however. Some hormones that mammals share with other vertebrates have taken on specialized functions in mammals associated with reproduction, including ovulation, pregnancy, and lactation (see chapter 10). These hormones include follicle-stimulating hormone (FSH), luteinizing hormone (LH), estrogen, testosterone, prolactin, progesterone, and oxytocin. The reaction to stress, involving adrenal gland hormones, has been studied extensively in some mammals, with particular regard to the effects of stress on reproduction and the immune system (see chapter 25). Hibernation and kidney function also involve specializations of the endocrine system that are unique to mammals (see chapter 9). In addition to the processes covered in other chapters, the endocrine system significantly affects mammalian physiology and behavior in other ways.

FUNCTIONAL ASPECTS

Because mammals are endothermic homeotherms, hormones are important for regulation of metabolic processes. The thyroid gland is the key modulator of metabolic activity. Through its hormone, thyroxin, and related compounds, the thyroid triggers increased cellular respiration in target tissues, resulting in additional heat production. Cold-adapted individuals have higher rates of thyroid output than those that have not been exposed to cold. The insulin-glucagon hormone system is responsible for maintaining the homeostasis of the blood carbohydrate levels. Insulin acts to lower blood glucose by promoting the conversion of glucose to glycogen and lipids. Glucagon has **antagonistic** (opposite) **actions** to insulin. It raises blood

glucose levels by promoting mobilization of fat reserves. An additional hormone that may be unique to mammals is calcitonin, which was not discovered until 1961. This fast-acting hormone, released from the parathyroid-thyroid complex, results in lower blood levels of calcium. This action is the opposite of that for parathyroid hormone, which raises blood calcium levels.

Hormones play key roles in many mammalian behavior systems (Nelson 2005). Some mammalian chemical communications systems (see chapter 21) rely on hormones. **Priming pheromones** (chemical substances that trigger generalized internal physiological events, such as the production and release of hormones) can function to regulate puberty in some rodents and are mediated by hormonal conditions in the donors (Vandenbergh 1983; Vandenbergh and Coppola 1986). The puberty-accelerating pheromone in male mouse urine depends on the testosterone level (Lombardi and Vandenbergh 1977). The puberty-delaying pheromone present in the urine of grouped females depends on adrenal glucocorticoids (Drickamer and Shiro 1984). Chemosignals also influence reproduction in larger mammals; for example, the urinary and vaginal compounds released by white-tailed deer (*Odocoileus virginianus*; Murphy et al. 1994) can act as sexual attractants and signal reproductive condition. Similarly, in Eld's deer (*Cervus eldi*), exposure of hinds to stag urine results in greater release of progesterone as measured in the hinds' feces, possibly increasing fertility (Hoscak et al. 1998).

Aggression in both male and female mammals is mediated in part by hormones. Higher levels of aggression in males are sometimes associated with higher levels of testosterone in some mammals, although this relationship is not true for all mammals under all conditions (Brain and Poole 1976; Barkley and Goldman 1977; Sapolsky 1997). The maternal aggression exhibited by female mammals around the time of parturition is mediated in part by prolactin (Svare and Mann 1983). Various aspects of behavior, including reproduction, are influenced by prenatal exposure to hormones and are called **organizational effects** (vom Saal 1979, 1989). In particular, in species with multiple births, female embryos positioned *in utero* between two male embryos are masculinized, which affects their behavior and reproductive success (Drickamer 1996). Other behavior systems are also under varying degrees of hormonal regulation.

Immune System

Organization and Function

The bodies of mammals have four lines of defense from foreign substances, which include bacteria, viruses, and nonliving materials. The first three defense mechanisms are nonspecific, occurring at any location where there is a threat. The outer barrier is composed of the skin and the linings of the digestive and respiratory tracts; these are mechanical barriers. A second line of defense involves the inflammatory reaction. A wound site swells and becomes red due to the white blood cells that accumulate to immobilize and then destroy bacteria or a small object, such as a splinter, that has penetrated the outer line of defense. The third defense system, or complement, involves proteins that are activated at wound sites. These proteins bring about holes or pores in the membranes of invading bacteria such that they swell and burst. The proteins also release chemicals that attract macrophages (a type of white blood cell) to the site to aid in destruction of the invader.

A fourth defense, the immune reaction, occurs when the body recognizes something as non-self, for example, a virus or bacterium, also called an antigen. Lymphocytes can produce specific antibodies that seek out and form complexes with the various types of antigens. The complex, in effect, results in the deactivation of the antigen, removing any potential deleterious effects on the animal from the foreign agent. These complexes are then engulfed and destroyed by other macrophages. When a mammal is vaccinated, it is given a small dose of a particular antigen, often in dead or attenuated form, so that the animal's immune system is triggered to produce antibodies. The immune system can "remember" this particular pathogen so that if the animal is later exposed to the live virus, a substantial immune response protects the animal. Due to the vaccination, the time to respond is shortened greatly, reducing the chance of a major infection from the bacterium or virus.

The immune system regulates certain body functions, most notably processes related to defenses against foreign materials that are threats to the body. The immune system of mammals is more complex than similar systems in other vertebrates and consists of several parts (figure 8.7).

The lymphatic system, which contains key elements of the immune system, comprises a series of one-way collecting vessels that serve to gather the fluids that have leaked from the circulatory system. As blood is squeezed through capillary networks, fluids pass out into the surrounding tissues. The lymphatic system returns those fluids to the closed circulatory system. A series of lymph nodes are located at junctions of vessels in the lymphatic system. The principal organs in the system are the thymus, spleen, and bone marrow. Lymph nodes function both as the site for production of the antibodies that comprise the defense mechanism portion of the immune system, and as locations where, in many cases, macrophages do their work by destroying antigen-antibody complexes.

Lymph nodes filter out dead cells, debris, and foreign materials in the lymphatic system as fluids are passed back toward the circulatory system. The thymus and spleen also cleanse the blood and lymph as fluids pass through these structures. Five types of white blood cells and their functions are (1) neutrophils, which serve to phagocytize, or consume, foreign particles, primarily bacteria; (2) several types of lymphocytes, which produce antibodies or

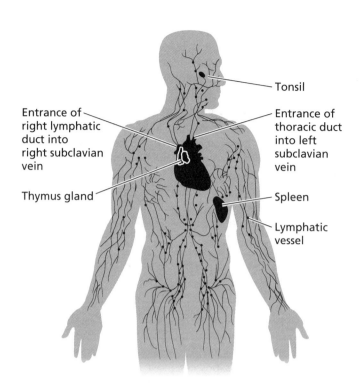

Figure 8.7 Mammalian immune system. The mammalian immune system consists of lymphatic ducts that return fluids to the circulatory system and a series of lymph nodes. Prominent among the lymph nodes are the tonsils, spleen, and thymus.

kill cells containing viruses; (3) monocytes, which convert to macrophages and act to consume bacteria and viruses; (4) eosinophils, which destroy and consume antigen-antibody complexes; (5) basophils, which congregate in tissues when there is an infection or inflammation and release histamine. Histamines are proteins that function to widen blood vessels with release of more fluids into the surrounding tissues, often leading to edema or swelling.

Biological Rhythms

Biological rhythms are patterns of activity that occur at regular intervals during the life of an animal. **Circadian rhythms** have a period of about 24 hours. **Circannual rhythms** have a period of about a year. **Ultradian rhythms** are those with a period of less than 24 hours. Because these cyclical patterns govern many of the activities of mammals, we examine them both in terms of their functional significance and the underlying mechanisms.

CIRCADIAN

Probably our most common perceptions about activity patterns in mammals relate to their cycle of daily activity.

Some mammals are **diurnal**, meaning they are active primarily during daylight hours and quiescent at night; others are **nocturnal**, exhibiting peak activity during hours of darkness and resting when there is daylight (Dunlap et al. 2003; Koukkari and Southern 2006). Some mammals, including types that are primarily diurnal or nocturnal, exhibit **crepuscular** activity, moving about more and feeding near sunrise and sunset. The primary cue for entrainment, or establishment and modification, of circadian rhythms (called a *Zeitgeber* or time-giver) is photoperiod. The photoperiod sets and changes the biological clock.

Activity patterns of about a day govern many aspects of the lives of mammals, just as they do for other animals. Members of the same mammalian species feed, sleep, mate, give birth, and disperse at about the same time of day. Various factors influence the timing of the activity phase. For mammals on which predation pressure is high, being nocturnal may reduce their chance of being detected. Alternatively, for some predatory mammals, such as insectivorous bats, peak activity may occur near dusk and just after nightfall, when large numbers of insects are active. Most mammals tend to give birth during the inactive phase of the daily cycle. Being able to move is necessary for avoiding predation (figure 8.8). For mammals that live in migratory herds, such as black wildebeest (*Connochaetes gnou*), giving birth during the inactive phase means that newborn calves have sufficient time within several hours after birth to become mobile enough to travel with the group. Nocturnal rodents that give birth during daylight hours in their burrows or nests can still forage during their normal nighttime activity phase. Obtaining sufficient resources just prior to parturition and at the onset of lactation may be critical for reproductive success.

Early mammals probably were nocturnal, and many of those living today are still nocturnal. Others have adopted a diurnal activity rhythm. Being active at different times of

the day has both advantages and disadvantages. Being active at night decreases the chances of heat stress and of being seen and preyed on. In addition, because the air is more humid, olfactory communication is enhanced. Social interactions involving visual communication are reduced at night. Competition for food may also be reduced with, for example, birds, which are primarily diurnal. Diurnal mammals, on the other hand, are better able to use vision to forage, and they can use complex visual communication. Suffering heat stress, being seen by potential predators, and competing for food can pose problems for diurnal mammals. They therefore tend to be larger than nocturnal mammals and live in larger, more complex social groups, on average, than mammals that are active primarily at night.

At least one mammalian evolutionary sequence may have revolved around circadian rhythms. Fossil history suggests that rodents evolved nearly concurrently with early primates from insectivore-like precursors and soon became dominant in the terrestrial habitat. Competition for food and nesting sites between rodents and early primates may have brought about two of the key changes that characterize primate evolution: a shift to a diurnal activity phase and adoption of a more arboreal existence (Simons 1972). Shifting activity to the trees may also have led to selective advantages for a more diurnal activity pattern because locomotion and foraging in the trees are safer during daylight. This is only one of several explanations for the evolution of primates as arboreal, diurnal animals, but it illustrates the possible importance of circadian activity patterns over the course of evolutionary time.

Although the mechanisms are not completely known, considerable progress has been made in the last decade in deciphering those mechanisms that regulate circadian rhythms in mammals. The timing mechanism appears to have three major components located in (1) the retina, (2) a portion of the thalamus in the brain called the intergeniculate nucleus, and (3) the suprachiasmatic nucleus (SCN) of the hypothalamus. Light striking the retina follows several pathways, two of which lead to the SCN and intergeniculate nucleus (Moore 1973; Harrington et al. 1985). One pathway from the intergeniculate nucleus then leads to the SCN. Animals with **ablation** (destruction by electrical or chemical techniques) of the SCN exhibit irregular cycles, implicating the SCN as the putative primary oscillator in mammals (Rusak and Zucker 1979; Moore 1982). Additional investigations that reveal spontaneous neural activity in the SCN, as well as the activity of neurotransmitters in this brain region, provide further evidence for the SCN's role as the primary pacemaker (Gillette et al. 1993; Glass et al. 1993). Recently, these findings were extended by the discovery that a chemical factor (TGF-X), which is released rhythmically by the SCN, inhibits locomotor activity, and disrupts circadian wake-sleep cycles (Kramer et al. 2001). Although the pineal gland does not appear to be a primary component of the timing mechanism, its removal (pinealectomy) in rats alters some aspects of circadian rhythms. These findings suggest that

Figure 8.8 Timing of parturition. Many mammals give birth to their young during the resting or sleep portion of their daily cycle. In this instance, a newly born wildebeest has a few hours to strengthen its wobbly legs so it can stand and move with the herd.

the pineal gland, via its secretion of melatonin, may connect various circadian processes based on the clock located in the SCN (Warren and Cassone 1995).

CIRCANNUAL

Circannual rhythms regulate a number of processes in mammals. As we will see in the next chapter, the annual process of hibernation is a key adaptation for some species that inhabit colder climates. The annual pattern of breeding in many mammals, particularly in temperate climates and at more northern latitudes, is also governed by a circannual rhythm. In tropical climates, the reproductive cycles of some mammals respond to rainfall and the resulting changes in vegetation as well as to population numbers of other animals. These resources provide the food necessary to reproduce and to support progeny when they are weaned. Some species of bats migrate each year between a summer breeding location and a winter hibernation location. The ultimate mechanisms behind circannual rhythms in mammals likely can be tied primarily to food resources and secondarily to climatic conditions.

Like circadian rhythms, the primary cue for annual patterns is photoperiod. Hormonal changes are cued by photoperiod, resulting in alterations of physiology, morphology, and behavior. For example, territorial behavior in golden jackals (*Canis aureus*) varies seasonally (Jaeger et al. 1996), and circannual patterns of reproductive hormones are related to shifts in testicular parameters in male southern pudu (*Pudu puda*; Reyes et al. 1997). For some mammals, including most small mammals in North America, an increasing photoperiod is correlated with these events, and reproduction begins. In other instances, a decline in photoperiod correlates with the fall mating

season, with young born the following spring, such as occurs with white-tailed deer or Rocky Mountain bighorn sheep (*Ovis canadensis*).

Seasonal changes in coat color (pelage dimorphism) are also triggered by circannual cycles. Shifts, for example in weasels (*Mustela* spp.) and snowshoe hares (*Lepus americanus*; figure 8.9), between a white winter coat that matches the snow-covered landscape and a darker coat in summer are critical to these animals' survival as predators or prey. The weasel's darker pelage of summer and white coat of winter each provide the camouflage necessary for stalking and preying on small mammals that constitute its diet. The snowshoe hare, on the other hand, needs its white winter coat as a means of preventing detection by predators. That same winter coat would stand out against a darker background of vegetation during summer months.

For some mammals, factors other than photoperiod can act as the primary cue that triggers events during an annual cycle. In the hotter, more humid climates of the southern United States, some species of small mammals, such as the beach mouse (*Peromyscus polionotus*), breed in the fall, when temperatures have declined from summer highs (Blair 1951; Drickamer and Vestal 1973). In other mammals, annual patterns of reproduction may be linked to certain plant compounds. For example, reproduction in montane voles (*Microtus montanus*) is triggered by a compound found in some fresh green plants (Negus and Berger 1977; Berger et al. 1987). The compound 6-methoxybenzoxazolinone (6-MBOA) may cue the start of reproduction in the spring, when fresh plant shoots begin to grow. Both the number and size of litters produced are influenced by 6-MBOA (Berger et al. 1987; table 8.1). Finally, social cues can be important with regard to hormones and seasonal patterns of reproduction. The social activities of conspecifics, often regulated by hormonal

A

B

Figure 8.9 **Pelage color changes.** Mammals that live in more northern latitudes often change their pelage color during the course of an annual molt. The snowshoe hare is white in winter and darker in summer, providing some degree of camouflage by nearly matching the color of the background. This annual shift in pelage color is regulated by external photoperiod and mediated by internal hormonal events.

events, may, for example, synchronize reproductive activity among members of a population of rhesus macaques (*Macaca mulatta*) (Vandenbergh 1969a; Vandenbergh and Drickamer 1974). The interactive roles of photoperiod and various environmental cues that influence hibernation have been the subject of a number of studies during the past decade in ground squirrels (Genus *Spermophilus*) and chipmunks (Genus *Tamias*) (Kawamichi 1996; Buck and Barnes 1999a,b; French 2000; Neuhaus 2000).

Photoperiod is the primary cue for circannual rhythms in most mammals because over many thousands of years, the annual cycle of changes in photoperiods is more stable and consistent than any other environmental cue. Changes in climatic conditions, for example, are more variable, and thus cueing on such events could lead to wasted energy. Through evolutionary time, photoperiod became the primary cue, and climate and changes in food resources (e.g., vegetation or the supply of insects) have come to be secondary cues. Some evidence for the effect of photoperiod on annual cycles comes from the work of Davis and Finnie (1975). Woodchucks (*Marmota monax*) from the northern hemisphere were sent to Australia. There, they entrained on some feature of the environment and eventually reversed their annual pattern, exhibiting hibernation at the appropriate time in the southern hemisphere, possibly cued by changes in photoperiod. Additional evidence for the importance of how changing photoperiod can affect circannual cycles comes from work by Heller and Poulson (1970) on two species of ground squirrels and four species of chipmunks. Individuals of several species of each genus were held under constant ambient and photoperiod conditions in the laboratory. They maintained circannual rhythms under these conditions even though various proximate cues (e.g., diet) were sometimes manipulated. The evidence is correlational, and a causal link between photoperiod and circannual rhythms in mammals has not yet been fully demonstrated.

Davis (1991) tested the possible effects of irradiance (light intensity) on the entrainment of circannual rhythms in California ground squirrels (*S. beecheyi*). Light intensity varies with the angle of the sun, and irradiance varies with latitude. Irradiance therefore could be an excellent predictor of entrainment of annual cycles. Davis held California ground squirrels in captivity inside a building but exposed them to the seasonal cycle of irradiance characteristic for Santa Barbara, California. The squirrels subsequently exhibited typical circannual patterns of weight gain and torpor. The possibility that irradiance is a key *Zeitgeber* for entrainment of circannual rhythms in mammals deserves further attention.

Hibernation is the primary phenomenon for which the underlying neural and hormonal correlates for circannual rhythms have been studied (Kayser 1965). The pineal gland plays a central role as a photoendocrine transducer in mammals (Reiter 1980; Gwinner 1986). Research has yet to reveal a critical role for this gland in circannual rhythms, however. Pinealectomies in ferrets (*Mustela putorius*) and golden-mantled ground squirrels (*Spermophilus lateralis*) did not eliminate circannual cycles (Herbert 1972; Zucker 1985), but the annual cycles of these animals were not synchronized with changes in photoperiod. Similarly, **lesions** (destroying an area of brain tissue via electrical or chemical means) in the central nervous system have not demonstrated that any particular structures are essential for circannual rhythms, although such treatments produce variations in the annual pattern. These studies have included lesions in brain areas, such as the SCN, that are essential to circadian rhythms (Zucker et al. 1983). Findings to date suggest couplings between the mechanisms mentioned earlier for circadian periodicities and those involved in annual patterns, but additional studies are needed to further elucidate the proximate mechanisms underlying circannual rhythms. One problem associated with research on these rhythms is the long time needed for data collection; waiting several years to obtain results from tests designed to determine whether a particular treatment has influenced rhythmic patterns is difficult and costly.

ULTRADIAN

Some mammals have regular cycles of activity with periods that are shorter than a day in length. Many microtine rodents exhibit 6–12 evenly spaced, short bouts of activity and rest during a 24-hour day (Daan and Slopsema 1978; Madison 1985; figure 8.10). Halle (1995) studied this phenomenon in tundra voles (*Microtus oeconomus*). Until recently, the primary methods for exploring ultradian rhythms involved either live trapping with trap checks made at very short intervals or radiotelemetry. Both of these methods, however, can possibly interfere with the ongoing activities of the animals. Halle's method used passage counters (Mossing 1975)—small tubes (7 cm wide by

Table 8.1   6-MBOA affects reproduction in pairs of montane voles (*Microtus montanus*) given either a control treatment or one of three doses of 6-MBOA*

Treatment (µg)	Total Number of Offspring	Sex Ratio (Male: Female)
Control	218	1.00:0.70
0.1	262	1.00:0.97†
1.0	272	1.00:1.08‡
10.0	285	1.00:1.02§

Source: Data from P. J. Berger et al., 1987, Effects of MBOA on Sex Ratio and Breeding Performance in Microtus montanus, *in* Biol. Reprod. 36:255–260.
* The affects differed significantly with respect to both the total number of pups produced and the sex ratio. The sex ratio differences were computed using chi-square analyses.
†Value differs from the control at p = .08.
‡Value differs from the control at p = .02.
§Value differs from the control at p = .04.

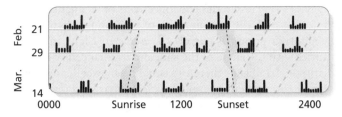

Figure 8.10 Ultradian rhythms. The activity rhythm of a female meadow vole (*Microtus pennsylvanicus*) weighing 30 g and fitted with a radiotransmitter collar. The animal was free-ranging in a field covered with 1 m of snow in Quebec, Canada. The vertical axis represents movement (in meters) at 15-minute intervals of detection. Spacing between the 3 days depicted reflects the time intervals between those days. The pair of dotted lines represents time of sunrise and sunset. Dashed lines represent ultradian periods of activity of about 4 hours duration and with an underlying circadian period of 23.85 hours/day.

5 cm long) placed in the runways used by these animals. A small gate in the center of the tube is attached to a microswitch. When a vole passes through the tube on its way along a runway, the switch closes, tripping an event recorder. The method does not permit individual identification but gives a better overall picture of undisturbed activity rhythms. The method also depends on the animals' using the runways, something done very regularly by voles but less often by other small mammals.

Halle (1995) found that root voles have about seven activity periods per day. Diurnal periods are 3.0–3.5 hours in duration, whereas nocturnal periods are 3.5–4.0 hours long. The voles were more active during daylight than at night but showed the highest levels of activity near sunrise and sunset. Changing seasons (shifts in daylength) resulted in changes in the ultradian patterns as the animals' internal clocks were reset by sunrise and sunset. The significance of ultradian rhythms in rodents most likely involves their metabolic needs and related foraging activities. Curiously, these ultradian activity bouts are synchronous within family groups (Mossing 1975; Webster and Brooks 1981), a phenomenon that has yet to be explained.

Ultradian rhythms are not restricted to rodents. Rhesus macaques exhibit regular short bouts of social activity between periods of more self-directed behavior (Maxim et al. 1976). Because these mammals are diurnal, the ultradian pattern was observed only during the active daylight portion of the daily rhythm. Periods for these ultradian patterns were generally 40–45 minutes in duration. The underlying mechanism for this ultradian rhythm in these monkeys may relate to an oscillator that affects cycles of rapid eye movement (REM) sleep (Kripke 1974).

With ongoing research on circadian rhythms rapidly producing results, it seems likely that attention may focus on ultradian rhythms in the future. Not only are new technologies likely to result in better detection of such rhythms, but we are also likely to find that such patterns have heretofore unrecognized functional significance. With an increased understanding of the proximate mechanisms for circadian rhythms, it may soon be possible to explore the underlying evolutionary causes of ultradian rhythms.

SUMMARY

Internal processes in mammals, like other vertebrates, are regulated by the nervous, endocrine, and immune systems. For many mammals, the senses of hearing and olfaction are most important. For others, vision is also key. Mammals hear and use sound over a wide range of wavelengths from 20 Hz to more than 20,000 Hz. Taste is of lesser importance for most mammals. Touch receptors are most highly developed in fossorial species.

The most obvious feature of the mammalian brain is the relative increase in the neocortex. In general, brain size has increased progressively both in absolute terms and relative to body mass for larger versus smaller mammals. Nerve cells in the cerebrum are associated with two properties that help to define the uniqueness of mammals: storage of information and associations made between events, both past and present. Mammals have more types of neurotransmitters than other vertebrates do.

The mammalian endocrine system is characterized by control from the pituitary gland and its connections with the hypothalamus. Specialized functions have evolved in mammals for hormone systems associated with reproduction, hibernation, kidney function, metabolic processes, and a number of behavioral systems. Chemosignals excreted in urine that affect reproductive physiology, aggression, and developmental effects involving intrauterine position are all examples of processes that are regulated, in part, by hormones.

The immune system forms part of the mammal's defense against external agents, including viruses and bacteria. Three nonspecific defense systems utilize the skin as a barrier, inflammation, and proteins known as complement. Specific antibodies are generated by specialized cells of the immune system to bind with antigens (anything the body recognizes as non-self) to form antigen-antibody complexes. The cells of the immune system phagocytize dead cells, foreign particles, and antigen-antibody complexes.

Biological rhythms, including circannual, circadian, and ultradian patterns, govern a number of life history features of mammals. Daily rhythms include cycles of activity and sleep, feeding behavior, mating, and the timing of births. The control of these rhythms involves the retina, the suprachiasmatic nucleus, and part of the thalamus. Circannual patterns are evident in hibernation, seasonal mating, and, for a few species of mammals—particularly bats and some cetaceans—migration. Changes in photoperiod are the most consistent cues for the establishment and maintenance of circannual patterns in mammals, although light intensity may be a good *Zeitgeber*. Other factors that affect annual cycles include climate, chemicals in plants, and social influences. Ultradian rhythms have received less attention than other mammalian rhythms. These short cycles, occurring within each daily pattern, are characteristic of some small mammals, as they awake, become active, feed, and return to rest at intervals of several hours.

SUGGESTED READINGS

Brown, R. E. 1994. An introduction to neuroendocrinology. Cambridge Univ. Press, New York.

Camhi, J. M. 1984. Neuroethology. Sinauer Assoc., Sunderland, MA.

Drickamer, L. C., S. H. Vessey, and B. Jakob. 2002. Animal behavior: Mechanisms, ecology, and evolution, 5th ed. Wm. C. Brown, Dubuque, IA.

Nelson, R. J. 2005. An introduction to behavioral endocrinology, 3rd ed. Sinauer Assoc., Sunderland, MA.

DISCUSSION QUESTIONS

1. Expansion of the cerebral hemispheres in mammals has resulted in more complex learning processes than those found in most other vertebrates. After consulting an appropriate book on the subject (see "Suggested Readings" in this chapter), summarize the differences in learning capacities between mammals and other vertebrates.

2. Using information from this and other chapters, summarize the integration of neural and hormonal mechanisms underlying these adaptational systems in mammals: (a) hibernation; (b) response to stress; and (c) female reproduction.

3. Although daylength is a primary cue for mammals with respect to seasonal activities and breeding in particular, other more proximate environmental cues can act as secondary triggers. Select any five species of mammals from different regions of the world and, after learning a little about their environmental conditions, suggest which environmental factors might be important secondary cues for each species.

4. Draw a triangle and label the three corners Nervous, Endocrine, and Immune. Using your knowledge from this textbook and general biology, write down the ways in which these three major body regulatory systems interact with one another.

5. What is the importance of the comparative method with regard to studies of mammalian nervous systems, endocrine systems, and biological rhythms? What are some of the possible advantages and disadvantages of using this method? (Excellent references for this method are Gittleman 1989 and Harvey and Pagel 1991.)

CHAPTER 9

Environmental Adaptations

Land mammals reside in environments characterized by extremes in temperature, precipitation, and altitude. Marine environments also vary greatly in terms of water pressure at different depths. Different terrestrial mammals may be faced with temperatures ranging from close to –65°C in the Arctic to 55°C in Death Valley, California. Daily fluctuations during winter can be extreme; for example, in Montana, the temperature in one day can plummet from 6°C to –49°C. Summer is no relief for mammals, and a hot spell in western Australia may reach 38°C and last for 162 continuous days! Although marine mammals live in water, they also experience temperature variation ranging from –2°C near the poles to 30°C near the equator. So how do mammals cope with such extremes? This chapter focuses on the diversity of mammalian adaptations that enhance survival in variable environments.

For many years, the terms **warm-blooded** and **cold-blooded** were used to conveniently divide animals into two groups: the vertebrates (typified by high, constant body temperature) and the invertebrates (typified by variable body temperatures). These terms are inaccurate and unscientific, however, as well as being unclear. Thermoregulatory terminology can be tricky and ambiguous, so it should be well defined (Hainsworth 1981; Bartholomew 1982; Hill and Wyse 1989; McNab 2002; Pough et al. 2005). The ability of mammals to colonize inhospitable environments is due largely to their ability to use **endothermy**—the maintenance of a relatively constant body temperature by means of heat produced from *inside* (*endo*) the body. The degree of endothermy varies in both space and time. No animal maintains the same body temperature over all parts of its body. The term **homeothermy,** the regulation of a constant body temperature by physiological means, is more precise than endothermy (Hill and Wyse 1989:99). The term **ectothermy** refers to the determination of body temperature primarily by sources outside (ecto) the body. This is not necessarily a passive system, and many ectotherms employ behavioral means to regulate their body temperature; for example, lizards will bask in the sun or rest on a warm rock. A similar term, **poikilothermy,** emphasizes the variation in body temperature under environmental conditions, as opposed to focusing on the mechanism by which the body temperature is maintained. Some examples of poikiotherms are clams, starfish, and many other invertebrates. Mammals and birds that are capable of dormancy—

hibernation, daily torpor, and estivation—are referred to as **heterothermic.** In such cases, regulation of body temperature may vary on different parts of the body (regional heterothermy) or at different times (temporal heterothermy). Many species of bats, for example, exhibit temporal heterothermy: They maintain constant body temperature while foraging but permit their body temperature to approach ambient temperature when they rest.

Maintaining an internally regulated body temperature offers numerous benefits for mammals. Both mammals and birds have high metabolic rates—at least eight times that of ectotherms. In terms of energy expenditure, maintaining such high body temperatures plus high levels of activity is costly, but it has the advantage of enhancing coordination of biochemical systems, increasing information processing, and speeding central nervous system functions. As a result, mammals have a refined neuromuscular system, thus enhancing their efficiency at capturing prey and escaping from predators. Mammals gain independence from temperature extremes in nature, can extend activity periods over a 24-hour period, and colonize many environments and ecological niches throughout the world. Mammals can match their thermoregulatory pattern to suit a given environment and thus take advantage of nutritional resources year-round.

Endothermy is thought to have evolved from an ectothermal condition probably two to three times in the Mesozoic era. Internal heat production probably evolved in the late Triassic within the group of mammal-like amniotes in response to selective pressure favoring sustained activity and temperature regulation. The evolution of endothermy in mammals has been the subject of considerable debate (Bennett and Ruben 1979; Farlow 1987; Hayes and Garland 1995; Grigg et al. 2004; Kemp 2006).

Heat Transfer Between a Mammal and the Environment

Most placental mammals residing in thermally benign environments maintain their body temperature between 36° and 38°C (Cossins and Bowler 1987). In general, the body temperature of birds is several degrees higher than that of mammals. Monotremes and marsupials, on the other hand, may have a core body temperature ranging from 30° to 33°C, but they thermoregulate quite well. To keep a constant body temperature, mammals must maintain a delicate balance between heat production (energy in) and heat loss (energy out)— thermodynamic equilibrium—to survive (Porter and Gates 1969). Heat is produced through metabolism of food or fat, cellular metabolism, and muscular contraction. The factors affecting the exchange of energy between a mammal and the environment are sunlight (solar radiation), reflected light, thermal radiation, air temperature and movement, and the pressure of water vapor of the air (figure 9.1). The properties of a mammal that influence the exchange of

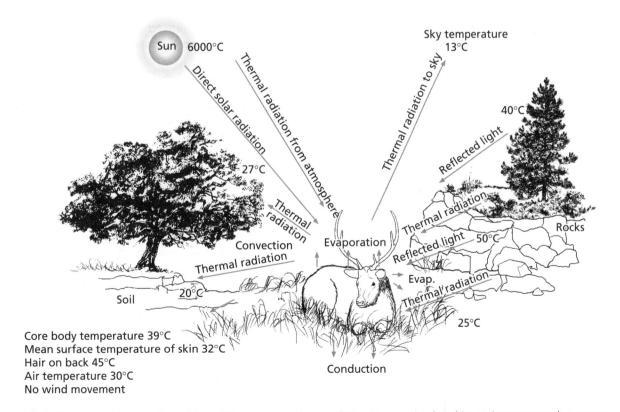

Figure 9.1 Energy exchange. Illustration of the energy exchanges between an animal and its environment under warm conditions. *Adapted from M. S. Gordon et al. (eds.), 1982,* Animal Physiology: Principles and Adaptations, *4th ed., Macmillan Publishing Company.*

energy are metabolic rate, rate of moisture loss, thermal conductance of fat or fur, absorptivity to radiation, and the size, shape, and orientation of the body. Most of the heat produced by mammals is lost to the environment passively by **radiation, conduction,** and **convection** (air movement) to a cooler environment and by the **evaporation** of water. Mammals must adjust their energy balance to meet different demands in their environment.

Temperature Regulation

As mentioned in chapter 8, mammals regulate body temperature by continuously monitoring outside temperatures at two locations: on the surface of the skin and at the hypothalamus. The hypothalamus, or mammalian "thermostat," is located in the forebrain below the cerebrum and operates by comparing a change in body temperature with a reference, or set point, temperature. Each species may have a different **set point,** or comfort zone, which is set at the hypothalamus. For most eutherian mammals, heat is generated by muscle contraction (shivering), brown fat (nonshivering thermogenesis), and activity of the thyroid gland. In marsupials and monotremes, heat production is due primarily to the activity of skeletal muscle (Augee 1978; Nicol and Andersen 1993). For mammals residing in variable environments, body temperature is maintained around a given set point, with each mammal's set point determined by such factors as insulation, behavioral postures, activity levels, and microclimatological regimes.

Adaptations to Cold

Most eutherian mammals maintain a core body temperature of about 38°C. Monotremes and marsupials tend to show lower body temperatures, with echidnas and platypuses exhibiting normal body temperatures ranging from 28° to 33°C (Augee 1978; Grigg et al. 1989, 1992). For each species, a range of environmental temperatures, referred to as the **thermoneutral zone,** occurs within which the metabolic rate is minimal and does not change as ambient temperature increases or decreases. The upper and lower limits of the thermoneutral zone are referred to as the upper and lower critical temperatures (figure 9.2). Decreasing environmental temperatures require that an animal increase its metabolic rate to balance heat loss. The temperature at which this becomes necessary is called the **lower critical temperature.** This temperature varies from species to species and is seasonally adjusted by the interplay of insulatory thickness, behavioral attributes, and integration with the hypothalamus. As environmental temperatures decrease, adjustments in **thermal conductance** (the rate at which heat is lost from the skin to the outside environment) and

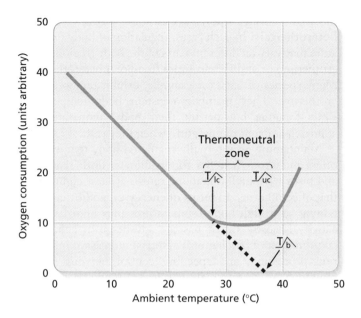

Figure 9.2 **Thermoneutral zone.** Relationship of oxygen consumption compared with ambient temperature in a hypothetical mammal. T_{lc} = lower critical temperature; T_{uc} = upper critical temperature; T_b = core body temperature. *Data from G. A. Bartholomew, 1982, Body temperatures and energy metabolism, in* Animal Physiology: Principles and Adaptations, *4th ed. (M. S. Gordon et al., eds.), Macmillan Company.*

metabolism are necessary if a species is to maintain euthermy. If this is not possible, death due to **hypothermia** (low body temperature) results. Thermal conductance (C) is expressed as the metabolic cost (MR), expressed in milliliters of oxygen per gram of body mass, for a given time interval per degree Celsius difference between body temperature (T_b) and environmental temperature (T_a)

$$C = MR/T_b - T_a$$

(McNab 1980b; Hill and Wyse 1989). Some mammals and birds from polar regions are so well insulated that they can withstand the lowest environmental temperatures on earth (about –70°C; figure 9.3) by simply increasing their resting metabolic rate (Scholander et al. 1950; Irving 1972). As we will learn later, however, some species can significantly decrease body temperature (undergo hibernation or torpidity) to combat seasonal decreases in food. To maintain body temperature within the thermoneutral zone when residing in cold environments, the individual must reduce the rate of heat loss to the environment. Mammals can achieve this reduction in many different ways: by being larger, possessing enhanced body insulation, changing peripheral blood flow, or modifying behavior (e.g., curling, **piloerection** [the fluffing of fur], nest building, and huddling in groups). By modifying the pelage, for example, an animal can reduce the gradient of heat flow. Do not forget that hair is a principal means of conserving energy (see chapter 6). Hair is a poor thermal conductor, so it greatly decreases the amount of body heat lost to the environment by decreasing thermal conductance (Speakman and Thomas 2003).

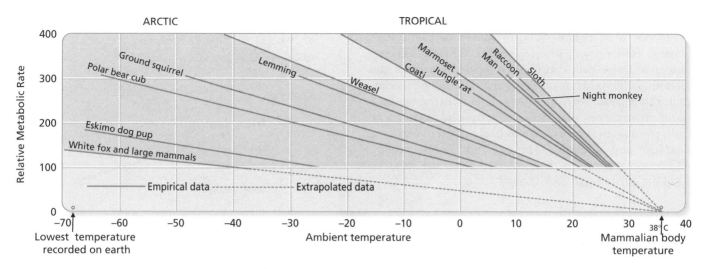

Figure 9.3 Metabolism and temperature. Metabolic rates of various mammals in relation to ambient temperature. The normal resting metabolic rate for each animal, in the absence of cold stress, is given the value 100%. Any increase at lower temperature is expressed in relation to this normalized value, making it possible to compare widely different animals. *Data from P. F. Scholander et al., 1950, Body insulation of some arctic and tropical mammals and birds,* Biological Bulletin *99:225–235.*

At the other end of the thermoneutral zone is the **upper critical temperature,** the temperature above which animals must dissipate heat by evaporative cooling to maintain a stable internal temperature. This zone is important to mammals residing in desert environments and is less variable than the lower critical temperature. Mammals employ many different mechanisms, such as seeking shelter in cooler underground burrows and restricting surface activity to hours of darkness, to avoid upper critical temperatures. Furthermore, evaporative cooling can be facilitated by pulmonary means, panting, sweating, spreading saliva, or using a form of dormancy called estivation. If a mammal cannot combat an increase in temperature by dissipating heat through evaporation or by muscular activity of panting, its body temperature will keep increasing, ultimately resulting in death. The mechanisms used to cope with heat are examined later in the chapter.

ENERGY BALANCE IN THE COLD

Winter is a potentially stressful time for northern mammals (and for mammals from extreme southern latitudes), and they employ a wide array of adaptations to cope with the food shortages and cold stress that the season brings. The key to survival in the cold is to maintain energy balance (Wunder 1978, 1984). Two major features determine the energy needs of mammals in the natural environment: the *physical environment* and the *activities* of the mammal (i.e., type and level of behavior, growth, or production of young; figure 9.4). To meet their energy needs, mammals must acquire energy—they must feed. Given a finite source of food during winter, mammals allocate energy initially to thermoregulation. Temperature regulation is essential for mammals to perform normal functions. For example, a mammal may freeze to death even with abundant food if it cannot assimilate that energy and produce heat fast enough

to balance heat loss (turnover capacity). Also, even a mammal with a high heat-generating capacity will perish if there is not enough food (energy) available in the environment to meet its needs. In sum, mammals residing in cold regions must have sufficient food available and sufficient turnover capacity to meet their thermoregulatory needs.

Avoidance and Resistance

Unlike many species of passerine (perching) birds inhabiting cold regions, most mammals do not undertake long migrations. They remain as residents during winter

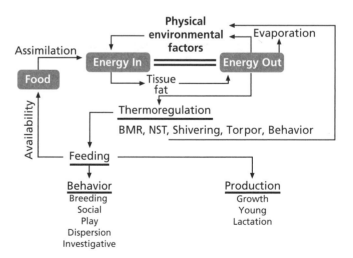

Figure 9.4 Energy balance. Conceptual model of avenues of energy balance for a small mammal, indicating a primary cascade for energy allocation. Lines represent both total energy flow and rate functions. *Adapted from B. A. Wunder, 1978,* Implications of a Conceptual Model, *Univ. Pittsburgh Press.*

months and have evolved mechanisms of *avoidance* and *resistance* to cold (table 9.1). Avoidance of cold entails energy conservation, whereas resistance requires energy expenditure—a debt that must be repaid (Speakman and Thomas 2003). Mammals rarely rely on a single mechanism to enhance survival in the cold but instead exhibit a suite of strategies that integrate behavioral, anatomical, and physiological specializations finely tuned to a specific habitat and lifestyle (Merritt and Zegers 2002). Thermoregulatory mechanisms vary according to taxon and the attributes of each species.

AVOIDANCE OF COLD

Body Size and Metabolism

Does a mammal's size enhance its energy conservation capacity in the cold? To answer this question, we must closely examine the relationship of body size and metabolism. Recall that in euthermic animals, heat production must equal heat loss, and heat loss is proportional to surface area. Resting, or basal, metabolic rate (BMR) is the volume of oxygen consumed per unit of time at standard temperature and pressure (STP) and is known for many species of mammals. McNab (1988) provides rates of basal metabolism and body masses for 320 species constituting 22 orders of mammals. MacMillen and Garland (1989) studied metabolic rates, body masses, climatic data, and latitudes for many species of rodents. Hill (1983) provides an excellent review of the thermal physiology and energetics of the Genus *Peromyscus*. It makes intuitive sense that larger animals require more food to maintain their body temperature and level of activity. Figure 9.5 (solid line) shows that larger animals consume more oxygen (they eat more food) than smaller animals—this is **total metabolic rate**. To compare oxygen consumption between animals of different sizes, metabolic rate is

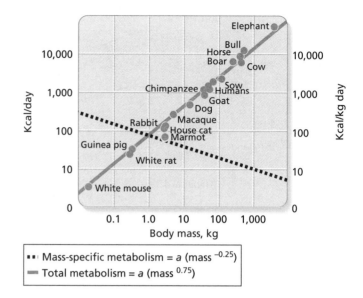

Figure 9.5 **Body size and metabolism.** Linear relationship comparing the log of body size in placental mammals with the log of total metabolism (solid line) and the mass-specific metabolism (broken line). a = proportionality constant (3.8) for placental mammals. *Data from M. Kleiber, 1932, Body size and metabolism, Hilgardia 6:315–353, 1932.*

adjusted for body size and expressed as **mass-specific metabolic rate** (rate of oxygen consumption per gram of body mass). Now it is apparent that the mass-specific metabolic rate decreases as body size increases (see figure 9.5, broken line). Given this relationship, we can estimate metabolic rate from the body mass of a species. The relationship of mass-specific and whole-animal metabolism is complex, and authors do not always agree on interpretation of this relationship (Packard and Boardman 1988; Hays and Shonkwiler 1996; Hays 2001; McNab 2002). The slope of the solid line in figure 9.5 has a value of about 0.75; that is, the metabolic rate scales to body mass to the 3/4 power—this relationship is generally known as Kleiber's law" (Kleiber 1932, 1961; Brody 1945; Brown and West 2000; Smil 2000; West and Brown 2005). Glazier (2005, 2006:326) contends that "the 3/4-power law is not universal, and should at most be regarded as a statistical rule or trend, rather than as an inviolable law." Although variation in the metabolic scaling exponent has been demonstrated both within and between species, this equation is quite useful in estimating the metabolic rate of mammals of known body mass and has important ecological and evolutionary consequences. The study of the consequences of a change in size, or in scale, is important in predicting the metabolic rates of mammals. Physiologists define **scaling** as the structural and functional consequences of a change in size or scale (Schmidt-Nielsen 1964; Calder 1984; Nagy 1987; McNab 1988, 2002; Withers 1992). As the size of an animal increases, volume and mass change more rapidly than does area. The change in proportions with increasing or decreasing body size is referred to as **allometry** and determines many patterns

Table 9.1 Winter survival mechanisms of mammals
Avoidance
Body size
Insulation
Appendages
Coloration
Modification of microclimatic regime
Communal nesting
Construction of elaborate nests
Foraging zones
Food hoarding
Reduction in level of activity
Reduction in body mass
Dormancy
Resistance
Increase in thermogenic capacity through BMR, NST, shivering

Abbreviations: BMR = basal metabolic rate; NST = nonshivering thermogenesis.

seen in body structure, metabolic rate, and heat flow. An excellent discussion of the relationship of body size and energy requirements is conducted by Speakman and Thomas (2003).

How do the above relationships of body mass and metabolism relate to the geographic distribution of mammals? As we will learn, several ecogeographic rules have been proposed to explain morphological variation on a geographic scale. The best known ecogeographic rule is **Bergmann's rule.** In 1847, Carl Bergmann contended that "on the whole . . . larger species live farther north and the smaller ones farther south" (Bergmann 1847 translated in James 1970). Mayr (1956:105) restricted Bergmann's rule to variation within species stating that "Races of warm-blooded animals from cooler climates tend to be larger than races of the same species from warmer climates" (Mayr 1956:105). Investigators testing the validity of Bergmann's rule have traditionally compared body size with latitude using latitude as proxy for temperature (Ashton et al. 2000). There is much debate concerning the interpretation and implications of Bergmann's research (Scholander 1955; Mayr 1963; Rosenzweig 1968; McNab 1971, 2002; Calder 1984; Lindstedt and Boyce 1985; Geist 1987; Ashton et al. 2000; Olcott and Barry 2000; Smith et al. 2002; Meiri and Dayan 2003; Ochocinska and Taylor 2003; Blackburn and Hawkins 2004; Rodriguez et al. 2006).

Bergmann's rule implies that some energetic advantage may be gained through a decreased surface-area-to-volume ratio. We can further generalize that the amount of heat loss depends on both the animal's surface area and the difference in temperature between its body surface and the surroundings. Large mammals, for example, have less heat loss than small mammals because of their large body mass relative to surface area. For example, visualize a mammal as a cube. With linear dimensions of 1 cm on each side, the cube has a surface of 6 cm^2 and a volume of 1 cm^3. Hence, the surface-area-to-volume ratio is 6:1. If you double the linear dimension of the cube, the total surface area increases to 24 cm^2, and the new volume is 8 cm^3. Now the surface-area-to-volume ratio is 3:1. By doubling the length, heat conservation is enhanced by reducing surface area relative to volume. Granted, mammals are not cubes, but arguments in support of an energetic interpretation of Bergmann's rule follow this logic: A larger mammal has less total surface area per unit volume and thereby benefits from a reduced rate of cooling. This interpretation sounds rather convincing, but there are problems. Small mammals know little about per-gram efficiency and are only concerned with total food requirements; a larger mammal clearly requires more food—a clear disadvantage for herbivores during winter. Interpretation of Bergmann's rule can be tricky. If we analyze body size of mammals distributed over a wide latitudinal range, we see mixed results. Rensch (1936) showed that 81% of North American species of mammals and 60% of European species of mammals were indeed larger at higher latitudes. However, McNab (1971) was critical of the work of Rensch (1936, 1938) because his data were

derived from a field guide rather than from measurements of individuals from different localities. McNab (1971) found that of 47 North American species examined, only 32% (15/47) followed the trend predicted by Bergmann's rule, which was thus invalid. Geist (1987) concurred, indicating that Bergmann's rule was invalid; although body size of large mammals initially increased with latitude, it reversed between 53° and 63°N; small body size occurs at the lowest and highest latitudes. However, Ashton et al. (2000) isolated problems with the analysis of McNab (1971) and found broad support for Bergmann's rule for all orders and most families of mammals examined (78/110 species examined; Ashton et al. 2000, tables 1–3). Temperature and latitude were strong predictors of body size variation (see Ashton et al. 2000 and Freckleton et al. 2003). In sum, an analysis of the validity of Bergmann's rule must integrate many factors associated with the biology of a species, such as pelage, behavior, temperature and water-related factors, size and type of food, primary plant production, and morphology, to mention just a few (Rosenzweig 1968; McNab 1971; Burnett 1983; Steudel et al. 1994; Yom-Tov and Geffen 2006).

For carnivorous mammals, the long, thin shape of weasels (Genus *Mustela*) has distinct disadvantages in cold climates. Their shape exposes a large surface area to the cold air, and their unique feeding requirements dictate that they pursue prey through small crevices and a labyrinth of subterranean runways, which is energetically very expensive. Their mobility would be greatly compromised by a dense, heavy pelage; thus, a short fur is essential for optimal agility. Compared with woodrats (Genus *Neotoma*), weasels have about a 15% higher surface-area-to-mass ratio—they cannot assume a spherical form by curling as woodrats can. As a result, cold-stressed weasels have metabolic rates 50–100% greater than less slender mammals of comparable size. So, being long and skinny may facilitate capturing prey, but weasels pay a high energy cost in the form of a rapid rate of heat loss. As a result, weasels must consume more food to maintain energy efficiency (Brown and Lasiewski 1972; King 1989, 1990; McNab 1989).

Insulation

For mammals that reside in cold environments, the most direct method of decreasing heat loss is to increase the effectiveness of insulation (see table 9.1). For mammals and birds, the best way to reduce thermal conductance is to possess fat, fur, or feathers between the body core and the environment (Withers 1992). Insulation value increases with the thickness of fur (figure 9.6) and is at maximum in such mammals as Dall sheep (*Ovis dalli*), wolves (*Canis lupus*), and arctic foxes (*Alopex lagopus*). The easiest way to understand the benefits of insulation is to think in terms of extending the limits of the lower critical temperature. The arctic fox, for example, has a lower critical temperature of –40°C, and at an environmental temperature of –70°C has elevated its metabolic rate only 50% above its

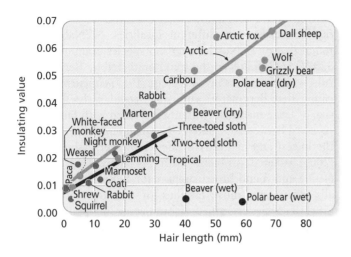

Figure 9.6 **The length of fur.** The insulating values of the pelts of arctic mammals (•) are proportional to the length of the hair. Pelts from tropical mammals (∞) have approximately the same insulating value as those of arctic mammals at short hair lengths, but long-haired tropical mammals like sloths have less insulation than arctic mammals with hair of the same length. Immersion in water greatly reduces the insulative value of hair, even for such semiaquatic mammals as the beaver and polar bear (μ). *Data from P. F. Scholander et al., 1950, Body insulation of some arctic and tropical mammals and birds, in* Biological Bulletin *99:225–235, 1950.*

normal metabolic rate! It is easy to understand that if an animal lowers its metabolic rate, it can get along with less food—a real plus if food is limited during winter months.

Some trade-offs are associated with insulation. The insulating value of fur is a function of its length and density; some small mammals such as lemmings (Genera *Dicrostonyx, Lemmus*) have very long fur relative to body mass. A thick coat on predators like weasels, however, would compromise their ability to forage in narrow crevices. Scholander (1955) contended that small mammals could not accumulate enough insulation in the form of fat or hair to cope with low winter temperatures without compromising agility. Research documenting the importance of pelage insulation in small mammals has focused primarily on the Families Muridae and Soricidae (Churchfield 1990). When these groups are faced with cold stress, heat loss was reduced by increasing pelage insulation by 11–19% during winter months. For small mammals, pelage changes may be useful in conserving energy and compensating for the large difference between body and environmental temperatures during winter (Bozinovic and Merritt 1992). As we will see later, however, small mammals that do not have a thick, long pelage take advantage of stable microclimatic regimes below ground and may employ social thermoregulation to cope with cold. On the other hand, larger mammals benefit greatly from insulation in the form of fat and fur. An excellent discussion of adaptation to cold by northern cervids is provided by Marchand (1996).

Large arctic mammals such as muskox (*Ovibos moschatus*), caribou (*Rangifer tarandus*), and wolves have long,

dense fur as well as thick layers of fat beneath the skin. The muskox, in particular, is equipped with an immense cloak of coarse guard hairs over 30 cm in length and an underfur of thick, dense, silky wool (figure 9.7). The hairs of the muskox are thicker at the tip than at the root so that they form an almost airtight coat. The animal's heat-losing extremities (tail and ears) are well hidden under the pelage. These heat-conserving adaptations permit the muskox to survive in the arctic tundra where winter temperatures commonly reach –40°C.

Insulating fur is a great advantage for thermoregulation in larger terrestrial mammals (Hudson and White 1985; Marchand 1996). Aquatic mammals have a slightly different strategy. For example, muskrats (*Ondatra zibethicus*), water shrews (*Sorex palustris*), beavers (*Castor canadensis*), and northern fur seals (*Callorhinus ursinus*) can function in waters at 0°C because of structural modifications of their fur that trap air. Their skin stays dry, shielded from the water by a layer of air. This layer of air not only helps with thermoregulation but assists with buoyancy. It is therefore easy to see that a polar bear (*Ursus maritimus*) with dry fur is far better off than one with wet fur. Fur is actually of limited value to an aquatic mammal. Body warmth is maintained because air is trapped

Figure 9.7 **An immense fur coat.** Musk ox are inhabitants of the arctic tundra of North America. They possess a massive dark brown coat composed mainly of long threads of high-quality wool. The outer coat is covered with long, coarse guard hairs. Heat-losing extremities are buried in the coat, an adaptation for survival in the arctic cold.

between hair and the skin; if this air is displaced by water, the fur loses most of its insulating value, which falls close to zero when the fur becomes wet.

Because the thermal conductivity of water is almost 25 times greater than that of air, marine mammals must protect themselves from heat loss, particularly when inactive (Schmidt-Nielsen 1997). Air trapped in dry underfur acts as an effective insulator for some groups. Marine mammals such as whales (chapter 17), walruses, and seals (chapter 16) possess a particularly effective insulator—**blubber.** Consisting of collagen and elastic fibers embedded in an incompressible matrix of adipocytes, blubber is located below the dermis of the skin (see figure 9.8). In addition to providing insulation, blubber acts to streamline the body (Koopman 1998), adjust buoyancy (Beck et al. 2000; Biuw et al. 2003), provide an energy storage site, support thermoregulation (Kanwisher and Sundnes 1966; Blix et al. 1979; Koopman 1998; Willis et al. 2005), and may provide a hydrodynamic function (Hamilton et al. 2004). The fat of marine mammals is entirely blubber; marine mammals have no visceral fat like other mammals. Thickness of blubber varies from 5 cm in small seals to 60 cm in the bowhead whale (*Balaena mysticetus*). In some whales, such as the gray whale (*Eschrichtius robustus*), blubber is also an energy reserve during migration. The blubber of seals is a better insulator than that of whales because it contains less fibrous connective tissue (Bonner 1990). During winter, the layer of blubber in a ringed seal (*Phoca hispida*) may account for 40% of its mass. Northern fur seals and other eared seals (Family Otariidae) are characterized by a dense covering of fur that traps air bubbles. Because they rely less on insulation in the form of blubber, they must remain active to maintain normal body temperature in frigid waters.

Activity on land, even with the air temperature at 12°C, can produce overheating, however, which may prove lethal. Marine mammals control overheating by remaining in the water, thus keeping their fur or skin moist; by panting; and also by waving their hind flippers, which are supplied with abundant sweat glands. Another way to control heating is by sleeping; in some species, sleeping reduces heat production by almost 25%. A southern elephant seal (*Mirounga leonina*) may reach 3,500 kg. This large, dark mass basking on pack ice in the Antarctic is subject to significant warming due to absorption of solar radiation. Blubber, permeated by blood vessels, is not passive insulation. For large elephant seals and Weddell seals (*Leptonychotes weddellii*) resting on pack ice, loss of heat is essential. In response to heat stress, these seals can increase the rate of blood flow through their skin by dilating special vessels that join arterioles directly to venules, bypassing the capillary beds. Because these bypasses (called arteriovenous anastomoses) are very near the surface of the skin, the increased flow of blood permits heat to be lost to the environment. **White adipose tissue** is typically considered an organ associated with the storage of lipid that acts to insulate mammals inhabiting cold environments. Recent research, however, indicates that white adipose tissue serves as an important endocrine organ associated with metabolic and physiological regulation (Trayhurn et al. 2000). White fat secretes a signal protein called leptin that is important in processes of energy balance, reproduction, and immunity (Zhang et al. 1994).

Appendages

The fur thickness of northern mammals may increase as much as 50% during winter months. Appendages such as legs, tail, ears, and nose, however, are potentially great heat "wasters." They cannot be well insulated, but to prevent heat loss from appendages, arctic land mammals permit temperatures at appendages to decrease, often approaching the freezing point. Appendages must be supplied with oxygen and kept from freezing, however, by circulating blood through them. This is done, without constant loss of heat from the extremity, by a process called **countercurrent heat exchange,** a form of peripheral heterothermy. This physiological mechanism shunts blood through a heat exchanger (the **rete mirabile,** or "miraculous net") that intercepts the heat on its way out and maintains the extremity at a considerably lower temperature than the core—the net result is a reduction of heat loss. As warm arterial blood passes into a leg, for example, heat is shunted directly from the artery to the vein and then carried back to the core of the body (figure 9.9). As a consequence, the appendages of many northern mammals are maintained at comparatively cold temperatures. Anatomical studies have shown that the foot pads of arctic canids possess a massive arteriovenous plexus through which blood flow to and heat loss from the foot pads is controlled (Henshaw et al. 1972). To keep the extremities soft and flexible at such low temperatures, the fats in the feet of northern mammals must have very low melting points, perhaps 30°C lower than ordinary body fats (Storey and Storey 1988). Countercurrent heat exchange is well developed in the tail of beavers (*Castor*

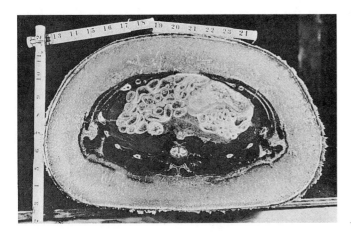

Figure 9.8 Blubber. Cross section of a frozen seal showing the thick layer of blubber. Of the total area in the photograph, 58% is blubber and the remaining 42% is muscle, bone, and visceral organs. The measuring stick is graduated in inches.

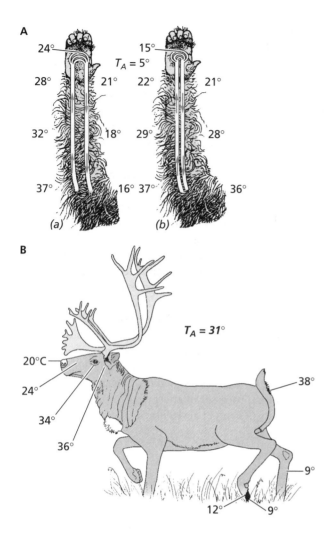

Figure 9.9 Countercurrent heat exchange. (A) A diagram representing the circulation in a limb of a mammal showing hypothetical temperature changes of the blood in the absence (a) and presence (b) of countercurrent heat exchange (°C). Arrows indicate direction of blood flow. In (b), the venous blood takes up heat (thus cooling the arterial blood) all along its path of return because even as it becomes warmer and warmer, it steadily encounters arterial blood that is warmer yet. (B) Regulation of external body temperature in the caribou (°C). Temperature regulation is accomplished in part by countercurrent heat exchange. An intricate meshwork of veins and arteries acts to keep the temperature of the legs near that of the environment so heat is not lost from the body. *(A) Data from R. W. Hill and G. A. Wyse, 1989,* Animal Physiology, *2nd ed., Harper and Row. (B) Data from J. F. Merritt, 1983,* Animals of the arctic, in Arctic Life: Challenge to Survive *(M. M. Jacobs and J. B. Richardson III, eds.), 1983, The Board of Trustees, Carnegie Institute.*

canadensis) and ears of Japanese hares (Cutright and McKean 1979; Ninomiya 2000). In air at low temperature, loss of heat from the tail may be reduced to less than 2% of the resting metabolic rate. Then, when faced with higher temperatures, over 25% of the heat produced at resting metabolism may be dissipated through the tail (Coles 1969).

Countercurrent heat exchange also occurs in the respiratory passages of mammals. Heat recovery (heat exchange) in the respiratory system is slightly different from the countercurrent heat exchange process discussed for the vascular system. For example, during inhalation, heat and water are added to air before it reaches the lungs. During exhalation, heat is typically lost to the environment due to warm air loss and evaporation of water. In the vascular system, heat exchange occurs at the same time between two channels, but in the respiratory system, the exchange occurs at different times in the same channel (Schmidt-Nielsen 1997). Caribou demonstrate the principle. When active, they produce a considerable amount of heat, which, due to their excellent insulation, may exceed their thermoregulatory requirements. Some must therefore be lost, or their body temperature would increase. When resting, however, caribou do not generate as much heat and must minimize heat loss to the environment. Caribou and other large mammals possess a flap of skin at the opening to the nasal passages. This flap of skin reduces the size of each opening to a small slit when the animal is inactive, thus minimizing heat loss from the lungs. When resting, the temperature of expired air is lower than body temperature. When exercising, the nostrils flare open, increasing the surface area of the mucous lining of the nose for heat exchange. Due to the increased rate of respiration during activity, the volume of air in the nasal passages has less time for heat exchange to occur. The temperature of expired air during exercise is therefore higher when the animal's heat production exceeds its requirements for temperature regulation. Thus, respiratory heat exchange for large arctic species serves a dual function: minimizing heat loss while at rest and increasing heat loss during periods of activity (Scholander 1957; Hainsworth 1981:223).

Conservation of heat in species occupying cold regions can also be enhanced by reducing the length of the exposed extremity. **Allen's rule** states that the appendages of endothermic animals are shorter in colder climates than those of animals of the same species found in warmer climates (Allen 1877). By reducing unnecessary body surface area, animals can reduce heat loss. It seems logical from the standpoint of thermal physiology that a long nose and tail, as well as long ears and legs, tend to add unnecessarily to the body surface and would be great heat wasters for cold-adapted species (Scholander 1955). For instance, arctic foxes have small, rounded, densely furred ears; a reduced muzzle; and short, stubby legs. Foxes living in deserts, such as swift foxes (*Vulpes velox*) of the southwestern United States or the fennec (*V. zerda*) of North Africa have large, elongated, thinly furred ears, and comparatively long legs (figure 9.10). For species residing in hot climates, long appendages are a valuable commodity because they aid in the dissipation of heat. In four species of macaques (Cercopithecidae: *Macaca*) residing in Asia, relative tail length generally decreases with increasing latitude in accord with Allen's rule (Fooden 1997; Fooden and Albrecht 1999). However, as with Bergmann's rule, the validity of this generalization is debatable, and there are exceptions. For example, the

A   **B**   **C**

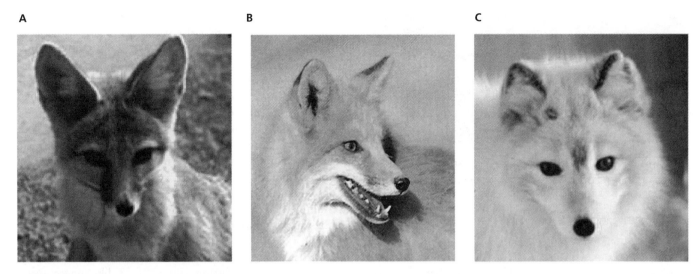

Figure 9.10 **Ears and heat loss.** The ears of foxes help to regulate body heat. (A) The swift fox (*Vulpes velox*) of the American southwest has relatively long ears that help to dissipate heat. (B) The red fox (*Vulpes vulpes*) of temperate America possesses intermediate-sized ears. (C) In contrast, the relatively short ears of the arctic fox (*Alopex lagopus*) help to conserve heat in the cold.

body form of bush dogs (*Speothos venaticus*) in the tropics of South America is not unlike that of arctic foxes. Furthermore, the length of ears and tail in hares (Genus *Lepus*) follows Allen's rule, but that for rabbits does not (Genus *Sylvilagus*; Stevenson 1986).

Coloration

Although northern mammals exhibit some fascinating anatomical specializations for coping with harsh winter climates, perhaps the most obvious trait defining them is their white pelage (figure 9.11). **Gloger's rule** states that "Races in warm and humid areas are more heavily pigmented than those in cool and dry areas" (Gloger 1833; Mayr 1970:200). Pigmentation in many mammals tends to be paler the closer the species habitat is to the arctic, with northern races (subspecies) of animals generally being lighter in color than their southern counterparts (Flux 1970; Johnson 1984). Coloration of pelage of many arctic mammals either remains white year-round or changes to white during winter. Those that remain white year-round include the polar bear (actually their hairs are transparent and pigmentless), arctic hares (*Lepus arcticus*) of the far north, and northerly forms of the caribou and gray wolf. It is noteworthy that arctic hares in the far north, such as those on Ellesmere and Baffin Islands and in Greenland (figure 9.12), do not change color seasonally but remain white during the short arctic summer. Because hares are so well adapted to winter, the cost in energy of changing to brown during the short summer exceeds the benefits of permanent adaptation to winter. A seasonal color change (dimorphism) is seen in collared lemmings (Genus *Dicrostonyx*), Siberian hamsters (*Phodopus sungorus*) snowshoe hares (*Lepus americanus*), mountain hares (*Lepus timidus*), ermines and weasels (Genus *Mustela*), and arctic foxes (*Alopex lagopus*), however. Species that undergo seasonal dimorphism accomplish the color

Figure 9.11 **A white fur coat.** The polar bear has a circumpolar distribution. It is one of the largest terrestrial carnivores in the world, feeding primarily on seals and fish. The thick white coat is composed of hollow hairs that trap warm insulating air and also act as an efficient solar collector for this bear when wandering from ice floe to ice floe in search of prey. Although their unique hairs convey a thermal advantage, polar bears enhance survival in cold by possessing an enormous thermal inertia (large body size) and large stores of subcutaneous fat.

changes in different ways. The hairs of ermines and weasels, for example, turn white along its entire length. This complete molt is characterized by each hair being lost and a new, white (winter) hair replacing the old, brown (summer) hair. In contrast, the snowshoe hares exhibit a different type of molt. Rather than going through a complete molt pattern, only the tips of the winter hairs of *L. americanus* are white; the bases remain gray. The timing of

Figure 9.12 Arctic hares in the far north. *Lepus arcticus* reside on the arctic tundra—the most northern distribution of all leporids. They normally show seasonal changes in coat color. Arctic hares inhabiting the far north, however, keep their white coats year-round, even though there is no snow cover in midsummer.

molt patterns for the Old World mountain hares is most strongly influenced by length of day (photoperiod) but is also correlated with average ambient temperature and duration of snow cover (Angerbjörn and Flux 1995). Furthermore, species of mammals that show seasonal color changes do not change color over their entire geographic range. For example, the North American long-tailed weasel (*M. frenata*) shows seasonal color changes north of about 40°N latitude, but south of this zone, they remain brown year-round. Individuals residing in the zone of overlap between these color groups show gradations in color from white to pied or brown (Hall 1951).

The specific cue (proximate factor) that triggers the onset of molt is decreasing daylength. In autumn, the optic nerve receives the stimulus, which is then transmitted to the hypothalamus. The factor that is ultimately responsible for establishing the duration of the winter coat is probably its thermal advantage, accounting for the close correlation between molting and mean annual temperatures (Flux 1970; Johnson 1984). Temperature and photoperiod, however, are not the sole determinants of white winter coloration. Transplantation experiments have demonstrated that heredity as well as temperature is involved in controlling pelage changes in mammals (Kliman and Lynch 1992). The molt cycle is geared to changes in the environment by way of changes in the hormonal system (namely, the endocrine glands: thyroid, pituitary, adrenal cortex, and pineal) as mediated by the brain and the hypothalamus. The seasonal cycles of molt and reproduction are closely related and are coordinated by the neuroendocrine system. In spring, in addition to changes in pelage color, density, and length, the brain signals the pituitary to secrete gonadotropins, which stimulate the gonads to prepare for the approaching breeding season. During spring, the hair follicles and testes of male weasels enlarge simultaneously; the testes begin to manufacture testosterone and sperm, and the hair follicles accumulate melanin and manufacture hair for a complete coat replacement (King 1989).

Why northern animals turn white in winter is not fully known. We assume that if the mechanism were not maintained by natural selection (if it did not confer an adaptive

advantage more often than a disadvantage), it would disappear. We assume that the color white acts to conceal both predators and prey (Stoner et al. 2003a, 2003b). This adaptation may conceal polar bears from a potential victim. While actively searching for prey, however, a weasel is surely easily detected, even with a white pelage. It is possible that the weasel's white color in snowy regions may allow it to blend with its background (cryptic coloration) and thus avoid predation by hawks, owls, and foxes. Many questions concerning the adaptive nature of color changes in northern mammals have yet to be answered (Walsberg 1983, 1991; Marchand 1996). For example, what can be the explanation for the blue color phase that commonly occurs in arctic foxes inhabiting the Pribilof Islands and many coastal areas of Alaska and Canada? If color acts to camouflage these animals, why are they slate blue during winter? If this blue phase is not adaptive, then how did it evolve and how is it maintained in the population? If cryptic coloration is important to conceal predator from prey, why don't gray foxes (*Urocyon cinereoargenteus*), fishers (*Martes pennanti*), and martens (Genus *Martes*) turn white in snowy regions?

The color white may also convey a thermal advantage. According to the laws of physics, black-colored animals lose heat by radiation faster than white-colored ones. Following Gloger's rule, we therefore expect to find white animals occurring in cold regions. Some investigators have misinterpreted the pertinent physical laws, however. Radiation of heat from an animal's body is in the form of infrared energy, which is unrelated to visible coloration. All animals are therefore considered to be thermodynamic "blackbodies," meaning that they absorb all incident radiation and reflect none. The color of the fur and underlying skin may be important to the amount of heat absorbed from solar radiation—its peak intensity is in the visible range. When exposed to direct solar radiation, dark-colored skin or fur absorbs more incident energy than light-colored skin or fur. The conservation of this heat energy depends on the length and thickness of the fur, not on its color. Many small mammals, such as voles and shrews, show increases in pelage density and length during winter (Khateeb and Johnson 1971; Bozinovic and Merritt 1992).

Animals living in northern regions may not only possess white fur but also have thicker, denser pelages. For example, mountain hares possess a white winter coat that is twice the length and thickness of their summer coat which provides an insulatory advantage in cold (Flux 1970). Walsberg (1991) examined coat insulation and solar heat gain in three species of subarctic mammals that shift between white winter pelages and darker summer pelages (*Lepus americanus*, *Mustela erminea*, and *Phodopus sungorus*). He contends that seasonal changes in coats of these species serve important roles in cryptic coloration, thermal insulation, and radiative heat gain. A white coat in winter does indeed result in a reduction in solar heat gain; however, this is not necessarily a result of increased heat reflectivity of the pelage or optical properties of the coat to solar radiation, but rather a product of increased coat insulation (Walsberg 1991). Generalizations concerning the adaptive significance of different biogeograpical "rules" or trends must take into account that the survivability of northern mammals may not result solely from forces that maximize heat conservation.

For some animals, the thermal advantage of hair is not necessarily contingent on its color, density, or length but rather on the anatomy of the hairs and color of the skin beneath it. For many members of the Family Cervidae, insulation depends on the air contained by the highly medullated guard hairs that provide insulation (Johnson and Hornby 1980). Research with polar bears indicates that a combination of their transparent, pigmentless hair and black skin may enhance their heat-conserving abilities (Grojean et al. 1980; Walsberg 1983; Tributch et al. 1990). Pigmentless hair traps and transmits to the skin 90% of sunlight in the invisible ultraviolet portion of the spectrum but only 10% of the visible light. Energy in the form of heat from the ultraviolet light is absorbed by the dark skin to help warm the body, while the visible light is reflected as white color. The hairs of polar bears act like optical fibers, with ultraviolet light entering at one end and bouncing along the inside of the hair shaft to reach the dark skin, where it is absorbed.

Modification of Microclimatic Regime

In northern regions, nonhibernating small mammals, particularly those of the rodent Families Cricetidae and Sciuridae (e.g., Genus *Glaucomys*), construct elaborate nests and engage in communal nesting (Cricetidae: Madison et al. 1984a; West and Dublin 1984; Wolff 1989; Bazin and MacArthur 1992; Sciuridae: Muul 1968; Stapp et al. 1991; Stapp 1992; Merritt et al. 2001). The greatest gain from huddling, however, should accrue to small mammals with a large surface-area-to-mass ratio and a limited capacity for increasing the insulation value of their pelage. Not just small mammals use huddling: 23 raccoons (*Procyon lotor*) were reported occupying one winter den (Mech and Turkowski 1966). During winter, female striped skunks (*Mephitis mephitis*) formed a communal nest with other females or with a single male (Wade-Smith and Verts 1982). Communal denning has adaptive value for winter survival and reproductive success, especially in northern latitudes. For example, in Alberta, an average of 6.7 striped skunks occupied communal dens with females more common in dens than males (Gunson and Bjorge 1979).

Both nest building and huddling conserve body heat through reductions in thermal conductance. Huddling in groups reduces each individual's exposed surface, thus reducing the cold stress and the metabolic requirement for heat production. Studies traditionally employed laboratory-based calorimetry to demonstrate energy savings of communal nesting during winter (Glaser and Lustick 1975; Martin et al. 1980; Casey 1981). More recent studies linked laboratory experimentation with field-derived data on nesting habits by employing radiotracers and radiotelemetry methods (Madison et al. 1984a; Andrews and Belknap 1986).

When muskrats huddle in a group, a major part of each individual's body surface is in contact with a neighboring animal (figure 9.13). Curling and retracting the extremities reduces heat loss, and microclimatic modification in the form of an elaborate, well-insulated nest below ground and snow cover (subnivean) adds greatly to energy savings. Sealander (1952) showed that at low temperatures, individuals of the Genus *Peromyscus* formed a "communal" group. Those at the bottom of the group enjoyed temperatures well above ambient levels, but by continually shifting position, each mouse in the huddle was periodically rewarmed and thus avoided hypothermia. Because heat loss by conduction or convection varies in direct proportion to the amount of surface area exposed, the energy savings of huddling can be easily calculated (Marchand 1996). Vickery and Millar (1984) provided a model for predicting the energy advantages and disadvantages of huddling. When applying their data on

A
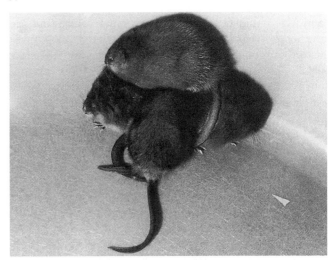

B

Figure 9.13 Social thermoregulation. Examples of (A) close and (B) loose aggregation responses of muskrats exposed to a temperature of 5°C.

Peromyscus, they found that huddling confers a distinct energy savings for mice subjected to ambient and nest temperatures well below thermoneutrality.

Many species of small mammals (principally rodents of the Subfamilies Arvicolinae and Cricetinae) form aggregations during winter to conserve energy (Merritt 1984). Rodents construct elaborate nests of grasses and herbs either under the litter, on the ground, or within a hollow tree or log. During the winter in northern regions, these nests are commonly located within the subnivean environment, which aids in insulating the nest site from fluctuating supranivean (above snow) temperatures. Radiotelemetry studies demonstrated that up to six muskrats may inhabit the same winter lodge in the marshes of Manitoba, Canada (MacArthur and Aleksiuk 1979). The resting metabolic rate of a group of four muskrats huddling in such a lodge during winter with an environmental temperature of –10°C shows up to a 13% energy savings over that of a single animal. In northern latitudes, aggregations of at least six muskrats are common. The physical structure and communal use of beaver lodges in southeastern Manitoba were assessed by Dyck and MacArthur (1993). Winter air temperatures outside the lodges reached a low of –41.4°C, but temperatures within the chambers of occupied lodges did not fall below 0°C. The mean monthly temperature of the nesting chamber consistently exceeded the mean monthly exterior air and water temperatures. The ameliorated microclimate within the lodges also facilitates periodic rewarming of foraging beavers, thus minimizing thermoregulatory costs during rest. Nests of taiga voles (*Microtus xanthognathus*) occupied by five to ten individuals remain between 7° and 12°C warmer than ground temperature within the subnivean environment and as much as 25°C warmer than supranivean temperatures (figure 9.14). Furthermore, nests were not completely vacated, so that foraging voles returned to a warm nest (Wolff 1980; Wolff and Lidicker 1981). Huddling by deer mice and voles may, for example, reduce energy requirements in the cold by as much as 16–36% (e.g., Gebczynska and Gebczynski 1971; Vogt and Lynch 1982; Andrews and Belknap 1986). For *Peromyscus leucopus*, a combination of torpor and huddling of three individuals within a nest at an ambient temperature of 13°C resulted in a daily energy savings of 74% compared with nontorpid, individual mice without a nest (Vogt and Lynch 1982). Aggregate nesting during winter is common for species of Genus *Peromyscus* (Madison et al. 1984b; Wolff and Durr 1986). The average daily metabolic rate of *P. maniculatus* is also lower in winter (Merritt 1984). When *P. maniculatus* and *P. leucopus* are syntopic (live within the same locality) in the Appalachian Mountains of Virginia, radiotelemetry studies have shown that *P. maniculatus* prefers nesting high in large, hollow trees, year-round, whereas *P. leucopus* uses both ground and tree nest sites in summer but shifts to underground nest sites in winter (Wolff and Durr 1986). Radiotelemetry studies have demonstrated that both species nest together during winter months (Wolff 1989).

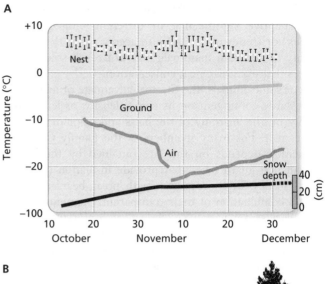

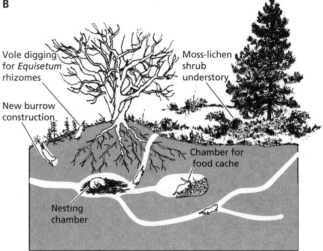

Figure 9.14 Nest temperatures of taiga voles. (A) Daily mean air temperature, ground temperature, and temperature in the nest of the taiga vole from 15 October to 1 December 1977, at a site 155 km northwest of Fairbanks, Alaska. The daily ranges of nest temperature are shown by the vertical bars. Increasing ground temperatures after 20 October were due to progressively deeper snow cover. (B) The communal winter midden-tunnel system and activities of a group of taiga voles. *Data from J. O. Wolff and W. Z. Lidicker, Jr., 1981, Communal inter nesting and food sharing in taiga voles, in* Behavioral Ecology and Sociobiology 9:237–240, 1981.

Most tree squirrels are solitary and euthermic during winter. Southern flying squirrels (*Glaucomys volans*), however, form "huddles" of up to 20 individuals (but groups fewer than 10 are more common) in hollow trees to conserve energy during winter. *Glaucomys volans* and *G. sabrinus* undergo periodic bouts of torpor during winter due to extended periods of food shortage and low temperature (Muul 1968). A group of six southern flying squirrels in New Hampshire, huddling within a wooden nest box and surrounded by temperatures of 6°C, reduced their energy expenditure by 36% (Stapp et al. 1991). This behavioral strategy, augmented by a nearby supply of hoarded nuts, enhanced survival during winter (Merritt et al. 2001).

Conservation of heat by a group of huddling animals is greatest when the nest is well insulated. Researchers have evaluated the thermal capacity of nests by calculating their shape, thickness of the nest wall (Wolfe 1970; Wolfe and Barnett 1977; Redman et al. 1999) and composition (King et al. 1964; Layne 1969). The resistance of nests to heat loss can also be measured quantitatively (Wrabetz 1980). Interspecific differences in nest-building behavior were correlated with available microhabitat and nest site preferences (Layne 1969). King and colleagues (1964) demonstrated a geographic correlation in the amount of nest material used by Genus *Peromyscus*. Northern forms used more nesting material under constant temperatures than did southern forms. Sealander (1952) demonstrated a seasonal difference in nest-building behavior: *P. leucopus* and *P. maniculatus* constructed more elaborate nests during winter than in summer, with winter nests conferring greater resistance to low temperatures. Pierce and Vogt (1993) employed outdoor enclosures and demonstrated that *P. leucopus* and *P. maniculatus* from northern New York constructed larger nests during winter compared with other times of the year. *P. maniculatus* constructed the largest nests.

Foraging Zones

Small mammals such as shrews and voles are active during midwinter and do not undergo physiological heterothermy (Wunder 1985; Merritt 1995). The thermal regime of the foraging zone, therefore, is crucial in dictating energy budgets of these winter-active mammals. The climatological regime of the foraging and nesting sites of shrews, voles, and mice has been examined during winter in Michigan, Ontario, and Pennsylvania (Pruitt 1957; Randolph 1973; Merritt 1986, respectively). In mixed deciduous forests, many small mammals forage in tunnels within soil covered by a rich layer of leaves. During winter, this foraging zone provides a stable, comparatively warm thermal regime. Snow covering the ground also provides additional insulation. Although ambient temperatures may reach –29°C in mid-January, the minimum temperature at the soil-leaf litter interface is about –4°C and 1°C within a subsurface tunnel (figure 9.15). Snow cover is an integral part of the life of small mammals (Merritt 1984; Marchand 1996). The presence of a sufficient depth of snow, called the **heimal threshold** (Pruitt 1957), insulates the subnivean environment against widely fluctuating environmental temperatures. High mortality rates among *Myodes gapperi* and *Peromyscus leucopus* during midwinter are attributable to a lack of snow cover to insulate the forest floor (Pruitt 1957; Beer 1961; Fuller et al. 1969). The period of autumn freeze is also crucial to the survival of small mammals due to great fluctuations in temperatures in their foraging zone. This period was found to produce increased mortality rates in small mammals studied in the Rocky

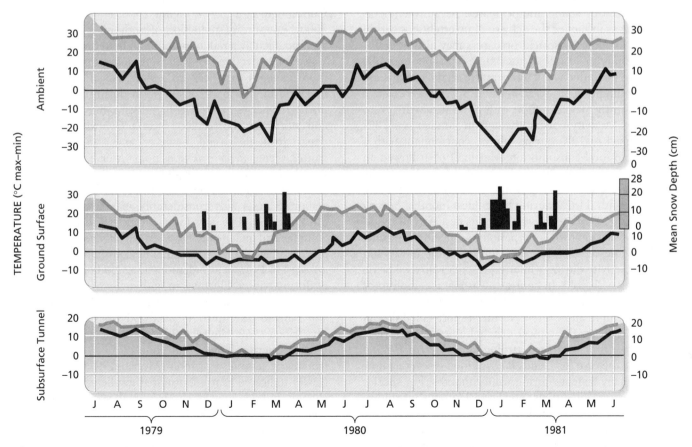

Figure 9.15 Thermal zone of small mammals. Maximum and minimum temperatures recorded on an Appalachian Mountain site from July 1979 to June 1981. Temperatures are recorded from 1.5 m above ground surface (ambient), at ground surface, and in a subsurface tunnel. Snow depth is shown by shaded bars. *Data from J. F. Merritt (ed.), 1984, Winter ecology of small mammals, in* Special Publication of Carnegie Museum of Natural History (10).

Mountains of Colorado (Merritt and Merritt 1978). Snow thickness in autumn was insufficient to insulate soil against fluctuating ambient temperatures, which may reach –20°C. Within the stable foraging zone, many species of small mammals establish caches of food to ensure that a predictable, quick energy source is readily available during winter, when food is scarce. Food hoarding is reported for 6 orders and 30 families of mammals, with most practitioners found in the Orders Rodentia and Carnivora. Because hoarding behavior is so important in the biology of Class Mammalia, a detailed discussion of this topic is provided in chapter 7.

Reduction in Level of Activity

Daily and seasonal temperature changes influence activity patterns of winter-active small mammals. Nonhibernating species residing in seasonal environments must secure adequate nourishment during winter to maintain a constant, high body temperature. Among winter-active mammals, shrews are characterized by rapid heat loss due to their large surface-area-to-volume ratio, high metabolic rate, and resulting high caloric requirements. We would therefore intuit them to be poor candidates for enduring cold stress. The northern short-tailed shrew (*Blarina brevicauda*) of eastern North America conserves energy by using cached food reserves and avoids cold environmental temperatures by restricting foraging and nesting to zones characterized by stable microclimates. Because temperature changes influence the activity of invertebrates within the soil, foraging by soricid predators may also be influenced. Churchfield (1982) used time-lapse photography to analyze the influence of temperature on the activity and food consumption of *Sorex araneus*. Shrews were active throughout the day and night with peaks in activity every 1–2 hours. Activity outside the nest in summer was 28% but dropped to 19% during winter, concomitant with a decrease in food consumption. Martin (1983) showed an annual activity range for *B. brevicauda* of 7–31% of the day. Daily activity during winter was greatly reduced, averaging only 11.6% during the coldest months. *Blarina* spends 80–90% of the day resting in its nest at a low metabolic rate, sleeping for long periods, and being intermittently highly active. The reduction in caloric intake and foraging activity during winter represents a survival tactic for coping with cold. In contrast, seasonal metabolism of *B. brevicauda* increases from a low in summer to a maximum in autumn and winter. Northern short-tailed shrews may depart from the typical metabolic profile of shrews due to their proclivity for hoarding food (Merritt 1986).

Unlike the case for Genus *Blarina*, food is severely limited for many rodents during winter months. For species of *Peromyscus* inhabiting northern latitudes, torpidity combined with communal nesting is important to conserving energy. For example, in western Kansas, *P. leucopus* spent 72% of the day in the nest during winter, compared with 28% of the day in summer (Baar and Fleharty 1976). During winter, when the cost of thermoregulation is

highest and food resources are lowest, available energy must be used to maintain metabolism. Energy loss can be minimized in part by curtailing locomotor activity. For voles (*Microtus pennsylvanicus*), cold temperatures during winter, especially at night, may select for more diurnal activity. During winter, voles demonstrate increased movements on warm days or forage below a mantle of snow. When there is no insulating blanket of snow, low temperatures stimulate the use of nests and inhibit activity in voles (Madison 1985). Hottentot golden moles (*Ambysomus hottentottus*) residing at high elevations (1,500 m) in the subtropical savannah of South Africa are exposed to cold and shortages of food during winter. These fossorial mammals cope with seasonal stressors by restricting activity to shorter, more intense periods, and by decreasing thermal conductance by increasing pelage insulation (Scantlebury et al. 2005).

Reduction of Body Mass

Small mammals, notably voles and shrews inhabiting seasonal environments, are reported to undergo a general decline in body mass during winter. This overwinter mass decline called **Dehnel's phenomenon** (Dehnel 1949) is thought to confer the adaptive advantage of decreasing caloric needs during winter when food resources are limited (Merritt and Merritt 1978; Hyvärinen 1984; Merritt 1984, 1995; Yaskin 1984; Churchfield 1990; McNab 1991; Churchfield et al. 1995; Taylor 1998; Hays and Lidicker 2000). For shrews, reduction of body mass and length is accompanied by a reduction of brain mass and skull depth, as well as a decline in most internal organs. As a result, these changes may lead to a decrease in mass-specific metabolism and food consumption during winter. Some northern species, namely collared lemmings (Nagy 1993) and northern short-tailed shrews (*Blarina brevicauda*), depart from this trend, however, and gain mass during winter. Genus *Peromyscus* does not conserve energy in this way during winter: Their ability to undergo torpor coupled with communal nesting, food hoarding, and use of elaborate nests aids considerably in conserving energy during winter. The cues for such fluctuations in mass are complex and include photoperiod, environmental temperature, and availability of food. The common shrew (*Sorex araneus*) in Eurasia also demonstrates body mass decreases during winter. Declines in body mass of shrews can be significant—*S. araneus* residing in Great Britain, Poland, and Finland lose up to 45% of body mass (Pucek 1965; Hyvärinen 1969; Churchfield 1982), and *Sorex cinereus* in North America showed a decline of 53% in body mass from early summer to winter (Merritt 1995). Churchfield (1990) indicated that a latitudinal gradient may exist for changes in body mass of shrews. Small size during winter confers an energy advantage by reducing food requirements. A small mammal eats less food and has a greater assimilation efficiency than a large one; consequently, it can reduce foraging time during cold and thus conserve energy. Although small body mass in winter decreases food requirements, it will reduce cold tolerance due to an increase in surface-area-to-volume ratio.

The survival mechanisms discussed have one thing in common, namely, they pertain to animals that are active during most of the winter and thus must forage to find food to maintain a high body temperature. Even small mammals such as Genus *Peromyscus* that exhibit short-term torpidity are quite thermolabile—because their body temperature decline is slight, they can quickly elevate their body temperature, forage on cached food, and then rewarm in a communal nest. As depicted in our model of the energy budget (see figure 9.4), thermoregulation is the highest priority: If a mammal cannot maintain euthermy, it cannot conduct other activities. But some small and medium-sized mammals residing in northern environments are quite lethargic during winter. They solve the problem of scarcity of food and low temperature by entering a prolonged and controlled state of dormancy called torpor, or hibernation.

Dormancy

Small mammals, such as bats and rodents, maintain high body temperature when active but are able to save energy by temporarily abandoning euthermia. **Dormancy** is defined as a period of inactivity characterized by a reduced metabolic rate and lowering of body temperature. **Torpor** is a form of dormancy characterized by a lowering of body temperature, metabolic rate, respiration, and heart rate. During the winter, torpor is referred to as hibernation, and during the summer, it is called estivation. Torpor that occurs daily is logically called daily torpor. These energy-conserving responses are sometimes grouped as forms of **adaptive hypothermia** (Bartholomew 1982). Unfortunately, "hibernation biologists" have difficulty agreeing on consistent terminology. The difference between these patterns should be treated as points along a continuum because one condition may shade imperceptibly into another. To understand these concepts, we must first establish some definitions. Our terms attempt to follow Bartholomew (1982), Lyman and colleagues (1982), French (1992), Geiser and Ruf (1995), Schmidt-Nielsen (1997), and Barclay et al. (2001).

The energy savings for mammals coping with cold depend on their drop in body temperature and length of time spent in a state of dormancy. Torpor is a type of dormancy in which body temperature, heart rate, and respiration are not lowered as drastically as in true hibernation. Body temperature declines markedly but not usually below 15°C. The lowest range of tolerable body temperatures during torpor is about 10° to 22°C (Wang and Wolowyk 1988). Patterns of torpor may extend for a period of hours or several days. Daily torpor is a response of small mammals to an immediate energy emergency. Examples of animals that undergo daily torpor are species of rodents (namely, Genus *Peromyscus*), many marsupials, "insectivores," bats, and some primates (Nestler et al. 1996). Torpor is a quite plastic condition and can provide significant energy savings for animals coping with cold

stress (Bartholomew 1982). Torpidity is certainly not a primitive physiological trait but is a unique feature adaptive for survivorship in seasonal environments. The ability of many rodents to undergo periodic bouts of torpor during winter is well known (Lyman et al. 1982; French 1992). This strategy aids in combating cold stress and scarcity of food during winter and is commonly accompanied by other energy-saving strategies. It is noteworthy that the ability to abandon homeothermy for torpor has never been shown in voles and most shrews, which must continue to forage under coldest conditions (Wunder 1985; Churchfield 1990).

Hibernation is defined as a profound dormancy in which the animal remains at a body temperature ranging from 2° to 5°C for periods of weeks during winter. Body temperature of mammals may reach as low as –6 °C without resulting in death due to freezing. Little brown bats (*Myotis lucifugus*) were exposed for 30 minutes to 6 hours to temperature extremes ranging from –3° to –9°C (Hurst and Wiebers 1967). Of three bats whose body temperature reached –6°C, two died during the experimental run and a third bat whose body temperature reached –6°C lived for 3 days following exposure. Bats subjected to –4°C lived for an indefinite period of time. More recently, Barnes (1989) measured core body temperatures as low as –2.9°C in hibernating arctic ground squirrels (*Spermophilus parryii*) held in outdoor burrows near Fairbanks, Alaska! The term hibernation is also referred to as seasonal torpor, deep hibernation, or true hibernation. Animals that undergo hibernation include ground squirrels, marmots, and hedgehogs. No mammal remains continuously in a dormant state during the entire period of hibernation, however, and the length of the period of hibernation varies with ambient temperature, body size, and species. We defer treatment of estivation (a form of shallow torpor typified by desert small mammals) for the later discussion of how mammals cope with heat.

Dormancy is exhibited by some seven orders of mammals (marsupials, "insectivores," elephant shrews, bats, primates, rodents, and carnivores) and at least 6 groups of birds (swifts, goatsuckers, hummingbirds, sunbirds, manakins, and colies). Representative birds and mammals and their sizes are given by Hudson (1978), Lyman and colleagues (1982), Geiser and Ruf (1995), and more recently Barclay et al. (2001). All mammals exhibiting adaptive hypothermia have a hypothalamic set point below which they do not allow body temperature to fall. For example, in very cold environments, for most species, the lowest temperature for hibernation may be regulated at just a couple degrees above freezing! For species in slightly warmer environments, body temperatures range slightly higher (5° to 10°C), and for species that employ torpor only for short-term emergencies, body temperatures range from 10° to 15°C. When challenged in such a way, these animals must arouse from torpor and restore their body temperature all the way to normal before reentering the state of hibernation. In terms of energy, this can be very costly.

The largest mammals to undergo true hibernation are marmots (Genus *Marmota*), which weigh about 5 kg. Hibernation is not possible or even necessary in very large mammals because of the large amount of energy necessary for arousal (Morrision 1960). Also, due to their size, large mammals can store sufficient energy in the form of internal fat to meet winter demands. Contrary to popular belief, bears do not exhibit "true" hibernation. Instead, they undergo a period of **winter lethargy** or shallow torpor with a decrease in body temperature of only 5° to 7°C below their active level of 38°C (Craighead et al. 1976; Rogers 1981; Watt et al. 1981; Hissa et al. 1994; Hissa 1997; Harlow et al. 2004). For example, as indicated by Hainsworth (1981:249), a grizzly bear (*Ursus arctos*) that weighs 386 kg requires 347,400 cal (calories) to raise its temperature 1°C (specific heat of animal tissue is about 0.9 cal/g • °C). If the grizzly bear were truly hibernating, we would assume that it must raise its body temperature from 5° to 37°C for arousal. The heat required to perform this arousal would be 11,116,800 cal! Just to maintain a basal metabolic rate, this bear would require about 6,100,000 cal/day. Also, many females are pregnant during the winter. During arousal, heat production increases above basal levels, and the amount of heat required in such a short period (daily) may not be feasible. Because large mammals can store more energy internally relative to their rate of energy use, a decrease in body temperature is less critical to them. Do bears exhibit periodical arousals during winter? Some investigators contend that bears do not engage in bouts of arousal during winter lethargy (Nelson et al. 1973; Nelson 1980; Hissa et al. 1994). In contrast, Harlow et al. (2004) suggest that bears engage in bouts of muscle activity during the winter denning period that results in retention of muscle strength without elevating their core body temperature and without arousing from shallow torpor (Harlow et al. 2004). Energetically speaking, it makes good sense to maintain a comparatively high body temperature for large mammals such as bears. Members of Order Carnivora do not exhibit true hibernation; however, as seen above, some species undergo inactivity and depression of body temperature during winter periods. For example, American badgers (*Taxidea taxus*), striped skunks (*Mephitis mephitis*), and raccoon dogs (*Nyctereutes procyonoides*) undergo inactivity during winter at northern latitudes (Harlow 1981; Korhonen and Harri 1984; Mutch and Aleksiuk 1977). Striped skunks enter spontaneous daily torpor with body temperature reaching 26°C—the lowest torpid body temperature recorded for any carnivore (Hwang et al 2007).

Mammals capable of adaptive hypothermia are found in the mammalian groups Subclasses Prototheria, Metatheria, and Eutheria. True hibernation has not been reported for members of the Orders Cetacea, Cingulata, Carnivora, Tubulidentata, Lagomorpha, Perissodactyla, and Artiodactyla. The short-nosed echidna (*Tachyglossus aculeatus*) and 15 species of marsupials undergo adaptive hypothermia (Geiser and Ruf 1995; Grigg and Beard 2000; Nicol and Andersen 2000). Marsupials capable of torpor range in size from Giles' planigale (Planigale gilesi, 8 g) to the western quoll (*Dasyurus geoffroii*, 1 kg). Of the Orders Afrosoricida, Erinaceomorpha, and Soricomorpha, only the hedgehogs (Genera *Erinaceus* and *Aethechinus*) and some tenrecs of Madagascar hibernate, as do the golden moles (Family Chrysochloridae). In Europe, hedgehogs have been a principal subject of hibernation studies. Shrews do not hibernate, but some may undergo shallow torpor (Fons et al. 1997). Two metabolic levels characterize the Family Soricidae (Vogel 1976; Nagel 1977; Genoud 1988). The Subfamily Crocidurinae (white-toothed shrews) is characterized by a low metabolic rate and low body temperature and undergoes torpor. In contrast, the Subfamily Soricinae (red-toothed shrews) exhibits high metabolic rates and elevated body temperatures and cannot undergo torpor (Churchfield 1990; Merritt 1995). The different metabolic levels of the two subfamilies represent evolutionary responses to different climates: White-toothed shrews are adapted to warmer, more southern latitudes, whereas the red-toothed shrews have evolved in more northerly latitudes.

Bats avoid cold by hibernating, migrating, or doing both. The picture of temperature regulation in the "microchiropterans" is more complicated, however (O'Farrell and Studier 1970; Lyman et al. 1982; Thomas 1995; Altringham 1996; Neuweiler 2000). Bats of the family Vespertilionidae (see chapter 13) are well-known hibernators, and most species occurring in the temperate zones spend the winter in caves for this purpose. Vespertilionid bats in temperate regions typically hibernate from October to April. In a hibernation site (hibernaculum), the ambient temperature may be near 5°C, and the bats are in deep hibernation, maintaining a body temperature of about 1°C above the ambient. This period is punctuated with occasional arousals during which the animal urinates, drinks, or changes location. Arousals occur every 1–3 weeks and last for only a few hours each time. In preparation for winter, temperate zone bats may establish body fat in autumn equal to a third of their body mass. In summer, shallow daily torpor (lasting for only a few hours) may occur during the day while bats are roosting. Body temperature rises again before feeding at dusk. Winter hibernation in bats differs from short-term torpor largely in the length of dormancy and the temperature decrease. The duration of hibernation for bats differs widely among species and within a species, depending on the geographic area. In the northeastern United States, for example, the little brown bat (*Myotis lucifugus*) hibernates for 6–7 months. Periods of hibernation for bats in warmer areas are considerably shorter. The larger "megachiropterans" are euthermic and maintain body temperature between 35° and 40°C (Ransome 1990; McNab and Bonaccorso 1995). Bartholomew and colleagues (1970), however, reported torpor in smaller species of fruit bats (*Nyctimene albiventer* and *Paranyctimene raptor*), and Coburn and Geiser (1996) described torpor in blossom bats (*Syconycteris australis*). When food was withheld, blossom bats remained in torpor from 1 to 10 hours at a body temperature of about 18°C. Torpor in primates is limited to the small lemurs of Madagascar. Mouse lemurs (Genus *Microcebus*), the smallest primates, range in mass from 29 to 63 g and exhibit an average body temperature during torpor of 24.9°C (Ortmann et al. 1996). Recently, Madagascar fat-tailed dwarf lemurs (*Cheirogaleus medius*) showed a wide daily fluctuation in body temperature of almost 20°C, closely tracking air temperature of their tree holes (Dausman et al. 2004).

The greatest number of hibernators are found in the Order Rodentia, specifically squirrels (Family Sciuridae). Rodents also show the greatest variation in length of dormancy. Within this group, we see a continuous integration between daily and seasonal torpor. Ground squirrels and marmots undergo periods of deep hibernation (Barnes 1989; Armitage et al. 1990; Ferron 1996; Buck and Barnes 1999a; Barnes and Buck 2000). Body temperature in marmots decreases from about 39°C during summer to between 2° and 8°C during deep hibernation. The lowest body temperature reported for hibernating marmots is about 4.0°C (Ferron 1996). Typically, the heartbeat slows from 100 to 15 bpm (beats per minute), and oxygen consumption falls to a tenth the normal rate. Marmots may only breathe once every 6 minutes when in deep hibernation; hibernating animals lose 30% of their weight by mobilizing body fat during winter. It is noteworthy that, with the exception of the woodchuck (*Marmota monax*), all species of the Genus *Marmota* are social and hibernate in groups (Arnold 1988).

Richardson's ground squirrels (*Spermophilus richardsonii*) of northwestern North America also undergo periods of deep hibernation. Adult squirrels may enter hibernation as early as mid-July and emerge 8 months later, in mid-March. During much of their long period of hibernation, body temperature is about 3° to 4°C and increases to about 38°C for their 4-month active phase. Like the torpor in many hibernators, torpor in ground squirrels during winter is interrupted by frequent intervals of rewarming to euthermy, and entry and arousal show a stepped progression in body temperature. Hibernation has been well studied in different ground squirrels, especially in golden-mantled (*Spermophilus lateralis*), Richardson's, arctic (*S. parryii*), and thirteen-lined (*S. tridecemlineatus*) ground squirrels. The "champion" of hibernators is the arctic ground squirrel; it will be discussed in detail later in this section.

Cricetides commonly survive cold by employing daily torpor coupled with communal nesting. Because their body temperatures are not as depressed as those of such deep hibernators as ground squirrels, they show a great deal of thermolability. All members of the Genus *Peromyscus* probably undergo some form of dormancy. Torpor occurs diurnally and lasts for less than 12 hours (Hudson 1978). For example, *Peromyscus* may be active on a warm day in mid-January in the north, but when ambient temperatures reach about 2° to 5°C, these small rodents may rapidly decrease body temperature to 13°C and undergo short-term torpor. The lability of body temperature in *P. leucopus* seems greater than that of other species of the genus (Hart 1971). Among the dipodids, hibernation (deep torpor) is reported for members of the Subfamily Zapodinae (Genera *Zapus* and

Napaeozapus; Brower and Cade 1966; Muchlinski 1980). In the Wasatch Mountains of Utah, the period of hibernation of *Z. princeps* ranges from early September to late July depending on elevation of the hibernacula (Cranford 1978). A summer active period of about 87 days spanned the period between snow melt in early summer and the beginning of the autumn snowfall season. Mice hibernated at an average depth of 59 cm, with no food caches. The mean soil temperature during hibernation was 4.6°C, and emergence from hibernation was cued by increasing soil temperature (figure 9.16). Birch mice (*Sicista betulina*) of northern and eastern Eurasia undergo daily torpor in response to cold. They decrease body temperature to about 4°C and wake spontaneously at night to feed. Most hamsters (namely, golden or Syrian and Turkish—*Mesocricetus auratus*, *M. brandti*) do not readily hibernate or exhibit torpor. On the other hand, Siberian hamsters (*Phodopus sungorus*), inhabitants of Siberia, undergo periodic daily torpor with body temperature near 19°C. Color change and torpor in these hamsters are quite variable; some individuals turn white in winter and exhibit torpor, whereas others remain brown and do not enter torpor. Dormice (Genera *Glis* and *Eliomys*) are also reported to undergo hibernation in response to food deprivation (Wilz and Heldmaier 2000).

The **cycle of dormancy** can be divided into three phases: entrance, period of dormancy, and arousal (Hudson 1973). An excellent example of the cycle of dormancy is that measured for the arctic ground squirrel (*Spermophilus parryii*; Barnes 1996; Boyer and Barnes 1999) and for

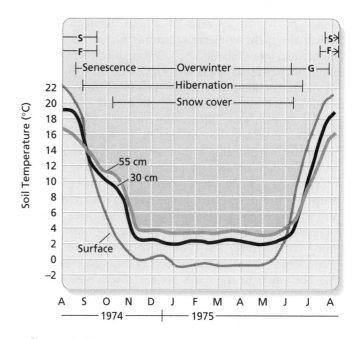

Figure 9.16 Soil temperatures in a montane environment. Daily average soil temperatures at the surface, 30 cm, and 55 cm depth at 2,900 m in Albion Basin, Alta, Utah, from 15 August 1974 to 15 August 1975. The lines are plotted through the weekly mean. Abbreviations: G = plant growth; F = flowering; S = plants setting seed. *Data from J. A. Cranford, 1978, Hibernation in the western jumping mouse (Zapus princeps), in* Journal of Mammalogy, *59:496–509.*

Richardson's ground squirrel (*S. richardsonii*; Wang 1978, 1979). Many mammals prepare for **entrance** to winter dormancy by putting on fat. For example, the woodchuck of North America begins putting on weight in midsummer. Just prior to hibernation, its weight is about 30% greater than in early summer. In autumn, after lining its hibernaculum with leaves and grasses, the obese woodchuck moves into its den, plugs the entrances, and curls up into a tight ball. The earthen plug is important for maintaining proper temperature and humidity in the den during winter. Increase in body mass among hibernators before dormancy can be impressive, reaching 80% in golden-mantled ground squirrels (*S. lateralis*), for example. During entrance, animals decrease heart rate, blood pressure, and oxygen consumption and finally exhibit a decline in body temperature. For some species, the environmental cues that signal preparation for hibernation are associated with the time of year and are induced by a combination of low temperature or lack of food. For others, entrance into dormancy may occur without an external stimulus. For such species, the duration of the daily light and the temperature cycle may synchronize to maintain the annual rhythm of dormancy. For arctic ground squirrels, entrance into hibernation begins in autumn with stepped periods of torpor alternating with periodic rewarming bouts reaching euthermy (figure 9.17A). The soil temperature during the period of entrance is about 0°C.

The **period of dormancy** is signified by a leveling off of the body temperature in late autumn. During the season of hibernation, arctic ground squirrels demonstrate 12 bouts of torpor with a duration of 1–3 weeks each. Bouts are separated by regular episodes of arousal (figure 9.17B) in which the body temperature returns to normal (that is, euthermic) levels (about 37°C). During hibernation, minimum body temperature falls to –2°C, whereas temperature of the surrounding soil is –10° to –15°C. **Arousal** from hibernation occurs in spring. To reduce energy expenditure during arousal, some hibernators employ passive rewarming. This is achieved by basking in the sun or practicing social thermoregulation and/or is facilitated by an increase in surrounding ambient temperature (Geiser et al. 2004).

Changes in body temperature of a specific episode of torpor of the arctic ground squirrel are illustrated in figure 9.17B. Each arousal episode includes three phases: (1) rewarming from torpor, (2) 24 hours of normal high body temperature, and (3) a slow cooling into torpor. Periods of arousal account for most of the energy used during dormancy. Energy costs associated with arousal from –2° to 37°C include (1) the cost of warming from hibernation to 37°C, (2) the cost of sustaining euthermy (37°C) for several hours, and (3) the cost of maintaining a metabolic rate above torpor as the body temperature slowly declines during reentry into torpor. For the entire period of dormancy, these three metabolic phases account for an average of about 83% of the total energy used by a squirrel (e.g., *Spermophilus richardsonii*; Wang 1979). Energy for arousal from hibernation is provided principally by nonshivering thermogenesis in brown adipose tissue augmented by shivering. This energy expense seems rather wasteful, and the

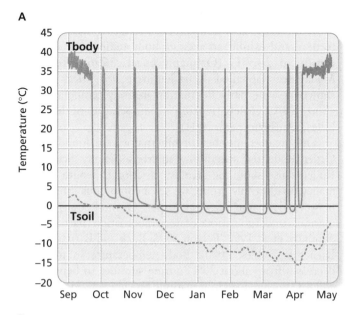

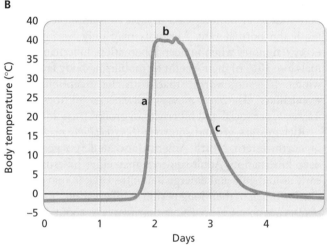

Figure 9.17 **Hibernation in the arctic ground squirrel.** (A) The period of seasonal hibernation of the arctic ground squirrel (*Spermophilus parryii*) residing in northern Alaska. Note the numerous bouts of torpor and arousal episodes and the extreme decrease in core body temperature occurring during deep torpor. (B) Each arousal episode includes three phases: rewarming, euthermia, and cooling. *B. B. Barnes and B. M. Barnes, 1999, Molecular and metabolic aspects of mammalian hibernation, BioScience 49: 713–724.*

function of such periodical arousals is not well understood. It is possible that periodic arousal permits hibernators to assess environmental conditions conducive to emergence.

RESISTANCE TO COLD

Increase in Thermogenic Capacity

Mammals employ many different tactics of energy conservation to avoid cold stress. If cold stress persists, however, mammals must resist it by processes that generate heat and thus require energy (see table 9.1). The major ways in which endotherms increase heat production are muscular activity and exercise, involuntary muscle contractions (shivering), and nonshivering thermogenesis (see figure 9.4). The most conspicuous mechanism by which endotherms increase heat production is by muscular activity from locomotion or shivering; however, the means need not be apparent for it to increase the rate of heat production (Bartholomew 1982; Korhonen et al. 1985; Kleinebeckel and Klussmann 1990; Schmidt-Nielsen 1997; May 2003; Hohtola 2004; Rose and Ikonomopoulou 2005). Shivering is well documented for marsupials, humans, canids, felids, selected rodents, lagomorphs, and ungulates as well as many species of birds. Small mammals residing in variable environments, however, employ a means of heat production called **nonshivering thermogenesis** that does not involve muscle contraction (Smith and Horwitz 1969; Jansky 1973; Girardier and Stock 1983; Trayhurn and Nicholls 1986; Cannon and Nedergaard 2004). **Brown adipose tissue,** the site of nonshivering thermogenesis, was first observed by Conrad Gesner in 1551 in the interscapular area of the Old World marmot (*Marmota alpina*). Brown fat, once referred to as the "hibernating gland," is found in all hibernating mammals and is the primary thermogenic tissue of cold-adapted small mammals (especially rodents and shrews). It is also well developed in newborn species of mammals, including humans. Within mammals, brown fat has been reported in many species spanning nine orders: Dasyuromorphia, Chiroptera, Afrosoricida, Soricomorpha, Rodentia, Lagomorpha, Artiodactyla, Carnivora, and Primates. The wide occurrence of nonshivering thermogenesis in small mammals was reviewed by Heldmaier (1971) and Jansky (1973). Monotremes do not possess brown adipose tissue (Hayward and Lisson 1992); however, researchers have isolated brown adipose tissue in some species of marsupials (Hope et al. 1997; Wallis 1979; Clements et al. 1998; Rose et al. 1999). Oliphant (1983) reported the presence of brown adipose tissue in ruffed grouse and chickadees; however, microanatomical studies based on multilocularity, increased vascularity, mitochondrial density, and cytochrome-c oxidase activity indicate that birds do not possess functional brown fat (Olson et al. 1988; Saarela et al. 1989, 1991; Brigham and Trayhurn 1994).

Unlike white adipose tissue, characterized by a single large droplet of fat with a peripheral nucleus (Pond 1978), brown fat contains many small droplets (multilocular) with a centrally located nucleus. In contrast to white adipose tissue, brown fat is highly vascular and well innervated. The cells contain many mitochondria, whereas the white fat cells have comparatively few. Brown fat is capable of a far higher rate of oxygen consumption and heat production than white fat is. The reddish brown color of brown fat is derived from iron-containing cytochrome pigments in the mitochondria, the essential part of the oxidizing enzyme apparatus of brown adipose tissue. White fat serves primarily as insulation and a storage site for food and energy. Brown fat, with its rich supply of mitochondria, serves as a miniature internal "blanket" that overlies parts of the systemic vasculature and becomes an active metabolic heater applied directly to the bloodstream (Wunder and Gettinger 1996). Deposits of brown

fat can be rather diffuse but are principally found in the interscapular, cervical, axillary, and inguinal regions in close proximity to blood vessels and vital organs (Hyvärinen 1994).

Temperature receptors in the skin sense cold and send impulses to the preoptic area of the hypothalamus—the "mammalian thermostat" located in the brain. Impulses are then relayed along the sympathetic nerves to the brown adipose tissue, where nerve endings release the neurohormone **norepinephrine.** At the brown adipose tissue, norepinephrine activates an enzyme (lipase) that splits triglyceride molecules into glycerol and free fatty acids. In the brown fat cell, the mitochondrial respiration is "uncoupled" from the mechanism of adenosine triphosphate (ATP) synthesis so that the energy of oxidation of the fatty acids is dissipated as heat instead of being used for ATP synthesis. A special protein, called **thermogenin,** is responsible for the uncoupling (Himms-Hagen 1985). When bats arouse from hibernation, the brown fat pad is much warmer than the rest of the body (Hayward and Lyman 1967; figure 9.18). The close proximity of Selzer's vein just beneath the interscapular brown fat permits rapid passage of warmed venous blood directly to the heart and brain with a minimum of heat loss.

A

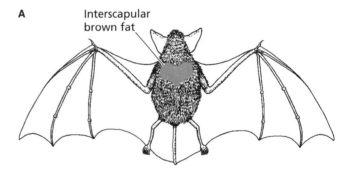

B

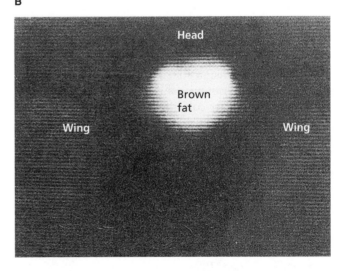

Figure 9.18 Brown fat. (A) Diagram showing the position of interscapular brown fat in the big brown bat (*Eptesicus fuscus*). (B) Photograph taken with heat-sensitive film indicating that the temperature of interscapular brown fat in the big brown bat is high during arousal from hibernation. *Adapted from Hayward and Lyman, 1967,* Mammalian Hibernation, *3rd ed., Elsevier.*

Research on northern small mammals has shown that dramatic increases in metabolic rate in cold are due to nonshivering thermogenesis. Nonshivering heat production usually tracks environmental temperatures, falling to the lowest rates in spring and summer, increasing in autumn, and peaking in winter. Typically, small mammals demonstrate a significant inverse relationship between nonshivering heat production and environmental temperature. Wunder and colleagues (1977) showed a 29% increase in oxygen consumption for prairie voles (*Microtus ochrogaster*) in winter compared with summer. Alaskan red-backed voles (*Clethrionomys rutilus*) exhibited a 96% increase in metabolism in winter. Although species of *Peromyscus* commonly employ many different survival adjustments (Merritt 1984), they are adept at increasing metabolism during winter. Maximum metabolism for *Peromyscus* from Iowa and Michigan ranged from a 24 to 70% increase over summer rates (Lynch 1973; Wickler 1980, respectively). Shrews typically show very high rates of nonshivering heat production. For example, for masked shrews (*Sorex cinereus*) residing in Pennsylvania, nonshivering thermogenesis in winter was almost twice that measured in summer—an increased capacity of 182% (Merritt 1995). Maximum metabolism, resting metabolic rate, and nonshivering thermogenesis were compared for 12 species of shrews from tropical and temperate regions (Sparti 1992). Maximum metabolism, and consequently improved cold tolerance, was pronounced in temperate species. Smith and Horwitz (1969) found a direct correlation between mass of brown fat and nonshivering thermogenesis. Like nonshivering thermogenesis, monthly changes in the mass of brown fat are inversely related to minimum environmental temperature for many species of voles, mice, and shrews (Merritt and Zegers 2002).

Adaptations to Heat

To survive in deserts, mammals must cope with a variety of demanding environmental challenges, such as intense heat during the day, cold nights, paucity of water and cover, and a highly variable food supply. Desert ecosystems are widespread and abundant—35% of the earth is covered with deserts. Mammals have successfully colonized desert ecosystems, as evidenced, for example, by the rich and diverse fauna of heteromyid and dipodid rodents (see chapter 18) in the deserts of North America and the Old World, respectively (Schmidly et al. 1993; Wilson and Reeder 2005). An excellent review of studies examining adaptations of mammals to desert environments is provided by Degen (1997).

The challenges faced by mammals in desert environments are even more severe than those encountered in cold regions; water as well as food is scarce, and the problem in temperature regulation is reversed. When we con-

sidered cold stress to mammals, we assessed the role of thermal radiation, conduction, convection, and other environmental influences on heat transfer between an individual and the environment, and then we focused on the many mechanisms employed to reduce heat flow to the environment (see figure 9.1). Recall that small mammals in cold regions reduced heat loss by up to 19% by increasing pelage insulation. Increased insulation reduced the gradient between the warm core of the body and the outside environment. In desert environments, however, the gradient between the internal and environmental temperatures is reversed. In some deserts, mammals may have to cope with air temperatures that reach 55°C and ground temperatures that exceed 70°C. Unlike arctic mammals concerned with conserving heat, desert mammals must dissipate heat or avoid it to maintain euthermy. Our discussion now focuses on the complex anatomical, physiological, and behavioral adaptations that enhance the survival of mammals in desert ecosystems.

Water is essential for survival. It constitutes 70% of the body mass of mammals, and water loss must be balanced by water gain. In mammals, **osmoregulation**—the maintenance of proper internal salt and water concentrations—is performed principally by the kidney. In addition to producing a concentrated urine, desert mammals cope with a lack of water by producing very dry feces. Evaporation, occurring mostly from the respiratory tract, is the major avenue of water loss but is also an important device for cooling. In this section, we examine evaporation across the skin and from respiratory passages, and the ways in which mammals keep cool in xeric (very dry) environments by employing evaporative cooling. Temperature regulation is also influenced by changes in insulation, appendages, metabolic rate, and body size. Mammals achieve water balance by eating succulent vegetation, drinking available water, and "metabolically" converting food into water (Nagy and Peterson 1980; Morton and MacMillen 1982; MacMillen and Hinds 1983). Desert mammals also conserve water by behavioral means such as selecting saturated burrows in which they reside during the heat of the day. Lastly, we will discuss a form of dormancy called estivation, which mammals residing in hot environments employ to survive heat and reduced food availability.

WATER ECONOMY

The Mammalian Kidney

Most of the elimination of excess water and soluble salts, urea, uric acid, creatinine, and sulfates occurs in the kidney. Mammalian **kidneys** are paired, bean-shaped structures located within the dorsal part of the abdominal cavity (figure 9.19). The kidney in cross section displays the following areas and structures. The outer **cortex** contains the renal corpuscles, convoluted tubules, and blood vessels. Masses of cortical tissue fill

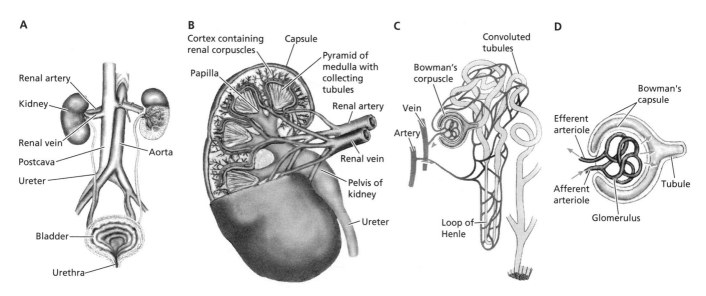

Figure 9.19 **The human excretory system.** (A) Ventral view of the entire system; (B) median section of one kidney; (C) relationship of Bowman's corpuscles, tubules, and blood vessels; (D) a single Bowman's corpuscle and adjacent tubule (shown also in cross section)—solid arrows show flow of blood, broken arrows show the excretory path. (C) and (D) are diagrammatic and much enlarged. *Adapted from T. I. Storer et al., 1979, General Zoology, 6th ed., McGraw-Hill.*

in between the pyramids of medullary tissue. The inner **medulla** is divided into triangular wedges called **renal pyramids.** Their broad bases are directed toward the cortex, and narrow apices **(renal papillae)** are oriented toward the center of the kidney, opening into the **calyx** and expanded **pelvis.** Ducts leading into the pelvis, the **ureters,** empty into the **urinary bladder,** which functions as a storage organ for urine. Another duct, the **urethra,** drains the bladder and carries its contents to the outside.

Nephrons (about 1.5 million in each kidney) are the functional units of the kidney and consist of a closed bulb, **Bowman's capsule** (or glomerular capsule), connected to a long coiled tube. Tubules of the various nephrons empty into collecting ducts that discharge into the pelvis of the kidney and then connect to the ureter. A microscopic mass of capillaries called the **glomerulus** is enclosed within the capsule. The capsule plus the inner glomerulus is called a **renal corpuscle.** Blood reaches the kidney via the large renal artery, a branch of the descending aorta. The blood arrives at the glomeruli by afferent arterioles. The blood is collected from the glomeruli by a number of venules and leaves the kidney by way of renal veins making their way to the inferior vena cava. Blood vessels enter a convoluted network within the glomerular capsule, then move through proximal and distal convoluted tubules and the **loop of Henle,** finally emptying into a branch of the renal vein. Exchange of substances takes place through active transport and osmosis almost exclusively between blood capillaries and nephrons.

The mammalian kidney performs many different roles including glomerular filtration, tubular reabsorption, and tubular secretion. These functions are well described by Hill and Wyse (1989) and Schmidt-Nielsen (1997). For mammals residing in desert environments, the ability to concentrate urine is paramount and closely tied to the function of the kidney (Fielden et al. 1990; Degen 1997). Because of the urine-concentrating ability of kidneys, mammals are able to produce urine that is hyperosmotic to that of blood plasma—up to 25 times the concentration of plasma. Understandably, the highest urine concentrations are found in mammals residing in desert habitats (Yousef et al. 1972; French 1993). The concentrating ability of the mammalian kidney in different species is closely related to the respective lengths of their loops of Henle and collecting ducts that transverse the renal medulla. The prominence of the medulla is commonly expressed as the relative medullary thickness (RMT), an important index of kidney adaptation (Sperber 1944). RMTs of mammals residing in arid areas are greater than those of mammals from more mesic environments (those with more moisture). The relationship between RMT and the kidney's maximum urine-concentrating capacity was first quantified by Schmidt-Nielsen and O'Dell (1961) and has proven very useful as a means of comparing kidney function in mammals. The anatomy of the papilla of the medulla can be compared visually for different species (figure 9.20). In desert-adapted small mammals, the papilla may extend beyond the margins of the renal capsule into the ureter. This extension is pronounced in small desert rodents (MacMillen and Lee 1969; Altschuler et al. 1979; Diaz and Ojeda 1999), shrews (Lindstedt 1980), and bats (Geluso 1978), suggesting that these species possess very powerful kidneys. In contrast, aquatic mammals, such as beavers, aquatic moles, water rats, and muskrats, have very short loops (shallow papillae) and produce a less concentrated urine.

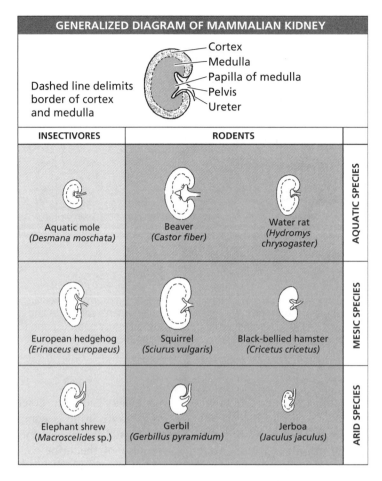

Figure 9.20 **The kidneys of selected mammals.** Aquatic species show little or no development of the papilla of the medulla. Genera *Desmana* and *Hydromys* lack the papilla. Genus *Castor* has two very shallow papillae. Mesic species have papillae. The papilla is especially well developed in arid species, so much so that it often penetrates well into the ureter (e.g., Genera *Macroscelides*, *Gerbillus*, and *Jaculus*). *Adapted from R. W. Hill and G. A. Wyse, 1989,* Animal Physiology, *2nd ed., Harper & Row.*

Urine and Feces

Comparative studies of renal function and morphology in mammals indicate a direct relationship between the ecological distribution of a species and its ability to conserve urinary water. The ability to concentrate urine in mammals is associated with long loops of Henle and tubules in the kidney that enhance the countercurrent exchange function. Species that reside in arid habitats tend to possess kidneys better adapted for water conservation. Representatives are found in the Orders Diprodontia (Family Macropodidae), Chiroptera, Cingulata (Family Dasypodidae), Rodentia (Families Muridae, Cricetidae, Octodontidae, Heteromyidae, Sciuridae), and Lagomorpha (Family Leporidae). Most mammals lose water by excretion in the urine and elimination in the feces. But the preceding groups have the ability to produce relatively dry feces and concentrated urine. For example, the average values for maximum urine concentration in desert heteromyids is superior to those of most mammals and comparable to those of other desert-adapted small mammals, such as dipodids from Asia and North Africa (see chapter 18) and murids from Australia and southern Africa

(French 1993). For example, the laboratory white rat can produce urine with twice the osmotic concentration that humans can achieve. The dromedary, and even dogs and cats, have a urine-concentrating equivalent to that of the white rat. As expected, the amount of water loss from feces is quite low for desert mammals. The feces of Merriam's kangaroo rat (*Dipodomys merriami*) are over 2.5 times as dry as those of white rats (834 versus 2,246 mg of water/g of dry feces). Furthermore, heteromyids commonly decrease fecal water loss by assimilating over 90% of the food they ingest. Desert rodents (e.g., kangaroo rats, sand rats, and jerboas) produce urine concentrations of 3,000–6,000 mOsm/L; Australian hopping mice (*Notomys alexis* and *N. cervinus*) can produce urine concentrations of over 9,000 mOsm/L (MacMillen and Lee 1969, 1970). Intuitively, we think of heteromyid rodents such as kangaroo rats and pocket mice as leaders in water conservation. Many other small mammals, however, such as pallid bats (*Antrozous pallidus*), canyon and house mice (*Peromyscus crinitus* and *Mus musculus*), and golden hamsters (*Mesocricetus auratus*), have evaporative water losses equivalent or greater to that of desert heteromyids. Interestingly, the spectacled hare-wallaby

(*Lagorchestes conspicillatus*) has the lowest mass-specific rate of water turnover reported for any mammal (Bradshaw et al. 2001)!

An additional way that desert rodents economize water loss is by producing a highly concentrated milk. The milk of *D. merriami* averages 50.4% water—a concentration comparable to that produced by seals and whales (Kooyman 1963). Furthermore, it has been demonstrated that in desert rodents, canids, and kangaroos, mothers reclaim water by consuming the dilute urine and feces of their young. This behavior may regain about one third of the water originally secreted as milk (Baverstock and Green 1975).

Diet

Since free drinking water is not available for desert mammals, they must obtain water from other sources, such as succulent plants or the body fluids of their prey or by consuming dry food. This requires subsisting on **metabolic water,** which is created in the cells by the oxidation of food, especially carbohydrates (Frank 1988; Hill and Wyse 1989; Degen 1997). Some desert mammals consume succulent plants for a source of water. Desert woodrats (*Neotoma lepida*) and cactus mice (*Peromyscus eremicus*) of southwestern North America consume large quantities of cactus (Genus *Opuntia*) as a source of both food and water. In addition, cactus is a staple in the diet of other xeric-adapted mammals, such as northern pocket gophers (*Thomomys talpoides*) inhabiting the dry short-grass prairies of Colorado (Vaughan 1967). Gophers and murids such as woodrats have evolved the ability to metabolize oxalic acid, an abundant compound of cactus that is toxic to other mammals. Some desert mammals depend on moist food such as cactus. Others rely primarily on dry seeds or halophytic plants (those that grow in salty soils), and their intake of water is quite minimal. Kangaroo rats, pocket mice of southwestern North America, and fat sand rats (*Psammomys obesus*) of the Saharo-Arabian deserts are able to subsist on dry food only and do not require water.

Halophytic plants of the Family Chenopodiaceae form a staple in the diet of many desert-dwelling small mammals but are, by definition, extremely high in salt concentrations (Degen 1997; Degen et al. 2000). Many small mammals that consume halophytes possess kidneys that produce highly concentrated urine. Fat sand rats obtain water from the leaves of the saltbush (*Atriplex halimus*). These gerbillid rodents are unusual in being diurnal and wholly herbivorous, whereas other members of the family are nocturnal and granivorous. Genus *Psammomys* scrapes off the outer surface of leaves with their teeth before consuming them—negligible amounts of leaf are scraped from moist plants and substantial amounts from dry plants (Kam and Degen 1992). Leaves possess up to 90% water but have high concentrations of salts and oxalic acid. To consume this plant material, the rat must produce urine with extremely high concentrations of salt as well as be able to metabolize large concentrations of oxalic acid. A

parallel development occurs in the chisel-toothed kangaroo rat (*Dipodomys microps*), an inhabitant of shrub habitats of western North America. This heteromyid harvests leaves of the chenopod (*A. confertifolia*) rather than foraging on seeds, as do most heteromyids. The epidermis of *Atriplex* leaves is high in electrolyte concentration, but the more internal parenchyma is low in electrolytes and high in starch. *D. microps* is able to consume the inner tissue by shaving off the peripheral epidermis, thus minimizing its consumption of salt. This kangaroo rat can perform such a task while conspecifics cannot because they possess lower incisors that are broad, flattened anteriorly, and chisel-shaped, permitting access to the plant's inner tissue. Because other sympatric kangaroo rats lack this adaptation, they cannot exploit this unique food resource and must rely on unpredictable seed crops (Kenagy 1972). Many other species exhibit tolerance for high levels of salt in their food or water (e.g., highland desert mice [*Eligmodontia typus*] of South America, fawn hopping-mice [*Notomys cervinus*] and Tammar wallabys [*Macropus eugenii*] of Australia, and western harvest mice [*Reithrodontomys megalotis*] of western North America, to mention just a few).

Desert carnivores and insectivores meet their moisture requirements by relying on their food rather than on free water. Southern grasshopper mice (*Onychomys torridus*) of hot, arid valleys and shrub deserts of southwestern North America consume primarily arthropods, including scorpions, beetles, and grasshoppers. Studies of water balance demonstrate that grasshopper mice can be maintained in the laboratory for more than 3 months on a diet of only fresh mouse carcasses. Grasshopper mice are able to survive the arid conditions of the desert because of their preference for animal foods high in water content (Schmidt-Nielsen 1964). Kit foxes, badgers (*Taxidea taxus*), coyotes (*Canis latrans*), desert hedgehogs (*Hemiechinus auritus*), fennecs, and the Australian dasyurid (*Dasycercus cristicauda*) are also able to subsist on a meat diet with minimal supplementation by free drinking water. Fennecs, xeric-adapted canids inhabiting the deserts of northern Africa, maintained water balance for a minimum of 100 days when fed only mice and no drinking water (Noll-Banholzer 1979).

In East Africa, many plains antelope are able to endure long periods of intense heat without drinking water. Two examples of nonmigratory ungulates, the eland (*Taurotragus oryx*) and oryx (*Oryx beisa*), are able to survive indefinitely without drinking water in an ecosystem typified by environmental temperatures reaching 40°C. A critical part of their ability to survive without drinking water is contingent on the fact that they can use metabolic water. Elands consume large quantities of the leaves of *Acacia*, which contain about 58% water. Oryx feed primarily on grasses and shrubs, a staple being the shrub *Diasperma*. Leaves of this shrub fluctuate in water content. During the day, when air temperature is high and humidity low, the leaves contain only 1% water; at night, they increase to 40% water due to decreased temperature and increased relative humidity. The oryx takes advantage of the variable water content of *Diasperma* by consuming it only late at night when water content is highest. During other hours, oryx opportunistically consume more succulent species of plants according to availability.

TEMPERATURE REGULATION

Evaporation

Metabolic processes, such as kidney function, all require energy. Changes in metabolic processes produce heat, and in desert environments, internal heat must be lost or an individual overheats and dies. **Evaporative cooling,** the major mechanism employed by mammals to reduce body temperature, is very effective as long as an animal has an unlimited supply of water. Evaporative cooling is relatively simple. When mammals cool by evaporation, they take advantage of a physical property of water's ability to absorb a great deal of heat when it changes state from a liquid to a vapor. In desert ecosystems, however, heat is intense and water scarce, so evaporative cooling is of limited utility except as a short-term response to a temperature crisis. In terms of thermal stress, it is clearly maladaptive for a kangaroo rat to venture into the desert sun. For such a small mammal to maintain normal body temperature under such circumstances, it would have to evaporate 13% of its body water per hour. This would be highly taxing, as most species die when they lose 10–20% of their body water. As we know from discussing how mammals cope with cold, many factors can modify the direct influence of environmental stressors. Although evaporative cooling requires some trade-offs, it represents a major line of defense for mammals combating heat.

We now focus on four major mechanisms of water loss known as **insensible,** or **transpirational, water loss.** This water loss occurs by diffusion through the skin and from the surfaces of the respiratory tract. It includes sweating, panting, saliva spreading, and respiratory heat exchange (Hill and Wyse 1989). Keep in mind that small mammals are limited by body size in the extent to which they can store and lose heat. As we will learn later, selection of cool and saturated microclimates represents a crucial strategy for conserving water by small mammals residing in xeric ecosystems.

Sweating

For many mammals, water loss occurs through the skin by way of sweat glands. There are two types of sweat glands: **apocrine** (found on the palms of the hands and bottom of the feet; they do not secrete for thermoregulation) and **eccrine** (distributed throughout the body; they secrete for evaporative heat loss). Water released from eccrine sweat glands evaporates from the surface of the skin, cooling it and the underlying blood. Sweating in response to overheating occurs only in primates and several species of

ungulates; it does not occur in rodents and lagomorphs. For humans working in a hot, dry environment, as much as 2,000 mL/h of water may be produced by eccrine sweat glands and lost by evaporation. Sweating appears to have evolved in mammals whose fur does not represent an appreciable barrier to surface evaporation, but the mechanism is not quite so simple. For animals that sweat, such as camels, the insulative barrier provided by the pelage takes on special significance.

Panting

Humans sweat to increase cooling by evaporation. In contrast, canids possess very few sweat glands and cool primarily by **panting**—a rapid, shallow breathing that increases evaporation of water from the upper respiratory tract. Panting is a common method of evaporative cooling for many carnivores and smaller ungulates, such as sheep, goats, and many small gazelles (Schmidt-Nielsen et al. 1970a). All mammals lose some heat as a result of evaporation of water from their respiratory passages. Inspired air is cooler and less humid than expired air; thus, heat is released from the evaporatory surface in both warming and humidifying the air. Water (and heat) is conserved during expiration when the warmed, moist expired air meets the cooler respiratory surfaces.

Sweating versus Panting

A major difference between sweating and panting is that the panting animal provides its own air flow over the moist surfaces, thus controlling the degree of evaporative cooling. A sweating animal has minimal control over the degree of evaporation. Another shortcoming of sweating is that sweat contains large amounts of salt. A profusely sweating human may lose enough salt in the sweat to become salt deficient. This is why we are reminded to drink lots of liquid and limit strenuous exercise outside on very warm days. In contrast, panting animals do not lose any electrolytes and do not become sodium-stressed. Panting does, however, have some drawbacks. The muscular energy associated with panting generates more heat than sweating, thus adding to the heat load. Second, the increased ventilation generated by panting can result in severe respiratory alkalosis—an elevation of serum pH attributable to excess removal of carbon dioxide.

Cool Brains

Panting has the major advantage of allowing an animal under sudden heat stress (e.g., a gazelle pursued by a cheetah) to maintain a high body temperature and yet keep its brain at a lower temperature. Taylor (1972) and more recently Mitchell et al. (1997) described this fascinating adaptation in artiodactyls. The brain is kept cooler than the body by the now familiar mechanism of countercurrent heat exchange. Arteries carrying warm blood from the heart toward the brain come into intimate contact with venous blood cooled by evaporation of water from the walls of the nasal passages within the **cavernous sinus**—a network of small vessels immersed in cool venous blood located in the floor of the cranial cavity where heat exchange occurs (figure 9.21). The venous blood from the nasal passages cools the warmer arterial blood heading toward the brain. As a result, the brain temperature may be 2° or 3°C lower than the blood in the core of the body (Baker 1979). Taylor and Lyman (1972) found that the small Thomson's gazelle of East Africa running for 5 minutes at a speed of 40 km/h exhibited a core body temperature of 44°C, but its brain was maintained at the cooler level of 41°C. Other devices may augment cooling of the brain due to panting. Cabanac (1986) described the cooling of the brain from venous blood returning from facial skin, exchanging with warm arterial blood within the cavernous sinus and influencing the temperature of the brain.

Saliva Spreading

When faced with heat stress, many rodents and marsupials spread saliva on their limbs, tail, chest, or other body parts. Grooming saliva assists in evaporative heat loss. This technique is less effective than sweating for evaporative cooling because the fur must be soaked with saliva before heat can be lost from the underlying surface of the skin. Furthermore, this technique is only effective for a short time. Because of their small size, most rodents have limited supplies of internal water to replenish the high rates of loss. Nonetheless, many rodents rely solely on saliva spreading for evaporative cooling. This mechanism is especially useful when heat stress is relatively short, for example, to prevent excessive hyperthermia while searching for cool refuge sites.

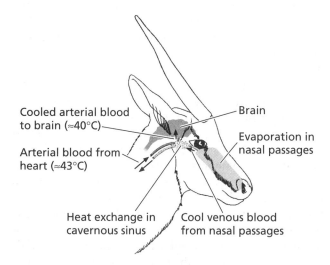

Figure 9.21 Cool brains. Schematic representation of the anatomical arrangement that promotes cooling of the brain in certain mammals. Dark vessels symbolize arterial blood flow, and light vessels, venous blood flow. In the cavernous sinus, the two flows are in intimate juxtaposition. *Adapted from C. R. Taylor and C. P. Lyman, 1972,* American Journal of Physiology *222:114–117.*

Secretion of saliva, like sweating and panting, is controlled by the hypothalamus. An increase in body temperature from a set point between 37° and 38.5°C is detected at preoptic tissues of the anterior hypothalamus and results in the activation of a salivary control center in the brainstem. Saliva then flows from the submaxillary and parotid glands (the salivary glands); saliva spreading thereby decreases or stabilizes the rising body temperature, which, in turn, signals the hypothalamus by a negative feedback mechanism. Certain nondesert rodents have been shown to increase evaporative heat loss to more than 100% of heat production, and at least half of this evaporative cooling is due to saliva spreading.

Respiratory Heat Exchange

Earlier, we discussed the importance of respiratory countercurrent heat exchange as a mechanism of conserving heat loss from respiratory passages in cold environments. Now, we see how water can be saved by evaporation in the lungs for animals residing in desert ecosystems. Many desert mammals, such as kangaroo rats, are able to cool expired air and thus reduce the amount of water lost to the environment. A brief discussion of the cycle of respiration helps elucidate this process of water conservation. As air is inhaled, it passes over moist tissues in the nasal passages where it is warmed and humidified. As the relatively dry air passes over the moist tissues of the nasal passages, these tissues are cooled due to evaporation, and heat is transferred from them to warm the inhaled air. During exhalation, the returning warm, saturated air from the lungs condenses on the cool walls of the nasal passages, thus conserving water. This countercurrent exchange—evaporation on inhalation and condensation on exhalation—conserves both water and energy. Kangaroo rats are extremely efficient at cooling expired air because of the unique morphology of their nasal passageways (Schmidt-Nielsen et al. 1970b). Compared with another desert inhabitant, the cactus wren (*Campylorhynchus brunneicapillus*), the nasal passages of the kangaroo rat are very narrow with a large wall surface, which enhances heat exchange between the air and the nasal tissues (figure 9.22). In birds such as the cactus wren, the passageway for air flow is wider and shorter and thus the surface area for contact is smaller. As a result, the nasal passageways of the cactus wren are less efficient at heat exchange than those of the kangaroo rat.

Depending on the ambient temperature and humidity, about 65–75% of the water vapor added to inspired air is recovered in the kangaroo rat's nasal passage during exhalation (figure 9.23). Kangaroo rats may inspire air at about 15° and 30°C that is 25% saturated with water vapor (these are arbitrarily selected air temperatures and humidity levels). In figure 9.23, the shaded bars indicate the amount of water vapor that must be added to 1 L of outside air as it is inhaled and brought to saturation at the body temperature of 38°C. The unshaded bar next to each shaded bar is the corresponding temperature of exhaled

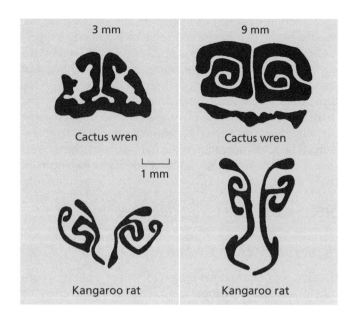

Figure 9.22 Comparative anatomy of nasal passages.
Cross sections of the nasal passageways of the cactus wren and kangaroo rat. The passages are wider and the wall area smaller in the bird than in the mammal of the same body size (about 35 g). In both animals, the profiles were obtained at a depth of 3 mm and 9 mm, respectively, from the external openings. *Adapted from K. Schmidt-Nielsen, 1972,* How Animals Work, *Cambridge Univ. Press.*

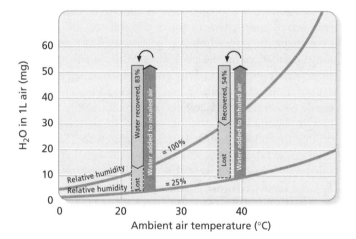

Figure 9.23 Respiratory water exchange. Diagram showing the recovery of water from exhaled air in the kangaroo rat. *Data from K. Schmidt-Nielsen et al., 1970, Counter-current heat exchange in the respiratory passages: Effect on water and heat balance,* Respiratory Physiology 9:263–276.

air under the selected conditions. If air is inhaled at 15°C, for example, it will be exhaled at 13°C. In this case, the cooling of the exhaled air causes the recondensation (recovery) of about 83% of the water that was added on inhalation; note that only about 17% of the water needed to saturate the respiratory air is lost by the animal. At 30°C, less water is recovered, but even so, 54% of the evaporated water is saved compared with what would be lost had the air been exhaled at body temperature. By

cooling expired air, kangaroo rats reduce the amount of water lost. As shown by the shaded bars in the figure, the amount of water lost and recovered depends on the saturation deficit between inspired air (25% humidity) and expired air (100% humidity) plus the air temperature of the inspired air. Expired air is always saturated, and thus its water content is directly related to its temperature.

Evaporation of water from the respiratory passages and through the skin (cutaneous) is a major avenue of water loss for small mammals residing in dry habitats (French 1993). For example, of the total water lost by *Dipodomys merriami*, 84% is due to evaporation in the respiratory tract, and only 16% is lost through the skin (Chew and Dammann 1961). The orange leaf-nosed bat (*Rhinonycteris aurantius*) of northern Australia has a rate of pulmocutaneous evaporation that is more than double that of other bats, and about seven times that measured in rodents of similar body mass (Baudinette et al. 2000; Tracy and Walsberg 2001). Desert-adapted species of rodents and many species of bats are able to reduce evaporative losses either by nasal heat exchange or by residing in humid microenvironments, namely subterranean burrows and caves, respectively. Living in such saturated microhabitats helps to reduce the vapor pressure deficit between the evaporating surface and the surrounding air.

Insulation

Earlier, we discussed how animals residing in cold environments used their pelage to decrease the flow of heat from the body to the environment. In contrast, animals adapted to hot climates use their fur to minimize the rate of heat they absorb from the environment. Dromedaries must cope with surface temperatures of 70°C and Merino sheep with temperatures as high as 85°C. Because the core body temperature of these ungulates is about 40°C when they are heat-stressed, the temperature gradient between the surface of the pelage and the body core is extremely steep. A shallower gradient in the opposite direction exists between core and skin if sweating occurs, as it does in camels. The maintenance of these two opposing temperature gradients is an essential component of the temperature regulation process of mammals residing in hot environments. Pelage insulation not only retards movement of heat from the environment to the skin but prevents large amounts of heat in the form of incident solar radiation from reaching the skin. The heat returns to the environment from the surface of the hair by convection and radiation (see figure 9.1). The reduction of surface temperature due to forced convection and air movement is significant and decreases the steepness of the thermal gradient across the pelage, thus reducing the heat load on the animal. In addition, the steepness of the gradient from the surface of the pelage to the skin is affected by the quality of the hair. For example, many mammals residing in arid regions have sleek, glossy, light-colored pelages that reflect many of the wavelengths of sunlight, reducing heating due to solar radiation.

Furthermore, the pelage of a mammal is not uniformly distributed. Guanacos (*Lama guanacoe*) reside in regions of South America characterized by intense solar radiation and high air temperatures. They possess gradations in pelage ranging from bare skin at the axilla, groin, scrotum, and mammary glands to thick pelage on the dorsum. These bare and sparsely furred areas, seen also in many desert antelopes, serve as **thermal windows** through which some of the heat gained from solar radiation can be lost by convection and conduction (Morrison 1966). In terrestrial mammals such as guanacos, African elephants (*Loxodonta africana*—Phillips and Heath 1992), and woodchucks (*Marmota monax*—Phillips and Heath 2001), thermal windows typically represent bare or sparsely haired thermoregulatory surfaces. These sites are mainly located in the appendages, which permit heat loss by increased peripheral blood flow (Williams 1990; Klir and Heath 1992). For amphibious pinnipeds, dissipation of heat is crucial when exposed to external thermal stress following haul-out. These mammals possess thermal windows distributed throughout their body surfaces (Oritsland et al. 1974; Mauck et al. 2003; Willis et al. 2005) that act as efficient dissipaters of heat during thermal stress. Such regions represent sites of convective heat loss associated with varying cutaneous blood flow. When seals leave the water, they are exposed to high ambient temperatures, and vasodilation of vessels leading to the periphery results in an increased blood flow to the skin. The numerous arteriovenous anastomoses of seals facilitate dissipation of heat via increased cutaneous blood flow (Bryden and Molyneau 1978)—this mechanism can be accurately tracked using infrared thermography. Recently, infrared thermography has been employed to isolate the appearance, distribution, and development of thermal windows on the head, trunk and extremities of harbour seals (*Phoca vitulina*), harp seals (*Phoca groenlandica*), and gray seals (*Halichoerus grypus*) (Mauck et al. 2003) (see figure 9.24). It is noteworthy that the functional significance of anatomical specializations such as thermal windows is commonly augmented by behavioral techniques. For example, otariids and walruses (*Odobenus rosmarus*) respond to changing environmental temperatures by adjusting posture and manipulating exposed surface area of their flippers to ambient air. These behavioral adjustments may be used to conserve or to dissipate heat in order to regulate energy balance (Fay and Ray 1968; Campagna and LeBoeuf 1988; Beentjes 1989, 2006). Horns, found in the Family Bovidae (see chapter 20), also serve as thermal windows for species residing in many different climates. Because horns are often richly vascularized, under certain circumstances they exhibit local vasodilation and serve as sites of heat loss, similar to bare patches on the coats of desert mammals.

Studies of the insulation of small desert rodents indicate a trend toward increased insulation. McNab and Morrison (1963) examined Genus *Peromyscus* from arid and mesic environments and concluded that desert subspecies showed a marked increase in insulation compared with their nondesert relatives. When the body temperature is less than the

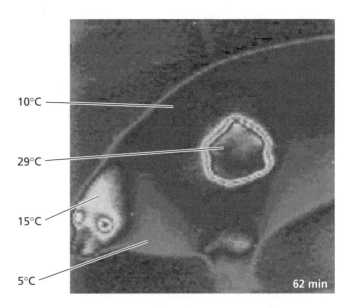

Figure 9.24   Thermal windows. Infrared thermogram of a harp seal (*Phoca groenlandica*) obtained 62 minutes after leaving the water. The large thermal window on the left side of the seal is clearly visible and exhibits temperatures up to approximately 24.0°C higher than those of the rest of the body surface. The thermal windows are separated from the surrounding areas by narrow but regular transition zones grading to lower temperatures. *Adapted from B. Mauck et al., 2003, Thermal windows on the trunk of hauled-out seals: hot spots for thermoregulatory evaporation*, Journal of Experimental Biology, *206:1727–1738.*

ambient temperature (e.g., during summer in desert ecosystems), it is advantageous for the mammal to have fur of a low conductance so as to slow the inward conduction of heat. The body temperature of desert rodents decreases at night; during the day, heat is a gradually stored in the fur, with a concomitant rise in body temperature.

Appendages

In the discussion of adaptations to cold, the physiological mechanism called peripheral heterothermy via countercurrent heat exchange was examined (see figure 9.9). These vascular arrangements, in which arteries and veins are closely juxtaposed, are widely reported to occur in the appendages of species that display regional heterothermy. The extent to which such a system can work for animals coping with heat stress depends on several properties of the countercurrent exchanger. Cooling of the outgoing arterial blood in the appendages is promoted by a high degree of contact between arteries and veins and by a relatively slow rate of blood flow through the exchanger. The arms of humans, the flippers of dolphins, and the limbs of tropical mammals, such as sloths, have a unique arrangement of countercurrent exchange. They possess two sets of veins: one superficial and distant to the major arteries, and the other deep and part of the exchange system. By changing the return of blood along these two venous systems, the mammal can emphasize heat dissipation or conservation in its extremity according to its thermal needs. The large ears of black-tailed jackrabbits (*Lepus californicus*) are excellent heat dissipators. Most excess heat generated during activity may be lost by dilation of the arteries in the ears (Hill et al. 1980). Thermal conductivity of the heat exchange system of jackrabbits is reported to be about ten times greater for animals at 23°C than at 5°C. Hart (1971) provides an excellent review of studies confirming the importance of appendages as dissipators of heat for desert mammals.

Metabolic Rate

Seed-eating rodents residing in desert ecosystems tend to exhibit low basal metabolic rates. Examples include heteromyids of North America (McNab 1979a; Hinds and MacMillen 1985; French 1993) and murids of Australia (MacMillen and Lee 1969, 1970) and Asia (Shkolnik and Borut 1969). French (1993) summarized the factors related to energy consumption in heteromyid rodents. They may use low metabolic rates to reduce overheating when they occupy a closed burrow system, or to cut pulmonary water loss in a dry environment. Metabolic rates of heteromyids are about a third less than those of other mammals when at rest. In addition, metabolic reductions are possible when the mammals undergo estivation during times of food scarcity.

Body Size

Earlier, we used the analogy of the mammal as a cube to help illustrate the energy implications of surface-area-to-volume ratios. Now we consider the implications of body size for mammals coping with heat rather than with cold. Animals living in hot environments gain heat from two sources: (1) the environment through conduction and convection plus radiation from ground and sun (see figure 9.1) and (2) metabolic heat gain. A mammal's **heat load** (sum of the environmental and metabolic heat gain) is therefore roughly proportional to its body surface area. Because small mammals have a much larger surface-area-to-volume ratio, they lose heat more readily than large mammals. Using the surface relationship, we can estimate the quantity of water required to dissipate a certain heat load. A small mammal such as a kangaroo rat would need to evaporate a great deal more water (relative to body size) to eliminate its heat load than would a large mammal (figure 9.25). To avoid this, as we shall see, small mammals escape the heat (curtail the heat load) by retreating to underground burrows during the day. In addition, desert mammals can decrease their body temperature to conserve water and decrease heat load.

DORMANCY

As we learned earlier, many mammals pass through unfavorable climatic periods by becoming inactive. A period of dormancy in reaction to cold is called torpor and hibernation.

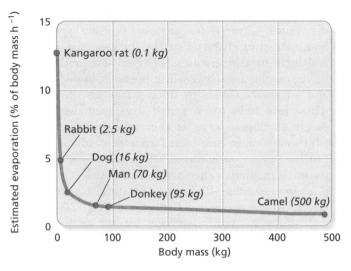

Figure 9.25 Evaporation in relation to body mass. Estimated evaporation compared with body mass of different mammals. For a mammal to maintain a constant body temperature under hot desert conditions, water must be evaporated in proportion to the heat load. Because of the larger relative surface area of a small animal, the heat load, and therefore the estimated evaporation in relation to the body size, increases rapidly in the small animal. The curve is calculated on the assumption that heat load is proportional to body surface. *Data from K. Schmidt-Nielsen, 1964,* Desert Animals: Physiological Problems of Heat and Water, *Oxford Univ. Press.*

A similar period in reaction to dry or hot conditions is called **estivation.** Estivation occurs in marsupials and insectivores but is most common among the rodents. It occurs at relatively high body temperatures, and the animals seem somewhat lethargic rather than torpid. If body temperatures are not measured, estivation may go unnoticed. For example, pigmy possums (*Cercartetus nanus*), an Australian marsupial, are able to eat and exhibit normal mobility at a body temperature as low as 28°C. Little pocket mice (*Perognathus longimembris*) exhibit a wide repertoire of behaviors while in estivation, ranging from eating at 23°C to shifting postures at temperatures as low as 6°C (figure 9.26). Typically, rodents capable of dormancy have narrow thermoneutral zones (28° to 35°C) and low basal metabolic rates. Estivation has not been as intensely studied as hibernation.

Among heteromyids, estivation is reported for all species of pocket mice and kangaroo mice (Genera *Perognathus, Chaetodipus,* and *Microdipodops*) but is poorly developed in the kangaroo rats (Genus *Dipodomys*; French 1993). Within heteromyids, two basic dormancy "profiles" occur: (1) those species that forage year-round and employ shallow torpor to cope with energy emergencies (e.g., species of Genus *Chaetodipus* that tolerate body temperatures of 10° to 12°C for less than 24 hours) and (2) those species of Genus *Perognathus* (e.g., *P. parvus*) that use supplies of cached seeds during dormancy that may last for up to 8 days at a body temperature as low as 2°C. A thorough discussion of dormancy in the Family Heteromyidae is provided by French (1993). One of the smallest rodents in the world, Genus *Baiomys* (adults = 9 g) of southwestern North America, undergoes bouts of estivation. If food and water are limit-

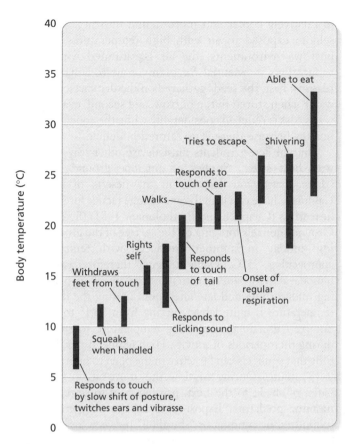

Figure 9.26 **Body temperature and behavior.** Range of minimal body temperatures for various patterns of behavior exhibited by the little pocket mouse during arousal from hibernation. *Data from G. A. Bartholomew and T. J. Cade, 1957, Temperature regulation, hibernation, and estivation in the little pocket mouse, Perognathus longimembris,* Journal of Mammalogy *38:60–72.*

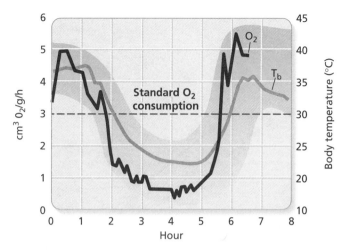

Figure 9.27 **Torpor in the cactus mouse.** Pattern of changes in oxygen consumption and body temperature in the cactus mouse during entry into and arousal from torpor at an ambient temperature of 19.5°C. The standard consumption line is for active mice at an ambient temperature of 20°C. The cycle of torpor was initiated by deprivation of food and water. *Data from R. E. MacMillen, 1965, Aestivation in the cactus mouse,* Comparative Biochemistry and Physiology *16:227–248.*

ed, these small mice readily decrease body temperature from 38° to 20°C. Periods of estivation are reported for species of Genus *Peromyscus* inhabiting hot regions. The cactus mouse (*P. eremicus*), an inhabitant of deserts of southwestern North America, undergoes torpor (estivation) in summer and winter in response to limited food and water (figure 9.27). Cactus mice remain in burrows during the driest part of the summer and enter torpor in response to food and water shortages at environmental temperatures of about 30°C. Under laboratory conditions, when deprived of food and water, *P. eremicus* enters torpor at an ambient temperature of 19.5°C, maintaining a body temperature of about 22°C for several hours. Body temperature does not drop below 15°C during torpor (MacMillen 1965). From the preceding discussion, it is apparent that rodents exploit many different adaptive patterns in dormancy, with the ultimate goal of saving energy in unpredictable environments.

Case Study of Estivation

Kangaroo mice can be used to illustrate the energy savings accrued by estivation. These small heteromyid rodents (10–14 g) occur in the Upper Sonoran sagebrush desert of western North America. Seeds, a staple in their diet, are gleaned from the sand and hoarded in underground burrows. Although environmental temperatures may range from 0° to 30°C, kangaroo mice rarely experience drastic changes in temperature due to their fossorial (burrow-digging) and nocturnal habits. In most deserts, seed production is quite seasonal and limited to very brief periods. As a result, strict homeothermy for granivorous rodents such as kangaroo mice would be a waste of energy. Kangaroo mice therefore cope with such environmental unpredictability by exhibiting great thermolability; duration and frequency of estivation is contingent on the availability of seeds and environmental temperature. For example, when food is in excess and environmental temperatures high, kangaroo mice exhibit homeothermy; if food is limited and temperatures unfavorable, they shift into heterothermy (estivation). Periods of estivation may last for only a few hours or for several consecutive days, and the more time they spend in dormancy, the less food they need. Brown and Bartholomew (1969) measured the influence of environmental (ambient) temperature on the energy needs (oxygen consumption) of homeothermic and torpid kangaroo mice (figure 9.28). At ambient temperatures of 5° to 25°C, the body temperatures of mice were 32° to 37°C when they were homeothermic, but during torpor, their body temperatures were just 1° to 3°C above ambient levels. As shown in the figure, the vertical distance between the "homeothermic" and "torpid" lines represents the energy savings per unit time of torpor at that temperature.

AVOIDANCE OF HIGH TEMPERATURES

Unlike large mammals, small mammals avoid extreme temperatures in desert ecosystems by adhering to fairly

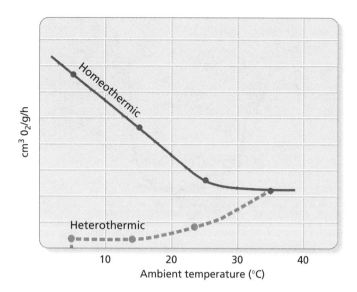

Figure 9.28 Torpor and euthermy in the kangaroo mouse. The effects of ambient temperature on the oxygen consumption of kangaroo mice when maintaining normally high body temperatures and when in torpor. The points are mean values. *Data from J. H. Brown and G. A. Bartholomew, 1969, Periodicity and energetics of torpor in the kangaroo mouse (Microdipodops pallidus),* Ecology 50:705–709.

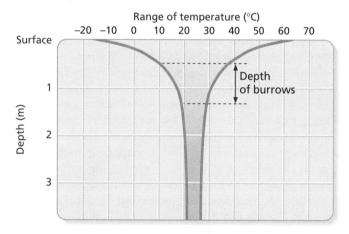

Figure 9.29 Soil temperature changes in deserts. Diagram showing the range of temperature compared with the depth of the soil. Temperature fluctuations are less extreme below the surface of the desert at depths typical of burrows of kangaroo rats. A temperature of 25°C represents an "average" burrow temperature. *Data from K. Schmidt-Nielsen, 1964,* Desert Animals: Physiological Problems of Heat and Water, *Oxford Univ. Press.*

definite periods of activity. With the exception of ground squirrels and chipmunks, all desert rodents of North America are nocturnal and fossorial. Small mammals optimize survival through avoiding extremes in temperature by residing in burrows below ground during the heat of the day. For a typical desert ecosystem, the temperature of the air in the burrow is mild compared with the extremes that occur on the surface (figure 9.29). Heteromyids search for seeds at night when it is cool and spend the day

in relatively cooler and more humid burrows. They are seldom exposed to air with high temperatures. Within burrow environments, the air is saturated with water vapor. This is essential for granivorous rodents for two reasons: first, the seeds gathered on the dry surface absorb water when stored in the burrow, and second, evaporative water loss is reduced considerably while the rodent is in its burrow because there is no saturation deficit.

Diurnal desert rodents must devise other ways to cope with heat stress. Studies of antelope ground squirrels (*Ammospermophilus leucurus*) in the deserts of southern California have elucidated some unique tactics for surviving heat stress (Chappell and Bartholomew 1981) (figure 9.30). Ground squirrels foraging on the surface of the ground during midday in summer are faced with temperatures approaching 75°C. Daily foraging therefore assumes a bimodal activity pattern, with most activity taking place during midmorning and late afternoon. During the day, body temperature is quite labile, varying from 36.1° to 43.6°C. Squirrels use fluctuating body temperature to store heat during their periods of activity. High temperatures limit the time that squirrels can be active in the open to no more than 9–13 minutes. During this time, they move rapidly from one patch of shade to the next, pausing only to seize food or monitor predators. Exposure to midday temperatures is minimized by running or "shuttling" between foraging and cooling sites. The body temperature of the antelope ground squirrel shows a pattern of rapid oscillations, rising while the squirrel is in the sun and dropping when it retreats to its burrow (Vispo and Bakken 1993; Hainsworth 1995; Degen 1997). In addition, ground squirrels employ their tail as a sort of parasol, or "heat shield." As shown in figure 9.31, the cape ground squirrel (*Xerus inauris*), an inhabitant of the hot arid regions of southern Africa, holds its wide, flat tail tightly over its back with the white ventral surface upward. In this position, the tail shades a large portion of the animal's back, thus lowering the effects of solar radiation on body temperature. On a hot day, to maintain their body temperature below 43°C, squirrels must retreat to burrows every few minutes. Burrows deeper than 60 cm usually have temperatures between 30° and 32°C. Ground squirrels and other rodents do not sweat or pant. Instead, they use this combination of transient heat storage and passive cooling in a deep, saturated burrow to permit activity during daylight hours when heat is extreme. As will be explained in the following section, antelope ground squirrels employ a strategy for coping with heat that is very similar to that of camels—saving water by allowing their body temperature to rise until the heat can be dissipated passively. The major difference between these squirrels and camels is a function of body mass. The large camel can store heat for an entire day and cool off at night, whereas the antelope ground squirrel goes through the same cycle many times during the day.

The Dromedary

The biology of camels (see chapter 20) has been reviewed by Gauthier-Pilters and Dagg (1981), Yagil (1985), and Kohler-

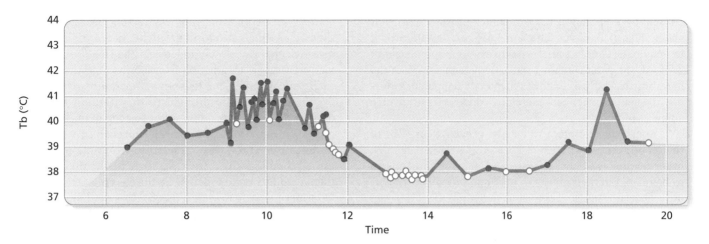

Figure 9.30 Activity and body temperature of the antelope ground squirrel. Short-term cycles of activity and body temperature of *Ammospermophilus leucurus*. Active (λ); inactive (μ). *Data from M. A. Chappel and G. A. Bartholomew, 1981, Standard operative temperatures and thermal energetics of the antelope ground squirrel (Ammospermophilus leucurus),* Physiological Zoology, *54:81–93.*

Rollefson (1991). The dromedary (*Camelus dromedarius*), touted as the "ship of the desert," occurs in semiarid and arid regions of the Old World. Camels take long journeys, some lasting from 2 to 3 weeks, with no opportunity to drink water. In the Sahara, dromedaries often remain without drinking water from October to May, existing solely on the water content in the forage. During such arduous trips, they are faced with severe problems of water conservation and temperature regulation. As a result, they have evolved unique mechanisms to cope with desert ecosystems.

One interesting mechanism employed by camels in coping with intense heat and lack of water of arid regions is their ability to vary body temperature as a device for saving water (Schmidt-Nielsen et al. 1957). Researchers have studied the daily cycle of body temperature in camels that received water and those that were dehydrated (figure 9.32). Camels given water show a very small

daily fluctuation in body temperature, ranging from 36°C in the early morning to a maximum of 39°C by midafternoon. If a camel is deprived of water, however, the daily variation in body temperature triples—falling to 34.5°C at night and climbing to 40.5°C during the day. This temperature variation is important because when camels are dehydrated, they can lose excess heat at night through radiation, conduction, and convection to a cooler environment, thus saving water! This results in a change in body temperature for dehydrated camels of about 6°C; for camels with water, the change is only half this (see figure 9.32). To determine how much water camels can conserve by storing heat during the day and losing it later that night by nonevaporative means, we use the following calculations. About 0.9 cal is required to increase each gram of tissue 1.0°C. A 500-kg camel that increases its temperature by 6°C can therefore store 2,700,000 cal of heat in its body. If this heat had to be lost

Figure 9.31 The heat shield. The cape ground squirrel (*Xerus inauris*) resides in hot deserts of southern Africa and uses its tail as a parasol. (A) The erected tail covers the entire dorsal surface of the sitting squirrel. (B) The tail is held over the back of a horizontally positioned squirrel shading its head and body from the sun. *Photograph from A. F. Bennett et al., 1984, The parasol tail and thermoregulatory behavior of the cape ground squirrel, Xerus inauris,* Physiological Zoology *57:57–62.*

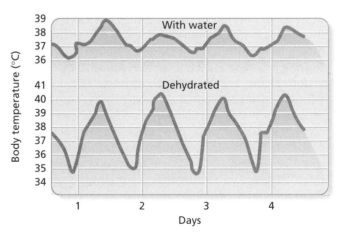

Figure 9.32 Heat storage and loss in the dromedary. Diagram showing the relationship of changes in body temperature compared to heat storage and loss in the dromedary. Camels store more heat when water availability is restricted. *Data from K. Schmidt-Nielsen et al., 1957, Body temperatures of the camel and its relation to water economy,* American Journal of Physiology *188:103–112.*

entirely by evaporation (each milliliter of water dissipates about 580 cal), however, a total of 4,655 mL of water would be required. Thus, a camel would have to evaporate about 4.5 L of water to maintain a stable body temperature at the nighttime level. By tolerating a high body temperature during the day, camels can therefore conserve considerable amounts of water.

Camels can tolerate a water loss greater than 30% of their body mass, whereas a 15% loss is lethal in most other mammals. Their toleration of dehydration appears to be related to their ability to maintain blood plasma volume near normal levels during dehydration. They can rehydrate quickly, taking in water equal to 30% of their body mass within a matter of minutes. Camels have an impressive ability to concentrate urine and absorb water from fecal material. Their kidneys can produce urine with a high chloride content; dromedary camels can drink a sodium chloride solution even more concentrated than seawater without ill effect.

Many of the camel's morphological, behavioral, and physiological adaptations enhance its survival during periods of heat. As a result of its thick pelage, much of the solar energy that strikes a camel's fur never reaches its skin. During summer, the surface of its fur can reach 70° to 80°C, but skin temperature stays at only 40°C. Abundant sweat glands release secretions onto the skin surface beneath the hair, and evaporative cooling aids temperature regulation. The combination of a thick pelage and high body temperature reduces the rate of heat gain to the core of the body, slows the rate of sweating needed, and thus conserves water. During the day, camels must resort to evaporative cooling through sweating and panting, coupled with deep, slow breathing to eliminate heat and reduce respiratory water loss. A thick fur coat protecting the evaporative surface of the bare skin from unrestrained access by environmental heat could also reduce the vaporization of sweat. So the camel has evolved an optimal thickness to its pelage: it is not so thick or dense as to interfere with dissipation of vaporized water from the skin to the atmosphere nor so thin as to provide inadequate protection. It seems counterintuitive that wearing a fur coat on a hot day may enhance thermoregulatory ability, but it works for camels. The deserts of northern Africa are cold in the winter, and camels have an extremely thick fur. In summer, camels shed their winter coat but retain hair of up to 6 cm long on the back and 11 cm on the hump. On the venter (abdomen) and surface of legs, the hair is only 1.5–2.0 cm long. As in guanacos and other large mammals, the lightly furred areas of camels act as thermal windows for heat loss.

The morphological peculiarities of camels couple with behavior to enhance water conservation and temperature regulation. Early in the morning, camels lie down on surfaces that have cooled overnight by radiation. They tuck their legs beneath the body with their ventral surface of short fur in contact with the cool ground. In this position, a camel exposes only its well-furred dorsum and flanks to the hot sun and places its lightly furred legs and venter in

contact with cool sand, thus conducting heat away from the body. Camels also form group huddles during the day, in which they lay closely together. The side-by-side contact prevents solar radiation from permeating the body of each animal and raising its body temperature.

The ability of the dromedary to withstand heat and dryness does not depend on water storage in its hump but rather on numerous physiological mechanisms aimed at water conservation. It is a popular misconception that the camel's hump is filled with water. Actually, the hump is filled with fat bound together by fibrous connective tissue. Contrary to myth, however, this fat does not help much as a source of energy for long travels in the desert or as a lightweight method of carrying water (Hill and Wyse 1989:156). To understand the costs and benefits of fat stored in the camel's hump, the amount of water gained and lost in breaking down (catabolizing) lipids or carbohydrates must be assessed. Consider the production of metabolic water in relation to respiratory water loss. For temperatures and humidities of a desert environment, camels actually lose more water across the lungs in obtaining oxygen to oxidize fats than they gain from the oxidation of those fats. In reality, the breakdown of fat to derive energy actually imposes a water deficit.

Other Desert Ungulates

Desert antelopes (e.g., oryx, gazelles, and elands) possess a number of adaptations for coping with heat and dehydration (Taylor 1972; figure 9.33). The pallid color and glossy fur of the desert antelope reflect direct sunlight, and the fur is excellent insulation against the heat. Heat is lost by convection and conduction from the venter, where the pelage is very thin. Unlike the dromedary, the antelope's coping mechanisms are independent of drinking water. Elands search for shade during the heat of day, whereas dromedaries remain in the full sun all day, with relief from heat not coming until evening. Elands and oryx both pant at high temperatures and can reduce their metabolic rates to conserve water. Furthermore, evaporation through the skin is reduced by 30% in elands and by 60% in oryx. Evaporative water loss through the respiratory tract is also reduced. In comparison to camels, oryx can reduce daytime evaporative water losses by suppressing sweating completely, even when very hot. The fat tissue in elands and dromedaries is localized in the hump rather than dispersed uniformly under the skin, where it would impair heat loss by radiation. The water requirements of antelopes are negligible, plus their ability to conserve water is greatly enhanced by their production of a concentrated urine and dry feces. The deserts of the Arabian Peninsula are characterized by unpredictable rainfall and summer ambient temperatures that exceed 45°C; such habitats are arguably the most inhospitable environments for mammals. Yet such habitats are home to Arabian oryx (*Oryx leucoryx*), Nubian ibex (*Capra ibex*), mountain gazelles (*Gazella gazella*) and sand

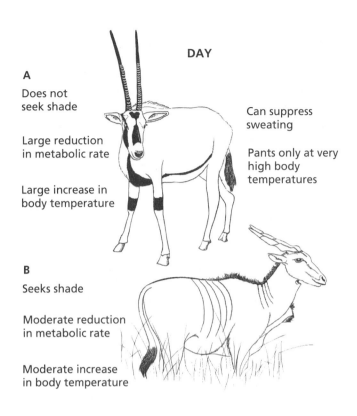
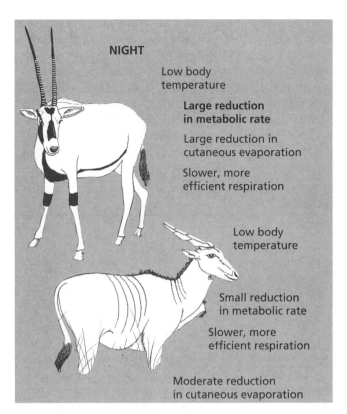

Figure 9.33 Water conservation in antelopes. Adaptations for water conservation in two nondrinking east African antelopes: (A) the oryx and (B) the eland. Figures on the left are for daytime; figures on the right are for night. *Adapted from C. R. Taylor, 1969, Scientific American.220:88-97,1969. Also adapted from M. S. Gordon, 1982, Animal Physiology: Principles and Adaptations, 4th ed., Macmillan Publishing Company.*

gazelles (*G. subgutturosa*). How do these ungulates cope with such extremes? Gazelles (*G. thomsonii* and *G. granti*) residing in arid savannahs of eastern Africa show very low values for evaporative water loss as a method of water conservation (Taylor 1970); however, their frugality

is five time higher than the value for sand gazelles (Ostrowski 2006)! Such low evaporative water loses indicates that sand gazelles have evolved a remarkable capacity to reduce water expenditures, a principal reason for their success in the deserts of Saudi Arabia.

SUMMARY

The success of mammals derives from their sophistication in form, function, and especially behavior. Mammals are widely distributed throughout the earth—ranging from tundra to deserts, each environment characterized by different extremes in climate. As a result of such strong selection pressures, mammals have evolved a unique suite of mechanisms to enhance survival. These adaptive strategies are diverse and highly sophisticated, with variation exhibited at the levels of ecosystem and species and between individuals.

Basic terminology in the physiology of environmental adaptation includes endothermy, homeothermy, euthermy, heterothermy, poikilothermy, and ambiguous terms such as warm- and cold-blooded. Most mammals maintain their body temperatures between 36° and 38°C. To do this, they must balance heat gain and heat loss against such forces as radiation, conduction, convection, and evaporation. The equilibrium maintained by mammals is achieved through interaction

with the environment by the mammalian thermostat—the hypothalamus—with its species-specific set points. Each species has unique coping mechanisms and peculiarities in its thermoneutral zone, thermal conductance, and metabolic rate. These characteristics and the important concept of countercurrent heat exchange help determine the distribution of species as well as such long-standing "biogeographic rules" as Bergmann's, Allen's, and Gloger's.

Mammals employ several mechanisms to defend against cold, including (1) behavioral mechanisms (communal nesting, nest construction, food hoarding, foraging dynamics, and activity patterns) and (2) anatomical and physiological mechanisms (body mass reduction during winter, decrease in thermal conductance, nonshivering heat production, and dormancy).

The challenges faced by mammals in desert environments are even more severe than those for species residing in the cold. Water is essential for the survival of all animals, so the role of water economy and the anatomy and physiology of the mammalian kidney are crucial to desert animals. Desert mammals conserve water by eating succulent plants and using metabolic water. Certain small mammals also have ingenious mechanisms for consuming halophytic plants. In addition to conserving water, xeric-adapted mammals regulate body temperature through evaporative cooling by sweating, panting, saliva spreading, and respiratory heat exchange.

As we have seen with mammals faced with cold, mammals residing in hot environments possess certain adaptive strategies that enhance their survival, including unique patterns in insulation, appendages, metabolic profiles, and body size. Just as animals in the cold use hibernation, desert mammals employ estivation to avoid heat and lack of water. Certain mammals epitomize the ability to survive in heat, namely heteromyid and dipodid rodents, dromedaries, and desert antelopes.

SUGGESTED READINGS

Barnes, B., and H. V. Carey (eds.). 2004. Life in the cold: Evolution, mechanisms, adaptation, and application. Pp. 1–593 *in* Twelfth International Hibernation Symposium. Biological Papers of the University of Alaska 27, Institute of Arctic Biology, University of Alaska, Fairbanks.

Degen, A. A. 1997. Ecophysiology of small desert mammals. Springer-Verlag, Berlin.

Folk, G. E., Jr., M. L. Riedesel, and D. L. Thrift. 1998. Principles of integrative environmental physiology. Austin and Winfield Publishers, San Francisco.

Gordon, M. S., G. A. Bartholomew, A. D. Grinnell, C. B. Jørgensen, and F. N. White (eds.). 1982. Animal physiology: principles and adaptations, 4th ed. Macmillan, New York.

Kemp, T. S. 2006. The origin of mammalian endothermy: A paradigm for the evolution of complex biological structure. Zoological Journal of the Linnean Society 147:473–488.

Koopman, H. N. 1998. Topographical distribution of the blubber of harbor porpoises (*Phocoena phocoena*). J. Mammal. 79:260–270.

McNab, B. K. 2002. The physiological ecology of vertebrates: a view from energetics. Comstock Publ. Assoc., Ithaca, NY.

Michener, G. R. 1992. Sexual differences in over-winter torpor patterns of Richardson's ground squirrels in natural hibernacula. Oecologia 89:397–406.

Pough, F. H., C. M. Janis, and J. B. Heiser. 2005. Vertebrate life, 7th ed. Prentice Hall, Upper Saddle River, NJ.

Schmidt-Nielsen, K. 1997. Animal physiology: Adaptation and environment, 5th ed. Cambridge Univ. Press, New York.

Speakman, J. R., and D. W. Thomas. 2003. Physiological ecology and energetics of bats. Pp. 430–490 in Bat Ecology (T.H. Kunz and M. B. Fenton, eds.). Univ. Chicago Press, Chicago.

Taylor, J. R. E. 1998. Evolution of energetic strategies of shrews. Pp. 309–346 *in* Evolution of shrews (J. M. Wójcik and M. Wolsan, eds.). Mammal Research Institute, Polish Academy of Sciences, Bialowieza, Poland.

Thomas, D. W. 1990. Winter energy budgets and costs of arousals for hibernating little brown bats, *Myotis lucifugus*. J. Mammal. 71:475–479.

Tomasi, T. E., and T. H. Horton (eds.). 1992. Mammalian energetics. Comstock Publ. Assoc., Ithaca, NY.

We refer you to the series of symposia on *Mammalian Hibernation* begun in 1959 by Charles P. Lyman and Albert R. Dawe. Symposia are held approximately every 3 years, each resulting in a volume of proceedings. The proceedings of the latest symposium, held from 25 July to 1 August 2004 in Seward, Alaska, were entitled *Life in the cold: Twelfth International Hibernation Symposium*. These volumes contain a wealth of information on hibernation biology spanning a scale from organismic to molecular and will help you keep pace with the changes in this interesting and dynamic field of biology.

DISCUSSION QUESTIONS

1. Distinguish among the following terms: homeothermy, ectothermy, poikiothermy, heterothermy.

2. Small mammals are common inhabitants of deserts and mountains. Describe the different adaptations they employ to maintain homeothermy in each environment.

3. Explain why it is advantageous for certain small mammals to abandon homeothermy during brief or extended periods of their lives.

4. Explain nonshivering thermogenesis. Track this heat-producing mechanism for a small mammal from its initiation in nature to its function in maintaining homeothermy.

5. Compare the advantages and disadvantages of the color white for animals residing in northern environments.

6. What are the benefits of having a larger body size in the Arctic? Compare the advantages and disadvantages of body size in terms of mass-specific metabolism and total energy needs.

7. Define countercurrent heat exchange. What are the advantages of this mechanism for a caribou of the Arctic and a sloth of Panama?

8. Define thermal conductance. Compare the benefits of a thick fur coat for a lemming in the Arctic compared with a dromedary in the Sahara Desert.

9. Compare the different types of dormancy in mammals: hibernation, torpor, and estivation. Provide a definition for each type and give examples of different mammals that fit into each category. What are the advantages and disadvantages of dormancy?

10. Describe the cycle of hibernation of a ground squirrel residing in northern Montana. Compare hibernation in this ground squirrel with estivation of a kangaroo mouse residing in the desert of California.

11. List the different water conservation strategies employed by desert mammals.

12. Discuss four major mechanisms of evaporative cooling (insensible water loss).

CHAPTER 10

Reproduction

Reproduction is the process by which organisms produce new individuals, passing on genetic material and thus maintaining the continuity of the species and of life. All mammals reproduce sexually by way of internal fertilization. Furthermore, all mammals are **dioecious**, that is, the sexes are separate, and a given individual normally has the sex organs and secondary sexual characteristics of only one gender.

As you will see in this chapter, mammals vary greatly in the structure and function of their reproductive system. The monotremes lay eggs and incubate their young, and as is the case for most reptiles and birds, monotremes have a **cloaca**—a common opening for the urinary and reproductive tracts. In addition, as with birds, only the left ovary is functional in monotremes; they have, however, true mammary glands but lack nipples (Hughes et al. 1975; Griffiths 1978; Griffiths et al. 1988; Grant 1995; Strahan 1995; Rismiller and McKelvey 2000).

The marsupials or pouched mammals, like the monotremes, possess a cloaca. The placenta in these animals is not very efficient due to the limited contact between the maternal and fetal blood supply. As a consequence, marsupials undergo a very brief gestation period and a prolonged period of nursing. The young are born in an altricial (poorly developed) state and require a long period of development in the pouch **(marsupium)** of the mother (Padykula and Taylor 1982; Dawson 1995; Tyndale-Biscoe 2005). The eutherian mammals, commonly called "placental mammals," have made a major advance in the evolution of their reproductive patterns. The key innovation was the appearance of a highly sophisticated placenta that facilitates efficient respiratory and excretory exchange between the maternal and fetal circulations. Unlike the marsupials, the eutherian mammals exhibit a longer gestation period and a shorter period of nursing or lactation. Furthermore, at birth, the young are more highly developed (precocial) than are young marsupials.

Mammals vary widely in the complexity of their reproductive patterns, but it is incorrect to interpret the seemingly more primitive reproductive patterns of the monotremes and marsupials as being less successful than those of eutherians. The success of a species can be evaluated only case by case and in the context of its specific ecological relationships. Many examples exist of successful groups of noneutherian mammals, one of which—the opossum—is well known. This marsupial originated in South America and has colonized environments as far north as Ontario, Canada, within only a short time. The great success of this species is demonstrated by its broad geographic distribution and high reproductive potential.

The variations in reproductive patterns of mammals seem endless. For example, gestation ranges from 10 to 14 days in some dasyurid marsupials to over 650 days in elephants. Lactation varies from only 4 days in pinnipeds such as the hooded seal (*Cystophora cristata*) to over 900 days in chimpanzees and orangutans (*Pan troglodytes* and *Pongo pygmaeus*; Hayssen et al. 1993). Most mammals give birth to between 1 and 15 young per litter, and intervals between births may range from 3 to 4 weeks in rodents to as long as 3 to 4 years or more in sirenians, elephants, and rhinos (Hayssen et al. 1993). Voles (Order Rodentia: Cricetidae) may hold the record for biotic potential: the meadow vole (*Microtus pennsylvanicus*) of the North American grasslands may produce up to eight or nine litters of five to eight young in a single year. As a result, a single female may potentially produce up to 72 offspring during one breeding season! Bailey (1924) reported that a captive female meadow vole produced 17 families in a single year. Naked mole-rats (*Heterocephalus glaber*, Family Bathyergidae) of Africa have the largest mean litter size of all wild mammals—up to 28 young (Sherman et al. 1999)! Variation in reproductive patterns may be explained by assessing environmental factors and the risk of raising young in highly seasonal environments. A thorough review of the seemingly endless reproductive strategies employed by mammals is presented by Hayssen et al. (1993).

To examine the variation in reproductive patterns in mammals, we divide this chapter into six parts. First, we concentrate on the role of the male and the female in the production of **gametes** (sperm produced by the testes of the male and ova produced by the ovaries of the female). The next five phases take place within the body of the female and include fertilization, implantation, gestation, parturition, and lactation. As we discuss each process, remember that each species of mammal has evolved a unique reproductive capacity for coping with the selective forces of the environment in which it resides.

The Reproductive Systems

THE MALE REPRODUCTIVE SYSTEM

The reproductive system of males comprises paired testes, paired accessory glands, a duct system, and a copulatory organ.

Testes

The **testes** (sing., *testis*) are the site of production of sperm (the male gametes) and the synthesis of male sex hormones, chiefly testosterone (figure 10.1). The paired testes of mammals are oval-shaped and may be suspended in a pouchlike, skin-covered structure called the **scrotum**. The position of the testes in mammals varies considerably. In many species (e.g., some bats and many rodents), the testes migrate from the body cavity into the scrotum during the breeding season and afterward are withdrawn into the inguinal canal. In some mammals (e.g., monotremes, some "insectivores," anteaters, tree sloths, armadillos, manatees and dugongs, all the seals except the walruses, whales, hyraxes, and elephants), the testes are always in

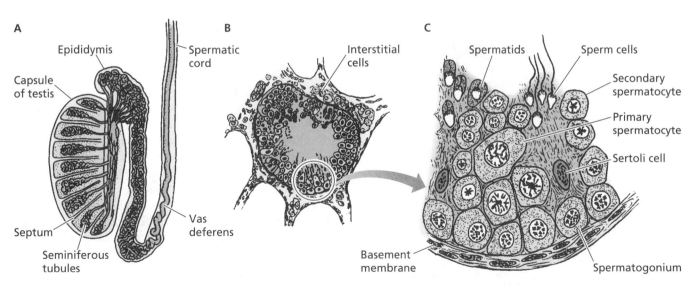

Figure 10.1 **Mammalian testis.** Section of the mammalian testis showing coils of (A) seminiferous tubules containing spermatozoans (B) in various stages of development. (C) The composite cross section of a seminiferous tubule shows spermatogonia near the basement membrane, progressing from spermatocytes to spermatids with mature spermatozoa (sperm cells) located near the lumen of the tubule. See Curtis (1975) and Hickman et al. (2004) for a description of the maturational process within the seminiferous tubule. *Adapted from Curtis, 1975, Biology, 2nd ed., Worth Publishers.*

the abdominal cavity. The testes remain in the scrotum permanently in most primates, artiodactyls, perissodactyls, and carnivores. It is possible that the temperature of the abdominal cavity is too high to permit development of sperm (**spermatogenesis**), and thus the scrotum allows for cooling of the testes. Temperature may not be the only critical factor for spermatogenesis, however, and a number of environmental factors interact to initiate and end spermatogenesis (van Tienhoven 1983; Hadley 1988).

Spermatogenesis, the production of spermatozoa by the process of meiosis, occurs within the **seminiferous tubules** of the testes. Within these tubules, spermatozoa mature by passing through several stages of development and are nurtured by **Sertoli cells** (see figure 10.1). Surrounding the tiny, coiled tubules within the testes are microscopic interstitial cells (**Leydig cells**), which are the site of secretion of the male hormone, **testosterone**. The secretions of Sertoli and Leydig cells provide an optimal microenvironment for spermatogenesis (Gordon 1982; Hadley 1988).

Ducts and Glands

After the sperm are mature, they must be transported to the exterior by a series of ducts (figure 10.2). On leaving the seminiferous tubules, sperm are collected in the **epididymis,** a highly coiled tube located on the surface of each testis. This tube serves both as a duct for passage of sperm and as a brief storage site where sperm and all glandular secretions are nourished prior to ejaculation. After the epididymis leaves the testis, its shape changes, becoming enlarged and straight to form the **ductus deferens** (vas

deferens), which passes to the urethra. Near the junction of the ductus deferens and the urethra, three different glands add secretions to the seminal fluid: the paired **seminal vesicles** (vesicular glands), a single **prostate gland,** and paired **bulbourethral glands** (Cowper's glands). Additional accessory glands, such as the paired **preputial glands** and **ampullary glands,** contribute their products to the seminal fluid (semen). Considerable variation in reproductive glands occurs among species of mammals, and not all species possess all the glands listed. In some animals, **coagulating glands** produce a substance that coagulates the secretions of the seminal vesicles. Following copulation, fluid from this gland contributes to formation of a **copulation plug,** which blocks the entrance to the female's vagina. Presence of this plug often indicates that a female has mated; it may persist for up to 2 days following copulation. The copulation plug is thought to be important in retaining the spermatozoa and preventing their loss from the female tract. Copulation plugs have been noted in many species of rodents and in some bats, insectivores, and marsupials (Austin and Short 1972a; Martin et al 2001). In addition to nourishing sperm, acting as a lubricant and, for some species, helping to form a copulation plug, glandular secretions enhance the alkalinity of the semen so as to protect the sperm from the acidic environment present in the male urethra and female vagina.

The Penis

Transfer of male sperm to the body of the female is facilitated by the highly vascular, erectile **penis**. The penis is composed of cylindrical bodies, the **corpora cavernosa,** which become filled with blood during sexual excitation, thus causing the organ to become erect. The sperm pass during coitus through the urethra, which is also used for urination. In some mammals (e.g., all carnivores, most primates, rodents, bats, and some insectivores), the tip of the penis (**glans penis**) may include a complex bony structure, the **os penis,** or **baculum** (figure 10.3). Mammalogists identify different species by this structure as well as use it as an index of approximate age of an individual (Burt 1960). The glans penis shows many different configurations, such as the bifurcated penis of some monotremes and marsupials and the corkscrew shape in the domestic pig. These configurations are compatible with the vagina of females of the same species. Furthermore, the penis of different mammals may be situated either anterior (most eutherian mammals) or posterior (marsupials, rabbits, hares, and pikas) to the scrotum. In mammals such as monotremes and marsupials, the penis is situated in a sheath within the cloaca, and in marsupials, it may be directed posteriorly (Biggers 1966).

THE FEMALE REPRODUCTIVE SYSTEM

The reproductive organs of females consist of a pair of ovaries, which produce the eggs and certain hormones; a pair of **oviducts** (Fallopian tubes), which act as the

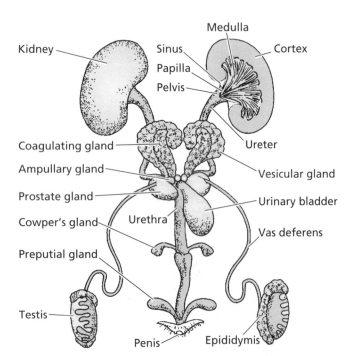

Figure 10.2 Male urogenital system. Composite diagram of the male urogenital system of the rat (*Rattus norvegicus*). *Adapted from DeBlase and Martin, 1981, A Manual of Mammalogy, 2nd ed., Wm. C. Brown Publishers.*

A

B

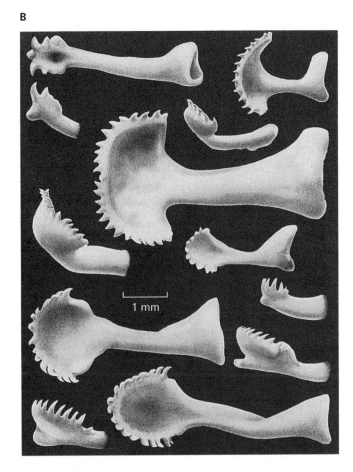

Figure 10.3 **The baculum.** Representative bacula of the Families (A) Mustelidae and Mephitidae and (B) Sciuridae.

channel for travel of the ova from the ovary to one or two enlarged **uteri** (sing., *uterus*), the site of embryonic development; and a **vagina,** the opening to the outside of the body. A **cervix** (pl., *cervices*) connects the uterus and vagina (figures 10.4 and 10.8).

The Ovary

The primary reproductive organ of the female is the **ovary.** In mammals, the ovaries are a pair of small, oval bodies that lie slightly posterior to the kidneys. They produce the female gametes, called **ova** (sing., *ovum*), and certain hormones. As within the testes of the male, chromosome reduction occurs in female gametes within the ovary. Immediately under the surface of the ovary is a thick layer of spherically grouped cells called **follicles** (figure 10.5), each of which encloses a single egg. At birth, large numbers of follicles are present in the female mammal (about 2 million in the ovaries of humans), but the number decreases steadily with age. The ovarian follicle is the source of three types of gonadal steroids: androgens (masculinizing), estrogens (feminizing), and progestins. The amounts of each class of steroid vary throughout the reproductive cycle of the female (Albrecht and Pepe 1990; Strauss et al. 1996).

Ovulation. As the egg matures and the surrounding follicle enlarges, they move closer to the surface of the ovary.

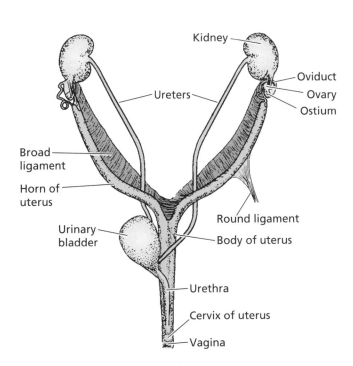

Figure 10.4 **Female urogenital system.** Composite diagram of the female urogenital system of the cat (*Felis catus*). *Adapted from W. J. Leach, 1961,* Functional Anatomy: Mammalian and Comparative, *McGraw-Hill.*

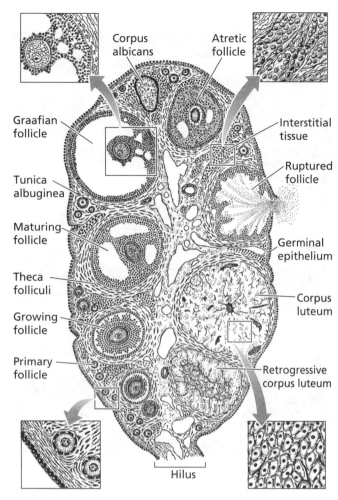

Figure 10.5 **Mammalian ovary.** Composite diagram of the mammalian ovary showing the progressive stages in the differentiation of a Graafian follicle shown on the left. The mature follicle may become atretic (*top*) or ovulate and undergo luteinization, as on the right. Insets show magnification of cellular components in each region of the ovary. *Adapted from C. D. Turner, 1966, General Endocrinology, 4th ed., W. B. Saunders Company.*

Ovulation occurs when the follicle bursts, releasing the egg, which penetrates the surface of the ovary and passes into the oviduct, where it may be fertilized if sperm are present. Some sperm may reach the site of fertilization (typically, the oviduct) within a few minutes, but for the most part, it takes several hours for **capacitation** (the physiological changes in the sperm that facilitate penetration of the covering of the egg) to take place (Austin and Short 1972c). For some species of mammals, however, the period between insemination and actual fertilization may be delayed for up to 12 months—a phenomenon called delayed fertilization (discussed in detail in "Reproductive Variations" in this chapter). If fertilization does not occur, the egg enters the uterus and degenerates.

The development of follicular cells is controlled by the action of **follicle-stimulating hormone (FSH)** and **luteinizing hormone (LH)** produced by the anterior pituitary gland (figure 10.6). Ovulation is induced by high

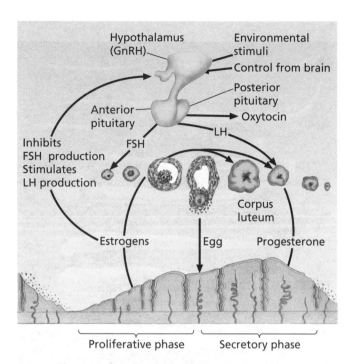

Figure 10.6 **Estrous cycle.** Hormones influence the estrous cycle of a mammal. FSH secreted by the anterior pituitary gland initiates the onset of follicular development. During the proliferative phase of the cycle, FSH and LH further the maturation of the follicle. Estrogens secreted by the follicular cells surrounding the egg cause thickening of the endometrium, further follicular growth, and inhibit the production of FSH. When the egg is mature, the enlarged follicle bulges from the surface of the ovary and bursts, releasing the egg (ovulation; see also figure 10.5). Ovulation, induced by high levels of LH, initiates the secretory phase of the cycle. The egg passes into the oviduct, where, if it is fertilized, it implants into the endometrium. Unfertilized eggs enter the uterus and degenerate. Under the influence of LH, the ruptured follicle fills with yellowish cells and is called the corpus luteum. The corpus luteum produces high levels of progesterone and estrogens that strongly inhibit GnRH secretion by the hypothalamus, thereby inhibiting secretion of FSH and LH necessary to permit maturation of new follicles. High levels of estrogens and progesterone stimulate the uterus to thicken, making final preparations for gestation. Progesterone also stimulates growth of the mammary glands. Abbreviations: FSH = follicle-stimulating hormone; LH = luteinizing hormone; GnRH = gonadotropin-releasing hormone. *Data from H. L. Gunderson, Mammalogy, 1976, McGraw-Hill.*

levels of LH. Under the influence of these hormones, the ruptured follicle fills with yellow follicular cells and is called the **corpus luteum** (pl., *corpora lutea*), or "yellow body." The corpus luteum continues functioning during early pregnancy and produces the female hormone **progesterone.** This hormone promotes growth of the uterine lining **(endometrium)** and makes possible the implantation of the fertilized egg (Stouffer 1987; Keys and Wiltbank 1988; Nelson 1995). Progesterone also stimulates development of the mammary glands. Inactive corpora lutea eventually degenerate into **corpora albicans,** or "white bodies," visible on the surface of the ovary. An examination of the number of corpora lutea and corpora albicans in the ovaries can reveal a great deal about the reproductive history of a particular female (Perry 1972). The number of corpora lutea is related to the number of ova that have been ovulated and can be used to estimate the number of embryos that were produced. The major female hormone secreted by the developing follicles is **estradiol** (a form of **estrogen**), under the control of the FSH (see figure 10.6). Estrogen is essential in promoting the proliferation of the endometrium. As the corpus luteum disintegrates, no more progesterone is produced, causing the "sloughing off" of the endometrium. Although this sloughing off and regrowth of the endometrium occurs cyclically in all mammals, it is most pronounced in primates. In some mammals, for example in dogs and cats, a false pregnancy, or **pseudopregnancy,** may occur. This results from the maintenance of the corpora luteum, without fertilization, beyond the time of normal regression. Mammals exhibiting a false pregnancy behave as if they are pregnant, even though they are not.

The Estrous Cycle. All female mammals except higher primates restrict copulation to specific periods of the sexual cycle. Periods of reproduction are controlled by hormones and by the nervous system and are regulated by environmental and social cues (Bronson 1989; see chapter 8). In nonprimates, a period of brief receptivity shortly before and after ovulation is called **estrus,** or **heat** (adj., *estrous*). In the majority of mammals, ovulation is **spontaneous,** that is, it occurs without copulation (figure 10.7). In contrast, if the egg is shed within a few hours following copulation, this situation is referred to as **induced** ovulation. Many investigators suggest, however, that these two strategies for ovulation cannot be separated and instead represent a continuum (Weir and Rowlands 1973). Some mammals known to be induced ovulators are rabbits, many carnivores, and some ground squirrels. Some desert rodents and montane voles (*Microtus montanus*) may also exhibit induced ovulation in response to the occurrence of green vegetation and associated nutritional factors in their diet (Beatley 1969; Reichman and Van De Graaff 1975; Negus and Berger 1977; Kenagy and Bartholomew 1985).

During the breeding season, the time span from one period of estrus to the next is called the **estrous cycle** (Gunderson 1976; Withers 1992). A species that has one such cycle per year, as is true for some carnivores, is termed **monestrous.** Other species, including rodents, rabbits, hares, and pikas, are **polyestrous,** meaning they have several cycles a year. In general, several stages may be seen during a given estrous cycle. As an individual moves from **anestrus** (the nonbreeding, quiescent condition) to **proestrus** (the beginning stage), the uterus begins to swell, the vagina enlarges, and ovarian follicles grow in response to several hormones—the animal is said to "come into heat." The ova are usually discharged from the ovary late in this estrous stage, which may be followed by **fertilization,**

A

Ovulation
Copulation
Fertilization
Implantation
Gestation
Parturition
Lactation

B

Copulation
Ovulation
Fertilization
Implantation
Gestation
Parturition
Lactation

C

Copulation
Delay
Ovulation
Fertilization
Implantation
Gestation
Parturition
Lactation

D

Ovulation
Copulation
Fertilization
Delay
Implantation
Gestation
Parturition
Lactation

Figure 10.7 Reproductive events. (A) *Spontaneous ovulation* is the usual sequence of reproductive events occurring in eutherian mammals. Several different groups show variation on this general sequence, including (B) *induced ovulation* exhibited by cats and some rodents, (C) *delayed fertilization* displayed by certain insect-eating bats, and (D) *delayed implantation* shown in a diverse array of groups. *Delayed development* exhibited by the Neotropical fruit bat and *embryonic diapause*, exhibited by almost all kangaroos, wallabies, rat kangaroos, pygmy possums, feathertail gliders and honey possums, are not shown in this figure.

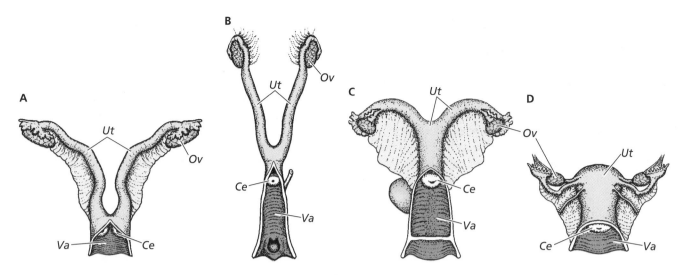

Figure 10.8 Types of uteri. The degree of fusion of the uterine horns varies among different groups of eutherian mammals. (A) A duplex uterus occurs in rodents, rabbits and hares, aardvarks, and hyraxes. Note the separate horns and distinct cervix of each. (B) Most carnivores and whales exhibit a bipartite uterus. (C) "Insectivores," most bats, primitive primates, pangolins, some carnivores, elephants, manatees and dugongs, and most ungulates possess a bicornuate uterus. (D) A simplex uterus is typified by higher primates, some bats, anteaters, tree sloths, and armadillos. *Abbreviations:* Ut = uterus, Ce = cervix, Ov = ovary, Va = vagina. *Adapted from E. M. Ramsey, 1982,* The Placenta: Human and Animal, *Praeger.*

the successful union of sperm and egg to form a zygote, and implantation, in which the embryo adheres to the uterine wall. Gestation represents the period of development within the uterus. The period of lactation (production of milk by the mammary glands) is reported to "block" the start of another estrous cycle; however, in certain mammals, parturition (birth) is followed immediately by another estrous cycle (postpartum estrus). Many members of the Family Cricetidae are known to ovulate within an hour after parturition. If, however, mating does not occur, the individual enters the normal condition of **metestrus,** followed by anestrus. Periods between mating in polyestrous species are termed **diestrus,** when the reproductive system recycles to the proestrous condition. Mammalogists can determine the different estrous periods by using the **vaginal smear technique,** which identifies specific cell types associated with a certain phase of the cycle (Kirkpatrick 1980). In humans and Old World primates, the ovarian and uterine cycles are different and are called menstrual rather than estrous cycles (Nadler 1975; Guyton 1986).

The Female Duct System and External Genitalia

The female duct system consists of paired oviducts (or Fallopian tubes), one or two uteri, one or two cervices, and the vagina (see figure 10.4). The external genitalia include the **clitoris,** the **labia majora,** and **labia minora.** The clitoris is the embryonic homologue of the penis, and in those species where males possess a baculum, females possess a small bone, the **os clitoris,** in the clitoris.

The architecture of the female reproductive tract also varies considerably in mammals. The monotremes and marsupials do not possess true vaginae (Chapter 11: figure

11.1). In the monotremes, the paired uteri and oviducts empty into a **urogenital sinus,** as do liquid waste products. This sinus passes into the cloaca. The female reproductive tract of marsupials is referred to as **didelphous,** meaning that the uteri, oviducts, and vaginal canals are paired (figure 11.1). During copulation, the forked penis of the male delivers sperm to the paired uteri of the female. In more advanced marsupials, such as the kangaroos, two lateral vaginae are used to conduct sperm, and a medial vagina functions as the birth canal.

In eutherian mammals, four principal uterine types occur that are based on the relationship of the uterine horns (figure 10.8). A **duplex** uterus is found in lagomorphs, rodents, aardvarks, and hyraxes and is characterized by two uteri, each with a cervix opening into the vagina. A **bipartite** uterus is typical of whales and most carnivores. Here, the horns of the uterus are separate but enter the vagina by a single cervix. The most widespread condition is called the **bicornuate** uterus, which is found in the "insectivores," most bats, primitive primates, pangolins, some carnivores, elephants, manatees and dugongs, and most ungulates. The uterine horns are Y-shaped, being separated medially but fused distally, where they form a common chamber, the body, which opens into the vagina through a single cervix. The last type of uterus is called **simplex,** in which all separation between the uterine horns is lacking, and the single uterus opens into the vagina through one cervix. This condition is seen in some of the bats, higher primates, and xenarthrans.

Implantation

Fertilization usually occurs shortly after ovulation in the oviduct. **Implantation** is the attachment of the embryo to

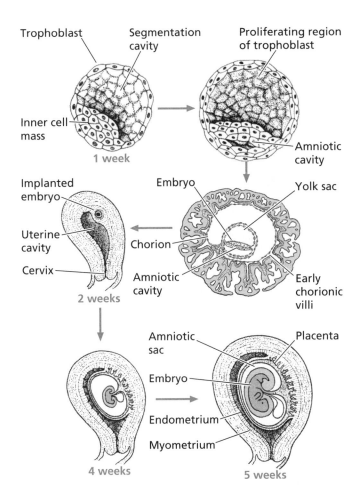

Figure 10.9 Human embryo. Diagram showing the early development of the human embryo and its extraembryonic membranes. *Adapted from Hickman et al., 2004,* Integrated Principles of Zoology, *12th ed., McGraw-Hill.*

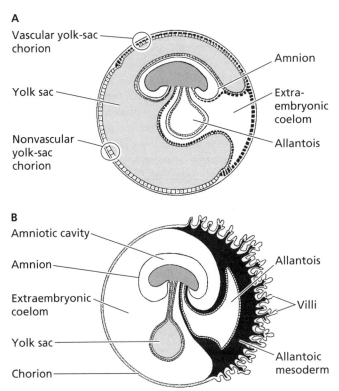

Figure 10.10 Embryo and extraembryonic membranes. Diagram showing the embryo and extraembryonic membranes of a (A) metatherian (kangaroo) and a (B) eutherian mammal. Note the greatly enlarged yolk sac of the metatherian that serves the nutritional needs of the developing embryo. In the eutherian mammal, extensive chorionic villi penetrate the endometrium and provide mechanical strength and increased surface area for rapid and efficient exchange of nutrients, gases, and waste products between the maternal and fetal blood supplies. *(A) Adapted from Torrey and Feduccia, 1979,* Morphogenesis of the Vertebrates, *John Wiley & Son; (B) Adapted from Balinsky, 1975,* Introduction to Embryology, *W. B. Saunders.*

the uterine wall. Cleavage of the **zygote,** or fertilized egg, occurs as it passes down the oviduct toward the uterus, propelled by ciliary action and rhythmic muscular contractions (peristalsis). Next, the blastocyst (the embryo at the 32- to 64-cell stage) implants on the wall of the uterus. The uterus of mammals is characterized by a thick muscular wall (the **myometrium**) surrounding the highly vascular endometrium—the specific site of implantation. The anterior pituitary gland (which produces FSH and LH) and the corpus luteum (which produces progesterone) collaborate to prepare the endometrium for implantation. The blastocyst differentiates into the embryo and the trophoblast (see figures 10.9 and 10.10). On contact with the lining of the uterus, the cells of the trophoblast region proliferate rapidly, producing enzymes that break down the **decidua**—the uterine mucosa in contact with the trophoblast. This permits the blastocyst to sink into the vascularized endometrium, and subsequently it is surrounded by a pool of maternal blood. At this point, the trophoblast thickens, extending thousands of fingerlike roots (villi) that penetrate the lining of the endometrium. The trophoblast further invades the decidua, eventually contributing to the

formation of the placenta. The more invasive the trophoblast, the larger the area of decidua through which both the embryo and placenta obtain nutrients, oxygen, water, ions, and hormones from the mother and discharge waste products. The degree of intimacy between the maternal and embryonic parts of the placenta varies in different mammals, and mammalogists employ these differences to help define various groups of mammals.

The Placenta. The placenta, a complex of embryonic and maternal tissues, performs several essential functions during pregnancy. It (1) physically anchors the fetus to the uterus, (2) transports nutrients from the circulation of the mother to the developing fetus, (3) excretes metabolites of the fetus into the maternal compartment, and (4) produces hormones that regulate the organs of the mother and fetus. Because both the placenta and embryo are genetically alien to the mother, uterine tissue should reject the embryo. This does not happen because the placenta has evolved the ability to suppress the normal immune response that the mother's body would ordinarily mount against it and the embryo. This ability to become an **allograft** (successful foreign transplant) is attributed to the production by the chorion of essential lymphocytes and proteins that block the mother's immune response by suppressing the formation of specific antibodies (Adcock et al. 1973; Strauss et al. 1996).

Although eutherians are often referred to as "placental" mammals, marsupials also have a placenta, albeit not one as efficient as that found in the eutherians. Most metatherians have only a choriovitelline placenta in which the yolk-sac cavity is greatly enlarged to form the placenta; it serves the nutritional needs of the developing embryo. Compared to eutherians, the placentae of most metatherians lack extensive villi and have a comparatively weak mechanical connection to the uterine lining (figure 10.11). The embryo of metatherians does not implant deeply in the endometrial mucosa but merely sinks into a shallow depression of the mucosa. The surface area for adhesion and absorption is increased by a slight wrinkling of the surface of the embryo that lies against the endometrium. Besides nutrition provided by the enlarged yolk sac, the developing embryo gains a limited amount of nutrition in the form of "uterine milk" derived from the uterine mucosa. The system of nourishment to the embryo from the mucosa is comparatively inefficient.

We can delimit four types of metatherian placentae based on the degree of apposition between the fetal and maternal tissues and the structure of the chorion and allantois (Austin and Short, 1976; Hughes, 1984; Mossman, 1987; Tyndale-Biscoe and Renfree, 1987; Dawson, 1995). Within the metatherians, the most common type of arrangement of fetal and maternal tissues is seen in eastern gray kangaroos (*Macropus giganteus*; Dawson, 1995). Here, the highly vascularized wall of the yolk-sac cavity is intimately connected with the maternal endometrial mucosa (figure 10.11). This kind of placenta is also found in other members of the Diprotodontia and in the Didelphimorphia (American opossums). The dasyurids show a variation on the latter condition in which the allantois and chorion are opposed early in development and then retreat without forming a placental structure. The koalas (*Phascolarctos cinereus*) and wombats (*Vombatus ursinus*) demonstrate an apposed, nonvascular, chorioallantoic placenta but still retain a vascular choriovitilline placenta. Here the large yolk-sac cavity retains its role as the primary surface for gas exchange and absorption of nutrients.

Lastly, the placenta of the bandicoots (Order Peramelemorphia: Family Peramelidae) show the greatest similarity to the eutherian placental condition. They possess a true "syncytialized" vascular chorioallantoic placenta yet still retain a vascular choriovitilline placenta (figure 10.11). In bandicoots, the allantois is large and highly vascular. The position of the embryo, close to the highly vascularized endometrial mucosa, promotes exchange of materials across allantoic membranes. However, the allantois of bandicoots lacks villi, thus compromising the surface area available for exchange of materials between fetal and maternal bloodstreams. Nutrition to the embryo is augmented by "uterine

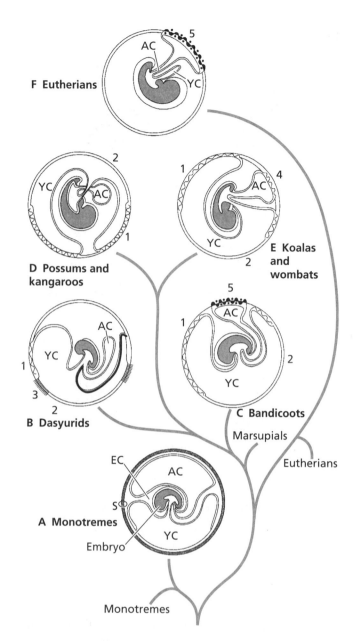

Figure 10.11 Types of placentation in marsupials and eutherians. Key: 1 = vascular choriovitilline placenta; 2 = nonvascular choriovitilline placenta; 3 = syncytialized choriovitilline placenta; 4 = apposed (nonvascular) chorioallantoic placenta; 5 = syncytialized (vascular) chorioallantoic placenta; AC = allantoic cavity; EC = extraembryonic coelom; S = shell; YC = yolk-sac cavity. *Adapted from Sharman (1976), Austin and Short (1976), and Pough et al (2005).*

milk" absorbed across the choriovitilline membranes. In sum, the arrangement of placental membranes in the bandicoots results in a less effective transfer of substances between the fetus and material circulations. The northern brown bandicoot (*Isodon macrou*) and the long-nosed bandicoot (*Perameles nasuta*) have one of the shortest gestation periods among mammals (ca. 12.5 days); however, young bandicoots are relatively procolial at birth. This fact suggests that the choriovitilline placenta is less efficient than the chorioallantoic placenta (Padykula and Taylor, 1982).

Compared to metatherians, all eutherian mammals possess a **chorioallantoic placenta,** which develops from the combination of the embryonic membranes, an outer chorion, and an inner vascularized allantois. The choriovitilline placenta is short-lived, and a complex "syncytialized," vascular chorioallantoic placenta is the functional one for most of gestation (figure 10.12). In this case, the blastocyst adheres to the endometrium and then sinks very deeply into it, forming a strong adhesion. This adhesion is enhanced by the rapid growth of **chorionic villi** (fingerlike projections of capillaries from the outermost embryonic membrane) that penetrate the endometrium. The uterus becomes highly vascularized at the point of attachment of the developing embryo. It is important to note that although the maternal and fetal circulations are in close contact, these two blood systems are not fused. Fetal blood does not circulate in the mother, and maternal blood does not circulate in the fetus. The complex, extensive network of villi provides not only mechanical strength but also an increased surface area for rapid and efficient exchange of nutrients, gases, and waste products between the maternal and fetal blood supplies. A human placenta is reported to possess about 48 km (30 miles) of villi (Bodemer, 1968).

The four types of placentas are classified based on the distribution of villi permeating the endometrial lining of the uterus (Ramsey 1982; figures 10.13 and 10.14). Lemurs, nonruminating artiodactyls, pangolins, and perissodactyls have a placenta that is **diffuse,** characterized by the villi scattered evenly throughout the uterus. A **cotyledonary** placenta, found in ruminating artiodactyls, is characterized by more or less evenly spaced patches of villi scattered within the uterus. Carnivores have a **zonary**

placenta, characterized by a continuous band of villi within the uterus. Many groups, including "insectivores," bats, most primates (including humans), some rodents, rabbits and hares, and pikas, have a placenta referred to as **discoidal,** in which villi occupy one or two disc-shaped areas within the uterus. A variation on the discoidal arrangement is the **cup-shaped discoid** found in most rodents.

Mammalian placentas are also categorized by an assessment of the speed and efficiency with which nutrients, oxygen, and waste material are exchanged between maternal and fetal blood supplies (see figure 10.13). Recall that as the blastocyst sinks into the endometrium and implantation proceeds, the chorionic villi grow farther into the endometrium. The uterus becomes greatly vascularized in the area of the implantation. Such an arrangement increases the surface area, allowing a more rapid interchange between the maternal and fetal circulation. Exchange rates depend on the number and thickness of capillary walls, connective tissue, and uterine and chorionic (fetal) epithelial tissues. Six arrangements are generally recognized (figure 10.14):

1. An **epitheliochorial** system is typified by having six tissue layers, with the villi resting in pockets in the endometrium. This system occurs in pigs, lemurs, horses, and whales.
2. The **syndesmochorial** system has one less layer because the epithelium of the uterus erodes at the

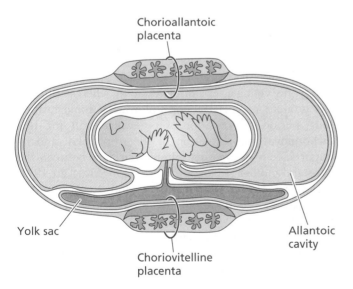

Figure 10.12 Mammalian placental structures. Two types of mammalian placental structures as seen in a transitional stage of an implanted embryo of a cat. Both a choriovitilline placenta and a chorioallantoic placenta are present at this stage. The chorioallantoic placenta grows outward and takes over the function of the choriovitilline placenta. *Adapted from Pough et al. (2005).*

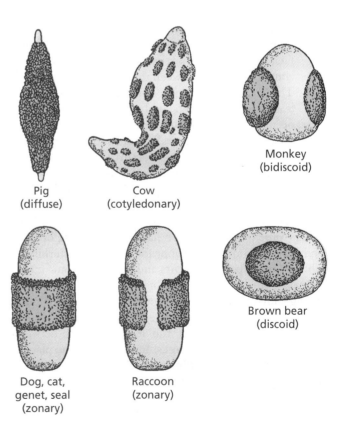

Figure 10.13 Placental villi. The shape and distribution of placental villi vary among different groups of mammals. *Adapted from E. M. Ramsey, 1982,* The Placenta: Human and Animal, *Praeger.*

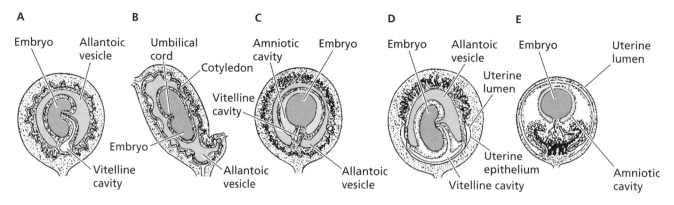

Figure 10.14 **Types of placentae.** Sectional views of different types of mammalian placentae: (A) diffuse, (B) cotyledonary, (C) zonary, (D) discoid, and (E) cup-shaped discoid. Views (A), (B), and (D) are sagittal, and views (C) and (E) are transverse sections. *Adapted from DeBlase and Martin, 1981,* A Manual of Mammalogy, *2nd ed., Wm. C. Brown Publishers.*

site of attachment, thus reducing the separation of the maternal and fetal bloodstreams. This system is found in ruminating artiodactyls, such as cows, sheep, and goats.

3. Even further erosion of the maternal tissue is exemplified by the endotheliochorial system of the carnivores. Here, the chorion of the fetus is in direct contact with the maternal capillaries.

4. In the **hemochorial** system seen in advanced primates (including humans), "insectivores," bats, and some rodents, no maternal epithelium is present, and the villi are in direct contact with the maternal blood supply.

5. In an **endothelioendothelial** system, maternal and fetal capillaries are next to each other, with no connective tissue between them.

6. The greatest destruction of placental tissues and least separation of fetal and maternal bloodstreams occur in the **hemoendothelial** system in which the fetal capillaries are literally bathed in the maternal blood supply. This system occurs in some rodents, rabbits, and hares (Austin and Short 1972b; Gunderson 1976; Ramsey 1982; Pough et al. 2005).

Many of you have observed the birth of pets or livestock (some of you may even have witnessed the birth of children) and noticed the appearance of the "afterbirth," which is the expelled placenta and extraembryonic membranes. Sometimes the afterbirth is expelled with the young shortly thereafter. Mammals such as pigs, lemurs, horses, and whales possess the epitheliochorial type of placenta, which provides the least intimacy between maternal and fetal membranes. In such cases, the placentas are termed **nondeciduous,** and due to the minimal attachment of villi in the uterine wall, they can pull away from the wall of the endometrium at birth without causing any bleeding. Mammals such as armadillos, sloths, anteaters, rodents, carnivores, and primates have more intimate associations between the endometrial and fetal bloodstreams. Such placentas are termed **deciduous** and are typified by the hemoendothelial system. As a result

of this close attachment, extensive erosion of the uterine tissue takes place at parturition. A portion of the uterine wall is actually torn away when the placenta separates at delivery. Bleeding occurs because some blood vessels and connective tissue of the uterine wall are shed with the placenta. Hemorrhaging following birth soon stops due to the collapse of the uterus by contractions of the myometrium (the smooth muscle layer of the uterus), which tends to constrict the blood vessels. The previous site of attachment of the fetus on the uterine wall is evidenced by a pigmented area called a **placental scar.** Placental scars form in rodents, shrews, carnivores, bats, rabbits, hares, and pikas, for example, and are useful in studying an individual's reproductive history (Lidicker 1973). Females that are **nulliparous** (never having been pregnant) show no evidence of placental scars, whereas a female that is termed **parous** has scars indicating prior parturition. A multiparous individual has placental scars of different ages, thus indicating that several fetuses or litters were produced.

The Trophoblast. The reproductive patterns of marsupials and eutherians are very different. Marsupials bear embryonic young after a brief gestation, whereas eutherians produce precocial young following a comparatively long period of gestation. One explanation for this dichotomy is the evolution of the trophoblast and its importance in preventing immunorejection between the fetus and mother (Lillegraven 1975, 1987; Luckett 1975; Eisenberg 1981; Tyndale-Biscoe and Renfree 1987; Vaughan et al. 2000). Recall that the **trophoblast** (see figures 10.9 and 10.15), the embryonic contribution to the placenta, greatly enhances fetal development by efficiently transferring nutrients and dissolved gasses between the fetal and maternal circulations. The trophoblast also prevents the immunorejection response from developing between the fetal and maternal circulatory systems. During early gestation in marsupials, membranes of the eggshell act as a barrier between antigen-bearing parts of the embryo and the uterine fluid, thus preventing immunorejection. Late in gestation, however, shell membranes are shed, leaving the embryo and maternal systems

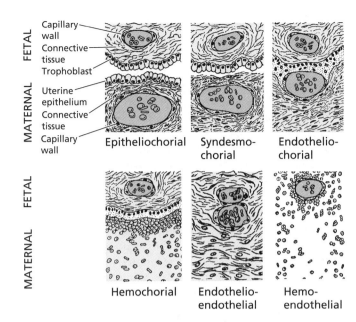

Figure 10.15 Internal structure of the placenta.
Diagram showing variation in the number of tissue layers separating maternal and fetal blood supplies in different systems of placental mammals. The number varies from one layer in a hemoendothelial system to six layers in an epitheliochorial system. *Adapted from E. M. Ramsey, 1982,* The Placenta: Human and Animal, *Praeger.*

vulnerable to immunological attack. To circumvent rejection, marsupials have evolved abbreviated gestation periods. In contrast, although eutherians lack shell membranes, early stages of the zygote are protected from mixing with maternal blood by the **zona pellucida**—a noncellular layer surrounding the zygote. Later in development, immunorejection is prevented by the trophoblast and the decidua. Throughout gestation, a close association of uterine and fetal tissues is maintained, yet the trophoblast provides a constant "line of defense" against the mixing of fetal and maternal tissues. As a result, the trophoblastic tissues have permitted eutherians to evolve longer periods of gestation, thus giving rise to the production of more precocial young and a decrease in the length of the energetically expensive period of lactation. The "appearance" of the trophoblast is of central importance in favoring the evolution of structural diversity and adaptive radiation of eutherian mammals.

Gestation and Parturition

The period of time from fertilization until the birth of the young is referred to as **gestation**. This period of intrauterine growth also varies greatly within mammals. The prototherian mammals (monotremes) lay eggs (oviparous) that hatch outside the body, so, by definition, gestation does not even occur in monotremes. The therian mammals (Infraclasses Metatheria and Eutheria), however, bear live

young (viviparous) and retain the embryo within the body. They therefore have a gestation period.

Marsupials bear altricial young following a brief gestation period, and lactation is protracted compared with similarly sized placental mammals (figure 10.16; Tyndale-Biscoe and Renfree 1987; Dawson 1995). Virginia opossums (*Didelphis virginiana*) may be familiar to many North Americans. This marsupial has a very short gestation—only about 12.5 days. Following this brief period of intrauterine development, from 4 to 25 embryo-like young (the size of honey bees) emerge from the birth canal. The young are altricial, and each weighs only about 0.13 g (0.005 oz)—an entire litter can fit into a teaspoon! Although altricial, the young have well-developed, muscular forelegs with sharp claws and can move hand over hand, up the hair of the mother's belly and into her pouch—traveling up to 50 mm (2 in.) in about 16.5 seconds. After reaching the mother's pouch, each pup takes hold of one of the 13 available nipples (figure 10.17), which then enlarges, forming a bulb within the mouth of the suckling young. They remain attached to the nipple until about 2 months later, when they are weaned. Although as many as 25 young are born, the entire litter cannot survive. Pouched young average about eight in number and, following weaning, ride on the mother's back for another month (McManus 1974).

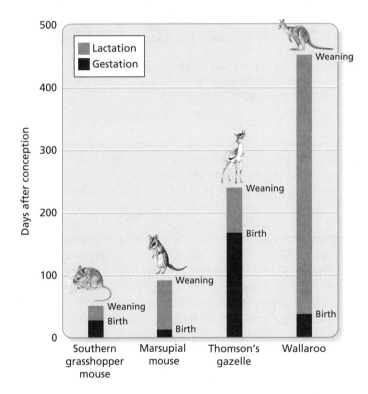

Figure 10.16 Gestation and lactation. Diagram comparing the periods of gestation and lactation of marsupial and placental mammals. Note that marsupials have a shorter interval of gestation and longer period of lactation than similarly sized species of placental mammals. *Data from Hickman et al., 1997,* Integrated Principles of Zoology, *10th ed., McGraw-Hill Companies, Inc., Dubuque, IA. All rights reserved.*

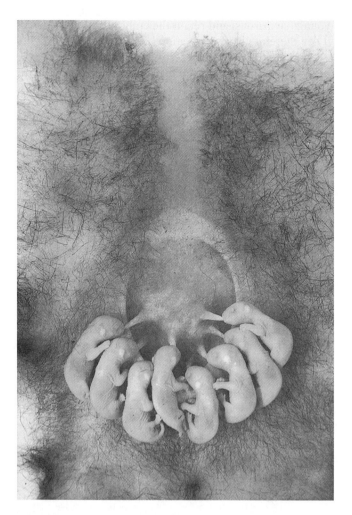

Figure 10.17 Young Virginia opossums. Photograph showing eight one-week-old Virginia opossums attached to the teats within the pouch of the mother.

The length of gestation, which varies according to taxonomic group, is correlated with body mass, the number of young per litter, and the degree of development of the young (Eisenberg 1981; Dawson 1995). A prime example of a mammal with a large body mass and the record holder for the longest gestation period (about 22 months) is the African elephant (*Loxodonta africana*). Although many rodents have very short gestation periods, a marsupial, the short-nosed bandicoot (Genus *Isoodon*) probably holds the record with a gestation of about 12.5 days. Bats are rather unusual among small mammals in displaying a comparatively long gestation period. Pregnancy generally lasts from 3 to 6 months, and most bats have only one young per litter. Some species, however, such as the red bat (*Lasiurus borealis*), produce litters of up to five young (Hayssen et al. 1993; Altringham 1996).

Body mass is not always the best predictor of gestation period. The blue whale (*Balaenoptera musculus*) is the largest animal that has ever lived. So it might be expected to have the longest gestation period. This huge mammal (about 130,000 kg; 143 tons), however, has a gestation period of only 10–12 months, similar to that of a horse.

The Order Cingulata (anteaters, sloths, and armadillos) are known to have longer gestation periods than predicted by body mass (McNab 1985). The two-toed sloth (*Choloepus didactylus*) of the Neotropical region weighs about 9 kg (20 lb) but has a gestation period of almost 11 months. This is about 4 months longer than the wapiti, or elk (*Cervus elaphus*), which weighs 22 times more (Hayssen et al. 1993; Vaughan et al. 2000). Other groups of mammals that do not exhibit longer gestation periods as body mass increases are the marsupials, whales and dolphins, higher primates, elephant shrews, and hystricomorph rodents (porcupines, guinea pigs, capybara, agoutis, pacas, cane rats and allies; Sadleir 1973).

Primates tend to exhibit long gestation periods compared with other mammals of similar body size. Some of the smaller primates, such as the lemurs, lorises, and marmosets, have gestation periods ranging from 60 to 200 days. Larger primates, such as gorillas, have gestation periods lasting about as long as that of humans (280 days). Newborn primates are precocial but require long periods of maternal care and lactation. These extended periods of care giving appear to be important in the evolution of socialization in higher primates. In addition to body mass, environmental factors such as latitude, ecosystem, and length of growing season influence the length of gestation period and litter size in mammals.

Reproductive Variations

For most mammals, fertilization occurs within several hours of insemination. The fertilized egg becomes implanted in the uterus, and development continues until birth. Thus, we define the gestation period as the interval between fertilization and parturition. Mammals may exhibit modifications that lengthen the fertilization or gestation period, including delayed fertilization, delayed development, delayed implantation, and embryonic diapause (Weir and Rowlands 1973; Gaisler 1979; Nowak 1994).

DELAYED FERTILIZATION

Many animals residing in seasonal environments employ special mechanisms to optimize survival. Bats of the Families Vespertilionidae and Rhinolophidae occupying temperate zones of the New and Old Worlds exhibit a reproductive tactic called **delayed fertilization** or **delayed ovulation** (Wimsatt 1945; Austin and Short 1972a; Oxberry 1979; Racey 1982; Nowak 1994; Bernard and Cumming 1997; Neuweiler 2000; see figure 10.7). Copulation takes place in September or October before hibernation commences. Follicular growth has occurred in the ovary, but ovulation does not happen at this time. The sperm are immotile and stored in either the uterus or the upper vagina, and then both sexes enter hibernation.

When the female emerges from hibernation in the spring, the eggs are ovulated, spermatozoa become motile, and fertilization takes place. It is noteworthy that some bats do copulate during the period of hibernation, especially species in the Genera *Myotis* and *Plecotus*. Implantation of the blastocyst occurs about a week after fertilization (Kunz 1982). The gestation period may last for 50–60 days, as in the silver-haired bat (*Lasionycteris noctivagans*) and little brown bat (*Myotis lucifugus*; Wimsatt 1945; Kunz 1982; Gustafson and Damassa 1985; Hayssen et al. 1993). Young are born in early summer, when insects are abundant. Delayed fertilization is advantageous for northern species of bats because it provides young with more time to build critical body mass in order to sustain their long period of hibernation.

Delayed fertilization is most common in temperate species of bats that undergo hibernation; however, mammalogists have found sperm storage in bats residing in tropical regions that apparently do not undergo true hibernation. Tropical species may store sperm as an adaptive strategy synchronized with the availability of food (Racey 1982).

DELAYED DEVELOPMENT

Delayed development differs from delayed fertilization in that the blastocyst implants shortly after fertilization, but development is very slow. A fertilized blastocyst implants in the uterus in summer but may have a 7-month period of development or gestation. Delayed development is reported for both micro- and megachiropterans (Racey 1982). *Artibeus jamaicensis* employs delayed development as a unique mechanism to synchronize the birth of young with the end of the dry season, when the availability of large fruits is at its peak (Fleming 1971; Racey 1982). The Old World insectivorous bat *Miniopterus australis* also exhibits delayed development (Richardson 1977). The reproductive delay in this species is reportedly in response to unpredictable availability of insects. Other species of bats reported to undergo delayed development are *Macrotus californicus* (Bradshaw 1961, 1962); *Haplonycteris fischeri* (Heideman 1988, 1989); *Otopteropus cartilagonodus* (Heideman et al. 1993); *Plenochirus jagori* (Heideman and Powell 1998); and *Rhinolophus rouxii*, *Cynopterus sphinx*, *Eptesicus furinalis*, and *Myotis albescens* (Racey 1982).

DELAYED IMPLANTATION

In **delayed implantation,** ovulation, copulation, fertilization, and early cleavage of the zygote up to the blastocyst stage occur normally (see figure 10.7). Development of the blastocyst is arrested, however, and each blastocyst floats freely in suspended animation in the reproductive tract until environmental conditions become favorable for implantation. Unlike delayed development, the blastocyst does not implant into the uterine wall but rather floats in

the reproductive tract. During this free-floating stage, the blastocyst is encased in a protective coat (called the zona pellucida) until the optimal time for its development. Eventually implantation occurs, and development proceeds normally (Enders 1963). Delayed implantation is either **obligate,** as in armadillos, in which the delay occurs as a normal, consistent part of the reproductive cycle, or **facultative,** as when rodents or "insectivores" are nursing a large litter or are faced with extreme environmental conditions. Delayed implantation is discussed in detail by Daniel (1970) and Mead (1989).

Although delayed fertilization and delayed development occur only in bats, delayed implantation is seen in many diverse groups, including "insectivores," rodents, bears, mustelids (weasels and allies), seals, armadillos, certain bats, and two species of roe deer (Genus *Capreolus*). Delayed implantation is a rule in the pinnipeds (Atkinson 1997; Laws et al 2003). An excellent review of delayed implantation in carnivores is provided by Mead (1989). Delayed implantation occurs in the Old World vespertilionids *Miniopterus schreibersi* and *M. fraterculus* (Kimura and Uchida 1983). The length of gestation in *M. schreibersi* varies geographically, and differences may in part reflect local environmental conditions (Racey 1982). For many species of bats, timing of implantation may be contingent on endocrine controls for different populations (see Nowak 1994; Neuweiler 2000).

Delayed implantation probably was first reported in the mid-1600s, as described in the "field notes" of William Harvey who joined King Charles I of England on hunting trips for European roe deer (see Gunderson [1976] for a discussion of the historical account). Many different mammals employ delayed implantation, and this reproductive strategy varies widely even within the same genus (Sandell 1984; King 1990). Siberian and European roe deer (*Capreolus capreolus* and *C. pygargus*) are the only ungulates that exhibit delayed implantation. Following the rut in July and August, implantation is delayed until December or January, and the young are born in the spring (April to June). The gestation period is between 264 and 318 days (Hayssen et al. 1993; Danilkin 1995; Sempéré et al. 1996). In North America, the black bear (*Ursus americanus*) exhibits delayed implantation. In Pennsylvania, for example, black bears mate during the summer, usually sometime from early June to mid-July. After the ova are fertilized, implantation of the blastocyst in the uterine wall is delayed for up to 5 months. Young are born in winter dens in mid-January, following an "actual" gestation period of about 60–70 days (Pelton 1982; Alt 1983; Hellgren et al. 1990; Hayssen et al. 1993). Mustelids also represent excellent examples of delayed implantation (King 1983; Mead 1989). Although great variation occurs among different species of "mustelids," the western spotted skunk (*Spilogale gracilis*) (Family Mephitidae) seems to represent a fairly typical example of delayed implantation (Mead 1968). Female spotted skunks enter estrus in September and mate in September or October. The zygotes undergo normal cleavage but stop at the blastocyst stage, at which time they float freely in the uterus for about 6.5 months. After implantation, gestation lasts only about a month, and young are usually born between April and June. The total period of pregnancy takes 210–260 days. Closely related to the spotted skunk is the fisher (*Martes pennanti*). Of all mammals, this mustelid has the most protracted delay to implantation. Fishers breed between March and April. Following a 9-month delay, blastocysts implant in January or February, and most births occur in March and early April (Wright and Coulter 1967; Powell 1993; Frost et al. 1997). The total pregnancy lasts between 327 and 358 days, about the same as that of the blue whale (Hayssen et al. 1993).

The adaptive advantage of delayed implantation is poorly understood (King 1984). Two closely related species of weasels residing in the same habitat may exhibit very different reproductive patterns. For example, in eastern North America, the long-tailed weasel (*Mustela frenata*) and the least weasel (*Mustela nivalis*) may occupy the same habitat, but the former demonstrates delayed implantation and the latter does not. Many mammalogists believe that by delaying the arrival of the young to spring, when hunting is easiest because small mammals are plentiful, the long-tailed weasel optimizes the likelihood of survival of its young. But why does the smaller least weasel not show such a delay?

The nine-banded armadillo (*Dasypus novemcinctus*) of the New World is another interesting example of variation in reproductive biology. Armadillos in the southern United States breed in July or August. Their copulatory position is rather unusual for a quadruped mammal—the female assumes a mating position on her back. Implantation is delayed until November, followed by a gestation period of about 4 months. In late February, the litter is born and comprises identical quadruplets, all of one sex—they all come from a single fertilized ovum—a phenomenon called **monozygotic polyembryony.** Because only one ovum was fertilized by a single sperm, all the young have exactly the same genetic makeup. Delayed implantation in this case seems to serve to time the birth of the litter for the spring flush of invertebrate food—the staple in the diet of armadillos (Lowery 1974).

EMBRYONIC DIAPAUSE

Many marsupial mammals undergo a condition called **embryonic diapause**; the blastocyst enters a state of dormancy during which division and growth of the cells may cease or continue at a slow pace until a signal is received from the mother (Tyndale-Biscoe, 2005). Embryonic diapause is similar to the process of delayed implantation discussed earlier for placental mammals. Among marsupials, embryonic diapause is reported to occur in almost all kangaroos, wallabies, rat-kangaroos, pygmy possums, feathertail gliders (*Acrobates pygmaeus*) and honey possums (*Tarsipes rostratus*). Western gray kangaroos (*Macropus fuliginosus*), Lumholtz's tree kangaroos (*Dendrolagus lumholtzi*), and musky rat kangaroos (*Hypsiprymnodon moschatus*) are not

known to exhibit embryonic diapause. Endocrine control of embryonic diapause has been intensively studied in the tammar wallaby (*Macropus eugenii*) (Sharman and Berger, 1969; Tyndale-Biscoe and Renfree, 1987; Gordon et al 1988; Tyndale-Biscoe, 2005).

Embryonic diapause is common in red kangaroos (*Macropus rufus*), residents of arid regions of Australia. Both red kangaroos and euros (= common wallaroos, *Macropus robustus*) are opportunistic breeders—reproduction is finely tuned to climatic changes. At a given time, an adult female red kangaroo simultaneously nourishes three young (called joeys): (1) one "weaned" joey running at the heel of the mother that occasionally suckles from an elongated teat from *outside* the pouch; (2) a nursing, pouch young attached to another teat *inside* the pouch, and (3) a tiny blastocyst lodged in one of the two uteri of the female (figure 10.18). For red kangaroos, the first pregnancy of the season is followed by a gestation period of about 33 days. The joey is born in a very altricial state yet makes its way to the pouch and attaches to a nipple. Under favorable conditions, adult females will breed continuously throughout the year. The presence of the suckling young causes the development of the new embryo to be arrested between the 70- and 100-cell stage. This period is referred to as embryonic diapause and may last about 240 days during which time the joey is growing in the pouch (Dawson 1995; Tyndale-Biscoe 2005). When the young at foot is about one-year old, it is fully weaned yet still associates with its mother. When the young in the pouch reaches 200 days of age, development of the blastocyst resumes, and it is born soon after the older joey has been permanently excluded from the pouch. At this point, the mother may again mate and become pregnant; however, because of suckling of the new pouch joey, the development of the

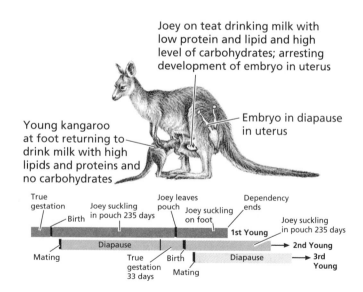

Figure 10.18 Embryonic diapause. Illustration showing embryonic diapause in the red kangaroo. The complex reproductive pattern of many kangaroos may result in having three young in different stages of development dependent on the mother at one time.

new embryo is arrested. If the pouch young dies, normal embryonic development of the blastocyst resumes almost immediately so that a new individual is born about a month later.

Red kangaroos are commonly faced with extreme drought and dwindling supplies of green herbage that greatly influence their reproductive activity. Young on the foot are first to feel the stress of drought. Their diet is changing from mother's milk to herbage; they exhibit a very high metabolism (twice as high as the mothers) and pronounced evaporative heat loss when exposed to high summer temperatures (Munn and Dawson 2001). Finally, during periods of drought in which forage is low in protein and water, young kangaroos cannot meet their energy needs and mothers are not able to provide sufficient milk for their rapidly growing young. At such times, young at the foot may perish; however, the mother still carries a small joey in the pouch. As the drought continues, her late stages of milk production will decline and the pouch young may die. The loss of the pouch young triggers the development of the uterine blastocyst. If the drought persists, she may produce a succession of young that she cannot nourish in the pouch. If the drought lasts for more than six months, red kangaroos and euros will eventually cease to breed altogether (Sharman 1963; Sharman and Clark 1967; Tyndale-Biscoe 2005). However, if rain occurs and plants grow, the milk production of the mother will increase, and the young will continue to develop to the end of the pouch life.

The reproductive plasticity in the form of embryonic diapause of red kangaroos and euros is truly remarkable—females breed well into a drought and can take immediate advantage of any improvement in climatological conditions shortly after they occur. This turnaround time is rapid; euros and red kangaroos are reported to have small young in their pouches very soon after the drought breaks (Tyndale-Biscoe 2005).

The suite of reproductive interactions outlined here is induced by the joey, whose suckling produces nervous stimulation of the hypothalamus. As a result, release of FSH and LH by the pituitary gland is inhibited, thereby reducing follicular development and subsequent ovulation. The production of estrogen and progesterone by the corpus luteum is suppressed by the increase in prolactin. Red kangaroos represent an excellent example of biological control of reproductive output—the suckling joey represents the proximate cause of embryonic diapause, yet the ultimate control is the selective pressures exerted by the environment.

Parturition

Before we begin our discussion of parturition, reviewing some terms concerning the developmental stages of young will be helpful. Newborn mammals can be categorized into two groups based on their degree of development. **Altricial** young are born hairless, blind, and essentially helpless. Rabbits, carnivores, and many rodents bear altricial young. **Precocial** young are born fully haired, with eyes open, and are able to get up and walk around shortly after birth. Good examples of precocial young are hares, many large grazing mammals (the ungulates), cetaceans, hyraxes, some rodents, and some primates.

Parturition, or birth, results when the fetus has completed its period of intrauterine growth. The mechanisms that act to terminate gestation and initiate parturition vary among species, are complex, and are not well understood (Gordon 1982; Pough et al. 1996). During the last part of gestation, the adrenal gland of the fetus begins to secrete **adrenocortical hormones,** such as **cortisol,** which initiates parturition. At this time, the placental secretions of estrogen increase and progesterone decreases. As a result of these hormonal changes, the placenta produces hormones called **prostaglandins,** which increase contractions of the uterus. The pressure of the fetus on the cervix then stimulates production of the hormone **oxytocin,** which is produced by the posterior pituitary gland. Oxytocin also increases uterine contractions and stimulates milk "letdown." **Relaxin,** a hormone produced by the corpora lutea, softens the ligaments of the pelvis so it can spread to allow the fetus to pass through the birth canal. After dilation of the vagina, rhythmic contractions of the uterus gradually force the fetus through the vagina to the outside. If the fetal membranes are not ruptured in the birth process, the mother tears them from the young, thus permitting the newborn to breathe. The mother usually severs the umbilical cord and consumes the placenta.

Lactation

DEFINITION AND PHYSIOLOGY

Lactation, the production of milk by the mammary glands, is the quintessential feature of Class Mammalia. Probably the most well-known characteristic of a mammal is its **mammae** or breasts—the word *mammal* is derived from the Latin word *mammalis*, meaning **breasts.** In females, these specialized skin glands produce milk to nourish the young; in males, they are usually rudimentary and nonfunctional. It is noteworthy, however, that lactation in males has been reported in a population of Dayak fruit bats (*Dyacopterus spadiceus*) from Malaysia (Francis et al. 1994).

Although the process is more complicated than we can cover here, lactation, like the estrous cycle, pregnancy, and parturition, is regulated by hormones as well as the presence of the suckling young. During pregnancy, high levels of the ovarian hormones estradiol and progesterone circulate in the blood and cause enlargement of the mammary glands, making them structurally ready to secrete milk. The actual production of milk does not take place until after parturition. Following parturition and the expulsion of the placenta, the estradiol concentration in the blood decreases. This decrease, in turn, signals the **anterior pituitary gland** to secrete the **lactogenic hormone,** or **prolactin.** Prolactin stimulates milk production but does not by itself cause the delivery, or letdown, of milk. Suckling by the newborn stimulates nerve receptors in the nipples. This information is transmitted to the hypothalamus and then to the posterior pituitary gland. This gland then releases oxytocin, the same hormone that is associated with uterine contractions during birth. Oxytocin stimulates the alveoli of the breasts to eject milk into the ducts, thus enabling the newborn to remove the milk by suckling. Prolactin is also important in inducing maternal behavior in females and, interestingly, sometimes in males.

COMPOSITION OF MILK

Milk contains fats, proteins (especially casein), and **lactose** (or milk sugar) as well as vitamins and salts. It provides nutrition for the growth of the newborn, transmits passive immunity, and may support the growth of symbiotic intestinal flora (Jensen 1995). The first product released by the mammary gland following birth is called **colostrum,** a protein-rich fluid containing antibodies that confer the mother's immunity to various diseases to the young. Other important milk proteins, such as **lysozyme** and **interleukin,** are present in milk throughout lactation. Lysozyme is reported to kill bacteria and fungi, protecting young animals from infection. The composition of milk varies greatly among groups of mammals (Jenness and Studier 1976; Oftedal 1984; Oftedal et al. 1987; Hayssen 1993; Kunz and Stern 1995). Mammals residing in northern environments have extremely high-fat, high-protein milk. The young of these species must increase weight rapidly to combat the cold after they are weaned and disperse. The milk of whales and seals, for example, may contain 40–61% fat and 11–12% protein—some milks are 12 times higher in fats and four times richer in protein than that of the domestic cow (Iverson 1993; Jensen 1995; Boness and Bowen 1996; Iverson et al. 1997; Mellish et al. 1999; Donohue et al. 2002). For example, the milk of the hooded seal (*Cystophora cristata*) contains about 61% fat and 11% protein, and as a result, pups may gain 20.5 kg in their short 4-day nursing period (Bowen et al. 1985). The extent to which young depend on milk for nutrition also varies greatly. Rodents, such as voles and mice, rely entirely on milk until they are weaned (Oswald and McClure 1990; Rogowitz 1998). In contrast, many young ungulates, such as deer (Genus *Odocoileus*), consume grass only a few days after birth, well before weaning occurs. Marsupials, bats, and primates have periods of lactation that are 50% longer than those of other mammals of similar body size (Hayssen et al. 1993). In many species of

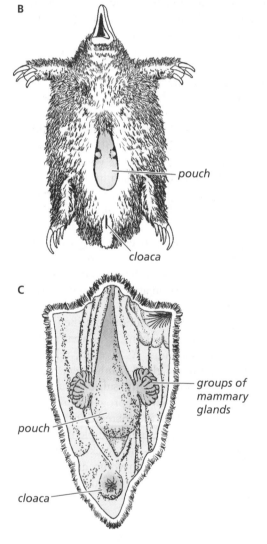

Figure 10.19 **An egg-laying mammal.** (A) Photograph of a young echidna between 2 and 4 months of age. (B) Lower surface of brooding female. (C) Dissection showing a dorsal view of the pouch, mammary glands, and two tufts of hair in the lateral folds of the mammary pouch from which the secretion flows. *Adapted from Nowak, 1991,* Walker's Mammals of the World, *5th ed., Johns Hopkins University Press.*

mammals, the composition of milk changes during the course of lactation. Milk produced during early stages of lactation and as weaning approaches often possesses more protein and less sugar than that delivered to young during the interim stages.

In addition to producing milk with different nutritive values, red kangaroos and wallaroos (Genus *Macropus*) may produce two milks of completely different characteristics at the same time (Dawson 1995; Tyndale-Biscoe 2005). During the earlier discussion of embryonic diapause in red kangaroos, we noted that the mother could simultaneously have two young, widely separated in age, both feeding on her milk. Because each joey has different requirements for nourishment, the mother must produce two different kinds of milk—two active teats are required for the nursing young. Oxytocin, the milk-letdown hormone, is elicited by different suckling stimuli. The first kind of milk is very dilute and is secreted for the newborn joey that exerts a slight sucking pressure. This young requires little nourishment as it is growing slowly and its metabolic rate is low. As this youngster develops and subsequently is at foot it requires a more concentrated milk and higher levels of protein to meet its energy demands for thermoregulation and to support its rapid growth rate. Early milk of kangaroos is more dilute than that of humans or cows milk, while late-stage milk of kangaroos is similar to that of seals.

All female mammals except the monotremes (platypuses and echidnas) have teats, or nipples, to facilitate the transfer of milk to the young, but their arrangement and number vary greatly. Female echidnas (Genera *Tachyglossus* and *Zaglossus*) typically lay a single egg that is transferred from the cloaca directly into the abdominal pouch, where it is incubated for about 10 days (figure 10.19). Because platypuses (*Ornithorhynchus anatinus*) do not have pouches, they keep their eggs (usually two, each about the size of a robin's egg) in a nest within a burrow, where they are incubated for about 2 weeks. Meanwhile, the embryo is nourished from the yolk. Monotremes lack true nipples, and after "hatching," the young (about the size of a raisin) suck milk that drains from the mother's mammary glands onto tufts of hair located on her abdomen, or in the pouch in the case of the echidna. In monotremes, the milk flows from pores in the skin rather than from nipples.

Marsupials generally have a circular arrangement of nipples in a pouch. The opossum has 13 nipples: 12 arranged in a U-shape, and the thirteenth located in the

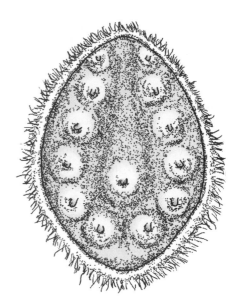

Figure 10.20 The marsupium. The opened pouch of the Virginia opossum has a horseshoe-shaped arrangement of mammae. *Adapted from J. F. Merritt, 1987,* Guide to Mammals of Pennsylvania, *Univ. of Pittsburgh Press.*

center (figure 10.20). The numbers of mammary glands or teats also vary among groups of mammals and are usually correlated with the size of the litter. The number of mammary glands ranges from 2 in many species of mammals to 29 in the tailless tenrec (*Tenrec ecaudatus*) of Madagascar. This latter species may be the most prolific mammal, with litters of as many as 32 young, but the average litter is between 15 and 20 (Louwman 1973; Eisenberg 1975; Nicoll 1983; Nicoll and Racey 1985). As mentioned earlier, naked mole-rats are a rival in maximizing fecundity (Sherman et al. 1999). Eutherian mammals typically have mammae arranged in two longitudinal ventral rows. Domestic pigs have large litters and, hence, have two rows of teats spanning all the way from near the forelimbs to between the hind limbs. Primates possess a pair of pectoral mammae, and horses have a pair of abdominal teats. The length of lactation varies greatly among mammals, ranging from only 4 days in hooded seals to more than 900 days in orangutans (*Pongo pygmaeus*). The record for "fastest developer" goes to the streaked tenrec (*Hemicentetes semispinosus*), whose young are weaned in about 5 days and begin breeding as early as 3–5 weeks of age (Hayssen et al. 1993; Sherman et al. 1999).

SUMMARY

Mammals reproduce sexually, and fertilization is internal. In the male reproductive system, the testes function to produce sperm (the male gametes) and to synthesize the male sex hormone (testosterone). Testes in mammals are oval-shaped and may or may not be suspended in a pouchlike sac called a scrotum. Sperm are produced in the seminiferous tubules of the testes and travel by a series of ducts. The ducts receive glandular secretions to the spermatozoa to form semen, which passes from the body via a canal (the urethra) into the highly vascular, erectile penis. The female reproductive system consists of a pair of ovaries that produce the ova and the female hormones, estrogen and progesterone.

Ovulation occurs when the follicle housing the ovum within the ovary bursts, releasing the egg into the oviduct. The development of follicular cells is stimulated by follicle-stimulating hormone and luteinizing hormone produced by the anterior pituitary gland. Follicular cells secrete estradiol (a form of estrogen) that causes a thickening of the endometrium in preparation for implantation of the egg. Ovulation is induced by high levels of luteinizing hormone. Stimulation of corpora luteum development is accomplished by luteinizing hormone. The corpus luteum, from the follicular cells, produces the hormone progesterone, which combines with estrogens to promote the growth of the uterine lining and makes possible the implantation of the fertilized egg. Progesterone also stimulates mammary gland development and maintenance of pregnancy.

Estrus, or heat, represents the period of receptivity to copulation shortly before and after ovulation. Ovulation may be spontaneous, as in most mammals, or induced, as in rabbits, many carnivores, and some ground squirrels. The female duct system consists of paired oviducts, one or two uteri, one or two cervices, and the vagina. Much variation occurs in this system among mammals. Four principal types of uteri based on the relationship of the uterine horns exist: duplex, bipartite, bicornuate, and simplex. Implantation occurs when the embryo attaches to the endometrium, or the lining of the uterine wall. The highly vascular endometrium passes nutrients and gases to and receives wastes from the extraembryonic tissues called the placenta. The relationship of the yolk sac to the developing embryo is the means of categorizing all mammalian placentas as either choriovitilline (most marsupials) or chorioallantoic (the bandicoots and all eutherian mammals). Placentas can also be classified by distribution of villi; speed and efficiency of exchange of nutrients, oxygen, and

waste materials; and the intimacy of the placenta with the uterine wall.

The period from fertilization until birth of the young is called gestation. The gestation period for marsupials is short and lactation is protracted, whereas the reverse is true for eutherian mammals. Typically, variation in the length of gestation is correlated with body mass, the number of young per litter, and the degree of development of the young. Larger animals tend to have longer periods of gestation. Modifications in the length of fertilization or gestation, however, are exhibited by many mammals. These reproductive variations include delayed fertilization (in bats of temperate zones), delayed development (in some bats in Neotropical areas), delayed implantation (in some "insectivores," rodents, bears, weasels and allies, seals, armadillos, and roe deer), and embryonic diapause (in kangaroos and some wallabies). Parturition, or birth, occurs when the fetus has completed its period of intrauterine growth. The degree of development of the newborn may be altricial or precocial. Processes associated with parturition are influenced by a variety of hormones, including oxytocin and relaxin.

The production of milk to nurse the newborn is called lactation and is aided by the hormones oxytocin and prolactin. Milk is composed of fats, proteins, and milk sugar. The composition of milk varies greatly among mammals, ranging from only 0.2% fat in black rhinos (*Diceros bicornis*) to as high as 61% fat in hooded seals (*Cystophora cristata*). Kangaroo mothers may produce two kinds of milk to nurse two joeys of differing ages. The number of mammae also varies greatly in mammals, ranging from 2 in many species to mammals to 29 in the tailless tenrec (*Tenrec ecaudatus*). Monotremes do not have true nipples or mammae. Their young suck milk that drains from mammary glands onto tufts of hair on the mother's abdomen.

SUGGESTED READINGS

Bodemer, C. W. 1968. Modern embryology. Holt, Rinehart and Winston, New York.

Dawson, T. J. 1995. Kangaroos: Biology of the largest marsupials. Comstock Publishing Assoc. Ithaca, NY.

Gordon, K., T. P. Fletcher, and M.B. Renfree. 1988. Reactivation of the quiescent corpus luteum and diapausing embryo after temporary removal of the sucking stimulus in the tammar wallaby. Journal of Reproduction and Fertility 83:401–406.

Hughes, R. L. 1984. Structural adaptations of the eggs and the fetal membranes of monotremes and marsupials for respiration and metabolic exchange. Pp. 389–421 in Respiration and metabolism of embryonic vertebrates (R. S. Seymour, ed.). Doordrecht, Junk.

Martin, R. E., R. H. Pine, and A. F. DeBlase. 2001. A manual of mammalogy with keys to the families of the world, 3rd ed. McGraw-Hill, Dubuque, IA.

Mossman, H. W. 1987. Vertebrate fetal membranes. Rutgers Univ. Press, New Brunswick, NJ.

Munn, A. J., and T. J. Dawson. 2001. Thermoregulation in juvenile red kangaroos (Macropus rufus) after pouch exit: higher metabolism and evaporative water requirements. Physiological and Biochemical Zoology 74:917–927.

Padykula, H. A., and J. M. Taylor. 1982. Marsupial placentation and its evolutionary significance. Journal of Reproductive Fertility, Suppl. 31:95–104.

Pough, F. H., C. M. Janis, and J. B. Heiser. 2005. Vertebrate life. 7th ed. Pearson Prentice Hall, Upper Saddle River, NJ.

Sharman, G. B. 1976. Evolution of viviparity in mammals. Pp. 32–70 in C. R. Austin and R. V. Short (eds.). Reproduction in mammals, volume 6. Cambridge Univ. Press, Cambridge, U.K.

Sharman, G. B., and P. J. Berger. 1969. Embryonic diapause in marsupials. Advances in Reproductive Physiology 4:211–240.

Sharman, G. B., and M. J. Clark. 1967. Inhibition of ovulation by the corpus luteum in the red kangaroo, Megaleia rufa. Journal of Reproduction and Fertility 14:129–137.

Tyndale-Biscoe, H. 2005. Life of marsupials. CSIRO Publishing, Collingswood, Australia.

Tyndale-Biscoe, H., and M. B. Renfree. 2005. Reproductive physiology of marsupials. Cambridge Univ. Press, Cambridge, UK.

DISCUSSION QUESTIONS

1. Trace the movement of spermatozoa from their origin and development in the seminiferous tubules of the testes to release at the penis.

2. Describe and give the function(s) of the following structures of the female reproductive system: ovaries, uterus, vagina, and mammary glands. How do these structures vary among monotremes, marsupials, and eutherian mammals?

3. Describe and distinguish the patterns in reproduction in monotremes, marsupials, and eutherian mammals. What aspects of reproduction are universal in Class Mammalia but not found in other vertebrates?

4. What reproductive attributes enhance survival of young residing in northern environments? List some such species, and state how the female makes physiological adjustments to enhance the survival of her young.

5. Delayed implantation is a common reproductive phenomenon for members of the Family Mustelidae (weasels and allies). The occurrence of this reproductive strategy varies widely among closely related species residing in similar environments. How can you explain this disparity, and how does it fit with the theory of natural selection?

6. Many mammals such as the opossum are known to produce more young that can be supported by the number of the mother's mammae. This attribute seems maladaptive. Explain this apparent waste of energy.

PART 3

Adaptive Radiation and Diversity

Part 2 detailed the great diversity in the structure and function of mammals. This section, including chapters 11 through 20, explores the life history characteristics, morphology, fossil history, and conservation of living orders and families of mammals. There are about 5,400 extant mammalian species currently recognized throughout the world —relatively few compared with over 9600 species of birds or the 40,000 species of fishes. The number of vertebrate species pale in comparison with the 104,000 species of mollusks, or the estimated tens of millions of species of insects.

Structurally and functionally, however, mammals are quite diverse. Consider the difference in size between the smallest and largest living species of birds. The largest, heaviest bird is the ostrich (*Struthio camelus*), which weighs about 135 kg as an adult. The smallest hummingbirds (Family Trochilidae) weigh about 3 g (or 0.003 kg). Although this is a difference of 4.5 orders of magnitude, consider the mass difference between the largest mammal, a blue whale (*Balaenoptera musculus*), at about 160,000 kg (160,000,000 g), and one of the smallest mammals, a pygmy shrew (*Sorex hoyi*) at about 3 g—a range of over 7 orders of magnitude! Size is just one aspect of a species that results from numerous selective factors operating through evolutionary time. As you read the chapters in this section, keep in mind not only the pronounced differences in structural and functional adaptations within the various mammalian orders but also the many examples of parallel or convergent evolutionary adaptations evident among phylogenetically distant groups. Environmental constraints force many otherwise unrelated taxa to adapt to similar problems in similar ways and, through time, to develop similar morphological features.

CLASSIFICATION

The Class Mammalia is divided into two subclasses of extant taxa, based primarily on reproductive characteristics: the Prototheria and the Theria. Prototherians are a small group of only four or five living species in the Order Monotremata. Therians make up the vast majority of mammals. They are divided into two Infraclasses: the metatherians and the eutherians. Metatherians are more commonly referred to as marsupials, whereas eutherians are typically called "placental" mammals. The term *placental* should be discouraged for this group as it implies that marsupials have no placenta. Marsupials do have a placenta, although it is not used as an accessory endocrine organ during prolonged gestation as in eutherians. Even the term *eutherian* is somewhat unfortunate, in its subtle implication of "progression" through marsupials to "good," (eu-) or more advanced, mammals. For lack of any reasonable alternative, however, we will use the terms placental and eutherian interchangeably. The approximately 330 extant species of marsupials, along with the monotremes, make up about 6% of the world's mammalian fauna. The remainder is made up of eutherians.

In this text we follow the classification of Wilson and Reeder (2005). The time period between the second (1993) and third (2005) editions of the Wilson and Reeder volumes saw numerous new mammalian species described. Taxonomic revisions have been based on a variety of genetic techniques as well as morphological approaches, and the number of recognized extant species has increased from about 4,600 in 1993 to over 5,400 as of 2005. More dramatically, the enormous amount of taxonomic work during this period has resulted in 5 new or proposed orders and 20 new (or resurrected) families, often from division of previous taxa. The following table is adapted from Wilson and Reeder (2005) and incorporates these proposed ordinal and familial revisions for extant taxa—several recently extinct families are not included. Where we discuss the various orders in chapters 11 through 20, both the revised and traditional approaches are presented, along with supporting literature and alternative taxonomic arrangements proposed by other authorities. In the following table, we only list those suborders we discuss that are not included in Wilson and Reeder (2005). For each family, the number of extant genera and species is given in parentheses, although for some families these figures should be viewed as close approximations. The conservation status of species noted in the test considered **endangered** (defined as in danger of extinction and whose survival is unlikely if causal factors continue) or **threatened** (i.e., **vulnerable**—likely to move into the endangered category in the near future if causal factors continue) are derived from various regional sources, as well as the IUCN Red Data Book (2006), on the internet at *www.redlist.org*.

CLASS MAMMALIA
Subclass Prototheria
    Order Monotremata
        Family   Ornithorhynchidae (1, 1)   Duck-billed platypus
                           Tachyglossidae (2, 4)   Echidnas
Subclass Theria
(Infraclass Metatheria)
    Order Didelphimorphia
        Family   Didelphidae (17, 87)   American opossums
    Order Paucituberculata
        Family   Caenolestidae (3, 6)   Shrew or rat opossums
    Order Microbiotheria
        Family   Microbiotheriidae (1, 1)   Monito del monte
    Order Dasyuromorphia
        Family   Myrmecobiidae (1, 1)   Numbat
                           Dasyuridae (20, 69)   Marsupial mice, Tasmanian devil
    Order Peramelemorphia
         Family   Peramelidae (6, 19)   Bandicoots and echymiperas
                           Thylacomyidae (1, 2)   Bilbies
    Order Diprotodontia
        Family   Phascolarctidae (1, 1)   Koala
                           Vombatidae (2, 3)   Wombats
                           Phalangeridae (6, 27)   Cuscuses, brushtail possums
                           Potoroidae (3, 8)   Potoroos, bettongs
                           Macropodidae (11, 65)   Kangaroos, wallabies, pademelons
       [*new*]   Hypsiprymnodontidae (1, 1)   Musky rat-kangaroo
                           Burramyidae (2, 5)   Pygmy possums

	Acrobatidae (2, 2)	Feathertail glider, possum
	Pseudocheiridae (6, 17)	Ringtail possums
	Petauridae (3, 11)	Striped and gliding possum
	Tarsipedidae (1, 1)	Honey possum

Order   Notoryctemorphia
    Family   Notoryctidae (1, 2)   Marsupial moles

(Infraclass Eutheria)

Order   Insectivora [some authorities now recognize the following three orders]
Order   Afrosoricida [*new*]
    Family   Tenrecidae (10, 30)   Tenrecs and otter shrews
                 Chrysochloridae (9, 21)   Golden moles
Order   Erinaceomorpha [*new*]
    Family   Erinaceidae (10, 24)   Hedgehogs and gymnures
Order   Soricomorpha [*new*]
    Family   Solenodontidae (1, 2)   Solenodons
                 Soricidae (26, 376)   Shrews
                 Talpidae (17, 39)   Moles, shrew-moles, and desmans
Order   Macroscelidea
    Family   Macroscelididae (4, 15)   Elephant shrews
Order   Scandentia
    Family   Tupaiidae (4, 19)   Treeshrews
    [*new*]   Ptilocercidae (1, 1)   Pen-tailed treeshrew

Order   Dermoptera
    Family   Cynocephalidae (2, 2)   Colugos
Order   Chiroptera
    [Suborder Megachiroptera]
    Family   Pteropodidae (42, 186)   Old World fruit bats
    [Suborder Microchiroptera]
    Family   Rhinopomatidae (1, 4)   Mouse-tailed bats
                 Craseonycteridae (1, 1)   Kitti's hog-nosed bat
                 Emballonuridae (13, 51)   Sac-winged or sheath-tailed bats
                 Nycteridae (1, 16)   Slit-faced bats
                 Megadermatidae (4, 5)   False vampire bats
                 Rhinolophidae (1, 77)   Horseshoe bats
    [*new*]   Hipposideridae (9, 81)   Trident and leaf-nosed bats
                 Noctilionidae (1, 2)   Fishing bats
                 Mormoopidae (2, 10)   Leaf-chinned bats
                 Phyllostomidae (55, 160)   New World leaf-nosed bats
                 Natalidae (3, 8)   Funnel-eared bats
                 Furipteridae (2, 2)   Smoky bats
                 Thyropteridae (1, 3)   Disk-winged bats
                 Myzopodidae (1, 1)   Sucker-footed bat
                 Vespertilionidae (48, 407)   Common bats
                 Mystacinidae (1, 2)   Short-tailed bats
                 Molossidae (16, 100)   Free-tailed bats
Order   Primates
    Family   Cheirogaleidae (5, 21)   Dwarf and mouse lemurs
                 Lemuridae (5, 19)   Lemurs
    [*new*]   Lepilemuridae (1, 8)   Sportive lemurs
                 Indriidae (3, 11)   Indrid lemurs, sifakas
                 Daubentoniidae (1, 1)   Aye-aye
                 Lorisidae (5, 9)   Lorises, potto
                 Galagidae (3, 19)   Bushbabies, galagos
                 Tarsiidae (1, 7)   Tarsiers
                 Callitrichidae (4, 26)   marmosets, tamarins

	Cebidae (6, 56)	New World monkeys
[*new*]	Aotidae (1, 8)	Night monkeys
[*new*]	Pitheciidae (4, 40)	Titis, sakis
[*new*]	Atelidae (5, 24)	Howler monkeys
	Cercopithecidae (21, 132)	Old World monkeys
	Hylobatidae (4, 14)	Gibbons, siamang
	Hominidae (4, 7)	Gorilla, chimpanzees, orangutan, humans

[Order Xenarthra—Some authorities now recognize the following two orders:]

Order Cingulata [*new*]

| Family | Dasypodidae (9, 21) | Armadillos |

Order Pilosa [*new*]

Family	Bradypodidae (1, 4)	Three-toed sloths
	Megalonychidae (1, 2)	Two-toed sloths
	Myrmecophagidae (2, 3)	True anteaters
[*new*]	Cyclopedidae (1, 1)	Silky anteater

Order Pholidota

| Family | Manidae (1, 8) | Pangolins |

Order Tubulidentata

| Family | Orycteropodidae (1, 1) | Aardvark |

Order Carnivora

Family	Felidae (14, 40)	Cats
	Herpestidae (14, 33)	Mongooses
[*new*]	Eupleridae (7, 8)	Madagascar mongooses
	Hyaenidae (3, 4)	Hyaenas
	Viverridae (15, 35)	Civets, genets
[*new*]	Nandiniidae (1, 1)	African palm civit
	Canidae (13, 35)	Dogs
	Ursidae (5, 8)	Bears, giant panda
	Mustelidae (22, 59)	Weasels, otters, badgers
[*new*]	Mephitidae (4, 12)	Skunks
	Odobenidae (1, 1)	Walrus
	Otariidae (7, 16)	Eared seals
	Phocidae (13, 19)	Earless seals
	Procyonidae (6, 14)	Raccoon, coati
[*new*]	Ailuridae (1, 1)	Red panda

Order Cetacea

Family	Balaenidae (2, 4)	Bowhead and right whales
	Balaenopteridae (2, 7)	Rorquals
	Eschrichtiidae (1, 1)	Gray whale
	Neobalaenidae (1, 1)	Pygmy right whale
	Delphinidae (17, 34)	Dolphins
	Monodontidae (2, 2)	Narwhal, beluga
	Phocoenidae (3, 6)	Porpoises
	Physeteridae (2, 3)	Sperm whales
	Platanistidae (1, 2)	Ganges and Indus River dolphins
[*new*]	Iniidae (3, 3)	Baiji, Franciscana
	Ziphiidae (6, 21)	Beaked whales

Order Rodentia

Family	Aplodontidae (1, 1)	Mountain beaver
	Sciuridae (51, 278)	Squirrels
[*new*]	Gliridae (9, 28)	Dormice
	Castoridae (1, 2)	Beavers
	Geomyidae (6, 40)	Pocket gophers
	Heteromyidae (6, 60)	Kangaroo rats and mice
	Dipodidae (16, 51)	Jerboas, birch mice, jumping mice
[*new*]	Platacanthomyidae (2, 2)	Spiny and soft-furred tree mice
[*new*]	Spalacidae (6, 36)	Zokors, bamboo rats, blind mole rats

[*new*]	Calomyscidae (1, 8)	Mouselike hamsters
[*new*]	Nesomyidae (21, 61)	Madagascar rats and mice
[*new*]	Cricetidae (130, 681)	Voles and mice
	Muridae (150, 730)	Old World rats and mice
	Anomaluridae (3, 7)	Scaly-tailed squirrels
	Pedetidae (1, 2)	Springhaas
	Ctenodactylidae (4, 5)	Gundis
	Bathyergidae (5, 16)	Mole-rats
	Hystricidae (3, 11)	Old World porcupines
	Petromuridae (1, 1)	Dassie rat
	Thryonomyidae (1, 2)	Cane rats
	Erethizontidae (5, 16)	New World porcupines
	Chinchillidae (3, 7)	Viscachas, chinchillas
	Dinomyidae (1, 1)	Pacarana
	Caviidae (6, 18)	Cavies, Patagonian "hare," guinea pigs, capybara
	Dasyproctidae (2, 13)	Agoutis, acouchis
[*new*]	Cuniculidae (1, 2)	Pacas
	Ctenomyidae (1, 60)	Tuco-tucos
	Octodontidae (8, 13)	Viscacha rats, coruro
	Abrocomidae (1, 10)	Chinchilla rats
	Echimyidae (21, 90)	Spiny rats
	Capromyidae (8, 20)	Hutias
	Myocastoridae (1, 1)	Nutria

Order   Lagomorpha
   Family

	Ochotonidae (1, 30)	Pikas
	Leporidae (11, 54)	Rabbits and hares

Order   Proboscidea
   Family

	Elephantidae (2, 2)	Elephants

Order   Hyracoidea
   Family

	Procaviidae (3, 4)	Hyraxes

Order   Sirenia
   Family

	Dugongidae (1, 1)	Dugong
	Trichechidae (1, 3)	Manatees

Order   Perissodactyla
   Family

	Equidae (1, 7)	Horses, asses, zebras
	Tapiridae (1, 4)	Tapirs
	Rhinocerotidae (4, 5)	Rhinoceroses

Order   Artiodactyla
   [Suborder Suiformes]
   Family

	Suidae (5, 19)	Pigs, warthogs
	Tayassuidae (3, 3)	Peccaries
	Hippopotamidae (2, 2)	Hippopotamuses

   [Suborder Tylopoda]
   Family

	Camelidae (3, 4)	Camels, llamas

   [Suborder Ruminantia]
   Family

	Tragulidae (3, 8)	Chevrotains
	Giraffidae (2, 2)	Giraffe, okapi
	Moschidae (1, 7)	Musk deer
	Cervidae (19, 51)	Deer
	Antilocapridae (1, 1)	Pronghorn antelope
	Bovidae (50, 143)	Antelope, sheep, goats, bison, cattle

TOTAL: Orders = 30; Families = 149; Genera = 1,222; Species = 5,413

Totals in Wilson and Reeder (2005) include some recently extinct taxa not included in this listing.
Twelve orders have a single family, and 23 families have a single species. In contrast, there are several very speciose families. For example there are 730 species in the rodent Family Muridae, 407 species of bats in the Family Vespertilionidae, and 376 species of shrews in the Family Soricidae.

CHAPTER 11

Monotremes and Marsupials

Monotremes and marsupials may be distinguished from eutherian mammals (or "placentals" as they are commonly called) on the basis of reproductive characteristics. Monotremes are the only mammals that are **oviparous**. Like birds and some other vertebrates, they lay eggs. Marsupials give birth to live young (they are **viviparous**) but are characterized by a very brief intrauterine gestation period. This results in reduced parental "investment," in terms of energy expenditure, in **neonates** (newborns) at birth. Maternal investment in lactation is high, however. In this chapter, we explore various aspects of form and function in these two lineages of noneutherian mammals.

Monotremata

As noted in chapter 4, mammals evolved from synapsid (reptilian) ancestors over a long time period. Monotremes (Subclass Prototheria) differ significantly from marsupials and eutherians (Subclass Theria) in their retention of various reptilian features. Extant monotremes are represented by two families. The Ornithorhynchidae, is **monotypic** (a group that includes a single taxon) and includes only the duck-billed platypus (*Ornithorhynchus anatinus*). The family Tachyglossidae includes the short-billed echidna (*Tachyglossus aculeatus*) and long-billed echidna (*Zaglossus bruijni*). Two additional species of long-billed echidnas in New Guinea, *Z. bartoni* and *Z. attenboroughi*, were recognized by Flannery and Groves (1998) based on characteristics of the claws and cranium, with the latter species described from a single specimen. On an evolutionary time scale, the monotremes probably owe their continued survival to relative geographic isolation from eutherians. They may have survived historically because they occupy ecological niches with little or no competition. Although McKenna and Bell (1997) placed the monotremes in two separate orders—Platypoda and Tachyglossa—most authorities retain the traditional approach.

The ordinal name Monotremata ("one opening") refers to the **cloaca**, a common opening for the fecal, urinary, and reproductive tracts. The most notable reptilian feature of monotremes is the structure of their reproductive tract and the fact that young hatch from small, somewhat rubbery-shelled eggs. The female reproductive tract includes separate uteri with large ovaries that produce large follicles during the breeding season (figure 11.1). The uteri have separate openings into the urogenital sinus that lead to the cloaca. After an egg is shed into the infundibulum, it passes to the fallopian tube, where fertilization occurs. The shell is deposited in the oviduct over a period of about 2 weeks and is composed of three identifiable layers (see Griffiths 1978, 1989 for details). Nutrient material is absorbed through the shell; thus, the eggs are permeable (not **cleidoic** or impermeable as is the case for birds). The early cleavage stages of the egg are **meroblastic**, that is, restricted to the anterior end (as opposed to holoblastic in which cleavage occurs throughout). The first cleavage furrow divides the germinal disk into two areas of unequal size. The second cleavage is perpendicular, so that at the four-cell stage, two large cells and two small cells (**blastomeres**) occur on the top of the yolk. Meroblastic eggs also occur in reptiles and birds.

Monotreme eggs are small, about 16 mm long and 14 mm wide. They are incubated for 10–11 days. During the final stages of incubation, a sharp egg tooth forms at the end of the snout of a developing young, which it uses to free itself from the egg, again as in birds and reptiles.

Neonatal monotremes are similar structurally to neonatal marsupials (Griffiths 1989). Development in both is rudimentary, although the forelimbs and shoulder muscles are well-developed. Once eggs hatch, as in all other mammals, the young are nursed. Monotremes have mammary glands, but unlike other mammals, they have no teats. Milk is secreted from pores on the belly of the platypus and from paired glandular lobes in the pouch of echidnas (see figure 10.19). The structure of the mammary glands is identical in monotremes and marsupials (Griffiths et al. 1973), and the process of lactation in monotremes is as "sophisticated and highly evolved" (Griffiths 1989) as in any mammal.

Besides laying eggs, monotremes exhibit several other reptilian features. The pectoral girdle (figure 11.2) has a corocoid, precoracoid, and interclavicle bone, as in primitive therapsid reptiles. Although they are homeotherms, the body temperature of monotremes is lower than that of therians, about 32°C. The chromosomes are unique among mammals in their mix of normal macrochromosomes and reptile-like microchromosomes and in their meiotic division. Details of the complex system of multivalent sex chromosomes during meiosis in the platypus remain problematic (Grützner et al. 2003, 2004; Ashley 2005, although see Rens et al. 2004). Sperm are **filiform** (thread-like) and reptilian in stucture, as is the anatomy of the testes (Carrick and Hughes 1978). The ultrastructure of spermatid development in the platypus was described by Lin and Jones (2000). Acrosomal development is

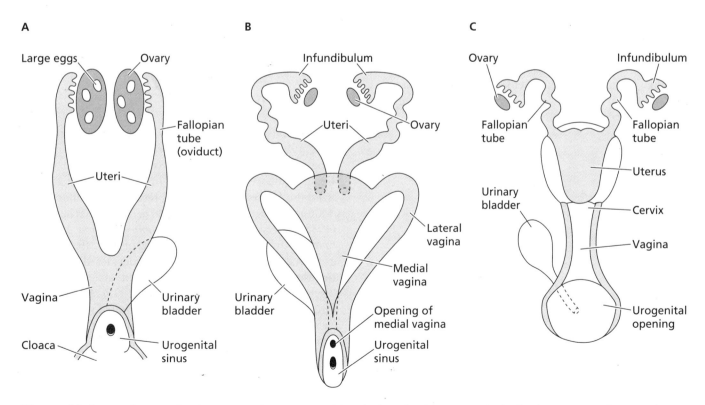

Figure 11.1 Female reproductive tracts. The structure of female reproductive tracts varies in the three groups of mammals: (A) prototherians (monotremes); (B) metatherians (marsupials); and (C) eutherians ("placentals"). *Adapted from MacFarland et al., 1985, Vertebrate Life, 2nd ed., Macmillan.*

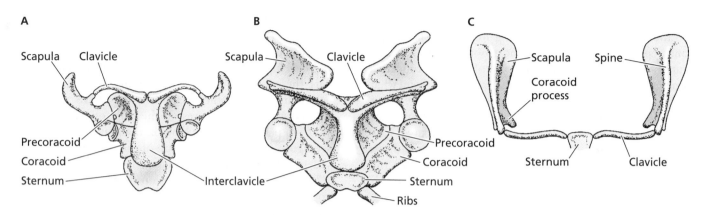

Figure 11.2 Reptile-like pectoral girdle in monotremes. Pectoral girdles in a (A) therapsid reptile, (B) short-beaked echidna, and (C) muskrat, a eutherian mammal. Monotremes have pectoral girdles very similar to ancient reptiles. *Adapted from L. H. Hyman, 1942, Comparative Vertebrate Anatomy, Univ. Chicago Press.*

reptilian, other features are the same as mammals, and some, including nuclear condensation, are unique to monotremes.

MORPHOLOGY

Although they are diverse in their outward appearance, the two monotreme families share several characteristics. The cranial features of monotremes are unique, with adults having indistinct sutures. Unlike therian mammals, the jugal bone is reduced or absent. The zygomatic arch is made up of the maxilla and squamosal bones. The dentary bone is greatly reduced, and adults are **edentate** (without teeth). With their elongated rostrum, lack of teeth, and high-domed cranium, monotreme skulls appear birdlike (figure 11.3). The **cochlea** (semicircular canals of the inner ear) of monotremes also are unique among mammals because they are not coiled.

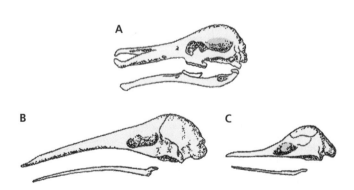

Figure 11.3 Monotreme skull morphology. Lateral view of skulls of the (A) duck-billed platypus, (B) long-beaked echidna, and (C) short-beaked echidna. All monotremes are edentate, although note the keratinized gum pads in the platypus. (Not to same scale.) *Adapted from T. E. Lawlor, 1979, Handbook to the Orders and Families of Living Mammals, 2nd ed., Mad River Press.*

Monotremes have several other distinctive features. Like marsupials, **epipubic bones** (attached to the pelvic girdle and projecting forward) occur in both sexes. Adult males have a large, hornlike medial spur on the ankle. In the platypus, this spur connects to a poison gland (see family Ornithorhynchidae). Details of the skull and post-cranial skeleton of the platypus and echidnas can be found in Grant (1989) and Griffiths (1989), respectively. Monotremes, like marsupials, have no **corpus callosum**, a bundle of nerve fibers that integrate the two hemispheres of the brain. Males have a baculum, permanently abdominal testes, and no scrotum. Monotremes exhibit a "mosaic evolution" given their relatively specialized mammalian features and numerous retained archaic reptilian characteristics.

FOSSIL HISTORY

Monotremes are of Mesozoic origin (Woodburne and Case 1996) and may have diverged from the therians as long as 230 million years ago (van Rheede et al. 2006). Fossil evidence of Australian ornithorhynchids extends from the Cretaceous and early Tertiary (Woodburne and Tedford 1975; Archer et al. 1978, 1992). Archer et al. (1985) described *Steropodon galmani*—the first Mesozoic mammal from the early Cretaceous of Australia based on a partial lower jaw. Originally considered to be an ornithorhynchid, it was placed in its own family Steropodontidae by Flannery et al. (1995). Another early Cretaceous monotreme, *Teinolophos trusleri*, was described by Rich et al. (2001, 2005). Characteristics of an ornithorhynchid from the mid-Miocene, *Obdurodon dicksoni*, were discussed by Musser and Archer (1998). The earliest tachyglossid, *Megalibgwilia ramsayi*, is from the early Miocene of Australia (Griffiths et al. 1991). The first evidence of a non-Australian fossil monotreme was a platypus, *Monotrematum sudamericanum*, from the early Paleocene of southern Argentina (Pascual et al. 1992, Pascual and Goin 2002). Musser (2003) provided an excellent review of the 60-million year fossil record of monotremes.

ECONOMICS AND CONSERVATION

Although certainly of intrinsic interest to mammalogists and the general public, monotremes have no economic importance. In the early days of Australian colonization, the platypus was hunted for its fur, but this was never a major industry. Most populations today are considered secure, although they are protected in all Australian states where they occur. The potential exists for habitat loss from stream erosion, reduced water quality, introduced species, and other factors (Grant and Temple-Smith 2003). The long-billed echidna once enjoyed a much more widespread range than it does today and is now considered endangered.

FAMILIES

Ornithorhynchidae

The semiaquatic, semifossorial platypus is the sole extant ornithorhynchid. It occurs near freshwater lakes and rivers at both high and low elevations along the east coast of Australia and throughout Tasmania, where it feeds on a variety of invertebrates, small fish, and amphibians. The platypus's physical appearance is so unique among mammals (figure 11.4) that the first specimen brought to London in 1798 was believed to be a hoax. Adult males average 50 cm in total length and 1700 g in body mass; females are smaller. Short, dense fur covers all but the bill, feet, and underside of the tail. The bill is distinctive and quite unlike that of a true duck. It is soft, pliable, and very sensitive and has nostrils at the tip. The bill is highly innervated both for tactile reception and to sense electric

fields generated by the muscle contractions of prey (Scheich et al. 1986; Proske et al. 1998; Manger and Pettigrew 1995). There are an estimated 40,000 "sensory mucous glands" in the bill that sense electric fields as a platypus moves its head from side to side while foraging (Proske and Gregory 2003). The small eyes and ears are situated in a groove extending from the bill. During a dive, this groove closes and the platypus relies on the sensitivity of the bill to locate prey.

Feet of the platypus are **pentadactyl** (five-toed), and the **manus** (forefoot) is webbed. Webbing is folded back when the platypus is on land. Nonetheless, terrestrial locomotion is energetically demanding (Fish et al. 2001). The long, sharp claws are used for burrowing. The spur on the hind limb connects to a large venom gland located in the thigh (figure 11.5). Spurs are strong, sharp, and about 12 mm long. The function of the spur is not entirely clear, but it may involve competition between males during the breeding season. People who have been accidentally envenomated in the hand experienced "immediate and intractable pain and marked swelling" (Fenner and Williamson 1996:438). Unlike echidnas, platypus neonates have three molariform teeth in each quadrant. These are shed before young emerge from the burrow. Continuously growing, keratinized pads in the gums take the place of teeth in adults. These pads grind food materials before they are swallowed.

Burrows are built in stream banks and generally are simple, although those constructed by females for nursing are 20–30 m long, with a nest chamber at the end. Unlike echidnas, the platypus has no pouch. Females curl their bodies around the eggs to incubate them; average litter size is two. As with birds, only the left ovary is functional. The young are not carried after hatching and remain in the burrow during the 4- to 5-month nursing period. Milk is secreted from the mammary glands onto distinct, protruding tufts of fur.

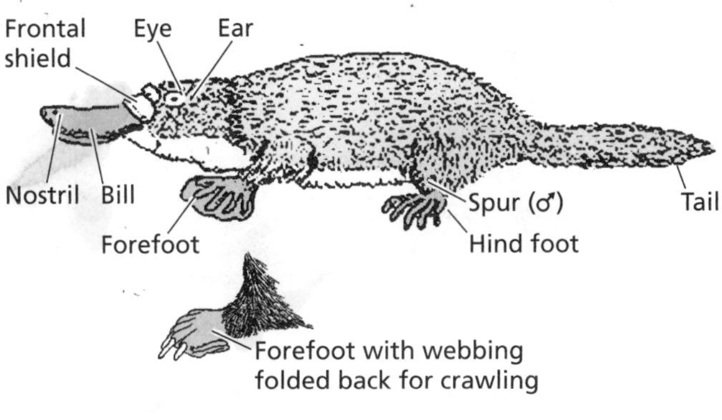

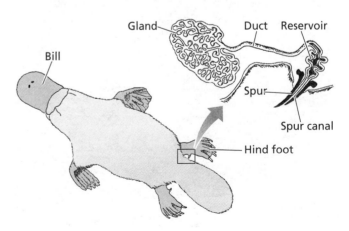

Figure 11.4 External features of the duck-billed platypus. The external anatomy of the duck-billed platypus is unique. *Adapted from D. W. Walton and B. J. Richards, 1989, Fauna of Australia, vol. 1B, Australian Government Publishing Service.*

Figure 11.5 Venom gland in the duck-billed platypus. Anatomy of the venom gland associated with the spur on the hind limb of the male platypus. *Adapted from D. W. Walton and B. J. Richards, 1989, Fauna of Australia, vol. 1B, Australian Government Publishing Service.*

Tachyglossidae

The short-beaked echidna (figure 11.6) occurs throughout Australia, New Guinea, and Tasmania in forest, scrub, and desert habitats. The long-beaked echidnas are restricted to forested highland areas of New Guinea. Maximum body mass is about 6 kg in *Tachyglossus* and 10 kg in *Zaglossus*. Both genera have a long "beak" that like the platypus contains electroreceptors, and a long, extensible tongue. Enlarged submaxillary salivary glands produce mucus that coats the tongue and makes it sticky. Guard hairs on the back and sides of the body are modified to form barbless spines up to 6 cm long. *Zaglossus bruijni* has much thicker hair and fewer spines than *T. aculeatus*. Large, scooplike claws on the feet enable echidnas to break into anthills and to burrow in a rapid, powerful manner. They are active anytime during the day or night but avoid extremes of temperature. Using its sticky tongue, *Tachyglossus* (meaning "rapid tongue") consumes ants, termites, and other insects, which are ground to a paste between the tongue and spiny palatal ridges. Their most common prey is ants of the genus *Iridomyrmex*. Short-beaked echidnas use monounsaturated fatty acids from these ants for energy during hibernation (Falkenstein et al. 2001). *Zaglossus* (meaning "long tongue") feed primarily on earthworms.

Unlike the platypus, the ankle spurs on male echidnas do not function, and they have no teeth at any stage of development. Also, echidnas have a pouch in which the eggs are incubated and hatched, again unlike the platypus. Echidnas in Australia mate from July through September. Gestation is about 23 days (Rismiller and McKelvey 2000).

The mammary glands converge to two small areas in the pouch, the milk areolae, where neonates cling to surrounding fur to nurse. Young remain in the pouch 45–50 days (Beard and Grigg 2000) until their spines begin to develop.

Marsupials

Marsupials (or metatherians) are often characterized by the female's abdominal pouch or **marsupium**, which gives rise to the common name of this group. This is a poor diagnostic feature, however, because not all marsupials have a marsupium, and as we have seen, a pouch occurs in echidnas. A pouch probably is a derived condition in marsupials (Tyndale-Biscoe and Renfree 1987; Harder et al. 1993); only about 50% of species have a permanent pouch. Marsupials are best distinguished from eutherians on the basis of their reproductive mode, specifically, the relatively small maternal energy investment in young prior to birth. In fact, no marsupials have litters that weigh more than 1% of the mother's body mass (Russell 1982). In contrast, small eutherians, such as rodents or insectivores, may have litters that weigh 50% of the mother's body mass. Maternal investment in lactation is much greater in marsupials (figure 11.7), however, so by the time young are weaned, total investment in a litter may be similar between marsupials and eutherians of similar body weight. Renfree (1993:450) noted that "Marsupials have, in effect, exchanged the umbilical cord for the teat." In addition to reproductive characteristics, marsupials differ from eutherians in many skeletal and anatomical features (table 11.1). The two groups also have different dental characteristics. Unlike eutherians, marsupials characteristically have a well-developed "stylar shelf" on the upper molars, as well as a "twinned" hypoconulid and entoconid in the lower molars (figure 11.8).

Figure 11.6 External morphology of the short-beaked echidna or spiny anteater. The beak is much shorter and straighter in the short-beaked echidna (*Tachyglossus aculeatus*) than in long-beaked echidnas (genus *Zaglossus*), and there are more spines in *Tachyglossus*.

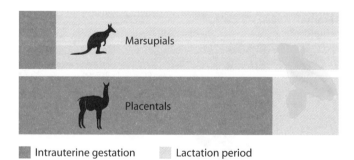

Figure 11.7 Lactation and gestation in therians. Relative lengths of gestation and lactation differ significantly in marsupials and eutherians. Most energy expenditure in marsupials occurs during lactation. The total amount of time that young are dependent varies on the size of the species. *Adapted from data in M. Archer and G. Clayton, 1984,* Vertebrate Zoogeography and Evolution in Australia, *Hesperian Press.*

Table 11.1 General skeletal and anatomical differences between metatherian (marsupial) and eutherian ("placental") mammals

Metatherians	Eutherians
Braincase small relative to body size; minimal development of neocortex; no corpus callosum	Braincase relatively large; greater complexity of neocortex; corpus callosum present
Auditory bullae usually absent, if present, formed primarily from alisphenoid bone	Auditory bullae present, formed from tympanic bone
Large vacuities often present in posterior part of palate	Palatal vacuities absent or small
Jugal bone large, jugal and squamosal bones articulate with dentary bone in mandibular fossa	Jugal bone does not articulate with dentary in mandibular fossa
Angular process of dentary inflected 90° (i.e., is perpendicular to the axis of the dentary) except in koala and honey possum	Angular process of dentary not inflected
Primitive dental formula 5/4, 1/1, 3/3, 4/4 = 50. Last premolar is the only deciduous tooth	Primitive dental formula 3/3, 1/1, 4/4, 3/3 = 44. Incisors, canines, and premolars are deciduous
Epipubic bones occur in both sexes	Epipubic bones do not occur
Female reproductive tract bifurcated (see text); tip of penis (glans) bifurcated	Reproductive tract and glans penis not bifurcated
Marsupium often present enclosing teats; opens either anteriorly or posteriorly	Marsupium not present
Scrotum anterior to penis, except in the mole *Notoryctes*; baculum never present	Scrotum posterior to penis; baculum sometimes present

As noted by Lee and Cockburn (1985), several life history characteristics of marsupials differ from those of eutherians, at least in matter of degree. Different characteristics should not be viewed as "shortcomings," however, but simply as different adaptive strategies. In each case, marsupials are consistently more conservative (less diverse) in their adaptive radiation than are eutherians. For example, marsupials generally have lower basal metabolic rates—about 70% of comparably sized eutherians.

Except for bandicoots, they also have slower postnatal growth. Relative brain size also is smaller in marsupials, especially when considering large species, as is the range of body size. For example, the difference in body mass between the smallest living marsupial species, the Pilbara ningaui (*Ningaui timealeyi*—adult body mass 2–9 g), and the largest, the red kangaroo (*Macropus rufus*—average mass of males 66 kg), is about four orders of magnitude. This is much less pronounced than the extremes noted in

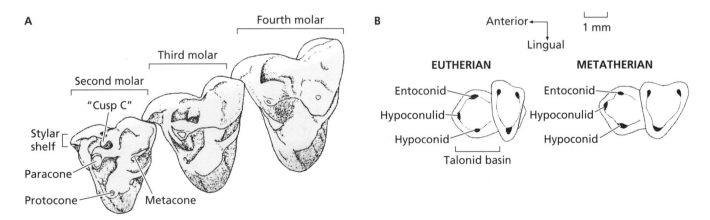

Figure 11.8 Marsupial molariform dentition. (A) Left maxillary of a marsupial containing molars 2–4, showing the characteristic stylar shelf on the labial (cheek) side, large size of the metacone relative to the paracone, and well-developed "stylar cusp C." (B) Lower molars, showing the hypoconulid closer to the entoconid ("twinned") in a marsupial (right). In a eutherian (left), the hypoconulid is equidistant between the entoconid and hypoconid. *Adapted from data in M. Archer and G. Clayton, 1984, Vertebrate Zoogeography and Evolution in Australia, Hesperian Press.*

eutherians; that is, seven orders of magnitude difference between pygmy shrews and blue whales. There are no marine marsupials, however, and therefore no parallel adaptation to that of cetaceans, so this type of comparison is questionable. Nonetheless, it illustrates the point of reduced morphological diversity in marsupials. Neither have marsupials developed true flight, like the bats. Consequently, marsupials have been unable to take advantage of the abundant feeding niches afforded by either plankton or night-flying insects. Likewise, there are no fossorial herbivores among the marsupials, and all the large marsupial carnivores are extinct. Marsupials are therefore more restricted than eutherians in their adaptive radiation and associated structural diversity. Nonetheless, there is a fascinating array of behavioral and morphological adaptations evident among the seven orders and approximately 330 currently recognized extant species of marsupials.

REPRODUCTIVE STRUCTURE

Females have a highly distinctive **bifurcated** (paired) reproductive tract. Two lateral vaginae are on either side of a medial vaginal canal, or sinus (figure 11.1), and the uterus is duplex. The two prongs of a male's bifid penis probably are compatible with the corresponding lateral vaginae during copulation. During copulation, sperm travel up the lateral vaginae. If fertilization occurs, the zygote(s) implant in the uterus (uteri). Following a short gestation period, **parturition** (birth) occurs through extension of the medial vaginal canal, not through the lateral vaginae.

The typical placenta of marsupials differs from that of eutherian mammals. Most marsupials have a **choriovitilline** (yolk-sac) placenta, in which the membranes are less developed and provide less nutrient exchange from the mother to the fetus (see figure 10.11 and associated text). Unlike the **chorioallantoic** placentae of eutherians, the choriovitilline placentae of marsupials have no **villi** (fingerlike projections of capillaries from the embryonic membrane). Because of the tremendous surface area they provide, villi enhance both nutrient exchange and the strength of fetal attachment. Maternal-fetal exchange is enhanced in marsupials, however, by wrinkling of the endometrium (wall of the uterus) after implantation. Among marsupials, only the bandicoots (order Peramelemorphia) and koala (family Phascolarctidae) have chorioallantoic placentae, but they still lack villi. Thus, gestation in marsupials is necessarily short because of relatively inefficient nutrient exchange and a weak structural connection between the fetus and the endometrium. The limited adaptive radiation of marsupials compared with eutherians, with a lack of hooves, flippers, or wings in metatheria, may be related to their limited intrauterine development time. Also, accelerated development of muscular, clawed forelimbs is necessary so that newborn marsupials can crawl from the vaginal opening to the nipples to nurse and grow. This necessarily

precludes the forelimbs from becoming more derived structures such as hooves, flippers, or wings.

An interesting feature of two New World families—didelphids and caenolestids—and one not found in Australasian marsupials (or other vertebrates), is paired sperm (figure 11.9). Although not paired in the testes, sperm become coupled at their heads in the epididymis (Tyndale-Biscoe and Renfree 1987). Paired sperm again separate once in the female's oviducts. Although this phenomenon has been known for over 100 years, its significance is uncertain. Bedford and coworkers (1984) suggested that such pairing may increase sperm survival and allow individuals to produce fewer sperm without lowering reproductive potential.

The scrotum is anterior to the penis in almost all marsupials. In the marsupial moles (Notoryctemorphia: Notoryctidae), the testes are abdominal; testes are only

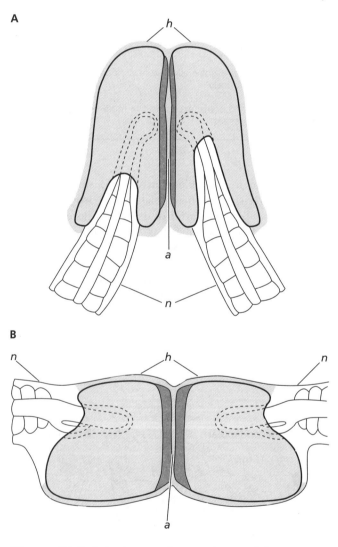

Figure 11.9 Paired sperm in some marsupials. Two families of New World marsupials have paired sperm, a feature unknown in any other mammals. Lateral view of the anterior ends of sperm pair typical of (A) a didelphid and (B) a caenolestid. Abbreviations: h = head; n = anterior portion of neck; a = acrosomes. *Adapted from H. Tyndale-Biscoe, 2005, Life of Marsupials, CSIRO Publishing.*

scrotal in wombats (Diprotodontia: Vombatidae) during the breeding season.

As noted earlier, not all marsupials have a marsupium. The numbat (*Myrmecobius fasciatus*), some New World opossums (Didelphidae), rat opossums (Caenolestidae), and many small, marsupial mice (Dasyuridae) have no pouch. In other didelphids and dasyurids, there is simply a fold of skin on either side of the teats. Pouches often are best developed in arboreal species and those that either burrow or jump. Pouches may open anteriorly or posteriorly (figure 11.10). Also, the condition and appearance of the pouch differs depending on the stage of the reproductive cycle. Woolley (1974) found no relationship between the structure of the marsupium and several life history characteristics. A marsupium, therefore, probably arose independently in various marsupial lineages (Kirsch 1977). The duration that young remain within the pouch, however, and the time of weaning depend on several factors, including maternal body mass, litter size, and number of litters per year.

Marsupial gestation periods are generally very short—12–13 days in some didelphids and bandicoots—and usually shorter than the interval between maternal estrous periods (Hsu et al. 1999). At parturition, neonates are tiny, often only a few milligrams, and never more than 1 g (figure 11.11); development of neonatal organ systems is just beginning. Despite their highly altricial (immature) features, the forelimbs and shoulder muscles of neonatal marsupials are well-developed, and the forefeet have deciduous claws. These features allow neonates to climb to a teat, whether in a pouch or not. Once attached, the teat swells, keeping the developing young in place during the early stages of the prolonged lactation period (see figure 10.17). During this teat attachment phase (Harder et al. 1993), the young do not voluntarily release the teat. This phase may be two thirds of the lactation period in larger species, but one third or less of that period in smaller species, especially those without a marsupium. During the next phase of lactation, the altricial young are left in the nest while the mother forages. This phase continues

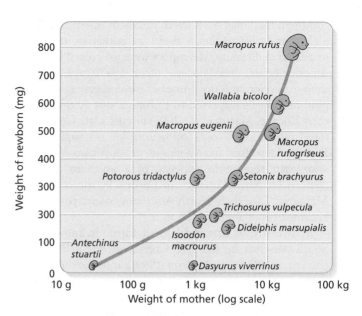

Figure 11.11 Small size of newborn marsupials. Body weight of neonatal marsupials is extremely small, never more than 1 g. An entire litter never weighs more than 1% of the mother's body mass. *Adapted from H. Tyndale-Biscoe, 2005,* Life of Marsupials, *CSIRO Publishing.*

until the young are weaned (Harder et al. 1993). Thus, "[m]arsupials have evolved a different but highly successful reproductive strategy when compared with eutherian mammals, in that their major reproductive investment is placed in lactation rather than gestation and placentation" (Tyndale-Biscoe and Renfree 1987:371).

ZOOGEOGRAPHY AND EARLY RADIATIONS

It is generally accepted that metatherians and eutherians derived from a common lineage, that is, they are sister taxa—more closely related to each other than either is to monotremes. An alternative view, called the Marsupionta hypothesis, suggests that marsupials are a sister taxon to monotremes. First proposed by Gregory (1947), the hypothesis has gained some support from molecular analyses (cf. Janke et al. 2002). However, most authorities (see Baker et al. 2004 for review) support the view that monotremes are basal to a therian clade of marsupials and eutherians as shown in figure 4.7.

Living marsupials occur today in North America (only one species north of Mexico); Central and South America; and Australasia, which includes Australia, Tasmania, New Guinea, and surrounding islands. How marsupials came to be where they are today (i.e., their historical zoogeography) remains controversial. Although Luo et al. (2003) describe a 125-million year old fossil specimen from China—*Sinodelphys szalayi*—with marsupial-like features, the oldest unequivocal fossil evidence of marsupials is from North America. A marsupial-like mammal, *Kokopellia juddi*, dates from the early Cretaceous of North America (Cifelli 1993) approximately 100 million years ago (mya).

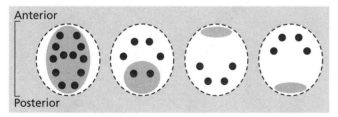

Figure 11.10 Types of pouches found in marsupials. Some species have no pouch. Others, such as the mouse opossums, antechinuses, and quolls, have (A) a fold of skin on either side of the teats. (B) Pouches found in the Virginia opossum, Tasmanian devil, and dunnarts enclose most, but not all the teats. The deepest pouches are found in arboreal or burrowing species. These may open either (C) anteriorly, as in possums and kangaroos, or (D) posteriorly, as in wombats and bandicoots. Open area of each pouch is shaded. *Adapted from D. Mcdonald, 1984,* Encyclopedia of Mammals, *Facts on File.*

Although they may have arisen in North America, marsupials eventually declined there as eutherian mammals increased in diversity. Marsupials were extirpated in North America by the mid-Miocene, about 15–20 mya. Prior to their extinction in North America, several genera of didelphid marsupials dispersed to Europe about 50 mya in the early Eocene. Martin et al. (2005) discuss a late Cretaceous (66 million years old) opossum-like marsupial from Europe. Described from a single molar, *Maastrichtidelphys meurismeti* represents evidence of a northern dispersal route from North America to Europe by the end of the Mesozoic. Like their North American counterparts, these became extinct about 20–25 mya.

A diversity of marsupial fossil forms is known from South America by the late Cretaceous/early Paleocene (Bonaparte 1990; de Muizon 1994), and marsupials still persist in South America. During most of this time, South America was isolated from North America. About 9 mya in the late Miocene, representatives of a few families moved along island arcs between the two continents (Marshall et al. 1982). About 2–5 mya, the Panamanian land bridge developed, and a major interchange of North and South American mammalian fauna occurred. Representatives of 17 North American families moved south during the late Pliocene and early Pleistocene. During the same period, representatives of 13 families moved from South America to the north (Webb 1985a). These transtropical migrations and resulting "intercontinental competition" caused many marsupials to go extinct, including large carnivores such as *Thylacosmilus* (figure 11.12). Only a few lineages of small insectivores and omnivores have remained. Of the marsupials that moved from South to North America during this faunal interchange, only the Virginia opossum (*Didelphis virginiana*) persists today.

Where exactly marsupials arose, or their pattern of dispersal, is unknown. Numerous alternative hypotheses have been proposed (see Marshall 1980 for an extensive review). Based on fossil discoveries, however, Marshall and associates (1990:479) suggest that marsupials originated in North America, but "share a single marsupial fauna with South America in the late Cretaceous." As already noted, they dispersed to Europe from North America. From South America, they dispersed to the Antarctic/Australian continent (figure 11.13). This probably occurred about 65 mya (Woodbourne and Case 1996; Goin et al. 1999), but certainly well before the opening of the Drake Passage between South America and Antarctica during the late Eocene. Antarctica and Australia then separated, isolating the ancestors of the extant Australian marsupial species. Springer et al. (1997) stated that marsupials entered Australia in the late Paleocene or early Eocene epoch. Compared with South American forms, Australian marsupials evolved in relative isolation from eutherians.

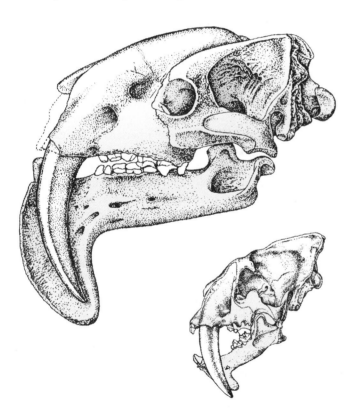

Figure 11.12 An extinct, large, carnivorous marsupial. The large, extinct, South American marsupial *Thylacosmilus atrox* (Thylacosmilidae), with a total length of the skull of about 23 cm, was very similar to the North American saber-toothed tiger *Smilodon* (inset—not to same scale), with skull length about 30 cm. *Adapted from D. Hundaker, 1997,* Biology of Marsupials, *Academic Press.*

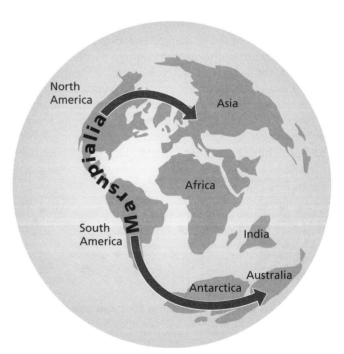

Figure 11.13 Early marsupial dispersal routes. Hypothesized origin and dispersal routes of ancestral marsupials during the early Cretaceous about 80 mya. Note juxtaposition of Antarctica and Australia. *Adapted from G. Marshall, 1980, in* Aspects of Vertebrate History *(L. L. Jacobs, ed.), Museum of Northern Arizona Press.*

ORDERS AND FAMILIES

All marsupials were once placed in a single order, the Marsupialia. Ride (1970), however, recognized four orders based on characteristics of the dentition and digits. Dentition in marsupials can be considered either **polyprotodont** (an unshortened mandible, lower incisors small and unspecialized) or **diprotodont** (a shortened mandible with first pair of lower incisors enlarged to meet upper incisors). Digits are either **didactylous** (unfused toes, each in their own skin sheath) or **syndactylous** (skeletal elements of the second and third toes fused, with both digits in a common skin sheath) (figure 11.14). These characteristics remain useful in broad descriptions of marsupial orders (table 11.2). Other dichotomous characters in addition to the lower incisors and digits of the hind foot have influenced the classification of marsupials. These include linear or v-shaped centrocrista on the occlusal surface of the

upper molars, the presence or absence of paired sperm (see figure 11.9), and a separate or continuous lower ankle joint pattern. Alternative taxonomic arrangements of marsupials, based on a variety of research techniques, have been proposed (see Woodburne and Case 1996; Springer et al. 1997). Taxonomic relationships among marsupial families proposed by Luckett (1994) and Phillips et al. (2006) are given in figure 11.15. Most classifications recognize both Australidelphia and Ameridelphia (North and South America) as major clades. Most extant marsupial orders were separate lineages by the early Paleocene if not earlier (Kirsch et al. 1997; Springer et al. 1997). As noted, we follow Wilson and Reeder (2005), who recognized 7 orders (see table 11.2) and 20 extant families. The first three orders we discuss (Didelphimorphia, Paucituberculata, and Microbiotheria) all have a single family and occur in the New World. The remaining four orders are Australasian in distribution.

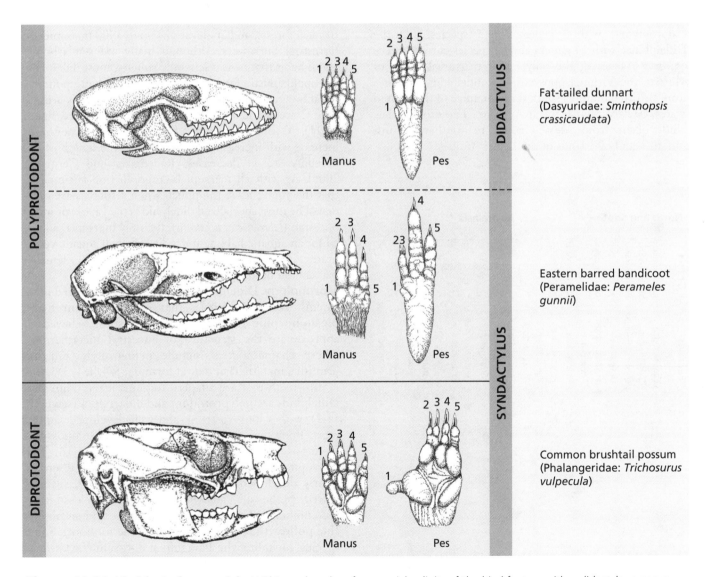

Figure 11.14 Hind feet of marsupials. Within each order of marsupials, digits of the hind feet are either didactylous or syndactylous. This may enhance the grasping ability of arboreal species. Dentition is either polyprotodont or diprotodont. *Adapted from H. Tyndale-Biscoe, 2005,* Life of Marsupials, *CSIRO Publishing.*

Table 11.2   Structure of the digits and dentition of the seven orders of marsupials discussed in this chapter*

| Digits | Dentition | |
	Polyprodont	Diprotodont
Didactylous	Didelphimorphia Microbiotheria Dasyuromorphia	Paucituberculata
Syndactylous	Peramelemorphia Notoryctomorphia†	Diprotodontia

* See Figure 11.15.
† There is debate over the characterization of both the dentition and digits in marsupial moles. Although clearly not diprotodont, the dentition is not typically polyprotodont. Likewise, whether syndactyly occurs or not is debatable.

Didelphimorphia

This order has been referred to as Ameridelphia (Marshall et al. 1990), reflecting its New World distribution. The single family Didelphidae includes 2 subfamilies; the Caluromyinae with 3 genera and 5 species, and the Didelphinae with 14 genera and 82 species currently recognized. As noted, the only extant marsupial north of Mexico, the Virginia opossum, is a didelphid. It ranges from British Columbia south through much of the United States, Mexico, and Central America. The other didelphids occur from Mexico south to southern South America, and on islands in the Lesser Antilles.

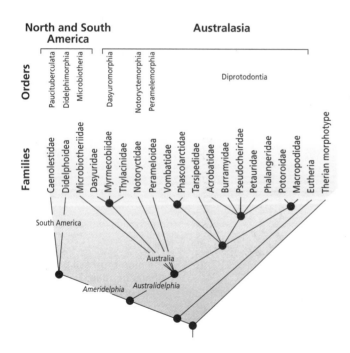

Figure 11.15 Taxonomic relationships of marsupials.
The Microbiotheriidae occur in South America but group with the Australian clade based on their phylogenetic relationships. *Data from W. P. Luckett, 1994,* Journal of Mammalian Evolution *2:225–283; Phillips et al., 2006,* Systematic Biology *55:122–137.*

Life history strategies of didelphids vary, as would be expected within the largest family of New World marsupials. They occur in almost all habitats from deserts to tropical forests, and at elevations up to 3400 m. Some are terrestrial burrowers, although many are semiarboreal and inhabit tree dens. Generally solitary, most didelphids are opportunistic feeders, their diet dependent upon seasonal forage availability. Gestation periods usually are less than 2 weeks; neonates weigh about 0.1 g (see figure 11.11). Within the genus *Didelphis*, mean litter sizes increase with increasing latitude, while the length of the breeding season decreases (Rademaker and Cerqueira 2006). As with all marsupials, reproduction is timed so that the young leave the pouch when resources are optimal. The most specialized didelphid is the yapok, or water opossum (*Chironectes minimus*), the only marsupial adapted for an aquatic habitat and a diet of small aquatic vertebrates. The hind feet are webbed, and the female's marsupium becomes watertight during dives.

Morphology. Didelphids are the most generalized marsupials (Marshall 1984); that is, they retain many **plesiomorphic** (ancestral) characteristics, believed to approximate the structure of ancestral metatherians. These characteristics include pentadactyly and the primitive metatherian dental formula (5/4, 1/1, 3/3, 4/4 = 50)—teeth that are adapted to generalized omnivory. Didelphids are polyprodont and didactylous (see table 11.2), with a long rostrum, and well-developed sagittal crest. Body mass ranges from 10 g in mouse opossums (genus *Marmosa*) to 2 kg in *Didelphis* (figure 11.16). The marsupium is well-developed in some genera, absent or poorly developed in others. Most have long, sparsely haired, prehensile tails (like the Virginia opossum) to accommodate their semiarboreal habits, and an opposable **pollex** (thumb or first digit on the forefoot). Some species, including the Patagonian opossum (*Lestodelphys halli*), which has the southernmost distribution of any didelphid, and several species of mouse opossums, have **incrassated** tails. That is, they store fat in the base of their tail for periods of torpor. As noted by Tyndale-

Figure 11.16 South American opossum. Note the long guard hairs and naked, prehensile tail in *Didelphis marsupialis*.

Biscoe (2005:137), "[c]ompared to the history of marsupials in Australia the adaptive diversity [of New World marsupials] has been very modest: South American marsupials did not give rise to specialized honey eaters, leaf eaters, gliders, browsers, or grazers; and the size range has also been modest."

Fossil History. This order, generally considered the most ancestral marsupial taxon, is known from the late Cretaceous in both North and South America. The group had radiated into 12 recognized genera by the Paleocene (Carroll 1988). A basal "didelphoid" from the early Paleocene, *Sznlinia gracilis*, from Bolivia was described by de Muizon and Cifelli (2001). Steiner et al. (2005) conducted genetic analyses on 19 species of didelphids and concluded the family originated in the mid-Eocene. Didelphimorphs occurred in Europe by the Eocene but died out there in the Miocene about 20 mya. They also were extirpated in North America in the Miocene but recolonized after the Panamanian land bridge formed during the Pleistocene.

Economics and Conservation. *Didelphis virginiana* is one of the few didelphids eaten by humans. It also is taken as a furbearer and occasionally by sport hunters in North America (Gardner and Sunquist 2003). In general, however, it is of minor economic importance for fur, for sport, or as a poultry depredator. Opossums are sometimes used in laboratory research. Loss of tropical habitat may adversely affect certain Central and South American didelphids. Several species are considered endangered, including the pygmy short-tailed opossum (*Monodelphis kunsi*), dryland mouse opossum (*Marmosa xerophila*), and the slim-faced slender mouse opossum (*Marmosops cracens*).

Paucituberculata

The single family, Caenolestidae, within this order contains three genera and six species of "shrew," or "rat,"

opossums. The four species of *Caenolestes*, including the recently described Andean caenolestid (*C. condorensis*), occur in dense vegetation in cold, wet, high-elevation forests and meadows of northwestern South America. The Peruvian shrew opossum or Incan caenolestid (*Lestoros inca*) is found in the Andes Mountains of southern Peru, whereas the Chilean shrew opossum or long-nosed caenolestid (*Rhyncholestes raphanurus*) is restricted to Chiloé Island, nearby Argentina, and south-central Chile at lower elevations than other members of the family. Caenolestids are primarily nocturnal, insectivorous or omnivorous, and terrestrial. With didelphids, caenolestids share the characteristic of paired spermatozoa within the epididymis—unknown in Australasian marsupials or eutherians.

Morphology. Caenolestids are small and shrewlike in appearance (figure 11.17). They have a long rostrum, small eyes, and the hind limbs are longer than the forelimbs. Adults weigh about 40 g. Total length is about 30 cm, half of which is the long, fully haired tail. They have no marsupium. Members of this order are didactylous and are the only New World marsupials that are diprotodont (see table 11.2). The dental formula is 4/3–4, 1/1, 3/3, 4/4 = 46–48. The procumbent first lower incisors have enamel only on the anterior surface, and the lower canines are

Figure 11.17 South American caenolestids. (A) The rat opossum, *Caenolestes obscurus*. (B) Note the diprotodont lower incisors, small posterior incisors, reduced size of the posterior molars, and characteristic preorbital vacuity. *Adapted from DeBlase and Martin, 1981*, A Manual of Mammalogy, *2nd ed., Wm. C. Brown Publishers.*

vestigial. It is generally believed that caenolestids are the only marsupials without a deciduous last premolar—although Luckett and Hong (2000) suggest they have one.

Fossil History. The earliest known fossil caenolestids are from the early Eocene. This family was extremely diverse during the Miocene, when they were the most abundant marsupials (Carroll 1988).

Economics and Conservation. Little is known about most of the species of caenolestids. None are considered endangered, although only a few *R. raphanurus* have been collected, and the species is considered vulnerable.

Microbiotheria

The single family in this order, Microbiotheriidae, contains only one extant species. The monito del monte—"little monkey of the mountains"—(*Dromiciops gliroides*) is nocturnal and arboreal and inhabits dense, humid forests of south-central Chile and adjacent Argentina. Its distribution is closely associated with beech (*Nothofagus*)/bamboo (*Chusquea*) forests (Hershkovitz 1999). Their distinctive, round nests are found in fallen logs, tree cavities, and thickets. Nests often are lined with leaves of water-resistant Chilean bamboo. Microbiotheriids forage primarily for invertebrates but occasionally may consume herbaceous material. Prior to hibernation, fat accumulates in the base of the incrassated, prehensile tail. Enough fat may be stored in a week to double an individual's body mass. Unlike the other two families of New World marsupials, *Dromiciops* does not have paired sperm (Tyndale-Biscoe 2005).

Morphology. These small (16–30 g), mouselike animals (figure 11.18) have a maximum head and body length of 13 cm. The well-furred, prehensile tail is also about 13 cm long. The fur is short and thick, and a well-formed pouch is evident in females. The soles of the hind feet have five distinct, transverse pads, or ridges. The dental formula is the same as in didelphids, with a total of 50 teeth. The skull of *D. gliroides* is noteworthy for its greatly inflated auditory bullae, unlike any other marsupial (figure 11.19). Giannini et al. (2004) discussed the postnatal ontogeny of the skull in *D. gliroides* based on 14 cranial dimensions. Previously considered a didelphid, this species was placed

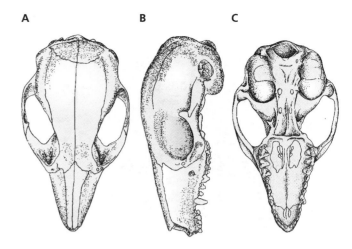

Figure 11.19 Skull of Dromiciops. (A) Dorsal, (B) lateral, and (C) ventral views of the skull of a monito del monte (*Dromiciops gliroides*). Note the five upper incisors in an arc-shaped pattern, large palatal vacuities, and inflated auditory bullae—highly unusual in a marsupial. *Adapted from L. G. Marshall, 1978,* Mammalian Species #99, *American Society of Mammalogists.*

in a separate order (Aplin and Archer 1987) based on tarsal morphology, serology, and karyotype.

Fossil History. Marshall et al. (1997) suggest that *Khasia cordillerensis*, from the early Paleocene of Bolivia, was a microbiotheriid. The family is known in South America from the late Oligocene fossil genus *Microbiotherium*. The phylogeny of *Dromiciops* remains unresolved among authorities. Most group microbiotheres among Australasian marsupials (see figure 11.15) and suggest that bidirectional dispersal between Antarctica and Australia occurred to account for current patterns (cf. Kirsch et al. 1991; Springer et al. 1998). Alternatively, Hershkovitz (1999:12) attributed similarities between *Dromiciops* and Australian marsupials to "parallelisms, convergences, or common retentions of primitive marsupial or therian characters."

Economics and Conservation. These harmless marsupials, called "colocolos" by natives, are of no economic value but are considered to bring bad luck if observed in or around homes. They currently are considered threatened.

Dasyuromorphia

These small to medium-sized Australasian marsupials include three families of carnivorous species; however, practically all extant species are in the family Dasyuridae. These include what are usually referred to as marsupial "mice." Given their characteristic foraging behavior, they might more properly be considered marsupial "shrews." With the recent extinction of the thylacine, or Tasmanian wolf (*Thylacinus cynocephalus*), the Tasmanian devil (*Sarcophilus harrisii*) is the largest living carnivorous marsupial. Analyses of dasyuromorphian phylogeny, based on cranial and dental characteristics, were done by Wroe et al. (2000).

Figure 11.18 Monito del monte. *Dromiciops gliroides*, the only member of the order Microbiotheria, has an unresolved phylogeny.

Morphology. Dasyuromorphians are polyprotodont and didactylous (see table 11.2). Canines are well-developed; several species, including the Tasmanian devil and the quolls (genus *Dasyurus*), have specialized **carnassial** (blade-like, shearing) molariform dentition. Tails are usually long, are often held erect, and are never prehensile.

Fossil History. Dasyuromorphian families were differentiated in Australia by the early Miocene. They probably were there much earlier, however, as the oldest fossils date from the Oligocene (Tedford et al. 1975).

Economics and Conservation. Aspects of conservation and economics are noted under the accounts of each family.

Thylacinidae. This monotypic family included the recently extinct Tasmanian wolf, or thylacine (figure 11.20). Recent fossil thylacines occurred throughout Australia and New Guinea, and the species was common in Tasmania prior to European colonization. The thylacine soon came into conflict with humans, however, because it preyed on domestic livestock among other large prey (Wroe et al. 2005). Thylacine populations rapidly declined throughout the 1800s and early 1900s because of predator control, habitat loss, and competition with the dingo (Johnson and Wroe 2003). The last thylacine died in the Hobart Zoo in 1934. Ironically, the species was given complete legal protection by the Tasmanian government 4 years later. The phylogenetic relationships of thylacinids and proposed affinities with extinct borhyaenids of South America have received much attention (Sarich et al. 1982; Krajewski et al. 1992). Thylacinids were quite diverse during the Miocene, and about 12 fossil species have been described. They ranged in body weight from 1 to 60 kg (Wroe 2001).

Myrmecobiidae. This monotypic family includes only the numbat (*Myrmecobius fasciatus*), which is highly specialized for **myrmecophagy** (a diet of ants and termites). They were extirpated from New South Wales by 1857 and South Australia by 1924 (Archer 1978). The species is now reduced to isolated populations in arid, scrub woodlands in southwest Western Australia. A combination of fox (*Vulpes vulpes*) control and numbat reintroduction has increased populations of *M. fasciatus* throughout much of western Australia (Friend and Thomas 2003). Numbats forage diurnally, which is unusual among marsupials. They are active about 5 hours a day as they search for prey in hollow logs. They spend the night in tree cavities and burrows (Cooper and Withers 2004, 2005). They are solitary except during the breeding season. The gestation period is two weeks. The usual litter size is four. They have no marsupium, and neonates cling to curly hair on the abdomen as they hang from the teats.

Numbats reach about 44 cm in total length, with the fully furred tail equal to the head and body length (figure 11.21). Adults weigh up to 700 g. The reddish or grayish brown pelage has seven transverse white and black stripes on the dorsum, and there is a black eye stripe. Like other myrmecophageous mammals, they have an elongated, tapered rostrum and tongue to take ants and termites. Their tongue can be extended 10 cm—close to half their head and body length. Numbats have small, degenerate, peglike teeth. They also have **supernumerary** (additional) cheekteeth, for a total number of teeth that may be 52—more than any other metatherian. Like many other marsupials, both sexes have a large presternal gland on the upper chest, probably used to mark territories (Friend 1989). The fossil record of this family dates only from the late Pleistocene, and relationships with other families remain questionable.

Dasyuridae. Like the New World didelphids, this large, diverse family of 15 genera and 61 species is the most generalized (ancestral) structurally and functionally of the Australasian marsupials. Based on DNA sequence analysis, Krajewski et al. (2000a) concluded that the family is monophyletic. Dasyurids range in size from the smallest marsupials, the tiny Pilbara ningaui, and other marsupial "mice," to the largest extant marsupial carnivore, the Tasmanian devil. Dasyurids occur throughout Australasia, where they occupy all terrestrial and semiarboreal habitats from deserts to high-elevation rainforests.

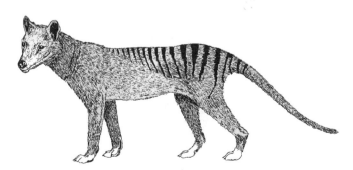

Figure 11.20 Recently extinct Tasmanian wolf. Very similar to eutherian canids in appearance and habits, the tylacine, or Tasmanian wolf (*Thylacinus cynocephalus*), was the largest recent marsupial carnivore, with a body mass of 15–35 kg. *Adapted from D. W. Walton and B. J. Richards, 1989,* Fauna of Australia, *vol. 1B, Australian Government Publishing Service.*

Figure 11.21 External features of a numbat. The numbat, or banded anteater, is distinctively striped. Like all species of ant and termite eaters, it has a long, extensible tongue.

Dentition totals 42–46 teeth and is specialized for a carnivorous or insectivorous diet. Jones and Barmuta (2000) discussed niche differences between three species of dasyurids, the Tasmanian devil, spotted-tailed quoll (*Dasyurus maculatus*) (figure 11.22), and eastern quoll (*D. viverrinus*), based on diet and habitat use. A marsupium usually is absent or poorly developed. In several species, males exhibit a mass die-off after mating the first time. The adaptive value of this semelparity may vary among the different species (Oakwood et al. 2001). Incrassated tails occur in several genera that inhabit deserts, including the dunnarts (*Sminthopsis*), the mulgara (*Dasycercus cristi-cauda*), and the pseudantechinuses (*Pseudantechinus*). Dasyurids primarily are nocturnal (although the mulgara also may be active during the day) and generally solitary. The fossil record extends to the early Miocene, during which time this group was rare. Krajewski et al. (2000b) provided a revised classification of dasyurids above the genus level.

Several dasyurids, including the southern dibbler (*Parantechinus apicalis*), the red-tailed phascogale (*Phascogale calura*), and several species of dunnarts, are endangered because of reduced distributions and population sizes.

Peramelemorphia

Bandicoots and bilbies (or rabbit-eared bandicoots) occur throughout Australasia. All are terrestrial omnivores, feeding on invertebrates, small vertebrates, and plant material. They occupy a variety of habitat types from arid deserts to tropical rainforests and jungle, often at high elevations.

Peramelemorphians are unusual among marsupials in having chorioallantoic placentae. Unlike eutherians, however, the placentae have no villi; thus, gestation periods in these species are no longer than in other metatherians. In fact, both the northern brown bandicoot (*Isoodon macrourus*) and the long-nosed bandicoot (*Perameles nasuta*) have gestation periods of 12.5 days—the shortest known among mammals.

There is continuing debate as to the affinity of peramelemorphians with each other and with other marsupial groups (Gordon and Hulbert 1989; Racey et al. 2001), and recognition of families has been in a state of flux. The bilbies (genus *Macrotis*) were placed in the family Peramelidae and then the Peroryctidae based on skull characteristics (Groves and Flannery 1990). Based on molecular data, bilbies are currently considered within the family Thylacomyidae. Also, the pig-footed bandicoot (*Chaeropus ecaudatus*), most likely now extinct, was placed in its own family (Chaeropodidae) by Westerman et al. (1999, 2001) based on mitochondrial DNA (mtDNA) analyses.

Morphology. Peramelemorphians have short, compact bodies, with a long, pointed rostrum. Extremes of body size occur in the thylacomyids (see later section). All are adapted for digging with strong claws on the second, third, and fourth digits of the forefeet. Hind limbs are larger than forelimbs, and the hind feet are long with a well-developed claw on an enlarged fourth digit (figure 11.23). Bandicoots are unique among marsupials in having a well-developed **patella** (kneecap) and no **clavicle** (collar bone) (Jones 1968).

Dentition is polyprotodont, with 5 pairs of upper incisors and 48 total teeth (46 teeth in some peroryctids). The canines are well-developed, and the molars are adapted for an omnivorous diet. The marsupium also is well-developed and opens posteriorly. There are eight teats, even though mean litter size usually is four. This allows for consecutive litters to be produced, as teats remain swollen after nursing the previous litter, and incoming neonates cannot attach to them.

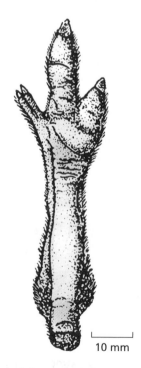

Figure 11.23 Hind foot of a bilby. The left hind foot of the greater bilby (*Macrotis lagotis*), showing the enlarged fourth digit and syndactylous second and third digits. *Adapted from D. W. Walton and B. J. Richards, 1989,* Fauna of Australia, *vol. 1B, Australian Government Publishing Service.*

Figure 11.22 Spotted-tailed quoll. An adult female with a juvenile.

Fossil History. The fossil record dates from the early Miocene of Queensland.

Economics and Conservation. Many species of Australian peramelids have suffered significant population declines because of habitat loss to domestic livestock and because of introduced predators. The western barred bandicoot (*Perameles bougainville*), once common on the mainland, now exists only on Bernier and Dorre islands off western Australia. In addition to the pig-footed bandicoot, the lesser bilby (*Macrotis leucura*) probably became extinct as recently as the 1960s. The greater bilby (*M. lagotis*) is endangered. The range of several species of bandicoots is greatly reduced from presettlement times.

Peramelidae. There are 6 genera and 19 species of bandicoots (figure 11.24) and echymiperas, one of which—the desert bandicoot (*Perameles eremiana*)—is probably extinct. Another bandicoot, *Microperoryctes aplini* from western New Guinea, was recently described by Helgen and Flannery (2004). Peramelids exhibit rather broad ecological flexibility (Gordon and Hulbert 1989), occurring in a variety of habitats, primarily in Australia. Bandicoots are smaller than bilbies. They range in size from the mouse bandicoot (*Microperoryctes murina*), with a maximum head and body length of 17 cm and tail of 11 cm, to the giant bandicoot (*Peroryctes broadbenti*), which reaches a total length of 90 cm and body mass of 5 kg. The five species of New Guinean spiny bandicoots or echymiperas (genus *Echymipera*) and the Ceram Island bandicoot (*Rhynchomeles prattorum*) lack the fifth upper incisor found in other peramelemorphs and have a total of 46 teeth. All species are nocturnal, terrestrial, and solitary. They are insectivorous or omnivorous and occur in grassland, shrub, and rainforest habitats. Both groups exhibit sexual dimorphism, with males being larger than females. Bandicoots have short, course pelage, often with stiff, quill-like guard hairs and relatively small ears and tail.

Thylacomyidae. Formerly considered to be in the family Peroryctidae, the bilbies now include two species. The distribution of the greater bilby (*Macrotis lagotis*) in Australia is much reduced from its historic range. As noted, the lesser bilby is probably extinct. Gibson and Hume (2000) concluded that the digestive strategy of the greater bilby is flexible to accommodate both plants and animals in their diet, which allows them to survive in the very arid regions to which they are now restricted (Gibson et al. 2002). Body mass of bilbies reaches 2.5 kg. Like the closely related peramelids, they have a chorioallantoic placenta and short gestation period (Flannery 1995). Bilbies have a long tail, 50–60% of the head and body length (figure 11.25) and longer, silkier pelage than bandicoots. Their long, rabbit-like ears reach beyond the tip of the snout and are folded over the eyes when individuals sleep. Bilbies are powerful burrowers; unlike bandicoots, they construct their own burrows. In the form of a deeply angled spiral, 2 m deep, burrows offer a refuge from desert heat during the day. Wild greater bilbies produce up to four litters per year, a much higher reproductive rate than in captive populations. However, longevity is much greater for captives (Southgate et al. 2000).

Diprotodontia

This diverse order of 11 families includes the familiar kangaroos, koala (*Phascolarctos cinereus*), wombats, and numerous other primarily herbivorous marsupials. As might be expected, given the 116 species in this order, adaptive radiation has been extensive. Species feed on insects, nectar, leaves, fruit, or they are omnivores. Many species are terrestrial, but some are arboreal. Three suborders were recognized by Kirsch et al. (1997). The wombats and koala are closely related and placed in the suborder Vombatiformes. The suborder Macropodiformes encompasses the families Potoroidae, Macropodidae, and the newly recognized, monotypic family Hypsiprymnodontidae (the musky

Figure 11.24 Southern brown bandicoot. Bandicoots such as *Isoodon obesulus* shown here have shorter, coarser pelage than rabbit-eared bandicoots.

Figure 11.25 Greater bilby. A greater bilby (or rabbit-eared bandicoot, *Macrotis lagotis*) has a longer tail and longer, silkier pelage than bandicoots.

rat-kangaroo). The remaining six families are included in the suborder Phalangeriformes, although this group is almost certainly not monophyletic (Kavanagh et al. 2004). **Morphology.** All species are diprotodont, and the second and third digits are syndactylous. In many arboreal diprotodonts, including the koala, ringtail possums (Pseudocheiridae), and cuscuses (Phalangeridae), the first two digits of the forefeet oppose the other three digits (figure 11.26), that is, they are **schizodactylous**. On the hind feet, the **hallux** (big toe) is also opposable. This is not the case in the terrestrial species, however. Dental adaptations include a large pair of lower incisors and three pairs of smaller upper incisors (although wombats have a single pair of upper and lower incisors). Upper canines are variable in size and shape; there are no lower canines.

Fossil History. The earliest fossil diprotodonts date from late Oligocene deposits of Australia.

Economics and Conservation. Many diprotodonts compete with domestic livestock on grazing lands. Other species have been hunted for meat or hides. Some species of wallabies and kangaroos have been seriously reduced in density and distribution following European settlement, with several recent extinctions being attributed to habitat loss or introduced predators.

Phascolarctidae. The familiar koala, with a superficial resemblance to a small bear (figure 11.27), is the only extant species in this family. Koalas occur in *Eucalyptus* woodlands throughout eastern and southeastern Australia. The koala is most closely related to wombats, and these families share a number of morphological characteristics. Both have a marsupium that opens posteriorly, vestigial tails (unusual in an arboreal species), and lack the first two premolars, among several other features. Unlike wombats, the koala has three upper incisors, and the den-

Figure 11.27 Koala. The familiar koala (*Phascolarctos cinereus*) spends a large part of each day resting in a tree.

tition is closed-rooted (i.e., not evergrowing). The angle of the dentary in koalas is not inflected, unlike any other marsupial except the honey possum (see table 11.1).

Koalas are sexually dimorphic, with males being 50% larger than females. With body mass from 6.5–12.5 kg, koalas are among the largest arboreal browsers and are on the ground only while moving between trees. They do not build a nest but simply rest in the forks of trees. Koalas are unusual among herbivores in their highly selective, specialized diet: leaves, stems, flowers, and even bark of numerous species of *Eucalyptus*. This is a very poor-quality forage, but koalas have reduced energy requirements because they are slow-moving and inactive up to 20 hours a day. Also, their alimentary tract has the largest cecum relative to body size of any mammal (see figure 7.14). Like the bandicoots and bilbies, the koala has a chorioallantoic placenta. But again, because there are no chorionic villi, gestation is only about 35 days. As one of the most recognizable marsupials, the koala is an important tourist attraction. Interestingly, hundreds of thousands of koalas were harvested annually as part of the fur trade in Queensland from about 1906 through 1927 (Hrdina and Gordon 2004). Although not considered either threatened or endangered, populations often are small (Sullivan et al. 2004), and the koala is not common.

Vombatidae. The two genera and three species of wombats are short-limbed, plantigrade, powerful burrowers (figure 11.28). Adult body mass is about 30 kg. The common wombat (*Vombatus ursinus*) is found in forested areas of southeastern Australia and Tasmania. The southern hairy-nosed wombat (*Lasiorhinus latifrons*) inhabits semiarid regions of south Australia, and the endangered northern hairy-nosed wombat (*L. krefftii*) is now restricted to the 3000-ha Epping Forest National Park in central Queensland. The fossil record dates from the early Miocene epoch.

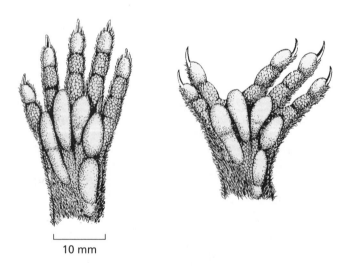

10 mm

Figure 11.26 Schizodactylous forefeet. In many arboreal diprotodonts, including the koala, ringtail possums, and cuscuses, the forefoot is schizodactylous. (A) digits closed; (B) digits open for grasping. *Adapted from D. W. Walton and B. J. Richards, 1989,* Fauna of Australia, *vol. 1B, Australian Government Publishing Service.*

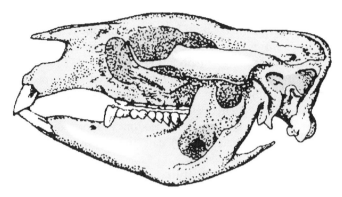

Figure 11.29 Skull of a wombat. In a common wombat (*Vombatus ursinus*), the single pair of upper and lower incisors—as in rodents and the aye-aye, a primate—is unique among marsupials. *Adapted from DeBlase and Martin, 1981, A Manual of Mammalogy, 2nd ed., Wm. C. Brown Publishers.*

Figure 11.28 Common wombat. Wombats are the largest herbivorous mammals that burrow.

Although they share several characteristics with the koala, wombats have several distinct features that reflect their terrestrial, grazing, semifossorial existence. Like rodents, they have a single pair of upper and lower incisors, no canines, and a reduced number of premolars (figure 11.29). Dentition is open-rooted and continuously growing, which is unique among marsupials. Wombats are nocturnal and consume primarily grasses and forbs (broad-leafed herbs). Unlike the koala, they have a poorly developed cecum. During hot, dry days, *Lasiorhinus* minimizes time on the surface of the ground by remaining in extensive, interconnected burrow systems. Benefits of burrows include protection from predators, fires, and harsh ambient conditions. Nonetheless, construction of burrows is energetically demanding, and wombats have a poor quality diet. The metabolic rate of wombats is extremely low, even for a marsupial (Evans et al. 2003). Most burrowing species are small; if large, they eat high-energy insects. Wombats are the largest burrowing herbivores (Shimmin et al. 2002; Finlayson et al. 2005). They are seasonal breeders (Taggart et al. 2005). The well-developed marsupium, with two teats, opens posteriorly. A single young is born after a gestation period of 21 days and leaves the pouch permanently by about 9 months of age.

Phalangeridae. This family includes 5 species of brushtail possums (genus *Trichosurus*), the scaly-tailed possum (*Wyulda squamicaudata*), and 4 genera and 21 species of cuscuses (Groves 2005). Brushtail possums and the scaly-tailed possum occur only in Australia, whereas the cuscuses are widespread throughout Australia, New Guinea, and

surrounding islands. The bear cuscus (*Ailurops ursinus*) is restricted to Sulawesi and nearby islands. The telefomin cuscus (*Phalanger matanim*) and the black-spotted cuscus (*Spilocuscus rufoniger*) are endangered. All phalangerids occur in scrub or heavily forested areas. The brushtail possum (*T. vulpecula*) was introduced into New Zealand beginning in the 1800s to establish a fur trade. Unfortunately, they have caused serious damage to native forests and orchards (Cowan 1995).

Phalangerids are medium-sized with large eyes, short snout, and soft, dense pelage. Adults range from 0.5–1.2 m total length and have a body mass of 1.1–4.5 kg. All are nocturnal, have long, prehensile tails, and are excellent climbers (figure 11.30). *Trichosurus* feed on leaves

Figure 11.30 Phalangerids vary in the amount of fur on the tail. (A) Small, nonoverlapping conical scales cover the distal part of the tail in *Wyulda*; (B) strongly prehensile tail in *Phalanger* has fur at the base only; (C) well-furred tail in *Trichosurus* with ventral friction pad. *Adapted from D. W. Walton and B. J. Richards, 1989, Fauna of Australia, vol. 1B, Australian Government Publishing Service.*

(**folivorous**), as well as flowers and nectar; *Phalanger* and *Wyulda* are omnivorous. The marsupium opens anteriorly; litter size is one or two. The gestation period is about 17 days, and young remain in the pouch for 4–7 months. Neonatal brushtail possums (*T. vulpecula*) enter the pouch about 2 minutes after birth, and attach to a teat 10–15 minutes later (Veitch et al. 2000). Fossil remains date from the late Oligocene. Based on mtDNA analyses, Ruedas and Morales (2005) date the origin of phalangerids to the same period.

Potoroidae. There are three genera and eight extant species of potoroids, including potoroos (*Potorous*) and bettongs (*Bettongia* and *Aepyprymnus*). Potoroids are small, secretive, densely furred animals. Like the macropodids, the hind limbs are larger than forelimbs, and the hind feet are large. The tail is weakly prehensile (bettongs carry nesting material with their tail); the upper canines are well-developed, and there is a large, carnassial premolar. The **sacculated** stomach (several chambers) is not as well-defined as in macropodids. Mean body mass of the musky rat-kangaroo is only 530 g, whereas the rufous bettong (*A. rufescens*) is about 3 kg. Potoroids are opportunistic omnivores or herbivores. Most include underground fungi as a large part of their diet (Lee and Cockburn 1985). They are essentially solitary and, with the exception of the musky rat-kangaroo, nocturnal. In the northern bettong, postpartum estrus occurs, followed by embryonic diapause. Pouch young permanently emerge at about 3.5 months of age and are weaned at 166–185 days old (Johnson and Delean 2001). No postpartum estrus has been detected in the musky rat-kangaroo, however (Lloyd 2001). The conservation status of several species of potoroids is of concern. Gilbert's potoroo (*Potorous gilbertii*) may be the most critically endangered marsupial in Australia. Population declines are due in large part to introduced predators. There is a single population in Two Peoples Bay National Park in western Australia. This population was discovered in 1994 after the species was presumed to be extinct (Sinclair et al. 2002). Gilbert's potoroo feeds predominantly on fungi (Nguyen et al. 2005). The burrowing bettong or boodie (*B. lesueur*) is endangered. Individuals were reintroduced to a site on the Australian mainland (Short and Turner 2000). The long-footed potoroo (*P. longipes*)—the first specimen of which was collected in 1968 (Seebeck and Johnston 1980)—is endangered, as are the woylie (*B. penicillata*), eastern bettong (*B. gaimardi*), and northern bettong (*B. tropica*). Two species—the desert rat-kangaroo (*Caloprymnus campestris*) and broad-faced potoroo (*P. platyops*)—are recently extinct.

Hypsiprymnodontidae. This monotypic family includes only the musky rat-kangaroo (*Hypsiprymnodon moschatus*), which occurs in northeastern Queensland, Australia. Formerly placed in either the family Potoroidae or the family Macropodidae, the species was placed in its own family by Burk and Springer (2000) based on sequences from three mtDNA genes. They suggested that *Hypsiprymnodon* diverged from macropodids in the early Oligocene.

Placement in its own family is consistent with several other genetic and morphological analyses (Kavanagh et al. 2004). Musky rat-kangaroos have a mean body weight of about 0.5 kg. Unlike macropodids, they do not hop bipedally but use all four legs. Neither do they exhibit embryonic diapause. They are generally terrestrial, foraging for fruits, seeds, and insects among leaf litter and downed logs on the forest floor. Populations have probably declined because of forest fragmentation.

Macropodidae. This is the largest marsupial family, with 11 genera and about 50 extant species. Body mass ranges from the 1-kg hare-wallabies (*Lagorchestes*) to 80-kg red kangaroos (*Macropus rufus*). Macropodids are browsing or grazing herbivores that occupy practically all terrestrial habitats (tree kangaroos, *Dendrolagus*, are semiarboreal) from deserts to rainforests throughout Australasia.

Their ecology and morphology parallel that of the eutherian artiodactyls. Macropodids have a large, sacculate stomach in which microorganism-aided digestion occurs. As in some artiodactyls, food is regurgitated for additional chewing and swallowed again. Macropodids are diprotodont, but, with the exception of the banded hare-wallaby (*Lagostrophus fasciatus*), the upper and lower incisors do not occlude. Canines are small or absent, and there is a diastema (figure 11.31). The molars are **hypsodont** (high-crowned), and as in elephants and manatees, **mesial drift** occurs (forward movement of the cheekteeth in the jaw, as worn anterior teeth drop out). Among marsupials, this serial replacement of cheekteeth, most pronounced in *Macropus*, occurs only in macropodids. Only the pygmy rock wallaby (*Petrogale concinna*), however, has supernumerary molars (more than the usual four) that are shed throughout life.

Macropodids ("big-footed") are characterized by their strong, well-developed hind limbs (figure 11.32) and large hind foot. Most species have a long, broad tail that acts as a balance during rapid (up to 50 km/h), bipedal, hopping locomotion. At slow speeds, the tail acts as an added limb, or "tripod" while sitting or foraging (figure 11.33). The

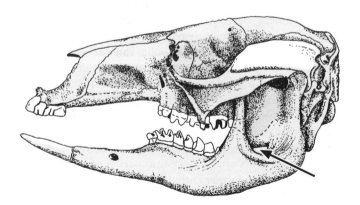

Figure 11.31 Wallaby skull. Skull of a wallaby shows the diprotodont dentition, and the diastema typical of herbivores. Note also the pronounced masseteric fossa of the mandible (arrow). *Adapted from DeBlase and Martin, 1981, A Manual of Mammalogy, 2nd ed., Wm. C. Brown Publishers.*

Figure 11.32 Red-necked wallaby. The enlarged hind limbs and reduced forefeet typical of macropodids are evident in this red-necked wallaby (*Macropus rufogriseus*).

enlarged hindquarters give macropodids a low center of gravity in the pelvic area, which helps individuals maintain balance while hopping.

The marsupium is large and opens anteriorly. Mammary glands are complex, and the physiology of lactation is very sophisticated, with pronounced differences in milk composition during lactation. Thus, a mother nursing a developing pouch young, and a young that has left the pouch but returns to nurse, produces milk of different nutrient composition from different teats (Green 1984). Gestation periods of macropodids are relatively long compared with those of other marsupials—close to the length of the estrous cycle. A single young is produced at a time, and

Figure 11.33 Tails in macropodids. Tails may serve a variety of functions in locomotion and foraging of macropodids. Tail may act as (A) a "tripod" while foraging; (B) a fifth limb in slow, "pentapedal" locomotion; or (C) as a counterbalance in rapid, bipedal hopping. *Adapted from D. W. Walton and B. J. Richards, 1989,* Fauna of Australia, *vol. 1B, Australian Government Publishing Service.*

embryonic diapause (see chapter 10) occurs in all species except the western gray kangaroo (*M. fuliginosus*).

European settlement has had little adverse affect on some of the larger species; in fact, livestock grazing has increased numbers and distribution. Conversely, some large and many smaller macropodids—including the bridled nailtail wallaby (*Onychogalea fraenata*), currently restricted to a single locality in central Queensland, Australia (Fisher et al. 2001), Calaby's pademelon (*Thylogale calabyi*), and several species of tree kangaroo—are considered endangered. Species that have gone extinct within the last 30–100 years include the toolache wallaby (*Macropus greyi*), central hare-wallaby (*L. asomatus*), eastern hare-wallaby (*L. leporides*), and the crescent nailtail wallaby (*O. lunata*).

Burramyidae. There are two genera and five species of pygmy possums—the smallest of the possums. Mean body mass of adults ranges from 7–50 g, and head and body length is only 5–12 cm. The mountain pygmy possum (*Burramys parvus*), considered endangered, is limited to terrestrial alpine areas above 1300 m elevation in southeastern Australia. Only four populations are known to exist within an area of 10 km². This species was known only from fossil remains until 1966, when a live animal was taken at Mount Hotham, Victoria (Mansergh and Broome 1994). Broome (2001) found that density, population structure, movements, and home range in a mountain pygmy possum population changed seasonally and were highly correlated with elevation. The remaining burramyids, in the genus *Cercartetus*, are all arboreal and occur in a variety of habitats in Australia and Tasmania. One species, the long-tailed pygmy possum (*C. caudatus*), also occurs in New Guinea. Tails are long and prehensile, the pouch opens anteriorly, and, like macropodids, females exhibit embryonic diapause.

Pygmy possums are nocturnal and omnivorous, consuming invertebrates, fruits, seeds, nectar, and pollen. They have long, extensible, "brushed" tongues with an extensive system of papillae (figure 11.34) that are especially well-developed in *Cercartetus*. Papillae may serve to increase the surface area for uptake of nectar and pollen. All burramyids enter torpor, but among them *B. parvus* is noteworthy because it is the only marsupial known to undergo prolonged periods of hibernation. Species of *Cercartetus* store fat at the base of the tail. Both these adaptations probably contribute to the relatively long life spans of these small marsupials. Based on mtDNA sequence data, Osborne and Christidis (2002) suggested that the two genera of burramyids diverged from each other during the Oligocene, which is consistent with fossil evidence (Brammall and Archer 1997, 1999) for this group.

Acrobatidae. This family, formerly included in the Burramyidae (Aplin and Archer 1987), includes only the feathertail glider (*Acrobates pygmaeus*), found in wooded habitats of eastern Australia, and the feather-tailed possum (*Distoechurus pennatus*), which occurs in disturbed forests, gardens, and rainforests of New Guinea (Flannery

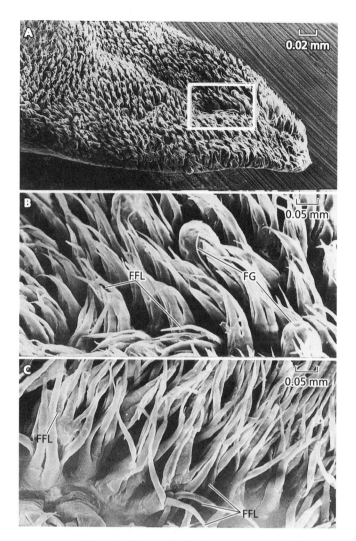

Figure 11.34 Fine structure of tongue papillae in a burramyid. Scanning electron micrographs of the papillae found on the tongue of the nectar-feeding eastern pygmy possum (*Cercartetus nanus*). (A) Dorsal surface of the tip of the tongue. (B) Finer detail of the tip. (C) Finer detail of the back of the tongue. Abbreviations: FFL = fine filiform papillae; FG = fungiform papillae. Papillae in burramyids probably increase the surface area for absorbing nutrients and move food more efficiently toward the esophagus. A similar adaptation occurs in the honey possum (see figure 11.37) and nectar-feeding bats.

Figure 11.35 Feathertail glider. Note the folded gliding membrane and flattened, feather-like tail in this small, gliding acrobatid.

1995). They are named for the long, stiff, featherlike hairs on the side of the tail. The feathertail glider (figure 11.35) is nocturnal and highly arboreal and, at 10–14 g body mass, probably the world's smallest gliding mammal. A furred **patagium** (gliding membrane) extends between the elbows and knees. The weakly prehensile tail aids both in climbing and as a rudder for gliding. The feather-tailed possum, however, is terrestrial and lacks a gliding membrane. Both species are primarily nectivorous and have long, brush-tipped tongues, similar to burramyids and tarsipedids, for taking nectar and pollen. The papillae on the tongue of *Acrobates* are longer and finer than those of *Distoechurus*. There is a general reduction in the size and number of teeth. Interestingly, although nectivorous, the

molars are bunodont for secondary feeding on insects. *Acrobates pygmaeus*, which may nest in groups of up to 20 individuals, has several litters per breeding season and exhibits embryonic diapause.

Pseudocheiridae. The 5 genera and 14 species in this family are closely related to the Petauridae, in which they were formerly included (Baverstock et al. 1990). Pseudocheirids occur in Australia, New Guinea, and a few surrounding islands (Flannery 1995). Most are slow-moving and inhabit forested areas; the rock ringtail possum (*Petropseudes dahli*) occurs on rocky slopes and outcrops. These arboreal, nocturnal species feed primarily on leaves, and many of their morphological features reflect this adaptation. Unlike petaurids, the molars of pseudocheirids are **selenodont** (cusps form crescent-shaped ridges) to finely chew leaves. The alimentary tract and very large cecum also are specializations for folivory. The arboreal habits of pseudocheirids are aided by the schizodactylous digits of the forefeet. The long, furred, prehensile tail is usually the same length as the head and body. A gliding membrane extends from the elbow to the ankle in *Petauroides*, not from the wrist as in *Petaurus* (figure 11.36 and following section). The marsupium opens anteriorly and encloses either two or four teats.

Petauridae. This family includes four species of striped possums (genus *Dactylopsila*), Leadbeater's possum (*Gymnobelideus leadbeateri*), and five species of wrist-winged gliders (*Petaurus*). Osborne and Christidis (2001) investigated mtDNA sequences within petaurids, and concluded that monophyly was not "well-supported." These arboreal gliders are named for the fact that the patagium extends from the wrist to the ankle (see figure 11.36). *Petaurus* is highly **convergent** (similar traits found in lineages that are not closely related) with the North American gliding squirrels (genus *Glaucomys*). All

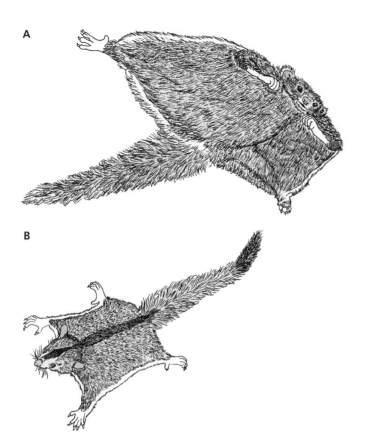

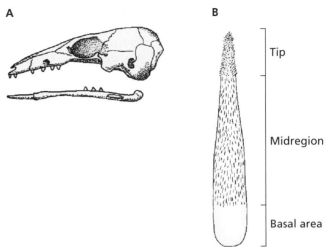

Figure 11.37 Features of the honey possum. (A) Except for the procumbent lower incisors, the honey possum has small, degenerate teeth, reflecting a soft diet. The dental formula is 2/1, 1/0, 1/0, 3/3 = 22, with the postcanine teeth not clearly differentiated. The palate is ridged to remove pollen and nectar from (B) the long, extensible, brush-tipped tongue. *Adapted from D. W. Walton and B. J. Richards, 1989,* Fauna of Australia, *vol. 1B, Australian Government Publishing Service.*

Figure 11.36 Gliding marsupials. The gliding membranes extend from the elbow to the ankle in (A) the greater glider (*Petauroides volans*), a pseudocheirid, and from the wrist to the ankle in (B) the sugar glider (*Petaurus breviceps*), a petaurid. *Adapted from D. W. Walton and B. J. Richards, 1989,* Fauna of Australia, *vol. 1B, Australian Government Publishing Service.*

petaurids are medium-sized (0.1–2.0 kg), with some type of dark dorsal stripe; long, well-furred, prehensile tails; and a well-developed marsupium that opens anteriorly. In some species, the pouch is partitioned into left and right compartments by a septum. Females usually give birth to a single young that remains in the pouch for 4 months. There are a total of 40 teeth; the diprotodont lower incisors are long and sharp, and the molars are bunodont.

Petaurids occur in forested areas of Australia, New Guinea, and surrounding islands. They are nocturnal and herbivorous or insectivorous. Arboreal locomotion is aided by the prehensile tail and opposable hallux. The endangered Leadbeater's possum was previously believed to be extinct, but populations were rediscovered in 1961 in Victoria, Australia. The mahogany glider (*P. gracilis*) and Tate's triok (*Dactylopsila tatei*) also are endangered.

Tarsipedidae. The unusual honey possum, or noolbenger (*Tarsipes rostratus*), is the sole member of this family. Average body mass of females is 12 g and males, about 9 g. Many of the adaptations of this tiny species are for its nectivorous diet. As noted by Bradshaw and Bradshaw (2001), the honey possum is the only ground-dwelling mammal that feeds exclusively on nectar and pollen. It has a long, pointed rostrum, and tubular mouth with extensi-

ble, brush-tipped tongue. The 22 small, peglike teeth are fewer in number than in any other marsupial (figure 11.37). The honey possum is nocturnal, highly arboreal, and occurs in shrubs and woodlands of southwestern Australia. The long tail is prehensile, and the hallux is opposable. Claws occur only on the two syndactylous digits of each hind foot. Instead, pads on the digits are used to grip branches.

The marsupium is well-developed. Honey possums exhibit delayed implantation, and pouch young occur throughout the year (Renfree et al. 1984). They are unique in giving birth to the smallest mammalian young; neonates weigh no more than 5 mg. Their sperm is about 0.3 mm long, the largest known among mammals. Although it shares certain characteristics with other marsupial families, the honey possum is unique in many respects, and probably represents the sole surviving member of an otherwise long-extinct lineage. There is no fossil history, and based on molecular analyses "…it has been difficult to determine the placement of *Tarsipes* relative to other diprotodontians" (Kavanagh et al. 2004:218).

Notoryctemorphia

This order encompasses a single family, the Notoryctidae, which includes the marsupial mole, *Notoryctes typhlops*. A second species, *N. caurinus*, has been described (see Johnson 1995). They represent the only completely fossorial marsupials and are widely distributed over much of western Australia in shrub-desert areas and sandy bottomland soils. Their diet consists of moth and beetle larvae and eggs, which are taken as individuals burrow through soil. Although the morphological characteristics noted in the following section are remarkably similar to those of eutherian talpids and chrysochlorids (see chapter 12), marsupial moles burrow differently. They essentially "swim" through the ground. Substrate collapses behind them and they leave no permanent tunnels. They also spend much more time foraging on the surface than eutherian moles and are active both day and night. Little is known of the reproductive biology of *Notoryctes*.

Morphology. Like eutherian moles, the body of *Notoryctes* is **fusiform** (torpedo-shaped, compact, and tapered) and adapted for a fossorial existence. Maximum head and body length of adults is about 14 cm, with a stubby tail 2–3 cm long. Adults weigh about 60 g. The iridescent pelage is long and silky. Other fossorial adaptations include strong forelimbs with greatly enlarged, scooplike claws on the third and fourth digits (figure 11.38). There is a thick, keratinized nasal shield used to push dirt, and the cervical vertebrae are fused for added rigidity. There are no pinnae. Located under the skin, the vestigial eyes are 1 mm in diameter and have no lens; the optic nerve also is greatly reduced.

Epipubic bones occur in both sexes but are reduced in size. Dentition is variable, with 40–44 teeth. The occlusal surface of the molars is **zalambdodont** (V-shaped), which is unusual among marsupials. During the breeding season, females have a well-developed marsupium that opens pos-

teriorly. The testes are never scrotal but, according to Johnson (1995), lie between the skin and the abdominal wall.

Fossil History. Notoryctids are known from the Miocene of Queensland. Their affinities with other marsupial lineages remain unresolved.

Economics and Conservation. Marsupial moles are eaten by Australian aborigines. They capture them by following the distinctive trails made by *Notoryctes* when traveling on the surface. Neither species has any impact on grazing lands, and both are considered endangered.

Figure 11.38 External features of a marsupial mole. The marsupial mole (*Notoryctes typhlops*) has large forefeet, reduced eye, no pinnae, and a thick, leathery nasal shield. *Adapted from DeBlase and Martin, 1981, A Manual of Mammalogy, 2nd ed., Wm. C. Brown Publishers.*

SUMMARY

Reproductive characteristics, as well as distinctive morphological and anatomical features, serve to differentiate monotremes and marsupials, both from each other and from eutherian mammals. Monotremes (prototherians), which include the duck-billed platypus and short- and long-beaked echidnas, have a cloaca, a single opening for the fecal, urinary, and reproductive tracts. Reproduction also is distinctive in these, the only oviparous (egg-laying) mammals. Eggs are small, with a semipermeable shell that allows for nutrient exchange from mother to the developing young prior to the eggs being layed. As in all mammals, neonatal monotremes are nursed. Monotremes have mammary glands but no teats. Like many marsupials, echidnas have a pouch (marsupium) in which eggs are incubated and hatchlings are nursed. The platypus has no pouch. Monotremes exhibit several features characteristic of reptiles, including a lower body temperature than most other mammals, microchromosomes, threadlike sperm, and uncoiled cochlea. The pectoral girdle retains coracoid, precoracoid, and interclavicle bones, as did primitive therapsid reptiles, the early ancestors of mammals.

The semiaquatic platypus is noteworthy in other features, including a unique bill and an ankle spur connected to a poison gland. The bill is highly innervated and responds to both tactile and electrical stimuli. During dives the platypus detects aquatic prey by sensing the weak electric fields generated by muscle contractions. Unlike the platypus, echidnas are strictly terrestrial. The short-beaked echidna feeds on ants and termites; the long-billed echidna, on earthworms; both species have large, scooplike forefeet for foraging and burrowing. Prototherians may have diverged from therians 200 mya, but the relationship of these groups remains unresolved. Most fossil evidence of monotremes is from Australia, with one early Paleocene specimen from southern Argentina. Extant monotremes probably owe their continued survival through evolutionary time to relative geographic isolation from eutherians. Today, monotremes occupy ecological niches with little competition.

The 7 orders and approximately 270 species of marsupials (metatherians) also differ in many respects from eutherians. Marsupials are named for the marsupium, or pouch, which varies in complexity among species. Not all marsupials, however, have a marsupium. The uterus is duplex, with lateral vaginae on either side of a medial vaginal canal. Like eutherians, marsupials have a placenta, but without villi. Thus, there is poor nutrient exchange from mother to fetus, and a weak structural connection between the fetus and the uterine wall. Both factors contribute to a very short intrauterine gestation period. Neonates are tiny and undeveloped; total mass of litters is generally less than 1% of maternal body mass. Most marsupial maternal reproductive investment, in time and energy expenditure, occurs during a prolonged lactation period. Marsupials exhibit many skeletal and anatomical differences from eutherians, including a relatively small braincase and no corpus callosum, reduced or absent auditory bullae, palatal vacuities, articulation of the jugal bone with the dentary, an inflected angular process, epipubic bones, and different dental characteristics. Generally, marsupials also have more conservative life history characteristics. They have a lower basal metabolic rate than eutherians and a smaller range in body sizes and occupy a narrower range of ecological niches. For example, no marsupials occupy niches comparable to eutherian bats, whales, or fossorial herbivores. All large marsupial carnivores are extinct. The oldest fossil marsupials are from North America, which is probably where they arose. With the exception of a single species, however, marsupials occur today only in South America and Australasia.

SUGGESTED READINGS

Augee, M. L. (ed.). 1992. Platypus and echidnas, Royal Zoological Society, New South Wales, Sydney.

Jackson, S. 2003. Australian mammals: Biology and captive management. CSIRO Publishing, Collingwood, Australia.

Jones, M., C. Dickman, and M. Archer (eds.). 2003. Predators with pouches: The biology of carnivorous marsupials. CSIRO Publishing, Collingwood, Australia.

Martin, R. 2005. Tree-kangaroos of Australia and New Guinea. CSIRO Publishing, Collingwood, Australia.

Strahan, R. (ed.). 1995. The mammals of Australia. Reed Books, Chatsworth, Australia.

Tyndale-Biscoe, H. 2005. Life of marsupials. CSIRO Publishing, Collingwood, Australia.

Wells, R. T., and P. A. Pridmore (eds.). 1998. Wombats. Surrey Beatty and Sons, Sydney.

DISCUSSION QUESTIONS

1. Contrast the various reptilian features of monotremes with those of marsupials. What are the differences and similarities?

2. The platypus uses its bill for locating aquatic prey through electroreception. Why do you think this complex adaptation developed in place of simply searching visually for prey?

3. In species with an enclosed marsupium, pouch young spend prolonged periods breathing air with a very high carbon dioxide concentration. How do the developing neonates accommodate this "poisonous" environment during the lactation period?

4. What are the advantages in the marsupial mode of producing "expendable neonates" with a prolonged lactation period, compared with the eutherian mode of prolonged gestation with placental involvement? What are the disadvantages?

5. The suggestion is often made that marsupials are "inferior" to eutherian mammals. What evidence would you present to either support or refute this argument?

6. As noted in the text, the creation of the Panamanian land bridge about 3 mya allowed dispersal and interchange of South American metatherians and North American eutherians. Were the marsupials generally outcompeted (what evidence can you give), and if so, why?

Afrosoricida, Erinaceomorpha, Soricomorpha, Macroscelidea, Scandentia, and Dermoptera

CHAPTER 12

The six orders discussed in this chapter have a confusing and chaotic taxonomic history. In our second edition, Order Insectivora included six families: solenodons, tenrecs, golden moles, hedgehogs and gymnures, shrews, and moles and desmans; however, this order was recently split into the three orders: Afrosoricida, Erinaceomorpha, and Soricomorpha (Wilson and Reeder 2005). The last two orders are sometimes combined as Lipotyphla. The elephant shrews or sengis (Order Macroscelidea), treeshrews (Order Scandentia), and colugos (Order Dermoptera) also represent mammalian groups whose members were included in the Order Insectivora at one time. Recent advances in molecular techniques have yielded many taxonomic changes (sensu Wilson and Reeder 2005); we have incorporated these changes throughout this edition.

Afrosoricida, Erinaceomorpha, and Soricomorpha

For the sake of discussion, and recalling traditional nomenclature, we will periodically refer to members of the three orders of Afrosoricida, Erinaceomorpha, and Soricomorpha as the "insectivores." Six diverse families comprise this rather primitive mammalian assemblage: hedgehogs, moonrats, and gymnures (Erinaceidae); tenrecs and otter shrews (Tenrecidae); golden moles (Chrysochloridae); shrews (Soricidae); moles, desmans, and shrew-moles (Talpidae); and solenodons (Solenodontidae). The evolutionary relationships of "Order Insectivora" (historically placed in the Suborder Lipotyphla) is the subject of considerable debate and is commonly referred to as a "wastebasket" taxon for families of uncertain affinities (Butler 1972; MacPhee and Novacek 1993; McKenna and Bell 1997). Many of their morphological characteristics are considered ancestral and probably represent characteristics common to the earliest mammals. As such, "insectivores" probably are near the ancestral stocks of many other orders of eutherians that have advanced, or more recently derived, characteristics. Phylogenetic relationships among and within families remain unresolved, although molecular analyses are providing additional insights. Recent molecular phylogenies indicate that the six families now placed in Afrosoricida, Erinaceomorpha, and Soricomorpha show a polyphyletic origin (Murphy et al. 2001a,b; Symonds 2005).

MORPHOLOGY

Unlike most mammalian orders, no key character or set of characters serves to identify "insectivores." This is one of the most anatomically diverse mammalian groups; each family has interesting adaptations for survival. Members of this group are generally small to medium-sized, pentadactyl, with generalized plantigrade locomotion, and long, somewhat pointed snouts. Pelage of adults often is made up only of guard hairs, sometimes modified as spines, as in hedgehogs and tenrecs. The **pinnae** (external ear) and eyes usually are small or absent; the eyes of golden moles are nonfunctional and without external openings. Primitive characteristics include a small braincase and a brain with smooth cerebral hemispheres. A ring-shaped tympanic bone is present instead of auditory bullae, and the anterior vena cavae are paired. In males, the testes are usually abdominal or within the inguinal canal; if external, the scrotum is anterior to the penis. A cloaca is present in some genera. The jugal bone is reduced or absent, and the pubic symphysis is reduced.

Another ancestral "insectivore" characteristic is their dentition. As noted in chapter 4, some "insectivores" retain tribosphenic molars, including tenrecids, chrysochlorids, solenodontids, and some soricids. Teeth are rooted, so they do not grow throughout life. The deciduous teeth are shed early and are seldom functional. The molars have four or five cusps and usually form a V-shaped (zalambdodont) or W-shaped (**dilambdodont**) occlusal pattern (figure 12.1). In many species, the total number of teeth is often the same as the primitive eutherian pattern: 3/3, 1/1, 4/4, 3/3 = 44.

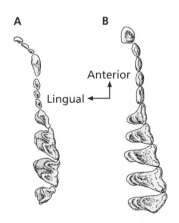

Figure 12.1 Representative occlusal surfaces in insectivores. (A) Occlusal surfaces of the left upper tooth row from a European mole (*Talpa europaea*) with a dilambdodont (W-shaped) cusp pattern. (B) Left upper tooth row from a giant otter shrew (*Potamogale velox*) with a zalambdodont (V-shaped) cusp pattern. *Adapted from A. Cabrera, 1925, Genera mammalium: Insectivora, Galeopithecia, Museo Nacional de Ciencias Naturales, Madrid, Spain.*

Given the morphological diversity within the currently recognized orders and the uncertain affinities of families, it is not surprising that Springer et al. (1997a) concluded that the order is not monophyletic. Using genetic analyses, they found the golden moles (see the subsequent section on Family Chrysochloridae) strongly associated with a group of taxa that includes elephant shrews, the aardvark (*Orycteropus afer*: Order Tubulidentata), and three orders discussed in chapter 19—elephants (Order Proboscidea), hyraxes (Order Hyracoidea), and sirenians. Similar relationships were reported by Mouchaty et al. (2000a), as well as by Stanhope et al. (1998) who proposed the new "Afrosoricida" order to include these taxa as well as the tenrecs (see the subsequent section on Family Tenrecidae and discussion by Bronner and Jenkins 2005).

FOSSIL HISTORY

Historically, we think of the fossil "insectivores" as a diversified assemblage with approximately 150 described genera. As noted by Butler (1972:254), this is because "any fossil eutherian not clearly related to one of the other orders is classifiable in the order Insectivora. The earliest fossil "insectivore,": the tiny, poorly known remains of *Batodon*, dates from the mid-Cretaceous period in North America, about 100 mya). The oldest members of clearly recognizable families—soricids and talpids—date from the Eocene epoch, about 50 mya (Harris 1998). Summaries of research focusing on paleontological relationships of New World and Old World shrews can be found in Wojcik and Wolsan (1998) and Merritt et al. (2005).

ECONOMICS AND CONSERVATION

Mammalian "insectivores" are of little economic importance today. Until the late 1800s, moles were trapped for their pelts, which were used for hats, apparel trim, and other purposes. Today, although they may damage lawns, fields, and gardens, moles have a relatively minor economic effect compared with rodents, for example. Hedgehogs, moles, and shrews take harmful insects as prey, but their effect is fairly negligible compared with insectivorous bats. Because many insectivores are found in tropical areas where habitats are often rapidly lost to logging and agriculture, several species are considered threatened or endangered (see the discussion of each family).

FAMILIES

Afrosoricida

Tenrecidae—Tenrecs and Otter Shrews. A great diversity of form and function of the Order Afrosoricida is reflected in the 30 species of the tenrecs and otter shrews (Family Tenrecidae—Bronner and Jenkins 2005). The 27 species of tenrecs are restricted to Madagascar, whereas the 3 species of otter-shrews are found in west-central Africa. Some authorities place otter-shrews in a separate family—the Potamogalidae. As suggested by their name, they are semiaquatic and closely resemble river otters (Carnivora: Mustelidae). The giant African water shrew (*Potamogale velox*) is the largest living insectivore, with a total length up to 640 mm. Using complete sequences of three mitochondrial DNA (mtDNA) genes, Stanhope et al. (1998) suggested that tenrecids (and golden moles) be placed in a new order, "Afrosoricidae," as noted previously. Recent work applying molecular methods to phylogenetic analyses has indicated that the Family Tenrecidae is part of a monophyletic African clade of mammals that represents one of four early eutherian radiations. This clade, the Superorder Afrotheria, includes the elephants, sea cows, hyraxes, aardvarks, elephant shrews, and golden moles (Springer et al 2004). Mouchaty et al (2000a) found that tenrecs were most closely related to the aardvark and African elephant (*Loxodonta africana*) based on analyses of 12 mitochondrial genes (figure 12.2).

The morphology of tenrecids defies a general description, as do their behavior and habitats. Rice tenrecs (Genus *Oryzorictes*) look like moles, are fossorial, and occur in marshy areas. Long-tailed tenrecs (Genus *Microgale*) resemble shrews and occupy thick vegetation and ground litter in a variety of habitat types. The web-footed tenrec (*Limnogale mergulus*) has a long, laterally flattened tail and webbed hind feet and looks like a small muskrat. It preys on fish, amphibians, and aquatic invertebrates in rivers, lakes, and marshes. This species is limited to stream habitat in eastern Madagascar and is active only at night (Benstead et al. 2001). The greater hedgehog tenrec (*Setifer setosus*) and

small Madagascar "hedgehogs" (*Echinops telfairi*) have sharp, barbed spines on the head, back, and sides like erinaceids (figure 12.3). Like hedgehogs, they have well-developed panniculus carnosus muscles and roll into a ball when threatened. Tenrecs have no auditory bullae and no jugal bone and, thus, an incomplete zygomatic arch (Cox 2006). Incisors and canines are usually small and unspecialized, and the upper molars are often zalambdodont.

Several species are heterothermic and enter torpor during the day or hibernate seasonally. Body temperatures generally are low, ranging from 30°–35°C while individuals are active. The streaked tenrec (*Hemicentetes semispinosus*) maintains a body temperature 1°C above ambient temperature while it hibernates during much of the winter (Stephenson and Racey 1994). Additionally, several species, such as the streaked tenrec and the long-tailed tenrecs, are believed to echolocate as part of their foraging activities. The common tenrec (*Tenrec ecaudatus*) has one of the largest litter sizes of any mammal, with as many as 32 young per litter (Eisenberg 1975). As with many mammalian groups in Madagascar, several species of tenrecs are reduced in density and distribution as a result of habitat loss or other factors. The web-footed tenrec (*Limnogale mergulus*) is endangered, as are several other species of tenrecs and otter-shrews.

Chrysochloridae—Golden Moles. The golden moles encompass 9 genera and 21 species (Bronner and Jenkins 2005). As noted, chrysochlorids appear to be more closely related to the aardvark and elephant shrews than to other "insectivore" families (Springer et al. 1997a; Stanhope et al. 1998; Emerson et al. 1999). Golden moles are distributed throughout central and southern Africa in forests, fields, and plains with soils suitable for burrowing. In these arid environments, free water is usually absent, and moles must exist without drinking water. Water requirements are greatly reduced by nocturnal foraging, torpidity, low metabolic rate, and very efficient kidney function (Seymour and Seely 1996; Fielden et al. 1990). Many of the same adaptations to enhance underground movement found in marsupial moles (Family Notoryctidae) and true moles (Family Talpidae) are seen in chrysochlorids—a good example of convergent evolution (figure 12.4). Golden moles have no pinnae; they have poorly developed eyes with fused eyelids covered with skin. The pelage moves equally well in any direction. It is a "metallic" or iridescent red, yellow, green, or bronze, depending on the species. A smooth, leather-like pad covers the nose, which golden moles use for pushing soil. The Namib desert golden mole (*Eremitalpa namibensis*) "swims" though the loose sand which immediately collapses behind it. Unlike talpids, the forelimbs of golden moles are under the body and do not rotate outward. They dig by forward extension of the limbs and scratching at the soil with large claws, especially those on the powerful third digit. For golden moles, as for talpids, burrow depth and construction depends on local soil characteristics (Seymour et al. 1998).

The principal prey of *E. namibensis* are dune termites (Genus *Psammotermes*) that reside in clumps of dune grass and ostrich grass found on scattered islands throughout

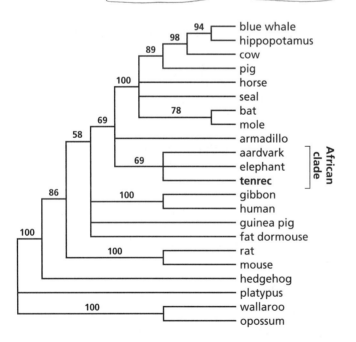

Figure 12.2 Phylogenetic position of tenrecs based on 12 mtDNA genes (from Mouchaty et al. 2000a). Numbers are bootstrap support values (500 replicates). Note the basal position of hedgehogs for eutherian taxa as well as the position of moles (Talpidae), another "insectivore" family.

A

B

C

Figure 12.3 Morphological diversity within the tenrecs. A great deal of morphological diversity is evident within the tenrecs, as seen in (A) a long-tailed tenrec (*Microgale dobsoni*); (B) a small Madagascar "hedgehog"; and (C) a giant otter shrew (*Potamogale velox*).

the vast sands of the Namib Desert. The moles make remarkably straight paths between clumps of grasses in search of food—sometimes moles cover 1400 m per night in search of prey (Fielden 1991) without visual aid. Recent work of Lewis et al. (2006) indicates that golden moles may use seismic cues to navigate, identify, and locate prey on or beneath the sand. Several authors have suggested that navigation is greatly enhanced in Namib and Cape golden moles (*E. namibensis* and *Chrysochloris asiatica*) by possessing disproportionately large auditory ossicles in their middle ear that serve as adaptations for detecting ground vibrations (Narins et al 1997; Mason and Narins 2002; Mason 2003; Willi et al 2006). The giant golden mole (*Chrysospalax trevelyani*) is endangered, as are several other chrysochlorids. Recent molecular methods applied to phylogenetic evidence indicates that golden moles are part of a monophyletic African clade of mammals (Superorder Afrotheria) that includes elephants, sea cows, hyraxes, aardvarks, elephant shrews, and the tenrecs mentioned above (Springer et al. 2004).

Erinaceomorpha

Erinaceidae—Hedgehogs and Gymnures. This family encompasses 10 genera and 24 species and includes hedge-

hogs, which have barbless spines on the back and sides, and gymnures, which lack spines (Hutterer 2005a). Erinaceids are an Old World family found in Africa, Europe, and Asia, including Sumatra, Borneo, and the Philippines. They inhabit many different habitats, including forests, grasslands, fields, and farmland. As their name suggests, desert hedgehogs (Genus *Paraechinus*) inhabit arid areas in North

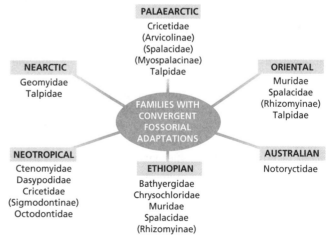

Figure 12.4 Fossorial mammals. Fossorial species from among several families occur in all faunal regions.

Africa, India, and Pakistan. Long-eared desert hedgehogs (Genus *Hemiechinus*) are found from northern Africa east to the Gobi Desert of Mongolia. Hedgehogs are mainly nocturnal and terrestrial, but some are semiarboreal. Adult *Hemiechinus auritus* weigh only 40–50 g, whereas the common European hedgehog (*Erinaceus europaeus*) reaches 1100 g. All are omnivorous, feeding on small vertebrates, eggs, fruits, and carrion in addition to invertebrates. Hedgehogs are well known for their ability to employ torpor as a mechanism for energy conservation in response to reduced energy availability and/or cold. Daily torpor may occur at any time of the year. Body temperature does not fall below 17°C, and the hedgehog may remain in a torpid state for periods of greater than 22 hours a time. During torpidity, their body temperature falls to within a few degrees of ambient (Geiser and Ruff 1995; Webb and Ellison 1998). See chapter 9 for information on dormancy in mammals. Basing their conclusion on mtDNA analyses, Mouchaty et al. (2000b:64) considered "the hedgehog as the most basal eutherian taxon." Murphy et al. (2001a), however, based on a much more extensive data set, did not support this conclusion.

A European hedgehog has approximately 5000 spines. Their defensive posture is typical for a mammal with spines or scales. When threatened, they roll into a tight ball with the spines directed outward (figure 12.5). This posture is aided by paired longitudinal "drawstring" muscles, the panniculus carnosus, on either side of the body. Hedgehogs have an interesting "self-anointing" behavior in which they spread large amounts of foamy saliva on the spines. This may be done as a sexual attractant during the breeding season, to reduce parasites, to clean the spines, or as additional protection from predators (Wroot 1984; Weldon 2004; D'havé et al. 2005). Several species, including the common European hedgehog, undergo true hibernation throughout the winter—the only insectivores that do. Desert hedgehogs commonly estivate (enter a dormant condition in the summer). Litter size is generally between four and six, and two litters a year may be produced. The young are altricial (immature). At birth, their short, soft spines have not yet broken through the skin, but the spines quickly grow in length after birth (figure 12.6) and harden within a few weeks.

Gymnures have no spines. They have omnivorous diets similar to those of hedgehogs and often are closely associated with **hydric** (wetland) habitats. The pelage is long and soft in the Philippines gymnure (*Podogymnura truei*) but very coarse and rough in the moonrat (*Echinosorex gymnurus*), which may reach 460 mm in head and body length and 2 kg in body mass. Moon rats are noted for their strong ammonia-like odor, which emanates from a pair of anal scent glands. The Philippines gymnure, Hainan gymnure (*Hylomys hainanensis*), and dinagat gymnure (*Podogymnura aureospinula*) are endangered.

Soricomorpha

Solenodontidae—Solenodons. The Family Solenodontidae is represented by one genus and four species

A

B

Figure 12.5 Hedgehog spines. (A) Lateral view of a long-eared hedgehog (*Hemiechinus auritus*). An adult hedgehog has about 5000 spines, each 2–3 cm long. The medulla of each spine is filled with air pockets to reduce weight. (B) The defensive posture of an African hedgehog (*Erinaceus frontalis*) is similar to that of armadillos and scaly anteaters.

(Hutterer 2005b). The family contains two extant species that are endangered and narrowly distributed: the Cuban solenodon (*Solenodon cubanus*) and the Haitian solenodon (*Solenodon paradoxus*) that resides in the Dominican Republic and Haiti. They are fossorial

Figure 12.6 Spines in a newborn hedgehog. Spines on these 2-day-old European hedgehogs (*Erinaceus europaeus*) erupt and harden after birth.

"insectivores" sheltering in caves, in crevices, and under logs; they construct extensive networks of tunnels reaching depths greater than 20 cm below ground (Eisenberg and Gozalez 1985). They inhabit forests of Cuba and Hispaniola to elevations of 2,000 m. Solenodons are one of the few native nonflying mammals that survived human settlement of the islands of the West Indies (MacPhee et al 1999). Solenodons are among the largest "insectivores," with total lengths approaching 600 mm. The pelage, naked tail, claws, and pinnae are all rather long, and the eyes are small. They have a pointed, highly flexible, and sensitive shrewlike snout used in capturing prey (figure 12.7). They inhabit rocky, brushy, or forested areas, move slowly in a zigzag path with a waddling gait, and are omnivorous. Solenodons emit high-frequency clicking sounds that probably function in echolocation. Along with

certain shrews and the platypus, they are among the few mammals that use a toxin. It is produced in the submaxillary glands, located at the base of the second lower incisor, which is large and deeply grooved to accommodate the toxic secretions (figure 12.8). Both species are considered endangered because of habitat loss and introduced predators, such as mongoose and feral cats, against which they are defenseless. Recovery of solenodon populations is also hampered by their low reproductive rate.

Combined gene sequences (13.9 kilobases) from *S. paradoxus* established that solenodons diverged from other lipotyphlan "insectivores" 76 mya in the Cretaceous period (Roca et al 2004). The described species in this family became extinct within the last several hundred years, with speculation that some species may have remained extant within the last 50 years. As noted by MacPhee et al. (1999:11), however, "there is no direct evidence at present that any species of *Nesophontes* even outlasted the close of the fifteenth century, let alone that of the nineteenth century."

Soricidae—Shrews. Shrews constitute the largest and most widely distributed family of insectivores. There are 26 genera and about 376 species, although many are of uncertain taxonomic status (Hutterer 2005b). This large family is placed in two subfamilies. The Subfamily Soricinae (the red-toothed shrews) includes three tribes found throughout much of the Nearctic, Palaearctic, and Oriental faunal regions (figure 12.11). The Subfamily Crocidurinae (the white-toothed shrews) is restricted to Old World faunal regions (see figure 12.9). Shrews are generally small. Body mass ranges from 3 g and 35 mm head and body length for adult pygmy white-toothed shrews (*Suncus etruscus*) and pygmy shrews (*Sorex hoyi*), two of the smallest mammals in the world, to 100 g and 150 mm head and body length for the musk shrew (*Suncus murinus*). Most species of shrews weigh 10–15 g and have a head and body length of about 50 mm. Legs are short, and the feet are unspecialized, except in the elegant water shrew (*Nectogale elegans*) and the North American water

Figure 12.7 Unusual insectivores—the solenodons. The small eyes and shrewlike snout are characteristic of the endangered Cuban solenodon. Prior to the arrival of Europeans, solenodons were among the largest predators within their restricted geographic area. They are quite defenseless against introduced mongooses and house cats, however. *Adapted from W. H. Flower and R. Lydekker, 1896,* An Introduction to the Study of Mammals Living and Extinct, *Adam and Charles Black.*

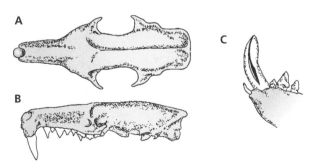

Figure 12.8 Use of toxin for prey capture by solenodons. (A) Dorsal and (B) lateral views of the skull of a Cuban solenodon. Note the incomplete zygomatic arch and enlarged first upper incisor. (C) The large second lower incisor is deeply grooved to accommodate secretions of toxin from the submaxillary gland below it. *Adapted from A. F. DeBlase and R. E. Martin, 1981,* A Manual of Mammalogy, *2nd ed., Wm. C. Brown Publishers.*

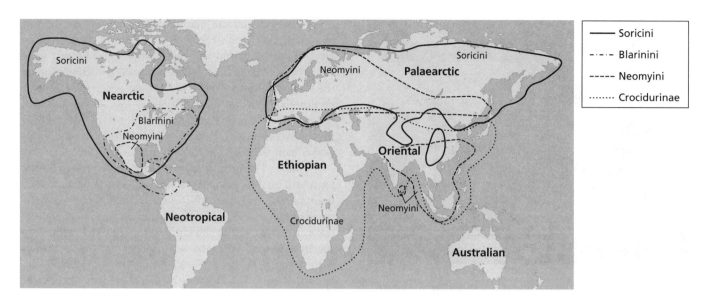

Figure 12.9 Worldwide distribution of shrews. Geographic distribution of red-toothed shrews (the Subfamily Soricinae, which includes three tribes: Neomyini, Soricini, and Blarinini) and the white-toothed shrews (Subfamily Crocidurinae). *Data from S. Churchfield, 1990,* The Natural History of Shrews, *Cornell Univ. Press.*

shrew (*Sorex palustris*), which are semiaquatic. Species associated with wet habitats have fimbriated hind feet. Shrews have small eyes and a long, pointed rostrum. Their pelage is short, dense, and usually dark-colored. In many species, lateral glands produce a musky odor that is most noticeable during the breeding season.

Shrew skulls have no zygomatic arch and no auditory bullae. The teeth are noteworthy in several ways. The first upper incisor is large, hooked, and has a posterior cusp that appears to be a distinct tooth (figure 12.10). The first lower incisor is long and procumbent (projecting horizontally forward). The upper molars are dilambdodont. The deciduous teeth are shed prior to parturition. In members of the Subfamily Soricinae, the tips of the teeth are a deep red (see figure 12.12). This color is caused by iron deposits (Dötsch and Koenigswald 1978) and is worn down and reduced as individuals age. Shrews with pigmented teeth are thus distinguished from the white-toothed shrews (Subfamily Crocidurinae), which do not have this characteristic. The groups also differ in biogeography, behavior, and physiology (Churchfield 1990; Merritt et al. 1994). Relationships among genera of African Crocidurinae were examined by Querouil et al. (2001, 2005), who used partial sequences of the mitochondrial 16S ribosomal RNA gene. Molecular phylogeny of Palearctic shrews from the Genera *Sorex*, *Crocidura*, *Neomys*, and *Suncus* were assessed using Restriction Fragment Length Polymorphism (RFLP) analysis of genome DNA and In Situ Polymerase Chain

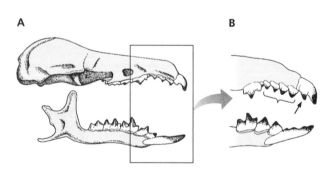

Figure 12.10 Characteristic features of shrew skulls. (A) A typical shrew skull. Note the lack of the auditory bullae and zygomatic arch. (B) The enlarged anterior portion of a red-toothed shrew showing the distinctively pigmented enamel, procumbent first lower incisor, secondary cusp on the first upper incisor (arrow), and unicuspids (bracket). The incisors function as tweezers picking up insect prey that are then passed to the sharp, multicusped posterior teeth for crushing and chewing. *Adapted from S. Churchfield, 1990,* The Natural History of Shrews, *Cornell Univ. Press.*

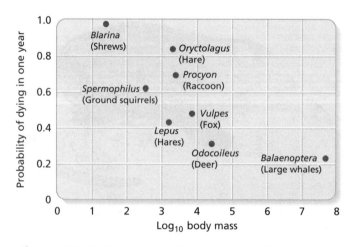

Figure 12.11 Mortality and body mass relationship. Inverse relationship between body size in a number of different mammals and the probability of dying within the first year of life. Note that the chance of a shrew such as *Blarina* living much beyond one year is very low. *Data from J. Eisenberg, 1981,* The Mammalian Radiations, *Univ. Chicago Press.*

A

B

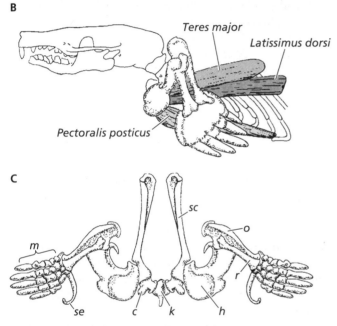

Teres major

Latissimus dorsi

Pectoralis posticus

C

sc

o

m

r

se   *c*   *k*   *h*

Figure 12.12 Morphological features of moles.
(A) The short hair, pointed snout, fusiform body shape, and large forepaws are evident in this eastern mole. (B) The massive forearm is supported and rotated by the teres major, latissimus dorsi, and pectoralis posticus muscles. (C) Forelimbs and highly modified pectoral girdle of the European mole. Note the large manus (*m*) with modified sesamoid bone (*se*), the small clavicle (*c*), keeled sternum (*k*), elongated scapula (*sc*), and elongated olecranon process (*o*) of the ulna. The humerus (*h*) is massive, is rectangular, and articulates with the radius (*r*). *Adapted from M. L. Gorman and R. D. Stone, 1990,* Natural History of Moles, *Cornell Univ. Press.*

Reaction (IS-PCR) methods (Bannikova and Kramerov 2005). This technique was useful in delineating two main lineages within the Palearctic Genus *Sorex* in addition to resolving other difficult phylogenetic questions arising from discordant morphological, karyological, biochemical, and mtDNA data associated with the phylogeny of shrews.

Shrews are mainly insectivorous, but many species are functional omnivores, consuming subterranean fungi. Shrews are found in many terrestrial habitats, usually with a heavy vegetative ground cover and abundance of invertebrates. This family is a good example of the ways body size constrains aspects of life history (Merritt 1995; Taylor 1998; McNab 2006). Shrews are often closely associated with water or moist habitats because they have a high respiratory water loss—a characteristic directly related to small size. The average life span of most shrews is only about a year (figure 12.11) because they have such high mass-specific metabolic rates, again associated with small size (Taylor 1998). Shrews are too small to hibernate or migrate, although some may enter torpor. As a result, most forage throughout the year, which exposes them to extremes of weather and predators, and affects life span. As noted by Merritt and Vessey (2000:235), "Daily activity patterns of shrews are controlled by metabolic requirements commensurate with their diminutive body mass and resultant high surface-to-mass ratios: they must forage often to avoid exhaustion of their energy stores. To remain homeothermic, shrews must partition a 24-h period into multiple bouts of foraging, rest, and sleep." Although several factors may affect activity in soricids, including feeding, photoperiod, precipitation, and temperature, few clear relationships are evident.

Many other aspects of these small insectivores are fascinating. Like solenodons, several species, including the short-tailed shrews and European water shrews (Genus *Neomys*), have salivary glands that secrete a toxin used to immobilize prey (See chapter 7) (Tomasi 1978; Churchfield 1990; Kita et al. 2004, 2005). Many species of soricine shrews use high-frequency sounds for interspecific communication, orientation, and prey detection (Tomasi 1979; Forsman and Malmquist 1988; Churchfield 1990). Echolocation in shrews is not as well-studied as in bats. Crocidurine shrews, and presumably soricine shrews as well, exhibit a behavior called "caravanning." Churchfield (1990:33) described this behavior, which involves a mother and her young, as follows: "Each youngster grasps the base of the tail of the preceding shrew so that the mother runs along with the young trailing in a line behind her." Numerous species of soricids are threatened or endangered.

Talpidae—Moles, Shrew-moles, and Desmans. Thirty-nine species of moles, desmans, and shrew-moles constitute this family; talpids are closely related to soricids (Mouchaty et al. 2000b; Hutterer 2005b). They are distributed throughout Europe, the Palaearctic region, and Asia, including Japan. In North America, they are found in southern Canada and most of the United States. The 17 genera of talpids are fairly diverse, ranging from moles, which are fossorial, to desmans, which are semiaquatic. Shrew-moles, the smallest talpids, are terrestrial. They are much more agile than moles, forage on the surface, and occasionally climb small shrubs. The star-nosed mole (*Condylura cristata*) of North America is unique among moles for its semiaquatic lifestyle and predilection for moist soils bordering streams and lakes. Star-nosed moles live rather considerably farther north than any other North American mole—it copes with cold encountered during foraging by possessing an elevated metabolic rate and high body temperature (Campbell et al. 1999).

The morphology of talpids reflects their fossorial life. The body shape is fusiform, with short, powerful limbs and short, smooth pelage. The pinnae are reduced or absent, and the eyes are minute. The sternum is keeled for enhanced pectoral muscle attachment, which is useful in digging. The forefeet are large and paddle-like with large claws. Because the radius articulates with the humerus, the forefeet are permanently rotated outward (figure 12.12) and produce the unique lateral digging motion of moles. The humerus is short and broad, again for enhanced muscle attachment, and articulates with both the clavicle and scapula to generate a great deal of force for digging. The olecranon process of the ulna also is very long for enhanced attachment of the triceps muscle. As a result, the digging muscles of the eastern mole (*Scalopus aquaticus*) can generate a force 32 times its own body mass (Arlton 1936; Hartman and Yates 2003).

The only evidence of moles that most people see is their tunneling activity and resulting molehills of excess excavated dirt. Burrowing is generally done in moist soils. Shallow tunnels 4–5 cm in diameter and a few centimeters below the surface are made when moles search for invertebrate prey. These are not as permanent as the complex, branching network of defended tunnels that extend to 150 cm deep and include the nest chamber (figure 12.13) (Gorman and Stone 1990; Hartman and Yates 2003). The Family Talpidae contains a diverse array of insectivores that exhibit some of the most unusual mechanosensory specializations found among Class Mammalia. Nearly all species of talpids possess small, domed mechanosensory organs called "Eimer's organs" located on their rhinarium (Catania 2000). With poorly developed eyesight and only a moderately developed sense of smell, the star-nosed mole (*C. cristata*) of eastern North America uses its impressive snout equipped with 22 projecting nasal appendages and long whiskers to search for prey below ground and in streams. We feature this fascinating mole and its equipment of "Eimer's organs" in chapter 7—feeding modes.

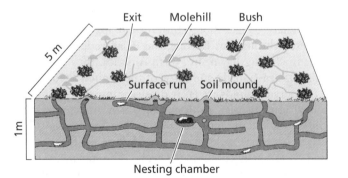

Figure 12.13 Mole burrow system. Example of depth and extent of burrow construction of a mole. *Data from A. V. Arlton, 1936, An Ecological Study of the Mole,* Journal of Mammalogy *17:349–371.*

Restricted to the Old World, desmans have a diet consisting of aquatic invertebrates and fish. The Russian desman (*Desmana moschata*) prefers ponds and marshes with thick vegetation. Conversely, the Pyrenean desman (*Galemys pyrenaicus*) inhabits streams and rivers with clear, fast-moving water. Head and body length is about 200 mm in the Russian desman and about 140 mm in the Pyrenean desman. The tail is laterally compressed in both species and equals the head and body length. The hind feet are webbed and fimbriated (having a fringe of hairs) for additional surface area. Both adaptations help desmans to swim rapidly. Their long, flexible snouts are very sensitive and are used to locate prey under water. Desmans echolocate to maneuver and locate prey (Richard 1973), like shrews and some tenrecs. Both species are considered vulnerable because of habitat loss, construction of dams and roads, and water pollution. Additionally, Russian desmans are harvested for their pelts. Pyrenean desmans are further reduced in number because of predation by introduced mink (*Neovison vison*). Several species of Asian moles and shrew-moles are endangered.

Macroscelidea

Macroscelididae—Elephant Shrews (= Sengis). The elephant shrews or sengis are in a single family (Macroscelididae), with 4 genera and 15 species (Schlitter 2005a). As noted previously, they are sometimes included in the Order Insectivora together with the treeshrews (Family Tupaiidae discussed in the next order) in the Suborder Menotyphla. Recent evidence of skeletal, cranial, and dental features, as well as molecular analyses, supports the current ordinal placement (Dene et al. 1980; Yates 1984). Recent molecular methods applied to phylogenetic evidence indicates that elephant shrews are part of a monophyletic African clade of mammals (Superorder Afrotheria) that includes elephants, sea cows, hyraxes, aardvarks, and golden moles and tenrecs mentioned previously (Springer et al. 2004).

Elephant shrews are restricted to central and eastern Africa from about 15°N latitude southward. An exception is the North African elephant shrew (*Elephantulus rozeti*), which occurs from Morocco to western Libya. Macroscelidids inhabit a diverse number of habitats, including desert, brushland, plains, forests, and rocky areas. They are strictly terrestrial, highly cursorial, and generally diurnal, except during hot weather, when they become increasingly nocturnal. Lovegrove et al. (2001) studied heterothermy in the North African elephant shrew and the eastern rock elephant shrew (*E. myurus*). During bouts of torpor, which never exceeded 24-hour periods, body temperatures declined to about 5°C and oxygen consumption was only 2% of the basal metabolic rate. Elephant shrews occurring in the southern African subregion (*Petrodromus tetradactylus, Elephantulus intufi,* and *E. brachyrhynchus*) reduce water and energy loss by avoiding extreme environmental temperatures; crepuscular activity and early morning sunbasking supplement heat gain in cold (Downs and Perrin 1995). Sengis feed on insects and other animal and plant material. Although quadrupedal, they move bipedally in an erratic fashion when alarmed. Sengis are easily alarmed: they are nervous animals and constantly twitch their ears and nose while making squeaking, chirping sounds. Although monogamy is found in less than 10% of mammals, all four genera of sengis are socially monogamous. Pairs of rufous sengis (*Elephantulus rufescens*) will construct and maintain a complex system of trails through the leaf litter—the males spend about 40% of their active daylight hours trail cleaning, compared to about 20% for female mates (Rathbun and Rathbun 2006). Scent-marking and sunbathing rolls are commonly performed on resting sites along trails. Sengis do not construct or use shelters or burrows but may temporarily use burrows of other animals (Rathbun and Redford 1981). Young sengis are highly precocial at birth—they will forage 1 day after birth (figure 12.14).

MORPHOLOGY

The common name of elephant shrews relates to their long, flexible, highly sensitive snout and large eyes and ears (figure 12.14). The pelage is long and soft. The hind legs are longer than the forelegs, which gives rise to the ordinal and family name ("long limbs"). The kangaroo-like hind legs allow for hopping—a bipedal form of locomotion when moving rapidly. Forelimbs are pentadactyl, whereas hind limbs have either four or five toes. Size varies: Head and body length and mass range from 95 mm and a maximum 50 g, respectively, in the short-eared elephant shrew (*Macroscelides proboscideus*) to 315 mm and about 400 g, respectively, in golden-rumped elephant shrews (Genus *Rhynchocyon*). Tail length is slightly shorter than head and body length. The skull has a complete zygomatic arch and auditory bullae, and the palate is noteworthy for its series of large openings (figure 12.15). All genera have functional incisors except *Rhynchocyon*. The molars in elephant shrews are quadrituberculate (four-cusped), and the occlusal surfaces are dilambdodont. As in treeshrews, a cecum is present.

FOSSIL HISTORY

The sengis represent a monophyletic radiation endemic to Africa. Just as all extant elephant shrews are endemic to Africa, all known fossil representatives of the order also are from Africa. Tabuce and colleagues (2001a) described an elephant shrew (*Nementchatherium senarhense*) from the Eocene epoch of Algeria. Other early elephant shrews include *Chambius kasserinensis* from the early to mid Eocene of Tunisia and *Herodotius pattersoni* from the late

Figure 12.14 Body morphology of elephant shrews (= sengis). (A). An adult rufous sengi (*Elephantulus rufescens*). Note the long proboscis, enlarged hind limbs, and kangaroo-like appearance of this rufous sengi. Convergence is evident in body form of sengis and several families of rodents, such as kangaroo rats (Rodentia: Heteromyidae), shown in figure 18.10, or jerboas (Rodentia: Dipodidae), in figure 18.15. (B). A 1-day-old rufous sengi (*E. rufescens*) illustrating the precocial nature of young.

Eocene of Egypt. Another Paleogene Macroscelidean fossil is *Metoldobotes stromeri*, which dates from the Oligocene epoch. Later fossil macroscelidids are quite variable structurally: Genus *Myohyrax* from the Miocene is similar to a hyrax (Hyracoidea: Procaviidae), and a Pliocene form (Genus *Mylomygale*) is similar to a rodent (Patterson 1965).

ECONOMICS AND CONSERVATION

The golden-rumped elephant shrew (*Rhynchocyon chrysopygus*) and the four-toed elephant shrew (*Petrodromus tetradactylus*) are harvested for meat in Kenya. The rufous elephant shrew (*Elephantulus rufescens*) is known to contract naturally occurring malaria and, with other species

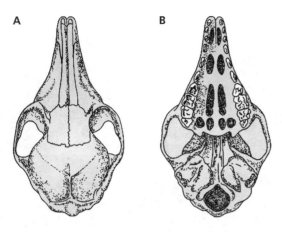

Figure 12.15 Elephant shrew (= sengi) skull characteristics. (A) Dorsal and (B) ventral views of the skull of the North African elephant shrew, showing the complete zygomatic arch, large auditory bullae, and numerous perforations in the palate. *Adapted from A. F. DeBlase and R. E. Martin, 1981, A Manual of Mammalogy, 2nd ed., Wm. C. Brown Publishers.*

in the genus, is used in medical research. Although an interesting group, elephant shrews are of little economic importance. Golden-rumped elephant shrews, black-and-rufous elephant shrews (*R. petersi*), and Somali elephant shrews (*E. revoili*) are endangered.

Scandentia

The Order Scandentia is represented by two families (Tupaiidae and Ptilocercidae), containing 4 genera and 19 species (Tupaiidae—treeshrews) and 1 genus and 1 species (Ptilocercidae—pen-tailed treeshrews) (Helgen 2005). Because they have long been considered the most primitive living primates, treeshrews have generated debate and controversy out of proportion to their size as a group. As such, they probably have a richer and more confused taxonomic history than any other mammalian family. An excellent review summarizing dental, skeletal, and anatomical evidence against considering treeshrews as primates is given by Campbell (1974). Anatomy and related systematics of the group are discussed in detail by Luckett (1980a). Treeshrews show a close relationship with primates (Sargis 2004) and demonstrate unique behaviorial traits of absentee maternal care and social monogamy (Emmons 2000). Most authorities, however, consider treeshrews to be a distinct lineage separate from either the primates or insectivores.

Treeshrews are restricted to the Oriental faunal region, ranging from India, southern China, and the Philippines southward through Borneo and the Indonesian islands. Throughout their range, treeshrews occur in forested habitats up to an elevation of 2400 m. They are diurnal, except for the pen-tailed treeshrew, and omnivorous,

feeding on different fruits and invertebrates. Species of treeshrews studied in Malaysia were primarily frugivorous concentrating on small, soft, avian-dispersed fruits. Although chiefly insectivorous, treeshrews may consume fruit to add extra calories or nutrients such as calcium to their high-protein diet (Emmons 1991). Vocalizations are limited, but all species scent-mark extensively. The common name is unfortunate because tupaiids are certainly not shrews. Also, the larger species such as the Philippines treeshrew (*Urogale everetti*), confined to the island of Mindanao, and the terrestrial treeshrew (*Tupaia [Lyonogale] tana*), of Borneo and Sumatra, spend most of their time on the ground.

MORPHOLOGY

Treeshrews superficially resemble squirrels, although they have a more slender snout (figure 12.16). Emmons (2000:17) noted that "The diurnal treeshrews have virtually no salient facial vibrissae, while the nocturnal *P. lowii* has a large spray of them." The limbs are equal in length, pentadactyl, and have long claws. Maximum total length is about 450 mm, half of which is the tail. Several characteristics of treeshrews resemble those of primates, including their large braincase, postorbital bar, scrotal testes, and structure of the carotid and subclavian arteries.

Figure 12.16 A typical tupaiid. This common treeshrew (*Tupaia glis*) superficially resembles a squirrel. Tupaiids have had a confused taxonomic history—at various times being placed in the Orders Primates and Insectivora, and currently within their own order, Scandentia. *Adapted from A. Cabrera, 1925, Genera Mammalium: Insectivora, Galeopithecia, Museo Nacional de Ciencias Naturales, Madrid, Spain.*

Unlike elephant shrews or sengis, with which they are sometimes grouped, treeshrews have tribosphenic molars and an unperforated palate. Like elephant shrews and several families of "insectivores," the occlusal surface of the upper molars is dilambdodont. The dental formula is 2/3, 1/1, 3/3, 3/3 = 38. The lower incisors are procumbent and used for grooming. Sargis (2001) discussed differences in the axial skeleton of the pen-tailed treeshrew compared to other tupaiids, and felt *P. lowii* was primitive to the more derived tupaiines.

FOSSIL HISTORY

The only definite fossil treeshrew (*Palaeotupaia sivalicus*) was described from mid-Miocene deposits in India (Chopra and Vasishat 1979). None of the previous reports of fossils referable to treeshrews, summarized by Jacobs (1980), provides unequivocal association with the family, however. Although not suggesting any phylogenetic relationship, Emmons (2000) noted the striking resemblance of treeshrews to Genus *Crusafontia*, a Jurassic period pantothere.

ECONOMICS AND CONSERVATION

Because of the continued loss of forest habitat throughout their range, six species are considered endangered or threatened. They do little damage to crops or plantations and are of no economic significance, although they may play a role in seed dispersal (Shanahan and Compton 2000).

Dermoptera

Cynocephalidae—Colugos. Order Dermoptera is represented by a single family (Cynocephalidae) containing two genera and two species (Stafford 2005). Dermopterans (literally, "skin-winged") commonly are called "flying lemurs" or colugos. The latter name certainly is preferable, however, because dermopterans do not fly and are not lemurs. Instead, they glide (**glissant**). Like the treeshrews, this order also has a confusing taxonomic history. Historically, colugos have been grouped taxonomically with the bats, "insectivores," and primates: McKenna and Bell (1997) considered dermopterans a suborder of the primates. This small order now contains a single family, Cynocephalidae, which has two genera (*Cynocephalus* and *Galeopterus*) each with one species. Using dental morphology, Stafford and Szalay (2000) argued that two genera should be recognized—the Philippine flying lemur (*C. volans*) and Sunda flying lemur (*G. variegatus*). In this text, we retain the traditional nomenclature, however.

The Philippine flying lemur is found in the southern Philippines, whereas the Sunda flying lemur occurs in Indochina, Malaya, Sumatra, Java, Borneo, and small islands nearby. These parapatric species are completely arboreal and inhabit lowland and upland forests and plantations. Colugos are primarily nocturnal, foraging on flowers, leaves, and fruits. Wischusen and Richmond (1998) reported that the mean duration of foraging bouts in *C. volans* was 9.4 minutes with 12 bouts per night being typical. Unlike other arboreal folivores—such as the koala (*Phascolarctos cinereus*) or three-toed sloth (*Bradypus variegatus*)—*C. volans* is a generalized forager, feeding on young leaves from many different tree species. Colugos den in tree cavities or hang upside down from branches during the day (with their head remaining upright, unlike bats). They move among trees very efficiently; colugos are the premier gliding mammals, and their morphological adaptations for this means of locomotion are pronounced.

MORPHOLOGY

The patagium of colugos is more extensive than in any of the gliding marsupials or rodents. It extends from the neck to the digits of the forelimbs, along the sides of the body and hind limbs, and encloses the tail (figure 12.17). Colugos weigh between 1 and 2 kg, with head and body length 340–400 mm and tail length 170–270 mm. *Cynocephalus variegatus* is the larger species. Despite the colugos' size, glide distances of over 100 m are common, with very shallow glide angles. Thus, while gliding 100 m, a colugo loses less than 10 m in elevation. Colugos are helpless on the ground but adept at climbing high into trees, aided by long, curved claws. Their brownish gray-and-white, mottled pelage camouflages them against tree trunks. Colugos have a keeled sternum, as do bats and other gliding species.

The dental formula is 2/3, 1/1, 2/2, 3/3 = 34. The first two lower incisors are procumbent and **pectinate**

Figure 12.17 Characteristic external features of colugos. The extent of the patagium is evident in this colugo.

("comblike" with 5–20 distinct prongs from a single root), a feature unique to dermopterans. These lower incisors are used to grate food and groom the fur. The upper incisors are small with distinct spaces between them, and the postorbital processes and temporal ridges are well developed (figure 12.18). Stafford and Szalay (2000) provided a detailed analysis of the craniodental morphology of the two species.

FOSSIL HISTORY

The extinct Family Plagiomenidae is placed in the Order Dermoptera and dates from the late Paleocene epoch in North America and early Eocene epoch in Europe. Marivaux et al. (2006) described newly discovered fossil dermopterans from different regions of South Asia (Thailand, Myanmar, and Pakistan) ranging from the late middle Eocene to the late Oligocene.

ECONOMICS AND CONSERVATION

Colugos are hunted for their fur and for food, but habitat loss to logging and farming is a greater threat to populations. *C. volans* is presently considered threatened.

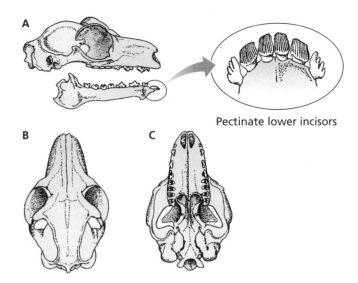

Figure 12.18 Colugo skull features. (A) Lateral, (B) dorsal, and (C) ventral views of the skull of a colugo with evident postorbital processes and temporal ridges. (Inset) Pectinate lower incisors, a characteristic feature of the two species in this unusual order. *Adapted from T. E. Lawlor, 1979,* Handbook to the Orders and Families of Living Mammals, *2nd ed., Mad River Press.*

SUMMARY

The Afrosoricida, Erinaceomorpha, Soricomorpha, Macroscelidea, Scandentia, and Dermoptera represent groups rich in diversity of form and function. The Afrosoricida, Erinaceomorpha, and Soricomorpha (formerly placed in Order Insectivora) include families that demonstrate convergent evolution—the talpids and chrysochlorids. The "insectivores" retain many primitive eutherian characteristics, including their brain anatomy, dentition, cranial morphology, postcranial structures, and cloaca. As such, they are thought to be ancestral to other mammalian orders. In contrast to their primitive characteristics, many "insectivores" have

developed highly specialized features, such as echolocation (tenrecs and shrews) and toxin in the saliva (solenodons and some shrews). Unlike many other orders, no key morphological character serves to identify the Afrosoricida, Erinaceomorpha, and Soricomorpha.

Elephant shrews (sengis), treeshrews, and colugos were also formerly included within the Order Insectivora. Each group is now placed in its own order, the Macroscelidea, Scandentia, and Dermoptera, respectively. The taxonomic history of these orders has been a mystery since the early 1800s and is one that is still being resolved today.

SUGGESTED READINGS

Churchfield, S. 1990. The natural history of shrews. Comstock Pub. Assoc., Ithaca, NY.

Crowcroft, W. P. 1957. The life of the shrew. Max Reinhardt, London.

Feldhamer, G. A., B. C. Thompson, and J. A. Chapman. 2003. Wild mammals of North America: Biology, management, and conservation, 2nd ed. Johns Hopkins Univ. Press, Baltimore.

George, S. B. 1986. Evolution and historical biogeography of soricine shrews. *Syst. Zool.* 35:153–162.

Gorman, M. L., and R. D. Stone. 1990. The natural history of moles. Comstock Pub. Assoc., Ithaca, NY.

Hartman, G. D., and T. L. Yates. 2003. Moles, Pp. 30-55, *in* Wild mammals of North America: Biology, management, and conservation (G. A. Feldhamer, B. C. Thompson, and J. A. Chapman, eds.), 2nd ed., Johns Hopkins Univ. Press, Baltimore.

Macdonald, D., and S. Norris (eds.). 2001. The new encyclopedia of mammals. Oxford Univ. Press, Oxford, UK.

Merritt, J. F., S. Churchfield, R. Hutterer, and B. I. Sheftel (eds.). 2005. Advances in the biology of shrews II. Special Publication of the International Society of Shrew Biologists 01, New York.

Nicoll, M. E., and G. B. Rathbun. 1990. African Insectivora and elephant-shrews: An action plan for their conservation. IUCN/SSC Specialist Group, IUCN, Gland, Switzerland.

Nowak, R. M. 1999. Walker's mammals of the world, 6th ed. Johns Hopkins Univ. Press, Baltimore.

Wojcik, J. M., and M. Wolsan (eds.). 1998. Evolution of shrews. Mammal Research Institute, Polish Academy of Sciences, Bialowieza.

DISCUSSION QUESTIONS

1. What do you think might be the functional significance of the pigmented teeth found in the soricid Subfamily Soricinae—the red-toothed shrews? Likewise, what is the adaptive significance of caravanning behavior in shrews?

2. The text discusses how body size influences the life history characteristics of shrews. What other factors would you expect also are affected by small body size? How does latitude enter the picture?

3. How does body size influence the available range of prey?

4. Why have so few species of mammals developed the use of toxins as part of their life history strategies?

5. The common tenrec may have the largest litters of any mammalian species. Given its extremely large numbers of young, what might you expect are its neonatal survival rate, dispersal of young from the natal area, and associated factors?

CHAPTER 13

Chiroptera

Bats are second only to rodents in the number of recognized species in the order. There currently are 18 recognized families encompassing 202 genera and 1,116 species of bats (Simmons 2005a). They are one of the more fascinating groups of mammals for both the general public and professional mammalogists. In terms of feeding, reproduction, behavior, and morphology, including structural adaptations for true flight, bats show a greater degree of specialization than any other mammalian order. Bats also exhibit a great range in body size, from the tiny hog-nosed bat (*Craseonycteris thonglongyai*) in Thailand, which weighs about 2 g, to fruit bats such as the large flying fox (*Pteropus vampyrus*) that weigh up to 1,200 g. The vast majority of bats, however, are relatively small, weighing from 10–100 g. This chapter examines the structural and functional adaptations of bats, as well as the continuing debate about their evolutionary history and the relationships among families.

Flight and echolocation have allowed bats to be widely distributed and fill many feeding niches. Most species are insectivorous, often taking insects in flight or gleaning them from foliage. Others are carnivorous, taking small vertebrates, such as frogs, mice, and occasionally other bats. A few species (especially the Family Noctilionidae) feed on fish (they are piscivorous). Many species are nectivorous, consuming pollen and nectar from flowers, and the fruit bats are frugivorous. Finally, the well-known vampire bats, which feed only on blood (and have contributed to many popular misconceptions), are sanguinivorous. Different feeding adaptations may occur within a given family. For example, all these feeding modes are seen among species of New World leaf-nosed bats (Family Phyllostomidae). These different feeding habits and foraging styles among bat species result in a greater variety of head shapes, dentition, and facial features than occurs in other mammalian orders.

Variation is also evident in the reproductive patterns of bats. Many species breed in the spring and exhibit the typical pattern of spontaneous ovulation (see chapter 10). Others, however, such as some of the common or vesper bats (family Vespertilionidae) and horseshoe bats (Family Rhinolophidae) in northern areas, exhibit **delayed fertilization**, in which mating occurs in the fall prior to migration or hibernation. Sperm is stored in the uterine tract, and ovulation and subsequent fertilization occur in spring. **Parturition** (birth) coincides with the emergence of insect prey. Other reproductive variations discussed in chapter 10 occur in different species of bats, including delayed implantation and delayed development (Racey 1982; Ransome 1990; Kunz and Pierson 1994).

As noted by Storz et al. (2000:152) "Female gregariousness at diurnal roosting sites has apparently facilitated polygynous mating systems in an ecologically diverse array of species." Mating systems that occur in various families of bats were placed in several subdivisions of the following three general categories by McCracken and Wilkinson (2000): (1) single male/multifemale groups, (2) multimale/multifemale groups, and (3) single male/single female monogamous groups. Ecological factors that affect these patterns include either male defense of females themselves, or male defense of the resources that females depend on.

As described later in the chapter, bats roost in many different places: in caves and hollow trees, under loose bark, in buildings, in understory vegetation and rock fissures, and in many other protected places. Some species roost alone, whereas others form aggregations that can vary from small to huge. The roosting habits of bats are adaptations that reflect the interrelationships of social structure, diet, flight behavior, predation risks, and reproduction of each species (Kunz 1982). Most species are strictly nocturnal (Speakman 1995), although diurnal activity is common in species of island-dwelling Old World fruit bats (Family Pteropodidae) and a few species of microchiropterans (Kunz and Pierson 1994).

Small mammals usually have several large litters per breeding season. Bats are unusual, however, in their limited reproductive potential. Although there are exceptions (usually in tropical species) bats most often produce a single litter per year, with only one or two young per litter. Thus, declining bat populations may be slow to rebound. Neonates typically weigh 20–30% of the maternal body mass (Kurta and Kunz 1987). Hayssen and Kunz (1996) examined the relationship between litter mass and several maternal morphological characteristics in over 400 species of bats from 16 families. They found different relationships in pteropodids from those occurring in other (microchiropteran) bats, suggesting different selection pressures on reproductive traits in the two suborders (see next section). Bats also have much longer life spans than is typical for small mammals, sometimes exceeding 30 years. In some species, prolonged life spans may be a function of reduced metabolic rate on both daily (torpor during the day) and seasonal (hibernation in winter) bases.

Because of the role bats fill as the major consumer of night-flying insects and as pollinators of plants, chiropterans are important in community structure and in economics. Despite this, and their ability to adapt to various environments, bat populations often are negatively affected

---

Table 13.1 Major differences between megachiropterans, the Old World fruit bats (Family Pteropodidae), and microchiropterans, the remaining 17 families of bats*

Microchiropterans	Megachiropterans
Echolocation for foraging and maneuvering; primarily insectivores	No echolocation, except in the genus *Rousettus*; primarily frugivores or nectivores
Tragus often well developed	No tragus; continuous inner ear margin
Nose or facial ornamentation often evident	No nose or facial ornamentation
No claw on second digit; second finger closely associated with third	Claw evident on second digit except in genera *Dobsonia, Eonycteris, Notopteris,* and *Neopterx*; second finger independent
Cervical vertebrae modified; head is flexed dorsally from main axis of body when roosting	Cervical vertebrae not modified; head held ventrally when roosting
Tail and tail membrane (uropatagium) often evident	Tail and uropatagium usually absent
Generally small body size; eyes generally small	Generally large body size; eyes usually large
Mandible with well-developed, long, narrow, angular process	Angular process of mandible absent, or broad and low if present
Postorbital process usually absent	Postorbital process well developed
Palate usually does not extend beyond last upper molars	Palate extends beyond last upper molars

* Additional differences are detailed by Koopman (1984) and Simmons (1995a).

by environmental perturbations, both natural and human-induced. As a result, many species throughout the world are in danger of extinction through loss of cave and riparian (riverbank) habitats, exposure to pesticides, and human exploitation.

Morphology

We divide the 18 families of bats described in this chapter into two suborders: the Megachiroptera and the Microchiroptera—although see the discussion under the Suborders and Families section later in this chapter for an alternative arrangement. Megachiropterans include a single family (Pteropodidae), the Old World fruit bats. The other 17 families of bats make up the Suborder Microchiroptera. As you might expect, given the names of these suborders, size is one of the obvious differences between the two groups (Hutcheon and Garland 2004). There is considerable overlap in size, however, because the largest microchiropterans are much larger than the smallest megachiropterans. Several other morphological differences between these suborders are given in table 13.1.

Bats are the only mammals that fly; consequently, many of their unique morphological features relate to flight. The ordinal name Chiroptera, derived from the Greek *cheir* (hand) and *pteron* (wing), refers to the modification of the bones of the hand into a wing, the primary adaptation for flight in bats (figure 13.1). The wing is formed from skin stretched between the arm, wrist, and

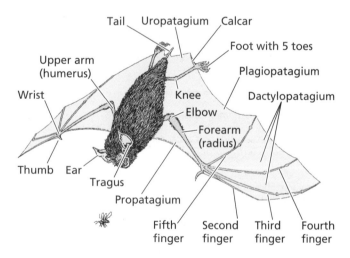

Figure 13.1 The major external features of bats.
Note the humerus, elongated forearm (radius), and fingers that form the wing. The claw on the thumb (digit one) occurs in all families, although it is rudimentary in smoky bats (Furipteridae) and short-tailed bats (Myzopodidae). Not all species have an external tail and uropatagium (see figure 13.2A). *Adapted from D. Macdonald,* Encyclopedia of Mammals, *Facts on File.*

finger bones. Several different muscle groups are used to keep the skin taut over the wing (Vaughan 1970a). Although the skin on the wings is very thin and appears delicate, it is fairly resistant to tears or punctures. The primary modification of the forelimb (wing) is its elongation, especially the forearm (radius and ulna), metacarpals, and fingers. The radius is greatly enlarged in bats—sometimes being twice as long as the humerus—and the ulna is very much reduced. Because of the demands of flight, movement is restricted to a single plane in the wrist, elbow, and knee joints of bats. The radius cannot rotate (as it does in humans), and the wrist (**carpal** bones) moves only forward and backward (flexion and extension). Thus, the wing gains strength and rigidity to withstand air pressures associated with flight. Wing membranes usually attach along the sides of the body in bats, although there are exceptions (see the section on the Family Mormoopidae in this chapter, for example).

The pectoral area also is highly developed and modified and contributes to a bat's ability to fly. The last cervical and first two thoracic vertebrae often are fused. Along with a T-shaped **manubrium** (the anterior portion of the sternum, or breast bone), the first two ribs form a strong, rigid pectoral "ring" to anchor the wings. In conjunction with the pectoral ring, the articulating **scapula** (shoulder blade) and proximal end of the humerus are highly modified for flight (Vaughan 1970b). Besides their primary function in flight, wings also aid bats in thermoregulation. The thin, vascularized wing membranes dissipate excess body heat generated during flight. Additional aspects of the wings will be considered when we examine the mechanics of how bats fly.

In many species, the **uropatagium** (the membrane between the hind limbs that encloses the tail—see figure 13.1) also aids bats in flying. Although not necessary for flight (some species have no uropatagium), this membrane may contribute to lift and help stabilize the body during turns and other maneuvers. Aerodynamic stability also is enhanced in bats by their body mass being concentrated close to their center of gravity. The size and shape of the uropatagium, and whether it completely encloses the tail, varies considerably among and within families (figure 13.2A). The **calcar**, a cartilaginous process that extends from the ankle, helps support the uropatagium.

The hind limbs of bats are small relative to the wings and are unique among mammals in being rotated 180° so that the knees point backward. This aids in various flight maneuvers, as well as in the characteristic upside-down roosting posture of bats. Bats roost head-down, hanging by the claws of the toes. A special locking tendon (Quinn and Baumel 1993; Simmons and Quinn 1994) allows them to cling to surfaces without expending energy.

As noted previously, the head and facial features of bats exhibit a great deal of diversity. Facial ornamentation, in the form of fleshy nose leafs, are conspicuous in New World leaf-nosed bats, horseshoe bats, the false vampire bats (Family Megadermatidae), and slit-faced

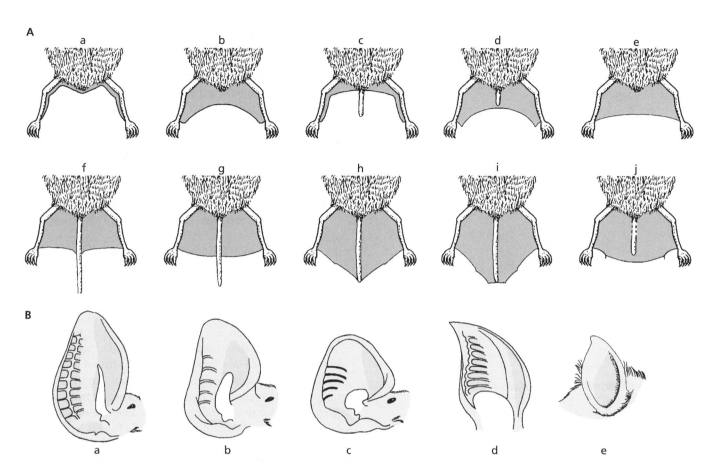

Figure 13.2 Variation in the uropatagium, tail, external ear, and tragus of representative families of bats. (A) (a) Fruit bats (Pteropodidae) in the genus *Pteropus*, as well as some New World leaf-nosed bats (Phyllostomidae); (b) New World leaf-nosed bats; (c) tube-nosed fruit bats in the genera *Nyctimene* and *Paranyctimene* (Pteropodidae); (d) New World leaf-nosed bats; (e) hog-nosed bat (Craseonycteridae), with extensive uropatagium but no external tail; (f) mouse-tailed bats (Rhinopomatidae) with long, thin tail free of the uropatagium; (g) free-tailed bats (Molossidae) with about one-half the tail free of the uropatagium; (h) common bats (Vespertilionidae), similar to tails in horseshoe bats (Rhinolophidae) and false vampire bats (Megadermatidae); (i) slit-faced bats (Nycteridae) with T-shaped tip of the tail; (j) sac-winged bats (Emballonuridae), fishing bats (Noctilionidae), and moustached bats (Mormoopidae) have a tail that protrudes through the uropatagium. (B) Representative sizes and shapes of the external ears and tragus in bats. (a–c) Variation in size and shape of the ear and tragus in vespertilionids; (d) antitragus in a rhinolophid; (e) continuous inner ear margin without tragus or antitragus in a pteropodid. *Adapted from Hill and Smith, 1984,* Bats: A Natural History, *British Museum of Natural History.*

bats (Family Nycteridae). Such ornamentation functions in the transmission of echolocation pulses because species in these families emit ultrasonic pulses through their nostrils.

"Blind-as-a-bat" is a totally misleading phrase. All bats have perfectly functional vision. Eyes are large in megachiropterans but relatively small in microchiropterans. In microchiropterans, however, most of which are nocturnal, visual acuity is secondary. Primary perception of the environment is through acoustic orientation, both through echolocation (see section on echolocation) and audible vocalizations important in various social interactions. The importance of hearing to microchiropterans is reflected in the tremendous variety evident in the size and shape of the pinnae (external ears) among species (Obrist et al. 1993; see figure 13.2B). Most microchiropterans have a **tragus**, a projection from the lower margin of the pinnae (see

figure 13.2B), that is important in echolocation. Some species also may have an **antitragus**, a small, fleshy process at the base of the pinnae (see figure 13.2B). An antitragus is especially evident in groups in which the tragus is absent or reduced, as in the horseshoe bats and free-tailed bats (Family Molossidae).

HOW BATS FLY

All aspects of the biology and natural history of bats are associated with their ability to fly. Compared with birds, flight in bats is slow but highly maneuverable. To comprehend how a bat flies, it is necessary to understand a few basic aerodynamic terms (see Norberg 1990 for detailed analyses of flight in vertebrates). A bat (like a bird or an

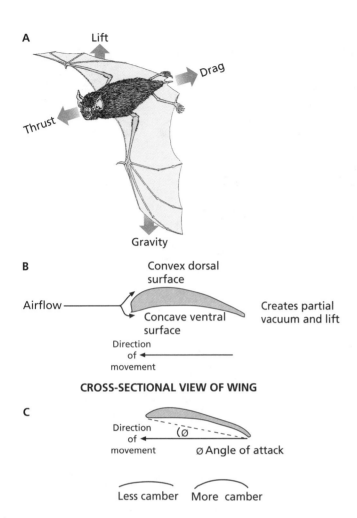

Figure 13.3 Aerodynamics of flight in bats. (A) In order to fly, the power producing forward thrust must overcome "backward" drag, while upward lift must be greater than the effects of "downward" gravity. Drag increases in proportion to the surface area of the wings and air speed, as well as increased camber and angle of attack. (B) Lift is created as air moves over the convex top of the wing as a bat moves forward. It has a greater distance to travel relative to the concave bottom of the wing. This produces less pressure above the wing and results in upward lift. A decrease in pressure with increased velocity of the air (or fluid) is known as Bernoulli's Principle. (C) The angle of attack is the angle of the wings relative to the horizontal plane in the direction of movement; camber is the degree of curvature of the wing.

airplane) is able to fly because it generates enough **lift** to overcome gravity and sufficient forward propulsive thrust to overcome **drag** (figure 13.3A). Lift and thrust are achieved in bats through the structure of the wings, in conjunction with the pectoral bones and muscles, the uropatagium, and hind limbs.

The dorsal surface of the wing in bats as in airplanes is convex and the ventral surface is concave, which causes air to move more rapidly over the wing than underneath it. This reduces the relative air pressure above the wing and results in lift (see figure 13.3B). In addition, to facilitate airflow, the surface of the wing is kept smooth and taut by

layered elastic tissue in the wing. Generally, the greater the **camber** (the extent of the front-to-back curvature of the wing), the more lift that can be produced. Likewise, the greater the **angle of attack** (see figure 13.3C), the more lift that can be generated. If the camber and angle of attack are too great, however, the smooth flow of air over the surface of the wing is disrupted and becomes turbulent; lift is greatly reduced or lost completely. Bats are capable of quickly adjusting both camber and angle of attack throughout the wing-beat cycle by use of several muscles in the wing, in conjunction with the movement of the wrist, thumb, fifth finger, and the hind limbs (Vaughan 1970c; Strickler 1978; Norberg 1990; Swartz et al. 1992).

As noted by Fenton (1985:16) "In flight the wing area proximal to the fifth digit provides lift, while the portion distal to the fifth digit acts as a variable pitch propeller providing thrust. Power comes from the contraction of nine pairs of flight muscles located on the chest and back." As a bat flies, thrust and lift are provided during the downstroke, in which the wings are fully extended and move in both a downward and forward motion. Little if any thrust or lift is developed during recovery, when the wings, partially folded to reduce drag, move upward and backward in preparation for the next downstroke. The chest muscles of bats power the downstroke and provide for lift and thrust. They are much larger than the back muscles, which function during the upstroke. In many species, the specialized scapula "locks" the movement of the humerus and stops the upstroke phase. See Vaughan (1970c) and Strickler (1978) for detailed analyses of the wing-beat cycle of bats and associated biomechanical factors.

The shape of a bat's wing is another factor that affects aerodynamic properties. When viewed from above, wing shape varies from short and broad, to long and thin in different families and species (figure 13.4). The proportion of wing length to width is called the **aspect ratio** and results from the relative lengths of the metacarpals and phalanges of the third, fourth, and fifth fingers. A high aspect ratio—long, narrow wings—adapts a species for sustained, relatively fast flight, as is seen in many free-tailed bats. These species would be expected to forage high above the ground, predominately in open habitats, free of obstructing vegetation. A low aspect ratio—short, wide wings—is seen in bats with slower, more maneuverable flight, as in the false vampire bats and the slit-faced bats. A low aspect ratio is associated with bats that forage more often in habitats with dense, obstructing understory vegetation. Besides aspect ratio, slow, maneuverable flight can be enhanced by low **wing loading**—the ratio obtained by the body mass of the bat divided by the total surface area of the wing. The lower the wing loading, the greater the potential lift and capacity for slow flight. In general, a direct relationship exists between the wing morphology and flight patterns of bats and their foraging and life history characteristics (Vaughan 1959, 1966; Findley et al. 1972; Lawlor 1973; Altringham 1996; Stockwell 2001).

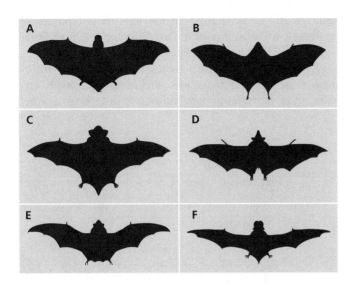

Figure 13.4 **Increasing aspect ratios of the wings in several different families of African bats.** The aspect ratio—calculated as the wingspan squared divided by the area of the wing—increases as wings become longer and thinner in different species of bats. Broader wings generally translate to slower, more maneuverable flight patterns. (A) Egyptian slit-faced bat (*Nycteris thebaica*) (Nycteridae); (B) heart-nosed bat (*Cardioderma cor*) (Megadermatidae); (C) Moloney's flat-headed bat (*Mimetillus moloneyi*) (Vespertilionidae); (D) straw-colored fruit bat (*Eidolon helvum*) (Pteropodidae); (E) Pel's pouched bat *Saccolaimus peli* (Emballonuridae); (F) Midas' free-tailed bat (*Mops midas*) (Molossidae). *Adapted from J. Kingdon, 1974*, East African Mammals, *vol. IIA, Univ. Chicago Press.*

Hearing can give the same amount of information as vision in terms of size, shape, texture, distance, and movement.

In echolocation, a bat generates and emits high-frequency, high-intensity sounds through the mouth or nose. These sound pulses are then transmitted through, scattered by, or reflected back by objects in their path. When reflected, the returning sound (echo), with altered characteristics, may be picked up by the bat. The characteristics of the returning echo are altered from the original sound pulse depending on the physical features of the object; distance, texture, movement, shape, and size. These features are all interpreted by the bat from the echo. Although simple in concept, the actual process is extremely complex, and the sensory apparatus of bats is very sophisticated. For example, the echo from a target 1 m away returns to the bat in 6/1000 of a second. Although 6/1000 second is very fast, bats are able to discriminate echo delays as short as 70 millionths of a second (Suga 1990). Many species also can determine the presence of objects as thin as 0.06 mm (0.002 in.) in diameter, the width of a human hair.

High-frequency sound pulses are particularly suitable for bat echolocation because objects about one wavelength in size reflect sound particularly well. For example, at a frequency of 30 kilohertz (kHz), the **wavelength** (distance from peak to peak in a sound wave; figure 13.5) is about 11 mm—roughly equivalent to the length of a small moth. Also, low-frequency sounds (long wavelengths) can "wrap around" small targets such as insects, and eliminate any echo. The negative aspect of high-frequency sound is that it is quickly absorbed by the atmosphere and thus has lim-

ECHOLOCATION

To find prey items and maneuver within their environment, most bats **echolocate**—that is, they emit high-frequency sound pulses and discern information about objects in their path from the returning echoes. Many of the structural characteristics of bats, including their seemingly bizarre facial features, relate to echolocation. As with the relationship between wing morphology, flight patterns, and life history features, many aspects of echolocation also are species-specific and relate to foraging. Most megachiropterans rely on vision and olfaction to locate food, avoid obstacles, and maneuver in their environment (although the rousette fruit bats—genus *Rousettus*—have a limited form of echolocation using low-frequency tongue clicks). Vision involves interpreting information about objects in the immediate environment from energy received as light waves produced by or reflected from those objects. Microchiropterans substantially augment visual senses with hearing, in the form of a highly sophisticated system of auditory echolocation. Hearing also involves interpreting information about objects in the environment from energy (received as sound waves) produced by or reflected from objects.

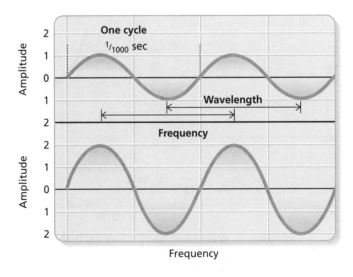

Figure 13.5 **Characteristics of a sound wave.** The upper sound wave represents a 1-kHz pure tone wave and is the same frequency as the lower one, but the amplitude is twice as great in the lower sound wave. Amplitude, or the intensity of sound, is measured in newtons per square meter. A newton equals 105 dynes (a dyne is the force necessary to move 1 gm of mass 1 cm/sec). *Adapted from Hill and Smith, 1984*, Bats: A Natural History, *British Museum of Natural History.*

ited range relative to lower frequency sound. If they were within the range of human hearing (and many species use signals that are), the high-intensity echolocation signals of some species would be very loud, including the signals from rhinolophids, molossids, emballonurids, rhinopomatids, and many vespertilionids. Bats avoid self-deafening from loud outgoing pulses by disarticulating the bones of the middle ear and dampening the sound. Other bats, including nycterids, phyllostomids, and megadermatids, are sometimes called "whispering bats." Their much lower intensity signals have a less effective range than species using signals of higher intensity. Interestingly, insects from at least six orders can hear ultrasound (Altringham 1996), and many night-flying insects have developed countermeasures to evade echolocating bats (Yager et al. 1990; Yager and May 1990; May 1991; Fullard et al. 1994; Fenton 1995).

The remarkable aspects of echolocation have been studied by numerous investigators (Simmons et al. 1975; Fenton 1985; Neuweiler 1990; Popper and Fay 1995) following Donald Griffin's pioneering work in the 1940s (see Griffin 1958). The sound pulses, produced by contraction of the cricothyroid muscles of the larynx (Novick and Griffin 1961; Novick 1977), generally are at frequencies above the range of human hearing, usually greater than 20 kHz. When a bat is taking insects in flight, the frequency and characteristics of echolocation pulses vary depending on whether the bat is in the search, approach, or terminal phase (figure 13.6). This variation involves changes in signal duration, loudness, and the nature and extent of frequency modulation and **harmonics** (integral multiples of fundamental frequencies that provide a broader scanning

ability). All these factors change in different phases of approach to a target and under different foraging situations, as well as among different species of bats.

As noted, bats perceive a great amount of detailed information from returning echoes, including size, shape, texture, and relative motion, in addition to distance to a target

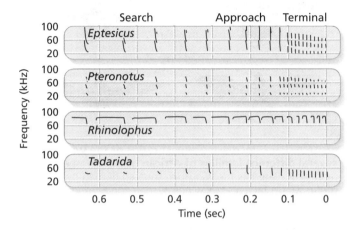

Figure 13.6 Increased pulse repetition at terminal phase. The increased pulse rate during the terminal phase of closing in on a prey item, referred to as the "feeding buzz," is evident in these reproductions of sonograms from four species of bats: big brown bat (*Eptesicus fuscus*) (Vespertilionidae); Wagner's mustached bat (*Pteronotus personatus*) (Mormoopidae); greater horseshoe bat (*Rhinolophus ferrumequinum*) (Rhinolophidae); and Brazilian free-tailed bat (*Tadarida brasiliensis*) (Molossidae). *Adapted from Hill and Smith, 1984,* Bats: A Natural History, *British Museum of Natural History.*

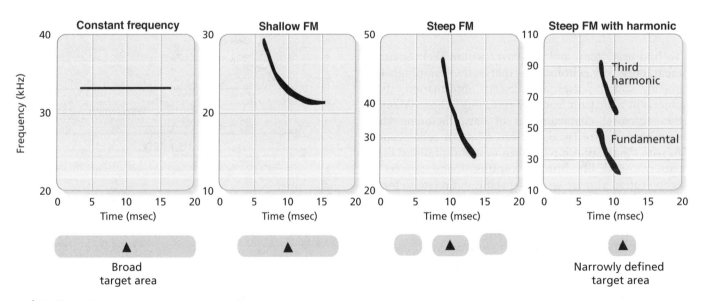

▲ = Target

Figure 13.7 Typical pattern of frequency changes in echolocation pulses that continuously narrows the target area perceived by a bat. Bats may use a narrow-band, long-duration, constant frequency (CF) pulse until a potential prey item (target indicated by ▲) is located. The target area (shaded) perceived by the bat becomes increasingly more narrowly defined as the bat shifts to broadband, shorter duration, echolocation pulses including a shallow frequency modulated (FM) sweep, a steep FM sweep, and finally a steep FM sweep with harmonics (which are integral multiples of fundamental frequencies) that pinpoint the target and provide fine details. *Data from M. B. Fenton, 1981,* American Scientist, *vol. 69.*

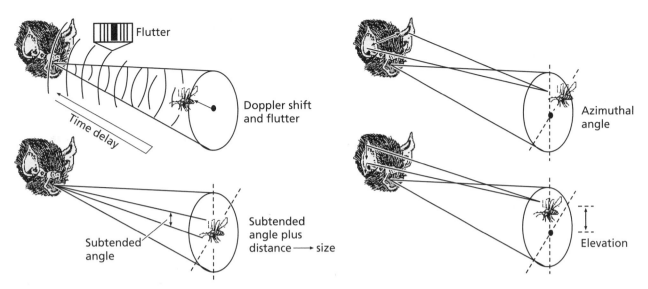

Figure 13.8 Information gained from returning echolocation signals. (Top left) Distance and relative speed of the target are indicated by the time delay and strength (in decibels; dB) of the returning pulse and by the Doppler shift. Flutters indicate the presence of rapidly beating insect wings. (Bottom left) The size of the target is reflected in the amplitude (subtended angle) of the returning signal in conjunction with the distance. (Top right) Time delay and amplitude differences between the ears convey information on azimuth (angular distance from center). (Bottom right) Interference patterns of the sound waves within the inner surface of the ear indicate elevation. *Adapted from N. Suga (1990),* Scientific American, *vol. 262.*

(figures 13.7 and 13.8). The constant-frequency (CF) component of the search phase, a pulse of constant pitch, is particularly useful for discriminating between moving objects, such as prey items, and stationary obstacles. Moving objects can be determined by their **Doppler shift**—a change in sound frequency of an echo relative to the original signal caused by the movement of one or both objects: the source (bat) or the target (insect). We observe a Doppler shift when a loud car races by us as we are standing still. The pitch of the engine noise sounds higher as it comes toward us because the sound waves are being "pushed" closer together. They pile up, the wavelength becomes shorter, and the sound frequency (pitch) is higher. As the car moves away from us, the sound waves are now being "stretched out" and become longer. The sound is now at a lower frequency (also, low-frequency sound travels farther than high-frequency sound). Because a bat and its insect prey are moving relative to each other, the returning echoes may be shifted in frequency. Many species of bats are able to compensate for this phenomenon.

Because bats cannot transmit and receive at the same time, outgoing pulses are often of short duration so as not to obscure the returning echoes. Bats that emit short calls and listen for echoes during the intervals are characterized as having "low-duty-cycle" echolocation. The alternative is "high-duty-cycle" echolocation that "... involves rapid emission of long constant-frequency calls, and takes advantage of the Doppler shift to avoid self-deafening, because call pulses and Doppler-shifted echoes are separated in frequency" (Simmons 2005b:165).

Broader band, frequency-modulated (FM) pulses, that go up or down, are better than CF pulses for determining the finer details of a target. Thus, the bat uses a series of FM pulses as it approaches closer to the target. Throughout the sequence, the rate of sound pulses increases (see figure 13.6), until it is extremely rapid during the terminal phase. This is because even small movements of the prey quickly create large angular differences in its relative position the closer the bat comes. The search/approach/termination sequence is repeated very rapidly. Although these generalizations of echolocation apply to all bats, there is substantial variation and diversity among families (table 13.2). The highly developed sensory apparatus, specialized neural anatomy, and sophisticated neural pathways of bats (Suga 1990) provide an amazing amount of detailed information about their surroundings.

Echolocation is not unique to bats; it occurs in other mammals, including some rodents, insectivores such as shrews and tenrecs, and seals (Fenton 1984a). Only in the toothed whales, however, might echolocation approach the level of sophistication found in bats.

EVOLUTION OF ECHOLOCATION AND FLIGHT

Echolocation and flight are two noteworthy characteristics of bats that are interrelated in the life history of all microchiropterans. As we will see in the next section (Fossil History), based on analyses of the articulation of the shoulder joint and enlarged scapula for flight muscles, and the enlarged cochlea region, Eocene bats were capable of flight and echolocation. Which one of these features came first has been a controversial question. It is generally assumed that the ancestors of these early bats were arboreal insectivores. There has been a great deal of

Table 13.2   Diversity of echolocation characteristics among families of bats

Characteristics of Echolocation Call	Family of Bat
1. Brief, broadband tongue clicks	Pteropodidae[*] (Genus *Rousettus*)
2. Narrowband; dominant fundamental harmonic	Some Vespertilionidae, Molossidae
3. Narrowband; multiharmonic	Craseonycteridae, Rhinopomatidae, Emballonuridae, Mormoopidae, Thyropteridae
4. Short broadband; dominant fundamental harmonic	Vespertilionidae (Genus *Myotis*)
5. Short broadband; multiharmonic	Megadermatidae, Nycteridae, Phyllostomidae, Mystacinidae, Vespertilionidae, Natalidae
6. Long, broadband	Myzopodidae
7. Constant frequency	Rhinolophidae, Hipposideridae, Noctilionidae

Adapted from Jones and Teeling (2006).
* Most pteropodids do not echolocate.

speculation concerning the evolution of these two features, however. Debate has centered on questions of what developed first—echolocation or flight—or whether these features may have developed at the same time. Currently, "flight first," "echolocation first," or "simultaneous development" are equally viable hypotheses (Simmons 2005b).

Fossil History

Fossil bats have been found on all continents except Antarctica (McKenna and Bell 1997). Based on occlusal surface patterns of the molars, bats are believed to have evolved from arboreal, shrewlike insectivores (Carroll 1988), although the time of divergence is unknown and no intermediate forms have been discovered. As shown in figure 13.9, fossils from eight extant families of bats have been found in early to late Eocene deposits. Fossil remains of bats from most other families date to the Oligocene or Miocene. The earliest fossil record of a bat is that of *Icaronycteris index*, from the early Eocene of North America (Jepson 1966). Many of the characteristics of *I. index* are considered primitive or unspecialized relative to modern bats. Primitive features include the total number of teeth (38, which is close to the primitive eutherian number of 44), lack of fusion in ribs and vertebrae, lack of a keeled sternum, and several other features noted by Jepson (1970). Nonetheless, other specialized

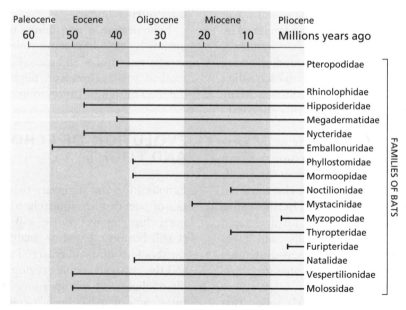

Figure 13.9 Fossil history of living (extant) families of bats. The fossil history of eight extant families of bats can be traced to the Eocene, with most of the remaining families documented from the Oligocene or Miocene. *Modified from Simmons (2005b).*

morphological features of modern microchiropterans did occur in *Icaronycteris* (figure 13.10), which was fully capable of flight. Several other early Eocene bat taxa include *Archaeonycteris, Hassianycteris, Palaeochiropteryx,* and *Tachypteron.* In their essential features, including flight and echolocation, bats have existed for over 50 million years. As summarized by Simmons (2005b:165), these Eocene bats "(1) were capable of powered flight, (2) used sophisticated echolocation like that of extant microchiropteran bats, (3) were taxonomically and ecologically diverse, and (4) were globally widespread, occurring on virtually all continents."

Megachiropterans include the fossil genera *Archaeopteropus* from the early Oligocene of Europe and *Propotto* from the early Miocene of Africa (Butler 1978). Given the differences noted in table 13.1, it has been suggested that bats are **diphyletic**; that is, megachiropterans and microchiropterans may have evolved independently, with megachiropterans being more closely related to the order Primates (Pettigrew 1986, 1995; Pettigrew et al. 1989). Based on numerous molecular and morphological studies, however, most investigators reject this hypothesis

and consider bats to be **monophyletic**, that is, derived from a common ancestral lineage (Gunnell and Simmons 2005). In fact, "most systematists agree that bat monophyly now represents one of the most strongly supported hypotheses in mammalian systematics" (Simmons 2005b:166).

Economics and Conservation

Bats are associated with both positive and negative economic effects. On the negative side, bats are associated with many bacterial, rickettsial, viral, and fungal diseases (Constantine 1970; Sulkin and Allen 1974). One of the fears people have is the association of bats with rabies. Rabies has been reported in numerous Old and New World species, but the percentage of infected individuals in any population generally is no more than 1–4% depending on the season and whether species are solitary or colonial (Constantine 1967; Brass 1994). Carnivores such as dogs, foxes, skunks, and raccoons are much more likely to be significant vectors of rabies than are bats (see chapter 27).

Vampire bats are directly responsible for losses of hundreds of millions of dollars to the livestock industry through disease transmission, including rabies. The distribution of vampire bats is restricted to New World tropical areas, however, as is their affect on livestock (see section on Family Phyllostomidae). Histoplasmosis is a fungal disease of the lungs sometimes associated with bat **guano** (fecal droppings) in caves and mines, but this health hazard is limited because few people frequent such places.

The benefits of bats far outweigh their negative attributes, however, and bats are one of the most economically beneficial mammalian groups. They are the only significant predator of nocturnal insects, consuming tremendous numbers of insect pests that would otherwise damage agricultural crops. For example, big brown bats (*Eptesicus fuscus*) feed on a variety of insects that often are vectors of plant diseases (Whitaker 1995). Bats also fill key ecological roles in many forest communities, especially as seed dispersers and plant pollinators (Fujita and Tuttle 1991; Fleming and Estrada 1993). Nearly 200 plant species of economic importance for food, timber, medicines, or fiber are pollinated by bats, including bananas, peaches, dates, and figs. Large accumulations of guano in caves have been mined in the past and used for fertilizer as well as an ingredient in gunpowder. Bats also are used in a variety of medical research programs, including studies of drugs, disease resistance, and navigational aids for the blind.

Population declines and range reductions of many species of bats throughout the world are due to cave closures and loss of foraging and roosting habitats, siltation or draining of riparian areas, insecticide accumulations,

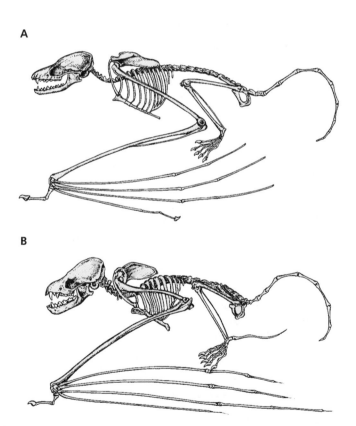

A

B

Figure 13.10 Similarities between the oldest fossil bat and modern bats. (A) Fossil skeleton of the bat *Icaronycteris* from the early Eocene. Note the development of the forelimb into a wing, with elongation of the phalanges, partial fusion of the radius and ulna, and dorsal position of the scapula. (B) Skeleton of a modern *Myotis* is similar in many respects to that of *Icaronycteris. Adapted from G. L. Jepsen, 1970, Bat origins and evolution in W. A. Wimsatt, ed.,* Biology of bats, *vol. I, Academic Press.*

and, in the case of large Old World fruit bats, consumption by humans. Many species have gone extinct in recent decades, especially endemic species of pteropodids (Kunz and Pierson 1994). In the United States, the gray bat and Indiana bat (*Myotis sodalis*) are endangered, as are some subspecies of big-eared bats (genus *Corynorhinus* [formerly *Plecotus*]). Hibernacula of these species, including caves and abandoned mines, often are gated to prevent disturbance of bats. A primary goal of conservation programs is to change people's perceptions of bats through education so that they might learn to appreciate these fascinating mammals rather than fear and persecute them. The organization Bat Conservation International has taken a leading role in these educational efforts.

Suborders and Families

The relationships of families of microchiropterans, and of megachiropterans to microchiropterans, are poorly

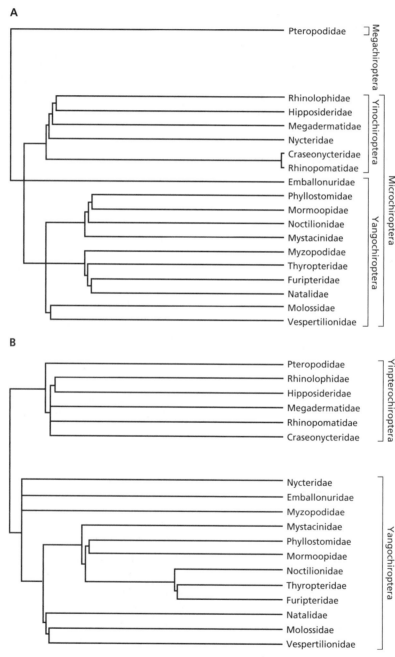

Figure 13.11 Alternative hypotheses of interfamily relationships in bats. Relationships of chiropteran families based on morphological data (A) do not correspond with those derived from molecular data (B). The suborders used in the text—Megachiroptera (for the family Pteropodidae) and Microchiroptera (for all other families)—are retained in morphologically based arrangements, with Yinochiroptera and Yangochiroptera as infraorders. In arrangements based on molecular data, the suborders are termed Yinpterochiroptera with the pteropodids grouped with several other families, and Yangochiroptera that encompasses the remaining families. *Modified from Simmons (2005b).*

understood, despite a long history of attempts at classification (see Smith 1980 for a review of early work). Despite advances in molecular techniques, these higher-level relationships remain unresolved (figure 13.11). The problem was stated by Simmons (1998:3): "Higher-level relationships of bats remain poorly understood despite recent advances in our understanding of relationships at lower taxonomic levels. Although most bat species can be assigned easily to one of several monophyletic families, there is little agreement concerning relationships among the latter taxa. This is not just a problem of resolution: previous hypotheses of interfamilial relationships are largely incongruent." The problem still remains, again as noted by Simmons (2005a:313): "No complete classification of bat families based on molecular data yet exists, and those complete classifications that are available...were based on morphology and are not at all congruent with the new molecular data." The fact that the phylogenetic affinities of bat families vary significantly based on molecular or morphological approaches is also discussed by Hutcheon and Kirsch (2004), Teeling et al. (2005), and Jones and Teeling (2006) among others. Simmons (2005a:313) felt that resolution is near and predicted that "a new consensus view of higher-level classification of bats that contradicts most traditional arrangements will soon emerge."

MEGACHIROPTERANS

Pteropodidae

Pteropodidae is the only family in the suborder Megachiroptera. Historically, this large assemblage of 42 genera and 186 species of fruit bats was divided into between 2 and 6 subfamilies based on feeding and morphological characteristics (Hill and Smith 1984, Corbet and Hill 1991, Koopman 1993; see Simmons 2005a). Fruit bats occur throughout tropical and subtropical areas of the Old World, usually in forested or shrubby areas that provide a predictable supply of fruit. They are distributed south and east of the Sahara Desert in Africa, east through India, southeast Asia and Australia, to the Caroline and Cook Islands in the Pacific.

Major features that distinguish the Old World fruit bats from the other families of bats (microchiropterans) were given in table 13.1. Pteropodids do not echolocate, the exception being the low-intensity tongue clicking of members of the genus *Rousettus*. The duration, amplitude, and peak frequency of echolocation in the Egyptian fruit bat (*R. aegyptiacus*) was described by Holland et al. (2004). Pteropodids have none of the distinctive facial features related to echolocation, such as a nose leaf or enlarged tragus, that are found in microchiropterans (figure 13.12). They navigate primarily using vision, and anatomically their large eyes are unusual. The choroid, which surrounds the retina, has numerous **papillae**, or projections. These extend into the retina and form undulating, uneven

Figure 13.12 **Facial features of a fruit bat, Family Pteropodidae.** The large eyes, lack of facial ornamentation, and simple ears with no tragus or antitragus of this Gambian epauleted fruit bat (*Epomophorus gambianus*) can be contrasted with most echolocating microchiropterans.

contours of unknown function (Suthers 1970). Most of the photoreceptors are rod cells, for black-and-white vision, although a few cone cells for color vision occur in some species, such as the Indian flying fox (*Pteropus giganteus*) (Pedler and Tilley 1969). Wang et al. (2004) investigated the opsin genes for color vision in the Philippine pygmy fruit bat (*Haplonycteris fischeri*) and the Ryukyu flying fox (*P. dasymallus*). They suggested these two species, which are most active at dawn and dusk, were sensitive to ultraviolet light.

Body size varies considerably among members of this family. The smaller species, such as the long-tongued fruit bat (*Macroglossus minimus*), pygmy fruit bat (*Aethalops alecto*), or spotted-winged fruit bat (*Balionycteris maculata*) weigh only about 15–20 g; the largest species in the genus *Pteropus* weigh up to 1,200 g, a difference of two orders of magnitude. Wingspans may reach 2.0 m in some species of *Pteropus* and *Acerodon*, the largest of any bat. Sexual dimorphism is evident in greater body size in males in the genera *Eonycteris*, *Epomops*, and *Epomophorus*. The hammer-headed fruit bat (*Hypsignathus monstrosus*) exhibits a greater degree of sexual dimorphism than any other species of bat, with males being nearly twice the body mass of females (Langevin and Barclay 1991). Male epauleted fruit bats (Genus *Epomophorus*) have light-colored tufts of fur on the shoulders ("epaulettes"), associated with glandular patches used to attract females during breeding.

Dental formulas also vary, with the teeth being specialized for a diet of fruit. Although some genera have 34 teeth, the more specialized nectar-feeders have a reduced number. Canines are always present, however. The molars

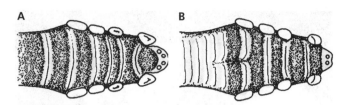

Figure 13.13 Ridges on the palatal surface of fruit bats. (A) Six ridges occur on the palate of epauletted bats (genus *Epomophorus*). (B) Additional flat, posterior serrated ridges are found in fruit bats of the genus *Epomops*. *Adapted from J. Kingdon, 1974,* East African Mammals, *vol. IIA, University of Chicago Press.*

Figure 13.14 Fruit bats roosting in a tree. Communal roosts are particularly common in the larger pteropodids (genus *Pteropus*).

generally have smooth surfaces, usually without cusps, but with shallow, longitudinal grooves. As noted in table 13.1, the palate extends beyond the last molar. In frugivores, the palate has ridges (figure 13.13) against which the tongue mashes ingested food. The juice is consumed and pulp and seeds discarded.

Fruit bats are primarily nocturnal, although some species may be diurnal. Large species may fly up to 100 km between roosting and foraging areas, depending on the local availability of fruit. Annually, species such as the straw-colored fruit bat (*Eidolon helvum*) may migrate 1,500 km to take advantage of seasonal food availability (Richter and Cumming 2006). Food resources are located primarily through olfaction. Pollen and nectar may make up a significant proportion of the diet in some species. Many of these species have bristle-like papillae on their long, extensible tongues to help collect pollen from flowers. Pteropodids thus serve valuable ecological functions in pollination (Andriafidison et al. 2006) and in seed dispersal (see McConkey and Drake 2006) of about 300 species of tropical plants. For example, one of the most common trees in India—*Ceiba pentandra*—is pollinated by the Indian flying fox, fulvous fruit bat (*Rousettus leschenaulti*), and the short-nosed fruit bat (*Cynopterus sphinx*) (Singaravelan and Marimutha 2004). Seeds usually are spit out or passed through the gut away from the parent tree, thus promoting dispersal. Shilton et al. (1999) examined retention time and viability of seeds in the gut of the short-nosed fruit bat. They suggested that ingested seeds potentially could be dispersed >300 km from the parent plant. Additionally, many pteropodids are folivores (eat leaves) because this is a food source that provides protein, carbohydrates, and minerals, especially calcium. Also, leaves are available throughout the year in the tropics and are a predictable resource temporally and spatially (Nelson et al. 2005).

Bats usually roost in trees or shrubs (figure 13.14), although a few species use caves or buildings. Roosts often are communal, especially in the larger pteropodids. The short-nosed fruit bat roosts in "tents" it constructs in the stems or fruit clusters of various tree species (Balasingh et al. 1995); most frequently the kitul palm (*Caryota urens*) in India (Bhat and Kunz 1995; Storz et al. 2000). Concentrations of up to a million straw-colored fruit bats

may occur at roosting sites (Kulzer 1969). Conversely, roosting dawn bats (genus *Eonycteris*) are solitary. Whether species are colonial or solitary is a function of several factors, including predation pressure, food availability, roosting site availability, and breeding pattern. In American Samoa, the Pacific flying fox (*Pteropus tonganus*) forms roosting colonies in isolated areas because of hunting pressure by people (Brooke et al. 2000). Gestation periods range from 100 to 125 days, although some pteropodids such as *E. helvum* (Mutere 1965) exhibit delayed implantation.

Where large congregations of fruit bats occur, they may become serious crop depredators. In Africa and on South Pacific islands, loss of roosting and foraging habitats and use of bats as food for people (see Wilson and Graham 1992) have severely reduced bat populations. A prime example is the Pacific flying fox population on the island of Niue, which is severely overharvested during a 2-month hunting season each year (Brooke and Tschapka 2002). The functional role of pteropodids in community ecology was discussed by Rainey et al. (1995) and

Richards (1995). Several species recently have become extinct, including the Nendo tube-nosed fruit bat (*Nyctimene sanctacrucis*), Palau Island flying fox (*Pteropus pilosus*), Guam flying fox (*P. tokudae*), and Panay giant fruit bat (*Acerdon lucifer*). Critically endangered species include the tube-nosed fruit bat (*N. rabori*), Comoro black flying fox (*P. livingstonii*), Rodrigues flying fox (*P. rodricensis*), Bulmer's fruit bat (*Aproteles bulmerae*), Negros naked-backed fruit bat (*Dobsonia chapmani*), and several others.

MICROCHIROPTERANS

Rhinopomatidae

The mouse-tailed bats are named for their long tails, which are nearly equal to the length of the head and body (see figure 13.2A [f]). They are the only living microchiropterans with a tail longer than their forearm. Adults weigh from 6 to 14 g and have dark dorsal and paler ventral pelage, with naked areas on the face, rump, and abdomen. The ears are large and connected across the forehead by a flap of skin, and the slitlike nostrils are valvular. The single genus, *Rhinopoma*, contains four species found throughout North Africa and the Middle East, to India and Sumatra. These insectivorous bats occur in arid regions where they roost in caves, cliffs, houses, and even Egyptian pyramids. Although mouse-tailed bats do not hibernate during the winter, they enter torpor and do not forage, living off stored body fat. A population of the greater mouse-tailed bat (*R. microphyllum*) in Iran fed almost entirely on beetles (Sharifi and Hemmati 2002). Females are monestrus and give birth to a single young in July or August after a gestation of about 123 days. Qumsiyeh and Jones (1986) summarized life history characteristics of the lesser mouse-tailed bat (*R. hardwickei*) and the small mouse-tailed bat (*R. muscatellum*). There is no fossil record for this family. They are considered primitive microchiropterans, however, based on several morphological features, including two phalanges on the second digit of the **manus** (hand or forefoot), unfused premaxillary bones, and an unmodified anterior thoracic region (Koopman 1984), as well as primitive chromosomal characteristics (Qumsiyeh and Baker 1985). MacInnes's mouse-tailed bat (*R. macinnesi*), distributed in northeast Africa, is considered threatened.

Emballonuridae

This family comprises 13 genera and 51 species of sac-winged or sheath-tailed bats. They enjoy a widespread distribution in tropical and subtropical habitats from Mexico to Brazil, sub-Saharan Africa, the Middle East, India, southeast Asia, Australia, and Pacific islands eastward to Samoa. Emballonurids range in body mass from 5–105 g. The total number of teeth varies from 30 to 34, and dentition is adapted for an insectivorous diet; molars have a distinct dilambdodont (W-shaped) occlusal pattern.

Figure 13.15 Roosts of proboscis bats. These sharp-nosed bats (*Rhynchonycteris naso*) are roosting in a hollow tree stump on the bank of the New River in northern Belize.

Emballonurids roost in caves, logs, houses, crevices, or the leaves of shrubby vegetation. Some species, such as ghost bats (Genus *Diclidurus*) are solitary, whereas others form groups up to 50 individuals, including the sharp-nosed bat (*Rhynchonycteris naso*; figure 13.15) and the white-lined bats (Genus *Saccopteryx*). The common names of the family refer to their distinctive morphological features. They are known as sac-winged bats because of glandular wing sacs on the ventral surface of the wings near the elbow (within the **propatagium**; see figure 13.1). The sacs are most prominent in males, and exude a red, odiferous substance that is probably important in pheromone production and attraction of females. Colonies of up to 60 individuals form in daytime roosts of greater sac-winged bats (*S. bilineata*—see figure 13.16). These colonies include a male and varying numbers of females. Scent from the sacs, as well as unusually complex audible vocalizations (songs), function in male courtship displays (Behr and von Helversen 2004, Davidson and Wilkinson 2004, Voigt et al. 2005). The family also is referred to as sheath-tailed bats because of the way the tail protrudes through the uropatagium (see figure 13.2A [j]) and slides up and down with movement of the hind limbs. Among emballonurids, the Seychelles sheath-tailed bat (*Coleura seychellensis*) is considered critically endangered as is Troughton's tomb bat (*Taphozous troughtoni*).

Figure 13.16 Greater sac-winged bat (*Saccopteryx bilineata*). This common emballonurid is one of four species in this New World genus.

Craseonycteridae

This monotypic family (Hill 1974) includes only the hog-nosed bat, *Craseonycteris thonglongyai*, known from a few locations in Kanchanaburi Province, western Thailand, and more recently from an area of southeast Myanmar (Bates et al. 2001). It is listed as endangered. It is the smallest species of bat and, based on body mass, among the smallest mammalian species in the world. Head and body length is about 30 mm, and body mass is about 1.5–2.0 g. Hog-nosed bats have large ears and tragus, and a distinctive plate on the nose. There is no external tail or calcar. Also known as the bumblebee bat, this species has a large uropatagium and broad wings adapted for slow, hovering flight. The bats roost in small caves where they are solitary. They feed on insects taken in flight or gleaned from foliage, as well as small spiders (Hill and Smith 1981). The echolocation characteristics of the species, and their relation to foraging, were described by Surlykke et al. (1993). This species is believed to be most closely related to the mouse-tailed bats (see figure 13.11).

Nycteridae

There is a single genus (*Nycteris*) and 16 species of slit-faced bats now recognized in this family. Fourteen species are distributed throughout sub-Saharan Africa and Madagascar; the Malayan slit-faced bat (*N. tragata*) occurs in Malaysia, Sumatra, and Borneo; and the Javan slit-faced bat (*N. javanica*) is on the East Indies islands of Java, Bali, and Kangean. Nycterids occur in habitats ranging from semiarid areas to savannas and tropical forests, where solitary individuals or small colonies roost in caves, trees, houses, or animal burrows. Nycterids are small to medium-sized bats weighing 10–30 g. Head and body length is 40–90 mm, with tail length up to 75 mm. They have large ears and a small tragus. The common name derives from the unusual longitudinal groove throughout the facial region (figure 13.17), which along with the external nose leaf, functions in their low-intensity echolocation calls, emitted through the nostrils. The uropatagium completely encloses the tail and is partially supported by the unique, T-shaped tip of the tail (see figure 13.2A [i]). Most species are insectivorous, but the large slit-faced bat (*N. grandis*) also feeds on frogs and other small vertebrates (Bayefsky-Anand 2005), including other bats (Fenton et al. 1981). Griffiths (1997) described a postlaryngeal chamber that is distinctive in the two Asian species compared to nycterids from Africa. No slit-faced bats are currently endangered.

Megadermatidae

The false vampire bats (the name reflects the historical, false belief that they feed on blood), encompass four genera and five species. They are found in tropical and savanna habitats throughout central Africa, India, southeast Asia, the East Indies, and Australia. These are moderate to large bats; the Australian giant false vampire or ghost bat (*Macroderma gigas*) has a wingspan of 60 cm and body mass up to 170 g (Breeden and Breeden 1967; Taylor 1984). Megadermatids have large ears that are united across the forehead; a **bifid** (divided) tragus; and a large, erect nose leaf. All species lack upper incisors, which is unusual for bats.

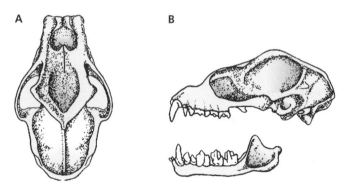

Figure 13.17 Skull of a slit-faced bat. (A) The dorsal aspect shows the deep concavity between the eye orbits. (B) Lateral view of the skull and mandible. *Adapted from G. S. Miller, 1907, Families and Genera of Bats, U.S. National Museum.*

They roost in caves, trees, buildings, and bushes. The four genera are variable in their feeding and roosting habits. The yellow-winged bat (*Lavia frons*) consumes insects and may be diurnal as well as nocturnal. Roosts may contain single individuals or groups of up to 100. The heart-nosed bat (*Cardioderma cor*) roosts in groups up to 80 individuals and feeds primarily on ground-dwelling beetles and scorpions (Csada 1996). The two species of Asian false vampire bats (Genus *Megaderma*) and *Macroderma gigas* feed on small vertebrates, including lizards, rodents, small birds, and other bats (Marimuthu et al. 1995). *Megaderma* may form very large colonies, whereas *Macroderma gigas*, a threatened species, roosts alone or in small groups. The range of *M. gigas* has contracted significantly and the species now occurs in distinct, isolated populations in northern Australia (Hoyle et al. 2001). There is a high amount of genetic subdivision due to limited gene flow between these populations (Worthington Wilmer et al. 1999).

Rhinolophidae

Traditionally, the horseshoe bats were divided into two subfamilies, the Rhinolophidae and the Hipposiderinae. Most authorities now consider these two taxa of Old World bats to constitute separate families based on several distinctive morphological and behavioral characteristics (table 13.3), and we consider the hipposiderids in the following section. Rhinolophids are widely distributed throughout Europe, Africa, the Middle East, Asia, Japan, the East Indies, and northern Australia. The single genus (*Rhinolophus*) and 77 species occur in a variety of habitats from deserts to tropical forests.

As might be expected given the large number of species, there is considerable morphological and behavioral variation in horseshoe bats. They are named for their distinctive facial ornamentation (figure 13.18), which ranges from structurally simple to very ornate. Generally, this includes a nose leaf above the nostrils, and below and on the sides of the nostrils a flap of skin shaped like a horseshoe. Between the horseshoe and nose leaf is a median projection called the **sella**. As noted in table 13.3, similar nose

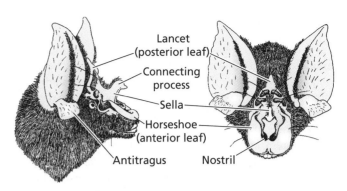

Figure 13.18 The distinctive nose leaf of a horseshoe bat. Note the anterior horseshoe shape, connecting stella, and posterior leaf in this broad-eared horseshoe bat (*Rhinolophus euryotis*). Horseshoe bats emit echolocation pulses through their nostrils, and this complex structure helps to beam the sound waves.

ornamentation occurs in the Family Hipposideridae, but without the sella. Nose ornamentation functions in directing the echolocation pulses emitted through the nostrils, not from the mouth (Robinson 1996). All horseshoe bats take insects while in flight (referred to as aerial hawking) in highly cluttered understory, using constant frequency echolocation calls (see table 13.2). Primary prey are small moths and beetles (Goiti et al. 2004; Pavey and Burwell 2004). Rhinolophids generally have large ears without a tragus. Ear length correlates to call frequency—species using high frequency have smaller ears—more so than in closely related hipposiderids (Huihua et al. 2003).

Rhinolophids roost singly or in small groups in caves, crevices, hollow trees, or houses. While roosting, they wrap their wings around themselves and fold up the uropatagium, thus appearing to be large cocoons. Species that hibernate exhibit delayed fertilization, mating in the fall, with ovulation and fertilization in the spring. Species from warmer regions ovulate and mate in the early spring. Nevertheless, rhinolophids commonly have a single young, with birth and lactation coinciding with the peak occurrence of insects in the spring. The convex horseshoe bat (*R. convexus*) that occurs in Malaysia, and Hill's horseshoe bat (*R. hilli*) in Rwanda, are critically endangered.

Table 13.3 Morphological and behavioral differences separating the horseshoe bats (Family Rhinolophidaae) and Old World leaf-nosed bats (Family Hipposideridae)

Rhinolophidae	Hipposideridae
Three phalanges in all toes of the feet except the first	Two phalanges in all toes of the feet
A median projection (sella) extends from the nose leaf (see text)	No median projection
No sac behind nose leaf	Males often with a sac behind nose leaf that can be everted and secretes a waxy material
Dentition: 32 teeth, with small lower premolar	Dentition: 28 to 30 teeth, no small lower premolar
Generally forage individually	Generally forage in small groups

Hipposideridae

As noted, the 9 genera and 81 species of leaf-nosed and roundleaf bats in this family were previously considered as a subfamily within the Rhinolophidae. Like rhinolophids, the hipposiderids are widely distributed throughout Asia, the East Indies, Australia, Africa, and the Middle East, where they take a diverse array of prey, primarily Coleoptera, Lepidoptera, and Diptera. For example, prey of Schneider's roundleaf bat (*Hipposideros speoris*) in Sri Lanka consisted of at least 27 families of insects taken among diverse habitats (Pavey et al. 2001a,b). Hiryu et al. (2005) mounted a small telemetry microphone on the head of great leaf-nosed bats (*H. armiger*) in Taiwan to investigate Doppler-shift compensation of their echolocation pulses. The bats changed frequency depending on flight speed such that the frequency of returning echoes remained constant. In Formosa, *H. armiger* used winter hibernacula that were warmer, with more stable temperatures, than non-hibernacula sites (Ho and Lee 2003). Lamotte's roundleaf bat (*H. lamottei*) is critically endangered, as is the Malayan roundleaf bat (*H. nequam*) and the Vietnam leaf-nosed bat (*Paracoelops megalotis*). Several other species are endangered.

Mormoopidae

This New World family formerly was considered a subfamily of the Phyllostomidae but gained family status based on work by Smith (1972). From examination of the mitochondrial cytochrome b gene and the nuclear recombination activating gene 2, Lewis-Oritt et al. (2001) recognized several subgenera within the genus *Pteronotus*. Simmons and Conway (2001) also described various subgenera after analysis of 209 morphological characters. The two genera and eight species of mormoopids are known as the mustached, ghost-faced, or naked-backed bats. They occur in semiarid to tropical forest habitats from the southwestern United States through Mexico, Central America, and the Caribbean, to northern South America. Mormoopids are small to medium-sized bats. All have a tail, and a tragus with a secondary fold of skin, best developed in the genus *Mormoops*. This group differs from phyllostomids in that it lacks a nose leaf and has very small eyes. The lips are enlarged and surrounded by numerous stiff hairs and, with a platelike growth on the lower lip, form a funnel into the mouth. Davy's naked-backed bat (*Pteronotus davyi*) and the big naked-backed bat (*P. gymnonotus*) actually have thick pelage on their back but appear naked because their wings meet middorsally, obscuring the fur (figure 13.19). All mormoopids are insectivorous; have long, narrow wings for rapid flight; and usually forage near water. They roost in warm, humid caves and are gregarious, sometimes forming large colonies.

Noctilionidae

Known as bulldog (figure 13.20A) or fishing bats, this family encompasses a single genus (*Noctilio*) and two species

Figure 13.19 **Davy's naked-backed bat (*Pteronotus davyi*).** The back only appears to be naked because the wings obscure the fur.

distributed in tropical lowlands from northern Mexico to northern Argentina, and on several Caribbean islands. The greater bulldog bat (*N. leporinus*) has a head and body length up to 13 cm. Males may weigh 70 g (Davis 1973) and have a 50-cm wingspan. Pelage color varies from yellow to brown, and apparently becomes darker in older, heavier individuals (Bordignon and de Oliveira-Franca 2004). These bats hunt at night and use their very long hind legs and large, rakelike feet with well-developed claws to scoop up small fish (figure 13.20B), crustaceans, and aquatic insects (Brooke 1994; Bordignon 2006). Prey are detected through echolocation near the surface of fresh or saltwater (Schnitzler et al. 1994). They also take terrestrial invertebrates. Small groups usually forage together. The lesser bulldog bat (*N. albiventris*) is smaller, 8–9 cm in head and body length, with maximum body mass about 40 g (Davis 1976). The hind limbs are not as well developed as in *N. leporinus*, and they apparently feed primarily on insects, again taken by echolocation. Both species roost in cliffs, caves, buildings, and hollow trees located near the water.

Phyllostomidae

This large New World family of leaf-nosed bats consists of 55 genera and about 160 species. Based on dentition,

A

B

Figure 13.20 A fishing bat. (A) From this close-up of the head, it is easy to see why they are also called bulldog bats. Note the long ears and well-developed tragus, with a comb-like lateral edge, and the lack of a nose leaf. (B) A noctilionid taking a small fish from the water.

nose- leaf structure, and skull features, Simmons (2005a) recognized seven subfamilies (table 13.4), including the vampire bats (Desmodontinae). There is disagreement, however, among authorities concerning the number of subfamilies.

Phyllostomids are widely distributed from the southwestern United States through Central America, the Caribbean, and northern South America. They occur from sea level to high elevations, in habitats ranging from desert to tropical forest. Nose-leaf ornamentation is present in most genera, although it is not as pronounced as in the Old World leaf-nosed bats (Family Hipposideridae). For those genera of phyllostomids without the nose leaf, such as vampire bats and many fruit-eating species, the upper lip usually has some type of plate or other promi-

Table 13.4 Subfamilies, number of genera, and number of species within the Phyllostomidae.

Subfamily	Number of genera	Number of species
Desmodontinae	3	3
Brachyphyllinae	1	2
Phyllonycterinae	2	5
Glossophaginae	13	32
Phyllostominae	16	42
Carolliinae	2	9
Stenodermatinae	18	67

Source: From Simmons (2005a).

nent outgrowth. The ears are sometimes connected at the base, and a tragus is present. A tail and uropatagium may or may not be present. There is great diversity in body size, ranging from small species such as the little white-shouldered bat (*Ametrida centurio*), with a head and body length of 4 cm, to the largest New World species, Linnaeus' false vampire bat (*Vampyrum spectrum*, not to be confused with megadermatids), with a head and body length of 13 cm and a wingspan of about 0.6 m.

Phyllostomids also exhibit a variety of feeding habits, probably the greatest diversity of any mammalian family. Many of the smaller Phyllostominae are insectivorous. Larger species in this subfamily, such as *V. spectrum*, the spear-nosed bat (*Phyllostomus hastatus*), and Peter's woolly false vampire bat (*Chrotopterus auritus*), are carnivorous, feeding on small vertebrates. Species in the Subfamilies Glossophaginae, Phyllonycterinae, Brachyphyllinae, Stenodermatinae, and Carolliinae primarily feed on nectar, pollen, or fruit.

The three species of true vampire bats in the subfamily Desmodontinae are sanguinivorous, that is, they consume only blood. Their dentition is specialized (figure 13.21), with large, sharp incisors for making an incision, and large canines. The cheekteeth are reduced in size, number, and complexity because blood does not require chewing. The common vampire bat (*Desmodus rotundus*) often preys on mammals, especially large **ungulates** (hoofed mammals), both domestic and wild. In contrast, the white-winged vampire bat (*Diaemus youngi*) and hairy-legged vampire bat (*Diphylla ecaudata*) primarily prey on birds.

Aspects of the biology of the common vampire bat are detailed in Brass (1994). Vampire bats make an incision about 1 cm long and 5 mm deep and 4 mm wide (Greenhall 1972) on the tail, snout, ears, feet, anus, or other area with sparse amounts of hair. Grooves on the sides and bottom of the tongue produce suction through capillary action while the bat laps up blood. Anticoagulant in the bat's saliva may keep the blood from coagulating for several hours (Hawkey 1966; Cartwright 1974). Nonetheless, blood loss is minimal, and prey usually are unaware that they have been attacked. A vampire bat probably consumes less than 8 L of blood a year, based on an average consumption of about 25 mL/day (Wimsatt

A

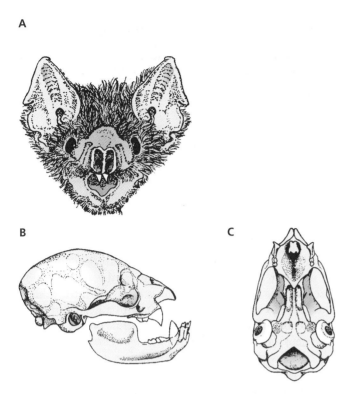

Figure 13.21 The common vampire bat (*D. rotundus*).
(A) Note the flat nose and relatively large eyes. (B) Lateral view and (C) ventral view of the skull show the specialized teeth. The incisors and canines are enlarged, with great reduction in the size and number of molars. There are a total of 20 teeth; other phyllostomids have 32–34 teeth. *Adapted from Hill and Smith, 1984,* Bats: A Natural History, *British Museum of Natural History.*

and Guerriere 1962). The open wound is subject to bacterial infections, parasites, and rabies, however. This can make livestock ranching impractical in areas of Central and South America within the range of vampire bats (Delpietro et al. 1993). Voigt and Kelm (2006) used stable carbon isotope analyses to document the preference of vampire bats for cattle instead of free-ranging ungulates. They suggested that livestock, which are usually fenced in, were a "more predictable resource" than free-ranging hosts.

Food sharing in vampire bats (Wilkinson 1984, 1990) is a fascinating social behavior that is a rare example of reciprocal altruism. Vampire bats will starve to death if they do not eat for about 3 days. Within roosting colonies of females (males roost individually), a bat that has eaten recently regurgitates blood to a roost-mate that is close to starvation. Food sharing even occurs between unrelated individuals but only in bats that have a close roosting association, that is, between those who can return the favor when necessary. Blood sharing was not observed in *D. youngi* (Schutt et al. 1999).

Phyllostomids roost in caves, trees, buildings, or the burrows of other animals, often with other species of bats. Single individuals, small groups, or large clusters may

occur. An example of a phyllostomid that forms exceptionally large colonies of 100,000 or more adults is the lesser long-nosed bat (*Leptonycteris curasoae*). Although this species forages cooperatively, Fleming et al. (1998) observed no allogrooming, communal nursing, or food sharing in a maternity colony. The tent-building bats (genus *Uroderma*), white bat (*Ectophylla alba*), and several species of fruit-eating bats (genus *Artibeus*; figure 13.22) construct a roost shelter by biting through the midrib of palm fronds, which causes the leaf to fold over and form a protective "tent." Kunz and McCracken (1996) suggested that the tent-making behavior may provide resources that aid in polygyny (males mating with more than one female). Eighteen species of tropical bats are known to make tents: 15 in the New World and 3 in the Old World (Kunz et al. 1994). The Puerto Rican flower bat (*Phyllonycteris major*) is recently extinct. The Guadeloupe big-eyed bat (*Chiroderma improvisum*), greater long-nosed bat (*Leptonycteris nivalis*), Thomas's yellow-shouldered bat (*Sturnira thomasi*), and Jamaican flower bat (*Phyllonycteris aphylla*) are endangered.

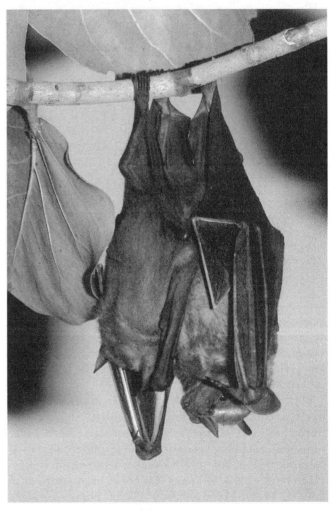

Figure 13.22 A great fruit-eating bat (*Artibeus lituratus*). Several species within this genus construct "tent" shelters from palm fronds.

Mystacinidae

The New Zealand short-tailed bats are unusual in several of their structural adaptations. Adapted for terrestrial locomotion and capture of prey on the ground to a greater degree than other bats, mystacinids have very sharp claws on the hind feet and thumb, each with a small, basal talon (figure 13.23). They also have a thick membrane along the side of the body. When an individual is not flying, the wings can be folded up, much like a sail. This makes the bats much more agile on the ground. The uropatagium can be "furled" in a similar manner. Short-tailed bats roost in caves, trees, and burrows in forested areas. Winter roosting activity in the lesser short-tailed bat (*Mystacina tuberculata*) was investigated by Sedgeley (2001) in an island population at the highest latitude of the species' distribution. Radio-monitored individuals roosted alone about 50% of the time, and communally about 50% of the time. Summer roosts were in large trees, with well-insulated cavities and stable temperatures (Sedgeley 2003). Lesser short-tailed bats may remain more active than other species of bats during cold temperatures because of their ability to forage on terrestrial invertebrates. The only species of bat known to burrow, they use their upper incisors to excavate. The dentition is typical for insectivorous bats, although the diet is diverse; in addition to insects they eat fruit, nectar, and arthropods. Jones et al. (2003) found that *M. tuberculata* echolocated to find prey during aerial foraging but echolocated only for orientation when on the ground. They probably use auditory and olfactory cues to locate prey in the leaf litter. Although both species were once widely distributed throughout New Zealand forests, *M. tuberculata*, the smaller of the two species with a body mass of 12–15 g, is now reduced to a small number of populations (Flannery 1987) and is threatened. The larger greater short-tailed bat (*M. robusta*; body mass of 25–35 g) has not been found since 1965 and probably is extinct.

Considering relationships among families of microchiropterans, the position of the mystacinids has been enigmatic, with little consensus between morphological and molecular analyses. However, recent analyses (Teeling et al. 2003, 2005; Simmons 2005b; Gunnell and Simmons 2005; Jones and Teeling 2006) suggest a close relationship of short-tailed bats and noctilionoid families (figure 13.11B).

Natalidae

The three genera and eight species of funnel-eared bats occur from northern Mexico to northern South America, and on many Caribbean islands. This family was formerly considered to have a single genus with three subgenera (*Natalus*, *Chilonatalus*, and *Nyctiellus*), but these were raised to full generic status based on work by Morgan and Czaplewski (2003). A proposed new species from Mexico, *N. lanatus*, was described by Tejedor (2005) based on morphological differences from the Mexican greater funnel-eared bat (*N. stramineus*). Natalids are small bats. Head and body length is 35–55 mm, and adults weigh from 4 to 10 g. The long, soft, dorsal pelage is yellow to reddish brown. The tragus is short, and the large ears are funnel-

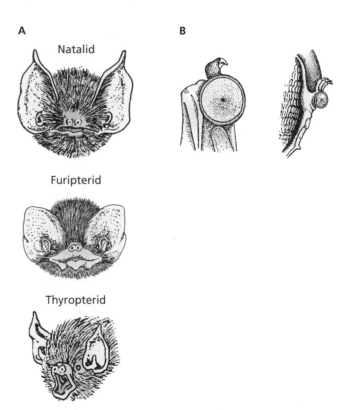

Figure 13.24 **Facial features and disk characteristics among three families of bats.** (A) (left) Mexican funnel-eared bat (*Natalus stramineus*), (center) smoky bat (*Amorphochilus schnablii*), and (right) Spix's disk-winged bat (*Thyroptera tricolor*). Note the characteristic three fleshy projections on the lower lip of the smoky bat. (B) Adhesive disks of a disk-winged bat are larger on the wrist (left) than on the foot (right).

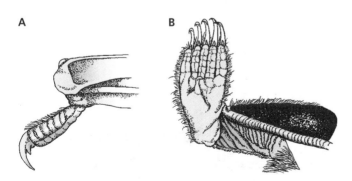

Figure 13.23 **Claws of the New Zealand short-tailed bat.** (A) The basal talon on the thumb is particularly evident. (B) Small talons also occur on the claws of the feet. Like flightless birds, these bats are highly adapted for terrestrial locomotion. *Adapted from G. E. Dobson, 1876, On Peculiar Structures in the Feet of Certain Species of Mammals Which Enable Them To Walk on Smooth Perpendicular Surfaces, Proceedings of the Zoological Society.*

like (figure 13.24), giving the family its common name. The forehead region is domed, and there is no nose leaf. Adult males have an unusual mass of glandular sensory cells, the "natalid organ," below the skin of the forehead (Hoyt and Baker 1980). The wings are long and slim (a high aspect ratio), and natalids have an erratic, butterfly-like flight as they consume insects in flight. These bats roost singly or in small clusters far back in caves and tunnels in lowland areas with high humidity. Morphological analyses suggest that natalids are most closely related to two other small New World families, the Furipteridae (the thumbless bats) and the Thyropteridae (the disk-winged or sucker-footed bats—see figure 13.11A). Molecular analyses (Dávalos 2005) place funnel-eared bats in close association with the Vespertilionidae and Molossidae, however.

Furipteridae

The smoky bats include two genera, each with a single species. The thumbless bat (*Furipterus horrens*) is found from Costa Rica to Peru and Brazil, whereas the smoky bat (*Amorphochilus schnablii*), a threatened species, occurs along coastal areas and interior valleys from western Ecuador to northwestern Chile. These bats are small; the body mass of *F. horrens* is only 3 g. They resemble funnel-eared bats in their general appearance (figure 13.24A), wing shape, inconspicuous nose leaf, ear shape, small tragus, and dense fur. Furipterids also are known as thumbless bats because this digit is so small and almost completely enclosed in the wing membrane that it appears to be absent. Smoky bats are insectivorous, and inhabit moist, low-elevation forested areas, where they roost in caves.

Thyropteridae

There are currently three species of disk-winged bats that are recognized. Peter's disk-winged bat (*Thyroptera discifera*) occurs from Nicaragua southward to northern South America. Spix's disk-winged bat (*T. tricolor*) is distributed from southern Mexico to Bolivia and southern Brazil. LaVal's disk-winged bat (*T. lavali*) is a rare species known from limited localities in northern South America. These bats are small; the body mass of *T. tricolor* is about 4 g. They are similar to the funnel-eared bats and smoky bats in features such as wing shape, high domed forehead, ear shape, and tragus (see figure 13.24A). The common name of these bats refers to their most striking anatomical feature: round, concave, suction disks at the base of the thumbs and on the soles of the feet (Wimsatt and Villa-R. 1970). The larger thumb disks are attached by a short stalk, or **pedicle** (see figure 13.24B). Disks are used to cling to stems, leaves, and other smooth, hard surfaces. A single disk is capable of supporting an individual's body mass. These species occupy humid forests, usually near water, where they feed on insects. They roost during the day, adhering by their disks, inside curled leaves of plants,

often the fronds of banana trees (Genus *Heliconia*). Several individuals, possibly a family group, sometimes share a single leaf (Findley and Wilson 1974), which usually is used as a roosting site for only a brief time (Vonhof and Fenton 2004). Characteristics of the echolocation calls have been described for *T. discifera* (Tschapka et al. 2000) and *T. tricolor* (Fenton et al. 1999). Echolocation pulses are low intensity for both species and include FM harmonics (see table 13.2). Solari et al. (2004), using morphological and genetic data, found that *T. tricolor* and *T. lavali* were sister taxa, with *T. discifera* the basal member of the genus.

Myzopodidae

This monotypic family includes the Old World sucker-footed bat (*Myzopoda aurita*), the only bat species **endemic** (naturally occurring in a limited geographic area) to Madagascar. Like the New World disk-winged bats (Thyropteridae), *M. aurita* has suction disks at the base of the thumbs and soles of the feet. There is no stalk associated with the thumb disk in *M. aurita*, however. The discs in myzopodids and thyropterids also differ anatomically (Schliemann and Maas 1978). Disks probably evolved independently in each family, another example of convergent evolution. A mushroom-shaped process at the base of the ear (figure 13.25) in this species is unique. *Myzopoda aurita* is insectivorous and probably restricted to rainforest habitats on the east coast of the island. The species is extremely rare; very few specimens have ever been collected, and little is known of its biology (Goepfert and Wasserthal 1995). It currently is considered threatened.

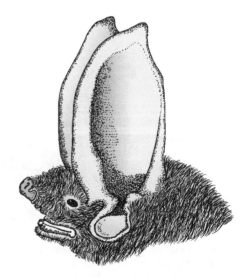

Figure 13.25 Old World sucker-footed bat. Notice the fleshy, mushroom-shaped structure at the base of the ear. Its function is unknown. *Adapted from O. Thomas, 1904,* On the Osteology and Systematic Position of the Rare Malagasy Bat Myzopoda aurita, *Proceedings Zoological Society.*

Vespertilionidae

There are more species of common bats than any other family of chiropterans. They enjoy a worldwide distribution in tropical, temperate, and desert habitats and are absent only from polar, high-elevation areas, and some oceanic islands. Simmons (2005a) placed the 48 genera and 407 species of vespertilionids in 6 subfamilies (table 13.5). Within the subfamily Myotinae, Genus *Myotis* (with 103 species) is the most widely distributed genus of bat, occurring worldwide.

As might be expected of so large a family, there is a great deal of diversity in size, pelage coloration, and morphological features. Head and body length ranges from 3 to 10 cm, and adult body mass from 4 to 50 g. The eyes are small, and a nose leaf is absent, except for a rudimentary one in the New Guinea big-eared bat (*Pharotis imogene*) and the long-eared bats, Genus *Nyctophilus*. The ears of some species are exceptionally large, up to 4 cm in *Corynorhinus* (figure 13.26). A well-developed tragus usually is present. The total number of teeth varies from 28 to 38, usually through loss of premolars. The number of molars always is 3/3, and the cusp pattern of upper molars

is dilambdodont. Species generally are insectivorous, although there are a few exceptions. For example, the fish-eating bat (*Myotis vivesi*) has adaptations similar to those of noctilionids, whereas the pallid bat (*Antrozous pallidus*) feeds on scorpions, beetles, other ground-dwelling prey, and sometimes fruit.

Caves are often favored roosting sites, although vespertilionids use many different types of shelter. Depending on the species, roosting may be solitary, in small clusters, or in large groups. Roosts may contain more than one species, although they generally do not form mixed clusters. Because insects are unavailable during winter, species in temperate regions either migrate or hibernate. Like other hibernators, bats arouse periodically throughout the winter. Although they may fly, they do not feed (Whitaker and Rissler 1993). Hibernating species may exhibit delayed fertilization, with females segregating into maternity colonies. In tropical species, breeding (without delayed fertilization) may take place throughout the year. Nevertheless, a single neonate is typical, although again there are exceptions. Some of the smaller species have two young; litters of up to four young may occur in the genus *Lasiurus*.

In North America, both the gray bat and the Indiana bat (Carter and Feldhamer 2005) are endangered. Numerous other species worldwide are threatened or endangered.

Table 13.5 Subfamilies, number of genera, and number of species within the Vespertilionidae*

Subfamily	Number of Genera	Number of Species
Vespertilionidae	38	238
Antrozoinae	2	2
Myotinae	3	106
Miniopterinae	1	19
Murininae	2	19
Kerivoulinae	2	23

From Simmons (2005a).
* Compare with Hill and Smith (1984) and Koopman (1993)

Molossidae

The 16 genera and 100 species of free-tailed bats are distributed in the Old World from southern Europe, Africa, Asia, and Australia, to the Fiji Islands. In the New World, they occur from southwestern Canada, through the United States and Caribbean, to Central America, and most of South America. Molossids occur in habitats from forests to deserts. They are medium-sized to large bats, with head and body length from 4 to 13 cm. The large ears meet on the forehead and point forward; there is a tragus but no nose leaf. The common name results from the tail extending well beyond the outer edge of the uropatagium (see figure 13.2A [g]). The family name derives from the Greek *mollossus*, or mastiff, a reference to the general doglike snout of these bats.

The hair in molossids is short and fine. In the two species of *Cheiromeles*, the hair is so short and sparse that they are called "naked bats."" Schutt and Simmons (2001) described unique characteristics of the thumb and calcar for the greater naked bat (*Cheiromeles torquatus*) and the lesser naked bat (*C. parvidens*). They suggested these structures may be related to a greater amount of quadrupedal locomotion in these species compared to other bats. Other free-tailed bats often have sensitive, erectile tufts, or crests, of hair between the ears, associated with glands. Crests are especially evident in male lesser mastiff bats (Genus *Chaerephon*). In the Genera *Tadarida* and *Molossus*, males have throat glands. Molossids have high aspect ratio wings (see figure 13.4F) and generally are swift aerial insectivores.

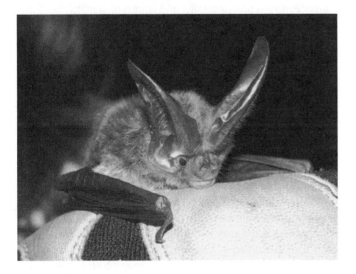

Figure 13.26 Large ears occur on some vespertilionid bats. Huge ears are evident on this Rafinesque's big-eared bat (*Corynorhinus rafinesquinii*).

Figure 13.27 Brazilian free-tailed bat (*Tadarida brasiliensis*). This common molossid may form aggregations of millions of individuals.

The dentition is typical of insectivorous bats; total number of teeth ranges from 26 to 32, depending on the species. Lee and McCracken (2005) analyzed fecal pellets from three colonies of Brazilian free-tailed bats (*Tadarida brasiliensis*; figure 13.27) in Texas and documented extremely diverse remains from 12 orders and 35 families of insects. Molossids do not hibernate, although species in north temperate regions, such as the mastiff bat (*Eumops perotis*) and Brazilian free-tailed bat, may migrate and go into torpor during the winter. They roost in caves, buildings, trees, and crevices and on foliage. The white-striped free-tailed bat (*T. australis*) in Australia selects older, larger trees with large cavities as roost sites (Rhodes and Wardell-Johnson 2006). They may be solitary, roost in small groups, or form extremely large aggregations. Brazilian free-tailed bats are especially gregarious; large numbers congregate in the southwestern United States. For example, an estimated 20 million *T. brasiliensis* have been reported from a single cave in central Texas. An estimated 50 million individuals once occupied one cave in Arizona (Cockrum 1969). These are the largest aggregations of any species of vertebrate. Although populations vary in their tendency to migrate and the routes they take, Russell et al. (2005) found no genetic structuring related to migratory groups. An interesting historical note is the experimentation done with *T. brasiliensis* during World War II in an attempt to use them to carry incendiary devices into enemy territory as "bat bombs" (Couffer 1992). This elaborate plan (called Project X-ray), which included dropping caged bats by parachute, was never implemented. Critically endangered molossids include Gallagher's free-tailed bat (*Chaerephon gallagheri*), the São Tomé free-tailed bat (*C. tomensis*), and Wroughton's bat (*Otomops wroughtoni*).

SUMMARY

Second only to rodents in the number of species in the order, bats represent unparalleled variety among the mammals. No order is more diverse than chiropterans in the number of feeding niches they fill. Most bats are insectivores; numerous other species are specialized to feed on fruit, nectar, or pollen. Other bats are carnivores, taking small terrestrial vertebrate prey, including other bats, and a few species feed on small fish or a diet of blood. Bats also exhibit diversity in their reproductive patterns and roosting ecology. Although many species exhibit spontaneous ovulation, vespertilionids in northern areas employ delayed fertilization in conjunction with hibernation or migration. Other species have delayed implantation or delayed development. Roosting ecology likewise varies among species. Bats may roost in large colonies, in smaller aggregations, in clusters of a few individuals, or be solitary. Roosting habits depend on interrelated characteristics of social organization, foraging ecology, predator avoidance, and reproduction.

Morphological diversity also characterizes the two suborders: the megachiropterans and the microchiropterans. Most bats are small—many less than 10 g body mass. The largest bats are the megachiropterans, Family Pteropodidae, that may reach 1,200 g and have wingspans close to 2 m. Size is only one of the differences between the suborders. Unlike microchiropterans, the facial characteristics of megachiropterans are simple. They have no nose leaf or other ornamentation, the eyes are relatively large, and the ears are without a tragus. Several other differences exist between the two suborders, including echolocation. Microchiropterans have a sophisticated system of acoustic orientation called echolocation. They sense their immediate environment and "locate" food or objects from the echoes returned from high-frequency sound pulses they generate and emit through either the mouth or nostrils. Although many similarities occur in the general aspects of echolocation in bats, species differ in the characteristics of sound pulse duration, timing, frequency modulation, intensity, and length of signals. Echolocation does not occur in megachiropterans. All these differences have led some authorities to suggest that the pteropodids were derived independently through evolutionary time, that is, bats are diphyletic. Recent genetic evidence suggests, however, that bats are monophyletic, that is, they evolved from a common ancestor, although relationships between families remain unresolved.

Despite the diversity found throughout the order, there is an obvious unifying characteristic; all bats have highly modified forearms and hands that form wings, and they are the only mammals that fly. Wings are the primary structures used to create the upward lift and forward thrust necessary for flight. Although there are common elements in wing structure for flight, there is diversity in the shape of wings. Wing shape and size, quantified in measures of aspect ratio and wing loading capacity, reflect the habitat, foraging characteristics, degree of maneuverability, and general life history characteristics of bat species.

Several species of bats worldwide are endangered or have recently gone extinct because of loss of roosting and foraging habitat, cave closure, insecticide accumulation, and other adverse influences. Despite the many positive economic benefits provided by bats, this fascinating and highly sophisticated group of mammals continues to suffer because of human ignorance, traditional misconceptions, and unfounded fears.

SUGGESTED READINGS

Adams, R. A., and S. C. Pedersen. 2000. Ontogeny, functional ecology, and evolution of bats. Cambridge Univ. Press, Cambridge.

Brigham, R. M., E. K. V. Kalko, G. Jones, S. Parsons, and H. J. G. A. Limpens (eds.). 2004. Bat echolocation research: tools, techniques, and analysis. Bat Conservation International, Austin, TX.

Crichton, E. C., and P. H. Krutzsch (eds.). 2000. Reproductive biology of bats. Academic Press, New York.

Hall, L., and G. Richards. 2000. Flying foxes: Fruit and blossom bats of Australia. Krieger Publishing Co., Malabar, FL.

Kunz, T. H., and M. B. Fenton (eds.). 2003. Bat ecology. Univ. Chicago Press, Chicago.

Zubaid, A., G. F. McCracken, and T. H. Kunz (eds.). 2005. Functional and evolutionary ecology of bats. Oxford Univ. Press, New York.

DISCUSSION QUESTIONS

1. Given the active dispersal abilities of bats, speculate as to why the large fruit bats (Pteropodidae) never dispersed to tropical areas of the New World. Likewise, why might the monotypic myzopodid have remained endemic to Madagascar, and the short-tailed bats (Mystacinidae) never dispersed beyond New Zealand?

2. What kinds of technological difficulties do you suspect may have limited early investigators in their studies of bat echolocation? What kinds of technological advances have allowed researchers to study bat echolocation in the field as well as the laboratory?

3. Does the fact that megachiropterans and microchiropterans differ so markedly in their use of echolocation and associated morphological adaptations argue against the idea that these two groups are monophyletic? What arguments can you make for and against echolocation as a key feature in this debate?

4. Vampire bats have highly specialized diets consisting only of blood. How might this specialization have arisen during the evolution of these bats? Why do you think vampire bats are restricted in their distribution to tropical areas? Given convergent evolution in other features of bats, why has sanguinivory never arisen in any Old World bats?

5. Generally, there is an inverse relationship in mammals between body size and litter size—small species have larger litters than large species. Nonetheless, most bats are small, but they also have small litters (usually one young). Why?

CHAPTER 14

Primates

Primates, one of the more ancient mammalian orders, probably originated in the Cretaceous, though its fossil record begins in the early Paleocene with representatives of the extinct Suborder Plesiadapiformes. Following radiations in the Paleocene and Eocene, primates dispersed into tropical areas of all continents except Australia by the mid-Tertiary. These radiations resulted in the two recognized suborders of living primates (Kay et al. 1997). The Suborder Strepsirhini consists of seven living families—Lemuridae, Lepilemuridae, Indriidae, Galagidae, Daubentoniidae, Cheirogaleidae, and Lorisidae (table 14.1)—and three extinct ones—Adapidae, Archaeolemuridae, and Palaeopropithecidae (Szalay and Delson 1979; Fleagle 1988; Groves 2001; Groves 2005). The Suborder Haplorhini consists of eight living families—Tarsiidae, Cebidae, Aotidae, Pitheciidae, Atelidae, Cercopithecidae, Hylobatidae, and Hominidae—and four extinct ones—Omomyidae, Parapithecidae, Oreopithecidae, and Pliopithecidae. The fossil evidence concerning the evolution of these primate groups has been reviewed by Kay and colleagues (1997). In this chapter, we first review the ordinal traits of primates, examine some general primate morphology and the fossil history for this order, and then note the economics and conservation of primates. Our major focus is on accounts of the families of living primates. We have more interest in the various primates than in any other group of mammals because we, as *Homo sapiens*, are part of this order.

Ordinal Characteristics

The name *primates* means "the first animals," which reflects an early (incorrect) and anthropocentric bias that gives special importance to the group that contains humans. Primates are characterized by several ordinal traits, although many primates are quite generalized and thus, in a sense, defy the sort of clear listing of specialized traits that can be made for many mammalian orders. In 1873, Mivart characterized the primates as "an **unguiculate, claviculate,** placental mammal with

Table 14.1 Living primates are divided into 2 suborders and 15 families comprising a total of 362 species

	Number of living species	Distribution
Suborder Strepsirhini		
Family Lemuridae	19	Madagascar, Comoros Islands
Family Cheirogaleidae	21	Madagascar
Family Lepilemuridae	8	Madagascar
Family Indridae	11	Madagascar
Family Daubentoniidae	1	Madagascar
Family Lorisidae	9	Central Africa, Southeast Asia, Sri Lanka
Family Galagidae	19	Africa
Suborder Haplorhini		
Family Tarsiidae	7	Indonesia, Philippines
Family Cebidae	56	Central and South America
Family Aotidae	8	Central and South America
Family Pithecidae	40	Central and South America
Family Atelidae	24	Central and South America
Family Cercopithecidae	132	Africa, Asia, Indonesia
Family Hylobatidae	14	Southeast Asia, China, Indonesia
Family Hominidae	7	Worldwide

Data from D. E. Wilson and D. M. Reeder, 2005, Mammal Species of the World, *Smithsonian Institution Press.*

orbits encircled by bone; three kinds of teeth at least at one time of life; brain always with a posterior lobe and a **calcarine fissure;** the innermost digits of at least one pair of extremities opposable; hallux with a flat nail or none; a well-marked cecum; penis pendulous; testes scrotal; always two pectoral **mammae.**" Although this list characterizes primates, none of these traits is unique to primates. What defines a primate has been the subject of considerable debate in recent decades (Schwartz et al. 1978; Luckett 1980a; Fleagle 1988). A major characteristic of primates is the cheekteeth. They are generally bunodont, having four sides like a square with four rounded cusps, and brachyodont, with a low crown.

Various investigators have extended this characterization by discussing the evolutionary trends that help to define the primates (Clark 1959; Macdonald 1984; Groves 2001).

1. The hands and digits have become refined, with increased mobility of the digits, nails replacing claws, and sensitive pads on the digits with friction ridges—which are important for grasping.
2. Both the absolute and relative brain sizes have increased, with elaboration of more **cerebral cortex.** A trade-off has occurred between increased dependence on sight, which is correlated with enlarged brain areas that are associated with vision, and decreases in brain areas associated with olfaction.
3. The muzzle region is shortened in primates, associated with a decline in the use of smell and a concomitant shift to binocular, stereoscopic, color vision.
4. Reproduction occurs at a slower rate, sexual maturity is delayed, and life spans are longer.
5. The diet has progressively shifted to greater reliance on fruits, seeds, and foliage, with a decline in the amount of animal matter consumed.
6. Social and mating systems have changed from ones based on overlapping male and female home ranges or territories to a diverse array of complex socio-spatial and breeding patterns.

Primates occupy a wide variety of habitats. Their geographical distribution is primarily tropical and subtropical, although there are some exceptions, such as the Japanese macaque (*Macaca fuscata*), which lives in areas that have considerable snow.

At one time, both the tree shrews (Order Scandentia) and the colugos (Order Dermoptera) were included in the same order with the primates. In past classifications, (Luckett and Szalay 1975; Fleagle 1988; Martin 1990), the primates were divided into prosimians and anthropoids. Here, we use the system proposed by Groves (2005; also Groves 2000; see figure 14.1). Although no investigators have used the emerging molecular technologies to attempt a phylogeny of all primates in a single analysis, numerous families and genera within the order have been analyzed. Ongoing research to refine the evolutionary tree for primate lineages includes the use of nuclear DNA

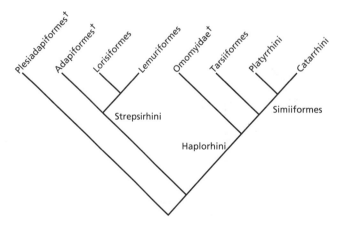

Figure 14.1 Primate relationships. Phylogeny of major primate groups proposed by Kay et al. (1997), with group names from Groves (2005). Extinct taxa are indicated with a dagger. The positions of Adapiformes and Omomyidae have been disputed. *Modified from Benton (2005).*

(Clisson et al. 2000), mitochondrial DNA (mtDNA) (Weinreich 2001), and DNA from ancient retroviruses that are inserted into the DNA (Johnson and Coffin 1999). Primate evolution and distribution have been affected by plate tectonics, climate, and chance (Fleagle and Gilbert 2006). The number of recognized species has increased dramatically in the past several decades, largely due to new techniques to distinguish among different groups, but also with the discovery of new species in Asia, Africa, and Southeast Asia.

Morphology

The primates can be distinguished by certain aspects of their skeletal morphology as well as characteristics of soft anatomy. All modern primates have a bony postorbital bar (figure 14.2) with the eyes generally directed forward. The snout, or muzzle, is reduced in most primates, as are the olfactory lobes in the brain; both traits reflect the relative diminution of the sense of smell in primates relative to their mammalian ancestral stock. The braincase surrounding the relatively large brain is enlarged. All primates share a petrosal-covered **auditory bulla,** the tympanic floor of which is derived from only the petrosal plate and ectotympanic bone. Primate jaws move mostly in the vertical plane, in contrast to many other mammals in whom considerable horizontal jaw movement takes place. The jaw symphysis became progressively more ossified in advanced primates. The cheekteeth of primates (figure 14.3) have been described as bunodont and brachyodont with the sides of the upper molar relatively filled out; the teeth are relatively complex and adapted primarily for grinding and secondarily for shearing (Schwartz 1986).

Primates are plantigrade, and usually pentadactyl. In some species, the hallux or pollex or both are reduced or absent. For most primates, the digits terminate in nails rather than claws.

Fossil History

The earliest radiation of primates in Europe and North America during the Paleocene and Eocene epochs produced a number of forms grouped together in the Plesiadapiformes (figure 14.4; Kay et al. 1997). These small, squirrel-like mammals possessed an elongated skull with the eye orbits coming together with the temporal region of the skull, not separated from the back of the skull by a bony plate as in most other primates (Fleagle 1988; Martin 1990). The radius and ulna in the forelimb and tibia and fibula of the hind limb were entirely separate, permitting rotation of the feet. The digits of the hands and feet were long enough to enable these primates to grasp the limbs and branches of trees. The terminal digits still had claws.

The Adapiformes, extinct relatives of today's lemurs, evolved during the early Eocene (Gebo et al. 2001). They ranged through the tropical and subtropical forests that

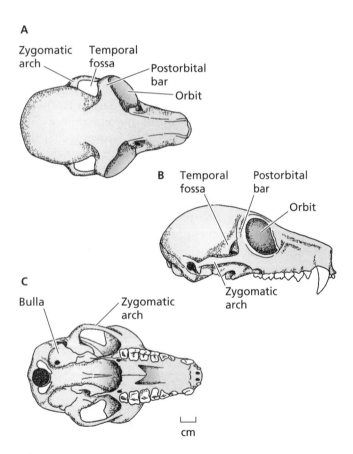

Figure 14.2 Strepsirhine skull. Three views (A) dorsal, (B) lateral, and (C) ventral of the skull of a strepsirhine (*Varecia variegata*). Key characteristics include the forward-facing eyes and the lateral postorbital bar. Strepsirhine primates have an orbital cavity confluent with the temporal fossa, in contrast to haplorhines, whose postorbital plate marks off the rear of the orbit. *Adapted from R. D. Martin, 1990*, Primate Origins and Evolution, *Princeton Univ. Press.*

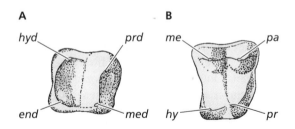

Figure 14.3 Primate teeth. The basic four-cusped crown pattern of primates: (A) lower left molar and (B) the upper right molar. Abbreviations: *end* = entoconid; *hy* = hypocone; *hyd* = hypoconid; *me* = metacone; *med* = metaconid; *pa* = paracone; *pr* = protocone; *prd* = protoconid. *Adapted from T. A. Vaughan, 1986*, Mammalogy, *3rd ed., W. B. Saunders Publishing.*

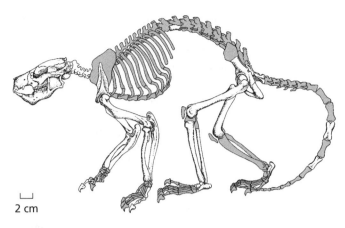

2 cm

Figure 14.4 Fossil primate. One of the earliest ancestral primates, *Plesiadapsis tricuspidens*. Note in particular the flexible spine and long tail. Skeletal portions shown with dark lines are known from fossils; portions shown in gray are reconstructed. *Adapted from I. Tattersall, 1970,* Man's Ancestors: An Introduction to Primate and Human Evolution, *John Murray.*

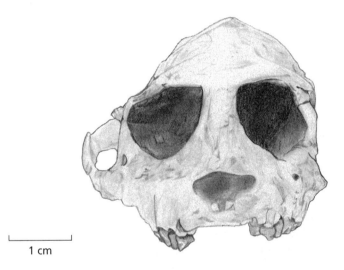

1 cm

Figure 14.5 Skull of *Parapithecus*, an anthropoidean primate from the Oligocene. *Drawing by Lisa Russell after E.L. Simon, 2001. Proceedings of the National Academy of Science USA 98: 7892-9897.*

then covered North America and Eurasia. Two genera in this group, *Smilodectes* and *Notharctus*, had a bony postorbital bar completely enclosing the orbit. No diastema occurred between the incisors and canines, resulting in a continuous tooth row. *Smilodectes* had eyes that were directed forward, making binocular, stereoscopic vision possible. The first digit of each foot was set apart from the others, suggesting a degree of opposability. These small primates were capable climbers, grasping branches as they moved through the trees. Other features associated with their mode of existence were a flexible back and long tail, useful for climbing and balance.

Another early haplorhine group, the Omomyidae, arose during the Eocene. *Tetonius* from North America and *Necrolemur* from Europe exemplify this group. These animals had large orbits located in a short, flat face; the large eyes suggest a nocturnal activity rhythm. Short jaws contained large canines and rather primitive molars. Fossil evi-

dence indicates that lemur-like primates (Lemuriformes) diversified in Asia, particularly on the Indian subcontinent (Marivaux et al. 2001). Fossils of one genus (*Afrotarsius*) from the Oligocene epoch are nearly identical to modern tarsiers in leg specializations that provide for the unique locomotion of members of this family of primates (Rasmussen et al. 1998).

Fossil remains of New World monkeys (Families Cebidae, Aotidae, Pitheciidae, Atelidae) are rare, making an assessment of platyrrhine origins difficult. However, these families evolved from African ancestors that somehow crossed the Atlantic Ocean and underwent adaptive radiation in South America in the Oligocene.

The chain of fossil evidence for the catarrhine primates of Asia and Africa (Old World primates) is more complete, extending from the Eocene to the Recent epochs (Fleagle 1988; Martin 1990). The fossil evidence from North Africa for one genus (*Parapithecus*) is relatively complete. This primate was characterized by a jaw that was only 5 cm in length but was deep, with a **condyle** located high on an **ascending ramus** to articulate with the skull. This latter trait is characteristic of modern cercopithecoid primates.

Paleontologists previously projected that the earliest primates likely diverged from their mammalian ancestors approximately 65 mya. In contrast, the evidence from DNA-based studies suggested a divergence at about 90 mya. Tavare et al. (2002) used a mathematical analysis of the fossil record to estimate an age of 81.5 mya for the most recent common ancestor of living primates. They accomplished this projection by entering three factors into their model. First, they used a figure of 235 as the minimum number of current primate species. Second, they added information on the morphological diversity of primates, both living and fossil. Lastly, they made the assumption that each species was extant for about 2.5 million years (my). The earliest primate is projected to have evolved in the tropics and moved northward over time. Both the dates obtained from the model and the projected tropical origin are subject to debate (Bower 2002).

The most widely accepted proposal for the development of traits that characterize primates is the arboreal theory (Martin 1979; but see Cartmill 1972). An arboreal ancestry explains the adaptations of the visual system, including the forward-facing eyes and enlarged visual cortex of the brain. It also helps explain many of the skeletal adaptations of the limbs and digits, which permit easy and safe movement in trees. The theory also accounts for hands gradually replacing the mouth for gathering and handling food.

Economics and Conservation

Humans are interested in primates for many reasons. In the Hindu religion, primates are considered sacred and so are protected (figure 14.6). Several species of macaques and langurs enjoy this special relationship with humans. In other cultures, primates are taken for food, as in some areas of West Africa and South America. In these instances, overhunting can reduce or eliminate particular species. Indeed, this has happened to the western red colobus monkey (*Colobus badius*) in Liberia, Cameroon, Ghana, and the Republic of the Congo (Gartlan and Struhsaker 1972; Leutenegger 1976; Oates et al. 2000). Still other primates have been and continue to be important for human medicine. Species such as *Macaca mulatta*, *Pan troglodytes*, and *Aotus trivirgatus*, among others, are used as animal models for various diseases. Probably the most widespread use of a nonhuman primate has been the testing of the Salk polio vaccine on large numbers of rhesus macaques. The capture and export of these monkeys from India reached alarming proportions by the 1960s and 1970s, eventually leading to a ban on their export. This resulted in the establishment of several colonies of rhesus monkey as breeding "farms" in places such as the Florida Keys and Morgan's Island, South Carolina, for the production of the stocks of monkeys needed for vaccine testing and other research uses.

The most significant threat to the continued existence of many primate populations comes from the destruction of their habitat by humans. In Indonesia and Brazil, and on the island of Madagascar and elsewhere, humans destroy vast areas of forest for lumber, other wood products, and fuel. In addition, major forested areas have been cleared in regions of Madagascar, Africa, and South America for agriculture to feed the ever-enlarging human population. These threats, and others such as the continuing warfare in areas of Rwanda inhabited by the western gorilla (*Gorilla gorilla*), may result in the demise of many species. The future holds little promise for a number of primate species.

The Ebola virus, which produces a hemorrhagic fever, has decimated some populations of gorillas and chimpanzees in the Congo River basin (Vogel 2006). This serious threat appears to be expanding, and initiating measures for protection are either impractical or expensive. Another recent threat to gorillas is the mining of coltan, a mineral from which tantalum is derived (National Public Radio 2001). Tantalum is used for coating components in a variety of modern electronic devices, including cell phones and computer parts. A high demand for tantalum has fueled a price increase from about $30/lb to more than $400/lb. The area where coltan is most readily mined is in the eastern Congo. The significant influx of miners has resulted in the destruction of gorilla habitat as well as the shooting of the animals for their meat. It is thus quite difficult to predict or even imagine the source of the next threat with regard to primate conservation.

Conservation efforts, specifically with respect to primates, include establishing reserves where the habitat should be left undisturbed. This has been done with some positive effect, for example, in Sulawesi, Indonesia, where reserves are designed to protect primate diversity that includes *Tarsius tarsier*, *Macaca nigra*, *M. tonkeana*, and several other species. Another type of conservation effort involving a limited number of species has been to develop captive breeding stocks of an endangered species for reintroduction into native habitat, as has been done with the golden lion tamarin (*Leontopithecus rosalia*) in Brazil (Magnanini et al. 1975; Kleiman 1976; Mittermeir and Cheney 1986; see chapter 29). This program has progressed to the point where golden lion tamarins placed into the wild in several locales, with suitable habitats of sufficient size that can be preserved, are now successfully reproducing, becoming viable populations. The outcome of both types of conservation measures will be known only several decades from now as researchers continue to monitor the diminishing population levels of a number of primate species.

Figure 14.6 Urban monkeys. A group of rhesus macaques (*Macaca mulatta*) forage on a street in a village in northern India. The monkeys are free to come and go because Hindu beliefs protect them.

Strepsirhine Primates

The living Strepsirhini (lemurs, lorises, and bushbabies) were formerly called Prosimians (figure 14.7). The seven families described here are concentrated on Madagascar but include species that live in Africa, Southeast Asia, and the Malay Archipelago. Several traits distinguish strepsirhine from haplorhine primates. One obvious external trait involves the **rhinarium,** an area of moist, hairless

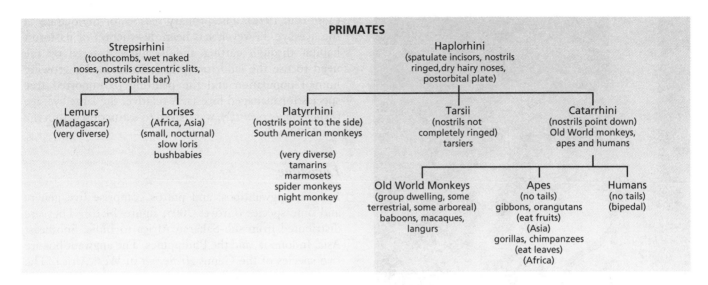

Figure 14.7 **Primate phylogeny and characteristics.** The tree diagram illustrates the relationships among primate groups and some key traits for each group. *Adapted from R. D. Martin, 1990,* Primate Origins and Evolution, *Princeton Univ. Press.*

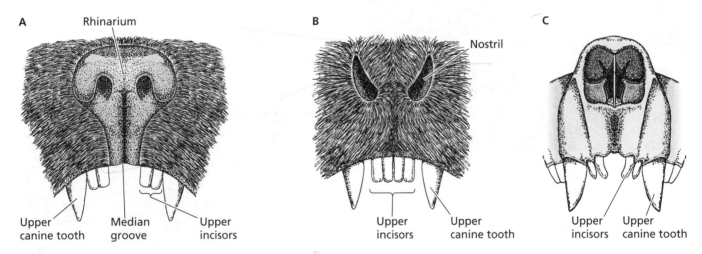

Figure 14.8 **Snout variation.** Snout of (A) a strepsirhine compared to that of (B) a haplorhine. Strepsirhines retain the ancestral rhinarium surrounding the nostrils. Hair obliterates the rhinarium in haplorhines. (C) Anterior view of the skull of a strepsirhine primate. *Adapted from R. D. Martin, 1990,* Primate Origins and Evolution, *Princeton Univ. Press.*

skin surrounding the nostrils (figure 14.8). The maximum dental formula for lemurs and lorises is 2/2, 1/1, 3/3, 3/3. Strepsirhines have a **bicornuate uterus** (uterus with two horns), which contrasts with the fused simplex uterus of the haplorhines. The noninvasive, **epitheliochorial placenta** in lemurs and lorises also distinguishes them from the haplorhines. In general, strepsirhines produce neonates that are smaller relative to the mother's body weight than those of haplorhines (Leutenegger 1973). Martin (1990) and Kay et al. (1997) concluded that both the Strepsirhini (lemurs and lorises) and Haplorhini (tarsiers and simian primates) are of monophyletic groups (see figure 14.7). Various molecular techniques, including immunodiffusion comparisons of albumins (Dene et al. 1976; Sarich and Cronin 1976), amino acid sequences (de Jong and Goodman 1988), and DNA hybridization data (Bonner et al. 1980) support this conclusion for the living

taxa, but cannot evaluate the placement of adapiforms among strepsirhines or omomyids among haplorhines. A summary of lemur biology (Gould and Sauther 2006) follows a somewhat different taxonomy than that used for this book, but the information regarding all aspects of lemur biology, evolution, and conservation is quite current and most useful for those wanting a more in-depth examination of the strepsirhine primates.

FAMILIES

Daubentoniidae

The sole species in this family is the aye-aye (*Daubentonia madagascariensis*; figure 14.9), which lives. Aye-ayes have a body length of about 400 mm and bushy tails that are

Figure 14.9 Daubentoniidae. The aye-aye uses its specially adapted digit to probe for insects.

longer than their bodies (ca. 600 mm). They have coarse dark brown or black fur. The face is short and broad with a tapered muzzle. The large eyes face forward.

Aye-ayes inhabit lowland rainforest. They are generally solitary, except for mothers with young, and have home ranges of approximately 5 ha (Petter and Petter 1967; Petter 1972). Females nurse their young with a single pair of inguinal mammae. Members of this species are nocturnal insectivores. The hands have particularly long fingers, and the middle finger bears a long, wirelike claw that is used for extracting insects from wood. They also use this elongated digit and claw to remove the pulp from fruits such as mangos and coconuts. The aye-aye possesses chisel-like incisors used for gnawing and chewing, much in the fashion of rodents.

The species originally inhabited regions of eastern and northwestern Madagascar (Petter 1962a, b; 1977); it is extirpated or nearly extirpated from all habitats except possibly Nosy Mangabe Island, where it was introduced in 1966 and 1967 (Petter and Petter-Rousseaux 1979). At one time, this species was protected by local custom because the people believed that anyone who harmed one would die. Conversely, they were then killed if found because they were thought to be evil and bring bad luck

(Tattersall 1972). The primary reason for the demise of the aye-aye, however, has been destruction of its forest habitat through cutting and burning, fostered by the need to use the land for agriculture to feed a growing human population and the planting of imported tree species for managed forests. A relative, the large aye-aye (*Daubentonia robustus*), was driven to extinction within the last 1000 years.

Lorisidae

Lorises, angwantibos, and pottos comprise five genera and nine species (Groves 2005; figure 14.10). They are distributed from sub-Saharan Africa to India, Southeast Asia, Indonesia, and the Philippines. The angwantibos are two species of the Genus *Arctocebus* in West Africa. The two species of slender loris (Genus *Loris*) are found in southern India and Sri Lanka. The three species of slow lorises (Genus *Nycticebus*) live in much of Southeast Asia and on neighboring islands. Groves (1998), using a series of morphometric measurements and discriminant function analyses, proposed additional species-level taxa for both genera of lorises, noting a taxonomic complexity and biogeographical differentiation not heretofore examined

Figure 14.10 Loridae. The slender loris is native to India and Sri Lanka. Dark fur accentuates the large eyes.

in these groups. The single species of potto (*Perodicticus potto*) is found in portions of West and Central Africa. The single false potto (*Pseudopotto martini*) is known from a single specimen from Cameroon. The Lorisidae formerly contained the galagos, which have been given separate familial status (Jenkins 1987; Groves 2001).

The smallest members of this family are the angwantibos and slender loris, being only 180–250 mm in head and body length and weighing 85–500 g. The slow lorises and pottos are larger, ranging from 200–400 mm in head and body length and weighing 1.0–1.4 kg. Tails are very short or absent in the Lorisidae, except for the potto, which has a tail about 65 mm in length. All lorisids have thick, woolly fur of darker colors, ranging from brown to gray and black with light underparts (Napier and Napier 1967). They have relatively large forward-facing eyes, with generally flattened faces, although some species have distinct muzzles that may be either pointed or short and rounded.

All lorisids are nocturnal and arboreal. Their habitat varies, depending on the species, from bamboo and evergreen forests to tropical rainforest and includes, for some species, logged areas and shrublands. They are slow but sure climbers, using a strong grip made possible by prehensile hands and feet that have the thumb (pollex) set at nearly a 180° angle from the remaining digits. It is thus termed pseudo-opposable. Many members of the family can climb well in a suspended position, below a limb, as well as locomoting on top of limbs. The index finger (second digit) is reduced in most species to a tubercle (knob), and the second toe is modified as a toilet or grooming claw in Genera *Loris*, *Nycticebus*, and *Perodicticus*. Lorisids are insectivorous, frugivorous, or both.

Lorisids live as single individuals or in pairs. Their spatial relations in the wild are poorly known. They generally have single births, though twinning occurs occasionally for all but the potto. Species within this family engage in marking behavior, using urine or anal glands. They also communicate using vocalizations and limited facial expressions.

Galagidae

In addition to this group being set aside as a separate family, its taxonomy has undergone considerable revision (Jenkins 1987; Groves 1989, 2001; Nash et al. 1989): what was one genus has been split into three genera, and a number of groups formerly given only subspecific status have been elevated to species-level taxa. The three genera are *Euoticus* (two species), *Galago* (14 species), and *Otolemur* (three species). As a group, these primates are known as bushbabies (figure 14.11). All members of this family live in Africa, inhabiting rainforest in West Africa and woodland savanna from Senegal to East Africa and down to southern Africa.

Members of the Family Galagidae are arboreal. A key feature of this group is their mode of locomotion, which involves leaping and bounding from branch to branch and between tree trunks. Two adaptations, well-developed

Figure 14.11 Galagonidae. Lesser bushbabies inhabit dry forests, savanna habitat with trees, and gallery forests from Senegal to Ethiopia and from Somalia to South Africa.

hind limbs and a long tail used for balance, aid in this form of locomotion. Bushbabies are nocturnal and have large eyes that allow them to see well at night. They range in size from the diminutive dwarf bushbaby (*Galago demidoff*), which has a head and body length of 120 mm, tail of 170 mm, and mass of only 60 g, to the much larger thick-tailed bushbaby (*Otolemur crassicaudatus*), which has a head and body length of 320 mm, tail length of 470 mm, and mass of 1.2 kg. Bushbabies are all pentadactyl, with the second toe modified as a toilet claw. Their diet varies from primarily insectivorous to omnivorous, including fruits, grains, and even small mammals. Bushbabies have a soft, woolly fur, ranging in color from gray to brown and russet brown, with lighter undersides.

Field study, limited primarily to *Otolemur crassicaudatus* and *Galago zanzibaricus*, suggests that galagids apparently live in groups of up to seven to nine individuals. For some species, mothers and infants may nest separately. Bushbabies build well-concealed nests in trees or use cavities in hollow trees. Single births or twins occur for all species studied. Mothers nurse their young with two to three pairs of mammae. Early in their infants' development, mothers carry their young with their canines, gripping them by the scruff of the neck. Bushbabies communicate via urine marking and vocalizations, notably the crylike sound made by young bushbabies, from which they get their common name. They also communicate visually with facial expressions and body postures.

Lemuridae

The true lemurs, Family Lemuridae, comprise five genera: *Eulemur* (11 species), *Hapalemur* (four species), *Lemur* (one species), *Prolemur* (one species), and *Varecia* (two species). Lemurids are found only on Madagascar (Tatersall 1982). Members of the group are diurnal or crepuscular and arboreal, but some species, *Lemur catta* for example, spend considerable time on the ground (figure 14.12). They are medium-sized, ranging from 260–450 mm in head and body length, with tails of 250–560 mm, and body mass of 2–4 kg. Phylogenetic relationships among three of the genera in this family were studied using restriction fragment DNA polymorphisms (Jung et al. 1992). The results support the conclusion that *Lemur* and *Hapalemur* diverged as a common stock from the ancestral group that led to our current *Eulemur* group. *Lemur* and *Hapalemur* later diverged, resulting in two genera. More recently, two investigations shed further light on the phylogeny of the Lemuridae (Wyner et al. 2000; Pastorini et al. 2002), though there is still considerable debate. Wyner et al. used morphology and multiple gene partitions, reporting that the genera *Lemur*, *Hapalemur*, and *Varecia* were all recovered as monophyletic in their analyses. In contrast, relationships within *Eulemur* were not as robust. Pastorini and colleagues, using mito-chondrial DNA reported four genera in this family, consisting of all but the genus *Prolemur* noted above. This last was elevated to genus status by Groves (2001).

Given their daytime activity phase, unique among strepsirhines, the eyes of lemurs are smaller than those of their close relatives. They also have a more prominent muzzle than most strepsirhine primates. Members of Genera *Eulemur*, *Lemur*, and *Varecia* have a diet consisting of fruits, flowers, and some vegetation, primarily leaves. That of Genus *Hapalemur* consists of bamboo shoots and reeds. The lower incisors and lower canines form a forward-projecting dental comb used in both autogrooming and allogrooming.

This is the most colorful of the strepsirhine families, with pelage colors ranging from gray and greenish gray to brown and reds of various hues. Almost all species have a distinct ruff of fur around the neck differing in color from that of the main body pelage. All species possess ear tufts. Some species, such as *Eulemur macaco*, exhibit **sexual dichromatism,** that is, males and females have distinctly different pelage colors. All members of the family, except *Lemur catta* (the ring-tailed lemur), locomote by clinging and leaping (chapter 6). Ring-tailed lemurs use considerable quadrupedal walking and climbing. Lemurids have hind limbs considerably longer and more developed than the forelimbs (see figure 14.12). They use their long tails for balance during their bounding movements and as they climb among tree limbs.

Socially, members of Lemuridae are more gregarious than other strepsirhines; *Hapalemur griseus* live in groups of 3–6 and several species of Genus *Eulemur* live in groups of up to 20 or more, including several adults of both sexes (Napier and Napier 1967). Single births are most common, but twins occur occasionally. Females nurse their young with a single pair of pectoral mammae. Young of Genera *Eulemur*, *Lemur*, and *Varecia* carry their infants on their abdomen at right angles to the main body axis. When they are somewhat older, the young may ride on their mother's back. The communication of lemurids includes considerable scent marking of branches and tree trunks, using urine, anal glands, and specialized sternal glands. *Lemur catta* possesses cutaneous arm glands, which are used to mark objects. Lemurids also have a modest vocal repertoire and use both facial expressions and body postures for visual communication. Human population and activities threaten several lemurids. *Prolemur simus*, the greater bamboo lemur, which inhabits a limited range of humid coastal forest in the east central region of Madagascar, is in immediate danger of extinction.

Lepilemuridae

A recent taxonomic revision has removed the Genus *Lepilemur* from the Family Lemuridae and placed it in a separate family of its own, Lepilermuridae, elevating what were seven subspecies of this genus to species level and splitting one species, so that there are now eight species in

Figure 14.12 Lemuridae. The ring-tailed lemur lives in relatively large social groups and is the only prosimian that spends a large proportion of its time in the terrestrial habitat. *Adapted from J. G. Fleagle, 1988,* Primate Adaptation and Evolution, *Academic Press.*

this monogeneric family (Groves 1989, 2001; Groves 2005). Members of the family are known by the common name sportive lemurs. They are restricted to the island of Madagascar, where they inhabit both dry deciduous forests and tropical rainforests. All are nocturnal and arboreal. They have dense, woolly fur that is usually shades of red, mixed with brown or gray. The undersides are pale gray or yellowish white. Some species have a lengthy spinal stripe from the head to the base of the tail. *Lepilemur* is medium-sized, with head and body length of 280–350 mm, tail ranging from 250–280 mm, and body mass of 0.5–1.0 kg (figure 14.13). The diet of sportive lemurs is quite different from that of most other strepsirhines (except possibly the lemurids); they primarily consume leaves but also eat bark, fruits, and flowers.

Lepilemurids locomote by vertical clinging and leaping (chapter 6), moving from one tree trunk to the next with occasional hops on the ground. As an adaptation for this mode of locomotion, the hind limbs are considerably longer than the forelimbs. *Lepilemur* also has a prehensile thumb, which is pseudo-opposable and capable of strong grips on vertical branches.

Socially, the sportive lemurs live solitary lives except for mothers with their infants. Some marking occurs with urine and the glands in the circumanal region. The primary means of communication is a relatively extensive vocal repertoire. Mothers give birth to single young and have been observed carrying their young in their mouths as they leap around. All species of this family are endangered to varying degrees, primarily due to habitat destruction on Madagascar.

Cheirogaleidae

The Cheirogaleidae comprise 5 genera and 21 species, all inhabiting Madagascar. Three of the genera—*Cheirogaleus* (seven species), *Allocebus* (one species), and *Phaner* (four species)—are all given the same common name, dwarf lemurs, whereas members of the fourth genus, *Microcebus* (eight species), are called mouse lemurs (figure 14.14).

Fossil evidence indicates that members of this family were once more widespread, occurring in areas of what is now Pakistan in south central Asia (Marivaux et al. 2001). The molecular phylogeny of Cheirogaleidae has been examined using mtDNA (Pastorini et al. 2001). Pastorini and colleagues used the aye-aye as an outgroup and samples from 26 individuals and analysis of the COIII gene to clarify relationships among taxa of this family (figure 14.15). The data support generic status for Genus *Mirza* as distinct from Genus *Microcebus*.

As the common names imply, all of these primates are small, the dwarf lemurs being 190–300 mm in head and body length, having tails the same length or slightly longer than the head and body length, and weighing from 300 g to slightly less than 1 kg. The mouse lemurs are even smaller, with head and body length of 130–170 mm, tail length of 170–280 mm, and body mass of about 60 g. Members of this family inhabit both wet and dry tropical forests. Because of the forest destruction throughout Madagascar, several of these species, particularly *Allocebus*

Figure 14.13 Lepilemuridae. Sportive lemurs (*Lepilemur mustelinus*) live in the deciduous and humid forests of Madagascar. They sleep during the day in tree hollows and feed at night on leaves supplemented with fruit, flowers, and bark.

Figure 14.14 Cheirogaleidae. Gray lesser mouse lemurs (*Microcebus murinus*) are among the smallest of the primates. They build nests and are primarily arboreal. *Adapted from J. G. Fleagle, 1988,* Primate Adaptation and Evolution, *Academic Press.*

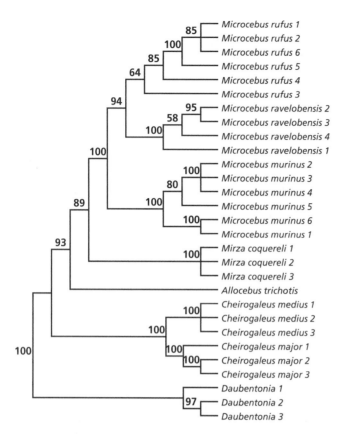

Figure 14.15 Cheirogaleidae systematics. Sequence analyses of the COIII gene from mtDNA was used to assess relationships among taxa in the Family Cheirogaleidae, using *Daubentonia* as an outgroup. The values above the bars are percentages obtained from 2500 replicates of a bootstrapping procedure, indicating the confidence in that node. *From Pastorini et al. (2001).*

trichotis, are thought to be nearly extinct in their natural habitat. The fur of Cheirogaleidae is dense and woolly, varying from gray to reddish brown to cinnamon, with lighter underparts. *Phaner* is characterized by a well-defined spinal stripe that bifurcates on the crown, where it joins dark eye rings. The eyes are large and forward-facing. Muzzles are clearly present in all members of the family, though how pronounced varies from one taxon to another.

All members of this family are arboreal and nocturnal. Locomotion of these small, agile primates is described as scurrying in short runs and darting squirrel-like from place to place. Mouse lemurs do much more leaping than dwarf lemurs. Genera *Cheirogaleus* and *Allocebus* are primarily frugivorous, Genus *Microcebus* is primarily insectivorous, and Genus *Phaner* consumes both insects and fruits.

Cheirogelids are solitary or found in pairs. Females give birth to one to three young per pregnancy and nurse them with two pairs of mammae, one inguinal and one pectoral. Communication apparently involves less scent marking than in most other strepsirhines, although more research is needed. Cheirogelids also communicate through vocalizations, facial expressions, and some body postures.

Indriidae

This family consists of three genera—*Avahi* (three species), *Indri* (one species), and *Propithecus* (seven species)—which occur only on Madagascar (figure 14.16). Much of the revision of this family, expanding greatly the number of species, was done by Groves (2001). All indrids are arboreal and nocturnal. Genus *Avahi* is the smallest, with head and body length of 260–300 mm, tail length of 550–700 mm, and a mass of about 1 kg. Genus *Propithecus* (sifakas) are intermediate in size, with head and body length of 400–600 mm, tail length of 450–600 mm, and body mass of 3.5–8.0 kg. Genus *Indri* is the largest strepsirhine primate, measuring 530–700 mm in head and body length, with a very short tail, and weighing 7–10 kg. The diet of all indrids consists of leaves, fruits, flowers, and bark. The families differ in their general habitat preferences; members of Genera *Avahi* and *Indri* frequent tropical rainforests, whereas *Propithecus* live in evergreen and deciduous forests (Macdonald 1984).

Locomotion of all members of this family involves vertical clinging and leaping between tree trunks and shrubs. In conjunction with this pattern of movement, the hind limbs are considerably more developed than the fore-

Figure 14.16 Indridae. Indri (*Indri indri*) is the largest of the prosimians. They live in the coastal rainforest in Madagascar and are endangered because of habitat destruction.

limbs. The hand is prehensile with the thumb pseudo-opposable. The lower incisors are arranged in a comblike configuration for grooming. The eyes are of moderate size and are directed forward, with a distinct but modest muzzle. The thick fur varies in color, depending on the species, from white and black to brown, maroon, reddish, and orange. There are no distinct sexual differences in either body size or color patterns within the different species in this family.

Indrids live in small social groups of three to six individuals, containing adults of both sexes and young. Most births are of single young, and mothers nurse their infants with a single pair of pectoral mammae. Young are generally carried crosswise to the body axis and under the abdomen as small infants, switching to riding on the backs of their mothers after a few weeks or months. Indrids, particularly Genus *Propithecus*, have distinct home ranges, which they defend against intruders. Communication is via scent marking, vocalizations, and visual expressions and postures. All of the taxa in this family are considered endangered or critically endangered.

Haplorhine Primates

The living haplorhine primates are divided into eight families (Groves 2005), which live in Africa, Asia, and Central and South America (see figure 14.7). In addition to the previously mentioned characteristics that distinguish haplorhines from strepsirhines, haplorhines have an invasive, **hemochorial** form of **placenta**; a postorbital plate, and spatulate incisors. The maximum dental formula is 2/2, 1/1, 2/2, 3/3 in anthropoid primates from Asia and Africa (Old World primates) and 2/2, 1/1, 3/3, 3/3 in those from Central and South America (New World primates). Distinctive differences in the visual and olfactory systems of Haplorhini also separate them from Strepsirhini (Fleagle 1988; Martin 1990).

Families

Tarsiidae

The tarsiers constitute a single genus (*Tarsius*) with seven species all inhabiting the islands of Indonesia, the Malay Archipelago, and the Philippines (Musser and Dagosto 1987). Fossil evidence indicates that tarsiids once inhabited Europe, North America, and mainland Asia. Discussions about whether *Tarsius* was best classified in the Strepsirhini or Haplorhini have recently been decided in favor of the latter categorization based on both fossil evidence and brain morphology (Groves 1998; Joffe and Dunbar 1998). Head and body length ranges from 95–140 mm, with tail from 200–260 mm, and body mass of 100–130 g (figure 14.17). A distinguishing trait of tarsiers is their ability to

Figure 14.17 Tarsiidae. Western tarsiers (*Tarsius bancanus*) live in pairs on Borneo, Bangka, and Sumatra. They inhabit rainforests, shrub areas, and plantations, where they consume insects and some vertebrates.

rotate their heads almost 180°, which is a function of the flexible neck vertebrae. The pelage is gray to gray-brown, with the face more ochre-colored in some species. A distinct tuft of fur occurs on the distal third of the tail.

All species are crepuscular and nocturnal. Their large eyes face forward, and they possess a reduced snout. The hind limbs, associated with the vertical clinging and leaping mode of locomotion, are considerably more developed than the forelimbs. Tarsiers bound rapidly from one tree trunk to another, sometimes covering distances of 2 m in one leap. They can also hop on the ground. Tarsiers inhabit primary and secondary rainforest, with some species also found in shrubland habitats. They are primarily carnivores, consuming insects, lizards, and spiders.

From studies of captive tarsiers, it appears that they are territorial, defending core areas (territories) within their overlapping home ranges. They spend considerable time patrolling and marking their territory boundaries. They live in pairs, although sometimes females and young are found alone. They give birth to single, precocial young, which are fully furred and capable of movement through the trees. Youngsters ride on the mother's back and also can be carried in her mouth.

Cebidae

In the classification of Groves (2005), the New World primates are divided into four families; this contrasts with the two-family scheme that was in place for many years. The four recognized families are Cebidae (marmosets, tamarins, capuchins, squirrel monkeys, and the callimico), Aotidae (night monkeys), Pitheciidae (titi monkeys and sakis), and Atelidae (howler monkeys, spider monkeys, muriqui, and woolly monkeys).

The Cebidae consist of 56 species, the largest group of New World primates, comprised of the marmosets, tamarins, capuchin monkeys, and squirrel monkeys, effectively combining major elements of the earlier Cebidae and Callithrichidae. The Genus *Callithrix* (marmosets; 21 species) live in areas from mid Central America southward to roughly the middle of South America. The marmosets are among the smallest primates, with a body length of 200 mm, tails somewhat shorter than the body, and a mass of less than 200 g. They inhabit the upper tree canopy where they feed by scraping tree bark to obtain gum and sap. They have claws on the ends of their digits rather than nails. They live in family groups of 3–15 individuals generally with a single male, several females, and the progeny. Marmosets exhibit both monogamy and polygyny. Infant care is shared among all members of the group. They scent-mark with urine around their home range, but are not thought to be territorial.

The Genus *Leontopithecus* (golden lion tamarins; four species) has red, gold, and yellow pelage. The golden lion tamarins measure up to 300 mm in body length and attain body mass of up to 1 kg (figure 14.18). They live in the rainforests of eastern Brazil, where they are arboreal. The diet consists primarily of insects obtained from under tree bark. They also consume small lizards and a variety of fruits. They live in small family groups, and parental care is shared.

The Genus *Saguinus* (tamarins; 17 species) live in tropical rainforests and open forests from Central America to the Amazon basin. Body lengths are from 180–300 mm, with tails of 250–440 mm, and body mass of 200–900g. Pelage colors vary from black, white, and brown mixes to some that are all black. Many taxa have some form of moustache around the mouth. They have longer lower canine teeth than the incisors—something that differentiates them from marmosets. They are omnivores, consuming plants, insects, fruits, and small invertebrates. Tamarins can live in social groups of up to 40 individuals comprised of several family groups, though more often they occur in single family units of 3–9 individuals. Tamarins are generally monogamous, though polygyny does occur. They give birth to twins, and the father is the primary caregiver.

The Genus *Callimico* is monotypic, the sole species being Goeldi's marmoset (*Callimico goeldii*). The traits and habits are similar to that of the closely related marmosets.

Figure 14.18 Callitrichidae. Golden lion tamarins (*Leontopithecus rosalia*) live in remnants of primary forest in Brazil. They have a varied diet that includes animal material, fruits, flowers, and gum from trees.

Members of the Genus *Cebus* (capuchins; eight species) inhabit a wide variety of habitats in wide geographic distribution from Honduras in Central America to the Amazon Basin and Paraguay. Bodies are brown with white chest and head, and a black cap; hence the name as they resemble an order of monks. The body is 300–550 mm with tails of the same length, and a body mass of up to 1.3 kg. The diet is omnivorous like the tamarins. They live in groups of 5–40 or more individuals consisting of related females, their progeny, and several males, one of which is dominant. They have single birth, and males do not participate in caring for the young. They use urine to mark territories, the boundaries of which are defended.

The Genus *Saimiri* (squirrel monkeys; five species) are found in tropical rainforests from Costa Rica south to central Brazil and Bolivia. The pelage is olive, yellow-orange, white, and black. The body is 250–350 cm in length, with tails of 350–420 mm, and a body mass of 750 g to 1 kg. They use their relatively long tails for balance and not for climbing. Squirrel monkeys are omnivores, consuming fruits, insects, bird eggs, nuts, and small vertebrates. They live in large social groups of up to 500 individuals, splitting up into smaller groups for daily foraging. They are more vocal than other cebids, using sounds as the primary means of communication. They do mark their tails and fur by rubbing them with urine. They are polygynous, and females are the only caregivers for the young.

Aotidae

The night monkeys (eight species) live in forests from Panama south to northern Argentina and Paraguay. They are one of the few nocturnal primates, and the only haplorhine to be active primarily at night. They use a great deal of vocal communication and lack color vision. They are monogamous, forming pair bonds, and live in small family groups of parents and their progeny. The diet consists primarily of leaves, but they also consume other plant parts.

Atelidae

The Atelidae were formerly a subgroup within the Cebidae, but they have been elevated to family status in the new classification scheme (Groves 2005). There are five genera, *Aloutta*, *Ateles*, *Brachyteles*, *Lagothrix*, and *Oreonax*. All members of the family have tails longer than their bodies, and the tails are prehensile, serving as a fifth hand. They are classified as **semibrachiators** as they swing by their arms from branch to branch and use the tail as a third arm to grab branches as they locomote.

The Genus *Aloutta* (howler monkeys; 10 species) inhabit the forests of Central and South America, feeding high in the canopy. The howler monkeys attain body lengths of 550–930 mm, with tails of about the same length. They can weigh up to several kilograms. They are folivores, consuming leaves and occasionally other plant parts. Their howling behavior, facilitated by an enlarged hyoid bone, enabling the expansion of a vocal sac, is among the loudest of any land mammal. Howler monkeys live in social groups with multiple males and females.

The Genus *Ateles* (spider monkeys; seven species) inhabit tropical forests from southern Mexico to central Brazil. Many members of this genus are endangered, primarily due to habitat destruction as population increases and there is greater demand for agricultural lands. They have a body length of about 600 mm, with tails of 890 mm, and body mass of up to 6.5 kg. The pelage colors include gold, black, and brown. They live in groups of 15–25 individuals, but often split up into smaller subgroups for daily foraging. They are polygynous, and females provide most of the infant care.

The Genus *Brachyteles* (woolly spider monkeys or muriquis; two species) live only in southeastern Brazil where they inhabit coastal Atlantic rainforests. As their name implies, they are closely related to both the spider and woolly monkeys. Muriquis have a body length of 400–600 mm and a tail that is longer than the body. They are folivores, but they also consume some fruits and flowers. They live in groups of about 8–40 individuals, are polygynous, and do not defend territories. Because of the nature of their habitat and the limited geographical distribution, muriquis are endangered.

The Genus *Lagothrix* (woolly monkeys; four species) are very similar to the spider and woolly monkeys with respect to size, diet, and social behavior. They inhabit forested regions of the Amazon Basin, Orinoco River region, and the eastern slopes of the Andes mountains. The pelage is pale gray and brown. They are 600–750 mm in body length, with tails that are slightly longer, and a body mass of up to 6 kg. They eat fruits and leaves.

Woolly monkeys live in social groups of up to 100 individuals, but split into smaller troops for foraging.

Oneonax flavicauda is a monotypic genus, the highly endangered yellow-tailed woolly monkey that inhabits very limited geographic regions of 1700–2500 m elevation in the Andes Mountains of Peru. Perhaps as few as 200–250 individuals of this species remain in the wild, making it the rarest of living primates. The yellow-tailed woolly monkeys weigh up to 5.7 kg for females and up to 8 kg for males. They live in montane cloud forests where their diet consists of fruits, along with some leaves, flowers, and buds. Their social structure is not well known because of their limited numbers. A combination of limited geographic distribution, habitat destruction, and the fact that they are slow to reach maturation make this a very tenuous situation for the yellow-tailed woolly monkey.

Pitheciidae

Members of this family were formerly a subgroup in the Cebidae. There are four genera, *Pithecia*, *Chiropotes*, *Cacajao*, and *Callicebus*. The saki monkeys are divided into two distinct genera, *Pithecia* and *Chiropotes*. The Genus *Pithecia* (five species) inhabit rainforests in northern and central South America. They attain a body length of 300–500 mm and have long bushy tails. Body mass can reach 2 kg. The pelage is brown, reddish brown, and gray. Some species lack fur on the face. Sakis are omnivores, consuming leaves, flowers, fruits, and sometimes small rodents. They use a varying series of vocal communications for contacting one another or warning of possible predation. They live in mixed sex groups of up to 30 individuals and are polygynous.

The closely related Genus *Chiropotes* (bearded sakis; five species) live in northeastern South America southward into northern Brazil. They have a body length of 300–500 mm, with long hairy tails and a body mass ranging from 2–4 kg. Bearded sakis are frugivores and also consume some other plant parts. They live in groups of 18–20 individuals with mixed sexes and are polygynous.

Titi monkeys of the Genus *Callicebus* (28 species) are among the most diverse primate groups. Many new species of titi monkeys have been recorded in the past several decades. The body size, tail size, and body mass are quite diverse in this group. The members of this genus live in forests ranging from Columbia southward through Brazil to northern Paraguay. Pelage is brown, black, or shades of reddish brown. Some species have distinct markings or stripes on the head, likely for visual communication. They also use vocal communication, defending a territory by using cries. Titis live in small family groups and exhibit monogamy, and males are the primary caregivers for infants and young.

The Genus *Cacajao* (uakaris; two species) live only in forests of the upper Amazon Basin. They range from 400–450 mm in body length and have tails that are much shorter, ranging from 150–180 mm, with a body mass of up to 3 kg. Members of the two species have long, coarse hairs on their body and heads that lack fur. In one species the head is black, and in the other it is scarlet. They occur in large groups of up to 100 individuals but are also found in smaller, family-sized groups.

Cercopithecidae

Also called "typical," or Old World, monkeys, the Cercopithecidae contains roughly a third of all primate genera (21) and about 40% of all known primate species (132). They are distributed throughout much of Africa and southern Asia, including the Malay Archipelago. Cercopithecids live as far north as northern Japan and as far south as southern Africa. Some species have cheek pouches for food storage, and males of all genera and female guenons have large canines. All have powerful muscles to provide good grinding action for the teeth, and sexually receptive females have **perineal swelling.**

The family can be divided into two groups (table 14.2). One group, the cercopithecine monkeys, consisting of 11 genera (73 species), predominates in Africa, although some species occur in Asia. Analyses conducted in the late 1990s using mtDNA (Zhang and Ryder 1998) and a combination of molecular, fossil, and biogeographic information (Stewart and Disotell 1998) reveal the history of Old World monkeys. The common ancestor of colobines and cercopithecines lived in Africa. The ancestors of today's Asian colobine groups migrated from Africa during the late Pliocene or early Pleistocene epoch. At about the same time, the ancestors of the macaques also spread from Africa to Asia, where considerable diversification of the group has occurred. At the generic level, mtDNA from eight species of *Macaca* has been tested using restriction endonucleases (figure 14.19). There appear to be at least four major groups within *Macaca* (Ya-Ping and Li-Ming 1993). We should also note that this genus is the most widespread of any nonhuman primate, extending from North Africa (*M. sylvanus*) to Japan (*M. fuscata*).

Cercopithecines live in a wide range of both forested and terrestrial habitats. The family ranges in size from the small talapoin monkey (Genus *Miopithecus*), with a head and body length of 340–370 mm, tail length of 360–380 mm, and body mass of 1.1–1.4 kg, to the largest members of the group, *Mandrillus* (drills and mandrills), with a head and body length of 800 mm, tail length of 70 mm, and body mass of 12.0–25.0 kg. Sexual dimorphism is pronounced in some genera (e.g., *Mandrillus*), intermediate in other genera (e.g., *Macaca*), and essentially absent in other genera (e.g., *Miopithecus*).

Coat colors include shades of brown, gray, green, and red as well as black and white. Some genera such as *Mandrillus* and *Cercopithecus* (guenons, cercopithecus monkeys) and some macaques (*Macaca*) have brightly colored patches of skin on the nose and face, the scrotum, or the rump. They also may have browridges or other patterns of hair in the head region that accentuate facial expressions or eye move-

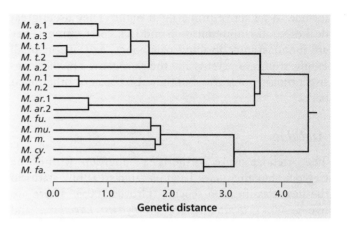

Figure 14.19 Macaque phylogenetic tree. Cladogram for eight species of Genus *Macaca*. Multiple numbers by a species abbreviation indicate multiple samples for that species. Abbreviations: M. mu = M. mulatta; M. cy. = M. cyclopis; M. fa. = M. fascicularis; M. n. = M. nemestrina; M. a. = M. assamensis; M. fu. = M. fuscata; M. t. = M. thibetana; M. ar. = M. arctoides. Data from Z. Ya-Ping and S. Li-Ming, 1993, *Phylogenetic Relationships of Macaques Inferred from Restriction Endonuclease Analysis of Mitochondrial DNA*, Folia Primatologica *60:7–17.*

ments. Cercopithecines are all quadrupedal, although some (e.g., Genus *Papio*, baboons) are better adapted for a terrestrial existence because they have relatively longer front limbs, whereas others (e.g., Genus *Cercocebus*, mangabeys) that are primarily arboreal have larger, more developed hind limbs. The guenons and mangabeys, as well as some less well-known genera such as *Allenopithecus* (Allen's swamp monkey) and *Lophocebus* (gray-cheeked mangabey), have a diet consisting primarily of leaves, whereas other genera, such as *Macaca* and *Papio*, eat more fruits and seeds and may, on occasion, catch and consume other animals.

Social organization, as well as other traits of cercopithecine monkeys, varies greatly (see table 14.2). As discussed in greater detail later (chapter 23), the varying forms of social organization in this taxon reflect a combination of evolutionary selection pressures related to finding resources, such as food and shelter, and to the threat of predation. Vocal and visual communication of cercopithecines have been the subject of considerable research, including studies of the differentiation of calls by vervet monkeys (*Chlorocebus aethiops*) in response to aerial versus ground predators (Seyfarth et al. 1980a, 1980b).

The second group of Cercopithecidae, the colobines, consists of 10 genera and 59 species (figure 14.20). Although found predominately in Asia, some colobines, notably in the Genera *Colobus* (black colobus monkeys), *Piliocebus* (red colobus monkeys), and *Procolobus* (olive colobus monkey), occur in Africa. Colobines are characterized by the absence of cheek pouches and the presence of a sacculated stomach and large salivary glands. The upper portion of the stomach has a nearly neutral pH; bacteria found here aid in fermentation and breakdown of

Table 14.2   Taxonomy and characteristics of the Cercopithecidae

Genus (no. species)	Distribution	Body Mass (kg)	Habitat	Mating/Social System(s)
Subfamily Cercopithecinae				
Allenopithecus (1)	Africa	4–8	Swamp forest	Groups of 10–30 of mixed sexes
Cercocebus (6)	Africa	6–10	Evergreen and rain forest	Groups of 20–40 of mixed sexes, polygynous
Cercopithecus (25)	Africa	2–9	Rainforest, montane forest, swamp forest	Groups of 10–50 of mixed sexes, mostly polygynous
Chlorocebus (6)	Africa	3–5	Wooded savanna	Groups of 10–50 of mixed sexes, polygonous
Erythrocebus (1)	Africa	4–13	Wooded savanna, open savanna	One-male groups
Lophocebus (3)	Africa	6–9	Evergreen forest	Groups of 20–40 of mixed sexes, polygynous
Macaca (21)	Africa, Asia	4–18	Montane forest, riverine forest, forest edge	Groups of 10–100+ individuals of mixed sexes, mostly polygynous
Mandrillus (2)	Africa	12–50	Rainforest	One-male units of 12–20, groups of 20–40 of mixed sexes
Miopithecus (2)	Africa	1–2	Swamp forest	Groups of 70–100 of mixed sexes, polygonous
Papio (5)	Africa	12–25	Savanna, rocky scrubland,	Groups of 10–80 of wooded savanna mixed sexes, one-male groups
Theropithecus (1)	Africa	14–21	Grassland	One-male units within large herds of 100+
Subfamily Colobinae				
Colobus (5)	Africa, Asia	5–15	Forest, wooded grassland	Groups of 4–15 of mixed sexes, polygynous, some territorial
Nasalis (1)	Asia	8–24	Mangrove, lowland rainforest	Groups of 12–20, polygynous, territorial
Piliocolobus (9)	Africa	3-11	Forest, savanna woodland	Groups of 5–20 of mixed sexes, polygynous, some territorial
Presbytis (11)	Asia	5–8	Rainforest, swamp forest	Groups of 12–30 of mixed sexes, polygynous, some territorial
Procolobus (1)	Africa	4-6	Forest	Groups of 5-15, polygonous
Pygathrix (3)	Asia	5–10	Rainforest, conifer forest	Groups of 15–60 of mixed sexes, polygonous
Semnopithecus (7)	Asia	5–24	Forest, scrub, cultivated	Groups of 15–50 of areas mixed sexes, polygynous, some territorial
Rhinopithecus (4)	Asia	5-10	Rainforest, conifer forest	Groups of 15-50, polygynous
Simias (1)	Indonesia	12-15	Rainforest	Groups of 12-20, polygynous, territorial
Trachypithecus (17)	Asia	4–14	Plantations, forest	Groups of 12–30 of mixed sexes, polygynous, some territorial

Data from D. E. Wilson and D. M. Reeder, 2005, Mammal Species of the World, *Johns Hopkins Univ. Press.*

Figure 14.20 Cercopithecidae. Proboscis monkeys (*Nasalis larvatus*) live in the mangrove swamps and lowland rainforests of Borneo. The function of the prominent nose is not fully understood.

leafy vegetation. Colobines are generally more slenderly built than cercopithecines. The molar teeth possess high pointed cusps, and the outside of the lower molars and insides of the upper molars are less convexly buttressed than in the cercopithecines. These differences in dentition and the more specialized stomach can be related to the diet of colobines, which for many genera consists of leafy vegetation. Not all colobines, however, are folivores. Many include fruits, seeds, soil, flowers, and other nonleafy materials in their diets, although few records exist of consumption of other vertebrates, and only a few species have been observed eating insects.

Colobines live in a variety of forest types, but they also occur in other habitats, including cultivated fields, rural and urban areas, and dry scrublands. Their coat colors are as varied as those of cercopithecines. Some of the common names for particular species exemplify the special pelage patterns that characterize many of them, including the golden snub-nosed monkey (*Rhinopithecus roxellana*), white-rumped black leaf monkey (*Trachypithecus francoisi*), and white-fronted sureli (*Presbytis frontata*). Most colobines are arboreal, although some, like the Hanuman, or northern plains gray langur (*Semnopithecus entellus*),

spend more than half their time on the ground. All colobines move by quadrupedal locomotion. They can walk bipedally along branches, supporting themselves by grasping other branches with their forelimbs, and they often leap between trees.

Colobines live in social groups ranging in size from occasional solitary individuals to aggregations of more than 100. Average size may be smallest in the Mentawi Islands sureli (*Presbytis potenziani*), which form groups of only three to four individuals. These monkeys are uniquely monogamous among colobines. Most other colobines live in groups that average 10–20 individuals with two or more adult males and adult females with a polygynous mating system.

Hylobatidae

The apes consist of the Families Hylobatidae and Hominidae. Goodman and colleagues (1990) have used genetic data (figure 14.21) to examine phylogenetic relationships among the apes. Their findings are similar to results obtained with DNA hybridization (Sibley and Ahlquist 1984). Note that Goodman and colleagues (1990; Goodman 1999) place the gibbons (Hylobatidae) in the Family Hominidae, whereas Groves (2005) place them in their own family, with 4 genera, *Hylobates* (7 species), *Bunopitheucs* (1 species), *Nomascus* (5 species), and *Symphalangus* (1 species).

Hylobatids inhabit evergreen rainforests and monsoon deciduous forests of southeast Asia and portions of the Malay Archipelago. The 10 species known as gibbons range from 450–650 mm in head and body length and have a body mass of 5–7 kg (figure 14.22). The siamang (*Sympphlagus syndactylus*) is larger, with head and body length of 750–900 mm and a body mass of 9–12 kg. Hylobatids do not have tails. Their fur is dense and shaggy, and pelage color ranges from white and gray to black

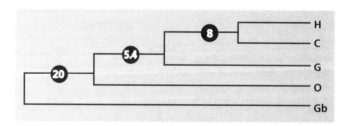

Figure 14.21 Phylogenetics of Hylobatidae and Hominidae. Maximum parsimony tree diagram for the ChrW(61561)ChrW(61544) region nucleotide sequence orthologues for the Families Hylobatidae and Hominidae. The number above each node is the difference in tree lengths between the tree shown and the nonparsimonious tree that adds the least length in breaking up the clade at that node. Abbreviations: *H* = humans, *C* = chimpanzees, *G* = gorillas, *O* = orangutans, *Gb* = gibbons. *Data from M. Goodman et al., 1990, Primate Evolution at the DNA Level and a Classification of Hominids,* J. Mol. Evol., *30:260–266.*

Figure 14.22 Hylobatidae. Pileated gibbons (*Hylobates pileatus*) inhabit tropical rainforest and semideciduous forest in southeast Thailand and western Kampuchea. Note especially the long arms used for brachiation through the arboreal habitat.

and brown. Distinct facial rings often differentiate the eyes, nose, or mouth, and some have tufts on the crown of the head.

One of the most distinguishing traits of the Hylobatidae is their true brachiation mode of locomotion. They also climb quadrupedally and walk, bipedally, on branches. Their arms are considerably longer than their legs. Their hands are prehensile with a fully opposable thumb used to grasp branches during their travels through the forest. About two thirds of the diet of gibbons is fruit, and much of the remainder is leaves. The diet of siamangs is about half leaves and half fruit.

Gibbons and siamangs live in family groups with a monogamous adult pair and several juveniles. A single young is produced every 2–3 years. Male siamangs provide considerable parental care. All species are highly territorial. Visual and vocal communications are infrequent within the family group. Most hylobatids engage in daily bouts of singing, which includes duets of great calls involving the adult male and female. These appear to serve at least two functions: (1) advertising their presence and territorial defense and (2) establishing and maintaining pair bonds.

Hominidae

The phylogeny of this last family has been debated and revised several times in recent decades (figure 14.23; Fleagle 1988; Groves 1989; Groves 2005). The Hominidae consists of four genera, *Gorilla* (two species), *Pan* (two species), *Pongo* (two species), and *Homo* (one species). These are the largest primates, and all lack tails. Most births are of single young, and young have the longest developmental periods of any primate (usually 2–3 years until weaning and several more years to attain sexual maturity). All species of Hominidae exhibit sexual dimorphism, with males larger than females. All hominids other than *Homo sapiens* are threatened or endangered.

The orangutans (*Pongo pygmaeus*—Bornean orangutan; *Pongo abelli*—Sumatran orangutan) are found in lowland and hilly tropical rainforests of Borneo and northern Sumatra (see figure 14.23A). Males have head and body lengths of about 970 mm and weigh 60–90 kg. Females have a head and body length of 780 mm and a body mass of 40–50 kg. They are covered by coarse, sparse, long hair that varies in color from reddish to orange, chocolate, or maroon. Orangs spend most of their lives in trees and locomote by brachiation. Their diet consists of about 60% fruit augmented by large quantities of leaves and shoots. They also consume insects, tree bark, eggs from bird nests, and small mammals. Orangs live solitary lives, except for sexual consortships and mothers with infants. Males have large home ranges that overlap the ranges of as many females as possible. Though most information on tool use pertains to chimpanzees, we now have data indicating that orangutans also use tools (van Schaik et al. 2003; Mulcahy and Call 2006). The most common vocalization is the long call of the male, usually given early in the morning. The call may function to attract mates, signal ownership of a particular area, or simply alert other orangutans of the male's location.

The two species of chimpanzee (*Pan troglodytes*, common chimpanzee, and *Pan paniscus*, pygmy chimpanzee or bonobo) live in central Africa in a variety of habitats, ranging from woodland savanna to deciduous and humid forests (see figure 14.23B). Common chimpanzees have a head and body length of 770–920 mm for males and 700–850 mm for females; males weigh about 40 kg and females about 30 kg. Pygmy chimpanzees, in spite of their name, are only slightly smaller than common chimpanzees. Morphological and molecular data used together in a cladistic analysis reveal that the two Pan species group together in a clade and that they are most closely related to the Genus *Homo* (Barriel 1997). Both species have long, coarse, sparse hair that is generally black but can turn gray

A   B   C

Figure 14.23 Hominidae. The great apes include (A) orangutans, (B) common chimpanzees, and (C) gorillas.

in older animals. Their diet consists primarily of fruits supplemented with leaves, seeds, flowers, and several other plant foods. In addition, common chimpanzees, but not bonobos, consume meat, sometimes caught through cooperative hunting (Goodall 1986).

Chimpanzees use tools in several contexts, such as to obtain food and water. Chimpanzees live in a complex community with approximately 12–100 individuals, splitting into smaller parties of 3–6 (common chimpanzees) or 6–15 (bonobos) that frequently change composition. Considerable evidence indicates that chimpanzees use stone tools (Mercador et al. 2002), save tools for future use (Mulcahy and Call 2006), and can be selective in terms of recruiting partners when collaboration is necessary, for example, to obtain food (Melis et al. 2006). Chimpanzees communicate through a full range of vocalizations and facial expressions as well as through demonstrative behavior, including throwing branches and rocks.

The gorillas (*Gorilla gorilla* and *Gorilla beringei*) are the largest primates, with males having a head and body length of 1600–1800 mm and a body mass of 140–180 kg, and females having a head and body length of 1400–1550 mm and body mass of 80–95 kg (see figure 14.23C). They live in tropical rainforests in widely separated populations in East and West Africa. The relatively dense fur is black to gray-brown. Older adult males have silver hairs on their backs. The diet of the gorillas in East Africa is almost entirely folivorous, consisting of leaves and stems, whereas West African gorillas eat proportionally more fruit. They live in relatively stable social groups of 5–30 individuals in East Africa, usually with one silver-backed male, several black-backed males, females, adolescents,

and younger gorillas; groups in West Africa are smaller, averaging about five individuals. Gorillas are primarily terrestrial, moving quadrupedally, often using the knuckles of the hand. They can climb trees, although this activity is more prevalent in younger animals. Some gorillas nest in trees at night. Males stand bipedally to perform a chest-beating display. They communicate by facial expressions and a series of barks, grunts, and other sounds. Their social behavior has been studied extensively by Schaller (1963) and Fossey (1983).

Humans (*Homo sapiens*) evolved from an ancestral African ape stock some time between 5 and 10 mya (Sarich and Wilson 1967a, b; Uzzell and Pilbeam 1971; Fleagle 1988; Martin 1990; Wood and Richmond 2000). In fact, evidence derived from mitochondrial and nuclear DNA (Templeton 2002), supports the idea that there were two waves of migration outward from Africa, separated by millions of years. The ancestral forms of hominids have been debated, but consensus has not been reached. New archeological data contribute to new interpretations several times each year (figure 14.24). Fossils include several forms of *Australopithecus*, several fossil forms of *Homo*, and a parallel line involving *Paranthropus*. One current hypothesis begins with *Australopithecus ramidus* (5–4 mya), then *Australopithecus afarensis* (4–2.7 mya), *Homo habilis* (2.2–1.6 mya), *Homo erectus* (2–0.4 mya), and *Homo sapiens* (400,000 years ago to the present). A slightly different version (figure 14.24), shows *Australopitheucus africanus* as being on the lineage to modern humans. There are numerous other possible schemes and only many years of additional data will provide a clearer picture of the path between early hominids and our species today.

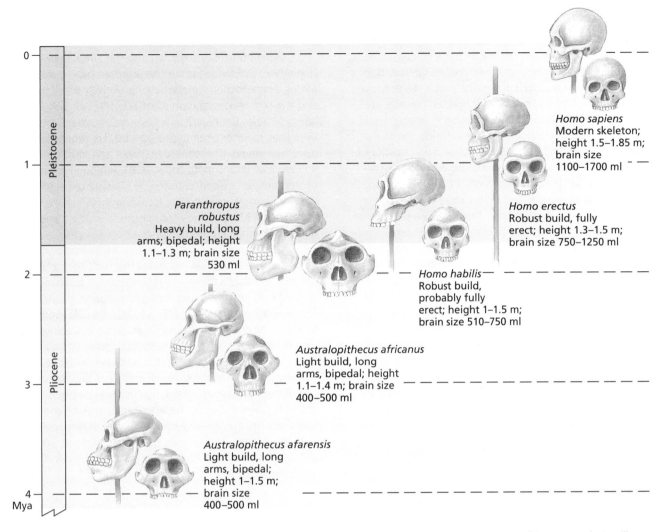

Figure 14.24 Human evolution. The pattern of evolution of the various hominids during the course of human evolution illustrates that (1) at times several forms coexisted and (2) a number of hominid forms have gone extinct during the evolution of this lineage.

New findings are continually altering and refining the perspective on various segments of the evolution of human ancestors. At one end of the spectrum is a finding from the Afar Depression in Eritrea of a cranium that dates to about 1 million years before the present and that has some characteristics that are typical of *H. erectus* and some that are typical of *H. sapiens* (Abbate et al. 1998). If confirmed, this finding would extend by several hundred thousand years into the past the origins of the *H. sapiens* lineage. That conclusion is contradicted by a summary of molecular data suggesting an origin of *H. sapiens* in Africa perhaps only several hundred thousand years ago (Jorde et al. 1998). At the other end of the spectrum is the finding west of Lake Turkana in East Africa of cranial fragments and part of a maxilla that have been dated to about 3.5 million years ago. These fossils are assigned to a new genus and species, *Kenyanthropus platyops* (Leakey et al. 2001). This fossil material adds a possible new lineage to the overall understanding of the hominid phylogeny at the border between the Pliocene and Pleistocene epochs.

Most recently, a possible new species of homonid, *Homo floresiensis*, was described from Flores Island in Indonesia (Brown et al. 2004; Morwood et al. 2005). This diminutive form is dated to 100,000 to just 12,000 years ago and appears to be an insular form exhibiting dwarfism. This and other findings that now occur on at least an annual basis are providing new information that helps to refine the tree of homonid evolution. Controversy and discussion occur for each new find as, for example, in the case of *Homo floresiensis*, where some investigators claim that the small head and body are due to microcephaly.

Humans are characterized by erect bipedalism; reduced sexual dimorphism relative to other hominids; a large brain, averaging about 1300 cm^3, contained in a large vaulted cranium; and a skeleton that is less robust and more gracile than those found in recent hominid ancestors. Human social organization is varied, including monogamy, polygyny, and in a few instances, polyandry. Young are born in a relatively altricial state, and there is an extended period of physical and behavioral development.

SUMMARY

Primates, the order of mammals that includes humans, comprises 15 families, 69 genera, and 376 species; considerable revision of the order occurred during the past 3–4 years, resulting in more families, genera, and species. Primates are largely tropical and subtropical in distribution, ranging through Africa, Asia, South and Central America, the Malay Archipelago, Japan, and Madagascar. Major characteristics of the group include refined hands and digits with nails replacing claws, binocular stereoscopic vision, a complete postorbital bar, a reduced muzzle, and slower rates of reproduction with increased developmental time. They exhibit a progression of sociospatial systems ranging from overlapping home ranges to a diverse array of social and mating systems. Cheekteeth are bunodont and brachydont. Primates, as a group, are generalists compared with most other mammal groups.

The first primates appeared about 70 mya and are grouped together as plesiadapids. The adapids probably evolved from the plesiadapids and are ancestral to modern lemurs. Another fossil group, Omomyidae, is ancestral to the tarsiers. *Parapithecus*, a fossil genus from Africa and Asia, is ancestral to today's cercopithecoid primates. The arboreal theory is the most accepted concerning the evolution of traits that characterize primates. Molecular techniques have been used to examine some phylogenetic relationships; ongoing investigations should help clarify questions concerning primate phylogeny.

Living primates are divided into the Strepsirhini and Haplorhini. Strephsirhines are characterized by a bicornuate uterus, epitheliochorial placentation, relatively small neonates, and a maximum dental formula of 2/2, 1/1, 3/3, 3/3. The group includes Daubentoniidae (aye-aye), Loridae (lorises), Galagidae (galagos), Lemuridae (lemurs), Lepilemuridae (sportive lemurs), Cheirogaleidae (dwarf and mouse lemurs), and Indridae (avahi, indri, and sifakas). Although the majority of strepsirhines are concentrated on Madagascar, some groups live in Africa and Asia, including the Malay Archipelago. Many of the Madagascan species are endangered due to habitat destruction.

Haplorhines are characterized by a fused simplex uterus, hemochorial placentation, neonates that are larger relative to the mother's size, differences in the rhinarium compared with sterpsirhines, and maximum dental formulae of 2/2, 1/1, 3/3, 3/3 (New World species) or 2/2, 1/1, 2/2, 3/3 (Old World species). The group includes Tarsiidae (tarsiers), Cebidae (marmosets, tamarins, lion tamarins, squirrel monkeys, and capuchin monkeys), Aotidae (night monkeys), Atelidae (spider monkeys, howler monkeys, woolly monkeys, and muriquis), and Pitheciidae (saki monkeys, bearded saki monkeys, uakaris. and titi monkeys), Cercopithecidae (cercopithecine and colobine monkeys), Hylobatidae (gibbons and siamang), and Hominidae (apes). Haplorhines, other than humans, are widely distributed throughout Africa, Asia, the Malay Archipelago, and Latin America.

SUGGESTED READINGS

Fleagle, J. G. 1988. Primate adaptation and evolution. Academic Press, New York.

Gould, L., and M. L. Sauther. 2006. Lemurs ecology and adaptation. Springer, New York.

Lehman, S. M., and J. G. Fleagle. 2006 Primate biogeography. Springer, New York.

Martin, R. D. 1990. Primate origins and evolution. Princeton Univ. Press, Princeton, NJ.

Richard, A. F. 1985. Primates in nature. W. H. Freeman, New York.

Szalay, F. S., and E. Delson. 1979. Evolutionary history of the primates. Academic Press, New York.

Wolfheim, J. H. 1983. Primates of the world, distribution, abundance, and conservation. Univ. of Washington Press, Seattle.

Wood, B., and B. G. Richmond. 2000. Human evolution: Taxonomy and paleobiology. J. Anat. 196:19–60.

DISCUSSION QUESTIONS

1. Primates are often characterized by a mixture of generalized and specialized traits. In terms of the morphology of six genera of living primates of your choice, write down those traits that you consider to be generalized and those that you think are specialized. Discuss your reasons for these classifications.

2. What relationships can you discern and describe between the physical size of different primates and their (a) diet, (b) habitat, and (c) social system?

3. Why do you think that virtually all primate distributions are limited to the tropics and subtropics?

4. Using your knowledge of molecular techniques to assess phylogenetic relationships, describe the studies you would carry out to investigate the relationships of the New World primates (Cebidae and Callitrichidae). How would you integrate the information you obtain with what is already known about phylogeny in these groups from morphological evidence on living primates and fossil history?

CHAPTER 15

Xenarthra [Cingulata and Pilosa], Pholidota, and Tubulidentata

The orders Xenarthra, Pholidota, and Tubulidentata share several characteristics related to a common feeding mode. Specifically, all are, or tend toward being, **myrmecophagous**, that is, they have diets composed predominantly of ants and termites. Because these insects form large colonies and are common in tropical and semi-tropical areas throughout the world, they offer an excellent potential energy source to various mammalian groups. Not only do the orders discussed in this chapter eat ants and termites, so do mammals within an array of other orders: monotremes (the echidnas), marsupials (the numbat, *Myrmecobius fasciatus*, and rabbit-eared bandicoots), and even some carnivores such as the aardwolf (*Proteles cristatus*) and the sloth bear (*Melursus ursinus*). Thus, several mammalian orders have ant-eating representatives distributed throughout all tropical or semitropical landmasses. Many other mammalian species also incorporate ants and termites as part of their diets.

Morphological features of each of the orders in this chapter are quite distinctive, reflecting adaptations for myrmecophagy. Most have long snouts and long, powerful, sticky tongues used to collect insects. Dentition is reduced or absent, and the coronoid process of the mandible is reduced, whereas the hyoid elements (bones and muscles of the tongue) are enlarged (Naples 1999). Strong, heavily clawed forepaws are used to dig into anthills and termite mounds. Small pinnae and valvular nostrils reduce susceptibility to biting from their prey. The integument also offers protection. The xenarthrans have either an armored carapace or are heavily furred. The pholidotes are covered with reptile-like scales, whereas the aardvark (Order Tubulidentata) has a tough, sparsely haired hide. The stomachs of individuals in all the orders are generally simple with thickened, muscular, often **keratinized** (a tough, fibrous protein) pyloric regions to aid in digestion and protect against the formic acid contained in many ant species. Also, the groups discussed in this chapter usually have low reproductive capacities—generally one young per litter—a strategy of small litter size and extended parental care (see chapter 25). These similar characteristics are an excellent example of convergent evolution (Reiss 2001) because, as noted, these diverse mammalian orders are not particularly closely related.

Xenarthra

This order, formerly referred to as Edentata (meaning "without teeth"), encompasses extant armadillos, anteaters (sometimes referred to as vermilinguas), and tree sloths found only in the Western Hemisphere. Xenarthrans are a morphologically diverse group; the "few living and numerous fossil representatives are characterized by quite disparate and highly derived (indeed, often bizarre) anatomical and ecological specializations (Gaudin et al. 1996:32). We will consider the xenarthrans from a more traditional standpoint as a single order encompassing five families: the two-toed sloths (Family Megalonychidae), three-toed sloths (Family Bradypodidae), anteaters (Family Myrmecophagidae), silky anteater (Family Cyclopedidae), and armadillos (Family Dasypodidae). A less traditional, although possibly more appropriate arrangement, is discussed in the following section.

Primarily restricted to tropical and semitropical habitats from Mexico southward throughout South America, only the nine-banded armadillo (*Dasypus novemcinctus*) occurs as far north as the southern United States. All xenarthrans have low metabolic rates (figure 15.1) and low body temperatures. Body temperature averages about 34°C, compared with 36° to 38°C in other mammals.

ALTERNATIVE XENARTHRAN TAXONOMY

Differences among xenarthrans are often so pronounced that alternative taxonomic systems have been proposed, such that "Xenarthra" may be better represented as two orders (Gardner 2005). The Order Cingulata includes the extant armadillos and extinct families of ground sloths (figure 15.2). The Order Pilosa encompasses two sister taxa, the two-toed tree sloths and the three-toed tree sloths in a clade with several associated extinct lineages. A second clade includes the true anteaters as well as the silky anteater (*Cyclopes didactylus*) in its own monotypic family. As noted by Gaudin (2003:36), "The twofold division of Xenarthra into Cingulata, including the armored glyptodonts and armadillos, and the Pilosa, including anteaters and sloths, has been corroborated by numerous studies."

Members of this order are limited geographically to warm regions due to their low basal metabolic rates and poor thermoregulatory abilities.

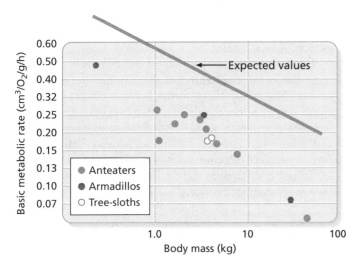

Figure 15.1 Lower metabolic rates in xenarthrans. The expected relationship of mass-specific basal metabolic rate of mammals (line), calculated as BMR = constant (body mass$^{-0.25}$), is much greater than the metabolic rates that occur in xenarthrans. *Data from B. K. McNab, 1985, Energetics, population biology, and distribution of xenarthrans, living and extinct, in* The Evolution and Ecology of Armadillos, Sloths, and Vermilinguas *(G. G. Montgomery, ed.), Smithsonian Institution Press.*

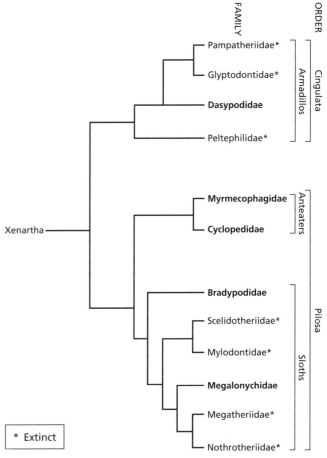

Figure 15.2 An alternative arrangement for the Xenarthra. Many recent studies suggest that this order would be better represented as two orders as noted here. *Redrawn and modified from Rose et al. (2005).*

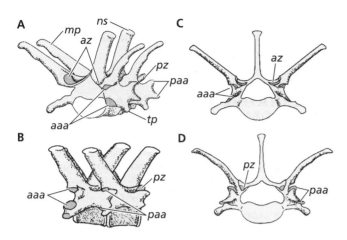

Figure 15.3 **Vertebrae of xenarthrans.** Third and fourth lumbar vertebrae in the nine-banded armadillo show the extra articular surfaces (shaded throughout). (A) Anterior three-quarters view. (B) Lateral view (anterior to the left). (C) Anterior view. (D) Posterior view. Abbreviations: az = anterior zygapophysis; mp = metapophysis; ns = neural spine; pz = posterior zygapophysis; tp = transverse process; aaa = anterior accessory intervertebral facets; paa = posterior accessory intervertebral facets. *Adapted from T. J. Gaudin and A. A. Biewener, 1992, The functional morphology of xenarthrous vertebrae in the armadillo* Dasypus novemcinctus *(Mammalia: Xenarthra),* Journal of Morphology *214:63-81.*

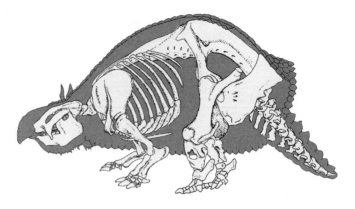

Figure 15.4 **The extinct glyptodont.** The extent of the armored carapace of this early Miocene xenarthran (*Propalaeohoplophorus*) is shown in outline, with a portion of the scute (scalelike) pattern shown in the center of the figure. Note the massive nature of the limbs and heavy, inflexible vertebrae, necessary to support the carapace, which made up 20% of the body mass. Total length approached 2 m. *Adapted from W. K. Gregory, 1951,* Evolution Emerging, *vol. 2, McMillan & Co.*

MORPHOLOGY

The older ordinal name *Edentata* is inappropriate because only the anteaters are "without teeth." The name *Xenarthra* refers to one of the distinguishing characteristics of the order—the presence of at least two accessory, or supplemental, "xenarthrous" intervertebral articulations or xenarthrales ("strange joint"), located primarily on the lumbar and some posterior thoracic vertebrae (figure 15.3). These give added rigidity to the axial skeleton (Gaudin and Biewener 1992; Gaudin 1999). Additional characteristics shared by this otherwise diverse group include loss of incisors and canines, and cheekteeth (if present) that are single-rooted and without enamel. In terrestrial species, the acromion and coracoid processes of the scapula are separate and well developed to enhance muscle attachment for digging. These are rudimentary, fused processes in most therian mammals. Also, the transverse processes of the anterior caudal vertebrae and the ischia are fused in all xenarthrans, except the silky anteater, now in its own family. Several other features of the skeleton and musculature define this order, including dermal ossicles in the skin, position of the infraorbital canal, and a secondary scapular spine, among others (Engelmann 1985; Gaudin 2003).

FOSSIL HISTORY

Using evidence from mitochondrial DNA, Hoss et al. (1996) concluded that xenarthran lineages diverged prior to the Paleocene epoch. The earliest known fossil xenarthrans are armadillos from the late Paleocene of South America, although the order probably dates from even earlier times. The fossil history of this group is rich and diverse, with over 150 described genera (Gaudin 2003). Extinct xenarthrans include giant armadillos (Family Glyptodontidae) that were over 3 m long with heavily armored head, back, and tail and that weighed over 1,800 kg (figure 15.4). The giant ground sloth (*Megatherium americanum*), another extinct xenarthran, weighed over 2,700 kg. Bargo et al. (2006) examined characteristics of the muzzle of the giant ground sloth and four other Pleistocene species of sloths to determine aspects of food intake.

The 14 genera and 31 species of xenarthrans living today represent less than a tenth of the known number of extinct genera. Several of these forms became extinct recently, eliminated by humans only a few hundred years ago. These include the ground sloth (*Mylodon listai*), whose bones, hide, and reddish hair have been found with human artifacts in a cave in southern Argentina. The Puerto Rican ground sloth (*Acratocnus odontrigonus*) and the lesser Haitian ground sloth (*Synocnus comes*) also survived until 400–500 years ago.

ECONOMICS AND CONSERVATION

Arboreal sloths are greatly affected by loss of tropical rainforest habitats. All three-toed sloth species (Genus *Bradypus*) are declining. The recently described pygmy three-toed sloth (*B. pygmaeus*) is critically endangered, and the maned sloth (*B. torquatus*) of Brazil is endangered. The giant armadillo (*Priodontes maximus*) is threatened because of habitat loss and hunting pressure, as are several other species of armadillos.

FAMILIES

Megalonychidae

This family includes 1 genus and 2 extant species of arboreal two-toed sloths and about 12 genera of recently extinct ground sloths. The two-toed sloths traditionally were included with the three-toed sloths in the Family Bradypodidae, and then in their own family, the Choloepidae. Based on a number of cranial characteristics (figure 15.5), as well as molecular analyses (Greenwood et al. 2001), authorities now include them as the only living members of the Family Megalonychidae (Webb 1985a; Wetzel 1985). The divergence of the two families and genera of sloths may date to the Eocene (Gaudin 2004). Hoffmann's two-toed sloth (*Choloepus hoffmanni*) occurs

from Nicaragua southward through Peru and central Brazil. Linnaeus's two-toed sloth (*C. didactylus*) is found east of the Andes Mountains in northern South America. As with all xenarthrans, members of this family have no incisors or canines, and the cheekteeth are usually 5/4. The anterior upper premolar is **caniniform** (shaped like a canine tooth; see figure 15.5) and separated from the rest of the molariform dentition by a diastema. These teeth

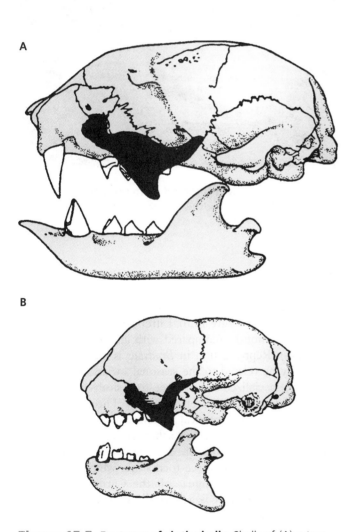

Figure 15.5 Features of sloth skulls. Skulls of (A) a two-toed sloth and (B) a three-toed sloth. Note the distinctive anterior caniniform premolar (a premolar that looks like a canine tooth) in the two-toed sloth. Both skulls have an incomplete zygomatic arch with the jugal bone (black), a flattened plate with upper and lower processes on the posterior edge. Also note the difference in the shape of the mandibles. (Both skulls are about half scale). *Adapted from E. R. Hall and K. R. Kelson, 1959,* Mammals of North America, *Ronald Press.*

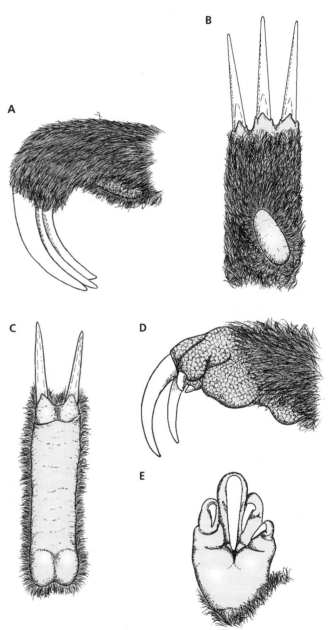

Figure 15.6 Large recurved claws in xenarthrans. Right front foot of several species showing similarity of the claws in each family. (A) Lateral view and (B) ventral view with toes spread of a three-toed sloth; (C) ventral view of a two-toed sloth; (D) lateral view of a giant anteater; and (E) ventral view of a southern tamandua, showing the large central digit. (Not to relative scale.) *Adapted from R. I. Pocock. 1924,* External characteristics of the South American edentates, Proceedings of the Zoological Society.

are sharp because of unique occlusion of the anterior surface of the lower caniniform premolar with the posterior surface of the upper one.

Two-toed sloths have two toes on the forefeet, each with a long (80–100 mm), sharp claw and three toes on the hind feet. *Choloepus* differs from three-toed sloths in the number of digits on the forefeet (figure 15.6) and in being somewhat larger and heavier. Their total length ranges up to 740 mm, and body mass reaches 8.5 kg. Unlike most mammals, which have seven cervical vertebrae, individual two-toed sloths may have five, six, or occasionally eight.

Two-toed sloths also are more active than three-toed sloths. They move to a different tree each day, and they have a broader range of feeding habits. They are almost entirely arboreal and folivorous (eat leaves) and spend most of their adult life hanging upside down from tree branches. They do, however, descend to the ground to defecate (at intervals of about 4 days in captivity) and apparently are capable swimmers. Their long, brownish gray pelage often has a greenish color because of green algae growing in it.

Two-toed sloths give birth to a single young; females do not reach sexual maturity until 3 years of age, males not until they are 4 or 5 years old. Taube et al. (2001) found gestation was 10 months for both species of *Choloepus* with a 15- to 16-month interval between births. There was little seasonality to reproduction, probably because of the more generalist diet of two-toed sloths compared to *Bradypus*. Two-toed sloths are very long-lived; captive individuals reach over 31 years of age. Both *Choloepus* and *Bradypus*, discussed in the following section, are poor thermoregulators. Their body temperature tracks the ambient temperature to a much greater extent than is the case in other mammals. This is especially evident in two-toed sloths, whose body temperature varies between 24° and 33°C, a factor that limits their geographic distribution to warm, tropical regions.

Bradypodidae

Three-toed sloths in the Genus *Bradypus* are distributed from Honduras south through northern Brazil. As with two-toed sloths, the four species in this family are arboreal folivores. They are active day and night, and have more narrowly restricted movement and feeding habits than two-toed sloths. Individuals may spend prolonged periods in the same tree. The teeth of three-toed sloths are cylindrical, with a central core of soft dentine, surrounded by harder dentine and then cementum. There is no enamel. The stomach has several compartments, and, as in other sloths, cellulose digestion is aided by microfauna. Foley et al. (1995) found that the pale-throated sloth (*B. tridactylus*) retained a large mass of digesta in the gut. Passage is slow because of low metabolic rate and slow fermentation. Three-toed sloths are smaller (mean about 60 cm total length and 4.5 kg in the brown-throated sloth, *B. variegatus*—figure 15.7) and generally more common than *Choloepus* in areas of sympatry. The heaviest bradypodids

Figure 15.7 **Brown-throated sloth (*Bradypus variegatus*).** The long forelimbs and coarse pelage are characteristic of sloths.

are maned sloths (*B. torquatus*), in which females are significantly larger than males (Lara-Ruiz and Chiarello 2005). Three-toed sloths have three well-clawed toes on the forefeet (see figure 15.6). Litter size is one, and gestation is 5–6 months (compared with about 10 months in *Choloepus*). Reproduction in *Bradypus* is clearly seasonal. This may be related to the seasonal variation in food of the more specialized diet of three-toed sloths compared to two-toed sloths (Taube et al. 2001). Three-toed sloths have eight or nine cervical vertebrae, which allows for greater flexibility in the neck; they can rotate their heads in a 270° arc. Anderson and Handley (2001) first described the pygmy three-toed sloth, which is endemic to Isla Escudo de Veraguas off the Caribbean coast of Panama. As noted, it is a critically endangered species. The maned sloth is endangered.

Myrmecophagidae

The two genera and three species in this family are truly edentate and highly specialized for myrmecophagy, as suggested by the family name. Anteaters occur in forested or savanna habitats from southern Mexico south into South America east of the Andes Mountains as far south as Paraguay. They have a long, tapered skull, with an elongated rostrum, especially in Genus *Myrmecophaga*, a long tongue, and very small mouth. The tongue of the giant anteater (*M. tridactyla*) has a maximum width of only 13 mm but can be extended up to 600 mm, a distance equal to the length of the skull (Naples 1999). It is anchored on the sternum (see figure 7.3) and is covered with a viscous secretion produced in the submaxillary glands. The tongue also has tiny, barblike spines directed posteriorly. Both the spines and the secretion aid in trapping ants. The giant anteater has coarse, shaggy gray hair with a dark diagonal stripe on the shoulders, and a bushy tail (figure 15.8A). Total length averages about 2 m, and they may weigh as much as 40 kg. They are entirely terrestrial and active throughout the day or night. The two species of lesser anteaters, or tamanduas, are intermediate in size, active day or night, and forage on the ground and in trees. They have coarse tan or brown pelage, and in the northern tamandua (*Tamandua mexicana*) and in southern specimens of the southern tamandua (*T. tetradactyla*) black fur forms a "vest" (figure 15.8B). Tamanduas have a prehensile tail that aids in climbing. All species have long, sharp, powerful claws for foraging, with the middle claw often enlarged (see figure 15.6D, E). Stomachs are simple, with the pyloric portion strengthened for digesting insects. As in sloths, litter size in anteaters is generally only one. Pereira et al. (2004) examined karyotypes of myrmecophagids. They reported 2n = 60 in the giant anteater, and 2n = 54–56 in the southern tamandua.

A

B

C

A

B

Figure 15.8 **Features of representative anteaters.** (A) Coarse hair, diagonal shoulder stripe, and long, tapered skull of the giant anteater (*Myrmecophaga tridactyla*). (B) Distinctive "vest" of the southern tamandua (*Tamandua tetradactyla*), a smaller, semiarboreal anteater.

Figure 15.9 **Representative armadillos.** (A) Andean hairy armadillo (*Chaetophractes nationi*); (B) six-banded armadillo (*Euphractus sexcinctus*); and (C) Llanos long-nosed armadillo (*Dasypus sabanicola*).

A

B

Figure 15.10 Distribution of the nine-banded armadillo.
(A) Range within the western hemisphere, and (B) the north-ernmost extent of the species range in the United States.
*(A) B. K. McNab, 1985, Energetics, population biology,
and distribution of xenarthrans, living and extinct, in* The
Evolution and Ecology of Armadillos, Sloths, and Vermilinguas
*(G. G. Montgomery, ed.), Smithsonian Institution Press; (B)
Feldhamer, Thompson, and Chapman (eds.), 2003,* Wild Mammals
of North America: Biology, Management, and Conservation,
Johns Hopkins Univ. Press.

Cyclopedidae

This newly recognized family is monotypic, with only a single genus and species. The silky anteater occurs from

southern Mexico to Brazil and Bolivia. The common name derives from the fine, soft, wooly pelage that is gray to yellowish in color with a darker dorsal band. Silky anteaters are small—about 430 mm in total length and 230 g in body mass. They are nocturnal, and almost entirely arboreal, even seeking termites and ants as well as shelter in trees. Like tamanduas, the silky anteater has a prehensile tail, which is naked on the ventral surface. The chromosome number, 2n = 64, is the highest among anteaters (Pereira et al. 2004).

Dasypodidae

The 9 genera and 21 species of armadillos (figure 15.9) occur in different habitats from the southeastern United States through Central America to the tip of South America. Three monophyletic subfamilies are currently recognized: Dasypodinae, Tolypeutinae, and Euphractinae (Delsuc et al. 2003; Gardner 2005). The long-nosed or nine-banded armadillo is the only xenarthran in North America (figure 15.10). It has extended its distribution significantly since the late 1800s through natural dispersal as well as by introduction to Florida (Taulman and Robbins 1996; Layne 2003); it is the most widely distributed of any xenarthran. Nine-banded armadillos in North America exhibit reduced genetic variability compared to those in South America (Huchon et al. 1999). Further distribution northward probably is limited by cold ambient temperatures, the high thermal conductance of armadillos, their inability to enter torpor, and the lack of food in winter (McNab 1985). Armadillos burrow extensively. The big hairy armadillo (*Chaetophractus villosus*) in Argentina makes short, simple burrows while searching for food or temporary shelter, whereas permanent home burrows are longer, deeper, and more complex (Abba et al. 2005).

A

B

Figure 15.11 Size comparison. Approximate relative sizes of (A) the giant armadillo (*Priodontes maximus*) and (B) the nine-banded armadillo (*Dasypus novemcinctus*). *Adapted from J. Eisenberg, 1989,* Mammals of the Neotropics, *Univ. Chicago Press.*

Armadillos feed opportunistically on a variety of invertebrates (Bolkovic et al. 1995), and also consume various amounts of vegetation and carrion. Bezerra et al. (2001) reported an instance of predation on a small mammal by the six-banded armadillo (*Euphractus sexcinctus*).

Armadillos have several unique morphological characteristics, the best known of which is the hard, armor-like carapace that gives the group its common name. Plates of ossified dermal "scutes" cover the head, back and sides, and tail in most species (Figures 15.9 and 15.11). Scutes are covered by nonoverlapping, keratinized epidermal scales. The bands of armor are connected by flexible skin. The outside of the legs also have some armored protection, but not the inside of the legs or the ventral surface. The tough skin in these areas is covered by coarse hair. In the United States, *D. novemcinctus* is commonly called the nine-banded armadillo because its carapace usually has nine movable bands, although band number varies throughout its geographic range.

The vertebrae in armadillos are modified for attachment of the carapace, which actually articulates with the metapophyses of the lumbar vertebrae (figure 15.12). Species vary in size from the tiny (100 g or less) pink fairy armadillo (*Chlamyphorus truncatus*) to the giant armadillo (see figure 15.11), which weighs up to 60 kg. Armadillos share the usual xenarthran reduction in dentition, having only peglike molariform teeth, which are small, open-rooted, homodont, and without enamel. The giant armadillo is exceptional among terrestrial mammals in having up to 100 small, somewhat vestigial cheekteeth. Although incisors form in embryonic *D. novemcinctus*, they degenerate and rarely persist by the time of birth.

Reproduction in *Dasypus* is also noteworthy. Delayed implantation occurs, as does **monozygotic polyembryony**

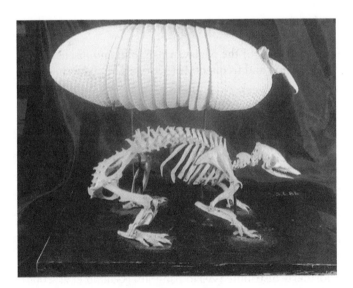

Figure 15.12 **Carapace and articulated skeleton of a nine-banded armadillo.** Note the distinctive metapophyses of the lumbar vertebrae for support of the carapace.

(several embryos from a single zygote) in several species. In *D. novemcinctus*, four young of the same sex are produced after division of a single fertilized ovum; in the southern long-nosed armadillo (*D. hybridus*), eight young normally are produced. From a common placenta, each embryo develops its own placenta, with no mixing of blood or nutrient material between embryos (Talmadge and Buchanan 1954). Eisenberg (1981:55) suggested that this is a "novel way of increasing the reproductive capacity" of *Dasypus* toward a more "r-selected" mode (Gleeson et al. 1994). Other aspects of obligate polyembryony were noted by Loughry et al. (2005). Enders (2002) described differentiation of the blastocyst and development of placentation in the nine-banded armadillo. This species has a hemochorial type of placenta (see figure 10.15) as do several other species of dasypodids (Adamoli et al. 2001). Finally, because *D. novemcinctus* contracts naturally occurring leprosy, it is a valuable model for a wide range of biomedical research projects (Storrs 1971; Pan American Health Organization 1978).

Pholidota

Pangolins, or scaly anteaters, are in a single family, Manidae, with a single genus (*Manis*) and eight species. As was true for the aardvark, pangolins were once placed in the Edentata based on morphological features associated with diet. Four species of pangolins are found in Africa south of the Sahara Desert. Patterson (1978) advocated placing these species in the Genus *Phataginus*. Based on analyses of numerous cranial characteristics, Gaudin and Wible (1999) concluded that the species of African pangolins are not monophyletic. Four other species of pangolins occur in Pakistan, India, Sri Lanka, southeast Asia, southern China, and Indonesia, including the Philippine or Palawan pangolin (*M. culionensis*), recently accorded species rank (see Schlitter 2005). The Palawan pangolin differs from the Malayan pangolin (*M. javanica*) in several skull characteristics and aspects of the scales (Gaubert and Antunes 2005). Habitats of pangolins include forests, savannas, and sandy areas, with distribution directly related to the occurrence of ants and termites, their primary prey (Heath 1992, 1995). Using data gathered from 15 ground pangolins (*M. temminckii*) in South Africa fitted with radiotransmitters, Swart et al. (1999) determined that ants, primarily *Anoplolepis custodiens*, formed 96% of their diet. The mean duration of feeding bouts by pangolins was 40 seconds, with the abundance and size of ants the primary factors determining the number of feeding bouts on a particular prey species. Some pangolins are terrestrial and are strong diggers, living in large, deep burrows, whereas others are arboreal, living primarily in trees and having semiprehensile tails. Even the terrestrial species are good climbers and occasionally may forage in trees. Pangolins may be active during the day but are primarily nocturnal. Occurring as solitary individuals or in pairs, they have limited vocal, visual, and auditory acuity. Olfactory communication plays a significant role in their behavior, however. Strong scent is produced from paired anal glands, and feces and urine are deposited along trails and trees. Like skunks (Carnivora: Mephitidae), they may eject an unpleasant smelling secretion from the anal glands. Gestation is about 140 days. Litter size is usually one, but occasionally twins are produced. The young are born with soft scales that do not harden until 2 days after birth. Newborn young cling to the female's back or tail, and, if threatened, the female curls up around the neonate.

MORPHOLOGY

The ordinal name means "scaly ones" and refers to the major diagnostic characteristic of this group. With the exception of the sides of the face, inner surface of the limbs, and the ventral surface, pangolins are covered with **imbricate** (overlapping) scales, somewhat like those of a pine cone (figure 15.13). Scales, which are composed of keratinized epidermis and are dark brown to yellow in color, serve the same protective function as the armor of armadillos or the spines of echidnas. A pangolin has the same number of scales throughout life; they grow larger as the individual grows. Skin and scales make up 20% or more of the body mass of most species, although, as might be expected, scales on arboreal species such as the tree pangolin (*M. tricuspis*) are lighter and thinner than those on terrestrial species. When alarmed, pangolins curl up in a ball, with the sharp-edged, movable scales directed outward. Pangolins may reach 1.6 m in length, with the tail comprising half the total. External ears do not occur in the four African species, but they are found in Asian pangolins.

Pangolins are truly edentate. Instead of teeth they use a long, powerful tongue for foraging. Sticky, viscous saliva is secreted onto the tongue by a large salivary gland in the chest cavity. Tongue muscles are enclosed in a sheath and pass over the sternum and anchor on the pelvis (Kingdon 1971). A pangolin's tongue is longer than its head and body length and is structurally similar to that of the giant anteater (Chan 1995; also see Reiss 2001). The skull is long and tapered, with a straight slender mandible and incomplete zygomatic arches (see figure 15.13). The morphology of the skull reflects the pangolin's diet, lack of teeth, and associated lack of any strong facial muscles for chewing. Pangolins are plantigrade and pentadactyl, with large, sharp, curved claws to break into ant and termite mounds, which they locate primarily through scent. As in other ant-eating species, pangolins have a stomach with a muscular, gizzard-like pyloric region for grinding ants and termites. Nisa et al. (2005) examined

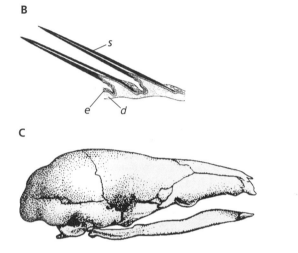

Figure 15.13 Pangolin characteristics. (A) Dorsal view of a long-tailed pangolin (*Manis tetradactyla*), showing its distinctive scales. (B) Diagram of a section through two scales, showing the scales (s), epidermis (e), and dermis (d). (C) Skull of a pangolin, with incomplete zygomatic arch and very simple mandible. *Adapted from K. Kowalski, 1976,* Mammals: An Outline of Theriology, *Polish Scientific Publishers, Warsaw.*

two stomachs of the Malayan pangolin and discussed the distribution and frequency of six types of endocrine cells.

FOSSIL HISTORY

Fossil material is meager, to an extent because pangolins lack teeth, which fossilize most easily, and all fossils are outside the current distribution of pangolins. Mid-Eocene pangolins with scales (*Eomanis krebsi* and *E. waldi*) are known from Europe, and pangolins occur in late Eocene deposits in Mongolia (Rose and Emry 1993). A fossil pangolin from the late Eocene of northern China, *Cryptomanis gobiensis*, represents the oldest and most northerly distributed Asian pangolin (Gaudin et al. 2006). It is similar anatomically to *Patriomanis americanus*, a late Eocene pangolin from western North America (Emry 2004).

ECONOMICS AND CONSERVATION

Native people eat African species (Carpaneto and Fusari 2000), and their scales are used for adornment. Scales also are used as good luck charms, with entire skins being very valuable (Kingdon 1971). In Asia, because powdered scales are believed to be of medicinal value, populations of the four species in this region are greatly reduced. For example, Wu et al. (2004) discuss threats to populations of the Chinese pangolin (*M. pentadactyla*) from overhunting and smuggling animals for food and medicinal uses.

Tubulidentata

This is the smallest eutherian order, with a single living member. The Family Orycteropodidae contains only one living member, the aardvark (*Orycteropus afer*), which occurs in Africa south of the Sahara Desert in habitats ranging from dry savanna to rainforests. As with pangolins, the aardvark is closely associated with the distribution of ant and termite mounds. An aardvark may excavate as many as 25 mounds a night (Taylor and Skinner 2000). The same mound may be visited repeatedly, with the aardvark foraging at previously excavated sites. Larvae, locusts, and even wild cucumbers (*Cucumis humifructus*) are also eaten. However, as with pangolins, the ant *Anoplolepis custodiens* is the most important prey species (Taylor et al. 2002). Aardvarks are solitary, secretive, and elusive, making them difficult to observe. They are predominately nocturnal (rarely crepuscular) and travel as much as 30 km a night as they forage. They locate prey with the aid of highly acute olfactory and auditory senses. Their burrow system is extensive; several openings occur throughout a small area. Females give birth to a single young after 7 months gestation.

Based on extensive molecular analyses of base pair sequences for almost all mammalian orders (Madsen et al. 2001; Murphy et al. 2001b), the aardvark occurs in a clade usually referred to as "Afrotheria" that includes elephants, manatees, hyraxes, tenrecs, and elephant shrews. More recent work has supported this view (see references in Holroyd and Mussell 2005). Other molecular analyses do not support the Afrotheria grouping, however. For example, Arnason and coworkers (1999) sequenced the entire mtDNA of the aardvark and concluded that Genus *Orycteropus* is most closely related to xenarthrans and a group that includes carnivores, ungulates, and whales. Unfortunately, the molecular, morphological, and fossil evidence has not resulted in a consensus as to the position of the aardvark relative to other mammalian orders.

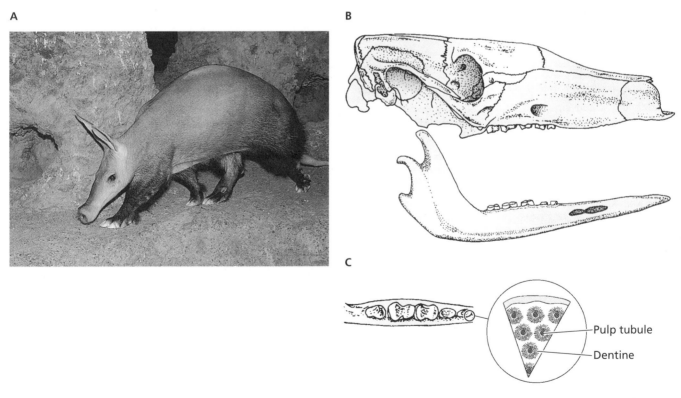

Figure 15.14 Features of the aardvark. (A) Piglike snout, large ears, and heavy claws of an aardvark. (B) Aardvark skull. (C) Diagrammatic representation of the occlusal surface of the toothrow of an aardvark with a section of the pulp tubules and surrounding dentine. *Adapted from DeBlase and Martin, 1981, A Manual of Mammalogy, 2nd ed., Wm. C. Brown Publishers, and from K. Kowalski, 1976, Mammals: An Outline of Theriology, Polish Scientific Publishers, Warsaw.*

MORPHOLOGY

The word *aardvark* is Afrikaans for "earth pig," and the species superficially resembles a pig (figure 15.14). It weighs up to 60 kg and is about 1.5 m in head and body length. The long, square snout is somewhat flexible, with tufts of hair that protect the nostrils during digging. The nasal septum has short, somewhat fleshy tentacles that probably serve an olfactory function. This supposition is further supported by the fact that the aardvark has more turbinate bones (scroll-like bones in the nasal passages) than any other mammal. Aardvarks have small eyes and large, erect ears; the latter is unusual in myrmecophagous species. The yellowish brown hide is very tough, sparsely haired, and insensitive to ant and other insect bites. The forefeet have four toes, and the hind feet five, each with a heavy, strong clawlike nail. Aardvarks can burrow rapidly as well as break into ant and termite mounds. As in anteaters and pangolins, the aardvark's skull is elongated (see figure 15.14), and it has a small tubular mouth with a long, sticky tongue. Aardvarks chew their food, however, and the dentition is fairly unusual. Incisors and canines are found only during the fetal stage, and adults typically have 20–22 cheekteeth. These teeth are composed of up to 1,500 pulp tubules surrounded by hexagonal prisms of dentine (see figure 15.14C) and account for the ordinal name ("tubule-toothed"). Teeth are open-rooted, without enamel, and are covered with cementum. Besides chewing teeth, aardvarks also grind ingested material in a muscular pyloric region of the stomach, similar to that of pangolins.

FOSSIL HISTORY

Like the fossil record of pangolins, that of the aardvark is fragmentary. The earliest known tubulidentate (Genus *Myorycteropus*) is from early Miocene deposits of east Africa. *Orycteropus gaudryi* dates from the late Miocene and, except for more cheekteeth, is similar to the extant species. A relatively unspecialized form, Genus *Leptorycteropus*, dates from the mid-Pliocene. Pleistocene remains are known from France, Greece, Turkey, India, and Madagascar (Patterson 1975; Lehmann et al. 2005).

ECONOMICS AND CONSERVATION

Aardvarks are eaten by natives, and the teeth are used as jewelry and good luck charms. Populations have been reduced throughout the range, although currently the species is not considered threatened.

SUMMARY

The orders discussed in this chapter offer an excellent example of phylogenetic convergence because many of the species "make a living" in similar ways. Although quite distinct morphologically, the aardvark, pangolins, and anteaters share a number of characteristics related to their myrmecophagous diets. The armadillos and sloths, along with the true anteaters, exhibit interesting structural diversity within a relatively small order. Named for the accessory vertebral articulations that give added rigidity to the axial skeleton, xenarthrans occur only in the New World and are restricted to warm regions because of their poor thermoregulatory abilities.

The pholidotes occur in both Africa and Asia. Called pangolins, or scaly anteaters, they are covered with keratinized, overlapping scales. These scales serve the same protective function that the carapace does in armadillos. Like the anteaters, pangolins are edentate.

The aardvark is not closely related to either xenarthrans or pholidotes, but like these groups, the aardvark's occurrence within its range is closely associated with ants and termites. The aardvark exhibits morphological and behavioral characteristics similar to other myrmecophagous species. Unlike anteaters and pangolins, however, aardvarks have cheekteeth.

SUGGESTED READINGS

Loughry, W. J., P. A. Prodohl, C. M. McDonough, and J. C. Avise. 1998. Polyembryony in armadillos. Am. Sci. 86:274–279.

Montgomery, G. G. (ed.). 1978. The ecology of arboreal folivores. Smithsonian Institution Press, Washington, D.C.

DISCUSSION QUESTIONS

1. Considering the symbiotic relationship between sloths and the algae that grow on their pelage, what are the benefits to both groups? Are there any negative consequences?

2. Discuss the relationship between low metabolic rates, poor thermoregulatory abilities, geographic distributions, and small litter sizes in xenarthrans.

3. Why might you expect pangolins that are primarily arboreal to have lighter, thinner scales than terrestrial species?

4. Why might you expect to find "altruistic" behavior among populations of the nine-banded armadillo rather than in some of the other species discussed in this chapter?

5. What is the significance of small eyes and ears in most myrmecophagous mammalian species?

Carnivora

A great deal of diversity exists among the 15 families and approximately 286 species now recognized in this order. Most of these species eat meat. Although it is easier to digest than vegetation, meat is much more difficult to locate, capture, kill, and consume. Thus, carnivores generally are thought of as the major group of mammalian predators because they feed primarily on animal flesh (including other mammals). Whereas numerous orders of mammals are herbivores, whales, insectivores, many bats, and a few rodent and marsupial species also consume animals. Carnivores occur naturally on all continents except Australia and have adapted to diverse niches in a variety of terrestrial and aquatic habitats.

Although most carnivores eat meat, this is not a defining characteristic of the order. Living carnivores share several morphological characteristics, including specialization of the teeth (see the Morphology section of this chapter). Carnivores are arranged into two suborders (Wozencraft 2005; Flynn and Wesley-Hunt 2005), based on the structure of their auditory bullae and carotid circulation (Flynn et al. 1988; Wozencraft 1989). The Suborder **Feliformia** (meaning "catlike") includes six families: the Felidae (cats), Herpestidae (mongooses), Hyaenidae (hyenas), Viverridae (civets) and the recently recognized Eupleridae (Madagascar mongooses) and the monotypic Nandiniidae (the African palm civet, *Nandinia binotata*) (see Gaubert et al. 2005). The other Suborder, **Caniformia** (meaning "doglike"), includes the Families Canidae (dogs), Ursidae (bears), Mustelidae (weasels), Procyonidae (raccoons), Mephitidae (skunks), the monotypic Ailuridae (the red panda, *Ailurus fulgens*) and three families of aquatic carnivores: Odobenidae (walrus), Otariidae (eared seals, including fur seals and sea lions), and Phocidae (earless, or true seals). The aquatic carnivores are referred to as **pinnipeds** (meaning "feather-footed"), based on the modification of their limbs into flippers. In the past, pinnipeds were considered either as a separate order of their own or as a suborder within the Order Carnivora, with the remaining members grouped in the Suborder Fissipedia (meaning "split-footed") because of the individual toes on each foot. Currently, a single order (figure 16.1) is considered most appropriate. Unlike whales, pinnipeds leave the water to rest, breed, and give birth. Also, the sea otter (*Enhydra lutris*) is essentially

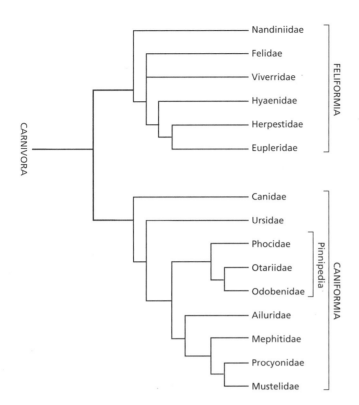

Figure 16.1 Phylogeny of modern carnivore families. Within the two traditional suborders, most authorities recognize two new families within the Feliformia—Nandiniidae and Eupleridae. Among the Caniformia, the red panda (family Ailuridae) groups most closely with the skunks (family Mephitidae), and the three families that comprise the pinnipeds form a clade. *Adapted and modified from Flynn et al. (2005) and Flynn and Wesley-Hunt (2005).*

Figure 16.2 Wolves are the primary predator of caribou. The primary mortality factor for young caribou in Alaska is predation by wolves.

aquatic. Several species of otters, and the polar bear (*Ursus maritimus*), spend much of their time in the water.

Smaller species of carnivores generally have larger litters and breed more frequently. Larger species, such as bears and the pinnipeds, have small litters (often a single young), and females may breed at intervals of several years. Induced ovulation or delayed implantation (see chapter 10) occur in certain species and are generally related to environmental or life history factors. Few carnivores have precocial young. Juveniles must learn to hunt successfully under a variety of circumstances, and an extended period of maturation and learning is often required before these skills are acquired and individuals disperse (Holekamp et al. 1997). This need to learn and adapt, as well as to develop a high degree of coordination and dexterity, is reflected in the high brain-to-body mass ratio of carnivores.

When hunting, individuals may be solitary, paired, or in small groups. Smaller carnivores, such as weasels, are generally solitary and restricted to taking smaller prey. Larger species, such as wolves, lions, hyenas, and African hunting dogs (*Lycaon pictus*), may hunt in packs. In addition to hunting smaller prey, packs are able to prey on larger, more dangerous species (figure 16.2) that may be

several times the size of individual members of the pack (Scheel and Packer 1991; Holekamp et al. 1997). It has been suggested (Kruuk 1972; Schaller 1972) that social behavior evolved primarily to improve hunting success and, through communication of an individual's dominant or submissive status, to reduce intraspecific competitive pressure within groups. Groups also may provide communal infant care, reduce predation, and defend territories from rival packs. Other investigators (Mills 1985; Packer et al. 1990) have found that cooperative hunting is just one of many factors that may favor group living. Caro (1994) provides a concise summary of factors affecting group living in carnivores.

Methods of hunting vary from concealment and a surprise pounce (stealth and "ambush" seen in many felids) to a stalk followed by a short, swift run (weasels) to a prolonged chase (wolves or hyenas). One of the functions of pelage in carnivores is concealment (as it is in potential prey species), and coloration often is related to hunting behavior. Generally, in cats and other groups in which concealment is critical, the pelage often has spots or stripes. Conversely, in predators such as canids, pelage is typically plain because concealment is less critical (figure 16.3). There are exceptions to both cases, however.

Species in the Order Carnivora exhibit several basic morphological and behavioral adaptations for searching out, capturing, and handling their prey so as to minimize the possibility of injury during the process. In this chapter, we explore general characteristics of carnivores as well as life history strategies within lineages that have evolved to deal with the diverse evolutionary developments between predators and their prey.

Morphology

The defining morphological characteristic of carnivores is the specialization of their fourth upper premolar (P^4) and

A

B

Figure 16.3 **Pelage coloration of carnivores.** Pelage coloration and pattern is often adapted to hunting technique. Species that hunt from concealment, including most members of the cat family, often have spots or stripes to help camouflage them—such as (A) the jaguar (*Panthera onca*). Species that do not hunt from concealment, such as (B) the gray fox (*Urocyon cinereoargenteus*), usually have pelage without stripes or spots.

Upper Carnassial tooth (*p⁴*)

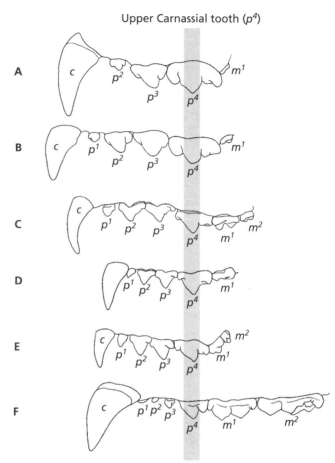

Figure 16.4 **Carnassial dentition.** The relative development of the carnassial dentition in various families of carnivores is represented by the vertical line through the maxillary carnassial tooth (P^4) in a (A) lion (Felidae); (B) hyena (Hyaenidae); (C) dog (Canidae); (D) marten (Mustelidae); (E) mongoose (Herpestidae); and (F) bear (Ursidae). Note the large canine tooth at the left (anterior) and the fact that the carnassial teeth are posterior (toward the back of the jaw). Mammalian dentition functions like a nutcracker, with the greatest force generated closest to the articulation of jaw and skull (the fulcrum point in a lever system). Thus, carnassial teeth are posterior in the dental arcade, rather than anterior, to better facilitate their crushing and shearing function. *Adapted from A. F. DeBlase and R. E. Martin, 1981, A Manual of Mammalogy, 2nd ed., Wm. C. Brown Publishers.*

first lower molar (m₁) as carnassial (shearing, or cutting) teeth. Carnassials are especially well developed in the predaceous felids, hyaenids, and canids but much reduced in the omnivorous ursids and procyonids (figure 16.4). Regardless of the relative development of carnassials, all carnivores have well-developed, elongated canine teeth.

Skulls are usually heavy, with strong facial musculature for crushing, cutting, and chewing flesh, ligaments, and bone. The relative development of facial muscle groups (see figure 7.6) and associated skull shape reflect different

life history patterns and relative use of canines and cheekteeth. Carnivores often have a deep, sharply defined, C-shaped **mandibular fossa** (the portion of the cranium that articulates with the mandible). This strong hinge joint, particularly evident in mustelids (figure 16.5), minimizes lateral movement of the mandible, as captured prey struggles, and permits only a vertical, or up-and-down, motion. Omnivorous carnivores, such as bears and raccoons, have a relatively flatter mandibular fossa that allows more lateral motion of the jaw as the animal chews.

The auditory bullae in carnivores (as in other eutherians) house the tympanic membrane and inner ear. Bullae are formed either entirely from the tympanic bone (derived from the reptilian angular bone) or from the tympanic and endotympanic bones. The structure of the

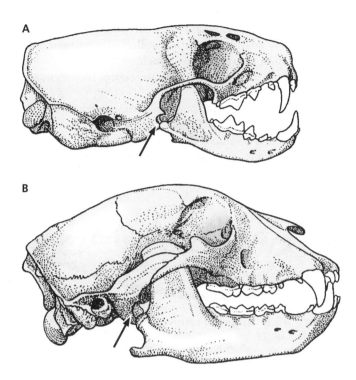

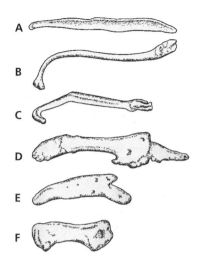

Figure 16.6 **Shapes and sizes of bacula.** The size and shape of the baculum varies considerably among carnivores. Lateral view of the baculum in a (A) canid (red fox [64 mm]); (B) procyonid (raccoon, [90 mm]); (C) mustelid (least weasel, [19 mm]); (D) herpestid (Egyptian mongoose, *Herpestes ichneumon* [18 mm]); (E) viverrid (common genet, *Genetta genetta* [6 mm]); and (F) felid (lion, [7 mm]). *Adapted from R. F. Ewer, 1973,* The Carnivores, *Weidenfeld/Nicolson.*

Figure 16.5 **Carnivore jaw articulation.** (A) The strongly C-shaped mandibular fossa of a mustelid restricts lateral movement of the lower jaw. (B) The fossa of a bear is flatter. *Adapted from T. E. Lawlor, 1979,* Handbook to the Orders and Families of Living Mammals, *2nd ed., Mad River Press.*

bullae is a criterion used to differentiate the two suborders of carnivores. In feliforms, both the tympanic and endotympanic bones form the bullae, with a septum occurring where the two meet. In caniforms, the bullae are formed almost entirely from the tympanic bone, and there is no septum. Ivanoff (2001) provided details of the auditory bullae of carnivores and discussed formation of the septum.

All but a few carnivore species have a well-developed **os baculum** (penis bone; figure 16.6). Although the function of the baculum is open to question, it may serve to prolong copulation in species with induced ovulation. Most carnivores have distinctive **anal sacs** associated with secretory anal scent glands. These occur on both sides of the anus and produce substances that function in defense and intraspecific communication. They are especially well developed in mephitids, mustelids, herpestids, and hyaenids. Skunks are well known for ejecting the anal gland secretion as a defensive mechanism. Weasels also have strong-smelling scent glands, but they are not used defensively. Anal sacs are relatively small in canids and felids and are absent in ursids and some procyonids.

Carnivores usually have well-developed claws on all digits. Even the unusual clawless otters (Genus *Aonyx*) have vestigial claws. In most felids and some viverrids, claws are retractile. This helps keep them sharp because they have less contact with the ground. Neither the pollex nor the hallux is opposable. The centrale, scaphoid, and lunar bones of the wrist are fused (figure 16.7) to form a scapholunar bone, which may add support for cursorial locomotion in some terrestrial species. As in ungulates, the clavicle

in carnivores is reduced or lost, which serves to increase the length of the stride and allow for faster running in cursorial species. Carnivores have an expanded braincase, but the postcranial skeleton of terrestrial species is generalized. Differences that occur among families often are only a matter of proportion. Differences in limb structure reflect locomotor adaptations: cursorial canids and felids are digitigrade, whereas ursids and procyonids are plantigrade. Specialized morphological adaptations of pinnipeds are noted in the Pinnipedia section of this chapter.

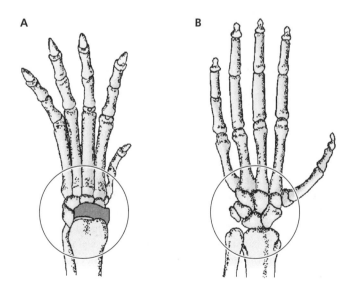

Figure 16.7 **Carnivore wrist structure.** (A) In carnivores, the centrale, scaphoid, and lunar bones of the wrist are fused (*shaded area*), whereas in other mammals, such as a (B) primate, they are not. *Adapted from D. Macdonald, 1984,* Encyclopedia of Mammals, *Facts on File.*

Carnivores range in size from a body mass of about 70 g in the least weasel (*Mustela nivalis*) to 800 kg in grizzly bears (*Ursus arctos*) and polar bears—over 11,000 times heavier. Secondary sexual dimorphism is often evident, with males being larger than females. This dimorphism is particularly pronounced in some pinnipeds. For example, male elephant seals (*Mirounga angustirostris*) may be six times larger than females. Evolution in size variation from small to large species within and among carnivore families reflects the size range of potential prey species available to them.

Because meat is easy to digest, carnivores have simple stomachs with an undeveloped cecum. Nonetheless, some specialized feeding habits have evolved. Besides those that eat only flesh, species are insectivorous, piscivorous, frugivorous, omnivorous, or almost completely herbivorous. Carbone et al. (1999) noted a direct relationship between the body mass of carnivores and the mass of their typical prey. They found that the maximum body size of carnivores that primarily prey on invertebrates was about 21 kg. Thus, diet choice and body size in carnivores are closely related. The structure of the dentition, including the carnassials, is correspondingly modified (see figure 16.4). Canids, felids, and mustelids subsist mainly on freshly killed prey. These families show correspondingly greater development in "tooth and claw;" they also have greater carnassial development and cursorial locomotion. In addition to live prey, canids, ursids, and hyaenids take a large amount of **carrion** (dead, often decaying animal matter). Because not all carnivores are strictly carnivorous (most ursids and procyonids are omnivorous), their diet varies depending on season and local availability of food. The giant panda bear (*Ailuropoda melanoleuca*) eats primarily bamboo shoots and roots and only occasionally eats animal matter (Wei et al. 1999).

Fossil History

The earliest known mammalian genus generally adapted for carnivory was *Cimolestes* (figure 16.8), from the late Cretaceous period, over 65 million years ago (mya). *Cimolestes* was small, about the size of a weasel, and is considered the basal group for both modern carnivores and an archaic group of terrestrial carnivores, the Creodonta. The creodonts (figure 16.9) extended from the late Cretaceous to the Miocene, when they became extinct, possibly through competition with modern carnivore lineages. Other early small carnivores were in the extinct families Viverravidae and Miacidae. These first recognizable carnivores, with P4/m₁ carnassial dentition, appeared in the late Paleocene and early Eocene. Unlike modern carnivores, the centrale, scaphoid, and lunar bones of the wrist in these early specimens were not yet fused. Although modern families of carnivores probably began to diverge by the late Eocene, most are recognizable only by the Oligocene, including canids, felids, viver-

rids, mustelids, and ursids (Flynn and Wesley-Hunt 2005). This adaptive radiation reflected corresponding diversification of prey groups, which was in turn related to the development of more diverse vegetative biomes in the early Cretaceous period.

The earliest ancestral pinnipeds, the enaliarctids, date from the late Oligocene and early Miocene (Ray 1976; Berta 1991). For example, the earliest walrus, *Proneotherium repenningi* (Demere and Berta 2001), dates from the early to

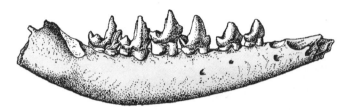

Figure 16.8 Ancestral carnivore. Lateral view of the lower jaw of the late Cretaceous *Cimolestes*. These early eutherian mammals may have been ancestral to both the creodonts (extinct by the Miocene epoch), and modern carnivores. Actual length of the jaw fragment is about 3 cm. *Adapted from R. L. Carroll, 1988*, Vertebrate Paleontology and Evolution, *W. H. Freeman.*

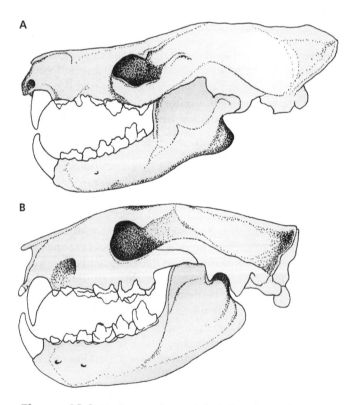

Figure 16.9 Early creodonts. (A) Skull and jaw of an early creodont, Limnocyon (Family Hyaenodontidae). (B) Unlike the modern lineage of carnivores, the carnassial teeth in creodonts, such as Oxyaena shown here (Family Oxyaenidae), involved the first or second upper molars and the second or third lower molars. Skull length is about 21 cm. (A) *From Flynn and Wesley-Hunt (2005).* (B) *Adapted from A. S. Romer, 1966,* Vertebrate Paleontology, *3rd ed., Univ. Chicago Press.*

mid Miocene, about 14–20 mya. Odobenids and otariids are closely related (see figure 16.1) and are often grouped in the Superfamily Otarioidea. They represent a sister group to the phocids.

Economics and Conservation

Many species of carnivores have close associations with humans. The dog (*Canis lupus familiaris*)—considered by Wozencraft (2005) as a subspecies of the wolf—was the first domesticated mammal species (see chapter 28). Dogs and domestic cats (*Felis catus*) have been popular pets throughout the world for thousands of years. Likewise, numerous species of carnivores have been important as furbearers throughout human history, including most species of felids and pinnipeds. Today, raccoons (*Procyon lotor*) are trapped for their pelts, as are foxes and many species of mustelids, including mink (*Mustela vison*), river otter (*Lontra [Lutra] canadensis*), marten (*Martes americana*), fisher (*M. pennanti*), and sable (*M. zibellina*)—see Feldhamer et al. (2003). Many other species are useful to the agricultural industry as predators on rodents and other agricultural pests.

Conversely, carnivores often are in conflict with human economic interests. Vilified as "red in tooth and claw," they are viewed by some people as viciously preying on innocent grass-eating livestock. As a result, the density and distribution of many species have been greatly reduced. The wolf (*Canis lupus*) is a prime example. It has been extirpated throughout much of its former range in the Old World. In the contiguous 48 United States, it occurs in only about 5% of its historic range (Paquet and Carbyn 2003). Similarly, in the last 200 years, the grizzly bear was practically extirpated in the United States (Schwartz et al. 2003). In response to changing sentiments toward conservation, however, many species, including the wolf, are beginning to recover (Mech 1995), often because of active reintroduction programs. Historically, large carnivores throughout the world also have been important as big-game trophies. This has resulted in serious overexploitation and range reduction of numerous species, as in the case of the tiger (*Panthera tigris*).

Suborders and Families

FELIFORMIA

Felidae

The cat family is distributed worldwide except for Australia, New Zealand, and surrounding islands; polar areas; Madagascar; Japan; and most oceanic islands.

Modern felids resulted from relatively recent divergence and speciation events in the late Miocene (Johnson et al. 2006). The classification of this family, especially regarding the Genus *Felis*, is controversial. Some authorities recognize only a few genera; however, Wozencraft (2005) considered 14 genera and 40 species within two subfamilies—Felinae and Pantherinae. All cats are characterized by a shortened rostrum (figure 16.10), well-developed carnassials, and large canine teeth that are "highly specialized for delivering an aimed lethal bite" (Ewer 1973:5). Felids kill their prey by suffocation or by biting the prey's neck so the canines enter between the vertebrae and separate the spinal cord. Loss or reduction in size of the other teeth is more evident in felids than in any other carnivores; total number of teeth is reduced to 28 or 30. The shortened toothrow and rostrum add to the force generated through the enlarged canines. The cheekteeth are bladelike and adapted for seizing and slicing meat from prey, rather than crushing bones, as occurs in hyaenids or canids.

Body masses range from about 2 kg in the black-footed cat (*Felis nigripes*) of Africa to 300 kg in tigers. Large felids are noted for their ability to roar (because flexible cartilage replaces the hyoid bone at the base of the tongue), whereas smaller species purr. Most cats are at least semiarboreal. They are digitigrade and have strongly curved, sharp claws to hold prey. Other than in the cheetah (*Acinonyx jubatus*), claws are protractable; in the resting position the claws are held "in." The dorsal surface of the tongue is covered by posterior-directed papillae that give the tongue a "sandpaper" feeling and may help to retain food in the mouth.

Felids prey almost exclusively on mammals and birds (Kruuk 1986), although the Asian flat-headed cat (*Prionailurus planiceps*) and fishing cat (*P. viverrinus*) consume fish, frogs, and even mollusks. Most felids are nocturnal. They are very agile and either stalk their prey or pounce from ambush. Their pelage is often spotted or striped, an adaptation to cryptic hunting behavior and

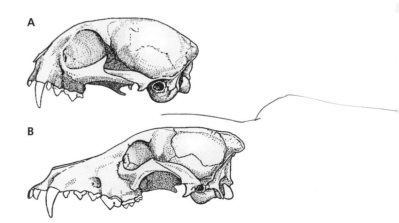

Figure 16.10 Shape of felid and canid skulls.
(A) A felid skull typically has a short, rounded rostrum compared with (B) the long rostrum typical of canids. *Adapted from T. E. Lawlor, 1979,* Handbook to the Orders and Families of Living Mammals, *2nd ed., Mad River Press.*

habitat. The diurnal cheetah, the fastest mammal in the world, relies on its ability to outrun prey over short distances. Species generally are solitary or form pairs; however, lions (*Panthera leo*) associate in prides that include up to 18 related females, their offspring, and several unrelated males (Packer 1986). The larger species are intolerant of other species; lions are known to kill leopards, and leopards kill cheetahs (Kingdon 1972). Most species of felids are endangered or threatened either because of overhunting, habitat loss, or the fur trade.

Hyaenidae

Hyaenids are native to the Old World in the Middle East, India, and Africa, where they are the most abundant large carnivore. They inhabit grassy plains and brushy habitats. Koepfli et al. (2006) suggest the hyaenids diverged in the mid-Oligocene, with living hyenas representing a small remnant of the nearly 70 described fossil species. The three genera and four extant species include the aardwolf (*Proteles cristatus*), which in the past has been placed in a separate family, the Protelidae, because of morphological and behavioral differences from hyenas. Hyenas are large carnivores; the spotted hyena (*Crocuta crocuta*) reaches 80 kg in body mass. Well-developed canines and cheekteeth make spotted hyenas highly competent hunters capable of killing large prey (Holekamp et al. 1997). They also are highly adapted to scavenging carcasses and feeding on carrion, using their large cheekteeth (figure 16.11A) to

crush bones. Although spotted hyenas generally are weaned when they are 1 year old, and morphological growth reaches a plateau at 20 months of age, their bite strength continues to increase until they are 5 years old (Binder and Van Valkenburgh 2000). Hyenas' scavenging specialization may allow them to minimize competition with sympatric canids. Hyenas regurgitate pellets of undigested material, including bone fragments, ligaments, hair, and horns. Holekamp et al. (1999) found that reproduction in the spotted hyena reflected seasonal variation in food availability. Unlike other hyaenids, the small, nocturnal aardwolf is myrmecophagous (feeds on ants and termites), which is unusual among carnivores. It primarily takes the snouted harvester termite (*Trinervitermes trinervoides*), which it locates through both sound and scent and by following aardvarks that have opened the mounds (Taylor and Skinner 2000). Reflecting these feeding habits, the dentition in *P. cristatus* is weakly developed, with smaller, more widely spaced cheekteeth than other hyaenids (figure 16.11B). The aardwolf spends the day in a den, which provides thermal stability to escape extremes of ambient temperatures and protection from black-backed jackels (*Canis mesomelas*) and other predators (Anderson and Richardson 2005).

Hyaenids are digitigrade and have nonretractile claws. The long neck and forelimbs and shorter hind limbs produce a characteristic sloping profile. Pelage is coarse, with a prominent mane in all but the spotted hyena, in which the mane is less apparent. This species is also noteworthy for its "laughing" vocalizations, one of several types of calls used in communication. Unlike most carnivores, females are slightly larger than males. Their external genitalia resemble that of males, such that it is very difficult to determine the sex of spotted hyenas in the field. As noted by Frank (1997:58), the "female has no external vagina; rather, the urogenital canal traverses the hypertrophied clitoris, which resembles a penis in size, shape and erectile ability" (figure 16.12). Various hypotheses have been proposed to explain the evolution of this female masculization, and debate continues (East and Hofer 1997; Frank 1997). All hyenas have **protrusable** (can be turned inside out) anal scent glands, which are dragged along the ground, depositing a pastelike substance to scent-mark territory (see figure 21.3). Hyaenids are unusual among carnivores in lacking a baculum. The brown hyena (*Hyaena brunnea*) has been reduced in numbers because of perceived livestock depredation. The aardwolf is also reduced in numbers, although neither are considered to be endangered.

Herpestidae

This Old World family is native to Africa, the Middle East, and Asia. The earliest known appearance of an extant herpestid, the slender mongoose (*Galerella sanguinea*) was from the late Miocene of Chad, Africa (Peigne et al. 2005). The 14 genera and 33 species of mongooses were included within the Family Viverridae (civets and

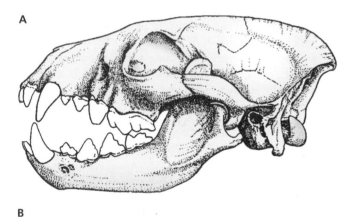

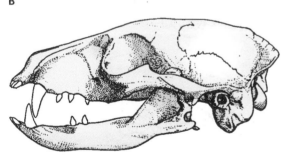

Figure 16.11 Hyaenid dentition. (A) The dentition of the striped hyena (*Hyaena hyaena*) is adapted for crushing bones, unlike (B) the reduced dentition of the myrmecophagous aardwolf. *Adapted from A. F. DeBlase and R. E. Martin, 1981, A Manual of Mammalogy, 2nd ed., Wm. C. Brown Publishers.*

Figure 16.12 Body morphology. Hyenas have a characteristic sloping profile. The high, strong shoulders provide leverage for individuals as they pull meat and hide from carcasses. This female has an erect pseudopenis.

Table 16.1 Morphological characteristics differentiating herpestids (mongooses) from closely related viverrids (civets and genets)

	Herpestidae (mongooses)	Viverridae (civets and genets)
Tail length	Less than head and body length	Equal to or greater than head and body length
Digits	Four or five; webbing reduced or absent	Five; webbing between toes
Claws	Nonretractile	Retractile
Ears	Short and round; no bursae on lateral margins	Long and pointed; bursae (pockets) on margins
Pelage	Usually uniform coloration	Usually spotted or striped
Behavior	Social, often forms groups; diurnal or nocturnal; terrestrial	Generally solitary; nocturnal; arboreal

genets) in the past. They differ from viverrids based on the structure of the auditory bullae and anal sacs, which are similar to those of hyaenids and mustelids. Anal sacs can be everted, exposing cutaneous glands (Kingdon 1972). Several other features distinguish herpestids from viverrids (table 16.1). Mongooses generally are small; body masses range from the dwarf mongoose (*Helogale parvula*) at 0.3 kg to the white-tailed mongoose (*Ichneumia albicauda*) at about 5.0 kg. Based on molecular data, Perez et al. (2006) found that herpestid genera grouped into two clades corresponding to the traditional subfamilies Herpestinae and Mungotinae.

Herpestids occur in a variety of terrestrial and semiarboreal habitats. They are feeding generalists, with associated generalized dentition (Ewer 1973). Some species are solitary, whereas others are highly social and form groups (Rood 1986). The meerkat (*Suricata suricatta*) is a herpestid that forms groups of family members in which individuals cooperate in activities such as rearing young and detecting predators. Meerkats have been extensively studied to test hypotheses of the evolutionary development of helping behavior by nonbreeders (O'Riain et al. 2000; Clutton-Brock et al. 2001), for example, in a social species. Also, Manser (2001) found that meerkats give acoustically different alarm calls depending on the type of predator encountered and the level of urgency of the threat. As in many other carnivore families, several diseases occur in herpestids, including rabies. The small Indian mongoose (*Herpestes javanicus*) was introduced to the West Indies and Hawaii, including several National Parks (Burde and Feldhamer 2003), where it is considered a pest. It preys on poultry and native fauna, especially nesting birds, and carries rabies. The Liberian mongoose (*Liberiictis kuhni*) is endangered, as is the marsh mongoose (*Herpestes palustris*).

Viverridae

The civets and genets occur only in the Old World. The 15 genera and 35 very diverse species occupy tropical and subtropical habitats in Europe, Africa, the Middle East, and Asia. In terms of taxonomy, Wozencraft (2005:548) felt "viverrids are one of the most problematic families of carnivores;" he recognized four subfamilies. Dental adaptations are diverse and reflect the variety of morphological adaptations in this family. In physical appearance and life history, viverrids resemble members of other carnivore families and fill many equivalent niches. For example, some viverrid genera are carnivorous and parallel mustelids or felids. Others are omnivorous like procyonids, some are frugivorous, and others are scavengers like hyenas. Species may be diurnal or nocturnal, occur singly or in groups, and may be terrestrial, semiaquatic, or arboreal. Civets retain several primitive features, including limbs that are generally short and unspecialized. The pelage usually is spotted or striped (see table 16.1).

Most civets have perianal glands, unique among carnivores. These are "composed of a compact mass of glandular tissue, typically lying between the anus and vulva or penis and opening into a naked or sparsely haired area, which may be infolded to form a storage pouch" (Ewer 1973:92). These glands produce a fluid called "civet," which is used for scent marking and functions in intraspecific communication among individuals. Civet has been economically important in perfume making for thousands of years. Several viverrids are endangered, including the

Malabar large spotted civet (*Viverra civettina*), the semi-aquatic otter-civet (*Cynogale bennettii*; Veron et al. 2006), and the crested genet (*Genetta cristata*).

Eupleridae

The seven genera and eight species of Malagasy (Madagascar) mongooses represent a monophyletic group resulting from a single radiation (Yoder et al. 2003). Species within this family were formerly considered within two other families. The ring-tailed mongoose (*Galedia elegans*), the brown-tailed mongoose (*Salanoia concolor*), the narrow-striped mongoose (*Mungotictis decemlineata*), and the Genus *Galidictis* were included within the Herpestidae. The Malagasy civet (*Fossa fossana*), fossa (*Crytoprocta ferox*), and falanouc (*Eupleres goudotii*) were formerly placed in the Viverridae. There are several endangered euplerids, including the giant-striped mongoose (*Galidictis grandidieri*), the narrow-striped mongoose, the fossa, and the falanouc. The remaining species are considered threatened. Found only in dry deciduous forests of western Madagascar, the narrow-striped mongoose is endangered because of habitat loss, human disturbance, and predation from domestic dogs (Woolaver et al. 2006).

Nandiniidae

As noted, this monotypic family includes only the African palm civet (*Nandinia binotata*), formerly included within the Viverridae but placed in its own family based on morphological and molecular criteria (Veron and Heard 2000; Yoder et al. 2003). They have dense, dark pelage with dark bands on the top of the tail. Palm civets occur throughout central Africa in forested habitats where they are primarily arboreal, herbivorous, and nocturnal. They also take insects and small vertebrates. Apparently, palm civets are easily tamed and keep houses free of rodent pests.

CANIFORMIA

Canidae

Canids include the wolves, coyotes, foxes, dingo, dholes, jackals, and the dog. The 13 genera and 35 species occur naturally throughout North and South America, Africa, Asia, and Europe. The dingo (*Canis lupus dingo*) was introduced to Australia, New Guinea, and parts of Asia 3,500–4,000 years ago (Corbett 1995). Habitats of canids range from hot, dry deserts to tropical rainforests to arctic ice. Canids generally take animal prey throughout the year; however, plant material may be taken seasonally by some species. Jackals often eat carrion, and the bat-eared fox (*Otocyon megalotis*) consumes a large amount of insects. Many canids are solitary. Four species—the wolf, African hunting dog, Asian dhole (*Cuon alpinus*), and bush dog (*Speothos venaticus*)—form packs and engage in cooperative

hunting (Moehlman 1986), although solitary wolves are also successful hunters (Thurber and Peterson 1993). Using DNA sequence data from 23 species, Bardeleben et al. (2005) examined phylogenetic relationships among canids.

Generally, canids have long limbs relative to head and body length and are adapted to pursue prey in open habitats. They range in body mass from 1 kg in the fennec (*Fennecus zerda*) to about 80 kg in the gray wolf. They are generally digitigrade and have nonretractile claws. The pollex and hallux are reduced. Unlike felid skulls, canid skulls characteristically have an elongated rostrum (see figure 16.10), with well-developed canines and carnassial teeth. Although there are a few exceptions, the typical dental formula of 3/3, 1/1, 4/4, 2/3 = 42 is close to the primitive eutherian number of 44. The earliest fossil remains of identifiable canids, Genus *Hesperocyon*, date from the early Oligocene of North America.

Several species of canids (e.g., foxes) are hunted for sport or trapped for their fur. In the United States, the coyote has been the target of state and federal predator control programs because of livestock depredations. The wolf is listed as endangered in the lower 48 United States and in India, Nepal, Pakistan, and Bhutan. The red wolf (*Canis rufus*), Simien jackal (*C. simensis*), and African hunting dog also are endangered species. The red wolf recently has been reintroduced in the United States (Phillips 1990), although authorities disagree as to whether this taxon is a distinct species, subspecies, or hybrid (see references in Wozencraft 2005). The Falkland Island wolf (*Dusicyon australis*) was driven to extinction in the 1870s.

Mustelidae

This large, diverse family of 22 genera and 59 species includes weasels, badgers, otters, and the wolverine (*Gulo gulo*). Traditionally, six subfamilies have been recognized, but Wozencraft (2005) recognized only two: the Lutrinae and Mustelinae. Mustelids are most commonly found throughout the Northern Hemisphere and are absent from Australia, Madagascar, the Celebes, and some other oceanic islands. Mustelids are highly specialized predators. They inhabit both terrestrial and arboreal habitats as well as fresh- and saltwater (figure 16.13).

Mustelids have long bodies with relatively short legs. They are digitigrade, pentadactyl, and have nonretractile claws. They range in size from the 30- to 70-g least weasel (*Mustela nivalis*), the smallest carnivore in the world, to the 55-kg wolverine (*Gulo gulo*). Males generally are about 25% larger than females. The carnassials are well developed, and no extant mustelid has more than one molar after the carnassial teeth. Nonetheless, the diverse dental adaptations within the mustelids reflect the varied life histories and diets in this family. The mandibular fossa is strongly C-shaped, which restricts the lateral movement of the mandible and allows little "give" for struggling prey (see figure 16.5).

Mustelids are noteworthy for their enlarged anal scent glands. The thick, powerful-smelling secretion (musk) is

Figure 16.13 Giant otter (*Pteronura brasiliensis*). Note the webbed forefeet of this mustelid that occurs in aquatic habitats throughout much of South America.

used for communication and defense. Some species, such as the marbled polecat (*Vormela peregusna*) and zorilla (*Ictonyx striatus*), also have a striking contrast in their black and white pelage (figure 16.14) as in skunks. This pattern probably serves as "warning coloration" to potential predators that their anal scent glands make them hazardous to capture.

Mustelids are monestrus, and many genera display both induced ovulation and delayed implantation. Although the actual gestation period is usually 1 or 2 months, because of delayed implantation, in some species the total period of pregnancy may be about 1 year. Mustelids constitute >50% of all mammalian species known to exhibit delayed implantation, which is prevalent in highly seasonal climates (Thom et al. 2004). A single litter per year is typical.

The black-footed ferret (*Mustela nigripes*) is critically endangered. Once believed to be extinct, it was reintroduced

(Svendsen 2003). It is currently listed as extinct in the wild. Several species of otters and weasels around the world also are endangered.

Mephitidae

Skunks formerly were considered to be mustelids. Based on molecular analyses, however, Dragoo and Honeycutt (1997) and Flynn et al. (2000) proposed that skunks, along with the stink badgers (Genus *Mydaus*), be considered as a separate Family Mephitidae. There are 4 genera and 12 species of mephitids. Two species of stink badgers occur in Indonesia and the Philippines, while 10 species of skunks in three genera—*Conepatus* (figure 16.15), *Mephitis*, and *Spilogale*—occur throughout a variety of habitats in the western hemisphere. Skunks are noteworthy for their noxious scent. As noted by Rosatte and Lariviere (2003:692), "The musk is an oily, yellow sulphur-alcohol compound known as butylmercaptan and it contains sulphuric acid." Skunks are highly omnivorous, taking insects, small vertebrates, and plant material. They are nocturnal and often solitary. From 20 to 30% of reported cases of rabies in North America each year are caused by skunks. Populations of the eastern spotted skunk (*Spilogale putorius*) have declined dramatically since the 1950s (Gompper and Hackett 2005).

Procyonidae

All 6 genera and 14 species in the raccoon family are restricted to the New World, where they typically inhabit forested temperate and tropical areas, usually near water. Body mass ranges from about 1 kg in the ringtail (*Bassariscus astutus*) and olingo (*Bassaricyon gabbii*) to 12 kg or more in raccoons (figure 16.16A); males usually weigh about 20% more than females. Procyonids typically have long, bushy tails (prehensile in the kinkajou [*Potos flavus*]), with alternating light and dark rings and obvious facial

Figure 16.14 Convergence in pelage characteristics. The zorilla, an African mustelid, has the same warning coloration of the more familiar striped skunk of the United States. *Adapted from J. Kingdon, 1977, East African Mammals, vol. IIIA, Univ. Chicago Press.*

Figure 16.15 Striped hog-nosed skunk (*Conepatus semistriatus*). This mephitid occurs throughout parts of Mexico, Central, and South America and shows the contrasting black and white pelage typical of skunks.

A

B

Figure 16.16 **Representative procyonids.** (A) The crab-eating raccoon (*Procyon cancrivorus*) closely resembles the North American raccoon. It occurs from Costa Rica to Uruguay. (B) The kinkajou is a highly arboreal member of the raccoon family with a prehensile tail.

markings. They are plantigrade; some have semiretractile claws, and all are adept at climbing trees. Procyonids have 40 teeth, except the kinkajou, which has 38. Dentition is generalized and adapted for an omnivorous diet, with fruit predominating in the kinkajou and olingos. The carnassials are fairly well developed only in the ringtail and cacomistle (*Bassariscus sumichrasti*). Kays and Gittleman (2001) discussed social organization in the kinkajou. Groups occur most often at denning and feeding sites, whereas most kinkajous (Figure 16.16B) observed were solitary. Conversely, the white-nosed coati (*Nasua narica*) is "the most social Neotropical carnivore" (Valenzuela and Ceballos 2000:811), with groups of up to 30 individuals. The raccoon (*Procyon lotor*), a popular game animal with both hunters and trappers, is one of the most commonly harvested furbearers in North America (Gehrt 2003). Populations are increasing in North America, and *P. lotor* has been introduced to numerous European countries as well as Japan. The raccoon also commonly contracts rabies (see chapter 27). Several species of raccoons and olingos are endangered.

Ursidae

Bears historically occurred throughout North America, the Andes Mountains of South America, Eurasia, and the Atlas Mountains of North Africa. Their habitats vary from tropical forests to polar ice floes. There are 5 genera and 8 species of ursids, including the giant panda (*Ailuropoda melanoleuca*)—once placed in its own family (Ailuropodidae).

Body mass in bears reaches 800 kg in grizzly and polar bears, the largest terrestrial carnivores. Sexual dimorphism is evident. Males are about 20% heavier than females in monogamous species such as the sun bear (*Helarctos malayanus*) and sloth bear (*Melursus ursinus*) and up to twice as large in polygamous species. All bears are plantigrade and pentadactyl and have nonretractile claws. The occurrence of epipharyngeal pouches, located near the pharynx and trachea, has been documented in most ursids, including the giant panda (Weissengruber et al. 2001). These pouches probably play a role in modification and amplification of vocalizations. Bears usually have 42 teeth. Canines are large, but the last upper molar is reduced, and carnassials are not well developed. Molars are broad, flat, and relatively unspecialized, reflecting an omnivorous diet. Only the polar bear is strictly carnivorous, feeding on fish and seals. The sloth bear, like the aardwolf, is highly myrmecophagous.

In northern areas, black bears (*Ursus americanus*), grizzlies, and Asiatic black bears (*U. thibetanus*) den in hollow trees, caves, or burrows. Polar bears use dens of ice and snow, and may abandon dens due to human disturbance (Durner et al. 2006). Bears may sleep through the winter, especially pregnant females, and live off their stored body fat. This process is called "winter lethargy" rather than hibernation (see chapter 9) because body temperature, heart rate, and other physiological processes are not reduced to the same extent as in true hibernation (Hellgren et al. 1990; Brown 1993). Bears are monestrus and exhibit delayed implantation. Young, often twins, are usually born between November and February while the female dens. In giant pandas, females move cubs to new dens three to four times before the cubs permanently leave the den at 3–4 months of age (Zhu et al. 2001). Den site selection and potential population size of black bears in an area is negatively impacted by the number of roads (Gaines et al. 2005).

Historically, bears have been hunted for their hides, meat, and fat. Most species have been eradicated through much of their ranges because of predation on domestic livestock, including the spectacled bear (*Tremarctos ornatus*—figure 16.17) in northern South America (Goldstein et al. 2006). Recently, large numbers of black bears have been illegally killed for their gallbladders, which are valuable in traditional Asian medicine. The giant panda is endangered; the Mexican subspecies of grizzly bear (*U. a. nelsoni*) is recently extinct.

Ailuridae

This monotypic family contains only the endangered red panda (*Ailurus fulgens*). The phylogenetic relationship of the red panda with other carnivores has always been

Figure 16.17 Spectacled bear (*Tremarctos ornatus*). This South American ursid is reduced throughout its range and currently is considered threatened.

Figure 16.18 Pinniped body shape. All pinnipeds have a streamlined, torpedo-shaped body for minimal resistance as they swim, as in this immature southern elephant seal (*Mirounga leonina*).

uncertain. In the past, it has been placed with raccoons (family Procyonidae), bears (family Ursidae), or with the giant panda in the family Ailuropodidae. Molecular and morphological data are not definitive, and authorities continue to debate placement of this highly enigmatic species. Also known as the lesser panda, *A. fulgens* occurs in the Himalayan Mountain areas of northern India, Nepal, Bhutan, Myanmar, and south-central China where it inhabits high-elevation forests and bamboo thickets. Red pandas, with a body weight of only 3–6 kg, feed primarily on bamboo and other vegetation, with occasional insects and small vertebrates. Han et al. (2004) and Zhang et al. (2006) described microhabitat selection of red pandas, and differences with the giant panda.

Pinnipedia

The last three families—the pinnipeds—exhibit specialized adaptations for an aquatic existence. Pinnipeds are not as totally adapted for life in the water as are whales, manatees, and the dugong, however. Whales (chapter 17) and sirenians (chapter 19) spend their entire lives in the water, but pinnipeds must "haul out" onto land or ice floes (where they are slow and vulnerable) to breed, give birth, or rest.

Being adapted to both terrestrial and aquatic conditions has caused all pinnipeds to be morphologically similar. Pinnipeds have a coarse pelage of guard hairs that helps protect them when they are out of the water. As in whales, body shape in pinnipeds is adapted to reduce turbulence and resistance (drag) as they swim through the water. Thus, bodies are fusiform, with no constriction in the neck region (figure 16.18). External genitalia are concealed in sheaths within the body contour, as are the teats, and external ears are reduced or lost. A subcutaneous layer of fat (blubber) provides energy, insulation, and buoyancy (see figure 9.8). It also serves to maintain a streamlined body shape, which enhances hydrodynamic properties and further reduces drag.

The limbs are relatively short, stout, and modified to form paddle-like flippers (figure 16.19). The forelimbs provide the propulsive force in otariids and odobenids; the hind limbs serve this function in phocids. These differences

are reflected in skeletal anatomy. Otariids and the walrus have enlarged cervical and thoracic (neck and upper chest) vertebrae that support the large muscle groups associated with the forelimbs. Because the hind limbs provide propulsion in phocids, the lumbar (lower back) vertebrae are relatively large (see figure 16.19).

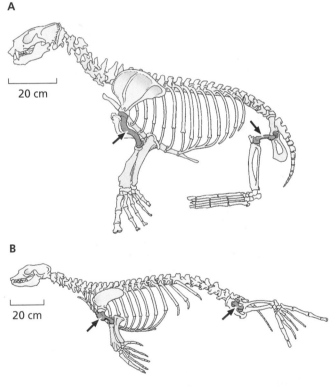

Figure 16.19 Pinniped skeletal characteristics. Lateral view of the skeleton of (A) a New Zealand fur seal (*Arctocephalus forsteri*), an otariid, and (B) a West Indian monk seal (*Monachus tropicalis*), a phocid. Note the very short, broad humerus and femur (arrows) and elongated foot bones in both. In otariids and the walrus, the cervical and thoracic vertebrae and scapula are enlarged to support the primary muscle groups that power the forelimbs for propulsion. The hind limbs propel phocids through the water, and the lumbar vertebrae are enhanced for muscle attachment. *Adapted from J. E. King, 1983, Seals of the World, British Museum of Natural History.*

Most pinnipeds have generalized feeding habits, and the primary function of the teeth is to grasp and hold prey rather than to chew. Teeth in most species approach homodonty, with the premolars and molars being similar and somewhat conical (figure 16.20). There are exceptions, however, in species that are more specialized feeders. For example, the crabeater seal (*Lobodon carcinophagus*) has distinctive cheekteeth that form a sieve to filter krill from the water. Unlike in terrestrial carnivores, the interorbital area in pinniped skulls is long and narrow, and the braincase is longer in proportion to the facial area (see figure 16.20).

The eyes of seals are relatively large and modified to focus underwater by means of a greater corneal curvature than occurs in the eyes of terrestrial mammals. On land, however, seals are quite nearsighted. Pinnipeds see effectively under conditions of reduced light. Like felids and some other terrestrial groups, they have a well-developed tapetum lucidum, a specialized membrane behind the retina. The tapetum lucidum increases light-gathering efficiency by reflecting back to the retina light that has passed through the retina but not been absorbed. All pinnipeds hear well underwater and are able to determine the direction from which sounds come. This ability is critical for locating prey, especially under reduced light conditions or in total darkness. King (1983) discussed several unique pinniped adaptations for directional hearing underwater. Work by Dehnhardt et al. (2001) on harbor seals (*Phoca vitulina*) suggested that pinnipeds are capable of locating and following for considerable distances the hydrodynamic trails left by fish and other potential swimming prey.

Many of the same physiological processes for diving seen in whales also occur in pinnipeds, including **bradycardia** (reduced heart rate) and shunting blood from peripheral areas to the brain and heart. Like whales, seals also have more oxygen-binding hemoglobin and myoglobin than do terrestrial mammals. The size of the spleen, and its importance in red blood cell storage, is related to diving capabilities in seals (Thornton et al. 2001; Cabanac 2002). Pinnipeds do not dive as deep or remain submerged as long as whales, however (DeLong et al. 1984; Lavigne and Kovacs 1988). Among pinnipeds, phocids dive deeper and remain submerged longer than do otariids. Aspects of dive duration and depth in pinnipeds no doubt relate to factors that include regional water conditions, seasonal variations in depth of prey species, and reproductive condition and molting in the seals (Frost et al. 2001). Methods used to investigate the diving ecology of seals, including time-depth recording equipment and analyses of data, are discussed by Gentry and Kooyman (1986), DeLong and Stewart (1991), and DeLong et al. (1992).

Pinnipeds had been harvested for thousands of years on a subsistence basis for fur, food, and oil, with little effect on populations. Large-scale commercial hunting and resultant overharvesting of many species began in the early 1800s, and many species were "commercially extinct" by the end of the century. Today, several species are still harvested, but most populations are secure. The most serious threats to pinnipeds include incidental drowning in fishing nets, habitat degradation, and environmental contamination of marine habitats (Reijnders et al. 1993).

Odobenidae

This monotypic family, once considered a subfamily of otariids, includes only the walrus (*Odobenus rosmarus*). Walruses have a circumpolar distribution in shallow arctic waters, where they remain near ice floes and rocky shorelines. Body mass of adult bulls is about 1,000 kg but can reach 1,600 kg. Typical of pinnipeds and most other carnivores, females are significantly smaller than males. The skin is thick and wrinkled, and the layer of blubber, which is usually 6–7 cm thick, can reach 15 cm. Certain morphological features of walruses are shared with otariids, whereas others are similar to phocids (table 16.2). In addition, walruses, like otariids, have naked ventral surfaces on all flippers, and the nails on the first and fifth digits of the hind flippers are rudimentary.

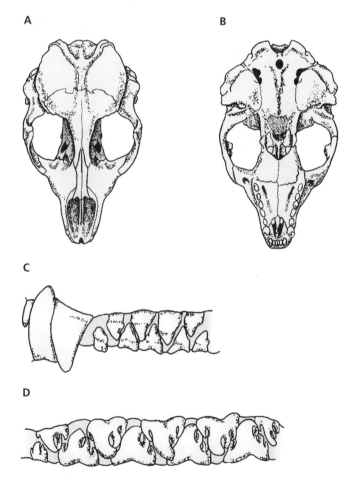

Figure 16.20 Pinniped skulls and cheekteeth. Dorsal view (A) and ventral view (B) of a phocid, the harbor seal (*Phoca vitulina*). Note the narrow interorbital regions, shortening of the nasal bones on the rostrum, and lack of a postorbital process in the phocid. The cheekteeth of otariids usually are single-cusped and peglike, as in (C) Steller sea lion (*Eumetopias jubatus*), whereas those of phocids are multicusped and particularly well developed in (D) the crabeater seal. *Adapted from M. Riedman, 1990,* The Pinnipeds: Seals, Sea Lions, and Walruses, *Univ. California Press.*

Table 16.2 Morphological characteristics differentiating the three families of pinniped carnivores*

	Phocidae (earless seals)	Odobenidae (walrus)	Otariidae (fur seals/sea lions)
External pinnae	No	No	Yes
Testes	Abdominal	Abdominal	Scrotal
Tip of tongue notched	Yes	No	Yes
Hind limbs rotate forward	No	Yes	Yes
Guard hairs with medulla	No	No	Yes
Underfur	Essentially absent	Essentially absent	Present in sea lions
Alisphenoid canal	Absent	Present	Present
Auditory bullae	Inflated	Small and flattened	Small and flattened
Transverse groove on upper incisors	No	No	Yes
Lower incisors present	Yes	No	Yes
Total number of teeth	26–36	18–24	34–38
Fused symphysis of lower jaw	No	Yes	No
Postorbital process	Absent	Absent	Present
Chromosome number	32–34	32	36

* For additional differences of the skull, middle ear, and inner ear, see Repenning (1972).

The significant feature in both male and female walruses is their tusks, which are enlarged upper canines (figure 16.21). These grow throughout life and can reach as long as 100 cm in males and 60 cm in females. The crown portion of the tusk is composed internally of dentine and externally of cementum. Other than a small cap at the end of the tooth that wears away during the first couple of years, there is no enamel. Tusks are used for defense, raking the bottom for mollusks, breaking through ice, hanging from ice floes while remaining in the water, and, among males, as weapons for establishing dominance hierarchies.

Walruses remain in shallow water, usually at depths of between 80 and 100 m (King 1983). Time-depth recorders have been used to describe diving patterns of pinnipeds, including walruses, in different parts of their range. In Bristol Bay, Alaska, Jay et al. (2001) followed four males for a month during the summer. Walruses were in the water almost 77% of the time, of which about 60% was spent diving. Dives generally were of short duration and shallow—

well within aerobic limits. Likewise, the mean depth of foraging dives for nine males on Svalbard, in the eastern Greenland Sea, was about 22 m (maximum = 67 m), with a mean duration of only 6 minutes (maximum = 24 minutes). The mean time spent in the water was 56 h, followed by 20 h on land (Gjertz et al. 2001). Walruses feed primarily on clams, other mollusks, and a variety of invertebrates taken from the muddy bottom. Ray et al. (2006) suggested that the feeding activities of walruses significantly impacted sediment, nutrient flux, and biological structure of thousands of square kilometers of the sea floor. Walruses are gregarious, often huddle together, and form groups of 100 to 1,000 individuals. Walruses are polygamous, mating during February and March on near-shore ice floes. Following a 3-month delayed implantation period and 12-month gestation, females haul out to give birth on land, usually to a single calf, in May or June. Thus, given the lactation period, females normally mate every 2–3 years.

Walruses are hunted for meat, oil, and hides, and the tusks (ivory) have been highly prized for artwork carvings (called scrimshaw) for hundreds of years. Although from 10,000 to 15,000 individuals are harvested annually from the Pacific population (Reijnders et al. 1993), walrus populations appear to be stable. Walruses date from the mid-Miocene, including fossil crania of *Pseudotaria muramotoi*, from northern Japan (Kohno 2006).

Otariidae

The eared seals include fur seals and sea lions in subpolar, temperate, or coastal waters of western North America, South America, Asia, southern Australia and New Zealand, and oceanic islands. They only inhabit marine communities, unlike phocids, which also occur in freshwater and estuarine communities. Several general characteristics of the 7 genera and 16 species of otariids have been noted previously and in table 16.2. Two subfamilies traditionally have

Figure 16.21 Walrus tusks. The large tusks of the walrus are canines.

been recognized: sea lions (Subfamily Otariinae), with blunt noses and little underfur, and fur seals (Subfamily Arctocephalinae), with pointed noses and abundant underfur. Based on mtDNA data, Wynen et al. (2001) concluded that the two-subfamily classification could not be supported. All otariids are highly dimorphic, with males being larger than females. For example, maximum body mass of male northern fur seals (*Callorhinus ursinus*) is five times greater than that of females; male southern sea lions (*Otaria byronia*) are twice the size of females.

Otariids feed on fish, cephalopods, and crustaceans. Northern fur seals in the north Pacific fed on 15 species of small squid, primarily *Watasenia scintillans* (Mori et al. 2001). Steller sea lions consume numerous species of fish throughout their range (Waite and Burkanov 2006; Womble and Sigler 2006). Duration of foraging trips in fur seals (*Arctocephalus pusillus*) in southeastern Australia increased from a mean of 3.7 days in summer to 6.8 days in winter (Arnould and Hindell 2001). Generally, otariids are much more gregarious than terrestrial carnivores (figure 16.22), and breeding colonies of up to a million individuals may occur within limited areas. All otariids breed on land in rocky, isolated areas that are inaccessible to potential predators. Phillips and Stirling (2001) described 11 distinct calls, grouped into investigative, threat, submissive, and affiliative categories, in South American fur seals (*Arctocephalus australis*). Otariid females must leave newborn pups to forage at sea. In the subantarctic fur seal (*A. tropicalis*), Charrier and coworkers (2001:873) found that "the mother's departure date is linked to the pup's ability to recognize her voice." This ability is crucial to survival of pups in a colonial species in which females nurse only their young. Kuhn et al. (2006) found that by the time they are weaned, California sea lion pups (*Zalophus californianus*) can store only about 53% of the oxygen that adults can prior to diving. As a result, pups cannot forage on resources in deeper water that are available to adults. Depth and duration of dives in the Australian sea lion (*Neophoca cinerea*) also increase with age (Fowler et al. 2006). Breeding males are polygynous and defend a territory with a group of 3–40 females. Most species are known to exhibit delayed implantation.

The most familiar species of otariid is the California sea lion, commonly displayed in circuses and zoos.

Figure 16.22 South American sea lions (*Otaria flavescens*). These otariids—like many pinnipeds—are highly gregarious.

Historically, all species were harvested for meat, hides, and oil from blubber. Today, relatively few individuals of most species are harvested for subsistence purposes; about 6,500 northern fur seals are taken annually (Reijnders et al. 1993). A remnant population of the Juan Fernandez fur seal (*Arctocephalus philippii*), once believed extinct due to overharvesting, is considered threatened. The Steller sea lion is endangered in the western Pacific. The Japanese sea lion (*Z. c. japonica*) is probably extinct.

Phocidae

The earless seals include 13 genera and about 19 recent species. Traditionally, two subfamilies have been recognized, the Phocinae (northern seals) and the Monachinae (southern seals and monk seals). Using a large mtDNA database, Davis et al. (2004) found "strong support" for groupings within these subfamilies. Phocids occur primarily in polar, subpolar, and temperate waters around the world; the monk seals (Genus *Monachus*) are the only pinnipeds that inhabit tropical areas. Besides in oceans, phocids occur in inland freshwater lakes and estuaries. The flippers of phocids are furred on all surfaces, and the nails are all the same size on the hind flippers.

Among pinnipeds, phocids are the most diverse in size. The smallest is the Baikal seal (*Pusa sibirica*), restricted to freshwater Lake Baikal, with a body mass of only 35 kg. The largest phocid (and the largest carnivore) is the northern elephant seal; male body mass is about 3,700 kg. Unlike otariids, phocids generally lack underfur (see table 16.2), and the cheekteeth are multicusped (see figure 16.20D). Phocids are not gregarious and do not form large breeding colonies. Almost all phocids breed on ice. Because they are clumsy and vulnerable on land, ice floes offer several benefits to breeding individuals. It is easier for seals to move on ice than on rocks, and they have quick access to the relative safety of deep water. Like other pinnipeds, they exhibit delayed implantation.

Fish and cephalopods form the diet of most species (Hammill and Stenson 2000; Holst et al. 2001). The leopard seal (*Hydrurga leptonyx*), the only pinniped that regularly feeds on warm-blooded prey, takes penguins and the young of other seals. As noted earlier, phocids dive deeper for prey and remain submerged longer than otariids and thus are the most aquatically adapted of the carnivores. Dive durations in southern elephant seals (*Mirounga leonina*) were longest in the morning and deepest at midday (Bennett et al. 2001). On land, however, phocids are less agile and mobile. They are unable to raise themselves on their front flippers and simply hunch their bodies, moving forward like inchworms.

Commercial sealing of many species is controversial, including the harvesting of young harp seals (*Phoca groenlandica*). The Mediterranean monk seal (*Monachus monachus*) and Hawaiian monk seal (*M. schauinslandi*) are endangered, and the Caribbean monk seal (*M. tropicalis*), last seen in the early 1950s, is most likely extinct.

SUMMARY

Carnivores are terrestrial or aquatic predators that usually consume other animals as a major part of their diet. Most morphological and behavioral characteristics of carnivores involve adaptations to enhance locating, capturing, killing, and consuming their prey without being injured in the process. Nonetheless, some specialized feeding habits have evolved. These include insectivory, as in the aardwolf and the sloth bear, scavenging on carrion by hyenas, omnivory in a number of species, and almost complete herbivory in the greater and lesser pandas.

All carnivores have digits with well-developed claws and dentition with enlarged canine teeth. The defining ordinal characteristic, however, is carnassial teeth-specialization of the fourth upper premolar (P^4) and first lower molar (m_1) for cutting and shearing. Carnassial dentition is especially well developed in highly predaceous families, such as felids, canids, and hyaenids, and less developed in more omnivorous groups, such as ursids and procyonids. Facial musculature is well developed. The articulation of the jaw with the cranium typically is hinged so as to reduce lateral motion as captured prey struggle to escape.

Size diversity is pronounced both within the order and within various families. The largest carnivores are the elephant seals. The largest terrestrial carnivores are the polar bear and grizzly bear, which are 11,000 times heavier than the smallest carnivore, the least weasel. With few exceptions, secondary sexual dimorphism is evident; males are often many times larger and heavier than females. Size of individual carnivores and method of hunting (solitary, paired, or in groups) generally relate to the size of the prey species the carnivore is able to capture. Although most species are solitary hunters, wolves, spotted hyenas, lions, and some others generally hunt in packs. Thus, they are able to prey on species that are several times larger than themselves. In addition to improved foraging efficiency, group membership may provide several additional benefits to individuals: communal infant care, reduced predation, and defense of feeding areas from rival packs.

The earliest carnivorous mammal, *Cimolestes*, from the late Cretaceous period, is considered the basal group for modern carnivores. An early group of carnivores, the creodonts, extended from the Cretaceous period to the Miocene. Creodonts are considered a sister group to the modern order. Most families of modern carnivores had developed by the Oligocene, concurrent with the adaptive radiation of herbivorous groups on which they preyed. Two suborders of modern carnivores are generally recognized. Six families occur within the Suborder Feliformia: felids, herpestids, hyaenids, viverrids, and the recently recognized Eupleridae and Nandiniidae. The Suborder Caniformia includes the canids, ursids, mustelids, procyonids, ailurids, mephitids, and three families of aquatic carnivores (the pinnipeds)—otariids, phocids, and an odobenid. Pinnipeds have distinctive morphological adaptations consistent with their aquatic life histories and in the past have been placed in their own order.

Carnivores have had a long association with humans. Dogs and cats are the most popular domestic pets throughout the world. Numerous other species of carnivores, mink and sable for example, are valuable in the fur industry. Conversely, most large predators, such as wolves, cougars, and bears, have been eliminated throughout much of their range because of conflict with human economic interests, often livestock depredation. Other species have been overexploited through trophy hunting or commercial harvest and are endangered as a result.

SUGGESTED READINGS

Bonner, N. 1994. Seals and sea lions of the world. Facts on File, New York.

Clark, T. W., M. B. Rutherford, and D. Casey, (eds.). 2005. Coexisting with large carnivores: Lessons from Greater Yellowstone. Island Press, Washington, DC.

Gittleman, J. L. (ed.). 1989. Carnivore behavior, ecology, and evolution, vol. I. Cornell Univ. Press, Ithaca, NY.

Gittleman, J. L. (ed.). 1996. Carnivore behavior, ecology, and evolution, vol. II. Cornell Univ. Press, Ithaca, NY.

DISCUSSION QUESTIONS

1. What are the adaptive benefits of delayed implantation to such carnivores as ursids and pinnipeds?

2. What adaptive benefits accrue to bears that enter winter lethargy in northern areas?

3. Sexual dimorphism often is evident among carnivores, with males being larger than females. Consider such factors as feeding and reproduction to hypothesize why size dimorphism might evolve.

4. The red panda has been included in the past in the families Ursidae, Procyonidae, and Ailuropodidae. Speculate as to why the red panda has had (and continues to have) such a confusing taxonomic history.

5. How might the striking black-and-white warning coloration of skunks and other species have evolved? How do potential predators learn to avoid these species?

CHAPTER 17

Cetacea

Whales are among the most fascinating and, in some ways, least understood of all mammals. The Order Cetacea (from the Greek word for "whale") is characterized by extremes. It includes the largest animals that have ever lived. Adult blue whales (*Balaenoptera musculus*) are heavier than the biggest dinosaurs (although probably not longer; see Gillette 1991). Whales have the loudest voices, they endure tremendous water pressure as they dive deeper than any other mammal, and certain species make some of the longest migrations. But much of the life history of whales remains unknown. Studying whales has often been difficult because of their relative inaccessibility and the wide-ranging movements of individuals. Only recently have advances in technology and molecular genetics allowed researchers to investigate the physiology, behavior, sound communication, social interactions, reproduction, phylogeny, and other aspects of whale biology more thoroughly. Conversely, because the whaling industry has provided carcasses for hundreds of years, probably more is known about the anatomy of whales than of many other large mammals.

In addition to the manatees and dugong (Order Sirenia; see chapter 19), whales are the only group of mammals that are entirely marine—they never leave the water. This chapter examines many of the interrelated morphological, physiological, locomotor, and behavioral adaptations of whales that have evolved in response to the demands of life in an aquatic environment.

Morphology

The 11 families of whales are divided into two well-defined extant suborders: the Mysticeti, or baleen whales, and the Odontoceti, or toothed whales. The primary feature of mysticetes is their baleen plates (see the section Mysticetes: Baleen and Filter-Feeding in this chapter), which take the place of teeth. Baleen is used to strain small marine organisms from the water for food. The teeth in odontocetes generally are homodont, simple, and peglike. Unlike most mammals, which are diphyodont (having two sets of teeth during a lifetime: deciduous and permanent), odontocetes are monophyodont. That is, they have a single set of teeth. Odontocetes also echolocate to orient within their environment and find prey; mysticetes do not. Several other differences exist between these two extant suborders (table 17.1). A third suborder of whales, the extinct Archaeoceti, is also recognized. These ancient whales had many features that were intermediate during the transition from primitive terrestrial mammals to fully aquatic whales.

Unlike fish, whose entire evolutionary history was in water, the adaptations of whales for life in the water are all secondary. That is, the morphological characteristics seen in whales today are all derived from those of ancestral land mammals that made a gradual transition from land to sea about 50 million years ago (see the section Fossil History in this chapter). As noted by Gingerich (2005:235), the Order Cetacea "is interesting from an evolutionary point of view, because it represents entry into and eventual mastery of a new aquatic adaptive zone markedly different from that of its terrestrial ancestors." Water, especially saltwater, provides a particularly buoyant environment. Thus, whales can be larger than any terrestrial mammal and still remain mobile because a buoyant environment they never leave supports them. Nonetheless, body size varies considerably among cetaceans; more than half of all whale species are the smaller dolphins and porpoises. Keep in mind, however, that even "small" whales are quite

large relative to the average size of most terrestrial mammals. Whales range in size from Heaviside's dolphin (*Cephalorhynchus heavisidii*), which weighs about 40 kg with a total length of 1.7 m, to the blue whale with a mass of 200,000 kg and a total length of 30 m (figure 17.1). No terrestrial mammals are the size of the large baleen whales because their skeletal support system would have to be so massive that they would be rendered immobile; there probably are metabolic and reproductive constraints

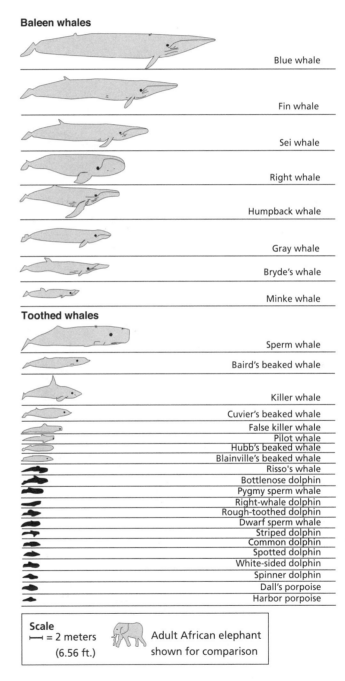

Figure 17.1 Relative sizes of baleen and toothed whales. The scale represents approximately 2 m. Adult African elephant is shown for comparison. *Data from R. T. Orr and R. C. Helm, 1989*, Marine Mammals of California, *rev. ed., Univ. California Press.*

Table 17.1 Primary differences between the two suborders of cetaceans: the mysticetes (baleen whales) and odontocetes (toothed whales)

Mysticeti	Odontoceti
Baleen present instead of teeth; teeth present in fetus lost before birth	Teeth present, usually homodont; exceed the primitive eutherian number in some species; monophyodont
Paired external nares ("blowholes"); located anterior to eye	A single external nare; located posterior to eye in all species except sperm whales
Facial profile convex, with no fatty "melon" present	Facial profile concave, with depression occupied by a "melon," or fatty organ
Skull symmetrical	Skull generally asymmetrical
Do not echolocate; auditory bullae (tympanoperiotic bones) attached to skull	Echolocate; tympanoperiotic bones not attached to skull
Nasal passages simple	Nasal passages with a complex system of diverticula
Mandibular condyle directed upward	Mandibular condyle directed posteriorly
Sternum consists of a single bone	Sternum consists of three or more bones

against such size as well. Several interrelated features of whales are associated with life in the water, including body and skull shape, thermoregulation, and the physiology of diving.

STREAMLINED BODY SHAPE

The general body shape of whales is the same whether they are large mysticetes or smaller odontocetes (figure 17.2A,B). Whales are fusiform (streamlined, or torpedo-shaped), which allows them to move forward through the water with less drag. Drag is affected by several factors, including the density and viscosity (resistance to flow) of water and the size, shape (cross-sectional area), and speed of the object moving through it. How these factors interact in fluid dynamics is suggested by the Reynolds number (Re), a dimensionless value. Of primary importance in determining Re is the product of the size and speed of an object moving through a fluid. Thus, a 0.3-mm-long zooplankton moving at 1 mm/s results in Re = 0.3. By comparison, a large whale moving at 10 m/s results in Re = 300,000,000. An excellent account of the mechanics of organisms in fluids, both water and air, is provided by Vogel (1994).

As a whale moves forward, water flows smoothly over it with minimal turbulence (referred to as **laminar flow**)

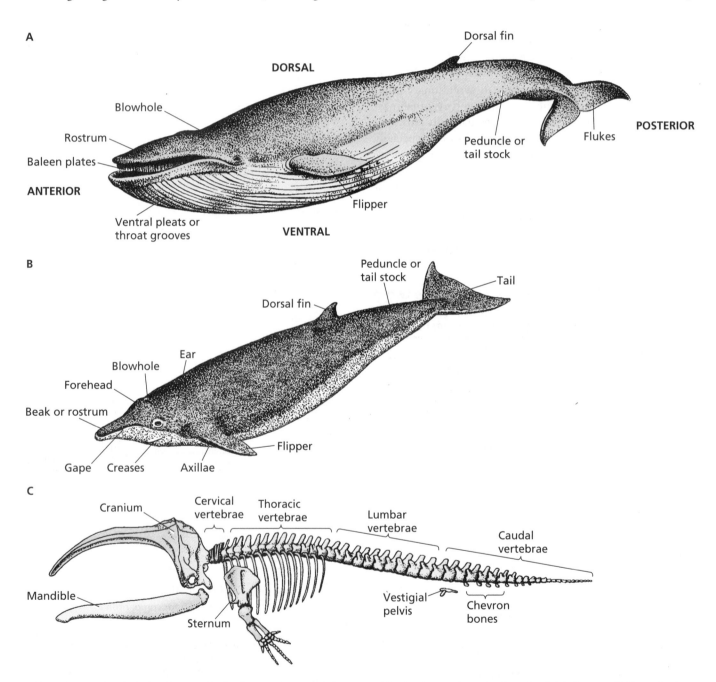

Figure 17.2 **Whale body morphology.** Lateral view of the general body plan and terminology for (A) baleen and (B) toothed whales. Whales have a streamlined body shape with few external appendages. (C) Skeleton of a mysticete whale. Note the vestigial hind limbs and chevron bones on the caudal vertebrae. *Adapted from T. A. Jefferson et al., 1993,* Marine Mammals of the World, *United Nations Environment Programme.*

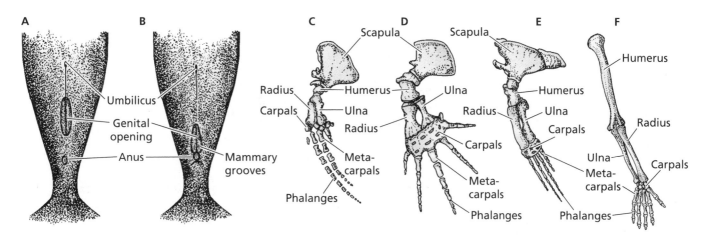

Figure 17.3 Genital grooves and forelimbs. Ventral view of the genital grooves in (A) a male whale and (B) a female whale. Forelimbs of whales have been modified to form paddle-like flippers: (C) a pilot whale (Genus *Globicephala*), (D) a blue whale, and (E) a right whale. Compared to a human arm (F), whales have a relatively short humerus, radius, and ulna. The phalanges have been lengthened and increased in number. Drawings are not to the same scale. *Adapted from P. G. H. Evans, 1987,* Natural History of Whales and Dolphins, *Facts on File.*

because of several structural features. Nothing protrudes from a whale's body that would increase drag and impede the smooth, continuous flow of water. Whales have no external ears or hind limbs, for example. Hind limbs have been reduced to small, vestigial innominate bones. These are not attached to the vertebral column (see figure 17.2C) but are imbedded in ventral muscle anterior to the anus (Slijper 1979). Likewise, the male has no scrotum, and testes are permanently abdominal. The penis is retractile and lies in a fold of skin (figure 17.3A). Streamlining is further achieved by almost complete absence of body hair. Insulation is provided by the subcutaneous blubber (the layer of fat under the skin). Thermoregulation is maintained by several physiological features noted in the section in this chapter on that topic.

The forelimbs of whales are modified as flexible flippers. The humerus, radius, and ulna are shortened, whereas the digits are greatly elongated by having additional **phalanges**, or finger bones (see figure 17.3C–E). The flippers do not provide power for forward thrust, however, but only act as stabilizers. The elongated tail that ends in a dorsoventrally flattened, horizontal fluke (see figure 17.2) generates power for locomotion. The **fluke** is supported only by connective tissue rather than by any skeletal element. Because water is denser than air and has much greater viscosity, the large surface area of the tail fluke moving up and down, which is termed **caudal undulation**, "pushes" against the water and propels a whale forward (Fish and Hui 1991; Fish 1992; also see Buchholtz 2001). The dorsal spinous vertebral processes provide sites for muscle attachment to raise the tail, and the **chevron bones** on the ventral portion of the anterior caudal vertebrae (see figure 17.2) provide attachment for the muscles that depress the tail. Another dimensionless value, the Strouhal number (St), is used to describe the swimming efficiency of cetaceans—as well as fish. As discussed by Rohr and Fish (2004), parameters that define St include stroke frequency, amplitude of the fluke, and mean forward velocity.

The simple postcranial skeleton of cetaceans is composed primarily of the backbone, which provides attachment sites for the large tail muscles. Limbs are not needed for support because the aquatic environment provides the needed buoyancy. Cervical vertebrae are reduced in size and are often fused; most whales have little mobility of the head.

SKULL STRUCTURE

Both suborders of whales have highly modified, **telescoped skulls** in which posterior bones of the cranium are compressed and overlap each other. The anterior, or rostral portion, of the skull is greatly elongated through extension of the nasal, premaxillary, maxillary, and frontal bones. These bones extend posteriorly such that they overlap the parietal bones (figure 17.4). Telescoping has accommodated the posterior displacement of the external **nares** (nostrils, or "blowholes") to the top of the skull, so that only this portion needs to be above water for a whale to breath. In addition, odontocetes have an asymmetrical skull structure, especially around the internal nares (figure 17.5). Asymmetry probably is related to echolocation and movement of the nares.

THERMOREGULATION

Whales face certain challenges living in an aquatic environment because of the physical properties of water. One of these is water's very high thermal conductivity: water absorbs heat from a warm body about 27 times faster than air does. Thus, whales must maintain their body temperature in frigid polar waters or the deep, cold waters of temperate and tropical areas. Loss of body heat represents loss of energy, which must be made up by increased food intake. Thus, it benefits individuals in cold water to minimize the amount of heat they radiate to the environment.

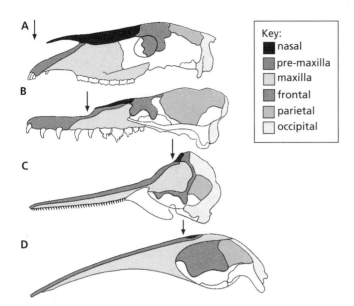

Figure 17.4 Telescoping of the skull in modern whales. Skulls of (A) a terrestrial mammal (the horse, *Equus caballus*); (B) an archaeocete whale (Genus *Basilosaurus*) with heterodont dentition; (C) a modern odontocete (common dolphin, *Delphinus delphis*) with homodont dentition; and (D) a modern mysticete (fin whale, *Balaenoptera physalus*). In modern whales the nasal, premaxilla, maxilla, and frontal bones have extended posteriorly to overlap the parietal bones. Arrows indicate the resultant posterior movement of the nares or nostrils. Skulls are not drawn to scale. *Adapted from E. J. Slijper, 1979, Whales, Cornell Univ. Press.*

Key:
- nasal
- pre-maxilla
- maxilla
- frontal
- parietal
- occipital

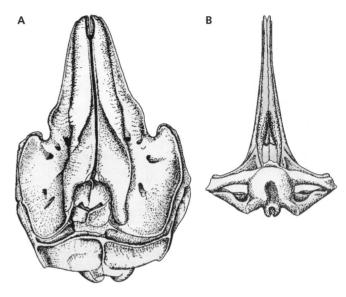

Figure 17.5 Odontocete verses mysticete skulls. Dorsal views of the skulls of (A) an odontocete, Risso's dolphin, and (B) a mysticete, the northern right whale. Note the asymmetrical skull typical of odontocetes. Skulls are not to the same scale. *Adapted from DeBlase and Martin, 1981,* A Manual of Mammalogy, *2nd ed., Wm. C. Brown.*

Unlike terrestrial mammals, whales have no insulating fur, nor can they burrow or construct nests to help maintain their body temperature against cold ambient conditions. Whales maintain thermal equilibrium in several ways. By virtue of their large size, they have a favorable surface-to-volume ratio (the "scale effect"). Even though they have a large surface area from which they lose heat, this area is small relative to their heat-producing body mass. Also, the subcutaneous layer of blubber acts to insulate whales. The thickness of blubber varies both seasonally and among species, from 5 cm in small dolphins to as much as 50 cm in the bowhead whale (*Balaena mysticetus*). When large whales are feeding at high latitudes, blubber may represent up to 70% of their body mass (Bonner 1989).

Although blubber forms a passive means of heat retention for thermoregulation, heat also may be actively retained in whales through their well-developed countercurrent heat exchange system. This system is associated with areas of the body with little blubber or muscle, such as the flippers, flukes, and head. In these areas, major arteries are surrounded by a closely associated network of veins. This *rete mirabile* ("wonderful net") acts as a heat exchanger (Bonner 1989). As the warm blood in the arteries moves from the body core outward toward the body surface, heat transfers to the adjacent colder blood in the returning venous system. Also, whales have somewhat of an advantage in thermoregulation compared with terrestrial species because large bodies of water generally maintain fairly constant temperatures.

PHYSIOLOGY OF DIVING

Whales must also face the challenges associated with diving. Because pressure increases by 1 atmosphere (atm; 14.7 lb/in.^{2}, or approximately 1 kg/cm^2) for every 10 m increase in depth, deep-diving species face tremendous physical as well as physiological demands. At a depth of 30 m, the pressure is 4 atm, or about 60 lb/in.^{2}; at 1,000 m, the water pressure is over 1,500 lb/in.^{2}. Whales have adapted in several ways to meet these challenges. They are able to survive the extreme pressures encountered during deep dives because their bones are relatively noncompressible. Their upper airways are fairly rigid as well and are supported by bundles of cartilage, although their lungs collapse at depth.

Like all mammals, whales must breathe air or they will die. Large species, however, are able to remain submerged without breathing for prolonged periods, up to 2 hours in sperm whales (see the section Physeteridae in this chapter) and bottlenose whales (*Hyperoodon ampullatus*; Benjaminsen and Christensen 1979). Whales can do so because they use oxygen more efficiently than terrestrial mammals. Land mammals use 4% of the oxygen they inhale with each breath, but whales use 12%. Oxygen is carried to the cells of the body more efficiently as well because the average **hematocrit** (number of erythrocytes per volume of blood) of whales is twice that of terrestrial mammals. Additionally, whales have two to nine times the amount of **myoglobin** (oxygen-binding protein in muscles) found in terrestrial mammals. Rapid gas exchange is facilitated by extra capillaries in the alveoli of the lungs. Whales also exhibit a characteristic "diving response," which includes bradycardia while submerged. They maintain normal arterial blood flow to the brain and heart but with vasoconstriction and reduced peripheral circulation. Nonetheless, constant blood pressure is maintained to vital organs throughout the dive—the combined result of increased blood pressure through vasoconstriction and decreased blood pressure because of bradycardia.

When humans dive while breathing compressed air, the increased water pressure causes nitrogen, which makes up about 79% of air, to dissolve in tissues and body fluids. If a diver ascends too quickly from a deep dive, the decreasing water pressure causes the nitrogen to come out of solution faster than it can be taken to the lungs and exhaled. The nitrogen forms bubbles in joints or other areas of the body, resulting in the "bends," or "decompression sickness," a condition that is not only painful but can be fatal. Whales dive deep and return to the surface rapidly but are able to avoid this problem. Cetacean lungs are relatively small compared with their body size. With increased depth and pressure, their lungs begin to collapse; at a depth of about 100 m, they are completely collapsed and contain no air. Any residual air is pushed to more rigid, cartilaginous portions of the trachea and bronchioles with little or no gas exchange. Thus, an increased invasion rate of nitrogen is no longer a problem with increased depth.

Although metabolic wastes accumulate during a dive, the respiratory center of the brain has a high tolerance to carbon dioxide buildup. Whales also tolerate high levels of lactic acid, produced by anaerobic respiration. They repay their "oxygen debt" when they surface and breath again, rapidly ventilating their lungs. This results in the characteristic "blow." This is a vapor cloud produced by the condensation of exhaled warm air contacting cooler outside air as well as spray from a small amount of water on the blowhole. Species sometimes can be identified by the shape, height, and direction of their blow.

It is interesting that the stomach structure of most whales is similar to that of ruminating artiodactyls in having three chambers (figure 17.6), although all whales are carnivores and do not ruminate. In whales, food is physically broken down by the highly muscular walls of the first chamber, or forestomach. No digestive enzymes are secreted in the first chamber, but they are in the second chamber, which is the main stomach. Numerous reticular folds allow the main stomach to greatly expand. The third chamber, or pyloric stomach, is named for its numerous pyloric glands (Slijper 1979; Langer 1996). Mysticetes and odontocetes have different morphological adaptations and feed in distinctly different ways on different types of prey (figure 17.7).

MYSTICETES: BALEEN AND FILTER-FEEDING

The baleen whales are named for their most characteristic feature. Baleen is composed of **keratin**, the same cornified protein material that makes up the horns of rhinoceroses and the fingernails of humans. Plates of baleen

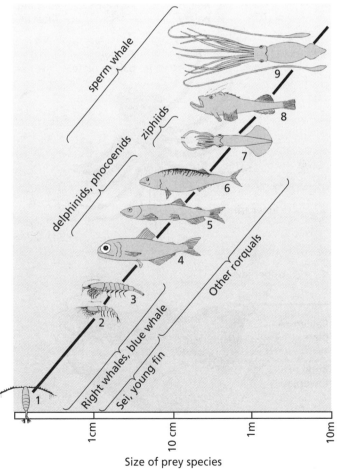

Figure 17.7 Size ranges of food items of baleen and toothed whales. Shown on a logarithmic scale are (1) calanoid copepods; (2–3) zooplankton; (4) lantern fish; (5) capelin; (6) mackerel; (7) small squid; (8) deep-water angler fish; and (9) large squid. *From D. E. Gaskin, 1982,* Ecology of Whales and Dolphins, *Heinemann Publishing.*

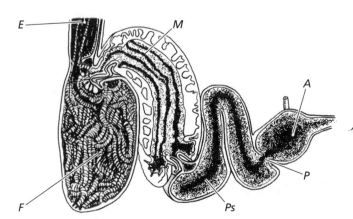

Figure 17.6 Whale stomachs. Several compartments are evident in the stomach of a bottlenose dolphin. (Abbreviations: E = esophagus; F = forestomach; M = main stomach; Ps = pyloric stomach; P = pylorus; A = ampulla of duodenum). *Adapted from E. J. Slijper, 1979,* Whales, *Cornell Univ. Press.*

hang in comblike fashion from the upper jaw only (figure 17.8). Depending on the species, from 200 to almost 500 plates occur on each side. Baleen plates in the front of the mouth are shorter; posterior baleen plates become progressively longer. The inner edge of each plate has a fringe of long, frayed filaments. These strainer-like filaments overlap each other from plate to plate and form a continuous filter. Baleen continues to grow throughout the life of an individual as the inner portion is continuously worn away. Again, depending on the species' feeding habits, baleen can be fairly short or up to 5.2 m long, as in the bowhead whale (figure 17.9). Fetal mysticetes actually have vestigial teeth that are lost before birth. Details of the formation and fine structure of baleen are given by Slijper (1979), Pivorunas (1979), and Bonner (1989).

Mysticetes feed on a variety of small marine organisms. **Plankton** (often referred to as krill, although not all plankton are krill) have limited locomotion and form large, floating aggregations. **Necton** is another category of slightly larger and more mobile marine organisms, including fish. Baleen whales feed on both plankton and necton, collectively referred to as **zooplankton** (see figure 17.7), as well as phytoplankton. Mysticetes generally do so by either "skimming" or "gulping." Skimming occurs in right whales (Family Balaenidae) and the gray whale (Family Eschrichtiidae). Individuals remain at or just below the surface. With their rostrum out of the water and mouth open, they move through large swarms of zooplankton. Water passes through the mouth and baleen, leaving a mass of zooplankton behind on the baleen. This material likely is scraped off by the tongue and swallowed.

The larger mysticetes (Family Balaenopteridae) are known as **rorquals**, which means "tube-throated." This term refers to the longitudinal grooves or pleats on the throat and chest that allow for great expansion of the oral cavity as the throat fills with water during gulping. In gulping, in contrast to skimming, individuals remain submerged and open their mouth. A huge oral cavity is created (figure 17.10A,B) as the lower jaw is distended (aided by a loosely joined, ligamentous mandibular symphysis; also see Lambertsen et al. 1995), the pleated throat is expanded, and the tongue is withdrawn into the large ventral pouch under the chest. Enormous amounts of water, filled with plankton and necton, flow into this oral cavity. Blue whales can gulp 16,000 gallons (60,640 liters, or about 64 tons!) of water at a time. This water is not swallowed but is expelled through the baleen and out the sides of the mouth as the throat and thoracic cavity contract (see figure 17.10C,D) and the tongue protrudes forward. The mass of zooplank-

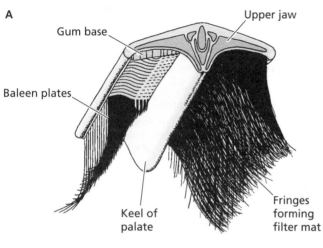

A

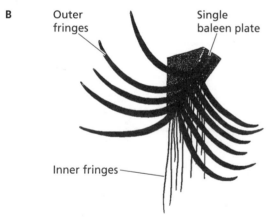

B

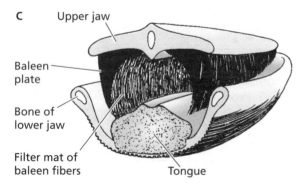

C

Figure 17.8 Baleen. (A) Arrangement of baleen in the upper jaw of mysticetes. Hundreds of plates may occur on each side of the mouth. (B) Example of a single plate. The size and number of plates vary among species. Growth is continuous as baleen is worn down from scraping by the tongue. (C) Transverse section through the head of a baleen whale showing the plates in place in relation to the large tongue. *Adapted from A. Pivorunas, 1979,* The Feeding Mechanisms of Baleen Whales, *American Scientist.*

Figure 17.9 Size and shape of baleen plates. Size and shape of plates are highly variable among different species. Representative plates of baleen from (A) minke; (B) sei; (C) Bryde's; (D) pygmy right; (E) gray; (F) humpback; (G) fin; (H) blue; (I) right; and (J) bowhead whales. *Adapted from P. G. H. Evans, 1987,* Natural History of Whales and Dolphins, *Facts on File.*

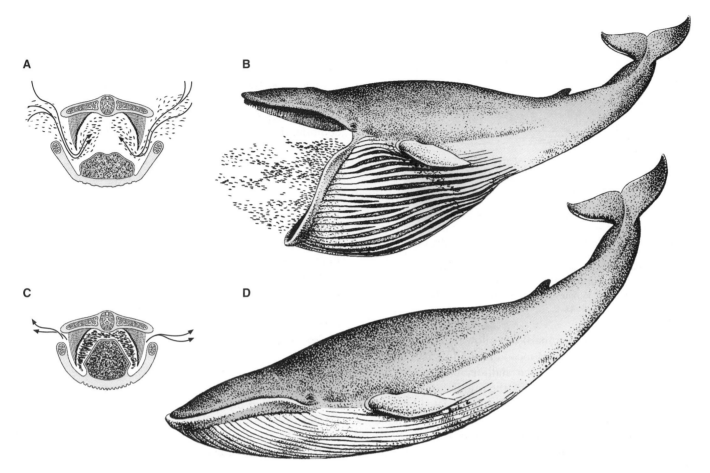

Figure 17.10 Gulping in baleen whales. (A) As the mouth opens, huge amounts of water pour in along with vast quantities of plankton and necton, as the (B) throat grooves allow for expansion of the oral cavity. (C) This water is then expelled through the filter-like baleen mat as (D) the throat contracts, trapping the food which is scraped off by the tongue and swallowed. *Adapted from N. Bonner, 1989,* Whales of the World, *Facts on File.*

ton trapped on the baleen is scraped off with the tongue and swallowed.

Unlike most other predators, baleen whales do not actively pursue their prey. Krill and other zooplankton passively float on ocean currents—they cannot attempt to out-run or outmaneuver whales. Thus, baleen whales do not need great speed or agility to capture them; locating concentrations of prey is all that is necessary. Baleen whales can become large at the expense of speed because increased size affords them thermoregulatory and other benefits.

In contrast, toothed whales are similar to more typical active predators: they must pursue and catch their prey. This necessitates a greater degree of agility and speed with associated restrictions on body size. Not surprisingly, of the approximately 67 species of odontocetes, two thirds are relatively small dolphins, river dolphins, and porpoises.

ODONTOCETES: DENTITION AND ECHOLOCATION

All odontocetes have teeth. The number varies among species, from a single pair of teeth in Genus *Mesoplodon* (see the section Ziphiidae in this chapter) to well over 100

pairs in some dolphins. Teeth are used to hold fish and squid, prey items that are found through echolocation. Opinion differs about where echolocation sound pulses are produced (see Morris 1986 for review). Purves (1967), Purves and Pilleri (1983), and Pilleri (1990) argued that echolocation sounds are produced in the larynx, as is true in bats and other mammals. In contrast, Norris (1968) suggested that the larynx is not involved but that echolocation pulses originate in the complex system of nasal sacs that occur in the forehead region of odontocetes. This idea is supported by most experimental evidence (see Au 1993 for review). As the sound is generated, it is reflected by the parabolic (dish-shaped) skull and focused through the oil-filled "melon" in the forehead (figure 17.11). The low-frequency echolocation sound pulses that whales emit are not as variable as those of bats, but the effective range of echolocation signals is much greater. The returning echoes are received via the relatively small, thin mandible. The mandible has an oil-filled sinus that channels the sound to the auditory bullae. The oil in the mandible and the melon is the same.

Because water conducts sound much better than air, whales receiving sound echoes could not achieve directionality or sensitivity unless they had special adaptations

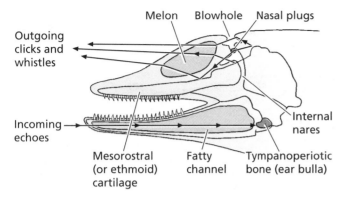

Figure 17.11 Echolocation in toothed whales. As in the dolphin shown here, sound in the form of clicks and whistles probably is produced in the complex nasal area. The sound is reflected from the skull and beamed out through the "melon." The returning echo moves through the oil-filled channel of the lower jaw to the ear bullae. *Adapted from P. G. H. Evans, 1987,* Natural History of Whales and Dolphins, *Facts on File.*

(not found in other mammals) that allow them to localize and discern the direction of incoming sound waves. They are able to do so because their auditory bullae, or **tympanoperiotic** bones (containing the middle ear), are not fused to the skull. Instead, the bullae are isolated by connective tissue and a system of sinuses filled with a unique foamed, mucous emulsion (Norris and Harvey 1974; Popper 1980). As noted by Au (1993:30): "Having the bullae physically isolated from the skull and therefore isolated from each other allows [them] to localize sounds received by bone conduction."

Fossil History

The morphological adaptations for aquatic life we described for whales are all secondarily derived. Whales evolved from terrestrial land mammals. Because extant whales are so distinctive morphologically and because intermediate fossil material is relatively rare, the phylogeny of cetaceans (figure 17.12) has been difficult to determine. Most paleontologists considered whales to be derived from within the extinct Order Condylarthra, which were archaic ancestral ungulates. Molecular evidence indicated that whales were most closely related to the Order Artiodactyla, specifically the hippopotamus family (Milinkovitch et al. 1998; Nikaido et al. 1999). Fossil and molecular evidence was reconciled with discovery of the Eocene whale remains *Artiocetus clavis* and *Rodhocetus balochistanensis* (Gingerich et al. 2001). Both these archaeocetes, in the Family Protocetidae, had an astragalus in the ankle—characteristic of artiodactyls (see chapter 20)—and provided strong support for a cetacean-artiodactyl relationship. Gingerich (2005:243) stated: "It is almost certain that the ancestor of the earliest archaeocetes would itself have been an artiodactyl."

Early groups of whales showed intermediate morphological stages during transition from terrestrial to fully aquatic mammals. Within the primitive Archaeoceti, the earliest known whale is *Pakicetus inachus* in the early Eocene Family Pakicetidae. It was small (less than 2 m long) and the ear structure was primitive, with none of the adaptations for deep diving or directional hearing underwater. In addition, the dentition was heterodont. Other archaeocete specimens also point to their intermediate structure between terrestrial Paleocene species and whales that were fully aquatic. Such a transitional specimen from the middle Eocene was *Basilosaurus isis*, which retained functional pelvic limbs and foot bones (Gingerich et al. 1990). Archaeocetes may have been amphibious during their transition from land to sea, returning to land after feeding in the water (Fordyce 1980). By the middle Eocene, however, whales were totally aquatic and highly adapted for life in the water.

Both modern suborders were clearly distinct by the early Oligocene. The earliest definite mysticete is *Aetiocetus* (Family Aetiocetidae) from the late Oligocene. Although identifiable as a primitive mysticete, it had teeth similar to archaeocetes rather than baleen. A wide range of primitive mysticetes from the Oligocene is known from the fossil record, including Genera *Phococetus*, *Llanocetus*, *Cetotheriopsis*, *Kekenodon*, and several others (Gingerich 2005). The Family Aetiocetidae, which appears to be the stem group from which mysticetes arose, became extinct

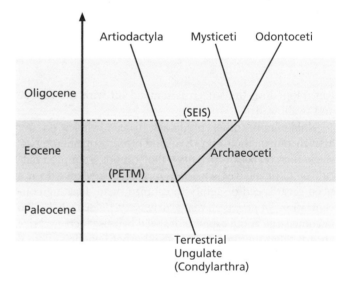

Figure 17.12 Proposed phylogeny of the Cetacea. The evolution of whales can be associated with changes in ocean temperatures. Archaeocetes derived from terrestrial ungulates in the early Eocene, concurrent with a global warming period called the Paleocene-Eocene Thermal Maximum (PETM). Likewise, modern suborders originated from the Archaeoceti concurrent with abrupt cooling events (small ephemeral ice sheets—SEIS) marking the Eocene-Oligocene boundary. *Adapted and modified from P. D. Gingerich, 2005,* Cetacea, *in* The Rise and Fall of Placental Mammals: Origins and Relationships of the Extant Clades *(K. D. Rose and J. D. Archibald, eds.), Johns Hopkins Univ. Press.*

by the late Oligocene. Another family from the Oligocene, the Agorophiidae, appears intermediate between the earlier archaeocetes and primitive true odontocetes. Based on their skull characteristics, even these Oligocene odontocetes probably echolocated (Wong 2002). Numerous genera of odontocetes also are known from the Oligocene fossil record, such as *Agorophius, Simocetus, Patriocetus,* and *Archaeodelphis* (Gingerich 2005). Most of the extant families of whales were recognizable by the Miocene. Williams (1998) provides a detailed summary of early cetacean families.

Economics and Conservation

Aboriginal whaling dates back thousands of years with Alaskan, Scandinavian, and other subsistence hunters using the meat and oil. These early hunting efforts were land-based and typically restricted to slower species inhabiting coastal waters. Using small boats, handheld harpoons (figure 17.13), and nets, hunters often tried to drive whales ashore to be killed. These efforts probably had a minimal effect on populations. The extent of whaling operations soon increased, however, as did their effect on populations. Basques were taking significant numbers of right whales during the first millennium, and by the 1500s and 1600s, most European countries had entered the whale trade using large ships that operated throughout the world. Whaling in New England began during the 1600s as well. By this time, exploitation of whales had moved from a subsistence level to a commercial enterprise because of the high economic value of the meat, oil, and "whalebone" (baleen). Commercial exploitation had a pronounced effect on whale populations, and significant declines began.

Technological advances in the whaling industry during the 1800s increased the efficiency of whaling even more, allowing additional species to be taken (figure 17.14). A modern, explosive harpoon gun was in use by 1864, concurrent with the replacement of sailing ships by steam-driven vessels. The early 1900s saw the development of floating factory ships. The stern slipway, in which whales were hauled directly aboard factory ships, was in use by 1925. This allowed whalers to operate more easily on the high seas. Many species were so severely depleted that they became "commercially extinct" and began to be protected from harvest.

Created in 1946, the International Whaling Commission (IWC) established harvest quotas on certain species and protected other species from any harvest, but it had no enforcement powers. Whale stocks, especially the larger species, continued to be overharvested due to the short-term economic interests of some countries. Today, the IWC, public opinion, political pressure, various governmental agencies, and nongovernmental conservation groups work toward conservation of whales and protection of ocean habitat (see chapter 29). In addition, the availability of alternative products to whale meat and oil has reduced the demand for whales. Nonetheless, some commercial whaling continues, generally for smaller species, by Japan, Russia, Norway, and occasionally other countries. Unfortunately, these commercial ventures are often conducted under the guise of "scientific collections." Also, continued subsistence hunting of bowhead whales by Inuit peoples remains highly controversial. Nonetheless, some large species, including the gray whale and even the blue whale, may be increasing in number. Other species, however, including bowhead and northern right whales, have not recovered. Slijper (1979), Tonnessen and Johnsen (1982), and Evans (1987) provide excellent summaries of the history of whaling.

Suborders

MYSTICETES

Baleen whales are large, with the females generally longer and heavier than the males. The smallest mysticetes, male pygmy right whales (Family Neobalaenidae; *Caperea marginata*), are about 6 m in length. During the summer, baleen whales generally feed in northern or southern polar latitudes, where they accumulate vast stores of subcutaneous fat (blubber). During the winter, they often migrate long distances to warmer, more equatorial areas. Feeding is greatly reduced during the winter because tropical waters contain less food, and whales subsist off their stored blubber. Baleen whales do not echolocate, although they produce a variety of sounds. These include "moans" and "grunts" in the low-frequency range, high-

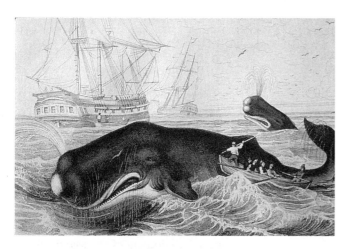

Figure 17.13 Early whaling. This print depicts a commercial arctic whaling operation in 1744. During this period, whalers used handheld harpoons and small boats, so whaling was necessarily land-based, with the dead whales being hauled onto shore to be processed.

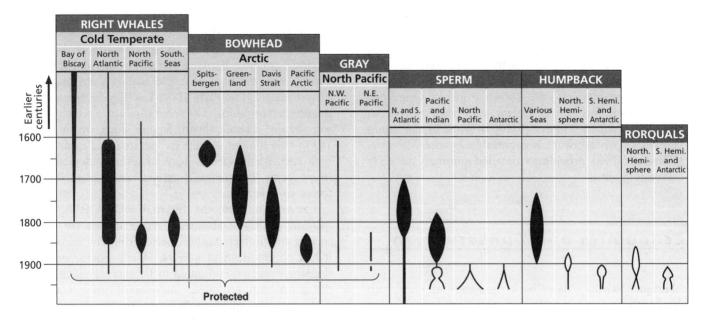

Figure 17.14 Schematic history of whaling. Solid bars indicate relatively primitive whaling operations, those from open boats with handheld harpoons; open bars indicate modern whaling operations. The width of the bars indicates only the development and decline of regional phases, not the relative numbers of each species taken. *Adapted from P. G. H. Evans, 1987,* Natural History of Whales and Dolphins, *Facts on File.*

er-frequency "chirps" and "whistles," and very high-frequency (as much as 30 kHz, or 30,000 cycles per second) pulsating clicks. Because these vocalizations are so loud and water is such an excellent conductor of sound, they may transmit for hundreds, and possibly thousands, of kilometers. Different sounds serve a variety of communication functions (Herman and Tavolga 1980), some of the most important of which may be identification of sex, social status, and location. See Evans (1987; his table 1.1) for a summary of the vocal characteristics of various cetacean species.

Although there are 4 extant families in this suborder, they include only 6 genera and 13 species. Thus, in terms of the number of species, baleen whales comprise only about 15% of living cetaceans. Sasaki et al. (2005) used mitochondrial DNA to investigate the phylogenetic relationships of baleen whales (Figure 17.15).

Balaenidae

We follow Mead and Brownell (2005), who include two genera and four species in this family. The bowhead whale, occasionally called the Greenland right whale, inhabits northern polar waters throughout the year. The black whale, or North Atlantic right whale (*Eubalaena glacialis*), is found in subpolar, temperate, and subtropical waters of the Northern Hemisphere, and the southern right whale (*E. australis*) is found in temperate and antarctic waters. The latter two species are considered conspecific by many authorities, although populations are genetically distinct. Formerly included within *E. glacialis*, many authorities consider the North Pacific right whale (*E. japonica*) a distinct species (Gaines et al. 2005). Warm

Gulf Stream waters may represent a thermal limit in the distribution of *E. glacialis* (Keller et al. 2006).

Balaenids are characterized by their stocky body shape (see figure 17.1); large head; a narrow, highly arched rostrum that accommodates very long, narrow baleen plates; a massive lower lip; lack of throat grooves; and no dorsal fin.

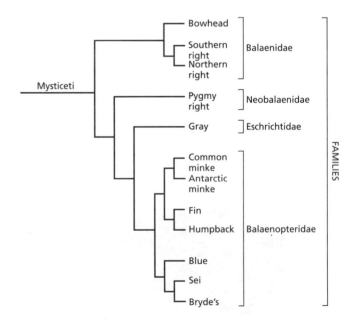

Figure 17.15 Proposed phylogeny of the Mysticeti. As noted in the text, 13 species of baleen whales are now recognized. The northern right whale in this figure does not differentiate between the North Pacific and North Atlantic right whales. *Modified from T. Sasaki et al., 2005, Mitochondrial phylogenetics and evolution of mysticete whales,* Syst. Biol. *54:77–90.*

Abundant **excrescences**, or **callosities**, occur on the head of *Eubalaena* (figure 17.16). These are rough, horny, white or yellow patches that are covered with barnacles and "whale lice" (small amphipod crustaceans). Three species of whale lice (Genus *Cyamus*) are known to occur on right whales. The largest excrescence, termed the "bonnet," is situated just anterior to the blowholes. The pattern of excrescences is specific to each whale and can be used to identify individuals. Although not known for certain, excrescences may function in aggressive interactions with conspecifics or predators (Reeves and Kenney 2003). Adult female North Atlantic right whales may mate with multiple males. Given the enormous size of the males' testes—about 1000 kg–sperm competition may be an important aspect of reproduction in this species (Mate et al. 2005).

Mysticetes feed by either skimming or gulping. Balaenids are highly specialized for skimming plankton at or just below the surface. Skimming is facilitated by the highly arched rostrum and 250–350 plates of baleen, up to 5.2 m long in bowhead whales, the longest of any mysticete (see figure 17.9). Skimming also is aided by the anterior separation of the baleen plates on each side of the jaw, which allows water to flow more easily into the open mouth (figure 17.17). This feeding behavior also explains why throat grooves found in the rorquals (see the following section), which feed by gulping, do not occur in balaenids.

These were the "right" whales to hunt because they were slow moving and found close to land. They had a great amount of blubber (and thus oil), meat, and baleen and therefore were economically valuable. Because of their high oil content, they have a tendency to float longer than other species after being killed. Bowhead and right whale populations were the first to be overexploited and dramatically reduced (see figure 17.14), and populations do not appear to be recovering, possibly because of the combined effects of predation by killer whales (*Orcinus orca*), hunting, net entanglement, tourism, loss of habitat, or inbreeding depression (Finley 2001; Fujiwara and Caswell 2001). Today, North Pacific and North Atlantic right whales remain endangered.

Neobalaenidae

This family includes only the pygmy right whale, which formerly was included in the Family Balaenidae. One of the least known cetaceans, it occurs only in temperate and cold waters of the Southern Hemisphere. The pygmy right whale is rarely observed and easily confused with minke whales. As suggested by the name, it is the smallest of the mysticetes—females are only about 6.5 m in total length, males about 6 m—about one third the size of the North Atlantic right whale. The flippers are small; there is a small **falcate** (curved toward the tail) dorsal fin and about 213–230 baleen plates per side. Like the closely related right whales, the pygmy right whale has a rostrum that is highly arched, with long, narrow baleen (Baker 1985). Unlike right whales, however, the pygmy right whale has two shallow throat grooves.

Eschrichtiidae

The gray whale (*Eschrichtius robustus*) constitutes a monotypic family. Males may reach a total length of about 13 m, females about 14 m. They have fairly slender bodies, broad flippers, and no dorsal fin, although the posterior third of the body has a series of low humps, or **crenulations**. There are two to five short throat grooves. Coloration is gray, with lighter mottled areas, and individuals are covered with an unusual number of barnacles and associated whale lice.

Gray whales feed during the summer in shallow waters of the North Pacific, specifically in parts of the Arctic Ocean, Bering Sea, and Okhotsk Sea. They roll on their sides and suck muddy sediment from the bottom to strain out a variety of worms, invertebrates, amphipods, and decapods (Moore et al. 2003). The 140–180 pairs of short baleen plates are narrow, stiff, and coarse. There is a gap in the baleen in the anterior part of the mouth. Small groups of gray whales are most common, although up to 150 individuals may congregate in good feeding areas.

Because their feeding areas freeze during the winter, in autumn gray whales migrate up to 18,000 km, one of the

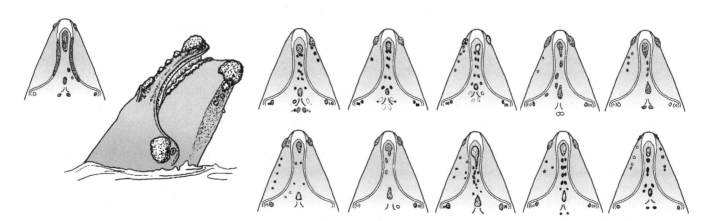

Figure 17.16 Excrescences on the head of right whales. Variation in the pattern of excrescences can be used to identify individuals, as in the dorsal view of the head of 11 right whales. *Adapted from N. Bonner, 1989,* Whales of the World, *Facts on File.*

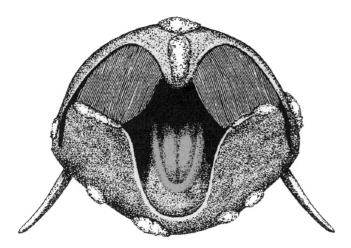

Figure 17.17 **Feeding by skimming.** Anterior view of a right whale feeding by skimming. The separation of the baleen plates in the front of the mouth is evident, as is the large tongue and lower lips. Note the excrescences. *Adapted from S. K. Katona et al., 1993, A Field Guide to Whales, Porpoises, and Seals from Cape Cod to Newfoundland, Smithsonian Institution Press.*

longest migrations of any mammalian species. They follow the Pacific coastline to calving grounds in Baja California or the Sea of Japan. They conceive during migration and give birth in January or February after a 13.5-month gestation period. A single calf is born in alternate years. Calves are born in shallow lagoons where the water is calmer and warmer than in the open ocean.

Gray whales swim slowly in shallow waters near shore, so they have always been easy prey for whalers. Populations probably were eliminated from the Atlantic Ocean by the late 1700s (Bryant 1995). The eastern Pacific population was protected since 1946, and gray whales appear to be increasing in numbers. The eastern population was removed from the endangered species list in 1994. The western population remains depleted, however, and its survival is uncertain (Weller et al. 2002).

Balaenopteridae

This family encompasses seven species of rorquals, or tube-throated whales. They all have parallel grooves, or pleats, along the throat and chest that allow expansion of the oral cavity during feeding. Unlike balaenids, the rostrum is not highly arched, the baleen plates are short, and they meet anteriorly. There are six species in the Genus *Balaenoptera*. All are long and slender, with short pectoral fins and dorsal fins that are posterior to the midpoint of the body. The five species are the blue whale, distributed throughout all oceans; the fin whale (*B. physalus*) and sei whale (*B. borealis*), also cosmopolitan in generally temperate waters; Bryde's whale (*B. edeni*), which occurs in tropical and warmer temperate areas; the small (7 m total length) common minke whale (*B. acutorostrata*), which is found in all oceans; and the Antarctic minke whale (*B. bonaerensis*), which occurs throughout waters of the Southern Hemisphere. The other species in the family is

the humpback whale (*Megaptera novaeangliae*), which enjoys a cosmopolitan distribution but is often more sedentary and found closer to shore than *Balaenoptera*. Humpbacks have a more robust body shape than other rorquals, and the flippers are very long, up to one third the body length (see figure 17.1). Numerous bumps cover the head. Each bump has a single, short, sensory hair projecting from it.

Balaenopterids feed on zooplankton by gulping; sei whales use both gulping and skimming. The larger species feed almost exclusively on krill, but the smaller species also may consume small fishes (see figure 17.7). Humpback whales are known to use a "bubble-netting" technique as well (Hain et al. 1982; Würsig 1988). In bubble-netting, the whale exhales a stream of air bubbles under water as it swims in a circle. This practice tends to congregate the plankton inside a curtain-like column of rising bubbles. The whale, with mouth open, then moves up through the column and consumes the concentrated plankton. Sometimes, coordinated groups of up to 20 humpbacks work together to concentrate prey (Jefferson et al. 1993). Rorquals feed during the summer, and they live off stored body fat during winter. Seasonal feeding and breeding activities of northern and southern populations of rorquals, like balaenids, are out of phase because they are in different hemispheres. As a result, they are genetically isolated, although they are considered to be conspecific.

All species of balaenopterids vocalize. For example, calls of blue whales have a seasonal pattern with peaks in late summer and early fall, and daily peaks at dawn and dusk. Peak calling correlates with the vertical migration of krill (Stafford et al. 2005; Wiggins et al. 2005). Humpback whales are well known for their complex, repetitive, long-lasting songs. Suzuki et al. (2006:1849) noted that a "song is a sequence of themes, where a theme consists of a phrase, or very similar phrases, repeated several times. A phrase is a sequence of several units." Songs are made up of complex series of multiple units. There are six basic songs with variations, and all the whales in a given region sing the same song. The song changes throughout the course of a season (Cerchio et al. 2001); all individuals in a region adapt to the changes. It is probable that songs function to help males compete for breeding females, as well as in social ordering of males (Darling and Berube 2001), because only mature males sing. Singing occurs during migration at higher latitudes as well as during the breeding season (Charif et al. 2001; Clark and Clapham 2004). Both Payne and Winn pioneered the study of songs in humpback whales (Payne and McVay 1971; Winn et al. 1981; see chapter 21.

Humpbacks, blue whales, and the other large rorquals were drastically overharvested throughout the early and mid 1900s. Blue, fin, and sei whales are endangered and are protected from whaling. Humpbacks have been protected since 1944, and populations are increasing; blue whales have been protected since 1965. Recovery has been slow, however, and most species almost certainly will never regain their former numbers.

ODONTOCETES

Toothed whales generally are smaller than baleen whales. Unlike male mysticetes, male odontocetes usually are larger than females. The forehead houses a complex system of nasal sacs and a fatty melon, both of which function in echolocation. All toothed whales have an asymmetrical skull, and their single external nare is often left of the center line. In all other respects, odontocetes exhibit bilateral symmetry. Skull asymmetry also functions in echolocation. Odontocetes echolocate to find food and to orient within their environment.

Delphinidae

The 17 genera and 34 species of dolphins are distributed practically worldwide in all but polar waters, including some tropical river systems in Asia and South America. Delphinids are the most varied family of cetaceans; they range in size from the smallest cetacean, Heaviside's dolphin (1.7 m in length and 40 kg), to the killer whale (9 m in length and 7,000 kg). Most delphinids have a rostrum that forms a beak and a falcate dorsal fin in the middle of the back. Delphinids vary from uniform gray or black to stripes, spots, or bands to the contrasting black-and-white patterns of the killer whale. Many morphological differences in this family are related to feeding. The total number of teeth varies among species (figure 17.18), from 2 to 7 pairs in Risso's dolphin (*Grampus griseus*) to over 120 pairs in the spinner dolphin (*Stenella longirostris*). Feeding habits of delphinids vary in terms of the size of prey species taken, distance from shore, and diving depth. Most dolphins, however, dive for short periods to depths of less than 200 m. Delphinids generally take small fishes and squid. Prey of the bottlenose dolphin (*Tursiops truncatus*) is highly variable; coastal and offshore dolphins take different prey (Reeves and Read 2003). In Florida, peak foraging occurred at dawn (Allen et al. 2001). Killer whales feed on fishes, seals, marine mammals, other cetaceans, birds, and even sharks.

Several **pelagic** (open ocean) species are gregarious and form **pods** (groups or schools) of up to 1,000 individuals, although typical group sizes are much smaller (see table 7.2 in Evans 1987). Species found closer to shore generally form smaller groups. Shark predation on bottlenose dolphins may have been a factor in the evolution of dolphin group formation, habitat use, and sociality (Heithaus 2001a, 2001b). Spinner dolphins and pantropical spotted dolphins (*Stenella attenuata*) are the species most often captured and drowned in purse-seine nets set for tuna. Special fishing techniques have been developed in attempts to minimize losses of dolphins (Macdonald 1984; Evans 1987). Other species of dolphins are taken for food in various parts of the world. Most species are insufficiently known to determine their conservation status, although Hector's dolphin (*Cephalorhynchus hectori*) is considered endangered.

Monodontidae

The two species in this family—the white whale or beluga (*Delphinapterus leucas*) and the narwhal (*Monodon monoceros*)

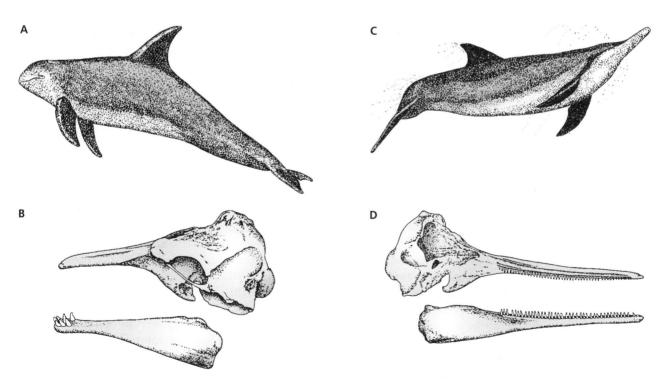

Figure 17.18 Variation in dolphin tooth number. (A) Risso's dolphin or grampus has (B) from 2 to 7 pairs of teeth, whereas (C) the spinner dolphin may have (D) a total of over 120 pairs of teeth. Note the lack of an external beak in Risso's dolphin. *Adapted from T. A. Jefferson et al., 1993,* Marine Mammals of the World, *United Nations Environment Programme.*

—are circumarctic in distribution. Both species occur in polar waters along the coast or pack ice. Both have somewhat robust bodies, rounded heads, and no dorsal fin (figure 17.19A, B). Belugas, generally found in shallow waters, are 3–5 m in length, and weigh up to 1,500 kg. Calves are gray at birth and become progressively paler as they mature; adults are white. The total number of teeth range from 14 to 44. Narwhals are similar in size to belugas, mottled grayish black dorsally, and paler on the ventral surface. Narwhals generally inhabit deeper waters than belugas. Dentition consists of one pair of upper teeth. In male narwhals, the left upper tooth develops into a long tusk (up to 3 m long and 10 kg) with a counterclockwise spiral (see figure 17.18C). The right tooth usually does not erupt. In females, neither of these teeth usually erupts. It has been suggested that the tusk is used to spear prey items or to focus echolocation signals. Most authorities agree, however, the tusks are just what they appear to be— weapons for establishing dominance breeding hierarchies among males (Hay and Mansfield 1989). Both species are gregarious, and aggregations over 1,000 individuals may form, although groups of 10–15 are more common (Jefferson et al. 1993). They feed at or near the bottom on a variety of fishes, small squid, and mollusks. Depth and duration of diving in belugas, as well as movement behavior, varies seasonally (Barber et al. 2001; A. R. Martin et al. 2001). Movement patterns of narwhals appear to be influenced by the dynamics and condition of sea ice (Laidre et al. 2004; Laidre and Heide-Jorgensen 2005). Inuits take both species for food; the beluga is considered threatened.

Phocoenidae

Porpoises formerly were considered a subfamily within the Delphinidae. Unlike dolphins, which generally have a beak, the three genera and six species of porpoises have no distinct beak. The total number of teeth range from 60–120 and are blunt with flattened crowns, in contrast to the sharply pointed, conical teeth in dolphins. Phocoenids generally are small and stocky, ranging in total length from 1.5 m in the California harbor porpoise, or vaquita (*Phocoena sinus*), to about 2.2 m in Dall's porpoise (*Phocoenoides dalli*). Most species have a short, triangular dorsal fin. They are distributed in Northern Hemisphere oceans and seas, including the Black Sea, and in Asian rivers. In the Southern Hemisphere, they occur in coastal waters of South America and around several island groups. The finless porpoise (*Neophocaena phocaenoides*) and species in the Genus *Phocoena* are found close to shore in bays and estuaries; Dall's porpoise occurs in deeper offshore waters. The spectacled porpoise (*Phocoena dioptrica*) is found in both inland and offshore waters of the Southern Hemisphere.

Dall's porpoises, like certain dolphins, are killed accidentally in fishing nets. Other species, including the harbor porpoise (*Phocoena phocoena*), are taken for food. The California harbor porpoise is limited to the northern portion of the Gulf of California, the most restricted range of any cetacean. It is considered critically endangered.

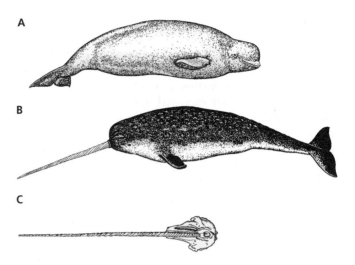

Figure 17.19 **Monodontids.** (A) Beluga and (B) narwhal whales, and (C) the spiral tusk of the narwhal. The dorsal portion of the cranium has been removed to show the root of the large, spiral left tusk of males and the typically unerupted right tusk. *Adapted from W. H. Flower and R. Lydekker, 1891,* An Introduction to the Study of Mammals, Living and Extinct, *Adam and Charles Black.*

Physeteridae

This family consists of the sperm whale (*Physeter catodon*) and two smaller species: the dwarf sperm whale (*Kogia sima*) and the pygmy sperm whale (*K. breviceps*). The latter two species are sometimes placed in their own family (Kogiidae). The sperm whale occurs in all but polar oceans. Males may reach 18 m in total length, the largest of any odontocete (see figure 17.1). The large, blunt head constitutes one third of its total length. Their flippers are small. Sperm whales have a small dorsal hump and series of posterior crenulations similar to gray whales. The thin mandible, with 18–25 pairs of teeth, is much shorter than the upper jaw, which contains sockets but no teeth.

Sperm whales have a huge, highly distinctive **spermaceti** organ (figure 17.20), several meters long. It may account for close to 10% of the total mass of an individual (Bonner 1989). The organ is composed mainly of two structures, the spermaceti organ or "case" and the "junk," both of which contain a waxy liquid, the spermaceti. In conjunction with the elaborate and convoluted nasal passages, the spermaceti organ serves as a lens to focus echolocation signals. Wahlberg et al. (2005) suggested: "The nasal complex of the sperm whale is nature's largest sound generator." Jaquet et al. (2001) found that sperm whales echolocate using "clicks" at 0.5- to 2-second intervals, and "creaks," which are very rapid clicks at intervals of 20 milliseconds (Miller et al. 2004). Clicks are emitted during deep foraging dives and are suited for long-range echolocation. Creaks are analogous to the "terminal buzz" of bats and are used when the whale closes in on prey. Repetitive series of clicks, called codas, are used for communication between individuals (Wahlberg et al. 2005; Zimmer et al. 2005). Sperm whales may reach depths of 3,200 m or

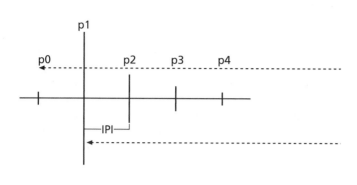

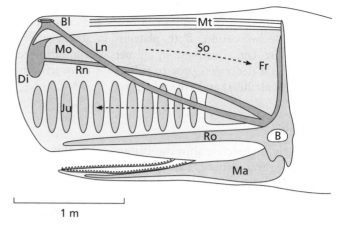

Figure 17.20 **Spermaceti organ of a sperm whale.** B = brain; Bl = blowhole; Di = distal air sac; Fr = frontal air sac; Ju = junk; Ln = left nare; Ma = mandible; Mo = "monkey lips"; Mt = muscle/tendon layer; Rn = right nare; Ro = rostrum; So = spermaceti organ. During a sperm whale echolocation click, a small amount of the energy produced "leaks" out anteriorly (represented by p0). Most of the energy is transmitted posteriorly into the spermaceti organ and bounces off the frontal air sac to be reflected out through the junk as the primary sound pulse p1. Continued reverberations at decreasing amplitude produce pulses p2–p4 at constant intervals producing the multipulsed clicks. *Adapted from W. M. X. Zimmer et al., 2005, Off-axis effects on the multipurpose structure of sperm whale usual clicks with implications for sound production,* J. Acoust. Soc. Am. *118:3337–3345.*

more, apparently diving straight up and down, and remain submerged for up to 2 hours (Leatherwood et al. 1983; Jefferson et al. 1993). Sperm whales feed mainly on large squid but also take a variety of fishes. Sperm whales exhibit complex spatial-temporal social behavior and form groups of up to 20 individuals (Christal and Whitehead 2001) with associated complex vocalizations (Rendell and Whitehead 2005). The evolution of group formation may be partially in response to predation pressure from killer whales (Pitman et al. 2001). Historically and recently, sperm whales have been one of the most important species in the whaling industry because of the vast amounts of spermaceti oil they contain (see Reeves and Read 2003). Sperm whales also are known for ambergris, a brownish, pliable, organic substance produced in the intestinal tract of some individuals. The function of ambergris remains unknown, although it may aid in digestion. Ambergris is extremely valuable as a perfume fixative ("worth its weight in gold"); pieces up to 421 kg have been found (Slijper 1979). Sperm whale hunting has been banned by the IWC since 1981.

Both pygmy and dwarf sperm whales are significantly smaller than *Physeter catodon*. They reach a maximum length of 3.4 and 2.7 m, respectively (see figure 17.1). Both species may be fairly common in tropical and temperate waters. Only the bottlenose dolphin is found stranded on beaches more often than pygmy sperm whales (Maldini et al. 2005). Pygmy and dwarf sperm whales are rarely seen at sea, however, and are poorly known. They feed on small squid, fishes, and crustaceans (Caldwell and Caldwell 1989).

Platanistidae

The taxonomic relationships of river dolphins have always been problematic. Previously, all river dolphins were included in this family, but molecular studies found that they were not a monophyletic group (Cassens et al. 2000; Hamilton et al. 2001; Nikaido et al. 2001). As such, some species have been placed in their own families (see Rice 1984; Heyning 1989) or in the Family Iniidae. Mead and Brownell (2005) considered the Family Platanistidae to include one genus and two species—the Ganges River dolphin, or susu (*Platanista gangetica*), and the Indus river dolphin, or bhulan (*P. minor*). These two species are small cetaceans 2–3 m long and found in estuaries, turbid rivers, and inland lakes in India, Pakistan, and Bangladesh. The Ganges River dolphin inhabits the Ganges-Brahmaputra River system. It often swims on its side with a foreflipper touching the bottom. The Indus River dolphin occurs in the Indus River of Pakistan. Platanistids are similar to other river dolphins in appearance, with long beaks and small eyes (figure 17.21).

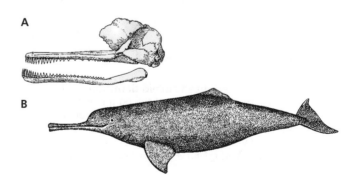

Figure 17.21 **The Ganges River dolphin.** (A) Lateral view of the skull shows the numerous homodont teeth and the distinctive maxillary crests overhanging the rostrum. There are variable numbers of teeth in the upper and lower jaws. (B) The robust body and extremely long rostrum are evident. *Adapted from T. A. Jefferson et al., 1993,* Marine Mammals of the World, *United Nations Environment Programme.*

Their eyes are without lenses, so they are virtually blind. They locate prey, generally small fishes and crustaceans, through echolocation. Both platanistids are negatively affected by dam construction, gill nets, decline in water quality, and decreased prey base because of overfishing (Smith et al. 2001) and are endangered.

Iniidae

Formerly included within the Platanistidae, this family includes three genera, each with a single species. The South American river dolphin, or boto (*Inia geoffrensis*), occurs in the Orinoco and Amazon River basins throughout northern South America. Also called the pink river dolphin, it echolocates using low-frequency whistles below 5 kHz (Ding et al. 2001). Adult males are significantly larger than females, suggesting intense competition between males related to mating opportunities (Martin and daSilva 2006). The Chinese river dolphin, or baiji (*Lipotes vexillifer*), is restricted to the Yangtze River system; the La Plata river dolphin, or franciscana (*Pontoporia blainvillei*), occurs in coastal marine waters of Argentina and southern Brazil. As noted, this group has a confused taxonomic history. The Chinese river dolphin is sometimes placed in its own monotypic Family Lipotidae, whereas the La Plata river dolphin is placed in the Family Pontoporiidae. The baiji is critically endangered (Reeves and Gales 2006).

Ziphiidae

The beaked whales occur in all oceans. This diverse family of 6 genera and 21 species of deep divers feeds on smaller species of squid (Santos et al. 2001a, 2001b) and fishes (see figure 17.7), apparently capturing them through suction (Heyning and Mead 1996). In the Atlantic Ocean, Sowerby's beaked whale (*Mesoplodon bidens*) and True's beaked whale (*M. mirus*) occur in colder waters to the north, whereas Blainville's beaked whale (*M. densirostris*) and Gervais' beaked whale (*M. europaeus*) are in warmer southern waters. MacLeod et al. (2006) discussed distribution patterns of ziphiids. Beaked whales are slender, with a pronounced beak and small dorsal fin. Total length ranges from 4–13 m. Female ziphiids may be larger than males, but there is no consistent sexual dimorphism in body lengths of species (MacLeod 2006). Although Shepherd's beaked whale (*Tasmacetus shepherdi*) may have up to 98 teeth, most ziphiids have only 1 or 2 teeth in each side of the lower jaw. These teeth usually occur only in males and are used primarily for intraspecific fighting. In the Genus *Mesoplodon*, the single pair of teeth is quite specialized among species (figure 17.22). MacLeod (2000) suggested that these sexually dimorphic teeth also may aid in recognition of morphologically similar sympatric species. Like the taxonomy of river dolphins, the taxonomy of ziphiids remains problematic; their natural history and conservation status are poorly known because they are uncommon and difficult to identify at sea.

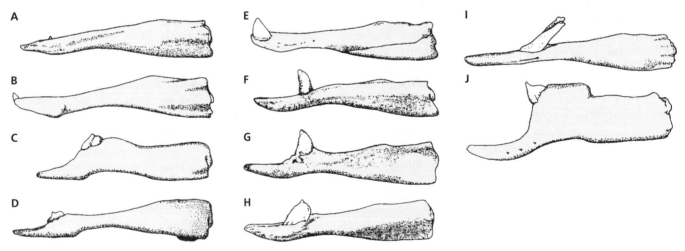

Figure 17.22 Specialized ziphiid dentition. Highly specialized and variable dentition occurs in the lower jaws of the ziphiid Genus *Mesoplodon*. Lateral view of the right mandible of (A) Gervais' beaked whale, *M. europaeus*; (B) True's beaked whale, *M. mirus*; (C) ginkgo-toothed beaked whale, *M. ginkgodens*; (D) Gray's beaked whale, *M. grayi*; (E) Hector's beaked whale, *M. hectori*; (F) Stejneger's beaked whale, *M. stejnegeri*; (G) Andrew's beaked whale, *M. bowdoini*; (H) Hubbs' beaked whale, *M. carlhubbsi*; (I) strap-toothed whale, *M. layardii*; and (J) Blainville's beaked whale, *M. densirostris*. *Adapted from T. A. Jefferson et al., 1993,* Marine Mammals of the World, *United Nations Environment Programme.*

SUMMARY

All whales exhibit similar morphological features because they are all adapted for life in the water. Compared with terrestrial mammals, whales are very large. Their large size can be accommodated because water provides a buoyant, supportive environment. Whatever their size, all whales have a streamlined, fusiform body.

There are no external ears, hind limbs, or fur to disrupt the smooth flow of water over the body as a whale moves. Forelimbs have been modified into paddle-like flippers, and a horizontally flattened tail fluke provides propulsion. The skull has also been modified by "telescoping," with resultant movement of the external nares to the top of the head. In place of fur, insulation is provided by a layer of subcutaneous fat—the blubber. Whales also have several physiological adaptations for diving. Compared with terrestrial mammals, they have a high hematocrit and more myoglobin, use oxygen more efficiently, and while submerged undergo bradycardia and reduced blood flow to peripheral areas of the body.

Despite these similarities, there are several differences in the two cetacean suborders. Mysticetes, or baleen whales, are named for the baleen plates that hang from the upper jaw in place of teeth. Baleen is used as a filter to strain small marine organisms (krill) from the water. Mysticetes, which account for only 15% of extant cetacean species, are the largest whales; they have sacrificed speed and agility for the benefits of increased body size. Their size is adaptive because their primary prey, krill, essentially is immobile. Species within the seven families of toothed whales, or Odontocetes, generally are smaller and more agile than baleen whales. They feed on fishes, squid, and larger prey items rather than on krill. Unlike mysticetes, odontocetes have an asymmetrical skull with a single external nare, a complex system of nasal sacs, and a large, bulbous forehead with an oil-filled melon. These structural features all relate to the ability of toothed whales to echolocate.

The specialized morphological characteristics of all whales are secondarily derived. That is, whales evolved from terrestrial land mammals, specifically early artiodactyls, which made a gradual transition from land to sea during the Eocene epoch. Intermediate stages during this transition are evident in several lineages of the Archaeoceti, an extinct suborder of whales. Both mysticetes and odontocetes were distinct by the early Oligocene.

Whales have been hunted for thousands of years for their meat, oil, and other valuable products. Subsistence hunting, which has a relatively minimal effect on populations, eventually gave way to large-scale commercial exploitation. Right whales and bowhead whales probably were the first species to experience significant population declines because of overharvesting. The International Whaling Commission was formed in 1946 to help conserve stocks of many species. Through the efforts of the IWC, governments, and many conservation agencies, most nations no longer harvest whales. Although populations of some species have begun to recover, others have not.

SUGGESTED READINGS

Clapham, P. J., S. B. Young, and R. L. Brownell, Jr. 1999. Baleen whales: Conservation issues and the status of the most endangered populations. Mammal Rev. 29:35–60.

Mann, J., R. C. Connor, P. L. Tyack, and H. Whitehead (eds.). 1999. Cetacean societies: Field studies of dolphins and whales. Univ. Chicago Press, Chicago.

Perrin, W. W., B. G. Würsig, and J. G. M. Thewissen (eds.). 2002. The encyclopedia of marine mammals. Academic Press, New York.

Reeves, R. R., and R. D. Kenney. 2003. Baleen whales: Right whales and Allies (*Eubalaena* spp). Pp.425–463 *in* Wild mammals of North America: Biology, management, and conservation. (G. A. Feldhamer, B. C. Thompson, and J. A. Chapman, eds.). Johns Hopkins Univ. Press, Baltimore.

Reeves, R. R., and A. J. Read. 2003. Bottlenose dolphin, harbor porpoise, sperm whale and other toothed cetaceans (*Tursiops truncates*, *Phocoena phocoena*, and *Physeter macrocephalus*). Pp. 397–424 *in* Wild mammals of North America: Biology, management, and conservation. (G. A. Feldhamer, B. C. Thompson, and J. A. Chapman, eds.). Johns Hopkins Univ. Press, Baltimore.

Reeves, R. R., B. S. Stewart, P. J. Clapham, and J. A. Powell. 2002. Guide to Marine Mammals of the World. Alfred A. Knopf, New York.

Reynolds, J. E., W. F. Perrin, R. R. Reeves, S. Montgomery, and T. J. Ragen (eds.). 2005. Marine mammal research: Conservation beyond crisis. Johns Hopkins Univ. Press, Baltimore.

DISCUSSION QUESTIONS

1. In contrast to whales, the seals and walrus, which spend the majority of their lives in water, are heavily furred. What life history aspects can you think of that make retention of fur adaptive in seals and the walrus but not in whales?

2. Can you offer any suggestions as to why humpback whales sing but apparently no other species of baleen whales do?

3. Sexual dimorphism is evident in whales. Why might females be larger than males in baleen whales, whereas the opposite situation occurs in toothed whales?

4. We discussed some of the ways that whales maintain their body heat in cold polar waters. Consider the other extreme: How do they dissipate heat while in tropical waters?

5. Despite conservation programs, many species of large whales have not recovered their former numbers. What negative factors affect oceans today that might mitigate against future recovery of whales?

CHAPTER 18

Rodentia and Lagomorpha

This chapter includes two structurally and functionally similar orders. The Order Rodentia is characterized by a single pair of upper and lower incisors and contains more species than any other mammalian order. The Order Lagomorpha—a much smaller group—is characterized by a second pair of small, round peg teeth posterior to the upper incisors. These orders are similar, however, in that individuals have specially adapted, enlarged incisors for gnawing and a diastema without canines between the incisors and the cheekteeth. Considering the origins and relationship of the two orders, there have been numerous studies based on morphology, paleontology, and molecular biology that offer support for a Superorder Glires encompassing rodents and lagomorphs (Meng and Wyss 2005). There also have been numerous studies that suggest no close relationship between these orders (see citations in Adkins et al. 2001). We have included rodents and lagomorphs in a single chapter because of their similarities in feeding and associated structural characteristics, and as noted by Carleton and Musser (2005:746) "Sister-group stature of Lagomorpha and Rodentia is decisively sustained in modern phylogenetic studies, both those drawing on morphological . . . and molecular data sources." Given the size of the Order Rodentia, we will examine an array of structural and functional adaptations that have enabled them to flourish throughout most of the world. At the same time, we will explore the considerable amount of convergence among unrelated rodent families that "make a living" in similar ways.

Rodentia

Rodents constitute the largest mammalian order, with 32 extant families currently recognized (Carleton and Musser 2005) and approximately 2272 species. About 42% of living mammalian species are rodents. Rodents enjoy a **cosmopolitan** (worldwide) distribution and are native everywhere except Antarctica, New Zealand, and a few oceanic islands. They have adapted very successfully to a wide range of terrestrial, arboreal, scansorial, fossorial, and semiaquatic habitats. Rodents are found in all biomes, often as **commensals** with humans. Also, they exhibit a diverse array of locomotor adaptations, including plantigrade, cursorial, swimming, fossorial, jumping, and gliding. The vast majority of rodents are small (20–100 g), although the largest, the capybara (*Hydrochoerus hydrochaeris*), may reach 50 kg.

Given the large number of rodent species, their degree of diversity and adaptability, and convergent evolutionary trends, it is not surprising that the systematic relationships of many families and subfamilies are complex and result in an array of suborders, superfamilies, and subfamilies. Despite the number of species and their widespread distribution and diversity, rodents are surprisingly uniform in several general morphological characteristics.

MORPHOLOGY

The diagnostic characteristic that defines all rodents is a single pair of upper and lower incisors. Their large incisors are open-rooted and evergrowing and are used for gnawing (the name *Rodentia* is derived from the Latin *rodere*, "to gnaw"). As gnawing quickly wears down the tips of the incisors, a chisel-like edge forms because the anterior side of each incisor is covered with enamel and wears more slowly than the posterior side, which lacks enamel (figure 18.1). In many species, the mouth can close behind the incisors, and the animal may have either internal or external cheek pouches for transporting food. Rodents also have a diastema, which is a gap between the incisors and cheekteeth that allows for maximum use of the incisors in manipulating food. Canine teeth are absent, and the number of molariform teeth is reduced. Molariform dentition may or may not be evergrowing, and many different occlusal cusp patterns occur (figure 18.2).

Rodents have dental formulae greatly reduced from the primitive eutherian number. A typical dental formula is 1/1,

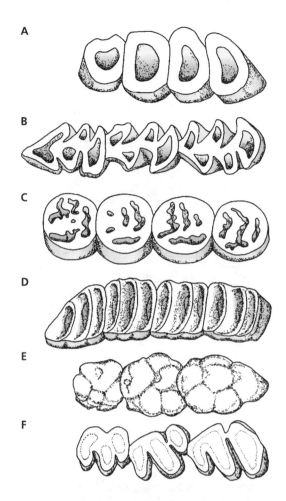

Figure 18.2 Rodent molariform occlusal surfaces. Many different patterns occur in the occlusal surfaces of molariform teeth of rodents. (A) Ring of enamel in a mole-rat; (B) prismatic in murids, voles, and lemmings; (C) flat with lakes of dentine surrounded by enamel in Old World porcupines; (D) transverse ridges in a chinchilla; (E) cuspidate pattern in many species; and (F) folded enamel in a murid. *Adapted from T. E. Lawlor, 1979,* Handbook to the Orders and Families of Living Mammals, *2nd ed., Mad River Press.*

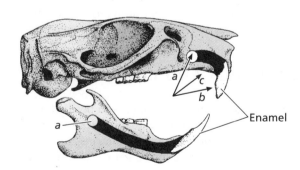

Figure 18.1 Skull of a rodent. Note the characteristic features of rodents, including a single pair of upper and lower incisors, which are open-rooted (*a*) and grow throughout an individual's life; the incisors' beveled edges (*b*); and the diastema (*c*) between incisors and cheekteeth. *Adapted from T. E. Lawlor, 1979,* Handbook to the Orders and Families of Living Mammals, *2nd ed., Mad River Press.*

0/0, 0/0, 3/3 = 16. Rodents never have more than two pairs of premolars, and no species has more than 22 teeth, except the silvery mole-rat (*Heliophobius argenteocinereus*), with 1/1, 0/0, 3/3, 3/3 = 28. Even then, not all the premolars are in place at the same time. Rodents generally are herbivorous or omnivorous, depending on the season and availability of food items. Coprophagy (reingestion of their own fecal pellets taken directly from the anus) has been reported in 11 families of rodents (Hirakawa 2001). Females have a duplex uterus, males have a baculum, and the testes may be scrotal only during the breeding season. The jaw musculature and skull structure associated with the dentition serve as important criteria for grouping rodents.

Despite confusion and lack of consensus in rodent classification, three groupings, first suggested by Simpson (1945), generally are accepted and useful for our purposes: **sciuromorph** (squirrel-shaped), **myomorph** (mouse-shaped), and **hystricomorph** (porcupine-shaped) rodents.

Each group is distinguished by skull structure and jaw musculature, specifically the origin of the masseter muscles (figure 18.3). These three groups are often given sub-ordinal status, although not all families readily fit into one of the groups.

Besides these cranial characteristics, all rodents, whether living or extinct, have one of only two types of lower jaw, depending on the insertion of the masseter muscle. They are either **sciurognathous,** with a relatively simple mandible that has insertion directly ventral to the molariform dentition, or **hystricognathous,** with a strongly deflected angular process and flangelike, or ridged, mandible for masseter insertion ventral and posterior to the teeth (figure 18.4). Carleton and Musser (2005) provided an excellent review of the higher-level classification of rodents. Numerous studies have examined the Order Rodentia for answers to questions of monophyly, the radiation of major lineages, and relationships among families. Although these questions will certainly continue to be examined, the large amount of recent molecular and morphological research "has made the case for rodent monophyly vastly more secure" (Carleton and Musser 2005:745).

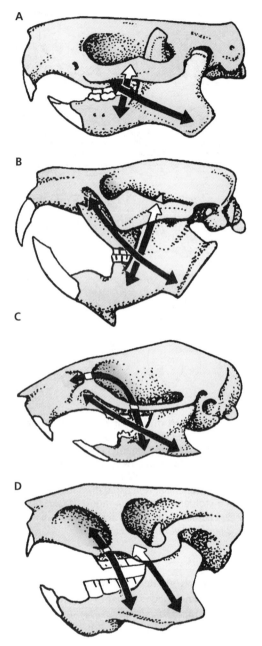

Figure 18.3 Masseter muscle complex in rodents. Diagrammatic representation of the origin and insertion of the masseter muscles in (A) an aplodontid-like, primitive protrogomorph form with the origin of the masseter muscle entirely on the zygomatic arch and not anterior on the rostrum. (B) Sciuromorphs have a small infraorbital foramen through which no masseter muscles pass. Origin of the middle muscle is anterior to the eye, and the deep muscle is beneath the zygomatic arch. (C) A slightly larger infraorbital foramen is found in myomorphs, with the deep portion of the masseter muscle passing through it and originating on the rostrum, whereas the middle masseter is anterior to the eye, as in sciuromorphs. (D) In hystricomorphs, the infraorbital foramen is very large; the deep masseter muscle passes through it and attaches anterior to the eye, and the middle masseter originates on the zygomatic arch. *Adapted from A. S. Romer, 1966, Vertebrate Paleontology, 3rd ed. Univ. Chicago Press.*

FOSSIL HISTORY

The reduced dentition, enlarged incisors, and diastema in all modern rodents appeared early in the fossil record. Rodent-like reptiles with these characteristics, the **Tritylodonts,** had developed by the late Triassic period, concurrent with the appearance of seed-bearing vegetation. Tritylodonts were succeeded by the multituberculates about 50 million years later, in the mid-Jurassic period. These early rodent-like, "prototherian" mammals also had enlarged incisors, no canines, and cheekteeth with multiple cusps (figure 18.5). Neither of these groups evolved into rodents, however. The oldest known members of the Order Rodentia, the sciuromorph Paramyidae (see figure 18.5C) represented by teeth from *Paramys atavus*, date from the late Paleocene in North America and Eurasia, as do fossil remains of Genus *Acritoparamys* (Meng and Wyss 2005). The paramyids were ancestral to several rodent families, and their closest living descendant, based on jaw and muscle structure and dentition, is the mountain beaver (Aplodontidae: *Aplodontia rufa*).

ECONOMICS AND CONSERVATION

Rodents had and continue to have both positive and negative effects on humans (figure 18.6). Several species are of economic importance as food for humans, and the fur of other rodents is valuable in the garment industry. North American rodents important as furbearers include the muskrat (*Ondatra zibethicus*) and beaver (*Castor canadensis*) as well as the introduced nutria (*Myocastor coypus*) (Baker and Hill 2003; Bounds et al. 2003; Erb and Perry 2003). The South American chinchilla (*Chinchilla lanigera*) is another important furbearer, with countless individuals bred in captivity throughout the world. On the negative side, numerous rodent species worldwide severely damage

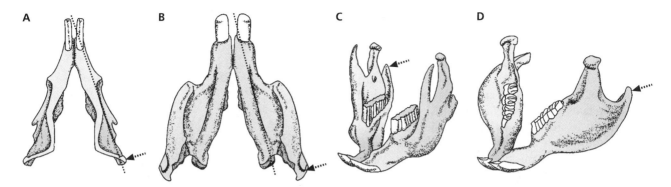

Figure 18.4 Sciurognath vs. hystricognath lower jaws. Ventral view of the mandible of (A) a sciurognathus rodent and (B) a hystricognathus rodent showing the relationship of the very long incisors (*dotted lines*) to the angular process (*arrows*). (C) In the dorsolateral view of the sciurognath, note the large coronoid process and surface directly below teeth for insertion of the masseter muscles. (D) In the hystricognath, the coronoid process is greatly reduced, and insertion of the masseter is on the large, bony flange of the angular process. *Adapted from R. J. G. Savage, 1986,* Mammal Evolution, *Facts on File.*

crops and grain stores. Rodents consume an average $30 billion worth of cash crops and cereal grains each year!

Historically, rodents, specifically European rats (*Rattus* spp.), were vectors for epidemics of bubonic plague that ravaged human populations periodically throughout the late Middle Ages, killing millions of people. Bubonic plague remains a problem throughout many less developed areas of the world. Rodents are vectors or reservoir hosts for a variety of other viral, bacterial, fungal, and protozoan infectious diseases, including murine typhus,

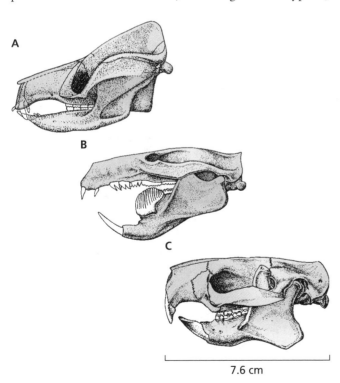

Figure 18.5 Rodent-like fossils. A mammal-like reptile (A) *Oligokyphus* (actual length about 7.6 cm), from the Jurassic period, with a diastema and enlarged incisors. These characteristics are also evident in a multituberculate mammal (B) *Ptilodus* (7.6 cm in length) from the Paleocene epoch. Compare with (C) *Paramys* (7.6 cm in length), from the oldest known rodent family, the Paramyidae. *Adapted from A. S. Romer, 1966,* Vertebrate Paleontology, *3rd ed., Univ. Chicago Press.*

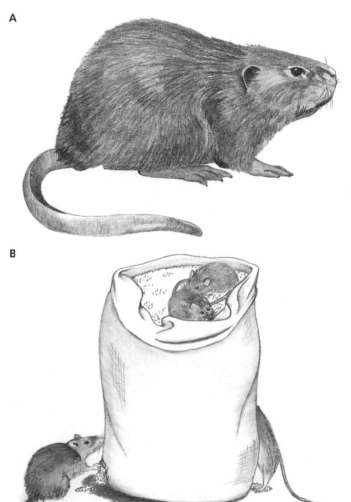

Figure 18.6 Economic effects of rodents. Rodents may have both significant positive and negative economic effects on humans. (A) Muskrats and several other species are important sources of fur in the United States, and millions are harvested yearly. (B) Norway rats (*Rattus norvegicus*) and numerous other rodent species cause billions of dollars in crop and property damage and are reservoirs for numerous diseases throughout the world. *Drawings by Lisa Russell.*

leptospirosis, listerosis, rickettsial diseases, Lassa fever, Q fever, histoplasmosis, Lyme disease, and hantaviruses (Cox 1979; see chapter 27).

Today, rodent populations are important for scientific purposes as well. Much of the work in population dynamics, animal behavior, physiology, and psychology has relied heavily on rodents as experimental animals. Researchers often use them in field or laboratory studies because of their rapid turnover rates, short gestation periods, large litter sizes, rapid sexual maturity, and easy handling. Many species of voles (Genus *Microtus*) or lemmings (Genus *Lemmus*) show predictable, 3–4-year population cycles throughout large geographic areas. Factors that cause these cycles, and how they operate, are still being studied to determine how external and internal influences affect population dynamics. Rodents such as guinea pigs (Genus *Cavia*), the house mouse (*Mus musculus*), and Norway rat (*Rattus norvegicus*) are important research animals in biological laboratories throughout the world.

SCIUROGNATH RODENTS

Families in this suborder generally have a small to medium-sized infraorbital foramen with either no passage of any part of the masseter muscle through it (or limited passage, as in the myomorphs). The springhares (Family Pedetidae), gundis (Family Ctenodactylidae), and scaly-tailed squirrels (Family Anomaluridae) are exceptions, however, because they have very large infraorbital foramina—more characteristic of hystricomorphs—for passage of much of the masseter muscle. All have typical sciurognath structure of the lower jaw, however. Thus, these three families may be considered as "hystricomorphous sciurognaths" and typify some of the difficulties involved in classifying rodents strictly on the basis of morphology.

Aplodontidae

This is a monotypic family of one extant genus and species. The mountain beaver (also called a boomer or sewellel) is endemic to the Pacific northwest from British Columbia to northern California. The species is of special interest because it is the most primitive living rodent, with a fossil history dating to the Paleocene. Mountain beavers (which are not limited to mountains and definitely are not beavers) occur in humid forested areas with a dense understory associated with heavy rainfall. They are restricted to wet areas, in part because their primitive kidneys cannot produce concentrated urine. Although *Aplodontia* is considered here as a sciuromorph, the masseter muscles originate entirely on the zygomatic arch (see figure 18.3); as such, mountain beavers can be considered the only living "protrogomorph" rodent (Wood 1965). They are stocky, have no external tail, and weigh up to 1.5 kg. The flattened skull is distinctive, appearing triangular from above, with flask-shaped auditory bullae and teeth that have unique projections (figure 18.7).

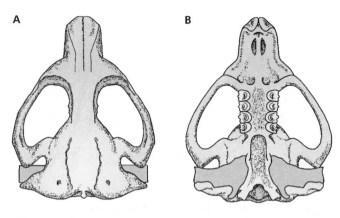

Figure 18.7 Mountain beaver skull. (A) Dorsal and (B) ventral views of a mountain beaver skull showing the triangular skull shape, flask-shaped auditory bullae (*shaded*), and labial projections on upper cheekteeth. Lingual projections occur on the lower cheekteeth. Actual size. *Adapted from T. E. Lawlor, 1979,* Handbook to the Orders and Families of Living Mammals, *2nd ed., Mad River Press.*

Mountain beavers are found in small colonies and burrow extensively, often near streams or in moist ground. Burrows may be fairly complex and can extend for 300 m. Mountain beavers also climb as part of their foraging and are capable swimmers. They are considered a pest and cause severe problems by damaging or consuming seedlings planted as part of forest regeneration programs (Feldhamer et al. 2003a).

Sciuridae

The squirrels are a large and diverse group of about 51 genera and 278 living species distributed worldwide except for Australia, Madagascar, southern South America, and some desert regions. Mercer and Roth (2003) discussed the worldwide diversification of sciurids in relation to major tectonic and other global changes. Traditionally, sciurids were divided into two subfamilies that reflected the diversity of habitats and behavioral characteristics of this large family: the tree squirrels and ground squirrels (Sciurinae) and the "flying" squirrels (Petauristinae). More recently, however, Thorington and Hoffmann (2005) recognized five subfamilies based on molecular studies. In addition to the Subfamily Sciurinae (with 20 genera), these subfamilies include the Ratufinae (with a single genus *Ratufa*), Sciurillinae (with a single genus *Sciurillus*), Callosciurinae (with 14 genera), and Xerinae (with 15 genera). Tree squirrels, including the well-known gray squirrel (*Sciurus carolinensis*) and the fox squirrel (*S. niger*), den in tree cavities or build nests among tree limbs. They generally are diurnal and arboreal. In boreal and cool temperate areas, most species hibernate. Ground squirrels and marmots are terrestrial and burrow, and many species both hibernate and estivate. Sciurids can be serious crop pests in many agricultural areas throughout the world. In North America, the conservation and management of the five species of prairie dogs (Genus

Cynomys)—all of which have declining populations—often are controversial and present challenges for resource agencies (Hoogland 2003).

Flying squirrels occur in both the Eastern and Western Hemispheres. They are distinguished by a furred patagium between the front and hind limbs, which in some species also includes the neck and tail. The patagium increases the surface area for extended gliding locomotion (figure 18.8) and enhances maneuverability. Members of the Genus *Petaurista*, for example, glide for distances up to 450 m and can turn 90 degrees in midair. In North America, the best-known "flying" squirrels are in the Genus *Glaucomys*. The scaly-tailed flying squirrels (see Family Anomaluridae in this chapter) have filled the gliding niche in Africa. In Australia, several species of marsupial phalangers (Family Phalangeridae) are also gliders. These cases of similar morphological adaptations and behaviors in similar habitats are good examples of evolutionary convergence.

Geomyidae

The 6 genera and approximately 40 species of pocket gophers are restricted to North and Central America from southern Canada through Mexico to extreme northern Colombia. They occur in a variety of habitats with soil conducive to burrowing. Geomyids are generally small, 120–440 mm in total length, including a tail 40–140 mm long. Morphology reflects their fossorial mode of existence. Visual and auditory acuity is reduced in favor of enhanced tactile and olfactory sensitivity. They have thickset, chunky bodies with small eyes and pinnae (figure 18.9). There are claws on the forefeet and hind feet for digging; those on the forefeet are larger. The incisors extend anteriorly so the lips can be closed behind them to keep dirt out of the mouth when the animal is digging. The pectoral girdle is highly developed for digging, with a short, powerful humerus and a keeled

Figure 18.9 Fossorial adaptations. This lateral view of a plain's pocket gopher (*Geomys bursarius*) shows the fusiform body shape of these fossorial species.

sternum for enhanced muscle attachment, similar to that in the three families of moles. The pelvic girdle is small and relatively undeveloped.

Gophers feed on subterranean parts of plants, primarily roots and tubers, although occasionally they leave their tunnels and forage aboveground. Food is brought back to storage chambers in large, externally opening, furred cheek pouches that extend from the mouth to the shoulder. Tunnels may be quite extensive and include chambers for shelter, nesting, food storage, and fecal deposits. Tunnels are most easily noted by a series of aboveground mounds. Pocket gopher mounds are differentiated from those of moles by having a fan rather than a conical shape and by the entrance hole not being in the center of the mound. Although gophers may damage croplands, gardens, and seedlings, they also aerate soil, increase water penetration into the soil, promote early successional plants, and enhance plant diversity and community structure (Sherrod et al. 2005). The Oaxacan pocket gopher (*Orthogeomys cuniculus*) and Queretaro pocket gopher (*Cratogeomys neglectus*) are critically endangered, and the Michoacan pocket gopher (*Zygogeomys trichopus*) is endangered.

Heteromyidae

The kangaroo rats, kangaroo mice, and pocket mice include 60 species in 6 genera. Three subfamilies are traditionally recognized (Alexander and Riddle 2005; Patton 2005). Heteromyids are distributed from southwestern Canada through the western United States and Central America to northwestern South America. Kangaroo rats (Genus *Dipodomys*) and kangaroo mice (Genus *Microdipodops*) take their name from their superficial similarity to marsupial kangaroos (figure 18.10). These external characteristics are adaptations for their distinctive, bipedal jumping locomotion. The spiny pocket mice (Genera *Liomys* and *Heteromys*) and pocket mice (Genera

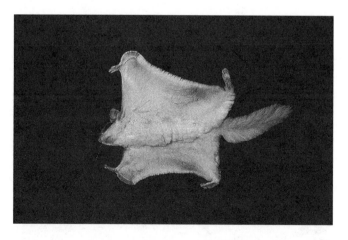

Figure 18.8 Gliding adaptations. The patagium and flattened tail, essential for maneuverability and extending time in the air, are clearly seen in this ventral view of a southern flying squirrel (*Glaucomys volans*).

A

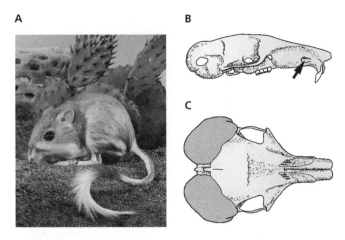

B

C

Figure 18.10 Kangaroo rat. (A) Long, tufted tail, elongated hind limbs, and kangaroo-like bipedal body shape of a desert kangaroo rat (*Dipodomys deserti*). (B) Lateral view of the cranium of *Dipodomys* showing the threadlike zygomatic arch and infraorbital foramen that pierces the rostrum (*arrow*), and (C) dorsal view showing very large, inflated auditory bullae (*shaded*).

Perognathus and *Chaetodipus*) are quadrupedal. Like the geomyids, all heteromyids have externally opening, furred cheek pouches, which they use to carry seeds back to their burrows for long-term storage. They may eat more perishable food items while foraging.

Heteromyid skulls are often quite distinctive. A characteristic feature of the family is that the infraorbital foramen pierces the **rostrum** (the anterior region of the cranium), and several genera have greatly inflated auditory bullae (see figure 18.10C). Large auditory bullae increase auditory acuity, which is of value in detecting potential nocturnal predators. They also aid in maintaining balance while heteromyids hop, often erratically, to elude those predators. Heteromyids are often found in arid regions; in the western United States, they are the rodents best adapted to dry, desert conditions. Kangaroo rats and pocket mice can exist for prolonged periods without free water. They survive on the moisture in food (mostly seeds) as well as metabolic water production. They also conserve water by producing a highly concentrated urine, being nocturnal, and remaining inactive in relatively humid burrows throughout the day. Spiny pocket mice occur in wetter, tropical habitats of Central and South America. Anderson and Timm (2006) described a new species of spiny pocket mouse (*Heteromys nubicolens*) from highland areas of Costa Rica. Several species of heteromyids are endangered.

Castoridae

This family contains only two extant species, both restricted to the Northern Hemisphere. The Canadian beaver (*Castor canadensis*) occurs in Alaska, Canada, and throughout much of the United States, and the European beaver (*C. fiber*) occurs throughout northern Eurasia. The family name comes from the two pairs of castor glands

and a pair of anal glands in both males and females. These glands produce material with strong pheromones that is deposited on logs and mud piles to demarcate territories.

Beavers are well adapted to aquatic habitats and prolonged periods in icy water. They are large (up to 35 kg) with long, heavy guard hairs overlaying fine, insulating underfur. Both body size and pelage enhance their heat retention capabilities. The familiar spatulate tail is used in swimming, construction of dams and lodges, and certain behaviors. Additional aquatic adaptations include a **nictating membrane** (a thin membrane that can be drawn over the eyeball), valvular nostrils and ears that close while under water, and fully webbed hind feet (figure 18.11). Beavers cut and feed on a variety of trees, but especially aspen. They eat the cambium, bark, leaves, and roots, and store sections of trees and twigs underwater for food during the winter. Haarberg and Rosell (2006) found that European beavers were central-place foragers, consistent with predictions of optimal foraging theory. Beavers can swim partially underwater with their mouths open, carrying cut branches, without water entering the lungs. They breathe through the nose, not the mouth. The epiglottis is above the soft palate, so as they breathe, air moves through the nostrils to the trachea and lungs. Also, the posterior part of the tongue is convex and blocks the pharynx, except when the animal swallows (Coles 1970).

Beavers have a tremendous influence on ecosystem structure and function through construction of dams, and they also have positive and negative economic effects (see Baker and Hill 2003). Dams are often very large and complex and may reach 600 m in length, although 25 m is about average. The depth and width of a watercourse influences dam-building behavior (Hartman and Tornlov 2006). Beavers impound large areas, creating ponds of still, often deep water. In the early 1800s, the quest for beaver pelts for the European market was a prime motivating factor in the

Figure 18.11 Beaver body morphology. Note the animal's fully webbed hind feet, thick pelage, and dorsoventrally flattened tail.

early exploration of western North America by trappers. Beaver populations were overexploited and extirpated throughout most of their range. Many current populations are the result of reintroductions to the former range.

Pedetidae

Two species are now recognized in this family, the South African springhare (*Pedetes capensis*) and the East African springhare (*P. surdaster*). Springhares occur in sandy soils or cultivated areas throughout semiarid regions. Total length reaches 900 mm, including a tufted tail up to 500 mm, and body mass ranges up to 4 kg. The skull is heavy with long nasal bones that extend beyond the premaxillary bones, and there is a large infraorbital foramen. As already noted, pedetids may be considered "hystricomorphous sciurognaths" based on their infraorbital structure so their phylogenetic position has been uncertain. The scaly-tailed flying squirrels and gundis also fall into this category.

Like North American kangaroo rats, springhares have large hind feet and muscular hind limbs modified for bipedal, jumping locomotion (figure 18.12). They exhibit erratic hopping when frightened or escaping predators. Springhares burrow and tunnel extensively, and if they sense danger as they are about to leave their burrow, they "spring" immediately into the air from the exit, a behavior that gave rise to the common name. Individuals use numerous different burrows and change them frequently (Peinke and Brown 2005). They breed throughout the year, but litter size is usually one. Springhares are nocturnal and herbivorous, feeding on a variety of cultivated crops. When foraging, they move slowly in a quadrupedal manner. They are hunted both because of the damage they do to crops and as a source of food by the Xhoisan people of South Africa.

Anomaluridae

The scaly-tailed flying squirrels are restricted to West and Central Africa, where they inhabit tropical forests. There are seven species in three genera, and with one exception—the Cameroon scaly-tailed squirrel (*Zenkerella insignis*)—all have a well-formed patagium that encloses the forelimbs, hind limbs, and tail. The fur is brightly colored and silky. The common name is derived from two rows of keeled scales found at the ventral base of the tail (figure 18.13). These scales probably add traction while climbing or landing on a tree trunk after a glide. Anomalurids are arboreal and den in tree cavities up to 35 m above ground. They are primarily nocturnal and herbivorous and may associate in colonies of up to 100 individuals. The taxonomic position of this group is uncertain; their resemblance to sciurid flying squirrels is due to convergence, not to phylogenetic relationship.

Ctenodactylidae

The four genera and five species of gundis are found in sparsely vegetated, rocky, semiarid regions of northern Africa from Morocco east through Somalia. They are similar to pikas (Order Lagomorpha: Ochotonidae), which are discussed later in this chapter. Their skulls have a large infraorbital foramen. They are unusual in that the paroccipital processes are very long and curve under and contact the auditory bullae. Gundis are up to 240 mm in head and body length, usually with very short tails, except in Speke's pectinator (*Pectinator spekei*). They have extremely dense pelage, which is groomed with the comb-like hind toes (the family name means "comb-toes"). Females have an unusual lateral pair of cervical mammae

Figure 18.12 Springhare. The reduced forelimbs, long, powerful hind legs, and long tail evident in these various postures of the springhare resemble those of the smaller kangaroo rat (heteromyid) and larger kangaroo (macropodid). *Adapted from D. Macdonald, 1984,* Encyclopedia of Mammals, *Facts on File.*

Figure 18.13 Convergence in gliding. The hairless, over-lapping scales at the base of the tail are apparent in this scaly-tailed flying squirrel. Note the similarity in patagium and body form between this African anomalurid and the North American flying squirrel in figure 18.8, a good example of phylogenetic convergence. *Adapted from D. Macdonald, 1984,* Encyclopedia of Mammals, *Facts on File.*

at the base of the forelimb, as well as a pair on the anterior part of the thorax. Gundis are herbivorous and are able to exist only on the water obtained in vegetation. They are diurnal and live in small colonies or family groups. Colonies vary in size depending on habitat and associated food availability. Nutt (2005) found most groups of the common gundi (*Ctenodactylus gundi*) in Tunisia were either multimale or multifemale, and she suggested cooperative breeding might occur. Runways and group defecation sites are often seen. They do not burrow but seek shelter among rocks during the hottest part of the day. When threatened or alarmed, gundis "play possum," a fear response in which they may remain motionless for up to 12 hours.

Muridae

This huge family has a very confusing taxonomic history, due in part to parallelism, convergence, and reversal throughout the evolutionary history of the group. Murids illustrate the dynamic nature of systematics as new, more sophisticated molecular techniques allow for more informed analyses of relationships. Musser and Carleton (2005) summarized the history of taxonomic research on this group. In the previous edition of this text, we considered the Family Muridae to encompass 1326 species in 17 subfamilies. Since then, numerous molecular studies suggested elevation of several of these subfamilies back to family status, including the Platacanthomyidae, Spalacidae, Calomyscidae, Nesomyidae, and Cricetidae,

which are discussed in the following sections. With the murids, these families now constitute the Superfamily Muroidea (table 18.1).

With the exception of Antarctica and some oceanic islands, murids are distributed worldwide, either naturally or through introduction, and have been introduced onto many formerly unoccupied islands. As might be expected of a family that includes over 13% of all living mammalian species, murids are found in a diverse array of habitats and show a wide range of locomotor adaptations. The usual dental formula is 1/1, 0/0, 0/0, 3/3 = 16. Molar structure within subfamilies and genera is quite variable. Molars may be rooted or rootless, cuspidate or prismatic (see figure 18.2). Tooth structure reflects the plant materials, invertebrates, or small vertebrates consumed. Likewise, few generalizations can be made concerning the diverse social and breeding behavior that occurs among murids.

Originally found throughout most of the Eastern Hemisphere, the two largest, most diverse genera (*Rattus*, with about 66 species, and *Mus*, with about 38 species) have been inadvertently introduced by humans throughout most of the world. These two genera include the ubiquitous Norway rat and the house mouse, two species commensal with humans almost everywhere. Most murids are nocturnal, but some genera, such as the striped grass mice (Genus *Lemniscomys*), are diurnal. Still others, such as the African grass rats (Genus *Arvicanthis*), water rats (Genus *Hydromys*), groove-toothed swamp rats (Genus *Pelomys*), and brush-furred mice (Genus *Lophuromys*), are active both day and night. Likewise, mean litter size in most species is usually 2–4, but it is 10–12 in the Natal multimammate mouse (*Mastomys natalensis*). Young are altricial except in the African spiny mice (Genus *Acomys*).

Cricetidae

With 140 genera and almost 700 species, the cricetids are almost as speciose as the murids. The Family Cricetidae encompasses the New World rats and mice. Again, the size and heterogeneity of this group make generalizations on form and function difficult. The largest genera include the well-known and ubiquitous voles (Genus *Microtus*), deer mice and white-footed mice (Genus *Peromyscus*), South American grass mice (Genus *Akodon*), and rice rats (Genus *Oryzomys*). Cricetids occur from the subarctic to the tropics. Their habitats, feeding habits, reproduction, and activity periods are as varied and diverse as the murids. Most species are terrestrial, but this group also includes the semiarboreal golden mouse (*Ochrotomys nuttalli*; figure 18.14A), the semiaquatic crab-eating rats (Genus *Ichthyomys*), the fossorial long-clawed mouse (*Notiomys edwardsii*), and the woodrats (Genus *Neotoma*), which occur in caves and rocky outcrops, scrublands, and forests (figure 18.14B). The typical body shape of voles and lemmings is short ears and tails; blunt, rounded snout; and fairly short legs. Again, most species are terrestrial,

Table 18.1 Families and subfamilies within the Superfamily Muroidea

Family/Subfamilies (number of genera, species)	Common Name(s)	Distribution
Calomyscidae (1, 8)	Mouselike hamsters	Middle East, Pakistan
Cricetidae (140, 697)		
Arvicolinae (28. 151)	Voles, lemmings, muskrats	Nearctic, Palaearctic, Northern Oriental
Cricetinae (7, 18)	Hamsters	Eastern Europe, Palaearctic, Oriental
Lophiomyinae (1, 1)	Crested rat	East Africa
Neotominae (16, 124)	Mice, woodrats, golden mouse	Nearctic
Sigmodontinae (84, 393)	New World rats and mice	Nearctic, Neotropical
Tylomyinae (4, 10)	Climbing rats	Neotropical
Nesomyidae (21, 61)		
Cricetomyinae (3, 8)	Pouched rats and mice	Africa south of Sahara
Delanymyinae (1, 1)	Delany's swamp mouse	Central Africa
Dendromurinae (6, 24)	African climbing mice, gerbil mice, fat mice, forest mice	Africa south of Sahara
Mystromyinae (1, 1)	White-tailed mouse	South Africa
Nesomyinae (9, 23)	Malagasy rats and mice	Madagascar
Petromyscinae (1, 4)	Rock mice, climbing swamp mouse	East central and Southwestern Africa
Platacanthomyidae (2, 2)	Spiny tree mouse, soft-furred tree mouse	Southwestern India, Southeast Asia
Muridae (150, 730)		
Deomyinae (4, 42)	Spiny mice, brush-furred rats	North Africa, Asia
Gerbillinae (16, 103)	Gerbils, sand rats	Ethiopian, Middle East, central Asia, India
Leimacomyinae (1, 1)	Buttner's forest mouse*	Togo (Africa)
Murinae (126, 561)	Old World rats and mice	Ethiopian, Palaearctic, Oriental, Australian, Nearctic
Otomyinae (3, 23)	Vlei rats, whistling rats	South and East Africa
Spalacidae (6, 36)		
Myospalacinae (2, 6)	Zokors	Northern China, Siberia
Rhizomyinae (2, 4)	Bamboo rats	Eastern India, China, Southeast Asia
Spalacinae (1, 13)	Blind mole rats	SE Europe, Middle East, NE Africa
Tachyoryctinae (1, 13)	Bamboo rats	East Africa

Source: *Data from Musser and Carleton (2005)*
* May be extinct.

although the heather voles (Genus *Phenacomys*) are arboreal, muskrats (Genera *Ondatra* and *Neofiber*) are aquatic, and the five species of mole voles (Genus *Ellobius*) are fossorial.

Platacanthomyidae

Formerly considered a subfamily within the murids, this family contains only two species; the spiny tree mouse (*Platacanthomys lasiurus*) of southwest India and the soft-furred tree mouse (*Typhlomys cinereus*) of China and Vietnam. Both species occur in forested habitats. Spiny tree mice have sharp spines throughout their dorsal fur with fewer, smaller spines on the ventral surface. The long tail ends in extended brushlike hairs. Populations may reach high enough densities to become pests. The pelage of *T. cinereus* is soft, dense, and without spines. This species occurs in high elevation deciduous forests.

Spalacidae

Another family formerly included among murids, the spalacids encompass 6 genera and 36 species of bamboo rats, zokors, and blind mole rats in 4 subfamilies (table 18.1). All are fossorial or semifossorial. The zokors, or

mole-rats (Genus *Myospalax*), are powerful diggers. They construct extensive burrow systems 2 m below the ground reaching up to 100 m long. Nonetheless, at night they may occasionally forage above ground. Blind mole rats (Genus *Spalax*) are mole-like in general body form except they do not have large, spatulate forefeet. Although they have small eyes, these are beneath the skin with no external openings. With little or no visual acuity, tactile sensitivity is enhanced by a line of short bristles on each side of the rounded snout. Blind mole rats inhabit different types of moist soils where they dig using their large incisors and powerful jaw muscles, as well as the snout. In the past, bamboo rats (Genus *Rhizomys*), the lesser bamboo rat (*Cannomys badius*), and the African mole-rats (Genus *Tachyoryctes*) have been grouped together in their own family, the Rhizomyidae.

Nesomyidae

This is a large family of 21 genera and 61 species found within six subfamilies (table 18.1). The pouched rats and mice (Genera *Beamys*, *Cricetomys*, and *Saccostomus*) within the Subfamily Cricetomyinae are named for their cheek pouches. They are found in various habitats where seeds, grains, and other materials are transported

A

B

Figure 18.14 North American cricetid rodents. (A) The golden mouse (*Ochrotomy nuttalli*) is a bright golden color and builds distinctive, softball-sized arboreal nests. (B) The eastern woodrat (*Neotoma floridana*) is a nocturnal species often found in rocky habitats. This species—also known as a packrat—is declining throughout much of the range.

back to burrows and nests in the pouches. Delany's swamp mouse (*Delanymys brooksi*) is a tiny climbing mouse found in swamps and marshes in higher elevation forests. The six genera within the Subfamily Dendromurinae include several species of climbing or tree mice (Genus *Dendromus*), with very long, semi-prehensile tails, and several species of fat mice (Genus *Steatomys*)—so named because they accumulate a great deal of fat that allows them to estivate during times with reduced food availability. The eight genera within the Subfamily Nesomyinae include ten species of tufted-tailed rats (Genus *Eliurus*) found in forested areas throughout Madagascar.

Calomyscidae

This family of a single genus (*Calomyscus*) includes eight species of mouse-like hamsters. They occupy a variety of habitats from forests to barren, rocky hills throughout the Middle East. In size, pelage, and general morphology, they are similar to Genus *Peromyscus*. Mouse-like hamsters may be active day or night depending on the season and feed primarily on seeds.

Dipodidae

This family includes 16 genera and 51 species of jerboas, birch mice, and jumping mice among 6 subfamilies (Holden and Musser 2005). Birch mice (Genus *Sicista*) and jumping mice (Genera *Zapus*, *Eozapus*, and *Napaeozapus*) formerly were included in their own family (Zapodidae). Dipodids exhibit a great deal of structural diversity based on the hind foot and associated life history, although in all species the tail exceeds head and body length. Birch mice have a simple, unmodified hind foot and are terrestrial or semiarboreal. These small

mice inhabit forests, meadows, and steppes from Europe east through central Asia. They have prehensile tails and are excellent climbers. Jumping mice have elongated hind feet used in ricochetal locomotion and a very long tail. The woodland jumping mouse (*Napaeozapus insignis*) can jump up to 2 m when alarmed. Pelage is brown on the dorsum, golden on the sides, and white on the venter. They occur in a variety of habitat types from Alaska throughout Canada and much of the United States. The Chinese jumping mouse (*Eozapus setchuanus*) is found in central China.

Jerboas are strongly bipedal and highly adapted for jumping, with the long, tufted tail used for balance. In most species, the three central metatarsals of the hind foot are fused, forming a cannon bone, and the first and fifth toes are lost. The strong hind legs are four times longer than the front legs. Hind foot length in these species is often half the head and body length. The largest, the great jerboa (*Allactaga major*), has a hind foot up to 98 mm long and is capable of leaping 3 m. The jerboas are distributed from the Sahara Desert east across southwestern and central Asia to the Gobi Desert. They are medium-sized rodents (figure 18.15), with head and body length ranging from about 35–260 mm. They have large eyes and ears and light, sandy-colored pelage; they inhabit arid, semidesert, and steppe regions, where they burrow in sandy or loamy soils. A vertical process on the jugal bone protects the eye when the head is used in digging, and a fold of skin closes the nostrils. Many of the same adaptations found in the Heteromyidae to conserve water also occur in jerboas. Dipodids are dormant or hibernate in burrows during the winter for prolonged periods, up to 9 months in some species. Likewise, jerboas may enter torpor in the summer during particularly hot, dry periods. Several species of jerboas and birch mice are endangered.

Figure 18.15 Convergence in desert rodents. Long hind feet and tail and reduced forelegs of the northern three-toed jerboa (*Dipus sagitta*) are similar to the pattern seen in desert rodents such as the kangaroo rats (figure 18.10) and the springhare (figure 18.12).

Figure 18.16 Fat dormouse. Also called the edible dormouse, *Glis glis* may weigh up to 200 g.

Gliridae

This group includes 9 genera and 28 species of dormice within three subfamilies. Many authorities consider the former family name—Myoxidae—as valid (see Holden 2005). The endangered desert dormouse (*Selevinia betpakdalaensis*), first described in 1939, is endemic to the small area north and west of Lake Balkhash in east Kazakhstan (central Asia). It burrows in sandy or clay soils and avoids desert heat by being nocturnal.

Dormice occur in Europe east to central and southern Asia, Africa, and southern Japan. They are small to medium in size, with head and body length 60–190 mm and tail length 40–165 mm. They are nocturnal, semiarboreal, and occupy forests, shrublands, residential areas, and rocky outcrops, where they forage for nuts, fruits, spiders, and insects. Juskaitis (2006) found that the fat dormouse (*Glis glis*) preyed on bird eggs, nestlings, and adult birds, as did the forest dormouse (*Dryomys nitedula*). They have a simple stomach without a cecum, suggesting a diet low in cellulose. Species may nest in tree cavities, attics, or burrows. Dormice put on weight in the fall and enter extended periods of hibernation (up to 7 months) or dormancy until spring. Fat dormice may not reproduce in years with poor mast crops and reduced resources (Fietz et al. 2005; Ruf et al. 2006). Many species are associated with residential areas, where they may become pests in gardens, orchards, or houses. The fat, or edible, dormouse (figure 18.16) in Europe is the largest member of the family and weighs up to 200 g. It is trapped for its fat and for use as food. The species has been considered a delicacy since ancient Roman times. The Chinese dormouse (*Chaetocauda sichuanensis*), Japanese dormouse (*Glirulus japonicus*), and Setzer's mouse-tailed dormouse (*Myomimus setzeri*) are endangered.

HYSTRICOGNATH RODENTS

Like sciurognaths, some uncertainty also surrounds the classification of hystricognath rodents. The mole-rats, discussed next, are a good example of the problems of placing families within this suborder. Although this group is clearly hystrico*gnathus* in its jaw structure, it has a small (i.e., nonhystrico*morphous*) infraorbital foramen. As noted previously, the anomalurids, pedetids, and ctenodactylids, which are considered "hystricomorphous sciurognaths," also provide examples of this difficulty in terms of clear subordinal associations.

Bathyergidae

Although the infraorbital foramen is small, with little passage of masseter muscle, jaws of bathyergids are hystricognathus with a strongly deflected angular process. This group also has an unusual crown pattern on the molars (see figure 18.2A). The 5 genera and 16 species of mole-rats are strictly fossorial. They occur in sandy soils in the hot, dry regions of Africa south of the Sahara Desert. All species burrow extensively, with tunnel length of the African mole-rat (*Cryptomys hottentotus*) reaching over 300 m. Tunnels are complex (Herbst and Bennett 2006), with numerous secondary branches and chambers for nesting, feeding, and defecation. Mole-rats feed primarily on underground bulbs and tubers of perennial plants. Several species of bathyergids are solitary—including the Namaqua dune mole-rat (*Bathyergus janetta*). Other species are found in pairs or small groups, as noted by Yeboah and Dakwa (2002) for the Togo mole-rat (*Cryptomys zechi*). Other species, however, form very large colonies of 80 or more individuals.

The naked mole-rat (*Heterocephalus glaber*) is of special interest. Whereas most bathyergids are fully furred, the naked mole-rat has bare, wrinkled skin with only a few tactile hairs present (figure 18.17). Along with its reduced metabolic rate, its lack of fur allows for easier dissipation of body heat while underground. Naked mole-rats have the most highly developed **eusocial** system known among mammals (see chapter 23), a colonial system similar to certain insects (Jarvis 1981; Honeycutt 1992). They cooperate in tunnel digging, predator defense, and reproduction. Individuals vary greatly in body size depending on their function in the colony—smaller workers versus larger nonworkers. The one reproductively active female (the "queen") is the largest individual and mates with one of very few reproductively active males in the colony. Other adults do not breed but take part in foraging, caring for young, and other cooperative functions. Nonbreeders are not sterile, however, and may disperse to establish new colonies or replace breeders that die. Individuals in a colony are very inbred (Reeve et al. 1990). Another eusocial bathyergid, the Damara mole-rat (*Cryptomys damarensis*), lives in colonies that average 16 individuals with a single breeding female. Because this species breeds with mole-rats from outside the colony, recognition is essential between colony and noncolony members. Jacobs and Kuiper (2000) found that recognition was by individually distinct cues rather than genetic relatedness. The same system seems to operate in naked mole-rats. Parag et al. (2006) related the degree of sociality and breeding patterns to baculum size and penile morphology in bathyergids. The genetic evolution of eusociality in mole-rats is a fascinating research area.

Hystricidae

The 3 genera and 11 species of Old World porcupines inhabit a variety of habitat types, including deserts, forests, and steppes. They occur throughout Africa, the Middle East, India, and Asia, including Indonesia, Borneo, and the Philippines. They are also found in Italy,

Figure 18.17 Naked mole-rats. Many morphological adaptations of these naked mole-rats are similar to those found in other fossorial species such as pocket gophers, but the wrinkled and nearly hairless skin is unique among terrestrial species.

where they probably were introduced thousands of years ago. Pelage is variable among the three genera. The long-tailed porcupine (*Trichys fasciculata*) has weak spines, or bristles, without quills and a long tail. Brush-tailed porcupines (Genus *Atherurus*) have short, soft spines on the head, legs, and ventral surface. Very long, flat, grooved spines are on the back. Crested porcupines (Genus *Hystrix*) have a short tail and long, hollow quills. These are grouped in clusters of five to six over the posterior two thirds of the body and produce a rattling sound as they move. Their head and shoulders are covered with long, stiff bristles. The tips of hystricid quills, unlike those of New World porcupines, do not have barbs. These species are herbivorous, terrestrial, and generally nocturnal. They are considered pests in portions of their range because of the damage they do to plantations, but they are also hunted or raised for food in many areas.

Erethizontidae

New World porcupines encompass 5 genera and about 16 species. The North American porcupine (*Erethizon dorsatum*) occurs throughout Canada and the northeastern and western United States, and the other species occur from southern Mexico south to northern Argentina. They are found in mixed coniferous forests, tropical forests, grasslands, and deserts. Despite their generally chunky, heavyset bodies, erethizontids are more arboreal than the Old World porcupines. The North American porcupine is both terrestrial and semiarboreal, whereas South American species (Genera *Coendou* and *Sphiggurus*) have prehensile tails (figure 18.18) and spend most of their time aboveground. Heavy spines with a barbed tip, embedded singly and not in clusters as they are in Old World porcupines, occur over much of the dorsum and sides of the body. The ventral surface has coarse, long hair and no spines. Contrary to popular belief, porcupines do not intentionally throw their quills at attackers. The quills are loosely embedded, however, especially in the tail, and can easily become detached in predators (Roze and Ilse 2003). Because of the barbed tip, the quill works continuously deeper into the wound and can even cause death. Nonetheless, porcupines have a variety of potential predators, with the fisher (*Martes pennanti*) the most adept at flipping porcupines over and attacking the unprotected ventral area. The pallid hairy dwarf porcupine (*Sphiggurus pallidus*), known only from two specimens from the West Indies, is considered to be extinct.

Petromuridae

The monotypic dassie rat (*Petromus typicus*) is restricted to southern Angola, Namibia, and northwest South Africa in rocky desert habitats. It is fairly small and squirrel-like, with soft, yellow-orange to brown pelage and no underfur. The skull is rather flat and the ribs flexible, which allows the dassie rat to squeeze through rocky crevices. Dassie rats are diurnal, feed on grasses and plant material, and are coprophagous. The crowns of the cheekteeth

A

B

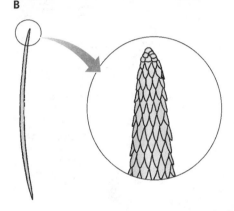

Figure 18.18 Porcupine quills. A. The familiar pelage of erethizontids with guard hairs modified as quills is seen in the Brazilian porcupine (*Coendou prehensilis*). Note the prehensile tail. (B) a diagrammatic enlargement of the overlapping barbs on the tip of a quill. *B. Adapted from G. A. Feldhamer, B. C. Thompson, and J. A. Chapman, 2003,* Wild Mammals of North America, *Johns Hopkins Univ. Press.*

are unusual; the labial side of the lower teeth and the lingual side of the upper teeth are raised above the rest of the occlusal surface. The family name Petromyidae also occurs in the literature for this family, which is most closely related to the Family Thryonomyidae.

Thryonomyidae

The two species of cane rats, or "grasscutters," occur throughout Africa south of the Sahara Desert. The greater cane rat (*Thryonomys swinderianus*) is a semiaquatic inhabitant of marshy areas, whereas the lesser cane rat (*T. gregorianus*) is found in drier upland areas. Both species have the typical chunky body shape of hystricomorphs, with the greater cane rat reaching a body mass of 10 kg. Their skulls are heavy, with a very large infraorbital foramen, and there are three grooves on the anterior surface of each upper incisor. Pelage is coarse and bristle-like. Both species may reach very high population densities and cause significant damage to cultivated crops and planta-

tions. van Zyl et al. (2005) described the gross anatomy and histology of the alimentary tract of the greater cane rat. Cane rats also are an important food source for native people; they are hunted or farmed, and large quantities of the meat are sold in markets. Van der Merwe and van Zyl (2001) examined growth rates of the greater cane rat on different experimental diets to determine when best to market them.

Chinchillidae

The three fairly diverse genera and six living species in this family are distributed in South America from the Andes Mountains of Peru south to southern Argentina. All have a large head with large eyes and ears, and very dense, soft pelage (figure 18.19A). The Argentine plains viscacha (*Lagostomus maximus*) lives in large colonies in scrub and grassland areas of Argentina, Paraguay, and Bolivia, where it creates large, extensive burrow systems. Size dimorphism is evident, with males twice as large as females. The three species of mountain viscachas (Genus *Lagidium*),

A

B

Figure 18.19 Representative chinchillids.
(A) Populations of chinchilla are greatly reduced in numbers throughout their native range. (B) A rabbit-like plains viscacha.

which look like rabbits (figure 18.19B) with long, strongly curved tails, form colonies in dry, rocky, rather barren areas at high elevation. The two species of chinchillas (Genus *Chinchilla*) also form colonies in these types of habitat. Populations of plains viscacha are greatly reduced because they are pests on crops and compete for range-land forage with domestic livestock. They also are taken for their fur and meat. Likewise, population densities of mountain viscacha have declined because they are hunted for their meat and fur. Chinchilla fur has been highly prized for centuries and is one of the most valuable of any mammal in the world. Both species are extirpated throughout much of their original ranges. The short-tailed chinchilla (*Chinchilla chinchilla*) is critically endangered; the long-tailed chinchilla (*C. lanigera*) is threatened. The Peruvian plains viscacha (*L. crassus*) is recently extinct.

Dinomyidae

The monotypic pacarana (*Dinomys branickii*) is a rare inhabitant of forested areas at higher elevations in the slopes and valleys from Colombia and Venezuela to Bolivia. Pacaranas look like huge guinea pigs (figure 18.20). The body is heavyset (about 15 kg) with coarse, brownish-black pelage and four lines of white spots on each side from the shoulders to the rump. Pacaranas are slow, lumbering, nocturnal herbivores, and their population levels apparently have never been high. Habitat destruction and consumption by humans have reduced the species even further, and they are endangered.

Caviidae

The guinea pigs, cavies, Patagonian "hares," and capybaras encompass 6 genera and about 18 species. They occur in the wild throughout South America, except for the Amazon Basin, in a variety of habitat types. Three subfamilies are currently recognized. The Subfamily Caviinae includes the guinea pigs (Genus *Cavia*) and

cavies (Genus *Galea*). The Subfamily Dolichotinae includes two species of Patagonian hares or maras (Genus *Dolichotis*), and the Subfamily Hydrochoerinae includes the capybaras and two species of cavies (Genus *Kerodon*; figure 18.21A). Capybaras were formerly in their own family (Hydrochoeridae) but are included within the caviids based on molecular data (Rowe and Honeycutt 2002). Members of the family share a short, robust body form with short limbs and ears and a vestigial tail. The two

A

B

Figure 18.21 Morphological differences in caviids. Morphological differences are evident when comparing (A) the rock cavy (*Kerodon rupestris*) and (B) the Patagonian "hare" or mara (*Dolichotis patagonum*).

Figure 18.20 A pacarana. This endangered rodent occurs in northern South America.

species of Patagonian hares are exceptions (figure 18.21B). They have longer ears and are adapted for cursorial locomotion by having long hind legs, a reduced clavicle, and reduced number of digits. The dentary bones of caviids have a prominent lateral groove. The upper cheekteeth converge anteriorly; thus, the toothrows form a V-shape. Caviids are probably the most abundant and widespread rodents in South America. Each genus is found in a different habitat, with only yellow-toothed cavies sympatric with other genera. The origin of the domestic guinea pig (*Cavia porcellus*) is uncertain, although it has been raised for consumption for thousands of years.

The subfamily name for capybaras (which means "water pig") suggests both their habitat affinity and general body form (figure 18.22). They are closely associated with water throughout their range, which extends from Panama south to Argentina east of the Andes Mountains. Capybaras are semiaquatic and have partially webbed feet. They swim and dive with agility and feed on aquatic plants in and around lakes, ponds, swamps, and other wet areas. They are heavyset with a large head and with small eyes and ears placed high on the head. Their pelage is short, coarse, and fairly sparse. Capybaras are the largest living rodents. Head and body length is about 1.3 m, and mean body mass is 28 kg, although individual body mass may reach 50 kg or more. The skull is distinctive with a very long paroccipital process, grooved dentary bones, and an upper third molar that is longer than the combined length of the other three cheekteeth. Capybaras are fairly gregarious, living in groups of 20 or more individuals. As a result, they often become serious agricultural pests. They are hunted either for food or to reduce crop damage. In Venezuela, capybaras are ranched for their meat and leather.

Dasyproctidae

This family is made up of about 11 species of agoutis (Genus *Dasyprocta*) and 2 species of acouchis (Genus *Myoprocta*). They occur from southern Mexico south to northern Argentina east of the Andes Mountains. Agoutis were introduced into Cuba and several other Caribbean islands. Dasyproctids are generally diurnal, solitary, burrowing herbivores and are important as predators and dispersers of the seeds of neotropical plants (Guimaraes et al. 2006). Likewise, as noted by Jorge and Peres (2005) population density and home range of red-rumped agoutis (*D. leporina*) were related to the spatial distribution of Brazil nut trees (*Bertholletia excelsa*). Both genera are similar in appearance with a coarse, glossy orange-brown to black pelage. The pelage is long, thick, and often of a contrasting color in the posterior area (the family name means "hairy-rumped"). Agoutis weigh up to 4 kg and inhabit forests, brushlands, and grasslands, always in association with water. Humans hunt them for food, and populations have declined in many areas. Acouchis are smaller, have a slightly longer tail, and are restricted to tropical rainforests of the Amazon Basin. Dasyproctids are sometimes considered to be in the same family as the pacas discussed in the following section. The Coiban agouti (*Dasyprocta coibae*) and the Ruatan Island agouti (*D. ruatanica*) are endangered.

Cuniculidae

The two species of pacas (Genus *Cuniculus*) occur from central Mexico south to Paraguay and Argentina in forested areas from lowlands to high elevations, usually near rivers or streams. Pacas have the typical robust hystricomorph form and weigh up to about 10 kg. Their coarse pelage is brown-black above with a paler venter and four rows of white spots on each side of the body. The skull is unique in that a portion of the zygomatic arch is enlarged and contains a large sinus (figure 18.23), possibly for amplification of vocalizations or tooth-grinding sounds. The dental formula is 1/1, 0/0, 1/1, 3/3 = 20. Pacas are nocturnal, terrestrial, burrowing herbivores that eat a variety of plants and fruits. They also may be serious

Figure 18.22 Capybaras. This South American species feeds on a variety of plant material, and in large groups can become a serious agricultural pest.

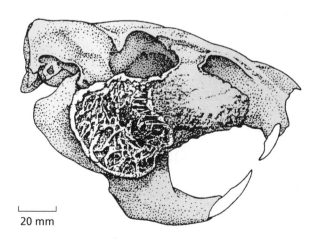

20 mm

Figure 18.23 Skull of the paca. The large, distinctive zygomatic arch of the lowland paca (*Cuniculus paca*) is evident. *Adapted from J. Eisenberg, 1989,* Mammals of the Neotropics, *Univ. Chicago Press.*

pests on crops, gardens, and plantations and sometimes are killed for this reason. More often, however, they are harvested for their excellent meat. Pacas are considered to be the best-tasting, most edible South American rodent. Hunting and habitat loss have greatly reduced population numbers in large parts of their range.

Ctenomyidae

This family is composed of about 60 species of tuco-tucos in the single Genus *Ctenomys*. They are distributed from central South America south to Tierra del Fuego and occur in habitat and soil types suitable for burrowing. These fossorial rodents are similar to North American pocket gophers in habits and appearance. Like geomyids, tuco-tucos have broad, thick incisors, small eyes, short, thick pelage, and enlarged claws. Unlike geomyids, however, they lack external cheek pouches. The family and genus name ("comb mouse") is derived from the stiff fringe of hairs around the soles of the hind feet and toes used to groom dirt from the fur. The burrow systems can be very extensive, and population densities may reach 200 per hectare. Burrow systems of male Mendoza tuco-tucos (*C. mendocinus*) are longer and more extensive than are those of females. As might be expected, Luna and Antinuchi (2006) found a direct relationship between the metabolic cost of digging and soil hardness in the Los Talas tuco-tuco (*C. talarum*). Tuco-tucos are quite vocal, and the "tloc-tloc" alarm call, which gave rise to the common name, can be heard when the animals are underground. Ctenomyids are diet generalists and consume a high proportion of the plant species available (Puig et al. 1999). Reingestion of fecal pellets in Pearson's tuco-tuco (*C. pearsoni*) occurs while the animals are resting or between feeding bouts (Altuna et al. 1998). Colonies may become serious pests in plantations because they destroy roots and girdle trees. Because their fossorial habits have resulted in relative isolation and disjunct distributions of species, *Ctenomys* is a diverse, rapidly evolving taxon (Castillo et al. 2005). Ctenomyids are closely related to the Family Octodontidae.

Octodontidae

The 8 genera and 13 species of octodontids occur in the Andean region of Peru, Bolivia, Chile, and Argentina. They occur in several habitat types from sea level to high elevation. Octodontids are small to medium-sized with thick, silky pelage, and they often have tufted tail tips. Unlike many other rodent families, the occlusal surfaces of the molars usually are simple, with a single lingual and labial fold forming a figure-8 pattern in most species (figure 18.24). This pattern gives rise to the family name. Habitats, life history, and feeding habits vary among species. The mountain degu (*Octodontomys gliroides*) and viscacha rat (*Octomys mimax*) primarily are surface-dwelling generalists. Degus (Genus *Octodon*), rock rats (Genus *Aconaemys*), and the red viscacha-rat

(*Tympanoctomys barrerae*) are semifossorial. The degu (*O. degus*) nests communally, and although adult females can recognize their own pups, they do not discriminate between related and unrelated pups during nursing (Ebensperger et al. 2006). The coruro (*Spalacopus cyanus*) is completely subterranean. This social octodontid digs tunnel systems up to 600 m long in which it stores large amounts of food (Begall and Gallardo 2000). Coruros maintain nomadic colonies of up to 26 individuals that move to a new area after depleting food resources at a site. Octodontids are of special interest because they exhibit the greatest span of chromosomal number ($2N = 38$–102) of any mammalian family. This range is because the red viscacha-rat is the only known mammalian tetraploid, with a chromosome complement of $4N = 102$. As a result, sperm in this species have a very large head with close to twice the nuclear DNA content of other mammalian species (Gallardo et al. 1999; Kohler et al. 2000). Mares and coworkers (2000) described two new genera and species of very rare ochtodontids in the salt flats in Argentina—the Chalchalero viscacha-rat (*Salinoctomys loschalchalerosorum*) and the golden viscacha-rat (*Pipanacoctomys aureus*).

Abrocomidae

Sometimes considered as a subfamily of the Families Octodontidae or Echimyidae, the chinchilla rats have the same geographic distribution as the octodontids. There are two genera and nine extant species of abrocomids. Chinchilla rats burrow among crevices in remote mountainous thickets and rocky areas. They are ratlike in appearance, with a pointed snout and long tail. They have long, soft, thick pelage but not of the same quality as true chinchilla fur. With 17 pairs of ribs, Bennett's chinchilla rat (*Abrocoma bennettii*) has more ribs than any other rodent.

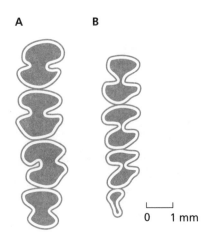

Figure 18.24 Molariform tooth shape. Left upper cheekteeth in (A) the coruro and (B) the rock rat (*Octomys mimax*) showing characteristic figure-8 shape in octodontids.

Echimyidae

This is a large family of about 21 genera and 90 extant species of spiny rats grouped into 4 subfamilies. They occur from Honduras south to central South America. Almost all species have spiny, sharp, stiff, bristle-like pelage. They appear ratlike with a pointed rostrum. Habitat preferences vary but are closely tied to availability of water. Some genera are semifossorial burrowers, others terrestrial, and some arboreal. Cordeiro et al. (2006) found Moojen's Atlantic spiny-rat (*Trinomys moojeni*) in Brazil bred throughout the year. Machado et al. (2005) determined that spiny-rats (Genus *Proechimys*) have karyotypes ranging from $2N = 14–62$. Several species of spiny rats in the Genera *Boromys*, *Brotomys*, and *Heteropsomys* from Cuba, Puerto Rico, and the Dominican Republic have been described from remains found with human artifacts and are very recently extinct (Guarch-Delmonte 1984).

Capromyidae

Eight genera and about 20 surviving species of hutias in 4 subfamilies exist in remote forested or rocky valleys and small islands in the West Indies. Hutias are chunky with short legs, a large head, and small eyes and ears. They range in head and body length from 200 to over 650 mm and in tail length from vestigial to long and prehensile. The coarse pelage varies from yellowish gray to black on the dorsum and is paler on the venter. Desmarest's hutia (*Capromys pilorides*) is an arboreal folivore, whereas others (Genus *Geocapromys*) feed on terrestrial plants. Hutias also consume small lizards.

Several genera and species of capromyids have become extinct during the last few hundred years in Cuba, Haiti, Puerto Rico, the Virgin Islands, and the Dominican Republic. Most extant species are considered threatened or endangered, not only because of habitat loss and over-harvesting for food but also from introduction of the mongoose (*Herpestes javanicus*) and feral house cats (*Felis catus*). In addition, the closely related giant hutias (Family Heptaxodontidae), which were also distributed throughout the Caribbean islands, have gone extinct in recent times, many since humans arrived on the islands.

Myocastoridae

The monotypic coypu, or nutria (*Myocastor coypus*), is often considered a member of either the Capromyidae or Echimyidae, but it is included in its own family based on morphological and serological differences. Coypus are native to South America from Chile and Argentina north to Bolivia and Brazil. The species is large (up to 10 kg body mass) and heavy-bodied, with rather coarse pelage and a long, round, sparsely furred tail (figure 18.25). Their webbed hind feet adapt them to aquatic habitats in and around fresh or brackish marshes or slow-moving streams. They make floating platforms that they use as feeding stations from which they consume aquatic vegetation.

Figure 18.25 Nutria morphology. Long, somewhat coarse pelage, webbed hind feet, and round tail are evident on this adult nutria.

Coypus have been widely introduced into Europe and North America. In the United States, most were introduced in the early to mid-1900s for fur farming. They escaped, or were released, and primarily occur in the Gulf Coast states and along the West Coast to Washington and the East Coast to Maryland. Populations also are established in several parts of southern Canada. Coypus fur is of major economic importance in South America, and in the United States, millions of pelts were harvested annually (Bounds et al. 2003) until the decline of the fur industry.

Lagomorpha

Lagomorphs (meaning "hare-shaped") include 11 genera and about 54 species of rabbits and hares in the Family Leporidae and 1 genus (*Ochotona*) and about 30 species of pikas in the Family Ochotonidae. Lagomorphs occur worldwide except for the southern portions of South America, Australia, and New Zealand, and islands such as Madagascar, the Philippines, and those in the Caribbean. A few species of leporids have been introduced into most of these places. The introduction of the European wild rabbit (*Oryctolagus cuniculus*) into Australia and New Zealand is a prime example of the potential pitfalls of introducing an **exotic** (a species outside its native range; see the section in chapter 27 on biological control).

MORPHOLOGY

All lagomorphs are small to medium-sized terrestrial herbivores. Coprophagy is common in lagomorphs, as it is in many rodents (see figure 7.11). Leporids reingest both soft and hard pellets, which enhances their ability to live on relatively low-quality vegetation (Hirakawa 2001). The diagnostic feature of the order is the occurrence of peg teeth, a second pair of small incisors without a cutting edge, immediately behind the larger, rodent-like first incisors (figure 18.26). A third pair of lateral incisors is lost before birth or immediately thereafter. The cutting edge of the primary incisors is notched in pikas but not in rabbits and hares (figure 18.27). The dental formula is 2/1, 0/0, 3/2, 2–3/3 = 26–28. The cheekteeth are hypsodont with two transverse enamel ridges, whereas rodents usually have several transverse ridges. The cheekteeth and incisors are open-rooted and evergrowing. In leporids, but not ochotonids, the rostral portion of the maxilla is **fenestrated** (having small, lattice-like perforations in the bone—see figure 18.26), and the frontal bone has a supraorbital process. Rabbits and hares generally have large ears and elongated hind limbs to accommodate their saltatorial locomotion. Rabbits have a well-known "cotton-ball" tail, but the tail in hares is longer. Pikas are rodent-like in appearance and have short limbs, small ears, and no tail. Although agile, they do not have the running ability of leporids. Unlike rodent feet, the feet of lagomorphs are fully furred. A cloaca is present, the uterus is duplex, and there is no baculum.

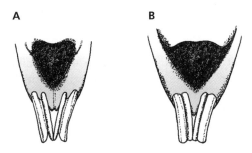

Figure 18.27 Lagomorph upper incisors. (A) Anterior view of the upper incisors of an ochotonid showing the characteristic notch on the cutting surface. (B) The upper incisors of a leporid do not have a notch. *Adapted from A. F. DeBlase and R. E. Martin, 1981,* A Manual of Mammalogy, *2nd ed., Wm. C. Brown Publishers.*

FOSSIL HISTORY

The earliest known lagomorphs, from the Family Eurymylidae, date from the late Paleocene of Mongolia and include stem taxa such as the Genera *Mimolagus* and *Mimotona* (Meng and Wyss 2005). Both Old and New World fossil leporids are known from the late Eocene, whereas fossil ochotonids are known from the mid-Oligocene in Asia. Both families were much more diverse in the Tertiary period than they are today, with 21 fossil genera of leporids and 23 fossil genera of ochotonids recognized (Carroll 1988). As noted, the phylogenetic relationship between lagomorphs and rodents continues to be debated among morphologists, paleontologists, and molecular biologists.

ECONOMICS AND CONSERVATION

Pikas are of limited economic importance, with a few Asian species considered to be agricultural pests. Leporids are often important as game species. Most species of cottontail rabbits (Genus *Sylvilagus*) are hunted; the eastern cottontail rabbit (*S. floridanus*) ranks first among game mammals in terms of the numbers taken and hours spent hunting them in North America. The snowshoe hare (*Lepus americanus*) historically has been an important furbearer, and several other species of hares are hunted for food and sport (Murray 2003).

Local populations of leporids may increase to such densities that they become serious crop depredators. Lethal methods such as poisoning and shooting, along with nonlethal methods such as fencing and repellents, are used in attempts to control local populations. In the western United States, ranchers and farmers occasionally have "rabbit drives" in which large numbers of people work together to round up and kill thousands of black-tailed jack rabbits (*L. californicus*) in an effort to reduce their density and resultant crop damage.

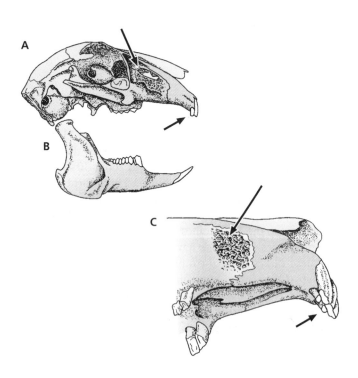

Figure 18.26 Peg teeth. Lateral view of the (A) cranium and (B) mandible of a leporid, the arctic hare. Although superficially rodent-like, note the peg teeth (*arrows*) and fenestrated rostrum, especially evident in the enlarged view (C). *Adapted from A. F. DeBlase and R. E. Martin, 1981,* A Manual of Mammalogy, *2nd ed., Wm. C. Brown Publishers.*

FAMILIES

Ochotonidae

Pikas commonly inhabit either steppe or forest areas, or steep, rocky (talus) slopes in alpine areas, where they live under and among the boulders. Distinct differences mark species within each habitat type. Pikas in talus areas do not burrow, are relatively asocial, and have low population densities and fecundity rates. Conversely, those in steppe or forest habitats dig burrows, are more social, and have higher population densities and fecundity. In both habitats, pikas are extremely vocal. Calls and "songs" are used to maintain distinct territories and social organization. These are enhanced with scent marking from apocrine glands on the cheeks. Given the acoustic characteristics of their alarm calls to warn of predators, Hayes and Huntly (2005) found the American pika (*O. princeps*) was less active during inclement weather with windy conditions that masked the calls. Pikas are well known for their "hay-making" activity (figure 18.28). An individual or breeding pair cuts and gathers vegetation throughout the summer and fall, cures it in piles in the sun and stores it in a traditional place within the territory. This cached hay is then used for winter food. These caches can weigh up to 5 kg and may directly affect the dynamics of local plant communities and sympatric species of herbivores (Smith et al. 1990). As in the American pika, Gliwicz et al. (2006) found that the northern pika (*O. hyperborean*) gathered and stored plants selectively rather than in proportion to their availability. The silver pika (*O. argentata*) and Kozlov's pika (*O. koslowi*)—both restricted to regions of China—are endangered. Based on severe declines in density and distribution, Wei-Dong and Smith (2005) recommended the Ili pika (*O. iliensis*) be listed as endangered as well. The Sardinian pika (*Prolagus sardus*) is recently extinct.

Leporidae

This family includes the rabbits and hares, which, although morphologically similar, differ in several ways. Rabbits usually build fairly well-constructed, fur-lined nests and give birth to altricial young. Hares do not construct nests but instead make shallow depressions on the ground called "forms" (figure 18.29), and they have precocial young. Neonatal hares are fully furred, have their eyes open, and are able to run a few hours after birth. Rabbits also have an interparietal bone in the skull, but hares do not. The karyotype of most rabbits is $2N = 42$; most hares have $2N = 48$. Based on a large molecular, cytogenetic, and morphological database, Robinson and Matthee (2005) strongly supported the monophyly of the leporids. Using mitochondrial DNA markers, Wu et al. (2005) described North American, Eurasian, and African species groups for the Genus *Lepus*.

Leporids occur in habitats from the snow and cold of the Arctic to deserts, grasslands, mountain areas, swamps,

Figure 18.28 North American pika. *Ochotona princeps* in typical rocky habitat, carrying cut vegetation.

and tropical forests. One of the adaptations of species in temperate or arctic areas, such as the arctic hare (*Lepus timidus*) or snowshoe hare (*L. americanus*), is pelage dimorphism. Coats are white in the winter and brown in the summer (see figure 8.9). Blending in with the habitat is beneficial because leporids are potential prey to a wide variety of predators, including lynx (Genus *Lynx*), coyotes (*Canis latrans*), red fox (*Vulpes vulpes*), marten (*Martes americana*), and fisher.

Like rodents, leporids have a high reproductive potential. They may have several litters per breeding season, with several individuals per litter. Postpartum estrus occurs, so that litters are produced in fairly rapid succession. Leporids also exhibit induced ovulation, further enhancing the probability that reproductively mature females will conceive. Similar to certain species of rodents that have 3–4 year cycles in population density,

Figure 18.29 Black-tailed jackrabbit form. This form is nothing more than a shallow, cleared depression—in this case, under sagebrush (*Artemisia tridentata*).

snowshoe hare populations exhibit 8–11 year cycles (see chapter 25). Density may change during this period by two orders of magnitude. Population increases result from increased birth rates and survival of young; the opposite occurs during population declines. The effect of food, predators, and other factors on causation of cycles has been investigated for decades (Krebs 1996).

In contrast to the high population densities of some leporids, six species of hares currently are considered to be endangered. These include the riverine rabbit (*Bunolagus monticularis*) in South Africa, the Tehuantepec jackrabbit (*Lepus flavigularis*) and volcano rabbit (*Romerolagus diazi*) in Mexico, the hispid hare (*Caprolagus hispidus*) in the foothills of the Himalayan Mountains, the Amami rabbit (*Pentalagus furnessi*) in the Ryukyu Islands of Japan, and the Sumatran striped rabbit (*Nesolagus netscheri*). Several species of cottontails also are endangered.

SUMMARY

Rodents represent about 42% of the living species of mammals in the world today. As might be expected of the largest mammalian order, the structural and functional characteristics of their locomotion and morphology vary greatly. The order is also highly adaptable in the variety of habitats occupied in an almost worldwide distribution. Despite their overall diversity, all rodents have a single pair of upper and lower chisel-shaped incisors, a diastema, and reduced numbers of molariform teeth. Convergence and parallelism of behavior and associated morphology is a key theme among rodent taxa. For example, behavior, structural similarity, and kangaroo-like bipedal locomotion are evident among some of the heteromyids, pedetids, dipodids, and certain murids in open, arid habitats. Likewise, the anomalurids and certain gliding sciurids show convergent adaptations, as do the fossorial geomyids, bathyergids, ctenomyids, and some murids.

Rodents influence our daily lives today as they have throughout history. The classification of rodents offers a continuing challenge and opportunity for worthwhile investigation, and the 32 rodent families discussed in this chapter may well be revised in the near future as additional phylogenetic evidence is gathered.

Although structurally and functionally similar in many respects to rodents, the lagomorphs constitute a much smaller and more restricted order in terms of habitats, morphology, and locomotion. As with rodents, a single key characteristic may be used to define the order. Both families of lagomorphs have peg teeth. These are a small pair of incisors posterior to the large, upper incisors. Several species of leporids are significant game animals, but the economic importance and effect on humans of the two families of lagomorphs are minor compared with those of rodents.

SUGGESTED READINGS

Anderson, P. K. 1989. Dispersal in rodents: A resident fitness hypothesis. Spec. Publ. No. 9, American Society of Mammalogists, Provo.

Barrett, G. W., and G. A. Feldhamer (eds.). 2007. Ecology, behavior, and conservation of the golden mouse: A model species for research. Springer Publishing, New York.

Genoways, H. H., and J. H. Brown. 1993. Biology of the Heteromyidae. Spec. Publ. No. 10, American Society of Mammalogists, Provo.

Kirkland, G. L., and J. N. Layne (eds.). 1989. Advances in the study of *Peromyscus* (Rodentia). Texas Tech. Univ. Press, Lubbock.

Tamarin, R. H. (ed.). 1985. Biology of the New World *Microtus*. Spec. Publ. No. 8, American Society of Mammalogists, Provo.

DISCUSSION QUESTIONS

1. Discuss the concept of kin selection and various factors that could contribute to eusocialty and reproductive "altruism" in species such as the naked mole-rat, including fossorial living (with associated relative protection from predators) or patchy distribution of resources.

2. How might breeding be suppressed in the majority of members in a colony of naked mole-rats, such that only the queen and one male breed?

3. Discuss the relationship between rabbits and hares in terms of the complexity of nest construction and whether neonates are altricial or precocial.

4. Convergence in rodents in terms of bipedal locomotion, fossorial adaptations, and gliding was noted in the summary. List two or three rodent taxa that show similar convergence in terms of adaptations for an aquatic niche.

5. What is the significance of coprophagy in mammals? Besides lagomorphs, what other mammalian orders exhibit coprophagy? Why are no carnivores coprophagous?

CHAPTER 19

Proboscidea, Hyracoidea, and Sirenia

The rationale for including in a single chapter elephants, the largest living terrestrial mammals; the rabbit-sized hyraxes; and the dugong and manatees, which never leave the water, may appear obscure. These three orders are often grouped together as "subungulates" based on their perceived evolutionary relationships (see Gheerbrant et al. 2005 for an excellent summary). Molecular studies have strengthened support for the phylogenetic affinity of these orders, often grouped in the clade "Afrotheria" (de Jong 1998; Madsen et al. 2001; Murphy et al. 2001). Along with the true ungulate Orders Perissodactyla and Artiodactyla (chapter 20), the orders in this chapter are derived from the primitive **Condylarthra**, a generalized ancestral order of land mammals that arose in the early Paleocene about 65 million years ago (mya). One of the recognizable groups within the Condylarthra was the Superorder **Paenungulata** ("near-ungulates"). By the early Eocene of Africa, about 54 mya, the Paenungulata had given rise to the Orders Proboscidea, Sirenia, and Hyracoidea and numerous extinct taxa, although classification of many of these fossil groups remains uncertain (Gheerbrant et al. 2005). Although the groups began to diverge in the Eocene and appear very different today, they share certain anatomical characteristics. None has a clavicle, the digits have short nails (no nails in the Amazonian manatee, *Trichechus inunguis*), and there are four toes on the forefeet (five in Asian elephants). Females have two pectoral mammae between the forelegs (hyraxes have two inguinal pairs as well) and a bicornuate uterus. Males have abdominal testes with no external scrotum and no baculum. All are nonruminating, herbivorous, hindgut fermenters. The symbiotic **microfauna** (ciliated protozoans and bacteria) that break down vegetation occur in an enlarged cecum. The ceca in hyraxes are particularly complex. Finally, the dugong, manatees, and elephants are unusual among mammals in their pattern of molariform tooth replacement, which is horizontal, not vertical as in other mammals. Proboscideans and sirenians flourished during the Oligocene and Miocene, but with only three extant families among them, they are today mere remnants of what were once very diverse and abundant groups. Because populations are declining throughout most of their range, the future of elephants, manatees, and the dugong is of concern to conservationists.

Proboscidea

Elephants today are represented by a single family, the Elephantidae, with two or three extant species. The Asian elephant (*Elephas maximus*) occurs south of the Himalayan Mountains in India, Sri Lanka, Indochina, and Indonesia and has been introduced to Borneo. There are three subspecies of Asian elephants: *Elephas maximus maximus* from Sri Lanka, *E. m. sumatrensis* from Sumatra, and *E. m. indicus* from mainland Asia (Shoshani 2005a). As in many other groups, the question about the validity of subspecies classification in elephants remains. Using DNA extracted from the dung of 118 Asian elephants from Sri Lanka, Bhutan, India, Laos, and Vietnam, Fernando et al. (2000) found very low genetic diversity. The only significant differentiation occurred between mainland populations and those in Sri Lanka. The African elephant (*Loxodonta africana*) is distributed throughout Africa south of the Sahara Desert. Likewise, Vidya et al. (2005) found almost no mtDNA variation in Asian elephants throughout southern India, suggesting past gene flow was extensive. Traditionally, two subspecies of African elephants have been recognized: the bush elephant (*Loxodonta africana africana*) in eastern, central, and southern Africa and the forest elephant (*L. a. cyclotis*) in central and western Africa. Because of sequence variation in four nuclear genes as well as differences in morphology and habitat, and because of limited gene flow, Roca et al. (2001) suggested that these two taxa be recognized as separate species. Likewise, Grubb et al. (2000) supported specific status for the forest elephant as *L. cyclotis* based on several morphological differences. Conversely, Debruyne (2003) argued to retain the subspecific status of the forest elephant. For ease of discussion, we will consider African elephants as a single species, acknowledging that this taxonomic question is unresolved.

African and Asian elephants exhibit several morphological differences, some of which are familiar to many people. African elephants are larger than Asian elephants and have much larger ears and tusks; the back is concave with the shoulder higher than the head, and the tip of the trunk has two lips (figure 19.1). The species have different occlusal surfaces on the cheekteeth (see figure 19.1) and different numbers of nails on the hind feet: four in Asian and three in African elephants. There are 19 pairs of ribs in Genus *Elephas*, 20 in Genus *Loxodonta*. Chromosome complement, however, is 2N = 56 in both species, with a high level of similarity in the structure of chromosomes (Houck et al. 2001).

Each species occurs in a variety of habitats, from grasslands and shrublands to forests. Regardless of habitat type, all elephants are closely tied to the availability of water. They are nonruminant herbivores; microbial action takes place in the cecum. During the wet season, herbaceous vegetation and grasses are eaten. Shrubs, leaves, and tree bark are taken in the dry season, with flowers and fruits eaten when available. Codron et al. (2006) found seasonal

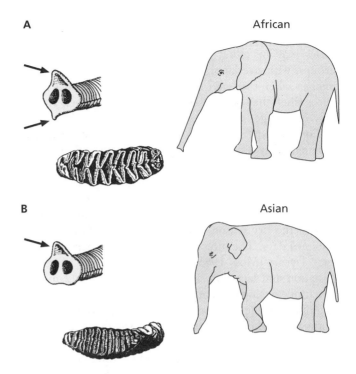

Figure 19.1 Characteristics of elephants. (A) An African elephant and (B) an Asian elephant. Note the different postures, with the shoulder above the head and a concave back in the African elephant. The Asian elephant has a larger, more bulbous skull, which is above the shoulder, and a convex back. The tip of the trunk has two lips (arrows) in African elephants and one lip in Asian elephants. The pattern of the laminar ridges on the occlusal surface of cheekteeth differs in African and Asian elephants. See text for other morphological differences. *Adapted from S. K. Eltringham, 1982,* Elephants, *Blandford Press.*

and spatial differences in diets of African elephants in Kruger National Park, South Africa. In agricultural regions, elephants can do extensive damage to cultivated crops. Large adults may consume up to 150 kg of vegetation a day, but because the digestive process is relatively inefficient compared with that of ruminants (see figure 7.9), half of this may pass through the gut undigested.

For the following reasons, large home ranges are needed to sustain groups of elephants: the amount of food necessary to maintain an individual elephant; the often dry, inhospitable habitat conditions; and the fact that aggregations of up to 50 individuals of mixed sex and various ages may form during portions of the year. As a result, home range size may be 1,600 km² during seasons when resources are scarce. Elephants may move up to 30 km a day to reach better habitat conditions. Extent and distance moved depend on the quantity and quality of food and water available. Elephant herds are capable of quickly degrading habitats and affecting the availability of resources to other herbivores.

Elephants are gregarious, with herds normally made up of family units of about 10 individuals (Laws 1974). Females and their young, led by a matriarch, remain in these units (Vidya and Sukumar 2005). Different units made up of related individuals often join temporarily to form "bond groups." Depending on the time of year, males may be either solitary or in temporary all-male groups, with overlapping home ranges. Adult males join females when the latter enter estrus. Juvenile males leave herds when they become sexually mature, between 10 and 17 years of age.

For both species, sexual maturity generally is reached in females between 9 and 12 years old, with peak fecundity from 25–45 years of age. Few, if any, females 50 years and older breed. Reproduction is tied to the wet season and the availability of food and water. Females are in estrus during the later part of the rainy season and first part of the dry season. Estrus is extremely brief, lasting only 2–4 days. The interval between estrous periods averages 4 years (Moss 1983) because of the long length of gestation and lactation. Following an average 22-month gestation, a single young is born (rarely twins) at the beginning of the wet season, when habitat conditions are optimal. Newborn African elephants weigh about 120 kg; newborn Asian elephants, 100 kg. They nurse for 3–4 years (with the mouth, not the trunk) and weigh 1,000 kg by 6 years of age. Although growth rate decreases by age 15, elephants grow throughout their lives. Males exhibit a reproductive period of 2–3 months each year called **musth**, during which hormone levels, sexual activity, and aggressive behavior increase. Greenwood et al. (2005) found that the ratio of two different molecular forms of the pheromone frontalin in Asian elephants elicited different responses in male and female conspecifics. Although males are physiologically capable of breeding by 10–15 years of age, most successful matings are by mature males 30–50 years old.

Elephants make a variety of vocalizations audible to humans, including trumpeting, growling, roaring, and snorting. Langbauer (2000) discussed 31 different types of calls in African elephants. Vocalizations vary depending on the size and composition of the group and the reproductive status of individuals (Payne et al. 2003). Elephants also communicate with one another through high-amplitude, extremely low-frequency sound, as low as 14–35 Hz (cycles per second) (Payne et al. 1986; Langbauer et al. 1991), well below the range of human hearing. Low-frequency sound allows groups or individuals up to 4 km apart to communicate and coordinate movements and helps males find females during their brief and unpredictable estrous periods. Low-frequency sound is used to communicate because it can travel over long distances and through obscuring vegetation much better than higher frequency sound. Also, a large animal like the elephant physically would have a difficult time producing high-frequency sound. Wood et al. (2005) focused on the low-frequency "rumble" call. They described two types of rumble associated with group feeding and a third type related to socializing and agitation. In female Asian elephants studied by Soltis et al. (2005), rumble vocalizations were exchanged in alternating sequences between individuals. Low-frequency

sounds also have a seismic component that may complement communication between individuals in a herd (O'Connell-Rodwell et al. 2006).

MORPHOLOGY

Elephants are the largest living terrestrial mammals. Aspects of anatomy, movement, dentition, and behavior relate to their size and associated long life span. The shoulder height of adult African bull elephants may reach 4 m, with maximum body mass up to 7,500 kg. Sexual dimorphism is evident, with body mass smaller in females. Because they grow throughout life, the oldest elephant in a group often is the largest. Asian elephants are smaller than African elephants. Mean body mass of male *Elephas* is about 4,500 kg; African forest elephants are smaller than bush elephants. The structure necessary to support such large mass gives rise to several modifications resulting in graviportal locomotion and a massive skeleton that makes up 15% of an individual's body mass, about twice that of most terrestrial mammals. The head is very large, in part to help support the trunk and ever-growing tusks. The bones surrounding the brain are thick but made less heavy by a series of air-filled pneumatic cavities, or sinuses (figure 19.2). The feet are broad, with the phalanges embedded in a matrix of elastic tissue to help cushion the weight (figure 19.3).

Throughout their evolutionary history, elephants may have benefited from increased body size in avoiding competition with the large herbivorous African perissodactyls that preceded them as well as the artiodactyls that arose later. Large body size confers other benefits in addition to reduced competitive pressure. It allows elephants to move greater distances in response to habitat conditions with relatively less energy expenditure. It also reduces predation pressure; only humans threaten adult elephants.

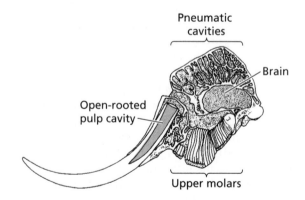

Figure 19.2 Elephant skull. Sagittal cross section of the skull of an African elephant, showing the extensive network of pneumatic cavities that help reduce the weight. Note the open-rooted pulp cavity of the tusk. The tusks grow throughout the life of the animal. *Adapted from D. Macdonald, 1984, Encyclopedia of Mammals, Facts on File.*

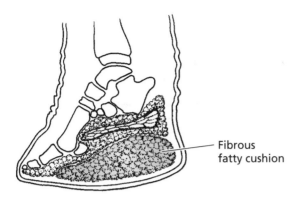

Figure 19.3 Forefoot of an African elephant. Although elephants are functionally plantigrade, they actually are digitigrade—walking on their toes. The bones of the feet, and the massive weight of the animal, are cushioned by fibrous, fatty connective tissue that uniformly distributes the weight over a broader area. *Adapted from B. Grzimek, 1990,* Animal Encyclopedia, *vol. 4, McGraw-Hill.*

Fibrous fatty cushion

Figure 19.4 African elephant. The large ears of African elephants not only function in auditory acuity, but provide a large, highly vascularized surface area from which to dissipate heat.

Along with benefits, large size also has its drawbacks. Elephants live in warm climates but have very few sweat or sebaceous glands. Heat dissipation without dehydration is a problem, especially in areas with limited water availability. Thus, elephants have very sparse body hair, a characteristic common to other large mammals in warm climates. Also, the wrinkled skin of elephants acts to hold water and facilitate its movement on the body surface, increasing the evaporative cooling effect (Lillywhite and Stein 1987). The large ears, especially of African elephants (figure 19.4), are highly vascularized, and when moved back and forth, they act as radiators to dissipate heat. Behavioral characteristics such as seeking shade and reducing activity during the middle of the day also help reduce heat load.

The trunk is the most recognizable feature of elephants and gives rise to the ordinal name. It is an elongated, flexible, muscular upper lip and nose, with the nasal canal throughout its length. Trunks were well developed in six fetuses of African elephants that were only 97–117 days old (Gaeth et al. 1999). The trunk has numerous uses. Because an elephant's head is so heavy, its neck is very short; thus, the animal cannot touch the ground with its mouth. The trunk is used to grasp food from the ground as well as from tall trees. It is extremely strong, yet the tip is sensitive enough to pick up small objects (such as peanuts in a zoo). Elephants drink by sucking up to 4 liters of water into their trunk at a time and squirting it into the mouth. Adults may drink 100 liters of water a day. Water is also sprayed from the trunk over the body to keep the animal cool. Mud and dust also can be sucked into the trunk and sprayed on the body for this purpose, as well as to reduce insect infestations. Also, elephants have an excellent sense of smell, and the trunk may be held upright in the air like a periscope to gain olfactory information from the surroundings.

The dental formula is 1/0, 0/0, 3/3, 3/3 = 26, and the teeth are highly specialized. The remaining upper incisor in extant elephants is the second (I2) and forms the characteristic tusk. The deciduous I2 is replaced by the permanent tooth between 6 and 12 months of age, when it is only about 5 cm long. Tusks (often collected by poachers as "ivory") are composed of dentine and calcium salts, with enamel only on the terminal portion. They are open-rooted (see figure 19.2) and grow throughout life. Tusks are largest in African bull elephants, reaching up to 3.5 m in length and 200 kg in weight in very old adults; 100–120 kg is more common. In female Asian elephants, the tusks are smaller and may not extend beyond the lower lip. Tusks are used in foraging, defense, and social displays.

The cheekteeth are large and hypsodont, with transverse **laminae** (ridges). These ridges are composed of dentine overlaid with enamel. Cementum occurs between the ridges. Posterior cheekteeth (equivalent to molars) are larger than those anterior (equivalent to premolars) and have a greater number of ridges. Asian elephants have more laminae than African elephants. Because the mandible is short and the cheekteeth long, only one upper and one lower molar (or parts of two) are active in each jaw at a time. Replacement of cheekteeth is horizontal from the back of the jaw—referred to as mesial drift. The new tooth moves forward as the worn anterior tooth is pushed out. The third and final molars begin to come in by about 30 years of age and last for the remainder of an individual's life (figure 19.5).

FOSSIL HISTORY

Although only two or three species of elephants survive today, this order was diverse and widespread throughout most of the Cenozoic era. Elephants occurred not only in Africa and Asia as today, but throughout the Pleistocene

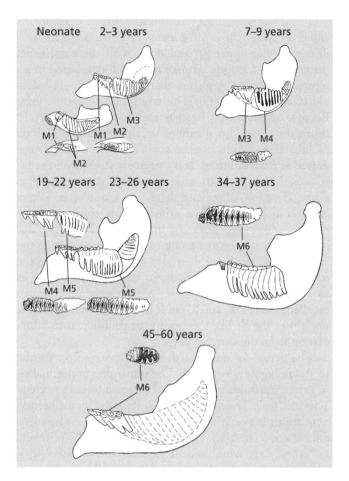

Figure 19.5 **Proboscidean cheekteeth.** Unlike the cheekteeth in most other species of mammals, those in elephants enter the jaw horizontally (mesial drift). Because all the molariform teeth are deciduous in elephants, M1–M3 are equivalent to premolars, and M4–M6 are equivalent to molars in other mammals. Progression of teeth is shown for the African elephant, with ages approximated, especially after 30 years of age. Note the larger size and greater number of laminar ridges in the posterior cheekteeth, M4–M6. *Data from J. Kingdon, 1979,* East African Mammals, *vol. IIIB* Large Mammals, *Univ. Chicago Press.*

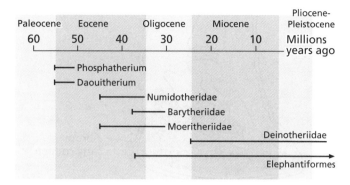

Figure 19.6 **Numerous lineages of proboscideans are recognized.** The only extant family, the Elephantidae, is included here within the Elephantiformes, along with the extinct Families Stegodontidae, Gomphotheriidae, Amebelodontidae, Mammutidae, and the early Genera Phiomia and Palaeomastodon. *Modified and adapted from Gheerbrant et al. (2005).*

in Europe and North America and even reached South America. The earliest fossil is *Phosphatherium escuilliei* from the early Eocene of Morocco (Gheerbrant et al. 2005). Extensive fossil remains exist of proboscideans throughout the Eocene, with several families recognized (figure 19.6). Two of these extinct families, the Moeritheriidae and the Deinotheriidae, diverged early. The moeritheriids (figure 19.7A), known from the Eocene and Oligocene of northern Africa, were only about a meter in height and probably amphibious. Deinotheriids occurred in Asia and Europe from the late Miocene to Pliocene. Referred to as "hoe-tuskers," they were elephant-size with large, downward-curving tusks in the lower jaw (see figure 19.7B). Three other extinct families (included as Elephantiformes in figure 19.6) have much closer affinities to today's elephants. The Family Gomphotheriidae, which was contemporary with moeritheriids and deinotheriids; the Family Mammutidae

(mastodons) from the early Miocene; and the Family Stegodontidae from the mid-Miocene all had the large body size and many of the characteristic specializations of elephants today. Gomphotheriids had a pair of tusks in both the upper and lower jaws. Mastodons probably survived until about 8,000 years ago. The North American mastodon (*Mammut americanum*) was contemporary with the arrival of humans on the continent. Skeletal material of mastodons was first collected in North America from the Hudson River in 1705 (Sikes 1971). Although members of the Family Stegodontidae survived until the late Pleistocene, only one proboscidean family remains extant.

The family that survives today, the Elephantidae, is recognizable from the late Miocene Genus *Stegotetrabelodon*. The extinct Genus *Primelephas*, from the late Miocene to early Pliocene, gave rise to the two genera still extant today as well as the extinct Genus *Mammuthus*, the mammoths that were contemporary with early humans. The woolly mammoth (*M. primigenius*; see figure 19.7D) was among the subjects of cave paintings by Paleolithic humans (figure 19.8). This species went extinct less than 4,000 years ago (Lister and Bahn 2000). Complete specimens of woolly mammoths have been found frozen in Siberian ice.

ECONOMICS AND CONSERVATION

The involvement of elephants in human culture, religious tradition, and history is extensive. Humans hunted mammoths as long as 70,000 years ago (Owen-Smith 1988). We are all aware of the central importance of elephants in circuses and as zoo animals since ancient times and of their role in the famous march of Hannibal over the Alps in 218 B.C.E. Asian elephants have been important as draft animals throughout many parts of Asia for over 5,000 years, although they can be considered "exploited captives" as opposed to domesticated species

A

B

C

D

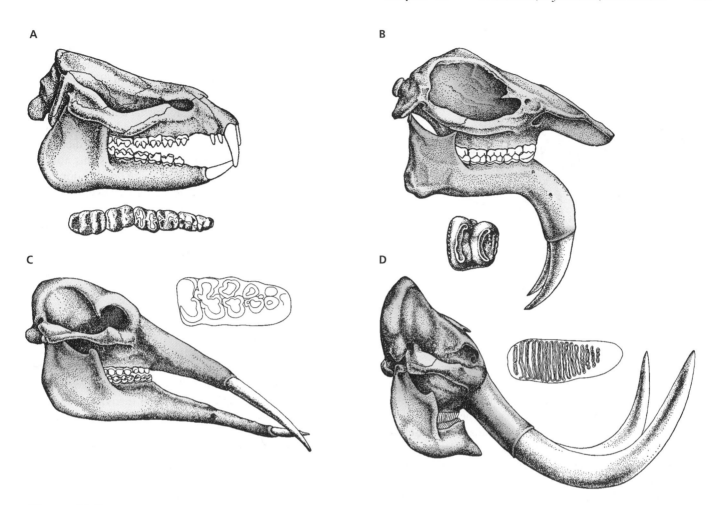

Figure 19.7 **Fossil skulls of early proboscideans.** (A) The late Eocene Genus *Moeritherium*, actual length about 33 cm, and (B) the Miocene Genus *Deinotherium*, actual length about 1.2 m. (C) Genus *Gomphotherium*, actual length about 1 m, occurred in the Miocene. (D) The actual length of the woolly mammoth skull is about 2.7 m. Note the changes in the molar cusp patterns of the early proboscideans, with the occlusal surface of mammoths being very similar to modern elephants. *Adapted from R. L. Carroll, 1988,* Vertebrate Paleontology and Evolution, *W. H. Freeman.*

(see chapter 28). Currently, the range of both species is much reduced, and the Asian elephant is endangered. The African elephant is considered either threatened or endangered.

Both African and Asian elephants remain important in terms of conservation efforts throughout their range. Population declines have been attributed to drought, loss of habitat associated with increased human population growth and desertification, and poaching for meat and ivory. When prices for ivory increased significantly during the 1970s and 1980s to over $100/kg (Douglas-Hamilton 1987), poaching escalated. At the same time, several African countries must cull elephants from areas where their population densities are too high relative to limited forage resources. The animals may destroy crops, increase soil erosion, and depress the resource base for other species (Owen-Smith 1988). Countries that cull elephants may depend on the legal harvest and sale of ivory to fund their wildlife management programs. Problems arise because legally harvested ivory cannot be distinguished from poached ivory. The argument has been made that legal trade promotes more illegal poaching by providing a

market, although Stiles (2004) believed this was not the case. Recent developments in molecular genetics may allow investigators to pinpoint the source population of ivory, as attempted by Wasser et al. (2004), and help alleviate this problem in the future. If two species of African elephants are recognized—the bush elephant (*Loxodonta africana*) and the forest elephant (*L. cyclotis*)—current management and conservation efforts would be affected accordingly.

Hyracoidea

Referred to in the Bible as "rock badgers," the hyraxes are composed of a single family, Procaviidae, with three genera and four species (Shoshani 2005b). The family name means "before the caviids (guinea pigs)" and points up the taxonomic confusion that has surrounded hyraxes. Previous authorities have recognized up to 12 species of hyraxes. Because of their superficial resemblance to

Figure 19.8 Paleolithic sketch. This drawing of a woolly mammoth is from the cave wall of Les Combarelles aux Eyzies, France. Not actual size.

rodents (figure 19.9), hyraxes were initially grouped with guinea pigs by taxonomists. Even the common name "hyrax" is unfortunate because it means "shrew mouse." Neither of these associations is accurate because, based on fossil and molecular evidence, hyracoids are most closely related to the other "subungulates," elephants and manatees. Nonetheless, the interordinal relationships of hyraxes "... are among the most debated questions of current mammalian phylogenetics" (Gheerbrant et al. 2005:93).

Hyraxes are distributed in central and southern Africa, Algeria, Libya, Egypt, and parts of the Middle East, including Israel, Syria, and southern Saudi Arabia. The rock hyrax (*Procavia capensis*) is the most widely distributed geographically and elevationally. They are found in rocky outcrops from sea level to 4,200 m elevation in Africa and the Middle East. The yellow-spotted rock

hyrax (*Heterohyrax brucei*) is found in similar rocky habitats in northeast to southern Africa. The two species of arboreal tree hyraxes (Genus *Dendrohyrax*) inhabit forested areas of Africa up to 3,600 m elevation. The terrestrial species are diurnal or crepuscular and form large colonies. Conversely, tree hyraxes are nocturnal and solitary.

All hyraxes are herbivorous and feed on a variety of vegetation. Grasses make up a large part of the diet of the rock hyrax; the hypsodont dentition grinds this abrasive material. The other hyraxes have more brachyodont dentition because they consume less abrasive vegetation. The southern tree hyrax (*D. arboreus*) in Parc National des Volcans, Rwanda, for example, feeds primarily on mature leaves of *Hagenia abyssinica* (Milner and Harris 1999). Although they do not ruminate, hyraxes have a unique digestive system involving one large cecum and a pair of ceca on the ascending colon (figure 19.10).

Colony size varies according to species, with the rock hyrax maintaining group sizes up to about 25 and the yellow-spotted hyrax up to 35. Rock hyraxes form social hierarchies with cooperative breeding. Unlike most species in which males have higher levels of testosterone than females, Koren et al. (2006) found adult females had testosterone levels equal to or greater than males. Hyraxes are poor thermoregulators. Individuals in a colony may huddle together to help conserve heat and maintain body temperature. Rock hyraxes bask in the sun during winter to help warm up, but during the summer they seek cool refuges in the rocks to escape lethal high temperatures and to reduce water loss (Brown and Downs 2005, 2006). As might be expected in colonial species, hyraxes are very

Figure 19.9 Hyrax characteristics. Hyraxes superficially resemble rodents but are not closely related to them.

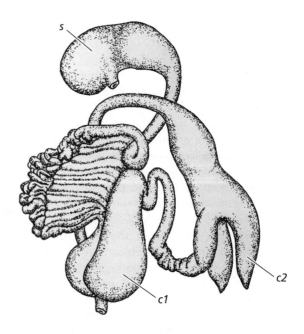

Figure 19.10 Unusual alimentary tract. The alimentary tract of a hyrax includes the stomach (*s*), a large cecum (*c1*) at the beginning of the large intestine, and a second, paired cecum (*c2*) at the end of the large intestine. *Adapted from B. Grzimek, 1990,* Animal Encyclopedia, *vol. 4, McGraw-Hill.*

vocal and make a variety of different types of sounds, including whistles, screams, croaks, and chatter.

Reproductive maturity in both sexes occurs at about 16 months of age. Females come into estrus once a year. Gestation is a relatively long 8 months, with litter size ranging from one to four. Neonates are precocial, with most births occurring during the wet season. Young males disperse from their natal area between the ages of 16 and 30 months and try to establish their own breeding territories.

MORPHOLOGY

Hyraxes are short, with compact bodies and very short tails. Pelage color is brown, gray, or brownish yellow. Total length varies from 32–60 cm and body mass from 1–5 kg; *P. capensis* is larger than *H. brucei*. There is no sexual dimorphism. Hyraxes from warm, arid regions have shorter, less dense pelage than tree hyraxes and those from high-elevation alpine areas. Hyraxes have a prominent middorsal gland surrounded by lighter colored hair (figure 19.11). The gland varies in size among species, being most noticeable in the western tree hyrax (*Dendrohyrax dorsalis*) and least so in the rock hyrax.

Because hyraxes inhabit rocky cliffs or move through trees, good traction for climbing and jumping has obvious adaptive value. This is achieved through specialized elastic pads on the soles of the feet. These pads are kept moist by secretory glands that make the feet similar to suction cups. The toes have short, hooflike nails except for the second digit on the hind feet, which has a claw used for grooming (figure 19.12).

Unlike elephants and manatees, the molariform dentition in hyraxes is not replaced horizontally. The dental formula for the permanent dentition is 1/2, 0/0, 4/4, 3/3 = 34. Deciduous canines may be retained in rare cases, but

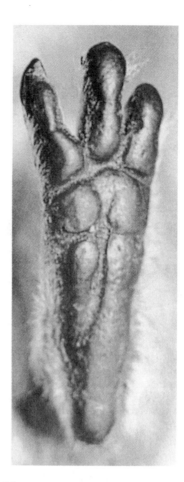

Figure 19.12 Hyracoid digits. The first and fifth toes are absent on the hind feet of a hyrax. The toes of the hyrax have hooflike nails, except for the second digit on the hind feet (pes). These have a claw that is used for grooming the fur. The soles have glandular pads that make them moist and increase adhesion to steep, rocky inclines.

there is usually a diastema between the incisors and cheekteeth (figure 19.13A). The upper incisors are long, pointed, and triangular in cross section and have a gap between them. They are evergrowing and stay sharp because the posterior sides do not have enamel. Gheerbrant et al. (2005; their table 7.4) discuss a number of other primitive as well as derived features of hyracoids.

FOSSIL HISTORY

Hyracoids first appear in the upper Eocene of North Africa and are very abundant and diversified at fossil sites in Morocco. They were the most common herbivore in early Tertiary African communities. One Eocene hyracoid—*Microhyrax lavocati*—described from fragmentary remains from Algeria (Tabuce et al. 2000, 2001) was very small, only about 3 kg. Several genera, some as large as modern tapirs, are placed in the extinct, paraphyletic Family Pliohyracidae, including *Megalohyrax eocaenus* (see figure 19.13B; Thewissen and Simons 2001). One of the largest fossil hyraxes is *Titanohyrax mongereaui*, which

Figure 19.11 Middorsal gland. A rock hyrax in typical habitat showing the prominent white fur surrounding the middorsal gland.

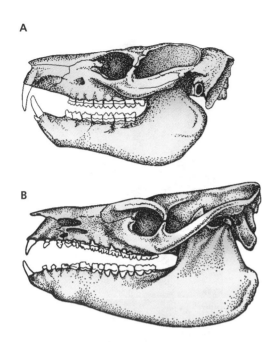

Figure 19.13 Hyracoid skulls. (A) The skull of a modern tree hyrax, with a well-defined diastema. Actual length about 12 cm. This can be contrasted with (B) an early Oligocene hyracoid Genus *Megalohyrax*. Actual length about 30 cm. *Adapted from A. S. Romer, 1966,* Vertebrate Paleontology, *3rd ed., Univ. Chicago Press.*

weighed about 800 kg. The earliest, most primitive fossil hyracoid is *Seggeurius amourensis* from Algeria. Several species within the Genus *Thyrohyrax* have been described from late Eocene and early Oligocene deposits in Egypt (DeBlieux et al. 2006). The first specimens attributable to modern procaviids are from the Miocene (Carroll 1988). By this time, only smaller forms of hyraxes survived. Larger species may have died out because they could not compete with ungulates.

ECONOMICS AND CONSERVATION

Tree hyraxes are hunted for both their meat and fur (from which blankets may be made). All tree hyraxes are affected by the loss of forest habitat, although none of the species of hyraxes are currently threatened or endangered. Because hyraxes in a colony defecate and urinate in a traditional place, massive caked deposits form on the rocks. This material has been used by natives as a decorative dye and by Europeans as a fixative in perfumes (Kowalski 1976).

Sirenia

This order is represented by two extant families. The Family Dugongidae has one extant species, the dugong (*Dugong dugon*), and the Family Trichechidae has three species of manatees in the Genus *Trichechus*. Like whales, sirenians never leave the water. Unlike whales, however, they are strictly herbivorous and represent the only mammalian marine herbivores. They inhabit coastal areas, estuaries, bays, and inland river systems in tropical regions, feeding on submerged and emergent vegetation. Dugongs feed on softer, less abrasive vegetation than manatees, however. Sirenian distribution is restricted to (1) relatively shallow coastal areas because the plants they depend on require sunlight and (2) tropical and subtropical regions with water temperatures = 20°C because of their low metabolic rates and poor thermoregulatory abilities. In contrast to terrestrial herbivores, sirenians have limited competition from other mammals within their feeding niche of shallow-water vegetation. The four extant species represent vestiges of an order that was abundant and diverse during the Tertiary period, with fossil remains of close to 20 genera known. The ordinal name is derived from the sirens, or sea nymphs, of mythology, and dugongs and manatees may have been the basis for the myth of mermaids.

MORPHOLOGY

Sirenians exhibit many of the same adaptations for life in the water as do cetaceans. They are large with a fusiform body shape (figure 19.14) and are devoid of fur except for very short, stiff bristles around the snout. In the West Indian manatee (*Trichechus manatus*), these modified vibrissae, or perioral bristles, are used for tactile exploration (Reep et al. 2001) and "in a prehensile manner in conjunction with elaborated facial musculature to bring plants into the mouth" (Marshall et al. 2000:649). There is no external ear, the nostrils are valvular and located on the top of the rostrum, the lips and snout are very flexible, and the tail is horizontally flattened. The forelimbs are paddle-like, and no external hind limbs are present. Only small, paired vestigial bones, suspended in muscle, represent the remains of the pelvis. The rostrum and lower jaw are deflected downward, especially in the dugong, to facilitate bottom feeding. The skeletal bones are very dense and massive (**pachyostotic**), an adaptation to increase body mass and overcome the buoyant effects of living in shallow-water habitats. The lungs are long and thin, extending for much of the length of the body cavity (figure 19.15). This helps to evenly distribute the buoyant effects of the air as an animal breathes. Unlike whales, sirenians have no dorsal fin.

FOSSIL HISTORY

The first fossils recognizable as sirenians, the Genus *Prorastomus*, are from the early Eocene, by which time aquatic adaptations were well underway. Nonetheless, the Eocene fossil sirenian *Pezosiren portelli*, although it spent most of its time in the water, had four well-developed legs

Figure 19.14 **Sirenian sizes.** The relative sizes of a (A) recently extinct Steller's sea cow, (B) manatee, and (C) dugong are shown in relation to a 6-foot-tall (2 m) person. Sirenians have a fusiform body shape, forelimbs modified into flippers, and no hind limbs. The most noticeable external differences between manatees and the dugong are in the shape of the tail and the extreme downward deflection of the rostrum in dugongs.

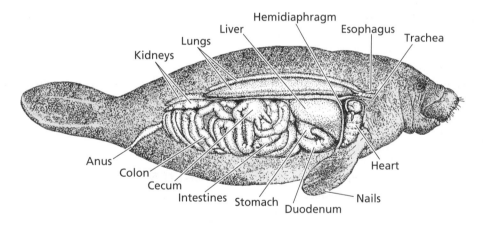

Figure 19.15 **Internal anatomy of a manatee.** Note the lungs lying in a horizontal position along the back. *Adapted from T. J. O'Shea, 1994, Manatees, Scientific American 271: 66–72.*

and was capable of terrestrial locomotion (figure 19.16). Middle and late Eocene remains of the Genus *Protosiren* occur in India, Europe, and North America. The oldest and smallest fossil from the genus—*P. eothene*—was described by Zalmout et al. (2003) from mid-Eocene deposits in Pakistan. Eocene fossil remains of sirenians are noteworthy in the retention of a fifth premolar. This is unique among Tertiary eutherian mammals (Carroll 1988), as sirenians were the last eutherians to retain five premolars (Domning et al. 1982; Gheerbrant et al. 2005). Better fossil evidence exists for the dugongids, including

Crenatosiren olseni from the late Oligocene of South Carolina (Domning 1997), than for the trichechids. Although dugongids are represented as early as the mid-Eocene by the Genus *Eotheroides*, there is no direct fossil evidence for the Genus *Dugong*. The early Miocene Genus *Dusisiren* (figure 19.17) is the earliest known ancestor of the recently extinct sea cow, Genus *Hydrodamalis* (see the section later in this chapter on Dugongidae). Fossil remains of trichechids date from the middle to late Miocene Genus *Potamosiren*. As mentioned, Tertiary sirenians were abundant, diverse, and widespread and reached

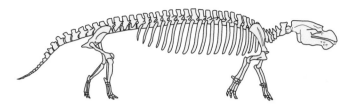

Figure 19.16 **Skeleton of *Pezosiren portelli*.** Unlike modern sirenians, this sirenian from the middle Eocene was capable of terrestrial locomotion. *Used with permission of Johns Hopkins Univ. Press.*

their peak diversity in the Miocene. However, as noted by Domning (1978), this diversity was somewhat confined by the lack of diversity in available marine feeding niches and of effective geographic barriers.

ECONOMICS AND CONSERVATION

There is reason for concern today about the immediate future of dugongs and manatees. Historically, hunting for meat, bones, hide, and fat has caused severe population reductions. Such reductions are not easily overcome by the very slow reproductive rate of this group. All four species are considered threatened, and in the United States, they are protected by the Marine Mammal and the Endangered Species acts. In other portions of their range, however, subsistence hunting continues. Hines et al. (2005) found the greatest threat to dugongs was mortality caused by incidental catch in fishing nets. In Florida, the West Indian manatee has been protected by the state since 1893. But animals are still lost because of poaching, being pinned by flood control gates and drowning, and being struck accidentally by boat propellers (figure 19.18). Manatees may be unable to detect and avoid oncoming boats because of poor hearing sensitivity at low frequencies (Gerstein et al. 1999). Loss of habitat to development presents a continuing threat to populations, as does destruction of submerged vegetation and reduced water quality. Commercial fishing, seismic surveys, and drilling for gas and oil also negatively affect populations. Because

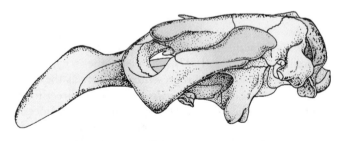

Figure 19.17 **Sea cow ancestor.** Lateral view of the cranium of *Dusisiren*, an early Miocene ancestor of the recently extinct sea cow. Note the strongly deflected rostrum, characteristic of dugongs today. *Adapted from R. L. Carroll, 1988, Vertebrate Paleontology and Evolution, W. H. Freeman.*

Figure 19.18 **Boat propeller wounds.** Contact with boat propellers is one of the primary mortality factors affecting manatee populations in Florida. Most manatees show evidence of scarring from propellers, with wounds such as these most often being fatal.

of their feeding habits, Amazonian manatees in South America have been used successfully to clear aquatic vegetation from inland canals (Allsopp 1960; although see Etheridge et al. 1985).

FAMILIES

Dugongidae

Dugongs are found in coastal areas of the Pacific Ocean throughout Micronesia, New Guinea and northern Australia, the Philippines, and Indonesia northward to Vietnam. In the Indian Ocean, they occur around Sri Lanka and India, and from the Red Sea south along the east coast of Africa to Mozambique. They attain a maximum body length of 4 m. Average mass is 420 kg, with a maximum of about 900 kg. Unlike manatees, in dugongs the tail fluke is notched (see figure 19.14), and the rostrum is much more strongly deflected downward (figure 19.19).

Although technically the adult dental formula is 2/3, 0/1, 3/3, 3/3, the anterior pair of upper incisors, all the lower incisors, and the canine are represented only by vestigial (**remnant**) alveoli. Thus, the dentition actually seen is 1/0,

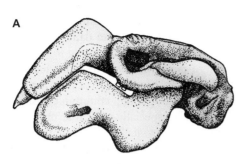

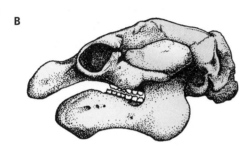

Figure 19.19 Sirenian skulls. Skull of a (A) dugong, showing the reduced dentition and strongly deflected rostrum, actually a greatly enlarged premaxilla bone. Note also that the anterior portion of the lower jaw (mandibular symphysis) is strongly deflected downward. Both are adaptations for bottom-feeding. (B) A West African manatee skull shows the greater number of cheekteeth than in dugongs. Teeth are replaced from the back of the jaws as they slowly move forward and as anterior teeth are worn out and lost. *Adapted from A. F. DeBlase and R. E. Martin, 1981,* A Manual of Mammalogy, *2nd ed., Wm. C. Brown Publishers.*

0/0, 3/3, 3/3. The upper incisor in males forms a short, thick tusk, but in females it does not erupt. The molars are cylindrical and cement-covered and move in horizontally from the back of the jaw as anterior teeth are worn away. In old adults, only one or two remain. These are open-rooted and grow throughout the life of the individual. Relative to their body size, the occlusal surface area of the cheekteeth of dugongs is very small compared to that of other herbivores. Lanyon and Sanson (2006:150) suggested that "…there may not have been great selection pressure acting on maintenance of an effective dentition in the dugong." In this regard, the dugong appears to be intermediate between manatees, with numerous complex and durable cheekteeth, and the closely related, recently extinct, Steller's sea cow (*Hydrodamalis gigas*), which was edentate.

Dugongs avoid freshwater much more than manatees. They generally are solitary or live in small groups. They forage by using their forelimbs to "walk" along the bottom, feeding in the substrate and pulling up plants with the deflected rostrum. Using time-depth recorders, Chilvers et al. (2004) monitored diving behavior of 15 dugongs at 3 sites in northern Australia. Mean duration of dives was <3 minutes, with mean maximum depth <5 m. There was significant variation in all dive parameters among individual dugongs. Unlike manatees, dugongs are restricted to the less abrasive sea grasses in the Families Hydrocharitaceae and Potamogetonaceae along coastal habitats, which have less silica content than true grasses (Family Gramineae). Loss of plant biomass from dugong grazing activity in sea-grass beds can exceed 50% of plant productivity (Masini et al. 2001). Sexual maturity is attained by both sexes between 7 and 14 years of age. A single young is born following a 12-month gestation.

The recently extinct Steller's sea cow was much larger than the dugong or manatees (see figure 19.14). Total length was about 7.5 m, greatest circumference was over 6 m, and body mass was about 5,000 kg, which was five to six times greater than sirenians today. This great size was an adaptation for the cold North Pacific waters they inhabited around the Commander Islands and Bering Island, where the species was first discovered in 1741. By then, its range had been greatly reduced, with a remnant population of 1,000–2,000. Sea cows had no teeth, but instead used rough plates in the mouth to forage on kelp. Slow moving in shallow water with few natural predators, they were easy prey for sailors seeking fresh meat and hides. Relentless slaughter followed their discovery, and Steller's sea cow was extinct by about 1768—only 27 years after its discovery.

Trichechidae

The West Indian manatee occurs from Florida south through the Caribbean Sea to northeast Brazil. Individuals have been reported as far north as Rhode Island (Odell 2003). The Amazonian manatee is found throughout the Amazon River Basin of South America, but it does not tolerate saltwater. The West African manatee (*Trichechus senegalensis*) is distributed in fresh- or saltwater rivers and estuaries, and along the west coast of Africa from Senegal to Angola. Cool winters, deep water, and strong currents are significant ecological barriers to dispersal (Domning 2005). Average body length is from 2.5–4 m, with body mass typically between 150 and 360 kg. Maximum body mass may approach 1,600 kg.

Unlike dugongs, manatees have a rounded, spatulate tail (see figure 19.14). Manatees have small nasal bones, unlike dugongs. They are also unusual among mammals in having six cervical vertebrae instead of seven. Adult dentition includes only cheekteeth; as in elephants, these are replaced consecutively from the rear of the jaw (mesial drift) as anterior molars are worn down and lost. Molars are brachyodont, bunodont, and close-rooted, and have enamel. Unlike elephants, manatees have four or five teeth in place at a time on each side of the jaws. Again in contrast to elephants, an indefinite number of molars, between 10 and 30, resulting from indefinite tooth germ formation, may move through each jaw quadrant during an individual manatee's lifetime. Teeth move forward as the bony interalveolar septa between them are constantly resorbed and redeposited. This "functional polyphyodonty" is an adaptation to the aquatic grasses and other abrasive, coarse, submerged vascular plants consumed by manatees, often in sand and mud, that quickly wear down teeth.

Manatees are usually seen as solitary individuals, paired, or in small groups. Hartman (1979) and Odell (2003) discussed their social ecology and behavior. In Florida, groups of several hundred may form in the winter as the animals congregate around warm-water discharge sites, but the primary social unit is a cow and her calf. Wind, water depth, abundance of food, water quality, and surrounding habitats are all factors that influence spatial distribution of West Indian manatees (Axis-Arroyo et al. 1998; Jimenez 2005). Seasonal movements and migratory behavior of manatees were described by Deutsch et al. (2003). The average one-way seasonal migration between northern warm season ranges and southern winter ranges was 280 km. Manatees tended to migrate when water temperatures fell below 20°C. Mean daily travel distance was only about 2.5 km, with males tending to move more than females. There is a great deal of individual variation in vocalizations of manatees. They vocalize for communication rather than navigation; communication between a cow and her calf is especially critical at night or in turbid waters (Sousa-Lima et al. 2002). Manatees are sexually mature between 3 and 4 years of age. A single calf is born after a gestation of about 13 months. Calves are weaned at 12–18 months of age. Thus, the reproductive potential of populations is limited (Koelsch 2001), and losses to boats and other causes may not be easily overcome.

SUMMARY

These three seemingly disparate orders arose from a common terrestrial ancestor and began to diverge by the early Eocene. Yet extant species continue to share several skeletal and anatomical features. They have a large cecum where vegetation is broken down by symbiotic microorganisms. Elephants, dugongs and manatees, and hyraxes all are hindgut fermenters. In this regard, their reduced digestive efficiency is similar to that of the perissodactyls. Similarities of anatomy, locomotion, dentition, and behavior in elephants and sirenians relate to size and their long life span. For example, replacement patterns (mesial drift) of the cheekteeth in the elephants and the manatees are similar. Teeth are replaced horizontally from the back of the jaw, moving forward slowly as older, anterior teeth are worn out and lost. This "functional polyphyodonty" is an adaptation for herbivory in these large, long-lived animals. The two groups differ in that elephants have only one or two molariform teeth that are functional at any one time. Manatees have several functional cheekteeth active at a time and produce an indefinite number of cheekteeth throughout their life. Dugongs have a much reduced number of functional cheekteeth. Elephants are terrestrial and have a graviportal structure to accommodate their bulk. Sirenians have the buoyancy of their water environment to help accommodate their large size. Proboscideans and sirenians are also similar in that they were much more diverse and widespread throughout the Tertiary than they are today, with numerous fossil genera recognized—as were the hyracoids. Today, the survival of elephants and sirenians is a concern as populations continue to decline in most areas. This is especially noteworthy for sirenians, as there are few other competitors for their feeding niche of shallow-water vegetation. Hyracoids are much smaller than either elephants or sirenians, and population densities appear to be secure.

SUGGESTED READINGS

Eltringham, S. K. 1991. The illustrated encyclopedia of elephants: From their origins and evolution to their ceremonial and working relationship with man. Salamandar Books, London.

Haynes, G. 1993. Mammoths, mastodonts, and elephants: Biology, behavior, and the fossil record. Cambridge Univ. Press, New York.

Langbauer, W. R., Jr. 2000. Elephant communication. Zoo Biology 19:425–445.

Reep, R. L., and R. K. Bonde. 2006. The Florida manatee: Biology and conservation. Univ. Press of Florida, Gainesville.

Spinage, C. 1994. Elephants. Poyser Natural History, London.

DISCUSSION QUESTIONS

1. Why do elephants use very low-frequency sound, whereas bats use very high-frequency sound? How might body size be a factor?

2. If today's environmental ethic and concern for endangered species were in vogue 250 years ago, do you think the Steller's sea cow would have gone extinct? Should we be concerned today for species with only remnant populations that may be "on their way out" anyway from an evolutionary standpoint?

3. What factors contribute to the low reproductive rate of both proboscideans and sirenians? How does this affect conservation efforts?

4. How does body size influence composition and group size in elephants? How about for hyraxes? What differences are evident?

5. Contrast the differences in the dentition of dugongs and manatees. What do these differences in families within the same order suggest about the evolutionary process in general?

6. What are the ultimate constraints on the upper limit of body size in terrestrial mammals? Think in terms of surface-area-to-mass ratio and skeletal support.

CHAPTER 20

Perissodactyla and Artiodactyla

The perissodactyls and artiodactyls are the modern ungulates, which are generally large, hoofed, terrestrial herbivores. The unifying characteristic of both orders is the structure of the limbs (figure 20.1). The term *ungulate* refers to mammals that walk on the tips of their toes, which end in thick, hard, keratinized hoofs. Ungulates often have a reduced number of toes and a lengthened foot such that the **calcaneum** (heel bone) does not articulate with the fibula. The limbs are restricted to movement in a single plane. Thus, ungulates are adapted for cursorial (running) locomotion. Although dentition varies among families, cheekteeth often are hypsodont, with complex occlusal surfaces. These and other morphological and life history characteristics of ungulates adapt many of them for existence on large, open expanses of land where they must be able to feed efficiently and outrun potential predators. The distribution and adaptations of large herbivores are dictated to a large extent by their forage resources. Conversely, many of the structural characteristics of forage plants are the result of coevolutionary pressures from large herbivores.

The two orders in this chapter include a great deal of structural diversity among taxa, the result of a rich, well-documented evolutionary history for both orders. The perissodactyls ("odd-toed" ungulates) today are a small order in terms of the number of extant species—a remnant of a group that flourished during the early to mid-Tertiary period. In contrast, modern artiodactyls ("even-toed" ungulates) encompass a diverse array of species. Today, ungulates are the most important group of mammals in terms of human commerce and economics. Many ungulate species have been translocated from their native areas and introduced throughout the world, where they are important as domesticated animals (chapter 28), for sport hunting, and for ecotourism. Many other species, however, are on the verge of extinction because of illegal poaching and habitat destruction.

Perissodactyla

The three families in this order are diverse in terms of their method of locomotion, life history, and morphology and initially may appear to have little in common. The Family Equidae includes the horses, zebras, and asses. The other two families are the more closely related Tapiridae (tapirs) and Rhinocerotidae (rhinoceroses) (Hooker 2005). All perissodactyls are large, terrestrial herbivores. These hindgut fermenters (that is, they do not ruminate) feed on fibrous vegetation that is often of poor quality. Like the artiodactyls, they share a common morphological feature—foot structure—that defines the order.

MORPHOLOGY

The ordinal name means "odd-toed" and refers to the main weight-bearing axis of each limb passing through the enlarged third digit, a condition called **mesaxonic**. Tapirs have four digits on the forefeet and three on the hind feet, whereas rhinos have three digits on all feet. The third digit is the only one remaining in equids (see figure 20.1). Perissodactyls have a deep groove in the proximal surface of the **astragalus** (ankle bone), which creates a pulley-like surface that limits the limbs to forward-backward movement. The skull in all species of perissodactyls is elongated by the lengthening of the rostrum. The cheekteeth are hypsodont and usually lophodont, adaptations that occur in large grazers to enhance the grinding of vegetation. Three upper incisors are retained in the equids and tapirids. Upper incisors are reduced in number or absent in the rhinos. Perissodactyls have a simple stomach, but there is an enlarged cecum at the junction of the small and large intestines, where the majority of the microorganism-aided breakdown of cellulose occurs. Food passes through the digestive system of a perissodactyl about twice as fast as through that of a ruminating artiodactyl. Because food is retained for less time, digestion is less efficient. For example, the digestive efficiency of a horse is only about 70% that of a cow. Perissodactyls compensate for reduced efficiency by consuming more food per unit of body mass. The enlarged cecum and colon provide storage and surface area for absorption of nutrients (see figure 7.9). Perissodactyls have a bicornuate uterus, diffuse placentation, and no baculum.

FOSSIL HISTORY

The perissodactyls (and artiodactyls) originated from the Condylarthra, the dominant mammalian herbivores of the early Paleocene (about 65 million years ago). Condylarths are considered to be the ancestors of many of the other lineages of large mammals, including proboscideans and sirenians. The oldest identifiable perissodactyl fossils are

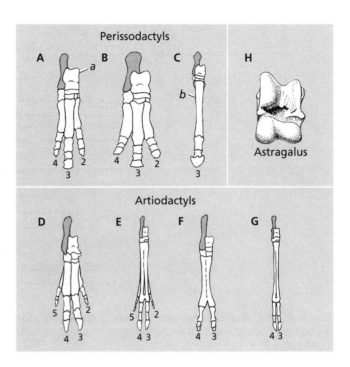

Figure 20.1 Hind foot structure of perissodactyls and artiodactyls. The hind feet of perissodactyls, including the (A) tapir, (B) rhinoceros, and (C) horse. In these "odd-toed" ungulates, the main axis of the limb passes through the enlarged third digit (i.e., they are mesaxonic). In equids, the third digit is the only one remaining (also see figure 20.4). Representative artiodactyls include the (D) pig, (E) deer, (F) camel, and (G) pronghorn. In these "even-toed" ungulates, the axis of the limb passes between the third and fourth digits (i.e., they are paraxonic). Note the vestigial second and fifth digits remain as "dew claws" in the pig and deer. For all examples, the heel bone (calcaneum) is shaded and articulates with the astragalus (a). (H) The grooved, pulley-like anterior surface of the astragalus in the caribou limits motion of the foot to a single plane. Note the fused metapodials (cannon bone, b) in C, E, F, and G. *Adapted from D. Macdonald, 1984,* Encyclopedia of Mammals, *Facts on File, and K. Kowalski, 1976,* Mammals: An Outline of Theriology, *Panstwowe Wydawnictwo Naukowe.*

from the early Eocene (50 mya). By this time, several lines of radiation are evident, and about 14 families are recognized (figure 20.2). During this period, perissodactyls far outnumbered the smaller, less diverse artiodactyls. By the end of the Oligocene (25 mya), however, 8 of the 14 families of perissodactyls were extinct. By the early Miocene, only the equids, tapirids, rhinocerotids, and the Chalicotheriidae remained. This last family included unusual ungulates with large forelimbs and short hind limbs adapted for standing semierect to feed on tall trees (Coombs 1983). Several genera had large, retractable claws instead of hoofs (figure 20.3).

Tapirs are among the more primitive extant large mammals, based on their four toes on the forefeet and brachyodont cheekteeth (Dawson and Krishtalka 1984). Originating in the early Eocene of North America, tapirs migrated both north into Asia and south into Central and South America. Tapirs were extirpated throughout most

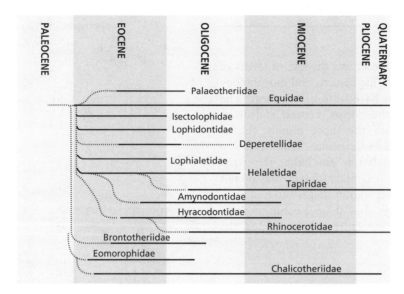

Figure 20.2 Geologic ranges of modern and extinct families of Perissodactyla. Several families of perissodactyls flourished in the early Eocene, but by the middle Miocene, only four families remained. Today, only 3 families are extant, which include only 6 genera and 16 extant species. *Adapted from R. L. Carroll, 1988*, Vertebrate Paleontology and Evolution, *W. H. Freeman.*

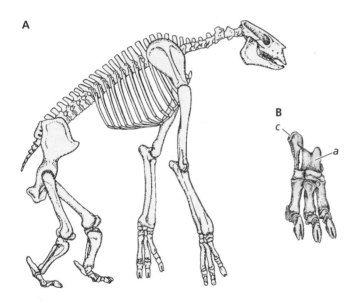

Figure 20.3 Example of a recently extinct perisso-dactyl. Chalicotheriids remained extant in Asia and Africa until the late Pleistocene epoch. (A) Skeleton of the large *Chalicotherium* from the Miocene epoch of Europe and (B) the clawed hind foot of *Moropus*. Note the calcaneum (c) and astragalus (a). *Adapted from R. L. Carroll, 1988*, Vertebrate Paleontology and Evolution, *W. H. Freeman.*

of North America by the late Pleistocene. For tapirs, like camels, the combination of migration and extirpation resulted in a discontinuous geographic distribution today. The current Genus *Tapirus* dates from 20 mya in the Miocene and has changed little since then. Other fossil tapirs are noted by Albright (1998) for remains in Texas and by Tong et al. (2002) in China.

Fossil evidence of the rhinocerotids dates from the late Eocene. Most of the genera extant today date from the Miocene (10–25 mya). They were extinct in North America

by the end of the Pliocene (2 mya), however, and never dispersed to South America. Rhinocerotids were abundant and widespread in the Old World until the late Pleistocene (about 60,000 years ago). The largest land mammal that ever lived was a rhinocerotid. *Indricotherium transouralicum* [*Baluchitherium grangeri*] was at least 5 m high at the shoulder. Although many estimates suggest the maximum body mass of *Indricotherium* was 30,000 kg or more, an upper estimate of 15,000–20,000 kg probably is more reliable (Economos 1981; Fortelius and Kappelman 1993).

The fossil history of equids is one of the best documented for any mammalian family (Froehlich 2002). This history shows increasing body size and skull proportions, increasing size and complexity of the cheekteeth, and reduction in the number of digits (figure 20.4). As noted by Hooker (2005:206), "Early equoids that have classically been referred to the genus *Hyracotherium* have long been regarded as the most primitive perissodactyls." Keep in mind, however, that the evolution of the horse was not a ladder-like, directed, progressive process, as suggested by figure 20.4, but a complex radiation of numerous divergent, overlapping lineages (figure 20.5). Equids passed most of their evolutionary history in North America, with migration to the Old World during the Miocene and to Central and South America in the Pliocene and Pleistocene. Equids left no Pleistocene descendants in the New World, however, becoming extirpated about 10,000 years ago. They were reintroduced to the New World by the Spanish conquistador Hernando Cortés in 1519.

ECONOMICS AND CONSERVATION

As a domesticated species, horses probably are second only to cattle in their importance in the development of

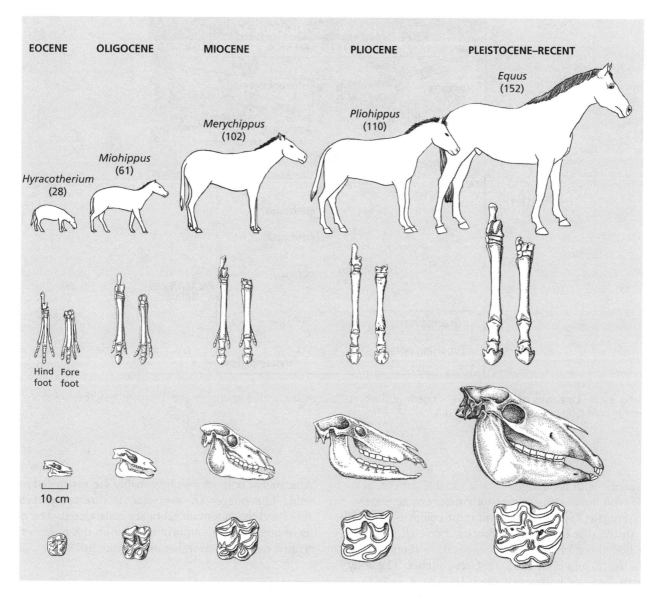

Figure 20.4 Evolution of the horse. The early Eocene *Hyracotherium* was a small, forest-dwelling browser. The forefeet had four toes, the skull had confluent temporal and orbital openings, and the brachyodont cheekteeth had simple enamel patterns. Note the progressive increase in size (shoulder heights in centimeters are in parentheses), with reduction in the number of digits, changed proportions of the skull with formation of a postorbital bar, and increasing complexity of the occlusal surface of the cheekteeth in the grazing Genus *Equus* that now inhabits open areas.

cultural and economic systems of humans. Domesticated in southern Ukraine and elsewhere about 5000 years ago (see chapter 28), horses have been introduced throughout most of the world and have been pivotal as an aid to travel, exploration, and warfare throughout history. There are over 150 recognized breeds of horses in the world today (Draper 1999). In contrast to the cosmopolitan distribution of domestic horses through introductions, several other species of equids are either recently extirpated in the wild or endangered. Likewise, tapirs are much reduced throughout their ranges because of hunting pressure and destruction of habitat. The plight of rhinoceroses is well known; all species have become prime targets of poachers for their horns and other body parts. Details of endangered species and conservation efforts are discussed in each family account.

FAMILIES

Equidae

Grubb (2005a) recognized seven extant species, all in the Genus *Equus* in addition to the recently extinct quagga (*E. quagga*). All have long, slender limbs, and only the third digit remains functional. Three upper and lower incisors occur in each quadrant, and the cheekteeth are large and hypsodont and have complex occlusal surfaces. Pelage color is variable in most equids, although the pattern of stripes in zebras depends on the species. Stripes are narrow and close together and extend down to the hooves in Grevy's zebra (*E. grevyi*). The mountain zebra (*E. zebra*) has broad stripes that do not extend to the ventral surface. Burchell's zebra (*E. burchellii*)—sometimes considered

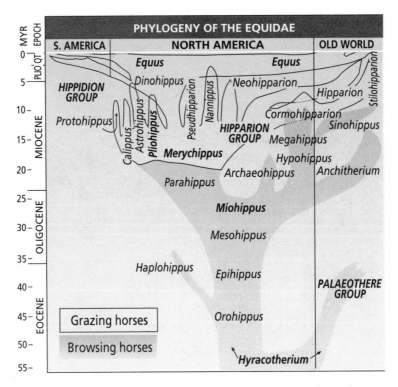

Figure 20.5 Lineages of the horse. Phylogeny of the horse is complex, with overlapping and divergent adaptive radiations. Genera in bold type are shown in Figure 20.4.

conspecific with the quagga—has a variable pattern, generally with broad stripes posteriorly that become narrower anteriorly. The neck mane on most equids is erect; it falls to the side only in the domestic horse (*E. caballus*). The body mass of wild equids ranges from about 250 kg in the ass (*E. asinus*) to 400 kg in Grevy's zebra. The body mass of domestic horses may reach 1000 kg.

The natural distribution of equids includes eastern Africa and central Asia from the Middle East to Mongolia. They inhabit short grasslands and desert scrublands and are never far from water. The basic social unit is the family group, generally 10–15 individuals made up of a highly territorial male, several females, and their offspring (Berger 1986). Young females leave the family group when they become sexually mature at about 2 years of age. Young males become sexually mature at the same age. They do not breed until they leave the family group and gain access to other females, which is by about 5 years of age. In Burchell's zebra, temporary aggregations of 100,000 individuals may form, depending on ecological conditions.

A single offspring is usual after a gestation period of about a year. Birth and subsequent mating 7–10 days later occur during the wet season, when vegetation is most abundant. Neonates are precocial. They begin to graze at about 1 month of age and are weaned at 8–13 months of age.

Several species of equids are endangered. These include Przewalski's horse (*Equus przewalskii*), although this may be a subspecies of the domestic horse (figures 28.8A, B) (see comments in Grubb 2005a:631), and the

African ass, both of which probably are extirpated in the wild. The onager (*E. hemionus*) is threatened. Grevy's zebra and the mountain zebra are endangered. The quagga of South Africa, uniform in color on the posterior and striped on the anterior, became extinct in 1872.

Tapiridae

There are four species in the single Genus *Tapirus*. The family has a discontinuous distribution: Baird's tapir (*T. bairdii*) occurs in Mexico, Central America, and northern South America; and two other species, the South American tapir (*T. terrestris*) and the mountain tapir (*T. pinchaque*), occur in northern South America. The Malayan tapir (*T. indicus*) occurs in Myanmar (Burma), Thailand, Malaya, and Sumatra. It has a chromosome complement of only $2N = 52$, whereas the New World species have a diploid number of $2N = 76$ or 80 (Houck et al. 2000). Tapirs have a chunky body with short legs and an elongated head with small eyes and ears (figure 20.6A). The nose and upper lip form a pronounced, flexible proboscis. Like an elephant's trunk, it is used to manipulate vegetation during feeding and, during locomotion, to gather olfactory information about the environment (Witmer et al. 1999). Mean head and body length is 180–250 cm, and body mass reaches as much as 300 kg. Pelage color in New World species is a uniform reddish brown to gray or black. The Malayan tapir is white on the trunk of the body and black on the head, shoulders, and

A

B

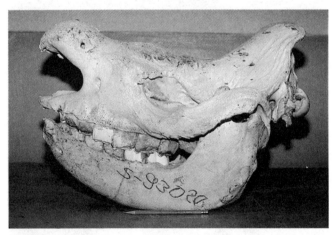

C

Figure 20.6 Representative perissodactyls. (A) South American tapir and (B) black rhinoceros. (C) The rhino skull shows the typical elongated nasal bones without any horn core or attachment site for the horns. Scale: pencil = 16 cm.

limbs. A short, bristly neck mane is characteristic of both Baird's and the South American tapir; the hide is very tough in all species.

Tapirs inhabit heavily forested areas. The mountain tapir lives at elevations of 2000–4500 m; the other species range up to 1200 m. All tapirs are nocturnal and feed on understory shoots, twigs, fruit, grass, and aquatic vegeta-

tion and occasionally on cultivated crops. Tapirs significantly affect plant ecology by dispersing seeds from a variety of plant species (Fragoso and Huffman 2000). All but the mountain tapir are associated with swamps, rivers, or other wet areas. They are good swimmers and feed or seek refuge in water. Novarino et al. (2004) found that the Malayan tapir preferred secondary forest habitat over primary or riparian forests.

Tapirs are generally solitary. Sexual maturity is reached at 3–4 years of age. Breeding occurs at any time during the year (Padilla and Dowler 1994). Usually, a single young is born after a gestation of about 395 days. Young have a reddish brown coat and are camouflaged with white spots and lines. They stay with their mother for 6–8 months, by which time the juvenile pelage is replaced with adult pelage. All species of tapirs suffer from loss of habitat due to logging, agriculture, and forest clearing and are declining in number and distribution. New World species are hunted for meat and hides. The mountain tapir is considered endangered, as is Baird's tapir. The other two species of tapirs are threatened.

Rhinocerotidae

There are four genera in this family, with five living species. Rhinoceroses are well known for their large, heavyset, graviportal structure (see figure 20.6B). They have small eyes and a prehensile upper lip that extends past the lower lip in black (*Diceros bicornis*) and Asian rhinos. The upper lip is used to gather vegetation. The white rhino (*Ceratotherium simum*) reaches 400 cm at the shoulder, with maximum body mass of 1700 kg. Body mass of adult male Indian rhinos (*Rhinoceros unicornis*) may be 2000 kg (Dinerstein 1991).

The family name refers to the rhino's horns, which have no bony core or keratinized sheath (figures 20.6C and 20.7A) but instead are a dermal mass of agglutinated, keratinized fibers (fused hairs). They are conical, often curve posteriorly, and may reach 175 cm in length in the white rhino. Asian rhinos have shorter horns. The anterior horn is positioned medially over the nasal bones. If two horns occur, the shorter, posterior one is over the frontal bones. Neither horn is attached to the bone, however, but to the skin over a roughened section of the skull bones. The nasal bones of the skull are large and project well above and anterior to the maxillae (see figure 20.6C).

Both white and black rhinos occur in sub-Saharan central and east Africa. The Indian rhino occurred in Pakistan and northern India, and the Javan rhino (*Rhinoceros sondaicus*) originally was in southeastern Asia from eastern India to Vietnam, Sumatra, and Java. The Sumatran rhino (*Dicerorhinus sumatrensis*) also originally was distributed throughout southeastern Asia, Sumatra, and Borneo. Based on molecular analyses, Asian and African lineages of rhinos are estimated to have diverged in the late Oligocene about 26 mya (Tougard et al. 2001). The geographic range of all species is greatly reduced to tropical and subtropical habitats due to human interfer-

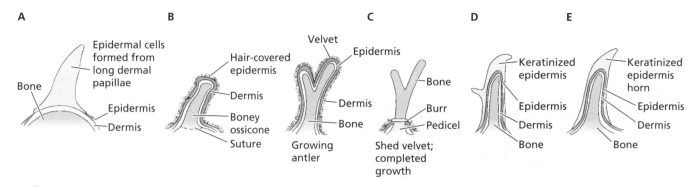

| A | B | C | D | E |

Figure 20.7 Head ornamentation in five ungulate families. (A) Rhinocerotidae, a perissodactyl; and four families of artiodactyls: (B) Giraffidae; (C) Cervidae; (D) Antilocapridae; and (E) Bovidae. *Adapted from H. Gunderson, 1976,* Mammalogy, *McGraw-Hill.*

ence, poaching, and habitat destruction. Depending on the species, rhinos occupy tropical rainforests, floodplains, grasslands, and scrublands. All are dependent on a permanent water supply for frequent drinking and bathing. Wallowing probably is necessary to help control body temperature (Owen-Smith 1975) and to reduce insect harassment. Rhinos forage on woody or grassy vegetation and occasionally fruits but prefer leafy material when available.

Aside from mother-and-offspring pairs, rhinos generally are solitary. Small groups of immature individuals may form in Indian and white rhinos. Females become sexually mature at 5 years of age and bear their first calves when 6–8 years of age. Gestation is about 8 months in the Sumatran rhino and about 16 months in the other species. Births, usually a single calf, occur at intervals of 2–4 years (Dinerstein and Price 1991). Young nurse for 1–2 years, although the white rhino begins to eat solid food by 1 week of age. Males generally do not breed before 10 years of age (Ryder 1993).

Populations of all species have declined during the last 150 years. As is the case with many large mammals, the quantity of forage required and low reproductive rates mitigate against recovery if populations are reduced. All species are considered to be endangered or critically endangered, with the Asian species near extinction. Rhinos are illegally harvested for their horns (figure 20.8), which, with other body parts, are valued in traditional Asian medicine for supposed aphrodisiac and medicinal properties. Horns also have been used traditionally for making dagger handles in the Middle East. White rhinos have been successfully translocated to parts of their former range in southern Africa.

Artiodactyla

Artiodactyls are much more selective feeders than perissodactyls, a factor in their greater adaptive radiation. In contrast to the 3 families and 16 extant species of perissodactyls, the artiodactyls include 10 living families, 89 genera, and approximately 240 species. Artiodactyls are distributed

almost worldwide, either naturally or through introduction. As might be expected in such a large group, there is tremendous diversity in body size and structure, and three suborders are recognized. The Suborder Suiformes, generally considered the least derived (most primitive) group, includes three families: the Suidae (pigs and warthogs), the Tayassuidae (peccaries), and the Hippopotamidae (hippopotamuses). The Suborder Tylopoda includes one family: the Camelidae (camels, llamas, and vicuña). Six extant families make up the Suborder Ruminantia, the most derived group: the Tragulidae (chevrotains, or mouse deer), the Giraffidae (giraffe and okapi), the Cervidae (deer), the

Figure 20.8 Poaching for horns. In the past, illegal rhino horn sold for tens of thousands of dollars a kilogram.

Moschidae (musk deer), the Antilocapridae (pronghorn), and the Bovidae (antelope, bison, goats, sheep, and many others). All ruminants, with the exception of chevrotains and musk deer, have some type of head ornamentation in the form of horns or antlers. Despite the vast array of species, artiodactyls share a common morphological characteristic that defines the order.

MORPHOLOGY

Like perissodactyls, the artiodactyls are defined by the structure of the foot. The main weight-bearing axis passes through the third and fourth digits, a condition termed **paraxonic.** The second and fifth digits are reduced and nonfunctional or absent. There is a definite trend toward cursorial locomotion in the more derived families. The Suiformes exhibit plantigrade locomotion, with unfused **metapodials** (metacarpals and metatarsals), in contrast to members of the Ruminantia, which are cursorial (specifically unguligrade, that is, walking on the tips of the toes), with metapodials fused to form a **cannon bone** (see figure 20.1C and E–G). The astragalus bone has a pulley-like surface above and below (see figure 20.1H). This "double pulley" system is above the distal portions of the limbs and allows great flexion and extension. At the same time, however, the astragalus limits distal limb motion so that it is parallel to the body. The clavicle is reduced or absent. Dentition varies in this order, with the cheekteeth ranging from bunodont and brachyodont to selenodont and hypsodont. The number of teeth varies, although in most families, the upper incisors and canines are reduced or absent. In some species, however, the canines form enlarged tusks as in pigs, peccaries, and musk deer. Other morphological features of artiodactyls are discussed by Theodor et al. (2005). Artiodactyls are diverse in their digestive anatomy, with simple, nonruminating stomachs occurring in suids and tayassuids, grading to much more complex, four-chambered ruminating stomachs in the more derived families. Size ranges from a maximum head and body length of 0.5 m in the mouse deer to almost 6 m in the giraffe and maximum body mass from 2.5 kg in the lesser mouse deer to 4500 kg in the hippopotamus. Using molecular evidence, Matthee et al. (2001 and references therein) concluded that the order Artiodactyla is not monophyletic. They found that hippopotamuses were more closely related to whales than to other artiodactyls (figure 20. 9). There are opposing views, however, based on morphological analyses. "Phylogenetic analyses of morphological data of extant mammals generally support sister-group relationship between artiodactyls and whales, but they have thus far not supported inclusion of whales in Artiodactyla" (Theodor et al. 2005:227). Characteristics specific to individual families are noted in the following sections.

FOSSIL HISTORY

The earliest ancestors of artiodactyls were the Condylarthra. The oldest recognized genus is the rabbit-sized *Diacodexis*

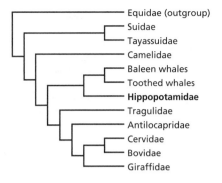

Figure 20.9 **Phylogeny of the Order Artiodactyla.** This phylogeny is based on one mitochondrial and eight nuclear DNA segments. Note the proposed relationship for hippos and whales. As noted in the text, morphological analyses do not support inclusion of whales within Artiodactyla. *Adapted from C. A. Matthee et al., 2001, Mining the Mammalian Genome for Artiodactyl Systematics,* Systematic Biology 50:367-390.

(figure 20.10) in the early Eocene (Theodor et al. 2005). The order then was relatively insignificant compared with the perissodactyls but radiated a great deal throughout the Eocene and Oligocene. Of the modern families, fossil evidence for the Ruminantia dates from the Eocene, as do the camels and tayassuids. Suids are known from the early Oligocene, and hippopotamids, from the early Miocene (figure 20.11).

ECONOMICS AND CONSERVATION

Most families of artiodactyls have had some type of economic importance to human civilizations for thousands of years. Domestic artiodactyls include pigs, camels, llamas, and cattle. Some species are no longer found in the wild but survive only in domestication (see chapter 28). Deer, the pronghorn antelope (*Antilocapra americana*), mountain sheep and goats, numerous African antelope, and other bovids all provide meat, hides, sport hunting,

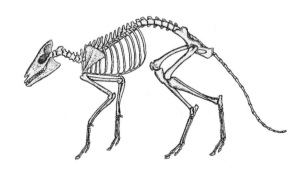

Figure 20.10 **An early artiodactyl.** The rabbit-sized *Diacodexis* from the early Eocene epoch was already highly adapted for cursorial locomotion. The hind limbs were elongated, as were the metapodials and the third and fourth digits. The astragalus restricted movement of the limbs to the vertical plane. *Adapted from K. D. Rose, 1982, Skeleton of Diacodexis,* Science 216:621-623.

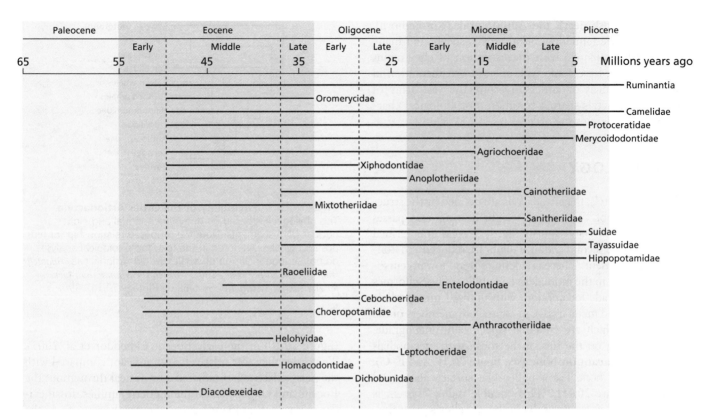

Figure 20.11 **Geological ranges of fossil and living lineages of artiodactyls.** Relationships among the families are uncertain and difficult to document. *Reproduced from Theodor et al., 2005, Artiodactyla, in* The Rise of Placental Mammals: Origins and Relationships of the Major Extant Clades *(K. D. Rose and J. D. Archibald, eds.), Johns Hopkins Univ. Press.*

and an economic base unparalleled by other mammalian groups. Many of these species are managed intensively on a sustained yield basis throughout much of their ranges.

FAMILIES

Suidae

The 5 genera and 19 living species of pigs have simple stomachs, bunodont cheekteeth, and large, ever-growing canines—the upper pair curving up and outward to form tusks (figure 20.12). Suids have short legs; heavyset bodies; thick skin with short, coarse pelage; small eyes; a relatively large head; and a prominent snout truncated at the end with a round, cartilaginous disk. Several layers of muscles are associated with the snout, which is used in rooting for food. Several species have large facial warts, most prominent in males (see figure 20.12A). Maximum body mass ranges from 10 kg in the pygmy hog (*Sus salvanius*) to 200 kg in the wild boar (*S. scrofa*). The native distribution of suids is Europe, Africa (except the Sahara Desert), and Asia, including Indonesia, Borneo, and the Philippines. They have been introduced to North and South America, Australia, and New Zealand, where both feral and domestic pigs now flourish.

Suids are gregarious and often forage in groups, although solitary individuals may occur, especially of the

desert warthog (*Phacochoerus aethiopicus*). Melis et al. (2006) related geographical variation in population densities of wild boars in western Eurasia to winter severity at higher latitudes. Their habitats include tropical forests, woodlands, scrubby thickets, grasslands, and savannas. Wild boars and the red river hog (*Potamochoerus porcus*) are omnivorous, whereas the babirusa (*Babyrousa babyrussa*; see figure 20.12B) of Sulawesi and nearby islands, the giant forest hog (*Hylochoerus meinertzhageni*), and the warthog are more specialized herbivores. The large head and mobile snout are used to root for food. Suids are sexually mature by 18 months of age, although males may not have access to females until they are 4 years old. Gestation is about 100 days in the pygmy hog, 115 days in domestic pigs, and about 175 days in the desert warthog. Litter size ranges from 1–2 in the babirusa and up to 12 in domestic pigs. Feral hogs are an important hunted species in California and throughout much of the southeastern United States (Sweeney et al. 2003). Unfortunately, they carry a number of diseases, including swine brucellosis, African swine fever, and pseudorabies. These can be transmitted to domestic pigs, creating potential problems for livestock operations. Rooting activities of feral hogs, in association with the highest reproductive rate among ungulates, can cause severe environmental impacts (Massei and Genov 2004), including damage in many national parks throughout the United States (Burde and Feldhamer 2005). The pygmy hog, found in the foothills

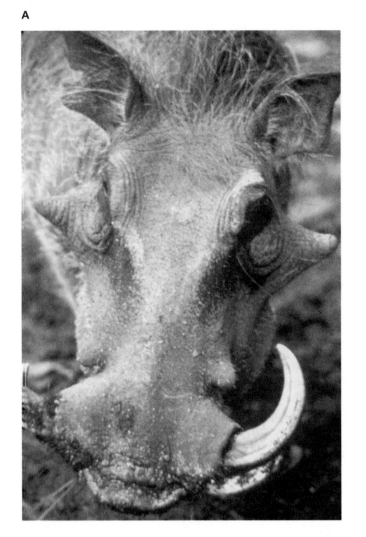

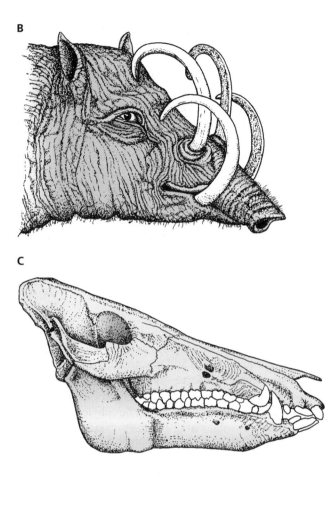

Figure 20.12 Representative suids. (A) Large facial warts, composed of dense connective tissue, are evident on this warthog. Warts may protect the facial area from an opponent's tusks during aggressive interactions involving ritualized head-to-head pushing contests. (B) Tusks are especially pronounced in the babirusa. The upper canines protrude through the skin of the rostrum and do not occlude with the lower canines. (C) The typical wedge shape of suid skulls is evident in the wild boar. Canine tusks are less pronounced than in the warthog and babirusa.

of the Himalayan Mountains, is critically endangered, as is the Visayan warty pig (*Sus cebifrons*). The Javan pig (*S. verrucosus*) is endangered.

Tayassuidae

Peccaries are the least specialized of the suiformes. The three genera, each with a single species, superficially resemble pigs. They have large heads and long, mobile, piglike snouts, but with thin legs and small hooves. Peccaries are smaller than pigs; they range up to 30 kg body mass. They have a total of 38 teeth and fewer tail vertebrae than pigs. Their tusk-like upper canines are small, sharp-edged, and point downward, again unlike those of pigs. Also, peccaries are found only in the New World, from the southwestern United States to central Argentina, where they occur in various habitats from desert scrublands to tropical rainforests. Competition may

occur with exotic feral hogs where they are sympatric with peccaries (Gabor and Hellgren 2000; Sicuro and Oliveira 2002). Peccaries are primarily diurnal herbivores. They root with their snouts as do pigs but occasionally take small vertebrates, invertebrates, eggs, fruit, and carrion. Beck (2006) noted that peccaries eat the fruit from at least 46 species of neotropical palms (Arecaceae). Through their feeding, dispersal, and trampling activities, peccaries have a great impact on the density and distribution of palms.

Peccaries generally form small groups of 5–15 individuals, although herds of the white-lipped peccary (*Tayassu pecari*) may number several hundred. Subherds often form with individuals commonly moving between various subherds (Keuroghlian et al. 2004). A rump gland is used in social communication. Breeding may occur throughout the year; in arid environments, it is affected by rainfall (Hellgren et al. 1995). The gestation period varies from

115 days in the collared peccary (*Pecari tajacu*) to 162 days in the white-lipped peccary. Litter size averages two. Female collared peccaries attain sexual maturity by about 8 months of age, males by 11 months old. Until recently, the Chacoan peccary (*Catagonus wagneri*) was known only from fossil evidence (Wetzel et al. 1975). The species is endangered by loss of habitat and possibly overharvesting. The collared peccary (figure 20.13) is hunted in Texas, New Mexico, and Arizona (Hellgren and Bissonette 2003), where populations generally are secure. Using nuclear and mtDNA, Gongora and Moran (2005) found Chacoan and white-lipped peccaries formed a clade distinct from the collared peccary (figure 20.14). Appropriate generic names, especially for the collared peccary, have generated controversy; both *Dicotyles* and *Tayassu* have been used in the past.

Hippopotamidae

The two species in this family differ greatly in size. The common hippopotamus (*Hippopotamus amphibius*) has a total length up to 4.5 m and maximum body mass of 4500 kg. The more primitive pygmy hippo (*Hexaprotodon* [*Choeropsis*] *liberiensis*) has a total length of only 2 m and body mass of about 250 kg. The phylogeny of hippopotamids is problematic, and the two extant species may not be closely related (Weston 2000). As noted, molecular evidence suggests that hippos may be closely related to whales (see figure 20.9), but the origin of the group remains unresolved, and debate continues (Boisserie et al. 2005; Boisserie and Lihoreau 2006). Both the common hippo and the pygmy hippo are essentially without hair except for a few bristles around the snout. Without temperature-regulating sweat glands, each species has glandular skin that exudes a pigmented fluid that appears red and gives rise to the misconception that they "sweat blood." The fluid helps protect against sunburn and may help keep wounds from becoming infected in the water. The cheek-

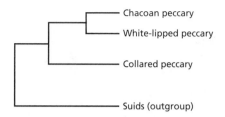

Figure 20.14 Relationship of peccaries (Family Tayassuidae). Based on molecular data, the Chacoan and white-lipped peccaries are more closely related than either is to the collared peccary. *Modified from Gongora and Moran (2005).*

teeth are bunodont, and *H. amphibius* has ever-growing, tusk-like lower canines and incisors, with the alveoli of the canines being anterior to those of the incisors (figure 20.15). Although the canines are not used in foraging, they are important in ritualized fighting to establish dominance, functioning as visual signals.

Both species are closely associated with rivers, lakes, and estuaries and are excellent swimmers and divers. Part of their dependence on water is because of rapid evaporative water loss through the epidermis. The hippopotamus can remain submerged for up to 30 minutes while walking along the bottom of lakes and rivers feeding on plants. In common hippos, the more aquatic of the two species, the eyes and nostrils are high on the head, allowing individuals to remain submerged throughout much of the day with only the eyes and nostrils above water.

The hippopotamus was originally distributed throughout most of Africa south of the Sahara Desert and along the Nile River. The pygmy hippo is restricted to the

Figure 20.13 Collared peccary. Morphologically similar to pigs, collared peccaries are a popular game mammal in parts of Texas, New Mexico, and Arizona.

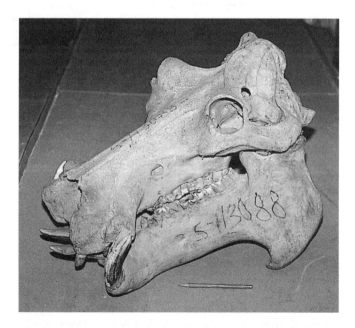

Figure 20.15 Skull of a hippopotamus. Note the eye orbits on the top of the cranium in the hippopotamus, the low-crowned cheekteeth, and the alveoli of the lower canines anterior to those of the incisors. Also note the very large mandible. Scale: pencil = 16 cm.

coastal regions of West Africa from Guinea to Nigeria. The aquatic specialization of common hippos and the geographic restriction of the more terrestrial pygmy hippopotamus may have been in response to pressure from other grazing artiodactyls, specifically bovids. The common hippo grazes on land for 5–6 hours each night, with grass as the primary food. Pygmy hippos also consume leaves and fruit. Hippopotamuses are not ruminants, but the stomach has septa and several blind sacs to slow the passage of food for more efficient digestion.

Common hippos are gregarious, forming herds of up to 40 individuals. Pygmy hippos are solitary or form pairs. Although there is a great deal of variability, males generally are sexually mature when 7 years old, females by 9 years of age. A single young is born after a gestation period of 200 days in pygmy hippos and 240 days in common hippos. Parturition may occur in the water. Calving intervals are about 2 years but may be affected by drought. Both species have been extensively overhunted, with populations further reduced by habitat destruction. The common hippopotamus is extirpated from the Nile River valley in Egypt. The pygmy hippo probably has never been common and currently is considered endangered. The Malagasy dwarf hippo (*Hippopotamus lemerlei*) and Malagasy pygmy hippo (*H. madagascariensis*) became extinct within the last 1000 years (Stuenes 1989). *Kenyapotamus*, from Miocene deposits in Kenya, is the oldest known hippopotamid.

Camelidae

This family has three genera and four species that, like the hippopotamuses, differ greatly in size. Head and body length in the one-humped, or dromedary, camel (*Camelus dromedarius*) and the two-humped Bactrian camel (*C. bactrianus*) approaches 3.5 m, with a body mass close to 700 kg. The nomenclature of the New World camelids is more problematic than their Old World relatives. The guanaco is considered to be the wild ancestor of the llama, and both are included in *Lama glama*. The domesticated alpaca, formerly referred as *Lama pacos*, is now considered synonymous with its presumed wild ancestor the vicuña (*Vicugna vicugna*) (see Kadwell et al. 2001; Grubb 2005b). The guanaco may reach 2.2 m in head and body length and 140 kg body mass. The vicuña is much smaller, about 55 kg. All camelids have a small head with long snout and cleft upper lip; a long, thin neck; and long legs with the metapodials fused to form a cannon bone. The toes, with nails on the upper surface, spread out as they contact the ground, with a broad pad acting to support the mass of the animal on soft, loose sand. Upper and lower canines are present, and the cheekteeth are selenodont. Only the outer, spatulate upper incisor is retained in adults. The vicuña is the only artiodactyl with ever-growing lower incisors. Camelids have a three-chambered, ruminating stomach and a short, simple cecum. An intriguing phenomenon in camelids involves their functional antibodies. Most vertebrates, and presumably all mammals, have

immunoglobulin molecules made up of four polypeptide chains. These molecules have two heavy and two light chains. Antibody molecules in camelids, however, are not tetrameric but instead are dimeric—they have only the two heavy chains—a unique adaptation (Nguyen et al. 2002; Su et al. 2002).

The dromedary camel may once have ranged throughout the Middle East but now survives only in domestication. It is the only domesticated mammalian species for which there is no information on wild or fossil forms (Corbet 1978). The Bactrian camel originally ranged throughout much of central Asia but now is restricted in the wild to the western Gobi Desert. Guanacos (figure 20.16) are found from sea level to 4000 m elevation in various habitats (Sosa and Sarasola 2005) in the Andes Mountains from southern Peru to Tierra del Fuego. Vicuñas occur in the grasslands of Peru, western Bolivia, northeast Chile, and northwest Argentina at elevations from 3700–4800 m (Arzamendia et al. 2006). All camelids are gregarious, diurnal, and herbivorous and are best suited to dry, arid climates. They can eat plants with a high salt content not tolerated by other grazers. They are well known for their ability to go long distances under difficult conditions, conserving water better than other large mammals (see chapter 9) and may lose up to 40% of their body mass through desiccation without harm (Gauthier-Pilters 1974). von Engelhardt et al. (2006) discussed physiological changes in camels during experimental water deprivation and rehydration. When camels are well-fed, their humps are firm and erect. The hump is a fat reservoir and shrinks, leaning to one side, when camels are nutritionally stressed. In camels, both limbs on either side of the body move in unison. This **pacing** locomotion allows for long strides with consequent lower energy expenditure. Wild camelids are social, forming groups of up to 30 individuals with different sex and age compositions depending on the season. A single young is born after a gestation of 300–320 days in Genus *Lama* and 365–440 days in Genus *Camelus*.

This family arose in North America in the early to mid Eocene (see figure 20.11) and was restricted to North America throughout most of the Tertiary period. It expanded to both Eurasia and South America during the Pliocene and became extirpated in North America in the late Pleistocene. Camelids have been introduced throughout the world for use as pack animals and for meat, wool, or milk. Both the llama and the dromedary camel have been domesticated for up to 5000 years (see chapter 28). In the wild, the Bactrian camel is critically endangered.

Tragulidae

This family is the most ancestral of the extant ruminants, that is, they have changed the least since the Oligocene, when they enjoyed a worldwide distribution. Today, they include three genera and eight species. The water chevrotain (*Hyemoschus aquaticus*) occurs in west-central Africa, and the Indian spotted chevrotain (*Moschiola meminna*)

Figure 20.16 **New World camelids.** (A) A guanaco (*Lama glama*), considered to be the wild ancestor of (B) the llama; (C) the vicuña (*Vicugna vicugna*). As noted in the text, relationships among New World camelids is confused, with little consensus—for example, see table 28.1 for alternative taxonomy.

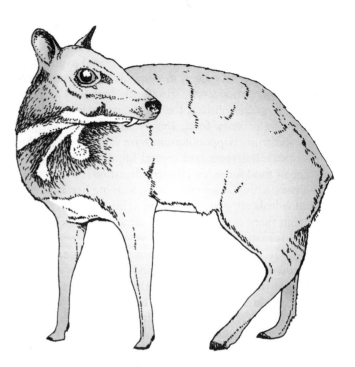

Figure 20.17 **A chevrotain, or mouse deer.** In these small artiodactyls, males have enlarged upper canines. *Original drawing by Kelly Paralis Keenan.*

(2004) recognized four additional species: the lesser mouse deer (*T. kanchil*), the Philippine mouse deer (*T. nigricans*), the Vietnam mouse deer (*T. versicolor*), and Williamson's mouse deer (*T. williamsoni*), known only from the type specimen. The Java mouse deer is the world's smallest artiodactyl and superficially resembles a tiny, chunky deer. Maximum head and body length is 50 cm, and body mass is 2.5 kg. Other species are slightly larger. Unlike deer, chevrotain males have no antlers and no facial or other body glands. The legs are thin with a cannon bone (except in the water chevrotain, the most ancestral form); dentition is selenodont and brachyodont; and there is a three-chambered, ruminating stomach. As in species of deer without antlers, tragulids have an enlarged, curved upper canine that extends below the upper lip (figure 20.17). Tragulids generally are solitary and nocturnal and spend most of the day hidden in brushy undergrowth, usually not far from water. Their spotted pelage helps camouflage them. They feed on grass, leaves, and fallen fruit and are in many ways the ecological equivalent of hares (Genus *Lepus*). Gestation in *Tragulus* is 4–6 months, and a single, precocial young is born each year. Tragulids are physically and sexually mature by 9 months of age. Hunting and habitat destruction have reduced populations of all species.

Giraffidae

This family includes both the giraffe (*Giraffa camelopardalis*) and the okapi (*Okapia johnstoni*). The giraffe occupies savannas, grasslands, and open woodlands throughout sub-Saharan Africa. The okapi is restricted to dense forested areas of the Republic of Congo (formerly Zaire). Both

lives in Sri Lanka and southern India. The third genus, *Tragulus*, traditionally encompassed two species—the Java mouse deer (*T. javanicus*) and the greater mouse deer (*T. napu*). In a recent revision, however, Meijaard and Groves species are characterized by long legs and neck that allow them access to forage out of reach of other large herbivores. In male giraffes, the top of the head may be 5.5 m off the ground, and the average shoulder height is 3 m (figure 20.18). Blood pressure to the head is maintained, in part, because the heart of an adult giraffe is large—over 2% of body weight—compared to about 0.5% in humans (Brook and Pedley 2002). Okapis are smaller, with a shoulder height about 1.7 m. Females are smaller than males in both species. These species are found where forage is available year-round at levels between 2 and 5.5 m. The teeth are small and brachyodont, which is unusual for a large herbivore. This suggests that they are selective feeders; forage selection is aided by their very long, muscular, prehensile tongue that, in the giraffe, can be extended almost 0.5 m. The long, muscular neck of both species is also used in intraspecific aggressive interactions of males as they establish dominance hierarchies. Interactions involve standing side by side and "necking"—pushing each other with intertwined necks and occasionally swinging the neck and using the horns to strike an opponent's head and neck. Extensive cranial sinuses help protect individuals from hammering blows. The horns are unique in being short, permanent, unbranched processes (**ossicones**) over the frontal and parietal bones (see figure 20.7B; figure 20.19). They are not extensions of the frontal bone but are distinct, fused to the cranium, and covered with hairy skin.

Both species have unusual pelage patterns. There is a great deal of variation in giraffe patterns and color shades, with white and melanistic individuals known. The okapi is reddish brown dorsally with distinctive striping on the legs and back of the thighs (figure 20.20). Okapis do not form herds but are solitary or live in small family groups (Bodmer and Rabb 1992). Group size in giraffes may be 30–50 but is more a reflection of seasonality (Le Pendu et

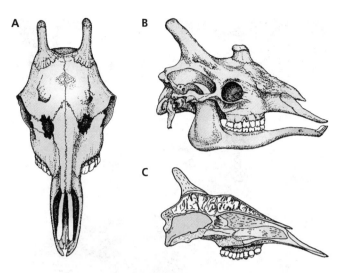

Figure 20.19 **Skull of a giraffe.** (A) Dorsal and (B) lateral views of the "horns" (ossicones) of a giraffe, showing distinct sutures on the frontal/parietal area. The horns are not extensions of the frontal bone as in bovids. (C) Sagittal section of a giraffe skull showing the extensive sinus cavities in the roof of the cranium. Sinuses allow blood to collect when the head is lowered so blood pressure does not increase to dangerously high levels. *Adapted from G. A. Bubenik and A. B. Bubenik, eds., 1990,* Horns, Pronghorns and Antlers, *Springer-Verlag.*

al. 2000) and forage availability than gregariousness. Gestation is variable in both species, from 427–488 days in the giraffe to a somewhat shorter period in the okapi (Kingdon 1979; Dagg and Foster 1982). Females of both species give birth to a single, precocial calf approximately every 2 years. Calf mortality is very high due to predation. Subsistence hunters have long taken giraffes for meat. Giraffes also are harvested illegally for their meter-long tail tufts, which are used to make bracelets for tourists. The okapi was not discovered by scientists until 1900 and has been protected by the government since 1933 (Hart and Hart 1988).

Moschidae

Formerly included in the Cervidae, the musk deer form a valid, monophyletic family (Su et al. 1999) with one genus (*Moschus*) and seven species (Grubb 2005b). Surprisingly, Hassanin and Douzery (2003) found this family more closely related to the Bovidae than the Cervidae based on analyses of nuclear and mtDNA. Musk deer are small, with head and body length between 80 and 100 cm and maximum body mass no more than 18 kg. The males do not have antlers, but as in the Chinese water deer (*Hydropotes inermis*) and tragulids, the upper canines are long and curved (figure 20.21). Musk deer are distributed from Siberia to the Himalaya Mountains in forested areas with dense understory vegetation. Wu et al. (2006) found that Siberian musk deer (*M. moschiferus*) selected for steep, rocky, higher elevation slopes near water and away from human disturbance. Musk deer are named for the musk gland, or "pod," that develops slight-

Figure 20.18 **The reticulated pattern of giraffes.** This pattern varies considerably both in shape and color. Giraffes feed selectively, and sex can be determined at a distance by feeding behavior. Males always stretch with the head extended upward, in contrast to females, which feed with their head down over the vegetation. Sexes therefore forage at different heights, reducing potential competition.

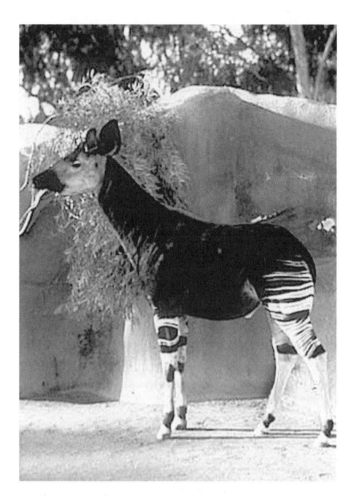

Figure 20.20 The okapi. This rare member of the giraffe family has a very distinctive coloration pattern.

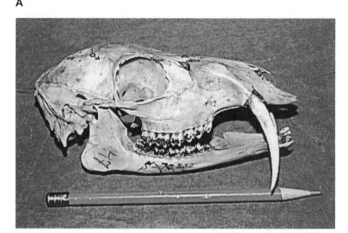

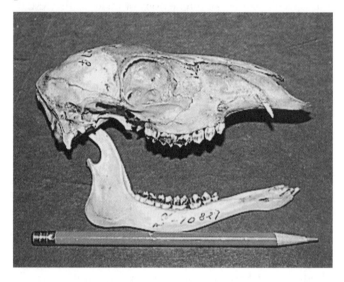

Figure 20.21 Musk deer skulls. (A) Male musk deer have no antlers, but enlarged, tusk-like upper canines are evident. (B) Upper canines are significantly smaller in female musk deer. Scale: pencil = 16 cm.

ly anterior to the genital area of males. In adults, this gland contains 18–32 g of reddish brown, gelatinous, oily material, probably important during the breeding season for attracting females and marking a territory. Musk is used both in fine perfumes and as an ingredient in Asian traditional medicine. As a result, hunting pressure on musk deer has always been exceptional. Musk deer have been produced on game farms for about 40 years to help ensure a reliable supply of musk and to reduce harvest pressure on wild populations.

Cervidae

The deer family includes 19 genera and 51 extant species grouped into 3 subfamilies (table 20.1). Alternative taxonomic arrangements have been proposed (Gilbert et al. 2006). Deer range in size from the southern pudu (*Pudu puda*), with a maximum body mass of 8 kg, to the moose (*Alces alces*) at 800 kg. Deer are widely distributed and are absent only from sub-Saharan Africa and Antarctica. They are not native to Australia or New Zealand but have been introduced there as well as to many other areas. The habitats occupied by deer are as varied as their size and include deciduous forests, marshes, grasslands, tundra, arid scrublands, mountains, and rainforests.

Deer exhibit sexual dimorphism, with males often being 25% larger in body mass and body dimensions than females. Male deer are also well known for their characteristic antlers, which are made of bone, are usually branched, and are supported on pedicels, which are raised extensions of the frontal bone (see figure 20.7C). Kierdorf and Kierdorf (2002) discussed the anatomical and histological development of pedicels from "antlerogenic periosteum." Antlers usually occur only in males. Exceptions include Chinese water deer, in which males do not have antlers, and caribou (*Rangifer tarandus*), in which both sexes have antlers. Antlers are deciduous; that is, they are shed each winter following the rut and regrown in the spring. Growing antlers are covered with haired, highly vascularized skin known as **velvet.** Antlers are the fastest growing tissue known other than cancer. Antler size and the number of **tines** (points) are a function of nutritional condition, genetic factors, and age. There is little correlation between individual age and the number of antler points,

Table 20.1 Subfamily designations within the deer family (Cervidae), with inclusive genera*

Capreolinae	Cervinae	Hydropotinae
Alces (2)	*Axis* (4)	*Hydropotes* (1)
Blastocerus (1)	*Cervus* (2)	
Capreolus (2)	*Dama* (1)	
Hippocamelus (2)	*Elaphodus* (1)	
Mazama (9)	*Elaphurus* (1)	
Odocoileus (2)	*Muntiacus* (11)	
Ozotoceros (1)	*Przewalskium* (1)	
Pudu (2)	*Rucervus* (3)	
Rangifer (1)	*Rusa* (4)	

*Number of species in each genus is in parenthesis. Taxonomy follows Grubb (2005b).

however. Different species of deer have a characteristic antler shape and size (figure 20.22), from single spikes in pudu, tufted deer (*Elaphodus cephalophus*), and brocket deer (Genus *Mazama*) to large, **palmate** (flattened) antlers with numerous tines in fallow deer (*Dama dama*), caribou, and moose. Antlers are important in mating behavior and success of males and, therefore, are a highly selected trait (Geist 1991).

Pelage color varies among and within species of deer. Most newborn cervids have white spots and lines on a darker background, providing excellent camouflage. Certain species maintain this pattern as adults, including fallow, sika (*Cervus nippon*), and axis deer (Genus *Axis*). Deer have no upper incisors; they browse or graze by cutting off forage between the lower incisors and a calloused upper pad. They are herbivores, and many of the temperate species change their diet depending on the season. Deer are gregarious, with species such as caribou forming herds of 100,000 or more. Visual, auditory, and olfactory senses are acute. Several glands may be present from which herd members gain reproductive and other information about each other. Glands may be located on the face (prelacrimal), between the toes (interdigital), or on the lower hind legs (tarsal or metatarsal glands). Males often fight each other, using their antlers, to establish territories on which to attract a group of females for breeding. Maintenance by males of female breeding groups depends on the tendency of females to remain grouped, and therefore defensible, and the extent to which estrus is synchronized. Gestation is generally 6–7 months. Among cervids, only the European roe deer (*Capreolus capreolus*) is known to exhibit delayed implantation. From 1–3 young are typical depending on habitat conditions and body condition of the female—called a doe, hind, or cow—depending on the species.

Deer have been hunted by humans for thousands of years for both meat and antler trophies. Today, many

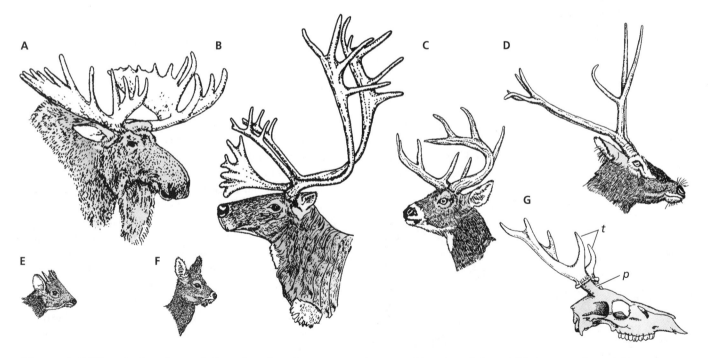

Figure 20.22 Relative size and diversity of antler structure in several species of deer. (A) Palmate antlers of the moose, the largest living cervid. (B) Large antlers of a male caribou. This is the only species of deer in which females also have antlers. (C) White-tailed deer (*Odocoileus virginianus*), a popular game species throughout North America. (D) Pére David's deer, being reintroduced to the wild. (E) The small, spike antlers of the pudu. (F) Male Chinese water deer lack antlers, but like musk deer (see figure 20.21) they have enlarged upper canines. (G) Skull of a sika deer showing the extension of the frontal bone (*p* = pedicel) on which antlers form and the typical points (*t* = tines) in this commonly introduced species. *Adapted from D. Macdonald, 1984, Encyclopedia of Mammals, Facts on File.*

species are domesticated to harvest both meat and antler velvet. Others have been introduced as free-ranging exotics to numerous countries outside their native range, often with unanticipated negative effects on native species (Feldhamer and Armstrong 1993). Overharvesting and habitat loss have resulted in several species becoming endangered, including the Visayan spotted deer (*Rusa alfredi*), Calamian deer (*Axis calamianensis*), Bawean deer (*A. kuhlii*), and Chilean guemal (*Hippocamelus bisulcus*). Pére David's deer (*Elaphurus davidianus*) is considered to be critically endangered. Extinct in the wild for several hundred years, the species has been reintroduced to parts of the original range.

Antilocapridae

This monotypic family includes only the pronghorn, which is endemic to western North America. Pronghorns have barrel-shaped bodies and long, thin legs. Shoulder height reaches about 1 m, and maximum body mass is 70 kg, with males being larger than females. Like deer, pronghorns have several different glands for olfactory communication. Both sexes have rump glands and interdigital glands; males have a gland below each ear and on the back. Chemical properties and possible functions of glands were discussed by Wood (2001, 2002). The horns consist of a keratinized sheath over a permanent, bony core extension of the frontal bone. Unlike members of the Family Bovidae, however, in pronghorns a new horn sheath grows each year under the old one, which splits and is shed following the breeding season. Horns are upright, with a posterior hook and a short, anterior branch, or prong (see figures 20.7D and 20.23). Horns on males may reach 250 mm in length. Horns may be absent on females; if present, they are much shorter and do not have the prong.

Pronghorns occur in open grasslands and semidesert areas, where they forage on grasses, forbs, and low shrubs, especially sagebrush (*Artemisia tridentata*). The pronghorn is the fastest New World mammal. Maximum speeds of up to 100 km/h probably are exaggerated, however (Byers 2003). Males are unusual in that they defend territories from March or April until after the rut, even though

breeding occurs only in the fall. Depending on her body condition, a female searches among territorial males for vigorous potential mates (Byers et al. 2006). However, rutting behavior is variable among populations (Maher 1991). The gestation period is about 250 days, and twins are common. During winter, large herds may form. Pronghorns migrate up to 160 km between distinct summer and winter ranges. The pronghorn is a popular game animal. In the 1920s, populations were very low but have recovered in the United States because of successful management programs. Populations of the Mexican subspecies (*Antilocapra americana peninsularis* and *A. a. sonoriensis*) are endangered, however. Population densities of Sonoran pronghorns are very small with associated loss of genetic diversity (Stephen et al. 2005). Also, predation rates on fawns and adults may be high (Bright and Hervert 2005).

Bovidae

The largest family of artiodactyls includes 50 genera and about 143 species. Indicative of the number of species and diversity of the group, five subfamilies traditionally have been recognized (Simpson 1945). More recently, Simpson (1984) proposed 10 subfamilies, Gentry (1990) proposed six, and Grubb (2001) divided bovids into nine subfamilies (table 20.2). We follow Grubb (2005b) who most recently recognized eight subfamilies of bovids (table 20.2). Designation of subfamilies and tribes within subfamilies is based on considerations of horn structure, cranial and skeletal features, behavior, genetics, feeding strategies, and other factors, all of which are interrelated. Generally, bovids have hypsodont and selenodont cheekteeth, with no upper incisors or canines. All species have four-chambered, ruminating stomachs. Size varies greatly, from the royal antelope (*Neotragus pygmaeus*) with a shoulder height of 250–300 mm and body mass to 2.5 kg, to the bison (*Bison bison*), several species of cattle (Genus *Bos*), and elands (Genus *Taurotragus*) with body mass approaching 1000 kg. All bovids have a pair of horns with the exception of the four-horned antelope (*Tetracerus quadricornis*). Horns are present on males and often on females. They have a bony core, which is an extension of the frontal bone (see figure 20.7E), covered with a keratinized sheath that is unbranched and rarely shed. Horns may be straight, spiraled, or curved (figure 20.24) and grow throughout life. Like antlers in deer, horns may function in defense against predators as well as in intrasexual fighting for access to breeding females. Variation in size and shape of horns reflects the fighting behavior of the species.

Absent naturally only from South America and Australia, domesticated bovids are distributed practically worldwide. Wild species are found primarily in Africa and Eurasia, where they occupy grasslands, savannas, scrublands, and forests. They also occur in harsher environments, including tundra, deserts, and swamplands. All are herbivores, with different subfamilies having different feeding strategies (figure 20.25). Feeding strategies of all extant bovids were categorized by Gagnon and Chew (2000) as frugivore, browser,

Figure 20.23 The pronghorn antelope. The black jaw patch, found only in males, is important in male-male and male-female behavioral interactions. Horns in females are short and without a prong.

Table 20.2 Subfamily designations of the Family Bovidae, according to various authors*

G. Simpson (1945)	C. Simpson (1984)	Gentry (1990)	Grubb (2001)	Grubb (2005b)
Antilopinae (13)	Aepycerotinae (1)	Alcelaphinae (5)	Aepycerotinae (1)	Aepycerotinae (1)
Bovinae (8)	Alcelaphinae (3)	Antilopinae (14)	Alcelaphinae (4)	Alcelaphinae (4)
Caprinae (13)	Antilopinae (8)	Bovinae (7)	Antilopinae (14)	Antilopinae (15)
Cephalophinae (2)	Bovinae (5)	Caprinae (10)	Bovinae (8)	Bovinae (9)
Hippotraginae (9)	Caprinae (11)	Cephalophinae (2)	Caprinae (10)	Caprinae (12)
	Cephalophinae (2)		Cephalophinae (2)	Cephalophinae (3)
	Hippotraginae (3)		Hippotraginae (5)	Hippotraginae (3)
	Neotraginae (7)		Peleinae (1)	Reduncinae (3)
	Reduncinae (2)		Reduncinae (2)	
	Tragelaphinae (3)			

* Number of species in each genus is in parenthesis. As is often true for large families, opinions diverge among authorities about the relationships among bovid genera. Future advances in molecular techniques and additional research in other areas will no doubt result in alternative classifications.

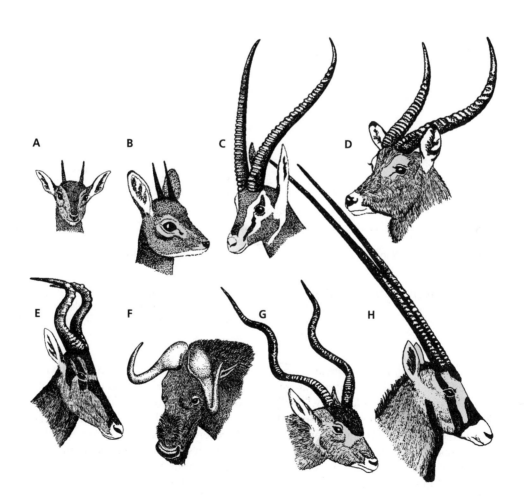

Figure 20.24 Variety of horn shapes and sizes in bovids. (A) suni (*Neotragus moschatus*); (B) klipspringer (*Oreotragus oreotragus*); (C) Grant's gazelle (*Gazella granti*); (D) waterbuck (*Kobus ellipsiprymnus*); (E) hartebeest (*Alcelaphus buselaphus*); (F) gnou (*Connochaetes gnou*); (G) addax (*Addax nasomaculatus*); and (H) oryx (*Oryx gazella*). Not to the same scale. *Adapted from* Horns, Pronghorns and Antlers (*G. A. Bubenik and A. B. Bubenik, eds.*), *Springer-Verlag.*

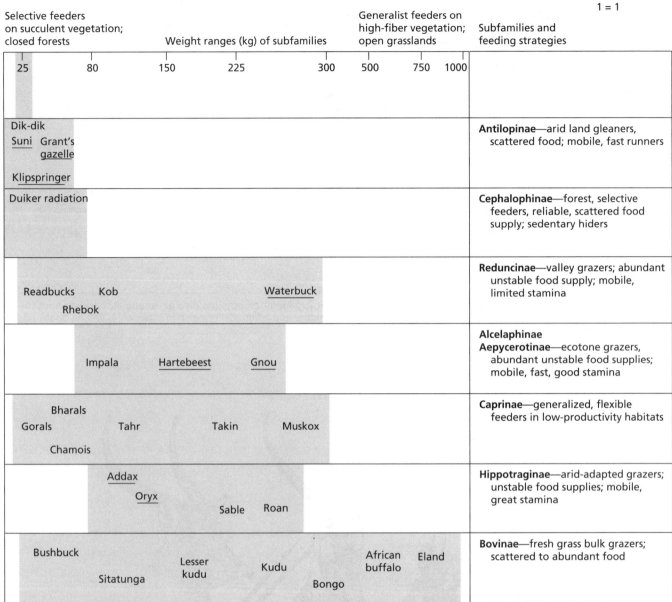

Figure 20.25 Feeding strategies of bovid subfamilies. Ecological niches vary, depending on the size of the species. Underlined names are species shown in figure 20.24. *Data from J. Kingdon, 1972,* East African Mammals, *vol. IIIB:* Large Mammals, *Univ. Chicago Press. Subfamilies; from P. Grubb (2005b),* Mammal Species of the World, *3rd ed. (D. E. Wilson and D. M. Reeder, eds.), Johns Hopkins Univ. Press.*

generalist, browser-grazer, variable grazer, or obligate grazer. Social systems are related to body size, feeding strategy, and predation pressure (Jarman 1974). Generally, small species are specialized feeders in dense, closed habitats and tend to be solitary or paired. Larger species are more gregarious and occupy more open areas, where they are generalist feeders on high-fiber (more cellulose) vegetation. Body sizes of bovids and group sizes are related to feeding modes, as is antipredator strategy (Brashares et al. 2000).

Cattle, sheep, and goats have been domesticated for over 5000 years and are an integral part of agricultural economies throughout the world (chapter 28). Many other species of bovids have been introduced or are hunted for

meat, hides, and sport. Numerous others throughout the world are adversely affected by habitat loss and overharvesting and are considered endangered. A previously unknown bovid, the saola (*Pseudoryx nghetinhensis*), was discovered in Vietnam (Dung et al. 1993). Recent molecular analyses place the saola in the subfamily Bovinae (Gatesy and Arctander 2000). Debate surrounding the validity of another recently described bovid from southeast Asia—the spiral-horned ox (*Pseudonovibos spiralis*) or Kting Voar ("wild cow that eat snake")—is well summarized by Brandt et al. (2001). Galbreath and Melville (2003) suggested this taxon not be recognized, however, until less problematic evidence is obtained.

SUMMARY

Both perissodactyls and artiodactyls are defined on the basis of limb structure. Perissodactyls are mesaxonic, with the main weight-bearing axis of the limbs passing through the third digit, whereas artiodactyls are paraxonic, with the axis of the limbs passing between the third and fourth digits. The superior digestive efficiency of ruminant artiodactyls probably led to the displacement of most lineages of perissodactyls early in the Cenozoic. The digestive efficiency of equids, for example, is only about 70% that of bovids. With the exception of the domestic horse, most species within the three surviving families of perissodactyls are severely depleted in numbers and geographic distribution. The order today represents a remnant of a much more diverse group that was extant in the early Tertiary period. In contrast, artiodactyls, especially the cervids and bovids, are widely distributed. Artiodactyls are very diverse in terms of both size (from the mouse deer to hippopotamus) and life history characteristics. Their success can be traced to superior digestive efficiency, especially the remarkable adaptation of rumination, which allows individuals to select large quantities of high-quality forage that is then processed in relative safety. Predation pressure and foraging behavior have led to major morphological developments in

many families of these two orders toward cursorial, unguligrade locomotion.

Horns and antlers occur in five of the families discussed in this chapter for defense against predators and as part of intraspecific social behavior. Rhinoceroses are the only perissodactyls with horns. These horns are unusual in that they are not made of bone and are not permanently attached to the skull bone. They are made of agglutinated keratin fibers and are attached to the skin over the nasal and frontal area. Giraffes also have an unusual horn structure. The horns are not extensions of the frontal bones but form from separate bones (ossicones) that fuse to the cranium. Antilocaprids and bovids have horns formed from keratinized sheaths over bony extensions of the frontal bone. These sheaths are deciduous in pronghorns but not in bovids. Cervids have antlers that, unlike horns, are made of bone and are branched, deciduous, and generally occur only on males. Horses and several species of artiodactyls are found worldwide, in many cases because of introductions. In contrast, other species such as camels and some bovids are no longer found in the wild but only in domestication. Still others are critically endangered because of habitat loss, poaching, or overharvesting.

SUGGESTED READINGS

Bubenik, G. A., and A. B. Bubenik (eds.). 1990. Horns, pronghorns, and antlers. Springer-Verlag, New York.
Cunningham, C., and J. Berger. 1997. Horn of darkness: rhinos on the edge. Oxford Univ. Press, New York.
Evans, J. W. 2000. Horses: A guide to selection, care and enjoyment. 3rd ed., Owl Books, Henry Holt and Co., New York.

Geist, V. 1998. Deer of the world: Their evolution, behaviour, and ecology. Stackpole Books, Mechanicsburg, PA.
Putman, R. 1988. The natural history of deer. Cornell Univ. Press, Ithaca, NY.
Rue, L. L., III. 2004. The encyclopedia of deer. Voyageur Press, Osceola, WI.

DISCUSSION QUESTIONS

1. Why is it adaptive for deer to shed their antlers every year following the rut, even though it means a considerable energy expenditure to regrow a new set beginning a few months later in the spring? What are the benefits of having deciduous antlers rather than keeping the same set of antlers throughout life?

2. Camels and tapirs both passed most of their evolutionary development in North America. Representatives of both families radiated both north and south during the Pliocene, so they currently are distributed in Central and South America as well as in Asia, even though they died out in North America during the Pleistocene. What reasons might explain why the highly mobile pronghorn antelope never left North America?

3. Consider figure 20.4 depicting the evolutionary development of the horse. Why will you never see a similar figure for rodents or bats?

4. What is the adaptive significance of antlers in female caribou? Consider the northern latitudes and the extreme climatic conditions where caribou occur. Also, keep in mind that males develop antlers before females each year and subsequently cast them in the winter before females do.

5. What kinds of characteristics have plants evolved in response to pressure from herbivores such as perissodactyls and artiodactyls?

6. Considering the length of their neck, what physiological mechanisms do giraffes use to maintain blood flow to their brain, or to regulate blood flow when bent down to drink?

PART 4

Behavior and Ecology

In this part, we tie the anatomical and physiological pieces together and consider what mammals do as they interact with their environment. The approach used in these chapters is to introduce the concepts and illustrate them with mammalian examples. The first four chapters (21–24) deal with animal behavior, a topic that is frequently taught as a separate course. Our emphasis here is on **behavioral ecology**, which deals with the animal's struggle for survival as it exploits resources and avoids predators and with how the animal's behavior contributes to its reproductive success. The first chapter (chapter 21) lays the groundwork for how individual mammals interact and communicate with other members of their species. In chapter 22, we deal with reproductive behavior—the problems of getting mates and caring for young—and then, in chapter 23, with living in groups. How mammals orient in the environment and find a place to live is the subject of chapter 24. Chapters 25 and 26 treat population and community ecology. These two chapters form a continuum that begins with single-species populations, expands in scope to include interactions among different species, and finally includes the structure and function of mixed-species communities.

CHAPTER 21

Communication, Aggression, and Spatial Relations

Behavior has been called the evolutionary pacemaker because an animal's behavior determines whether it can survive and reproduce (Wilson 1975). Natural selection acts on individuals interacting with the environment and with other organisms. Studying the behavior of mammals is particularly challenging because of the great diversity of forms and lifestyles in the class, ranging from solitary species such as northern short-tailed shrews (*Blarina brevicauda*) to highly social forms such as lions (*Panthera leo*) and chimpanzees (*Pan troglodytes*). A second difficulty in studying mammals is that most species are small, nocturnal, and often secretive (Eisenberg 1981), making them much harder to observe than, say, insects or birds. In this chapter, we explore the ways mammals communicate; then we consider how they interact and distribute themselves in space.

Mammals have a wide range of sensory and neural abilities, centered on the neocortex, a structure that has taken over many functions from other, more primitive parts of the brain of other vertebrates (chapter 8). Evolution of this structure is generally credited with giving mammals an ability greater than their reptilian ancestors had to modify their behavior in response to environmental circumstances (i.e., to learn).

Communication

Just as nerves and hormones (chapter 8) convey messages from one part of the body to another, **displays** are behavior patterns that convey messages from one individual to another. Displays carry an encoded **message** that describes the sender's state. The recipient of the message makes **meaning** of the message. Mammals send messages to members of their own species and to other species through a diversity of sounds, colors, odors, and postures. A **signal** is the physical form in which the message is coded for transmission through the environment.

WHAT IS COMMUNICATION?

When a male lion (or a domestic house cat [*Felis catus*], for that matter) sniffs a bush, breathing through open mouth and flared nostrils, then backs up, turns, raises its tail, and sprays urine on the bush, much more than simple excretion is going on. Wilson (1975) defined **biological communication** as an action on the part of one organism (the sender) that alters the probability of occurrence of behavior patterns in another organism (the receiver) in a fashion adaptive to either one or both of the participants. The word **adaptive** implies that the signal or response enhances survival and reproduction and that it is partly under genetic control and influenced by natural selection.

Who benefits from communication: the sender, the receiver, or both? Behavioral ecologists generally argue that the sender must benefit (Slater 1983), and some have compared animal communication with advertising, a method of sending a message in which the sender manipulates the behavior of others. The receiver may benefit or may be harmed. Signals become exaggerated, or **ritualized**, so that the sender can control the behavior of the receiver with a minimum of wasted energy (Dawkins and Krebs 1978; Krebs and Dawkins 1984).

Signalers may send misleading information. **Deceit** is relatively rare in most species, probably because natural selection favors those individuals that ignore or devalue fake signals. An example of deceit in mammals is the behavior of young male northern elephant seals (*Mirounga angustirostris*). They may enter the harems of territorial males, look and act like females so as to not draw the attention of the dominant male, and attempt to sneak copulations. Females, however, respond to any mounting attempt by vocalizing loudly, thereby alerting the harem master, which then intervenes (Cox and Le Boeuf 1977).

PROPERTIES OF SIGNALS

Discrete versus Graded

Animal signals can be divided into two general types: discrete and graded. **Discrete**, or digital, signals are sent in a simple either/or manner. For example, equids (zebras, horses, and asses) communicate hostility by flattening their ears and friendliness by raising their ears (figure 21.1). **Graded**, or analog, signals, on the other hand, are more variable and communicate motivation by their intensity. Thus, equids indicate the intensity of hostile or friendly emotions by the degree to which the mouth opens: the more open the mouth, the more intense the signal. Note, however, that the mouth-opening pattern is the same for both hostility and friendliness.

Distance and Duration

Although mammalian species typically have only 20–40 different displays (Wilson 1975), signals vary in several ways that increase their information content. The distance a signal travels may vary. Low-frequency sounds, such as those made by elephants and some species of whales, travel many kilometers. Visual displays, at the other extreme, usually operate over much shorter distances. The duration of a signal may also vary. Alarm signals, such as those produced by ground squirrels, have a localized, short-term effect and thus a rapid fade-out time. Bright pelage on primates and other male adornments, such as antlers on cervids, may last the entire breeding season.

Composite Signals, Syntax, and Context

Two or more signals can be combined to form a **composite signal**. In the zebra (*Equus* spp.) example, ear position, with the ears forward (friendly) or backward (hostile), is coupled with the degree to which the mouth is open, providing additional information on motivation (see figure 21.1).

Animals possibly convey additional information by changing the **syntax**—the sequence of signals. In human speech, syntax is very important. Contrast the different meanings when the order of words is changed in the following phrases: "The bear eats Steve" versus "Steve eats the bear." Little evidence exists that nonhuman animals use syntax naturally. Laboratory-reared chimpanzees, however, use syntax as they communicate with one another and with their human companions (Savage-Rumbaugh 1986; Savage-Rumbaugh and Brakke 1990).

The same signals can have different meanings depending on the **context**—that is, on what other stimuli are

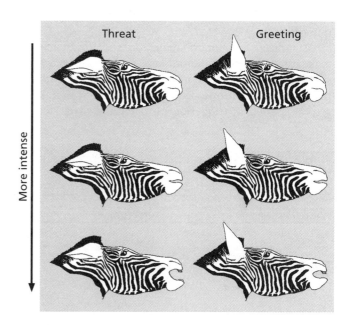

Figure 21.1 Facial signals in zebras. Ears convey a discrete signal. They are either laid back as a threat or pointed upward as a greeting. The mouth conveys a graded signal; degrees of openness indicate the intensity of hostility or friendliness. *Redrawn from E. Trumler, 1959, "Das Rossigkeitsgesicht" und ähnliches Ausdrucksverhalten bei Einhufern. Zeit. für Tierpsychol. 16:478–488.*

impinging on the receiver. For example, the lion's roar can function as a spacing device for neighboring prides, an aggressive display in fights between males, or a means of maintaining contact among pride members.

MODES OF COMMUNICATION

We know that mammals have several sensory systems, including visual, auditory, and olfactory, by which they get information about their environment. Mammals use different **sensory channels** to send signals; the choice of channel depends on the animal's environment and the type of information being sent. Table 21.1 summarizes some of the properties of signals traveling via different channels.

Odor

From an evolutionary standpoint, one of the earliest channels of communication was chemical; odor is used throughout the animal kingdom, with the exception of most bird species. Pheromones are chemical signals that elicit responses in other individuals, usually of the same species. Most pheromones are involved in mate identification and attraction, spacing mechanisms, or alarm. Chemical signals have probably evolved and become widespread for several reasons: such signals can transmit information in the dark, can travel around solid objects, can last for hours or days, and are efficient in terms of the cost of production (Wilson 1975). Because pheromones must diffuse through air or water, however, they are slow to act and have a long fade-out time. Compared with other channels, odors also are somewhat limited in the number of signals that can be sent at one time.

As indicated in many studies on rodents, mammals make extensive use of pheromones. Two general classes of substances, which differ in effect, have been identified: **priming pheromones** that produce a generalized response, such as triggering estrogen and progesterone

production that leads to estrus, and **signaling pheromones** that produce an immediate motor response, such as the initiation of a mounting sequence. Bronson (1971) suggested that pheromones in mice have the following functions:

Priming Pheromones	Signaling Pheromones
Estrus inducer	Fear inducer
Estrus inhibitor	Male sex attractant
Adrenocortical activator	Female sex attractant
Aggression inducer	Aggression inhibitor

Examples of priming pheromones are substances in male house mouse (*Mus domesticus*) urine that speed up sexual maturation in young females (Lombardi and Vandenbergh 1977). Other urinary products stimulate the adrenal cortex and block implantation of embryos. Different signaling pheromones either decrease or increase aggression. It is unclear how many different chemicals are involved; different functions may be served by the same pheromone (Drickamer 1989). The sources of these products include urine, feces, the sexual accessory glands, and a number of specialized skin glands. For example, Belding's ground squirrels (*Spermophilus beldingi*) (figure 23.6) produce distinct odors from at least five specialized glands: the mouth area, the back, the ears, the foot pads, and the anus. Each of these products contains information that uniquely identifies a particular individual, creating a singular bouquet that can be used in social situations (Mateo 2006).

Mule deer (*Odocoileus hemionus*) and other cervids produce pheromones from the tarsal and metatarsal glands on the hind legs, from the tail, and from urine. The leg is rubbed against the forehead, then the forehead on twigs, thus transmitting odors to the twigs (Muller-Schwarze 1971; figure 21.2). It turns out, however, that the primary source of the tarsal gland scent is urine and that the gland has to be recharged about once a day with urine, a behavior pattern referred to as rub-urination (Sawyer et al. 1994). Apparently fat-soluble pheromones present in the

Table 21.1	General properties of the major sensory channels of communication used by mammals			
	Sensory Channels			
Signal Property	**Olfactory**	**Auditory**	**Visual**	**Tactile**
Range	Long	Long	Medium	Short
Transmission rate	Slow	Fast	Fast	Fast
Travel around objects	Yes	Yes	No	No
Night use	Yes	Yes	Little	Yes
Fade-out time	Slow	Fast	Fast	Fast
Locate sender	Difficult	Varies	Easy	Easy
Cost to send signal	Low	High	Medium	Low

Data from John Alcock, 1989, Animal Behavior: An Evolutionary Approach, *4th ed., Sinauer Associates.*

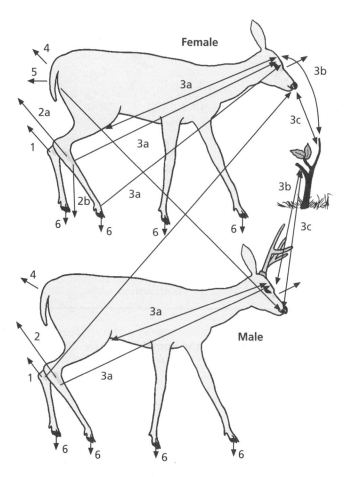

Figure 21.2 Sources of scents used in intraspecific communication in the mule deer. The scents of the following are transmitted through the air: tarsal gland (1), metatarsal gland (2a), tail (4), and urine (5). When the animal lies down, the metatarsal gland marks the ground (2b). The tarsal gland, scented with urinary pheromones, is rubbed against the forehead (3a), and the forehead is rubbed against twigs (3b). Marked objects are sniffed and licked (3c). The interdigital glands (6) deposit scent on the ground. *Redrawn from D. Muller-Schwarze, 1971, Pheromones in Black-Tailed Deer, in* Animal Behavior *19:141–152.*

urine dissolve in sebum from the tarsal gland and convey information about individual identity and social status. Olfactory investigation in ungulates is characterized by **flehmen**, a retraction of the upper lip exhibited soon after sniffing the anogenital region of another or while investigating freshly voided urine (Estes 1972). This behavior pattern is especially common in males during the breeding season but also occurs in females.

Many of the pheromones produced by mammals function as a means of establishing territories or home ranges, much as does bird song. The advantage of pheromones is that the odor may last for many days. Because these substances are often associated with the urinary and digestive systems, eliminative behavior is often highly specialized. For example, spotted hyena (*Crocuta crocuta*) clans mark the boundaries of their territories by establishing latrine areas. Clan members defecate in a particular area and then paw the ground to deposit additional scents from digital

glands. The feces, which are loaded with digested bone, turn white and become quite conspicuous. Hyenas also engage in "pasting." Both sexes possess two anal glands that open near the rectum. When pasting, the hyena straddles long stalks of grass; as the stems pass underneath, the animal everts its anal gland and deposits a strong-smelling yellowish substance on the grass stems (Kruuk 1972; Drea et al. 2002; figure 21.3).

Sound

Much more information about immediate conditions can be transmitted faster by sound than by chemicals. Sound can be produced by a single organ, can travel around objects and through dense vegetation, and can be used in the dark. Information can be conveyed by both frequency and amplitude modulation. The best frequency for a particular animal to use seems to depend on the environment. Brown and Waser (1984) demonstrated that there is a **"sound window"** for blue monkeys (*Cercopithecus mitis*) living in the forests of Kenya and Uganda. At around 200 hertz (Hz), sounds are attenuated very little and are relatively unaffected by background noises. This frequency corresponds to that of the "whoop-gobble" call produced by the adult male, which has a special vocal sac. Blue monkeys also seem to be much more sensitive to sounds in this range than the more terrestrial rhesus monkey (*Macaca mulatta*). Howler monkeys (*Alouatta* spp.) in the Neotropical rain forest also signal to groups many kilometers away with low-frequency calls, in the 40- to

Figure 21.3 Pasting as a form of territorial marking. Spotted hyena pasting by depositing anal gland secretions on a grass stalk.

100-Hz range. Animals with smaller home ranges, for example, squirrel monkeys (*Saimiri sciureus*), use higher-frequency sounds, which dissipate rapidly after striking leaves and branches. Such calls serve to maintain contact among group members or neighboring troops without attracting predators from a distance.

Ultrasonic sounds (frequencies above those audible to humans; above 20 kHz) are used by a variety of animals, particularly mammals. The distress calls of young rodents and some of the vocalizations of dogs and wolves are well above the range of human hearing, as are the echolocation sounds of most bats. Although bat echolocation is used mainly to locate food and other objects, communication also occurs between predator and prey. Some noctuid moths, for example, do not produce sounds themselves, but they possess tympanic membranes on each side of the body that receive sonar pulses from bats (Roeder and Treat 1961). Depending on the location and intensity of sound stimulation, the moth may take evasive action—fly away in the opposite direction, dive, or desynchronize its wing beat to produce erratic flight.

Underwater sound has properties somewhat different from those of sound transmitted though the air, traveling much farther and faster. Marine mammals produce clicks, squeals, and longer, more complex sounds incorporating many frequencies. The short-duration sounds function in echolocation. Baleen whales (Suborder Mysticeti) produce lower and longer sounds than do the toothed whales (Suborder Odontoceti), such as dolphins. Payne and McVay (1971) analyzed the sounds of the humpback whale (*Megaptera novaeangliae*) and determined that their calls are varied and occur in sequences of 7–30 minutes in duration and are then repeated (figure 21.4). The songs have a great deal of individuality, and each whale adheres to its own for many months before developing a new one.

What is the function of humpback whale songs? Researchers now know that only males sing and that songs are sung in the winter breeding areas, so the assumption is that male songs are directed at potential mates, at male competitors, or at both. All males within a region sing very similar songs, and the songs change progressively during the season and over years (Payne and Payne 1985). Surprisingly, within-season changes in one area, near Hawaii, were similar to changes in a second area, off the coast of Mexico, some 5,000 kilometers away. Because it is unlikely that males would move such distances within a season, these findings suggest that there are rules governing change that are similar in both places (Cerchio et al. 2001). Detailed observations of singing behavior and responses to it have been hampered by researchers' inability to pinpoint which males are singing and to observe the responses of distant animals. Special hydrophone arrays are making it possible to pinpoint the singers, and digital acoustic tags can be attached to individuals to remotely record behavior of recipients (Miller et al. 2000).

Some species of terrestrial mammals produce very low (infrasonic) frequency sounds. Desert rodents such as kangaroo rats have greatly inflated auditory bullae, making them sensitive to low-frequency sounds and vibrations

Figure 21.4   Song of the humpback whale. The song of a humpback whale can be broken up into units, phrases, themes, songs, and song sessions. Each whale sings its own variation of the song, which may last up to a half an hour. *Data from R. S. Payne and S. McVay, 1971, Songs of the Humpback Whales, in* Science, 17:585–597.

such as those produced by snakes and other predators. Banner-tailed kangaroo rats (*Dipodomys spectabilis*) defend their territories by foot drumming (Randall 1984), producing sounds in the 200- to 2,000-Hz range. Both Asian elephants (*Elephas maximus*) and African elephants (*Loxodonta africana*) use very low frequency "rumbles," in the range of 14–35 Hz, to communicate over distances of several kilometers (Payne et al. 1986; Poole et al. 1988). Such infrasonic sounds have very high pressure levels (over 100 decibels [db]) and suffer little loss due to the environment. These calls seem to be used mainly for long-distance coordination of group movements and location of mates.

Vision

Visual displays enable the receiver to locate the signaler precisely in space and time and to identify very subtle signals, but they must usually operate in daylight and over short distances. Social and diurnal mammals such as many primates rely heavily on visual cues (chapter 14). Many cursorial, or running, mammals have white rump or tail patches; in the presence of predators, the hairs are erected and the tail waved, making a conspicuous display. Observers have suggested several functions for this **flagging behavior**, such as:

1. Distracting the predator from other members of the group
2. Warning other group members
3. Confusing the predator when many group members are displaying
4. Signaling the predator that it has been detected
5. Eliciting premature pursuit

In the case of (4) or (5), the predator may go off in search of less-alert prey or may be lured into an unsuccessful pursuit (Smythe 1977). Others have argued that rump patches have little to do with defense against predators; rather they function in intraspecific social communication (Guthrie 1971). See Hasson (1991) for a review of communication between predators and prey.

Another example of a pelage feature whose function is disputed is the male lion's mane (figure 21.5). Darwin (1871) suggested that it serves as protection of the vulnerable head and neck region against attacks of rival males. Schaller (1972) suggested instead that it signals the male's quality as a prospective mate, with longer and darker manes indicating superiority. A series of observations and experiments with stuffed lions by West (2005) specifically addressed this issue. She found no evidence to support Darwin's hypothesis that the mane protected males from attacks. Females preferred males with dark manes, while rival males avoided them. Males with the darkest manes were generally dominant and tended to get first access to estrous females. The length of a male's mane was not related to dominance or access to mates, however. As with any signal, there is a cost to producing it. West demonstrated that males with the darkest manes suffered more from the heat than males with lighter colored manes; having higher body temperatures lowers sperm counts. An additional cost was that dark-maned males ate less, as indicated by belly size.

Touch

In most primates and in some gregarious carnivores, grooming is an important social activity and functions not only to remove ectoparasites but also as a "social cement" in the reaffirmation of social bonds (figure 21.6). Females of a number of primate species groom individuals of higher dominance rank more than those of lower rank. There also seems to be competition among group members to groom those at the top of the hierarchy. Females also

Figure 21.6 Female rhesus monkey grooming offspring. In addition to removing ectoparasites and other foreign matter from the skin and hair, grooming acts as "social cement," solidifying social bonds among group members. Although most grooming occurs between close relatives, it increases between unrelated males and females during the mating season.

groom kin more often than nonkin (Schino 2001). Long-term grooming patterns exist between some nonrelatives (Sade 1965; Smuts 1985). Widespread use of tactile stimuli occurs during copulation. In many rodents, stimulation of the rump region of a female in estrus produces concave arching of the back and immobility (lordosis). In mammals such as cats and rabbits, vaginal stimulation induces ovulation (chapter 10).

One of the most unusual structures among mammals is the snout of the star-nosed mole (*Condylura cristata*). It consists of 22 fleshy, mobile appendages. Scanning electron microscope images reveal that these rays are densely populated with mechanosensory organs called "Eimer's organs" that are very sensitive to touch (Catania 1999). When microelectrode recordings were made in moles' brains following stimulation of the rays, the receptive field for the star was a huge area in the somatosensory cortex (figure 21.7; Catania and Kass 1996). One clear function of these structures is in locating and identifying food items, especially important for underground living.

Electric Field

At present, no evidence exists that mammals use electric fields for communication as do some species of fish, but several recent studies suggest that some species can detect the weak electric fields produced by prey animals. The

Figure 21.5 Male lion with mate. Males with darker manes tend to be dominant and to be preferred by females as mates.

Anatomical proportions

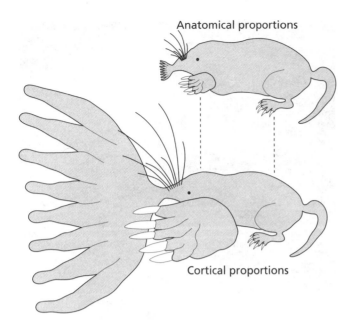

Cortical proportions

Figure 21.7 Cortical representation of body parts in the somatosensory cortex of the brain of the star-nosed mole. (A) The actual proportions of the various body parts. (B) The relative size of each body part as it is represented in the somatosensory cortex. In addition to the huge area represented by the star, note also the large representation of the forelimbs. The latter could be linked to the use of forelimbs in the excavation of tunnels. *Redrawn from K. C. Catania and J. H. Kass. 1996, The Unusual Nose and Brain of the Star-Nosed Mole, in BioScience 46:578–586.*

platypus (*Ornithorhynchus anatinus*) uses sensors on its bill to detect the electric field produced by earthworms (Scheich et al. 1986), and the star-nosed mole has detectors in the tentacles of the star that may serve the same function (Gould et al. 1993). As we pointed out in the preceding section, however, the main neural connections are to the somatosensory cortex, suggesting that touch is the more important sense (Catania and Kaas 1996).

FUNCTIONS OF COMMUNICATION

Although we have briefly reviewed some of the properties of signals and the channels typically used by mammals, we still need to explain what these signals are used for. All communication ultimately functions to increase fitness, but here we consider its more immediate, or proximate, functions (Wilson 1975; Smith 1984).

Group Spacing and Coordination

Cebus monkeys of the South American rain forest forage in dense vegetation. A group of 15 may spread out over an area 100 m in diameter as they search the treetops for fruits. In addition to the sound of moving branches, an observer hears a continual series of "contact" calls from the different members of the group. An individual that becomes isolated utters a "lost" call, which is much loud-

er than the contact calls. Marler (1968) suggested that primates use the following types of spacing signals:

1. Distance-increasing signals, such as branch shaking, which may result in another group's moving away
2. Distance-maintaining signals, such as the dawn chorus of howler monkeys, which regulates the use of overlapping home ranges
3. Distance-reducing signals, for example, the contact or lost calls of *Cebus* monkeys
4. Proximity-maintaining signals such as occur during social grooming within groups

Recognition

Species recognition is important prior to mating to avoid infertile matings between members of closely related species (Mateo 2006). Mammals, however, communicate information that is more individualized than that needed for species recognition alone. Recognition of kin has now been demonstrated in a number of mammalian species—in some cases, even in the absence of interactions with kin early in life (Holmes and Sherman 1983). Kin recognition enables social species to behave nepotistically and thus increase their own inclusive fitness (chapter 23). It may also allow individuals to avoid possible costs of inbreeding incurred by mating with close relatives.

How might kin identify each other even if they have never interacted? One possibility is **phenotype matching**, a process in which the individual uses as a referent, or template, kin whose phenotypes are learned by association. The referent is then compared with the stranger; if the two are similar, then the stranger is treated like a relative (Holmes and Sherman 1983). Belding's ground squirrels recognized littermates later in life, even when separated from each other as neonates (Holmes 1988).

Cues used to assess genetic relatedness may come from genes in the major histocompatibility complex (MHC), a cell recognition system that is used by the immune system to identify self and nonself. In rodents, and possibly other mammal species, genetic differences in this region produce urinary odor cues that can be used by whole organisms to recognize genetic similarities in other individuals (Yamazaki et al. 1976; Brown and Eklund 1994). House mice and some other mammals, including humans, may use this information to choose genetically dissimilar individuals as mates. Such choices would lead to production of heterozygous offspring that would be more resistant to disease. Choosing genetically dissimilar mates would also result in less inbred offspring (Penn and Potts 1999).

Although humans have a relatively poorly developed sense of smell, fragrances such as perfumes have been used for thousands of years to boost sexual attractiveness. It now appears that human preferences for perfume are correlated with their MHC type: persons that have similar MHC types also have similar preferences for perfumes (Milinski and Wedekind 2001). Thus, perfumes may serve to amplify the body odors that reveal a person's immunogenetics.

Reproduction

Sexually receptive females or males may communicate to advertise their identity or condition, court a member of the opposite sex, form a bond, copulate, or perform post-copulatory displays. As described earlier, these behavior patterns are part of species identification, but they may also communicate an individual's reproductive condition. An example of such behavior is the roaring of red deer (*Cervus elaphus*) stags (figure 21.8). Males with high roaring rates obtain more copulations than do males with low roaring rates (Clutton-Brock et al. 1982).

Aggression and Social Status

In social groups in which members of a species are in close proximity, it is sometimes in an individual's best interest to compete with and fight with others for possession of a resource, be it food, space, or access to another individual. Physical combat is expensive in terms of energy and carries the risk of death or injury, even for winners. To avoid unnecessary energy expenditure and risk of injury, social species have evolved displays that communicate information about an individual's mood and about how the animal is likely to behave in the near future (figure 21.9). As a result of previous encounters, the animal may be dominant or submissive to another, and its behavior is thus predictable. Presented with a limiting resource, the submissive individual will yield to the dominant without an overt fight. Many species compete for resources only infrequently; thus, aggressive interactions make up only a small proportion of their behavioral repertoire.

Alarm

Mammals use a variety of signals to alert group members to danger. Although vocalizations used for territory

Figure 21.9 Threat by female rhesus monkey. The recipient of the threat is out of the picture to the upper right. Note that the older of the two offspring is also threatening.

defense (i.e., "songs") are often complex and vary greatly within and among closely related species, alarm calls are likely to be simple, hard to locate, and differ little among species. Members of species that live together and are endangered by the same predators benefit mutually by minimizing divergence among alarm vocalizations (Marler 1973). For example, Marler was unable to tell the difference between the "chirp" alarm calls of African blue monkeys and red-tailed monkeys (*Cercopithecus ascanius*), whereas the male songs differed greatly between the two species. Grivet monkeys (*Chlorocebus aethiops*) communicate **semantically** in that they use different signals to warn about different objects in their environment. Group members climb trees when they hear alarm calls given in response to leopards, look up when they hear eagle

Figure 21.8 Roaring by red deer stag. During the rut on the Isle of Rhum, Scotland, the male roars to defend his harem of females against other males.

alarms, and look down when they hear snake alarms (Struhsaker 1967; Seyfarth et al. 1980a). Young grivets give alarm calls in response to a variety of animals and often not to dangerous ones; however, their ability to classify predators and give appropriate alarm calls improves with age (Seyfarth et al. 1980b).

Mice and rats excrete a substance in their urine when they are electrically shocked, attacked by another mouse, or otherwise stressed. This substance acts as an alarm pheromone and may cause others to avoid the area (Rottman and Snowden 1972).

Hunting for Food

A selection pressure in favor of group living is increased efficiency in finding food, which involves both communication about its location and cooperation in securing it. This attribute is exemplified in the African wild dog (*Lycaon pictus*), a canid distantly related to domestic dogs and wolves. Just prior to hunting prey that often consists of large ungulates, members of the pack engage in an intense greeting ceremony, or "rally," consisting of a frenzy of nosing, lip-licking, tail-wagging, and circling (Creel and Creel 1995). The rally ensures that pack members are alert and ready to hunt in a coordinated fashion (figure 21.10). Social carnivores may even communicate information about what type of prey they are about to hunt. For example, Kruuk (1972) noted that hyenas hunting zebra sometimes passed by prey they had hunted on other occasions. The hyenas' behavior made it apparent even to the human observer that they were interested only in hunting zebra.

Giving and Soliciting Care

A wide variety of signals is used between parent and offspring or among other relatives in the begging and offering of food, as can be observed with domestic cats and dogs. Baby mice, when they are chilled, emit high-

frequency sounds that are inaudible to humans but can be heard by adult mice, who can assist them.

Soliciting Play

Play consists of behavior patterns that may have many different functions in the adult, such as sex, aggression, and exploration. The play bow in canids (figure 21.11) is communication about play and informs others that the motor patterns that follow are not the real thing (Bekoff 1977). Play behavior occurs in a wide variety of mammals and is rare or absent in other taxa. The function of play itself has been debated. Observers usually agree that they can recognize play when they see it, but they have had great difficulty ascribing definitions or functions to it. They most often suggest that the function of play is to practice motor skills and behavior patterns used later in adult life. Play may also serve the immediate function of enhancing muscle development and coordination (Fagen 1981).

Figure 21.11 Soliciting play in domestic dogs. The play bow performed by the dog on the right communicates that the behavior patterns that follow are play.

Aggression and Competition

DEFINITIONS

Aggression is behavior that appears to be intended to inflict noxious stimulation or destruction on another organism (Moyer 1976). The word *aggression* emphasizes offensive behavior. Behavioral ecologists take a more functional approach and consider aggression as a form of resource **competition**, in which rivals are actively excluded from some limited resource, such as food, shelter, or mates (Archer 1988).

Another term often used in the same way as aggression is **agonistic behavior** (not to be confused with agnostic behavior). This term includes all aspects of conflict, both attack (offensive) and escape (defensive). It includes threats, submissions, chases, and physical combat, but

Figure 21.10 Greeting by wild dogs. Wild dogs "rally" before setting out on a hunt.

it specifically excludes predatory aggression, which is included in ingestive behavior (Scott 1972).

The most important forms of agonistic behavior (Wilson 1975; Moyer 1976) include

Territorial: exclusion of others from some physical space

Dominance: control, as a result of a previous encounter, of the behavior of a conspecific

Sexual: use of threats and physical punishment, usually by males, to obtain and retain mates

Parental: attacks on intruders when young are present

Parent-offspring: disciplinary action by parent against offspring, usually associated with weaning

COMPETITION FOR RESOURCES

Most agonistic behavior involves competition for some limited resource—namely, food; water; access to a member of the opposite sex; or space, such as sites for nesting, wintering, or refuge from predators. Competition can be divided into two forms: **exploitation**, in which individuals use resources and deprive others of them without directly interacting, and **interference**, in which organisms interact so as to reduce one another's access to or use of resources. In the tropics, many species of birds and mammals compete for fruits of various kinds. Fruit-eating bats forage by night, whereas fruit-eating birds, primates, and many rodents forage by day. In this instance, bats reduce the resource for the other taxa by exploitative competition with no direct interaction.

More often, competing individuals are not passive but interfere with others seeking the same resource. Some may establish territories and defend resources against others, or they may establish dominance and control the access of others to the resources, as happens during the breeding season when dominant male deer drive subordinate males away from territories containing females. The importance of competition in structuring populations and communities of mammals will become evident in Chapters 25 and 26.

Spatial Relations

HOME RANGE

Most organisms spend their lives in a relatively restricted part of the available habitat and learn the locations of food, water, and shelter in the vicinity. This area, which is used by an animal in its day-to-day activities and in which the animal spends most of its time, is its **home range**. It may be difficult to determine the actual boundaries of the home

range because an animal or group may occasionally wander some distance away to a place it never revisits. Such excursions are rare, however, and identifying the boundaries of an individual's home range is generally straightforward.

The size of the home range depends on the size of the animal as well as on the quality of resources the home range contains. Its size is an approximate function of the mass of the animal and its metabolic rate (McNab 1963). For mammals, $A = 6.76W^{0.63}$, where A (the area) equals the expected home range in acres and W equals the mass in kilograms. Thus a 20-g mouse should have a home range of about 0.57 acres, or 0.23 hectares (ha); this figure is within the observed range. The productivity of the habitat is also important: white-footed mice range farther in less-productive habitats; males also range farther than females within the same habitat as they search for mates. An area that is defended against conspecifics would probably have a smaller expected value for the equation. This relationship between home range size and body mass suggests that ultimately what determines home range size of most mammals is the food supply.

The area of heaviest use within the home range is the **core area**. This location may contain a nest, sleeping areas, water source, or feeding site. As with home range, the designation of a core area is somewhat arbitrary but useful in understanding the behavior and ecology of different species or the same species in different habitats or at different population densities. Figure 21.12 illustrates home ranges and core areas of baboons in Africa.

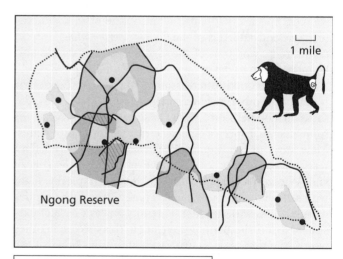

Figure 21.12 Home ranges and core areas of groups of baboons in Nairobi Park, Kenya. Although home ranges of these nine groups overlap extensively, core areas overlap little. *Data from I. DeVore, 1965,* Primate Behavior: Field Studies of Monkeys and Apes, *Holt, Reinhart & Winston.*

TERRITORY

An area occupied more or less exclusively by an individual or group and defended by overt aggression or advertisement is a **territory**. To demonstrate territory, an individual, mated pair, or group must have exclusive use of some space and also must exhibit defense of that area.

Territorial animals spend much time patrolling the boundaries of their space, vocalizing, visiting scent posts, and making other displays. Such behavior would seem to take more time and energy than would simple exploitative competition. In fact, however, these displays often evolved so as to require relatively little energy, for once the territory is established, the neighbors are conditioned and need only occasional reminders to keep out. The cost of defense, then, could be less than the benefit of having exclusive use of a resource.

The key to when an animal should establish a territory seems to be **economic defendability** (Brown 1964), meaning that the costs (energy expenditure, risk of injury, etc.) are outweighed by the benefits (access to the resource). Such things as the distribution of the limited resource in space and whether or not the availability of the resource fluctuates seasonally are important factors. A limited resource—say, food—that is uniformly distributed in time and space is most efficiently used if members of the population spread themselves out through the habitat, possibly defending areas. A resource that is clumped in space and that is unpredictable might favor overlapping home ranges, colonial living, or possibly nomadism. The quality of food may also be important; high-energy food sources are more likely to be defended than low-energy sources. Species that construct elaborate burrow systems (e.g., ground squirrels of the Genus *Spermophilus*) or food caches (e.g., woodrats of the Genus *Neotoma*) should also be more likely to defend territories (Eisenberg 1981).

Although food is often assumed to be the resource that is defended, other resources may be critical and, thus, the focus of competition (Maher and Lott 2000). Males of some mammal species defend territories to acquire mates. Male arctic ground squirrels (*Spermophilus parryii*) defend territories against other reproductively active males (Lacey and Wieczorek 2001). Although females typically mate with several males, the male on whose territory she resides usually mates with her first. Paternity analysis has demonstrated first male sperm precedence in this species; in other words, the first male to mate with an estrous female sires her offspring (Lacey et al. 1997).

A mating system involving a peculiar type of territory is the **lek**. In a lek, the only resource that the organism defends is the space where mating takes place. Feeding and nesting occur away from the site. Lek, or arena, systems are characterized by promiscuous, communal mating; the males are likely to have evolved elaborate ornaments such as horns or antlers. The same area typically is used year after year. Males arrive early in the breeding season and, through highly ritualized agonistic

Figure 21.13 Male topi on lek. Some males of this species of African antelope congregate on small territories close to those of other males in the breeding season, while others defend dispersed territories or none at all.

behavior, stake out their plots. Certain territories or displaying males appear to be more attractive to females than neighboring ones, in the sense that some males do much more breeding than others. Females move through the areas in which the males are displaying, mate with one or more males, and then leave. While on the lek, the males do little or no feeding; they spend all their time and energy patrolling the boundaries, displaying to other males, and attempting to attract females into their area. Although lek mating systems are rare among mammals, they do occur in a wide range of species, from some African antelope species such as topi (*Damaliscus korrigum*; Balmford 1991; figure 21.13) to African hammer-headed bats (*Hypsignathus monstrosus*; Bradbury 1977).

DOMINANCE

When individuals live in a social group, as in many species of primates, carnivores, and ungulates, access to resources may be determined through dominance interactions rather than territoriality. One individual is dominant to another if it controls the behavior of the second individual (Scott 1966). In another sense, dominance is a prediction about the outcome of future competitive interactions (Rowell 1974). If four or five mice that are strangers to one another are put together, several outcomes are possible. Most likely a single mouse becomes a despot, being dominant to the rest, while all the subordinates are more or less equal.

Another possibility is a linear hierarchy, or pecking order, often seen in primates such as baboons (*Papio* spp.)

and macaques (*Macaca* spp.) and in spotted hyenas, in which A dominates B, B dominates C, and so on.

Sometimes triangular relationships form in primates, in which individual A dominates B, who dominates C, who, in turn, dominates A, creating a circular relationship.

Also in primates, especially in chimpanzees, coalitions may affect dominance, such that A dominates B or C alone, but C and B together dominate A.

Among nonhuman primates, third parties may intervene, sometimes supporting the aggressor and sometimes the recipient of the aggression. The term **alliance** is used when two individuals repeatedly form coalitions. Alliances have been documented not only in nonhuman primates but also in bottlenose dolphins (*Tursiops truncatus*; Connor et al. 1992).

Much variation occurs in the intensity of dominance and the frequency of reversals, in which the subordinate individual wins an encounter with a dominant individual. Among closely related species and subspecies, one may see a clear-cut hierarchy in one species and not in the other. For example, captive African green monkeys (*Chlorocebus sabaeus*) from West Africa show little or no dominance hierarchy, even when access to some highly prized food is limited; however, the closely related grivet monkey (*C. aethiops*) from East Africa has a pronounced linear hierarchy. Furthermore, species that are territorial in the wild may, when crowded in captivity, change to a dominance hierarchy. Sometimes the dominance rank order is a function of the resource for which the animals are competing; for example, one individual may have first access to water, another to a favored breeding site.

Costs and Benefits of Dominance

 Studies of mammals living in groups indicate that dominant animals are well fed and healthy. Subordinates may be malnourished or diseased and thus suffer higher mortality rates. Christian and Davis (1964) reviewed data for mammals showing that low-ranking individuals, frequent losers in fights, have higher levels of adrenal glucocortical hormones than do dominants. These hormones elevate blood sugar and prepare the animal for "fight or flight."

The cost of this elevation in hormones is a reduction in antigen-antibody and inflammatory responses—the body's defense mechanisms—and a reduction in levels of reproductive hormones.

Field studies of baboons generally support these findings: In groups and stable dominance hierarchies with few rank reversals, dominant males had lower concentrations of adrenal cortical hormones than did subordinates (Sapolsky 1990). Furthermore, although basal testosterone levels were similar in high- and low-ranking males, those levels plummeted in low-ranking males under the stress of being darted and held captive for several hours. High-ranking males actually showed an increase in testosterone under such conditions (Sapolsky 1991).

Dominance is often directly correlated with reproductive success, as in northern elephant seals, in whom the highest-ranking males do virtually all of the mating (Haley et al. 1994). In Japanese macaques (*Macaca fuscata*), both behavioral and genetic data demonstrated that high-ranking males monopolized most mating and paternity, although one third of the young were sired by nontroop males (Soltis et al. 2001). In some systems, the highest-ranking male may be so busy displaying that lower-ranking males copulate with the females. In many species, the male dominance hierarchy is age-graded, with younger, lower-ranking males working their way up the hierarchy as older males die off or leave the group. Thus, low rank does not necessarily mean low lifetime reproductive success. Much-needed studies on lifetime reproductive success began to be published only in the last few decades (Clutton-Brock 1988).

Popular literature has emphasized the importance of dominance and aggression in "keeping the species fit" because the survivors of fierce battles are the most vigorous and therefore "improve" the species by passing those traits on. The simplest and most likely explanation of aggression and dominance, however, is that dominant individuals benefit. As with other traits, an optimum level of aggression exists, depending on the individual's particular social and physical environment. Individuals that are too aggressive are selected against, as are those that are too passive. For example, female olive baboons (*Papio anubis*) show a linear dominance hierarchy, and dominant females have priority of access to scarce resources. Although high-ranking females have shorter interbirth intervals and higher infant survival rates than low-ranking females, they also suffer more miscarriages and long-term infertility (Packer et al. 1995). These results suggest that qualities essential to achieving high rank may also carry reproductive costs.

In his studies of cooperatively breeding carnivores, Creel has (2005) found that dominant individuals actually have higher levels of stress hormones than subordinates. Comparing the dwarf mongoose (*Helogale parvula*), the African wild dog, and the gray wolf (*Canis lupus*), he found that in one or both sexes dominant group members had higher levels of circulating glucocorticoid hormones than did subordinates. In some instances these increases were associated with higher levels of aggression. Thus, our earlier assumption that lower-ranking individuals are always under more stress needs to be reevaluated.

Extreme Forms of Aggression: Infanticide and Siblicide

The hanuman langur monkey (*Semnopithecus entellus*) lives in a variety of habitats in India. In some areas, social groups of langurs contain only one adult male, 5–10 females, and their young. Several observers reported instances in which a new male came into the group, chased out the old male, and killed some or all of the infants. One interpretation of such behavior is that the group is socially disorganized by the change in males, and the usual restraints on overt aggression are absent. Once the new male has established himself, he ceases his attack on the young. From the standpoint of the species or group, such behavior is maladaptive and should be rare, which, in fact, it is. A second interpretation of these events is that **infanticide** is adaptive, at least for the new male, because he removes the offspring of the presumably unrelated male and causes the females to come into estrus sooner to bear his own offspring (Hrdy 1977a; 1977b). This view is consistent with the notion that social behavior results from the action of natural selection on the individual.

When new male lions first enter a pride in an attempted takeover, they kill the smaller cubs and evict the older cubs and subadults (Pusey and Packer 1987). Infanticide by incoming males accounts for about one fourth of cub mortality (Packer et al. 1990); females therefore group their young into crèches to protect them from nomadic males. Males benefit by killing cubs because they eliminate the offspring of competing males and bring the females into estrus sooner.

Infanticide is also performed by a variety of rodent species and may explain why females defend territories (Wolff, 1993). Under crowded conditions rodents may even eat their young, usually those already dead. Such behavior could be interpreted as social pathology, which results from crowded conditions and leads to maladaptive behavior. Note, however, that under crowded conditions or when food is scarce, the young would probably not survive anyway, so the parents save their reproductive investment for more propitious times by consuming their young. Under adverse conditions, rodents and lagomorphs resorb developing embryos in the uterus, with the same effect as cannibalism. In kangaroos, most of the development of the young takes place in the pouch, so females can terminate "pregnancy" simply by throwing the young out of the pouch.

Such killing of young is not limited to adults. Among spotted hyenas, the young fight vigorously while in the den, forming dominance relationships within hours. One sibling often kills the other, referred to as siblicide, especially when the two are of the same sex (Frank et al. 1991). When food supplies are adequate, however, both cubs typically survive and become close allies (Smale et al. 1995). For additional papers on infanticide and siblicide in mammals, see Hausfater and Hrdy (1984), Parmagiano and vom Saal (1994), Ebensperger (1998), and Dobson et al. (2000).

SUMMARY

Communication may be defined as an action on the part of one organism that alters the behavior of another organism in a fashion adaptive to either the sender alone or both the sender and the receiver. Behavior patterns that are specially adapted to serve as social signals are termed displays. Some definitions of communication imply that both sender and receiver must benefit and that evolution moves toward maximization of information transfer. An alternative view argues that communication is a means by which the sender manipulates the receiver, who may benefit or may be harmed; the purpose of a display is to persuade, not to inform.

Social signals, which vary in fade-out time, effective distance, and duration, convey information by being discrete or graded. They may be combined to form composite signals, and the order in which they appear may affect the information transmitted (syntax).

Channels of communication include odor (mainly via pheromones), sound, touch, and vision. Communication functions in group spacing and coordination; species, population, mate, and kin recognition; reproduction; agonism and social status; alarm; hunting for food; giving and soliciting care; and soliciting play.

Agonistic behavior, defined as social fighting among conspecifics, includes all aspects of conflict, such as threat, submission, chasing, and physical combat, but excludes predation. Aggression emphasizes overt acts intended to inflict damage on another and may include predation, defensive attacks on predators by prey, and attacks on inanimate objects. Agonistic behavior involves competition, one form of which is exploitation, as individuals passively use up limited resources. A second form is interference, in which individuals actively defend resources.

Much conflict behavior involves the social use of space. The home range is the area used habitually by an individual or group; the core area is the zone of heaviest use within the home range. Territory is the space that is used exclusively by an individual or group and is defended. Individual territories occur most often when a needed resource is predictable and uniformly distributed in space. Some territories include food and water supply, nest site, and mates; other territories contain only one defended resource, for example, a place where only mating takes place, such as in the lek.

Species that live in more or less permanent groups usually develop dominance hierarchies in which individuals control the behavior of conspecifics on the basis of the results of previous encounters. Dominance hierarchies are not always determined by fighting ability; age, seniority, maternal lineage, and alliances with friends have been shown to be important factors in some mammals. Although dominant animals often have the highest reproductive success, there are costs to being dominant, especially in highly social species.

Agonistic behavior is usually highly ritualized and communicative in nature; killing or wounding is infrequent. Recent observations interpret overt aggression such as infanticide as an expression of genetically selfish behavior on the part of an individual rather than as a maladaptive response to abnormal conditions.

SUGGESTED READINGS

Archer, J. 1988. The behavioural biology of aggression. Cambridge Univ. Press, New York.

Bradbury, J. W., and S. L. Vehrencamp. 1998. Principles of animal communication. Sinauer Associates, Sunderland, MA.

Halliday, T. R., and P. J. B. Slater (eds.). 1983. Animal behavior, vol. 2: Communication. W. H. Freeman, New York.

Harcourt, A. H., and F. B. M. de Waal (eds.) 1992. Coalitions and alliances in humans and other animals. Oxford Univ. Press, New York.

Harper, D. G. C. 1991. Communication. Pp. 374–397 in Behavioural ecology: an evolutionary approach (J. R. Krebs and N. B. Davies, eds.), 3d ed. Blackwell, Boston.

Hauser, M. D. 1996. The evolution of communication. MIT Press, Cambridge, MA.

DISCUSSION QUESTIONS

1. Distinguish among each of the following terms: *communication, message, meaning, signal,* and *display,* giving examples of each.

2. Table 21.1 illustrates the properties of different signal channels in terms of propagation through the environment. Construct a second table that shows the primary channels used by each order of mammal. Do you see any trends in the results based on the evolutionary history (phylogeny) or ecology of the groups?

3. Contrasting views have been put forth about the evolution of communication systems in animals: (1) displays evolve so as to maximize information transfer between sender and receiver (i.e., both sender and receiver benefit), and (2) displays evolve so as to maximize the ability of the sender to manipulate the behavior of the receiver. Give examples of mammalian communication systems illustrating both of these views.

4. This chapter presented examples of how aggressive, dominant individuals have greater reproductive success than low-ranking individuals. Why doesn't this trend lead to the evolution of higher and higher levels of aggression?

CHAPTER 22

Sexual Selection, Parental Care, and Mating Systems

Chapters 8 and 10 dealt with how hormones and environmental factors interact to influence reproduction (proximate causation). This chapter seeks to explain why natural selection has led to the reproductive patterns observed in nature (ultimate causation). Why do males in some species of mammals fight vigorously for access to females, but those in others fight little and form long-term pair bonds with females? Why is the male northern elephant seal (*Mirounga angustirostris*) more than three times bigger than the female, the male silver-backed jackal (*Canis mesomelas*) about the same size as the female, and the male roof rat (*Rattus rattus*) smaller than the female (Ralls 1976)? In this chapter, we deal with questions that address the ultimate evolutionary forces that have led to sexual dimorphism in some species of mammals. We will attempt to determine why males of some species provide little care for young but those of other species share parental responsibility equally with the female, and why males and females of some species form pair bonds but in others one sex has multiple mates.

Anisogamy and the Bateman Gradient

Many organisms reproduce asexually during one part of their life cycle and reproduce sexually at other times, but mammals are less flexible and rely exclusively on sexual reproduction. In mammals, the two sexes are anatomically different and are produced in about equal numbers. Female mammals produce few large, sessile, and energetically expensive eggs, and males produce many small, motile, and energetically cheap sperm. This difference in gamete size, referred to as **anisogamy**, sets the stage for a host of differences in the reproductive behavior patterns of males and females (Trivers 1972). The relatively small number of expensive female gametes is likely to be a limited resource for which males compete. A male's reproductive success is likely to be a function of how many different females he can inseminate, whereas a female's success depends on how many eggs and young she can produce. Several features of mammalian reproduction further promote differences in reproductive behavior between the sexes: Fertilization occurs in the oviducts, so it is certain that the female is the biological parent of offspring born to her, but paternity is not as certain. Also, only female mammals gestate and lactate, thus limiting the investment males can make in offspring.

One of the results of these differences is that the sexes are likely to follow different reproductive strategies. Bateman (1948) demonstrated in laboratory populations of fruit flies (*Drosophila melanogaster*) that nearly all the females mated. Some males, however, mated several times, and others failed to mate at all. In other words, the variance in copulatory success was higher for males than for females. Males that copulated most also sired the most offspring, the **Bateman gradient** (Andersson 1994). Females, on the other hand, needed only to mate once to produce the maximum number of offspring.

When reproductive success is determined over an animal's entire lifetime, the Bateman gradient becomes evident, especially in species where male-male competition for access to grouped females is intense. For example, male lions (*Panthera leo*; Packer et al. 1988; figure 22.1) and male northern elephant seals (Haley et al. 1994) have a much higher variance in lifetime reproductive success than do the females of those species.

Sex Ratio

The sex ratios of populations of most species tend to be about 50:50 at birth or hatching but may deviate significantly among adults. The **operational sex ratio** considers only the reproductively active members of the population. Deviations in this ratio can have a large influence on the mating system, as members of the abundant sex compete for access to members of the scarcer one.

According to a model based on maternal condition proposed by Trivers and Willard (1973), the Bateman gradient sets the stage for another possible deviation from a 1:1 sex ratio. The model assumes that mothers in the best physical condition produce healthier offspring that are better able to compete for mates or other resources. Because males are more likely than females to compete for additional mates, the reproductive success of males is highly variable, and males should require greater maternal investment. For example, male Antarctic fur seal pups (*Arctocephalus gazella*) are heavier at birth, grow faster, and weigh more at 60 days than their sisters (Goldsworthy 1995). Trivers and Willard (1973) argue that reproductive success of sons should be high if their mothers are in good condition but low, perhaps zero, if their mothers are in poor shape. Female offspring are likely to breed anyway, regardless of their mother's condition. Citing data from mammals such as mink, deer, seals, sheep, and pigs, these investigators predicted that a female would produce male offspring if she were in good condition and female offspring if she were in poor condition.

Some studies lend support to Trivers and Willard's model. Dominant red deer (*Cervus elaphus*) hinds (i.e., females) have access to the best feeding sites and are able to invest more in their offspring via lactation (Clutton-Brock et al. 1984; figure 21.8). Male offspring in this species grow faster than do females and would seem to benefit more from this greater investment. As adults, males compete intensely to control harems. A female that produces a successful son can achieve more than twice the reproductive success of a female producing a daughter. As predicted by Trivers and Willard (1973), a positive correlation exists between dominance rank of hinds and the percentage of sons produced (figure 22.2). This relationship disappeared at high population density, however (Kruuk et al. 1999). In a field study of common opossums (*Didelphis marsupialis*) in Venezuela, females receiving supplemental food produced more sons than daughters (Austad and Sunquist 1987). Furthermore, in a sample of old females in poor condition, the sex ratio was skewed toward female offspring, as predicted. Not all evidence supports the hypothesis, however. In European roe deer (*Capreolus capreolus*), the situation is reversed in that large, high-quality mothers invest more in daughters than in sons (Hewison et al. 2005). One possible reason is that mating success among males is not that much more variable than for females, so daughters may be more valuable than sons.

Males may also benefit from manipulating the sex of their offspring. Continuing with the example of red deer, males vary widely in fertility (Gomendio et al. 2006). More fertile males have faster-swimming sperm and fewer abnormal sperm than do less fertile males. More fertile males also have larger and more elaborate antlers, preferred by females, that may be inherited by their sons.

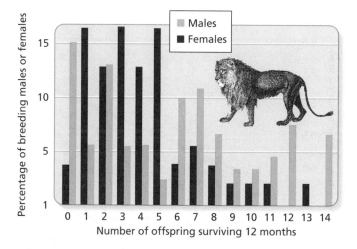

Figure 22.1 Lifetime reproductive success for African lions. Female reproductive success varies little; most females produce 1–5 surviving offspring. Male reproductive success varies widely; males sire 0–14 offspring. *Data from C. Packer et al., 1988, Reproductive Success of Lions, in* Reproductive Success *(T. H. Clutton-Brock, ed.), Univ. Chicago Press.*

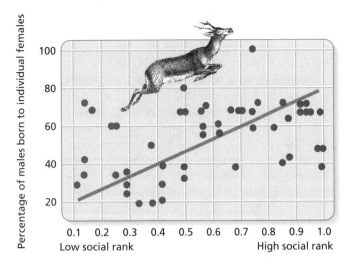

Figure 22.2 Red deer birth sex ratios. Birth sex ratios produced by individual red deer females differing in social rank over their life spans. Measures of maternal rank were based on the ratio of animals the subject threatened or displaced to animals that threatened or displaced it. High-ranking females tended to have sons. *T. H. Clutton-Brock, S. D. Albon, and F. E. Guinness, 1984, Maternal Dominance, Breeding Success and Birth Sex Ratios in Red Deer,* Nature *308:358–360.*

When females of similar condition were maintained in enclosures with ample food and then were inseminated with sperm from males of varying fertility, those inseminated by the most fertile males produced the most sons (Gomendio et al. 2006; figure 22.3), consistent with the Trivers and Willard hypothesis.

Models different from that of Trivers and Willard have been proposed to explain biased sex ratios in other mammals. In studies of a South African strepsirhine primate, the brown greater galago (*Otolemur crassicaudatus*), Clark

(1978) noted that the population sex ratio was male-biased. Female offspring tended to remain near the mother's home range, but males dispersed, the usual patterns for mammals (see chapter 24). Daughters therefore compete with their mothers and sisters for food. By producing fewer daughters, Clark (1978a,b) argued, local resource competition is reduced. A troop of hamadryas baboons (*Papio hamadryas*) living in Amboseli National Park, Kenya, has been studied for many years by Altmann and colleagues (1988). Dominant females produce more daughters than sons, whereas subordinate females produce more sons than daughters, the opposite result predicted by Trivers and Willard. Altmann and colleagues (1988) point out that daughters, who remain in their natal group for life, share the social rank of their mothers and can benefit from their mother's high rank. Presumably, high-ranking families get larger and even more powerful as more and more daughters are born. On the other hand, as with sons of brown greater galagos, baboon sons usually leave the natal troop and thus do not benefit by their mother's rank. In these baboons, the best strategy for dominant females is to produce daughters, whereas for subordinate females, it is to produce sons.

Some species of mammals breed cooperatively such that offspring from previous litters help parents rear the next crop of young. If one sex helps more than the other, parents would benefit by producing more of that sex. For example, alpine marmots (*Marmota marmota*) live in colonies with a territorial pair, subordinates, yearlings, and young of the year. Subordinates warm juveniles, groom them, and cover them with hay, thus increasing their survival. In a field study in the French Alps, the overall sex ratio of marmots born into the population was male-biased (Allaine et al. 2001). Furthermore, juvenile survival increased with the number of subordinate males in the hibernaculum, but not with the number of females.

A

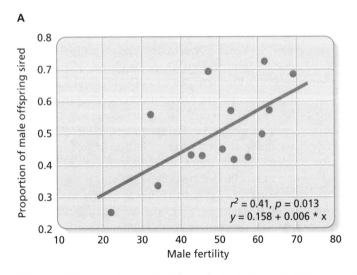

B

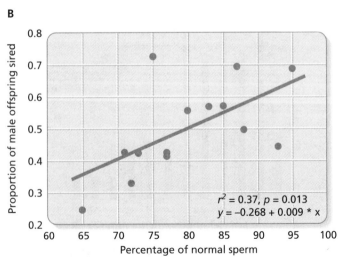

Figure 22.3 Red deer birth sex ratio, father's fertility, and father's percentage of normal sperm. The proportion of male offspring born to females artificially inseminated with sperm from males varying in fertility and percentage of normal sperm is shown here. The sex ratio of offspring was skewed toward males with sperm from males of high fertility and a high percentage of normal sperm. *Data from M. Gomendio et al., 2006, Male Fertility and Sex Ratio at Birth in Red Deer,* Science *314:1445–1447.*

These findings support the helper repayment model, in which the helping sex becomes less costly to produce because it repays the parents by helping to rear offspring. The parents can thus afford to produce more of that sex.

It is not clear how such adjustments of sex ratios occur at the proximate level. Some evidence points to the timing of insemination within the estrous cycle, such that more of one sex is conceived earlier and more of the other sex later in the cycle, as in white-tailed deer (*Odocoileus virginianus*; Verme and Ozoga 1981), Brown [Norway] rats (*Rattus norvegicus*; Hedricks and McClintock 1990), golden hamsters (*Mesocricetus auratus*; Huck et al. 1990), and humans. A number of studies, including those in humans, suggest that levels of sex hormones present at the time of conception affect offspring sex ratios (James 1996). In many species of mammals, the sex ratio at conception is male-biased. Intrauterine mortality rates are typically higher for male than for female embryos, and these rates are probably highest among mothers in poor condition. In this way, the maternal condition hypothesis could be explained, because females in poor condition would lose their male embryos and possibly come into estrus again. After birth, other mechanisms may operate to select for one sex or the other.

Sexual Selection

The success of an individual is measured not only by the number of offspring it leaves but also by the quality or probable reproductive success of those offspring. Thus, it becomes important who its mate will be. Darwin (1871) introduced the concept of **sexual selection**, a process that produces anatomical and behavioral traits that affect an individual's ability to acquire mates. He believed that sexual selection was a different process than natural selection because the former could produce traits that might reduce an individual's survival and thus oppose natural selection. Sexual selection can be divided into two types: one in which members of one sex choose certain mates of the other sex (**intersexual selection**) and a second in which individuals of one sex compete among themselves for access to the other sex (**intrasexual selection**). One result of either mechanism is that the sexes come to look different, that is, they are **dimorphic**.

INTERSEXUAL SELECTION

Often, individuals of one sex (usually the males) "advertise" that they are worthy of an investment; then members of the other sex (usually the females) choose among them. Most naturalists after Darwin discounted the importance of mate choice in evolution, but it has recently become a popular topic of study. Fisher (1958) developed a model for how exaggerated traits could evolve. Suppose some trait that appears in males (e.g., antlers) is preferred by

females, for whatever reason. Assuming that the trait is to some extent heritable, these females are likely to produce sons with that trait and daughters that prefer that trait when choosing a mate of their own. Further development of the trait proceeds in males, as does the preference for it in the females, resulting in a **runaway selection** process. Traits produced by runaway sexual selection are arbitrary and are not linked to a male's fitness, except by female choice.

Sexually selected traits may become so exaggerated that survival of the males is reduced. Counterselection in favor of less ornamented males should then occur, and the effects of natural and sexual selection would be brought to a steady state. An often-cited example of runaway sexual selection is the extinct Irish elk (*Megaloceros giganteus*). The enormous antlers of Irish elk—much larger than those of the extant North American elk or wapiti (*Cervus elaphus canadensis*) shown in figure 22.4—apparently functioned for social display, but little evidence supports the notion that heavy antlers directly caused their extinction (Barnosky 1985).

As alternatives to Fisher's runaway selection hypothesis are several "good genes," or **indicator models**, which assume that the trait favored by females in some way indicates male fitness. One is the **handicap** hypothesis proposed by Zahavi (1975). Agreeing that sexual selection can produce traits that are detrimental to survival, he adds that they are both costly to produce and linked to superior qualities in the males. Important to this hypothesis is the notion of **"truth in advertising"**: the male's handicap must be honest and linked to overall genetic fitness. Only in this way do females benefit by picking a male with this handicap. Antlers of deer could be viewed as a handicap because any male that can bear the cost of such structures and still survive must be fit indeed. One problem with this model is that when a male with a handicap is picked by a female, not only are his favorable traits passed on to his offspring but so are the genes for the handicap, which should be selected against.

Although Darwin and Fisher thought of sexual selection as a process distinct from natural selection, some

Figure 22.4 Male North American elk with antlers.
A sexually selected trait.

biologists argue that the two are inseparable. Kodric-Brown and Brown (1984) claim that most sexually selected traits are aids, rather than handicaps, to survival. Thus, a male deer with a big set of antlers may be dominant over other males and may have better access to a food supply. The researchers further argue that the sexually selected trait must be a reliable indicator of the male's condition. Thus, the trait must not be entirely genetically fixed but rather must be influenced by environmental conditions.

Evidence for an indicator role for a sexually selected trait comes from a hypothesis put forth by Hamilton and Zuk (1984). Working with birds, they argued that sexually selected traits have evolved to reveal an animal's state of health, specifically whether or not it is free from disease or parasitism. They predicted that species in which the males have showy plumage would be more prone to infestation with blood parasites. Comparisons of plumage showiness among museum specimens confirmed their prediction. They also predicted that within such species, brightness of plumage is linked to overall condition, such that brighter males are relatively disease-free and are therefore preferred as mates. Some experimental evidence, mostly from birds, supports their model, but the idea is still controversial, and alternative explanations have been proposed. So far, there are few tests of the Hamilton-Zuk hypothesis in mammals. In humans, Low (1990) found no direct evidence that sexual selection was linked to pathogen stress, although societies with high levels of pathogen stress tended to be polygynous, with some males having more than one mate.

In addition to the size or color of sexually selected traits, the symmetry of paired traits may also indicate fitness. **Fluctuating asymmetry** refers to random deviations from bilateral symmetry in paired traits (Andersson 1994). This means, for example, that when a paired trait, such as horns or canines, is measured for length, thickness, or some other attribute, the right and left sides may differ. These deviations are thought to reflect the inability of the organism to maintain developmental homeostasis (i.e., symmetry) in the presence of environmental variation and stress. Greater asymmetry is associated with low food quality and quantity, habitat disturbance, pollution, disease, and genetic factors such as inbreeding, hybridization, and mutation. For example, the stress of vegetation removal caused an increase in asymmetry of mandible development in least shrews (*Sorex cinereus*; Badyaev et al. 2000). Northern elephant seals suffered a severe loss of genetic diversity after overhunting by humans at the end of the 19th century. After this genetic "bottleneck," there was an increase in asymmetry of a number of bilateral skull characteristics (Hoelzel et al. 2002).

Symmetrical individuals are likely to be dominant and preferred as mates. Thus, male fallow deer (*Dama dama*) with symmetrical antlers were dominant over those with asymmetrical antlers (Maylon and Healy 1994), and male oribi (*Ourebia ourebi*) with symmetrical horns had larger harems than did asymmetrical males (Arcese 1994).

Fluctuating asymmetry may indicate fitness in both sexes. Among gemsbok (*Oryx gazella*), both males and females with asymmetrical horns were in poorer condition and lost more aggressive encounters than did those with symmetrical horns (Møller et al. 1996). Furthermore, symmetrical males were more often territorial breeders than were asymmetrical males, and symmetrical females more often had calves than did asymmetrical females.

Fluctuating asymmetry is most pronounced in sexually selected traits. For example, canine teeth are used as weapons in male-male competition in a variety of primates. In general, males (but not females) from species subject to the strongest sexual selection (as indicated by size dimorphism and male-male competition) showed the highest asymmetry in canines (Manning and Chamberlain 1993). No relationship was found between canine asymmetry and body mass or diet type. The connection between fluctuating asymmetry and environmental stress is suggested by the marked increase in asymmetry of canines (but not of premolars) of lowland gorillas (*Gorilla gorilla*) during the 20th century (Manning and Chamberlain 1994). Such an increase is consistent with environmental degradation associated with increasing rates of deforestation during that century.

INTRASEXUAL SELECTION

Intrasexual selection involves competition within one sex (usually males), with the winner gaining access to the opposite sex. Competition may take place before mating, as with ungulates such as deer (Family Cervidae) and African antelope (Family Bovidae). Typically, males live most of the year in all-male herds; as the breeding season approaches, they engage in highly ritualized battles, using their antlers or horns. The winners of these battles gain dominance and do most of the mating. Antlers are better developed in those cervid species in which males compete strongly for large groups of females (Clutton-Brock et al. 1982).

It is often difficult to determine which type of sexual selection is operating to produce an observed effect because members of both sexes may be present during courtship. As an example, note that the antlers of deer demonstrate the effects of both female choice and male-male competition. Females may incite competition among males and thus maintain some control over the choice of mate. For example, female northern elephant seals vocalize loudly whenever a male attempts to copulate. This behavior attracts other males and tests the dominance of the male attempting to mate (Cox and Le Boeuf 1977). In response to the female's sounds, the dominant harem master drives off low-ranking, potentially inferior mating partners.

Sperm Competition

Competition among males to sire offspring does not necessarily cease with the act of copulation. Females of some species may mate with several males during a single estrous period, creating the possibility of **sperm competition**, that is, a situation in which one male's sperm fertilize a disproportionate share of eggs (Parker 1970). Rather than being thought of as sperm actually fighting it out to gain access to eggs, sperm competition can be considered a selection pressure leading to two opposing types of adaptation in males: those that reduce the chances that a second male's sperm will be used (first-male advantage) and those that reduce the chances that the first male's sperm will be used (second-male advantage; Gromko et al. 1984).

Adaptations of first males include mate-guarding behavior and the deposition of copulatory plugs, both of which reduce the chance of sperm displacement by a second male. In mammals, adaptations of second males are probably restricted to dilution of the first male's sperm by frequent ejaculation of large amounts of sperm from a second male. Female Rocky Mountain bighorn sheep (*Ovis canadensis*; figure 22.5) usually mate with more than one male during estrus. Dominant rams seek to guard estrous females from forced copulations by subordinate males but are not always successful. If a subordinate male does achieve a copulation, the dominant male immediately copulates with that female himself, probably reducing the chances that the subordinate male's sperm will fertilize the egg (Hogg 1988).

Copulatory plugs occur in rodents (Dewsbury 1988), bats (Fenton 1984b), and some primates (Strier 1992). Although the plugs in domesticated guinea pigs (*Cavia porcellus*) appear to block subsequent inseminations (Martan and Shepherd 1976), those in deer mice (*Peromyscus maniculatus*) are ineffective (Dewsbury 1988). In the latter species, the plugs probably function to retain the sperm within the female's reproductive tract. Dewsbury (1984) found that in the muroid rodents he tested, the last male to mate or the male ejaculating most often sired most of the offspring.

Sperm competition has been suggested as an explanation for the relatively large number of deformed sperm in mammals—up to 40% in humans, for example. Baker and Bellis (1988) argue that these deformed sperm play a "kamikaze" role, staying behind to form a plug to inhibit passage of sperm from a second male. Meanwhile, a small number of "egg-getter" sperm proceed to the oviduct to attempt fertilization. The kamikaze sperm hypothesis has been criticized on several grounds: selection should favor use of seminal fluids rather than sperm to form the plug, and there should be more deformed sperm in ejaculates from species in which the female is likely to mate with several males. No such relationship seems to exist, however (Harcourt 1989). Clearly, more studies are needed to test this hypothesis adequately.

In primates, the amount and quality of sperm that the male produces are related to the type of mating system. In gorillas and Bornean orangutans (*Pongo pygmaeus*), the winners of male-male competition have relatively free access to females. In common chimpanzees (*Pan troglodytes*), however, several males may attempt to mate with a female in estrus. Møller (1988) argues that, in this case, competition takes place in the female's fallopian tubes and that the male with the most and best sperm fertilizes the egg. In fact, chimps have larger testes than other apes and produce a high-quality ejaculate—greater numbers of and more motile sperm.

Among woolly spider monkeys (*Brachyteles arachnoides*), males sometimes form aggregations, taking turns copulating with an estrous female at intervals of about 20 minutes (Milton 1985; Strier 1992). Copious amounts of ejaculate are produced, and a plug forms in the female's vagina. This plug is rather easily removed by subsequent males, so it may not function as a deterrent. As might be predicted from the previous discussion, these monkeys have extraordinarily large testes. Rather than competing aggressively for access to females, these males appear to engage in sperm competition (Milton 1985).

Among mammals in general, a similar relationship exists: Species in which only one male has access to one or more females have smaller testes than do species in which more than one male have access to females (Kenagy and Trombulak 1986; figure 22.6). Some notable exceptions do occur, however. The supposedly monogamous southern grasshopper mouse *Onychomys torridus*) has extremely large testes (nearly five times the predicted size), in spite of the presumed lack of sperm competition. Lions, on the other hand, have rather small testes, in spite of the observation that one male may copulate more than 50 times in 24 hours with the same female, and a female often mates with more than one pride male (Kenagy and Trombulak 1986).

Brown and colleagues (1995) have suggested that factors other than sperm competition among males may explain the differences in relative testes mass among primates. They argue that the size and length of the female's vagina may determine how much sperm is needed to effect fertilization: a larger or longer vagina requires more ejaculate. They note that the vagina of a female gorilla is approximately 10 cm long, but that of a chimpanzee is about 17 cm. The latter is made even longer by

Figure 22.5 Sperm competition in bighorn sheep.
Dominant rams copulate at high frequency immediately after a "sneak" copulation by a subordinate ram.

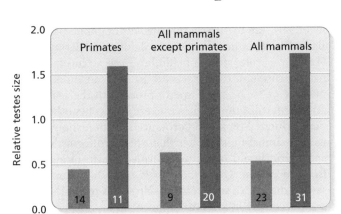

Figure 22.6 Mean relative testes size of mammals in relation to mating system. Sample sizes are at the bottom of each column. Species in which more than one male mates with a female have relatively large testes. *Data from G. J. Kenagy and S. C. Trombulak, 1986, Size and Function of Mammalian Testes in Relation to Body Size,* Journal of Mammalogy *67:1–22.*

the genital swelling during estrus. The factors that affect vagina size and shape are yet to be explored.

Postcopulation Competition

Following conception, male-male competition may take a different form. In some, but not all, species of mice, the **Bruce effect** operates early in pregnancy: a strange male (or his odor) causes the female to abort and become receptive (Bruce 1966). Although the effect has been reported in some strains of lab mice (*Mus musculus*) and in some species of the genera *Peromyscus* and *Microtus*, others point out that the phenomenon has not been demonstrated in natural populations and is therefore most likely a laboratory artifact (Mahady and Wolff 2002; Wolff 2003). Among northern plains gray langurs (*Semnopithecus entellus*), strange males may take over a group, driving out the resident male. The new male may then kill the young sired by the previous male (Hrdy 1977a). Females who have lost their young soon become sexually receptive, and the new resident male can inseminate them. Infanticide by adult males thus may be viewed as a second-male adaptation. Similar findings have been made for lions—coalitions of males typically kill cubs on taking over a pride (Packer 1986).

Parental Investment

Sexually reproducing species face a number of reproductive decisions, although not necessarily conscious ones:

How much of the resources available should be spent on reproduction versus continued growth and survival? Of those resources spent on reproduction, how should they be allocated to individual offspring? Once young are born, what may the parents do to improve their offspring's chances to survive? Although most animals provide no care for their offspring, one parent or both provide at least some care in all species of mammals. We define **parental investment** as any behavior pattern that increases the offspring's chances of survival at the cost of the parent's ability to rear future offspring (Trivers 1972). Because an egg requires a greater investment of energy than does a sperm, male and female parental strategies are expected to differ. Because eggs are likely to be limited in number, males are expected to compete for the opportunity to fertilize them and thus are subject to sexual selection, as described in the previous section. Figure 22.7 shows that the cost of parental investment rises more quickly for females than for males. The optimum number of offspring is thus lower for females than for males. A female is likely to mate, but given her already large investment, her ability to invest further may be limited; thus, she is choosy about which male fertilizes her eggs. Males, on the other hand, try to inseminate as many females as possible. Mammalian mating systems in which the male mates with more than one female should be the most common (Trivers 1972).

WHICH SEX INVESTS?

Even though only one sperm fertilizes an egg, millions are generally required in each ejaculation to ensure fertilization

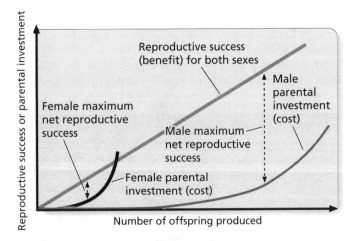

Figure 22.7 Parental investment and reproductive success as a function of number of offspring produced. Because the parental investment (cost) for females usually rises more steeply than for males, the optimum number of offspring (highest reproductive success at lowest cost) for females is less than for males. Males, consequently, must seek more than one mate to attain maximum net reproductive success. Broken lines indicate maximum female and male net reproductive success. *Adapted from R. L. Trivers, 1972, Parental Investment and Sexual Selection, in* Sexual Selection and the Descent of Man *(Bernard Campbell, ed.), Aldine de Gruyter.*

of even a single egg (Dewsbury 1982). Also, the number of times most males can ejaculate within a certain period is limited. Remember also that sperm competition in males of some species leads to selection for increased sperm production. Thus, a male's investment in sperm is not necessarily trivial, and he too can be expected to be somewhat choosy about his mate.

Factors other than anisogamy may affect the contribution of each sex to parental care. Certain taxonomic groups of animals are predisposed to a particular pattern: Male mammals typically mate with more than one female, and the male contributes little to raising the offspring. Trivers (1972) reasons that confidence of parentage might explain which sex cares for the young. In species in which internal fertilization occurs and in which sperm competition could take place, the male has no way of "knowing," consciously or otherwise, whether his sperm fertilized the egg. He might therefore be inclined to desert the females with which he has copulated and seek additional mates. The female, on the other hand, is certain of her genetic relationship to the offspring, so she invests further in current offspring. Another reason why females might stay with the young, suggested by Williams (1975), is that parental care evolved in the sex that is most closely associated physically with the embryos. In mammals, that sex is always female.

Because gestation and milk production are restricted to the female in mammals, the male can do relatively little to provide direct care for the young. In mammals whose young are relatively advanced (precocial) at birth, the opportunities for male investment are even lower. In these species, males compete for multiple mates more than they do in species whose young are immature (altricial) at birth and in which the male can share more evenly the investment with the female (Zeveloff and Boyce 1980).

Competition among males of some mammalian species is intense. For example, mating of northern elephant seals takes place on land in dense colonies; males establish dominance hierarchies, and only the high-ranking males breed (Le Boeuf 1974; Le Boeuf and Reiter 1988; Haley et al. 1994). Typically, less than one-third of the males copulate at all, and the top five males do at least 50% of the copulating (figures 22.8 and 22.9). The males make no investment in offspring beyond the sperm, as evidenced by the fact that males may trample pups as they strive to inseminate females. The males probably have no way of knowing which young are their own because pups are born a year after copulation.

At the opposite extreme are species in which the sexes share more or less equally in care of young, as in canids such as silver-backed jackals (Moehlman 1983). Both parents defend the territory and hunt cooperatively. The males play a crucial role because, in cases in which the father disappears, the female and her offspring die (Moehlman 1986). Among tamarins and marmosets (Family Callitrichidae), females produce twins. Adult males typically show strong interest in the infants soon after birth and carry them as much or more than the

Figure 22.8 Extreme sexual dimorphism in elephant seals. Among a herd of females, two males fight to establish dominance. Males differ strikingly from females; are about three times larger; possess an enlarged snout, or proboscis; and have cornified skin around the neck. In this highly polygynous species, males invest nothing in their offspring other than DNA from sperm.

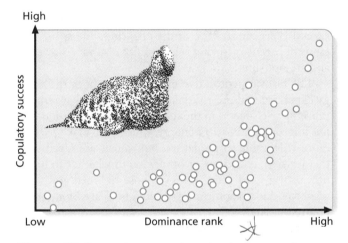

Figure 22.9 Dominance and sex in elephant seals. Copulatory success as a function of dominance in elephant seals for 72 males is shown here. *Data from M. P. Haley et al., 1994, Size, Dominance and Copulatory Success in Male Northern Elephant Seals, Mirounga angustirostris, in* Animal Behaviour *48:1249–1260.*

female does (Vogt 1984). Young from previous years also carry young. The cost of carrying two infants is substantial. In laboratory tests, an adult common marmoset (*Callithrix jacchus*) carrying two newborn suffered a 17% decrease in leaping ability (Schradin and Anzenberger 2001). Under natural conditions, this cost could lead to decreased foraging ability and increased vulnerability to predators. By sharing the task, the costs and risks can be spread among family members.

What are the proximate causes of male care of young? Hormonal changes, such as increased prolactin and decreased testosterone, can lead to paternal care (reviewed by Jones and Wynne-Edwards 2001). In the California mouse (*Peromyscus californicus*), exposure to the pregnant female's urine, as well as other cues from her, enhances paternal behavior. These cues also suppress infanticidal aggression by the male in other rodent species. But in meadow voles (*Microtus pennsylvanicus*), contact with the pups also is necessary. Male Campbell's desert hamsters (*Phodopus campbelli*), a species that lives in the cold deserts of Siberia, participate in all aspects of care except lactation, even assisting in the delivery of the young. In lab experiments, the presence of the pregnant female was not necessary for males to show extensive paternal care toward the pups, suggesting that the response is hard-wired or triggered by some other cue (Jones and Wynne-Edwards 2001). Evidence is accumulating that receptors in the forebrain for the hormone vasopressin are causally linked to increased paternal care (Hammock and Young 2005).

PARENTAL CARE AND ECOLOGICAL FACTORS

The extent of parental care varies widely among different mammalian species. Parental care in primates may last for several years, up to 25% of the offspring's life span. The kinds of parent-offspring relationships also vary, however, and are not simple to describe.

The term **reproductive effort** denotes both the energy expended and the risk taken for breeding that reduces reproductive success in the future. Finding mates and caring for young take extra energy, which may make parents more vulnerable to predators. Individuals are faced with the decision, conscious or otherwise, of whether to breed now or wait until later. If they choose to breed now, they need to decide if some effort should be spared for another attempt later.

Several environmental factors influence the investment parents make in their young after birth. For example, species adapted to stable environments have a tendency to have a larger body size, develop more slowly, enjoy a longer life span, and bear young at repeated intervals (**iteroparity**) rather than all at once (**semelparity**). Semelparity is very rare in mammals (see chapter 25). Typically, iteroparous individuals occupy a home range or territory. These stable conditions favor production of small numbers of young that receive extensive care and thus have a low mortality rate. Such species are sometimes said to be **K-selected**, in reference to the fact that populations are usually at, or near, *K*, the **carrying capacity** of the environment (Pianka 1970). Intraspecific competition is likely to be intense, and the emphasis is on producing few, high-quality offspring rather than large numbers (e.g., gorillas).

Species that are adapted to fluctuating environments have high reproductive rates, rapid development, and small body size and provide little parental care. Their populations tend to be controlled by physical factors, and their mortality rate is high. Such species are said to be **r-selected**, where *r* refers to the reproductive rate of the population (Pianka 1970). For example, meadow voles are considered to be r-selected: They mature at an early age and have larger, more frequent litters than other, similarly sized rodent species.

In environments in which survival of offspring is low and unpredictable, parents may be expected to "hedge their bets" and put in a small reproductive effort each season. The predictions of **bet-hedging** thus seem to contradict those of *r*- and *K*-selection. In unstable, unpredictable environments, *r*-selection should operate, yielding high reproductive rates. Bet-hedging theory, however, predicts low reproductive rates and the spreading of reproductive effort across many breeding seasons. These and other problems have led many researchers to reject the whole concept of *r*- and *K*-selection (Stearns 1992), but some mammalogists still find it a useful way to classify patterns of mammalian life history. For a further discussion of life history traits, see chapter 25.

Prolonged dependency and extensive parental care are also favored when a species, such as many larger felids and canids, depends on food that is scarce and difficult to obtain. Much effort is spent searching for prey, and in some species, cooperation is needed for the kill. During the prolonged developmental period, the young benefit from considerable learning through observation of parents and play.

The Old World monkeys (Family Cercopithecidae) and the great apes (Family Hominidae) have the longest period of dependency. Typical of these species is an infancy of 1.5–3.3 years and a juvenile phase of 6–7 years, making up nearly a third of the total life span (figure 22.10). The prolonged dependency in these large-brained species may be related to their complex social life, in which they must recognize and remember interactions with many individuals across long spans of time.

Figure 22.10 Extended family of rhesus monkeys. The 3-year-old son at left will soon leave the group, but will probably not breed for several more years. The 4-year-old daughter at right will remain in the group for life and has had her first infant.

Parent-Offspring Conflict

Anyone who has raised children or grown up with a sibling has observed frequent disagreement between parent and child. It is not unusual to see a mother rhesus monkey (*Macaca mulatta*) swat her 10-month-old infant or raise her hand over her head, thereby pulling her nipple from its mouth. Frequently, the infant responds by throwing a temper tantrum. Much of the conflict can be viewed as a disagreement over the amount of time, attention, or energy the mother should give to the offspring (i.e., the infant wants more than the parent wants to give).

Such conflict in humans is often interpreted as being maladaptive and related to psychological problems of parent or child or to some negative cultural influence. Work with nonhuman primates, however, led primatologists to the idea that the conflict is a natural part of the weaning process and is necessary for the infant to become an independent and functioning member of the social unit (Hansen 1966). Nevertheless, it is still unclear from an evolutionary standpoint why the infant should resist the weaning process.

Trivers (1974) put forth the hypothesis that conflict arises because natural selection operates differently on the two generations. To maximize her lifetime reproductive success, the mother should invest a certain amount of time and energy in current offspring and then wean the young and invest in new young. The mother's optimum investment is a trade-off between investment in current young and effort toward future offspring. The current offspring, however, profits from continued care until the cost to its mother is twice the benefit (since the offspring shares half its mother's genes). At this point the offspring's own fitness starts to decline also because the offspring benefits to some extent by having its mother produce future, related offspring. Trivers' (1974) approach, based on the coefficient of relationship and kin selection (discussed further in chapter 23), makes intuitive sense but has not been rigorously tested.

Among species of birds, where there is typically more than one young in the nest, begging displays often appear to be extravagant and costly, as would be predicted if offspring were trying to manipulate their parents into providing more food than the parents could afford. Begging behavior as a form of parent-offspring conflict has been less studied in mammals, however. In gray seals (*Halichoerus grypus*), the single pup emits begging calls more often when hungry than when satiated, and the mother responds to these begging calls by presenting her nipples (Smiseth and Lorentsen 2001). Begging therefore seems to be an honest signal of need. Given the lack of predation on the isolated islands on which these seals breed, begging is not likely to be very costly. Thus, in this species, there is little indication that begging has become extravagant and costly, as predicted if parent-offspring conflict were at work.

Conflict is not always limited to parents and offspring; siblings may aggressively compete with one another over the distribution of parental care. Pigs (*Sus scrofa*) are born with razor-sharp teeth that they use in fights with siblings to gain access to favored teats for nursing (Fraser 1990; Fraser and Thompson 1991). Among spotted hyenas (*Crocuta crocuta*), litters are usually twins, and pups are also born with fully erupted canines and incisors (figure 22.11). Fighting begins immediately after birth, and if littermates are of the same sex, one is often killed (Frank et al. 1991). Spotted hyenas are unusual in that young are precocial, and both sexes have high levels of circulating androgens at birth.

Mating Systems

Anisogamy prevails in all mammals, with females investing more in each egg than males invest in each sperm. This difference sets the stage for male-male competition for access to females and for attempts by males to mate with more than one female, a condition referred to as **polygyny**. Polygyny results in greater variation in reproductive success for males than for females; for each male that fertilizes the eggs from a second female, another male is likely to fertilize none, as was shown for lions at the beginning of this chapter (see figure 22.1). Sexual selection tends to act more strongly on males than on females among polygynous species; however, not all species are polygynous.

In trying to evaluate the adaptive significance of differences in mating systems, researchers must look at ecological factors as well as historical ones. For example, group size may be related to predator pressure and food distribution. In the open plains, where large predators are present and food is widely distributed and often clumped, omnivorous primates and grazing mammals such as ungulates live in large groups in which mating with several

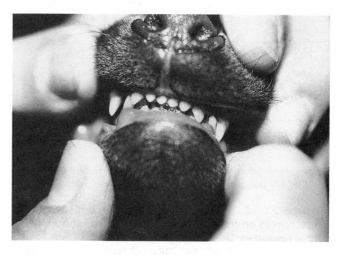

Figure 22.11 Dentition of spotted hyena on day of birth. Canines are 6–7 mm long; incisors are 2–4 mm long. Teeth are used in combat with siblings.

members of the opposite sex is likely for both males and females. In densely forested areas, where communication over long distances is difficult, small family units and monogamy are more common; the same is true where food is uniformly distributed. Thus, the spatial distribution of resources (food, nest sites, or mates) influences the type of mating system (figure 22.12).

Classifications of mating systems traditionally have been based on the extent to which males and females associate (bond) during the breeding season. **Monogamy** refers to an association between one male and one female at a time and includes an exclusive mating relationship between pair members. **Polygamy** incorporates all multiple-mating, nonmonogamous mating systems. The following multiple mating systems are subsets of polygamy. Polygyny is association between one male and two or more females at a time. **Polyandry** is association between one female and two or more males at a time. **Promiscuity** refers to the absence of any prolonged association and multiple mating by at least one sex.

Emlen and Oring (1977) developed the following new ecological classification of mating systems based on the ability of one sex to monopolize or accumulate mates. Although developed primarily for birds, the system seems applicable to mammals as well.

MONOGAMY

In monogamous systems, neither sex is able to monopolize more than one member of the opposite sex. Monogamy is relatively rare in mammals. It is found in

less than 5% of mammal species (Kleiman 1977). Facultative (i.e., optional) monogamy may occur when densities are low and the home range of a male overlaps only that of a single female. Obligate monogamy typically occurs when investment from the male is necessary for the female to rear offspring. When the habitat contains scattered, renewable resources or scarce nest sites, monogamy is the most likely strategy. If there is no opportunity to monopolize mates, an individual benefits from remaining with its initial mate and helping to raise the offspring. The formation of long-term pair bonds also seems advantageous because less time is needed to find a mate during each reproductive cycle. Predation risk is another factor promoting monogamy. Some species live in small social units and behave secretively in order to reduce the chances of being eaten.

Monogamy is reported in most mammalian orders, with the bulk of examples coming from Orders Primates, Carnivora, and Rodentia (Kleiman 1977). Among primates, the forest-dwelling marmosets and tamarins are all monogamous, with extensive male care in the form of carrying of infants. The larger carnivores, especially canids such as jackals, foxes, and wolves, are nearly all monogamous; males carry, feed, defend, and socialize offspring (Moehlman 1986). Among the Rodentia, the prairie vole (*Microtus ochrogaster*) is generally monogamous (Getz and Carter 1980), as is the California mouse. In the latter species, males and females form long-term pair bonds and occupy overlapping ranges distinct from other pairs, and males spend as much time in the nest with young as do females. Ribble (1991) analyzed the parentage of 82 offspring from 22 families of *P. californicus* using DNA fingerprinting and found no instances of mixed paternity among litters. The male and female associated with each litter were, in all cases, the biological parents.

POLYGYNY

In polygynous systems, individual males monopolize more than one female. In **resource defense polygyny**, males defend areas containing the feeding or nesting sites critical for reproduction, and a female's choice of a mate is influenced by the quality of the male and of his territory. Territories that vary sufficiently in quality may cross the **polygyny threshold**—the point at which a female may do better to join an already mated male possessing a good territory than an unmated male with a poor territory (Orians 1969).

Many mammalian species are probably facultatively polygynous. In habitats where feeding or nesting resources cannot be monopolized by males, monogamy is likely, whereas habitats with clumped resources that are defensible would favor polygyny (see figure 22.12). Thus, variation in mating systems of a single species would be expected.

Female defense polygyny may occur when females are gregarious for reasons unrelated to reproduction, as

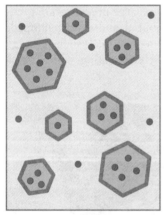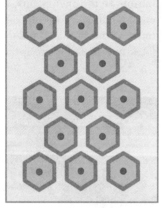

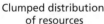

Clumped distribution
of resources

Uniform distribution
of resources

High potential
for polygamy

Little potential
for polygamy

Figure 22.12 The influence of the spatial distribution of resources on the ability of individuals to monopolize those resources. Dots are resources and hexagons are defended areas. Uniform distribution of resources on the right offers little opportunity for monopolization; monogamy is the likely mating system here. *Adapted from L. C. Drickamer et al., 2002,* Animal Behavior, *5th ed., McGraw-Hill.*

when females herd for protection against predators or gather around resources such as food or nesting sites. Some males monopolize females and exclude other males from their harems. In many species of seals (Families Otariidae, Phocidae), the females haul out on land to give birth, and they mate soon after. The females are gregarious because there are limited numbers of suitable sites, and the males monopolize the females for breeding. Intense competition among males results in marked sexual dimorphism and a large variance in male reproductive success, as already described for northern elephant seals (see figure 22.9; Haley et al. 1994).

If males are not involved in parental care and have little opportunity to control resources or mates, **male dominance polygyny** may develop. If female movements or concentration areas are predictable, the males may concentrate in such areas and pool their advertising and courtship signals. Females then select a mate from the group of males. These areas are called leks, and males congregate and defend small territories within them in order to attract and court females. Leks do not contain resources (food or nesting sites) but are purely display sites for mate choice and copulation. Females select a mate, copulate, and then leave the area and rear their young on their own. Often, older or more dominant males occupy the preferred territories or have the most attractive displays and thus do most of the copulating.

Male hammer-headed fruit bats (*Hypsignathus monstrosus*) from central and western Africa display at traditional sites along riverbanks (Bradbury 1977). Each territory is about 10 m apart, and the males emit a loud clanking noise to attract females. Once chosen, a male copulates with the female and resumes calling immediately. Some males, for whatever reason, have much higher success than others at attracting females, and in one year, 6% of the males achieved 79% of the copulations. As might be expected, this species shows extreme sexual dimorphism, with males nearly twice the size of females. Each male has a huge muzzle that ends in flaring lip flaps and a large larynx associated with the clanking sound they make.

Leks also are seen in several species of ungulates, such as Uganda kob (*Kobus kob*; Buechner and Roth 1974) and topi (*Damaliscus korrigum*; Gosling and Petrie 1990). In topi (figure 21.13), Gosling and Petrie (1990) found that the largest and reproductively most successful males defend single territories, while the smaller males cluster in leks. In another study of topi, however, males defending clustered territories near the center of the lek got the most matings (Bro-Jørgensen and Durant 2003). Those males were also larger and older, but they suffered more wounds than males at the periphery. In Uganda kob, a dual male strategy also exists, and males on single territories are less successful than those in leks (Balmford 1991). The reproductive success of lekking male fallow deer varies widely, with a few males having spectacular success while most males do no breeding at all (Appolonio et al. 1992). Competitively inferior males follow a low-risk strategy and defend single territories; in these, they can count on

getting a few copulations but spend less time and energy fighting and displaying.

In the absence of territory or dominance, **scramble polygyny** may operate, as males try to obtain copulations. Where females are widely dispersed, as with 13-lined ground squirrels (*Spermophilus tridecemlineatus*), males become highly mobile during the breeding season. Rather than actively competing with other males for territory or dominance, the most successful breeders are those males that cover the most area in search of estrous females (Schwagmeyer 1988).

POLYANDRY

In polyandrous systems, females monopolize more than one male. Because female investment in gametes and offspring exceeds that of males, polyandry is expected to be rare, especially in mammals, where only females gestate and lactate. We have already pointed out that in most mammal species, females provide the bulk of parental care, with males seeking new mates and investing little in offspring. Under certain circumstances, polyandry might be expected if food availability at the time of breeding is highly variable or if breeding success is very low due to high predation on the young. True polyandry should result in role reversal, with large females competing for smaller mates, female ornamentation via sexual selection, male parental care, and female dispersal in search of mating opportunities. Such a pattern does not occur in mammals, although there are dozens of species of birds in which, except for egg-laying, the male does all the parenting alone.

In many species of mammals, however, females mate with more than one male, and sometimes littermates have more than one father, for example in mice of the genus *Peromyscus* (Birdsall and Nash 1973; Xia and Millar 1991). Such systems, however, are usually referred to as promiscuous rather than polyandrous because males also mate with multiple females, males provide little or no care of offspring, and no lasting bond exists between the partners. Several species of larger canids show most of the features of true polyandry. Although mainly monogamous, the African wild dog (*Lycaon pictus*) sometimes exhibits polyandry; females are occasionally mated by several males, males provide extensive care of pups, and females are the dispersing sex (Moehlman 1986; see figure 21.10).

THE NEUROENDOCRINE BASIS OF MATING SYSTEMS

The neuroendocrine control of mating systems is now beginning to be understood. At the level of the central nervous system, two hormones, oxytocin and vasopressin, have been implicated in the regulation of behavior patterns associated with monogamy, such as formation of long-term pair bonds, defense of nest and mate, and high levels of paternal behavior. In prairie voles, receptors for these hormones in the forebrain differ in number between males and females and between monogamous and polygynous individuals (Young et al. 1998). Young et al. (2001) hypothesize that oxytocin and vasopressin affect affiliation and social attachment in monogamous species by way of reward pathways in the brain.

Vasopressin is necessary and sufficient for aggression against intruders and for pair bonding (Winslow et al. 1993). Receptors for vasopressin are present at higher levels in the forebrain of monogamous prairie voles than in polygynous species such as the meadow vole. A gene, *V1aR*, has been identified that controls the number of vasopressin receptors in the forebrain. Lim et al. (2004) were able to transfer extra copies of this gene into the forebrains of meadow voles. These individuals then behaved more like prairie voles, forming monogamous pair bonds. One factor affecting the activity of the *V1aR* gene is the presence of a microsatellite in the regulatory portion of the gene: The longer the satellite, the more active the gene. As expected, meadow voles have shorter microsatellites in this region than do prairie voles. Additionally, prairie voles with long microsatellites form stronger pair bonds with their mates and males lick and groom their pups more than do those with short ones (Hammock and Young 2005).

SUMMARY

In mammals, as in almost all other vertebrates, sexual reproduction is the only option. The gametes of the two sexes differ anatomically, with the female producing few large, nonmotile eggs and the male producing many small, motile sperm. In most species, males attempt to mate with more than one female and vary greatly in their reproductive success, whereas females usually mate only once per breeding bout and are often reproductively successful. In some species of mammals, parents may adaptively adjust the sex ratio of their offspring to produce the sex that will have the highest reproductive success.

Sexual selection affects anatomy and behavior at the time of mating. Intersexual selection involves choices made between males and females, with the females usually making the choice. Sexual selection may lead to the evolution of elaborate secondary sexual characteristics, particularly in males. The evolution of such traits is thought to have come about by either:

1. A process of runaway selection in which some arbitrary trait evolves because females prefer it and so produce sons with the trait and daughters that prefer the trait. The trait thus continues to evolve until countered by natural selection;

2. A linkage of the trait with superior genes in the male so that females selecting males with the trait obtain more fit mates.

In intrasexual selection, competition takes place within one sex (usually the male), with the winner gaining access to the opposite sex. Competition may take place before copulation, or after copulation in the form of sperm competition, blocked implantation, or infanticide.

Parental investment in offspring is generally greater by females than males, as males seek additional mates and leave the female to care for young; however, in some species, males provide extensive care. Species living in harsh, unpredictable environments or suffering high juvenile mortality rates tend to have high reproductive rates and provide little care compared with those adapted to stable environments.

Offspring often seem to want more attention or food than the mother is willing to give and resist attempts by the mother to wean them. Conflicts may reflect the fact that parents share only half their genome with an offspring. What is best for the parent is not necessarily what is best for the offspring.

Most mammals have polygynous mating systems in which males are able to monopolize more than one female. In such systems, males are usually large, may be more ornamented than the smaller females, and provide little if any care of offspring. In a few species, males cluster on leks, defending a small territory that functions solely to attract females for mating. In some orders, especially Carnivora, food is difficult to obtain, and both parents share in caring for offspring—processes that lead to monogamy. Polyandry, in which females monopolize more than one male, is rare in mammals. It has been reported in some of the larger canids, where females enlist the help of several males to raise offspring. Complete sex role reversal, with males as the sole parent, is unknown in mammals. The hormones oxytocin and vasopressin, acting via receptors in the forebrain, have been implicated in the control of mating systems and parental care.

SUGGESTED READINGS

Andersson, M. 1994. Sexual selection. Princeton Univ. Press, Princeton, NJ.

Clutton-Brock, T. H. 1991. The evolution of parental care. Princeton Univ. Press, Princeton, NJ.

Daly, M., and M. Wilson. 1983. Sex, evolution, and behavior, 2nd ed. Willard Grant, Boston.

Krebs, J. R., and N. B. Davies, eds. 1997. Behavioural ecology: An evolutionary approach, 4th ed. Blackwell Science, Inc., Malden, MA.

Trivers, R. L. 1972. Parental investment and sexual selection. Pp. 136–179 *in* Sexual selection and the descent of man, 1871–1971 (B. Campbell, ed.). Aldine, Chicago.

Trivers, R. L. 1974. Parent-offspring conflict. Am. Zool. 14:249–264.

Wolff, J. O., and P. W. Sherman (eds.). 2007. Rodent societies: An ecological and evolutionary perspective. Univ. Chicago Press, Chicago.

DISCUSSION QUESTIONS

1. Differences between the sexes are usually assumed to have come about via sexual selection. Can you think of selection pressures other than competing for mates that might produce such differences?

2. As you have seen in this chapter, polygyny is common in mammals. In birds, however, monogamy is the rule (Greenwood 1980). Discuss the possible reasons for this taxonomic difference in mating systems.

3. Discuss the possible relevance to humans of Trivers' concept of parent-offspring conflict.

4. Why is polyandry so uncommon in mammals? What sorts of ecological and phylogenetic circumstances might favor polyandry?

5. Using the figure shown here, describe the general relationship between testes weight and body weight, and then explain why some genera are above the line and some below it. What can you surmise about human mating systems from this graph (Harcourt et al. 1981)?

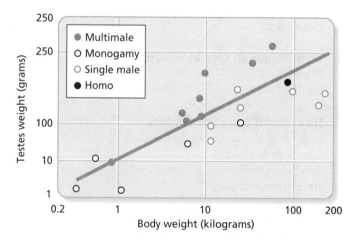

Paired testes weight (g) versus body weight (kg) for different primate genera. *Data from A. H. Harcourt et al., 1981, Testis Weight, Body Weight, and Breeding System in Primates,* Nature 293:55–57.

CHAPTER 23

Social Behavior

A **society** is a group of individuals of the same species that is organized in a cooperative manner, extending beyond sexual and parental behavior. In this chapter, we first explore several examples of cooperative social behavior in mammals; then we discuss the evolutionary costs and benefits of social behavior in general. Finally, we consider the way natural selection has brought about the evolution of social systems in mammals. For overviews of sociality in different groups of mammals, refer to the chapters in part 3.

Sociality has evolved independently in many groups of animals, ranging from invertebrates to primates. The complexity of social behavior might be expected to increase as organisms move from simple to more sophisticated, yet by some criteria, just the opposite is true: Some invertebrates, such as the Portuguese man-of-war (Genus *Physalia*), form colonies of individuals that cooperate much more extensively than mammals do (Wilson 1975).

Examples of Cooperative Social Behavior

As was explained in the previous chapter, male mammals typically provide little care for their young. Instead they seek additional mates in polygynous relationships (Eisenberg 1981). The predominant role of the mother in rearing young is no doubt due, in part, to the fact that only female mammals gestate and lactate. Milk is the "cement" of the young mammal's first social relationship (Wilson 1975). Most mammalian social systems are organized matrilineally; mothers and offspring may stay together, and groups are thus composed of mothers, daughters, sisters, aunts, and nieces. Because of the prevalence of polygyny and the associated tendency of male mammals to disperse as they reach sexual maturity (chapter 24), adult males are usually unrelated to other adults in the group. Complex social organization has evolved in some species in nearly all mammalian orders, but especially among carnivores, cetaceans, and primates. Within each order, the most highly social species tend to be large-sized and have large brains, and the highly social terrestrial species tend to forage aboveground in open habitats in the daytime (Wilson 1975). The cooperation that characterizes social living takes one or more of the following forms that extend beyond parental care.

ALARM CALLING

When a Belding's ground squirrel (*Spermophilus beldingi*) sees a predator such as a weasel (*Mustela* sp.), it may emit a shrill alarm call, alerting other nearby ground squirrels to run for cover (Sherman 1977). The caller, however, is now more likely to fall prey to the predator. Behavior that appears to be costly to the individual but beneficial to others is said to be **altruistic**. How such behavior evolves has been the focus of much controversy, as will be seen later in this chapter.

COOPERATIVE REARING OF YOUNG

Although cooperative rearing of young—that is, individuals other than the young's mother nursing or providing food for the young—is not common, it does occur in social carnivores and some rodents. Among carnivores, lionesses (*Panthera leo*) share the nursing of cubs in the pride (Packer et al. 1992; figure 23.1), and subordinate wolves (*Canis lupus*) regurgitate food for the alpha female and her litter (Mech 1970). In the banded mongoose (*Mungos mungo*) of Africa, several females breed synchronously, giving birth in a communal den and nursing each other's offspring (Rood 1986). Although the dwarf mongoose (*Helogale parvula*) is monogamous, it lives in packs of about 10 individuals. Breeding, like breeding in wolves,

Figure 23.1 Communal nursing in lions. Females nurse young born to other pride members, a trait that may have evolved via kin selection because pride females are closely related.

Figure 23.2 A family of dwarf mongooses. All members of a group help to raise the young, even though the offspring are usually produced by a single, dominant pair.

is suppressed among subordinate females. These nonreproductive members guard the den and bring insects to the young (Rood 1980; figure 23.2). Subordinate females occasionally do become pregnant, but most of their offspring die, probably due to infanticide by the dominant female. Subordinate females may also become "pseudo-pregnant," a condition in which the female is in a hormonal state of pregnancy, although no embryo is present. She may then lactate and nurse the dominant female's young (Rood 1980; Creel et al. 1992).

In another cooperatively breeding carnivore that has become a popular zoo species, the meerkat (*Suricata suricatta*) young produced by the dominant female are reared by up to 30 helpers. Helpers engage in babysitting (guarding the natal burrow for the first month of pup life) and also in bringing food (Clutton-Brock et al. 2000). Helpers do not seem to specialize in particular tasks (Clutton-Brock et al. 2003), nor do they help in proportion to their relatedness to those offspring (Clutton-Brock et al. 2001).

Rodents, such as house mice (*Mus musculus*) and white-footed mice (*Peromyscus leucopus*), may form communal nests in which several litters of different ages are present (Wolff and Durr 1986). In some cases, females nurse their own young and those of other females simultaneously (Jacquot and Vessey 1994; figure 23.3). Lactation doubles a female's energy need (Millar 1978), so such costly and seemingly altruistic behavior demands an explanation. For reviews of nonoffspring nursing in mammals in general, see Packer et al. (1992), and for rodents in particular, see Hayes (2000).

COALITIONS AND ALLIANCES

Most of the species that live in large groups, from herds of wildebeests to schools of dolphins, have evolved in open habitats, such as savannas or oceans. Such groups

Figure 23.3 **Nonoffspring nursing in the white-footed mouse.** The older pups' mother is the sister of the nursing female.

Figure 23.4 **An alliance between two male olive baboons.** The two males on the right are challenging the male on the left in the foreground.

may consist of members of both sexes and associated young; they may persist throughout the year or only during the breeding season. The degree of cooperation seen in such groups varies widely from none, except between mothers and offspring, to complex coalitions and alliances among both related and unrelated members.

In the large, multimale groups found in many Old World monkeys and apes, cooperation in caring for young does not extend to nursing or giving food to offspring other than one's own. Group membership is usually highly stable, however, and dominance hierarchies are prominent in some species. Males and females may play different social roles within the group. In rhesus monkeys (*Macaca mulatta*), the highest-ranking males act as "control" animals (Bernstein and Sharpe 1966). These males protect the group against serious extragroup challenges and reduce intragroup conflict by intervening in fights among group members (Vessey 1971).

Learning and early experience play a large part in determining the social structure in primates. Although **kinship**, the sharing of a common ancestor in the recent past, is important in structuring primate social systems, coalitions and "friendships" among nonrelatives are also evident, as in hamadryas baboons (*Papio hamadryas*; Smuts 1985; figure 23.4). Chimpanzees (*Pan troglodytes*), among other nonhuman primates, form coalitions in competitive situations, such as when two individuals cooperate to defeat or take some resource away from a third. They also reciprocate, meaning that after individual A intervenes on behalf of individual B, individual B is later likely to intervene on behalf of individual A (de Waal 1992). In laboratory settings, chimpanzees given various tasks to get a food reward "know" when they need to recruit a collaborator and keep track of which collaborators are more effective, choosing them for future tasks (Melis et al. 2006). Such complex social relationships require the ability to keep track of large numbers of individuals, and this ability is related to

brain size. Among primate species, there is a positive correlation between the relative size of the neocortex and both group size and the complexity of social networks (Kudo and Dunbar 2001).

Bottlenose dolphins (*Tursiops truncatus*) live in large groups that vary in membership. Males form stable, first-order alliances of two or three in order to herd estrous females and keep them away from other males. Two first-order alliances may combine to form a second-order alliance of five or six males. Second-order alliances are able to take females from first-order alliances and are also effective at protecting females from other alliances (Connor et al. 1992).

EUSOCIALITY

The epitome of social organization, seen most often in insects of the Order Hymenoptera, is referred to as **eusociality**. Three traits characterize this pattern: (1) cooperation in the care of young; (2) reproductive castes, with nonreproductive members caring for reproductive nestmates; (3) overlap between generations such that offspring assist parents in raising siblings (Wilson 1975). Jarvis (1981) first demonstrated mammalian eusociality in the naked mole-rat (*Heterocephalus glaber*; figure 23.5). She captured 40 members of a colony from their burrow system in Kenya and studied them for 6 years in an artificial burrow system in the laboratory. Only one female in the colony ever had young; mother and young were fed but not nursed by male and female adults of the worker caste; members of this caste were not seen to breed. Another caste of nonworkers assisted in keeping the young warm; males of this caste bred with the reproducing female (or "queen"). This species fits all of the criteria for eusociality just mentioned. Another distantly related species, the Damara mole-rat (*Cryptomys damarensis*), also appears to be eusocial (Jarvis et al. 1994).

Figure 23.5 Naked mole-rat colony. The large, pregnant female in the center gave birth to 28 pups the next day. The other adults feed her, care for the young, or maintain the burrow system.

Why Mammals Live in Groups

It is often assumed that living in complex social groups is somehow superior to living a more solitary life, yet costs and benefits are associated with each. Most of the benefits of sociality listed in the following section can be related to two ecological factors: predation pressure and resource distribution (Alexander 1974). Keep in mind that these advantages may not, by themselves, have led to the evolution of sociality; rather, they may have become secondarily advantageous once sociality had already evolved via one of the other selective pressures.

BENEFITS

Protection from Physical Factors

White-footed mice frequently form communal nests in winter (Wolff and Durr 1986); huddling has been shown to conserve significant amounts of energy (Hill 1983). This benefit leads to the formation of aggregations, as with large clusters of bats, but not necessarily to organized social groups.

Protection Against Predators

Detection of and communication about danger are more rapid when individuals are in groups, and predator deterrence may be enhanced by mobbing and group defense. According to the "many eyes" hypothesis, individuals in large groups spend less time watching for predators and so can spend more time in other activities such as feeding. Examples include prairie dogs (*Cynomys* spp.; Hoogland 1979b) and long-tailed ground squirrels (*Spermophilus undulatus*; Carl 1971). In the latter study,

observers could approach within 3 m of isolated individuals but no closer than 300 m to grouped individuals before waves of alarm calls swept through the colony. Musk oxen (*Ovibos moschatus*) and other ungulates form a defensive perimeter in response to wolf attacks (Gunn 1982); adults form a line or circle, keeping themselves between the predator and dependent young.

Painted hunting dogs, also called African wild dogs (*Lycaon pictus*), are cooperative breeders and live in packs of up to 20 adults (see figure 21.10). Only the alpha pair breeds; the other adults are reproductively suppressed and help raise the pups (Creel and Creel 1995). An important function of helpers is to protect the pups from predators such as lions when the rest of the pack is off hunting prey. There is a cost to this behavior, however, because smaller hunting parties have greater difficulty making kills and defending the carcass against competitors, and there are fewer adults available to regurgitate food to the pups. Courchamp and colleagues (2002) concluded that the minimum pack size is 5 adults. Above this number, the pack can afford to leave a baby-sitter behind and still hunt effectively. Below this number, the pack is not likely to survive because of the trade-off between pup guarding and hunting. The observation that there is a minimum group size below which populations decline is referred to as the **Allee effect** (Courchamp et al. 1999). This species is critically endangered, so understanding the determinants of pack size is especially important.

Finding and Obtaining Food

Living in groups may make it easier to find and obtain food. Wolves (Mech 1970) and lions (Schaller 1972) are able to capture large species of prey—such as moose (*Alces alces*) in the case of wolves and African buffalo (*Syncerus caffer*) in the case of lions—that would be nearly impossible to capture alone. That said, researchers have disagreed about whether getting food is the main determinant of group size in social carnivores. Caraco and Wolff (1975) showed that for lions feeding on Thomson's gazelle, two was the optimum pride size. The fact that lions live in much larger groups than two suggests that other benefits of grouping, such as defense of territory, are paramount (Packer et al. 1990).

In their studies of painted hunting dogs, the Creels (1995, 1997) analyzed data from more than 900 hunts and 400 kills and concluded that when both costs and benefits of cooperative hunting were considered, the observed pack size of 10 was close to the predicted optimum of 12–14 adults. Thus, obtaining food seems to be a major selective force shaping group size in hunting dogs.

Although herbivores are thought to group mainly for defense against predators, in bighorn sheep (*Ovis canadensis*) the locations of feeding areas and migration routes are remembered by older members of the band. This information is transmitted to subsequent generations via tradition (Geist 1971).

Group Defense of Resources

Lion prides are territorial, defending space containing food resources against other prides (Schaller 1972). Similarly, wolf packs defend space against neighboring packs (Mech 1970). Interpack fights to the death have been observed in painted hunting dogs, and the largest pack always wins (Creel and Creel 1995).

Assembling Members for Location of Mates

As seen in chapter 22, hammer-headed fruit bats (*Hypsignathus monstrosus*; Bradbury 1977) and fallow deer (*Dama dama*; Appolonio et al. 1992) breed on leks, where males defend small territories and display to attract females for copulation.

Division of Labor Among Specialists

This feature of advanced social behavior is found in eusocial insects but is rare among mammals. Recent studies of naked mole-rats, discussed previously, suggest that only one female is reproductively active in the colony and that the other adults specialize in tasks such as maintaining the tunnel system or nurturing the young (Sherman et al. 1991).

Richer Learning Environment for Young

This advantage is frequently suggested as important for mammals in general and primates in particular. Dependence on learning provides for greater behavioral plasticity, but it requires a long period of physiological and psychological dependence. Large-brained and highly social species, such as dolphins and primates, spend as much as 25% of their lives dependent on parents or other relatives.

COSTS

Only a few studies have directly attempted to assess the possible disadvantages of sociality. The following are several obvious costs of living in groups:

Increased Intraspecific Competition for Resources

In prairie dog colonies, the amount of agonistic behavior per individual increases as a function of group size. Also, black-tailed prairie dogs are more highly social and have higher rates of aggression than do the less social white-tailed prairie dogs (Hoogland 1979a). Among frugivorous (fruit-eating) species of primates, the average distance traveled in search of food each day increases as a function of group size, potentially setting an upper limit on group size (Janson and Goldsmith 1995). This relationship does not hold for leaf-eating

species, perhaps because there is less competition for leaves than for fruits.

Increased Chance of Spread of Diseases and Parasites

Ectoparasites, such as fleas and lice, are more numerous in larger and denser prairie dog colonies than in smaller ones (Hoogland 1979a). Fleas transmit bubonic plague, epidemics of which periodically decimate prairie dog colonies. Hence, members of dense colonies are more at risk (see chapter 27).

Interference with Reproduction

Parental care misdirected to nonoffspring and killing of young by nonparents exemplify this cost. Female white-footed mice with young are aggressive toward strange adults; in the absence of the mother, pups are usually killed by intruders (Wolff 1985). One of the factors influencing dispersal by male lions is defense of their cubs against infanticide by new coalitions of males (Pusey and Packer 1987); small cubs are almost invariably killed when new males take over a pride. Brazilian free-tailed bats (*Tadarida brasiliensis*) roost in caves in dense colonies containing millions of bats (see chapter 13). Mothers returning from a night's foraging for insects have to find their own infant among the thousands present. Most of the time, they find their own young by vocalization, but 17% of the time mothers suckle someone else's offspring (McCracken 1984).

How Social Behavior Evolves

INDIVIDUAL VERSUS GROUP SELECTION

A basic element of sociality is cooperation. Individuals work together, often sacrificing personal gain, to achieve a common goal that benefits the social group. How such altruistic behavior could evolve was not a topic of great concern for biologists prior to the 1960s. It was generally assumed that groups with cooperating individuals would be more successful than those without cooperators, a type of **group selection**. This thinking culminated in the publication of a book by V. C. Wynne-Edwards (1962). In that book and a sequel (Wynne-Edwards 1986), he argued that the evolutionary significance of social behavior is that organisms can track resources in the environment more efficiently. Intraspecific competition became ritualized into contests whose intensity was proportional to the supply of the limiting resource. Because the result of such competition leads to reduced reproductive success of those participating, Wynne-Edwards believed that natural selection must be acting at the level of the group. Note that individuals must sacrifice personal reproduction for the good of the group.

Not everyone, however, believed that group selection was an important force for evolution. R. A. Fisher (1958), in his book *Genetical Theory of Natural Selection*, pointed out that his fundamental theorem referred strictly to "the progressive modification of structure or function only in so far as variations in these are of advantage to the *individual* [emphasis added]." His theorem offered no explanation for the existence of traits that would be of use to the species to which an individual belongs. Much earlier, Darwin (1859) had stated that "if it could be proved that any part of the structure of any one species had been formed for the exclusive good of another species, it would annihilate my theory." If Fisher and Darwin were correct, then explanations for the evolution of social behavior had to be sought based on natural selection at the level of the individual and its offspring, rather than at the group or species level.

The publication of *Adaptation and Natural Selection* by Williams (1966) led to a rapid shift in thinking among researchers working on social behavior. Williams argued that when considering any adaptation, researchers should assume that natural selection operates at that level necessary to explain the facts, and no higher—usually at the level of parents and their young. The argument is based on evidence that natural selection at the group or population level is so weak that it is almost always outpaced by selection of individual phenotypes. In other words, genes promoting altruistic behavior are swamped by genes favoring selfish behavior. Group selection is not impossible: If some genes decrease individual fitness but make it less likely that a group, population, or species will become extinct, then group selection will influence evolution (Williams 1966). Most behavioral ecologists today follow Williams' rule of parsimony (derived from Occam's razor), which adopts the simplest theory explained by the facts and argues that group selection is weak and should be invoked only when lower-level selection has been ruled out. More realistic models of group selection continue to be developed (Wilson 1980; Wilson and Dugatkin 1997), however, and conditions favoring evolution by means of group selection may not be as restrictive as was once thought. For further discussion of the controversy, see the books by Brandon and Burian (1984) and Sober (1984).

THE SELFISH HERD

One way to explain gregarious behavior in mammals at the individual level is to suppose that it is a form of cover-seeking, in which each individual tries to reduce its chances of being caught by a predator. Hamilton (1971) suggested how this might work, using a hypothetical lily pond in which some frogs and a predatory water snake live. The snake stays on the bottom of the pond most of the time and feeds at a certain time of day. Because it usually catches frogs in the water, the frogs climb out on the edge before

the snake starts to hunt. They do not move inland from the rim because of even more threatening terrestrial predators. The snake surfaces at some unpredictable place and catches the nearest frog. What should each frog do to minimize the chances of its being eaten? Hamilton demonstrated mathematically that it should jump around the rim, moving into the nearest gap between two other frogs. The end result would be an aggregation.

Such an example could be also applied to herds of ungulates; when each individual behaves selfishly and moves into the center of the herd to minimize its chances of being picked off, the end result is a tightly packed group. Although the selfish herd might explain some aggregations, it does not explain the defensive groupings of musk oxen. Adult musk oxen put themselves between their young and the wolves rather than trying to selfishly minimize the chances of being attacked.

KIN SELECTION

To explain the evolution of cooperative behavior, Hamilton (1963, 1964) presented a theory incorporating both the gene and the individual as units of selection. In its simplest form, this **kin selection** theory suggests that if a gene that causes some kind of altruistic behavior appears in the population, the gene's success depends ultimately not on whether it benefits the individual carrying the gene but on the gene's benefit to itself. If the individual that benefited by the act is a relative of the altruist and therefore more likely than a nonrelative to be carrying that same gene, the frequency of that gene in the gene pool increases. The more distant the relative, the less likely it is to carry that gene, so if the gene is to spread, the ratio between the benefit to the recipient and the cost to the altruist must be greater. This relationship, called Hamilton's rule, can be expressed algebraically as follows:

$$b/c > 1/r$$

where b is the benefit to the recipient, c is the cost to the altruist, and r is the coefficient of relationship, that is, the proportion of genes shared by the two participants by way of descent from a common ancestor. In full siblings, who share half their genes, $r = 1/2$, and therefore b/c must exceed 2 for altruistic genes to spread. In other words, if an individual more than doubles the fitness of a sibling through an altruistic act that causes that individual to leave no offspring, genes promoting that behavior could spread through the population. The more distant the relative, the lower r is, and the higher the benefit-to-cost ratio (b/c) must be. Thus, for first cousins, $r = 1/8$, so b/c must exceed 8. Of course, the control of an altruistic behavior pattern most likely involves more than one gene, but it is assumed that the same principles apply.

It now becomes necessary to consider an individual's fitness as including both a direct component, measured by the reproductive success of its own offspring, and an indirect component, measured by the reproductive success of its relatives other than its own offspring. These two components make up an individual's **inclusive fitness**.

Hamilton's idea about the evolution of cooperative behavior through kin selection stimulated considerable research because it remained to be seen if cooperative acts are in some way distributed to relatives according to the degree of relatedness. Such studies, however, require information about paternal and maternal relationships that is usually not available to the observer of a natural population. It is particularly difficult to determine paternity in mammals because of internal fertilization. Molecular techniques such as DNA fingerprinting (see chapter 3), however, now make it possible to determine parentage in a variety of species.

Another problem is the recognition factor. How do animals recognize kin in order to "correctly" distribute their altruistic acts? Perhaps the simplest and most common way is by familiarity. When given a choice in a laboratory arena, young northeast African spiny mice (*Acomys cahirinus*) prefer to huddle with littermates rather than nonlittermates (Porter et al. 1981). Siblings separated soon after birth and raised by foster mothers treated each other as strangers, but unrelated young raised together responded to each other as siblings. Some species of mammals can recognize relatives with whom they have never associated. A possible mechanism for this is called phenotype matching, in which the individual uses its own kin as a "template" to compare with strangers who might be kin (Holmes and Sherman 1983). Such a mechanism may explain instances when siblings recognize each other even when reared apart. Thus, golden-mantled ground squirrel (*Spermophilus lateralis*) juveniles preferred to play with littermates over nonlittermates, supporting the familiarity mechanism, but siblings reared apart preferred each other as playmates over nonsiblings reared apart (Holmes 1995).

A third mechanism of kin recognition that does not require learning is the presence of "recognition" genes that enable individuals to recognize others with those same genes and to behave cooperatively toward those individuals. This mechanism has been called the "green beard effect" because such a trait could be selected for if genes controlling it confer not only the green beard but also the preference for green beards on others. Cues used to assess genetic relatedness may come from genes in the major histocompatibility complex (MHC), a cell recognition system used by the immune system to distinguish self from nonself. In rodents, and possibly other mammal species, genetic differences in this region produce urinary odor cues that can be used to recognize genetic similarities or differences in other individuals (Yamazaki et al. 1976; Brown and Eklund 1994). These cues have been shown to affect mate choice in wild house mice (Potts et al. 1991). For a review of kin recognition in rodents, see Mateo (2003).

One of the most thorough tests of kin selection was Sherman's study of alarm calls in Belding's ground squirrels (Sherman 1977; figure 23.6). He found that

Figure 23.6 Belding's ground squirrel giving an alarm call. This conspicuously calling squirrel, a lactating female, is more likely to be attacked by a predator than is a noncaller. Nearby ground squirrels benefit because they can remain hidden or take cover. Females with mothers, sisters, or offspring in the colony are most likely to call.

whenever a terrestrial predator (such as a weasel or a coyote) was seen, the calling squirrel stared directly at the predator while sounding the alarm. Sherman suggested many hypotheses to explain this behavior, two of which we discuss here. First, the predator may abandon the hunt once it is seen by the potential prey (in this hypothesis, the caller is behaving selfishly). Second, other ground squirrels in the area may benefit from the warning, even though the caller may be harmed (in this hypothesis, the caller is behaving altruistically).

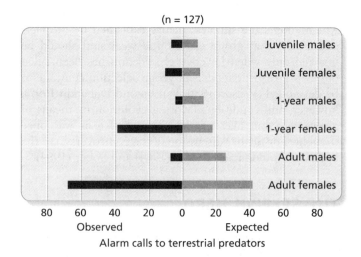

Figure 23.7 Expected and observed frequencies of alarm calls by Belding's ground squirrels to terrestrial predators. Expected frequencies are those that would be predicted if the animals called in proportion to their frequency of occurrence in the population. Calls to terrestrial predators are given disproportionately by females. *Data from P. W. Sherman, 1985, Alarm Calls of Belding's Ground Squirrels, in* Behav. Ecol. Sociobiol. *17:313–323.*

Sherman demonstrated that callers attract predators and are more likely to be attacked after calling, and thus they are not behaving selfishly.

Because he kept records on mothers and offspring, Sherman knew that the males leave the natal area several months after birth and that the females are philopatric, remaining near their place of birth to breed. He also found that adult and yearling females are much more likely to call than would be expected by chance and that males are less likely to do so. Furthermore, females with female relatives in the area (such as mothers or sisters, but not necessarily with offspring) call more frequently in the presence of a predator than those with no female relatives living nearby (figure 23.7). Sherman concluded that the most likely function of the alarm call is to warn family members.

Working with primates, Massey (1977) tested the kin selection model in rhesus monkeys. She found that within an enclosed group, monkeys aid each other in fights in proportion to their degree of relatedness. In observations of the same species in a free-ranging situation, Meikle and Vessey (1981) found that when males leave the natal group, they usually join groups containing older brothers. They associate with brothers in the new group, aid each other in fights, and avoid disrupting each other's sexual relationships, in contrast to their behavior toward nonbrothers.

Bertram (1976) estimated that males in a pride of lions are related on the average almost as closely as half siblings ($r = 0.22$), and females are related as closely as full cousins ($r = 0.15$). Lions cooperate in several ways, including hunting, driving out intruders, and caring for young. He found that males show tolerance toward cubs at kills and seldom compete for females in estrus and that cubs suckle communally from the pride females. Based on the degree of relatedness among pride members, Bertram concluded that kin selection is in part responsible for these cooperative behaviors. He was not able, however, to evaluate its importance compared with other selective pressures.

Black-backed jackals (*Canis mesomelas*) live in the brushland of Africa, where monogamous pairs defend territories, hunt cooperatively, and share food. Frequently, offspring from the previous year's litter help rear their siblings by regurgitating food for the lactating mother as well as for the pups (Moehlman 1979; figure 23.8). The number of pups surviving is directly related to the number of helpers (figure 23.9). One-and-a-half extra pups survived, on the average, for each additional helper—a yield of one pup per adult involved. Only one-half pup per adult survived when just the parents were involved. Given the fact that helpers are as closely related to their siblings as they would be to their own offspring ($r = 0.5$), yearling jackals gain genetically by aiding their parents. In addition to increasing their indirect fitness by aiding siblings, helpers may increase their direct fitness by gaining experience in rearing young, becoming familiar with the home territory, and possibly gaining a

portion of the home territory, all of which would increase their own chances of breeding successfully at a later time (Moehlman 1983, 1986).

Kin selection may also explain the nursing of nonoffspring by female mammals, such as in lions (figure 23.1). Recall that r between adult females in the pride is about 0.15. Among rodents, female white-footed mice are occasionally observed nursing young from two different-aged litters (figure 23.3). In some instances, the mother of the older pups is the sister of the nursing female (Jacquot and Vessey 1994), consistent with kin selection. Among gray mouse lemurs (*Microcebus murinus*) in Madagascar, adult

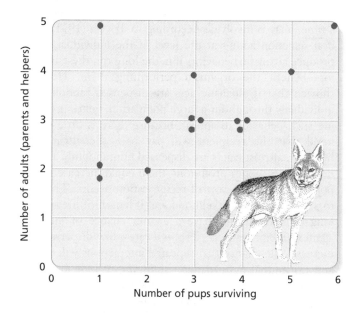

Figure 23.8 Black-backed jackal helper. The helper, probably from the previous year's litter, is about to regurgitate food to the pups.

Figure 23.9 Black-backed jackal pup survival as a function of the number of adults in the family. Young from previous years may assist their parents in rearing the current litter. Each helper increases the number of surviving pups by an average of about 1.5. Helpers may, therefore, increase their inclusive fitness more by rearing siblings than by attempting to breed on their own. *Data from P. D. Moehlman, 1979, Jackal Helpers and Pup Survival,* Nature 277:382–383.

females typically form small daytime sleeping clusters. Based on genetic analysis, these groups consist of close female relatives that regularly groom and nurse related offspring other than their own and adopt related dependent young after their mother's death (Eberle and Kappeler 2006).

Cooperation Among Nonkin

A critical feature of a theory such as Hamilton's is that it is testable and falsifiable. Although the theory cannot be confirmed simply by finding cases of nepotism, it can be falsified in specific cases by observing cooperation among nonrelatives in a natural population. For example, McCracken and Bradbury (1977) demonstrated that the degree of relatedness of colonial greater spear-nosed bats (*Phyllostomus hastatus*) is too low to explain their coloniality on the basis of kin selection.

Recall that in lions, the males that take over prides are frequently related; their cooperation can be explained by kin selection. Additional studies of these same prides, however, suggest that at least some of the coalitions are among unrelated males, so some additional explanation is needed (Packer and Pusey 1982).

RECIPROCAL ALTRUISM

Individuals may cooperate and behave altruistically if there is a chance that they will be the recipients of such acts at a later time. Such a situation, called **reciprocal altruism**, is similar to mutualism (see the following section) except that a time delay is involved. According to Trivers (1971), natural selection acting at the level of the individual could produce altruistic behaviors if in the long run these behaviors benefit the organism performing them. He first showed that if altruistic acts are dispensed randomly to individuals throughout a large population, genes promoting such behavior disappear, because there is little likelihood that the recipient will pay back the altruist. If, however, altruistic acts are dispensed nonrandomly among nonrelatives, genes promoting them could increase in the population if some sort of reciprocation occurs. The factors that affect that likelihood are (l) length of life span—long-lived organisms have a greater chance of meeting again to reciprocate; (2) dispersal rate—low dispersal rate increases the chance that repeated interactions will occur; and (3) mutual dependence—clumping of individuals, as occurs when avoiding predation, increases the chances for reciprocation. In any social system, nonreciprocators (cheaters) can be expected, but as cheating increases, altruistic acts become less frequent.

A few field studies have suggested the importance of reciprocal altruism. Working with olive baboons (*Papio anubis*) in Africa, Packer (1977) studied coalitions among males in which two presumably unrelated males joined forces against a third. If that third male was in consort with a female in estrus, one of the attackers might gain

access to her. The pair of males tended to maintain the previously established coalition, and the next "stolen" female would be taken over by the other member of the male pair.

A different sort of reciprocity has been demonstrated in common vampire bats (*Desmodus rotundus*) in Costa Rica by Wilkinson (1984; figure 13.21). At night, these bats feed on blood, primarily from cattle and horses, and then return to a hollow tree to roost during the day. Wilkinson marked nearly 200 bats that roosted in 14 trees and spent 400 hours observing them in their roosts. He recorded 110 cases of blood sharing, where one bat ate blood that was regurgitated by another bat. Not surprisingly, most of these exchanges were between mothers and offspring. In most of the other feedings, he was able to determine both the coefficient of relationship between the pair and an index of association based on how often the pair had been together in the past. Wilkinson demonstrated that both relatedness and association contributed significantly to the pattern of exchange (figure 23.10). Close relatives and associates were fed more often than would be expected by chance.

For reciprocity to persist, (l) the pairs must persist long enough to permit reciprocation; (2) the benefit to the receiver must exceed the cost to the donor; and (3) donors must recognize cheaters (those that do not reciprocate) and not feed them. Through additional studies on captive animals, Wilkinson demonstrated that vampire bats meet these conditions. In this example, both kin selection and reciprocal altruism contribute to the cooperative behavior pattern of blood sharing.

Behavioral ecologists have used game theory to predict the circumstances in which unrelated individuals might cooperate. In a game called the Prisoner's Dilemma, two players have a choice of cooperating with each other or defecting in a series of moves. Depending on the relative costs and benefits of cooperating and defecting, it sometimes pays to follow a tit-for-tat strategy in which one player cooperates on the first move and then does whatever the other player does on his or her move (Axelrod and Hamilton 1981). For example, lions cooperate in the defense of the pride territory against strangers. When recordings of the roaring from strange females were played within the pride territory, several resident females usually advanced to check it out (Heinsohn and Packer 1995). Some lionesses were leaders and advanced boldly, but others were laggards, hanging back and letting the other member of the pair incur the risk. According to the tit-for-tat strategy, on subsequent pairings of the same leader and laggard, the leader should defect and also become a laggard. In fact, although leaders were more hesitant to advance when paired with known laggards, they did so regardless, contradicting the predictions of the model. Clearly, the relationships among the pride members are more complicated than those assumed by the model. Possibly laggards excel at hunting or in nursing cubs and so are "forgiven" for being bad at pride defense.

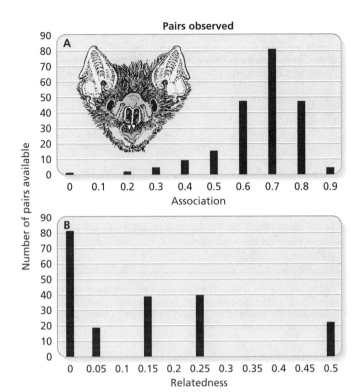

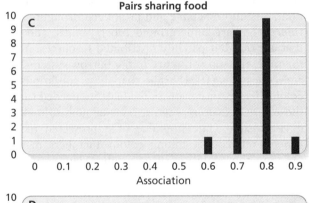

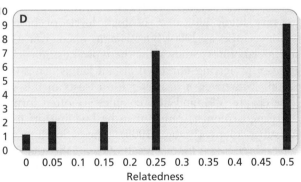

Figure 23.10 Reciprocal blood sharing in vampire bats. Graphs (A) and (B) show the frequency of pairs observed, based on the degree of association and relatedness. Graphs (C) and (D) show the frequency of blood sharing, excluding mother-young pairs. Both degree of association and relationship independently predict blood sharing, implicating both reciprocal altruism and kin selection in the evolution of this behavior. *Data from G. S. Wilkinson, 1984, Reciprocal Food Sharing in the Vampire Bat,* Nature *308:181–184.*

Although "pure" altruism, where no reciprocity appears to exist, is predictably rare, infant humans and young chimpanzees readily assist humans to achieve goals in a variety of situations in the laboratory when no immediate benefit is received (Warneken and Tomasello 2006). In one such task, young chimps handed out-of-reach objects to the human experimenter, but only when it was clear that the experimenter was trying to get the object and not when the objects were dropped deliberately.

MUTUALISM

In some situations, both individuals benefit from the relationship, and there is no apparent cost to either. An example might be the huddling of mice in cold weather described earlier (Wolff and Durr 1986). Such interactions, referred to as **mutualism**, could explain some cases of cooperation and might be included with the concept of the selfish herd. Another example might be the lion that helps another bring down a wildebeest; both benefit immediately by feeding on the prey. These cases are also similar to reciprocal altruism in the long run, except that there is no delay in returning the favor. Little research has actually been done to test the importance of mutualism in the evolution of social behavior; it is difficult to distinguish mutualism from selfish behavior because there is never a cost to the individuals involved. Usually, however, mammalogists look for examples of apparent altruism, in which the costs to the actor seem to outweigh the benefits, and tend to ignore cases of mutualism.

PARENTAL MANIPULATION OF OFFSPRING

In the discussion of parent-offspring conflict in chapter 22, we pointed out that because offspring share only half of each parent's genes, the interests of the parents may differ from those of their offspring. This difference is manifested by the offspring's demanding more investment from the parents than the parents are willing to give (Trivers 1974). An initial assessment might be that offspring should win such conflicts because they are the ones that must survive and pass the genes on to future generations. But Alexander (1974) argued that parents should win in the long run. Suppose that some offspring have genes that give the offspring a competitive advantage over siblings to the extent that they reduce the parents' lifetime reproductive success. For example, a highly competitive baby mouse that pushes all of its siblings out of the nest would receive more food from its parents and probably increase its direct fitness, although its indirect fitness would suffer. The parents' fitness would also suffer, however, because they would only raise one young that year. When the young mouse grew up and had young of its own, it would pass those competitive traits on, and its fitness would be less than that of a mouse with less competitive young that could coexist in the nest. Genes that will be favored are those that cause offspring to behave so as to maximize the lifetime reproductive success of the parent. Alexander pointed out that the parents thus "manipulate" the offspring to the parents' advantage. He listed the following types of behavior as examples of **parental manipulation**:

- Limiting the amount of parental care given to each offspring so that all have an equal chance to survive and reproduce
- Restricting parental care or withholding it entirely from some offspring when resources become insufficient for an entire brood
- Killing some offspring or feeding some offspring to others
- Causing some offspring to be temporarily or facultatively sterile helpers at the nest
- Causing some offspring to become permanent (obligately sterile) workers or soldiers

The last type of behavior, the extreme form of parental manipulation, is rare or nonexistent in mammals, with the possible exception of the mole-rats discussed previously.

Alexander's reasoning was criticized by Dawkins and Carlisle (1976) because the argument can be turned around and viewed from the perspective of the offspring: Parents that allow themselves to be manipulated to provide more care than they otherwise would leave more offspring themselves. It becomes a difficult task to sort out whether parents or offspring win in such situations. Although parental manipulation is not inevitable, it needs to be considered as an explanation for some types of cooperation in mammals.

ECOLOGICAL FACTORS IN COOPERATION

Emlen (1982, 1984) developed an "ecological constraints" model to explain the evolution of cooperative breeding in birds and mammals. He considered environmental factors that restrict the chances for individuals to breed independently. One condition that might favor staying at home and helping the parents or others rear offspring is a stable, predictable environment. In these cases, unoccupied territories are absent or rare; young have little chance of dispersing and breeding on their own (table 23.1, #1). An example of this might be the helpers among jackals discussed previously.

Because helping has just been explained on the basis of habitat saturation in stable environments, it is ironic to find that cooperative breeding is most common in arid regions of Africa and Australia, where rainfall is highly variable and unpredictable. In this case, Emlen argues that the same behavior results for different reasons (table 23.1,

Table 23.1 Ecological constraints that severely limit personal reproduction

Type of Constraint	Cause of Constraint
1. Breeding openings are nonexistent	Species has specialized ecological requirements; suitable habitat is saturated, and marginal habitat is rare (stable environments)
2. Cost of rearing young is prohibitive	Unpredictable season of extreme environmental harshness (fluctuating, erratic environments)

Result: Grown offspring postpone dispersal and are retained in the parental unit. The population becomes subdivided into stable, social, kin groups.

From S. T. Emlen, 1982, The Evolution of Helping, I. An Ecological Constraint Model, in American Naturalist 119:29–39. Copyright ©1982 by the University of Chicago Press, Chicago, IL. Reprinted with permission.

#2). The African mole-rat belongs to the Family Bathyergidae, all but two species of which are solitary (see also chapter 18). The two eusocial species are distantly related; both occupy arid regions of Africa where rainfall is unpredictable and prolonged droughts are the rule. Foods are underground tubers, bulbs, and corms that are rich but patchily distributed resources. Most of the time, the soil is too dry to dig burrow systems. Jarvis and colleagues (1994) argue that these conditions favor coloniality because many individuals are needed to dig on those rare occasions when it rains enough to make digging possible. The presence of a protected nest and genetic relatedness among colony members may then promote the evolution of eusociality (Honeycutt 1992; figure 23.11).

Genetic similarity among members of naked mole-rats is very high, indicative of high levels of inbreeding (Honeycutt et al. 2000). Surprisingly, when given a choice in the laboratory, they show the typical mammalian pattern of preferring unrelated distant kin as mates (Ciszek 2000). Moreover, a disperser form exists in this species (see chapter 24), and new colonies are often formed by unrelated colonists rather than by budding off from an existing colony (Braude 2000). These findings suggest that inbreeding is more a result of coloniality than a cause of it and that ecological factors may be paramount.

Hystricognath rodents (e.g., guinea pigs, chinchillas, capybaras) occupy a variety of habitats in South America and display a wide range of social behavior. In a comparative study of 76 of these species, Ebensperger and Cofré (2001) tested several hypotheses to explain the evolution of communal living. They concluded that group living was more related to reducing the cost of burrow construction than to predatory risk or the need for extended parental care.

The ecological constraints model may explain why some animals stay home and refrain from breeding, but

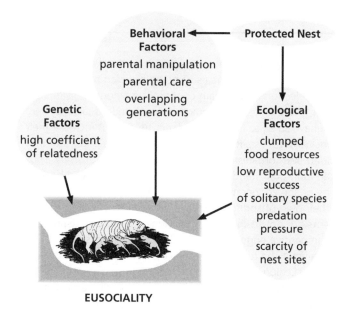

Figure 23.11 Genetic, ecological, and behavioral factors promoting the evolution of eusociality. All the factors shown here may play a role in the evolution of colonial living in the naked mole-rat. *R. L. Honeycutt, 1992, Naked Mole-Rats,* American Scientist *80:43–53.*

why should they bother to help? We have already mentioned the kin selection argument, whereby indirect fitness is enhanced. But often nonrelatives or very distant relatives are aided, as with the dwarf mongooses, in which nonrelatives guard and feed young of the dominant pair (see figure 23.2). Reciprocity may be involved: Rood (1983) observed an instance in which a female was later assisted by the unrelated young she had helped rear.

Comparative Studies

In trying to understand the mechanisms involved in the evolution of social behavior, it is necessary to consider the type of environment in which the population lives. Several attempts have been made to relate mating system and social structure to habitat and distribution of resources. The relationship between feeding behavior and group size is evident in the great diversity of African antelope (see figure 20.25). Jarman (1974) classified these herbivores on the basis of five feeding styles, ranging from selective species that feed on only a few highly nutritious parts of localized plants to unselective species that feed primarily on grasses and forage of low nutritive value. The selective species are small, solitary, monogamous, and monomorphic, and they defend small territories. The least selective species are large, gregarious, polygamous, and sexually dimorphic, and they occupy large home ranges. Jarman argued that feeding style is an important determinant of group size; both group size and the pattern of movement over the home range affect reproductive strategies and social behavior.

One problem with such studies is that the role of phylogeny, or the evolutionary history of the group, is usually ignored. Using phylogenetically based statistical techniques, Brashares and colleagues (2000) reevaluated Jarman's (1974) conclusions and generally confirmed them. One exception was that the positive relationship between body mass and group size reported by Jarman disappeared when phylogenetic relationships were factored in.

Several attempts have been made to classify primate societies on the basis of habitat and niche variables. The first was by Crook and Gartlan (1966), with a later version by Eisenberg and colleagues (1972). At one end of a continuum are nocturnal, arboreal, forest-dwellers that feed mostly on insects. They tend either to be solitary, such as some lemurs (Genus *Microcebus*), or to live in small, monogamous groups (e.g., tamarins of the Genus *Saguinus*). At the other end are diurnal, terrestrial, plains-dwellers that feed mostly on seeds, tubers, or other nutritious plant parts. They tend to live in large, multimale groups (e.g., baboons of the Genus *Papio*).

Some species may have a particular mating and social system not so much because of present circumstances but because it has been retained from their ancestors (Brooks and McLennan 1991). Both morphological and molecular data and the methods of cladistics can be used to prepare a phylogenetic tree. Various traits, such as mating system (monogamy versus polygamy), presence or absence of sexually selected traits, and type of parental care, can be independently mapped on the tree in an effort to understand how often and in what sequence the traits evolved.

One such study addresses the question of the function of **concealed ovulation** in primates (Sillén-Tullberg and Møller 1993). As we saw in chapter 10, most female mammals undergo estrus, becoming sexually attractive and receptive to males around the time of ovulation. In some primates, however, including humans, no obvious behavioral or morphological changes are associated with ovulation. Concealed ovulation has been thought to have two possible functions: One is a female tactic to promote paternal care. The male in a monogamous mating system is "forced" to stay around to impregnate the female and guard her against other males. A second hypothesis is that in a polygynous system with concealed ovulation, no male is certain that he is the father, so males are less likely to harm any infants in the group on the chance that they are their offspring. In other words, concealed ovulation may function to conceal ovulation from either the male in a pair or from all the males in the group. The analysis showed that ovulatory signs tend to disappear under polygyny but rarely under monogamy, suggesting that the second hypothesis is correct and that concealed ovulation functions to confuse paternity and thereby to reduce the chances of infanticide by males.

SUMMARY

A society may be defined as a group of individuals of the same species, organized in a cooperative manner that extends beyond sexual behavior. Types of cooperation include alarm calling, cooperative rearing of young, formation of coalitions and alliances, and eusociality, with the formation of reproductive castes. The potential benefits of social behavior include protection from physical elements, predator detection and defense, group defense of resources, and division of labor. The costs are increased competition, spread of contagious diseases, and interference in reproduction.

Most early researchers of animal social systems assumed that traits favoring cooperation could be selected for even when they were detrimental to individuals. Research on the evolution of social behavior in the last two decades has demonstrated that many types of cooperation can be explained at the level of parents and their young: Organisms cooperate for genetically selfish reasons rather than for the good of the group.

Aggregations may occur as individuals attempt to reduce their own chances of being selected by a predator. Kin selection may explain cooperation among individuals that share genes by common descent. Individuals may behave so as to lower their direct fitness but increase their inclusive fitness. Reciprocal altruism may be important in long-lived organisms that do not disperse extensively. Thus, an individual stands a good chance of being "paid back" if he or she cooperates with others. Parents may manipulate their own offspring to behave in ways beneficial to the parents but not to the offspring. Genes giving an offspring a competitive edge against its siblings could be selected against if the reproductive success of the parent is lowered.

Ecology plays a large role in determining the observed types of social behavior, as evidenced by comparative studies of closely related species that occupy different habitats. It is also necessary to consider the evolutionary history of these species, however, because some social traits may have evolved in response to selective pressures in the distant past.

SUGGESTED READINGS

Alexander, R. D. 1974. The evolution of social behavior. Ann. Rev. Ecol. Syst. 5:325–383.

Alexander, R. D., and D. W. Tinkle (eds.). 1981. Natural selection and social behavior. Chiron Press, New York.

Clutton-Brock, T. 2002. Breeding together: Kin selection and mutualism in cooperative vertebrates. Science 296:69–72.

Harcourt, A. H., and F. B. M. de Waal. 1992. Coalitions and alliances in humans and other animals. Oxford Univ. Press, New York.

Krebs, J. R., and N. B. Davies, eds. 1997. Behavioural ecology: An evolutionary approach, 4th ed. Blackwell Science, Oxford.

Nowak M. A. 2006. Five rules for the evolution of cooperation. Science 314:1560–1563.

Rubenstein, D. I., and R. W. Wrangham (eds.). 1986. Ecological aspects of social evolution. Birds and mammals. Princeton Univ. Press, Princeton, NJ.

Solomon, N. G., and J. A. French. 1996. Cooperative breeding in mammals. Cambridge Univ. Press, Cambridge.

Wolff, J. O., and P. W. Sherman (eds.). 2007. Rodent societies: An ecological and evolutionary perspective. Univ. Chicago Press, Chicago

DISCUSSION QUESTIONS

1. How might you argue that colonial invertebrates such as the Portuguese man-of-war have a more "perfect" society than does a group of chimpanzees?

2. Distinguish among the following types of natural selection: individual, kin, and group. Give possible examples of each.

3. Woodchucks (*Marmota monax*) live in hay fields and fencerows in the eastern United States. They live a solitary life and are aggressive toward one another. They breed each year, and the young disperse by 1 year of age. Yellow-bellied marmots (*M. flaviventris*) live at intermediate elevations in the Rockies. They are colonial but moderately aggressive, breeding annually but occasionally skipping a year. Juveniles disperse in the second year. Olympic marmots (*M. olympus*) inhabit meadows at high elevations on the West Coast. They are highly colonial, breed in alternate years, and are highly tolerant of others; juveniles do not disperse until their third year. Design a series of observations and experiments to explain the large differences in social behavior of these closely related species. Compare your approach with that of David Barash (1974), who made detailed comparisons of these species.

CHAPTER 24

Dispersal, Habitat Selection, and Migration

Most mammalian movement occurs in a relatively small area, such as when an individual acquires resources within its home range or marks and defends the area against conspecifics. These local movements are sometimes referred to as **station keeping.** They may include relatively long round trips, as, for example, when a mouse harvests seeds and returns them to a cache. We dealt with station keeping in chapter 21. On a larger scale, **ranging** behavior includes forays outside the home range, usually in search of suitable habitat or mating opportunities. According to Dingle (1996), this type of movement includes **natal dispersal,** movement from the natal site to a site where reproduction takes place. Movement on the largest scale is **migration,** persistent movement across different habitats in response to seasonal changes in resource availability and quality. It involves special physiological changes during which the animal does not respond to the presence or absence of local resources. In this chapter, we first treat natal dispersal, discussing both proximate and ultimate reasons for it, and then treat the process of habitat selection, or the finding of a place to live. Finally, we consider migration and **homing,** the process of getting back to a home range or nest site.

Dispersal from the Place of Birth

In many species of animals, members of one sex disperse from the place of birth before breeding, whereas members of the other sex are **philopatric,** breeding near the place where they were born. Among mammals, usually the males disperse, but the opposite is true in birds (Greenwood 1980). Natal dispersal means leaving the site of birth or social group (emigration), traversing unfamiliar habitat, and settling into a new area or social group (immigration). Although Dingle (1996) considers dispersal a type of ranging, others have grouped it under migration or treated it separately (McCullough 1985). Moving away from known ground is risky because the individual is unfamiliar with the location of food and shelter and is no longer in the presence of familiar neighbors and relatives. Recall from chapter 23 that cooperative behavior can evolve by both kin selection and reciprocal altruism. Both of these behavior patterns require that individuals remain in the vicinity of relatives or associates, however. Given these costs, there must be considerable benefits for dispersal behavior to be so widespread.

CAUSES OF DISPERSAL

The causes of dispersal can be understood at several different levels. At the proximate level, we wish to know the immediate reasons why an individual leaves the natal area. For instance, a male might be forced out by its parents or other residents, or it might respond involuntarily to increases in testosterone levels associated with sexual maturation. At the ultimate level, researchers wish to know the long-term, evolutionary causes of dispersal. For instance, individuals that fail to disperse may have lower reproductive success because their offspring are inbred and therefore less viable. Natural selection would then favor dispersers. For species living in fluctuating environments, individuals that colonize new habitats might be favored.

The ultimate cause of dispersal from the natal site has been argued by many to be the avoidance of inbreeding. The costs of inbreeding, referred to as **inbreeding depression,** have been documented in many laboratory and zoo populations (Ralls et al. 1979) but only recently studied in natural populations. Inbreeding depression manifests itself through reduced reproductive success and survival of offspring from closely related parents compared with offspring of unrelated parents. It is caused by increased homozygosity of the inbred offspring and the resulting expression of deleterious recessive alleles. For example, African lion (*Panthera leo*) males typically leave the natal pride at sexual maturity and attempt to breed with females from other prides. Males from a small, inbred population, however, showed lower testosterone levels and more abnormal sperm than did males from a large, outbred population (Wildt et al. 1987). Presumably, these inbred prides are prone to extinction. When both inbred and outbred white-footed mice (*Peromyscus leucopus*) were released back into the natural habitat, the inbred stock survived less well than the outbred stock, although differences between the two stocks had not been great in the laboratory environment (Jiménez et al. 1994).

If one or the other sex disperses, the chance of matings between related individuals lessens. Among black-tailed prairie dogs (*Cynomys ludovicianus*), young males leave the family group before breeding, whereas females remain. Also, adult males usually leave groups before their daughters mature (Hoogland 1982). Among primates such as vervet monkeys (*Chlorocebus aethiops*) and hamadryas baboons (*Papio hamadryas*), males usually leave the natal group at, or shortly after, sexual maturation. They usually transfer to a neighboring group with age peers or brothers (figure 24.1). Several years later, they may again transfer alone to a third group. Cheney and Seyfarth (1983)

argued that this pattern of nonrandom followed by random movement minimizes the chances of mating with close kin. Packer (1977) reported that a male baboon that failed to disperse at sexual maturity and then mated with relatives sired offspring with lower survival rates than the offspring of outbred males. Thus, generally, a real cost in reduced fitness seems to be associated with inbreeding.

The effects of inbreeding depend on past population size and mating patterns. Populations with a long history of outbreeding tend to show the most severe effects once inbreeding takes place. This is because recessive mutations accumulate in the population during outbreeding without ill effect but are more likely to be present in both parents and thus passed on to offspring when inbreeding takes place. Populations that have survived episodes of inbreeding in the past, however, such as the cheetah (*Acinonyx jubatus*), may tolerate current inbreeding with few ill effects, in part because the deleterious recessive alleles already have been selected out of the population (Shields 1982; see chapter 29).

A second cause of dispersal from the natal site may be the reduction of competition with conspecifics for food, shelter, or mates (Dobson 1982; Moore and Ali 1984). As we pointed out in chapter 22, gestation and lactation by only females means that males tend to provide little or no direct care for offspring. Most species of mammals are polygynous, that is, males mate with more than one female. Males may be forced to disperse as they compete for access to females. Although the inbreeding-avoidance hypothesis predicts that one sex should disperse, it does not predict which sex should disperse; the competition

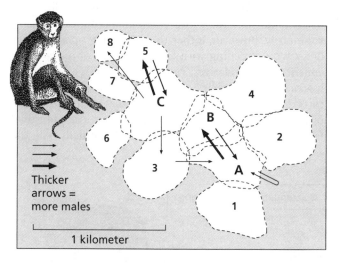

Figure 24.1 Group transfer by natal and young adult male vervet monkeys from three social groups between March 1977 and July 1982. Arrows indicate direction of movement. Letters indicate ranges of the three main study groups; numbers indicate ranges of regularly censused groups. Only groups with ranges adjacent to the main study groups are shown. Males usually transferred to neighboring groups with brothers or age peers. *Data from D. L. Cheney and R. M. Seyfarth, 1983, Nonrandom Dispersal in Free-ranging Vervet Monkeys: Social and Genetic Consequences,* The American Naturalist *122:392–412.*

hypothesis predicts that males should be the dispersing sex in polygynous species. According to Greenwood (1980), in such systems the reproductive success of males is limited by the number of females with which they can mate, and males are likely to range farther than females as they search for mates. Females, on the other hand, are limited by resources (food and nesting sites) that can best be obtained and defended by staying at home. Among group-living mammals, females typically form the stable nucleus, and the males attempt to maximize their access to them, frequently moving from one group to another.

Hamilton and May (1977) proposed a somewhat different competition model in which animals disperse so as to avoid local resource competition with close relatives and thus avoid lowering their indirect fitness. In a new habitat, they are likely to be competing with nonrelatives and therefore would suffer no such cost. These models are not contradictory, but more research is needed on both the proximate and ultimate causes of dispersal. The role of dispersal in population regulation is discussed in chapter 25.

EXAMPLES OF DISPERSAL

Lions

Dispersal in lions follows the typical mammalian pattern, in that females usually remain in or near their natal pride, whereas males always leave, usually before 4 years of age, to become nomads or to form coalitions that take over new prides (Pusey and Packer 1987; figure 24.2). Competition with other males seems to be an important factor because departures most often occur when a new coalition of males takes over the pride. Some males appear to leave voluntarily, however, either to find mating opportunities or to avoid breeding with kin. Coalitions of males often consist of close relatives. A coalition controls a pride until it is ousted by another coalition a few years later or leaves to take over another pride. In all cases, males leave the pride before their daughters start mating. An additional factor is that new coalitions of males usually kill the young cubs in the new pride (see chapter 21). Thus, breeding males must remain in a new pride long enough to ensure the survival of their cubs. Pusey and Packer (1987) concluded that male–male competition, mate acquisition, protection of young cubs, and inbreeding avoidance all play roles in the evolution of dispersal patterns of lions.

Belding's Ground Squirrels

The question of why individuals disperse was addressed at several different levels of analysis by Holekamp and Sherman (1989). Belding's ground squirrels (*Spermophilus beldingi*) also follow the typical mammalian pattern in that females remain in the natal area for life, whereas males disperse as juveniles. The proximate causes of male dispersal seem to be the prenatal effects of testosterone on the male embryo during a critical period in development in the mother's uterus (organizational effects of hormones; see chapter 8) and the attainment of a critical body mass after birth. Testosterone seems less important later in life (activational effects); castration of males just prior to natal dispersal did not prevent dispersal. Holekamp and Sherman were not able to test the inbreeding-avoidance and competition hypotheses directly but concluded that inbreeding avoidance was the more likely means by which dispersal increased fitness (table 24.1).

INBREEDING VERSUS OUTBREEDING

If inbreeding depression were the only factor involved in natal dispersal, individuals might be expected to disperse as far as possible from relatives. Such is not usually the case, however. According to Shields (1982), most species that have been adequately studied are relatively philopatric and remain close to the place of birth. He cites apparent cases of **outbreeding depression,** in which matings between members of different populations within a species yield less-fit offspring. Members of a population may possess adaptations to local conditions that are

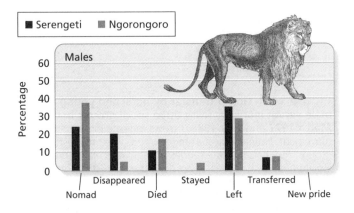

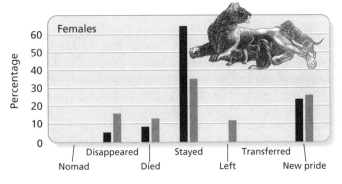

Figure 24.2 The fate of subadult lions by 4 years of age at two African sites. Note the differences between the sexes in dispersal pattern. *Data from Anne E. Pusey and C. Packer, 1987, The Evolution of Sex-Biased Dispersal in Lions,* Behaviour, *101:275–310.*

Table 24.1   Why juvenile male Belding's ground squirrels disperse

Levels of Analysis	Summary of Findings
Physiological mechanisms	Dispersal by juvenile males is apparently caused by organizational effects of male gonadal steroid hormones. As a result, juvenile males are more curious, less fearful, and more active than juvenile females.
Ontogenetic processes	Dispersal is triggered by attainment of a particular body mass (or amount of stored fat). Attainment of this mass or composition apparently also initiates a suite of locomotors and investigative behaviors among males.
Effects on fitness	Juvenile males probably disperse to reduce chances of nuclear family incest.
Evolutionary origins	Strong male biases in natal dispersal characterize all ground squirrel species, other ground-dwelling sciurid rodents, and mammals in general. The consistency and ubiquity of the behavior suggest that it has been selected for directly across mammalian lineages.

From Kay E. Holekamp and Paul W. Sherman, 1989, Why Male Ground Squirrels Disperse, American Scientist, *77:232–239. Copyright © 1989. Reprinted by permission of* American Scientist, Journal of Sigma Xi, The Scientific Research Society.

lost through outbreeding. Thus, two areas might differ slightly in temperature, humidity, or types of food available. If each population is genetically adapted to these conditions, then they would be better off mating with individuals with those same adaptations.

A certain degree of inbreeding may be advantageous. In sexually reproducing organisms, the loss of alleles in the offspring can be reduced by half if the parents are related. Furthermore, according to Shields (1982), coadapted gene complexes are less likely to be disrupted in matings between relatives. Finally, kin selection (chapter 23), which results in the evolution of cooperative behaviors, can operate only when relatives are in close proximity. More cooperative behavior is predicted within philopatric species and within the philopatric sex. Indeed, among Belding's ground squirrels, females are philopatric and engage in altruistic alarm calling. Males who disperse away from relatives do not make alarm calls (Sherman 1981; see also chapter 23).

The preceding arguments may predict some sort of "optimal" inbreeding strategy in which matings between very close relatives (siblings, or parents and offspring) are avoided but matings with more distant relatives are favored. In laboratory tests, female white-footed mice in estrus preferred males who were first cousins over nonrelatives or siblings (Keane 1990). Heavier pups and larger litters resulted from matings between first cousins than between individuals of other degrees of relationship. Based on electrophoretic analysis of blood enzymes in natural populations, however, little if any mating between relatives seems to occur in this species (Wolff et al. 1988).

FEMALE DISPERSAL

Although males are the dispersing sex in most species of mammals, exceptions exist in several orders. Among lagomorphs, pikas (*Ochotona princeps*) live in relatively isolated patches of talus (rock debris) on mountain slopes (see chapter 18). Most juvenile pikas remain close to their birth site for life. Individuals occasionally disperse both within and between patches of talus, however. Of those that moved more than 100 m, females moved more than males at sites in Alberta (Millar 1971) but not in Colorado (Smith 1987). Similarly, among banner-tailed kangaroo rats (*Dipodomys spectabilis*) in Arizona, most juveniles stayed home, sharing all or part of the maternal home range. Of those that did disperse, females had a tendency to move farther (Jones 1987). This species is solitary, living in dispersed mounds that are "inherited" from the previous occupant. Other examples of species in which females are the dispersing sex include the chimpanzee (*Pan troglodytes*), African wild dog (*Lycaon pictus*), and white-lined bats (*Saccopteryx* spp.). These groups have widely differing social systems, and it is not yet clear why the typical mammalian pattern is reversed (Greenwood 1983).

Habitat Selection

Generally, plants depend on natural agents such as currents of air or water or other organisms for dispersal. The result is an opportunistic dissemination of plant individuals; few ever reach environments conducive to survival and reproduction. In contrast, the well-developed locomotive abilities of mammals allow them to play more active roles in finding places to live. **Habitat selection** can be defined as choosing a place to live, which does not necessarily imply a conscious choice or that individuals make a critical evaluation of the entire constellation of factors confronting them. More often the choice is an "automatic" reaction to certain key aspects of the environment.

If a species of mammal occupies an area and reproduces there, all its needs are met and it can compete with other species successfully. Any one of several factors can prevent a species from occupying particular habitats, however. One obvious factor is dispersal ability. Oceanic islands such as New Zealand and Hawaii provide much habitat suitable for mammals; yet, other than marine mammals, only a few species of bats have made it there on their own. Many species such as rats (*Rattus* spp.) and domestic ungulates have been introduced by humans and have thrived.

Sometimes behavior patterns keep species from occupying apparently suitable habitats. The white-footed mouse lives in woodlands, but it is rarely found in adjacent fields, which are inhabited by a closely related species, the deer mouse (*Peromyscus maniculatus bairdii*). In fact, *P. leucopus* seems perfectly capable of living in fields also. Why should a species not take advantage of suitable habitat? One possibility is that the habitat is not actually suitable, perhaps because of competition, predation, or other, undetected, factors. It is also possible that such habitats may not have been suitable in the past; if organisms responding to certain environmental cues in previous optimal habitats left more offspring, their genetically influenced behavior patterns would become widespread and persist. New environments, although suitable, might not contain those cues and therefore not be used. White-footed mice prefer the vertical structure that is present in woodlands but generally lacking in field habitats (M'Closkey 1975). This mouse orients toward large trees (Joslin 1977b) and avoids areas with high densities of woody stems (Barry and Francq 1980).

Even if an individual can and "wants" to get to a place, other factors may prevent its becoming established. These factors include predators, parasites, disease agents, **allelopathic** agents (plant toxins or antibiotics), or competitors. Although it is difficult to demonstrate conclusively that one species prevents an area from being colonized by another, experiments and observations lend support to this idea (see chapter 26).

Risk of predation and competition may restrict habitat use. Hairy-footed gerbils (*Gerbillurus tytonis*) live in vegetated islands in a sea of sand in the Namib Desert of southwestern Africa (Hughes et al. 1994). Individuals preferred sites around bushes or grass clumps over open areas and were more active on new-moon nights than on full-moon nights. They also gave up feeding at seed trays sooner in open areas and on full-moon nights (figure 24.3). These differences were likely caused by greater risk of predation in open areas and when the moon was full. When four-striped grass mice (*Rhabdomys pumilio*), a close competitor of the gerbil, were removed, gerbils increased foraging activity, especially in the grass clumps.

Finally, a species may be absent from an area because of abiotic (nonbiological) factors. Each organism has a range of tolerances for a variety of physical and chemical factors, and much of its behavior is directed toward staying within these limits. Temperature and moisture are the main factors that limit the distribution of life on Earth, but physical factors, such as light, soil structure, or fire, and chemical factors, such as pH and nutrients, may be

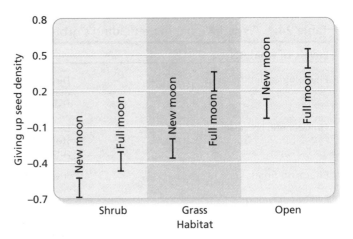

Figure 24.3 **The effect of predation risk on habitat use in hairy-footed gerbils.** Values along the y-axis are giving up seed densities (GUDs) (log10 scale) in food trays containing seeds mixed with sand that were placed in different habitats. High values mean that gerbils left the trays when many seeds were still present. Bars denote 95% confidence intervals. Gerbils gave up at higher seed densities in open areas and on full-moon nights. *Data from J. J. Hughes et al., 1994, Predation Risk and Competition Affect Habitat Selection and Activity of Namib Desert Gerbils,* Ecology, *75:1397–1405.*

important as well. Adaptations of mammals to some of these conditions were discussed in chapter 9.

DETERMINANTS OF HABITAT PREFERENCE

Genes and Environment

What roles do genetics and learning play in habitat choice? If two animals reared from birth in identical environments differ in habitat preference when they are tested as adults, the conclusion is that those differences result from hereditary factors.

Wecker (1963) conducted one of the classic studies on habitat selection in mammals on deer mice (*Peromyscus maniculatus*). This species, a common North American rodent, is divided into many geographically variable subspecies; two are the larger, long-eared, long-tailed forest form (*P. m. gracilus*) and the smaller, short-eared, short-tailed grassland (prairie) form (*P. m. bairdii*). The grassland subspecies does well in forest-like laboratory conditions where its food preference and temperature tolerance are similar to those of the forest subspecies. Thus, experimenters assumed that the avoidance of forests in the prairie deer mouse is a behavioral response (Harris 1952).

Wecker's objective was to assess the genetic basis of this behavior and to test the importance of "habitat imprinting" (Thorpe 1945). He constructed an enclosure halfway in a forest and halfway in a grassland, released the grassland subspecies in the middle, and recorded their locations. The animals he tested were of three basic types: (1) wild-caught in grassland; (2) offspring of wild-caught parents, reared in a laboratory; and (3) reared in a laboratory for 20 generations. Both wild-caught mice and their offspring selected the grassland half of the enclosure, whatever their previous experience. Laboratory stock and their offspring showed no preference, whether or not they had been raised in forest conditions. Laboratory stock reared in a grassland enclosure until after weaning, however, showed a strong preference for the grassland when tested later (table 24.2).

Wecker reached the following conclusions:

1. The choice of grassland environment by grassland deer mice is predetermined genetically.
2. Early grassland experience can reinforce this innate preference but is not a prerequisite for subsequent habitat selection.
3. Early experience in forest or laboratory is not sufficient to reverse the affinity of this subspecies for the grassland habitat.
4. Confinement of these deer mice in the laboratory for 12 to 20 generations results in a reduction of the hereditary control over the habitat selection response.
5. Laboratory stock retains the capacity to "imprint" on early grassland experience but not on forest.

Wecker also suggested that learned responses, such as habitat imprinting, are the original basis for the restriction of this subspecies to grassland environments; genetic control of this preference is secondary.

In a series of laboratory experiments designed to explore the importance of early experience on bedding preference in inbred house mice (*Mus musculus*), Anderson (1973) raised animals either on cedar shavings or on a commercial cellulose material. When he tested them later, he found that the mice preferred the bedding on which they had been raised, although females raised on cellulose "drifted" toward cedar shavings in subsequent tests. Naive mice preferred cedar shavings. Both this experiment and the one by Wecker illustrate the complex interaction of genes and environment in the development of habitat preferences in mammals.

Tradition

Inherited tendencies and imprinting may be involved in restricting habitat choice to a small part of the potential range, but **tradition**—behavior passed from one generation to the next through the process of learning—may also be important. Such behavior seems to be important in the movements of many species of ungulates.

Mountain sheep (*Ovis canadensis*) live in unisexual groups; females are likely to stay in the natal group but may switch to another female group when they are between 1 and 2 years of age (Geist 1971). Mothers do not tend to chase their young away at weaning, as most other mammals do. Young rams desert the natal group at 2 years of age and join all-ram bands. Females follow an older, lamb-leading female, whereas males follow the largest-horned ram in the band. When rams mature, they are followed by younger rams and pass on to them their habitat preferences and migration routes.

Until the nineteenth century, mountain sheep occupied a much larger range in North America and Asia than they do today. Measures enacted to protect sheep, mainly hunting regulations, have done little to increase their numbers. Formerly inhabited parts of the range that appear intact often have not been recolonized, and transplants to suitable areas often have been unsuccessful. In contrast, deer (*Odocoileus* spp.) and moose (*Alces alces*) have recolonized areas rapidly and have reached population densities higher than ever (Geist 1971).

Why have sheep failed to extend their range, but moose and deer have done so? Geist pointed out that deer and moose, which are relatively solitary, establish ranges by individual exploration after being driven out of the mother's range; sheep, in contrast, transmit home-range knowledge from generation to generation and often associate with group members for life.

Understanding the niches of these ungulate species can help explain the sheep system, which may seem a rather poor adaptation to the environment. Moose are associated over much of their range with early successional communities that follow forest fires, where moose are a "pioneer species" that disappears when the climax coniferous forest

Table 24.2   Habitat selection by deer mice as a function of hereditary background and early experience*

Number of Mice Tested	Hereditary Background	Early Experience	Habitat Preference
12	Grassland	Grassland	Grassland
13	Laboratory	Grassland	Grassland
12	Grassland	Laboratory	Grassland
7	Grassland	Forest	Grassland
13	Laboratory	Laboratory	None
9	Laboratory	Forest	None

Data from S. C. Wecker, 1963, The Role of Early Experience in Habitat Selection by the Prairie Deer Mouse, Peromyscus maniculatus bairdi, *Ecological Monographs 33:307–325, Ecological Society of America.*

*The outdoor test enclosure was forest on one side and grassland on the other. Preference was measured by the percentage of time, amount of activity, and depth of penetration by mice in each side of the enclosure.

regenerates. Moose therefore must continually colonize these newly formed early successional habitats. Each spring when her new calf is born, the cow drives away her yearling, which may wander some distance before establishing its new range.

Sheep habitats consist of stable, long-lasting, climax grass communities that exist in small patches. Geist argued that, given the distance between patches and the ease with which wolves can prey on sheep, the best strategy for the sheep is to stay on familiar ground. Sheep also have at least two and as many as seven seasonal home ranges that may be separated by 30 km or more. These areas are visited regularly by the same sheep year after year at the same time. Knowledge of the location of these ranges and the best times to visit them is transmitted from one generation to the next. Because new habitats rarely become available, there is no advantage to an individual's dispersing and attempting to colonize other areas.

Tradition may also play a role in habitat use by mantled howling monkeys (*Alouatta palliata*), which are Neotropical folivores. Many trees within their home ranges produce leaves that contain secondary compounds that make them unpalatable or even toxic to the monkeys (Glander 1982). Glander (1977) postulated that older group members provide a reservoir of knowledge about the location and timing of seasonal foods that reduces the chances of the group's encountering toxic leaves.

THEORY OF HABITAT SELECTION

Theoretical approaches to the problem of habitat selection are in a rather early stage of development, and no single general theory is generally accepted (Rosenzweig 1985). Several different approaches have been used. One is to think of habitats as patches, or areas of suitable habitat interspersed among areas of unsuitable habitat, and to apply optimal foraging theory, first developed by MacArthur and Pianka (1966; see chapter 7). This theory enables researchers to predict which habitat patches an animal should select and when it should leave one habitat and move to another so as to get the greatest benefit for the least cost. This economic model incorporates such factors as the availability of resources in various patches and the costs of getting from one patch to another. Although the resource is usually assumed to be food (i.e., energy), nest sites or mates are other possibilities. For a review of optimal foraging theory, see Stephens and Krebs (1986).

A second approach is the **ideal free distribution,** which predicts how individuals distribute themselves so as to have the highest possible fitness (Fretwell and Lucas 1970). It assumes that animals have complete and accurate knowledge about the distribution of resources (ideal) and that they are passive toward one another and can go to the best possible site (free). Individuals settle in habitats so that the first arrivals get the best resources. As density increases, less desirable areas are occupied, and animals spread themselves out so that all have the same fitness in the absence of intraspecific competition. One obvious result of such a distribution is that rich habitats will have more individuals than poor ones.

If intraspecific competition occurs via dominance or territory (see chapter 21), a despotic distribution develops, with some individuals monopolizing the best resources (Fretwell 1972). Another variation on the ideal free distribution is the ideal preemptive distribution (Pulliam and Danielson 1991). Potential breeding sites differ in quality—that is, in the expected reproductive success of their occupants—and individuals choose the best unoccupied sites. These best sites are thus preempted and are no longer available to others. Several studies have shown that individuals in the preferred habitat have higher fitness, as measured by reproductive success and survival, than those in less preferred habitat. For instance, Grant (1975) found that meadow voles (*Microtus pennsylvanicus*) in the preferred grassland habitat had higher survival and reproductive success than those in the less preferred woodland.

Habitat selection also affects the growth rates and densities of populations, a topic considered in chapter 25. It may even lead to the formation of new species. Although the usual models of speciation require the formation of geographic barriers to gene flow, reproductive isolation could evolve if members of two parts of a population came to prefer different microhabitats in the same region. If the preferences were heritable and individuals mated assortatively (i.e., with others having similar habitat preferences), a barrier to gene flow would be created that could eventually lead to the formation of new species (Rice 1987). Such a mechanism has yet to be demonstrated in mammals, however.

Migration

Of all the movements that animals make, perhaps none has generated so much interest and controversy as migration. Migratory behavior differs from station-keeping movements associated with resources and maintenance of home ranges or territories. Migration takes an animal out of its home range and habitat type. It is triggered by proximal cues, such as photoperiod, that are linked to ultimate factors, such as a shortage of resources. Migration may also occur in response to endogenous rhythms (Dingle 1996). Earlier definitions of migration require that the individual make a round trip, but one-way movements are included in most recent definitions. In migrating, mammals may use all three means of vertebrate locomotion: flying, swimming, and walking.

BATS

As the only mammals with true flight, bats might be expected to show migratory behavior comparable to the only other flying vertebrates, the birds. Such is not the case for most species. As we pointed out in chapter 13, bats have evolved relatively slow, maneuverable flight, necessary to catch insects on the wing. The wings are even used to trap insects. Bat wings are thin airfoils of high camber (curvature) that produce high lift at slow speeds but produce excessive drag at the high speeds necessary for long-distance migration. Rather than migrate, many species use hibernation, an energetically less-costly way of dealing with cold temperatures and lack of food.

In spite of these flight constraints, some species of bats do migrate, most often to and from caves and other shelters used as hibernation sites (Griffin 1970). Information about bat migration comes from the seasonal appearances and disappearances at roosting sites coupled with recoveries of banded individuals (Fenton and Thomas 1985). Radiotelemetry has been used in a few instances, mostly to study foraging trips and roosting behavior (Fenton et al. 1993). In a study involving the banding of more than 73,000 little brown bats (*Myotis lucifugus*), individuals migrated more than 200 km from hibernation caves in southwestern Vermont. They generally moved southeast into Massachusetts and neighboring states for the summer (Davis and Hitchcock 1965). The endangered Indiana bat (*M. sodalis*) migrates from hibernation caves in Kentucky and southern Indiana as far north as Michigan (Barbour and Davis 1969).

Swifter flying species such as the hoary bat (*Lasiurus cinereus*) and free-tailed bat (*Tadarida brasiliensis*) move even greater distances. Hoary bats migrate from summer ranges in the Pacific Northwest, as far north as Alaska, south into central California and Mexico for the winter. In winter, they are not found above 37°N latitude, a limit probably set by the distribution of flying insects (Griffin 1970). Free-tailed bats seem to have both migratory and nonmigratory populations. Those in southern Oregon and northern California are year-round residents, but those in the southwestern United States migrate south into Mexico for the winter (Dingle 1980). For instance, a population in the Four Corners area (where Colorado, Utah, New Mexico, and Arizona meet) has a well-established flyway through the Mexican states of Sonora and Sinaloa west of the Sierra Madre Oriental Mountains. The routes of some southwestern U.S. populations have yet to be identified.

A newer technique to study bat migration is the use of stable hydrogen isotope ratios of bat hair (Cryan et al. 2004). The mean annual hydrogen isotope ratio of local precipitation and groundwater changes with latitude and elevation and is incorporated into the hair, making it possible to determine the general location where the hair was grown. For example, a male hoary bat captured in Mexico in September had a ratio consistent with a region north of the Canadian border, some 2,000 km distant, confirming earlier conclusions about migration patterns in this species (Cryan et al. 2004). Advantages of this method are that there is no need to band or radio track animals and one can get hair samples from museum specimens.

Long-distance migration is not restricted to bat species in temperate regions. Seasonal shifts in rain patterns trigger migration in some species of African bats (Fenton and Thomas 1985). For instance, West African fruit bats of the family Pteropodidae migrate distances of 1500 km each year, following rains into the Niger River basin (Thomas 1983). One of the most spectacular migrations of fruit bats is the influx of 5–10 million straw-colored fruit bats (*Eidolon helvum*) into Kasanka National Park in Zambia in response to food supply (Richter and Cumming 2006).

CETACEANS

Most species of baleen whales (Suborder Mysticeti) spend summer months at high latitudes, feeding on plankton in the highly productive antarctic and arctic waters. As winter approaches in each area, whales migrate to warmer subtropical and tropical waters. Food supply does not drive this migration; tropical waters are relatively unproductive, and, in fact, whales do not feed during migration or at their wintering grounds. Instead, they rely on fat deposits. The benefit of moving to warmer water is likely the energy savings from reduced heat loss, especially for calves. Calves are born in the tropical breeding grounds, and lactating females with their newborn calves move to feeding areas at higher latitudes as spring approaches. Breeding cycles of species that breed inshore, such as the humpback whale (*Megaptera novaeangliae*) and the gray whale (*Eschrichtius robustus*), are fairly well known, but little is known about the breeding habits of the offshore species, such as the blue whale (*Balaenoptera musculus*) and fin whale (*B. physalus*; Dingle 1980).

California gray whales spend their summers feeding in the North Pacific and Arctic Oceans. In autumn, they migrate south to subtropical breeding grounds off the coast of Baja California (Orr 1970). Humpback calves are born in September, and lactating females begin the trek back north with their calves in the spring, usually after males and newly pregnant females have already left (Dingle 1980).

PINNIPEDS

Many species of seals and sea lions (chapter 16) migrate thousands of kilometers from island breeding and molting areas to oceanic feeding areas. Island breeding sites are chosen because they are relatively free of predators. Northern elephant seals (*Mirounga angustirostris*) breed on island rookeries off California, migrate to foraging areas in the North Pacific and Gulf of Alaska, and later return to the islands to molt. Using geographic location–time–depth recorders, Stewart and DeLong (1995) found that seals travel linear distances of up to 21,000 km during the 250 to 300 days they are at sea. Each individual makes two round-trip migrations per year, returning to the same foraging areas during postbreeding and postmolt movements. Males migrate farther north than females, where they feed off the Alaskan coast (figure 24.4).

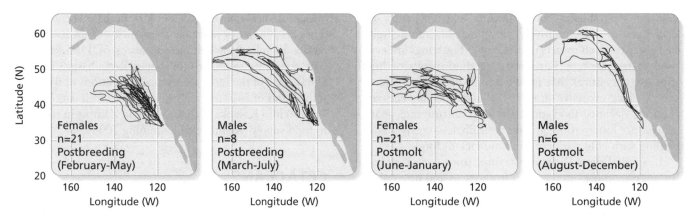

Figure 24.4 Seasonal migratory tracks of northern elephant seals in the eastern North Pacific. Each seal makes two migrations per year, returning to land for breeding and again for molting. These seals were marked on San Miguel Island, California. *Data from B. S. Stewart and R. L. DeLong, 1995, Double Migrations of the Northern Elephant Seal, Mirounga angustirostris,* Journal of Mammalogy, *76:196–205.*

Data-recording tags can do double duty by providing information about ocean temperature and salinity (Pala 2006). Nine elephant seals tagged in 1998–1999 covered an average of 4634 km over 67 days, with a mean dive duration of 20 min and mean depth of 428 m (Boehlert et al. 2001). In so doing, they provided a wealth of information about ocean conditions at different depths. These and other data have been added to the World Ocean Database.

UNGULATES

Large ungulates migrate long distances as well. The best-studied northern species is the barren-ground caribou (*Rangifer tarandus*). Herds migrate north to calving grounds above the timberline in spring and return south in winter, covering distances of more than 500 km (Orr 1970; figure 24.5). More recently, the movements of individuals

have been monitored by satellite tracking (Craighead and Craighead 1987). Migrations seem to be made up of series of fairly straight movements that are little influenced by landmark features such as rivers (Bovet 1992; figure 24.6).

The mass migrations of wildebeests (*Connochaetes taurinus*) in East Africa are spectacular (figure 24.7). The

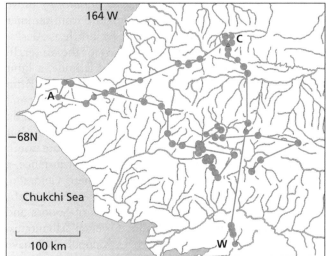

Figure 24.6 Migration route of an adult female caribou in northwest Alaska in 1984. The line connects successive satellite fixes (filled circles). The caribou left the winter range (W) on 15 May and arrived on the calving ground (C) on 30 May, where she calved on 5 June, and stayed until 16 June. She then moved to the herd's aggregation area (A), where she stayed on 4 and 5 July. She spent the summer traveling east, with occasional one- to two-week stays in localized areas. The last fix was on 7 October, while she was moving toward the winter range. Due to the hydrographic features of the area (thin lines = rivers), the caribou's route was probably as often across as it was parallel to valleys. *Data after Craighead and Craighead (1987), and Fairbanks of the World Map (1:2500000) of the USSR Main Admin. of Geodesy and Cartography, Moscow, 1973; in J. Bovet, 1992, Mammals, in* Animal Homing *(F. Papi, ed.), Chapman and Hall.*

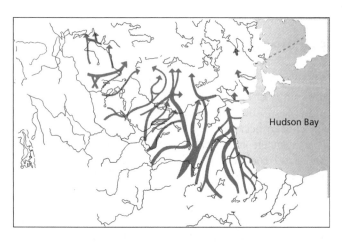

Figure 24.5 Spring migration routes of barren-ground caribou. In spring, herds move north to calving grounds above the timberline; in autumn, they return south to the shelter of forests. *Data from H. L. Gunderson, 1976,* Mammalogy, *McGraw-Hill.*

Figure 24.7 Wildebeests migrating in the Serengeti. The mass migration of wildebeests is associated with seasonal rainfall and the consequent growth of grasses.

Serengeti population spends the wet season, usually December through April, in the southeastern Serengeti plains of Tanzania, where short grasses are lush and calving takes place. Large migratory herds form at the beginning of the dry season, in May and June, as millions of animals move, sometimes in single file, northwest toward Lake Victoria. In July and August, near the end of the dry season, herds move northeastward into the Masai-Mara of Kenya and return south to the breeding grounds between November and December. These patterns vary considerably, however, depending on the timing of rainfall (Dingle 1980).

Other ungulates engage in elevational migration; elk (*Cervus elaphus*) and mule deer (*Odocoileus hemionus*) move into high-elevation summer ranges that are relatively free of snow and then return to milder winter ranges at lower elevations (McCullough 1985). Mountain sheep follow the same routes each year, climbing up into isolated patches of lush grazing meadow as the snow melts in spring and returning to lower elevations in autumn and winter.

Homing

Whether migrating thousands of kilometers or simply foraging within the home range, most species of mammals return to a home range, nest site, or den, a process called **homing.** For example, mule deer have shown site fidelity for both summer and winter home ranges, with straight-line distances between seasonal ranges of as much as 115 km (Thomas and Irby 1990). Of 30 deer radio-tracked for more than a year, 29 returned to the same home ranges in both summer and winter. Similar results have been obtained for elk, white-tailed deer (*Odocoileus virginianus*), moose, and mountain sheep (Bovet 1992). Nonmigratory species must also find their way home each time they return to a nest or den after foraging or searching for mates.

ORIENTATION AND NAVIGATION

How do mammals find their way home? Most of what is known about the process of homing comes from studies of homing pigeons, but enough work has been done with mammals to suggest that they use many of the same mechanisms. Most research has been done on rodents and bats. One technique for studying homing is to displace the animal from its home range and record such variables as the direction it heads when released (the "vanishing bearing"), the time it takes to get home, whether it makes it home at all, and, in some cases, the route it takes to get back home. The roles of various senses can be studied via manipulations (e.g., use of blindfolds).

In one of the earliest studies, in which deer mice were released at distances well beyond their home range (Murie and Murie 1931), some individuals made it back home. Three possible mechanisms were suggested: (1) the mice were sufficiently familiar with the terrain; (2) the mice had some sort of homing instinct, or sense of direction; or (3) the results were due to chance. These possibilities form the basis for much of the later work on homing mechanisms (Joslin 1977a).

As might be expected, homing success decreases as a function of the distance the individual is displaced (Bovet 1992; figure 24.8). Homing success increases as a function of home-range size, even within the same species. For instance, house mice with large home ranges homed over greater distances than did those with small home ranges (Anderson et al. 1977).

Several studies have added sensory deficits to displacement, with mixed results. Making white-footed mice **anosmic** (eliminating the sense of smell) had little effect on homing ability, whereas blinding them produced negative

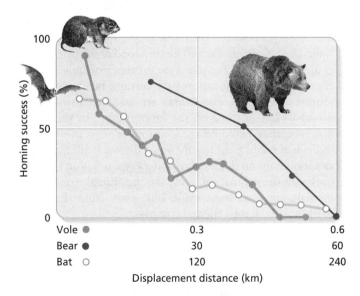

Vole ●	0.3	0.6
Bear ●	30	60
Bat ○	120	240

Displacement distance (km)

Figure 24.8 Relationship between homing success and displacement distance. Meadow voles (*Microtus pennsylvanicus*), n = 460; black bears (*Ursus americanus*), n = 112; Indiana bats (*Myotis sodalis*), n = 700. *Data after Robinson and Falls, 1965; MacArthur, 1981; Hassell, 1960; in J. Bovet, 1992,* Mammals in Animal Homing *(F. Papi, ed.), Chapman and Hall.*

effects (Cooke and Terman 1977; Parsons and Terman 1978). Blindfolding bats produced little deficit over short distances but reduced success over long distances (>32 km; Mueller 1966); at longer distances vision appears to play a role in the orientation and navigation of bats (Williams and Williams 1967, 1970; Williams et al. 1966).

Path Integration

When a small mammal takes a circuitous trip away from home, it is able to head directly back home, even in the absence of external cues. Hamsters (*Mesocricetus auratus*) fed in the center of an arena are able to get back to their peripheral nestbox with food, even when tested in the dark and in the absence of other external cues (Etienne et al. 1988). Somehow, signals generated during locomotion allow an individual to update its position relative to its starting point. Such a navigational process is termed **path integration,** or dead reckoning. It involves internal, or egocentric, spatial localization in that the animal needs no external referent.

One problem with this type of navigation is that if the animal is accidentally shifted off course, as from wind or current, it may not be able to compensate and get back home without some external referent. Path integration also accumulates more and more error the greater the number of turns in the route. Thus, it is unlikely that such a mechanism would be reliable for long-distance homing (Etienne et al. 1996).

Distant Landmarks

Several laboratory studies have demonstrated that rodents can locate food or shelter based on their relationship to distant visual cues, a mechanism sometimes referred to as **piloting** (Bovet 1992). Landmarks would seem best for locating objects within the home range, such as when gray squirrels (*Sciurus carolinensis*) locate a food cache (McQuade et al. 1986). Piloting requires the existence of some sort of cognitive, or mental, map of the terrain, but it does not require a compass. Landmarks are assumed to be most useful within or close to the home range. In principle, however, they could be used on longer trips as well. An individual could head toward some distant landmark such as a mountain on the way out and away from it on the way back. On the return trip, however, the animal would have to maintain a constant angle 180° away from the landmark, that is, have a mental compass.

Sun Compass

Another way to maintain direction is to use the sun as a compass and maintain a constant angle to it while traveling. Of course, the sun is not fixed in the sky, so an individual must have some sort of internal clock that enables it to compensate for the movement of the sun across the sky (about 15°/h). One way to test for the existence of a

sun compass is to "clock-shift" the animal in the laboratory by delaying the onset of the light–dark cycle and then test it in the field. For instance, if the test animal is shifted 6 hours in the laboratory, it should head 90° off course when tested. Thirteen-lined ground squirrels (*Spermophilus tridecemlineatus*) were tested in an outdoor arena 100 m west of their home cages with only the sky visible (Haigh 1979). When released in the arena, they burrowed in the direction of the home cages. These same individuals were then clock-shifted 6 hours in the laboratory. When retested, most shifted their burrowing direction 90° in a clockwise direction, as predicted (figure 24.9). The ability to use the sun as a compass with time compensation has also been demonstrated in several other species of small rodents (Bovet 1992). Big brown bats (*Eptesicus fuscus*) use the post-sunset glow to travel from their roost to favorite foraging areas (Buchler and Childs 1982). They depart the roost in the evening at a colony-specific angle to the glow that is independent of landmarks. Other celestial cues such as the stars are used by migrating birds but have not been demonstrated in mammals.

Magnetic Compass

Less clear is the extent to which mammals obtain directional information from geomagnetic cues, although many other organisms do. Frequent reports of otherwise healthy whales that strand themselves on beaches suggest that these whales have made navigational errors in areas where magnetic minima intersect the coast (Klinowska 1985). Strandings of male sperm whales around the North Sea occur most often during times of low sun spot activity, as part of an 11-year cycle (Vanselow and Ricklefs 2005).

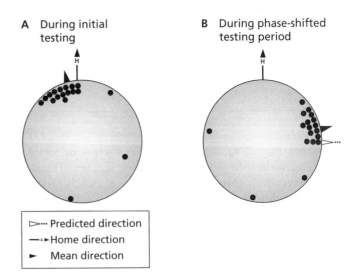

Figure 24.9 Digging responses of thirteen-lined ground squirrels in outdoor enclosures. *Data from G. R. Haigh, 1979, Suncompass Orientation in the Thirteen-lined Ground Squirrel,* Spermophilus tridecemlineatus, Journal of Mammalogy *60:629–632.*

Variations in solar flux are associated with changes in the Earth's magnetic field that may disorient migrating individuals. When aerial sightings of fin whales off the northeast U.S. coast were plotted on maps, no association could be drawn between location and measurements such as bottom depth or slope. Sightings of migrating, but not feeding, animals, however, *were* associated with areas of low geomagnetic-field intensity and gradient in autumn and winter (Walker et al. 1992). On the other hand, sightings of short-beaked saddleback dolphins (*Delphinus delphis*) off the coast of southern California were related to bottom topography but not to magnetic patterns (Hui 1994).

When white-footed mice were displaced 40 m from their home areas in woods and released in a circular arena in adjacent fields, exploratory and escape behavior was concentrated in the homeward direction. A second group of mice was treated exactly the same, except that a magnetic field opposite of that of Earth's was established in the transport tube. Those mice concentrated their activity in the opposite direction from home, suggesting that they had a magnetic sense and used geomagnetic fields as a compass cue (August et al. 1989). Similar results have been found in several other species of rodents (Bovet 1992) and apparently even in humans (Baker 1987).

Among subterranean mammals, visual cues are lacking, so geomagnetic cues could be especially important. Blind mole-rats (*Spalax ehrenbergi*) use path integration over short distances but rely on geomagnetic forces over longer distances, based on laboratory experiments (Kimchi et al. 2004). Relatively little is known about the physiological mechanisms involved in this sensory ability; see Lohmann and Johnsen for a review (2000). As with other animals, the primary receptors in mammals are believed to be magnetite-based. Working with another species of mole-rat (*Fukomys anselli*), Wegner et al. (2006) found that applying a local anesthetic to the cornea of the eye disrupted the mole-rats' ability to orient to the nest site, although the ability to detect light and dark was not impaired. Thus, these authors tentatively conclude that the magnetic receptors are in the cornea.

If a mammal is going to get home in unfamiliar terrain without the use of landmarks, it needs a map as well as a compass to know its location in relation to home. For instance, it might be able to use the magnetic isoclines to get information about longitude. It would, however, need another gradient along a second axis in order to get information about latitude, to fix itself in two-dimensional space. Such a grid-based bicoordinate navigation system has not yet been demonstrated conclusively in any animal.

FACTORS AFFECTING SPATIAL ABILITY

Mammals, among other animals, are assumed to have a cognitive map, that is, a representation of geometric relationships among a home site, terrain surrounding the home site, goals to be visited, and the terrain surrounding those goals. It has long been known that the hippocampus, a large forebrain structure (chapter 8), is involved in spatial learning in mammals (O'Keefe and Nadel 1978), so the hippocampus is a likely seat of such a map. Although rats with lesions in this region can still learn single associations in order to recognize a goal, they are unable to form relational representations involving multiple stimuli (Eichenbaum et al. 1990).

Sex differences in spatial ability exist in a variety of mammals: males perform better than females. This difference was first documented in humans, but has now been shown in other species. In a number of rodent species, males outperform females on spatial tasks such as the radial arm maze and the Morris water maze. In the water maze, the animal must learn to navigate to a hidden submerged platform in a round tank filled with opaque water (Galea et al. 1996).

These differences have been linked to mating systems such that the difference is greatest in species with polygynous or promiscuous mating systems, and especially for those with scramble competition for mates (chapter 22). When two species of *Microtus* were compared, one with a promiscuous mating system (*pennsylvanicus*) and the other with a monogamous mating system (*pinetorum*), males of the former, but not the latter, species performed better than females on spatial tasks (Gaulin and Fitzgerald 1986). This difference was also reflected in their brains: males of the polygynous species had hippocampi that were 11% larger than those of females, while in the monogamous species the difference was only 2% (Jacobs et al. 1990). Additional experiments with *M. pennsylvanicus* reveal that females prefer males with better spatial ability and that such males have larger home ranges and visit more females during enclosure tests (Spritzer et al. 2005a,b). Surprisingly, males with better spatial abilities did not sire more offspring than males with poorer spatial abilities (Spritzer et al. 2005a).

Sex differences in spatial ability also depend on the reproductive status of the animal. In both meadow voles and deer mice, the difference was evident only in adult, breeding animals, implicating sex hormones in the effect (Galea et al. 1996).

SUMMARY

Movements are made by mammals at several different levels of spatial scale. At the smallest scale—station keeping—the animal obtains and defends resources in its home range. Ranging involves forays outside the home range in search of new habitat or mating opportunities and may include natal dispersal. Migration is the persistent movement across different habitats without regard to presence or absence of resources. Among mammals, migration usually consists of a round trip.

Within a species' range, individuals may disperse or remain in the natal area to breed. In mammals, males typically disperse, but females are philopatric. By dispersing, the chances of matings between close relatives are reduced, and inbreeding is minimized. Mammals may also disperse to reduce competition for resources among relatives and neighbors. Most mammal species are polygynous; females defend resources needed for reproduction, and males range over a larger area in search of mates. Although there are demonstrated costs to extreme inbreeding, outbreeding may also incur genetic costs; some studies show an optimal degree of inbreeding, in which first cousins are favored as mates.

Habitat selection refers to the choice of a place in which to live. Organisms may fail to colonize an otherwise suitable area because of the inability to get there, as evidenced by the lack of mammals on oceanic islands. Behavior patterns may restrict mammals to a fraction of the habitat that they would presumably be adapted to occupy. Other factors such as competitors, parasites, predators, diseases, or physical and chemical factors can restrict a species' distribution.

Experiments with mammals have been conducted to determine the roles of genes and experience in habitat selection. Selection of the "correct" habitat is under some degree of genetic control, as has been demonstrated by studies in which animals have been reared in isolation and later tested in various habitats. Early experience can modify later choices, however. Tradition, the transmission of knowledge of habitats from one generation to the next, is thought to be important in some ungulate species.

Theoretical approaches to habitat selection include optimal foraging models, in which individuals choose patches (habitats) and stay in them so as to maximize the gain of some resource. The ideal free distribution assumes that, in the absence of competition, individuals settle in the best habitats first, with later arrivals moving into suboptimal sites.

Migration in mammals has been well documented in bats, cetaceans, pinnipeds, and ungulates. Many bats with temperate distributions migrate from hibernacula, often caves, to summer roosts and then return the following autumn. Baleen whales move from high-latitude feeding grounds to tropical seas to breed. Pinnipeds such as elephant seals spend most of their time at sea as they migrate from pelagic feeding grounds to island sites that are used for both breeding and molting. Ungulates such as barren-ground caribou migrate between northern breeding and southern wintering grounds. Elk and deer make vertical migrations as they spend summers at high elevations and then return to lower elevations in winter. Wildebeests in East Africa engage in mass migrations in association with seasonal rainfall patterns and the growth of grasses.

Migrants often return to the same home ranges they occupied previously, and nonmigrants traversing their home ranges in search of resources return to a nest or den, a process called homing. Individuals that are experimentally displaced from their home ranges often traverse unfamiliar terrain and return home. Short-distance homing can occur by path integration, in which the animal stores and integrates information from the outward trip. Landmarks may be used for trips within familiar ground, a process called piloting. Some species of mammals use the sun as a compass and are able to compensate for the apparent movement of the sun across the sky. The ability to detect and use geomagnetic cues has also been demonstrated in a few species of mammals. The area in the brain most involved in spatial ability is the hippocampus. Males in species that are polygynous have better spatial ability and larger hippocampi than males of monogamous species.

SUGGESTED READINGS

Bovet, J. 1992. Mammals. Pp. 321–361, *in* Animal homing (F. Papi, ed.). Chapman and Hall, New York.

Chepko-Sade, B. D., and Z. T. Halpin (eds.). 1987. Mammalian dispersal patterns. The effects of social structure on population genetics. Univ. of Chicago Press, Chicago.

Dingle, H. 1996. Migration: the biology of life on the move. Oxford Univ. Press, New York.

Krebs, C. J. 2001. Ecology: The experimental analysis of distribution and abundance, 5th ed. Benjamin Cummings, San Francisco.

Rankin, M. A. (ed.). 1985. Migration: mechanisms and adaptive significance. Marine Science Institute, Port Aransas, TX.

Swingland, I. R., and P. J. Greenwood. 1983. The ecology of animal movement. Clarendon Press, Oxford, England.

Multiple Authors. 2006. Migration and Dispersal. Special Section. Science, 313:775–800.

DISCUSSION QUESTIONS

1. In contrast to mammals, most species of female birds disperse farther from the natal site to breed than do males. Why might this be so? Consult the references by Greenwood (1980, 1983).

2. Review the life history of a mammalian species of your choice. Try to explain the dispersal patterns of each sex in light of our discussion of both proximate and ultimate causation. Include ecological as well as genetic factors in your answer.

3. Although tradition is claimed to play a role in the habitat preference of some animals, such as mountain sheep (Geist 1971), firm data are lacking. Design an experiment to demonstrate the role of tradition in habitat choice.

4. African mole-rats (*Cryptomys hottentotus*) are colonial, subterranean rodents that dig the longest underground burrow systems of any mammal. Burrow systems are linearly arranged, with the main tunnel often more than 200 m long. The main tunnel is usually oriented in a north–south direction. Burda and colleagues (1990) explored the possible role of the geomagnetic field as a cue for underground orientation, considering that these animals are virtually blind. Family groups were provided with nesting material in a light-proof arena. They were then exposed to the local geomagnetic field (control) or to experimental fields produced by Helmholtz coils surrounding the arena. The symbols in the figure denote the location of nests along the walls of the arena. What conclusions can you draw from these results? What additional experiments might be necessary?

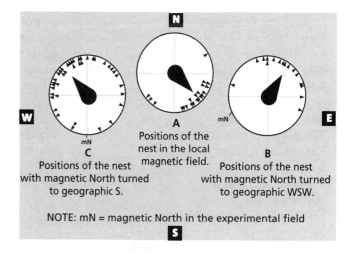

Arrows represent the mean vectors for nest location.
Data from H. Burda et al., 1990, Magnetic Compass Orientation in the Subterranean Rodent Cryptomys hottentotus *(Bathyergidae)* Experientia, *46:528–530.*

CHAPTER 25

Populations and Life History

Populations are groups of organisms of the same species, present at the same place and time. Their boundaries may be natural, or they may be defined arbitrarily by an investigator. A central problem for mammalian ecologists for decades has been to understand and predict the population dynamics of mammals. Predicting outbreaks of pest species or declines of endangered species relates directly to humans' quality of life. In this chapter, we consider processes that affect populations of mammals; that is, how populations grow and how they are regulated in nature. We also deal with life history tactics of individuals, exploring trade-offs that mammals make as they allocate resources to reproduction versus survival.

Population Processes

Populations of mammals increase via births, or **natality;** they decrease via deaths, or **mortality;** or they change in numbers via movements. One-way movement into a population is **immigration,** whereas one-way movement out of a population is **emigration.** The term **migration** refers to movements (usually round-trip) that are repeated each year, shown most spectacularly by some species of bats, ungulates, cetaceans, and pinnipeds (see chapter 24).

RATE OF INCREASE

Populations of all organisms have a great capacity for increase, perhaps best illustrated by the human population of the world that is presently increasing at nearly 2% per year and doubling in less than 40 years. The English economist Thomas Malthus was perhaps the first to realize that populations had the potential to increase exponentially but the means to support them did not (Malthus 1798). Arithmetical growth takes place by adding a constant amount during each time interval, but **exponential** growth occurs in proportion to the number already present. Populations increase exponentially because as each individual matures, it begins to breed along with the rest of the population. Similarly, in a savings account in which the interest is compounded, the interest begins earning interest along with the principle. The more often the interest is compounded, the faster the savings grow. When breeding occurs continuously, the population grows as if interest were compounded instantaneously, requiring the use of calculus. The difference between arithmetical and exponential growth can be seen clearly in figure 25.1. Darwin (1859) wrote, in *On the Origin of Species*, "There is no exception to the rule that every organic being naturally increases at so high a rate, that, if not destroyed, the earth would soon be covered by the progeny of a single pair." Darwin pointed out that even one of the slowest breeding animals, the elephant, would increase from a single pair to nearly 19 million in just 750 years if unchecked.

The rate of growth of a population undergoing exponential growth at a particular instant in time is a differential equation:

$$\frac{dN}{dt} = rN \qquad (1)$$

where N = the number of individuals in the population; t = the time interval; and r = the growth rate per individual, or the intrinsic rate of natural increase. r can be calculated as the individual birth rate minus the individual death rate (sometimes written as $r = b - d$). If a closed population is assumed, then immigration and emigration are zero.

In words, formula (1) can be read as

The rate of change in population size	=	The contribution of each individual to population growth	×	The number of individuals present in the population

If the number of individuals is plotted against time, the result is a curve that becomes increasingly steep during exponential growth (see figure 25.1). For example, reindeer (*Rangifer tarandus*) were introduced on several of the Pribilof Islands in the Bering Sea off Alaska. Four males and 21 females were released on 106-km² St. Paul Island in 1911. In the absence of hunting or predators, numbers increased rapidly, reaching a peak of about 2000 in 1938 (Scheffer 1951; figure 25.2). The population then crashed to only 8 in 1950 because the habitat was overgrazed. Smaller species of mammals, with shorter generation times, can have even higher growth rates. Under laboratory conditions, vole populations can double in as little as 79 days (Leslie and Ranson 1940)!

LIFE TABLES

Not all individuals in a population are identical, of course; they differ in sex, age, and size. These differences affect natality, mortality, and movement rates. Mortality is summarized in populations by means of **life tables,** originally developed by human demographers to provide data on life expectancy for life insurance companies. In his classic study of wolves (*Canis lupus*) of Mount McKinley, Alaska, Murie (1944) collected 608 Dall's sheep (*Ovis dalli*) skulls and estimated the age at death based on the size of their horns. For instance, 121 sheep were less than 1 year old at death, 7 were between 1 and 2 years, and so forth. The oldest sheep were between 13 and 14 years. From these data, Deevey (1947) constructed a life table, shown in table 25.1. Note that mortality was high in the first year, declined to low levels from about 2 to 8 years, then increased once more. From life tables can be computed the expectation of further life (e_x), the amount of time an individual can expect to live once it survives to a particular age (the fifth column in table 25.1).

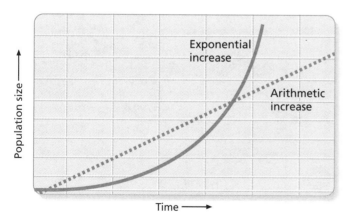

Figure 25.1 Arithmetic and exponential growth curves. The exponential curve increases more slowly at first but then accelerates past the arithmetic curve, which increases at a steady, incremental pace throughout. *Adapted from W. P. Cunningham and B. W. Saigo, 1997,* Environmental Science: A Global Concern, *4th ed., William C. Brown Publishers.*

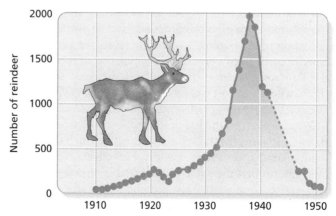

Figure 25.2 Population growth of a herd of reindeer. Reindeer were introduced on one of the Pribilof Islands, Bering Sea, in 1911 and increased exponentially until 1938, after which numbers declined rapidly. *Data from V. B. Scheffer, 1951, The rise and fall of a reindeer herd,* Sci. Mon. *73:356–362.*

Table 25.1 Life table for the Dall mountain sheep constructed from the age at death of 608 sheep in Mount Mckinley (now Denali) National Park.

Age Interval, Years, x	Number Dying During Age Interval, d_x	Number Surviving at Beginning of Age Interval, n_x	Number Surviving as a Proportion of Newborn, l_x	Expectation of Further Life, e_x
0–1	121	608	1.000	7.1
1–2	7	487	0.801	7.7
2–3	8	480	0.789	6.8
3–4	7	472	0.776	5.9
4–5	18	465	0.764	5.0
5–6	28	447	0.734	4.2
6–7	29	419	0.688	3.4
7–8	42	390	0.640	2.6
8–9	80	348	0.571	1.9
9–10	114	268	0.439	1.3
10–11	95	154	0.252	0.9
11–12	55	59	0.096	0.6
12–13	2	4	0.006	1.2
13–14	2	2	0.003	0.7
14–15	0	0	0.000	0.0

Based on data in Murie (1944), quoted by Deevey (1947).

The data for life tables are usually collected in two different ways. **Static**, sometimes called vertical, life tables are generated from a cross section of the population at a specific time. Individuals are assigned an age group, and mortality rates are recorded for each age group. These are used by life insurance actuaries to calculate insurance risks for humans. The sheep life table (see table 25.1) is a type of static life table in which the age at death was observed from skulls. Static life tables require the assumption that the population is stationary and that the birth and death rates of each group are constant, both rather unlikely circumstances. **Cohort**, sometimes called horizontal, life tables are generated by following a group, or cohort, of individuals of similar age from birth to death; these data are much harder to obtain, especially for mammals that are often highly mobile and long-lived. An example of a cohort life table for a relatively short-lived species comes from white-footed mice (*Peromyscus leucopus*), few individuals of which lived more than 1 year (Goundie and Vessey 1986). In this population, spring-born individuals lived an average of just 10 weeks after weaning. Cohort life tables for a long-lived species come from a provisioned population of rhesus monkeys (*Macaca mulatta*) on islands off the coast of Puerto Rico, where a cohort was followed for 18 years (Meikle and Vessey 1988). Another island example comes from red deer (*Cervus elaphus*) on the Isle of Rhum, Scotland, where calves born in 1957 were followed for 9 years, by which time 92% had died (Lowe 1969). Cohort life tables require the assumption that the age class followed is representative of the entire population; this assumption is rather unlikely because mortality rates are apt to change over time.

SURVIVORSHIP CURVES

Life tables can also be used to construct **survivorship curves** to show graphically the pattern of mortality across different age groups. The number of survivors (l_x, the fourth column in table 25.1) is plotted against age. Often the number of survivors is expressed as a proportion of the total, as in table 25.1. It is best to plot the $\log_{10}$ of survivors, so that a straight line is obtained when mortality rates are constant across all age categories. Thus, if there were 1000 individuals to begin with and the mortality rate was 50% per year, an arithmetic plot would show a sharply declining curve, but a log plot of the same data would show a straight line. Different types of survivorship curves are illustrated in figure 25.3. Murie's sheep generally show a type I curve: Once they survive the rather high mortality rate in the first year, the rate is nearly zero until 8 or 9 years, when it rises sharply and the curve breaks steeply down. Most other large mammals also show type I survivorship curves (figure 25.4). White-footed mice, and probably many other small mammals, show a type II curve,

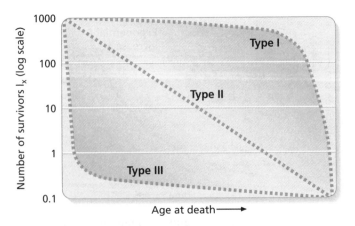

Figure 25.3 Hypothetical survivorship curves. Type I represents low mortality rates early in life, with most individuals dying at an old age. Type II represents constant mortality rates at all ages. Type III represents high mortality rates early in life. Most mammal species show type I or type II curves. *Adapted from C. J. Krebs, 1994,* Ecology, *4th ed., HarperCollins.*

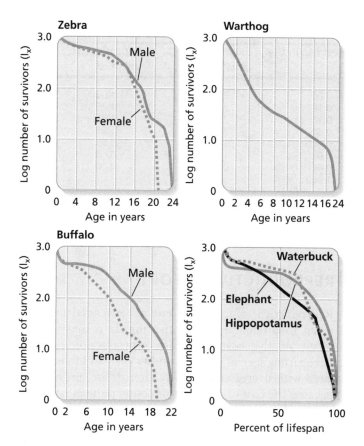

Figure 25.4 Survivorship curves of African ungulates. After the first few years, most resemble type I curves, with the exception of the warthog, which resembles a type II curve. *Data from E. R. Pianka, 1994,* Evolutionary Ecology, *5th ed., HarperCollins.*

in which mortality rates are more or less constant at all ages (Schug et al. 1991), similar to the warthog curve in figure 25.4. Type III curves, in which mortality rates are highest at early stages of development, are characteristic of invertebrates and fish but probably are not seen in mammals because of the larger investment female mammals make in their offspring through gestation and lactation (chapter 22).

AGE STRUCTURE

Birth rates as well as death rates vary with age because very young and very old individuals do not breed in most species of mammals. The **age structure** of a population is determined in a static fashion by calculating the proportion of the total population made up of individuals of various ages. This information can then be graphed, as shown in figure 25.5. For humans with a type I survivorship curve, distributions with triangular, "pinched" shapes indicate a high potential for increase because there are many individuals just reaching reproductive age and few old individuals. Much of the human population lives in less-developed countries with age structures similar to those shown in figure 25.5C. Stationary populations and those with slow growth rates, typical of developed countries, tend to be rectangular in shape, with

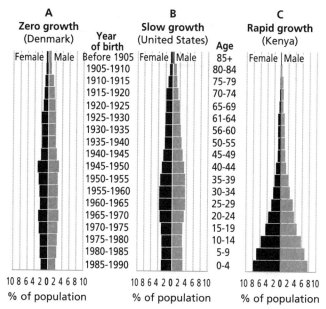

Figure 25.5 Age-structure diagrams for human populations. (A) Zero growth is shown for Denmark, (B) slow growth for the United States, and (C) rapid growth for Kenya. *Data from U.S. Census Bureau, The United Nations, and Population Reference Bureau.*

more even percentages in each age group. Short-lived mammals with type II survivorship curves tend to have triangular distributions even when stationary because mortality rates are typically high and independent of age.

It is also useful to add to the life table the number of offspring produced by an average female of age x during that age period; this figure is called **fecundity** and is written as m_x (table 25.2). Multiplying this column times l_x gives the realized fecundity for each age class. Summing the $l_x m_x$ column gives the net reproductive rate R_0 for the population. **Reproductive value,** which is the sum of an individual's current reproductive output and its expected future output at age x, can also be estimated. A female's reproductive value increases with age once she passes through the juvenile stage, peaks early in adulthood, then declines in old age.

The ratio of males to females and the type of mating system can affect population growth. As we pointed out in chapter 22, many species of mammals have a polygynous mating system in which a small number of males do most of the mating. Although the sex ratio at birth tends to be 50:50 (chapter 22), males typically suffer higher mortality rates than do females. This leads to female-biased sex ratios among adult breeders. For this reason, most measures of reproductive rates are calculated for females rather than for both sexes.

Life History Traits

Life history traits, including size at birth, litter size, age at maturity, and degree of parental care, directly influence the life table schedules of fecundity and survival. Natural selection recognizes only one currency: successful offspring.

Table 25.2 Hypothetical life table*

Age, x	Survivorship, l_x	Fecundity, m_x	Realized Fecundity, l_x, m_x	Expectation of Life, e_x	Reproductive Value, v_x
0	1.0	0.0	0.00	3.40	1.00
1	0.8	0.2	0.16	3.00	1.25
2	0.6	0.3	0.18	2.67	1.40
3	0.4	1.0	0.40	2.50	1.65
4	0.4	0.6	0.24	1.50	0.65
5	0.2	0.1	0.02	1.00	0.10
6	0.0	0.0	0.00	0.00	0.00
Sums		2.2	1.00 (R_0)		

*Compare columns with table 25.1. Fecundity (m_x) has been added so that net reproductive rate (R_0) and reproductive value (v_x) can be calculated.

Although all organisms have presumably been selected to maximize their own lifetime reproductive success, mammals vary widely in the relative amounts of energy they expend on activities that enhance fecundity versus those that enhance survival. Much of life history theory examines trade-offs between reproduction and survival.

Aspects of reproduction such as when to breed and how many young to produce are subject to natural selection. For example, red squirrels (*Tamiasciurus hudsonicus*) in Yukon, Canada, have begun breeding earlier each year in response to increasing spring temperatures, advancing 18 days in 10 years. Some of this shift is a plastic response to increased food availability, but the rest is a heritable response to selection (Réale et al. 2003).

SEMELPARITY VERSUS ITEROPARITY

As we pointed out in chapter 22, some organisms, such as annual plants, insects, and the Pacific salmon, reproduce only once during their entire lifetimes. These "big bang" breeders are called semelparous (from the Latin *semel*, meaning "once," and *pario*, meaning "to beget"). They exert a huge effort during their reproductive episode and then die. Some marsupial mice (*Antechinus* spp.) from Australia are semelparous in that the males reach sexual maturity, mate during a brief period, then physically decline and die shortly thereafter (Cockburn et al. 1985). Several species of New World marsupials in the family Didelphidae also seem to be at least partially semelparous. Males of the Brazilian gracile opossum (*Gracilinanus microtarsus*) breed when they are one year old. Survival rates decrease sharply at this time, and only a few males survive to breed the following year (Martins et al. 2006). Semelparity seems to be favored when extensive preparation for breeding is necessary or favorable environmental conditions are ephemeral or uncertain. Most other mammal species are iteroparous (from the Latin *itero*, meaning "to repeat"), that is, able to breed more than once during their lifetimes.

REPRODUCTIVE EFFORT

Reproductive effort is the energy expended and risk taken to produce current offspring. It is measured in terms of the cost to future reproduction. Among iteroparous species, high fecundity early in life is correlated with decreased fecundity later in life, presumably because the cost of early reproduction reduces later survival and reproduction. In long-lived species with type I survivorship curves, such as many ungulates (see figure 25.4), reproductive effort is likely to increase with age as reproductive value declines. In other words, older animals, with little time left to live, should put maximum effort into current breeding. Long-lived species of mammals are expected to show a gradual increase in effort with age, as measured by the number or mass of offspring. Short-lived species, especially those with high extrinsic mortality rates (i.e., not related to breeding effort), should put maximum effort into early reproduction because they are not likely to live long enough to suffer any negative consequences of high reproductive effort early in life.

Although the increased energy expenditure and risk associated with current reproduction have been documented in many species of mammals, only a few studies have shown that these costs reduce later survival and reproductive success. Lactation is much more costly for mammals than is gestation; lactation doubles the energy requirements of a mouse (Millar 1978), and it delays ovulation in most species examined (Short 1983). Bighorn sheep ewes suffer a cost of reproduction in having higher parasite loads when lactating than when they are not (Festa-Bianchet 1989). Among red deer, in years when winter food was low, females that successfully reared an offspring had lower survival rates than those that failed to get pregnant or lost a young before weaning (Clutton-Brock et al. 1989; figure 25.6). Parental care after weaning can also be costly. In red-necked wallabies (*Macropus rufogriseus*), mothers that associate closely with their offspring are more likely to lose their next infant (Johnson 1986).

A

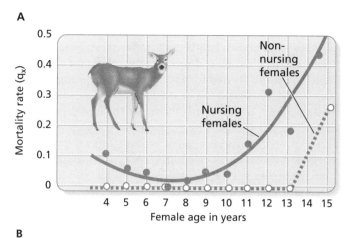

B

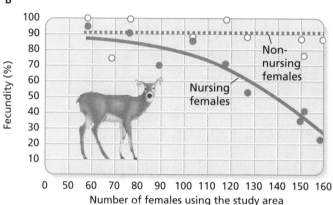

Figure 25.6 The cost of reproduction in female red deer and the effect of density on fecundity.
(A) Nursing females have a higher rate of mortality at all ages. (B) Nonnursing females have a higher probability of calving in the next year than nursing females do at high densities. *Data after T. H. Clutton-Brock et al., 1983, The costs of reproduction to red deer hinds, J. Anim. Ecol. 52:367–383.*

SENESCENCE

Inspection of the life table shown in table 25.2 indicates that fecundity increases until the fourth age interval and then declines. Most mammals experience **senescence**, that is, their fecundity gradually decreases and their mortality rate increases, resulting from deterioration in physiological function with age. Assuming that survival is advantageous to an individual of any age, researchers need to discover what causes senescence and why natural selection does not eliminate it.

The "rate-of-living" theory of aging links senescence to metabolic rate (Austad and Fischer 1991). According to this theory, aging is viewed as an inevitable consequence of physiological processes in which biochemical errors and toxic metabolic by-products accumulate. Life span should be negatively related to the rate at which physiological processes occur. Thus, organisms with low metabolic rates should live longer than those with high metabolic rates (Austad and Fischer 1991).

A second, evolutionary theory of aging argues that selection acts more strongly on traits expressed at an early age, when most individuals in the population are still alive. Selection gets weaker as organisms age because fewer individuals are around on which it can act. Also, some genes enhance fitness at an early age, and they would be selected for but might act incidentally to reduce fitness later on (Stearns 1992). According to this theory, species that suffer high extrinsic rates of mortality, as from predation or other environmental hazards, should senesce more quickly, as should those that invest heavily in reproduction at an early age. Thus, senescence would not be expected to begin until after sexual maturity, and it should show up earlier in populations with higher mortality rates. There is little evidence of senescence in most natural populations of mammals because few individuals live long enough to show it. Predation, disease, starvation, or other factors tend to kill individuals well short of their potential life spans. The situation is different in zoos or other captive populations, where unlimited food is available, diseases are controlled, and predators are absent; there, senescence is seen frequently.

One way to test between these theories is to compare groups of mammals that have different metabolic rates. Marsupials have basal metabolic rates about 70% that of similar-sized eutherian mammals. Yet life spans of captive marsupials are only about 80% that of comparable eutherians, the opposite of the situation that would be predicted by the rate-of-living theory (Austad and Fischer 1991). Bats, on the other hand, have metabolic rates comparable to those of other eutherians, yet relative to their body size, they are the longest-lived group of mammals. Why do bats live so long? Austad and Fischer (1991) argue that this difference is attributable to flight; bats are protected from predation and can roost in protected sites. They emphasize that birds are also relatively long-lived, as are species of gliding mammals (squirrels, marsupials, and dermopterans). These examples therefore tend to support an evolutionary theory of aging that links life span to levels of environmental hazard.

r- AND *K*-SELECTION

The evolution of life history traits is linked to the growth rates of populations and their environments. In temperate and arctic regions, populations are frequently reduced by density-independent factors, mainly climate. Populations in such areas often have high intrinsic rates of natural increase (*r* from equation 1, which was discussed earlier), including early maturity, high fecundity, and little parental care of offspring. Such opportunistic species are said to be *r*-selected. Conversely, populations of tropical species living in more constant climates are thought to be influenced by density-dependent factors, mainly competition. They often have low intrinsic rates of increase, including later maturation, low fecundity, and much parental care of offspring. Such populations are said to be *K*-selected because they remain at the carrying capacity (*K* from equation 2, which is discussed in more detail later; MacArthur and Wilson 1967). Pianka (1970) listed a variety of life history traits associated with these types of species in addition to those mentioned earlier.

In the eastern United States, two common small mammal species are the white-footed mouse and the meadow vole (*Microtus pennsylvanicus*). The white-footed mouse is widely distributed in relatively stable woodland habitats. Individuals mature at about 60 days, and the litter size is four to five. The meadow vole, on the other hand, inhabits fields. Fields typically undergo succession to woods unless maintained by fire or other disturbances and so are relatively temporary habitats. *Microtus* typically has litter sizes of eight or more and reaches sexual maturity in as little as 30 days, giving it a high intrinsic rate of increase. Relative to *Peromyscus*, *Microtus* would thus be considered *r*-selected.

A consistent relationship between environmental fluctuations and life history traits has been difficult to demonstrate. Thus, many apparently *r*-selected species occur in tropical climates, and many *K*-selected species occur in temperate and arctic regions. Furthermore, many species tend to be intermediate between these extremes, making them difficult to classify. Finally, there is no evidence that natural selection acts on life history traits in this way (Stearns 1992). In spite of these problems, the terms *r*- and *K*-selection continue to be used by some ecologists studying life history patterns (Boyce 1984; Krebs 2001).

BET-HEDGING

An alternative approach to understanding the relationship between life history traits and environmental conditions is to consider how an organism should expend reproductive effort when faced with unpredictable outcomes. When survival of offspring varies widely from year to year and is unpredictable, parents might be selected that have smaller litter sizes and spread breeding across several seasons, a strategy referred to as bet-hedging (Stearns 1976).

The predictions of bet-hedging contradict those of *r*- and *K*-selection in some cases. In unstable, unpredictable environments, *r*-selection, which favors high reproductive rates, should be important. Bet-hedging theory, however, suggests that the opposite would be favored: low reproductive rates, greater parental care, and reproductive effort spread across many breeding seasons.

FAST AND SLOW LIFE HISTORIES

A more recent approach to understanding the diversity of life history patterns is to place them on a continuum, with species that mature early and have high reproductive rates and short generation times (fast) at one end, and those that mature late and have low reproductive rates and long generation times (slow) at the other end (Promislow and Harvey 1990). Oli (2004) used the ratio of the fertility rate to the age at first reproduction to quantify the tempo of life for 138 populations of mammals. At one end are species with high fertility rates and a short time to first reproduction (a high ratio) and at the other end are those with a low fertility and longer time to first reproduction (a low ratio). This ratio can be used to classify populations as fast, medium, or slow, with about one third of the populations in each group. The relationship is independent of body size or phylogeny, so it is not simply a matter of large animals such as elephants having slow life histories and small animals such as mice having fast life histories. Overall, "fast" mammal species reach sexual maturity earlier, have higher population growth rates, die younger, and have lower survival rates than "slow" mammal species.

Population Growth and Regulation

LIMITING FACTORS

Although a host of factors, such as food, water, and shelter, must be present for any population of mammals to survive, typically only a single factor limits the growth of a population at any one time. This idea originated with Liebig (1847), a plant physiologist who studied the effects that nutrients, such as nitrogen, had on plant growth. The axiom is referred to as Liebig's **law of the minimum.** When Liebig's idea is extended to plant and animal populations, one factor will be the limiting factor. Tests for which factor limits a population's growth can be conducted by manipulating the amount of one factor while holding others constant and recording changes in the population. If food is the limiting factor, an increase in the population will result if the food supply is augmented.

A number of studies have shown that adding food leads to increased reproduction and to higher population size in mammals. For example, Holling (1959) studied the population response of some small mammals to different amounts of a staple food, cocoons of the European pine sawfly (*Neodiprion sertifer*). Shrews of the genus *Sorex* increased more or less linearly with increasing food supply up to a population of 25 shrews per acre; then the population leveled off (figure 25.7). At that point, some resource other than sawflies became limiting. Likewise, deer mice increased in numbers up to about eight per acre, then remained constant despite the food supply (Holling 1959). The abundance of sawflies seemed to be unrelated to the population density of another genus of shrew (*Blarina*). Thus, some other factor was limiting the population of that species. Although Liebig's law explains how populations are limited in some instances, factors may interact in complex ways to regulate numbers, as we explain later in this chapter.

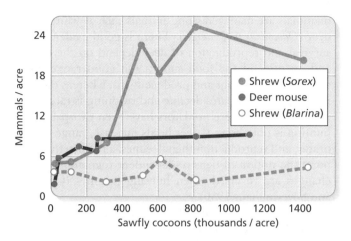

Figure 25.7 Numerical response of small mammals to increasing food supply. Populations of the shrew (genus Sorex) and deer mouse (genus Peromyscus) increase in size, up to a point, as food supply increases. Shrews of the genus Blarina, however, show no response to increased food supply (sawfly cocoons), so their numbers must be limited by some other resource. *Data from C. S. Holling, 1959, The components of predation as revealed by a study of small mammal predation of the European pine sawfly,* Can. Entomol. *91:293–320.*

LOGISTIC GROWTH

As shown by the reindeer example in figure 25.2, populations of mammals or anything else do not increase indefinitely. As density rises, the presence of other organisms reduces the birth rate, increases the death rate, or triggers emigration, and population growth slows or stops. One model for such growth, the **logistic equation,** yields a sigmoid, or S-shaped, curve. Numbers increase slowly at first, then increase rapidly, as shown by equation 1. Then, however, the growth rate begins to slow as numbers approach an upper limit, or asymptote. The upper limit is often called the carrying capacity, or the equilibrium density. The carrying capacity is typically set by the amount of food or some other factor that is limiting the population at that time (figure 25.8). Equation 1 can be modified to give the equation for logistic growth where *K* = the carrying capacity.

$$\frac{dN}{dt} = rN\left(1 - \frac{N}{K}\right) \qquad (2)$$

The term in parentheses is a density-dependent term that ranges from 0 to 1. When *N* is very small in relation to *K*, *N/K* is close to 0, so the term in parentheses is close to 1. When the term in parentheses is 1, equation 2 becomes the same as equation 1, and the population increases exponentially. As *N* approaches *K*, *N/K* approaches 1, and the term in parentheses approaches 1 minus 1, which is 0. At that point, the whole right part of the equation goes to 0 and so the growth rate, shown on the left side, is also 0, and population growth stops. If *N*

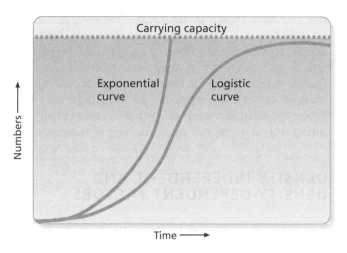

Figure 25.8 Comparison of exponential and logistic population growth curves. The rate of logistic growth declines gradually to zero as the carrying capacity is approached. *Adapted from W. P. Cunningham and B. W. Saigo, 1997,* Environmental Science: A Global Concern, *4th ed., William C. Brown Publishers.*

should exceed *K*, the growth rate would be negative, and the population would decline. In this simple model, the population growth rate declines linearly with increasing population size, but the relationship is curvilinear (Sibly et al. 2005).

Laboratory populations of mice often show logistic growth when a few pairs are used to start a colony and food, water, and nesting material are supplied freely (Terman 1973). Some natural populations remain at equilibrium for long periods of time, presumably at the carrying capacity. For instance, sheep were introduced on the island of Tasmania in the early 1800s. Numbers showed a logistic growth pattern, reaching an asymptote of about 2 million and remaining there for many years (Davidson 1938; figure 25.9). Often, however, populations show an exponential increase followed by rapid decline, as in the

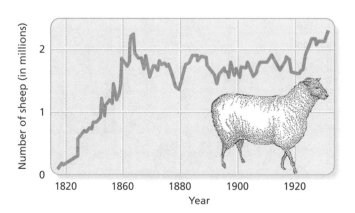

Figure 25.9 Logistic growth of sheep population. Number of sheep on the island of Tasmania following their introduction in the early 1800s. *Data from J. Davidson, 1938, On the growth of the sheep population in Tasmania,* Trans. Roy. Soc. South Aus. *62:342–346.*

reindeer herd shown in figure 25.2. Many populations seem to fluctuate, increasing and decreasing without any apparent pattern. Short-lived mammals that breed seasonally, such as mice, may show annual fluctuations, with year-to-year variation in the summer peaks and winter troughs, as seen for white-footed mice (figure 25.10). These populations tend to increase exponentially in spring and early summer, often crashing in autumn.

DENSITY-INDEPENDENT AND DENSITY-DEPENDENT FACTORS

The factors in nature that keep populations from increasing indefinitely are traditionally divided into two types: those that are density-independent and those that are density-dependent. The type depends on whether the factor's effect on the population is or is not a function of population density. Density-independent factors are such things as storms, fires, floods, or other climatic conditions that destroy individuals without regard to how many are present. Density-dependent factors include food supply, shelter, predators, competitors, parasites, and disease. Their effect on the population is stronger in crowded than in sparse populations, and they can be expressed mathematically by the $1 - K/N$ term in equation 2.

If population regulation is defined as the maintenance of numbers within certain limits and not simply as random fluctuations, density-dependent factors become regulatory agents. Only when a factor increases its negative effect on population growth as population increases can an equilibrium of numbers be maintained (Krebs 2001). Mammals, as warm-blooded organisms, are buffered against weather relative to most other taxa and thus may be less influenced by density-independent factors. The best examples of the role of density dependence from natural populations of

mammals are managed game animals such as white-tailed deer (*Odocoileus virginianus*). There is good evidence that survival and reproduction of deer depend on the quality and quantity of the food supply. On good range, most females get pregnant and produce twins, whereas on poor range, pregnancy rates decline and twinning is rare (table 25.3). In areas where predators have been extirpated and hunting is not allowed, deer herds increase, range deteriorates, and reproduction then declines, a pattern that supports the notion of density-dependent regulation. These changes are reversed when hunting resumes (Cheatum and Severinghaus 1950).

Red deer on the Isle of Rhum, Scotland, show a number of density-dependent changes in fecundity and juvenile mortality (Clutton-Brock et al. 1985). The density of females on the Rhum study site was negatively correlated with fecundity and survival for both mothers and offspring (see figure 25.6). Especially important was winter mortality of calves, which was very high in years of high density. Presumably, forage was scarce in those years. Male calves were especially affected, with over 70% mortality when female density was highest. As noted in chapter 22, males seem to be more sensitive to their mother's condition than females, perhaps because males grow faster in this sexually dimorphic species.

Equation 2 assumes a linear decline in population growth rate with density, producing the logistic growth curve shown in figure 25.8, but in reality the density-dependence function may be convex or concave. In a synthesis of long-term population data from 79 species of mammals, Sibly et al. (2005) found that the relationship is usually a concave curve, meaning that population growth rate slows rapidly as populations begin to increase, then remains low as the population slowly approaches the carrying capacity (*K*). The relationship is not influenced by body size or taxonomy. One implication of this result is that populations of mammals probably spend most of their time above the carrying capacity.

REGULATION OF POPULATIONS

Intraspecific competition received much attention as a regulatory factor in the past because it is the only one that is "perfectly" density-dependent. In other words, the intensity of competition among members of the same population is likely to vary directly with the size of the population; as the population increases, so does competition for the limiting resource. Unsuccessful competitors may leave the area, die, or have lower reproductive success. Other biotic regulatory factors, such as predation and interspecific competition, include other species of organisms that, in turn, are affected by other agents; thus, these biotic regulators are not perfectly density-dependent. For example, predation involves a complex relationship between predator and prey. Although wolves eat caribou, they are not necessarily the regulators of the caribou population. Because the wolf population is affected by

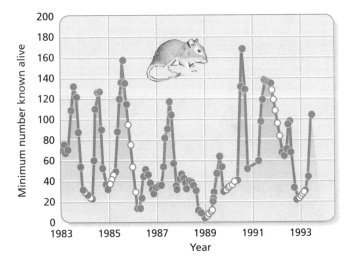

Figure 25.10 Fluctuations of white-footed mice in a small, 2-hectare woodlot in northwest Ohio. Open blocks denote missing data estimates. *Unpublished data from S. H. Vessey.*

Table 25.3 Reproductive parameters of white-tailed deer in five regions of New York, 1939–1949

Region*	Percent of Females Pregnant	Embryos per Female	Corpora Lutea per Ovary†
Western (best range)	94	1.71	1.97
Catskill periphery	92	1.48	1.72
Catskill central	87	1.37	1.72
Adirondack periphery	86	1.29	1.71
Adirondack central (worst range)	79	1.06	1.11

Data from E. L. Chaetum and C. W. Severinghaus, 1950, *Variations in Fertility of White-Tailed Deer Related to Range Conditions*, Trans. North Am. Wildl. Conf., *15:170–189.*
*Arranged by decreasing suitability of range.
†Note the decline in fecundity with decreasing quality of range.

factors other than the numbers of caribou, such as diseases or the abundance of alternative prey species, it will not be able to track the caribou population exactly. An outbreak of disease might reduce the size of the wolf pack and allow the caribou to "escape" control and begin to destroy grazing land. Intraspecific competition among the caribou, however, varies directly with the population density and the limiting resource.

Behavioral-Physiological Factors

During cyclical peaks in abundance, snowshoe hares (*Lepus americanus*) are seen under practically every bush; during these population peaks, a disturbance as slight as a hand clap was reported to be enough to send them into convulsions and coma, followed by death, apparently due to hypoglycemic shock (Green et al. 1939). Christian (1950) proposed that mammalian populations could be regulated by shock disease caused by exhaustion of the adrenal gland, following prolonged psychological stress from agonistic interactions at high population levels. This negative-feedback loop, which depends on intraspecific competition, is perfectly density-dependent. The idea grew from Selye's (1950) work on the **general adaptation syndrome** (GAS) in which nonspecific stressors, such as heat, cold, or defeat in a fight, produce a specific physiological response. Adrenocorticotrophic hormone (ACTH) released from the anterior pituitary, under control of the hypothalamus, stimulates production of glucocorticoids by the adrenal gland. These hormones, such as cortisone, function mainly to elevate blood glucose to prepare the body for fight or flight.

The phenomenon of death due to adrenal exhaustion turned out to be an extreme case, and Christian (1978) subsequently modified his hypothesis after a series of laboratory experiments. There is, in fact, a rise in adrenocortical output in response to increasing population density. These hormones, along with corticotropin-releasing factor, ACTH, and others reduce direct fitness in numerous ways. For example, the body's two main defense mechanisms—the immune and the inflammatory responses—are inhibited by these hormones. Such changes obviously increase the likelihood of morbidity or mortality. At the same time, growth and sexual maturation are inhibited by increased adrenocortical output, as are spermatogenesis, ovulation, and lactation (Rivier et al. 1986). Some of these

effects on reproduction persist even into subsequent generations, in spite of a reduction in population density. Field data supporting these findings came from studies of a variety of mammals such as sika deer (*Cervus nippon*), brown [Norway] rats (*Rattus norvegicus*), house mice (*Mus musculus*), and woodchucks (*Marmota monax*; Christian 1978). There was no evidence, however, of increased adrenal gland size with population density in lemmings (Krebs 1963) or meadow voles (To and Tamarin 1977). Part of the problem is that the stressor is not density itself but the agonistic behavior associated with competition for limited resources. Thus, in laboratory mice, as little as 2 minutes per day exposure to a trained "fighter" mouse produces a pronounced stress response, lowering the body's defense against parasites (Patterson and Vessey 1973). Clearly, this negative-feedback loop works under some conditions, but its generality in natural systems remains to be demonstrated.

Work with pheromones has augmented some of these findings. Substances released in the urine produce effects on conspecifics without a physical encounter. In some species of mice, the smell of strange male urine blocks pregnancy by preventing implantation, a phenomenon referred to as the Bruce effect (Bruce 1966); in crowded or unstable populations, the birth rate could thus be lowered. This effect may be a laboratory artifact, however. Field studies on several species of voles (*Microtus* spp.) have failed to demonstrate the Bruce effect (de la Maza et al. 1999; Mahady and Wolff 2002).

House mice avoid the urine of a mouse recently defeated by another mouse, and urine from stressed mice produces an adrenocortical response in naive mice (Bronson 1971, 1979). Females grouped together produce substances voided in urine that delay sexual maturation in other females (Drickamer 1974a). Six volatile organic compounds produced under the influence of the adrenal gland have been isolated by Novotny and colleagues (1986). Various combinations of these substances, when added to the urine of adrenalectomized females or to plain water, restore the delay effect.

The same delay in the onset of estrus has been produced in the laboratory by exposing young mice to urine of females taken from high-density populations in the field (Massey and Vandenbergh 1980). In an experimental study, field populations were increased by adding new mice to simulate a population explosion (Coppola and Vandenbergh 1987). Urine from resident females before the explosion had little delay effect on maturation of young mice in the laboratory, but after the explosion, urine from these same females produced a pronounced delay. In contrast, odor from mature males produces the opposite effect, accelerating the sexual maturation of young females (Vandenbergh 1969b). Although most of these effects are best viewed as adaptations of individuals to gain reproductive advantage, they may act incidentally to regulate population size.

Behavioral-Genetic Factors

A competing hypothesis proposed that natural selection favors different genotypes at high and low population levels. While working with voles (*Microtus* spp.) that have 3- to 4-year population cycles, Chitty (1960) noticed that populations continued to decline even under seemingly favorable environmental conditions. Voles from declining populations were highly aggressive, intolerant, and bred poorly; voles from increasing populations were mutually tolerant and rapid breeders. He postulated that a change took place in the quality of animals in a declining population. Through natural selection, the proportion of aggressive individuals increased, and even though they could compete in crowded conditions, their reproductive rates were low, and the population declined (figure 25.11). It seems unlikely that a change in gene frequency could take place over a span of only a few years, but dispersal of animals of one genotype can produce a rapid change in the gene frequency in the rest of the population. Myers and Krebs (1971) reported that the frequency of one allele in the blood serum of voles was significantly different in dispersers than in residents. The behavioral changes reported in vole populations have not been linked to specific genes, however. Evidence from the laboratory shows that the age of puberty can be shifted in house mice after only a few generations of artificial selection (Drickamer 1981). The low birth rates observed in peak populations in the wild could be due, in part, to selection for late-maturing individuals.

When snowshoe hares were captured from both peak and trough populations in the Yukon, Canada, and raised in large outdoor pens, the descendants of the peak populations had reduced reproductive rates and went extinct, while descendants of trough populations reproduced well and survived for the entire 16-year breeding program (Sinclair et al. 2003). The authors conclude that the presence of different phenotypes at different times in the 10-year hare cycle could provide a causal mechanism for the cycle.

Tamarin (1980) emphasized the role of dispersal itself as a regulating mechanism. If dispersal is blocked by a fence or is blocked naturally, as on islands, regulation fails and high populations and depleted resources result—the **fence effect** (Krebs et al. 1973). Fences may be social as well as physical (Hestbeck 1982, 1988). When neighboring densities are low, animals are able to disperse into them. If the neighboring densities are high, dispersal may be blocked because the surrounding habitat is occupied by hostile competitors. Blocked dispersal could thus lead to high population densities, followed by population crashes as resources are used up.

One of the noncycling vole species is the Mediterranean pine vole (*Microtus duodecimcostatus*). The reason it does not cycle may be the presence of dispersal sinks adjacent to main populations (Paradis 1995). Sink populations exist in low-quality habitats only by virtue of immigration from source populations in high-quality habitats. Dispersal between sources and sinks seems to stabilize populations, keeping them from cycling.

Several other mechanisms have been proposed that incorporate both genetics and aggressive behavior. One is that individuals behave less aggressively toward kin than nonkin. As a population increases, the likelihood of non-relatives' interacting increases, leading to an increase in aggression, followed by population decline (Charnov and Finerty 1980). A second idea links increased genetic heterozygosity with increased aggression. Several studies have shown that heterozygous individuals are more aggressive than homozygous individuals. As a population increases, so does dispersal and outbreeding, leading to an increase in heterozygosity. The resultant increase in aggressiveness could force a decline via increases in mortality rates and decreases in reproduction (Smith et al. 1978). So far, little hard evidence exists to support either of these hypotheses.

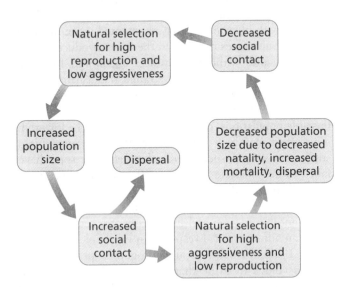

Figure 25.11 Chitty's model of population regulation in mammals. In response to increased population size and social contact, aggressive individuals with low reproductive rates are selected, leading to a decline in population. Modifications of the model emphasize the role of dispersal of certain phenotypes in causing the decline. This model predicts genetic changes in the population as different genotypes are favored at high versus low densities. *Data from D. Chitty, 1967, The natural selection of self-regulatory behavior in animal populations,* Proceedings of the Ecological Society of Australia, 2:51–78; *and C. Krebs, 1964, The lemming cycle at Baker Lake, Northwest Territories, during 1959–1962, in* Arctic Institute of North America Technical Paper No. 15.

Traits Favoring Population Self-Regulation

Wolff (1997) developed a model that predicts whether a species of mammal has the potential to be regulated by intrinsic (behavioral) or extrinsic factors (figure 25.12). Intrinsic factors include female territoriality, dispersal, and suppression of breeding by conspecifics. One reason females defend territories is to protect their offspring from infanticide, according to Wolff (1993). Species with altricial young that are raised in a den or burrow site are especially vulnerable to infanticide by strange females. We have already seen from the previous models that dispersal can regulate populations, but only if the dispersal rate increases with population density. According to Wolff's model, suppression of breeding by females is adaptive if male relatives fail to disperse or if there is a threat of infanticide. In the former instance inbreeding would lower fitness; in the latter instance reproductive effort would be wasted. Species most likely to fit these criteria are rodents, terrestrial carnivores, insectivores, rabbits, and prosimian primates. Other groups, such as ungulates, cetaceans, bats, and marsupials, typically do not have female territoriality or altricial young that remain in a nest. Because the conditions favoring intrinsic regulation of populations are so restrictive, Wolff argues that most species of mammals are probably regulated by extrinsic factors.

Evolution of Population Regulation

Proponents of self-regulation have not resolved two important problems. First, most of the data demonstrating behavioral and physiological changes in response to increasing population size are from laboratory populations, in which densities are often unrealistically high and emigration is usually prevented. The results, therefore, cannot be extrapolated to natural populations. In natural populations, furthermore, many extrinsic factors come into play, and no single mechanism can be implicated. Second, if such self-regulatory mechanisms exist, do they evolve specifically for that purpose, and, if so, what is the unit of selection?

It is probably true that self-regulation is adaptive for a population or species that has the potential for destroying

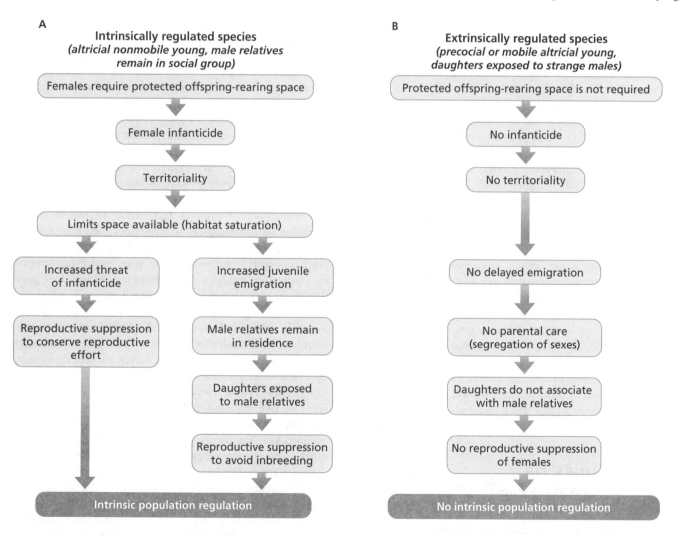

Figure 25.12 Behavioral and life history traits contributing to intrinsic (behavioral) (A) and extrinsic (B) regulation of mammalian populations. Rodents, insectivores, terrestrial carnivores, rabbits, and prosimian primates tend to show traits in pathway A, while ungulates, marine mammals, bats, and marsupials tend to show traits in pathway B. *Data from J. O. Wolff, 1997, Population regulation in mammals: An evolutionary perspective,* J. Anim. Ecol., *66:1–13.*

or using up its resources, but many of the mechanisms discussed here seem to be maladaptive for the individuals involved. An animal that is genetically programmed to respond to increased fighting by cutting back on reproduction will lose out, genetically speaking, to another individual that can keep on reproducing. How could a system evolve in which an animal responds to aggression by reducing its defense mechanisms and therefore, by definition, becomes less fit?

One answer is by group selection, as Wynne-Edwards (1962, 1986) argued. Groups or populations that avoid overexploitation of the environment survive at higher rates and may later colonize the habitats left vacant by imprudent groups that became extinct. Although group selection is considered by most evolutionary biologists to be theoretically possible and several mechanisms for it have been proposed (Lewontin and Dunn 1960; Wilson 1980), the rate of change in gene frequency would usually be too slow compared with the rate of change through individual selection. To maintain a trait that is adaptive for the group but not for the individual, group extinction would have to be more rapid than individual extinction—and that does not seem to be the case. More than three decades of study of diverse organisms have failed to produce strong support for the group selection hypothesis. Most population biologists, therefore, favor arguments based on individual selection and reject Wynne-Edward's (1962, 1986) hypothesis.

Arguments based on kin selection (see chapter 23 offer one way around the problem of group versus individual selection. Many social groups are composed of close relatives. Thus, behavioral and physiological traits that reduce direct fitness but increase indirect fitness, in this case by increasing the likelihood of survival of a kinship group, could be selected for. In other words, it might benefit an individual, evolutionarily speaking, to sacrifice its immediate reproductive output if the potential survival of the group as a whole, which contains its relatives, is enhanced.

A second, more widely accepted answer to the question of self-regulation is that these "mechanisms" are nothing more than the consequence of an individual's behaving in its own best interest. Reproductive restraint in the face of overcrowding may be the best individual strategy; a delay in maturation or the production of fewer young can save energy for a better time. The greatest genetic representation in the next generation is not synonymous with having the most offspring. Litter size in small mammals has evolved to produce the number of offspring that is most likely to survive and reproduce; too large a litter could result in survival of few or no young.

FIELD STUDIES OF POPULATION REGULATION

Descriptive and Correlational Studies

Given the large number of factors that might act singly or in combination to regulate mammal populations, it is perhaps not surprising that much disagreement exists about their importance in different populations. Food and predators have received much attention; the role of interspecific competition is discussed in the next chapter. Descriptive studies, sometimes spanning many years, occasionally demonstrate correlations among various factors and population density. Trees such as oak and hickory tend to produce large crops of seeds at irregular intervals, a phenomenon referred to as masting. The abundance of mast (primarily acorns) in autumn is positively correlated with winter survival, winter breeding, and subsequent population densities of white-footed mice (Ostfeld et al. 1996a; Wolff 1996). Thus, mast appears to be a limiting factor for these mice, at least in some years. In correlational studies, however, it is difficult to show a cause-and-effect relationship. In this example, some other variable, also correlated with abundance of mast, might actually be the limiting factor.

American red squirrels depend on highly variable crops of pine cones that do not mature until autumn, *after* the squirrels have committed to reproduction for that year. Somehow the squirrels anticipate the size of the upcoming seed crop by giving birth to larger litters and more often breed as yearlings in advance of a large seed crop that autumn (Boutin et al. 2006).

The role of predators in regulating prey populations has also been the focus of many correlational studies. For instance, in southern Sweden, field voles (*Microtus agrestis*) and wood mice (*Apodemus sylvaticus*) are heavily preyed on by generalist predators: common buzzards (*Buteo buteo*), red foxes (*Vulpes vulpes*), and domestic cats (*Felis [catus] sylvestris*; Erlinge et al. 1983). Analysis of pellets, scats, and prey remains for 2 years showed that predators consumed as many mice and voles as were produced during each year! Predators can show two types of responses to changes in the availability of prey. A **numerical response** means increases in the numbers of predators (see figure 25.7). This, of course, takes time because the generation time of predators usually exceeds that of prey, and it is common for changes in predator numbers to lag behind their prey. Predators may also show a functional response, changing the numbers of a particular prey item in their diets. In the Swedish study, predators showed a **functional response,** switching from eating other prey to eating mice and voles as numbers of these two species increased, and thus exerted a density-dependent effect on mouse and vole populations. These researchers conclude that generalist predators are effective regulators of small mammals in southern Sweden and that they keep the vole populations from cycling. Vole populations do cycle every 3–4 years farther north, where these generalist predators are lacking.

Population densities of mammalian predators in the Order Carnivora are closely linked to the biomass of their prey, irrespective of body mass: 10,000 kg of prey support about 90 kg of predator (Carbone and Gittleman 2002). The number of carnivores per unit of prey productivity is related to carnivore mass at a power of -0.75. Thus, carnivores are tied not only to prey size but also to prey biomass. The relationship predicts predator density across more than three orders of magnitude and can thus be used in conservation efforts (Carbone and Gittleman 2002).

Field Experiments

Cause-and-effect relationships can be best revealed by experimental manipulations. One of the simplest experiments is to add food to a population and monitor changes in birth, death, and movement rates as well as changes in home range size. Many studies have been done on populations of mammals, with mixed results. Typical responses to supplemental food are decreased home range size, increased body mass, and increased breeding activity (Boutin 1990). Food addition, however, usually does not prevent major declines in populations, such as in voles that undergo multiannual fluctuations, indicating that other factors must be involved.

Other field experiments have manipulated predators. European rabbits (*Oryctolagus cuniculus*), introduced into Australia, have become serious pests there, sparking interest in the possible role of predators in keeping rabbit numbers in check. In New South Wales, rabbits have several predators, including red foxes and domestic cats. Predators were removed from some plots, and their stomach contents analyzed for rabbit remains (Pech et al. 1992). When rabbits were at low density, predators kept numbers in check, but if predators were removed, rabbits increased in numbers, as one might expect. When predators were allowed back into those areas, however, rabbit populations continued to increase. In other words, the rabbits "escaped" control by the predators. A different predator-prey system in Chile suggests a somewhat similar effect of predators on their mammalian prey. The degu (*Octodon degus*) is a medium-sized caviomorph rodent weighing about 150 g. Its main predator is the culpeo fox (*Lycalopex culpaeus*). On study plots where foxes and aerial predators such as owls were excluded, degus tended to achieve higher densities than on control plots, where predators were allowed access (Meserve et al. 1996). Survival rates of degus and other species of small mammals also were higher on predator-free plots.

Several attempts have been made to study the combined effects of food and predators in what is called a factorial design. At least four plots are needed: one with added food, one with predators excluded, one with both added food and predators excluded, and one left alone as a control. Each plot is usually replicated at least once, for a minimum of eight. For rodent studies, various types of fences and nets are used to exclude predators. In one such study, prairie vole (*Microtus ochrogaster*) densities increased in pens where high-quality food was added and in those where predators were excluded (Desy and Batzli 1989). Highest vole densities were achieved in those pens with both added food and predators excluded. In this study, the effects of increased food and reduced predation were additive (figure 25.13). Food operated mostly by influencing reproduction, whereas predators affected survival of adults

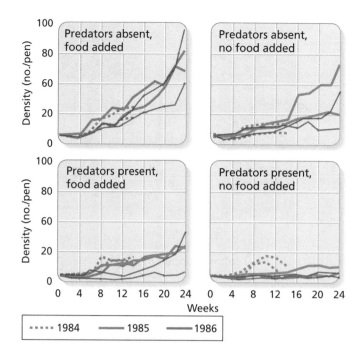

Figure 25.13 The role of food and predators in regulating vole populations. Population density of voles in four treatments (two replicates each year). *Data from E. A. Desy and G. O. Batzli, 1989, Effects of food availability and predation on prairie vole demography: A field experiment,* Ecology, *70:411–421.*

and young. For an example in which the effects of food and predators were nonadditive, see the study by Krebs and colleagues (1995) on snowshoe hares discussed in the following section.

CYCLES

Many populations of mammals fluctuate rather widely, presumably in response to environmental changes. True cycles, those with constant periods, are rare in nature, and none is known in the southern hemisphere or the tropics (Sinclair and Gosline 1997). Mammals are unusual in having a number of species that cycle, ranging from 3–4 years in smaller species such as certain voles and lemmings (*Lemmus* spp.) to 10 years or more in larger species such as lynx (*Lynx canadensis*) and snowshoe hares. The existence of cycles has fascinated and puzzled mammalogists and ecologists ever since Charles Elton (1924) introduced the subject. Cycle amplitude and length tend to increase with latitude (Hanski et al. 1991). Both the causes of these cycles and their apparent synchrony in time and space demand explanation.

As an example of a cyclic species, Siberian brown lemmings (*Lemmus sibiricus*) at Barrow, Alaska, have cycles of about 4 years (figure 25.14). During one of the increases, breeding starts in the fall, and by spring, large numbers of lemmings are present, many of which fall prey to snowy owls (*Nyctea scandiaca*), arctic foxes (*Alopex lagopus*), weasels (*Mustela* spp.), and other predators. During the

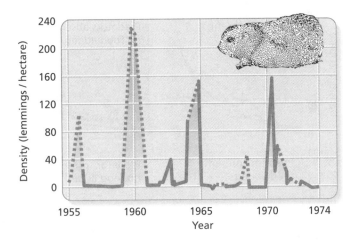

Figure 25.14 **Lemming population cycles.** Estimated lemming densities in the coastal tundra at Barrow, Alaska, for a 20-year period. *Data from E. O. Batzli et al., 1980, The herbivore based trophic system, in* An Arctic Ecosystem: The Coastal Tundra at Barrow, Alaska *(J. Brown et al., eds.), Dowden, Hutchinson and Ross.*

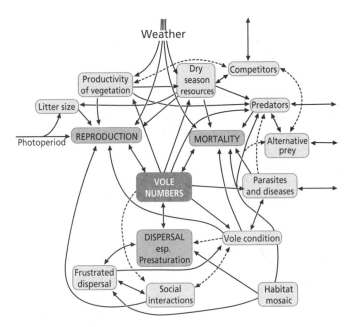

Figure 25.15 **Factors affecting vole populations.** Schematic representation of factors known (*solid arrows*) or suspected (*dashed arrows*) to influence numbers of California voles. *Data from W. Z. Lidicker, Jr., 1988, Solving the enigma of microtine "Cycles,"* J. Mammal. *69:225–235.*

summer, the population crashes to a low level, where it remains for 1–3 years.

Many agents, including extrinsic factors such as changes in food quantity or quality, parasites, disease, and predators, and intrinsic factors such as physiological stress from crowding and changes in gene frequency, have been proposed to explain these cycles (Pianka 1994). Many researchers have focused on one or two factors in an effort to explain cycles. For instance, Hanski and colleagues (1991) have argued that small, specialist predators such as weasels, which are major predators at high latitudes in Scandinavia, increase the amplitude of the cycles. These authors contend that larger, generalist predators, which are more important at lower latitudes, tend to stabilize rodent populations.

In the high arctic tundra of northeast Greenland, the Nearctic collared lemming (*Dicrostonyx groenlandicus*) is the main source of food for no less than four vertebrate predators, making it one of the world's simplest predator-prey communities (Gilg et al. 2006). Lemmings follow a 4-year cycle, with peak densities more than 100 times trough densities. The most specialized predator, the short-tailed weasel (*Mustela erminea*), shows a delayed numerical response such that maximum stoat predation rates precede the lowest lemming densities and maintain low prey densities for at least two successive years. Additionally, minimum predation rates by stoats occur the winter preceding lemming peaks, leading the authors to conclude that stoats are the main drivers of the cycle.

Others have argued that single-factor explanations are not likely to be universal and that multifactorial explanations are needed to explain multiannual fluctuations. In these models, extrinsic and intrinsic factors act synergistically and sequentially to produce cycles (Lidicker 1988; figure 25.15). A problem with these models is that they

are complex and difficult to test. Note the large number of factors in Lidicker's model for the California vole (*Microtus californicus*).

Most spectacular is the snowshoe hare cycle in the boreal forests and tundra of North America. Among the estimates of abundance of hares and their mammalian predators are the numbers of furs brought in by trappers to the Hudson Bay Company (MacLulich 1957). Each cycle is about 10 years in length, and the lynx and hare populations are highly synchronized, with the peaks in lynx numbers following those of the hare by a year or two (figure 25.16). One hypothesis for the cause of these cycles examines time delays in the interaction of three **trophic** (feeding) levels: the winter food supply of hares, the number of hares, and the number of predators (primarily lynx and great horned owls, *Bubo virginianus*). A second hypothesis includes only two trophic levels: hares and their predators.

Mathematical models and laboratory experiments demonstrate that cycles can occur in the absence of any regular environmental fluctuations. The logistic equation (equation 2) produces population cycles for certain values of r, the intrinsic rate of natural increase. When the response of population growth to density is time-delayed, either damped or increasing oscillations result. This result has been confirmed in laboratory populations of invertebrates, but field tests are lacking. Mathematical models of predator-prey interactions also predict stable cycles under certain conditions, perhaps explaining the lynx-hare cycles. Keith (1983) concluded, however, that food shortage during the winter initiates the decline in hare density,

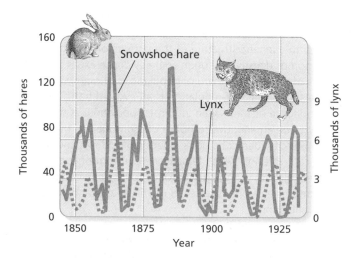

Figure 25.16 Population cycles of the lynx and the snowshoe hare. Comparison of population cycles of both species in the Hudson Bay region of Canada, as indicated by fur returns to the Hudson Bay Company. *Data after MacLulilch, 1937 in R. E. Ricklefs, 1990,* Ecology, *4th ed., W. H. Freeman.*

Whatever the immediate causes of hare cycles, the observation that cycles frequently are synchronized in time and space also requires an explanation. Could some continent-wide synchronizer be affecting climate and thus altering the hare's food supply over a wide region? One candidate suggested early on were peaks in sunspot activity that have about an 11-year mean. The role of sunspots was later discounted because the mean of the hare cycles was closer to 10 years (Keith 1963). This theory has been resurrected, however, based on more recent analyses (Sinclair et al. 1993). These researchers found a correlation between dark marks observed in tree rings from logged spruce trees and the abundance of hares in the Yukon. The dark marks represent stunted growth resulting from browsing by hares on the apical shoots of spruce. In peak hare years, after the more preferred foods of birch and willow are already consumed, hares feed on the less preferred spruce. Dark marks also correlate with the fur records from trapping between 1751 and 1983 in other parts of Canada. Both dark marks in tree rings and hare numbers are correlated in turn with sunspot activity, all with a 10-year periodicity. Sinclair and colleagues (1993) further suggest that solar variability is somehow connected with weather patterns, which directly or indirectly (e.g., through precipitation and food supply) affect hare numbers.

If a global synchronizer such as sunspot cycles is operating in the hare cycle, cycles might be expected to be synchronized among boreal forests within similar climatic regions of both the Old and New World. One study of the Finnish mountain hare (*Lepus timidus*) found that hare density cycles in Finland and Canada were not synchronized (Ranta et al. 1997), suggesting either that sunspots are not involved or that the effect of solar activity on weather is region-specific (Sinclair and Gosline 1997).

The topic of population cycles continues to be vexing. After decades of work by mammalian ecologists from many countries, including some studies of voles spanning more than 25 years, researchers know much about what are not the causes of cycles. A more complete understanding of their causes awaits additional statistical tests of large data sets and further experiments.

with predators playing a secondary role. Lynx density seems to depend on hare density, so the lynx is food-limited, but whether hares are limited only by predators or by both food and predators is unclear (Keith 1987).

In an experiment designed to test the role of food and predators on hare density, supplemental food and mammalian predator abundance were manipulated on 1-km² plots in the Yukon (Krebs et al. 1995). Mammalian predators were excluded from some plots by means of electric fences. Hare density during the cyclic peak and decline doubled when predators were excluded and tripled when food was added. When predator exclusion was combined with food addition, hare density increased 11-fold. Thus, both food supply and predation seem important in population cycles of hares, and they seem to interact in a synergistic, or nonadditive, way. These results tend to support the three-trophic-level hypothesis.

SUMMARY

Populations increase via natality and immigration and decrease via mortality and emigration. All species of mammals are capable of exponential increase in numbers, resulting in an increasingly steep growth curve; however, they rarely do so in nature because of the operation of limiting factors (food, predators, disease, and others).

Life tables are used to compute expectation of further life, survivorship curves, net reproductive rate, and reproductive value. Most species of large mammals show type I survivorship

curves, with low mortality rates until old age. Small mammals more often show constant mortality rates at all ages, giving a type II curve.

Life history traits affect fecundity and survival as reflected in the life table schedules. Most mammals are iteroparous, breeding repeatedly during their lives. Reproductive effort, the energy put into current reproduction that reduces future survival and reproduction, is predicted to increase with age in long-lived species. In short-lived species subject to heavy rates of predation, reproductive effort is high at sexual maturity. Senescence, a decline in physiological function with age, seems to be a consequence of natural selection acting more strongly on traits favoring survival and reproduction early in life than on traits favoring survival to old age.

Species adapted for severe, fluctuating climates are sometimes said to show r-selected traits, with early maturation and large litter sizes. Those adapted to stable habitats are said to be *K*-selected, with larger body size and slower reproductive rates. On the contrary, bet-hedging theory predicts that individuals should reproduce at lower rates for a longer period when survival of offspring is low and unpredictable. Mammal species can be placed on a fast-slow continuum, with those that mature early and have high reproductive rates and short generation times (fast) at one end, and those that mature late and have low reproductive rates and long generation times (slow) at the other end.

Some populations of mammals show logistic growth, with densities gradually approaching the limit that can be supported by the environment. Density-dependent factors, whose negative effects on growth increase with density, are likely to be more important than density-independent factors in regulating such populations.

Several mechanisms have been proposed by which populations of mammals could regulate their own numbers, by means of either behavioral-physiological responses to increased density or behavioral-genetic mechanisms, in which the genetic structure of the population changes as a function of density. Although some evidence exists for both types of responses in natural populations of mammals, none shows that self-regulation is an evolved trait. The reproductive restraint seen in dense populations is more likely to be an adaptive response of individuals in the population to crowded conditions. Species of rodents and terrestrial carnivores, in which females defend territories to protect altricial young and reproduction is suppressed by the presence of opposite-sexed relatives, are candidates for self-regulation.

A number of studies have attempted to correlate food supply and predators with population growth, suggesting that both are important. A smaller number of field experiments have tested for interactions among food, predators, and competitors in regulating population density.

At high latitudes, densities of a few species cycle, typically every 10 years for larger mammals such as the lynx and hare and every 3–4 years for smaller mammals such as voles and lemmings. Both food and predators have been implicated in hare cycles, and there is some evidence that cycles are synchronized by large-scale climate fluctuations linked to solar activity.

SUGGESTED READINGS

Krebs, C. J. 2001. Ecology: The experimental analysis of distribution and abundance, 5th ed. Benjamin Cummings, San Francisco.

Pianka, E. R. 1994. Evolutionary ecology, 5th ed. HarperCollins, New York.

Ricklefs, R. E., and G. Miller. 1999. Ecology, 4th ed. W. H. Freeman, New York.

Roff, D. 2001. Life history evolution. Sinauer Associates, Sunderland, MA

Stearns, S. C. 1992. The evolution of life histories. Oxford Univ. Press, New York.

DISCUSSION QUESTIONS

1. Statements suggesting that many factors contribute to regulating animal populations seem to run counter to Liebig's law of the minimum. Is there any way to resolve this apparent contradiction?

2. Studies of litter size in white-footed mice have shown that litters of six produce the most surviving offspring, yet the average litter size is closer to five. Why would a female mouse produce less than the optimum number of offspring in a particular litter? For a discussion of this topic, see Morris (1992).

3. Understanding factors affecting the size of populations is a central problem in ecology and mammalogy. Why has social behavior been implicated in so many of the theories of population regulation?

4. The data in the following table refer to mice that were exposed 2 min/day either to trained fighter mice, to trained nonfighter mice, or to no mice for days 1–14. On day 8, each mouse was fed equal numbers of tapeworm eggs. On day 22, the mice were killed, their adrenal and thymus glands weighed, and the intestinal tapeworms counted and weighed. Adrenal gland weight is positively related to adrenocortical hormone secretion. The thymus is involved in the immune response. Discuss how these data may relate to population regulation and behavior.

Effect of fighter mice exposure on mean host weights, organ weights, and tapeworm number

	No Mice	Groups Exposed To Trained Nonfighter Mice	Trained Fighter Mice
Group size	10	10	7
Adrenal weight (mg)	4.39	4.42	5.88*
Adrenal weight (mg)/body weight (g) × 100	13.1	13.1	16.6*
Thymus weight (mg)	37.0‡	35.0	21.9*
Body weight (g)	33.7	34.0	35.8
Worm number/mouse	11.6§	15.4§	75.3†
Worm weight/mouse weight (mg)	6.37	4.04	82.28†

Source: Data from Patterson, M. A. and S. H. Vessey, 1973, Tapeworm (Hymenolepis nana) *infection in male albino mice: Effect of fighting among the hosts,* J. Mammal. *54:784–786.*
*Significantly different from other two groups, $p < .01$.
†Significantly different from other two groups, $p < .001$.
‡Mean based on nine weights.
§Three of these mice had no worms.

CHAPTER 26

Community Ecology

In this chapter, we focus on the interactions among species and the effect those interactions have on both the living and nonliving features of their environment—the subject of **community ecology.** Interactions among species affect the abundances of populations (chapter 25) and contribute to natural selection among phenotypes, thus influencing the evolution of coexisting species. These interactions ultimately explain why there are so many different species of mammals. Populations of different species interacting in an area at the same time make up a **biological community.** A more inclusive term is **ecosystem,** which includes the biotic (living) components (i.e., the community) plus the abiotic (nonliving) components. Communities and ecosystems are not simply random assemblages of species but seem to show similar patterns of assemblages across different continents. Understanding these so-called assembly rules is a central problem in ecology, and the study of mammalian communities has played a significant role in this research. **Community structure** encompasses patterns of species composition and abundance, temporal changes in communities, and relationships among locally coexisting species. At the simplest level, community ecology can be studied by looking at interactions between two species, which we do first in this chapter. Then we explore more complex and larger-scale patterns of mammalian diversity. Our goal is to explain the way single-species populations are integrated into larger biological levels of organization.

Mammalogists have played important roles in an understanding of the interactions among populations. One of the first North American community ecologists was C. Hart Merriam, a mammalogist who developed the concept of life zones. Working in the San Francisco Mountains of northern Arizona, Merriam (1894) concluded that temperature defined floral and faunal zones, forming elevational bands up the sides of mountains and latitudinal bands from the equator to the north and south poles. Early researchers viewed communities as interrelated units responding collectively to abiotic factors (reviewed by Mares and Cameron 1994). The degree to which communities can be considered discrete units rather than random assemblages of populations has been hotly debated by ecologists, as we shall explain later in this chapter.

The Ecological Niche

Each species of mammal occupies one or more habitats, has certain physical and chemical environmental tolerances, and performs a specific functional role in the community. Joseph Grinnell, who worked with birds and mammals in the early 1900s, was one of the first ecologists to use the term *niche*, by which he meant the habitat a species occupies as a function of its physiological and behavioral attributes (Grinnell and Swarth 1913). The term was defined somewhat differently by Elton (1927) to include an organism's functional role in the community in terms of its trophic (feeding) level (Mares and Cameron 1994). Today, most ecologists include both the distributional and functional components, considering a niche to be the total of adaptations of a species to a particular environment (Pianka 1994). Ecologists can never know all the parameters necessary to define a niche completely, so they emphasize those factors that potentially limit distribution and abundance and those for which organisms compete.

Most studies of mammals provide information potentially useful in determining niche attributes of individual species. Laboratory studies can determine physiological tolerances of a species to such conditions as heat, cold, and moisture (see chapter 9). Measuring its niche in the field is more difficult, however, because biotic and abiotic factors often interact to produce unpredictable results. Competitors, predators, and symbionts affect the distribution, population dynamics, and niche of most species.

Species Interactions and Community Structure

INTERSPECIFIC COMPETITION

Individuals of the same species may compete for limited resources such as food or shelter, leading to density-dependent changes in population growth (see equation 2 from chapter 25). Different species may also compete for those resources, affecting rates of population growth of both species. Equation 2 from chapter 25 can be modified to include the effect of competing species 2 on species 1 by adding a term called the coefficient of competition (a_{12}):

$$\frac{dN}{dt} = r_1 N_1 \left(1 - \frac{N_1}{K_1} - \frac{a_{12} N_2}{K_1}\right) \quad (1)$$

where the subscripts 1 and 2 refer to values for species 1 and species 2. Notice that the density-dependent term in parentheses now contains a term including the number of individuals of species 2. When $a = 1$, the interspecific competitor, species 2, has the same effect on population growth of species 1 as another member of species 1, and interspecific competition is intense. As a declines,

approaching zero, the last term of the equation approaches zero, and the effect of species 2 on species 1 can be ignored. A similar equation can be written to describe the effect of species 1 on species 2, where the coefficient of competition is a_{21}. This equation is too simple to actually model competition in nature, but it serves as a starting place to develop hypotheses and design experiments to further understand this important ecological process.

Interspecific competition also affects the habitat and food preferences of different species. In the absence of competition, a species occupies its **fundamental niche,** the full range of conditions and resources in which the species can maintain a viable population. In the presence of competing species, however, the species may be restricted to a narrower **realized niche,** where some habitats or resources are not available because competitors occupy them (Hutchinson 1957). If the competitors are strong enough, a realized niche may become so small that local extinction results. This line of reasoning has led to the **competitive exclusion principle:** If two competing species coexist in a stable environment, they do so as a result of differentiation of their realized niches (Ricklefs and Miller 1999). If no differentiation takes place, then one species will exclude the other. Although this principle makes intuitive sense and has been widely accepted by ecologists, it is difficult to confirm experimentally.

One way to visualize interspecific competition is to look at only one niche component at a time, picking the one that is limiting for the competitors. Consider several species of desert rodents feeding on different-sized seeds from various plants. It is relatively easy to determine which types of seeds are stored for later consumption because seeds can be collected from the cheek pouches of trapped animals. The feeding niches of the rodents can be inspected by plotting seed size on one axis and the utilization (consumption) rate on the other. In the hypothetical example shown in figure 26.1, each of the three species shown has a roughly normal, or bell-shaped, utilization curve with a mean, or average. Niche width (w), measured by the standard deviation around the mean, and the distance between utilization peaks (d) indicate the amount of interspecific competition. When d is large and w is small, little overlap occurs, and the coefficient of interspecific competition is small (figure 26.1A). When d is small and w is large, overlap is great, and the coefficient is large. In this case, d might increase or w might decrease, leading to reduced competition. Another outcome is that one species would be eliminated from the area. If possible, more than one niche dimension needs to be examined because coexisting species may overlap extensively in one niche dimension but only slightly in another (Schoener 1974).

How different do two species need to be to coexist in the same general habitat? The ratio d/w is a measure of niche overlap, and when this ratio is low (i.e., less than 1) coexistence is theoretically possible. However, differences in carrying capacity among the competing species, environmental fluctuations, and the importance of other niche parameters may all affect the outcome.

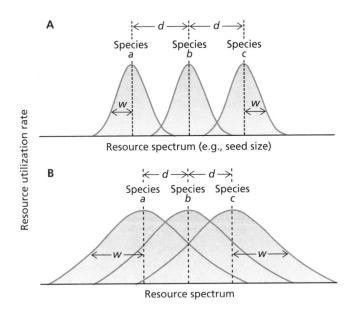

Figure 26.1 **Resource utilization curves for three species coexisting along a one-dimensional resource spectrum.** The term *d* is the distance between adjacent curve peaks; *w* is the standard deviation of the curves. (A) Narrow niches with little overlap (*d* > *w*), that is, relatively little interspecific competition; (B) broad niches with great overlap (*d* < *w*), that is, relatively intense interspecific competition. *Data from M. Begon et al., 1986,* Ecology, *Sinauer Associates.*

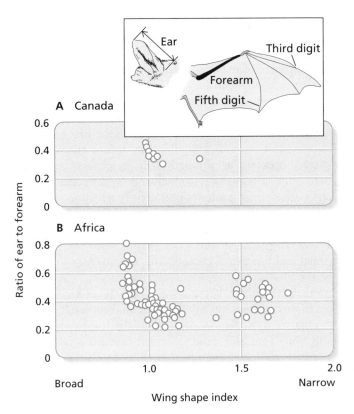

Figure 26.2 **Niche space of insectivorous bats of (A) southeastern Ontario, Canada, and (B) Cameroon, Africa.** The horizontal axis is the ratio of the lengths of the third and fifth digits of the hand and the vertical axis is the ratio of ear length to forearm length. *Data from M. B. Fenton, 1972,* The structure of aerial-feeding bat faunas as indicated by ears and wing elements, *Can. Journal Zoology 50:287–296.*

One way to indirectly evaluate a niche is to assess morphological traits related to important functions, such as feeding. The number of species packed into a niche space can be estimated by comparing morphological distances between nearest neighbors. Fenton (1972) used several dimensions to define morphological space among different species of bats. One was the ratio of ear length to forearm length; the higher the ratio, the larger the ears and the more important echolocation is presumed to be in the foraging behavior. A second was the ratio of third and fifth digits of the wing bones, yielding an index of wing shape. A high ratio on this dimension indicates a long, thin wing shape, whereas a low ratio indicates a short, broad wing shape (called aspect ratio—see figure 13.4). The former favors high-speed flight, whereas the latter favors low-speed flight and greater maneuverability. Plotting both ratios together provides an insight into the niche space occupied by different species (figure 26.2). In the temperate zone communities in Ontario, bat species are small insectivores, and the points are clustered fairly closely. In the tropics, bats fill many other roles, such as fruit-, nectar-, fish-, and even bat-eaters, and the scatter of the points reflects the wide range of wing and ear shapes. Other studies of bats (Findley and Black 1983; Schum 1984) have confirmed that external morphology and diet are correlated and that variation in morphology increases with the number of species present.

The role of interspecific competition in determining niche parameters and the species composition of communities has been controversial. Although competition is assumed to be a driving force, it is possible that competition played a major role in adaptation of the species in the past, but it is hardly observable today. Another problem is that competition may occur only during occasional "crunches," when resources become limiting for a brief period (Wiens 1977) or when species newly come together. At other times, a particular resource may not be limiting. Nevertheless, these crunches may be crucial in determining niche breadth and the distribution of species across the landscape.

Several factors provide indirect evidence of the importance of competition in natural communities of mammals. One is the distribution of species, both in the presence and the absence of presumed competitors. For instance, two species of chipmunk, the cliff chipmunk (*Tamias dorsalis*) and the Uinta chipmunk (*T. umbrinus*), are found on isolated mountaintops in the Great Basin area of Nevada. When each species is present alone, it occupies a wide range of elevations, but when together, one species occupies lower elevations and the other higher (Hall 1948; figure 26.3). The assumption is that in the presence of competition, each species reduces niche overlap. Conversely, each species shows **competitive release,** expanding its range in the absence of competitors.

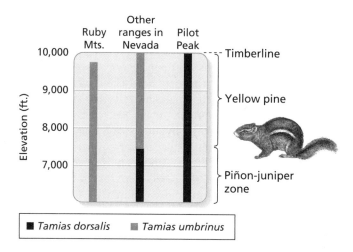

Figure 26.3 The distribution of two species of chip-munk in the Great Basin area of Nevada. Note that where only one species is present, each occupies the entire slope, but if both are present, one occupies higher elevations and the other lower elevations. *Data from E. R. Hall, 1948,* Mammals of Nevada, *Univ. California Press.*

Introductions by humans provide indirect evidence of competition, especially when mainland species are introduced on islands. The result is often extinction of the island species, presumably due to competitive exclusion. Thus, introduction of the placental canids, the dingo (*Canis lupus dingo*) and red fox (*Vulpes vulpes*), into Australia was a factor in the extinction of the marsupial fauna such as the Tasmanian wolf (*Thylacinus cynocephalus*) and the Tasmanian devil (*Sarcophilus harrisii*). Based on fossil evidence, the disappearance of both the Tasmanian wolf and devil from southern Australia coincided with the arrival of the dingo to that region. Both marsupial species survived on Tasmania because the island was cut off from the mainland before the dingo got to the southwestern mainland of Australia.

In the United States, introduced species of deer have had a negative effect on native species. Sika deer (*Cervus nippon*), native to Japan and the East Asian mainland, were introduced into Maryland's eastern shore in 1916. During the 1970s and 1980s, the proportion of white-tailed deer (*Odocoileus virginianus*) harvested declined sharply, from 75% to 35%, and that of sika deer showed a corresponding increase, from 25% to 65% (Feldhamer and Armstrong 1993). Such negative correlations do not, however, demonstrate a cause-and-effect relationship.

Character Displacement

Other observational evidence for competition comes from comparisons of morphology and behavior of individuals in the presence or absence of close competitors. When the ranges of two similar species overlap (i.e., their distributions are **sympatric**), the species tend to differ more from each other than where their ranges do not overlap (i.e., their distributions are allopatric). This shifting of traits in areas where both species are present has been referred to as character displacement. In northern Europe, the Eurasian pygmy shrew (*Sorex minutus*) and the common shrew (*S. araneus*) overlap extensively. Allopatric populations, however, occur on islands off the coasts of England and Sweden. Pygmy shrews have significantly smaller lower jaws in sympatric populations than in allopatric populations (Malmquist 1985; figure 26.4). In this study, however, jaw measurements of the larger, common shrew did not shift. Skull size is related to prey size, that is, individuals with larger skulls can eat larger prey than those with smaller skulls. The inference is that in zones of overlap, one or both of the species diverge in prey size selection, leading to reduced competition and coexistence of both species.

Two species of bats, the long-eared myotis (*Myotis evotis*) and the southwestern myotis (*M. auriculus*), coexist at several sites in New Mexico where their ranges overlap. When Gannon and Rácz (2006) measured distances between various landmarks on mandibles from museum specimens, they found that the mandibles of the two species differed more in sympatry than in allopatry, consistent with the idea that the species shifted diets in regions where both species were present. Based on the differences in mandible shape, the researchers surmise that in sympatry *M. evotis* specializes on hard-bodied beetles, while *M. auriculus* specializes on soft-bodied moths.

Although usually restricted to closely related species, character displacement may occur among more distantly related taxa on a community-wide basis. Three sympatric

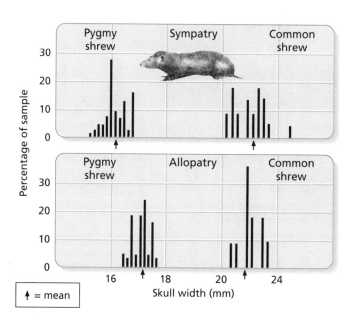

Figure 26.4 Character displacement in shrews. The pygmy shrew and the common shrew are allopatric on islands off the coast of Sweden but sympatric on the mainland. Note that the measure of skull width is smaller for the pygmy shrew in sympatry. *Data from M. G. Malmquist, 1985, Character displacement and biogeography of the pygmy shrew in northern Europe,* Ecology *66:372–377.*

small cat species in Israel are strongly sexually dimorphic, and the sexes can be treated separately for comparison (Dayan et al. 1990). The average canine diameters for the six "morphospecies" were remarkably evenly spaced, consistent with the idea that niche partitioning of prey size occurs among species and between sexes. These cats are exclusively carnivorous, and canine teeth are used to kill prey. The canines are well adapted to wedge between the vertebrae of the prey's neck, severing the spinal cord or hind brain. Larger prey would clearly require thicker canine teeth, making canine diameter a good predictor of the size of prey captured. Dayan and colleagues (1990) argue that the evenly spaced sizes indicate character divergence to reduce competition for prey. Although competi-

tion for food is the presumed explanation, it is also possible that sexual selection has played a role in canine size. Males of these species may use canines for display or fighting with other males, as do many polygynous species of primates. Thus, canine size may be affected by male-male competition for access to females as well as by size of prey.

Character displacement may also exist among mustelids. McNab (1971) hypothesized that prey size combined with character displacement governs body size for three species of North American weasels (*Mustela* spp.); sympatric species of weasels were forced by competition to maintain minimum size differences. Harvey and Ralls (1985), who examined condylobasal measurements of skull length in North American weasels, found no evi-

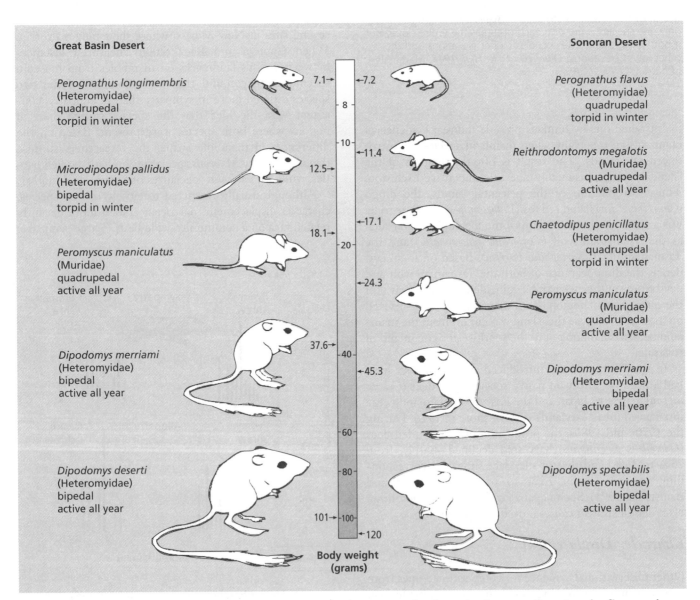

Figure 26.5 **Convergence in structure between a six-species community from the Sonoran Desert and a five-species community from the Great Basin Desert.** Numbers are average weights (grams). Note the similarities in body size, form, taxonomic affinity, and other characteristics between species occupying similar positions in each community. Also notice the displacement in body size in *Peromyscus maniculatus* and *Dipodomys merriami* (both are larger in the Sonoran Desert) to compensate for the different numbers and sizes of coexisting species. *Data from J. H. Brown, 1975, Geographical ecology of desert rodents, in* Ecology and Evolution of Communities *(M. L. Cody and J. M. Diamond, eds.), Belknap Press of Harvard Univ. Press.*

dence of character displacement, however, concluding that most effects were due to latitudinal gradients in body size. In a still more recent study using the diameter of the upper canine, almost no overlap occurred among different sympatric species (Dayan et al. 1989). Size ratios were nearly evenly spaced, as was true in the cat example. Equal size ratios between species adjacent in a size ranking were considered by Dayan and colleagues as evidence for community-wide character displacement as a result of competition, supporting McNab's (1971) original hypothesis.

Limiting Similarity

One of the ways two or more species can coexist is to evolve different body sizes, thus essentially dividing the area into different feeding niches and avoiding competition. A ratio in length of 1:1.3 was suggested for the minimum difference between two closely related competitors to permit coexistence (Hutchinson 1959), prompting a number of investigators to compare body size at the community level. Numerous species of seed-eating rodents occupy the deserts of the southwestern United States. When Brown (1975) compared the rodent communities in the Sonoran Desert with those of the Mojave and Great Basin Deserts, he noticed that the distributions of body sizes of different species were remarkably similar (figure 26.5). Brown also observed that species of similar size tended to "replace" one another in different habitats. Thus, the 7-g silky pocket mouse (*Perognathus flavus*) in the Sonoran Desert was replaced by the 7-g little pocket mouse (*P. longimembris*) in the Great Basin Desert, and the 11-g western harvest mouse (*Reithrodontomys megalotis*) was replaced by the 12-g pale kangaroo mouse (*Microdipodops pallidus*). In addition, when the same species was present in both deserts, it was displaced in size, filling a gap (e.g., the deer mouse, *Peromyscus maniculatus*, was 24 g in the Sonoran Desert but only 18 g in the Great Basin Desert).

Removal Experiments

Observational studies can provide only indirect, correlational evidence of the importance of competition in structuring ecological communities. One way to provide direct evidence of interspecific competition is to remove one of the competitors and monitor the response of the other species. In one such study, three sympatric species of small mammals—the white-footed mouse (*Peromyscus leucopus*), the golden mouse (*Ochrotomys nuttalli*), and the northern short-tailed shrew (*Blarina brevicauda*)—were livetrapped, marked, and released in a pine plantation in Tennessee (Seagle 1985). A number of vegetation characteristics were recorded at each capture site to determine the microhabitat preferences for each species. When white-footed mice were removed from one grid, golden mice demonstrated competitive release by shifting their microhabitat preference from open forest to areas with more fallen logs, denser canopy development, and denser understory.

Four species of chipmunks (*Tamias* spp.) are found along an altitudinal-vegetational gradient on the eastern slope of the Sierra Nevada, California. The chipmunk species are contiguously allopatric, that is, their ranges are adjacent but not overlapping, and each is restricted to its own vegetational community along the gradient (figure 26.6). Chappell (1978) conducted a series of observations and experiments to reveal the physiological and behavioral factors that might produce this distribution. For example, captures of the least chipmunk (*T. minimus*) increased in wooded areas at higher elevations, where the yellow-pine chipmunk (*T. amoenus*) was removed. Behavioral observations confirmed that *T. amoenus* was dominant to *T. minimus*, suggesting that competitive exclusion was responsible for the failure of *T. minimus* to colonize the wooded zones. Where *T. minimus* had been removed, however, *T. amoenus* failed to invade. The failure of the socially dominant *T. amoenus* to move down slope into the sagebrush zone in the absence of *T. minimus* is probably due to physiological constraints: *T. amoenus* is not adapted to tolerate the hot, dry conditions at lower elevations. In this example, both behavioral and physiological factors affect the distribution of species.

Not all removal studies provide clear-cut results. Along the Mediterranean coast of Israel is a narrow strip of sand dunes, of recent origin, perhaps only a few hundred years old. Two species of seed-eating gerbilline rodents occupy the area, and they show considerable overlap in diet and microhabitat preference (Abramsky and Sellah 1982). The larger species, Tristram's jird (*Meriones tristrami*), colonized the dunes from the north. It can live on sand as well as other soil types. The smaller species, Anderson's gerbil

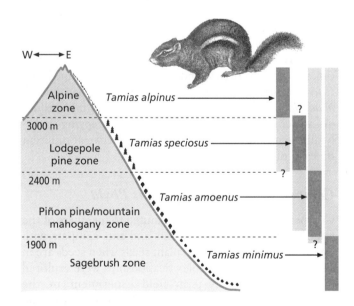

Figure 26.6 Distribution of four species of chipmunk on the eastern slope of the Sierra Nevada, California. Dark portions of bars denote realized niches; entire bars show fundamental niches. *Data from M. A. Chappell, 1978, Behavioral factors in the altitudinal zonation of chipmunks* (Eutamias), Ecology *59:565–579.*

(*Gerbillus* [*allenbyi*] *andersoni*) colonized from the south, and it is found only on sand. When alone, *M. tristrami* occupies sand. Where the two species are sympatric, however, *G. andersoni* occupies sand and *M. tristrami* occupies soil types other than sand. The exclusion of *M. tristrami* from the habitat it prefers when alone (sand) has been interpreted as resulting from interspecific competition. When *G. andersoni* was removed from an area containing both species, it was expected that *M. tristrami* would expand its habitat and occupy the sand areas, showing competitive release. No such effect was observed, however, suggesting that competition was not responsible for the different substrates occupied. One interpretation of these results is that these species did, in fact, compete in the past and that habitat selection is under genetic control. Thus, genetic changes in the population of *M. tristrami* in areas of overlap with *G. andersoni* might have led to its failure to occupy sand in the absence of the other species. This phenomenon has been referred to as "the ghost of competition past."

In some cases, two very similar species are found in the same habitat. In the Appalachian Mountains, two species of the genus *Peromyscus* frequently coexist. White-footed mice and deer mice are so similar in appearance in some regions that molecular techniques are needed to tell the two species apart with certainty. They eat the same foods (Wolff et al. 1985) and defend territories against members of their own species and the other species (Wolff et al. 1983). The two species may occupy different microhabitats because *P. maniculatus* shows more arboreal activity than does *P. leucopus* (Harney and Dueser 1987). Based on reciprocal removal of each species on different plots, neither species seemed to encroach on the other's microhabitat, however, suggesting little in the way of competitive interactions (Harney and Dueser 1987). By checking live traps in the middle of the night, Drickamer (1987) captured *P. leucopus* more often before 1:00 a.m. and *P. maniculatus* more often after 1:00 a.m. Using live traps with timers on them, Bruseo and Barry (1995) confirmed that *P. leucopus* becomes active earlier each night than does *P. maniculatus*. Thus, these two similar species may reduce intraspecific competition and coexist by occupying slightly different microhabitats and by being active at different times of night.

Competition Among Different Phyla

Competition is not necessarily limited to members of the same taxonomic group. Rodents and ants compete with one another as they prey on plant seeds when seeds are in short supply. These complex relationships were explored by means of a long-term field experiment in the Chihuahuan Desert in southeastern Arizona (Brown et al. 1986). Twenty-four plots were fenced off as part of a 10-year study. Each plot was 50 m by 50 m, and three general manipulations were performed: (1) exclusion of some or all rodent species using different-sized gates in the fences; (2) exclusion of ant species by means of poisoning

colonies; and (3) the addition of seeds. Rodent densities increased in the plots with added food, suggesting that seeds were a limiting factor. Excluding large granivorous rodents (kangaroo rats, *Dipodomys* spp.) led to increases in small granivorous rodents, suggesting competition among rodents for seeds. When all rodents were excluded, large-seeded plants were favored because the rodents preferred eating the larger seeds; small-seeded plants then declined, as did the ants that fed on them. Thus, although rodents and ants seemed to be competing for seeds, they actually had a mutualistic relationship: By their feeding on seeds of different plants, both large- and small-seeded plants coexisted, as did the ants and rodents that fed on them (figure 26.7).

PREDATION

In chapter 25 we discussed predation from the perspective of prey population dynamics and whether predators can regulate the numbers of prey. Here we take a broader perspective and consider the role of predators in structuring communities. One type of effect is referred to as a **trophic cascade**, in which predators depress populations of herbivores to the point that plant biomass increases.

Keystone Predators and Trophic Cascades

Some species appear to play critical roles in the community, such that "not all species are created equal." Although mammals typically account for only a small

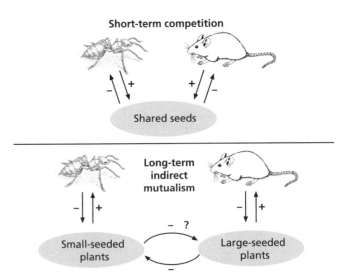

Figure 26.7 Interactions between granivorous rodents and ants. In the short term, the two taxa compete if they overlap in their feeding on limited seeds, but in the longer term, the two taxa can have indirect mutualistic effects on each other if they feed differentially on different plant species that also compete with each other. Plus signs refer to positive interactions; negative signs to negative interactions. *Adapted from J. H. Brown et al., 1986, Experimental community ecology: The desert granivore system, in* Community Ecology (*J. Diamond and T. J. Case, eds.*), *Harper & Row.*

fraction of the biomass and energy flow in most ecosystems, they sometimes regulate the structure and dynamics of the entire community. **Keystone predators** are those that exert a controlling force in the community, affecting abundance and distribution of many other species and sometimes actually increasing the species diversity in the community.

For example, sea otters (*Enhydra lutris*) prey extensively on sea urchins along the coast of the western United States. Fur traders decimated otter populations from 1741 to 1911, except for a few remnant populations in Alaska and central California. Since being protected, sea otter populations have become reestablished over much of their original range. In areas where sea otters are present, sea urchins are sparse and algae communities (mainly kelp) thrive, evidence of a trophic cascade. In areas where sea otters are absent, sea urchins become abundant and eliminate much of the kelp and associated marine species (Estes et al. 1978).

When sea urchins were experimentally removed from plots on Torch Bay, Alaska, where no otters were present, a complex association of several species of kelp developed. Similar effects were seen when otters were transplanted into areas with high densities of sea urchins. The sea urchins were quickly depleted by the otters, and the kelp community developed several years later (Duggins 1980; table 26.1). Sea otters, therefore, exert a profound effect on community structure and fit the definition of keystone predators.

Sea otter numbers declined once again over much of Alaska's west coast in the 1990s, at rates as high as 25% per year. The evidence points to increased predation by killer whales (*Orcinus orca*) as the cause (Estes et al. 1998). As predicted by the decline in sea otters, sea urchin biomass increased eightfold while kelp density declined 12-fold in the same decade. Killer whales typically feed mainly on seals in this region, but seal populations have plummeted as a result of declining fish stocks. In response, killer whales appear to have shifted to sea otters as prey.

Herbivores can also act as keystone species. The African elephant (*Loxodonta africana*) prefers forest edge, woodland, and brushland. It feeds on a diet of browse (twigs, shoots, and leaves) supplemented by grass. While foraging on bark, elephants destroy the shrub understory and girdle trees, all of which leads to loss of woodland and the creation of grassland (figure 26.8). Once forest canopy is opened up, fire accelerates the process. In large parts of sub-Saharan Africa, the vegetation has shifted from woodland to grassland as a result of elephants. This change ultimately works to the disadvantage of the elephants, however, because they need woody species for browse, but it works to the advantage of grazing ungulates (Laws 1970).

More than one species may play a keystone role in the same community. Species occupying a particular trophic level can be subdivided into **guilds,** or groups of species that exploit a common resource base in a similar fashion. In the Chihuahuan Desert, several species of kangaroo rats form a seed-eating guild (Brown and Heske 1990). When these rodents were excluded from this desert shrubland, a chain of events occurring over 12 years changed it to desert grassland. In the absence of kangaroo rats, annual plants with large seeds increased in number, raising the vegetative cover to the point where ground-feeding birds could no longer forage on grass seeds. This led to the increase in grasses. Both selective foraging by genus *Dipodomys* on large seeds and soil disturbance by foraging, caching, and burrowing tend to maintain the shrub desert. Of the 15 species of rodents that live in the Chihuahuan Desert, Brown and Heske (1990) considered the kangaroo rats to form a **keystone guild** of large-seed eaters.

Table 26.1 Sea otters as a keystone species: Changes in sea urchin and kelp densities in response to the reintroduction of sea otters

	Mean Density (no./m² ± SD)*		
	Torch Bay (no otters) $N = 80$	**Deer Harbor (otters present < 2 y)** $N = 24$	**Surge Bay (otters present < 10 y)** $N = 80$
Urchins			
Strongylocentrotus franciscanus	6±7 (79%)	0.03±0.03 (4%)	0
S. purpuratus	4±13 (31%)	0.08±0.06 (8%)	0
S. droebachiensis	6±14 (65%)	0.2±0.1 (12%)	0
Kelps			
Annuals	3±7 (33%)	10±5 (100%)	2±5 (28%)
Laminaria groenlandica	0.8±5 (6%)	0.3±0.6 (21%)	46±26 (99%)
Samples with no kelp	(64%)	(0%)	(0%)

Data from D. O. Duggins, 1980, Kelp beds and sea otters: An experimental approach, Ecology 61:448.
*Numbers in parentheses are the percent of these quadrats in which the species or group were observed. N = number of 1.0 m² quadrats.

Figure 26.8 Elephants as a keystone species. By destroying shrubs and trees, elephants promote grassland, which in turn favors grazing ungulates.

Predator Removals and Introductions

A straightforward way to test the role of predation in regulating communities is to remove the predators. Researchers predict that herbivores, normally kept in check by predators, would explode and plant biomass would decrease, a trophic cascade. Such experiments are ongoing in South America as areas are flooded by dams designed to provide hydroelectric power. In one such project in Venezuela, 4300 square kilometers were flooded, forming many islands (Terborgh et al. 2001). Large predators, with their large home ranges, quickly disappeared from these islands. A result was an explosion of herbivores, mainly howler monkeys, rodents of several species, iguanas, and leaf-cutter ants, that reached densities of 10 to 100 times those on the mainland. This outcome demonstrates the importance of top-down control of ecosystems in which predators keep herbivore populations sufficiently in check such that food is not limiting. In a few decades, these hyperabundant herbivores are predicted to reduce the once species-rich forests to a small group of herbivore-resistant plants.

Another example of a trophic cascade comes from the introduction of predators on islands. In the late 19th and early 20th centuries, arctic foxes (*Alopex lagopus*) were introduced to more than 400 islands in the Aleutian archipelago to bolster the collapsing fur trade (Croll et al. 2005). Several islands remain fox free, however, and now serve as controls in this unwitting experiment. Breeding seabird densities are now almost 2 orders of magnitude higher on the fox-free islands than on fox-infested islands. The reduction of seabirds by fox predation has led to a reduction in bird guano input. The resulting loss of nutrient input from the ocean has, in turn, led to a transformation from tall grasses to a dwarf shrub/forb community on the fox-infested islands (Croll et al. 2005).

Risk of Predation

Food partitioning is one way in which communities are structured, but risk of predation may also be important. Rodents in the Great Basin Desert in Nevada partition the microhabitat so that kangaroo rats and kangaroo mice forage in the open, whereas pocket mice and deer mice forage near bushes. The former have hyperinflated auditory bullae, associated with high auditory acuity, and elongated hind legs, associated with bipedal locomotion. These adaptations aid in the detection of, and escape from, predators such as long-eared owls (*Asio otus*), coyotes (*Canis latrans*), kit foxes (*Vulpes macrotis*), and gopher snakes (*Pituophis melanoleucus*). Dice (1945) showed that owls find prey more easily under moonlight than starlight. In a field experiment, illumination simulating moonlight was added to some grids by means of lanterns, and shadows were created at some sites by means of parachute canopies (Kotler 1989). In general, illumination led to reduced foraging in open microhabitats. Food in the form of bird seed was added to some bush sites and to other open sites. The bipedal species responded positively to additional food in both bush and open sites, whereas the quadripedal species responded only to food in the bush sites. Kotler (1989) concluded that both risk of predation and resource availability interacted to affect foraging behavior and habitat choice. Thus, both predators and distribution of resources need to be considered when trying to predict the species composition of a community.

Among African carnivores, lions and hyenas not only compete with cheetahs, they also prey on them. Playback studies show that cheetahs actively moved away from calls of lions and hyenas (Durant 2001). They also stopped hunting and moved away from high concentrations of prey (gazelles) to avoid lions and hyenas.

MUTUALISM

Interactions involving mammals and members of other taxa can be mutualistic, in which both species benefit either directly or indirectly. Examples of direct mutualism, in which both species are in contact, include seed dispersal by rodents, pollination of many species of plants by bats (chapter 13), and digestion of cellulose by the endosymbionts within the rumens of their ungulate hosts (chapters 7 and 20). Mutualistic relationships between birds and mammals include one between the African honey guide (*Indicator indicator*) and the African honey badger (*Mellivora capensis*). The bird vocalizes and "leads" the badger to a bee's nest. The badger opens up the nest and feeds on honey, while the honey guide consumes wax (Vaughan 1986). Oxpeckers (*Buphaga* spp.) eat the ticks they remove from large African ungulates to the benefit of both bird and mammal. Unfortunately, as domestic cattle replace native species of ungulates, oxpeckers have declined in number, in part because cattle are treated with pesticides that then eliminate the oxpecker's food supply.

Indirect mutualism involves positive effects without direct contact between the species. An example is the ant-rodent mutualism in the Arizona desert discussed earlier (see figure 26.7). Commensal relationships, in which one species benefits and the other is more or less unaffected, include that between cattle egrets (*Bubulcus ibis*) and cattle (genus *Bos*; Heatwole 1965). Cattle egrets feed on insects stirred up by the cattle. Another example of commensalism is the moving substrate provided for barnacles attached to the whale's skin.

PLANT-HERBIVORE INTERACTIONS

Herbivorous mammals, especially ungulates, are important components of many ecosystems, especially grassland communities. The effect of grazing on the plant community has been viewed by ecologists both as a predator-prey relationship, in which grazers benefit and plants are harmed, and as a mutualistic one, in which numerous species of grazers and plants have coevolved adaptations in response to one another. The most spectacular concentration of grazing animals in the world occurs in the Serengeti-Masai-Mara ecosystem in Tanzania and Kenya, where hundreds of thousands of blue wildebeests (*Connochaetes taurinus*), Thomson's gazelles (*Eudorcas* [*Gazella*] *thomsonii*), Burchell's zebras (*Equus burchellii*), and African buffalos (*Syncerus caffer*) concentrate, along with lesser numbers of more than 20 other species (figure 26.9, see also figure 7.10). In the wet season, usually ending in May, huge numbers of migrating wildebeests pass

Figure 26.9 Thomson's gazelles grazing in the Masai-Mara, Kenya. Previous grazing by zebras and wildebeests stimulates new plant growth, improving forage for gazelles. See also figure 7.10.

through the area and reduce the aboveground green biomass by more than 80% and plant height by more than 50% (McNaughton 1976). Within several weeks, however, a dense lawn of green leaf forage appears in the heavily grazed areas, providing ideal grazing for Thomson's gazelles. Grazing by the wildebeests stimulates growth of senescent grasses, which provides forage of higher nutrient content and digestibility for the gazelles. This phenomenon has been called grazing facilitation, in which the feeding activity of one herbivore species increases the food supply for a second. No evidence exists that grazing ever increases overall plant production or fitness, however (Belsky 1986).

COMMUNITY ASSEMBLY RULES

Most ecologists now agree that ecological communities are not random collections of species; certain combinations of species seem to occur more frequently than would be expected by chance. Competition, predation, and mutualism all play roles in structuring mammalian communities and may act in concert. Several researchers have developed rules for assembling communities based on the competitive use of resources. One such rule, suggested by Fox (1989), is that species entering a community are drawn from a different group, or guild, until each group is represented, and then the series repeats. Communities lacking a representative from one group but having more than one from another group are less stable and more subject to invasion. Fox evaluated this rule for three groups of marsupials and for rodents in eastern Australia. When tested against a null model of chance assembly, significantly more sites fit the rule. One species from each of three sizes of shrews (small, medium, large) was usually present at a number of sites in western Kentucky and Tennessee, rather than several species from the same size group (Feldhamer et al. 1993). Other communities of shrews in North America seem to follow a similar rule (Fox and Kirkland 1992).

Seed-eating desert rodents of the family Heteromyidae, from North America, and of the subfamily Gerbillinae, from Europe and Asia, show convergence in morphology (Ben-Moshe et al. 2001). Skull and tooth measurements among the species show a nonrandom pattern of evenly dispersed means, indicative of community-wide character displacement. This difference is related to the sizes of seeds that each species prefers. Such convergence of community structure on different continents indicates that general rules govern the structure of ecological communities and guilds.

Findley (1989) asked the somewhat more general question of whether morphological differences between **syntopic** (present at the same time and place) species of rodents are greater than would be the case if the species were drawn randomly from the available pool of colonists. Using skull measurements of the rodents from a number of sites in New Mexico and Sonora, Mexico, he found that morphological differences in size, but not shape, among coexisting species were greater than expected by chance. He concluded that this nonrandom pattern may have resulted from interactions such as competition or from species-specific responses to resources that were differentially distributed.

Other investigators have failed to find evidence that competition for resources structures communities. Five feeding guilds of phyllostomid bats were identified from northeastern Brazil by means of fecal samples from field-caught specimens: foliage-gleaning insectivores, nectarivores, frugivores, sanguinivores, and omnivores (Willig and Moulton 1989). A number of morphological measurements were made from each species to see if community members were assorted by size. The observed species compositions were no different from what would be found if species were selected from the available species pool at random, suggesting that competition is not structuring the bat community. Similarly, in a comparison of bat communities from different regions of Venezuela, it was concluded that although competition is important at the population level, it was not necessarily important at the community level (Willig and Mares 1989). The failure to find a pattern could be due to the spatial heterogeneity of the habitats in tropical South America combined with the great dispersal abilities of bats.

Community Function

ENERGY FLOW AND COMMUNITY METABOLISM

One way to view the interconnectedness of communities is to trace the flow of energy through different trophic levels. Although often presented as linear food chains for simplicity, most communities are really interconnected in a weblike fashion. **Energy,** defined as the ability to do work, enters the community as electromagnetic light energy from the sun. Converted by photosynthetic plants into chemical energy (in the form of sugars), it is then available to animals. Under natural conditions, green plants convert less than 1% of the light energy available to them. In an old-field community in Michigan studied by Golley (1960), meadow voles (*Microtus pennsylvanicus*) consumed about 2% of the plant material available to them, and least weasels (*Mustela nivalis*) ate about 31% of the available voles (figure 26.10). In this linear food chain, so little energy was converted into weasels that a higher carnivore dependent on weasels for food could not be supported. Of course, many other organisms and energy pathways existed within this community, and most of the plants were eaten by herbivorous insects. A more complex food web (figure 26.11) depicts trophic interaction in the alpine tundra in the central Rocky Mountains. Actually

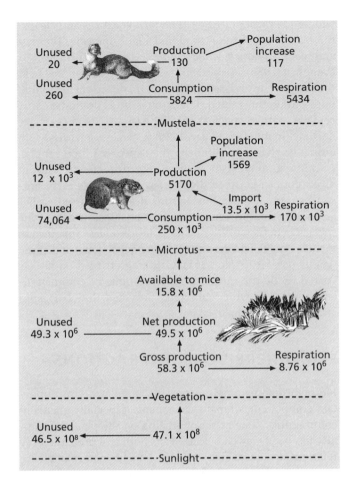

Figure 26.10 Energy flow diagram of a portion of a food web in an old-field community in southern Michigan. Numbers are in Calories per hectare per year. *Data from F. B. Golley, 1960, Energy dynamics of a food chain of an old-field community, in* Ecological Monographs, *30:187–206.*

measuring the flow of energy in such a system would be extremely difficult. Food webs are traditionally thought of as being relatively static, relying on the persistence of species making up the community. More recent notions consider food webs as dynamic and flexible, accommodating a certain amount of change in species composition (de Ruiter et al. 2005).

Pathways of energy flow through a desert scrub ecosystem were documented by Chew and Chew (1970). Thirteen species of small and medium-sized mammals were present. Small mammals, both herbivores and granivores, played a small role in terms of energy, converting only 0.016% of the primary, aboveground production to mammal tissue that was then available to carnivores.

The production efficiency of small mammals was measured for populations in nine ecosystem types in Europe and North America (Grodzinski and French 1983). Productivity refers to the addition of new tissue in the form of growth of individuals plus new individuals from reproduction. Production was divided by respiration to give production efficiency. Efficiency was lowest for shrews (0.7%) and highest for herbivores (3.4%).

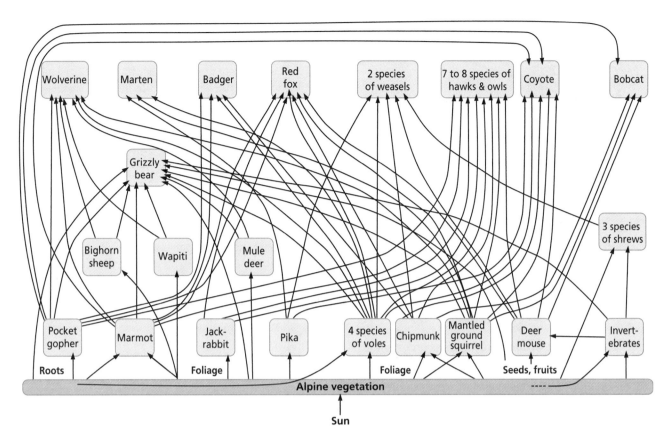

Figure 26.11 Food web in the alpine tundra community of the Beartooth Plateau. Insectivorous and herbivorous birds are not included. *Data from R. S. Hoffmann, 1974, Terrestrial vertebrates, in* Arctic and Alpine Environments *(J. D. Ives and R. G. Barry, eds.), Harper & Row.*

Efficiencies were low for all these small mammals compared with poikilotherms because of the high respiratory cost of homeothermy in small mammals.

Although mammals typically do not account for a large proportion of energy flow in communities, their biomass may be quite high, resulting in a large "standing crop." Furthermore, the total effect of mammals on vegetation may be much greater than the amount assimilated. Common voles (*Microtus arvalis*) in agricultural fields in Poland destroyed as much as 13 times more vegetation than they actually used for energy requirements (Grodzinski et al. 1977). Mammals affect the vegetation in many ways such as by cutting, trampling, burrowing, and nesting in it. Finally, because of their large size, long life, and high activity, some species of mammals play dominant roles in the community by influencing vegetation and other animals, as we have described with elephants, sea otters, and other keystone predators.

COMMUNITY DEVELOPMENT

Ecological succession is the replacement of species in a habitat through a regular progression of stages leading ultimately to a stable state, the **climax** community. Species occupying early successional communities tend to be *r*-selected, with high dispersal rates, rapid growth rates, and high reproductive rates (chapter 24). Those in later stages are more likely to be *K*-selected, with lower dispersal rates, delayed maturation, and lower reproductive rates.

Old-field succession, in which fields are monitored for varying numbers of years since abandonment from agriculture, is often studied. The diversity of species typically increases during succession, although it may decline somewhat at the climax. In one study in Minnesota, a census of small mammals was taken for 18 fields that ranged in age from 2 to 57 years since abandonment from agriculture (Huntly and Inouye 1987). The youngest fields were dominated by short-lived, introduced Eurasian plant species. In middle-aged and older fields, native species of prairie grasses dominated. Woody shrubs were found only in fields older than 50 years and were never common. Of the six species of small mammals trapped, white-footed mice were caught in fields of all ages, whereas meadow voles and masked shrews (*Sorex cinereus*) tended to be caught in older fields. Generally, the number of small mammal species increased with the age of the field; this increase was associated with a striking increase in plant nitrogen, a measure of primary productivity (figure 26.12). Thus, mammal communities change as the vegetation changes.

Researchers assume that mammals respond to changes in vegetation, but mammals may affect the vegetation and thus the course of succession. Voles can extensively influence

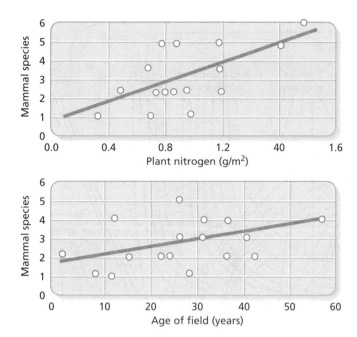

Figure 26.12 Effect of secondary succession on small mammal communities. Total number of species of small mammals in fields as a function of field age (years since abandonment from agriculture) and nitrogen content of vegetation (g/m²). *Data from N. Huntly and R. S. Inouye, 1987, Small mammal populations of an old-field chronosequence,* Journal of Mammalogy *60:739–745.*

vegetation, especially during peak years (Batzli and Pitelka 1970; Grodzinski et al. 1977). In the Serengeti, grazing by large herds of ungulates has a profound effect on the vegetation by increasing energy and nutrient flow rates (McNaughton 1985). McNaughton argued that mammals and plants both have coevolved traits, resulting in interdependence rather than a relationship in which animals gain and plants lose.

Community Patterns

ISLAND BIOGEOGRAPHY

It has often been observed that island habitats have fewer species than do comparable mainland sites. MacArthur and Wilson (1967) developed a model that predicts a dynamic equilibrium of the number of species on islands. Although the species identity changes through time, the total number remains constant as new species colonize and resident species become extirpated. Immigration rates by new species are affected by island size; smaller islands are smaller targets and therefore have lower colonization rates. Extinction rates are higher on small islands, in part because population sizes are smaller. Also important is the distance of the island from the colonizing pool of species

on the mainland; islands farther away have lower immigration rates (figure 26.13). Thus, small islands equilibrate at fewer species than large islands, and distant islands equilibrate at fewer species than near islands.

It makes sense that the number of species should increase as the area being sampled increases. Larger areas typically have more habitats, which reduces the chances that an individual species will become extirpated. The relationship, called the **species-area relationship,** can be described by the equation

$$S = cA^z \quad (2)$$

where S = number of species, c = a constant measuring the number of species per unit area, A = area being sampled, and z = a constant measuring the slope of the line relating S and A. The exponent z is typically in the range of 0.2–0.3 for mammals (Mártin and Goldenfeld 2006).

The application of island biogeography theory to mammalian communities has met with mixed results. Compared with birds and insects, mammals are relatively poor dispersers across water. For instance, Lawlor (1986)

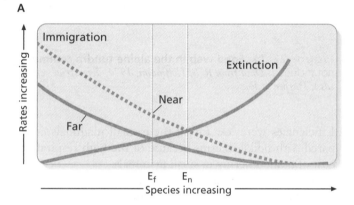

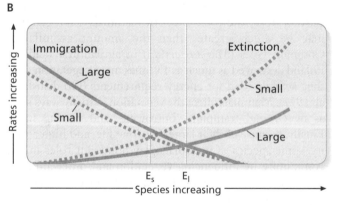

Figure 26.13 Equilibrium numbers of species on islands as a function of immigration and extinction rates. In (A), an island far from a colonizing source or mainland should equilibrate with fewer species, E_f, than an otherwise identical near island, E_n. In (B), a large island equilibrates with more species, E_l, than a small one, E_s, at the same distance from the mainland. Equilibrium numbers are shown where immigration and extinction curves intersect. *Data from Robert H. MacArthur and Edward O. Wilson, 1967,* The Theory of Island Biogeography, *Princeton Univ. Press.*

compared isolated, oceanic islands with so-called land-bridge islands, those that were connected to the mainland at the end of the Ice Age. The species-area curve for oceanic islands is nearly flat (z is low relative to landbridge and mainland areas). Oceanic islands have fewer species than predicted from equilibrium theory, probably because colonization rates are so low for mammals.

Island size and distance from the mainland were evaluated in the Thousand Islands region of the St. Lawrence River in New York (Lomolino 1986). For species such as the red fox and raccoon (*Procyon lotor*) to be present, islands had to be above a critical size. For other species, such as deer mice, both island size and distance from the mainland were important factors in determining presence or absence of a species (figure 26.14).

The Great Basin of North America consists of a vast "sea" of sagebrush desert interspersed at irregular intervals by isolated mountain ranges (islands). The upper slopes of these mountains are well vegetated, with cool, mesic (moist) conditions. The mammal species on these mountaintops (above approximately 2500 m) are derived from the boreal faunas of both the Sierra Nevada Mountains to the west and the Rocky Mountains to the east. As can be seen from figure 26.15, more species were found as the area of the mountaintop "island" increased, showing a linear relationship on log scale and approaching the number found in the saturated "mainland" areas (Brown 1971). The "islands" close to the "mainland," however, do not tend to have more species than those farther away, and the rate of colonization by new species was

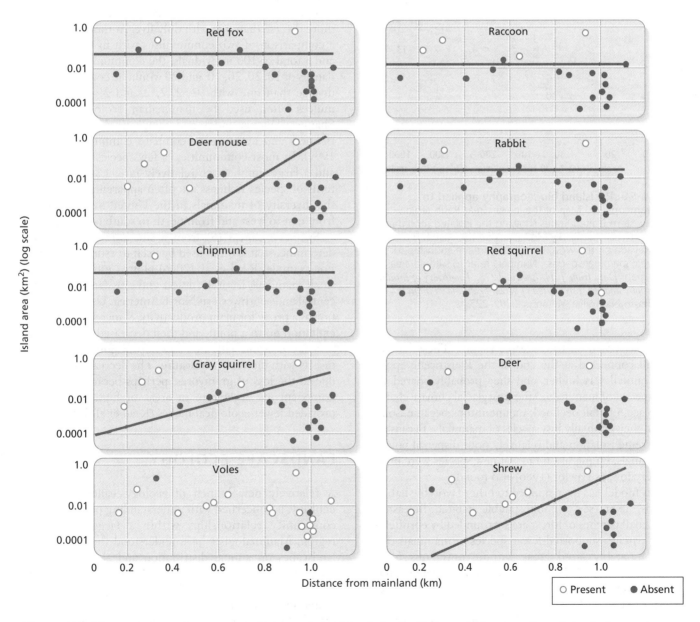

Figure 26.14 Occurrence of 10 species of mammals on islands in the Thousand Islands region of the St. Lawrence River in New York as a function of island size and isolation. Many species occur only on islands above a certain critical size (*horizontal lines*). Other species, such as the deer mouse, are affected by both island size and distance from the mainland. *Data from M. V. Lomolino, 1986, Mammalian community structure in islands,* Biological Journal of the Linnean Society 28:1–21.

A

B

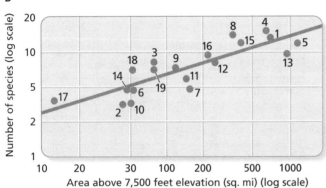

Figure 26.15 Island biogeography applied to mountaintops. (A) Map of the lower Great Basin region of the western United States showing the isolated mountain ranges between the Rocky Mountains on the east and the Sierra Nevada on the west. (B) Species-area relationship for the boreal mammal species. Numbers refer to sample areas on the map. *Data from J. H. Brown, 1978, The theory of insular biogeography and the distribution of boreal birds and mammals,* Great Basin Naturalist Memoirs 2:209–227.

much lower than on true oceanic islands. These habitats were all connected at the end of the Pleistocene epoch, when rainfall was higher, and they probably shared the same species of mammals. After postglacial climate change took place, rainfall declined, mountaintops became isolated, and they gradually lost species of mammals. Because of the extreme isolation and relatively poor dispersal powers of mammals, immigration rates of new species were probably very low, as Lawlor (1986) also found.

This model has been applied to other favorable habitat "islands" in a "sea" of inhospitable habitat. In eastern Iowa, small patches of forest are surrounded by cornfields. These patches supported fewer species of mammals than did comparably sized areas within contiguous forest (Gottfried 1979; 1982). Smaller and more isolated patches had fewer species than larger and less isolated patches, as expected.

The species area relationship shown in equation [2] depends mainly on the statistical properties of distances between conspecific individuals; it best describes communities where the individuals within each species cluster together, and where there are many rare species but only a few common species, conditions that are common to most ecological communities. The relationship does not directly involve processes such as competition, immigration, or the effects of landscape variability (Mártin and Goldenfeld 2006).

SPECIES DIVERSITY

Because of the ever-increasing encroachment of humans into natural communities throughout the world, there is much interest today in **biodiversity**. Biodiversity has two main components: **species richness,** which is simply the number of species in an area, and **evenness,** the relative abundance of individuals within each species. Thus, a community that is made up of ten species of mammals is more diverse than one with only five. At the same time, in a comparison of two communities with five species each and a total of 100 individuals, the community with abundances of 20, 20, 20, 20, and 20 would be considered more diverse than one with 92, 2, 2, 2, and 2. Two diversity indices often used by mammalian ecologists are the Shannon-Wiener index and Simpson's index. Both use richness and evenness to compare communities (Krebs 1999). In most communities, a few species are abundant and a large number are relatively rare. Large-scale patterns of species richness are often apparent; for example, the diversity of mammals in the United States increases from east to west and from north to south (see chapter 5).

Species richness (S) from a site is relatively easy to determine, so it is often used by itself to estimate diversity. For example, Read et al. (2006) plotted estimates of S of small mammals from 43 desert and grassland sites across central and southwestern North America. Using precipitation as a proxy for plant productivity, S increased with precipitation up to a point, and then decreased. The increase in S is consistent with increases in plant forage, seeds, and insects with increasing moisture. The decrease was mostly due to the loss of granivores, perhaps because the dense, homogeneous vegetation in areas of high precipitation provided fewer ecological niches (Read et al. 2006).

LANDSCAPE ECOLOGY

A relatively new branch of ecology, called **landscape ecology,** is concerned with understanding population and community relationships within a large geographic region. Mammalogists have understood the influence of landscapes on animal distribution and abundance since the pioneering work of Merriam (1894), but only recently have methods been developed to quantify the effects of spatial scale on community structure. An important application of this approach is in the design of nature reserves, discussed more fully in chapter 29. Problems such as movements of individuals through habitat patches differing in size and shape, responses to habitat fragmentation, design of connecting corridors, and patterns of dispersal across habitat patches are modeled. This discipline explores the behavior of individuals, populations, and communities as they respond to different features of the environment at different levels of spatial scale (Forman and Godron 1986; Merriam and Lanoue 1990). New techniques include computer simulations based on fractal geometry, satellite imagery, and computerized geographic information systems (GIS).

Landscape ecology has great potential for helping researchers predict the effects of habitat fragmentation, one of the major causes of species extinctions. Classical island biogeography theory focuses on species richness as well as patterns of colonization and extinction as a function of habitat patch size and isolation, and it assumes that the surrounding matrix is homogeneous (e.g., water). For these and other reasons, it has met with mixed results when applied to terrestrial systems, as noted earlier in this chapter. Landscape ecology, on the other hand, incorporates information about how landscape patterns of many types influence reproduction and dispersal of local populations. Such studies can be done on natural landscapes or on experimental plots where the arrangement of habitat elements can be manipulated (Diffendorfer et al. 1995).

At the population level, it is important to understand how mammals living in isolated habitat fragments move through less suitable areas to get to more suitable ones (Merriam 1995). Studying how animals move about their home ranges and territories (chapters 21 and 24) is not sufficient; ecologists must also study the spatial structure of the landscape mosaic and how dispersing animals move through it. Local interbreeding populations, called **demes,** can be connected to other demes via dispersal, forming **metapopulations** (Hanski 1996). It therefore becomes necessary to define the demographic unit, or population, under study. Researchers know that immigration and emigration are important population forces (chapter 24), but they often have little idea of what the boundaries of a population actually are. For example, in a long-term study of white-footed mice in a small, isolated 2-hectare woodlot in Ohio, it had been assumed that the population was essentially defined by the boundaries of the woodlot. Livetrapping mammals in the surrounding agricultural fields, however, revealed that mice were making extensive use of cropland at certain times of the year (Cummings and Vessey 1994). In eastern Ontario, near the northern edge of its range, this forest-dwelling species has adapted to living in cornfields year-round, where densities were similar to those in wooded areas (Wegner and Merriam 1990). Densities were quite low compared with other studies, and the spatial scale over which the mice moved was unusually large.

Landscape ecology can also be viewed at the community level and above, where patches of the same community types are studied in a spatially explicit way (Lidicker 1995). Properties of landscapes that can be studied at this level include spatial configuration (the dispersion of patches), edge-to-area ratios (sizes and shapes of patches),

and connectedness (links among patches of the same community type).

MACROECOLOGY

At a still larger scale is the study of **macroecology,** which explores patterns of body mass, population density, and geographic range at the scale of whole continents (Brown and Maurer 1989; Brown 1995). It involves the nonexperimental (i.e., nonmanipulative) investigation of relationships among populations, including patterns of abundance, body size, metabolic rates, geographic distribution, and diversity. Such large-scale studies require massive amounts of data on distributions, densities, and body sizes of representative species in each community before meaningful statistical patterns can be seen. Most of these data have already been collected for mammals, however, at least in temperate regions. The goal of macroecology is to identify and explain emergent statistical patterns in terms of general processes that can then be applied to unstudied areas. One statistical pattern already discussed is the species-area curve (see equation [2]), which shows how the number of species increases with the area sampled (e.g., see figure 26.15). The species-area relationship can be studied at all spatial scales, from the smallest microhabitats to entire continents, and it has provided much insight into how communities are assembled.

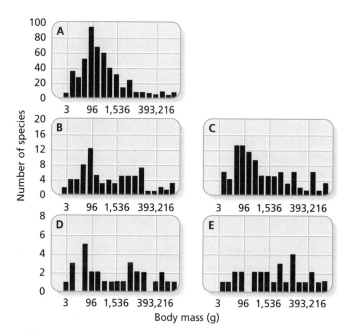

Figure 26.16 Frequency distributions of body masses (log scale) among species of North American land mammals. Distribution for the entire continent, for land mammals within (A) biomes, (B) northern deciduous forest, and (C) desert; and for land mammals within small patches of relatively uniform habitat within each of these biomes—(D) Powdermill Reserve and (E) Rio Grande Bosque. *Data from J. H. Brown and B. A. Maurer, 1989, Macroecology: The division of food and space among species on continents,* Science, *243:1145–1150.*

Other patterns involve attributes of individual organisms. When the number of species of terrestrial mammals in North America is plotted against average body mass, one finds that there are many more small species than large ones, and a strong peak in mass occurs between 50 and 100 g (Brown and Maurer 1989; figure 26.16A). When smaller spatial scales are used, however, comparing deciduous forest, desert, or still smaller patches of relatively uniform habitat, the distributions become flatter, giving approximately equal numbers of species in each size category (see figure 26.16B–E). Why are there so many small species, and why does the pattern shift as the scale changes? One possible explanation for this shift is interspecific competition, which reduces the number of similarly sized species in the same habitat. Large species could be relatively rare at the continental scale due to energy constraints; smaller populations are more likely to go extinct. Finally, Brown and Maurer (1989) hypothesize that a greater number of small species exist at the continental scale because they have smaller geographic ranges and replace one another more frequently across the landscape than do large species. These small species are specialized because they need high-quality food to maintain their relatively high metabolic rates.

SUMMARY

Biological communities consist of interacting populations of organisms in a prescribed area. Communities can be studied at the simplest level by looking at interactions between two species and then expanding the analysis to include more complex interactions and large-scale patterns of distribution and abundance.

Central to the study of any species is the idea of ecological niche, an organism's habitat and functional role in the community. Interspecific competition affects both the distribution and abundance of species. The competitive exclusion principle states that complete competitors cannot coexist indefinitely.

Evidence that interspecific competition is an important force in structuring mammalian communities has been difficult to obtain. Distributions of species with and without presumed competitors sometimes show competitive release. The divergence of traits sometimes seen in areas where close competitors coexist, called character displacement, is taken as evidence of competition. The best evidence of competition comes from removal experiments, in which species are experimentally removed and the response of the presumed competitor is monitored. Such experiments have produced mixed results.

Predators also may be important forces in structuring communities. Keystone predators are those that control the distribution and abundance of many other members of the community, often by limiting a particular species of prey. Trophic cascades exist when predators reduce herbivore densities to the point that plant growth increases. The existence of keystone predators and trophic cascades is often discovered by removal experiments. Guilds are groups of species exploiting a common resource base in a similar fashion. Guilds may also act as keystone species complexes. Risk of predation may also act to structure communities; prey species may modify their foraging behavior and habitat choice to reduce predation risk.

Mutualistic relationships, in which both species benefit, are common but have not been well studied in mammals. Interactions among grazing ungulates and forage plants may be mutualistic in some systems because some species of ungulates make more edible forage available for others.

Several attempts have been made to devise rules for assembling communities and to test the general hypothesis that communities are not simply random collections from the pool of available species. In some cases, such as small mammals in Australia, shrews in North America, and rodents in the deserts of North America, the composition is nonrandom, suggesting that competitive interactions have resulted in the inclusion of only those species sufficiently different from those already present to allow coexistence. Studies of other groups, however, especially Neotropical bats, fail to show anything but a random assortment of species based on size and shape.

Several studies have documented the flow of energy through the mammalian component of ecosystems. Herbivorous and granivorous mammals consume relatively little of the energy available to them and thus provide little to higher trophic levels. Some species of mammals play important roles by controlling other components of the ecosystem.

Mammals show changes in both species composition and abundance in response to ecological succession. In old-field succession, the number of species of small mammals tends to increase, possibly in response to higher quality food. Grazing ungulates influence succession by their effects on forage plants.

Studies of islands, varying in size and isolation from mainland habitats, have shown that communities consist of species in dynamic equilibria that result as new species colonize and resident species go extinct. Although some studies of mammals support the idea, habitat islands that are highly isolated have fewer species than predicted, perhaps because mammals are relatively poor dispersers.

Landscape ecology is the study of the distribution of individuals, populations, and communities across different levels of spatial scale. This approach holds much promise for a better understanding of how components of communities respond to fragmented habitats and for the design of nature reserves. On a still larger scale, macroecology explores patterns of body mass, population density, and geographic range at a continental scale.

SUGGESTED READINGS

Brown, J. H. 1995. Macroecology. Univ. Chicago Press, Chicago.

Lidicker, W. Z., Jr. 1995. Landscape approaches in mammalian ecology and conservation. Univ. Minnesota Press, Minneapolis.

Morris, D. W., Z. Abramsky, B. J. Fox, and M. R. Willig (eds.). 1989. Patterns in the structure of mammalian communities. Texas Tech Univ. Press, Lubbock.

Ricklefs, R. E., and G. L. Miller. 1999. Ecology, 4th ed. W. H. Freeman, New York.

DISCUSSION QUESTIONS

1. Character displacement is usually considered to be a result of competition among closely related species in sympatry. For what other reasons might species show character displacement?

2. Which species in figure 26.1B is most likely to be eliminated, and why?

3. Suppose that you conduct a removal experiment to test for competition between two sympatric species of mammals. What dependent measures would you use to demonstrate that competition is or is not occurring at your site? If you find no effects of removal of one species on the other, what conclusions can you draw?

4. Which of the following paragraphs best describes mammalian communities? Muster as much support as possible for your position. You may wish to consult an ecology textbook, such as Krebs (2001), Ricklefs and Miller (1999), or Begon et al. (2005).

   a. The distribution of individual species in space and time suggests that each [species] responds to its own unique set of requirements independently of the effects on other species. Community boundaries are best understood as being arbitrarily defined units, more for the convenience of the investigator, than as highly integrated levels of organization.

   b. Communities include closely integrated species with complementary functional roles. Predictable patterns of species distribution and guild composition reflect the close coevolution of species in response to interspecific competition.

PART 5

Special Topics

The final three chapters of this book, which make up part 5, cover topics of increasing importance to mammalogists. In chapter 27, we discuss general characteristics of parasites and their effects on mammalian hosts. We also investigate host–parasite coevolution and specificity. Finally, we consider diseases that occur in mammals, including those with a long history of human impact such as plague and rabies, and several of more recent interest, such as Lyme disease, hantaviruses, and severe acute respiratory syndrome (SARS) which are transmitted from nonhuman mammals to people.

The second special topic, discussed in chapter 28, is the domestication of mammals. The biological and cultural processes of mammalian domestication that began 15,000 years ago continue today. Domestication has had a significant effect on human history. Nonetheless, relatively few species of mammals have been domesticated, and we will explore the reasons why.

The final chapter of the text deals with one of the most important aspects of current mammalogy—conservation. In chapter 29, we discuss various factors and their interactions that contribute to the decline of mammalian populations throughout many parts of the world, as well as potential solutions. A better understanding of these issues and how they affect mammalian fauna will ultimately lead to better and more effective conservation decisions.

CHAPTER 27

Parasites and Diseases

Parasitism is a form of interaction, or **symbiosis** ("living together"), between two species; one species (the parasite) benefits at the expense of the second species (the host). Parasites are similar to predators in that they both benefit at the expense of another species. Unlike most parasites, however, predators directly remove individuals from the prey population. Parasites usually do not kill their hosts, although there are exceptions, and effects on prey populations often may be subtle and difficult to determine.

Parasites include a vast array of diverse life-forms; there are more species of parasites in the world than nonparasites. Nonetheless, parasites share several general characteristics: they are smaller than their host, usually are physiologically dependent on the host, spend either their entire life (permanent parasites) or part of it (temporary parasites) in or on the host, and derive essential nutrients from the host. Parasites can be categorized into two general groups: microparasites and macroparasites. **Microparasites** include viruses, bacteria, and fungi—disease agents often not thought of as parasites—and protistans. They usually are microscopic and have rapid regeneration rates within their hosts. **Macroparasites** include the flatworms or platyhelminths (tapeworms and flukes); nematodes (roundworms); acanthocephalans (thorny-headed worms); and arthropods (ticks, fleas, lice, flies, and mites). They are larger than microparasites and have longer generation times; generally, most do not reproduce entirely within the host.

Parasites can be categorized in other ways as well. They are either **obligate,** meaning that they must spend at least part of their life cycle as a parasite, or **facultative,** meaning that they are organisms that are not normally parasitic but become so. **Endoparasites** occur within the body of the host, whereas **ectoparasites** occur either on or embedded in the host's body surface. Janzen (1985) noted the following generalizations about animal ectoparasites, several of which also apply to endoparasites:

1. They are subject to almost no predation while on the host.
2. The chemical content of their food is relatively constant among mammalian hosts.
3. Close contact is common, and the host can chemically identify the parasite.

4. Hosts play a major role in the interhost movement of ectoparasites.
5. There are often large numbers of parasites on a host.
6. A large percentage of the parasite's life cycle is spent on or near the host.

In this chapter, we will discuss the effects of parasites on mammalian host populations, host–parasite coevolution, and the different types of parasites. We also will focus on mammalian parasites and the associated human **diseases** (clinical conditions that can be observed or measured) they may impart. More specifically, we will consider several **zoonoses** (from the Greek words *zoo*, meaning "animals," and *noses* meaning "diseases"), which for our purposes are defined as diseases transmitted from nonhuman mammals to people. Mammals sometimes are a primary reservoir, or source, of the infective organism of human zoonoses (Marcus 1992). Reservoirs may enable a disease to persist in an area at moderate to low levels, which is termed an **enzootic** phase. When various factors amplify the occurrence and distribution of an enzootic disease, it becomes **epizootic.** Arthropod parasites often serve as **vectors** (carriers) of viruses, bacteria, or other microparasites to humans and other mammalian species; these microparasites are the **etiological,** or causal, agents of the disease. We will concentrate on a few common or historically important zoonoses as well as on some relatively recent diseases with which mammalogists should be familiar. Understanding mammalian parasites and diseases is important because of the risks they present in mammalian studies and the roles they play in the evolution and life histories of species (Childs 1995).

Parasite Collection

Mammalian ectoparasites can be collected from live or dead specimens in the field or laboratory. Snap-trapped small mammals can be placed in individual plastic bags that are then sealed to keep parasites with the host. Animals can be examined under a dissecting microscope for ectoparasites on or embedded in the skin or clinging to the fur, such as mites. The eyes, ears, lips, and genital area should be searched as well. Parasites can be removed with a needle. Another technique for removing ectoparasites from small mammals is to immerse them in water containing detergent. Whatever removal method is used, ectoparasites can be preserved in 70% alcohol. Certain ectoparasites, such as fleas and larger mites, may leave dead animals as the animals cool. This is less of a problem with live-trapped mammals.

Collection of endoparasites is through necropsy in a laboratory. Depending on the objectives of the study, all visceral organs, as well as the brain, should be examined. Stomach and intestinal contents can be emptied into a tray, or these organs can be slit open and the walls examined for embedded individuals. Parasitology laboratory manuals, for example Dailey (1996), should be consulted for details of specific techniques on examination, recovery, preparation, and identification of both micro- and macroparasites. Whitaker (1968) summarized methods of parasite collection and preservation for *Peromyscus* that are applicable to other small species. Keys to identification of parasite groups are available; however, identification to family level or below often is difficult and should be done by a specialist familiar with the group.

Effects of Parasites on Host Populations

DIRECT MORTALITY EFFECTS

Parasites and diseases occur in all mammalian species. Although parasites usually do not kill their **definitive hosts** (where the parasite reaches sexual maturity), they may or may not have an obvious adverse effect on the host. Yuill (1987) suggested three circumstances, however, in which parasitism can be a mortality factor (table 27.1). An intermediate host may also die if its death enhances transmission of the parasite to a definitive host. Also, mortality can occur when parasites associate with **accidental hosts,** that is, those in which they normally are not found. For example, rinderpest, a viral disease normally found in livestock, is highly pathogenic in wild African ungulates and has caused several epizootic outbreaks (Barrett and Rossiter 1999; Kitching 2000). Likewise, *Baylisascaris procyonis*, a nematode normally

Table 27.1 Circumstances in which parasites may act as mortality factors in mammalian host populations

Circumstance	Example
1. When death of the host facilitates transmission of the pathogen	Rabies (see text)
2. A "generalist" pathogen establishes transmission cycles involving many host species	The screwworm fly (*Cochliomyia hominivorax*) in deer
3. The pathogen "wanders" over wide enough geographic area and a long enough time period such that infected local host populations are not seriously threatened	The yellow fever virus in howler monkeys (Genus *Alouatta*) and marmosets (Genus *Callithrix*)

From T. M. Yuill, 1987, Diseases as components of mammalian ecosystems: mayhem and subtlety, Canadian Journal of Zoology 65:1061–1066.

found in raccoons (*Procyon lotor*), is pathogenic in woodrats (Genus *Neotoma*). Current population declines of woodrats in the northeastern United States may be related, at least in part, to *Baylisascaris* infection. White-tailed deer (*Odocoileus virginianus*) are the definitive host for meningeal worm (*Parelaphostrongylus tenuis*), another nematode. Although it has little effect on whitetails, in other cervids, especially moose (*Alces alces*), meningeal worm (figure 27.1) can cause fatal neurological disease. The parasite is a serious management consideration in translocating white-tailed deer outside their native range (Comer et al. 1991; Samuel et al. 1992).

BIOLOGICAL CONTROL

When a mammalian population reaches such a high density that it is considered a pest, an unusual pathogen can purposely be introduced as a **biological control** (using predators or parasites to reduce host population density). As noted by Roberts and Janovy (1996), successful biological control agents have several characteristics:

1. High host-searching capacity.
2. A limited but wide enough range of hosts to maintain the parasite population.
3. A life cycle that is either shorter than or synchronized with the life cycle of the pest population.
4. Ability to survive in all habitats occupied by the host.
5. Easy replication so that large numbers are available to introduce.
6. Ability for rapid control of the pest population.

Biological control in mammals is rare and usually produces additional unforeseen problems (Spratt 1990). The European wild rabbit (*Oryctolagus cuniculus*) in Australia is one of the few examples of "successful" control. Introduced into Australia in the 1840s, wild rabbit populations increased dramatically. They soon devastated native flora and became a serious agricultural pest as well. In the 1950s, myxomatosis virus was introduced to control rabbit populations. Initially, control was successful as over 99% of infected rabbits died. Resistance to the virus developed in certain rabbits, however, and these individuals became the primary breeding nucleus (Yuill 1987). Also, attenuated, or nonlethal, strains of the virus developed, and eventually, rabbit populations once again reached pest proportions. Another attempt at control was made more recently by introducing a different pathogen—rabbit hemorrhagic disease (RHD) virus. In some regions of Australia, rabbit populations have declined by 95%. It is hoped that rabbit control can be sustained with RHD virus in conjunction with other control methods, including fumigation, hunting, and destruction of warrens (Drollette 1997; Cooke 2002).

NONMORTALITY EFFECTS

Effects of parasitism on individual definitive hosts and mammalian populations may be subtle and inconspicuous. The impact a parasite has is a function of its numbers; the sex, age, and overall condition of the host; the season of the year; and other variables. Parasitism can increase energy costs and decrease energy gained in hosts (Yuill 1987). This may result in decreased movement or reproduction (Smith et al. 1993), reduced host growth, altered behavior, or reduced survival of offspring (Munger and Karasov 1994). All these factors can have negative consequences on host population dynamics, although Ostfeld et al. (1996) found that heavy tick infestation had no effect on the fitness of white-footed mice hosts. Many factors have received little attention by mammalogists, however, in part because they are very difficult to document in field studies. Grenfell and Gulland (1995) summarized observational, quantitative, and experimental studies done on the effects of micro- and macroparasites on mammalian reproduction and survival.

Coevolution of Parasites and Mammalian Hosts

Throughout the nearly 250-million-year history of mammals, micro- and macroparasites have been adapting to mammals, while mammals in turn have been adapting as hosts. Host–parasite **coevolution** was defined by Kim (1985a:670) as "reciprocal evolutionary change in interacting species in a parasite community involving both the parasite species versus the host and the parasite species versus other parasites." Relationships (congruence) between host and parasite phylogenies may be difficult to assess, however, and may represent "coaccommodation" rather than coevolution (Brooks and McLennan 1991; Page 1993). Nonetheless, many mammal–parasite assem-

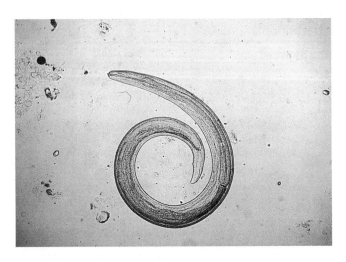

Figure 27.1 Meningeal worm. This nematode is a serious mortality factor in moose and other cervids. The primary host, white-tailed deer, is not affected by the parasite.

blages are believed to have coevolved (Waage 1979; Price 1980; Futuyma and Slatkin 1983; Dittmar et al. 2006), with the result that many parasites are closely associated with a particular group of mammals; that is, they show pronounced specificity.

PARASITE SPECIFICITY

Most species of parasite probably descended from free-living ancestors. Through evolutionary time, they developed a close, often obligatory, association with hosts such that generalist parasites are relatively rare. As noted by Adamson and Caira (1994:S85), "Parasites are specialists, not just of hosts, but of microhabitat or tissue site . . . Parasite specificity is not fundamentally different from specificity, or niche restriction, in free-living organisms, except that the niche of parasites is part of another organism."

Evolution can shape parasite specificity in three interacting ways (Adamson and Caira 1994). Initially, specificity is a vestige of the microhabitat and feeding mode of the parasite's free-living ancestors and the manner in which the parasite–host relationship arose. A second factor is host ecology, especially important in passively transmitted parasites that have little effect on the physiological or immune systems of their host. These types of parasites (e.g., pinworm nematodes or trichomonadid protists) occur in hosts that are ecologically rather than phylogenetically similar. The third factor in host–parasite specificity can involve a number of coevolutionary adaptations, including

1. Synchrony with host phenology to maximize the transmission cycle of the parasite.
2. Hatching or excystment cues, as in mammalian schistosomes.
3. Migration cues, as occur in digeneans (trematode flatworms).
4. Specific cell surface receptors, which are important in a variety of protists, including *Giardia* (Lev et al. 1986), *Leishmania* (Alexander and Russell 1992), and *Plasmodium* (Braun Breton and Pereira da Silva 1993).
5. Ability to evade host immunological responses.
6. Competitive exclusion, which reduces realized niches such that different parasites do not occur in the same host.

Adaptations of parasites include the means to find an individual host, attach to it, reproduce, and subsequently disperse either eggs, juveniles, or adults. Additional physiological and morphological adaptations involve the structure of mouth parts and the apparatus for feeding and digestion (Kim 1985b). Such a high degree of host specificity may evolve that a parasite may become species-specific, that is, not be able to survive on taxa other than its host.

A prime example of parasite–host specificity is the association of parasitic mites of the Family Myobiidae with their mammal hosts (Fain 1994). Two subfamilies of myobiids occur only on New World marsupials (chapter 11). The Subfamily Archemyobiinae occurs only on opossums (Family Didelphidae), with a single species of mite found on the monito del monte (Family Microbiotheriidae: *Dromiciops australis*), whereas the Subfamily Xenomyobiinae parasitizes only shrew opossums (Family Caenolestidae). A third myobiid subfamily, the Myobiinae, is divided into two tribes. The tribe Australomyobiini occurs only on Australian marsupials. More specifically, the Genus *Australomyobia* is found on marsupial "mice" (Family Dasyuridae), and the Genus *Acrobatobia* lives only on pygmy possums (Family Burramyidae). The tribe Myobiini includes all the genera and species of mites that parasitize eutherian mammals. For example, mammals within the former Order Insectivora (the current Orders Afrosoricidae, Erinaceomorpha, and Soricomorpha—see chapter 12) host 16 genera and 67 species of myobiid mites that are highly species-specific (Fain 1994). Likewise, the 21 genera and 230 species of myobiid mites that occur on bats are highly specific. Each genus of mite is generally restricted to one family or subfamily of bat, with many species of mite specialized for a single host genus. Numerous rodent species also are parasitized by myobiid mites (figure 27.2). Again, the "concordance between the radiations of the Myobiidae and that of their rodent hosts is remarkable" (Fain 1994:1281). In North American chipmunks (Genus *Tamias*), there also is a high degree of concordance between host species and ectoparasites (Jameson 1999). Coevolution also has been shown in pocket gophers (Order Rodentia: Geomyidae) and their associated chewing lice (Reed and Hafner 1997). A direct relationship exists between body size of host gophers and their parasitic lice (Morand et al. 2000). This relationship results from the close association between the diameter of the guard hairs in gophers and the size of the rostral groove by which lice attach to those hairs (Reed et al. 2000). Studies of these types of relationships allow comparison of the rates of speciation and evolution in host and parasite.

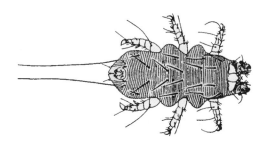

Figure 27.2 Host–parasite specificity. The myobiid mite *Myobia musculinus* is a very common fur mite of house mice (*Mus musculus*). *Adapted from A. C. Oudemans, 1912, Acarologische Aanteekeningen, Gravenhage Ber. Ned. Ent. Ver. 3.*

Mammalian Parasites and Diseases

Numerous species of parasites and diseases are associated with mammals. We will note examples of mammalian parasites from each major group, including parasites and diseases that affect humans. Space does not permit a discussion of most zoonoses; for more detailed information on zoonoses and other mammal-related diseases, see Acha and Szyfres (1989), Gorbach et al. (1998), and the *CRC Handbook Series on Zoonoses* (Beran 1994).

PROTISTANS

Since the 1980s, the classification of these single-celled eukaryotes has been fairly chaotic. Up to 3 kingdoms and 19–45 phyla have been proposed (Marquardt et al. 2000). Locomotion is by flagella, cilia, or pseudopodia (figure 27.3), and they reproduce asexually, sexually, or both. Thousands of species are free-living. Many other parasitic species cause no adverse effects for their hosts. For example, trypanosomes are widely distributed protistan parasites that usually are benign in mammals. A number of protistan species are pathogenic, however. For example, trypanosomiasis (*Trypanosoma evansi*) occurs in many mammals, including elephants, dogs, horses, camels, and deer, and is fatal within several weeks if untreated. Nagana, caused by *Trypanosoma brucei brucei*, infects wild African ruminants but apparently causes no disease. It is pathogenic, however, in a variety of livestock species and has kept 4 million square miles of African grazing land out of livestock production. Toxoplasmosis (*Toxoplasma gondii*) is widespread in over 200 mammalian species, including

humans, and can be fatal in numerous species (Sanger 1971). Theileriosis (Genus *Theileria*) occurs primarily in numerous species of ruminant artiodactyls throughout the world as well as in rodents, primates, xenarthrans, and the aardvark (*Orycteropus afer*). *Theileria parva* is a tick-borne protist that infects red blood cells and is highly pathogenic in cattle. Babesiosis is another protistan disease carried by ticks, and it infects rodents, carnivores, and ungulates. There are >100 reported species in the Genus *Babesia*. Although babesiosis can be fatal, animals often do not show clinical disease. Mortality rates from *Babesia bigemina*, however, can approach 90% in cattle.

Some protistans are highly pathogenic to humans. A good example of the diseases they can cause is African sleeping sickness, which occurs throughout central Africa and is caused by two subspecies of *Trypanosoma brucei*: *T. b. rhodesiense* and *T. b. gambiense*. The vectors are several species of tsetse fly in the Genus *Glossina*. In people infected with *T. b. gambiense*, sleeping sickness progresses from headache and fever to muscular and neurological involvement, and eventually to coma and death. People infected with *T. b. rhodesiense* usually die before these symptoms develop. As many as 20,000 new cases of sleeping sickness occur annually, of which 50% prove fatal. Past epidemics have killed hundreds of thousands of people (Maguire and Hoff 1992). If not fatal, infection often results in permanent brain damage.

A closely related trypanosome protozoan, *T. cruzi*, causes American trypanosomiasis, or Chagas' disease. Reservoirs include over 100 mammalian species, particularly domestic cats and dogs, as well as bats, armadillos, and rodents. Several species within three genera of reduviid bugs ("assassin bugs") are the vectors. This disease affects 12 million to 19 million people and is a leading cause of cardiovascular death in Central and South America. In the United States, *T. cruzi* occurs predominately in the South, where it may be more prevalent than previously believed (Roberts and Janovy 1996).

Leishmaniasis is a complex of zoonotic diseases common to tropical and subtropical regions and is caused by some species in the Genus *Leishmania*. The reservoirs of these diseases, which are transmitted through the bite of sandflies (Family Psychodidae), are canids and rodents. Ashford et al. (1992) estimated at least 400,000 new cases of leishmaniasis annually worldwide. In humans, *Leishmania donovani* causes Dum Dum fever, also called kala-azar. After an incubation period of several weeks, respiratory or intestinal infection and hemorrhage occur. If untreated, kala-azar is usually fatal within 2–3 years.

Many campers and hikers are familiar with giardiasis. Found worldwide, it is the most prevalent protistan parasite in humans (figure 27.4). *Giardia duodenalis*, with reservoirs in beavers, dogs, and sheep, is picked up in water. It is highly contagious and can cause severe intestinal disorders ("beaver fever") but is not fatal. Giardiasis is common in developing countries with poor water and sewage treatment facilities, but it also occurs in industrialized nations. Most people with the parasite show no symptoms, however.

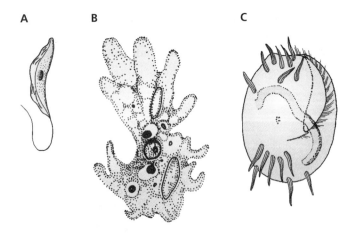

Figure 27.3 Protistans. These parasites can be grouped by their means of locomotion: flagella, pseudopodia, or cilia. A single flagellum occurs in (A) *Trypanosoma*, the protozoan genus responsible for African sleeping sickness and Chagas' disease. (B) This amoeba has pseudopodia ("false feet"). (C) Cilia occur in this free-living *Euplotes*. *Adapted from L. S. Roberts and J. Janovy, 1996,* Foundations of Parasitology, *Wm. C. Brown Publishers.*

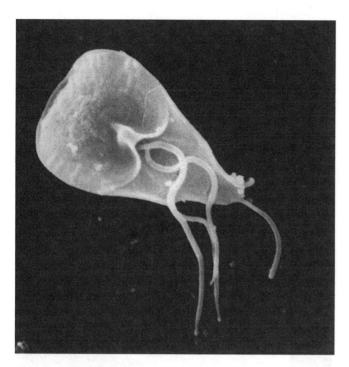

Figure 27.4 *Giardia lamblia.* This flagellated protozoan causes giardiasis.

Finally, several species of the Genus *Plasmodium* infect mammals, including rodents, nonhuman primates, and humans. Species of *Plasmodium* cause several varieties of malaria. Worldwide, 1.5 billion people are exposed to the disease, and up to 3 million die each year.

PLATYHELMINTHS

Two classes within the Phylum Platyhelminthes are entirely parasitic and common in humans and other mammals. The Class Trematoda includes both visceral and blood flukes. The Class Cestoidea includes the tapeworms. Both groups have indirect life cycles. That is, adult parasites occur in the primary (definitive) host and produce eggs, which are usually expelled. The larvae enter one or more intermediate hosts and pass through several growth stages before once again entering a primary host. In mammals, the dwarf tapeworm (*Vampirolepis* [*Hymenolepis*] *nana*) and taenids such as *Taenia solium* are exceptions in that they can hatch and remain within the host.

Trematodes are usually dorsoventrally flattened, leaf-shaped parasites found in a wide variety of wild and domestic mammals throughout the world. Most trematodes require two intermediate hosts; at least one, a snail, is required for development. Hundreds of species of trematodes have been reported from numerous mammalian species. Many trematodes cause significant economic losses in livestock, such as the large liver flukes, *Fasciola hepatica* and *F. gigantica*, common in many species of wild and domestic herbivores. These flukes also can

infect humans. More than 30 species of *Paragonimus* infect mammals, primarily carnivores. Many are zoonotic and cause pulmonary disease in millions of people (Goldsmith et al. 1991)—*P. westermani* is most common in people. All trematodes that parasitize humans belong to the Subclass Digenea. About 17 species of digenetic trematodes can infect the blood, liver, intestines, lungs, or brain of humans. For example, five different species of blood fluke cause schistosomiasis. This disease was known as early as 5000 years ago in ancient Egypt. Today, an estimated 200 million people are infected with schistosomiasis in parts of Africa, the Middle East, Southeast Asia, and South America, and a million die each year. The cycle of transmission to humans generally does not involve other mammals as intermediate hosts, although *Schistosoma mansoni* and *S. japonicum* occur in numerous mammalian species that may act as reservoirs.

Cestodes are the highly specialized, ribbonlike tapeworms found in the intestine of numerous mammalian species, including humans. Two orders of tapeworms infect people: Pseudophyllidea and Cyclophyllidea. Almost all species require two intermediate hosts to complete their life cycle. Juvenile tapeworms encyst in animals consumed as food by humans. A common pseudophyllidean tapeworm is *Diphyllobothrium latum*. It occurs in numerous fish-eating carnivores, including canids, felids, mustelids, phocids, otariids, and ursids as well as an estimated 9 million people worldwide (Hopkins 1992), where it causes pernicious anemia. Most tapeworms in wild mammals are cyclophyllideans. This order includes the Family Taeniidae, the most important medically for humans. For example, infection from *Taenia saginata* results from eating undercooked beef (figure 27.5). Usually, a single worm is present, which can attain a length of 25 m, although 3–5 m is typical. Likewise, consumption of undercooked pork may result in infection from *Taenia solium* (figure 27.6). Individual *T. solium* can survive in human hosts up to 25 years. Ingestion of eggs by people leads to cysticerosis, which can be fatal. The tapeworm Family Hymenolepididae contains many species that infect mammals. One, the dwarf tapeworm, is the most common human tapeworm, although pathogenicity is rare. It is unique in that an intermediate host is not necessary (Roberts and Janovy 1996).

Dogs and other carnivores are definitive hosts for the tapeworm *Echinococcus granulosus*. Herbivores, generally ruminants, serve as intermediate hosts. When infected, individual herbivores are more prone to predation by carnivores. The parasite thus "promotes" its own successful transmission. When eggs of *E. granulosus* are accidentally ingested by humans, they can cause a serious disease—hydatidosis. If growth of the juvenile parasite occurs in the central nervous system or heart, surgical removal may be necessary to prevent severe disability or death. Canids, cats, and rodents, especially arvicolines, can also be infected with *E. multilocularis*. In humans, this parasite can cause alveolar hydatid disease. Untreated, it is fatal in about 75% of cases.

A

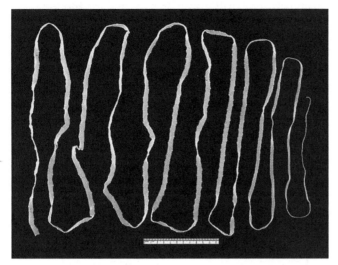

B

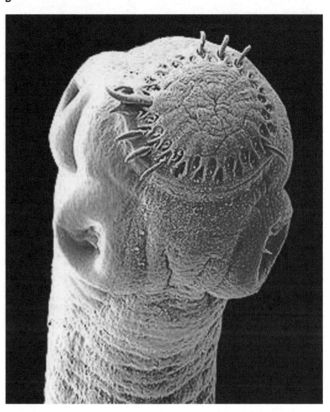

Figure 27.5 Tapeworms. (A) The tapeworm *Taenia saginata* may grow to be 25 m long. Tapeworms consist of a head, or scolex, that attaches to the intestinal lining of the host; a germinal center where body segments, or proglottids, form; and the body segments themselves. Several different species of tapeworms occur in humans. (B) Head of *T. pisiformis.*

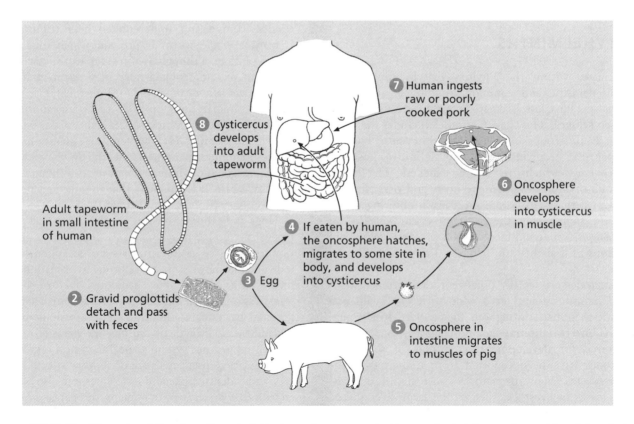

Figure 27.6 The life cycle of *Taenia solium.* Individual parasites can persist 25 years in their human host. *Adapted from L. S. Roberts and J. Janovy, 1996,* Foundations of Parasitology, *Wm. C. Brown Publishers.*

NEMATODES

These cylindrical, unsegmented worms are tapered at both ends. They usually have a direct life cycle, that is, no intermediate host is required. Generally called roundworms, there may be 500,000 species worldwide. Numerous species of free-living nematodes, as well as thousands of parasitic species, infect many mammalian hosts. Nematode infection occurs most often in tropical and subtropical developing countries with poor sanitation facilities. "Some of the most dreaded, disfiguring, and debilitating diseases of humans are caused by nematodes" (Roberts and Janovy 1996:385).

Whipworms, in the Family Trichuridae (Order Trichurida), include about 70 described species in the Genus *Trichuris* that are found in mammals. One of these, *T. trichiura*, infects an estimated 800 million people worldwide, although most cases are asymptomatic. Many mammalian species also harbor the liver nematode *Capillaria hepatica* (Family Capillariidae), although it usually is found in rodents. This nematode has been used as a biological control agent to reduce house mouse (*Mus musculus*) populations in Australia, apparently with little success (Singleton et al. 1995). Infections in humans are very rare but can be fatal.

A familiar human disease caused by a nematode parasite, and one of the most widespread, is trichinellosis (previously called trichinosis). Five species of *Trichinella* (Family Trichinellidae) are carried by wild carnivores and rodents. Campbell (1988) recognized four types of infection cycles, one domestic and three wild (figure 27.7). People are most likely to become infected with trichinellosis by eating undercooked pork carrying encysted larvae. Living juveniles may invade muscle tissue (figure 27.8), where they produce an immune response. An estimated 150,000 to 300,000 new cases of trichinellosis occur each year in Europe and the United States. Most cases are asymptomatic; no more than about 150 people yearly have heavy enough infections to produce clinical symptoms that are reported. In severe cases, however, death may result from respiratory or cerebral involvement or heart failure. Arctic explorers have died from eating infected polar bear meat.

The Order Strongylida includes numerous intestinal nematodes in domestic and wild mammals as well as humans. For example, two species of hookworms in the Family Ancylostomidae occur in over a billion people worldwide: *Necator americanus*, the most common, and *Ancylostoma duodenale*. Although most infected people are asymptomatic, heavy infections lead to malnutrition, heart involvement, and death in up to 60,000 people a year. Wild ruminants and domestic livestock also are parasitized by nematodes in the Family Trichostrongylidae. These include *Haemonchus contortus*, found in the abomasum, and several species of *Ostertagia* and *Trichostrongylus*, which can lead to extreme economic losses in livestock. Large intestinal roundworms (Family Ascarididae) such as *Ascaris lumbricoides* can reach 46 cm in length. It is estimated that 25% of the world's population harbors *A. lumbricoides* (Freedman 1992), with about 20,000 deaths yearly due to intestinal blockage. Lungworms, including

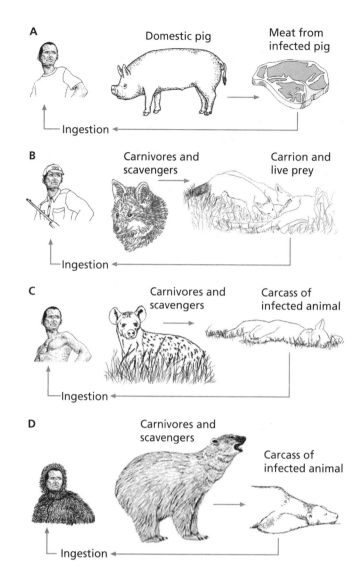

Figure 27.7 ***Trichinella*** **transmission.** Domestic and sylvatic ("wild") zoonotic cycles involve humans as accidental hosts. (A) Domestic cycle of transmission, often caused by *T. spiralis*; (B) temperate zone sylvatic cycle (*T. britori*); (C) torrid zone sylvatic cycle (*T. nelsoni*); (D) frigid zone sylvatic cycle (*T. nativa*). *Adapted from L. S. Roberts and J. Janovy, 1996, Foundations of Parasitology, Wm. C. Brown Publishers.*

Dictyocaulus filaria, *Protostrongylus rufescens*, and *Muellerius capillaris*, are found in a variety of ungulates, carnivores, and other mammals. Infection with these parasites may be seriously debilitating, often leading to pneumonia; heavy infections may be pathogenic. Lungworms can be a serious management consideration in big game species, including bighorn sheep (*Ovis canadensis*; Arnett et al. 1993). *Toxocara canis* is another ascaridid found worldwide in canids. Humans are an accidental host in which juvenile nematodes "wander." This results in a disease called visceral larva migrans. Usually, symptoms are not serious, but depending on the number of parasites and the organ infected, death can result.

Seven nematode species, collectively referred to as filarial worms, infect the bloodstream of over 650 million people. They often cause debilitating diseases, including elephantiasis and river blindness, and result in up to

Figure 27.8 Trichinellosis. Juvenile *Trichinella spiralis* encysted in human muscle. This nematode parasite causes trichinellosis, usually referred to as trichinosis. *Photo from Centers for Disease Control and Prevention.*

50,000 deaths per year. Filarial worms require an arthropod intermediate host. *Dirofilaria immitis*, heartworm, is a major nematode in dogs and other carnivores.

ACANTHOCEPHALANS

Compared to parasitic platyhelminths and nematodes, the thorny-headed worms are less common in mammals. They are named for the hooks that cover the proboscis (figure 27.9), which imbeds in the small intestine and holds the worm in place. Acanthocephalans have been reported from a wide range of mammalian species (DeGiusti 1971), but most often they are found in carnivores. Five species, including *Macracanthorhynchus hirudinaceus*, have been reported in people (Schmidt 1971). They are of minor importance in human health, however.

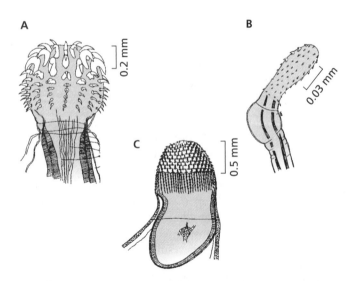

Figure 27.9 Thorny-headed worms. A variety of different types of hooks occur on the proboscids of these representative acanthocephalans: (A) *Sphaerechinorhynchus serpenticola*, (B) *Pomphorhynchus yamagutii*, and (C) *Owilfordia olseni*. *Adapted from L. S. Roberts and J. Janovy, 1996,* Foundations of Parasitology, *Wm. C. Brown Publishers.*

Pigs commonly carry *M. hirudinaceus*, and heavy infections can lead to death.

ARTHROPODS

Arthropods include ticks, fleas, mites, flies, mosquitoes, and lice (figure 27.10). Arthropods are the largest phylum of animals, with over a million described species. They are **metameric** (segmented) and have a chitinous exoskeleton. They may serve as either intermediate or definitive hosts for protistans, platyhelminths, and nematodes. Although most species are of no medical importance, some arthropods are mammalian parasites that are among the most significant vectors in the transmission of zoonoses (table 27.2).

Vector-Borne Zoonoses

PLAGUE

Plague has killed more people and has had a greater effect on human history than any other zoonotic disease (Twigg

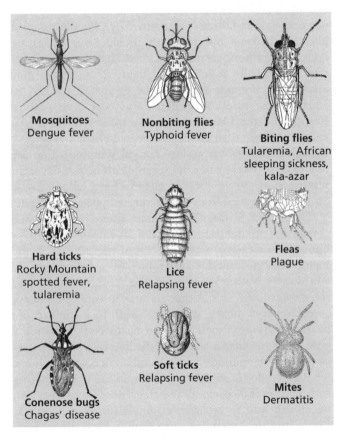

Figure 27.10 Arthropods. Different types of arthropods may serve as vectors, transmitting pathogens for human diseases. *Adapted from B. J. Bogitsh and T. C. Cheng, 1990,* Human Parasitology, *W. B. Saunders.*

Table 27.2 Representative orders and genera of parasitic arthropods associated with mammals

Order	Common Name	Representative Genera of Mammalian Parasites
Class Insecta		
Dermaptera	Earwigs	*Hemimerus*
Mallophaga	Chewing lice	*Bovicola, Haematomyzus, Heterodoxus, Trichodectes*
Anoplura	Sucking lice	*Haematopinus, Pediculu,*[1] *Phthirus*
Hemiptera	True bugs	*Cimex, Leptocimex, Rhodnius,*[2] *Triatoma*[2]
Coleoptera	Beetles	*Amblyopinus, Platypsyllus*
Siphonaptera	Fleas	*Ctenocephalides, Pulex, Tunga, Xenopsylla*[3]
Diptera	Flies, gnats, mosquitoes	*Aedes,*[4] *Anopheles,*[5] *Culex,*[6] *Cnephia, Cuterebra,*[7] *Glossina,*[8] *Psychoda, Simulium, Tabanus*
Lepidoptera	Moths	*Arcyophora, Calpe, Lobocraspis*
Class Arachnida		
Ixodida	Hard ticks	*Amblyomma,*[9,10] *Argas, Boophilus, Dermacentor,*[9] *Haemaphysalis, Hyalomma, Ixodes,*[11] *Ornithodoros, Rhipicephalus*[9]
Mesostigmata	Mites	*Echinolaelaps, Halarachne, Liponyssus, Ornithonyssus, Pneumonyssus*
Prostigmata	Mites	*Cheyletiella,*[12] *Dermodex, Leptotrombidium, Psorergates, Trombicula*[13]
Astigmata[14]	Mites	*Chorioptes, Psoroptes, Sarcoptes*

Taxonomy from L. S. Roberts and J. Janovy, 1996, Foundations of Parasitology, 5th ed. Wm. C. Brown Publishers.

Notes: 1, Transmits epidemic typhus, trench fever, and relapsing fever; 2, vector for Chagas' Disease; 3, vector for plague along with numerous other genera; 4, transmits yellow fever and dengue fever; 5, transmits malaria; 6, transmits encephalitis; 7, bot flies; 8, transmits sleeping sickness; 9, transmits Rocky Mountain spotted fever; 10, transmits tularemia; 11, transmits Lyme disease; 12, causes mange; 13, chiggers; 14, causes mange

1978; Lewis 1993). The etiological (infectious) agent is the bacterium *Yersinia pestis*, and fleas are the primary vector. There are more than 1,500 species of fleas, most of which are probably capable of plague transmission (Poland et al. 1994; Krasnov et al. 2006). Female oriental rat fleas (*Xenopsylla cheopis*) are the best-documented plague vectors (figure 27.11). They transmit bacteria as they feed on an infected animal, often black rats (*Rattus rattus*) or brown [Norway] rats (*R. norvegicus*). In the United States, many other rodents carry plague, especially ground squirrels (*Spermophilus* spp. and *Cynomys* spp.) and the deer mouse (*Peromyscus maniculatus*; Gage et al. 1995). Over 200 mammalian species are known to be naturally infected with plague (Poland et al. 1994).

In about 12% of infected fleas, bacteria multiply rapidly and within 9–25 days fill their gut. These are called "blocked" fleas. When they try to feed on humans or other mammalian species, blood enters their gut and picks up bacteria. Because their gut is blocked, blood is regurgitated back into the wound, and plague bacteria enter the host (figure 27.12). Blocked fleas probably can transmit plague bacteria for about 14 days before they starve to death. Humans also may pick up the bacterium through flea fecal material rubbed into wounds.

There are three clinical types of plague: bubonic, septicemic, and pneumonic. Bubonic plague is the most common; bacteria concentrate in the lymph nodes of the armpits and groin, where, after a 1–8 day incubation period, they cause extreme swelling (figure 27.13). These swollen areas are called **buboes**—thus, the name bubonic plague. Because of internal bleeding and necrosis of tissue under the skin, these areas turn black. Historically, bubonic plague is referred to as the Black Death. Untreated, the human mortality rate for bubonic plague initially was about 75%. Today, 25–50% of untreated cases are fatal (Roberts and Janovy 1996). Of the three types of plague, however, bubonic is the least infectious.

Occasionally, plague bacteria do not concentrate in lymph nodes but instead infect the entire bloodstream. This results in septicemic plague. Pneumonic plague results when the bacteria in human hosts move to the lungs. The lungs fill with a frothy, bloody fluid—a hemorrhagic bronchiopneumonia. Droplet-borne bacteria are highly infectious and may be spread directly from person to person without the need of intermediate fleas. The human mortality rate for untreated septicemic or pneumonic plague is 100%, and following onset of symptoms, life expectancy is only 1–8 days.

In the last 1500 years, three plague **pandemics** (large-scale outbreaks over wide geographic areas) have been recorded. The first began in Arabia and spread to Egypt in a.d. 542 and to the Roman Empire and Europe between

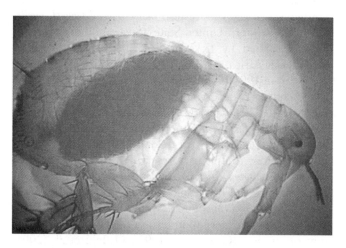

Figure 27.11 Oriental rat flea (*Xenopsylla cheopis*). The gut filled with blood can be clearly seen.

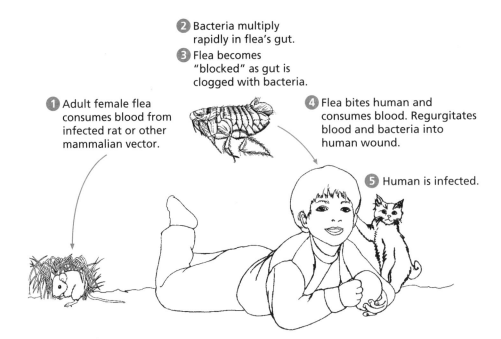

1 Adult female flea consumes blood from infected rat or other mammalian vector.

2 Bacteria multiply rapidly in flea's gut.

3 Flea becomes "blocked" as gut is clogged with bacteria.

4 Flea bites human and consumes blood. Regurgitates blood and bacteria into human wound.

5 Human is infected.

Figure 27.12 Bubonic plague. The common vector-borne plague cycle involves "blocked" fleas and a mammalian reservoir.

558 and 664 (Twigg 1978a; Acha and Szyfres 1989). An estimated 100 million people died. The second pandemic, the Black Death of Europe, probably began in Asia in the 1340s, spread across Europe, and reached England in 1348. Over 25 million people died as periodic **epidemics** (outbreaks affecting many people in an area) continued throughout Europe until the mid-1600s. Following this pandemic, it took 200 years for Europe to regain its 1348 population level (Twigg 1978). In the last plague pandemic, which broke out in 1894 and continued until the 1930s, 10 million people died throughout Southeast Asia, South Africa, and South America.

Plague is not just of historical interest, however. Several areas throughout the world continue to harbor the disease, with 1000 to 3000 cases reported each year. Plague is endemic in rodent populations in localized focal areas in western North America, southern Africa, the Middle East, China, and Southeast Asia (figure 27.14). Although human deaths still occur from plague each year, it is treatable with a variety of antibiotics if

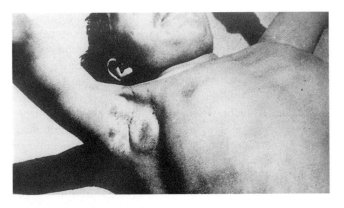

Figure 27.13 Plague symptoms. A bubo (lymphadenitis) such as shown here typically occurs in people infected with bubonic plague.

properly diagnosed. However, given the ability of microorganisms to mutate rapidly and become immune to antibiotics, and the ability of people to travel anywhere in the world in a matter of hours, the potential exists for new outbreaks of plague.

LYME DISEASE

Named for Lyme, Connecticut, where it was first recognized in the United States in 1977, this disease is increasingly common, with 23,305 cases reported in the United States in 2005 (figure 27.15). The etiological agent is the spirochete (corkscrew-shaped) bacterium *Borrelia burgdorferi*. The vectors are ticks of the *Ixodes ricinus* complex, found throughout North America, Europe (where the disease was originally described), and Asia (Lane et al. 1991). The disease correlates closely with distribution of the deer tick (*I. scapularis*) in the northeastern and midwestern United States and the western black-legged tick (*I. pacificus*) in the western United States. Several other species of *Borrelia* carried by ticks or lice cause related diseases throughout Africa, Asia, and North America (Burgdorfer and Schwan 1991). Numerous mammalian species, primarily rodents, are reservoirs for Lyme disease throughout the world. The main reservoirs in central and eastern North America are white-footed mice (*Peromyscus leucopus*) and white-tailed deer (Ostfeld 1997). In the western United States, the primary reservoir is the dusky-footed woodrat (*Neotoma fuscipes*). In Europe, the main rodent reservoirs are the bank vole (*Clethrionomys glareolus*) and wood mice (*Apodemus* spp.). Rodents probably are not affected by the bacterium in terms of decreased reproduction or survival (Gage et al. 1995). Ostfeld et al. (2006) attempted to predict the extent of occurrence of Lyme disease based on key determinants such as climate, density of

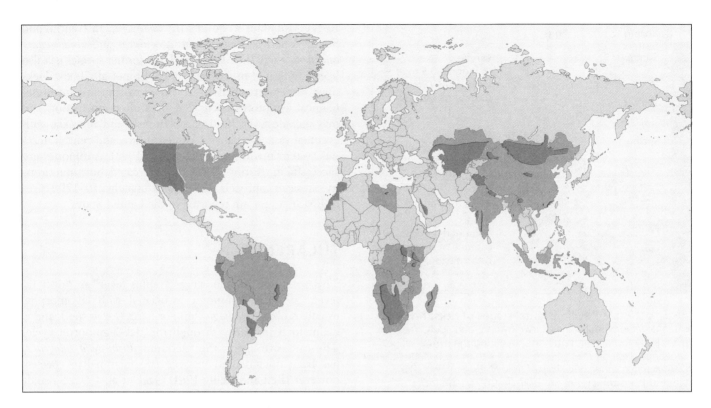

Figure 27.14 Worldwide distribution of plague. Countries reporting plague (orange) and probable plague focal areas (gray). *Data from Centers for Disease Control and Prevention.*

rodents and deer, and abundance of acorns—a primary food resource. They found risk in a given year was best predicted by the density of mice and chipmunks the previous year, and the abundance of acorns two years earlier. Reliability of such predictive models may be best at local rather than landscape scales (Schauber et al. 2005).

Ticks likely transmit the spirochete to humans through saliva but only after 24 hours or more of attachment (Piesman et al. 1991; White 1993). Ticks pass through three developmental stages: larva, nymph, and adult (figure 27.16). They are most likely to transmit Lyme disease during the nymph stage. They actively feed during this developmental stage but are unlikely to be noticed because of their small size (about 1 mm). Although larval ticks also feed and are even smaller (0.5 mm), they rarely

carry the infection. Adult ticks can also transmit Lyme disease. Because of their relatively large size, however, they are much more likely to be noticed and removed in less than the 24 hours necessary to transmit infection.

Symptoms of Lyme disease may include fatigue, fever, muscle and joint pain, and a characteristic bull's-eye-shaped skin rash. This rash (erythema migrans), seen 3–30 days after infection, lasts 2–3 weeks. When diagnosed early, Lyme disease may be successfully treated by antibiotics. Unfortunately, positive diagnosis often is difficult because the diversity of symptoms varies from patient to patient. For example, a rash does not form in about 20–40% of cases (Barbour and Fish 1993). Conversely, a rash may result from an allergic reaction to the tick saliva rather than infection from Lyme disease. Clinically variable, Lyme disease can

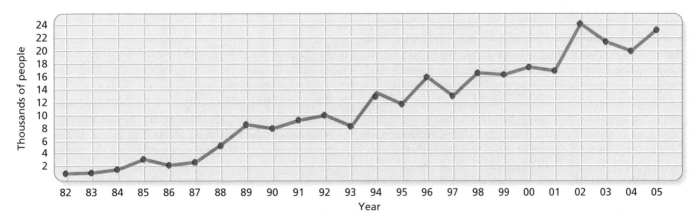

Figure 27.15 Lyme disease. The number of cases of Lyme disease reported yearly in the United States continues to increase. *Data from Centers for Disease Control and Prevention.*

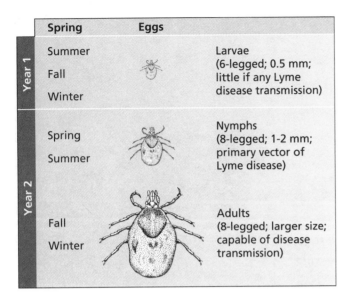

Year 1	Spring	Eggs	
	Summer		Larvae (6-legged; 0.5 mm; little if any Lyme disease transmission)
	Fall		
	Winter		
Year 2	Spring		Nymphs (8-legged; 1-2 mm; primary vector of Lyme disease)
	Summer		
	Fall		Adults (8-legged; larger size; capable of disease transmission)
	Winter		

Figure 27.16 Developmental stages of ticks. Ticks have three developmental stages: larva, nymph, and adult. They are most likely to transmit Lyme disease during the nymph stage. Drawings illustrate relative (not actual) size of each stage for *Dermacentor andersoni. Adapted from H. Brown, 1969,* Basic Clinical Parasitology, *3rd ed., Appleton-Century-Crofts.*

produce either acute or chronic disease. If untreated, the mortality rate is low (probably <5%), but infection can be disabling, leading to severe arthritis; nervous system involvement, including numbness, pain, or meningitis; and occasionally, cardiac arrhythmia (White 1993).

ROCKY MOUNTAIN SPOTTED FEVER

Rocky Mountain spotted fever (RMSF) is another well-known tick-borne zoonosis caused by a rickettsial bacterium, *Rickettsia rickettsii*. First reported in 1896 from Idaho as "black measles," RMSF is broadly distributed throughout North America, Mexico, Central America, and parts of South America. Other rickettsial diseases in the United States and worldwide include rickettsialpox, murine typhus, louse-borne typhus, Q fever, and monocystic ehrlichiosis (Gage et al. 1995). *Rickettsia rickettsii* is transmitted transovarially; that is, female ticks pass the infection to their offspring. Nonetheless, the bacterium probably would not be maintained without the small mammals that serve as "amplifying" hosts (Gage et al. 1995) because *R. rickettsii* negatively affects the survival and reproduction of the ticks that carry it (McDade and Newhouse 1986).

Numerous species of rodents and lagomorphs serve as hosts for the ticks that carry RMSF, as do the opossum (*Didelphis virginiana*), carnivores, and deer (McDade and Newhouse 1986). The name of this disease actually is a misnomer, as it occurs from western Canada through the United States and Central America to Brazil (the Centers for Disease Control and Prevention frequently call it "tick-borne typhus"). In the southeastern United States, the American dog tick (*Dermacentor variabilis*) is the primary vector. In the western United States, the usual vector is the

Rocky Mountain wood tick (*D. andersoni*). In Central and South America, *Rhipicephalus sanguineus* and *Amblyomma cajennense* carry RMSF, although the former species has also been implicated in Arizona (Nicholson et al. 2006). Ticks must remain attached 10–20 hours to transmit infection. Clinical symptoms occur 2–4 days after inoculation and may include fever, headache, skin rash, and anorexia, with eventual vascular damage, kidney failure, and central nervous system involvement (Clements 1992). Antibiotics are successful in treating RMSF; left untreated, human mortality rates can approach 30%. Approximately 250–1200 cases of RMSF a year are reported in the United States.

TULAREMIA

This zoonosis, also known as "rabbit fever" or "deer-fly fever," is most commonly associated with lagomorphs, usually rabbits (*Sylvilagus* spp.) or hares (*Lepus* spp.), and is maintained primarily through the tick-lagomorph cycle. The causative agent, the bacterium *Francisella tularensis*, has been documented in over 100 mammalian species, however (Bell and Reilly 1981; Gage et al. 1995; Sjöstedt 2005), including voles (*Microtus* spp. and *Clethrionomys* spp.), beavers (*Castor canadensis*), and muskrats (*Ondatra zibethicus*). Tularemia occurs worldwide in the Northern Hemisphere from above the Arctic Circle to 20°N latitude. Humans can contract the disease in several ways (figure 27.17): through direct contact with infected animals, from biting flies (Family Tabanidae) or ticks—primarily the lone star tick (*Amblyomma americanum*), American dog tick, and Rocky Mountain wood tick—or from water contaminated by urine from infected animals. The incubation period to onset of symptoms is 3–5 days. Six types of tularemia have been described; all begin with fever, chills, muscle and joint pain, and malaise and lead to respiratory involvement. The seriousness of the disease varies among individuals depending on the route of infection, bacterial load, and the particular strain of *F. tularensis* acquired (Hopla and Hopla 1994). Tularemia is successfully treated with antibiotics. In the United States, where 100–200 cases a year are reported, overall mortality rate of untreated cases is about 7%.

Nonvector Zoonoses

The remaining zoonotic diseases we discuss do not involve arthropod vectors. The first, rabies, has an ancient history, whereas the hemorrhagic fevers and prion diseases have been recognized as a concern only relatively recently.

RABIES

Rabies is a disease of great antiquity. The Eshnunni Code of Mesopotamia (pre-2200 B.C.) called for fining

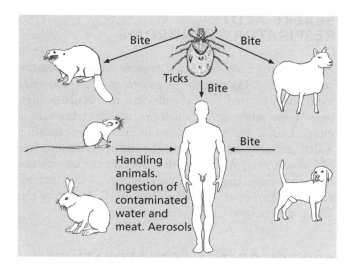

Figure 27.17 **Tularemia.** The several possible modes of transmission for tularemia from mammalian reservoirs to people. *Adapted from Acha and Szyfres, 1989,* Zoonoses & Communicable Diseases, 2nd edition, *Pan American Health Organization, Washington, DC.*

the owner of a rabid dog that killed someone. The Greek philosopher Democritus was the first to describe cases of rabies in 500 B.C. The causative agents of rabies and related diseases are distinct molecular strains of RNA-viruses in the Family Rhabdoviridae, Genus *Lyssavirus* (the Greek *lyssa* means "madness"). *Lyssavirus* occurs almost worldwide in a variety of mammalian hosts. Rabies virus is transmitted to humans most commonly through bite wounds or cuts and less commonly through mucous membranes or from inhalation. After inoculation, the virus infects nerve tissue. It eventually reaches the spinal cord and moves to the brain. Multiplying rapidly in the brain, the infection moves out along peripheral nerves throughout the body. In humans, symptoms generally appear 30–90 days after exposure, with fever, headache, and unusual tactile sensations. A number of other effects may follow, including apprehension, agitation, disorientation, hypersalivation, and paralysis. Once symptoms appear, death generally occurs in 100% of cases within a week or sooner (Fishbein 1991; Krebs et al. 1995).

Carnivores are the primary reservoir for the maintenance and transmission of rabies worldwide. Domestic dogs are the main carrier in developing countries; 22 types of rabies virus have been identified just from dogs. In North America, primary hosts are raccoons (*Procyon lotor*), skunks (most often the striped skunk, *Mephitis mephitis*), foxes (usually the red fox, *Vulpes vulpes*), and various species of bats (figure 27.18). In Europe, the red fox is the main vector, with the raccoon dog (*Nyctereutes procyonoides*) increasingly important in Eastern Europe (Holmala and Kauhala 2006). Rabid vampire bats in Central and South America, principally *Desmodus rotundus* (see chapter 13), primarily affect the cattle industry, but they may also infect humans (Lopez et al. 1992). Rodents and lagomorphs also are capable of transmitting rabies but are not considered natural reservoirs. They

represent less than 1% of reported cases in the United States, almost all of which involve marmots (*Marmota monax*; Childs et al. 1997).

Rabies infection in humans is relatively rare in developed countries. In the United States, only 37 cases were diagnosed between 1981 and 1998, and 12 of these were acquired in foreign countries (Krebs et al. 2000). No one in the United States died from rabies in 2000. Controlling rabies, however, costs $300 million in the United States each year (Dobson 2000), including operating diagnostic laboratories in all 50 states. It remains a significant health problem in less-developed countries (Krebs et al. 1995). Over 55,000 people a year die of rabies worldwide, primarily in Asia and Africa. Much research is directed toward developing vaccines to reduce the incidence of rabies in wildlife populations. Although untreated rabies in humans is almost always fatal, there have been rare exceptions. Brass (1994) documented three cases of people surviving rabies, but all suffered subsequent neurological impairment. Presumably, other mammalian carriers also succumb to infection.

HEMORRHAGIC FEVERS

Hemorrhagic fever actually encompasses a number of different viral diseases involving four families of viruses: flaviviruses, arenaviruses, bunyaviruses, and filoviruses. These widely distributed, negatively-stranded RNA viruses are often spread directly from rodent hosts to humans. The arenaviruses (Family Arenaviridae) include species that cause a variety of human diseases, including lymphocytic choriomeningitis, which is rarely serious. Lassa fever, however, infects 100,000–300,000 people yearly in West Africa with about 5,000 deaths annually. Argentine, Bolivian, Brazilian, and Venezuelan hemorrhagic fevers also occur (Childs et al. 1995). Untreated, these diseases are fatal in 10% to 30% of cases.

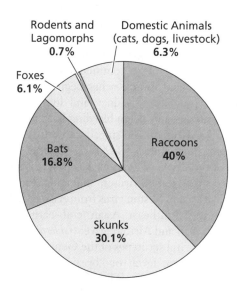

Figure 27.18 **Rabies in the United States.** The graph shows the distribution of 7437 reported cases of rabies in the United States through the year 2001. *Data from Centers for Disease Control and Prevention.*

Viruses in the Family Bunyaviridae include the Genus *Hantavirus*. In Europe and Asia, hantaviruses cause several zoonotic diseases, collectively referred to as hemorrhagic fever with renal syndrome, in up to 200,000 people a year. First recognized in the United States in 1993, hantavirus pulmonary syndrome (HPS) occurs primarily in the Southwest and secondarily in the Midwest and Southeast. By 2007, however, it was identified in 31 states as well as Canada, Panama, Bolivia, Brazil, Argentina, Chile, Paraguay, and Uruguay. Symptoms of HPS, like all viral hemorrhagic fevers, include fever, muscle aches, headache, and cough. Rodents are the reservoir of the virus, primarily the deer mouse, the cotton rat (*Sigmodon hispidus*), white-footed mouse, and rice rat (*Oryzomys palustris*) in North America. Worldwide, more than 27 different strains of hantavirus currently are recognized, each generally associated with a single rodent species, suggesting a long history of coevolution (Yates et al. 2002). The contagion is spread, often in dry, dusty areas, through infected rodent saliva or excreta inhaled as aerosols, or directly through broken skin (including bites) or contact with mucous membranes.

Hantavirus may be maintained and transmitted among wild rodent populations by individuals biting each other; arthropod vectors are not involved in transmission. Apparently, infection in rodents does not affect their survival or overall fitness. Unlike previous researchers, Biggs et al. (2000) found a significant relationship between the prevalence of hantavirus and the density of deer mice on two study sites. Domestic pets are not known to carry hantavirus. Few cases of HPS have been identified compared with other zoonoses—460 in the United States as of 2007. However, given the high mortality rates (35%) and the large number of researchers who work with wild rodents, mammalogists should be aware of the potential hazards and observe measures to minimize exposure.

One or more of the filoviruses cause Ebola hemorrhagic fever (McCormick and Fisher-Hoch 1994). Highly pathogenic, Ebola virus was first noted from outbreaks in 1976 in Zaire (Republic of Congo) and Sudan that resulted in over 600 deaths. Another outbreak occurred in Kikwit, Zaire, in 1995 (250 deaths), and Uganda in 2000–2001 (224 deaths). Periodic outbreaks in Africa continue to occur. The typical fever, headaches, and muscle aches begin 4–16 days after infection and progress to kidney and liver involvement. Eventually, patients may begin bleeding both internally and externally. The virus can be spread easily from person to person; the mortality rate is 50–90%. The natural reservoirs and hosts of the Ebola virus, as well as how it is spread, are unknown. Ebola-related viruses, including the Marburg virus from Africa and Reston virus from Asia, have been isolated from the Southeast Asian crab-eating macaque (*Macaca fascicularis*) and African grivet (*Chlorocebus aethiops*).

Four distinct viral serotypes of the Genus *Filovirus* cause dengue hemorrhagic fever, or "breakbone fever," which does not involve a vector. The virus is carried directly between humans by mosquitoes of the Genus *Aedes*, especially *A. aegypti*, although wild monkeys can be involved as mammalian reservoirs. Hundreds of thousands of cases are reported yearly, with a mortality rate of about 5%.

SEVERE ACUTE RESPIRATORY SYNDROME

First reported from Asia in 2003, severe acute respiratory syndrome (SARS) quickly spread to over 24 countries worldwide. Caused by a coronavirus that leads to high fever, body aches, and pneumonia, the mortality rate is close to 10%. The natural mammalian reservoir host for SARS originally was believed to be the masked palm civet (*Paguma larvata*). More recently, however, three species of horseshoe bats have been documented as the natural reservoir (Lau et al. 2005; Li et al. 2005), including the Chinese horseshoe bat (*Rhinolophus sinicus*).

SPONGIFORM ENCEPHALOPATHIES

Finally, a group of transmissible, progressive, neurodegenerative diseases that afflict mammals (table 27.3) results from a very different method of infection from those discussed previously. The infectious agents in each are **prions**—small, modified forms of cellular protein thought to be associated with synaptic function in neurons. Prion diseases, or spongiform encephalopathies, are characterized by large vacuoles (open areas) that occur in the cortex and cerebellum of the brain. These produce loss of motor control, dementia, paralysis, and eventually death. One of these diseases is bovine spongiform encephalopathy, or "mad cow" disease—one of several prion diseases. Another, of more urgent concern in North America, is chronic wasting disease (CWD). The only known hosts for CWD are mule deer (*Odocoileus hemionus*), white-tailed deer, and elk (*Cervus elaphus*). First recognized in 1978, symptoms of infection in cervids include weight loss, excessive salivation and urination, and tremors (Williams et al. 2002). Infected animals probably die within a few months. Endemic to the tricorner region of Colorado, Wyoming, and Nebraska, CWD has now been documented in 11 states and 2 Canadian provinces (Belay et al. 2004).

Table 27.3 Prion diseases

Disease	Natural Host
Scrapie	Sheep and goats
Transmissible mink encephalopathy (TME)	Mink
Chronic wasting disease (CWD)	Mule deer; elk
Bovine spongiform encephalopathy (BSE)*	Cattle
Kuru	Humans
Creutzfeldt-Jakob disease (CJD)	Humans
Gerstmann-Sträussler-Scheinker syndrome (GSS)	Humans
Fatal familial insomnia	Humans

From S. B. Prusiner, Prion biology, Prion Diseases of Humans and Animals, *S. Prusiner, J. Collinge, J. Powell, and B. Anderton (eds.), Ellis Horwood.*
*"Mad cow" disease.

In other infectious agents, genetic information is transmitted through nucleic acids. Besides being proteins, prions are unique disease agents because they appear to be *both* infectious and hereditary, with the disease mechanism believed to be a spontaneous change in protein structure.

Although transmission to humans is probably rare, evidence suggests that people can be infected by ingestion of prion-infected animal products. This is a very active area of current research (DeArmond and Prusiner 1995; Dalsgaard 2002).

SUMMARY

Parasitism is a symbiotic relationship between a parasite and a host. Parasites are relatively small, often microscopic; derive their nutrients from their host; and spend all or a part of their lifetime in or on the host. Depending on its size, life history characteristics, and host attachment, a given species of parasite is either a micro- or macroparasite and an endo- or ectoparasite. Microparasites include viruses, bacteria, fungi, and protistans. Macroparasites include platyhelminths, nematodes, acanthocephalans, and arthropods. Members of each of these groups are parasitic on mammals.

Many parasites do not kill their definitive host and may produce only subtle adverse effects. Parasites are more often pathological to intermediate and accidental hosts or to heavily infected definitive hosts. The influence of parasites on the dynamics of mammalian host populations is often subtle and inconspicuous. At the population level, a number of variables may affect overall mortality rate or reproductive potential of host populations. Parasites are rarely used for biological control of mammalian populations, but some attempts have been made.

Certain species of parasites are often closely associated with a particular mammalian taxon, resulting in a high degree of host specificity. These closely associated host–parasite assemblages may have coevolved; that is, reciprocal changes took place in the behavior, physiology, and morphology of both host and parasite.

Some parasitic species serve as vectors of disease-causing organisms. Organisms that cause disease are called etiological agents. Mammalian species serve as reservoirs for many diseases that can infect humans. Several zoonoses are discussed, representative of viral, bacterial, protistan, platyhelminth, and arthropod parasites. Many of these zoonotic diseases, including rabies and hemorrhagic fevers, have a high human mortality rate. Throughout history, for example, plague has killed hundreds of millions of people. Other important zoonotic diseases have been discovered relatively recently. These include Lyme disease, the Ebola and hantaviruses, and protein-based prion diseases, all fertile areas for basic and applied research.

SUGGESTED READINGS

Collinge, S. K,. and C. Ray (eds.). 2006. Disease ecology: community structure and pathogen dynamics. Oxford University Press, New York.

Goodman, J. L., D. T. Dennis, and D. E. Sonenshine (eds.). 2005. Tick-borne diseases of humans. ASM Press, Washington, DC.

Majumdar, S. K., J. E. Huffman, F. J. Brenner, and A. I. Panah (eds.). 2005. Wildlife diseases: landscape epidemiology, spatial distribution and utilization of remote sensing technology. Pennsylvania Academy of Science, Easton.

Padovan, D. 2006. Infectious diseases of wild rodents. Corvus Publishing, Anacortes, WA.

Wobeser, G. A. 2006. Essentials of disease in wild animals. Blackwell Publishing, Oxford.

DISCUSSION QUESTIONS

1. Investigate current control methods for one of the zoonoses discussed in the text. Specifically, what efforts are possible for breaking the transmission cycle, enhancing host resistance, destroying the infectious agent, or destroying the vector?

2. What role does the mobility of mammalian hosts play in coevolutionary development of host–parasite assemblages? Why might host–specific relationships be more likely to develop in pocket gophers than in white-tailed deer?

3. We noted that relatively few studies have investigated the effects of parasitism on the dynamics of host populations. Discuss some of the practical difficulties in designing such a study.

4. Over 100 years ago, Robert Koch, a German medical bacteriologist, developed a set of procedural steps for identifying the causative agent of a disease. These procedures, Koch's Postulates, are still used today. Try to determine necessary steps for identifying the causative agent of a disease, then look up Koch's procedures in a microbiology textbook.

Domestication and Domesticated Mammals

sMost people have seen cows and horses grazing in fields and are familiar with barnyard animals such as goats and pigs. You may have a dog or cat as a pet. These domesticated mammals, living in close association with humans, are "different" in certain respects from the wild ancestors from which they are derived. In this chapter, we will discuss the domestication of mammals and the benefits to humans associated with it. Considering the number of mammalian species, relatively few have been **domesticated**, that is, "bred in captivity for purposes of economic profit to a human community that maintains total control over its breeding, organization of territory, and food supply" (Clutton-Brock 1999:32). This definition is useful for our purposes; Bökönyi (1989) notes several other definitions used by researchers.

However domestication is defined, several questions arise concerning domesticated animals. Why domesticate at all? What was the process of domestication among early humans? How did it begin? Why did people choose to stop killing wild animals and begin to herd and propagate them? When did various species begin to be domesticated and where? Why were so few species of mammals "selected" for domestication? What behavioral or physiological characteristics lend a species to successful domestication? These are all interesting questions because of their effect on human history and because, in addition to mammalogy, they encompass anthropology, archaeology, and related disciplines. As we explore these questions, we will find that domestication includes both biological and cultural processes that continue today. Some mammalian species have been domesticated for thousands of years, and even today people are evaluating the potential of wild species for eventual domestication.

Why Domesticate Mammals?

From the dawn of human existence, our ancestors survived as hunters, gatherers, and scavengers. As the glaciers receded 10,000–12,000 years ago and the worldwide climate changed, humans began to make a transition from small groups of nomadic hunter-gatherers to a more sedentary lifestyle. This change allowed for the eventual establishment of discrete, larger human population centers. A major "cultural revolution" necessarily accompanied the transition from a nomadic to a sedentary lifestyle, however. This revolution involved the need for a sustainable food resource base to maintain permanent settlements (Reed 1984). Dependable, sustainable energy sources were provided by both cultivated food plants and domesticated animals. Thus, a major behavioral shift in humans gradually took place from simply hunting and killing animals to keeping and raising them. Eventually, domesticated animals provided numerous benefits to human society, economics, and culture.

Archaeological evidence suggests that agriculture, the cultivation of cereal grains (or "plant domestication"), began about 9,000 years ago (Reed 1977, 1984). Several criteria were necessary for agriculture to begin in an area (Sauer 1969). Primarily, food could not be chronically scarce; people had to have the time and resources to experiment with cultivation. They probably were already sedentary and had an alternative food source, such as fish. Whatever the circumstances, the rise of agriculture was preceded by the domestication of mammals. The first domesticated animals raised purely for food were sheep and goats as early as 10,000–11,000 years ago. This development was critical because food storage in those days was a problem. Goats and sheep provided "live stock," or a "walking larder" (Clutton-Brock 1989), that could accompany small groups of people as they moved. One mammalian species was domesticated even earlier than sheep

and goats, however. This was the dog, first domesticated about 15,000 years ago (figure 28.1), but not primarily to provide food (Savolainen et al. 2002; Gentry et al. 2004). Instead, dogs were valuable companions and probably assisted in hunting. They also scavenged human debris around camps and thus helped reduce intrusion by unwanted predators, warned of approaching predators, and as they do today, no doubt provided mutual affection. Other mammalian species eventually were domesticated not only for food and milk but also for a variety of other purposes. They provided hides for shelter or for use as carrying containers; wool or fur for clothing; dung for fuel; and transportation for people and products. They also were important in sport, war, or simply for prestige. Domestication of animals dictated the partial or complete separation of breeding stock from their wild ancestors. The process of domestication began human attempts to control nature. Substituting artificial selection for natural selection was an innovation in human history of the magnitude of controlling fire or developing tools (Davis 1987).

How Domestication Began

The process of domestication began with Paleolithic people and continued over thousands of years. Domestication probably was not a conscious decision on the part of early humans. That is, initially people probably were unaware of what was happening as ephemeral associations between

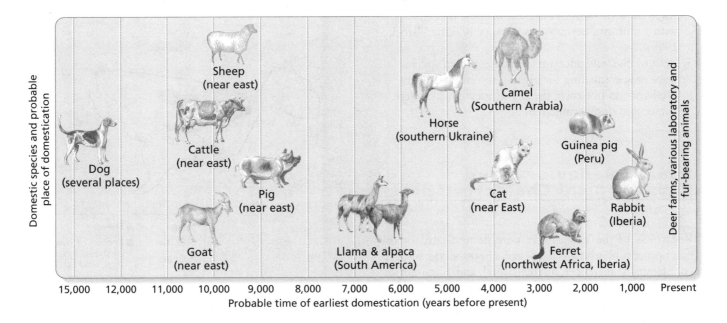

Figure 28.1 Important domesticated mammals. The estimated times and places of origin of domesticated mammals are based on archaeological evidence. *Data from D. F. Morey, 1994,* The early evolution of the domestic dog, *American Scientist 82:336–347; and other sources.*

animals and people became more long-term and involved. Nonetheless, over a 5,000-year period beginning about 10,000 years ago, humans changed from nomadic hunters and gatherers to sedentary farmers and herders in more densely populated areas. Domesticated mammals made this change possible and allowed people to become non-food-producing specialists, such as artisans, scholars, and soldiers. This revolution began with the development of dogs from wolves (see the section on the dog in this chapter) because wolves and humans hunted the same prey and were necessarily in close association. Zeuner (1963) summarized the probable stages of mammal domestication, all of which can be easily visualized in the transition from wolves to dogs, but they also apply to other species.

Probable Stages of Mammal Domestication

1. Initially, the contacts between people and wild species were loose, with free breeding, uncontrolled by humans.
2. Through time, individual free-ranging animals were confined in and around human settlements, with breeding eventually occurring in captivity.
3. Breeding of confined individuals became more selective and organized by humans, probably with occasional crossing with wild individuals, to obtain certain desired behavioral and morphological characteristics.
4. Human economic considerations led to greater selectivity for various desirable properties, with the resultant formation of different **breeds**, that is, animals with a uniform, heritable appearance (Clutton-Brock 1999).
5. Because the wild ancestors of domesticated species were now competing for grazing resources or were predators on livestock, these wild species often were persecuted or exterminated.

Shared Behavioral and Physiological Characteristics

Regardless of the species that were domesticated or the exact process by which it happened, almost all domesticated mammals share certain behavioral and physiological characteristics that lend them to domestication. Because the characteristics are rather specific, relatively few mammalian species have been successfully domesticated over many thousands of years. These characteristics were first summarized over 140 years ago by Galton (1865) and more recently by Diamond (1999). Specifically, mammalian species suitable for domestication are expected to be

1. Adaptable in terms of diet and environmental conditions
2. Highly social, with behavior based on a juvenile period with strong social bonding and eventual development of a dominance hierarchy
3. Easily maintained as a resource for products as well as for food when necessary
4. Easily bred in captivity, with rapid growth rates, because living offspring are one of the most important products
5. Closely herded, not adapted for instant flight, and thus easy for herders to tend, especially important for livestock such as sheep, goats, and pigs

These criteria necessarily preclude individuals that are strongly territorial and resist close association (cats are an exception). They also limit the potential number of domesticates. Archaeological evidence and the range overlap of ancestral species suggest that domestication of most mammals began in the Fertile Crescent area of the Middle East (figure 28.2) and possibly simultaneously in China and eastern Asia (Isaac 1970; Davis 1987). Recent genetic investigations based on mtDNA data for most domesticated mammals point to multiple domestication events—and resultant genetic lineages—that occurred in a number of different locations (Bruford et al. 2003; Gentry et al. 2004; Dobney and Larson 2006). Diamond (1999) provided a well-documented analysis of *why* most mammalian domestication arose in this part of the Old World—ultimately an

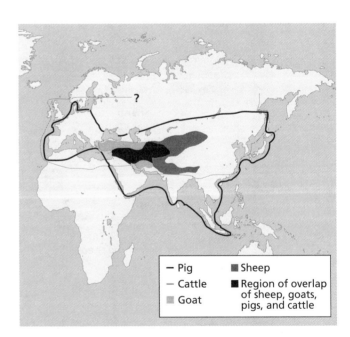

Figure 28.2 Areas of origin. The hypothesized area of origin of domesticated sheep, goats, pigs, and cattle overlaps the region of early agriculture and the ranges of the ancestral species. As noted in the text, multiple domestication events likely occurred for most species throughout these geographic areas. *Data from Erich Isaac, 1970,* Geography of Domestication, *Prentice-Hall.*

argument of geographic determinism—and the subsequent consequences on human history.

Mammals That Have Been Domesticated

Relatively few mammalian species have been domesticated. Of the approximately 5,400 species of extant mammals recognized today, less than 1% have been domesticated (table 28.1). Most mammalian species are small and of little use as domesticates. Nonetheless, of the almost 150 species of large herbivores (over 50 kg), only 5, all from Eurasia, have been domesticated and spread beyond their original range—horse, cow, sheep, goat, and pig (Diamond 1999).

Different stages in the development of domesticated mammals can be associated with evolving human social and economic development, including the concept of private ownership of animals. Zeuner (1963) grouped domesticated mammals into several temporal or functional stages as follows:

Temporal

Preagriculture	Dog, goat, sheep, reindeer
Early agriculture	Cattle (including water buffalo, yak, and banteng), pig

Functional

Transport and labor	Horse and ass, camels and llama, elephants
Pest destruction	Cats, ferrets
Provide consumer products	Deer ranches, African bovids, fur farms

Rats, mice, and rabbits could be included within an additional functional category, namely medical research. These groups are useful as we briefly review the major groups of domesticated mammals. As noted in previous chapters, some species of domesticated mammals are no longer found in the wild, and their ancestry is uncertain. For other domesticated species, their wild ancestors are clearly evident. It is also important to remember that "the intuitive notion that each modern domestic animal (when discussed as a global population) is descended solely from a single wild species is almost certainly incorrect" (Dobney and Larson 2006).

Table 28.1 Taxonomic grouping of domesticated mammals and their wild ancestors*

Domestic Form	Wild Ancestor
Lagomorphs	
Rabbit (*Oryctolagus cuniculus*)	European wild rabbit (*Oryctolagus cuniculus*)
Rodents	
Guinea pig (*Cavia porcellus*)	South American cavy (*Cavia aperea*)
Laboratory mouse (*Mus* [*domesticus*] *musculus*)	House mouse (*Mus musculus*)
Laboratory rat (*Rattus norvegicus*)	Norway rat (*Rattus norvegicus*)
Chinchilla (*Chinchilla lanigera*)	Chinchilla (*Chinchilla lanigera*)
Golden hamster (*Mesocricetus auratus*)	Syrian hamster (*Mesocricetus auratus*)
Carnivores	
Dog (*Canis* [*familiaris*] *lupus*)	Wolf (*Canis lupus*)
Ranched fox (*Vulpes vulpes*)	Red fox (*Vulpes vulpes*)
Ferret (*Mustela furo*)	Polecat (*Mustela putorius*)
Ranched mink (*Mustela vison*)	Wild mink (*Mustela vison*)
Cat (*Felis* [*catus*] *sylvestris*)	Wild cat (*Felis sylvestris*)
Perissodactyls	
Horse (*Equus caballus*)	Wild horse (*Equus ferus*)
Donkey (*Equus asinus*)	Wild ass (*Equus africanus*)
Artiodactyls	
Pig (*Sus* [*scrofa*] *domesticus*)	Wild boar (*Sus scrofa*)
Dromedary camel (*Camelus dromedarius*)	Unknown (*Camelus* sp.)
Bactrian camel (*Camelus bactrianus*)	Bactrian camel (*Camelus ferus*)
Llama (*Lama glama*)	Guanaco (*Lama guanicoe*)
Alpaca (*Lama* [*pacos*] *glama*)	Vicuña (*Vicugna vicugna*)
Domestic buffalo (*Bubalus bubalis*)	Indian water buffalo (*Bubalus arnee*)
Domestic cattle (*Bos taurus*)	European, Asian, N. African aurochs (*Bos primigenius*)
Zebu (*Bos frontalis*)	Indian aurochs (*Bos gaurus*)
Yak (*Bos grunniens*)	Wild yak (*Bos mutus*)
Bali cattle (*Bos javanicus*)	Banteng (*Bos javanicus*)
Domestic sheep (*Ovis aries*)	Mouflon (*Ovis orientalis/musimon/vignei*)
Domestic goat (*Capra hircus*)	Bezoar goat (*Capra aegagrus*); markhor (*C. falconeri*)

Data from A. Gentry et al., 2004, The naming of wild species and their domestic derivatives, *J. Archaeological Science* 31:645–651; J. Clutton-Brock, 1999, A Natural History of Domesticated Animals, *Cambridge University Press;* G. B. Corbet and J. Clutton-Brock, 1984, Appendix: Taxonomy and nomenclature, in Evolution of Domesticated Animals (*I. L. Mason, ed.*), pp. 434–438, Longman; S. J. M. Davis, 1987, The Archaeology of Animals, *Yale Univ. Press;* and others.

*Species such as elephants and reindeer, which may be considered "exploited captives," are not included. There is little consensus among authorities on scientific names applied to domesticated mammals.

DOG

The dog (*Canis [familiaris] lupus*) was the first domesticated mammal (see figure 28.1). Based on morphological, behavioral, and genetic criteria (Clutton-Brock and Jewell 1993), dogs clearly originated from wolves (*C. lupus*). They are an exception to the usual reasons mammals were domesticated, that is, as a source of food or raw materials. Instead, dogs originally provided a mutually beneficial hunting team with Paleolithic people at the end of the Pleistocene epoch (about 10,000–12,000 years ago) and development of a mutually beneficial association is easy to envision. Both wolves and early humans formed small hunting groups and stalked the same prey throughout similar environments around the world. As opportunistic scavengers, wolves and humans often hunted in the same areas. These wolves were tolerated by humans, and their pups may have been cared for and occasionally tamed. Soon their utility was evident because they warned against predators or other enemies and helped to

locate game animals (Isaac 1970). As generations passed, certain individual, tame wolves that had the requisite temperament and behavior (and were not eaten by humans during times of food shortage) began to breed and leave offspring with increasingly more submissive, placid dispositions. Through time and in different parts of the world, these lines of wolves eventually evolved into "dogs." Savolainen et al. (2002) concluded from mtDNA data that dogs probably originated in East Asia about 15,000 years ago. With continued artificial selection, different breeds evolved. Several breeds are recognized from ancient Egypt and early Roman times (Olsen 1985; Clutton-Brock and Jewell 1993). Through continued artificial, selective breeding, many types of dogs exist today (figure 28.3). In 2006, the American Kennel Club officially recognized over 150 breeds in 7 groups—sporting, nonsporting, working, hound, toy, terrier, and herding. All are the same species and reflect the variety that artificial selection has produced over a 10,000- to 12,000-year period.

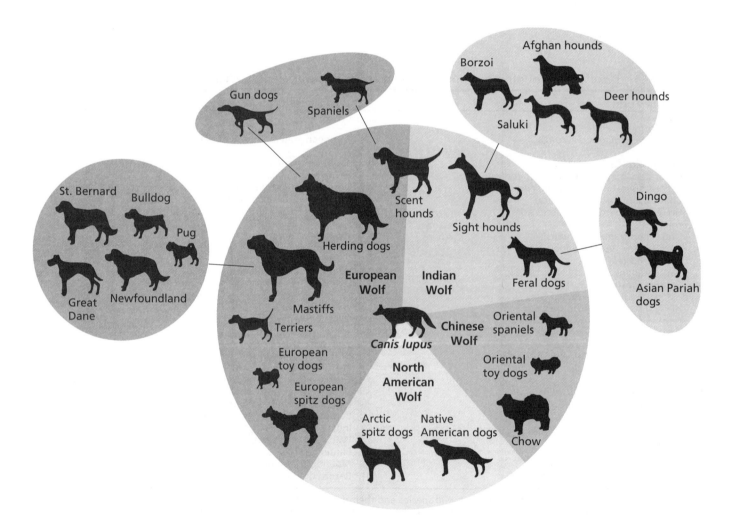

Figure 28.3 Different dog breeds. The numerous breeds of dog that occur in the world today (all descended from the wolf) are the products of thousands of years of artificial selection under human control. *Data from J. Clutton-Brock and P. Jewel 1993, Origin and domestication of the dog, in* Miller's Anatomy of the Dog, *3rd ed., W. B. Saunders.*

GOATS AND SHEEP

Originating in central and western Asia, goats and sheep were the first mammals domesticated as livestock. They provided humans with a sustained resource base for food, milk, hides, and other products as early as 10,000 years ago. Domestication was favored by their environmental and feeding adaptability, social behavior, and ease of herding. Domestication of early wild stock no doubt occurred either when young wild animals were reared by and imprinted on humans or when wild flocks were confined near water and eventually became habituated to people (Clutton-Brock 1999). Within 2,000 years, several morphological changes were evident in domesticated stock, including smaller size with shorter limbs, changes in the shape of horns (figure 28.4), and hornless females. Another useful morphological change in domestic sheep involved the pelage; it no longer shed, as occurred in wild species. Domestic sheep could be sheared, and their wool was not lost in the field.

Authorities do not agree on the systematics of sheep and goats. Grubb (2005) recognized eight species of goats, several of which are considered subspecies by other authors; some authorities recognize only two species. Domestic goats (*Capra hircus*) are derived from the ancestral bezoar goat (*C. aegagrus*), the markhor (*C. falconeri*), or both (Pidancier et al. 2006—see table 28.1). Domestic sheep (*Ovis aries*) probably originated from early mouflon (*O. musimon/orientalis/vignei*) or the closely related argali (*O. ammon*). Hiendleder et al. (2002:893) concluded that "a large number of wild and possibly ancestral species and subspecies exist."

CATTLE

Economically, cattle are the most important livestock (table 28.2). Domesticated cattle probably arose from the

Table 28.2 Estimated numbers of domesticated mammals worldwide as of 2003*

Perissodactyla	Numbers
Horses	55,470,000
Asses	40,328,000
Mules	12,806,000
Artiodactyla	
Cattle	1,371,117,000
Sheep	1,024,040,000
Pigs	956,017,000
Goats	767,930,000
Buffalo	170,661,000
Camels	19,074,000

Data from Food and Agriculture Organization of the United Nations, 2003, vol. 57.
*These figures vary yearly.

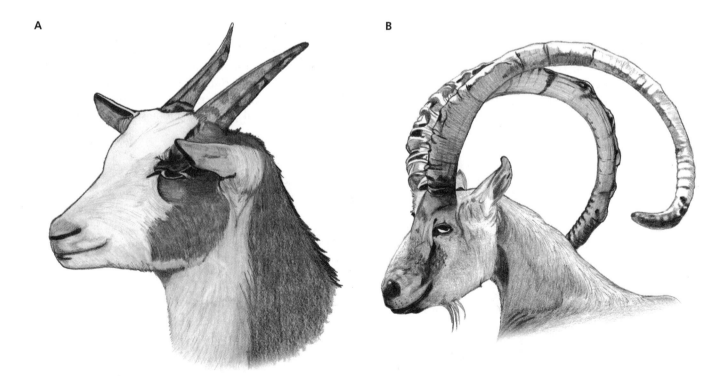

Figure 28.4 Domestication changes morphology. Among other characteristics, domestication results in changes in horn shape. (A) Domestic goats have horns that are either straight or twisted, whereas (B) their wild ancestors have large, scimitar-shaped horns. *Drawings by Lisa Russell.*

aurochs, or giant wild ox (*Bos primigenius*), about 10,000 years ago. Male wild oxen were very large, up to 2 m in shoulder height (figure 28.5), and reportedly very fierce. Widely distributed throughout much of the Northern Hemisphere, the aurochs became extinct in 1627, with the death of the last individuals in Poland. There has been much speculation as to the process of domestication of cattle from the wild progenitor. Given the size and fierceness of aurochsen (plural of *aurochs*, as in *ox* and *oxen*), they could not have been easy to capture or restrain. They probably were attracted to, and enticed to remain near, settlements by humans making salt and water available. Because aurochsen would have trampled crops and attracted unwelcome predators, they eventually might have been driven into fenced areas where they were easier to house and handle. Over time and through artificial selection resulting in smaller sizes (thus, easier to handle), distinctive breeds of "cattle" eventually emerged. Götherström et al. (2005:2345) believed cattle originated in numerous geographical areas and the process was "far more complex than previously believed." They concluded (p. 2344) that "Hybridization and backcrossing with wild ancestors may thus have been a common phenomenon during domestication, with the prevailing direction of such events being crosses of wild males with domestic females, meaning that it is not detected by analysis of DNA." Several breeds of cattle existed by the time of ancient Egypt, and today about 800 breeds are recognized.

Given thousands of years of domestication, interbreeding, and crossbreeding with wild species, it should not be surprising that the taxonomy of cattle is uncertain. They now are considered to encompass several species within the Family Bovidae, Subfamily Bovinae (see chapter 20). These are included in two groups: humpless cattle (*Bos taurus*), the most common species of domesticated mammals throughout the world; and humped cattle, which include the distinctive zebu (*B. [indicus] taurus*), with its large muscular hump, large hanging ears, and narrow face (figure 28.6). Recent molecular analyses (Bradley et al. 1996; Troy et al. 2001) have indicated distinct mtDNA lineages in *Bos taurus* and *B. indicus*. It is believed that zebu were domesticated independently, coming from the Asian form of aurochsen, *Bos primigenius namadicus*. There are several other species of cattle, none of which occurs beyond the range of its ancestral species. These include the domestic yak (*B. grunniens*), which today serves a variety of purposes in the high elevations of the Himalaya Mountains. Yaks interbreed with *B. taurus*, although only the female offspring are fertile. Wild yaks (sometimes considered a distinct species, *B. mutus*) are now very rare. Cattle also include the Asian mithan, or gayal (*B. frontalis*), which is not truly domesticated. The mithan is used today, as it was historically, primarily for sacrificial purposes (Clutton-Brock 1999). The mithan is believed to be descended from the gaur (*B. [gaurus] frontalis*), the largest extant bovid. The banteng (*B. javanicus*) is another domesticated bovid bred in Malaysia, Sumatra, Borneo, and Java. Like the yak, all these other bovines are, or have been, actively crossbred with domestic cattle. There also are several species of domesticated water buffalo (Genus *Bubalus*). The domestic buffalo (*B. bubalis*) is the most common species of cattle in much of Asia.

PIG

The ancestor of domestic pigs (*Sus [scrofa] domesticus*) was the wild boar (*S. scrofa*), and possibly other closely related wild suids. Wild boars were widely distributed throughout most of the Old World. They may have been domesticated first in the Middle East and western Asia about 9,000 years ago with multiple independent centers of domestication (Larson et al. 2005). Pigs are ideal for domestication because of their diverse feeding habits. It is easy to envision the initial process of domestication. Pigs could

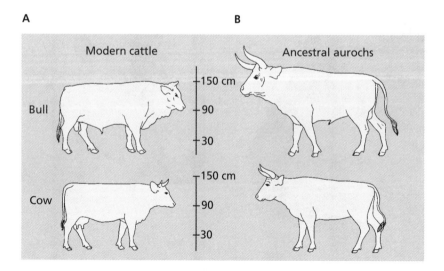

Figure 28.5   Ancestors of cattle. Relative sizes of (A) representative modern cattle and (B) aurochsen. There are almost 800 breeds of cattle today, with much variation in size. *Data from S. J. M. Davis, 1987,* The Archaeology of Animals, *Yale Univ. Press.*

Figure 28.6 Modern cattle. Cattle are often placed in two general categories, those with humps, such as this zebu, and those without humps, such as most breeds found in North America.

Figure 28.7 Breeds of domestic pigs. More than 100 breeds of domestic pigs are recognized throughout the world today. The Yorkshire breed shown here is generally all white. Originally bred in England, it is still referred to there as the Large White.

have thrived by scavenging on discarded food remains in association with early human settlements, before they were eventually confined. Also, piglets are easily tamed and acclimated to humans. Historically, groups of domesticated pigs were allowed to roam loose, watched by a swineherd, or housed in a "pig sty" or pen. Numerous breeds of pigs are now recognized (figure 28.7). Pigs have served as food and as sacrificial animals in many cultures. Pigs may have been considered unclean by ancient Egyptians, and later by some Middle Eastern cultures, and religious sanctions were raised against eating pork. Such prohibitions must have developed only after pigs had been domesticated and eaten for thousands of years.

HORSE AND DONKEY

The horse (*Equus caballus*) was one of the last common livestock species to be domesticated (see figure 28.1), from 5,000 to 6,000 years ago in the southern Ukraine (Bökönyi 1984). Authorities suggest many more distinct ancestral lineages were involved in the domestication of the horse than in most other domesticated mammalian species (Vilá et al. 2001; Jansen et al. 2002). From this region, domesticated horses spread throughout Europe and Asia. As noted by Clutton-Brock (1999), this species is the least changed from the ancestral stock. Although originally used as a source of meat, the primary purpose of domesticated horses soon became carrying people and goods long distances as quickly as possible. Horses revolutionized transportation on land. For the first time in history, humans could move themselves and their goods to other places faster than their own legs could carry them. With the introduction of chariots around 2000 b.c., horses also provided a revolutionary military advantage. "Without the horse Alexander the Great and Genghis Khan could not have made their Asian conquests. There could have been no European Crusades to the Holy Land, and the Spanish followers of Columbus could not have destroyed the civilizations of the Aztecs and the Incas in the Americas" (Clutton-Brock 1996:85). Horses were already well adapted for these functions, so little artificial selection was necessary. Thus, it is very difficult to distinguish domestic from wild horse remains at archaeological sites, and the ancestor of the domestic horse is uncertain. Most authorities consider horses to be derived from the feral horse, or tarpan (*E. [ferus] hemionus*). Nevertheless, about 150 breeds of horses are recognized today. The domestic horse has a long, flowing mane that is not shed. This is the result of a mutation not found in wild horses, which have a short, erect mane (figure 28.8) that is shed annually.

The origin and taxonomy of the donkey or domestic ass (*E. asinus*) also are uncertain and somewhat confusing. The only livestock species with an African ancestry, they probably were derived from the African wild ass (*E. africanus*) at about the same time as the horse. Beja-Pereira et al. (2004) determined the domestic donkey arose from at least two distinct wild populations, probably in northeast Africa. Asian species and subspecies of wild equids also occur, including the onager, or kulan (*E. hemionus*), and the kiang (*E. kiang*), which have been interbred with each other for millennia.

CAMELS, LLAMA, AND ALPACA

As discussed in chapter 20, the taxonomy of the Family Camelidae is uncertain because some of the domesticated species no longer exist in the wild. Camelids in the Old and New World always have been valued for their strength and endurance in harsh environments. As a result, as is true of horses, little artificial selection has taken place away from wild forms, and wild remains often cannot be differentiated

A

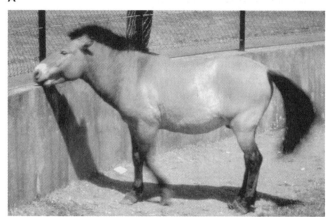

B

Figure 28.8 Horses. (A) Przewalski's horse (*Equus* [*przewalskii*] *caballus*), probably extinct in the wild, retains the characteristic short, erect mane of wild ancestral equids, unlike (B) the domestic horse.

from domesticated ones at archaeological sites. Little is known of the ancestors of domesticated camelids, due in part to early hybridization within both Old and New World groups (Stanley et al. 1994). The one-humped (dromedary) camel (*Camelus dromedarius*) may have been domesticated in southern Arabia about 5,000 years ago, primarily as a pack animal and for riding (Bulliet 1975). Two-humped Bactrian camels (*C. bactrianus*), primarily draft and pack animals, may have been domesticated near present-day Iran about the same time.

Domestication of New World camelids, the llama (*Lama glama*) and alpaca (*L. pacos*), probably occurred earlier, 6,000–7,500 years ago, in high-elevation regions of the Andes Mountains (Pieres-Ferreira et al. 1976; Novoa and Wheeler 1984). As noted in chapter 20, genetic evidence (Kadwell et al. 2001) has shown that the alpaca is descended from the vicuña (*Vicugna vicugna*), whereas the llama is a domesticated guanaco (*L. guanicoe*). Native llamas and alpacas flourished until the introduction of European livestock, which followed the Spanish conquest of the Inca Empire in 1532. Even today, llamas and alpacas are relegated to marginal lands unsuitable for introduced livestock, and their fleece is of relatively poor quality compared to that of the vicuña.

CAT

Domestic cats (*Felis catus*) represent the exception to the general characteristics noted earlier that define a "suitable" domesticant. As carnivores, their diet is not particularly adaptable. Neither are they highly social, nor do they provide a resource for valuable products. Despite thousands of years of association with humans, cats are little changed from the ancestral wild cat (*F. sylvestris*) of Eurasia. Most domestic cats can easily revert to a feral state if necessary. It is unknown when cats were first domesticated; their association with humans may be very ancient. Initially, cats may have entered permanent human settlements in the Fertile Crescent of the Middle East by following their primary prey—rodents. Through time and

continued association with people, cats may have progressed from a commensal, rodent-catching status to eventual domestication (Baldwin 1975). Again, it is impossible to assess whether remains at many early archaeological sites are wild or domestic cats; however, cats are known to have been fully domesticated by at least 3,000–4,000 years ago in ancient Egypt. The Egyptians considered cats to be sacred animals (Serpell 2000), and it was a capital offense to kill one. When they died, they were mummified. Enormous numbers of mummified cats have been excavated (one collection shipped to England weighed 19 tons!). Today, of course, cats compare with dogs as the most common household pets. There is no consensus on the number of recognized breeds among various cat associations in the United States; however, there are at least seven breeds of longhair cats and ten of shorthair.

ELEPHANTS AND REINDEER

Other species of mammals, such as elephants (chapter 19) and reindeer, may be considered "exploited captives." These were defined by Clutton-Brock (1999:130) as "mammals whose breeding remains more under the influence of natural rather than artificial selection" regardless of the length of association with humans. A degree of overlap and subjectivity certainly exists in these two categories. These species may be considered tamed and more tractable in that their flight distance in response to humans is reduced to zero, but they are not truly domesticated. As noted by Dobney and Larson (2006:262), "a strict wild/domestic dichotomy . . . prevents a deeper appreciation of those animals whose lives are spent somewhere in between."

Elephants

The difference between tamed and domesticated is especially evident in elephants. Because of their size, strength, and unique appearance, both the African (*Loxodonta africana*) and the Asian elephant (*Elephas maximus*) have been important economically, culturally, and historically for thousands of years: in the ivory trade, as beasts of burden, in zoos and circuses, for ceremonial purposes, and in early warfare. However, neither species has been bred in captivity for many generations nor subjected to artificial selection. This is partly because the reproductive behavior of elephants is difficult to control. As noted in chapter 19 females do not reach sexual maturity until they are 9 to 12 years old, gestation is 22 months with several years between estrous periods, and a single calf is produced that is dependent on its mother for a prolonged period. Also, in elephants, as in horses and camels, artificial selection has not taken place because the characteristics of use to humans—size, strength, and endurance—already exist naturally.

Humans have been taming elephants for about 4,000 years, originally for use in warfare. Alexander the Great used elephants in his conquests; he first learned of them in his campaigns in India and then incorporated them into his armies. For several hundred years thereafter, and well before the famous march of Hannibal across the Alps beginning in 218 B.C., elephants were used in numerous ancient battles (Scullard 1974). Elephants were used against cavalry, to trample infantry, and to break down walled encampments. Elephants can be unreliable and difficult to control, however, especially in the noise and confusion of a battlefield. They "are usually finely balanced between fight and flight, and . . . do not distinguish readily between friend and foe. In many ancient battles the elephants charged the enemy and then retreated through their own lines, causing havoc to friend and foe alike" (Douglas-Hamilton 1984:195).

The Indian elephant has been used as a beast of burden for several thousand years (Carrington 1959). Their principal function was in removing logs from forest tracts. Given the amount of food elephants consume and the fact they can work hard for only a few hours a day, they are relatively inefficient and expensive to maintain. Thus, their role in the timber industry today is decreasing because machinery is more efficient and because the decrease in wild populations makes it more difficult to replace tame individuals that are injured or die. The use of elephants as a domesticated mammal is rapidly becoming negligible.

Reindeer

Of all the species of cervids (deer), reindeer (*Rangifer tarandus*) are most likely to be associated with humans. Very gregarious and nonterritorial, they naturally form large herds. Nonetheless, they are considered exploited captives because they continue to live in semiwild conditions with little if any controlled breeding. Although reindeer may have been semidomesticated (or more properly "herded") very early, no direct evidence exists as to when this first occurred. Also, it is unknown whether they were first domesticated in one area or independently in several different areas. Nonetheless, the adaptations of reindeer that allow them to survive the extremes of harsh northern environments have been directly responsible for the habi-

tation of northern regions of Scandinavia and Russia by humans, who depend greatly on reindeer. Reindeer provide meat and hides and, secondarily, transportation. Their milk is not an important product, however. In the Western Hemisphere, domesticated reindeer have been introduced to the Seward Peninsula of Alaska, the lower Mackenzie Valley of northwest Canada, and parts of coastal Greenland (Skjenneberg 1984).

Morphological Effects of Domestication

The preceding discussion points out that, of the common domesticated mammals, almost all exhibit distinct morphological, physiological, reproductive, and behavioral changes from their wild ancestors (table 28.3). Several generalizations are possible about these changes (O'Regan and Kitchener 2005). The most obvious change is reduction in size. Most breeds of dog are smaller than ancestral wolves, and cattle are smaller than aurochsen. This trend may have resulted from intentional selection toward smaller individuals that were easier to handle or herd, or from dietary changes associated with captivity. Skulls are often shorter, and dentition is reduced. Cranial capacity also is reduced in domesticated mammals, and they have relatively smaller brains than their wild counterparts. Changes in horn size, shape, and growth rate also have occurred; these especially are evident in the goat. Domesticated species also have a greater variety of coat colors and patterns. In wild counterparts, this variability might be maladaptive and reduce survival.

Many of these changes, viewed together, reflect **neoteny** (the retention and persistence of juvenile characteristics in adults). Neoteny results in several behavioral changes in response thresholds in domesticated species (Ratner and Boice 1975; Price 1984). Domesticates have slower reactions, reduced flight distance, and less percep-

Table 28.3 Some common characteristics of domesticated mammals*

Trait	Domesticated Mammals
Dwarf or giant varieties	All
Piebald coat coloration	All
Wavy or curly hair	Sheep, dogs, donkeys, horses, pigs, goats
Rolled tails	Dogs, pigs
Shortened tails; few vertebrae	Dogs, cats, sheep
Floppy ears	Dogs, cats, pigs, horses, sheep, goats, cattle
Changes in reproductive cycle	All

Modified from Dobney and Larson, 2006, Journal of Zoology 269:261–271.
*See text for additional traits.

tion of the immediate environment (Hemmer 1990). These characteristics lead to more docile and easily handled individuals. Domestication often leads to greater reproductive potential as well. Compared with wild species, domesticates usually have earlier puberty and larger litter sizes and breed more often throughout the year. All these effects lead through time to distinctive breeds, whether in dogs, cats, pigs, or cattle. Although breeds are similar to subspecies in wild animals, the distinctiveness of breeds is maintained by human-influenced reproductive isolation rather than by geographical isolation, as is the case in subspecies.

Domestication: Human Artifact or Evolutionary Process?

Domestication may be viewed as somewhat of an "artificial" process, against the "natural order" of things, and a condition "imposed" on certain species by humans. Budiansky (1992) has suggested, however, that domestication be viewed instead as an evolutionary process driven by natural selection. This process has resulted in a symbiosis between humans and certain other species—that is, a symbiosis similar to other mutually beneficial relationships between species, such as the one occurring between ants and aphids. Humans obviously benefit from domestic animals, but the domesticants also benefit. They receive a reliable food supply, shelter, protection from predation pressure, and increased reproductive potential. Domestication may be viewed as an inevitable evolutionary consequence of unpredictable, changing environmental conditions at the end of the Pleistocene epoch. These conditions brought humans and domesticants closer together and created a new ecological niche, namely, association with humans (Morey 1994). In this view, natural variation among individual animals in a population resulted in the physiological, morphological, and behavioral changes eventually seen in domesticated species. Thus, they retained more curiosity, a less-developed species-specific sense of recognition, and care-soliciting behavior as the result of selection pressure rather than human preference.

Certainly prehistoric people worked toward domesticating species such as dogs, goats, and sheep. The question, however, is one of cause and effect. As noted by Morey (1994:346), "The issue lies with the presumption that the eventual result—highly modified animals under conscious human subjugation—explains the process that started those animals toward that end." Thus, it is possible that domesticated mammals selected humans as much as humans chose them. However it happened, placing domestication in the context of an evolutionary process, as opposed to considering it a strictly human sociocultural phenomenon, is an intriguing idea.

Current Initiatives

Although reindeer have been exploited for food and transportation for thousands of years, other species of cervids are just beginning to be considered for domestication. Deer farming—husbandry in penned conditions—to produce venison and antler velvet is a growing practice throughout the world (Hudson et al. 1989; Haigh and Hudson 1993). Often, exotic deer species— most commonly red deer (*Cervus elaphus*), fallow deer (*Dama dama*), and sika deer (*C. nippon*; figure 28.9)—are preferred by game farmers. The increasing number of game farms and possible conflicts with native wildlife are a serious concern to wildlife managers (Feldhamer and Armstrong 1993; Wheaton et al. 1993). The potential for commercial exploitation of many other large ungulates, primarily bovids, is currently being explored in the United States and throughout the world. Likewise, debate continues in Australia about raising kangaroos as livestock. Foxes, mink (*Mustela vison*), and chinchillas are raised on farms for fur production. Likewise, several rodent species, including the capybara (*Hydrochoerus hydrochaeris*) and paca (Genus *Agouti*; see chapter 18), are being produced on farms as food sources (Robinson and Redford 1991). Various strains of rats and mice, important in medical research, occur only in laboratory colonies.

Today, there are billions of individual domesticated mammals throughout the world, as livestock (see table 28.2) and as pets. They play critical roles in the lives of humans—socially, culturally, and economically. Unfortunately, the demands of livestock for pasture and grazing lands and the needs of an increasing human population often mean that less habitat and fewer resources are available for the wild mammals of the world. The extent to which wild mammals will remain a viable part of ecosystems depends on current and future conservation efforts—the subject of the next chapter.

Figure 28.9 Deer farms. There are thousands of deer farms throughout the United States and the world, such as this one in Ireland where Japanese sika deer are raised.

SUMMARY

All domestic mammals descended from wild ancestral species. This process has been occurring for thousands of years and for a variety of reasons—companionship, food, milk, fur, beasts of burden, sport, or prestige—and it continues today. Relatively few of the approximately 5,400 currently recognized mammalian species have been domesticated, however, because the behavioral and physiological attributes of species necessary for successful domestication are stringent. They must be social, have an adaptable diet, and be easily herded and bred under a variety of environmental conditions. Thus, only five species of large, herbivorous mammals have been domesticated and spread worldwide: horse, cow, goat, sheep, and pig. Once domesticated, species change morphologically and behaviorally from their wild ancestors in several ways. Body size often becomes smaller, pelage increases or decreases according to climate, the skull is shortened and dentition reduced, and submissive behavior patterns increase. Generally, domesticated species retain behavioral and morphological characteristics similar to juveniles, a condition referred to as neoteny. Neotenates have slower reaction times, reduced flight distance, and less environmental perception, all of which make domesticated animals easier to handle and herd. It has been suggested that the initiation of domestication was not entirely a human innovation but an inevitable consequence of the evolutionary process.

Domestication is artificial selection for characters beneficial to humans, and it results in recognizable breeds. This process is similar to evolution, including development of subspecies during natural selection of wild stock. Unlike subspecies, breeds are not necessarily geographically restricted. Domestication probably began 15,000 years ago between prehistoric humans and wolves, which led to domestic dogs. The first livestock species were sheep and goats. They were domesticated 10,000 years ago, before the development of agriculture. Sheep and goats provided the "walking larder" that helped make possible the major transition of humans from small groups of nomadic hunter-gatherers to permanent, sedentary, larger agricultural population centers. This transition in turn led to significant changes in the structure and function of human civilization. Domestication of many species was first centered in the Fertile Crescent area of the Middle East, but the process certainly arose independently in other regions of Asia, Europe, South America, and Africa. The process of domesticating new species continues today, with several species of deer held on game farms, a situation that is not without controversy. Billions of individual livestock animals provide a critical economic base for much of human society today and have contributed to human social and cultural development for thousands of years.

SUGGESTED READINGS

Clutton-Brock, J. 1992. Horse power: A history of the horse and donkey in human societies. Harvard Univ. Press, Cambridge, MA.

Draper, J. 1999. Horse breeds of the world. Hermes House, London.

Herre, W., and M. Röhrs. 1990. Animals in captivity: Domestic mammals. Pp. 570–599 *in* Grzimek's encyclopedia of mammals, vol. 5. (S. P. Parker, ed.). McGraw-Hill, New York.

Mason, I. L. (ed.). 1984. Evolution of domesticated animals. Longman, New York.

Trut, L. N. 1999. Early canid domestication: the farm—fox experiment. Am. Sci. 87:160–169.

Zeder, M., D. Bradley, E. Emshwiller, and B. Smith (eds.). 2006. Documenting domestication: New genetic and archaeological paradigms. Univ. California Press. Berkeley.

DISCUSSION QUESTIONS

1. Given the presence of pronghorn antelope, bighorn sheep, bison, and other large, social ungulates, speculate on why the native North Americans never domesticated any livestock species. Likewise, why was the kangaroo never domesticated in Australia?

2. Although livestock raising is now common in Africa, what factors might mitigate against livestock in Africa, especially several thousand years ago?

3. The question of private ownership of wildlife is central to deer farming. In addition to this issue, what other social, ethical, and biological questions and problems surround the commercialization of deer and other "wild" species?

CHAPTER 29

Conservation

The gray wolf (*Canis lupus*; figure 29.1A) once roamed over most of North America. Its range was reduced over the past two centuries so that it was almost gone from the lower 48 states (see figure 29.1B). Extirpation of the wolf over much of its range occurred primarily through hunting and trapping because wolves were considered dangerous and possible predators on livestock. Bison (*Bison bison*) numbered in excess of 60 million in North America about 1860. Three decades later, however, fewer than 200 were left in the wild, and shortly thereafter, the only remaining bison were in reserves. Again, human pressure was the major factor accounting for the dramatic decrease in numbers.

Madagascar (figure 29.2) is home to more species of strepsirhine primates than any other location on Earth. During the past 50–75 years, however, a tremendous decline in numbers has occurred for most of these species. A number of the strepsirhine primates found only on Madagascar are on the verge of extinction (see chapter 14). Of the 60 species of primates that inhabit Madagascar, 49 (82%) are currently listed as endangered. The primary driving force behind this severe decline is habitat destruction. Many of the forests that were home to these primates have been cut down so that the land could be used for agriculture and the wood used for fires and construction.

A striking example of what may be in store for other mammals is the recent declaration that the baiji (*Lipotes vexillifer*), the Chinese river dolphin, became the first Cetacean to go extinct in the modern era (Revkin 2006). The baiji once roamed the Yangtse River, using its sonar to capture fish. The growing human population along the Yangtse almost certainly doomed this dolphin. While a formal declaration of the baiji's demise will take some years, those who recently spent six weeks in search of baiji, finding none, feel that the species is now functionally extinct.

Because we are mammals and some of our mammalian relatives, for example, wolves, elephants, nonhuman primates, and pandas, are among the featured players in the current conservation drama, it seems logical that we conclude this text with a consideration of the plight of mammals in the modern world. In so doing, we will consider briefly the manner in which an integrated knowledge of the physiology, ecology, and behavior of mammals can be critical in making effective conservation decisions. What factors are responsible for the

A

B

Figure 29.1 Wolves. (A) The wolf is a predator and scavenger, living alone or in small packs that hunt as a unit. (B) Its range once encompassed much of North America but now is restricted to only the northern portion of the continent.

continuing decline in some mammal populations? Why are some species on the verge of extinction while others may be increasing their numbers? What are some of the solutions to these problems?

Nature of the Problem

Conservation has a history almost as long as some of the problems it seeks to resolve. As humans have increased their standard of living, they have overused and degraded their environment. Today, some groups of humans still live

in close association with their immediate surroundings. Reverence for the land and all of its plant and animal resources, once true for all humans, has declined over the centuries. The land and associated resources are considered expendable commodities, to be used up like any other commodity rather than conserved for future generations (Leopold 1966).

In North America and other parts of the world, a conservation ethic was reborn and gained momentum during the past century (Owen 1971). Prior to that time, a few farsighted individuals, for example Thomas Jefferson, noted that humans were despoiling their surroundings and needed to be concerned about the rate at which they were using up their natural resources. Establishment of Yellowstone

Figure 29.2 Prosimian habitat loss. Madagascar, an island of just over 587,000 km², is home to the richest prosimian primate fauna in the world. The loss of forested habitat, as shown here, has lead to sharp declines in the populations of many of these primates. The forested area, which once covered 62,000 km², has been reduced to 26,000 km².

National Park (1872) in the American West and the Adirondack Forest Preserve in upstate New York (1885) were among the first actions taken to preserve wilderness areas on a large scale.

The first modern event in conservation in North America was probably the White House Conference of 1908, organized and led by President Theodore Roosevelt. This conference resulted in the establishment of the National Conservation Commission, headed by Gifford Pinchot. The natural resources inventory, completed by this commission, resulted in the withdrawal of more than 200 million acres (80 million hectares) from further settlement. People like Roosevelt, Pinchot, and John Muir stand out for their commitment to conservation and their ability to balance the need for better resource management with the needs of increasing civilization. Together, they greatly increased the numbers of national parks and national forests.

President Franklin Roosevelt's establishment of the Works Progress Administration (WPA) and Civilian Conservation Corps (CCC) resulted both in employment for millions during the Great Depression and afterward, and in a prodigious number of natural resource conservation and related endeavors in virtually every state in the country. Projects carried out by CCC members remain today with Americans in many of their state and national parks, forests, and preserves. Also under Franklin Roosevelt, a Natural Resources Board was appointed. In 1934, the board completed the second full inventory of American natural resources. Other key agencies founded at this time include the Soil Conservation Service and the Tennessee Valley Authority, both of which continue to play roles in American conservation and resource management. The Wildlife Restoration Act of 1937 provided for exploration of conservation and resource problems, and for finding and implementing various means of remediating these problems.

The most recent wave of interest in conservation can be traced to the 1960s. This interest was spurred by such books as Rachel Carson's *Silent Spring* (1962) and Paul and Anne Ehrlich's *The Population Bomb* (1968). Legislative actions in the United States included the Clean Air Act (1964), National Wilderness Act (1964), Endangered Species Act (1973), Marine Mammal Protection Act (1972), and bills establishing numerous additional national parks, monuments, wilderness areas, and scenic rivers. The general awakening to environmental concerns such as the greenhouse effect along with global warming, pollution, and diminishing supplies of nonrenewable natural resources that began in the 1960s led to the first Earth Day in 1970. The movement has continued to gain momentum. The need to maintain biodiversity on a global scale has become an acknowledged concern for people in many countries. Biodiversity is the summation of all of the living plants, animals, and other organisms that characterize a particular region or country or the entire earth. With this brief background, we are now ready to explore in more detail some of the specific problems affecting mammals around the globe.

The list of factors that have led to declining numbers for many mammal species, and extinction for some, is lengthy, and the exact factors differ from one situation to another. To explain the decline in numbers for any particular species, its extirpation from certain regions, or even its extinction, it is necessary to understand the interactions of multiple factors. One factor, global warming, is now recognized as a critical factor affecting all aspects of our environment. The possible effects of global warming on mammal populations are hard to judge, but clearly changing water temperatures will affect sea mammals, higher land temperatures and longer warm seasons will result in changes for breeding seasons and foods for many mammals. We will not deal with global warming in a direct manner in what follows, but we note its critical, emerging importance.

HUMAN POPULATION

Early humans most likely lived in small subsistence groups, possibly consisting of one or several related families. Regional population density was quite low, but even then, exponential growth occurred. By about 1850, the total world population reached 1 billion people (figure 29.3). As of 2007, the population exceeds 6.6 billion individuals, and it continues to increase exponentially. Predictions for the future vary, with a conservative estimate at more than 8 billion by 2020. The continued explosion of the human population worldwide results primarily from an unchecked birth rate, and secondarily from somewhat lower death rates for infants and adults. Some areas, such as Europe, have approached a near stable population size. Other regions, including Latin America and much of Africa, have the highest annual growth rates of any of the world's regions, exceeding 2% per year.

The major consequence of the ever-increasing human population is a need for additional land and other resources. There is tremendous pressure to provide enough food, necessitating clearing more and more land for agriculture. Forests are cut down to provide both land for growing crops and the wood for fuel and construction (Simmons 1981). Other critical habitats that are home to many mammals are transformed for human use, such as wetlands, which are drained and filled. The accumulation of wastes, both human and those incidental to human activities (industrial and commercial), is an increasing problem. Armed conflicts that occur constantly around the world are, in large measure, a product of the growth of the human population and the competition for resources. The consequences for mammalian wildlife can be quite direct. The conflict in Rwanda resulted in the loss of some of the endangered eastern gorillas (*Gorilla beringei*). Several dozen species and subspecies of mammals have gone extinct since 1900 (figure 29.4), many due to direct or indirect human influences.

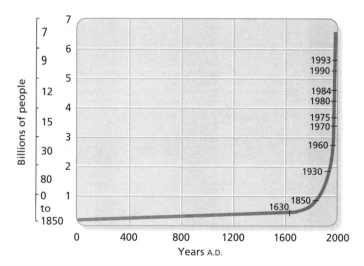

Figure 29.3 Human population growth. The human population began to enter the exponential growth phase, as shown in this graph, even before the Industrial Revolution in the first half of the 19th century, when the total number of individuals was about 1 billion. Since then, the population has doubled almost three times. *Data from Agency for International Development, Washington, DC, 1968*, Food vs. People.

HABITAT DESTRUCTION

Habitat destruction, caused primarily by human activities, is manifested in several forms. Probably the most publicized type of habitat destruction involves forests, particularly tropical rainforests. Over 45% of the original tropical rainforest of the world has been destroyed, including a 45% forest loss in Thailand between 1961 and 1985, all of the primary rainforest in Bangladesh, 85% of the forest in the Ivory Coast in West Africa, and more than 35% of the forest in Brazil (Lean et al. 1990). Much of this destruction has occurred since 1940. In addition, considerable forested areas in the temperate zones have been eliminated. Small mammals are most affected by forest destruction. In both Neotropical and African rainforests, the diversity and abundance of bats is diminished when the forest habitat is disturbed (Fenton et al. 1998; Medellin et al. 2000). Road construction and timber extraction both resulted in changes in rodent community structure in an African tropical forest (Malcolm and Ray 2000).

Cutting of forests occurs for several reasons. One of these, provision of land for additional agriculture, has been mentioned. For example, vast areas of forest are cut down each year in a number of regions of Africa and Latin America to grow crops and provide pastures for livestock (figure 29.5). Because most tropical soils are of relatively poor quality, a given area is productive for crops for only a few years. The land is then abandoned, and people move on to repeat the process in another nearby locale.

Over time, vast areas of what were once forests of various types are destroyed. Most of the nutrients in these forest ecosystems are in the vegetation. When the area is logged or burned, the nutrients are lost, leaving an ecological desert. The secondary growth that replaces the original forest after the land is abandoned is a poor substitute for the original native forest. Opportunistic and exotic plants can come to dominate in such situations, leaving almost nothing of the original plant and animal communities. Leveling forests for agriculture does not occur only in less developed nations, however. In Australia, the need for grazing land for livestock is magnified by the low density of forage. One cow can require up to 80 hectares/year to provide sufficient food. Many eucalyptus forests are leveled using ball-and-chain arrangements stretched between large bulldozers that move across the landscape.

A number of mammals are endangered or threatened as the result of deforestation. Endangered species are those likely to go extinct as a result of human activities and natural causes in all or a major portion of their range. Threatened (some listings use the term *vulnerable* instead) species are those that are likely to become endangered in the near future; they may be at risk for extinction. A variety of species of lemurs on Madagascar, where more than 80% of the original forest has been removed by humans, are endangered; black-and-white ruffed lemurs (*Varecia variegata*) and indris (*Indri indri*) are examples. Critically endangered are diademed sifakas (*Propithecus diadema*) and

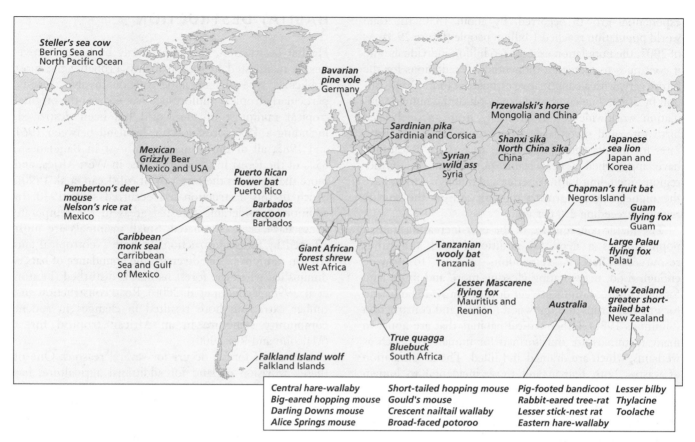

Central hare-wallaby	Short-tailed hopping mouse	Pig-footed bandicoot	Lesser bilby
Big-eared hopping mouse	Gould's mouse	Rabbit-eared tree-rat	Thylacine
Darling Downs mouse	Crescent nailtail wallaby	Lesser stick-nest rat	Toolache
Alice Springs mouse	Broad-faced potoroo	Eastern hare-wallaby	

Figure 29.4 **Mammalian extinctions.** The worldwide distribution of mammals that have become extinct in the past 400 years. *Data from G. Caughley and A. Gunn. 1996,* Conservation Biology in Theory and Practice, *Blackwell Science Publications.*

Figure 29.5 **Deforestation and agriculture.** (A) In a number of the world's countries, slash-and-burn agriculture or similar practices result in deforestation of plots to provide land for grazing livestock or growing crops to support an increasing human population. Because of the generally poor quality of tropical soils, however, crops are successful for only a few years, and then the process is repeated on another plot. (B) In other regions, forests have been converted to rice paddies.

the aye-aye (*Daubentonia madagascariensis*). Woolly lemurs (*Avahi laniger*) and ringtail lemurs (*Lemur catta*) are quite vulnerable due to habitat loss (Green and Sussman 1990; Mittermeier et al. 1994). Similarly, some New World primates are on the verge of disappearing because of the loss of forest habitat. These include the

southern muriqui (*Brachyteles arachnoides*) and the golden-lion tamarin (*Leontopithecus rosalia*), about which we will have more to say later in the chapter, when we deal with several specific case histories. Finally, among the primates, all three of the great apes, chimpanzees (Genus *Pan*), the gorilla, and the orangutan (Genus *Pongo*) are endangered due primarily to the destruction of their forest habitats.

Nonprimate mammals are also threatened by deforestation. The giant pandas (*Ailuropoda melanoleuca*) are diminished throughout much of their range in China, due, in part, to destruction of their bamboo forest food resource (figure 29.6). The volcano rabbit (*Romerolagus diazi*), which inhabits the slopes of the volcanoes Popocatepetl and Ixtacihuatl and nearby ridges in Mexico, has diminished greatly in numbers due to hunting pressure and the clearing of its pine forest habitat for agriculture. Several subspecies of sika deer (*Cervus nippon*) are declining due to reduction of their forest habitats in Japan, in Taiwan, and on the Asian mainland. The three South American tapirs (*Tapirus* spp.), the only living New World representatives of the Order Perissodactyla (not including the horse and burro, which were introduced), live in a variety of forest habitats from lowlands to elevations above 3500 m. In spite of their generalized requirements, these species are diminishing rapidly over much of their range as habitat is encroached on by the expanding human population. Subspecific populations of some mammals, such as the Florida panther (*Puma concolor coryi*), Asiatic lion (*Panthera leo persica*), and western giant eland (*Taurotragus derbianus derbianus*), are listed by various conservation agencies as threatened because of the loss of forest habitat.

Forests are not the only habitat altered by humans. Most other habitat types, including grasslands, prairies, savannas, swamps, marshes, and other wetlands, all have been usurped by humans for their own use, with negative consequences for certain mammal species. Much of the historical range of the American bison on the Great Plains is now used for agriculture. Though the great herds were killed for several reasons, their prairie habitat was turned into vast agricultural monocultures. Several species of prairie dogs (*Cynomys* spp.) likewise have been reduced. Their habitat has been taken for agriculture, and they have been poisoned or shot as pests (Miller et al. 1994b). The drastic reduction of the black-footed ferret (*Mustela nigripes*) is directly related to declines of prairie dogs, their primary prey. In China, the extensive cultivation of wetlands for rice and other grain crops led to the virtual extinction in the wild of Pére David's deer (*Elaphurus davidianus*) over 3500 years ago (figure 29.7). Finally, the loss of relatively minor but nonetheless significant habitats such as mangroves in many tropical and subtropical regions has contributed to the reduction of numbers for species such as manatees (*Trichechus* spp.) and the dugong (*Dugong dugon*; Hartman 1979; see chapter 19).

HABITAT DEGRADATION

Habitat degradation may be due to (1) changes in habitats due to acid rain, introduction of toxic wastes, synthetic chemicals, oil spills, and so on; (2) introduction of predators and other alien or nonnative species, such as domestic dogs (*Canis familiaris*) and cats (*Felis* [*catus*] *sylvestris*), wild rats (*Rattus* spp.), or mongooses (*Herpestes* spp.) into particular habitats; or (3) the fragmentation of habitats, which can occur in a variety of ways. We will examine each of these briefly with a few mammalian examples.

The effects of acid rain have been most noticeable in the northeastern portion of the United States and neighboring areas of Canada as well as in parts of Europe. While individual species may not have been threatened or extirpated, the habitat alterations associated with acid rain have undoubtedly had adverse effects on numerous mammals that depend on those habitats. These effects have yet to be adequately measured. The effects of industrial wastes and by-products, such as mercury, pesticides, and

Figure 29.6 Habitat destruction. Giant pandas (*Ailuropoda melanoleuca*) are declining precipitously throughout much of their range in China because of the loss of the bamboo forest habitat that is their home and source of food.

Figure 29.7 Value of captive herds. Peré David's deer once roamed much of northeastern China but ceased to exist in the wild about 3500 years ago, when its swamp habitat was taken over for cultivation by local people. However, unlike some other animals that have been driven to extinction in such scenarios, these deer were maintained in herds in captivity. Although never domesticated, sizable herds of Peré David's deer live in many parks and reserves today.

polychlorinated biphenyls (PCBs), on various habitats also remain largely unknown. Carnivores, higher on the food chain, are most likely to suffer from such pollution because such dangerous chemicals are concentrated as they move through the food chain from producers to top-level carnivores. Populations of numerous species of bats worldwide have declined, and it is possible that increased exposure to chemicals has played a role in these declines. The effects of oil spills on populations of marine mammals such as sea otters and porpoises have been documented for each maritime accident, such as the *Exxon Valdez* disaster in Alaska (Bowyer et al. 1995). Although a species' existence does not appear to be threatened by such events, the ecological integrity of large regions is often affected for many years.

Another type of damaging effect results from the influences of some of the chemicals that humans discharge into the environment. **Endocrine disrupters** are chemicals that mimic the effects of hormones or otherwise interfere with normal functioning of the endocrine system within an organism. Many of the livestock food additives and pesticides used in agriculture contain such synthetic chemical compounds. The negative effects of these chemicals with respect to endocrine functions in reproduction and development are only now being detected (Colburn et al. 1996). The effects include infertility, abnormal development, cancer, and others just now being discovered.

The introduction of predators, which has sometimes been included in the topic of biological invasions (Drake et al. 1989), has radically affected many mammalian populations and continues to pose problems for a number of species. In these instances, humans act as dispersal agents, moving into a previously undisturbed area and bringing with them their domestic dogs and cats. On many Caribbean islands and in the Hawaiian Islands, humans introduced mongooses (Family Herpestidae) in an attempt to control rodent populations; the rats had arrived earlier with the humans and their cargoes. One factor in the decline of the hutias (*Plagiodontia aedium* and *P. araeum*) on Hispaniola probably was the introduction of the small Asian mongoose (*Herpestes javanicus*) for rodent control. Early aboriginal colonists brought the dingo (*Canis lupus*) with them to Australia. One of the affects of the dingo was likely the extinction of the thylacine, or Tasmanian wolf (*Thylacinus cynocephalus*) on mainland Australia. Evidence also exists that humans, finding that Tasmanian wolves posed a threat to their sheep, aided in its demise, particularly on the island of Tasmania. The introduction of dogs and cats has played a critical role in the drastic decline in numbers of the numbat, or banded anteater (*Myrmecobius fasciatus*), in Australia; it is no longer found in New South Wales or South Australia.

A third type of habitat degradation is fragmentation. **Habitat fragmentation** is the process by which a previously continuous area of similar habitat is reduced in area and divided into smaller parcels. This process is best known because of its effect on migratory songbirds in eastern deciduous forests of the United States (Robinson et al.

1995). The amount of forested land in northeastern and Midwestern United States has actually increased in the past 100–150 years. The effect of this additional forested land is diminished, however, because much of it is fragmented and not in large, contiguous habitat, and some is comprised of opportunistic species, not leading to the original climax community. Fragmentation may in fact significantly benefit some mammalian populations. Clearly, the fragmented landscape of much of the central Atlantic states and Midwest has benefited the white-tailed deer (*Odocoileus virginianus*). Interspersing good cover with food resources has lead to sizable populations in most states. A study on habitat fragmentation in southeastern Australian forests provides additional information regarding effects on mammalian communities (Lindemayer et al. 2000). Continuous eucalypt forests supported communities that were richer in mammalian species than did patchy eucalyptus forests, which in turn were richer than exotic pine plantations. Habitat fragmentation can interact synergistically with other factors that negatively impact mammals, increasing the overall effect. Synergistic factors include grazing, invasions by exotic species, and hunting activities (Hobbs 2001; Peres 2001).

Not all fragmentation involves forests. Roadways have been tested as possible barriers for movements by small mammals (Oxley et al. 1974; Swihart and Slade 1984). Data for several rodent species indicate that construction of highways can be a barrier to genetic exchange. Morphological features diverged in populations of bank voles (*Clethrionomys glareolus*) from different sides of a highway (Sikorski and Bernshtein 1984). Furthermore, Oxley and colleagues (1974) reported that roads and highways clearly influenced the tendencies of white-footed mice (*Peromyscus leucopus*) and eastern chipmunks (*Tamias striatus*) to cross barriers during their daily movements (figure 29.8). Although the distances they moved should have led to considerable road crossing, minimal numbers of both species actually ventured across the "barrier."

The effects of road corridors on mammal communities are receiving considerable attention in recent years. Both the highway itself, acting as a barrier, and the collateral effects of the road corridor are being examined (Forman and Deblinger 2000). Excluding Alaska and Hawaii, 22% of the land area of the United States is ecologically impacted by roads (Forman 2000). Roads may actually lead to increased diversity due to edge effects, but this effect can in turn negatively impact the normal fauna in a locale. Furthermore, roads lead to the potential for introduction of exotic species, and with the human activity that inevitably comes with a road, considerable disruption occurs in the lives of the animals living in that area (Trombulak and Frissel 1999).

In addition to these forms of habitat degradation, much habitat is lost to grazing domestic livestock. Most affected are ungulates, such as bison and antelope, over much of the western United States, along with other mammals (Fleischner 1994). Many of these effects can be indirect in that the habitat is modified. The nature of the vegetation

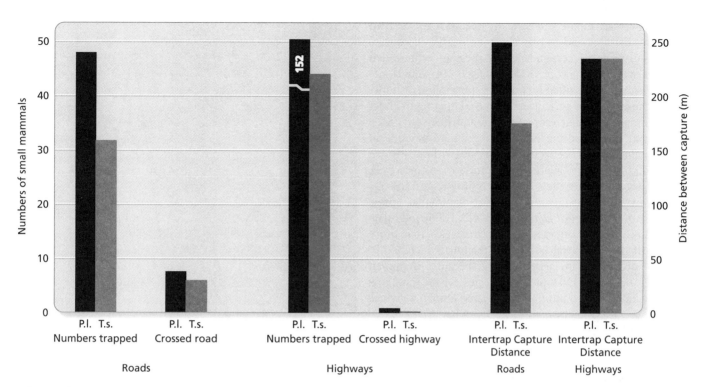

Figure 29.8 **Roads as possible barriers to genetic exchange.** Relatively few white-footed mice (*Peromyscus leucopus* = P. l.) and eastern chipmunks (*Tamias striatus* = T. s.) cross country roads and highways. This is true even though the ranges of their movements, as determined from data on intertrap capture distances, indicate that they travel more than sufficient distances to cross such "barriers." For the four pairs of histograms on the left, use the left axis; for the two pairs of bars on the right, use the right axis. *Data from D. J. Oxley, M. B. Fenton, and G. R. Carmody, 1974, The effects of roads on small mammals, Journal of Applied Ecology 11:51–59.*

on the grasslands is altered, and riparian (riverbank) habitats, with their high levels of vegetative diversity, are trampled and perhaps irreparably altered. One example of small mammals adversely affected by various agricultural practices are the kangaroo rats (*Dipodomys* spp.): many of these species are now endangered, particularly in California (Price and Endo 1989; Price and Kelly 1994).

In a number of countries in Africa, humans graze their cattle on savannas or other similar grasslands, where the cattle compete with a variety of wild ungulates. A related problem is that cattle carry diseases such as rinderpest that can be transmitted to some of the wild ungulates, often decimating local or regional populations. Similar phenomena occur in areas of India and Southeast Asia. Disease has impacted other mammals, including several species of prairie dogs in the western United States. Both black-tailed prairie dogs (*Cynomys ludovicianus*) and Gunnison's prairie dog (*C. gunnisoni*) are periodically decimated in localized areas by outbreaks of plague (*Yersinia pestis*) carried by fleas. Plague was likely introduced to the west coast about 1900 and has spread eastward.

In the past several decades it has become apparent that another form of habitat degradation involves fire, primarily in forested areas. Fires in several regions, such as Sumatra (Kinnaird and O'Brien 1998) and Brazil (Nepstad et al. 1998; Schulze 1998) are now well documented. What is not fully known yet is the impact of these fires on mam-

mals; additional studies in the first part of the 21st century will be needed on the direct and indirect effects on mammalian populations and communities. Yet another recent development is the finding that habitat degradation and animal loss occurs during wars between nations and civil disturbances within countries (Dudley et al. 2002).

A final, more specialized instance of habitat degradation concerns human activities in and around caves. Caves are a limited resource, and thus bats tend to be concentrated in them, putting the bats in peril should the caves be damaged or disturbances to the habitat be persistent. Many caves serve as roosting sites for bats, particularly during hibernation (see chapter 13). When humans repeatedly enter caves, disturb the bats, or despoil the habitat in some manner, bats may not return. Bat species whose habitat has been degraded in this manner and that have therefore suffered declines in North America include the gray bat (*Myotis grisescens*), the Indiana bat (*M. sodalis*), and Townsends' big-eared bat (*Corynorhinus townsendii*). Cave closure can also adversely affect bats. Furthermore, such closure may be a factor in the decline of woodrats (*Neotoma* spp.) in the eastern United States.

SPECIES EXPLOITATION

Another major threat to mammals is the use of firearms to hunt them and traps to capture them. Some hunting and trapping are legitimate and certainly necessary for survival of native cultures and to provide some degree of control for populations of certain species, such as white-tailed deer, beaver, and raccoons (*Procyon lotor*). In some locations, mammals are killed by trapping, poisoning, or hunting because they are agricultural pests, competing in effect with humans for their crops (Burton and Pearson 1987). Regulated sport hunting is also a legitimate practice in many countries and occurs in many locales without adverse effect on mammal populations. In some nations, however, hunting occurs either illegally or with little regard to conservation and management of the mammalian prey. Examination of the *Red Data Book* (IUCN 2006) and related volumes on endangered mammals reveals that more than half the terrestrial nonrodent mammals included in the list are there, in large measure, because of human hunting pressure. Examples of species whose numbers have declined due to hunting and related human activities include a variety of primates (blue monkey *Cercopithecus mitis*; northern plains gray langur, *Semnopithecus entellus*), cervids (brow-antlered deer, *Cervus eldii*; several subspecies of sika deer), camelids (guanaco, *Lama guanicoe*; vicuña, *Vicugna vicugna*), and felids (leopard, *Panthera pardus*; tiger, *Panthera tigris*). Some species of mammals, such as sable antelope (*Hippotragus niger*) and North American beaver (*Castor canadensis*), have been killed or trapped for their horns or pelts. Species such as the sable antelope are endangered as a consequence of past hunting. Others, such as the beaver, were exploited for many decades (see chapter 2) but have recovered to such substantial numbers in most locations that they are often considered pests. Trophy hunting, particularly for ungulates with impressive horns or antlers, has taken an additional toll on some species, particularly in Africa and Asia. Trophy hunting has proven valuable in at least one location, Zambia, where funds from managed trophy hunting are channeled back to those whose resources have been restricted by management practices; the system appears to be functioning quite well (Lewis and Alpert 1997).

A comparison of the relative abundance of mammals weighing more than 1 kg in two areas of Amazonian forest—one area with frequent hunting pressure and a second area hunted only infrequently—revealed significant effects (Bodmer et al. 1997). There were significant differences between the two locales; the overall conclusion was that species with long-lived individuals and with longer generation times were more vulnerable to extinction than were species with short life spans and shorter generation times. This finding was confirmed and extended in a meta-analysis of data on extinctions (Cardillo et al. 2005). Extinction risk for small (< 3 kg) mammals is driven largely by environmental factors, but in larger mammals there is, in addition, the added risk of intrinsic factors.

Some marine mammals have been devastated by hunting pressure. Most notable of these are the numerous species of whales that were killed in large numbers over several centuries (figure 29.9; see also figure 17.14). Both baleen and toothed whales have been hunted for a variety

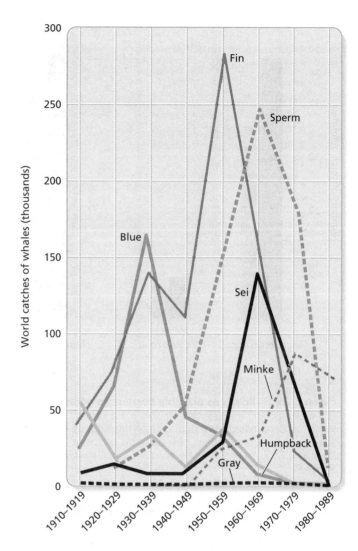

Figure 29.9 World catches of whales. The rate of capture for many types of whales increased sharply in the early part of the 20th century and continued to rise, with the use of newer technologies, until about the 1970s. Then a combination of virtual disappearance of some species and the implementation of whaling quotas followed by bans on hunting allowed some populations to recover. *From Geoffrey Lean et al., 1990,* Atlas of the Environment. *Copyright © Banson Marketing Ltd., London.*

of purposes, most notably for their oil, which was used for fuel and in lamps until petroleum oil was discovered. Other whale products include meat for human and animal consumption, bone meal, spermaceti (a waxy solid used in ointments and cosmetics), ambergris (a waxy material from the intestines used as a fixative in perfumes), and extracts from the liver and endocrine glands. Such inventions as harpoon guns and steam-powered ships hastened the demise of many types of whales. Whaling continued at a steady pace with the use of factory ships and sonar devices until the last quarter of the 20th century. International treaties now protect all whales from further hunting, although several nations persist in taking some whales each year and are attempting to remove the whale hunting ban for some species.

Sea otters (*Enhydra lutris*) are another example of an overexploited sea mammal. Beginning in the late 1700s, these marine mammals were trapped in large numbers for their pelts. By the 20th century, they had diminished to very low numbers. Recovery of sea otter populations in the northern Pacific Ocean was made possible via protection and reintroduction. In addition, numerous porpoises and dolphins have been ensnared in nets used for catching fish and other seafood, where they drown because they cannot reach the surface to breathe.

Some mammal populations have declined, in part, because they make good pets. Especially affected have been some species of small South American primates, such as Goeldi's marmoset (*Callimico goeldii*), and related species. Some African primates are also kept as pets; Diana monkeys (*Cercopithecus diana*) are a good example. For several decades, numerous primates have been exported for use in medical testing and experiments. At the peak, some 30,000–40,000 wild rhesus macaques (*Macaca mulatta*) were exported annually from India, largely for use in testing the Salk polio vaccine. Primates are often ideal subjects for medical testing purposes because of their close evolutionary relationship to humans. In the mid-1970s, the Indian government wisely put an end to these exports. Researchers now rely instead on monkey farming for the production of some primates of medical importance.

A final, more subtle, process involves human domestication of wild mammals. Some stocks of formerly wild ungulates, probably the progenitors of stocks of domestic animals, have become extinct or nearly extinct because they were interbred with domestic forms over many generations. An example is the banteng (*Bos javanicus*) of southeast Asia, which interbred with domestic cattle.

CULTURAL AND RELIGIOUS ISSUES

In a number of cultures, particularly in Africa and Asia, ivory is believed to have special powers. Carved and decorated, it is valued as an art object in many human societies. The limited supply of ivory, primarily from elephants (*Elaphus maximus* and *Loxodonta africana*) and walruses (*Odobenus rosmarus*), has put a premium on this commodity. The result has been extensive poaching of wild elephants (most often in Africa), primarily for the removal of the tusks (Eltringham 1979). Elephant feet are sometimes made into tables for the tourist trade and export. This problem has received much attention in the popular and scientific press in the past decade, although no clear solutions seem to be available (Cohn 1990; Milner-Gulland and Mace 1991; Hoare and DuToit 1999). Similarly, the horns of rhinoceroses are valued for their use in traditional Asian medications and as an aphrodisiac. They are also prized as sheaths and handles for knives in some Middle Eastern cultures. Such practices led to poaching of these animals throughout much of their range, to the point where fewer than 5000 black rhinoceroses (*Diceros bicornis*) now exist and the white rhinoceros (*Ceratotherium simum*)

is endangered. One practice initiated in the last few years to save rhinoceroses involves dehorning them, thus removing the prize for which they are sought by poachers. The efficacy of this practice is much debated, and possible behavioral side effects may result from it (Berger and Cunningham 1994; Berger et al. 1994; Rachlow and Berger 1997; Berger and Cunningham 1998).

In North America, black bears (*Ursus americanus*), and sometimes other bear species, have been killed illegally to remove their gallbladders. This organ is thought to have medicinal or aphrodisiac powers by some people in the Near and Far East. Another mammalian organ thought to have medicinal properties are pinniped penises. Modern molecular techniques, that is, sequencing of the cytochrome-b gene, are now being used to assess which pinnipeds are involved in this illegal trade (Malik et al. 1997). These and other cultural practices that involve particular mammals as delicacies or ceremonial foods have contributed to declining numbers for a variety of species.

ISLANDS AS SPECIAL CASES

Isolated in varying degrees from nearby land masses, islands of varying sizes—from small 1- to 2-hectare islets to larger islands, such as New Guinea—have, through geological time, tended to develop unique faunas and floras. The entire process has been well described in terms of island biogeography (MacArthur and Wilson 1967). The models developed by MacArthur and Wilson related the fauna and flora of islands to a series of factors. The size, age, and distance of an island from either a continent or other islands can influence the number of species present through differential rates of colonization and extinction (see chapter 5).

On islands, some of the factors affecting the decline of certain mammals are exacerbated. Mammals that are endemic to specific islands or groups of islands face special problems. Of the 83 documented cases of mammalian extinctions since the end of the 15th century, 51 (61.5%) have been of mammals living on islands (Ehrlich and Ehrlich 1981; Reid and Miller 1989).

The introduction of a predator or competitor can severely harm an endemic mammal species. We already mentioned the effects of mongooses introduced on Caribbean and Hawaiian islands for rodent control and of dingos accompanying humans as they migrated to Australia. Brown rats (*Rattus norvegicus*) and roof rats (*R. rattus*), introduced to Madagascar possibly as long as 2000 years ago, decimated much of the native small rodent fauna (Goodman 1995). The introduction of the European rabbit (*Oryctolagus cuniculus*) to Australia created intense competition with other small mammals. The solenodons on Hispaniola (*Solenodon paradoxus*) and Cuba (*S. cubanus*) have both been driven nearly extinct by a combination of predation from introduced dogs and cats and loss of their forested habitat. In New Zealand, several bat species, most notably the long-tailed wattled bat (*Chalinolobus tuberculatus*) and short-tailed bat (*Mystacina tuberculata*), have been nearly eliminated by introduced predators that include dogs, cats, weasels (*Mustela* spp.), and rats. Since these island-based endemics do not have other source populations from which recolonization can occur, their demise is hastened if humans do not take special conservation measures.

In recent years, some island situations have served as model systems for ways to attempt to eradicate invasive species (Krajick 2005). One system that serves as a highlights scenario is the Channel Islands off the coast of California. On Santa Cruz Island, feral domestic swine posed a significant threat to plants that are, in many cases, unique to this island or to just a few islands in the group. An eradication campaign aimed at the several thousand pigs on the island has worked remarkably well; shooting the pigs is quite effective. With major reductions in feral pigs, the native vegetation rebounded rapidly, so that by 2006 the original vegetation was back up to 90% coverage.

Another mammalian example involves eradicating up to 150,000 feral goats from Isabela Island in the Galapagos. In addition to shooting goats, two additional strategies are employed to attain near complete elimination of the goats. Some 600 goats fitted with radio collars are permitted to roam, attracting the hard-to-find goats to be shot. Also, hormonally sterilized nannies are turned loose to attract billies that can be killed. These efforts are winning, and normal animal and plant populations are rebounding on Isabela Island. On a worldwide basis, through eradication procedures, particularly on islands, and using different methods depending on the prey, at least seven different mammalian invader species have been removed from islands—a total of almost 600 eradications as of 2005.

Fragments of what were once larger habitats and the various parks and reserves of different sizes that have been established are considered by some to be ecological islands (Newmark 1995). The rates of extinction and recolonization in the various national parks in the western United States fit the predictions from models of island biogeography and related theory. These sorts of formulations—specifically regarding the sizes of the reserves and their possible connections via corridors—deserve careful consideration when conservation plans are made for various mammals (Soulé and Wilcox 1980). New fields of study, such as landscape ecology, provide a broad framework within which particular conservation strategies can be evaluated.

HYBRIDIZATION

One problem, fully recognized only in recent years, concerns the possible extinction of species through hybridization (Avise and Hamrick 1996; Rhymer and Simberloff 1996). The problem can arise in at least three different ways. One way is through the introduction of an exotic species, either accidentally or deliberately (Drake et al. 1989). Exotic species are those that are not native to a particular locale or region, and their introduction can result

in **introgression,** the mixing of the gene pools. The interbreeding of feral house cats (*Felis catus*) with the wild cat (*Felis sylvestris*) in remote areas of Scotland and a similar problem involving the African wild cat (*F. libyca*) in southern Africa exemplify this problem (Stuart and Stuart 1991; Hubbard et al. 1992). Ongoing attempts to reintroduce the red wolf (*Canis rufus*) in eastern North America may be doomed, in part, because evidence from tooth morphology (Nowak 2002) and mitochondrial DNA indicate that red wolves already have significant genes from both gray wolves and coyotes (*C. latrans*; Wayne and Jenks 1991). Given the increasing coyote populations in most areas where red wolves could be released, the probability is high that introgression will occur wherever attempts are made to reestablish red wolves.

A second way hybridization occurs is when habitat modification results in the possibility of two previously separated species meeting and interbreeding. This process can take any of the following forms. Local habitat change, such as clearing brush around natural or artificial ponds, has resulted in the hybridization of several species of tree frogs (Genus *Hyla*; Cade 1983). Regional habitat change, such as lands that have been cleared for agriculture or the reforestation process that occurs when such lands are abandoned, can apparently result in hybridization for some species. The ongoing mixing of the blue-winged warbler (*Vermivora pinus*) and golden-winged warbler (*V. chrysoptera*) in the eastern and central United States is an example (Gill 1980). Although clear examples have yet to be found for mammals, these problems are still a concern.

A third possibility involves humans' own conservation attempts and the introgression of gene pools that can result from such activities. When humans attempt to save a species or subspecies by crossbreeding it with a closely related species or subspecies or when they attempt to enhance a stock of wild animals by introducing animals from a distant population, often of a different subspecies, two issues arise. First, outbreeding depression can occur when new genotypes produced by crossing stocks are inferior to the original native stock and when the new stock is at a disadvantage with respect to adaptation to local conditions. Mammalian examples of outcrossing for conservation reasons include the Florida panther (*Puma concolor coryi*), wisent or European bison (*Bison bonasus*), and wood bison (*Bison bison athabascae*; Fisher et al. 1969; Fergus 1991). In the first case, animals from the Texas subspecies (*Puma concolor cougar*) were introduced into Florida, and in the latter two cases, stocks were enhanced by introduction of American plains bison (*Bison bison*). The second issue is that humans are, to varying degrees, transforming the original gene pool of a species or subspecies, thus opening up the myriad issues surrounding both the definition of a species and what exactly should be conserved. The magnitude of the possible negative effects is unclear in all cases, but what is certain is that introgression of gene pools occurs, and the original species gene pool is altered forever.

A fourth, related consideration involves the escape or release of animals bred for domestic use, such as mink or fox for furs. In many cases, artificial selection within these captive stocks alters the genetics, as for example with larger litter sizes, shifts in pelage coloration, disease resistance, or behavioral temperament. When captive-bred animals find their way back into wild populations and interbreed, the effects on extant populations could be quite detrimental.

Solutions to Conservation Problems

Just as a variety of factors contribute to the decline of many populations and species of mammals, so are there many solutions to the problems, some more general and others particular to specific situations. The single most important factor contributing to the demise of the other mammalian species is the growing human population. For a number of reasons, religious and cultural in nature, this problem unfortunately may be the most difficult one to attack. Mammalian species conservation is most critical precisely where growth rates of the human population are at their highest levels—in Africa; Latin America; and southeast Asia, including Indonesia, all of the Malay Archipelago, and the Philippine Islands. If even the more modest of the projections for human population growth are correct, then it seems likely that many dozens of threatened or endangered species will, in fact, become extinct before the middle of this century. No programs that humans develop will have a chance to succeed without control of the human population.

To protect the remaining wildlife, several measures have received renewed emphasis. First, there is a continuing move to establish and maintain reserves, where native mammals, as well as the plants and other animals that constitute their communities, have an opportunity to survive. National governments and nongovernmental agencies have aided in the establishment of new reserves as well as the addition of lands to existing reserves. This is a continuing process. In many states within the United States and in a number of other countries, preservation of areas for hunting and trapping has provided a stimulus and funding for wildlife areas. In most instances, these reserves provide habitat for many species, not just those being hunted.

Second, in conjunction with the establishment of these reserves, national and international laws concerning hunting, trading, exporting, and other practices that directly affect endangered mammals must be enforced. Many countries have laws designed to protect endangered mammals. Human population pressure, however, drives a need for more food and space for agriculture. Funding for enforcement and education of the populace concerning the laws is lacking, and corruption allows poaching and related activities to flourish in some countries. Many nations, including the United States, are feeling increasing pressure from development interests to permit exploration for and

mining of minerals and other resources and other uses in areas that are designated reserves or wilderness preserves (figure 29.10). On top of this are efforts to develop agricultural land and natural areas for human habitation. Such activities invariably have negative consequences for wildlife, including mammals. Until these sorts of situations are changed, establishment of reserves will only protect mammals in a limited way.

A third major initiative, now being pursued in some countries, involves educating the local residents concerning the benefits to be gained from protecting organisms with which they share their environment. If residents can understand, for example, that large mammals living in their region are a potential economic resource through ecotourism, then there may be hope for saving some of these mammals. Ecotourism provides the funds needed to protect mammals and the habitat those mammals need to survive. Unfortunately, ecotourism sometimes has the negative consequence of overrunning an area with tourists. Cheetahs, for example, are followed by hoards of tourists in minibuses, which interferes with their attempts to stalk and kill prey. The economics of conservation issues need to be examined further, but the basic principle that local people can benefit from conserving rather than killing and consuming the wildlife and the habitat must become firmly established.

Finally, a series of international treaties and agreements has potential benefits for many endangered mammals. The more than 80 national signatories to the Commission on International Trade in Endangered Species (CITES) have agreed to regulate the import and export of many animal species. Trade in products such as ivory and rare furs is regulated or banned altogether. In a similar manner, starting in 1946, treaties promulgated by the International Whaling Commission (IWC) governing the hunting of whales eventually provided for the virtual cessation of whale hunting. Note, however, that many

Figure 29.10 Problems of development. Though designated as a national park, the Tauro National Park in Australia is used for cattle grazing. In other countries, reserves and parks have been degraded by the intrusion of development interests seeking oil and minerals.

nations ceased whaling simply because the whale populations had already become so greatly diminished as to be economically unprofitable. A combination of agreements like these and cooperative multinational efforts may save some of the presently endangered species of mammals. For others, however, it is most likely already too late. Organizations such as the International Union for the Conservation of Nature (IUCN), CITES, and the IWC, as well as conservation groups such as the World Wildlife Fund and the Nature Conservancy, provide a basis for obtaining the cooperation of all "shareholders" of the Earth's riches that can lead to productive and successful conservation programs.

In some locations, perhaps most notably in Europe, a combination of factors results in the reemergence of some mammals long missing from major areas (Enserink and Vogel 2006). Thus, a variety of carnivores, including bears, lynx, and wolves, are increasing in numbers in many areas of Europe. What happened to result in these changes? First, local groups and entire nations are working together to conserve and restore habitats that are conducive to the reappearance of these carnivores. Second, some of the mammals have adapted; where these formerly would shun contact with or living near humans, some carnivores have changed their behavior so that they are more tolerant of civilization. Third, in some cases, hunting laws and practices have changed to guarantee greater production for populations of carnivores.

As the nations of the world become more aware of the problems, plans for conserving and managing resources are being developed. A Species Survival Commission (SSC) exists for each mammalian order. For some endangered species, species survival plans (SSPs) have been created. These plans cover the most critically endangered species and involve breeding animals in captivity for reintroduction into their natural habitat (table 29.1). A desperation measure being employed with some mammals includes careful planning of a captive breeding program that will lead to the eventual reintroduction of some animals in the wild, with the hope that they will either reestablish an extirpated population or augment a critically low one, thus resulting in overall species survival. One program is succeeding with the reintroduction of gray wolves to habitat they formerly occupied in the Greater Yellowstone Ecosystem in the western United States. In this instance, the animals were captured in Canada and transplanted to the park. Such programs are generally expensive and labor-intensive (Fritts 1993; Mech 1995). Wolves now occur in at least four and possibly several more states. This is an example of a successful reintroduction program and one that has led to other similar attempts at mammalian translocations to attempt to reestablish populations of, for example black-footed ferrets, lynx, and elk to locales where they were extirpated.

Considerable discussion is taking place, even among knowledgeable conservation biologists, concerning whether it is more sensible and practical to invest in saving species that are on the verge of extinction via SSPs and reintroductions or to spend funds on mammals (or other animals) that are still living in the wild in sufficient numbers, although threatened or endangered. Another current debate concerns whether conservationists should be working on a species-by-species basis or instead working to preserve habitats (Lindemayer et al. 2002). In the latter

---

Table 29.1   Species survival plan*

1. *Designation of species:* A critically endangered species is selected prior to extinction.
2. *Appointment of species coordinator:* An individual agrees to take charge of the SSP for the designated species and must then coordinate all activities relative to the SSP.
3. *Organization of propagation group:* The propagation group consists of representatives of the institutions that hold captive members of the species and have filed a memorandum of participation. It is the main advisory body for the SSP for the species in question.
4. *Appointment of studbook keeper:* One member of the propagation group is responsible for compiling this vital book of records on the captive members of the species.
5. *Compilation of studbook:* This document is the collection of essential information on the identity, date and place of birth, date and cause of death, parents, and other characteristics (e.g., temperament) of each captive member of the species. It is becoming common practice to obtain DNA fingerprint information for many species. This book serves as the basis for all plans for captive breeding, and the information it contains is the heart of the process.
6. *Formulation of master plan:* Each individual in the captive population is evaluated with respect to its potential breeding status, relationship to others that might be used for breeding,

health, and other characteristics. Each individual is assigned to one of five categories: (a) breed, (b) hold without breeding, (c) transfer to another location, (d) surplus, or (e) research. These decisions are made for the population at each participating institution and then in coordination with the other cooperating institutions. In this way, optimum use can be made of the existing genetic variability to attempt to retain that variability in subsequent generations.
7. *Review of master plan:* The master plan is reviewed by the Wildlife Conservation and Management Committee of the American Association of Zoological Parks and Aquariums.
8. *Update of master plan:* Each SSP requires an annual report summarizing all activities for the previous year, which is then published in the *International Zoo Yearbook*. These reports contain estimates of the existing genetic variability and the minimum viable population projected for the species.
9. *Compilation of husbandry handbook:* This handbook constitutes a record of information on how to care for and manage a particular species that has an SSP. Recording detailed information on diet, behavior, diseases, and many other aspects of the species' biology can aid others working with the same or similar species. A thorough survey of all of the available literature and other information on the species should be included.

*A variety of factors is considered when a species survival plan is put together for a particular mammal. The basic components of an SSP are listed with a synopsis of the procedure at each stage.

case, entire community structures may be preserved. A law to protect endangered or threatened habitats could be superior to the Endangered Species Act because it would protect all species, not just those currently listed as endangered. Some biologists have taken the notion of habitat restoration to a new level (Donlan et al. 2006). The idea is to restore most of the long lost megafauna in North American ecosystems; a process they term Pleistocene Rewilding. The group of 12 scientists promotes the idea with plans to begin with selected species; a large tortoise, wild horses, followed by elephants and lions. Bowles and Whelan (1995) provide details concerning both the planning and applied aspects of recovery of threatened and endangered species.

An issue that is important to all efforts to save mammals is a **population viability analysis** (PVA), which is the determination of how many animals are needed in a population to prevent it from going extinct (Schoenwald-Cox et al. 1983; Soule 1987; Ellner et al. 2002; Reed et al. 2002). Small populations are more likely to be lost than are large ones (figure 29.11), as, for example, was the case for bighorn sheep (*Ovis canadensis*; Berger 1990). When a population is reduced below the minimum viable population due to habitat loss, habitat degradation, or other human-related activities, loss of a particular population or extinction on a broader scale can occur rapidly (Soule 1987). An analysis of population viability in West Indian manatees (*Trichechus manatus*) reveals that 10% decreases in either the rate of adult survival or fecundity would likely result in extinction of this species (Marmontel et al. 1997). Maintaining viable populations depends on enforcement of laws regarding human boating activities, the primary cause for manatee mortality.

Related to the idea of PVA is the concept of genetic variability. Genetic variability is critical for populations to retain the capacity to adapt to a changing environment (Avise and Hamrick 1996). If insufficient genetic variability remains, the probability of extinction increases. Thus, conservation biologists working with mammals must consider the sizes of existing populations when trying to determine if they are viable or susceptible to local extinction. Small populations can also suffer rapid declines due to short-term fluctuations in birth or death rates, or they may be affected by predation, food supplies, natural catastrophes, and other similar natural events.

Advances in DNA-related technologies are being applied to conservation problems to address the issue of genetic variation. By using available forms of DNA fingerprinting, it is possible to mate captive animals so as to maximize their genetic differences, aiding in maintaining overall heterozygosity of the species' gene pool. The potential for variation in phenotypic traits borne by those genes is thought by many to be a critical component of the process of successfully reintroducing animals to their native habitats from captive, often small, breeding populations. Inbreeding depression is a cause for concern in all captive breeding programs and must be monitored and analyzed carefully (Ralls et al. 1988; Kalinowski et al. 2000).

Case Studies

The following three examples illustrate some of the conservation problems faced by mammalogists and how they are being solved. Each has resulted in some hopeful signs of eventual recovery.

ARABIAN ORYX

Arabian oryx (*Oryx leucoryx*; figure 29.12) once roamed over virtually all of the Arabian Peninsula. Hunting, enhanced by the use of motor vehicles (accompanying the development of oil reserves) and modern weapons technology (commencing with the active involvement of European powers in the region during World War I), reduced their numbers to a few, very small, scattered populations by the 1960s (Grimwood 1988). The last wild Arabian oryx was probably killed in 1972. Fortunately, some farsighted individuals had begun a program to save the species, commencing with the capture of some wild oryx in 1962. Captive populations already existed in Arabia. New captive breeding programs were established first in the United States and then at several zoos in Europe. A critical problem, faced in the initial stages of this effort, was that few animals were available for breeding, perhaps only four or five! Additional stocks of Arabian oryx were discovered in private collections at several locations in the area of the Arabian Peninsula as the project progressed. Because maintaining genetic diversity could be critical to the successful survival of such animals reintroduced into native habitat, a key goal of captive

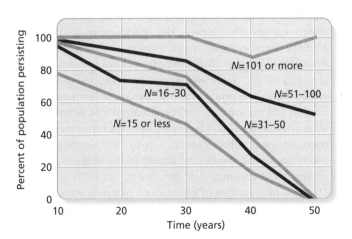

Figure 29.11 Minimum viable population size.
The relationship between original population size and the percentage of populations persisting over time. Most populations with about 100 bighorn sheep survived for 50 years, whereas smaller populations did not. *Data from R. B. Primack, 1993, Essentials of Conservation Biology, Sinauer Associates.*

Figure 29.12 Arabian oryx. These modest-sized ungulates, up to 1.4 m long and weighing about 55 kg, formerly inhabited much of the Arabian Peninsula. Conservation efforts under way today have succeeded in reestablishing Arabian oryx populations at several locations.

breeding programs like the one for the oryx is to maintain careful records on all known specimens. To accomplish this, a studbook (a record of all of the captive oryx, their locations, breeding status and, where known, their genealogy) was established.

By the 1970s, the breeding program was quite successful. The first reintroductions to the Middle East involved oryx established at the New Shaumari Reserve in Jordan, where they remain today. The primary site selected for reintroduction of Arabian oryx to the wild was at Yalooni in Oman. The site has rocky outcrops, temperatures that range from summer highs of 47°C to winter lows of 7°C, and minimal rainfall. Foggy seasons also occur, however, when moisture from the Arabian Sea is deposited on the landscape, providing for periods of relatively lush vegetation. In 1980, oryx were moved from captive breeding locations in the United States to small enclosures at the release site. Release into the wild occurred in stages. After confinement in small enclosures for several months, the oryx were then maintained in a larger, 1-km² enclosure for a period of up to 2 years. Finally, in 1982, a herd of ten animals was released into the wild. Herds set for release comprised specific numbers of males and females, adults and young, based on knowledge of the species and its reproductive patterns. Such knowledge was obtained through prior information on wild Arabian oryx and careful observation of the animals in captivity. The latest data indicate that there are now more than 250 wild oryx. The interest in preserving this species resulted in the establishment of captive breeding populations in Bahrain, Qatar, United Arab Emirates, Oman, Jordan, and Saudi Arabia. Regular meetings of the Coordinating Committee for Conservation of the Arabian Oryx provide international communication, cooperation, and a united effort to push for more reintroductions to establish populations in native habitat throughout the original range of the species. The most recent data

from the IIUCN reveal up to six wild populations resulting from reintroductions, and in excess of 1000 animals in these populations.

Education and local interest play vital roles in the success of such projects. The local people, the Harasis, act as wardens in protecting the oryx, as gamekeepers in managing the stock, and as science aides in participating in scientific observations of the animals. Several subsequent releases have taken place, and the small herds have readily established territories. At this time, the project appears to be a success.

WHALES

Numerous species of whales have been hunted for centuries to obtain a variety of products (see chapter 17). With the advent of modern power ships and technologies for tracking, killing, and processing whales, populations of many species plummeted (table 29.2). As is true for many conservation efforts, issues surrounding whale conservation become entangled in economic and political considerations. Whale hunting dates back thousands of years, when Basques first organized large-scale hunts and before that peoples in other locations in areas throughout the world likely used whales as a source for food, oil, and other needs. Initially, the Basques exploited North Atlantic right whales (*Eubalaena glacialis*) in the Bay of Biscay. As they captured and used most of the whales available there, they moved farther and farther northward and out into the Atlantic Ocean. Whaling expanded during the 17th through 19th centuries, with major whaling fleets mounted by the Dutch, British, and Americans. Soon these fleets discovered the rich supply of whales in the Pacific Ocean. Later, into the 20th century, Japan, Germany, Norway, Korea, Russia, and other nations joined in the hunt for whales.

Techniques used for hunting and harpooning whales and bringing them to the side of the ship for processing improved greatly during the 19th century. Herman Melville immortalized many aspects of this in his novel *Moby-Dick*. By the 20th century, technological advances changed all the processes. Large powered ships replaced sailing vessels, weapons such as explosive harpoons were used to kill the whales, and eventually large factory ships were launched, all of which resulted in a rapid increase in the number of whales taken. Up to 30,000 whales were taken each year beginning about 1930; at least 330,000 whales were captured and processed between 1910 and 1966. As populations of one species of whale were depleted, the factory ships and their crews moved to another species. By the early 1800s, North Atlantic right whales and bowhead whales (*Balaena mysticetus*) had been overharvested. Blue whales (*Balaenoptera musculus*), the largest mammals on Earth, became scarce next, followed in order by fin (*B. physalus*), sei (*B. borealis*), and common minke (*B. acutorostrata*) whales (Burton 1980; Burton and Pearson 1987). Some controls were attempted but proved

Table 29.2 Population estimates of several species of whales indicating the precipitous declines for some during the 20th century

Species (Common Name)	Estimated Population Sizes		
	Pre-1900	1970	Present
Balaenoptera musculus (blue whale)	200,000	5,000	10,000
B. acutorostrata (minke whale)	350,000	350,000	350,000
B. borealis (sei whale)	250,000	45,000	50,000
B. physalus (fin whale)	500,000	84,000	120,000
Eubalaena glacialis (right whale)	320,000	3,000	3,000
Physeter catodon (sperm whale)	2,000,000	487,000	1,000,000
Megaptera novaeangliae (humpback whale)	100,000	3,000	12,000
Balaena mysticetus (bowhead whale)	65,000	5,000	8,000

Data from P. B. Best, 1988, Right whales, (Eubalaena australis) *of Tristan da Cunha: A clue to the "non-recovery" of depleted stock?* Biological Conservation 46:23–51; S. D. Kraus, 1990, Rates and potential causes of mortality in North Atlantic right whale (Eubalaena australis), Marine Mammal Science 6:278–291; R. B. Primack, 1993, Essentials of Conservation Biology, *Sinauer Associates.*

Note: Bans on whaling may have come just in time. Some species have already made a measurable recovery.

generally ineffective, in large measure because whaling was so profitable. Whaling itself and those employed to use its products formed important components of the economies of the nations involved in the hunt. For political reasons, governments were hesitant to move forward with plans to slow down the slaughter.

The IWC, formed in 1946, was largely ineffective. Its members tended to ignore the evidence provided by scientific advisers, who indicated that there would be dire consequences if restraint and long-term recovery plans were not initiated. Breakthroughs occurred in the early 1980s. In 1981, the commission voted to ban hunting sperm whales and the next year voted for a moratorium on all commercial whaling to take effect in 1985. Most nations complied, but several, most notably Russia, Norway, Peru, Japan, Korea, and Iceland, have all continued to do some whaling, though often under the guise of "scientific collecting." At this time, there is growing pressure to resume commercial whaling, particularly of such species as minke whales, which apparently have survived in sufficient numbers or have undergone significant recovery during the moratorium.

The IWC and scientists working with them provide estimates of existing populations of many whale species. Unfortunately, due to a lack of resources, some of these estimates date back to the late 1980s, and few are more recent than 5 years ago. These data do seem to indicate that for several whale species, the moratorium on harvesting whales has been effective; numbers for minke whales and pilot whales indicate some degree of recovery. For most other species, where data are available, there is little evidence of significant recovery. One problem that came to light in recent years is based on the use of genetic information to estimate historical numbers of whales for some species (Roman and Palumbi 2003). The IWC projects estimates for historical populations for humpback whales and fin whales of 20,000 and 40,000, respectively. Using those numbers and the criterion that a whale species must attain 50% of the historical populations size, both of these species would be eligible for harvesting. However, using

genetic information, the true historical populations were more nearly 200,000 for humpback whales and perhaps 360,000 for fin whales. Obviously, correct estimates of the historical populations are necessary for proper conservation and, in particular, for establishing the potential for resuming and harvesting of whales. Additional chapters in the story of whale hunting remain to be written with the hopeful prospect that species whose stocks were so decimated may eventually recover.

GOLDEN-LION TAMARIN

Only about 2% of the original contiguous Atlantic coastal forest in Brazil that is the home for *Leontopithecus rosalia* remains today. Factors leading to the near extinction of this species included habitat destruction from forest removal, to provide lumber and to create open lands for cultivation and cattle grazing, the human traffic in golden-lion tamarins as pets, and the development of lands for upscale human housing along the coast of Brazil.

As the population dwindled to a few hundred wild tamarins and another 100 in zoos by about 1970, the first steps in the recovery for this species were taken. A national reserve, Poco das Antas Biological Reserve, was established in 1974. The year before, an official studbook for golden-lion tamarins was initiated, and the coordination of the recovery project was assigned to the National Zoological Park in Washington, D.C. (Kleiman et al. 1990). The cooperation that developed in the ensuing decades between scientists at zoological parks in several countries and the authorities and conservation biologists in Brazil can serve as a model for similar species preservation efforts.

Progress proceeded in several steps. First, the captive populations were not breeding well and were in fact declining. Information on behavior and reproductive physiology was collected. In tamarin groups, only the dominant pairs generally breed; reproduction of other individuals is suppressed. In addition, female golden-lion tamarins are very aggressive toward one another. Initial steps thus involved forming appropriate breeding groups (see Kleiman and Rylands 2002). This approach enhanced reproduction and the possibility that sufficient numbers could be produced to eventually plan for releases in native habitat. With the establishment of additional breeding centers, the captive population grew to nearly 400 individuals by the mid 1980s. Today there are four different international management and recovery teams working together with the Brazilian government on different species of lion tamarins (*Leontopithecus* spp.; Kleiman and Mallinson 1998), with the collective goal of preserving the Atlantic rainforest where the tamarins and a number of other endangered species live.

A key feature of such programs is to maintain breeding while attempting to maximize the remaining genetic diversity. Any time that a program starts with a limited stock of animals, as in this instance, only a portion of the total genetic diversity of the species will be represented. To foster diversity, a careful breeding program was established, using the information in the studbook. In this way, particular pairs could be bred to maintain the genetic heterozygosity. This plan often meant that tamarins were shipped between breeding centers at different zoos. The time needed for quarantine and adjustment under such circumstances adds to the logistical difficulties of such programs.

At the same time, additional research was taking place on the few tamarins living in the wild in Brazil. Steps were taken in Brazil to protect areas where these animals still remained. Furthermore, a campaign of public education concerning the tamarins heightened people's awareness of the need for their conservation and habitat preservation. In 1983, reintroductions to natural habitat in Brazil began with captive-bred golden-lion tamarins. Despite some losses and various setbacks, it appears that the program is succeeding overall in establishing new populations in former habitat and augmenting existing populations.

An update on the success of the reintroduction of golden-lion tamarins reveals that there are now more than 1000 living in the wild. As recently as 2004–2005, the IUCN changed the status from critically endangered to endangered. Some reintroductions continue, but the overall limitation in terms of the existing, suitable habitat has become a limiting factor. Two related efforts are now underway to address this problem. First, reforestation efforts are gaining momentum in some locales. However, this is a long-term process, with mature forests of the type needed for the tamarins taking perhaps half a century or more. Given that the recovery effort is now some 30 years in duration, the reforestation efforts may occur in a workable time frame. Second, habitat corridors are in development, connecting some fragmented swatches of coastal rainforest. While this effort may not do much to increase the total amount of available habitat, it should lead to more opportunities for dispersal and heightened interbreeding of localized populations. A new threat has emerged in the form of a predator, the tayra (*Eira barbara*), that gains access to the nests of gold-lion tamarins, killing entire groups. This is a more serious type of predation than the situation wherein a hawk or snake would capture and kill one or two members of a group (Smithsonian Institution 2004).

SUMMARY

Conservation of mammals, something that humans have done in various forms for many centuries, has only really become an area of strong interest during the past 100 years. During that time, several waves of interest in conservation-related activities have occurred, most notably during the last 25 years. The passage of many new environmental laws, both in the United States and abroad, specifically including the Endangered Species Act, have heightened our awareness of the fragility of the environment and the need for renewed vigilance.

The decline of a mammal species is likely due to a series of interrelated factors. The seemingly uncontrolled exponential growth of the human population is the single most critical factor in the decline of wild mammals and other animals and plants. Virtually all environmental problems can be traced directly or indirectly to human population growth. Habitat destruction—most commonly deforestation but also the disappearance or degradation of savannas, marshes, and other habitats—poses a serious threat to a wide variety of species. Habitat degradation can take several forms, including the release of pollutants, introduction of predators, and fragmentation. Exploitation of some species through hunting has been a major factor in their decline. Finally, cultural practices, such as those associated with ivory or the use of particular animal parts as traditional medicines or aphrodisiacs, can lead to declining populations. Islands, with their endemic fauna and lack of source populations for recolonization, present special problems with respect to conservation of mammals.

Various solutions are being attempted to overcome these problems. Solutions must be found soon to control human population growth, or most other measures taken for species conservation will be of little value. More reserves are being established, and existing reserves are being enlarged. Laws governing hunting, trapping, and trading in mammals and mammalian products need to be enforced. International cooperation in program development and financing is necessary to ensure the success of mammalian conservation plans. Examples of successful projects include the cessation of whale harvests and the reintroduction of captive-bred mammals such as the Arabian oryx and golden-lion tamarin. These examples provide different perspectives on how conservation biologists are approaching the issue of saving mammals from possible extinction.

SUGGESTED READINGS

Conservation Biology. This journal has excellent articles on mammals in each issue.

Hoyt, J. A. 1994. Animals in peril. Avery, Garden City Park, NY.

Leopold, A. 1966. A Sand County almanac. Oxford Univ. Press, New York.

Nash, R. 1967. Wilderness and the American mind. Yale Univ. Press, New Haven, CT.

Primack, R. B. 1993. Essentials of conservation biology. Sinauer Assoc., Sunderland, MA.

Soulé, M. E. 1987. Viable populations for conservation. Cambridge Univ. Press, New York.

Tudge, C. 1992. Last animals at the zoo. Oxford Univ. Press, Oxford, England.

DISCUSSION QUESTIONS

1. Examine a week of issues of a major city newspaper and compile a list of articles and topics that relate to conservation. Given your knowledge of the conservation field from material in this and other courses, how balanced do you feel the coverage is of the various conservation issues? How many articles could have (but did not) refer to the human population increase as a serious problem? How many articles referred to the human population problem at all?

2. In addition to the topics we covered in this chapter, what other factors pose threats to mammals?

3. When reintroductions are done, such as the golden-lion tamarin in Brazil or the gray wolf in Yellowstone National Park, what measures must be taken to provide for the best possible success rate? Can you suggest some ways to accomplish these measures?

4. One of the biggest sources of conflict facing those who work on species conservation is economic development. From your own background and course work, what examples can you provide of ways in which corporations and conservationists have been able to cooperate, providing for the interests of both groups?

5. List the four or five most important conservation issues in the world today. How would you recommend solving these problems?

Glossary

A

ablation The destruction of tissue by electrical or chemical techniques.

abomasum The fourth and last chamber of the stomach of a ruminant; often called the "true stomach."

accidental host A host in which a parasite does not normally occur.

active dispersal Ecological dispersal events in which individuals move by terrestrial locomotion or flight.

adaptive Making an individual more fit to survive and reproduce in comparison with other individuals of the same species.

adaptive hypothermia A group of energy-conserving responses of mammals and birds characterized by the temporary abandonment of homeothermy.

adrenocortical hormones Hormones produced by the cortex of the adrenal gland.

age structure The proportion of individuals in a population in different age classes.

agenesis A tooth missing from the dental complement that is normally found in a species.

aggression Behavior that appears to be intended to inflict noxious stimulation or destruction on another organism.

agonistic behavior Behavior patterns used during conflict with a conspecific, including overt aggression, threats, and retreats.

agouti Hair pigments that exhibit mixtures or banding from pheomelanin and eumelanin.

albinism The condition in which all hairs are white because of an absence of pigment, caused by a genetic mutation.

alignment In molecular evolution, the definition of homologous sites within homologous DNA or protein sequences; homologous sites are typically placed in the same column (hence "aligned") of a data matrix, the rows of which are the individual sequences.

Allee effect The unfavorable consequences of undercrowding, as when individuals fail to breed because group numbers are below some critical density.

allelopathic The direct inhibition of one species by another using noxious or toxic chemicals.

Allen's rule The biogeographic "rule" that states that extremities of endothermic animals are shorter in colder climates than those of animals of the same species found in warmer climates.

alliance A long-term association between two or more individuals in which they cooperate against a third party, as in dominance interactions.

allograft A piece of tissue or an organ transferred from one individual to another individual of the same species; a successful foreign transplant.

allopatric Occurring in different places, usually referring to geographical separation of populations. Compare to **sympatric**.

altricial Neonates that are born in a relatively undeveloped condition (eyes closed and with minimal fur present) and require prolonged parental care; as opposed to precocial.

altruistic Behavior that reduces personal fitness for the benefit of others.

alveolus The socket in a jaw bone for the root(s) of a tooth.

ambergris A form of excrement of sperm whales.

ambulatory locomotion Walking; usually associated with a plantigrade foot posture.

amphibious mammals Those that spend time in both terrestrial and aquatic habitats.

amphilestids A family of triconodont mammals that was extant for about 50 million years from the mid-Jurassic to the early Cretaceous periods.

ampullary glands Small, paired accessory reproductive glands in some male mammals that contribute their products to the semen.

anal sacs A paired glandular area in carnivores that produces secretory substances.

anestrus The nonbreeding, quiescent condition of the reproductive cycle of a mammal.

angle of attack The angle of the wings of a bat relative to the ground (i.e., the horizontal plane in the direction of movement).

angora Continuously growing, long, flowing hair that may or may not be shed.

anisogamy The condition in which the female gamete (ovum) is larger than the male gamete (sperm).

annual molt A rapid process each year during which most hairs are replaced in species living at temperate and northern latitudes.

Anomodontia A suborder of the Therapsida. These mammal-like amniotes were primarily herbivorous and were extinct by the late Triassic period.

anosmic Without the sense of smell.

antagonistic actions Having opposite effects, as hormones on target tissues.

anterior pituitary gland The anterior lobe of the pituitary gland.

antitragus A small fleshy process on the ventral margin of the pinnae of bats, often those that lack a tragus.

antlers Paired processes that are found only on the skull of cervids (deer), made entirely of bone, branched, and shed yearly.

apocrine sweat glands A type of sweat gland found on the palms of the hands and bottom of the feet of mammals; highly coiled structures located near hair follicles (sudoriferous glands).

apomorphic Characters that are derived or are of more recent origin.

appendicular skeleton The portion of the postcranial skeleton that consists of the pectoral and pelvic girdles, arms (forelimbs), and legs (hind limbs).

aquatic mammals Those that live most of the time in water but come onto land periodically for certain activities such as breeding and parturition.

arboreal Living or moving about in trees.

Archaeoceti An extinct order of whales that had features intermediate between terrestrial mammals and fully marine species.

arousal The third, or final stage, of the cycle of dormancy; characterized by shorter periods of dormancy and increases in body temperature; usually occurs in late winter.

articular bone A bone in the mandible of lower vertebrates and primitive mammals; becomes the malleus bone in the middle ear of modern mammals.

ascending ramus The upward curving branch of the jaw that articulates with the base of the skull; the vertical portion of the dentary bone.

aspect ratio The ratio of the length of a wing to its width; short, wide wings have a low aspect ratio.

association cortex Regions of the cortex that are not specifically identifiable as sensory or motor cortex, where information is processed and integrated across the various sensory modalities.

astragalus One of the ankle bones; in ungulates, its pulleylike surface limits motion to one plane.

atlas The first cervical vertebra; it articulates with the occipital condyles of the skull anteriorly and with the axis posteriorly.

auditory bulla The auditory (hearing) vessicle with the tympanic floor derived from only the petrosal plate and ectotympanic bone.

autocorrelation As used in radiotelemetry, locations or fixes that are too close together in time, that is, dependent on the previous location.

awns The most common guard hairs on mammals, having an expanded distal end with firm tips and a weak base. They exhibit definitive growth and usually lie in one direction, giving pelage a distinctive nap.

axial skeleton The skull, vertebral column, ribs, sternum, hyoid aparatus, and laryngeal skeleton of a vertebrate.

axis The second cervical vertebra; it articulates anteriorly with the atlas.

B

background extinction The permanent loss of a species from the biosphere due to normal ecological processes associated with gradual environmental change.

baculum A penis bone found in certain mammalian orders (also called **os penis**).

baleen The fringed plates of keratinized material that hang from the upper jaw of mysticete whales. Baleen grows throughout the life of an individual and is used to filter small marine organisms from the water for food.

basal metabolic rate The rate of energy conversion in a resting animal with no food in its intestine, at an ambient temperature that causes no thermal stress.

Bateman gradient The large variation in the reproductive success of males compared to females in relation to the number of mates they obtain.

behavioral ecology The study of an animal's struggle for survival as it exploits resources and avoids predators, as well as how an animal's behavior contributes to its reproductive success.

Bergmann's rule The biogeographic "rule" that races of warm-blooded species living in warmer climates are smaller than races from cooler climates.

bet-hedging Spreading of risk, reducing the chance of catastrophic failure to survive and reproduce. Bet-hedging leads to more frequent but less intense bouts of reproduction.

bicornuate Type of uterus in eutherian mammals that has a single cervix and the two uterine horns fused for a part of their length (found in insectivores, most bats, primitive primates, pangolins, some carnivores, elephants, manatees, dugongs, and most ungulates).

bifid Divided into two equal parts.

bifurcated Paired; with two corresponding halves.

binomial nomenclature A system for naming all organisms in which there are two names, one for the genus and the other for the species.

biodiversity The living plants, animals, and other organisms that characterize a particular region or country or the entire earth.

biogeography The study of the patterns of distribution of organisms, including both living and extinct species.

biological classification The grouping of organisms into ordered categories according to their attributes, reflecting their similarities and consistent with their evolutionary descent.

biological communication An action on the part of one organism (the sender) that alters the probability of occurrence of behavior patterns in another organism (the receiver) in a fashion adaptive to either one or both of the participants.

biological community An association of interacting populations, usually defined by the nature of their interaction in the place in which they live.

biological control Using a species of organism to reduce the population density of another species in an area through parasitism or predation.

biological species concept (BSC) A definition of species in which actual or presumed reproductive isolation is the sole criterion for recognizing a species.

biome A broad ecosystem characterized by particular plant life, soil type, and climatic conditions.

bipartite A type of uterus in eutherian mammals that is almost completely divided along the median line, with a single cervical opening into the vagina (found in whales and most carnivores).

blastomeres Early cleavage cells.

blubber A thick subcutaneous layer of fat below the dermis that prevents heat loss from the body core and provides energy, insulation, and bouyancy to whales, seals, and walruses.

body hair The outer layer of hair or fur; also called guard hair.

bootstrapping A process involving construction of replicate sets of data using the parameters of the original data to provide for subsequent analyses.

Bowman's capsule A part of the kidney; the invaginated distal portion of the uriniferous tubule, which contains the glomerulus (also called renal capsule).

brachiation Using the forelimbs to swing from branch to branch.

brachyodont Cheekteeth with low crowns found in certain omnivorous mammals such as primates, bears, pigs, and some rodents; as opposed to hypsodont.

bradycardia A reduced heart rate associated with diving or torpor.

breasts Milk-producing glands unique

to mammals; see **mammae**.

breed Individuals of the same species that have a uniform, heritable appearance.

bristles Firm, generally long hairs that exhibit angora growth, such as in the manes of horses or lions. Bristles function in communication, augmenting or accentuating facial expressions or body postures.

brown adipose tissue A special type of fat packed with mitochondria; the site of nonshivering thermogenesis in eutherian (placental) mammals.

Bruce effect In mice, the effect of a strange male, or his odor, that causes a female to abort and become receptive.

buboes Swollen lymph nodes.

bulbourethral glands In male mammals, the small paired glands that secrete mucus into the urethra at the time of sperm discharge (also called Cowper's glands).

bunodont Low-crowned teeth that have rounded, blunt cusps used primarily for crushing.

C

caching The handling of food for the purpose of conserving it for future use; synonymous with food hoarding.

calcaneum The heel bone; largest and most posterior of the ankle bones.

calcar A process that extends medially from the ankle of bats and helps support the uropatagium.

calcarine fissure A shallow groove or sulcus separating convolutions of the brain; located on the internal walls of the cerebral hemispheres.

callosities Hardened, thick areas on the skin; for example, rough patches or outgrowths found on certain species of whales (same as excrescences) or ischial callosities in cercopithecid primates.

calyx A small collecting area for the renal pyramids found within the mammalian kidney.

camber Considered in cross section, the amount of curvature in a wing.

caniform Doglike.

canine Unicuspid tooth posterior to the incisors and anterior to the premolars. If present, there is never more than one canine tooth in each quadrant.

caniniform Canine-shaped.

cannon bone Metapodials that are fused to form a single long bone in many unguligrade species.

capacitation The physiochemical changes in the spermatozoa that enable them to penetrate the protective covering of cells surrounding the oocyte.

cardiac muscle Heart muscle.

carnassial Bladelike, shearing cheekteeth found in most carnivores; the last upper premolar and the first lower molar in extant mammals. Most highly developed in felids and canids.

carnivorous (carnivory) Consuming a diet primarily of animal material (characterizing meat-eating members of the Order Carnivora, marsupial dasyurids, and others).

carpal A group of bones of the forefoot distal to the radius and ulna that form the wrist.

carrion Animal matter after it is dead and often decaying.

carrying capacity (K) The number of individuals in a population that the resources can support; the equilibrium density reached in a logistic growth curve.

caudal (vertebrae) The most posterior vertebrae; the number varies with tail length.

caudal undulation Vertical motion of the tail of a marine mammal to produce forward momentum.

cavernous sinus A network of small vessels immersed in cool venous blood located in the floor of the cranial cavity and important in heat exchange in certain carnivores and artiodactyls.

cellulolytic Cellulose-splitting enzymes.

cementum A layer of bony material that covers the roots of teeth and helps keep them in place. In many species, cementum annuli (rings) are evident that correlate with the age of the individual.

center of origin The location where a particular taxon arose.

centers of diversity Geographic regions that have a large number of species per unit area.

cerebral cortex The upper main part of the brain, consisting of two hemispheres.

cervical (vertebrae) The most anterior vertebrae, numbering seven in most mammals.

cervix The tip of the uterus that sometimes projects into the vagina.

character displacement Divergence in the characteristics of two otherwise similar species where their ranges overlap; thought to be caused by the effects of competition between the species in the area of overlap.

characters In systematic biology, all heritable features of organisms.

cheekteeth Dentition that is posterior to the canines, that is, premolars and molars.

chevron bones Found on the ventral part of the caudal vertebrae of marine mammals; sites for attachment of the muscles that depress the tail.

chorioallantoic placenta A type of placenta found in eutherians and peramelemorphians (bandicoots and bilbies) composed of two extra-embryonic membranes: an outer chorion and an inner vascularized allantois. Highly vascularized villi enhance nutrient exchange and mechanical connection to the uterine lining.

chorionic villi Fingerlike projections of capillaries from the outermost embryonic membrane that penetrate the endometrium; they increase exchange between maternal and fetal systems.

choriovitelline placenta A type of yolk-sac placenta found in all marsupials except bandicoots; lacks villi and has a weak mechanical connection to the uterine lining.

circadian rhythms Activity patterns with a period of about 24 hours.

circannual rhythms Seasonal activity patterns with a period of about one year.

cladistic biogeography An approach to historical biogeography that seeks to reconstruct the relationships of areas based on the distribution and phylogenies of species occuring in those areas.

cladistics A process by which unweighted, nonmetric characters (traits) are used to organize organisms exclusively into taxa on the basis of joint descent from a common ancestor.

cladogram A simple representation of the branching pattern or phyletic lineage, which does not attempt to represent rates of evolutionary divergence.

clavicle The bone connecting the scapula and sternum.

claviculate Having a clavicle, the bone connecting the scapula and sternum.

claws Structures that occur at the ends of the digits, grow continuously, and consist of two parts, a lower or ventral subunguis, which is continuous with the pad at the end of the digit, and an upper or dorsal unguis.

cleidoic Eggshells that are impermeable to air or nutrient exchange, as are found in birds.

climax community The end point of a successional sequence; a community that has reached a steady state.

climbing An often arboreal locomotion involving the use of the forelimbs and hind limbs to grasp branches and other objects to move about the habitat.

cline A pattern of monotonic change in some character among individuals of a species along a geographic transect through the species' range.

clitoris The small erectile body at the anterior angle of the female vulva; homologous to the penis in the male mammal.

cloaca A chamber into which the digestive, reproductive, and urinary systems empty and from which the products of these systems leave the body.

closed-rooted Teeth that do not grow throughout the life of an individual, as opposed to open-rooted teeth.

coagulating glands Anterior prostate glands of some mammals; upon ejaculation, the secretion of these glands, when mixed with the secretions of the seminal vesicles, sometimes forms a viscous substance that constitutes a "copulatory plug" left in the vagina.

cochlea The spirally coiled tubular cavity of the inner ear containing necessary organs for hearing.

coevolution In parasite-host relationships, reciprocal evolutionary changes that result in close association of certain species of parasites with certain host species.

cohort life table The age-specific survival and reproduction of a group of individuals of the same age, recruited into the population at the same time and followed from birth to death. Compare to **static life table**.

cold-blooded Pertaining to any animal whose body temperature remains close to that of ambient temperature. Includes all animals except birds and mammals; poikilothermy, ectothermy.

colostrum A special type of protein-rich mammalian milk secreted during the first few days before and after birth of young; contains antibodies that confer the mother's immunity to various diseases to the young.

commensal(ism) Different organisms living in close association with each other; one is benefited and the other is neither benefited nor harmed; in close association with humans.

community ecology The study of the interactions among species and the effect those interactions have on both the living and nonliving features of their environment.

community structure Patterns of species composition and abundance, temporal changes in communities, and relationships among locally coexisting species.

comparative phylogeography A research program that uses genealogical concordance among multiple codistributed species to infer common historical processes that have shaped the geographic history of regional biotas.

competition The attempt of two or more organisms (or species) to use the same limited resource.

competitive exclusion principle The hypothesis that two or more species cannot coexist on a single resource that is limiting to both.

competitive release The expansion of range of one species that sometimes occurs when a competing species is removed.

composite signal In communication, a signal that is made up of more than one signal.

concealed ovulation The absence of any behavioral or morphological changes associated with estrus.

conduction The movement of heat from regions of high temperature to regions of low temperature.

Condylarthra A diverse lineage of Paleocene herbivores, a generalized ancestral order, from which arose several orders including proboscideans, sirenians, cetaceans, perissodactyls, and artiodactyls.

condyle A rounded process at the end of a bone, providing for an articulation with a socket of another bone. Occipital condyles provide articulation between the skull and the vertebral column.

cones Retinal receptors for color vision.

conservation Is the care, protection, and management of natural re-sources. The term also refers to the preservation of resources from waste and to the restoration of natural resources.

conspecific Individuals or populations of the same species.

context In communication, stimuli other than the signal that are impinging on the receiver and might alter the meaning of a signal.

continental drift The movement over geological time of the large land masses of the earth's surface as a result of plate tectonics.

convection Movement of heat through a fluid (either liquid or gas) by mass transport in currents.

convergence (convergent) (1) The presence of a similar character for two taxa whose common ancestor lacked that character. (2) The evolution of similar morphologies by distantly related lineages that inhabit regions with similar ecological, geological, and climatic conditions.

convergent evolution The evolutionary origination of similar derived characters in species whose most recent common ancestor did not possess such characters; derived similarities due to convergence are not homologous, they are homoplastic.

coprophagy Feeding upon feces (as in shrews, lagomorphs, rodents); also called refection.

copulation plug A plug of coagulated semen formed in the vagina after copulation; found only in certain species of mammals.

core area The area of heaviest use within the home range.

corpora albicans The degenerated corpus luteum formed after the birth of the fetus or after the egg fails to implant in the uterus (also called "white bodies").

corpora cavernosa A mass of spongy tissue surrounding the male urethra within the penis.

corpus callosum The bundle of nerve fibers that integrates the left and right hemispheres of the brain in eutherian mammals.

corpus luteum An endocrine structure that forms from the remnants of the ovarian follicle after ovulation. The corpus luteum ("yellow body") secretes progestins that support the uterine lining for blastocyst implantation.

corridor route A faunal interchange where there is minimal resistance to the passage of animals between two geographic locations.

cortex (1) The structure that surrounds the medulla and makes up most of the hair shaft. (2) The area of the kidney that contains the renal corpuscles, convoluted tubules, and blood vessels.

cortisol An adrenal cortical hormone.

cosmopolitan Essentially worldwide geographic distribution.

cotyledonary placenta A type of chorioallantoic placenta in which the villi are grouped in well-shaped rosettes separated by stretches of smooth chorion.

countercurrent heat exchange An arrangement of blood vessels that allows peripheral cooling particularly of appendages and at the same time maintains an adequate blood supply without excessive heat loss.

countershading Having ventral body pelage that is more lightly colored than the dorsal surface.

cranium The upper portion of the skull including the bones that surround the brain.

crenulations A series of low humps or ridges on the back of gray whales.

crepuscular Activity concentrated near sunrise and sunset.

crown The portion of a tooth that projects above the gum, composed of enamel and dentine.

crypsis Pattern of camouflage in which the color of the pelage matches that of the substrate. Most commonly found in small, terrestrial mammals.

cryptic coloration A pelage pattern that matches the general background color of the animal's habitat.

cup-shaped discoid placenta The type of chorioallantoic placenta that is a variation on the discoidal arrangement of the placenta.

cursorial A type of locomotion in which at least a portion of the time is spent running.

curation The task of assembling and maintaining a systematic collection, including the database of information associated with each specimen.

cycle sequencing A modification of the Sanger chain-termination DNA sequencing protocol that uses the polymerase chain reaction to amplify template molecules and incorporate dideoxy nucleotides; can be performed with smaller amounts of template than are required for standard Sanger sequencing.

cusp A projection or point on the chewing surface of a tooth. Molariform teeth have several cusps, including a protocone, metacone, and so on.

cuticle The thin, transparent, outer layer of the hair; forms a scalelike pattern on the surface.

cycle of dormancy The three phases of dormancy: entrance, period of dormancy, and arousal.

Cynodontia The diverse group of theriodont therapsid reptiles from which mammals evolved.

D

data matrix In phylogenetics, a summary of the variation in characters from which a phylogeny is estimated; typically, species are the rows, characters are the columns, and character states are the cell values.

deceit In communication, the sending of misleading information.

decidua The uterine mucosa in contact with the trophoblast.

deciduous placenta The type of placenta in which a portion of the uterine wall is torn away at parturition.

definitive hairs Hairs that attain a particular length and are shed and replaced periodically.

definitive host A host in which a parasite reaches sexual maturity.

Dehnel's phenomenon The observation that reduction in the body size of the common shrew (*Sorex araneus*) in autumn is accompanied by shrinkage in the size of the skull.

delayed development A condition found in some neotropical bats and characterized by a reduced growth rate of the embryo following implantation. Differs from delayed fertilization in that the blastocyst implants

shortly after fertilization, but development is very slow.

delayed fertilization An adaptation of certain species of hibernating bats where mating occurs in late summer or autumn. Following copulation, sperm floats free in the uterine tract during the winter. Ovulation then occurs in the spring with subsequent fertilization and implantation.

delayed implantation The postponement of embedding of the blastocyst in the uterine epithelium for several days or months.

deme A local population within which individuals theoretically mate more or less at random.

dendrogram A treelike diagram of the relationships in a phylogeny.

dentary bone The single bone of the lower jaw or mandible in mammals.

dentine Hard, dense, calcareous (containing calcium carbonate), acellular material under the enamel in a tooth; mesodermal in origin.

dermis The part of the skin that consists of connective tissue and is vascularized; located beneath the epidermis.

diaphragm A muscular partition that separates the thoracic (chest) and abdominal cavities.

destructive sampling Permanent removal of some or all tissue associated with a standard museum specimen (e.g., study skin, skeleton, spirit specimen) for biochemical analysis.

diastema A gap between adjacent teeth, for example, between incisors and cheekteeth in rodents, lagomorphs, artiodactyls, and perissodactyls.

didactylous Often used to refer to marsupial orders in which the digits are unfused, each with its own skin sheath; as opposed to syndactylous.

didelphous Pertaining to the female reproductive tract of marsupials in which the uteri, oviducts, and vaginas are paired.

diestrus The final stage of the estrous cycle; cornified cells are rare and some mucus may be present in a vaginal smear; progesterone levels increase, reach a peak, and then decline.

diffuse placenta The type of chorioallantoic placenta in which the villi are expansive and distributed over the entire chorion.

digastric digestive system See **foregut fermentation**.

digitigrade A foot posture associated with cursorial locomotion; the metacarpals and metatarsals are elevated off the substrate, such that only phalanges make contact with the ground.

dilambdodont Tooth cusps and associated ridges arranged in a W-shaped pattern.

dimorphic Having more than one form, size, or appearance; usually referring to the difference between males and females of a species.

dioecious Pertaining to an organism in which male and female reproductive organs occur in different individuals.

diphyletic A group whose members are descended from two distinct lineages.

diphyodont Two sets of teeth during a lifetime. In the typical mammalian pattern, deciduous teeth ("milk" or "baby" teeth) are followed by permanent counterparts.

diprotodont Dentition in marsupials, specifically the Paucituberculata and Diprotodontia, in which the lower jaw is shortened and the single pair of lower incisors is elongated to meet the upper incisors; as opposed to polyprotodont.

discoidal placenta The type of chorioallantoic placenta in which the villi are limited to one or two disc-shaped areas.

discontinuous distribution A condition that occurs when a widespread species becomes restricted or split into isolated geographic locations.

discrete In communication, signals that are all or none.

disease The deviation from a normal, healthy state; illness with specific causes and symptoms.

disjunct distribution A gap between the ranges of sister-species or larger monophyletic groups.

dispersal (1) Movements that occur within the lifetime of the individual, as, for example, when it leaves its natal site. (2) Long-term movement patterns involving species in a historical zoogeographic sense.

displays Behavior patterns that convey messages from one individual to another.

disruptive coloration Patterns of stripes or colors as part of the fur that stand out from the basic background fur pattern.

diurnal Active primarily during daylight hours and quiescent at night.

DNA-DNA hybridization Measures similarity between strands. The DNA is heated and the strands separate but remain intact. When the strands are allowed to cool, they collide with one another by chance, and complementary base pairs again link together. When DNA from a single species has been used, cooling results in homoduplexes. When DNA samples from two species are heated and cooled together, heteroduplexes are formed. Strands are separated and tested for thermal stability by slowly raising the temperature and assessing the mixture for DNA strand disassociation.

DNA fingerprinting A process in which DNA is cleaved by restriction enzymes that result in repetitive units of 16-64 base pairs in length. The fragments are run on gels and visualized by the Southern blot method. The band patterns produced in this manner are unique to each individual.

DNA sequencing The exact listing of the order of the four nucleotide bases (A [adenine], G [guanine], T [thymine], C [cytosine]) within a section of DNA.

Docodonta An order of late Jurassic mammals known only from the remains of complex tooth and jaw fragments.

domesticated Individuals or species that are bred in captivity to benefit a human community that controls breeding, territory, and food supply.

Doppler shift The apparent change in sound or light frequency caused by movement of the source or the receiver.

dormancy A period of inactivity in which an animal allows its body temperature to approximate ambient temperature.

drag As an object moves through a medium, the resistive force from friction with resulting loss of momentum. Friction results from resistance to movement of water or air and the size, shape, and speed of the object moving through it.

Dryolestidae A diverse family of early omnivorous mammals in the Order Eupantotheria that were extinct by the mid-Cretaceous period.

ductus deferens The tube that carries sperm from the epididymus to the cloaca or urethra in male mammals (also called vas deferens).

duplex A type of uterus in which the right and left parts are completely unfused and each has a distinct cervix; found in lagomorphs, rodents, aardvarks, and hyraxes.

E

eccrine sweat glands Sweat glands with separate ducts that lead to the body surface, through which water is forced outward. Found throughout the body, they are important as a means of evaporative cooling in mammals.

echolocate Emit high frequency sound pulses and gain information about the surrounding environment from the returning echoes.

ecological biogeography The subdiscipline of biogeography that seeks to explain current species in terms of interaction between organisms and the ecosystems in which they occur.

ecological dispersal Movements that occur within the lifetime of the individual, as, for example, when it leaves its natal site.

ecological succession The replacement of populations in a community through a more or less regular series to a stable end point (see **climax community**).

economic defendability A state in which the defense of a resource yields benefits that outweigh the costs of defending it.

ecosystem The interacting biotic (living) components (i.e., the community) plus the abiotic (nonliving) components in a defined area.

ectoparasites Parasites that occur on or embedded in the body surface of their host.

ectothermy The maintenance of body temperature primarily by sources outside (**ecto**) the body; cold-blooded, poikilothermy.

edentate Without teeth. True anteaters, pangolins, and the monotremes are all edentate.

Eimer's organs The sensitive tactile organs located on the snouts of moles and desmans.

embryonic diapause A period of

arrested development of an embryo at the stage of the blastocyst (70- to 100-cell stage); found in some kangaroos and wallabies.

emigration The movement of individuals out of a population.

enamel The outer portion on the crown of a tooth. Dense, acellular, and ectodermal in origin, it is the hardest, heaviest, most friction-resistant tissue in vertebrates.

endangered species Species that are likely to go extinct in all or a major portion of their range as a result of human activities and natural causes.

endemism The biogeographic phenomenon in which a species or larger clade occurs only in a restricted area.

endocrine disrupters Chemical compounds that mimic the effects of hormones, facilitating or inhibiting processes normally regulated by the endocrine system; particularly important during prenatal and early postnatal development.

endocrine glands Specialized groups of cells that produce chemical substances that are released into the bloodstream.

endometrium The inner lining of the uterus in which blastocysts implant during gestation.

endoparasites Parasites that occur inside the body of their host.

endotheliochorial placenta The arrangement of the chorioallantoic placenta in which the chorion of the fetus is in direct contact with the maternal capillaries.

endothelioendothelial placenta The arrangement of the chorioallantoic placenta in which the maternal and fetal capillaries are next to each other with no connective tissue between them.

endothermy The maintenance of a relatively constant body temperature by means of heat produced from inside (endo) the body; emphasis is on the mechanism of body temperature regulation; homeothermy, warm-blooded.

energy The ability to do work.

entoconid One of the accessory cusps found in the lingual portion of the talonid of lower molars.

entrance In hibernation, the first stage of the cycle of dormancy; characterized by decrease in heart rate, reduction in oxygen consumption, and decrease in body temperature.

enucleate Without a nucleus. Red blood cells (erythrocytes) in adult mammals are enucleate.

enzootic A disease affecting animals only in certain areas, climates, or seasons.

epidemic A severe disease outbreak affecting many people, often over a widespread area.

epidermis The outer layer of the skin, which consists of three layers: the outer stratum corneum, the middle stratum granulosum, and the inner stratum basale.

epididymus A coiled duct that receives sperm from the seminiferous tubules of the testis and transmits them to the ductus deferens.

epipubic bones Paired bones extending anteriorly from the pelvic girdle. Seen in early reptiles, they occur in monotremes and almost all marsupials.

epitheliochorial placenta The arrangement of the chorioallantoic placenta typified by having six tissue layers, with the villi resting in pockets in the endometrium; the least modified placental condition.

epizootic Rapid, widespread disease; epidemic.

erythrocytes Red blood cells.

estivate (estivation) A period of several days or longer in summer during which a mammal allows its body temperature to approximate ambient temperature.

estradiol See **estrogen**.

estrogen Any of the C_{18} class of steroid hormones, so named because of their estrus-generating properties in female mammals; produced by developing ovarian follicles, under the control of follicle-stimulating hormone. Biologically important estrogens include estradiol, estrone, and estriol.

estrous cycle A sequence of reproductive events, including hormonal, physiological, and behavioral, that typically occur at regular intervals in a female mammal; generally divided into four stages: proestrus, estrus, metestrus, and diestrus.

estrus The period during which female mammals will permit copulation (adj., **estrous**); specifically, when ovulation occurs; detected via vaginal cytology or behavior; also called **heat**.

etiologic The specific causal agent in a disease.

eumelanin Pigment mixtures that provide shades of black and brown.

Eupantotheria An order of early mammals with tribosphenic teeth from which more advanced, therian lineages evolved by the mid to late Cretaceous period.

eusocial (eusociality) A social system involving reproductive division of labor, that is, castes, and cooperative rearing of young by members of previous generations.

euthemorphic An often square molar with four major cusps (protocone, hypocone, paracone, and metacone). Various modifications occur in different mammalian groups.

evaporation The conversion of liquid into vapor.

evaporative cooling Cooling due to absorption of heat when water changes state from a liquid to a vapor. Heat is absorbed from the surface at which the change of state occurs and is carried away with the water vapor produced.

evenness The relative abundance of individuals within each species in a community.

event-based methods Analytical approaches in historical biogeography that use phylogeny and process models to reconstruct evolutionary events such as speciation, extinction, dispersal, or vicariance within the history of a single monophyletic group

excrescences Hardened, thick areas on the skin; rough patches or outgrowths found on certain species of whales (same as callosities).

exemplars In systematic biology, species selected for inclusion in a phylogenetic study to represent larger, presumably monophyletic groups.

exotic Nonnative; a species introduced to an area in which it does not occur naturally.

exploitation competition A type of competition in which organisms passively use up resources; also called scramble competition. Contrast with **interference competition**.

exponential (growth) In reference to rates of increase (or decrease) in population size in which the number present (N) is raised to a power; accelerating population growth as rate of change depends on the number of organisms present.

extinction The loss of a species, which is often a natural process and the ultimate fate for all species.

F

facultative An organism that is not dependent on establishing a parasitic relationship but can do so if the opportunity arises.

facultative delayed implantation A form of delayed implantation in which the delay occurs because the female is nursing a large litter or faces extreme environmental conditions.

falcate Curved or hooked.

faunal interchange Active long-term species dispersal movements.

fecundity The number of offspring produced during a unit of time.

feliform Catlike.

female defense polygyny A mating system in which males control access to females directly by competing with other males.

female reproductive system Reproductive organs of females consisting of a pair of ovaries, a pair of oviducts, one or two enlarged uteri, a vagina, and a cervix.

fence effect The tendency of populations to reach high densities when surrounded by a fence or natural barrier.

fenestrated Describing an area of a skull with light, feathery, latticelike bone structure.

feral Wild or free-ranging individuals or populations that were once domesticated.

fertilization The penetration of an egg by a sperm with the subsequent combination of paternal and maternal DNA.

field metabolic rate The rate of energy use in an animal engaging in normal activities under natural conditions.

filiform Thin and threadlike in shape.

filter route A faunal interchange where only certain species move between land masses because of some type of barrier.

fimbriation (fimbriated) Stiff fringe of hairs between the toes that aid species, such as shrews, in locomotion.

flagging behavior Alarm signaling, as with the use of the tail or rump patch.

flehmen A retraction of the upper lip exhibited soon after sniffing the anogenital region of another or while investigating freshly voided urine.

flight (powered) Volant locomotion in the aerial environment through the use of wings that beat through the use of energy.

fluctuating asymmetry Random deviations from bilateral symmetry in paired traits, such as horns.

fluke The horizontal, dorsoventrally flattened distal end of a whale's tail.

focal animal sampling The recording of all occurrences of specified behavior patterns or interactions of a selected individual or individuals during a period of prescribed length.

folivorous (folivory) Consuming a diet of leaves and stems (characterizing koalas, gliders, ringtail possums, sloths, pandas, dermopterans, various megachiropterans, primates, and rodents).

follicle A small cavity or pit; in the reproductive system of mammals a group of follicles enclose a single egg immediately under the surface of the ovary.

follicle-stimulating hormone (FSH) A hormone produced by the anterior pituitary gland that stimulates development of ovarian follicles and secretion of estrogens; stimulates spermatogenesis; stimulates Leydig cell development and testosterone production in males.

food hoarding The handling of food for the purpose of conserving it for future use; synonymous with caching or storing.

foramen magnum An opening at the base of the skull that provides the pathway for the spinal cord to enter the brain.

foramina Openings in bone, for example, in the facial region for the eyes and at the base of the skull for the spinal cord.

foregut fermentation The digestive process characterized by mammals that possess a complex, multichambered stomach with cellulose-digesting microorganisms that enable them to derive nutrients from highly fibrous foods. Also called rumination; digastric digestive system.

fossorial Digging under the ground surface to find food or create shelter.

frugivorous (frugivory) Consuming a diet of fruit (characterizing pteropodid and phyllostomid bats, phalangerids, and primates such as indrids, lorisids, cercopithecids, colobines, pongids).

functional response The change in the rate of consumption of prey by a predatory species as a result of a change in the density of its prey. Compare to **numerical response**.

fundamental niche The full range of conditions and resources in which the species can maintain a viable population. Compare to **realized niche**.

fur The most common underhair; consists of closely spaced, fine, short hairs.

fusiform A cigar- or torpedo-shaped body form tapered at both ends.

G

gait A pattern of regular oscillations of the legs in the course of forward movement.

gamete A mature, haploid, functional sex cell (egg or sperm) capable of uniting with the alternate sex cell to form a zygote.

general adaptation syndrome (GAS) The situation in which nonspecific stressors, such as heat, cold, or defeat in a fight, produce a specific physiological response.

geographic information systems (GIS) Computer software programs used to store and manipulate geographic information.

gestation The length of time from fertilization until birth of the fetus.

ghost lineage The existence of a clade prior to the earliest occurrence of its fossil representatives, inferred from the oldest fossils of its sister group.

glans penis The head or distal end of the penis.

gliding The aerial locomotion involving the use of a membrane (patagium) to provide lift, but without any active power.

glissant The gliding locomotion found in colugos, "flying" squirrels, and other species where patagia provide extended surface areas.

global positioning system (GPS) The use of satellite-based radio signals and a receiver to obtain the latitude and longitude at any location on the earth's surface.

Gloger's rule The biogeographic "rule" that states, "Races in warm and humid areas are more heavily pigmented than those in cool and dry areas."

glomerulus The minute, coiled mass of capillaries within a Bowman's capsule of the mammalian kidney.

graded In communication, signals that are analog or varying continuously.

granivorous (granivory) Consuming a diet of primarily fruits, nuts, and seeds.

graviportal A mode of locomotion in large species in which the limbs are straight and pillarlike and adapted to support great mass. Species such as elephants and hippopotamuses have their legs directly under the body, sacrificing agility for support.

group selection A selection that operates on two or more genetic lineages (groups) as units; broadly defined, this includes kin and interdemic selection.

guano Bat fecal droppings that often accumulate in large amounts where colonies roost.

guard hair The outer layer of hair or fur (overhair); comprised of three types: awns, bristles, and spines.

guilds Groups of species that exploit a common resource base in a similar fashion.

gumivorous (gumivory) Consuming a diet of exudates from plants such as resins, sap, or gum (characterizing marmosets, mouse lemurs, petaurid gliders, and Leadbeater's possum).

H

habitat destruction A condition in which the expanding human population exerts strong negative effects on a variety of habitats in an effort to meet agricultural and industrial needs; habitats are altered such that some or all of the original fauna and flora can no longer exist in the community.

habitat fragmentation A condition in which the continuous area of similar habitat is reduced and divided into smaller sections because of roads, fields, and towns.

hair Cylindrical outgrowths from the epidermis composed of cornified epithelial cells; a unique feature of mammals.

hair horns The "horns" of rhinoceroses. These consist of agglutinated, kertinized fibers similar to hairs that develop above the nasal bones. They are not homologous to the true horns of bovids.

hallux The first (most medial) digit of the pes (hind foot); the big toe in humans.

handicap In sexual selection, the hypothesis that apparently deleterious sexual ornaments possessed by males are attractive to females because they indicate that the males bearing them have such vigor that they can survive even with the handicap.

haplotypes Alleles at a locus defined by their DNA sequences or restriction site maps.

harmonics Integral multiples of a fundamental sound frequency.

heat See **estrus**.

heat load The sum of the environmental and metabolic heat gain.

Heimal threshold The depth of snow required to insulate the subnivean (below the snow) environment against fluctuating environmental temperatures.

hematocrit The number of red blood cells per unit volume of blood.

hemochorial placenta The arrangement of the chorioallantoic placenta that lacks maternal epithelium; villi are in direct contact with the maternal blood supply.

hemoendothelial placenta The arrangement of the chorioallantoic placenta in which the fetal capillaries are literally bathed in the maternal blood supply. This arrangement shows the greatest destruction of placental tissues and least separation of fetal and maternal bloodstreams.

herbivorous Animals that consume plant material.

heterodont Teeth that vary in form and function and generally include incisors, canines, premolars, and molars.

heterothermic (heterothermy) Animals that at times exhibit high and well-regulated body temperatures (homeothermic) and at other times exhibit body temperatures that are close to that of the environment (ectothermic).

hibernation A form of adaptive hypothermia characterized by pro-

found dormancy in which the animal remains at a body temperature ranging from 2 to 5°C for periods of weeks during the winter season.

hindgut fermentation A digestive system in which food is completely digested in the stomach and passes to the large intestine and cecum, where microorganisms ferment the ingested cellulose; also called a monogastric system.

historical biogeography The subdiscipline of biogeography that deals with past changes in species ranges, as well as the causes and evolutionary consequences of those changes.

homeothermy The regulation of a constant body temperature by physiological means regardless of external temperature; endothermy, warm-blooded.

home range The area in which an animal spends most of its time engaged in normal activities.

homing The process of returning to a home range, nest site, or den.

homologous In evolutionary biology, an adjective that applies to the same character in two or more organisms or species that was inherited from a common ancestor that also had the character.

homodont Teeth that do not vary in form and function; often peglike in structure, as in toothed whales and some xenarthrans.

homoplasy In systematic biology, a term that describes the occurence of similar, derived characters in two species due to convergent evolution; a mistaken hypothesis of homology.

hooves Large masses of keratin (the unguis) completely surrounding the subunguis; found in the ungulates (perissodactyls and artiodactyls); a specialized variation of claws.

hormones Chemical substances that are released into the bloodstream or into body fluids from endocrine glands and that affect target tissues.

horns Processes in bovids, formed from an inner core of bone, that extend from the frontal bone of the cranium, are covered by a sheath of keratinized material, and are derived from the epidermis.

hydric Wetland habitats; as opposed to xeric (very dry) areas.

hyoid apparatus A series of bones

located in the upper throat region that are modified remnants of the gill arches in ancestral fish. The hyoid supports the base of the tongue and larynx.

hypocone A cusp that is posterior to the protocone and lingual (toward the tongue) in upper molars. It is labial (toward the cheek) in lower molars (where it is called a hypoconid). The addition of this cusp often forms quadritubercular molars.

hypoconid One of the accessory cusps found in the labial portion of the talonid of lower molars.

hypoconulid One of the accessory cusps found in the posterior portion of the talonid of lower molars.

hypodermis The innermost layer of the integument, consisting of fatty tissue; the base of each hair follicle is located in this layer, along with vascular tissues, parts of sweat glands, and portions of the dermal sensory receptors.

hypothalamic-pituitary portal system The vascular connection between the hypothalamus and pituitary gland.

hypothalamus The part of the midbrain, located below the thalamus, that contains collections of neuron cell bodies (nuclei). Sensory input to the hypothalamus comes from other brain regions and from cells within the hypothalamus that monitor conditions in blood that passes through the region. The hypothalamus is the key brain region for body homeostasis; the mammalian "thermostat."

hypothermia A condition in which the temperature of the body is subnormal.

hypsodont Cheekteeth with high crowns, often with complex folding ridges; as opposed to brachyodont.

hystricognathous A mandible in certain rodents; in ventral view, the angular process is lateral to the alveolus of the incisor (as opposed to sciurognathus).

hystricomorph Rodents in which the infraorbital foramen is greatly enlarged.

I

ideal free distribution The distribution of individuals among resource patches of different quality that equalizes the net rate of gain of each individual; assumes that organisms are free to move and have complete knowledge about patch quality.

imbricate Overlapping; as in fish scales, for example.

immigration The movement of individuals into a population.

implantation The attachment of the embryo to the uterine wall of the female mammal.

inbreeding depression The reduced reproductive success and survival of offspring from closely related parents compared to offspring of unrelated parents. It is caused by increased homozygosity of the inbred offspring and the resulting expression of deleterious recessive alleles.

incisors Usually unicuspid teeth anterior to the canines that are used for cutting or gnawing.

inclusive fitness The sum of an individual's direct and indirect fitness. Direct fitness is measured by reproductive success of one's own offspring (descendant relatives), and indirect fitness is measured by the reproductive success of one's nondescendant relatives.

incrassated Thickened or swollen.

incus The second of the three bones of the middle ear in mammals (ossicles); derived from the quadrate bone.

indicator models Models of sexual selection that assume that the trait favored by females in some way indicates male fitness.

induced ovulation Ovulation that occurs within a few hours following copulation; the act of copulation serves as a trigger for ovulation.

infanticide The killing of young.

infrasound Sound frequencies of less than 20 Hz.

insectivorous (insectivory) Consuming a diet of insects, other small arthropods, or worms.

insensible water loss The mechanism by which water is lost by diffusion through the skin and from the surfaces of the respiratory tract; also called transpirational water loss.

in situ hybridization Experimental procedure in which nucleic acid probes are chemically bound to chromosome spreads such that specific DNA sequences can be localized on the chromosomes.

integument The outer boundary layer between an animal and its environment; the skin.

interference competition A form of competition in which organisms defend or otherwise control limited resources; also called contest competition. Contrast with **exploitation competition**.

interleukin Any of several compounds that are produced by lymphocytes or monocytes and function especially in regulation of the immune system.

intersexual selection The selection of characteristics of one sex (usually males) based on mate choices made by members of the other sex (usually females).

intrasexual selection The selection of characteristics of the sexes based on competition among them (usually males) for access to members of the other sex (usually females).

introgression The mixing of gene pools.

island rule Large mammals on islands tend to be smaller than their mainland relatives, whereas small mammals on islands tend to be larger than their mainland relatives.

iteroparity (iteroparous) The production of offspring by an organism in successive bouts. Compare to **semelparity (semelparous)**.

K

K-selection Selection favoring slow rates of reproduction and growth, characteristics that are adapted to stable, predictable habitats. Compare to **r-selection**.

karyotype The characteristic number and shapes of the chromosomes of a species.

keratin A tough, fibrous scleroprotein found in epidermal tissues (in hard structures such as hair and hooves, for example).

keratinized Made of keratin.

keystone guild A group of species exploiting a common resource and controlling the distribution and abundance of many other members of the community.

keystone predator Species that control the distribution and abundance of many other members of the community, often by limiting a particular species of prey.

kidneys Paired, bean-shaped structures in mammals located within the dorsal part of the abdominal cavity; the principal organ that regulates the volume and composition of the internal fluid environment.

kin selection The selection of genes due to an individual's assisting the survival and reproduction of nondescendant relatives who possess the same genes by common descent.

kinship The possession of a common ancestor in the not-too-distant past.

krill Small marine organisms fed on by baleen whales.

L

labia majora Two large lateral folds of skin that border and cover the vulva area.

labia minora Two lateral folds of skin that cover the vaginal opening and are largely covered by the labia majora.

lactation The production of milk by mammary glands.

lactogenic hormone A hormone, namely prolactin (PRL), produced by the anterior lobe of the pituitary gland that induces lactation and maintains the corpora lutea in a functioning state in mammals (originally called luteotropic hormone, LTH).

lactose A 12-carbon sugar present in the milk of mammals.

lambdoidal crest A bony ridge at the rear of the cranium.

laminae Ridges on teeth that may have distinct cusps.

laminar flow The smooth movement of air or water over a surface with a minimum of turbulence.

landscape ecology The study of the distribution of individuals, populations, and communities across different levels of spatial scale.

laparoscopy The use of fiber optic techniques to perform a laparotomy.

laparotomy An internal examination of the female reproductive tract to determine the condition of the ovaries and uterus.

law of the minimum The idea that only a single factor limits the growth of a population at any one time.

lek An area used, usually consistently,

for communal courtship displays.

lesion Wound; an area of tissue destroyed via electrical or chemical means.

Leydig cells The interstitial cells between the seminiferous tubules in the testes that produce androgens in response to luteinizing hormone secreted by the anterior pituitary gland.

life table A summary by age of the survivorship and fecundity of individuals in a population.

lift The upward force created as air moves over the top of a wing.

locomotion A form of movement. Running, jumping, gliding, swimming, or flying are all types of locomotion that occur in various mammals.

logistic equation The mathematical expression for a sigmoid (S-shaped) growth curve in which the rate of increase decreases in linear fashion as population size increases.

loop of Henle A long, thin-walled kidney tubule present only in mammals and some birds. The concentrating ability of the mammalian kidney is closely related to the length of the loops of Henle and collecting ducts.

lophodont An occlusal pattern in which the cusps of cheekteeth form a series of continuous, transverse ridges, or lophs, as in elephants.

lophs Elongated ridges formed by the fusion of tooth cusps.

lower critical temperature The temperature at which an animal must increase its metabolic rate to balance heat loss.

lumbar (vertebrae) The lower back vertebrae that number from four to seven in mammals; sometimes partially or entirely fused.

lutenizing hormone (LH) A hormone produced by the anterior pituitary gland that stimulates corpora lutea development and production of progesterone in females.

lysozyme A crystalline enzyme-like protein present in tears, saliva, milk, and many other animal fluids that is able to destroy bacteria by disintegration.

M

macroecology The patterns of body mass, population density, and geo-

graphic range at a continental scale.

macroparasites Parasites that are larger and have longer generation times than microparasites; they usually do not reproduce entirely within or on the host.

male dominance polygyny A mating system in which males compete and acquire dominance ranks that influence their access to females, with higher-ranking males obtaining more mates.

male reproductive system Reproductive organs of males consisting of paired testes, paired accessory glands, a duct system, and a copulatory organ.

malleus The first of the three bones of the middle ear in mammals (ossicles). The "hammer" connects the tympanic membrane (eardrum) and the incus; derived from the reptilian articular bone.

mammae Milk-producing glands unique to mammals (sing., **mamma**); see **breasts**.

mammalogy The study of animals that constitute the class Mammalia, a taxonomic group of vertebrates in the Kingdom Animalia.

mammary glands (mammae) Milk-producing, hormone-mediated glands, unique to mammals; similar to apocrine glands in development and structure.

mandible The lower jaw, consisting of paired dentary bones that meet anteriorly at the mandibular symphysis and articulate posteriorly with the squamosal bones of the cranium to form the jaw joint.

mandibular fossa A part of the cranium with which the mandible (lower jaw) articulates.

manubrium The long, lever-type arm of the malleus that attaches to the tympanic membrane (eardrum); also the anterior (uppermost) portion of the sternum.

manus The forefoot; together, the carpals, metacarpals, and phalanges.

marine mammals Mammals that spend their entire lives in the ocean and never come onto land.

mark-recapture study An experimental procedure for studying the characteristics of wild populations; individuals are captured, marked, released, and (perhaps) captured again at a later time. Recapture rates can be used in conjunction with other data and assumptions to estimate population parameters.

marsupium An external pouch formed by folds of skin in the abdominal wall. Found in many marsupials and in echidnas, the marsupium encloses mammary glands and serves as an incubation chamber.

masseter One of three main masticatory muscles of mammals that functions to close the mouth by raising the mandible. Pronounced in herbivorous mammals, the masseter aids with the horizontal movement of the jaw.

mass extinction The sudden, catastrophic, and simultaneous extinction of many species on a global scale.

mass-specific metabolic rate The rate of energy necessary per gram of body mass; refers to energy demands within the tissues of an animal.

maxilla One of a pair of large bones that form part of the upper jaw, carrying teeth; it also forms portions of the rostrum, hard palate, and zygomatic arch.

meaning In communication, how the recipient of a message interprets that message.

medulla (1) The central portion or shaft of the hair. (2) The internal area of the kidney divided into triangular wedges called renal pyramids.

melanin A group of brown pigments produced in certain dermal chromatophores (melanophores). These pigments absorb ultraviolet radiation. They are injected into growing hair cells to give pelages their distinctive colors.

melanism A condition in which an animal is generally all black, due to a genetic mutation.

meroblastic The type of egg cleavage in reptiles, birds, and monotremes in which only part of the cytoplasm is cleaved due to a large amount of yolk.

mesaxonic Having a weight-bearing axis of a limb pass through the third digit, as in perissodactyls.

mesial drift Molariform dentition that is replaced horizontally rather than vertically. As anterior teeth wear out, they move forward and are replaced from the rear of the jaw by posterior teeth; occurs in macropodids, ele-

phants, and manatees.

message In communication, information about the state of the sender.

metabolic rate Energy expenditure measured in kilojoules per day.

metabolic water Water produced by aerobic catabolism of food; also known as oxidation water.

metacone A cusp that is posterior to the protocone. It is labial (toward the cheek) in upper molars and lingual (toward the tongue) in lower molars (where it is called a metaconid).

metameric Segmented; repeated body units.

metapodials A general term for both the metacarpal and metatarsal bones.

metapopulation A set of local populations or demes linked together via dispersal.

metestrus The third stage of the estrous cycle: leucocytes appear among the cornified epithelial cells in a vaginal smear, corpora lutea are fully formed, and progesterone levels are high.

microfauna Symbiotic ciliated protozoans and bacteria in the forestomach or cecum of herbivores that break down cellulose and other plant materials.

microparasites Parasites that are microscopic and have rapid regeneration times generally within the host.

microsatellite markers Tandem repeats of short DNA sequences, usually multiples of two to four bases.

migration A persistent movement across different habitats in response to seasonal changes in resource availability and quality. In mammals, these typically are round-trip movements, and the individual returns to the same breeding and wintering areas each year.

mitochondrial DNA (mtDNA) All or part of the circular DNA molecules found within mitochondria; in mammals, mtDNA is just over 16,000 bases long and contains 2 ribosomal RNA genes, 13 protein-coding genes, 22 transfer RNA genes, and one large noncoding ("control") region.

molar A nondeciduous cheektooth with multiple cusps that is posterior to the premolars.

molecular clock A model of sequence evolution for a particular gene or protein in which the rate of substitu-

tion is the same in all lineages of organisms of a particular group.

molecular cytogenetics Application of molecular methods, particularly in situ hybridization of nucleic acids, to the study of chromosomes.

molting The seasonal replacement of definitive hair and sometimes angora hair.

monestrous Having a single estrous period or heat each year.

monogamy (monogamous) A mating system in which a single male and female pair for some period of time and share in the rearing of offspring.

monogastric system See **hindgut fermentation**.

monophyletic group (clade) A group consisting of an ancestral species, all of its descendants, and nothing else.

monotypic Having only one member in the next lower taxon. For example, the aardvark is a monotypic order. It has only one family, one genus in that family, and one species in the genus.

monozygotic polyembryony Reproductive process in some armadillos in which a single zygote splits into separate zygotes and forms several identical embryos all of the same sex.

morphometrics The measurement of the characteristics of organisms, including such features as the skeleton, pelage, antlers, or horns.

mortality Death, usually expressed as a rate.

mosaic evolution A pattern wherein the different components of an existing structure evolved at different rates through evolutionary time.

multiparous Describing a female that has had several litters or young; often with evidence of placental scars of different ages.

Multituberculata An order of herbivorous mammals with large lower incisors and molariform teeth with numerous large cusps that extended for 120 million years from the late Jurassic period to the late Eocene epoch.

musk Secretions from scent glands found in mustelids and a variety of other mammalian species.

musth The reproductive period in male elephants.

mutualism (mutualistic) A mutually beneficial association between different kinds of organisms.

mycophagous (mycophagy) Consuming a diet of fungi (characterizes many sciurids, murids, and the marsupial Family Potoroidae).

mycorrhiza The mutualistic association of the mycelium of a fungus with the roots of a seed plant.

myoglobin A protein in the muscles that binds oxygen.

myometrium The thick muscular wall surrounding the highly vascular endometrium in mammals.

myomorph Rodents in which the infraorbital foramen is small to moderate in size.

myrmecophagous (myrmecophagy) Feeding primarily on colonial insects such as ants and termites. Many mammalian families are primarily or secondarily myrmecophagous.

Mysticeti A modern order of baleen whales; species that have two external nares and a symmetrical skull and do not echolocate.

N

nails A specialized variation of claws that evolved in primates to facilitate better gripping ability and precision in object manipulation by the hands and feet. Only the dorsal surface of the end of each digit is covered by the nail.

nares External nostrils or "blowholes" in whales.

natal dispersal More or less permanent movements from the natal site to a site where reproduction takes place.

natality Birth, usually expressed as a rate.

nectarivorous (nectarivory) (also nectivorous, nectivory) Consuming a diet of nectar; found in about six genera of bats and marsupial honey possums.

necton Larger marine organisms with movements independent of waves and currents.

neonates Newborn animals.

neotony The retention of juvenile characteristics in an adult.

nephrons Functional units of the mammalian kidney; consisting of the Bowman's capsule and a long, unbranched tubule running through the cortex and medulla and ending in the pelvis.

niche The role of an organism in an ecological community, involving its way of living and its relationships with other biotic and abiotic features of the environment.

nictating membrane A thin membrane that functions as a "third eyelid" in certain species.

nocturnal Exhibiting peak activity during hours of darkness and resting when there is daylight.

nondeciduous placenta A type of placenta that separates easily into embryonic and maternal tissue at parturition, resulting in little or no damage to the uterine wall.

nonshivering thermogenesis Means of heat production in mammals that does not involve muscle contraction.

norepinephrine A catecholamine found in sympathetic postganglionic neurons of mammals that stimulates production of heat by brown adipose tissue (also called noradrenaline).

nulliparous A female that has never given birth; shows no evidence of placental scars or pregnancy.

numerical response A change in the population size of a predatory species as a result of a change in the density of its prey. Compare to **functional response**.

numerical taxonomy A system in which individuals are organized into taxa based on unweighted estimates of overall similarity.

nunatak Refugia found within ice sheets during periods of glaciation; pockets of variable size that were not covered by the advancing glaciers.

nutritional condition A measure of how well a particular individual at a particular time has assimilated the nutrients necessary for normal metabolism.

nutritional requirements The types and minimum amounts of nutrients needed by an individual, or characteristic of a species, to meet the metabolic costs of normal activity.

O

obligate delayed implantation A form of delayed implantation in which the delay occurs as a normal, consistent part of the reproductive cycle, as in armadillos.

obligate parasites Organisms that must spend at least part of their life cycle as a parasite.

observability (1) Extent to which habitat permits regular, direct observation. (2) The fact that when species are watched, individuals may be seen for different portions of the time period depending on age, sex, or dominance status.

occlusal The surfaces of upper and lower teeth that contact each other during chewing. Occlusal surfaces of teeth have one or more cusps.

Odontoceti A modern order of toothed whales; species with a single external nare, asymmetrical skull, and echolocation.

omasum The muscular third chamber of the stomach of a ruminant.

omnivorous (omnivory) Consuming both animal and vegetable food (characterizing most rodents, bears, raccoons, opossums, pigs, and humans).

open-rooted Teeth that grow throughout the life of an individual, as opposed to closed-rooted teeth.

operational sex ratio The number of reproductively active males and females in a population, expressed as a proportion.

organizational effects Morphological, physiological, and behavioral differences in adults resulting from prenatal exposure to hormones that alter the developmental trajectories of various cells and tissues.

os baculum See **os penis**.

os clitoris A small bone present in the clitoris in some mammal species. Homologous to the baculum (**os penis**) in males.

os penis A bone in the penises of certain mammals (also called a **baculum**).

os sacrum Fused sacral vertebrae in mammals.

osmoregulation The maintenance of proper internal salt and water concentrations; this function is performed principally by the kidneys in mammals.

ossicles (auditory) The three bones (the malleus, incus, and stapes) of the middle ear in mammals that transmit sound waves from the tympanic membrane (eardrum) to the inner ear.

ossicones Short, permanent, unbranched processes of bone that form the horns in giraffes.

outbreeding depression A condition that occurs when new genotypes produced by crossing stocks are inferior to the original native stock; the new stock is at a disadvantage with respect to adaptation to local conditions, possibly because of the breaking up of coadapted gene complexes.

outgroup Any group (taxon) used for comparisons in a phylogenetic analysis; the outgroup cannot contain any members that are part of the study group.

ova Eggs shed from an ovary (singular: **ovum**)

ovaries The female gonads; the site of egg production and maturation (sing., **ovary**).

oviducts The ducts that carry the eggs from the ovary to the uterus (also called Fallopian tubes).

oviparous Able to reproduce by laying eggs, as in monotremes; unlike therian mammals, which are viviparous.

ovulation The releasing of an egg by the ovary into the oviduct.

oxytocin A hormone produced by the posterior pituitary gland that causes rhythmical contractions of the uterus during parturition and enhances milk "letdown."

P

pachyostotic Describing bones that are very dense.

pacing A gait in which both legs on the same side are raised together.

Paenungulata A group within the generalized ancestral Order Condylarthra from which evolved elephants, dugongs, manatees, and hyraxes.

palmate Flattened or weblike.

palynology Studies of pollen preserved in bogs and other moist places. Because plants that are characteristic of a region at a given time reflect the existing climate, knowledge of flora requirements can be used to describe past climate conditions.

panbiogeography A method of defining ancestral biotas by plotting disjunct distributions of many groups on a map and connecting them by lines to form "tracks".

pandemic A large-scale disease outbreak over a wide geographic area.

Pangaea A large landmass formed by all the continents about 200 million years ago prior to their drifting apart.

panting A method of cooling characterized by very rapid, shallow breathing that increases evaporation of water from the upper respiratory tract, as occurs in canids and small ungulates.

papillae Small, protruding projections.

paracone A cusp that is anterior to the protocone. It is labial (toward the cheek) in upper molars and lingual (toward the tongue) in lower molars (where it is called a paraconid).

paraxonic Having a weight-bearing axis of a limb pass through the third and fourth digits, as in artiodactyls.

parental investment Any investment in offspring that increases its chances of survival and reproduction at the expense of the parents' ability to invest in other offspring.

parental manipulation The selective providing of care to some offspring at the expense of other offspring so as to maximize the parents' reproductive success.

parous Describing female mammals that are pregnant or show evidence of previous pregnancies (e.g., possess placental scars).

parsimony analysis of endemicity (PAE) A numerical method for identifying areas of endemism and the relationships among them based on the presence or absence of species at specific localities across a larger region.

parsimony (rule of) The practice of adopting the simplest explanation for an observation consistent with the facts; in taxonomy, determining which cladogram best represents the evolution of a particular group. The tree with the fewest steps or branching points (character states) is generally accepted as the best representation of the phylogeny.

parturition The process of giving birth in mammals.

passive dispersal Movements in which the dispersing organisms have no active role.

passive integrated transponder (PIT) A technology used for radiotracking; PIT tags include an integrated circuit programmed with a unique identification code that is transmit-

ted when the tag is in proximity to a transceiver.

patagium An integumentary membrane stretching from the body wall to the limbs or tips of digits; it serves as the airfoil in gliding mammals and the wing in bats.

path integration A process by which an organism uses internal spatial localization to return from an outward-bound trip; also called ideothetic or dead reckoning.

pectinate A comblike structure with several prongs or projections in a row.

pectoral girdle Bones of the shoulder region providing for articulation of the forelimbs; the scapula and clavicle or only the scapula form the shoulder joint in most mammals.

pedicel (pedicle) (1) A short supporting stalk or stem. (2) In deer, the extension of the frontal bone on which the antlers occur.

pelage All the hairs on an individual mammal.

pelagic The open ocean; away from coastal areas.

pelvic girdle The bones of the hip region, providing for articulation of the hind limbs and consisting of the paired ilia, ischia, and pubic bones.

pelvis A large cavity within the mammalian kidney; the renal pelvis empties into the ureter.

Pelycosauria One of two orders within the reptilian Subclass Synapsida. Pelycosaurs had more primitive characteristics than the other order, the Therapsida.

penis The male copulatory organ through which sperm are deposited in the female reproductive tract and urine leaves the body.

pentadactyl Having five digits. The hands and feet of humans are pentadactyl, as are those of insectivores.

Peramuridae A family of Jurassic mammals that probably gave rise to the lineage of advanced therians, that is, mammals of metatherian-eutherian grade.

perineal swelling The swelling and sometimes reddening of tissues in the anogenital region of some primates in estrus produced by the actions of estrogen.

period of dormancy The second stage of the cycle of dormancy characterized by leveling off of body temperature; usually occurs in early winter; see **dormancy**.

phalanges Bones of the fingers and toes; the distal-most bones in the manus and pes.

phenotype matching A mechanism by which kin may recognize one another; individuals use as a reference kin whose phenotypes are learned by association.

pheomelanin (xanthophylls) Pigment mixtures that produce various shades of red and yellow.

pheromones Airborne chemical signals that elicit responses in other individuals, usually of the same species.

philopatric Living and breeding near the place of birth.

phylogenetic biogeography An approach to understanding the historical biogeography of a group of species by combining a reconstructed phylogeny with the "peripheral isolation" model of speciation.

phylogenetic classification The principle that only monophyletic groups should be recognized with formal taxonomic names.

phylogenetics (phylogeny) The evolutionary history of various groups of living organisms.

phylogenetic species concept (PSC) A definition of species in which a species is the smallest diagnosable cluster of individual organisms within which there is a parental pattern of ancestry and descent.

phylogeography The study of biogeographic history, including living forms, fossils, geology, and molecular tools.

phylogram A tree diagram attempting to represent the degree of genetic divergence among the taxa represented by the lengths of the branches and the angles between them.

piloerection Fluffing of the fur.

piloting The use of familiar landmarks to locate food or shelter.

pinnae External ears that surround the auditory meatus and channel soundwaves to the tympanic membranes (eardrums); not found in many marine and fossorial mammals.

Pinnipedia (pinnipeds) Literally, "feather-footed"; aquatic carnivores that include the seals and walrus.

piscivorous (piscivory) Consuming a diet composed primarily of fish (characterizes bulldog bats [*Noctilio*]).

pituitary gland The master gland of the endocrine system; located below the hypothalamus.

placenta A highly vascularized endocrine organ developed during gestation from the embryonic chorion and the maternal uterine wall (endometrium); connects to the umbilical cord through which nutrient and waste exchange occurs between mother and fetus.

placental scar A pigmented area on the uterine wall formed from prior attachment of a fetus.

plankton Floating plant and animal life in lakes and oceans; movements are primarily dependent on waves and currents.

plantigrade Walking on the soles of the hands and feet.

plate tectonics The theory that the earth's crust, including the surfaces of continents and the ocean floors, is made up of a series of geological plates.

plesiomorphic Characters that are ancestral or that appeared earlier.

pods Groups, schools, or herds of animals; specifically applied to whales.

poikilothermy Pertaining to animals whose body temperature is variable and fluctuates with that of the environment; includes all animals except birds and mammals; cold-blooded, ectothermy.

pollex The first (most medial) digit of the manus (forefoot); the thumb in humans.

polyandry A mating system in which females acquire more than one male as a mate.

polyestrous Pertaining to species that exhibit several periods of estrus or heat per year.

polygamy (polygamous) A mating system in which both males and females mate with several members of the opposite sex.

polygyny A mating system in which some males obtain more than one mate and females provide most of the care of offspring.

polygyny threshold The point at which a female will benefit more by joining an already mated male possessing a good territory rather than an unmated male on a poor territory.

polymerase chain reaction (PCR) A procedure for preparing large amounts of DNA from small amounts of sample.

polymorphism Character variation among individuals within a species.

polyprotodont Dentition in several orders of marsupials in which the lower jaw is equal in length to the upper jaw and the lower incisors are small and unspecialized; as opposed to diprotodont.

populations Groups of organisms of the same species, present at the same place and time.

postjuvenile molt A molt that starts soon after weaning.

precocial Born in a relatively well-developed condition (eyes open, fully furred, and able to move immediately) and requiring minimal parental care, for example, snowshoe hares, deer, porcupines, and many bovids; as opposed to altricial.

prehensile Possessing digits or tail able to grasp branches and other objects.

premolars Cheekteeth that are anterior to the molars and posterior to the canines. Unlike molars, there are both deciduous and permanent premolars.

preputial glands Modified sebaceous glands that in males of some species contribute to the formation of the semen and in others secrete a scent used for marking.

priming pheromones Chemical communication substances that produce generalized internal physiological responses, such as the production and release of hormones.

prions Small, modified proteins thought to be the disease agents in spongiform encephalopathies, including "mad cow" disease.

procumbent Projecting forward more or less horizontally, as teeth in shrews, horses, and prosimian primates.

proestrus The beginning stage of the estrous cycle when nucleated cells are present in a vaginal smear and when estrogen, progesterone, and lutenizing hormone levels reach their peak.

progesterone A steroid hormone produced in small quantities by the follicle and in larger quantities by the corpus luteum; promotes growth of the uterine lining and makes possible the implantation of the fertilized egg.

prolactin (PRL) A hormone produced by the anterior pituitary gland that has many actions relating to reproduction and water balance in mammals. For example, PRL promotes corpus luteum function in ovaries and stimulates milk production.

promiscuity A mating system in which there is no prolonged association between the sexes and in which multiple matings by both sexes occurs.

propatagium The anterior portion of a bat's wing that extends from the shoulder to the wrist.

prostaglandins Lipid-based hormones that communicate between cells over a short distance; involved in several aspects of reproductive function such as increased contractions of the uterus.

prostate gland A mass of muscle and glandular tissue surrounding the base of the urethra in male mammals; at the moment of sperm release it secretes an alkaline fluid that has a stimulating effect on the action of the sperm.

protein electrophoresis A method that uses the characteristic migration distance of various proteins in an electric field to identify and compare individuals; sometimes called allozyme analysis.

protein immunology The cross-reactivity of the homologous (original) antigen (protein) and a heterologous antigen (protein from a different, related species) to provide an estimate of the degree of genetic relationship between the two species.

protocone The primary cusp in a tribosphenic molar at the apex of the trigon. It is lingual in upper molars and labial in lower molars (where it is called the protoconid).

protrusible Capable of being turned inside out; bulging or jutting out.

provincialism The phenomenon in which areas of endemism for a large number of species overlap, resulting in biogeographic provinces

pseudopregnancy Any period when there is a functional corpus luteum and buildup of the endometrial uterine layer in the absence of pregnancy (synonymous with luteal phase).

pterygoideus One of three main masticatory muscles of mammals that functions to close the mouth by raising the mandible; important in stabilizing and controlling the movement of the jaw.

pulp cavity The part of the tooth below the gumline that contains nerves and blood vessels to maintain the dentine.

Q

quadrate bone A bone in the posterior part of the mandible of lower vertebrates; becomes the incus in modern mammals.

quadritubercular (quadrituberculate) Describing a square or rectangular cheektooth with four major cusps: protocone, paracone, metacone, and hypocone.

R

r-selection Selection favoring rapid rates of reproduction and growth, especially among species that specialize in colonizing short-lived, unstable habitats. Compare to **K-selection**.

radiation Energy transmitted as electromagnetic waves (e.g., ultraviolet, visible, and infrared).

radioimmunoassay (RIA) A method for assessing hormone levels on small blood samples involving a radioactively labeled antibody mixed with blood samples from an animal in a competitive binding assay.

radiotelemetry A method for determining the location and movements of an animal by using a transmitter affixed to the individual, the signals from which are monitored with an antenna and a receiver from known points in the study area.

random amplified polymorphic DNA (RAPDs) Involves the use of restriction enzymes with short primers in conjunction with PCR. The resulting DNA fragments are 200 to 2000 base pairs long and appear as a series of bands on a gel. The primers are not locus-specific, and thus the amplified loci are said to be anonymous.

ranging Movements that include forays outside the home range, usually in search of suitable habitat or mating opportunities.

Rapoport's rule The latitudinal breadth of species ranges tends to be larger for species at higher latitudes than for species at lower latitudes.

realized niche The range of conditions and resources in which the species can maintain a viable population when in the presence of competitors or predators. Compare to **fundamental niche**.

reciprocal altruism The trading of altruistic acts by individuals at different times, that is, the payback to the altruist occurs some time after the receipt of the act.

reciprocal monophyly A pattern of genealogical relationship among haplotypes sampled from two distinct groups, such that all haplotypes from each group are more closely related to one another than to any haplotype from the other group.

refugia Small geographical areas that preserve ancestral biodiversity during periods of envirnomental change.

regionalization The scientific process of identifying biogeographic provinces.

relaxin A hormone produced by the corpora lutea that acts to soften the ligaments of the pelvis so it can spread and allow the fetus to pass through the birth canal.

renal corpuscle A unit of the mammalian kidney, located in the cortical portion of the mammalian kidney, that is composed of Bowman's capsule and the glomerulus.

renal papillae Narrow apices of the cortex of the mammalian kidney. Because the papillae are composed of long loops of Henle, their prominence gives an indication of the number and length of such loops.

renal pyramids Triangular wedges of the medulla of the mammalian kidney.

reproductive effort The energy expended and risk taken to reproduce, measured in terms of the decrease in ability of the organism to reproduce at a later time.

reproductive value The sum of an individual's current reproductive output and its expected future output at age x.

reservoir A source that maintains a disease agent in nature.

resource defense polygyny A mating system in which males control access to females indirectly by monopolizing resources needed by females.

restriction fragment-length polymorphisms (RFLPs) Fragments of DNA that have been isolated and cut with one or more restriction enzymes. They are placed on a gel for electrophoresis and stained to permit viewing of the fragments sorted by size.

rete mirabile A complex mass of intertwined capillaries specialized for exchange of heat or dissolved substances between countercurrent flowing blood (also called miraculous net, marvelous net, wonderful net).

reticulum The second of the four compartments of the stomach of ruminants; a blind-end sac with honeycomb partitions in its walls.

rhinarium An area of moist, hairless skin surrounding the nostrils.

ribs Bones attached to the thoracic vertebrae on the dorsal surface and, in most cases, to the sternum on the ventral surface. The rib cage, or thoracic basket, surrounds and protects the vital internal organs.

ritualized Describing behavior patterns that have become modified through evolution to serve as communication signals.

rods Retinal receptors for black-and-white vision.

root The portion of a tooth that is below the gum and fills the alveolus.

rorquals Literally, "tube throated;" large baleen whales with longitudinal grooves on their throats that allow for expansion as they fill with water during feeding.

rostrum The anterior portion of the face or cranium.

rumen The first and largest compartment of the four-part stomach of ruminants.

ruminant artiodactyl A member of the Order Artiodactyla that "chews its cud," or ruminates (e.g., cervids, bovids, antilocaprids, giraffids).

ruminate (rumination) To chew the cud; see **foregut fermentation**.

runaway selection Selection for ornaments (usually in males) that happens due to the genetic correlation and the resulting positive feedback relationship between the trait and the preference for the trait.

rut The mating season in cervids and other artiodactyls.

S

sacculated A stomach with more than one chamber and symbiotic microorganisms for cellulose digestion present in the first chamber(s). The stomach of certain herbivores, whales, and marsupials is sacculated.

sacral (vertebrae) In most mammals, vertebrae that are fused to form the **os sacrum**, to which the pelvic girdle attaches.

sagittal crest The bony midline ridge on the top of the cranium formed by the temporal ridges.

salivary amylase A potent digestive enzyme produced by the salivary glands.

saltatorial locomotion Jumping and ricocheting. Jumping involves the use of all four feet; ricocheting involves propulsion provided only by the two hind limbs.

sanguinivorous (sanguinivory) Feeding on a diet of blood (e.g., vampire bats).

scaling Structural and functional consequences of a change in size or in scale among animals.

scan sampling The recording of the current activity of all or selected members of a group at predetermined intervals.

scansorial Adapted for climbing.

scapula A part of the pectoral girdle; the shoulder blade.

scent glands Modified sweat or sebaceous glands that produce substances used for a wide variety of functions in mammals.

schizodactylous Grasping digits in which the first two most medial oppose the remaining three.

sciurognathous Describing the mandible in certain rodents; in ventral view, the angular process is in line with the alveolus of the incisor (as opposed to hystricognathous).

sciuromorph Rodents in which the infraorbital foramen is relatively small.

scramble polygyny A mating system in which males actively search for mates without overt competition.

scrotum A bag or pouch of skin in the pelvic region of many male mammals that contains the testicles.

seasonal molt The change of the pelage more than once each year.

sebaceous glands Structures associated with hair follicles that secrete oils to keep the hair moist and waterproof.

sectorial Cutting or shearing teeth.

selenodont A cusp pattern in molariform teeth of goats, sheep, cows, and deer in which the lophs form cresent-shaped ridges or "half-moons" on the grinding surface.

sella A median projection of the nose leaf of horseshoe bats in the Subfamily Rhinolophinae.

sem antic Of or relating to the meaning of signals; specifically used to denote the use of different alarm signals to warn about different predators.

semelparity (semelparous) The production of offspring by an organism once in its life. Compare to **iteroparity (iteroparous)**.

semen A product of the male reproductive system that includes sperm and the secretions of various glands associated with the reproductive tract (also called seminal fluid).

semibrachiators An animal that moves by swinging from branch to branch.

seminal vesicles The swollen portion of a male reproductive duct in which sperm are stored and that secretes a fluid useful in the transmission of sperm during copulation (also called vesicular glands).

seminiferous tubules The long, convoluted tubules of vertebrate testes in which sperm cells are produced and undergo various stages of maturation or spermatogenesis.

senescence The gradual deterioration of function in an organism with age, leading to increased probability of death.

sensory channels The physical modality used for signaling, for example, odor or vision.

Sertoli cells Cells that line the seminiferous tubules and that surround the developing sperm, which they nourish.

set point A "reference" temperature in the hypothalamus; analogous to a thermostatic control.

sexual dichromatism The presence of distinctly different pelage colors in males and females.

sexual dimorphism A difference in the

sexes in form, such as size; males are often larger than females, although the opposite occurs in some species.

sexual selection Selection in relation to mating; composed of competition among members of one sex (usually males) for access to the other sex and choice of members of one sex by members of the other sex (usually females).

shaft The central structure of a hair, comprised of the inner medulla, surrounded by the cortex, and covered by a thin outer cuticle.

signal The physical form in which a message is coded for transmission through the environment.

signaling pheromones Airborne chemical signals that produce an immediate motor response, such as the initiation of a mounting sequence.

simplex A type of uterus in eutherian mammals in which all separation between the uterine horns is lacking; the single uterus opens into the vagina through one cervix (found in some bats, higher primates, xenarthrans).

sister groups Two monophyletic groups that are each other's closest relatives.

smooth muscle An involuntary muscle.

society A group of individuals belonging to the same species and organized in a cooperative manner. Usually assumed to extend beyond sexual behavior and parental care of offspring.

somatic muscle Muscle derived from embryonic somites; somatic muscles orient the body in the external environment.

sound window The use of frequencies for communication that are transmitted through the environment with little loss of strength (attenuation).

speciation The evolutionary process by which new species are derived from ancestral species.

species-area relationship The equation describing the increase in number of species as a function of the area sampled.

species dispersal The extension of a species range into a previously unoccupied area.

species range The geographic area over which individuals of a particular species occur at a given point in time.

species richness The number of species in an area.

sperm competition A situation in which one male's sperm fertilize a disproportionate number of eggs when a female copulates with more than one male.

spermaceti An organ found in the head of certain species of toothed whales. It contains a waxy liquid that may function in diving physiology and echolocation.

spermatogenesis A series of cell divisions and chromosome and cytoplasmic changes involved in the production of functional spermatozoa, beginning with the undifferentiated germinal epithelium.

spines Stiff, enlarged guard hairs that exhibit definitive growth.

spirit specimens Museum specimens that consist of carcasses, or portions of carcasses, preserved in alcohol; many spirit specimens are initially fixed in formalin.

spontaneous ovulation Ovulation that occurs without copulation.

stapes Also called the "stirrup;" the last of the three middle ear bones (ossicles) found in mammals. In other vertebrates, this is the only ossicle (the columella) in the middle ear.

static life table A life table generated from a cross section of the population at a specific time. Compare to **cohort life table**.

station keeping Local movements of an animal within its home range, as it acquires resources or marks and defends its territory.

sternum A series of bony elements along the midventral line of the thoracic region that articulate with the ribs and (in some species) the clavicle; a part of the axial skeleton.

striated muscle A voluntary or skeletal muscle.

study skin A museum specimen that consists of the preserved integument of an individual mammal; study skins of small mammals are usually stuffed with cotton and dried in a flattened, linear pose.

subspecies A taxonomic category below the level of species; subspecies are usually named to recognize discrete polymorphism across the geographic range of a species.

subterranean Living underground for all activities.

subunguis The lower or ventral por-

tion of the claw, which is continuous with the pad at the end of the digit.

supernumerary An additional tooth or teeth in a position where they do not normally occur in a species.

survivorship curve (l_x) The proportion of newborn individuals alive at age x, plotted against age.

sweat glands Epidermal glands that lie deep in the dermis but are connected to the surface of the skin by a coiled tube. They produce a watery secretion (sweat).

sweepstakes route A dispersal route in which some unusual occurrence carries an organism or group of organisms across a dispersal barrier into a previously unoccupied area.

symbiosis Two species living together in which one benefits and the other may benefit (mutualism), be unaffected (commensalism), or harmed (parasitism).

Symmetrodonta An early order of therian mammals with tribosphenic teeth that include small carnivores or insectivores from the late Triassic period.

sympatric Occurring in the same place; usually referring to areas of overlap in species distributions. Compare to allopatric.

synapomorphy A sharing of a derived trait by two or more taxa.

Synapsida One of four subclasses of reptiles, this is the subclass from which mammals evolved.

syndactylous Having digits in which the skeletal elements of the second and third toes are fused and share a common skin sheath, as in the marsupial Orders Peramelemorphia and Diprotodontia; as opposed to didactylous.

syndesmochorial placenta The arrangement of the chorioallantoic placenta that possesses one less layer than the epitheliochorial condition.

syntax The information provided by the sequence in which signals are transmitted.

synteny The localization of homologous genes on the same chromosome in different species.

syntopic Being present at the same time and place.

T

talonid The basin or heel in lower

molariform teeth posterior to the trigonid that occludes with the protocone of the upper molar.

tapetum lucidum A reflective layer lying outside the receptor layer of the retina that causes the eye shine when light strikes the retina at night. This structure aids in night vision by reflecting light that has passed through the receptor layer back toward the retina.

taxonomic key An arrangement of the traits of a group of organisms into a series of hierarchical, dichotomous choices.

taxonomic revision A redefinition of species boundaries within a group of closely related species, usually based on a consideration of all or most of the available specimens (and perhaps genetic data) representing the group.

taxonomy A description of species and the process of classifying them into groups that reflect their phylogenetic history.

taxon sampling The rationale for choosing which species to include in a phylogenetic study.

telescoped skulls Compressed posterior bones and elongated anterior bones in the cranium of modern whales, with associated movement of the nares to the top of the skull.

temporalis One of three main masticatory muscles of mammals that functions to close the mouth by raising the mandible. Pronounced in carnivorous mammals, the temporalis assists in holding the jaws closed and aids in the vertical chewing action.

territory An area occupied exclusively and defended by an animal or group of animals.

testes Oval glands in males, often in the scrotum, that produce sperm.

testosterone A steroid hormone secreted by the testes, especially in higher vertebrates; responsible for the development and maintenance of sexual characteristics and the normal production of sperm.

Therapsida One of two orders within the reptilian Subclass Synapsida. These mammal-like amniotes eventually gave rise to mammals.

Theriodontia One of two suborders within the Order Therapsida, the mammal-like amniotes. Primarily

carnivorous, theriodonts encompassed several diverse lineages.

thermal conductance The heat loss from the skin to the outside environment.

thermal windows Bare or sparsely furred areas of certain mammals that reside in regions characterized by intense solar radiation and high air temperatures (e.g., guanacos and many desert antelopes). They function as sites through which some of the heat gained from solar radiation can be lost by convection and conduction.

thermogenin A mitochondrial protein responsible for heat production by brown adipose tissue due to uncoupling oxidative phosphorylation.

thermoneutral zone A range in environmental temperatures within which the metabolic rate of an animal is minimal.

thoracic (vertebrae) Articulating with the ribs; from 12 to 15 pairs in mammals.

threatened species Those that are likely to become endangered in the near future (some listings use the term vulnerable instead).

tine A point or projection on an antler.

torpor A form of adaptive hypothermia or dormancy in which body temperature, heart rate, and respiration are not lowered as drastically as in hibernation.

total metabolic rate The total quantity of energy necessary to meet energy demands of an animal.

tradition A behavior pattern that is passed from one generation to the next through the process of learning.

tragus A projection from the lower margin of the pinnae of many microchiropteran bats that functions in echolocation.

transpirational water loss The mechanism by which water is lost by diffusion through the skin and from the surfaces of the respiratory tract; also called insensible water loss.

tribosphenic Molars with three main cusps (the trigon) arranged in a triangular pattern. Cusp patterns of many modern mammalian groups are derived from this pattern.

Triconodonta An order of small, carnivorous mammals characterized by

molars that had three cusps in a row. The lineage extended for 120 million years until the late Cretaceous period.

trigon(id) The three cusps (protocone, paracone, and metacone) of a tribosphenic molar. The suffix **-id** is applied to the mandibular dentition; main cusps are the protoconid, paraconid, and metaconid.

Tritylodonts A lineage of rodentlike reptiles that existed for about 50 million years from the late Triassic to the mid-Jurassic periods.

trophic Pertaining to food or nutrition.

trophic cascade Indirect effects of predators on plant biomass. Also referred to as top down effects

trophoblast The outer layer of the blastocyst in mammals; attaches the ovum to the uterine wall and supplies nutrition to the embryo as part of the placenta.

truth in advertising In sexual selection, the hypothesis that a male's ornaments or behavior are reliable indicators of his overall genetic fitness.

turbinal (turbinate) bones Structures found within the nasal area that increase the surface area for reception of chemical cues and that secrete mucus to aid in filtering small particles from incoming air.

tympanoperiotic The auditory bullae and middle ear apparatus of whales; not fused to the skull so that the direction of incoming sound waves can be determined.

tympanum A membrane (eardrum) at the interior end of the external auditory meatus; connects to the ossicles of the middle ear.

type specimens One or more specimens, usually part of a systematic collection, that constitute the type of a species; under the International Code of Zoological Nomenclature, the type is the "name bearing" specimen associated with a particular species.

U

ultradian rhythms Activity rhythms with a period of less than 24 hours.

ultrasound Sound frequencies greater

than 20,000 kHzzz.

underfur Fine, short hairs that occur beneath guard hairs in the pelage of a mammal.

unguiculate Having nails or claws instead of hooves.

unguis The upper or dorsal portion of the claw, which is a scalelike plate that surrounds the subunguis.

ungulates Mammals with hooves; perissodactyls and artiodactyls.

unguligrade A running locomotion with only the hooves (tips of the digits) on the ground; characteristic of ungulates.

unicuspid Teeth with a single cusp. Canine teeth are unicuspid, as are premolars in many species.

upper critical temperature The temperature at which an animal must dissipate heat to maintain a stable internal temperature.

ureters Ducts within the mammalian kidney that drain the renal pelvis.

urethra The tube through which urine is expelled from the urinary bladder.

urinary bladder The organ that stores urine in mammals.

urogenital sinus A common chamber for the reception of products from the reproductive and urinary systems. In mammals, it is found in monotremes and marsupials.

uropatagium The membrane between the hind legs of bats that encloses the tail; also called the interfemoral membrane.

uteri In female mammals a muscular expansion of the reproductive tract in which the embryo and fetus develop; opens externally through the vagina (sing., **uterus**).

V

vagina The part of the female reproductive tract that receives the male penis during copulation.

vaginal smear technique A procedure for monitoring different stages of the estrous cycle by observing changes in the types of cells lining the vaginal canal.

valvular Describing nostrils or ears that can be closed when an animal is under water.

variable number tandem repeats (VNTRs) Individual loci where alleles are composed of tandem repeats that vary in terms of the number of core units.

vectors Any agent or carrier that transmits a disease organism.

velli Very short, fine hairs sometimes referred to as "down" or "fuzz."

velvet Haired and highly vascularized skin covering growing antlers.

vestigial Reduced; remnant; atrophied.

vertebrae A series of bony elements that form the spinal column of the axial skeleton, stretching from the base of the skull to the tail.

viable population analysis Determination of the number of animals needed in a population to prevent it from going extinct within a given time period (often 1000 years).

vibrissae Long, stiff hairs with extensive enervation at the base of the follicle that are found on all mammals except humans.

vicariance The splitting of an ancestral species range into two or more smaller ranges by the development of a dispersal barrier, resulting in interrupted gene flow, genetic differentiation, and perhaps speciation of isolated populations.

villi Fingerlike projections of capillaries from the outermost embryonic membrane that penetrate the endometrium; increases exchange between maternal and fetal systems (same as chorionic villi).

viviparous Able to give birth to live young. Therian mammals are viviparous; prototherians are oviparous.

volant Having powered flight.

voucher specimen A museum specimen (e.g., study skin, skull) that serves permanently to identify the source of a tissue, protein, or DNA sample.

W

warm-blooded Characterized by having a constant body temperature, independent of environmental temperature. Typified by birds and mammals only; endothermy, homeothermy.

wavelength The distance from one peak to the next in a sound (or light) wave.

white adipose tissue The major fatty tissue of mammals that functions in body insulation, mechanical support, and buoyancy and as an energy reserve.

wing loading In bats (and birds), the body mass divided by the total surface area of the wings.

winter lethargy A period of winter dormancy in which body temperature of the animal decreases only about 5 to 6°C from euthermy, as in black bears (*Ursus americanus*).

wool Underhair that is long, soft, and usually curly.

X

xanthophylls Pigments that produce mixtures of red and yellow (pheomelanin).

Z

zalambdodont Tooth cusps that form a V-shape.

Zeitgeber "Time giver"; environmental cues that serve to set and adjust biological clocks.

zona pellucida A noncellular layer surrounding the zygote.

zonary placenta The type of chorioallantoic placenta in which the villi occupy a girdlelike band about the middle of the chorionic sac.

zoogeography The study of distributions of animals, including mammals.

zoonoses Diseases transmitted from vertebrate animals (nonhuman mammals) to people.

zooplankton Animal material including both plankton and necton fed on by baleen whales.

zygomatic arch Bony structure that surrounds and protects the eye and serves as a place of attachment for jaw muscles.

zygote A diploid cell resulting from the union of the male and female gametes.

References

Abba, A. M., D. E. Udrizar Sauthier, and S. F. Vizcaino. 2005. Distribution and use of burrows and tunnels of *Chaetophractus villosus* (Mammalia: Xenarthra) in the eastern Argentinian pampas. Acta Theriol. 50:115–124.

Abbate, E., A. Albianelli, A. Azzaroli, M. Benvenuti, B. Tesfamariam, P. Bruni, N. Cipriani, R. J. Clarke, G. Ficcarelli, R. Macchiarelli, G. Napoleone, M. Papini, L. Rook, M. Sagri, T. M. Tecle, D. Torre, and I. Villa. 1998. A one-million-year-old Homo cranium from the Danakil (Afar) Depression of Eritrea. Nature 393:458–460.

Abe, H., R. Matsuki, S. Ueno, M. Nashimoto, and M. Hasegawa. 2006. Dispersal of *Camellia japonica* seeds by *Apodemus speciosus* revealed by maternity analysis of plants and behavioral observation of animal vectors. Ecological Research 21:732–740.

Abraham, G. E., F. S. Manlimos, and R. Gazara. 1977. Radioimmunoassay of steroids. Pp. 591–999 *in* Handbook of radioimmunoassay (G. E. Abraham, ed.). Marcel Dekker, New York.

Abramsky, Z., and C. Sellah. 1982. Competition and the role of habitat selection in *Gerbillus allenbyi* and *Meriones tristrami*: a removal experiment. Ecology 63:1242–1247.

Acha, P. N., and B. Szyfres. 1989. Zoonoses and communicable diseases common to man and animals, 2nd edition. Pan Am. Health Org., Sci. Publ. No. 503. Washington, DC.

Adamoli, V. C., P. D. Cetica, M. S. Merani, and A. J. Solari. 2001. Comparative morphologic placental types in dasypodidae (*Chaetophractus villosus*, *Cabassous chacoensis*, *Tolypeutes matacus* and *Dasypus hybridus*). Biocell 25:17–22.

Adams, G. P., P. G. Griffin, and O. J. Ginther. 1989. In situ morphologic dynamics of ovaries, uterus, and cervix in llamas. Biol. Reprod. 41:551–558.

Adams, G. P., C. Plotka, C. Asa, and O. J. Ginther. 1991. Feasibility of characterizing reproductive events in large nondomestic species by transrectal ultrasonic imaging. Zoo Biol. 10:247–259.

Adams, L., and S. Hane. 1972. Adrenal gland size as an index of adrenocortical secretion rate in the California ground squirrel. J. Wildl. Dis. 8:19–23.

Adams, R. A. 2003. Bats of the Rocky Mountain West. Univ. Press of Colorado, Boulder.

Adamson, M. L., and J. N. Caira. 1994. Evolutionary factors influencing the nature of parasite specificity. Parasitol. 109:S85–S95.

Adcock, E. W., F. Teasdale, C. S. August, S. Cox, G. Meschia, F. C. Battaglia, and M. A. Naughton. 1973. Human chorionic gonadotropin: its possible role in maternal lymphocyte suppression. Science 181:845–847.

Adkins, R. M., E. L. Gelke, D. Rowe, and R. L. Honeycutt. 2001. Molecular phylogeny and divergence time estimates for major rodent groups: evidence from multiple genes. Mol. Biol. Evol. 18:777–791.

Adler, E., M. A. Hoon, K. L. Mueller, J. Chandrashekar, N. J. P. Ryba, and C. S. Zucker. 2000. A novel family of mammalian taste receptors. Cell 100:693–702.

Agresti, A. 1996. An introduction to categorical data analysis. John Wiley and Sons, New York.

Albrecht, E. D., and G. J. Pepe. 1990. Placental steroid hormone biosynthesis in primate pregnancy. Endocrin. Rev. 11:124–150.

Albright, L. B. 1998. A new genus of tapir (Mammalia: Tapiridae) from the Arikareen (earliest Miocene) of the Texas Coastal Plain. J. Vert. Paleontol. 18:200–217.

Alexander, J., and D. G. Russell. 1992. The interaction of Leishmania species with macrophages. Adv. Parasitol. 31:175–254.

Alexander, L. F. and B. R. Riddle. 2005. Phylogenetics of the New World rodent family Heteromyidae. J. Mammal. 86:366–379.

Alexander, R. D. 1974. The evolution of social behavior. Annu. Rev. Ecol. Syst. 5:325–383.

Alexander, R. McN. 1993a. The energetics of coprophagy: a theoretical analysis. J. Zool. Lond. 230: 629–637.

Alexander, R. McN. 1993b. The relative merits of foregut and hindgut fermentation. J. Zool. Lond. 231:391–401.

Alexander, R. M., and H. C. Bennet-Clark. 1977. Storage of elastic strain energy in muscle and other tissues. Nature 265:114–117.

Allaine, D., F. Brondex, L. Graziani, J. Coulon, and I. Till-Bottraud. 2001. Male-biased sex ratio in litters of Alpine marmots supports the helper repayment hypothesis. Behav. Ecol. 11:507–514.

Allen, E. G. 1938. The habits and life history of the eastern chipmunk (*Tamias striatus lysteri*). Bull. N. Y. State Mus. 314:1–122.

Allen, J. A. 1877. The influence of physical conditions in the genesis of species. Radical Rev. 1:108–140.

Allen, M. C., A. J. Read, J. Gaudet, and L. S. Sayigh. 2001. Fine-scale habitat selection of foraging bottlenose dolphins *Tursiops truncatus* near Clearwater, Florida. Mar. Ecol. Progr. Ser. 222:253–264.

Allsopp, H. 1960. The manatee: ecology and use for weed control. Nature 188:762.

Alt, G. L. 1983. Timing and parturition of black bears (*Ursus americanus*) in northeastern Pennsylvania. J. Mammal. 64:305–307.

Altenbach, J. S. 1979. Locomotor morphology of the vampire bat, *Desmodus rotundus*. Spec. Pub., Am. Soc. Mammal. 6:1–137.

Altenbach, J. S. 1989. Prey capture by the fishing bats *Noctilio leporinus* and *Myotis vivesi*. J. Mammal. 70:421–424.

Altmann, J. 1974. Observational study of behavior: sampling methods. Behaviour 48:227–265.

Altmann, J., G. Hausfater, and S. A. Altmann. 1988. Determinants of reproductive success in savannah baboons, *Papio cynocephalus*. Pp. 403–418 *in* Reproductive success (T. H. Clutton-Brock, ed.). Univ. Chicago Press, Chicago.

Altringham, J. D. 1996. Bats: biology and behaviour. Oxford Univ. Press, Oxford, England.

Altschuler, E. M., R. B. Nagle, E. J. Braun, S. L. Lindstedt, and P. H. Krutzsch. 1979. Morphological study of the desert heteromyid kidney with emphasis on the genus *Perognathus*. Anat. Record 194:461–468.

Altuna, C. A., L. D. Bacigalupe, and S. Corte. 1998. Food-handling and feces reingestion in *Ctenomys pearsoni* (Rodentia, Ctenomyidae). Acta Theriol. 43:433–447.

Amrine-Madsen, H., M. Scally, M. Westerman, M. J. Stanhope, C. Krajewski, and M. S. Springer. 2003. Nuclear gene sequences provide evidence for the monophyly of Australidelphian marsupials. Mol. Phylogenet. Evol. 28:186–196.

Anderson, L. T. 1973. An analysis of habitat preference in mice as a function of prior experience. Behaviour. 47:302–339.

Anderson, M. D. and P. R. K. Richardson. 2005. The physical and thermal characteristics of aardwolf dens. S. Afr. J. Wildl. Res. 35:147–153.

Anderson, P. K., G. E. Heinsohn, P. H. Whitney, and J. P. Huang. 1977. *Mus musculus* and *Peromyscus maniculatus*: homing ability in relation to habitat utilization. Can. J. Zool. 55:169–182.

Anderson, R. M. 1965. Methods of collecting and preserving vertebrate animals, 4th edition, rev. Bull. Nat. Mus. Can. 69:1–199.

Anderson, R. P., and C. O. Handley, Jr. 2001. A new species of three-toed sloth (Mammalia: Xenarthra) from Panama, with a review of the genus *Bradypus*. Proc. Biol. Soc. Wash. 114:1–33.

Anderson, R. P., and C. O. Handley, Jr. 2002. Dwarfism in insular sloths: biogeography, selection, and evolutionary rate. Evolution 1045–1058.

Anderson, R. P., and R. M. Timm. 2006. A new montane species of spiny pocket mouse (Rodentia: Heteromyidae: *Heteromys*) from northwestern Costa Rica. Am. Mus. Novitates 3509:1–38.

Anderson, S., and T. L. Yates. 2000. A new genus and species of Phyllotine rodent from Bolivia. J. Mammal. 81:18–36.

Andersson, M. 1994. Sexual selection. Princeton Univ. Press, Princeton, NJ.

Andrades-Miranda, J., L. F. B. Oliviera, C. A. Lima-Rosa, A. P. Nunes, N. I. T. Zanchin, and M. S. Mattevi. 2001. Chromosome studies of seven species of *Oligoryzomys* (Rodentia: Sigmodontinae) from Brazil. J. Mammal. 82:1080–1091.

Andriafidison, D., R. A. Andrianaivoarivelo, O. R. Ramilijaona, M. R. Razanahoera, J. MacKinnon, R. K. B. Jenkins, and P. A. Racey. 2006. Nectarivory by endemic Malagasy fruit bats during the dry season. Biotropica 38:85–90.

Andrews, P., and E. M. O'Brien. 2000. Climate, vegetation, and predictable gradients in mammal species richness in southern Afr. J. Zool. Lond. 251:205–231.

Andrews, R. V., and R. W. Belknap. 1986. Bioenergetic benefits of huddling by deer mice (*Peromyscus maniculatus*). Comp. Biochem. Physiol. 85A:775–778.

Angerbjörn, A. 1986. Gigantism in island populations of wood mice (*Apodemus*) in Europe. Oikos 47:47–56.

Angerbjörn, A., and J. E. C. Flux. 1995. *Lepus timidus*. Mamm. Species 495:1–11.

Animal Care and Use Committee. 1998. Guidelines for the capturing, handling, and care of mammals as approved by the American Society of Mammalogists. J. Mammal. 79:1416–1431.

Aplin, K. P., and M. Archer. 1987. Recent advances in marsupial systematics with a new syncretic classification. Pp. xv–lxxii *in* Possums and opossums: studies in evolution (M. Archer, ed.). Royal Zoological Society, New South Wales, Sydney.

Appolonio, M., M. Festa-Bianchet, F. Mari, S. Mattioli, and B. Sarno. 1992. To lek or not to lek: mating strategies of male fallow deer. Behav. Ecol. 3:25–31.

Aranoff, S. 1993. Geographic information systems: a management perspective. WDL Publications, Ottawa, Ontario.

Arbogast, B. S., and G. J. Kenagy. 2001. Comparative phylogeography as an integrative approach to historical biogeography. J. Biogeogr. 28:819–825.

Arbogast, B. S., R. A. Browne, and P. D. Weigl. 2001. Evolutionary genetics and Pleistocene biogeography of North American tree squirrels (*Tamiasciurus*). J. Mammal. 82:302–319.

Arcese, P. 1994. Harem size and horn symmetry in oribi. Anim. Behav. 48:1485–1488.

Archer, J. 1988. The behavioural biology of aggression. Cambridge Univ. Press, New York.

Archer, M. 1984a. Earth-shattering concepts for historical zoogeography. Pp. 45–59 *in* Vertebrate zoogeography and evolution in Australasia. Hesperian Press, Carlisle, Australia.

Archer, M. 1984b. Introduction. Pp. xi–xxiv *in* Vertebrate zoogeography and evolution in Australasia. Hesperian Press, Carlisle, Australia.

Archer, M. 1978. The status of Australian dasyurids, thylacinids, and myrmecobiids. Pp. 29–43 *in* The status of endangered Australian wildlife (M. J. Tyler, ed.). Royal Zoological Society, South Australia.

Archer, M., and A. A. Bartholomai. 1978. Tertiary mammals of Australia: a syntopic review. Alcheringa 2:1–19.

Archer, M. and J. A. W. Kirsch. 1977. The case for the Thylacomyidae and Myrmecobiidae, Gill 1872, or why are marsupial families so extended? Proc. Linnean Soc. New S. Wales 102:18–25.

Archer, M., M. D. Plane, and N. S. Pledge. 1978. Additional evidence for interpreting the Miocene *Obdurodon insignis*, Woodburne and Tedford 1975, to be a fossil platypus (Ornithorhynchidae: Monotremata) and a reconsideration of the status of Ornithorhynchus agilis De Vis, 1895. Aust. Zool. 20:9–27.

Archer, M., T. F. Flannery, A. Ritchie, and R. E. Molnar. 1985. First Mesozoic mammal from Australia—an early Cretaceous monotreme. Nature

318:363–366.

Archer, M., P. Murray, S. J. Hand, and H. Godthelp. 1992. Reconsideration of monotreme relationships based on the skull and dentition of the Miocene *Obdurodon dicksoni* (Ornithorhynchidae) from Riversleigh, Queensland, Australia. Pp. 75–94 in Mammalian phylogeny (F. Szalay, M. Novacek, and M. McKenna, eds.). Springer-Verlag, New York.

Arita, H. T., P. Rodriguez, and E. Vázquez-Domínguez. 2005. Continental and regional ranges of North American mammals: Rapoport's rule in real and null worlds. J. Biogeogr. 32:961–971.

Arletazz, R. 1996. Feeding behaviour and foraging strategy of free-living mouse-eared bats, *Myotis myotis* and *Myotis blythii*. Anim. Behav. 51:1–11.

Arlton, A. V. 1936. An ecological study of the mole. J. Mammal. 17:349–371.

Armitage, K. B., J. C. Melcher, and J. M. Ward, Jr. 1990. Oxygen consumption and body temperature in yellow-bellied marmot populations from montane-mesic and lowland-xeric environments. J. Comp. Physiol. B Biochem. Syst. Environ. Physiol. 160:491–502.

Armstrong, R. B. 1981. Recruitment of muscles and fibers within muscles in running animals. Symp. Zool. Soc. Lond. 48:289–304.

Arnason, U., A. Gullberg, and A. Janke. 1999. The mitochondrial DNA molecule of the aardvark, *Orycteropus afer*, and the position of the Tubulidentata in the eutherian tree. Proc. Roy. Soc. Lond. Ser. B 266:339–345.

Arnett, E. B., L. R. Irby, and J. G. Cook. 1993. Sex- and age-specific lungworm infection in Rocky Mountain bighorn sheep during winter. J. Wildl. Dis. 29:90–93.

Arnold, W. 1988. Social thermoregulation during hibernation. J. Comp. Physiol. B Biochem. Syst. Environ. Physiol. 158:151–156.

Arnould, J. P. Y., and M. A. Hindell. 2001. Dive behaviour, foraging locations, and maternal-attendance patterns of Australian fur seals (*Arctocephalus pusillus doriferus*). Can. J. Zool. 79:35–48.

Arroyo-Cabrales, J., R. R. Hollander, and J. Knox Jones, Jr. 1987. *Choeronycteris mexicana*. Mamm. Species. 291:1–5.

Arzamendia, Y., M. H. Cassini, and B. L. Vila. 2006. Habitat use by vicuña (*Vicugna vicugna*) in Laguna Pozuelos Reserve, Jujuy, Argentina. Oryx 40:198–203.

Ashford, R. W., P. Desjeux, and P. deRaadt. 1992. Estimation of population at risk of infection and number of cases of leishmaniasis. Parasitol. Today 8:104–105.

Ashley, T. 2005. Chromosome chains and platypus sex: kinky connections. Bioessays 27:681–684.

Ashton, D. G. 1978. Marking zoo animals for identification. Pp. 24–34 in Animal marking, (B. Stonehouse, ed.). Univ. Park Press, Baltimore.

Ashton, K. G., M. C. Tracy, and A.de Queiroz. 2000. Is Bergmann's rule valid for mammals? Am. Nat. 156: 390–415.

Astúa de Moraes, D., R. T. Santori, R. Finotti, and R. Cerqueria. 2003. Nutritional and fibre contents of laboratory-established diets of Neotropical opossums (Didelphidae). Pp. 229–237 in Predators with pouches: the biology of carnivorous marsupials (M. Jones, C. Dickman, and M. Archer, eds.). CSIRO Publishing, Collingwood, Australia.

Atherly, A. G., J. R. Girton, and J. F. McDonald. 1999. The science of genetics. Saunders College Publishing, Fort Worth, TX.

Atkinson, S. 1997. Reproductive biology of seals. Reviews of Reproduction 2: 175–194.

Au, W. W. L. 1993. The sonar of dolphins. Springer-Verlag, New York.

Audubon, J. J., and J. Bachman. 1846–1854. The viviparous quadrupeds of North America, 3 vols. V. G. Audubon, New York.

Augee, M. L. 1978. Monotremes and the evolution of homeothermy. Aust. Zool. 20:111–119.

August, P. V., S. G. Ayvazian, and J. G. T. Anderson. 1989. Magnetic orientation in a small mammal, *Peromyscus leucopus*. J. Mammal. 70:1–9.

Austad, S. N., and K. E. Fischer. 1991. Mammalian aging, metabolism, and ecology: evidence from the bats and marsupials. J. Gerontol. 46:B47–53.

Austad, S. N., and M. E. Sunquist. 1987. Sex ratio manipulation in the common opossum. Nature 324:58–60.

Austin, C. R., and R. V. Short (eds.). 1972a. Reproduction in mammals, book 1: germ cells and fertilization. Cambridge Univ. Press, Cambridge, UK.

Austin, C. R., and R. V. Short (eds.). 1972b. Reproduction in mammals, book 2: embryonic and fetal development. Cambridge Univ. Press, Cambridge, UK.

Austin, C. R., and R. V. Short (eds.). 1972c. Reproduction in mammals, book 4: reproductive patterns. Cambridge Univ. Press, Cambridge, UK.

Austin, C. R., and R. V. Short (eds.). 1976. Reproduction in mammals, book 6: the evolution of reproduction. Cambridge Univ. Press, New York.

Austin, C. R., and R. V. Short 1984. Hormonal control of reproduction, 2d edition. Cambridge Univ. Press, New York.

Avise, J. C. 1994. Molecular markers, natural history and evolution. Chapman and Hall, New York.

Avise, J. C. 1996. Toward a regional conservation genetics perspective: phylogeography of faunas in the southeastern United States. Pp. 431–470 in Conservation genetics: case histories from nature (J. C. Avise and J. L. Hamrick, eds.). Chapman & Hall, New York.

Avise, J. C. 2000. Phylogeography: the history and formation of species. Harvard Univ. Press, Cambridge, MA.

Avise, J. C. 2003. Phylogeography: the history and formation of species. Harvard Univ. Press, Cambridge, MA.

Avise, J. C. 2004. Molecular markers, natural history, and evolution, 2d edition. Sinauer Associates, Sunderland, MA.

Avise, J. C., and R. M. Ball. 1990. Principles of genealogical concordance in species concepts and biological taxonomy. Oxford Surv. Evol. Biol. 7:45–67.

Avise, J. C., and J. L. Hamrick (eds). 1996. Conservation genetics: case histories from nature. Chapman and Hall, New York.

Avise, J. C., J. Arnold, R. M. Ball, E. Bermingham, T. Lamb, J. E. Neigel, C. A. Reeb, and N. C. Saunders. 1987. Intraspecific phylogeography: the mitochondrial DNA bridge between population genetics and systematics. Annu. Rev. Ecol. Syst. 18:489–522.

Avise, J. C., and R. M. Ball, Jr. 1990. Principles of genealogical concordance in species concepts and biological taxonomy. Pp. 45–67 in Oxford Surveys in Evolutionary Biology, vol. 7 (D. Futuyma and J. Antonovics, eds.). Oxford Univ. Press, Oxford, UK.

Avise, J. C., R. A. Lansman, and R. O. Slade. 1979. The use of restriction endonucleases to measure mitochondrial DNA sequence relatedness in natural populations. I. Population structure and evolution in the genus *Peromyscus*. Genetics 92:279–295.

Axelrod, R., and W. D. Hamilton. 1981. The evolution of cooperation. Science 211:1390–1396.

Axis-Arroyo, J., B. Morales-Vela, D. Torruco-Gomez, and M. E. Vega-Cendejas. 1998. Factors associated with habitat use by the Caribbean manatee (*Trichechus manatus*), in Quintana Roo, Mexico (Mammalia). Rev. Biol. Trop. 46:791–803.

Ayala, F. J. 1986. On the virtues and pitfalls of the molecular evolutionary clock. J. Heredity 77:226–235.

Azcon, R., J. M. Ruiz-Lozano, and R. Rodriguez. 2001. Differential contibution of arbuscular mycorrhizal fungi to plant nitrate uptake (15N) under increasing N supply to the soil. Can. J. Bot. 79:1175–1180.

Baar, S. L., and E. D. Fleharty. 1976. A model of the daily energy budget and energy flow through a population of the white-footed mouse. Acta Theriol. 21:179–193.

Bachman, G. C. 1994. Food restriction effects on the body composition of free-living ground squirrels, Spermophilus beldingi. Physiol. Zool. 67:756–770.

Badgely, C., and D. L. Fox. 2000. Ecological biogeography of North American mammals: species density and ecological structure in relation to environmental gradients. J. Biogeogr. 27:1437–1467.

Badyaev A. V., K. R. Foresman, and M. V. Fernandes. 2000. Stress and developmental stability: vegetation removal causes increased fluctuating asymmetry in shrews. Ecology 81: 336–345.

Bailey, V. 1924. Breeding, feeding, and other life habits of meadow mice (*Microtus*). J. Agri. Res. 27:523–535.

Bailey, W. J., J. L. Slightom, and M. Goodman. 1992. Rejection of the "flying primate" hypothesis by phylogenetic evidence from the ε-globin gene. Science 256:86–89.

Baird, S. F. 1859. General report on North American mammals. J. B. Lippincott, Philadelphia.

Baker, A. N. 1985. Pygmy right whale, *Caperea marginata* (Gray, 1846). Pp. 345–354 in Handbook of marine mammals, vol. 3: the sirenians and baleen whales (S. H. Ridgway and R. Harrison, eds.). Academic Press, New York.

Baker, B. W., and E. P. Hill. 2003. Beaver, *Castor canadensis*. Pp. 288-310 in Wild mammals of North America: biology, management, and conservation (G. A. Feldhamer, B. C. Thompson, and J. A. Chapman, eds.). Johns Hopkins Univ. Press, Baltimore, MD.

Baker, C. S., S. R. Palumbi, R. H. Lambertsen, M. T. Weinrich, J. Calambokidis, and S. J. O'Brien. 1990. Influence of seasonal migration on geographic distribution of mitochondrial DNA haplotypes in humpback whale. Nature 34:238–240.

Baker, M. A. 1979. A brain-cooling system in mammals. Sci. Am. 240:114–123.

Baker, M. L., J. P. Wares, G. A. Harrison, and R. D. Miller. 2004. Relationships among the families and orders of marsupials and major mammalian lineages based on recombination activating gene-1. J. Mammalian Evol. 11:1–16.

Baker, R. J., and R. D. Bradley. 2006. Speciation in mammals and the genetic species concept. J. Mammal. 87:643–662.

Baker, R. J., C. S. Hood, and R. L. Honeycutt. 1989. Phylogenetic relationships and classification of the higher categories of the New World bat family Phyllostomidae. Syst. Zool. 38:228–238.

Baker, R. J., M. J. Novacek, and N. B. Simmons. 1991. On the monophyly of bats. Syst. Zool. 40:216–231.

Baker, R. R. 1987. Human navigation and magnetoreception: the Manchester experiments do replicate. Anim. Behav. 35:691–704.

Baker, R. R., and M. A. Bellis. 1988. 'Kamikaze' sperm in mammals? Anim. Behav. 36:936–938.

Bakker, R. T. 1971. Dinosaur physiology and the origin of mammals. Evolution 25:636–658.

Balasingh, J., J. Koilraj, and T. H. Kunz. 1995. Tent construction by the short-nosed fruit bat, *Cynopterus sphinx* (Chiroptera: Pteropodidae) in southern India. Ethology 100:210–229.

Balcomb, K. C., III, and M. A. Bigg. 1986. Population biology of three resident killer whale pods in Puget Sound and off southern Vancouver Island. Zoo Biol. Monogr. 1:85–95.

Baldi, R., C. Campganga, S. Pedraza, and B. J. LeBouef. 1996. Social effects of space availability on the breeding behaviour of elephant seals in Patagonia. Anim. Behav. 51:717–724.

Baldwin, B. H., B. C. Tennant, T. J. Reimers, R. G. Gowan, and P. W. Concannon. 1985. Circannual changes in serum testosterone concentrations of adult and yearling woodchucks (*Marmota monax*). Biol. Reprod. 32:804–812.

Baldwin, J. A. 1975. Notes and speculations on the domestication of the cat in Egypt. Anthropos. 70:428–448.

Baldwin, R. A., A. E. Houston, M. L. Kennedy, and P. S. Liu. 2004. An assessment of microhabitat variables and capture success of striped skunks (*Mephitis mephitis*). J. Mammal. 85:1068–1076.

Balmford, A. P. 1991. Mate choice on leks. Trends Ecol. Evol. 6:87–92.

Bannikova, A. A., and D. A. Kramerov. 2005. Molecular phylogeny of palearctic shrews inferred from RFLP and IS-PCR data. Pp. 87–98 in Advances in the biology of shrews II (J. F. Merritt, S. Churchfield, R. Hutterer, B. I. Sheftel, eds.). Special Publication of the International Society of Shrew Biologists 01, New York.

Baquero, R. A., and J. L. Telleria. 2001. Species richness, rarity and endemicity of European mammals: a biogeographical approach. Biodiv. Conserv. 10:29–44.

Barash, D. P. 1974. The evolution of marmot societies: a general theory. Science 185:415–420.

Barber, D. G., E. Saczuk, and P. R. Richard. 2001. Examination of beluga-habitat relationships through the use of telemetry and a geographic information system. Arctic 54:305–316.

Barbosa, A., and J. Benzal. 1996. Diversity and abundance of small mammals in Iberia: peninsular effect or habitat suitability? Z. Saugertierk. 61:236–241.

Barbour, A. G., and D. Fish. 1993. The biological and social phenomenon of Lyme disease. Science 260:1610–1616.

Barbour, R. W., and W. H. Davis. 1969. Bats of America. Univ. Kentucky Press, Lexington.

Barclay, R. M. R., and R. M. Brigham. 1994. Constraints on optimal foraging: a field test of prey discrimination by echolocating insectivorous bats. Anim. Behav. 48:1013–1021.

Barclay, R. M. R., C. L. Lausen, and L. Hollis. 2001. What's hot and what's not: defining torpor in free-ranging birds and mammals. Can. J. Zool. 79:1885–1890.

Bardeleben, C., R. L. Moore, and R. K. Wayne. 2005. A molecular phylogeny of the Canidae based on six nuclear loci. Mol. Phylogenet. Evol. 37:815–831.

Bargo, M. S., N. Toledo, and S. F. Vizcaino. 2006. Muzzle of South American Pleistocene ground sloths (Xenarthra, Tardigrada). J. Morphol. 267:248–263.

Barkley, M. S., and B. D. Goldman. 1977. The effects of castration and silastic implants of testosterone on intermale aggression in the mouse. Horm. Behav. 9:32–48.

Barnes, B. M. 1989. Freeze avoidance in a mammal: body temperatures below 0°C in an arctic hibernator. Science 244:1593–1595.

Barnes, B. M. 1996. Relationships between hibernation and reproduction in male ground squirrels. Pp. 71–80 in Adaptations to the cold: Tenth International Hibernation Symposium (F. Geiser, A. Hulbert, and S. Nicol, eds.). Univ. New England Press, Armidale, Australia.

Barnes, B. M., and L. Buck. 2000. Hibernation in the extreme: burrow and body temperatures, metabolism, and limits to torpor bout length in arctic ground squirrels. Pp. 65–72 in Life in the cold: Eleventh International Hibernation Symposium (G. Heldmaier and M. Klingenspor, eds.). Springer-Verlag, Berlin.

Barnosky, A. D. 1985. Taphonomy and herd structure of the extinct Irish elk, Megaloceros giganteus. Science 228:340–344.

Barrantes, G. E., and L. Daleffe. 1999. Allozyme genetic distances and evolutionary relationships in marsupials of North and South America. Acta Theriol. 44:233–242.

Barrett, T., and P. B. Rossiter. 1999. Rinderpest: the disease and its impact on humans and animals. Adv. Virus Res. 53:89–110.

Barriel, V. 1997. Pan paniscus and Homonid phylogeny: morphological data, molecular data and "total evidence." Folia Primat. 68:50–56.

Barry, R. E., Jr., and E. N. Francq. 1980. Orientation to landmarks within the preferred habitat by Peromyscus leucopus. J. Mammal. 61:292–303.

Bartholomew, G. A. 1982. Body temperature and energy metabolism. Pp. 333–406 in Animal physiology: principles and adaptations (M. S. Gordon, G. A. Bartholomew, A. D. Grinnell, C. B. Jørgensen, and F. N. White, eds.), 4th edition. Macmillan, New York.

Bartholomew, G. A., W. R. Dawson, and R. C. Lasiewski. 1970. Thermoregulation and heterothermy in some of the smaller flying foxes (Megachiroptera) of New Guinea. Z. Vergl. Physiol. 70:196–209.

Barton, R. A., and P. H. Harvey. 2000. Moasic evolution of brain structure in mammals. Nature 405:1055–1058.

Bateman, A. J. 1948. Intra-sexual selection in Drosophila. Heredity 2:349–368.

Bateman, J. A. 1959. Laboratory studies of the golden mole and mole-rat. Afr. Wildl. 13:65–71.

Bates, P. J. J., T. New, K. M. Swe, and S. S. H. Bu. 2001. Further new records of bats from Myanmar (Burma), including Craseonycteris thonglongyai Hill 1974 (Chiroptera: Craseonycteridae). Acta Chiropt.

3:33–41.

Batzli, G. O. 1985. Nutrition. Pp. 779–811 in Biology of New World Microtus (R. H. Tamarin, ed.). Spec. Pub. No. 8, American Society of Mammalogists.

Batzli, G. O. 1994. Special feature: mammal-plant interactions. J. Mammal. 75:813–815.

Batzli, G. O., and I. D. Hume. 1994. Foraging and digestion in herbivores. Pp. 313–314 in The digestive system in mammals: food, form, and function (D. J. Chivers and P. Langer, eds.). Cambridge Univ. Press, Cambridge, UK.

Batzli, G. O., and F. A. Pitelka. 1970. Influence of meadow mouse populations on California grassland. Ecology 51:1027–1039.

Baudinette, R. V., S. K. Churchill, K. A. Christian, J. E. Nelson, and P. J. Hudson. 2000. Energy, water balance and the roost microenvironment in three Australian cave-dwelling bats (Microchiroptera). J. Comp. Physiol. B 170:439–446.

Baverstock, P., and B. Green. 1975. Water recycling in lactation. Science 187:657–658.

Baverstock, P. R., M. Archer, M. Adams, and B. J. Richardson. 1982. Genetic relationships among 32 species of Australian dasyurid marsupials. Pp. 641–650 in Carnivorous marsupials (M. Archer, ed.). Roy. Zool. Soc. NSW, Mossman, Australia.

Baverstock, P. R., M. Krieg, J. Birrel, and G. M. McKay. 1990. Albumin immunologic relationships of Australian marsupials 2: the Pseudocheiridae. Aust. J. Zool. 38:519–526.

Bayefsky-Anand, S. 2005. Effect of location and season on the arthropod prey of Nycteris grandis (Chiroptera: Nycteridae). Afr. Zool. 40:93–97.

Bazin, R. C., and R. A. MacArthur. 1992. Thermal benefits of huddling in the muskrat (Ondatra zibethicus). J. Mammal. 73:559–564.

Beard, L. A., and G. C. Grigg. 2000. Reproduction in the short-beaked echidna, Tachyglossus aculeatus: Field observations at an elevated site in south-east Queensland. Proc. Linnean Soc. New S. Wales 122:89–99.

Bearder, S. K., and R. D. Martin. 1980. Acacia gum and its use by bush-babies, Galago senegalensis (Primates:Lorisidae). International J. Primatology 1:103–128.

Bearzi, M. 2006. California sea lions use dolphins to locate food. J. Mammal. 87:606–617.

Beatley, J. C. 1969. Dependence of desert rodents on winter annuals and precipitation. Ecology 50:721–724.

Beck, C. A., W. D. Bowen, and S. J. Iverson. 2000. Seasonal changes in buoyancy and diving behaviour of adult grey seals. J. Exp. Biol. 203:2323–2330.

Beck, H. 2006. A review of peccary-palm interactions and their ecological ramifications across the Neotropics. J. Mammal. 87:519–530.

Bedford, J. M., J. C. Rodger, and W. G. Breed. 1984. Why so many mammalian spermatozoa—a clue from marsupials? Proc. Roy. Soc. Lond. Ser. B 221:221–233.

Beentjes, M. P. 1989. Haul-out patterns, site fidelity and activity budgets of male Hooker's sea lions (Phocarctos hookeri) on the New Zealand mainland. Marine Mammal Science 5:281–297.

Beentjes, M. P. 2006. Behavioral thermoregulation of the New Zealand sea lion (Phocarctos hookeri). Marine Mammal Science 22:311–325.

Beer, J. R. 1961. Winter home ranges of the red-backed mouse and white-footed mouse. J. Mammal. 42:174–180.

Begall, S., and M. H. Gallardo. 2000. Spalacopus cyanus (Rodentia: Octodontidae): an extremist in tunnel constructing and food storing among subterranean mammals. J. Zool. Lond. 251:53–60.

Begon, M., J. L. Harper, and C. R. Townsend. 1996. Ecology: individuals, population, and communities, 3rd edition. Blackwell Science, Cambridge, MA.

Begon M., C. R. Townsend, and J. L. Harper. 2005. Ecology: from individuals to ecosystems, 4th edition. Blackwell Publishing, Malden, MA.

Behr, O., and O. von Helversen. 2004. Bat serenades—complex courtship songs of the sac-winged bat (Saccopteryx bilineata). Behav. Ecol. Sociobiol. 56:106–115.

Beidleman, R. G., and W. A. Weber. 1958. Analysis of a pika hay pile. J. Mammal. 39:599–600.

Beja-Pereira, A., P. R. England, N. Ferrand, S. Jordan,

A. O. Bakhiet, M. A. Abdalla, M. Mashkour, J. Jordana, P. Taberlet, and G. Luikart. 2004. African origins of the domestic donkey. Science 304:1781.

Bejan, A., and J. H. Marden. 2006. Constructing animal locomotion from new thermodynamics theory. Am. Sci. 94:342–349.

Bekoff, M. 1977. Social communication in canids: evidence for the evolution of a stereotyped mammalian display. Science 197:1097–1099.

Belay, E. D., R. A. Maddox, E. S. Williams, M. W. Miller, P. Gambetti, and L. B. Schonberger. 2004. Chronic wasting disease and potential transmission to humans. Emerging Infectious Diseases. 10:977–984.

Belk, M. C., and M. H. Smith. 1996. Pelage coloration in oldfield mice (Peromyscus polionotus): antipredator adaptation? J. Mammal. 77:882–890.

Bell, J. F., and J. R. Reilly. 1981. Tularemia. Pp. 213–231 in Infectious diseases of wild mammals, 2nd edition (J. W. Davis, L. H. Karstad, and D. O. Trainer, eds.). Iowa State Univ. Press, Ames.

Bell, R. H. V. 1971. A grazing ecosystem in the Serengeti. Sci. Am. 225:86–93.

Bell, W. J. 1991. Searching behavior. Chapman and Hall, New York.

Belovsky, G. E. 1978. Diet optimization in a generalist herbivore: the moose. Theoret. Pop. Biol. 14:105–134.

Belovsky, G. E. 1984. Herbivorous optimal foraging: a comparative test of three models. Am. Nat. 124:97–115.

Belovsky, G. E., and O. J. Schmitz. 1994. Plant defenses and optimal foraging by mammalian herbivores. J. Mammal. 75:816–832.

Belsky, A. J. 1986. Does herbivory benefit plants? A review of the evidence. Am. Nat. 127:870–892.

Benjaminsen, T. and I. Christensen. 1979. The natural history of the bottlenose whale Hyperoodon ampullatus. Pp. 143–164 in Behavior of marine animals, vol. 3: cetaceans (H. E. Winn and B. L. Olla, eds.). Plenum Press, New York.

Ben-Moshe, A., T. Dayan, and D. Simberloff. 2001. Convergence in morphological patterns and community organization between old and new world rodent guilds. Am. Nat. 158:484–495.

Bennett, A. F., and J. A. Ruben. 1979. Endothermy and activity in vertebrates. Science 206:649–654.

Bennett, A. F., R. B. Huey, H. J. Alder, and K. A. Nagy. 1984. The parasol tail and thermoregulatory behavior of the cape ground squirrel, Xerus inauris. Physiological Zoology 57:57–62.

Bennett, D. K. 1980. Stripes do not a zebra make, part I: a cladistic analysis of Equus. Syst. Zool. 29:272–287.

Bennett, K. A., B. J. McConnell, and M. A. Fedak. 2001. Diurnal and seasonal variations in the duration and depth of the longest dives in southern elephant seals (Mirounga leonina): possible physiological and behavioural constraints. J. Exp. Biol. 204:649–662.

Bennett, S., L. J. Alexander, R. H. Crozier, and A. G. Mackinlay. 1988. Are megabats flying primates? Contrary evidence from a mitochondrial DNA sequence. Aust. J. Biol. Sci. 41:327–332.

Benson, S. B. 1933. Concealing coloration among some desert rodents of the southwestern United States. Univ. Calif. Publ. Zool. 40:1–70.

Benstead, J. P., K. H. Barnes, and C. M. Pringle. 2001. Diet, activity patterns, foraging movement, and responses to deforestation of the aquatic tenrec Limnogale mergulus (Lipotyphla: Tenrecidae) in eastern Madagascar. J. Zool. 254:119–129.

Benton, M. J. 1997. Vertebrate palaeontology, 2nd edition. Chapman and Hall, London.

Benton, M. J. 2005. Vertebrate palaeontology, 3rd edition. Blackwell Science Publishers, Malden, MA.

Beran, G. W. (ed.). 1994. CRC handbook series on zoonoses, 2nd edition. CRC Press, Boca Raton, FL.

Berger, J. 1986. Wild horses of the Great Basin: social competition and population size. Univ. Chicago Press, Chicago.

Berger, J. 1990. Persistence of different-sized populations: an empirical assessment of rapid extinctions in bighorn sheep. Conserv. Biol. 4:91–98.

Berger, J., and C. Cunningham. 1994. Phenotypic alterations, evolutionarily significant structures, and rhino conservation. Conserv. Biol. 8:833–840.

Berger, J., and C. Cunningham. 1998. Natural variation in horn size and social dominance and their importance to the conservation of black rhinocerous. Conserv. Biol. 12:708–711.

Berger, P. J., N. C. Negus, and C. N. Rowsemitt. 1987. Effect of 6-methoxybenzoxazolinone on sex ratio and breeding performance in *Microtus montanus*. Biol. Reprod. 36:255–260.

Berger, J., C. Cunningham, and A. A. Gawuseb. 1994. The uncertainty of data and dehorning black rhinos. Conserv. Biol. 8:1149–1152.

Bergmann, C. 1847. Über die Verhältnisse der Wärmeökononie der Thiere zu ihren Grösse. Göttinger Studien 1:595–708.

Bernard, R. T. F., and G. S. Cumming. 1997. African bats: evolution of reproductive patterns and delays. Q. Rev. Biol. 72:253–274.

Berndtson, W. E. 1977. Methods for quantifying mammalian spermatogenesis: a review. J. Anim. Sci. 44:818–833.

Berner, R.A. 1997. The rise of plants and their effects on weathering and atmospheric CO2. Science 276:544–545.

Bernstein, I. S., and L. G. Sharpe. 1966. Social roles in a rhesus monkey group. Behaviour 26:91–104.

Berry, R. J. 1970. The natural history of the house mouse. Field Stud. 3:219–262.

Berta, A. 1991. New Enaliarctos (Pinnipedimorpha) from the Oligocene and Miocene of Oregon and the role of "Enaliarctids" in pinniped phylogeny. Smithsonian Contrib. Paleobiol. 69:1–33.

Bertram, B. C. R. 1976. Kin selection in lions and in evolution. Pp. 281–301 in Growing points in ethology (P. P. G. Bateson and R. A. Hinde, eds.). Cambridge Univ. Press, New York.

Bertram, B. C. R., and J. M. King. 1976. Lion and leopard immobilization using C1-744. East Afr. Wildl. J. 14:237–239.

Bewick, T. 1804. A general history of quadrupeds. G. and R. Waite, New York.

Bezerra, A. M. R., F. H. G. Rodrigues, and A. P. Carmignotto. 2001. Predation of rodents by the yellow armadillo (*Euphractus sexcinctus*) in Cerrado of the central Brazil. Mammalia 65:86–88.

Bhat, H. R., and T. H. Kunz. 1995. Altered flower/fruit clusters of the kitul palm used as roosts by the short-nosed fruit bat, *Cynopterus sphinx* (Chiroptera: Pteropodidae). J. Zool. Lond. 235:597–604.

Bibb, M. J., R. A. Van Etten, C. T. Wright, M. W. Walberg, and D. A. Clayton. 1981. Sequence and organization of mouse mitochondrial DNA. Cell 26:167–180.

Bick, Y. A. E., and W. D. Jackson. 1967. DNA content of monotremes. Nature 215:192–193.

Biewener, A., R. M. Alexander, and N. Heglund. 1981. Elastic energy storage in the hopping of kangaroo rats (*Dipodomys spectabilis*). J. Zool. Lond. 195:369–383.

Biggers, J. D. 1966. Reproduction in male marsupials. Symp. Zool. Soc. Lond. 15:251–280.

Biggs, J. R., K. D. Bennett, M. A. Mullen, T. K. Haarmann, M. Salisbury, R. J. Robinson, D. Keller, N. Torrez-Martinez, and B. Hjelle. 2000. Relationship of ecological variables to Sin Nombre virus antibody seroprevalence in populations of deer mice. J. Mammal. 81:676–682.

Binder, W. J., and B. Van Valkenburgh. 2000. Development of bite strength and feeding behaviour in juvenile spotted hyenas (*Crocuta crocuta*). J. Zool. Lond. 252:273–283.

Birdsall, D. A., and D. Nash. 1973. Occurrence of successful multiple-insemination of females in natural populations of deer mice (*Peromyscus maniculatus*). Evolution 27:106–110.

Birky, C. W. 1991. Evolution and population genetics of organelle genes: mechanisms and models. Pp. 112–134 in Evolution at the molecular level (R. K. Selander, A. G. Clark, and T. S. Whittam, eds.). Sinauer Associates, Sunderland, MA.

Birney, E. C., and J. R. Choate (eds.). 1994. Seventy-five years of mammalogy, 1919–1994. Special Pub. No. 11, Am. Soc. Mammal.

Biuw, M., B. McConnell, C. J. A. Bradshaw, H. Burton, and M. Fedak. 2003. Blubber and buoyancy: monitoring the body conditions of free-ranging seals using simple dive characteristics. J. Exp. Biol. 206:3405–3423.

Bjornhag, G. 1994. Adaptations in the large intestine allow small animals to eat fibrous foods. Pp. 287–309 in The digestive system in mammals: food, form, and function (D. J. Chivers and P. Langer, eds.). Cambridge Univ. Press, Cambridge, UK.

Black, H. L. 1974. A north temperate bat community: structure and prey populations. J. Mammal. 55:138–157.

Blackburn, D. G. 1991. Evolutionary origins of the mammary gland. Mammal Rev. 21:81–96.

Blackburn, D. G., V. Hayssen, and C. J. Murphy. 1989. The origins of lactation and the evolution of milk: a review with new hypotheses. Mammal Rev. 19:1–26.

Blackburn, T. M., and B. A. Hawkins. 2004. Bergmann's rule and the mammal fauna of northern North America. Ecography 277:15–724.

Blacket, M. J., S. J. B. Cooper, C. Krajewski, and M. Westerman. 2006. Systematics and evolution of the dasyurid marsupial genus *Sminthopsis*: II. The Murina species group. J. Mammal. Evol. 13:125–138.

Blair, W. F. 1951. Population structure, social behavior, and environmental relations in a natural population of the beach mouse (*Peromyscus polionotus leucocephalus*). Contr. Lab. Vert. Biol. Univ. Mich. 48:1–47.

Blix, A. S., H. J. Grav, and K. Ronald. 1979. Some aspects of temperature regulation in newborn harp seal pups. Am. J. Physiol. 236:R188–197.

Bloch, J. I., K. D. Rose, and P. D. Gingerich. 1998. New species of Batodonoides (Lipotyphla, Geolabididae) from the early Eocene of Wyoming: Smallest known mammal. J. Mammal. 79:804–827.

Bloedel, P. 1955. Hunting methods of fish-eating bats, particularly *Noctilio leporinus*. J. Mammal. 36:390–399.

Bloemendal, H. 1977. The vertebrate eye lens. Science 197:127–138.

Bodemer, C. W. 1968. Modern embryology. Holt, Rinehart and Winston, New York.

Bodmer, R. E., and G. B. Rabb. 1992. Okapia johnstoni. Mammal. Species No. 422:1–8.

Bodmer, R. E., J. F. Eisenberg, and K. H. Redford. 1997. Hunting and the likelihood of extinction of Amazonian mammals. Conserv. Biol. 11:460–466.

Boehlert, G. W., Costa, D. P., Crocker, D. E., Green, P., O'Brien, T., Levitus, S., and Le Beouf, B. J. 2001. Autonomous pinniped environmental samplers: Using instrumented animals as oceanographic data collectors. Journal of Atmospheric and Oceanic Technology 18: 1882–1893.

Bogdanowicz, W., M. B. Fenton, and K. Daleszczyk. 1999. The relationship between echolocation calls, morphology and diet in insectivorous bats. J. Zool. Lond. 247:381–393.

Boisserie, J. R., and F. Lihoreau. 2006. Emergence of Hippopotamidae: new scenarios. C. R. Palevol. 5:749–756.

Boisserie, J. R., F. Lihoreau, and M. Brunet. 2005. The position of Hippopotamidae within Cetartiodactyla. Proc. Nat. Acad. Sci. USA 102:1537–1541.

Bökönyi, S. 1984. Horse. Pp. 162–173 in Evolution of domesticated animals (I. L. Mason, ed.). Longman, New York.

Bökönyi, S. 1989. Definitions of animal domestication. Pp. 22–27 in The walking larder: patterns of domestication, pastoralism, and predation (J. Clutton-Brock, ed.). Unwin Hyman, London.

Bolkovic, M. L., S. M. Caziani, and J. J. Protomastro. 1995. Food habits of the three-banded armadillo (Xenarthra: Dasypodidae) in the dry Chaco, Argentina. J. Mammal. 76:1199–1204.

Bonaparte, J. F. 1990. New late Cretaceous mammals from the Los Alamitos Formation, northern Patagonia. Natl. Geogr. Res. 6:63–93.

Bonato, V., K. G. Facure, and W. Uieda. 2004. Food habits of bats of subfamily Vampyrinae in Brazil. J. Mammal. 85:708–713.

Boness, D. L., and W. D. Bowen. 1996. The evolution of maternal care in pinnipeds. BioScience 46:645–654.

Bonner, N. 1989. Whales of the world. Facts on File, New York.

Bonner, T. I., R. Heinemann, and G. J. Todaro. 1980. Evolution of DNA sequence has been retarded in Malagasy lemurs. Nature 286:420–423.

Bonner, W. N. 1990. The natural history of seals. Facts on File, New York.

Bookhout, T. A. (ed.). 1996. Research and management techniques for wildlife and habitats, 5th edition, rev. Wildlife Society, Bethesda, MD.

Boonstra, R., C. J. Krebs, S. Boutin, and J. M. Eadie. 1994. Finding mammals using far infra-red thermal imaging. J. Mammal. 75:1063–1068.

Bordingnon, M. O. 2006. Diet of the fishing bat Noctilio leporinus (Linnaeus) (Mammalia, Chiroptera) in a mangrove area of southern Brazil. Revista Brasileira de Zoologia 23:256–260.

Bordingnon, M. O., and A. de Oliveira-Franca. 2004. Fur colour variation in fishing bat *Noctilio leporinus* (L., 1758) (Mammalia, Chiroptera). Revista Brasileira de Zoologia 6:181–189.

Bounds, D. L., M. H. Sherfy, and T. A. Mollett. 2003. Nutria, *Myocastor coypu*. Pp. 1119–1147 in Wild mammals of North America: Biology, management, and conservation. (G. A. Feldhamer, B. C. Thompson, and J. A. Chapman, eds.). Johns Hopkins Univ. Press, Baltimore.

Boulieré, F. 1964. The natural history of mammals, 3rd edition, rev. Alfred A. Knopf, New York.

Bourliére, F. 1970. The natural history of mammals. Alfred A. Knopf, New York.

Boutin, S. 1990. Food supplementation with terrestrial vertebrates: patterns, problems, and the future. Can. J. Zool. 68:203–220.

Boutin, S., Wauters, L. A., McAdam, A. G., Humphries, M. M., Tosi, G. & Dhondt, A. A. 2006. Anticipatory reproduction and population growth in seed predators. Science 314:1928–1930.

Bovet, J. 1992. Mammals. Pp. 321–361 in Animal homing (F. Papi, ed.). Chapman and Hall, New York.

Bowen, W. D., O. T. Oftedal, and D. J. Boness. 1985. Birth to weaning in 4 days: remarkable growth in the hooded seal, *Cystophora cristata*. Can. J. Zool. 63:2841–2846.

Bower, B. 2002. Heads up: problem solving pushed bright primates toward bigger brains. Science News 161:166.

Bowles, M. L., and C. J. Whelan. 1995. Restoration of endangered species. Cambridge Univ. Press, New York.

Bown, T. M., and M. J. Kraus. 1979. Origin of the tribosphenic molar and metatherian and eutherian dental formulae. Pp. 172–181 in Mesozoic mammals: the first two-thirds of mammalian history (J. A. Lillegraven, Z. Kielan-Jaworowska, and W.A. Clemens, eds.). Univ. California Press, Berkeley.

Bowyer, R. T., J. W. Testa, and J. B. Faro. 1995. Habitat selection and home ranges of river otters in a marine environment: effects of the Exxon Valdez oil spill. J. Mammal. 76:1–11.

Boyce, M. S. 1984. Restitution of r-selection and K-selection as a model of density-dependent natural selection. Annu. Rev. Ecol. Syst. 15:427–447.

Boyer, B. B., and B. M. Barnes. 1999. Molecular and metabolic aspects of mammalian hibernation. BioScience 49:713–724.

Bozinovic, F., and J. F. Merritt. 1992. Summer and winter thermal conductance of Blarina brevicauda (Mammalia:Insectivora:Soricidae) inhabiting the Appalachian Mountains. Ann. Carnegie Mus. 61:33–37.

Bradbury, J. W. 1977. Lek mating behavior in the hammer-headed bat. Z. Tierpsychol. 45:225–255.

Bradley, D. G., D. E. McHugh, P. Cunningham, and R. Loftus. 1996. Mitochondrial diversity and the origins of African and European cattle. Proc. Natl. Acad. Sci. USA 93:5131–5135.

Bradley, R., and R. C. Lowson. 1992. Bovine spongiform encephalopathy: the history, scientific, political and social issues. Pp. 285–299 in Prion diseases of humans and animals (S. Prusiner, J. Collinge, J. Powell, and B. Anderton, eds.). Ellis Horwood, New York.

Bradshaw, C. J. A., R. J. Barker, R. G. Harcourt, and L. S. Davis. 2003. Estimating survival and capture probability of fur seal pups using multistate mark-recapture models. J. Mammal. 84:65–80.

Bradshaw, F. J., and S. D. Bradshaw. 2001. Maintenance nitrogen requirement of an obligate nectarivore, the honey possum, Tarsipes rostratus. J. Comp. Physiol. B 171:59–67.

Bradshaw, G. V. R. 1961. Le cycle des reproduction

des *Macrotus californicus* (Chiroptera, Phyllostomatidae). Mammalia. 25:117–119.

Bradshaw, G. V. R. 1962. Reproductive cycle of the California leaf-nosed bat, *Macrotus californicus*. Science 136:645–646.

Bradshaw, R. H., R. F. Parrott, J. A. Goode, D. M. Lloyd, R. G. Rodway, and D. M. Broom. 1996. Behavioral and hormonal responses of pigs during transport—effect of mixing and duration of journey. Anim. Sci. 62:547–554.

Bradshaw, S. D., K. D. Morris, and F. J. Bradshaw. 2001. Water and electrolyte homeostasis and kidney function of desert-dwelling marsupial wallabies in western Australia. J. Comp. Physiol. B 171:23–32.

Brain, P. F., and A. E. Poole. 1976. The role of endocrines in isolation-induced intermale fighting in albino laboratory mice, II: Sex steroid influences in aggressive mice. Aggr. Behav. 2:55–76.

Brammall, J., and M. Archer. 1997. A new Oligocene-Miocene species of *Burramys* (Marsupialia, Burramyidae) from Riversleigh, north-western Queensland. Memoirs of the Queensland Museum 41:247–268.

Brammall, J., and M. Archer. 1999. Living and extinct petaurids, acrobatids, tarsipedids and burramyids (Marsupialia): relationships and diversity through time. Aust. Mammal. 21:24–25.

Brändli, L., L.-J. Lawson Handley, P. Vogel, and N. Perrin. 2005. Evolutionary history of the greater white-toothed shrew (*Crocidura russula*) inferred from analysis of mtDNA, Y, and X chromosome markers. Mol. Phylogenet. Evol. 37: 832–844.

Brandon, R. N., and R. M. Burian, ed. 1984. Genes, organisms, populations. MIT Press, Cambridge, MA.

Brandt, J. H., M. Dioli, A. Hassanin, R. A. Melville, L. E. Olson, A. Seveau, and R. M. Timm. 2001. Debate on the authenticity of *Pseudonovibos spiralis* as a new species of wild bovid from Vietnam and Cambodia. J. Zool. Lond. 255:437–444.

Brannon, M. P. 2000. Niche relationships of two syntopic species of shrews, *Sorex fumeus* and *S. cinereus*, in the southern Appalachian mountains. J. Mammal. 81:1053–1061.

Brashares, J. S., T. Garland, Jr., and P. Arcese. 2000. Phylogenetic analysis of coadaptation in behavior, diet, and body size in the African antelope. Behav. Ecol. 11:452–463.

Brass, D. A. 1994. Rabies in bats: natural history and public health implications. Livia Press, Ridgefield, CT.

Braude, S. 2000. Dispersal and new colony formation in wild naked mole-rats: evidence against inbreeding as the system of mating. Behav. Ecol. 11:7–12.

Braun, J. K., and M. A. Mares. 2002. Systematics of the Abrocoma cinerea complex (Rodentia: Abrocomidae), with description of a new species of Abrocoma. J. Mammal. 83:1–19.

Braun Breton, C., and L. H. Pereira da Silva. 1993. Malaria proteases and red blood cell invasion. Parasitol. Today 9:92–96.

Breeden, S., and K. Breeden. 1967. Animals in eastern Australia. Australasian Publ., Sydney.

Bremer, K. 1992. Ancestral areas: a cladistic reinterpretation of the center of origin concept. Syst. Biol. 41:436–445.

Brigham, R. M., and P. Trayhurn. 1994. Brown fat in birds? A test for the mammalian BAT-specific mitochondrial uncoupling protein in common poorwills. Condor. 96:208–211.

Briggs, J.C. 1995. Global biogeography. Elsevier, Amsterdam, Netherlands.

Bright, J. L., and J. J. Hervert. 2005. Adult and fawn mortality of Sonoran pronghorn. Wildl. Soc. Bull. 33:43–50.

Britten, R. J. 1986. Rates of DNA sequence evolution differ between taxonomic groups. Science 231:1393–1398.

Britten, R. J., D. E. Graham, and B. R. Neufield. 1974. Analysis of repeating DNA sequences and a speculation of the origins of evolutionary novelty. Methods Enzymol. 29:363–418.

Brody, S. 1945. Bioenergetics and growth. Reinhold, New York.

Bro-Jørgensen J., and S. M. Durant. 2003. Mating strategies of topi bulls: getting in the centre of attention. Animal Behaviour 65: 585–594.

Bronner, G. N. 1995. Cytogenetic properties of nine species of golden moles (Insectivora: Chrysochloridae). J. Mammal. 76:957–971.

Bronner, G. N., and P. D. Jenkins. 2005. Order Afrosoricida. Pp. 71–81 *in* Mammal species of the world: a taxonomic and geographic reference (D. E. Wilson and D. M. Reeder, eds.). 3d ed., vol. 1, Johns Hopkins Univ. Press, Baltimore, MD.

Bronson, F. H. 1971. Rodent pheromones. Biol. Reprod. 4:344–357.

Bronson, F. H. 1979. The reproductive ecology of the house mouse. Q. Rev. Biol.54:265–299.

Bronson, F. H. 1989. Mammalian reproductive biology. Univ. Chicago Press, Chicago.

Bronson, F. H., and B. E. Eleftheriou. 1964. Chronic physiological effects of fighting in mice. Gen. Comp. Endocrinol. 4:9–14.

Brook, B. S., and T. J. Pedley. 2002. A model for time-dependent flow in (giraffe jugular) veins: uniform tube properties. J. Biomech. 35:95–107.

Brooke, A. P. 1994. Diet of the fishing bat, *Noctilio leporinus* (Chiroptera: Noctilionidae). J. Mammal. 75:212–218.

Brooke, A. P., and M. Tschapka. 2002. Threats from overhunting to the flying fox, *Pteropus tonganus* (Chiropters: Pteropodidae) on Niue Island, South Pacific Ocean. Biol. Conserv. 103:343–348.

Brooke, A. P., C. Solek, and A. Tualaulelei. 2000. Roosting behavior of colonial and solitary flying foxes in American Samoa (Chiroptera: Pteropodidae). Biotropica 32:338–350.

Brooks, D. R., and D. A. McLennan. 1991. Phylogeny, ecology, and behavior. Univ. Chicago Press, Chicago.

Broome, L. S. 2001. Density, home range, seasonal movements and habitat use of the mountain pygmy possum *Burramys parvus* (Marsupialia: Burramyidae) at Mount Blue Cow, Kosciuszko National Park. Aust. Ecol. 26:275–292.

Brower, J. E., and T. J. Cade. 1966. Ecology and physiology of *Napaeozapus insignis* (Miller) and other woodland mice. Ecology 47:46–63.

Brown, C. H., and P. M. Waser. 1984. Hearing and communication in blue monkeys (*Cercopithecus mitis*). Anim. Behav. 32:66–75.

Brown, C. L., C. E. Rupprecht, and W. M. Tzilkowski. 1990. Adult raccoon survival in an enzootic rabies area of Pennsylvania. J. Wildl. Dis. 26:346–350.

Brown, G. 1993. The great bear almanac. Lyons and Burford, New York.

Brown, J. H. 1971. Mammals on mountaintops: non-equilibrium insular biogeography. Am. Nat. 105:467–478.

Brown, J. H. 1975. Geographical ecology of desert rodents. Pp. 315–341 in Ecology and evolution of communities (M. L. Cody and J. M. Diamond, eds.). Belknap Press of Harvard Univ. Press, Cambridge, MA.

Brown, J. H. 1995. Macroecology. Univ. Chicago Press, Chicago.

Brown, J. H., and G. A. Bartholomew. 1969. Periodicity and energetics of torpor in the kangaroo mouse, *Microdipodops pallidus*. Ecology 50:705–709.

Brown, J. H., and D. W. Davidson. 1977. Competition between seed-eating rodents and ants in desert ecosystems. Science 196:880–882.

Brown, J. H., and A. C. Gibson. 1983. Biogeography. C. V. Mosby, St. Louis.

Brown, J. H., and B. A. Harney. 1993. Population and community ecology of heteromyid rodents in temperate habitats. Pp. 618–651 in Biology of the Heteromyidae (H. H. Genoways and J. H. Brown, eds.). Spec. Pub., Am. Soc. Mammal. 10:1–719.

Brown, J. H., and E. J. Heske. 1990. Control of a desert-grassland transition by a keystone rodent guild. Science 250:1705–1708.

Brown, J. H., and R. C. Lasiewski. 1972. Metabolism of weasels: the cost of being long and thin. Ecology 53:939–943.

Brown, J. H., and M. V. Lomolino. 1998. Biogeography, 2nd edition. Sinauer Associates, Sunderland, MA.

Brown, J. H., and B. A. Maurer. 1989. Macroecology: the division of food and space among species on continents. Science 243:1145–1150.

Brown, J. H., and G. B. West. 2000. Scaling in biology. Oxford Univ. Press, New York

Brown, J. H., and D. E. Wilson. 1994. Natural history

and evolutionary ecology. Pp. 377–397 *in* Seventy-five years of mammalogy, 1919–1994 (E. C. Birney and J. R. Choate, eds.). Spec. Pub. No. 11, Am. Soc. Mammal.

Brown, J. H., O. J. Reichman, and D. W. Davidson. 1979. Granivory in desert ecosystems. Annu. Rev. Ecol. Syst. 10:201–227.

Brown, J. H., D. W. Davidson, J. C. Munger, and R. S. Inouye. 1986. Experimental community ecology: the desert granivore system. Pp. 41–61 *in* Community ecology (J. Diamond and T. J. Case, eds.). Harper and Row, New York.

Brown, J. L. 1964. The evolution of diversity in avian territorial systems. Wilson Bull. 76:160–169.

Brown, J. L., and A. Eklund. 1994. Kin recognition and the major histocompatibility complex: an integrative review. Am. Nat. 143:435–461.

Brown, K. J., and C. T. Downs. 2005. Seasonal behavioural patterns of free-living rock hyrax (*Procavia capensis*). J. Zool. 265:311–326.

Brown, K. J., and C. T. Downs. 2006. Seasonal patterns in body temperature of free-living rock hyrax (*Procavia capensis*). Comp. Biochem. Physiol. Part A Mol. Int. Physiol. 143:42–49.

Brown, L., R. W. Shumaker, and J. F. Downhower. 1995. Do primates experience sperm competition? Am. Nat. 146:302–306.

Brown, P., T. Sutikna, M.J. Morwood, R.P. Soejono, R.P., Jatmiko, E.W. Saptomo, and R.A. Due. 2004. A new small-bodied hominin from the late Pleistocene of Flores, Indonesia. Nature 431:1055–1061.

Brown, P. T., T. W. Brown, and A. D. Grinnell. 1983. Echolocation, development, and vocal communication in the lesser bulldog bat, Noctilio albiventris. Behav. Ecol. Sociobiol. 13:287–298.

Brown, R. D. (ed.). 1983. Antler development in Cervidae. Caesar Kleberg Wildlife Research Institute, Kingsville, TX.

Brown, R. E., and D. W. Macdonald. 1985. Social odours in mammals, 2 vols. Clarendon Press, Oxford, England.

Brown, S. G. 1978. Whale marking techniques. Pp. 71–80 *in* Animal marking (B. Stonehouse, ed.). Univ. Park Press, Baltimore.

Bruce, H. M. 1966. Smell as an exteroceptive factor. J. Anim. Sci. Suppl. 25:83–89.

Bruford, M. W., D. G. Bradley, and G. Luikart. 2003. DNA markers reveal the complexity of livestock domestication. Nature Review Genetics 4:900–910.

Brundin, L. 1966. Transantarctic relationships and their significance as evidenced by midges. K. Sven. Vetenskapsakad. Handl. Ser. 4 11:1–472.

Brundin, L. 1988. Phylogenetic biogeography. Pp. 343–369 *in* Analytical biogeography (A.A. Myers and P.S. Giller, eds.). Chapman & Hall, New York.

Bruns, V., H. Burda, and M. J. Ryan. 1989. Ear morphology of the frog-eating bat (*Trachops cirrhosus*, Family: Phyllostomidae): apparent specializations for low-frequency hearing. J. Morph. 199: 103–118.

Bruns Stockrahm, D. M., B. J. Dickerson, S. L. Adolf, and R. W. Seabloom. 1996. Aging black-tailed prairie dogs by weight of eye lenses. J. Mammal. 77:874–881.

Bruseo, J. A., and R. E. Barry, Jr. 1995. Temporal activity of syntopic Peromyscus in the central Appalachians. J. Mammal. 76:78–82.

Bryant, P. J. 1995. Dating remains of gray whales from the eastern North Atlantic. J. Mammal. 76:857–861.

Bryden, M. M., and G. S. Molyneux. 1978. Arteriovenous anastomoses in the skin of seals. II. The California sea lion (*Zalophus californianus*) and the northern fur seal (*Callorhinus ursinus*)(Pinnipedia:Otariidae). Anatomical Record 191:253–260.

Buchholtz, E. A. 2001. Vertebral osteology and swimming style in living and fossil whales (Order: Cetacea). J. Zool. Lond. 253:175–190.

Buchler, E. R., and S. B. Childs. 1982. Use of post-sunset glow as an orientation cue by the big brown bat (*Eptesicus fuscus*). J. Mammal. 63:243–247.

Buck, C. L., and B. M. Barnes. 1999a. Annual cycle of body composition and hibernation in free-living arctic ground squirrels. J. Mammal. 80:430–442.

Buck, C. L., and B. M. Barnes. 1999b. Temperatures of hibernacula and changes in body composition of arctic ground squirrels over winter. J. Mammal. 80:1264–1276.

Buckland, S. T., D. R. Anderson, K. P. Burnham, J. L. Laake, D. L. Borchers, and L. Thomas. 2001. Introduction to distance sampling. Oxford Univ. Press, New York.

Budiansky, S. 1992. The covenant of the wild: why animals choose domestication. W. Morrow, New York.

Buechner, H. K.,. and H. D. Roth. 1974. The lek system in Uganda kob. Am. Zool. 14:145–162.

Buffon, G. L. L. 1858. Buffon's natural history of man, the globe, and of quadrupeds. Hurst, New York.

Bulliet, R. W. 1975. The camel and the wheel. Harvard Univ. Press, Cambridge, MA.

Burda, H., S. Marhold, T. Westenberger, R. Wiltschko, and W. Wiltschko. 1990. Magnetic compass orientation in the subterranean rodent *Cryptomys hottentotus* (Bathyergidae). Experientia. 46:528–530.

Burde, J. H., and G. A. Feldhamer. 2003. Mammals of the National Parks: conserving America's wildlife and parklands. Johns Hopkins Univ. Press, Baltimore.

Burde, J. H., and G. A. Feldhamer. 2005. Mammals of the National Parks: Conserving America's Wildlife and Parklands. Johns Hopkins Univ. Press, Baltimore.

Burgdorfer, W., and T. G. Schwan. 1991. Borrelia. Pp. 560–566 in Manual of clinical microbiology, 5th edition (A. Balows, W. J. Hauslei, K. L. Herrmann, H. D. Isenberg, and H. J. Shadomy, eds.). American Society of Microbiology, Washington, DC.

Burk, A., and M. S. Springer. 2000. Intergeneric relationships among Macropodoidea (Metatheria: Diprotodontia) and the chronicle of kangaroo evolution. J. Mammalian Evolution 7:213–237.

Burnett, C. D. 1983. Geographic and climate correlates on morphological variation in *Eptesicus fuscus*. J. Mammal 64:437–444.

Burt, W. H. 1960. Bacula of North American mammals. Misc. Pub. Mus. Zool., Univ. Mich. 113:1–75.

Burt, W. H., and R. P. Grossenheider. 1980. A field guide to mammals of North America north of Mexico, 3rd edition. Macmillan, Boston.

Burton, J. A., and B. Pearson. 1987. Collins guide to the rare mammals of the world. Collins, London.

Burton, R. 1980. The life and death of whales. Andre Deutsch, Worcester, England.

Bush, M. 1996. Methods of capture, handling, and anesthesia. Pp. 25–40 in Wild mammals in captivity (D. G. Kleiman, M. E. Allen, K. V. Thompson, and S. Lumpkin, eds.). Univ. Chicago Press, Chicago.

Bush, M. B. 1994. Amazonian speciation: a necessarily comples model. J. Biogeogr. 21:5–17.

Butcher, E. O. 1951. Development of the piliary system and the replacement of hair in mammals. An. N. Y. Acad. Sci. 53:508–516.

Butler, P. M. 1972. The problem of insectivore classification. Pp. 253–265 in Studies in vertebrate evolution (K. A. Joysey and T. S. Kemp, eds.). Winchester Press, New York.

Butler, P. M. 1978. Insectivora and Chiroptera. Pp. 56–68 in Evolution of African mammals (V. J. Maglio and H. B. S. Cooke, eds.). Harvard Univ. Press, Cambridge, MA.

Butler, P. M. 1992. Tribosphenic molars in the Cretaceous. Pp. 125–138 in Structure, function, and evolution of teeth (P. Smith and E. Tchernov, eds.). Freund Publishing House, London.

Byers, J. A. 1997. American pronghorn. Univ. Chicago Press, Chicago.

Byers, J. A. 2003. Pronghorn, Antilocapra americana. Pp. 998–1008 in Wild mammals of North America: biology, management, and conservation (G. A. Feldhamer, B. C. Thompson, and J. A. Chapman, eds.). Johns Hopkins Univ. Press, Baltimore.

Byers, J. A., A. A. Byers, and S. J. Dunn. 2006. A dry summer diminishes mate search effort by pronghorn females: evidence for a significant cost of mate search. Ethology 112:74–80.

Cabanac, A. J. 2002. Contracted spleen in seals, estimates of dilated organs, and diving capacity. Polar Biol. 25:1–4.

Cabanac, M. 1986. Keeping a cool head. New Physiol. Sci. 1:41–44.

Caccone, A., and J. R. Powell. 1989. DNA divergence among hominoids. Evolution 43:925–942.

Cade, T. J. 1983. Hybridization and gene exchange

among birds in relation to conservation. Pp. 288–309 in Genetics and conservation (C. M. Schoenwald-Cox, S. M. Chambers, B. MacBryde, and L. Thomas, eds.). Benjamin Cummings, Menlo Park, CA.

Cain, A. J. 1954. Animal species and their evolution. Princeton Univ. Press, Princeton, NJ.

Cain, S. A. 1944. Foundations of plant geography. Harper, New York.

Calder, W. A. 1984. Size, function, and life history. Harvard Univ. Press, Cambridge, MA.

Caldwell, D. K., and M. C. Caldwell. 1989. Pygmy sperm whale *Kogia breviceps* (de Blainville, 1838): dwarf sperm whale *Kogia simus* Owen, 1866. Pp. 235–260 in Handbook of marine mammals, vol. 4: river dolphins and the larger toothed whales (S. H. Ridgway and R. Harrison, eds.). Academic Press, New York.

Camhi, J. M. 1984. Neuroethology. Sinauer Associates, Sunderland, MA.

Campagna, C., and B.J. LeBoeuf. 1988. Thermoregulatory behaviour of southern sea lions and its effect on mating strategies. Behaviour 107:72–90.

Campbell, C. B. G. 1974. On the phyletic relationships of the tree shrews. Mammal Rev. 4:125–143.

Campbell, K. L., I. W. McIntyre, and R. A. MacArthur. 1999. Fasting metabolism and thermoregulatory competence of the star-nosed mole, *Condylura cristata* (Talpidae: Condylurinae). Comp. Biochem. Physiol. 123A:293–298.

Campbell, W. C. 1988. Trichinosis revisited—another look at modes of transmission. Parasitol. Today 4:83–86.

Cannon, B., and J. Nedergaard. 2004. Brown adipose tissue: function and physiological significance. Physiological Review 84:277–359.

Cantoni, D. 1993. Social and spatial organization of free-ranging shrews, *Sorex coronatus* and *Neomys fodiens* (Insectivora, Mammalia). Anim. Behav. 45:975–995.

Caraco, T., and L. L. Wolf. 1975. Ecological determinants of group sizes of foraging lions. Am. Nat. 109:343–352.

Carbone, C., G. M. Mace, S. C. Roberts, and D. W. Macdonald. 1999. Energetic constraints on the diet of terrestrial carnivores. Nature 402:286–288.

Carbone, C., and J. L. Gittleman. 2002. A common rule for the scaling of carnivore density. Science 295:2273–2276.

Carbone, C., G. M. Mace, S. C. Roberts, and D. W. Macdonald. 1999. Energetic constraints on the diet of terrestrial carnivores. Nature 402:286–288.

Cardillo, M., G. M. Mace, K. E. Jones, J. Bielby, O. R. P. Bininda-Emonds, W. Sechrest, C. D. L. Orme, and A. Purvis. 2005. Multiple causes of high extinction risk in large mammal species. Science 309:1239–1241.

Carl, E. A. 1971. Population control in arctic ground squirrels. Ecology 52:395–413.

Carleton, M. D. 1973. A survey of gross stomach morphology in New World Cricetinae (Rodentia, Muroidea), with comments on functional interpretations. Misc. Publ. Mus. Zool., Univ. Michigan. 146:1–43.

Carleton, M. D. 1984. Introduction to rodents. Pp. 255–265 in Orders and families of recent mammals of the world (S. Anderson and J. K. Jones, Jr., eds.). John Wiley and Sons, New York.

Carleton, M. D. 1985. Macroanatomy. Pp. 116–175 in Biology of New World Microtus (R. H. Tamarin, ed.). Spec. Pub. No. 8, American Society of Mammalogists.

Carleton, M. D., and G. G. Musser. 1984. Muroid rodents. Pp. 289–446 in Orders and families of recent mammals of the world (S. Anderson and J. K. Jones, Jr., eds.). John Wiley and Sons, New York.

Carleton, M. D., and G. G. Musser. 2005. Order Rodentia. Pp. 745–752 in Mammal species of the world: a taxonomic and geographic reference. 3rd edition (D. E. Wilson and D. M. Reeder, eds.). Johns Hopkins Univ. Press, Baltimore.

Caro, T. M. 1994. Cheetahs of the Serengeti plains. Univ. Chicago Press, Chicago.

Carpaneto, G. M., and A. Fusari. 2000. Subsistence hunting and bushmeat exploitation in central-western Tanzania. Biodivers. Conserv. 9:1571–1585.

Carrick, F. N., and R. L. Hughes. 1978. Reproduction

in male monotremes. Aust. Zool. 20:211–231.

Carrington, R. 1959. Elephants: a short account of their natural history, evolution, and influence on mankind. Basic Books, New York.

Carroll, R. L. 1988. Vertebrate paleontology and evolution. W. H. Freeman, New York.

Carson, R. 1962. Silent spring. Fawcett, Boston.

Carter, T. C., and G. A. Feldhamer. 2005. Roost tree use by maternity colonies of Indiana bats and northern long-eared bats in southern Illinois. Forest Ecol. Mgmt. 219:259–268.

Cartmill, M. 1972. Arboreal adaptations and the origin of the Order Primates. Pp. 97–122 in Biology of the primates (R. H. Tuttle, ed.). Aldine-Atherton, Chicago.

Cartmill, M. 1985. Climbing. Pp. 73–88 in Functional vertebrate morphology (M. Hildebrand, D. M. Bramble, K. F. Liem, and D. B. Wake, eds.). Harvard Univ. Press, Cambridge, MA.

Cartwright, T. 1974. The plasminogen activator of vampire bat saliva. Blood 43:317–326.

Casey, T. M. 1981. Nest insulation: energy savings to brown lemmings using a winter nest. Oecologia. 50:199–204.

Cassens, I., S. Vicario, V. G. Waddell, H. Balchowsky, D. Van Belle, W. Ding, C. Fan, R. S. Lal Mohan, P. C. Simoes-Lopes, R. Bastida, A. Meyer, M. J. Stanhope, and M. C. Milinkovitch. 2000. Independent adaptation to riverine habitats allowed survival of ancient cetacean lineages. Proc. Natl. Acad. Sci. USA 97:11343–11347.

Castillo, A. H., M. N. Cortinas, and E. P. Lessa. 2005. Rapid diversification of South American tuco-tucos (*Ctenomys*; Rodentia, Ctenomyidae): contrasting mitochondrial and nuclear intron sequences. J. Mammal. 86:170–179.

Catania, K. C. 1995. A comparison of the Eimer's organs of three North American moles: the hairy-tailed mole (*Parascalops breweri*), the star-nosed mole (*Condylura cristata*), and the eastern mole (*Scalopus aquaticus*). J. Comp. Neurol. 354:150–160.

Catania, K. C. 1999. A nose that looks like a hand and acts like an eye: the unusual mechanosensory system of the star-nosed mole. J. Comp. Physiol. A 185:367–372.

Catania, K.C. 2000. Epidermal sensory organis of moles, shrew-moles, and desmans: a study of the family Talpidae with comment on the function and evolution of Eimer's organ. Brain, Behavior, and Evolution 56:146–174.

Catania, K. C., and J. H. Kaas. 1996. The unusual nose and brain of the star-nosed mole. BioScience 46:578–586.

Catesby, M. 1748. The natural history of Carolina, Florida, and the Bahama Islands, 3 vols. B. White, London.

Catzeflis, F. M., J. Michaux, A. W. Dickerman, and J. A. W. Kirsch. 1993. DNA hybridization and rodent phylogeny. Pp. 159–172 in Mammal phylogeny, vol. 2: placentals (F.S. Szalay, M.J. Novacek, and M.C. McKenna, eds.). Springer-Verlag, New York.

Cawthorn, J. M. 1994. A live-trapping study of two syntopic species of *Sorex*, *S. cinereus* and *S. fumeus*, in southwestern Pennsylvania. Pp. 39–43 in Advances in the biology of shrews (J. F. Merritt, G. L. Kirkland, Jr., and R. K. Rose, eds.). Special Pub. No. 18, Carnegie Museum of Natural History, Pittsburgh.

Cerchio, S., J. K. Jacobsen, and T. F. Norris. 2001. Temporal and geographical variation in songs of humpback whales, *Megaptera novaengliae*: synchronous change in Hawaiian and Mexican breeding assemblages. Anim. Behav. 62:313–329.

Chakraborty, R., and K. K. Kidd. 1991. The utility of DNA typing in forensic work. Science 254:1735–1739.

Chaline, J., P. Mein, and F. Petter. 1977. Les grandes lignes d'une classification évolutive des Muroidea. Mammalia 41:245–252.

Chambers, L. K., G. R. Singleton, and C. J. Krebs. 2000. Movements and social organization of wild house mice (*Mus domesticus*) in the wheatlands of northwestern Victoria, Aust. J. Mammal. 81:59–69.

Champoux, M., and S. J. Suomi. 1994. Behavioral and adrenocortical responses of rhesus macaque mothers to infant separation in an unfamiliar environment. Primates 35:191–202.

Chan, L. K. 1995. Extrinsic lingual musculature of

two pangolins (Pholidota: Manidae). J. Mammal. 76:472–480.

Chappell, M. A. 1978. Behavioral factors in the altitudinal zonation of chipmunks (*Eutamias*). Ecology 59:565–579.

Chappell, M. A., and G. A. Bartholomew. 1981. Standard operative temperatures and thermal energetics of the antelope ground squirrel *Ammospermophilus leucurus*. Physiol. Zool. 54:81–93.

Charif, R. A., P. J. Clapham, and C. W. Clark. 2001. Acoustic detections of singing humpback whales in deep waters off the British Isles. Mar. Mammal Sci. 17:751–768.

Charnov, E., and J. Finerty. 1980. Vole population cycles: a case for kin-selection? Oecologia 45:1–2.

Charnov, E. L. 1976. Optimal foraging: the marginal value theorem. Theoret. Pop. Biol. 9:129–136.

Charrier, I., N. Mathevon, and P. Jouventin. 2001. Mother's voice recognition by seal pups. Nature 412:873.

Chatterjee, H. J. 2006. Phylogeny and biogeography of gibbons: a dispersal-vicariance analysis. Int. J. Primatol. 27:699–712.

Cheatum, E. L. 1949. The use of corpora lutea for determining ovulation incidence and variations of fertility in white-tailed deer. Cornell Vet. 39:282–291.

Cheatum, E. L., and C. W. Severinghaus. 1950. Variations in fertility of white-tailed deer related to range conditions. Trans. N. Am. Wildl. Conf. 15:170–189.

Chen, X., C. R. Dickman, and M. B. Thompson. 1998. Diet of the mulgara, *Dasycercus cristicauda* (Marsupialis:Dasyuridae), in the Simpson Desert, central Australia. Wildl. Res. 25:233–242.

Cheney, D. L., and R. M. Seyfarth. 1983. Nonrandom dispersal in free-ranging vervet monkeys: social and genetic consequences. Am. Nat. 122:392–412.

Cheng, J., Z. Xiao, and Z. Zhang. 2005. Seed consumption and caching on seeds of three sympatric tree species by four sympatric rodent species in a subtropical forest, China. Forest Ecol. Mgmt. 216:331–341.

Chew, R. M., and A. E. Chew. 1970. Energy relationships of the mammals of a desert scrub (*Larrea tridentata*) community. Ecol. Monogr. 40:1–21.

Chew, R. M., and A. E. Dammann. 1961. Evaporative water loss of small vertebrates, as measured with an infrared analyzer. Science 133:384–385.

Childs, J. E. 1995. Special feature: zoonoses. J. Mammal. 76:663.

Childs, J. E., J. N. Mills, and G. E. Glass. 1995. Rodent-borne hemorrhagic fever viruses: a special risk for mammalogists? J. Mammal. 76:664–680.

Childs, J. E., L. Colby, J. W. Krebs, T. Strine, M. Feller, D. Noah, C. Drenzek, J. S. Smith, and C. E. Rupprecht. 1997. Surveillance and spatiotemporal associations of rabies in rodents and lagomorphs in the United States, 1985–1994. J. Wildl. Dis. 33:20–27.

Chilvers, B. L., S. Delean, N. J. Gales, D. K. Holley, I. R. Lawler, H. Marsh, and A. R. Preen. 2004. Diving behaviour of dugongs, *Dugong dugon*. J. Exp. Mar. Biol. Ecol. 304:203–224.

Chittenden, H. 1954. The American fur trade, 2 vols. Stanford Univ. Press, Stanford, CA.

Chittenden, H. M. 1986. The American fur trade of the Far West, 2 vols. Univ. Nebraska Press, Lincoln.

Chitty, D. 1960. Population processes in the vole and their relevance to general theory. Can. J. Zool. 38:99–113.

Chitty, D., and M. Shorten. 1946. Techniques for the study of the Norway rat *Rattus norvegicus*. J. Mammal. 27:63–78.

Chopra, S. R. K., and R. N. Vasishat. 1979. Sivalik fossil tree shrew from Haritalyangar, India. Nature 281:214–215.

Choudhury, A. 2001. An overview of the status and conservation of the red panda *Ailurus fulgens* in India, with reference to its global status. Oryx 35:250–259.

Christal, J., and H. Whitehead. 2001. Social affiliations within sperm whale (*Physeter macrocephalus*) groups. Ethology 107:323–340.

Christian, J. J. 1950. The adreno-pituitary system and population cycles in mammals. J. Mammal. 31:247–259.

Christian, J. J. 1963. Endocrine adaptive mechanisms and the physiologic regulation of population growth. Pp. 189–353 *in* Physiological mammalogy, vol. I: mammalian populations (W. V. Mayer and R. C. Van Gelder, eds.). Academic Press, New York.

Christian, J. J. 1978. Neurobehavioral endocrine regulation of small mammal populations. Pp. 143–158 *in* Populations of small mammals under natural conditions (D. P. Snyder, ed.). Spec. Pub. Ser. 5. Pymatuning Lab Ecology, Univ. Pittsburgh.

Christian, J. J., and D. E. Davis. 1964. Endocrines, behavior, and population: social and endocrine factors are integrated in the regulation of growth of mammalian populations. Science 146:1550–1560.

Churchfield, S. 1982. The influence of temperature on the activity and food consumption of the common shrew. Acta Theriol. 27:295–304.

Churchfield, S. 1990. The natural history of shrews. Cornell Univ. Press, Ithaca, NY.

Churchfield, S., J. Hollier, and V. K. Brown. 1995. Population dynamics and survivorship patterns in the common shrew *Sorex araneus* in southern England. Acta Theirol. 40:53–68.

Churchfield, S., J. Barber, and C. Quinn. 2000. A new survey method for water shrews (*Neomys fodiens*) using baited tubes. Mammal Rev. 30:249–254.

Cifelli, R. L. 1993. Early Cretaceous mammal from North America and the evolution of marsupial dental characters. Proc. Natl. Acad. Sci. USA 90:9413–9416.

Cifelli, R.L. 2001. Early mammalian radiations. J. Paleontol. 75:1214–1226.

Cifelli, R. L., and S. K. Madsen. 1998. Triconodont mammals from the medial Cretaceous of Utah. J. Vert. Paleontol. 18:403–411.

Ciszek, D. 2000. New colony formation in the "highly inbred" eusocial naked mole-rat: outbreeding is preferred. Behav. Ecol. 11:1–6.

Clapham, P. J., and C. A. Mayo. 1987. Reproduction and recruitment of individually identified humpback whales, *Megaptera novaeangliae*, observed in Massachusetts Bay, 1979–1985. Can. J. Zool. 65:2853–2863.

Claridge, A. W., J. M. Trappe, S. J. Cork, and D. L. Claridge. 1999. Mycophagy by small mammals in the coniferous forests of North America: nutritional value of sporocarps of *Rhizopogon vinicolor*, a common hypogeous fungus. J. Comp. Physiol. B 169:172–178.

Claridge, A. W., and S. J. Cork. 1994. Nutritional value of hypogeal fungal sporocarps for the long-nosed potoroo (*Potorous tridactylus*), a forest-dwelling mycophagous marsupial. Aust. J. Zool. 42:701–710.

Claridge, A. W., and T. W. May. 1994. Mycophagy among Australian mammals. Aust. J. Ecol. 19:251–275.

Clark, A. B. 1978. Sex ratio and local resource competition in a prosimian primate. Science 201:163–165.

Clark, C. W., and P. J. Clapham. 2004. Acoustic monitoring on a humpback whale (*Megaptera novaeangliae*) feeding ground shows continual singing into late spring. Proc. Roy. Soc. Biol. Sci. Ser. B 271:1051–1057.

Clark, J. D., J. E. Dunn, and K. G. Smith. 1993. A multivariate model of female black bear habitat use for a geographic information system. J. Wildl. Mgmt. 57:519–526.

Clark, W. E. L. 1959. History of the primates, 5th edition. Univ. Chicago Press, Chicago.

Clarke, M. R. 1978a. Structure and proportions of the spermaceti organ in the sperm whale. J. Mar. Biol. Assoc. U.K. 58:1–17.

Clarke, M. R. 1978b. Bouyancy control as a function of the spermaceti organ in the sperm whale. J. Mar. Biol. Assoc. U.K. 58:27–71.

Clements, F., P. Hope, C. Daniels, I. Chapma, and G. Wittert. 1998. Thermogenesis in the marsupial *Sminthopsis crassicaudata*: effect of catecholamines and diet. Aust. J. Zool. 46: 381–390.

Clements, M. L. 1992. Rocky Mountain spotted fever. Pp. 1304–1312 *in* Infectious diseases (S. L. Gorbach, J. G. Bartlett, and N. R. Blacklow, eds.). W. B. Saunders, Philadelphia.

Clisson, I., M. Lathuilliere, and B. Cronau-Roy. 2000. Conservation and evolution of microsatellite loci in primate taxa. Am. J. Primatol. 50:205–214.

Clutton-Brock, J. (ed.). 1989. The walking larder: patterns of domestication, pastoralism, and predation.

Unwin Hyman, London.

Clutton-Brock, J. 1992. The process of domestication. Mammal Rev. 22:79–85.

Clutton-Brock, J. 1996. Horses in history. Pp. 83–102 *in* Horses through time (S. L. Olsen, ed.). Roberts Rinehart Publ., Boulder, CO.

Clutton-Brock, J. 1999. A natural history of domesticated animals, 2nd edition. Cambridge Univ. Press, New York.

Clutton-Brock, J., and P. Jewell. 1993. Origin and domestication of the dog. Pp. 21–31 *in* Miller's anatomy of the dog, 3rd edition (H. E. Evans, ed.). W. B. Saunders, Philadelphia.

Clutton-Brock, T. H. 1988. Reproductive success. Univ. Chicago Press, Chicago.

Clutton-Brock, J., and P. Jewell. 1993. Origin and domestication of the dog. Pp. 21–31 *in* Miller's anatomy of the dog, 3rd edition (H. E. Evans, ed.). W. B. Saunders, Philadelphia.

Clutton-Brock, T. H., F. E. Guinness, and S. D. Albon. 1982. Red deer: behavior and ecology of two sexes. Univ. Chicago Press, Chicago.

Clutton-Brock, T. H., S. D. Albon, and F. E. Guinness. 1984. Maternal dominance, breeding success, and birth sex ratios in red deer. Nature 308:358–360.

Clutton-Brock, T. H., M. Major, and F. E. Guinness. 1985. Population regulation in male and female red deer. J. Anim. Ecol. 54:831–846.

Clutton-Brock, T. H., S. D. Albon, and F. E. Guinness. 1989. Fitness costs of gestation and lactation in wild mammals. Nature 337:260–262.

Clutton-Brock T. H., P. N. M. Brotherton, M. J. O'Riain, A. S. Griffin, D. Gaynor, L. Sharpe, R. Kansky, M. B. Manser, and G .M. McIlrath. 2000. Individual contributions to babysitting in a cooperative mongoose, *Suricata suricatta*. Proc. Roy. Soc. Biol. Sci. Ser. B 267: 301–305.

Clutton-Brock, T. H., P. N. M. Brotherton, M. J. O'Riain, A. S. Griffin, D. Gaynor, R. Kansky, L. Sharpe, and G. M. McIlrath. 2001. Contributions to cooperative rearing in meerkats. Anim. Behav. 61:705–710.

Clutton-Brock T. H., Russell A. F., and Sharpe L. L. 2003. Meerkat helpers do not specialize in particular activities. Anim. Behav. 66: 531–540.

Coburn, D. K., and F. Geiser. 1996. Daily torpor and energy saving in a subtropical blossom-bat, *Syconycteris australis* (Megachiroptera). Pp. 39–45 *in* Adaptations to the cold: Tenth International Hibernation Symposium (F. Geiser, A. J. Hulbert, and S. C. Nicol, eds.). Univ. New England Press, Armidale, NSW, Australia.

Cockburn, A. 1991. An introduction to evolutionary ecology. Blackwell Sci. Publ., London.

Cockburn, A., M. P. Scott, and D. J. Scotts. 1985. Inbreeding avoidance and male-biased natal dispersal in *Antechinus* spp. (Marsupialia: Dasyuridae). Anim. Behav. 33:908–915.

Cockrum, E. L. 1962. Introduction to mammalogy. Ronald Press, New York.

Cockrum, E. L. 1969. Migration in the guano bat, *Tadarida brasiliensis*. Univ. Kansas Mus. Nat. Hist. Misc. Publ. No. 51:303–336.

Codron, J., J. A. Lee-Thorp, M. Sponheimer, D. Codron, R. C. Grant, and D. J. DeRuiter. 2006. Elephant (*Loxodonta africana*) diets in Kruger National park, South Africa: spatial and landscape differences. J. Mammal. 87:27–34.

Cohn, J. P. 1990. Elephants: remarkable and endangered. BioScience 40:10–14.

Colburn, T., D. Dumanoski, and J. P. Myers. 1996. Our stolen future. Plume Penguin Books, New York.

Cole, A. J. (ed.). 1969. Numerical taxonomy: proceedings of the colloquium in numerical taxonomy held in the Univ. St. Andrew, September 1968. Academic Press, London.

Coles, R. W. 1969. Thermoregulatory function of the beaver tail. Am. Zool. 9:203a.

Coles, R. W. 1970. Pharyngeal and lingual adaptations in the beaver. J. Mammal. 51:424–425.

Coley, P. D., and J. A. Barone. 1996. Herbivory and plant defenses in tropical forests. Annu. Rev. Ecol. Syst. 27:305–335.

Colwell, R. K., and D. C. Lees. 2000. The mid-domain effect: geometric constraints on the geography of species richness. Trends Ecol. Evol.

15:70–76.

Comer, J. A., W. R. Davidson, A. K. Prestwood, and V. F. Nettles. 1991. An update on the distribution of *Parelaphostrongylus tenuis* in the southeastern United States. J. Wildl. Dis. 27:348–354.

Conner, D. A. 1983. Seasonal changes in activity patterns and the adaptive value of haying in pikas (*Ochotona princeps*). Can. J. Zool. 61:411–416.

Connor, E. F., and E. D. McCoy. 1975. The statistics and biology of the species–area relationship. Am. Nat. 113:791–833.

Connor, R. C., R. A. Smolker, and A. F. Richards. 1992. Dolphin alliances and coalitions. Pp. 415–443 in Coalitions and alliances in humans and other animals (A. H. Harcourt and F. B. M. de Waal, eds.). Oxford Univ. Press, New York.

Connour, J. R., K. Glander, and F. Vincent. 2000. Postcranial adaptations for leaping in primates. J. Zool. Lond. 251:79–103.

Conroy, C. J., and J. A. Cook. 2000. Molecular systematics of a holarctic rodent (Microtus: Muridae). J. Mammal. 81:344–359.

Constantine, D. G. 1967. Bat rabies in the southwestern United States, Public Health Rep. 82:867–888.

Constantine, D. G. 1970. Bats in relation to the health, welfare, and economy of man. Pp. 319–449 in Biology of bats, vol. II (W. A. Wimsatt, ed.). Academic Press, New York.

Contoli, L. 2000. Rodents of Italy species richness maps and forma Italiae. Hystrix 11:39–46.

Contreras, L. C., J. C. Torres-Mura, A. E. Sportorno, and L. I. Walker. 1994. Chromosomes of *Octomys mimax* and *Octodontomys gliroides* and relationships of octodontid rodents. J. Mammal. 75:768–774.

Cook, J. A., and T. L. Yates. 1994. Systematic relationships of the Bolivian tuco-tucos, genus *Ctenomys* (Rodentia: Ctenomyidae). J. Mammal. 75:583–599.

Cook, R. C., J. G. Cook, and L. D. Mech. 2004. Nutritional condition of northern Yellowstone elk. J. Mammal. 85:714–722.

Cooke, B. D. 2002. Rabbit haemorrhagic disease: field epidemiology and the management of wild rabbit populations. Rev. Sci. Tech. OIE 21:347–358.

Cooke, J. A., and C. R. Terman. 1977. Influence of displacement distance and vision on homing behavior of the white-footed mouse (*Peromyscus leucopus noveboracensis*). J. Mammal. 58:58–66.

Coombs, M. C. 1983. Large mammalian clawed herbivores: a comparative study. Trans. Am. Philosoph. Soc. 73:1–96.

Cooper, C. E., and P. C. Withers. 2004. Influence of season and weather on activity patterns of the numbat (*Myrmecobius fasciatus*) in captivity. Aust. J. Zool. 52:475–485.

Cooper, C. E., and P. C. Withers. 2005. Physiological significance of the microclimate in night refuges of the numbat *Myrmecobius fasciatus*. Aust. Mammal. 27:169–174.

Cooper, S. M. 1991. Optimal hunting group size: the need for lions to defend their kills against loss to spotted hyaenas. Afr. J. Ecol. 29:130–136.

Coppola, D. M., and J. G. Vandenbergh. 1987. Induction of a puberty-regulating chemosignal in wild mouse populations. J. Mammal. 68:86–91.

Corbet, G. B. 1978. The mammals of the Palaearctic region: a taxonomic review. British Museum of Natural History, London.

Corbet, G. B. 1988. The mammals of the Palearctic region: a taxonomic review. Cornell Univ. Press, Ithaca, NY.

Corbet, G. B., and J. E. Hill. 1991. A world list of mammalian species, 3rd edition. British Museum of Natural History Publ., London.

Corbett, L. K. 1995. The dingo in Australia and Asia. New South Wales Univ. Press, Sydney.

Cordeiro, J., A. Dirceu, and S. A. Talamoni. 2006. New data on the life history and occurrence of spiny rats *Trinomys moojeni* (Rodentia: Echimyidae), in southeastern Brazil. Acta Theriol. 51:163–168.

Cork, S. J., and G. J. Kenagy. 1989a. Nutritional value of hypogeous fungus for a forest-dwelling ground squirrel. Ecology 70:577–586.

Cork, S. J., and G. J. Kenagy. 1989b. Rates of gut passage and retention of hypogeous fungal spores in two forest-dwelling rodents. J. Mammal. 70:512–519.

Corneli, P. S., and R. H. Ward. 2000. Mitochondrial

genes and mammalian phylogenetics: increasing the reliability of branch length estimation. Mol. Biol. Evol. 17:224–234.

Cossins, A. R., and K. Bowler. 1987. Temperature biology of animals. Chapman and Hall, New York.

Costa, L.P. 2006. The historical bridge between the Amazon and Atlantic forest of Brazil: a study of molecular phylogeography with small mammals. J. Biogeogr. 30:71–86.

Cote, S. D., M. Festa-Bianchet, and K. G. Smith. 1998. Horn growth in mountain goats (*Oreamnos americanus*). J. Mammal. 79:406–414.

Cotgreave, P., and P. Stockley. 1994. Body size, insectivory and abundance in assemblages of small mammals. Oikos 71:89–96.

Cott, H. B. 1966. Adaptive coloration in animals. Methuen, London.

Coues, E. 1877. Monographs of North American Rodentia. No. 1—Muridae. Smithsonian Institution, Washington, DC.

Couette, S., G. Escarguei, and S. Montuire. 2005. Constructing, bootstrapping, and comparing morphometric and phylogenetic trees: a case study of New World monkeys (Platyrrhini, Primates). J. Mammal. 86:773–781.

Couffer, J. 1992. Bat bomb: World War II's other secret weapon. Univ. of Texas Press, Austin.

Courchamp, F., T. Clutton-Brock, and B. Grenfell. 1999. Inverse density dependence and the Allee effect. TREE 14:405–510.

Courchamp, F., G. S. A. Rasmussen, and D. W. Macdonald. 2002. Small pack size imposes a trade-off between hunting and pup-guarding in the painted hunting dog *Lycaon pictus*. Behav. Ecol. 13:20–27.

Cowan, P. E. 1995. Brushtail possum. Pp. 68–98 in The handbook of New Zealand mammals (C. M. King, ed.). Oxford Univ. Press, Auckland, New Zealand.

Cowles, C. J., R. L. Kirkpatrick, and J. O. Newell. 1977. Ovarian follicular changes in gray squirrels as affected by season, age, and reproductive state. J. Mammal. 58:67–73.

Cowlishaw, G., and J. E. Hacker. 1997. Distribution, diversity and latitude in African primates. Am. Nat. 150:505–512.

Cox, C. R., and B. J. Le Boeuf. 1977. Female incitation of male competition: a mechanism in sexual selection. Am. Nat. 111:317–335.

Cox, F. E. G. 1979. Ecological importance of small mammals as reservoirs of disease. Pp. 213–238 in Ecology of small mammals (D. M. Stoddart, ed.). Chapman and Hall, London.

Cox, P. G. 2006. Character evolution in the orbital region of the Afrotheria. J. Zool. 269:514–526.

Coyne, J. A., and H. A. Orr. 2004. Speciation. Sinauer Associates, Sunderland, MA.

Cracraft, J. 1983. Species concepts and speciation analysis. Pp. 159–187 in Current ornithology, vol. 1 (R.F. Johnston, ed.). Plenum Press, New York.

Cracraft, J. 1988. Deep-history biogeography: retrieving the historical patterns of evolving continental biotas. Syst. Zool. 37:221–236.

Cracraft, J. 1989. Speciation and its ontology: the empirical consequences of alternative species concepts for understanding patterns and processes of differentiation. Pp. 28–59 in Speciation and its consequences (D. Otte and J.A. Endler, eds.). Sinauer Associates, Sunderland, MA.

Cracraft, J. 1991. Patterns of diversification within continental biotas: hierarchical congruence among the areas of endemism of Australian vertebrates. Aust. Syst. Bot. 4:211–227.

Cracraft, J., and K. Helm-Bychowski. 1991. Parsimony and phylogenetic inference using DNA sequences: some methodological strategies. Pp. 184–220 in Phylogenetic analysis of DNA sequences (M.M. Miyamoto and J. Cracraft, eds.). Oxford Univ. Press, New York.

Craighead, D. J. and J. J. Craighead. 1987. Tracking caribou using satellite telemetry. Nat. Geog. Res. 3:462–479.

Craighead, J. J., J. R. Varney, F. C. Craighead, and J. S. Sumner. 1976. Telemetry experiments with hibernating black bears. Pp. 357–371 in Bears—their biology and management (M. R. Pelton, J. W. Lentfer, and G. E. Folk, Jr., eds.) International Union for Conservation of Nature and Natural Resources, Morges, Switzerland.

Crandall, L. S. 1964. The management of wild mammals in captivity. Univ. Chicago Press, Chicago.

Cranford, J. A. 1978. Hibernation in the western jumping mouse (*Zapus princeps*). J. Mammal. 59:496–509.

Craw, R.C. 1988. Continuing the synthesis between panbiogeography, phylogenetic systematics and geology as illustrated by empirical studies on the biogeography of New Zealand and the Chatham Islands. Syst. Zool. 37:291–310.

Creel, S. 1997. Cooperative hunting and group size: assumptions and currencies. Anim. Behav. 54:1319–1324.

Creel, S. 2005. Dominance, aggression, and glucocorticoid levels in social carnivores. J. Mammal. 86:255–264.

Creel, S., and N. M. Creel. 1995. Communal hunting and pack size in African wild dogs, *Lycaon pictus*. Anim. Behav. 50:1325–1339.

Creel, S., N. Creel, D. E. Wildt, and S. L. Monfort. 1992. Behavioural and endocrine mechanisms of reproductive succession in Serengeti dwarf mongooses. Anim. Behav. 43:231–246.

Crisci, J.V., L. Katinas, and P. Posada. 2003. Historical biogeography: an introduction. Harvard Univ. Press, Cambridge, MA.

Crockett, C. M. 1996. Data collection in the zoo setting, emphasizing behavior. Pp. 545–565 in Wild mammals in captivity (D. G. Kleiman, M. E. Allen, K. V. Thompson, and S. Lumpkin, eds.). Univ. Chicago Press, Chicago.

Crockett, C. M., C. L. Bowers, G. P. Sackett, and D. M. Bowden. 1993. Urinary cortisol responses of longtailed macaques to five cage sizes, tethering, sedation, and room change. Am. J. Primatol. 30:55–74

Croft, D. A., L. R. Heaney, J. J. Flynn, and A. P. Bautista. 2006. Fossil remains of a new, diminutive *Bubalus* (Artiodactyla: Bovidae: Bovini) from Cebu Island, Philippines. J. Mammal. 87:1037–1051.

Croizat, L. 1964. Space, time, form: the biological synthesis. Caracas, published by the author.

Croll, D.A., and B. R. Tershy. 2002. Filter feeding. Pp. 428–432 in Encyclopaedia of marine mammals (W. F. Perrin, B. Wursig, and J. G. M. Thewissen, eds.). Academic Press, San Diego.

Croll, D. A., J. L. Maron, J. A. Estes, E. M. Danner, and G. V. Byrd. 2005. Introduced predators transform subarctic islands from grassland to tundra. Science 307: 1959–1961.

Crompton, A. W. 1972. The evolution of the jaw articulation of cynodonts. Pp. 231–251 in Studies in vertebrate evolution (K. A. Joysey and T. S. Kemp, eds.). Winchester Press, New York.

Crompton, A. W., and K. Hiiemae. 1969. How mammalian molar teeth work. Discovery 5:23.

Crompton, A. W., and F. A. Jenkins, Jr. 1979. Origin of mammals. Pp. 59–73 in Mesozoic mammals: the first two-thirds of mammalian history (J. A. Lillegraven, Z. Kielan-Jaworowska, and W. A. Clemens, eds.). Univ. California Press, Berkeley.

Crompton, A. W., and Z. Luo. 1993. Relationships of the Liassic mammals *Sinoconodon*, *Morganucodon oehleri*, and *Dinnetherium*. Pp. 30–44 in Mammal phylogeny: Mesozoic differentiation, Multituberculates, Monotremes, early Therians, and Marsupials (F. S. Szalay, M. J. Novacek, and M. C. McKenna, eds.). Springer-Verlag, New York.

Crook, J. H., and S. S. Gartlan. 1966. Evolution of primate societies. Nature 210:1200–1203.

Cryan, P. M., M. A. Bogan, R. O. Rye, G. P. Landis, and C. L. Kester. 2004. Stable hydrogen isotope analysis of bat hair as evidence for seasonal molt and long-distance migration. J. Mammal. 85:995–1001.

Csada, R. 1996. Cardioderma cor. Mammal. Species No. 519:1–4.

Cumming, D. H. M., R. F. DuToit, and S. N. Stuart. 1991. African elephants and rhinos: status survey and conservation action plan. IUCN/SSC African Elephant and Rhino Species Group, Gland, Switzerland.

Cummings, J. R., and S. H. Vessey. 1994. Agricultural influences on movement patterns of white-footed mice (*Peromyscus leucopus*). Am. Mid. Nat. 132:209–218.

Currah, R. S., E. A. Smreciu, T. Lehesvirta, M. Niemi, and K. W. Larsen. 2000. Fungi in the winter diets of northern flying squirrels and red squirrels in the boreal mixedwood forest of northeastern Alberta. Can. J. Botany 78:1514–1520.

Currie, D.J. 1991. Energy and large-scale patterns of animal- and plant-species richness. Am. Nat. 137:27–49.

Curtis, H. 1975. Biology, 2nd edition. Worth Publ., New York.

Cutright, W. J., and T. McKean. 1979. Countercurrent blood vessel arrangement in beaver (*Castor canadensis*). J. Morphol. 161:169–176.

Daan, S., and S. Slopsema. 1978. Short-term rhythms in foraging behaviour of the common vole, *Microtus arvalis*. J. Comp. Physiol. A127:215–227.

Dagg, A. I., and J. B. Foster. 1982. The giraffe: biology, behavior, and ecology. R. E. Krieger, Huntington, NY.

Dailey, M. D. 1996. Meyer, Olsen, and Schmidt's essentials of parasitology, 6th edition. Wm. C. Brown, Dubuque, IA.

Dalquest, W. W. 1950. The genera of the chiropteran family Natalidae. J. Mammal. 31:436–443.

Dalsgaard, N. J. 2002. Prion diseases. An overview. APMIS 110:3–13.

Daniel, J. C., Jr. 1970. Dormant embryos of mammals. BioScience 20:411–415.

Danilkin, A. A. 1995. Capreolus pygargus. Mamm. Species 512:1–7.

Darling, J. D., and M. Berube. 2001. Interactions of singing humpback whales with other males. Mar. Mammal Sci. 17:570–584.

Darlington, P. J., Jr. 1957. Zoogeography: the geographical distribution of animals. John Wiley and Sons, New York.

Darwin, C. 1871. The descent of man, and selection in relation to sex. D. Appleton, New York.

Darwin, C. R. 1859. On the origin of species. Dent, London.

da Silva, M. N. F., and J. L. Patton. 1998. Molecular phylogeography and the evolution and conservation of Amazonian mammals. Mol. Ecol. 7:475–486.

Dausmann, K. H., J. Glos, J U. Ganzhorn, and G. Heldhaier. 2004. Hibernation in a tropical primate. Nature 429:825–826.

Dávalos, L. M. 2005. Molecular phylogeny of funnel-eared bats (Chiroptera: Natalidae), with notes on biogeography and conservation. Mol. Phylo. Evol. 37:91–103.

Davenport, T. R. B., W. T. Stanley, E. J. Sargis, D. W. De Luca, N. E. Mpunga, S. J. Machaga, and L. E. Olson. 2006. A new genus of African monkey, *Rungwecebus*: morphology, ecology, and molecular phylogenetics. Science 312:1378–1381.

David-Gray, Z. K., J. Gurnell, and D. M. Hunt. 1999. Estimating the relatedness in a population of grey squirrels, *Sciurus carolinensis*, using DNA fingerprinting. Acta Theriol. 44:243–251.

Davidow-Henry, B. R., and J. K. Jones, Jr. 1988. Notes on reproduction and postjuvenile molt in two genera of pocket gophers, *Cratogeomys* and *Thomomys*, in Texas. Tex. J. Sci. 40:459–461.

Davidson, J. 1938. On the growth of the sheep population in Tasmania. Trans. Roy. Soc. S. Aust. 62:342–346.

Davidson, S. M., and G. S. Wilkinson. 2004. Function of male song in the greater white-lined bat, *Saccopteryx bilineata*. Anim. Behav. 67:883–891.

Davis, C. S., I. DeIisle, I. Sterling, D. B. Sinoff, and C. Strobeck. 2004. A phylogeny of the extant Phocidae inferred from complete mitochondrial DNA coding regions. Mol. Phylo. Evol. 33:363–377.

Davis, D. E. 1991. Seasonal change in irradiance: a Zeitgeber for circannual rhythms in ground squirrels. Comp. Biochem. Physiol. 98A:241–243.

Davis, D. E., and E. P. Finnie. 1975. Entrainment of circannual rhythm in weight of woodchucks. J. Mammal. 56:199–203.

Davis, D. E., and F. B. Golley. 1963. Principles in mammalogy. Reinhold Publ., New York.

Davis, S. J. M. 1987. The archaeology of animals. Yale Univ. Press, New Haven, CT.

Davis, S. J. M., and F. R. Valla. 1978. Evidence for domestication of the dog 12,000 years ago in the Natufian of Israel. Nature 276:608–610.

Davis, W. B. 1973. Geographic variation in the fishing bat, *Noctilio leporinus*. J. Mammal. 54:862–874.

Davis, W. B. 1976. Geographic variation in the lesser noctilio, *Noctilio albiventris* (Chiroptera). J. Mammal. 57:687–707.

Davis, W. H., and H. B. Hitchcock. 1965. Biology and migration of the bat, *Myotis lucifugus*, in New England. J. Mammal. 46:296–313.

Dawkins, R., and T. R. Carlisle. 1976. Parental investment, mate desertion, and a fallacy. Nature 262:131–133.

Dawkins, R., and J. R. Krebs. 1978. Animal signals: information or manipulation? Pp. 282–309 *in* Behavioural ecology: an evolutionary approach (J. R. Krebs and N. B. Davies, eds.). Sinauer Associates, Sunderland, MA.

Dawson, M. R., and L. Krishtalka. 1984. Fossil history of the families of recent mammals. Pp. 11–57 *in* Orders and families of recent mammals of the world (S. Anderson and J. K. Knox, Jr., eds.). John Wiley and Sons, New York.

Dawson, T. J. 1983. Monotremes and marsupials: the other mammals. Edward Arnold, London.

Dawson, T. J. 1995. Kangaroos: biology of the largest marsupials. Comstock Publishing, Ithaca, NY; Cornell Univ. Press, Ithaca, NY.

Dayan, T., D. Simberloff, E. Tchernov, and Y. Yom-Tov. 1989. Inter-and intraspecific character displacement in mustelids. Ecology 70:1526–1539.

Dayan, T., D. Simberloff, E. Tchernov, and Y. Yom-Tov. 1990. Feline canines: community-wide character displacement among the small cats of Israel. Am. Nat. 136:39–60.

Dearing, M. D. 1997. The function of hay piles of pikas (*Ochotona princeps*). J. Mammal. 78:1156–1163.

DeArmond, S. J., and S. B. Prusiner. 1995. Etiology and pathogenesis of prion diseases. Am. J. Pathol. 146:785–811.

DeBlieux, D. D., M. R. Baumrind, E. L. Simons, P. S. Chatrath, G. E. Meyer, and Y. S. Attia. 2006. Sexual dimorphism of the internal mandibular chamber in Fayum Pliohyracidae (Mammalia). J. Vert. Paleo. 26:160–169.

Debruyne, R. 2003. Differenciation morphologique et moleculaire des Elephantinae (Mammalia, Proboscidea): statut systematique de l'elephant d'Afrique de foret, *Loxodonta africana cyclotis* (Matschie, 1900). Ph.D. Dissertation, Museum National d'Histoire Naturelle, Paris, France.

DeCesare, N. J., and D. H. Pletscher. 2006. Movements, connectivity, and resource selection of Rocky Mountain bighorn sheep. J. Mammal. 87:531–538.

Deevey, E. S., Jr. 1947. Life tables for natural populations of animals. Q. Rev. Biol. 22:283–314.

Degen, A. A. 1997. Ecophysiology of small desert mammals. Springer-Verlag, Berlin, Germany.

Degen, A. A., M. Kam, I. S. Khokhlova, and Y. Zeevi. 2000. Fiber digestion and energy utilization of fat sand rats (*Psammomys obesus*) consuming the chenopod Anabasis articulata. Physiol. Biochemi. Zool. 73:574–580.

DeGiusti, D. L. 1971. Acanthocephala. Pp. 140–157 *in* Parasitic diseases of wild mammals (J. W. Davis and R. C. Anderson, eds.). Iowa State Univ. Press, Ames.

Dehnel, A. 1949. Studies on the genus Sorex. Ann. Univ. M. Curie-Sklodowkka, Sect. C4:17–102.

Dehnhardt, G., B. Mauck, W. Hanke, and H. Bleckmann. 2001. Hydrodynamic trail-following in harbor seals (*Phoca vitulina*). Science 293:102–104.

de Jong, W. W. 1998. Molecules remodel the mammalian tree. Trends Ecol. Evol. 13:270–275.

de Jong, W. W., and Goodman, M. 1988. Anthropoid affinities of Tarsius supported by a-crystallin sequences. J. Hum. Evol. 17:575–582.

Delahay, R. J., J. A. Brown, P. J. Mallinson, P. D. Spyvee, D. Handoll, L. M. Rogers, and C. L. Cheeseman. 2000. The use of marked bait in studies of the territorial organization of the European badger (*Meles meles*). Mammal Rev. 30:73–87.

de la Maza H. M., J. O. Wolff, and A. Lindsey 1999. Exposure to strange adults does not cause pregnancy disruption or infanticide in the gray-tailed vole. Behav. Ecol. Sociobiol. 45: 107–113.

Delaney, M.J., and M.J.R. Healey. 1967. Variation in long-tailed field mouse (*Apodemus sylvaticus*) in south-west England. J. Zool. 1152:319–332.

DelGiudice, G. D., U. S. Seal, and L. D. Mech. 1987. Effects of feeding and fasting on wolf blood and urine characteristics. J. Wildl. Mgmt. 51:1–10.

DelGiudice, G. D., R. A. Moen, F. J. Singer, and M. R. Riggs. 2001. Winter nutritional restriction and simulated body condition of Yellowstone elm and bison before and after the fires of 1988. Wildl. Monogr. 147:1–60.

DeLong, R. L., and B. S. Stewart. 1991. Diving patterns of northern elephant seal bulls. Mar. Mammal Sci. 7:369–384.

DeLong, R. L., G. L. Kooyman, W. G. Gilmartin, and T. R. Loughlin. 1984. Hawaiian monk seal diving behavior. Acta Zool. Fenn. 172:129–131.

DeLong, R. L., B. S. Stewart, and R. D. Hill. 1992. Documenting migrations of northern elephant seals using daylength. Mar. Mammal Sci. 8:155–159.

Delpietro, H. A., N. Marchevsky, and E. Simonetti. 1993. Relative population densities and predation of the common vampire bat (*Desmodus rotundus*) in natural and cattle-raising areas in north-east Argentina. Prev. Vet. Med. 14:13–20

Delsuc, F., M. J. Stanhope, and E. J. P. Douzery. 2003. Molecular systematics of armadillos (Xenarthra: Dasypodidae): contribution of maximum likelihood and Bayesian analyses of mitochondrial and nuclear genes. Mol. Phylogenet. Evol. 28:261–275.

DeMatteo, K. E., I. J. Porton, D. G. Kleiman, and C. S. Asa. 2006. The effect of male bush dog (*Speothos venaticus*) on the female reproductive cycle. J. Mammal. 87:723–733.

de Meijere, J. C. 1894. Über die Haare der Säugetiere besonders über ihre Anordung. Gegenhaur's Morphol. Jahrb. 21:312–424.

De Mendonca, P. G. 1999. Impact of radio-collars on yellow-necked mice (*Apodemus flavicollis*, Mammalia, Rodentia). Mammal Review 29:129–134.

Demere, T. A., and A. Berta. 2001. A reevaluation of *Proneotherium repenningi* from the Miocene Astoria formation of Oregon and its position as a basal odobenid (Pinnipedia: Mammalia). J. Vert. Paleontol. 21:279–310.

Demers, M. N. 1997. Fundamentals of geographic information systems. John Wiley and Sons, New York.

De Muizon, C. 1994. A new carnivorous marsupial from the Paleocene of Bolivia and the problem of marsupial monophyly. Nature 370:208–211.

De Muizon, C., and R. L. Cifelli. 2001. A new basal "didelphoid" (Marsupialia, Mammalia) from the early Paleocene of Tiupampa (Bolivia). J. Vert. Paleontol. 21:87–97.

Dene, H., M. Goodman, and W. Prychodko. 1976. An immunological examination of the systematics of Tupaioidea. J. Mammal. 59:697–706.

Dene, H., M. Goodman, W. Prychodko, and G. Matsuda. 1980. Molecular evidence for the affinities of Tupaiidae. Pp. 269–291 *in* Comparative biology and evolutionary relationships of tree shrews (W. P. Luckett, ed.). Plenum Press, New York.

D'Eon, R. G., and R. Serrouya. 2005. Mule deer seasonal movements and multiscale resource selection using global positioning system radiotelemetry. J. Mammal. 86:736–744.

Derocher, A.E. 1999. Latitudinal variation in litter size of polar bears: ecology or methodology? Polar Biol. 22:350–356.

Derocher, A. E., M. Anderson, and Ø. Wiig. 2005. Sexual dimorphism of polar bears. J. Mammal. 86:895–901.

de Ruiter P. C., V. Wolters, J. C. Moore, and K. O. Winemiller. 2005. Food web ecology: playing Jenga and beyond. Science 309: 68-71.

Descalzi, M., G. N. Cameron, and J. W. Jacobson. 1998. Measuring similarity among hispid cotton rats (*Sigmodon hispidus*) of know relatedness with DNA fingerprinting. J. Mammal. 79:594–603.

Dessauer, H. C., C. J. Cole, and M. S. Hafner. 1996. Collection and storage of tissues. Pp. 29–47 *in* Molecular systematics, 2nd edition (D. M. Hillis, C. Moritz, and B. K. Mable, eds.). Sinauer Associates, Sunderland, MA.

Desy, E. A., and G. O. Batzli. 1989. Effects of food availability and predation on prairie vole demography: a field experiment. Ecology 70:411–421.

Detwiler, L. A., and R. Rubenstein. 2000. Bovine spongiform encephalopathy: an overview. ASAIO J. 46:S73–S79.

Deutsch, C. J., J. P. Reid, R. K. Bonde, D. E. Easton, H. I. Kochman, and T. J. O'Shea. 2003. Seasonal movements, migratory behavior, and site fidelity of West Indian manatees along the Atlantic coast of the

United States. Wildl. Monogr. 151:1–77.

de Waal, F. B. M. 1992. Coalitions as part of reciprocal relations in the Arnhem chimpanzee colony. Pp. 233–258 in Coalitions and alliances in humans and other animals (A. H. Harcourt and F. B. M. de Waal, eds.). Oxford Univ. Press, New York.

Dewsbury, D. A. 1982. Ejaculate cost and male choice. Am. Nat. 119:601–610.

Dewsbury, D. A. 1984. Sperm competition in muroid rodents. Pp. 547–571 in Sperm competition and the evolution of animal mating systems (R. L. Smith, ed.). Academic Press, New York.

Dewsbury, D. A. 1988. A test of the role of copulatory plugs in sperm competition in deer mice (Peromyscus maniculatus). J. Mammal. 69:854–857.

D'Havé, H., J. Scheirs, R. Vergagen, and W. DeCoen. 2005. Gender, age and seasonal dependent self-anointing in the European hedgehog Erinaceus europaeus. Acta Theriol. 50:167–173.

Diamond, J. 1999. Guns, germs, and steel: the fates of human societies. W. W. Norton, New York.

Diaz, A. 1998. Comparison of methods for measuring rabbit incidence on grasslands. Mammalia 1998: 205–212.

Diaz, G. B., and R. A. Ojeda. 1999. Kidney structure and allometry of Argentine desert rodents. J. Arid Environ. 41:453–461.

Dice, L. R. 1939. Variation in the cactus mouse, Peromyscus eremicus. Contr. Lab. Vert. Biol. Univ. Mich. 8:1–27.

Dice, L. R. 1945. Minimum intensities of illumination under which owls can find dead prey by sight. Am. Nat. 79:385–416.

Dice, L. R. 1947. Effectiveness of selection by owls of deer mice (Peromyscus maniculatus) which contrast in color with their background. Contrib. Lab. Vert. Biol. Univ. Mich. 50:1–15.

Dickman, C. R., and C. Huang. 1988. The reliability of fecal analysis as a method for determining the diet of insectivorous mammals. J. Mammal. 69:108–113.

Dickman, C. R., A. S. Haythornthwaite, G. H. McNaught, P. S. Mahon, B. Tamayo, and M. Letnic. 2001. Population dynamics of three species of dasyurid marsupials in arid central Australia: a 10-year study. Wildl. Res. 28:493–506.

Dietz, R.S. 1961. Continent and ocean basin evolution by spreading of the sea floor. Nature 190:854–857.

Diffendorfer, J. E., N. A. Slade, M. S. Gaines, and R. D. Holt. 1995. Population dynamics of small mammals in fragmented and continuous old-field habitat. Pp. 175–199 in Landscape approaches in mammalian ecology and conservation (W. Z. Lidicker, Jr., ed.). Univ. Minnesota Press, Minneapolis.

Dilkes, D. W., and R. R. Reisz. 1996. First record of a basal synapsid ("mammal-like reptile") in Gondwana. Proc. Roy. Soc. Lond. 263:1165–1170.

Dinerstein, E. 1991. Sexual dimorphism in the greater one-horned rhinoceros (Rhinoceros unicornis). J. Mammal. 72:450–457.

Dinerstein, E., and L. Price. 1991. Demography and habitat use by greater one-horned rhinoceros in Nepal. J. Wildl. Mgmt. 55:401–411.

Ding, W., B. Wursig, and S. Leatherwood. 2001. Whistles in boto, Inia geoffrensis, and tucuxi, Sotalia fluviatilis. J. Acoust. Soc. Am. 109:407–411.

Dingle, H. 1980. Ecology and evolution of migration. Pp. 1–101 in Animal migration, orientation, and navigation (S. A. Gauthreaux, Jr., ed.). Academic Press, New York.

Dingle, H. 1996. Migration: the biology of life on the move. Oxford Univ. Press, New York.

Dittmar, K., M. L. Porter, S. Murray, and M. F. Whiting. 2006. Molecular phylogenetic analysis of nycteribiid and streblid bat flies (Diptera: Brachycera, Calyptratae): implications for host associations and phylogeographic origins. Mol. Phylogenet. Evol. 38:155–170.

Dobney, K., and G. Larson. 2006. Genetics and animal domestication: new windows on an elusive process. J. Zool. 269:261–271.

Dobson, A. 2000. Raccoon rabies in space and time. Proc. Natl. Acad. Sci. USA 97:14041–14043.

Dobson, F. S. 1982. Competition for mates and predominant juvenile male dispersal in mammals. Anim. Behav. 30:1183–1192.

Dobson, F. S., R. K. Chesser, J. L. Hoogland, D. W. Sugg, and D. W. Foltz. 1998. Breeding groups and

gene dynamics in a socially structured population of prairie dogs. J. Mammal. 79:671–680.

Dobson, F. S., Chesser R. K., and Zinner B. 2000. The evolution of infanticide: genetic benefits of extreme nepotism and spite. Ethol. Ecol. Evol. 12: 131–148.

Dobzhansky, T. 1937. Genetics and the origin of species. Columbia Univ. Press, New York.

Domning, D. P. 1978. Sirenia. Pp. 573–581 in Evolution of African mammals (V. J. Maglio and H. B. S. Cooke, eds.). Harvard Univ. Press, Cambridge, MA.

Domning, D. P. 1997. Fossil Sirenia of the West Atlantic and Caribbean region, VI: Crenatosiren olseni (Reinhart, 1976). J. Vert. Paleontol. 17:397–412.

Domning, D. P. 2001. The earliest known fully quadrupedal sirenian. Nature 413:625–627.

Domning, D. P. 2005. Fossil Sirenia of the West Atlantic and Caribbean region. VII. Pleistocene Trichechus manatus Linnaeus, 1758. J. Vert. Paleontol. 25:685–701.

Domning, D. P., G. S. Morgan, and C. E. Ray. 1982. North American Eocene sea cows (Mammalia: Sirenia). Smithsonian Contrib. Paleontol. 52:1–69.

Dondini, G., and S. Vergari. 2000. Canivory in the greater noctule bat (Nyctalus lasiopterus) in Intaly. J. Zool. Lond. 251:233–236.

Donlan, C. J., J. Berger, C. E. Bock, J. H. Bock, D. A. Burney, J. A. Estes, D. Foreman, P. S. Martin, G. W. Roemer, F. A. Smith, M. E. Soule, and H. W. Greene. 2006. Pleistocene rewilding: an optimistic agenda for the 21st century. Am. Nat. 168:660–681.

Donoghue, M. J., and B. R. Moore. 2003. Toward an integrative historical biogeography. Integ. Comp. Biol. 43:261–270.

Donohue, M. J., D. P. Costa, E. Goebel, G. A. Antonelis, and J. D. Baker. 2002. Milk intake and energy expenditure of free-ranging northern fur seal, Callorhinus ursinus, pups. Physiol. Biochemi. Zool. 75:3–18.

Dötsch, C., and W. Koenigswald. 1978. Zur rotfärbung von soricidenzähnen [On the reddish coloring of soricid teeth]. Z. Säugetierk. 43:65–70.

Douglas-Hamilton, I. 1984. African elephant. Pp. 193–198 in Evolution of domesticated animals (I. L. Mason, ed.). Longman, New York.

Douglas-Hamilton, I. 1987. African elephants: population trends and their causes. Oryx 21:11–24.

Dowling, T. E., C. Moritz, J. D. Palmer, and L. H. Rieseberg. 1996. Nucleic acids III: analysis of fragments and restriction sites. Pp. 249–320 in Molecular systematics, 2nd edition (D. M. Hillis, C. Moritz, and B. K. Mable, eds.). Sinauer Associates, Sunderland, MA.

Downs, C. T., and M. R. Perrin. 1995. The thermal biology of three southern African elephant-shrews. J. Thermal Biol. 20:445–450.

Doyle, T. J., R. T. Bryan, and C. J. Peters. 1998. Viral hemorrhagic fevers and hantavirus infections in the Americas. Emerging Inf. Dis. 12:95–110.

Dragoo, J. W., and R. L. Honeycutt. 1997. Systematics of mustelid-like carnivores. J. Mammal. 78:426–443.

Drake, J. A., H. A. Mooney, F. diCastri, R. H. Groves, F. J. Kruger, M. Rejmánek, and M. Williamson. 1989. Biological invasions. John Wiley and Sons, New York.

Draper, J. 1999. Horse breeds of the world. Hermes House, London.

Drea, C. M., S. N. Vignieri, H. S. Kim, M. L. Weldele, and S. E. Glickman. 2002. Responses to olfactory stimuli in spotted hyenas (Crocuta crocuta): II. Discrimination of conspecific scent. J. Comp. Psychol. 116: 342–349.

Drickamer, L. C. 1974. Social rank, observability, and sexual behaviour of rhesus monkeys (Macaca mulatta). J. Reprod. Fert. 37:117–120.

Drickamer, L. C. 1974a. Sexual maturation of female house mice: social inhibition. Dev. Psychobiol. 7:257–265.

Drickamer, L. C. 1981. Selection for age of sexual maturation in mice and the consequences for population regulation. Behav. Neur. Biol. 31:82–89.

Drickamer, L. C. 1987. Influence of time of day on captures of two species of Peromyscus in a New England deciduous forest. J. Mammal. 68:702–703.

Drickamer, L. C. 1989. Pheromones: behavioral and

biochemical aspects. Pp. 269–348 in Advances in comparative and environmental physiology (J. Balthazart, ed.). Springer-Verlag, Berlin.

Drickamer, L. C. 1996. Intra-uterine position and anogenital distance in house mice: consequences under field conditions. Anim. Behav. 51:925–934.

Drickamer, L. C., and B. Shiro. 1984. Effects of adrenalectomy with hormone replacement therapy on the presence of sexual maturation-delaying chemosignal in the urine of grouped female mice. Endocrinology 115:255–260.

Drickamer, L. C., and B. M. Vestal. 1973. Patterns of reproduction in a laboratory colony of Peromyscus. J. Mammal. 54:523–528.

Drollette, D. 1997. Wide use of rabbit virus is good news for native species. Science 275:154.

Ducrocq, S., Y. Chaimanee, V. Suteethorn, and J. J. Yaeger. 1998. The earliest known pig from the Upper Eocene of Thailand. Palaeontology 41:147–156.

Ducroz, J.-F., M. Stubbe, A. P. Saveljev, D. Heidecke, R. Samjaa, A. Ulvi ius, A. Stubbe, and W. Durka. 2005. Genetic variation and population structure of the Eurasian beaver Castor fiber in eastern Europe and Asia. J. Mammal. 86:1059–1067.

Dudley, J. P., J. R. Ginsberg, A. J. Plumptre, J. A. Hart, and L. C. Campos. 2002. Effects of war and civil strife on wildlife and wildlife habitats. Conserv. Biol. 16:319–329.

Dugatkin, L. A. 1997. Cooperation among animals: an evolutionary perspective. Oxford Univ. Press, New York.

Duggins, D. O. 1980. Kelp beds and sea otters: an experimental approach. Ecology 61:447–453.

Dumont, E.R. 2003. Bats and fruit: an ecomorphological approach. Pp. 398–429 in Bat ecology (T.H. Kunz and M. B. Fenton, eds.). Univ. Chicago Press, Chicago.

Dung, V. V., P. M. Giao, N. N. Chinh, D. Tuoc, P. Arctander, and J. J. MacKinnon. 1993. A new species of living bovid from Vietnam. Nature 363:443–445.

Dunlap, J. C., J. Loros, and P. J. DeCoursey. 2003. Chronobiology: biological timekeeping. Sinauer Associates, Sunderland, MA.

Dunstone, N., and M. L. Gorman (eds.). 1993. Mammals as predators. Oxford, Clarendon Press.

Durant, S. M. 2001. Living with the enemy: avoidance of hyenas and lions by cheetahs in the Serengeti. Behav. Ecol. 11:624–632.

Durner, G. M., S. C. Amstrup, and K. J. Ambrosius. 2006. Polar bear maternal den habitat in the Arctic National Wildlife Refuge, Alaska. Arctic 59:31–36.

Duval, B. D., E. Jackson, and W. G. Whitford. 2005. Mesquite (Prosopis glandulosa) germination and survival in black-gram (Bouteloua eriopoda) grassland: relations between microsite and heteromyid rodent (Dipodomys spp.) impact. J. Arid Environ. 62:541–554.

Dyck, A. P., and R. A. MacArthur. 1993. Seasonal variation in the microclimate and gas composition of beaver lodges in a boreal environment. J. Mammal. 74:180–188.

Dyhrepoulsen, P., H. H. Smedegaard, H. H. Roed, and E. Korsgaard. 1994. Equine hoof function investigated by pressure transducers inside the hoof and accelerometers mounted on the first phalanx. Equine Vet. J. 26:362–366.

East, M. L., and H. Hofer. 1997. The peniform clitoris of female spotted hyaenas. Trends Ecol. Evol. 12:401.

Ebensperger, L. A. 1998. Strategies and counterstrategies to infanticide in mammals. Biological Reviews 73: 321–346.

Ebensperger, L. A., and H. Cofré. 2001. On the evolution of group-living in the New World cursorial hystricognath rodents. Behav. Ecol. 12:227–236.

Ebensperger, L. A., M. J. Hurtado, and I. Valdivia. 2006. Lactating females do not discriminate between their own young and unrelated pups in the communally breeding rodent, Octodon degus. Ethology 112:921–929.

Eberhard, I. H., J. McNamara, R. J. Pearse, and I. A. Southwell. 1975. Ingestion and excretion of Eucalyptus punctata D.C. and its essential oil by the koala, Phascolarctos cinereus (Goldfuss). Aust. J. Zool. 23:169–179.

Eberle, M., and P. M. Kappeler. 2006. Family insurance: kin selection and cooperative breeding in a solitary primate (*Microcebus murinus*). Behav. Ecol. Sociobiol. 60: 582–588.

Ecke, D. H., and A. R. Kinney. 1956. Aging meadow mice, *Microtus californicus*, by observation of molt progression. J. Mammal. 37:249–254.

Economos, A. C. 1981. The largest land mammal. J. Theoret. Biol. 89:211–215.

Eder, T. 2005. Mammals of California. Lone Pine Publishing, Auburn, WA.

Edwards, A. W. F., and L. L. Cavalli-Sforza. 1963. The reconstruction of evolution. Ann. Hum. Genet. 27:105–106.

Edwards, S. V., and P. Beerli. 2000. Variance in coalescence time and comparative phylogeography. Evolution 54:1839–1854.

Eisenberg, J. F. 1981. The mammalian radiations. Univ. Chicago Press, Chicago.

Ehrlich, P., and A. Ehrlich. 1981. Extinction. Random House, New York.

Ehrlich, P. R., and A. H. Ehrlich. 1968. The population bomb. Amereon, Mattituck, NY.

Eichenbaum, H., C. Stewart, and R. G. M. Morris. 1990. Hippocampal representation in place learning. Neurosci. 10:3531–3542.

Eisenberg, J. F. 1963. The behavior of heteromyid rodents. Univ. Cal. Publ. Zool. 69:111–114.

Eisenberg, J. F. 1975. Tenrecs and solenodons in captivity. Internl. Zoo Yearb. 15:6–12

Eisenberg, J. F. 1978. The evolution of arboreal herbivores in the Class Mammalia. Pp. 135–152 *in* The ecology of arboreal folivores (G. G. Montgomery, ed.). Smithsonian Institution Press, Washington, DC.

Eisenberg, J. F. 1981. The mammalian radiations. Univ. Chicago Press, Chicago.

Eisenberg, J. F., and G. Gozalez. 1985. Observations on the natural history of *Solenodon cubanus*. Acta Zool. Fennici 173:275–277.

Eisenberg, J. F., N. Muckenhirn, and R. Rudran. 1972. The relationship between ecology and social structure in primates. Science 176:863–874.

Elgmork, K. 1982. Caching behavior of brown bears (*Ursus arctos*). J. Mammal. 63:607–612.

Elias, B. A., L. A. Shipley, R. D. Sayler, and R. S. Lamson. 2006. Mating and parental care in captive pygmy rabbits. J. Mammal. 87:921–928.

Ellegren, H. 1991. Fingerprinting birds' DNA with a synthetic polynucleotide probe (TG)n. Auk. 108:956–958.

Ellner, S. P., J. Fieberg, D. Ludwig, and C. Wilcox. 2002. Precision of population viability analysis. Conserv. Biol. 16:258–261.

El-Rabbany, A. 2006. Introduction to GPS: the global positioing system, 2nd edition. Artech House Publ., Norwood, MA.

Elrod, D. A., E. G. Zimmerman, P. D. Sudman, and G. A. Heidt. 2000. A new subspecies of pocket gopher (*Geomys*) from the Ozark Mountains of Arkansas with comments on its historical biogeography. J. Mammal. 81:852–864.

Elton, C. 1924. Periodic fluctuations in the numbers of animals: their causes and effects. Brit. J. Exptl. Biol. 2:119–163.

Elton, C. S. 1927. Animal ecology. Sidgwick and Jackson, London.

Eltringham, S. K. 1978. Methods of capturing wild animals for purposes of marking them. Pp. 13–23 *in* Animal marking (B. Stonehouse, ed.). Univ. Park Press, Baltimore.

Eltringham, S. K. 1979. The ecology and conservation of large African mammals. Univ. Park Press, Baltimore.

Emerson, G. L., C. W. Kilpatrick, B. E. McNiff, J. Ottenwalder, and M. W. Allard. 1999. Phylogenetic relationships of the order Insectivora based on complete 12S rRNA sequences from mitochondria. Cladistics 15:221–230.

Emerson, S. B. 1985. Jumping and leaping. Pp. 58–72 *in* Functional vertebrate morphology (M. Hildebrand, D. M. Bramble, K. F. Liem, and D. B. Wake, eds.). Harvard Univ. Press, Cambridge, MA.

Emlen, S. T. 1982. The evolution of helping, I: An ecological constraints model. Am. Nat. 119:29–39.

Emlen, S. T. 1984. Cooperative breeding in birds and mammals. Pp. 305–339 *in* Behavioural ecology: an evolutionary approach, 2nd edition (J. R. Krebs and N. B. Davies, eds.). Sinauer Assoc., Sunderland, MA.

Emlen, S. T., and L. W. Oring. 1977. Ecology, sexual selection, and the evolution of mating systems. Science 197:215–223.

Emmons, L. H. 1991. Frugivory in treeshrews (*Tupaia*). Am. Nat. 138:642–649.

Emmons, L. H. 1999. A new genus and species of abrocomid rodent from Peru (Rodentia: Abrocomidae). Am. Mus. Nov. 3279:1–14.

Emmons, L. H. 2000. Tupai: a field study of Bornean treeshrews. Univ. California Press, Berkeley.

Emry, R. J. 2004. The endentulous skull of the North American pangolin, *Patriomanis americanus*. Bull. Am. Mus. Nat. Hist. 30:130–138.

Enders, A. C. (ed.). 1963. Delayed implantation. Univ. Chicago Press, Chicago.

Enders, A. C. 2002. Implantation in the nine-banded armadillo: how does a single blastocyst form four embryos? Placenta 23:71–85.

Engelmann, G. F. 1985. The phylogeny of the Xenarthra. Pp. 51–64 *in* The evolution and ecology of armadillos, sloths, and vermilinguas (G. G. Montgomery, ed.). Smithsonian Institution Press, Washington, DC.

Enserink, M., and G. Vogel. 2006. The carnivore comeback. Science 314:746–749.

Entwistle, A. C., P. A. Racey, and J. R. Speakman. 1996. Habitat exploitation by a gleaning bat, *Plecotus auritus*. Philosophical Trans. Roy. Soc. Lond. Ser. B 351:921–931.

Erb, J., and H. R. Perry, Jr. 2003. Muskrats, *Ondatra zibethicus* and *Neofiber alleni*. Pp. 311–348 *in* Wild mammals of North America: biology, management, and conservation. (G. A. Feldhamer, B. C. Thompson, and J. A. Chapman, eds.). Johns Hopkins Univ. Press, Baltimore.

Erickson, C. J. 1991. Percussive foraging in the aye-aye, *Daubentonia madagascariensis*. Animal Behav., 41:793–801.

Erickson, C. J. 1994. Tap-scanning and extractive foraging in aye-ayes, *Daubentonia madagascariensis*. Folia Primatologica 62:125–135.

Erlinge, S., G. Göransson, L. Hansson, G. Högstedt, O. Liberg, I. N. Nilsson, T. Nilsson, T. von Schantz, and M. Sylvén. 1983. Predation as a regulating factor on small rodent populations in southern Sweden. Oikos 40:36–52.

Essop, M. F., E. H. Harley, and I. Baumgarten. 1997. A molecular phylogeny of some Bovidae based on restriction-site mapping of mitochondrial DNA. J. Mammal. 78:377–386.

Estes, J. A., N. A. Smith, and J. F. Palmisano. 1978. Sea otter predation and community organization in the western Aleutian Islands, Alaska. Ecology 59:822–833.

Estes, J. A., M. T. Tinker, T. M. Williams, and D. F. Doak. 1998. Killer whale predation on sea otters linking oceanic and nearshore ecosystems. Science 282:473–476.

Estes, R. D. 1972. The role of the vomeronasal organ in mammalian reproduction. Mammalia 36:315–341.

Etheridge, K., G. B. Rathbun, J. A. Powell, and H. I. Kochman. 1985. Consumption of aquatic plants by the West Indian manatee. J. Aquat. Plant Mgmt. 23:21–25.

Etienne, A. S., R. Maurer, and F. Saucy. 1988. Limitations in the assessment of path dependent information. Behaviour. 106:81–111.

Etienne, A. S., R. Maurer, and V. Seguinot. 1996. Path integration in mammals and its interaction with visual landmarks. J. Exp. Biol. 199:210–209.

Evans, M., B. Green, and K. Newgrain. 2003. The field energetics and water fluxes of free-living wombats (Marsupialia: Vombatidae). Oecologia 137:171–180.

Evans, P. G. H. 1987. The natural history of whales and dolphins. Facts on File, New York.

Ewer, R. F. 1968. Ethology of mammals. Plenum Press, New York.

Ewer, R. F. 1973. The carnivores. Weidenfeld and Nicolson, London.

Excoffier, L., P. E. Smouse, and J. M. Quattro. 1992. Analysis of molecular variance inferred from metric distances among DNA haplotypes: application to human mitochondrial DNA restriction data. Genetics 131:479–491.

Fagen, R. 1981. Animal play behavior. Oxford Univ. Press, New York.

Fain, A. 1994. Adaptation, specificity, and host–parasite coevolution in mites (*Acari*). Int. J. Parasitol. 24:1273–1283.

Falkenstein, F., G. Kortner, K. Watson, and F. Geiser. 2001. Dietary fats and body lipid composition in relation to hibernation in free-ranging echidnas. J. Comp. Physiol. B 171:189–194.

Faria, K. de C., and E. Morielle-Versute. 2006. Genetic relationships between Brazilian species of Molossidae and Phyllostomidae (Chiroptera, Mammalia). Genetica 126:215–225.

Farlow, J. O. 1987. Speculations about the diet and digestive physiology of herbivorous dinosaurs. Paleobiology 13:60–72.

Farris, J. S. 1983. The logical basis of phylogenetic analysis. Pp. 7–36 *in* Proceedings of the First Meeting of the Willi Hennig Society (N. H. Platnick and V.A. Funk, eds.). Columbia Univ. Press, New York.

Fay, F. H., and C. Ray. 1968. Influence of climate on the distribution of walruses, *Odobenus rosmarus* (Linnaeus). I. Evidence from thermoregulatory behavior. Zoologica 53:1–18.

Feldhamer, G. A., and W. E. Armstrong. 1993. Interspecific competition between four exotic species and native artiodactyls in the United States. Trans. N. Am. Wildl. Nat. Res. Conf. 58:468–478.

Feldhamer, G. A., R. S. Klann, A. S. Gerard, and A. C. Driskell. 1993. Habitat partitioning, body size, and timing of parturition in pygmy shrews and associated soricids. J. Mammal. 74:403–411.

Feldhamer, G. A., B. C. Thompson, and J. A. Chapman (eds.). 2003. Wild mammals of North America: biology, management, and conservation, 2nd edition. Johns Hopkins Univ. Press, Baltimore.

Feldhamer, G. A., J. A. Rochelle, and C. D. Rushton. 2003a. Mountain beaver, *Aplodontia rufa*. Pp. 179–187 *in* Wild mammals of North America: biology, management, and conservation. (G. A. Feldhamer, B. C. Thompson, and J. A. Chapman, eds.). Johns Hopkins Univ. Press, Baltimore.

Feldman, M. W., J. Kumm, and J. Pritchard. 1999. Mutation and migration models of microsatellite evolution. Pp. 98–115 *in* Microsatellites: evolution and applications (D. B. Goldstein and C. Schlötterer, eds.). Oxford Univ. Press, Oxford.

Felsenstein, J. 1983. Parsimony in systematics: biological and statistical issues. Annu. Rev. Ecol. Syst. 14:313–333.

Felsenstein, J. 1985. Confidence limits on phylogenies: an approach using the bootstrap. Evolution 39:783–791.

Felsenstein, J. 2004. Inferring phylogenies. Sinauer Associates, Sunderland, MA.

Felsenstein, J. 2006. PHYLIP (Phylogenetic Inference Package), version 3.6. Distributed by the author. Department of Genome Sciences, Univ. Washington, Seattle.

Fenner, P., and J. Williamson. 1996. Platypus envenomation. Pp. 436–440 *in* Venomous and poisonous marine animals: a medical and biological handbook (J. A. Williamson, P. J. Fenner, J. W. Burnett, and J. F. Rifkin, eds.). Univ. New South Wales Press, Sydney.

Fenton, M. B. 1972. The structure of aerial-feeding bat faunas as indicated by ears and wing elements. Can. J. Zool. 50:287–296.

Fenton, M. B. 1984a. Echolocation: Implications for ecology and evolution of bats. Q. Rev. Biol. 59:33–53.

Fenton, M. B. 1984b. Sperm competition? The case of vespertilionid and rhinolophid bats. Pp. 573–587 in Sperm competition and the evolution of animal mating systems (R. L. Smith, ed.). Academic Press, New York.

Fenton, M. B. 1985. Communication in the Chiroptera. Indiana Univ. Press, Bloomington.

Fenton, M. B. 1995. Constraint and flexibility—bats as predators, bats as prey. Pp. 277–289 in Ecology, evolution, and behaviour of bats (P. A. Racey and S. M. Swift, eds.). Oxford Univ. Press, Oxford, England.

Fenton, M. B., D. Audet, D. C. Dunning, J. Long, C. B. Merriman, D. Pearl, D. M. Syme, B. Adkins, S. Pedersen, and T. Wohlgenant. 1993. Activity patterns and roost selection by *Noctilio albiventris* (Chiroptera: Noctilionidae) in Costa Rica. J. Mammal. 74:607–613.

Fenton, M. B., D. H. M. Cumming, I. L. Rautenbach, G. S. Cumming, M. S. Cumming, G. Ford, R. D. Taylor, J. Dunlop, M. D. Hovorka, D. S. Johnston, C. V. Portfors, M. C. Kalcounis, and Z. Mahlanga. 1998. Bats and the loss of tree canopy in African woodlands. Conserv. Biol. 12:399–407.

Fenton, M. B., and D. W. Thomas. 1985. Migrations and dispersal of bats (Chiroptera). Pp. 409–424 in Migration: mechanisms and adaptive significance (M. A. Rankin, ed.). Marine Science Institute, Port Aransas, TX.

Fenton, M. B., D. W. Thomas, and R. Sasseen. 1981. Nycteris grandis (Nycteridae): an African carnivorous bat. J. Zool. 194:461–465.

Fenton, M. B., J. Rydell, M. J. Vonhof, J. Eklof, and W. C. Lancaster. 1999. Constant-frequency and frequency-modulated components in the echolocation calls of three species of small bats (Emballonuridae, Thyropteridae, and Vespertilionidae). Can. J. Zool. 77: 1891–1900.

Fergus, C. 1991. The Florida panther verges on extinction. Science 251:1178–1180.

Fernando, F., M. E. Pfrender, S. E. Encalada, and R. Lande. 2000. Mitochondrial DNA variation, phylogeography, and population structure of the Asian elephant. Heredity 84:362–372.

Ferraris, J. D., and S. R. Palumbi. 1996. Molecular zoology: advances, strategies, and protocols. Wiley-Liss, New York.

Ferron, J. 1996. How do woodchucks (Marmota monax) cope with harsh winter conditions? J. Mammal. 77:412–416.

Festa-Bianchet, M. 1989. Individual differences, parasites, and the costs of reproduction for bighorn ewes (Ovis canadensis). J. Anim. Ecol. 58:785–795.

Fielden, L. J. 1991. Home range and movements of the Namib Desert gold mole, Eremitalpa granti namibensis (Chrysochloridae). J. Zool. Lond. 223:675–686.

Fielden, L. J., M. R. Perrin, and G. C. Hickman. 1990. Water metabolism in the Namib Desert golden mole, Eremitalpa granti namibensis (Chrysochloridae). Comp. Biochem. Physiol. 96A:227–234.

Fietz, J., M. Pflug, W. Schlund, and F. Tataruch. 2005. Influences of the feeding ecology on body mass and possible implications for reproduction in the edible dormouse (Glis glis). J. Comp. Physiol. Biochem. Syst. Env. Physiol. 175:45–55.

Findley, J. S. 1969. Biogeography of southwestern boreal and desert mammals. Univ. Kansas Misc. Publ. Mus. Nat. Hist. 51:113–128.

Findley, J. S. 1989. Morphological patterns in rodent communities of southwestern North America. Pp. 253–263 in Patterns in the structure of mammalian communities (D. W. Morris, Z. Abramsky, B. J. Fox, and M. R. Willig, eds.). Texas Tech Univ. Press, Lubbock.

Findley, J. S. 1993. Bats: a community perspective. Cambridge Univ. Press, Cambridge, UK.

Findley, J. S., and H. Black. 1983. Morphological and dietary structuring of a Zambian insectivorous bat community. Ecology 64:625–630.

Findley, J. S., and D. E. Wilson. 1974. Observations on the Neotropical disk-winged bat, Thyroptera tricolor Spix. J. Mammal. 55:562–571.

Findley, J. S., and D. E. Wilson. 1982. Ecological significance of chiropteran morphology. Pp. 243–260 in Ecology of bats (T. H. Kunz, ed.). Plenum, New York.

Findley, J. S., E. H. Studier, and D. E. Wilson. 1972. Morphologic properties of bat wings. J. Mammal. 53:429–444.

Finlayson, G. R., G. A. Shimmin, P. D. Temple-Smith, K. A. Handasyde, and D. A. Taggart. 2005. Burrow use and ranging behaviour of the hairy-nosed wombat (Lasiorhinus latifrons) in the Murraylands, South Australia. J. Zool. Lond. 265:189–200.

Finley, K. J. 2001. Natural history and conservation of the Greenland whale, or bowhead, in the northwest Atlantic. Arctic 54:55–76.

Firestone, K. B., M. S. Elphinstone, W. B. Sherwin, and B. A. Houlden. 1999. Phylogeographical population structure of tiger quolls Dasyurus maculatus (Dasyuridae: Marsupialia), an endangered carnivorous marsupial. Molec. Ecol. 8:1613–1625.

Fish, F. E. 1992. Aquatic locomotion. Pp. 34–63 in Mammalian energetics: interdisciplinary views of metabolism and reproduction (T. E. Tomasi and T. H. Horton, eds.). Cornell Univ. Press, Ithaca, NY.

Fish, F. E., and C. A. Hui. 1991. Dolphin swimming—a review. Mammal Rev. 21:181–195.

Fish, F. E., P. B. Frappell, R. V. Baudinette, and P. M. MacFarlane. 2001. Energetics of terrestrial locomotion of the platypus (Ornithorhynchus anatinus). J. Exp. Biol. 204:797–803.

Fishbein, D. B. 1991. Rabies in humans. Pp. 519–549 in The natural history of rabies (G. M. Baer, ed.). CRC Press, Boca Raton, FL.

Fisher, D. O., and I. P. F. Owens. 2000. Female home range size and the evolution of social organization in macropod marsupials. J. Anim. Ecol. 69:1083–1098.

Fisher, D. O., S. P. Blomberg, and S. D. Hoyle. 2001. Mechanics of drought-induced population decline in an endangered wallaby. Biol. Conserv. 102:107–115.

Fisher, J., N. Simon, and J. Vincent. 1969. Wildlife in danger. Verlag, New York.

Fisher, R. A. 1958. Genetical theory of natural selection. Dover Publ., New York.

Flannery, T. 1984. New Zealand: a curious zoogeographic history. Pp. 1089–1094 in Vertebrate zoogeography and evolution in Australasia. Hesperian Press, Carlisle, Australia.

Flannery, T. 1995. Mammals of New Guinea. Cornell Univ. Press, Ithaca, NY.

Flannery, T. F. 1987. An historic record of the New Zealand greater short-tailed bat, Mystacina robusta (Microchiroptera: Mystacinidae) from the South Island, New Zealand. Aust. Mammal. 10:45–46.

Flannery, T. F., and C. P. Groves. 1998. A revision of the genus Zaglossus (Monotremata, Tachyglossidae), with description of new species and subspecies. Mammalia 62:367–396.

Flannery, T. F., M. Archer, T. H. Rich, and R. Jones. 1995. A new family of monotreme from the Cretaceous of Australia. Nature 377:418–420.

Fleagle, J. G. 1988. Primate adaptation and evolution. Academic Press, New York.

Fleagle, J. G., and C. C. Gilbert. 2006. Biogeography and the primate fossil record: the role of tectonics, climate, and chance. Pp 375–418 in Primate biogeography (S. Lehman and J. G. Fleagle, eds.). Springer, New York.

Fleischner, T. L. 1994. Ecological costs of livestock grazing in western North America. Conserv. Biol. 8:629–644.

Fleming, T. H. 1970. Notes on the rodent faunas of two Panamanian forests. J. Mammal. 51:473–490.

Fleming, T. H. 1971. Artibeus jamaicensis: delayed embryonic development in a neotropical bat. Science 171:402–404.

Fleming, T. H. 1973 Numbers of mammal species in North and Central American forest communities. Ecology 54:555–563.

Fleming, T. H. 1974. The population ecology of two species of Costa Rican heteromyid rodents. Ecology 55:493–510.

Fleming, T. H. 1982. Foraging strategies of plant-visiting bats. Pp. 287–325 in Ecology of bats (T. H. Kunz, ed.). Plenum, New York.

Flemming, T. H. 1988. The short-tailed fruit bat: a study in plant-animal interactions. Univ. Chicago Press, Chicago.

Fleming, T. H. 1993. Plant-visiting bats. Am. Sci. 81:460–467.

Fleming, T. H. 1995. The use of stable isotopes to study the diets of plant-visiting bats. Symp. Zool. Soc. Lond. 67:99–110.

Fleming, T. H., and A. Estrada (eds.). 1993. Frugivory and seed dispersal: ecological and evolutionary aspects. Adv. Veg. Sci. 15:1–392.

Fleming, T. H., and V. J. Sosa. 1994. Effects of nectarivorous and frugivorous mammals on reproductive success of plants. J. Mammal. 75:845–851.

Fleming, T. H., A. A. Nelson, and V. M. Dalton. 1998. Roosting behavior of the lesser long-nosed bat, Leptonycteris curasoae. J. Mammal. 79:147–155.

Flower, W. H., and R. Lydekker. 1891. An introduction to the study of mammals living and extinct. Adam and Charles Black, London.

Flux, J. E. 1970. Colour change of mountain hares (Lepus timidus scoticus) in northeast Scotland. J. Zool. 162:345–358.

Flynn, J. J., and G. D. Wesley-Hunt. 2005. Carnivora. Pp. 175–198 in The rise of placental mammals: origins and relationships of the major extant clades (K. D. Rose and J. D. Archibold, eds.). Johns Hopkins Univ. Press, Baltimore.

Flynn, L. J., N. A. Neff, and R. H. Tedford. 1988. Phylogeny of the Carnivora. Pp. 73–115 in The phylogeny and classification of the tetrapods, vol. I: Mammals (M. J. Benton, ed.). Clarendon Press, Oxford.

Flynn, J. J., M. A. Nedbal, J. W. Dragoo, and R. L. Honeycutt. 2000. Whence the red panda? Mol. Phylogenet. Evol. 17:190–199.

Flynn, J. J., J. A. Finarelli, S. Zehr, J. Hsu, and M. A. Nedbal. 2005. Molecular phylogeny of the Carnivora (Mammalia): assessing the impact of increased sampling on resolving enigmatic relationships. Syst. Biol. 54:317–337.

Focardi, S., P. Marcellini, and R. Montanaro. 1996. Do ungulates exhibit a food density threshold? A field study of optimal foraging and movement patterns. J. Anim. Ecol. 65:606–620.

Fogel, R., and J. M. Trappe. 1978. Fungus consumption (Mycophagy) by small animals. Northwest Sci. 52:1–31.

Foley, W. J., and C. McArthur. 1994. The effects and costs of allelochemicals for mammalian herbivores: an ecological perspective. Pp. 370–391 in The digestive system in mammals: food, form, and function (D. J. Chivers and P. Langer, eds.). Cambridge Univ. Press, New York.

Foley, W. J., W. V. Engelhardt, and P. Charles-Dominique. 1995. The passage of digesta, particle size, and in vitro fermentation rate in the three-toed sloth Bradypus tridactylus (Edentata: Bradypodidae). J. Zool. Lond. 236:681–696.

Folk, G. E., Jr., and M. A. Folk. 1980. Physiology of large mammals by implanted radio capsules. Pp. 33–43 in A handbook of bioteletry and radio tracking (C. J. Amlaner, Jr., and D. W. Macdonald, eds.). Pergamon Press, Oxford.

Fons, R., S. Sender, T. Peters, and K. D. Jürgens. 1997. Rates of rewarming, heart and respiratory rates and their significance for oxygen transport during arousal from torpor in the smallest mammal, the Etruscan shrew Suncus etruscus. J. Exp. Biol. 200:1451–1458.

Fooden, J. 1972. Breakup of Pangea and isolation of relict mammals in Australia, South America, and Madagascar. Science 175:894–898.

Fooden, J. 1997. Tail length variation In Macaca fascicularis and M. mulatta. Primates 38:221–231.

Fooden, J., and G. H. Albrecht. 1999. Tail-length evolution In fascicularis-group macaques (Cercopithecidae:Macaca). Int. J. Primatol. 20:431–440.

Fordyce, R. E. 1980. Whale evolution and Oligocene southern ocean environments. Paleogeogr. Paleoclimatol. Paleoecol. 31:319–336.

Forman, R. T., and M. Godron. 1986. Landscape ecology. Wiley, New York.

Forman, R. T. T. 2000. Estimate of the area affected ecologically by the road system of the United States. Conserv. Biol. 14:31–35.

Forman, R. T. T., and R. D. Deblinger. 2000. The ecological road-effect zone of a Massachusetts (U.S.A.) suburban highway. Conserv. Biol. 14:36–46.

Forsman, K. A., and M. G. Malmquist. 1988. Evidence for echolocation in the common shrew, Sorex araneus. J. Zool. Lond. 216:655–662.

Fortelius, M. and J. Kappelman. 1993. The largest land mammal ever imagined. Zool. J. Linnean Soc. 108:85–101.

Fossey, D. 1983. Gorillas in the mist. Houghton Mifflin, Boston.

Fossey, D., and J. Harcourt. 1977. Feeding ecology of free-ranging mountain gorilla (Gorilla gorilla beringei). Pp. 415–447 in Primate ecology: studies of feeding and ranging behaviour in lemurs, monkeys, and apes (T. H. Clutton-Brock, ed.). Academic Press, London.

Foster, J. B. 1964. Evolution of mammals on islands. Nature 202:234–235.

Fowler, S. L., D. P. Costa, J. P. Y. Arnould, N. J. Gales, and C. E. Kuhn. 2006. Ontogeny of diving behaviour in the Australian sea lion: trials of adolescence in a late bloomer. J. Anim. Ecol. 75:358–367.

Fox, B. J. 1989. Small-mammal community pattern in Australian heathland: a taxonomically-based rule for species assembly. Pp. 91–103 in Patterns in the structure of mammalian communities (D. W.

Morris, Z. Abramsky, B. J. Fox, and M. R. Willig, eds.). Texas Tech Univ. Press, Lubbock.

Fox, B. J., and G. L. Kirkland, Jr. 1992. North American soricid communities follow an Australian small mammal assembly rule. J. Mammal. 73:491–503.

Fragoso, J. M. V., and J. M. Huffman. 2000. Seed-dispersal and seeding recruitment patterns by the last Neotropical megafaunal element in Amazonia, the tapir. J. Trop. Ecol. 16:369–385.

Francis, C. M., E. L. P. Anthony, J. A. Brunton, and T. H. Kunz. 1994. Lactation in male fruit bats. Nature 367:691–692.

Frank, C. L. 1988. Diet selection by a heteromyid rodent: role of net metabolic water production. Ecology 69:1943–1951.

Frank, L. G. 1997. Evolution of genital masculinization: why do female hyaenas have such a large 'penis'? Trends Ecol. Evol. 12: 58–62.

Frank, L. G., S. E. Glickman, and P. Licht. 1991. Fatal sibling aggression, precotial development, and androgens in neonatal spotted hyenas. Science 252:702–704.

Fraser, D. 1990. Behavioral perspectives on piglet survival. J. Reprod. Fertil. Suppl. 40:355–370.

Fraser, D., and B. K. Thompson. 1991. Armed sibling rivalry by domestic piglets. Behav. Ecol. Sociobiol. 29:1–15.

Fraser, F. C., and P. E. Purves. 1960. Hearing in cetaceans. Bull. Br. Mus. Nat. Hist. 7:1–140.

Freckleton, R. P., P. H. Harvey, and M. Pagel. 2003. Bergmann's rule and body size in mammals. Am. Nat. 161:821–825.

Freedman, D. O. 1992. Intestinal nematodes. Pp. 2003–2008 in Infectious diseases (S. L. Gorbach, J. G. Bartlett, and N. R. Blacklow, eds.). W. B. Saunders, Philadelphia.

Freeland, W. J., and D. H. Janzen. 1974. Strategies in herbivory by mammals: the role of plant secondary compounds. Am. Nat. 108:269–289.

Freeman, P. W. 1979. Specialized insectory: beetle-eating and moth-eating molossid bats. J. Mammal. 60:467–479.

Freeman, P. W. 1981. Correspondence of food habits and morphology in insectivorous bats. J. Mammal. 62:166–177.

Freeman, P. W. 1988. Frugivorous and animalivorous bats (Microchiroptera): dental and cranial adaptations. Biol. J. Linnean Soc. 33:249–272.

Freeman, P. W. 1995. Nectarivorous feeding mechanisms in bats. Biol. J. Linnean Soc. 56:439–463.

French, A. R. 1992. Mammalian dormancy. Pp. 105–121 in Mammalian energetics (T. E. Tomasi and T. H. Horton, eds.). Comstock Publ. Assoc., Ithaca, NY.

French, A. R. 1993. Physiological ecology of the Heteromyidae: economics of energy and water utilization. Pp. 509–538 in Biology of the Heteromyidae (H. H. Genoways and J. H. Brown, eds.). Spec. Pub. Am. Soc. Mammal. 10:1–719.

French, A. R. 2000. Interdependence of stored food and changes in body temperature during hibernation of the eastern chipmunk, Tamias striatus. J. Mammal. 81:979–985.

French, N. R., B. G. Maza, H. O. Hill, A. P. Aschwanden, and H. W. Kaaz. 1974. A population study of irradiated desert rodents. Ecol. Monogr. 44:45–72.

Fretwell, S. D. 1972. Populations in a seasonal environment. Princeton Univ. Press, Princeton, NJ.

Fretwell, S. D., and H. L. Lucas. 1970. On territorial behaviour and other factors influencing habitat distribution in birds, I: Theoretical development. Acta Biotheoret. 19:16–36.

Freudenberger, D. O., I. R. Wallis, and I. D. Hume. 1989. Digestive adaptations of kangaroos, wallabies, and rat-kangaroos. Pp. 179–189 in Kangaroos, wallabies, and rat-kangaroos (G. Grigg, P. Jarman, and I. D. Hume, eds.). Surrey Beatty and Sons, Chipping Norton, NSW, Australia.

Friend, J. A. 1989. Myrmecobiidae. Pp. 583–590 in Fauna of Australia: Mammalia, vol. IB (D. W. Walton and B. J. Richardson, eds.). Australian Government Publ. Service, Canberra.

Friend, J. A. 1995. Numbat, Myrmecobius fasciatus. Pp. 160–162 in The mammals of Australia (R. Strahan, ed.). Reed Books, Chatswood, Australia.

Friend, J. A., and N. D. Thomas. 2003. Conservation

of the numbat (Myrmecobius fasciatus). Pp. 452–463 in Predators with pouches: the biology of carnivorous marsupials (M. Jones, C. Dickman, and M. Archer, eds.). CSIRO Publishing, Collingwood, Australia.

Fries, R., A. Eggen, and G. Stranzinger. 1990. The bovine genome contains polymorphic microsatellites. Genomics 8:403–406.

Fritts, S. H. 1993. Controlling wolves in the greater Yellowstone area. In Ecological issues in reintroducing wolves into Yellowstone National Park (R. S. Cook, ed.). National Park Service (Scientific Monograph NPS/NRYELL/NRSM-93/22), Washington, DC.

Froehlich, D. J. 2002. Quo vadis Eohippus? The systematics and taxonomy of the early Eocene equids (Perissodactyla). Zool. J. Linnean Soc. 134:141–256.

Frost, H. C., W. B. Krohn, and C. R. Wallace. 1997. Age-specific reproductive characteristics in fishers. J. Mammal. 78:598–612.

Frost, K. J., M. A. Simpkins, and L. F. Lowry. 2001. Diving behavior of subadult and adult harbor seals in Prince William Sound, Alaska. Mar. Mammal Sci. 17:813–834.

Fujita, M. S., and M. D. Tuttle. 1991. Flying foxes (Chiroptera: Pteropodidae): threatened animals of key ecological and economic importance. Conserv. Biol. 5:455–463.

Fujiwara, M., and H. Caswell. 2001. Demography of the endangered North Atlantic right whale. Nature 414:537–541.

Fullard, J. H., J. A. Simmons, and P. A. Saillant. 1994. Jamming bat echolocation: the dogbane tiger moth Cynia tenera times its clicks to the terminal attack calls of the big brown bat Eptesicus fuscus. J. Exp. Biol. 194:285–298.

Fuller, M. R. 1987. Applications and considerations for wildlife telemetry. J. Raptor. Res. 21:126–128.

Fuller, W. A., L. L. Stebbins, and G. R. Dyke. 1969. Overwintering of small mammals near Great Slave Lake, northern Canada. Arctic 22:34–55.

Futuyma, D. J., and M. Slatkin (eds.). 1983. Coevolution. Sinauer Associates, Sunderland, MA.

Gabor, T. M., and E. C. Hellgren. 2000. Variation in peccary populations: landscape composition or competition by an invader? Ecology 81:2509–2524.

Gaeth, A. P., R. V. Short, and M. B. Renfree. 1999. The developing renal, reproductive, and respiratory systems of the African elephant suggest an aquatic ancestry. Proc. Natl. Acad. Sci. USA 96:5555–5558.

Gage, K. L., R. S. Ostfeld, and J. G. Olson. 1995. Nonviral vector-borne zoonoses associated with mammals in the United States. J. Mammal. 76:695–715.

Gagnon, M., and A. E. Chew. 2000. Dietary preferences in extant African Bovidae. J. Mammal. 81:490–511.

Gaines, C. A., M. P. Hare, S. E. Beck, and H. C. Rosenbaum. 2005. Nuclear markers confirm taxonomic status and relationships among highly endangered and closely related right whale species. Proc. Roy. Soc. Biol. Sci. Ser. B 272:533–542.

Gaines, W. L., A. L. Lyons, J. F. Lehmkuhl, and K. J. Raedeke. 2005. Landscape evaluation of female black bear habitat effectiveness and capability in the North Cascades, Washington. Biol. Cons. 125:411–425.

Gaisler, J. 1979. Ecology of bats. Pp. 281–342 in Ecology of small mammals (D. M. Stoddart, ed.). Chapman and Hall, London.

Galbreath, G. J. 1982. Armadillo, Dasypus novemcinctus. Pp. 71–79 in Wild mammals of North America: biology, management, and economics (J. A. Chapman and G. A. Feldhamer, eds.). Johns Hopkins Univ. Press, Baltimore.

Galbreath, G. J., and R. A. Melville. 2003. Pseudonovibos spiralis: Epitaph. J. Zool. Lond. 259:169–170.

Galea, L. A. M., M. Kavaliers, and K. P. Ossenkopp. 1996. Sexually dimorphic spatial learning in meadow voles Microtus pennsylvanicus and deer mice Peromyscus maniculatus. J. Exp. Biol. 199:195–200.

Gallardo, M. H., J. W. Bickham, R. L. Honeycutt, R. A. Ojeda, and N. Kohler. 1999. Discovery of tetraploidy in a mammal. Nature 401:341.

Galton, F. 1865. The first steps toward the domestication of animals. Trans. Ethnol. Soc. Lond. 3:122–138.

Gannon W. L., and G. R. Racz 2006. Character displacement and ecomorphological analysis of two long-eared Myotis (M. auriculus and M. evotis). J. Mammal. 87: 171–179.

Gardner, A. L. 1977. Feeding habits. Pp. 293–350 in Biology of bats of the New World Family Phyllostomatidae, part II. (R. J. Baker, J. Knox Jones, Jr., and D. C. Carter, eds.). Spec. Pub.No. 13, Museum of Texas Tech Univ., Lubbock.

Gardner, A. L. 1982. Virginia opossum, Didelphis virginiana. Pp. 3–36 in Wild mammals of North America: biology, management, and economics (J. A. Chapman and G. A. Feldhamer, eds.). Johns Hopkins Univ. Press, Baltimore.

Gardner, A. L. 2005. Order Cingulata. Pp. 94–99 in Mammal species of the world: a taxonomic and geographic reference. (D. E. Wilson and D. M. Reeder, eds.) 3rd edition, 2 vols. Johns Hopkins Univ. Press, Baltimore.

Gardner, A. L., and M. E. Sunquist. 2003. Virginia opossum, Didelphis virginiana. Pp. 3–29 in Wild mammals of North America: biology, management, and conservation. G. A. Feldhamer, B. C. Thompson, and J. A. Chapman (eds.). Johns Hopkins Univ. Press, Baltimore.

Gartlan, J. S., and T. T. Struhsaker. 1972. Polyspecific associations and niche separation of rain-forest anthropoids in Cameroon, West Africa. J. Zool. Lond. 168:221–266.

Gaston, K. J., T. M. Blackburn, and J. I. Spicer. 1998. Rapoport's rule: time for an epitaph? Trends Ecol. Evol. 13:70–74.

Gatesy, J., and P. Arctander. 2000. Hidden morphological support for the phylogenetic placement of Pseudoryx nghetinhensis with bovine bovids: a combined analysis of gross anatomical evidence and DNA sequences from five genes. Syst. Biol. 49:515–538.

Gatesy, J., and M. A. O'Leary. 2001. Deciphering whale origins with molecules and fossils. Trends Ecol. Evol. 16:562–570.

Gaubert, P., and A. Antunes. 2005. Assessing the taxonomic status of the Palawan pangolin Manis culionensis (Pholidota) using discrete morphological characters. J. Mammal. 86:1068–1074.

Gaubert, P., W. C. Wozencraft, P. Cordeiro-Estrela, and G. Veron. 2005. Mosaics of convergences and noise in morphological phylogenies: what's in a viverrid-like carnivoran? Syst. Biol. 54:865–894.

Gaudin, T. C. 2003. Phylogeny of the Xenarthra (Mammalia). Senckenbergiana biologica 83:27–40.

Gaudin, T. C. 2004. Phylogenetic relationships among sloths (Mammalia, Xenarthra, Tardigrada): the craniodental evidence. Zool. J. Linnean Soc. 140:255–305.

Gaudin, T. J. 1999. The morphology of the xenarthrous vertebrae (Mammalia: Xenarthra). Fieldiana, Geology 41:1–38.

Gaudin, T. J. 1999a. Pangolins. Pp. 855–857 in Encyclopedia of paleontology (R. S. Singer, ed.). Fitroy Dearborn Publishers, Chicago.

Gaudin, T. J. 1999b. Xenarthrans. Pp. 1347–1353 in Encyclopedia of paleontology (R. S. Singer, ed.). Fitroy Dearborn Publishers, Chicago, IL.

Gaudin, T. J., and A. A. Biewener. 1992. The functional morphology of xenarthrous vertebrae in the armadillo Dasypus novemcinctus (Mammalia: Xenarthra). J. Morphol. 214:63–81.

Gaudin, T. J., and J. R. Wible. 1999. The entotympanic of pangolins and the phylogeny of the Pholidota (Mammalia). J. Mammal. Evol. 6:39–65.

Gaudin, T. J., J. R. Wible, J. A. Hopson, and W. D. Turnbull. 1996. Reexamination of the morphological evidence for the cohort Epitheria (Mammalia, Eutheria). J. Mammal. Evol. 3:31–79.

Gaudin, T. J., R. J. Emry, and B. Pogue. 2006. A new genus and species of pangolin (Mammalia, Pholidota) from the Late Eocene of Inner Mongolia, China. J. Vert. Paleo. 26:146–159.

Gaulin, S. J., and R. W. FitzGerald. 1986. Sex differences in spatial ability: an evolutionary hypothesis and test. Am. Nat. 127:74–88.

Gauthier-Pilters, H. 1974. The behaviour and ecology of camels in the Sahara, with special reference to nomadism and water management. Pp. 542–551 in The behaviour of ungulates and its relation to management (V. Geist and F. Walther, eds.). IUCN Publ. No. 24, Morges, Switzerland.

Gauthier-Pilters, H., and A. Dagg. 1981. The camel, its evolution, ecology, behavior, and relations to man. Univ. Chicago Press, Chicago.

Gebczynska, Z., and M. Gebczynski. 1971. Insulating properties of the nest and social temperature regulation in *Clethrionomys glareolus* (Schreber). Acta Zool. Fennici. 8:104–108.

Gebo, D. L., M. Dagasto, K. C. Beard, and T. Qi. 2001. Middle Eocene primate tarsals from China: implications for haplorhine evolution. Am. J. Phys. Anthro. 116:83–107.

Gehrt, S. D. 2003. Raccoon, *Procyon lotor* and allies. Pp. 611–634 *in* Wild mammals of North America: biology, management, and conservation. 2nd edition (G. A. Feldhamer, B. C. Thompson, and J. A. Chapman, eds). Johns Hopkins Univ. Press, Baltimore.

Geiser, F. 2003. Thermal biology and energetics of carnivorous marsupials. Pp. 238–253 *in* Predators with pouches: the biology of carnivorous marsupials (M. Jones, C. Dickman, and M. Archer, eds.). CSIRO Publications, Collingwood, Australia.

Geiser, F., and T. Ruf. 1995. Hibernation versus daily torpor in mammals and birds: physiological variables and classification of torpor patterns. Physiol. Zool. 68:935–966.

Geiser, F., R. L. Drury, G. Kortner, C. Turbill, C. R. Pavey, and R. M. Brigham. 2004. Passive rewarming from torpor in mammals and birds: energetic, ecological and evolutionary implications. Pp. 51–62 *in* Life in the cold: revolution, mechanisms, adaptation, and application (B. M. Barnes and H. V. Carey, eds.). Twelfth International Hibernation Symposium. Biological Papers of the Univ. Alaska, 27. Institute of Arctic Biology, Univ. Alaska, Fairbanks.

Geist, V. 1966. Validity of horn segment counts in aging bighorn sheep. J. Wildl. Mgmt. 30:634–635.

Geist, V. 1971. Mountain sheep: a study in behavior and evolution. Univ. Chicago Press, Chicago.

Geist, V. 1987. Bergmann's rule is invalid. Can. J. Zool. 65:1035–1038.

Geist, V. 1991. Bones of contention revisited: did antlers enlarge with sexual selection as a consequence of neonatal security strategies? Appl. Anim. Behav. Sci. 29:453–469.

Geluso, K. N. 1978. Urine concentrating ability and renal structure of insectivorous bats. J. Mammal. 59:312–323.

Genoud, M. 1988. Energetic strategies of shrews: ecological constraints and evolutionary implications. Mammal Rev. 18:173–193.

Genoways, H. H., and J. H. Brown (eds.). 1996. Biology of the Heteromyidae. Spec. Pub. No. 10, American Society of Mammalogists.

Gentry, A. W. 1990. Evolution and dispersal of African Bovidae. Pp. 195–227 *in* Horns, pronghorns, and antlers: evolution, morphology, physiology, and social significance (G. A. Bubenik and A. B. Bubenik, eds.). Springer-Verlag, New York.

Gentry, A., J. Clutton-Brock, and C. P. Groves. 2004. The naming of wild animal species and their domestic derivatives. J. Archaeological Sci. 31:645–651.

Gentry, R. L., and G. L. Kooyman. 1986. Methods of dive analysis. Pp. 28–40 in Fur seals: maternal strategies on land and at sea (R. L. Gentry and G. L. Kooyman, eds.). Princeton Univ. Press, Princeton, NJ.

Gerard, A. S., and G. A. Feldhamer. 1990. A comparison of two survey methods for shrews: pitfalls and discarded bottles. Am. Mid. Nat. 124:191–194.

Gerstein, E. R., L. Gerstein, S. E. Forsythe, and J. E. Blue. 1999. The underwater audiogram of the West Indian manatee (*Trichechus manatus*). J. Acoust. Soc. Am. 105:3575–3583.

Getz, L. L., and C. S. Carter. 1980. Social organization in *Microtus ochrogaster* populations. The Biologist 62:56–69.

Gheerbrant, E., D. P. Domning, and P. Tassy. 2005. Paenungulata (Sirenia, Proboscidea, Hyracoidea, and relatives). Pp. 84–105 *in* The rise of placental mammals: origins and relationships of the major extant clades (K. D. Rose and J. D. Archibald, eds.). Johns Hopkins Univ. Press, Baltimore.

Giannini, N. P., F. Abdala, and D. A. Flores. 2004. Comparative postnatal ontogeny of the skull in *Dromiciops gliroides* (Marsupialia: Microbiotheriidae). American Museum Novitates 3460:1–17.

Gibbs, H. L., P. J. Weatherhead, P. T. Boag, B. N. White, L. M. Tabak, and D. J. Hoysak. 1990. Realized reproductive success of polygynous red-winged blackbirds revealed by DNA markers. Science 250:1394–1397.

Gibson, L. A., and I. D. Hume. 2000. Digestive performance and digesta passage in the omnivorous greater bilby *Macrotis lagotis* (Marsupialia: Peramelidae). J. Comp. Physiol. B 170:457–467.

Gibson, L. A., I. D. Hume, and P. D. McRae. 2002. Ecophysiology and nutritional niche of the bilby (*Macrotis lagotis*), an omnivorous marsupial from inland Australia: a review. Comp. Biochem. Physiol. Part A Mol. Integrative Physiol. 133:843–847.

Gilbert, C., A. Ropiquet, and A. Hassanin. 2006. Mitochondrial and nuclear phylogenies of Cervidae (Mammalia, Ruminantia): systematics, morphology, and biogeography. Mol. Phylogenet. Evol. 40:101–117.

Giles, R. H. 1971. Wildlife management techniques, 3rd edition, rev. Wildlife Society, Washington, DC.

Gilg, O., B. Sittler, B. Sabard, A. Hurstel, R. Sane, P. Delattre, and L. Hanski. 2006. Functional and numerical responses of four lemming predators in high arctic Greenland. Oikos 113: 193–216.

Gill, F. B. 1980. Historical aspects of hybridization between blue-winged and golden-winged warblers. Auk. 97:1–18.

Gillette, D. D. 1991. *Seismosaurus halli gen.* et sp. nov., a new sauropod dinosaur from the Morrison Formation (Upper Jurassic/Lower Cretaceous) of New Mexico, USA. J. Vert. Paleontol. 11:417–433.

Gillette, M. U., S. J. DeMarco, J. M. Ding, E. A. Gallman, L. E. Faiman, C. Liu, A. J. McArthur, M. Medanic, D. Richard, T. K. Tcheng, and E. T. Weber. 1993. The organization of the suprachiasmatic pacemaker of the rat and its regulation by neurotransmitters and modulators. J. Biol. Rhythms. 8:S53–S58 (suppl.).

Gilmore, D. P. 1969. Seasonally reproductive periodicity in the male Australian brush-tailed possum. J. Zool. 157:75–98.

Gingerich, P. 1987. Early Eocene bats (Mammalia: Chiroptera) and other vertebrates in freshwater limestones of the Willwood Formation, Clark's Fork Basin, Wyoming. Contrib. Mus. Paleontol. Univ. Michigan 27:275–320.

Gingerich, P. D. 2005. Cetacea. Pp. 234–252 *in* The rise of placental mammals: origins and relationships of the extant clades (K. D. Rose and J. D. Archibald, eds.). Johns Hopkins Univ. Press, Baltimore.

Gingerich, P. D., B. H. Smith, and E. L. Simons. 1990. Hind limbs of Eocene Basilosaurus: evidence of feet in whales. Science 249:154–156.

Gingerich, P. D., M. ul Haq, I. S. Zalmout, I. H. Khan, and M. S. Malkani. 2001. Origin of whales from early artiodactyls: hands and feet of Eocene Protocetidae from Pakistan. Science 293:2239–2242.

Girardier, L. and M. J. Stock (eds.). 1983. Mammalian thermogenesis. Chapman and Hall, New York.

Giraudoux, P., B. Pradier, P. Delattre, S. Deblay, D. Salvi, and R. Defaut. 1995. Estimation of water vole abundance by using surface indices. Acta Theriol. 40:77–96.

Gittleman, J. D. 1989. Carnivore behavior, ecology, and evolution. Comstock Publ. Assoc., Ithaca, NY.

Gittleman, J. D. 1996. Carnivore behavior, ecology, and evolution, vol. 2. Comstock Publ. Assoc., Ithaca, NY.

Gittleman, J. L. 1989. The comparative approach in ethology: aims and limitations. Persp. Ethol. 8:55–83.

Gjertz, I., D. Griffiths, B. A. Kraft, C. Lydersen, and O. Wiig. 2001. Diving and haul-out patterns of walruses, *Odobenus rosmarus*, on Svalbard. Polar Biol. 24:314–319.

Glander, K. E. 1977. Poison in a monkey's Garden of Eden. Nat. Hist. 86 (March):35–41.

Glander, K. E. 1982. The impact of plant secondary compounds on primate feeding behavior. Yearb. Phys. Anthro. 25:1–18.

Glantz, S. A. 1992. Primer of biostatistics, 3rd edition. McGraw-Hill, New York.

Glaser, H., and S. Lustick. 1975. Energetics and nesting behavior of the northern white-footed mouse, *Peromyscus leucopus noveboracensis*. Physiol. Zool. 48:105–113.

Glass, J. D., U. E. Hauser, W. Randolph, S. Ferriera, and M. A. Rea. 1993. Suprachiasmatic nucleus neurochemistry in the conscious brain: correlation with circadian activity rhythms. J. Biol. Rhythms 8:S47–S52 (suppl.).

Glazier, D. S. 2005. Beyond the "3/4-power law": variation in the intra- and interspecific scaling of metabolic rate in animals. Biol. Rev. Cambridge Philosophical Society 80:611–662.

Glazier, D. S. 2006. The 3/4-power law is not universal: evolution of isometric, ontogenetic metabolic scaling in pelagic animals. Bioscience 56:325–332.

Gleeson, S. K., A. B. Clark, and L. A. Dugatkin. 1994. Monozygotic twinning: an evolutionary hypothesis. Proc. Natl. Acad. Sci. USA 91:11363–11367.

Glennon, M. J., W. F. Porter, and C. L. Demers. 2002. An alternative field technique for estimating diversity of small-mammal populations. J. Mammal. 83:734–742.

Gliwicz, J., S. Pagacz, and J. Witczuk. 2006. Strategy of food plant selection in the Siberian northern pika, *Ochotona hyperborea*. Arctic Antarctic Alpine Res. 38:54–59.

Gloger, C. L. 1833. Das Abändern der Vögel durch Einfluss des Klimas. Breslau, Breslau.

Goepfert, M. C., and L. T. Wasserthal. 1995. Notes on echolocation calls, food, and roosting behaviour of the Old World sucker-footed bat, *Myzopoda aurita* (Chiroptera: Myzopodidae). Z. Saeugetierk. 60:1–8.

Goin, F. J. 2003. Early marsupial radiations in South America. Pp. 30–42 *in* Predators with pouches: the biology of carnivorous marsupials (M. Jones, C. Dickman, and M. Archer, eds.). CSIRO Publishing, Collingwood, Vic., Australia.

Goin, F. J., J. A. Case, M. O. Woodburne, S. F. Vizcaino, and M. A. Reguero. 1999. New discoveries of "opposum-like" marsupials from Antarctica (Seymour Island, Medial Eocene). J. Mammal. Evol. 6:335–365.

Goiti, U., J. R. Aihartza, and I. Garin. 2004. Diet and prey selection in the Mediterranean horseshoe bat *Rhinolophus euryale* (Chiroptera, Rhinolophidae) during the pre-breeding season. Mammalia 68:397–402.

Goldingay, R. L., and R. P. Kavangah. 1991. The yellow-bellied glider: a review of its ecology and management considerations. Pp. 365–375 *in* Conservation of Australia's forest fauna (D. Lunney, ed.). Royal Zoological Society, Sydney.

Goldsmith, R., D. Bunnag, and T. Bunnag. 1991. Lung fluke infections: paragonimiasis. Pp. 827–831 *in* Hunter's tropical medicine, 7th edition (G. T. Strickland, ed.). W. B. Saunders, Philadelphia.

Goldspink, G. 1981. The use of muscles during flying, swimming, and running from the point of view of energy saving. Symp. Zool. Soc. Lond. 48:219–238.

Goldstein, D. B., and D. D. Pollock. 1997. Launching microsatellites: a review of mutation processes and methods of phylogenetic inference. J. Hered. 88:335–342.

Goldstein, I., S. Paisley, R. Wallace, J. P. Jorgenson, F. Cuesta, and A. Castellanos. 2006. Andean bear-livestock conflicts: a review. Ursus 17:8–15.

Goldsworthy, S. D. 1995. Differential expenditure of maternal resources in Antarctic fur seals, *Arctocephalus gazella*, at Heard Island, southern Indian Ocean. Behav. Ecol. 6:218–228.

Golley, F. B. 1960. Energy dynamics of a food chain of an old-field community. Ecol. Monogr. 30:187–206.

Golley, F. B. 1961. Effect of trapping on adrenal activity in Sigmodon. J. Wildl. Mgmt. 25:331–333.

Gonyea, W. J., and R. Ashworth. 1975. The form and function of retractable claws in the Felidae and other representative carnivorans. J. Morphol. 145:229–238.

Goodall, J. 1986. The chimpanzees of Gombe. Harvard Univ. Press, Cambridge, MA.

Goodman, M. 1999. The genomic record of humankind's evolutionary roots. Am. J. Hum. Genet. 64:31–39.

Goodman, M., D. A. Tagle, D. H. A. Fitch, W. Bailey, J. Czelusniak, B. F. Koop, P. Benson, and J. L. Slighton. 1990. Primate evolution at the DNA level and a classification of hominids. J. Mol. Evol. 30:260–266.

Gomendio M., A. F. Malo, A. J. Soler, M. R. Fernandez-Santos, M. C. Esteso, A. J. Garcia, E. R. S. Roldan, and J. Garde. 2006. Male fertility and sex ratio at birth in red deer. Science 314:1445–1447.

Gompper, M. E., and H. M. Hackett. 2005. The long-term, range-wide decline of a once common carnivore: the eastern spotted skunk (*Spilogale putorius*). Anim. Cons. 8:195–201.

Gongora, J., and C. Moran. 2005. Nuclear and mitochondrial evolutionary analyses of collared, white-lipped, and Chacoan peccaries (Tayassuidae). Mol. Phylogenet. Evol. 34:181–189.

González, S., F. Álvarez-Valin, and J.E. Maldonado. 2002. Morphometric differentiation of endangered pampas deer (*Ozotoceros besoarticus*), with description of new subspecies from Uruguay. J. Mammal. 83:1127–1140.

Goodchild, M. F. 2003. Geographic infrmation science and systems for environmental management. Annu. Rev. Envir. Res. 28:493–519.

Goodman, M. 1978. Protein sequences in phylogeny. Pp. 141–159 *in* Molecular evolution (F. J. Ayala, ed.). Sinauer Associates, Sunderland, MA.

Goodman, S. M. 1995. *Rattus* on Madagascar and the dilemma of protecting the endemic rodent fauna. Conserv. Biol. 9:450–453.

Gorbach, S. L., J. G. Bartlett, and N. R. Blacklow (eds.). 1992. Infectious diseases. W. B. Saunders, Philadelphia.

Gorbach, S. L., J. G. Bartlett, and N. R. Blacklow (eds.). 1998. Infectious diseases, 2nd edition. W. B. Saunders, Philadelphia.

Gordon, G., and A. J. Hulbert. 1989. Peramelidae. Pp. 603–624. *in* Fauna of Australia: mammalia, vol. IB (D. W. Walton and B. J. Richardson, eds.). Australian Government Publishing Service, Canberra.

Gordon, K., T. P. Fletcher, and M.B. Renfree. 1988. Reactivation of the quiescent corpus luteum and diapausing embryo after temporary removal of the sucking stimulus in the tammar wallaby. J. Reprod. Fertility 83:401–406.

Gordon, M. S. 1982. Animal physiology: principles and adaptations, 4th edition. Macmillan, New York.

Gorman, M. L., and R. D. Stone. 1990. The natural history of moles. Comstock Publ. Assoc., Ithaca, NY.

Gosling, L. M., and M. Petrie. 1990. Lekking in topi: a consequence of satellite behaviour by small males at hotspots. Anim. Behav. 40:272–287.

Gotelli, N. J., and A. M. Ellison. 2004. A primer of ecological statistics. Sinauer Associates, Sunderland, MA.

Götherström, A., C. Anderung, L. Hellborg, R. Elburg, C. Smith, D. G. Bradley, and H. Ellegren. 2005. Cattle domestication in the Near East was followed by hybridization with aurochs bulls in Europe. Proc. Roy. Soc. Lond. B 272:2345–2350.

Gottfried, B. M. 1979. Small mammal populations in woodlot islands. Am. Mid. Nat. 102:105–112.

Gottfried, B. M. 1982. A seasonal analysis of small mammal populations on woodlot islands. Can. J. Zool. 60:1660–1664.

Gould, E. 1983. Mechanisms of mammalian auditory communication. Pp. 265–342 *in* Advances in the study of mammalian behavior (J. F. Eisenberg and D. G. Kleiman, eds.). Spec. Pub. No. 7, American Society of Mammalogists.

Gould, E., W. McShea, and T. Grand. 1993. Function of the star in the star-nosed mole, *Condylura cristata*. J. Mammal. 74:108–116.

Gould, L., and M. L. Sauther. 2006. Lemurs, ecology and adaptation. Springer, New York.

Goundie, T. R., and S. H. Vessey. 1986. Survival and dispersal of young white-footed mice born in nest boxes. J. Mammal. 67:53–60.

Grand, T., E. Gould, and R. Montali. 1998. Structure of the proboscis and rays of the star-nosed mole, *Condylura cristata*. J. Mammal. 79:492–501.

Granjon, L., J.-F. Cossn, E. Quesseveur, and B. Sicard. 2005. Population dynamics of the multimammate rate *Mastomys huberti* in an annually flooded agricultural region of central Mali. J. Mammal. 86:997–1008.

Grant, P. R. 1975. Population performance of *Microtus pennsylvanicus* confined to woodland habitat and a model of habitat occupancy. Can. J. Zool. 53:1447–1465.

Grant, T. 1995. The platypus. Univ. New South Wales Press, Sydney.

Grant, T. R. 1989. Ornithorhynchidae. Pp. 436–450 *in* Fauna of Australia: mammalia, vol. IB (D. W. Walton and B. J. Richardson, eds.). Australian Government Publ. Service, Canberra.

Grant, T. R., and P. D. Temple-Smith. 2003. Conservation of the platypus, *Ornithorhynchus anatinus*: threats and challenges. Aquat. Ecosys. Health Mgmt. 6:5–18.

Grassman, L. I., Jr., M. E. Tewes, N. J. Silvy, and K. Kreetiyutanont. 2005. Ecology of three sympatric felids in a mixed evergreen forest in north-central Thailand. J. Mammal. 86:29–38.

Graur, D., and W.-H. Li. 2000. Fundamentals of molecular evolution, 2nd edition. Sinauer Associates, Sunderland, MA.

Green, B. 1984. Composition of milk and energetics of growth in marsupials. Pp. 369–387 *in* Physiological strategies in lactation (M. Peaker, R. G. Vernon, and C. H. Knight, eds.). Academic Press, New York.

Green, G. M. G. and R. W. Sussman. 1990. Deforestation history of the eastern rain forests of Madagascar from satellite images. Science 248:212–215.

Green, J. S., F. F. Knowlton, and W. C. Pitt. 2002. Reproduction in captive wild-caught coyotes (*Canis latrans*). J. Mammal. 83:501–506.

Green, K., M. K. Tory, A. T. Mitchell, P. Tennant, and T. W. May. 1999. The diet of the long-footed potoroo (*Potorous longipes*). Aust. J. Ecol. 24:151–156.

Green, R. G., C. L. Larson, and J. F. Bell. 1939. Shock disease as the cause of the periodic decimation of the snowshoe hare. Am. J. Hyg. 30:83–102.

Greenbaum, I. F., and C. J. Phillips. 1974. Comparative anatomy of and general histology of tongues of long-nosed bats (*Leptonycteris sanborni* and *L. nivalis*) with reference to infestation of oral mites. J. Mammal. 55:489–504.

Greenhall, A. M. 1972. The biting and feeding habits of the vampire bat, *Desmodus rotundus*. J. Zool. Lond. 168:451–461.

Greenhall, A. M., and U. Schmidt (eds.). 1988. Natural history of vampire bats. CRC Press, Boca Raton, FL.

Greenhall, A. M., G. Joermann, and U. Schmidt. 1983. *Desmodus rotundus*. Mamm. Species 202:1–6.

Greenhall, A. M., U. Schmidt, and G. Joermann. 1984. *Diphylla ecaudata*. Mamm. Species 227:1–3.

Greenhall, A. M., and U. Schmidt (eds.). 1988. Natural history of vampire bats. CRC Press, Boca Raton, FL.

Greenhall, A. M., and W. A. Schutt, Jr. 1996. *Diaemus youngi*. Mamm. Species 533:1–7.

Greenwood, A. D., J. Castresana, G. Feldmaier-Fuchs, and S. Paabo. 2001. A molecular phylogeny of two extinct sloths. Mol. Phylogenet. Evol. 18:94–103.

Greenwood, D. R., D. Comeskey, M. B. Hunt, and E. J. Rasmussen. 2005. Chemical communication: chirality in elephant pheromones. Nature 438:1097–1098.

Greenwood, P. J. 1980. Mating systems, philopatry, and dispersal in birds and mammals. Anim. Behav. 28:1140–1162.

Greenwood, P. J. 1983. Mating systems and the evolutionary consequences of dispersal. Pp. 116–131 *in* The ecology of animal movement (I. R. Swingland and P. J. Greenwood, eds.). Clarendon Press, Oxford.

Gregory, W. K. 1947. The monotremes and the palimpsest theory. Bulletin of the American Museum of Natural History 88:1–52

Grenfell, B. T. and F. M. D. Gulland. 1995. Ecological impact of parasitism on wildlife host populations. Parasitol. 111:S3–S14.

Griffin, D. R. 1958. Listening in the dark. Yale Univ. Press, New Haven, CT.

Griffin, D. R. 1970. Migrations and homing of bats. Pp. 233–265 *in* Biology of bats (W. A. Wimsatt, ed.). Academic Press, New York.

Griffiths, M. 1978. The biology of the monotremes. Academic Press, New York.

Griffiths, M. 1989. Tachyglossidae. Pp. 407–435 *in* Fauna of Australia: mammalia, vol. IB (D. W. Walton and B. J. Richardson, eds.). Australian Government Publ. Service, Canberra.

Griffiths, M., M. A. Elliott, R. M. C. Leckie, and G. I. Schoefl. 1973. Observations on the comparative anatomy and ultrastructure of mammary glands and on the fatty acids of the triglycerides in platypus and echidna milk fats. J. Zool. 169:255–279.

Griffiths, M., F. Kristo, B. Green, A. C. Fogerty, and K. Newgrain. 1988. Observations on free-living, lactating echidnas, *Tachyglossus aculeatus* (Monotremata: Tachyglossidae), and sucklings. Aust. Mammal. 11:135–143.

Griffiths, M., R. T. Wells, and D. J. Barrie. 1991. Observations on the skulls of fossil and extant echidnas (Monotremata: Tachyglossidae). Aust. Mammal. 14:87–101.

Griffiths, T. A. 1997. Phylogenetic position of the bat *Nycteris javanica* (Chiroptera: Nycteridae). J. Mammal. 78:106–116.

Griffling, J. P. 1974. Body measurements of black-tailed jackrabbits of southeastern New Mexico with implications of Allen's rule. J. Mammal. 55:674–678.

Grigg, G., and L. Beard. 2000. Hibernation by echidnas in mild climates: hints about the evolution of endothermy? Pp. 5–19 *in* Life in the cold: Eleventh International Hibernation Symposium (G. Heldmaier and M. Klingenspor, eds.). Springer-Verlag, Berlin.

Grigg, G. C., L. A. Beard, and M. L. Augee. 1989. Hibernation in a monotreme, the echidna *Tachyglossus aculeatus*. Comp. Biochem. Physiol. 92A:609–612.

Grigg, G. C., L. A. Beard, T. R. Grant, and M. L. Augee. 1992. Body temperature and diurnal activity pattern in the platypus, *Ornithorhynchus anatinus*, during winter. Aust. J. Zool. 40:135–142.

Grigg, G. C., L. A. Beard, and M. L. Augee. 2004. The evolution of endothermy and its diversity in mammals and birds. Physiol. Biochem. Zool. 77:982–997.

Grimwood, I. 1988. "Operation oryx": The start of it all. Pp. 1–8 *in* The conservation and biology of desert antelopes (A. Dixon and D. Jones, eds.). Helm, London.

Grinnell, J., and H. S. Swarth. 1913. An account of the birds and mammals of the San Jacinto area of southern California, with remarks upon the behavior of geographic races on the margins of their habitats. Univ. Calif. Pub. Zool. 10:197–406.

Grodzinski, W., and N. R. French. 1983. Production efficiency in small mammal populations. Oecologia 56:41–49.

Grodzinski, W., M. Makomaska, R. Tertil, and J. Weiner. 1977. Bioenergetics and total impact of vole populations. Oikos 29:494–510.

Grojean, R. E., J. A. Suusa, and M. C. Henry. 1980. Utilization of solar radiation by polar animals: an optical model for pelts. Appl. Optics 19:339–346.

Gromko, M. H., D. G. Gilbert, and R. C. Richmond. 1984. Sperm transfer and use in the multiple mating system of *Drosophila*. Pp. 371–426 *in* Sperm competition and the evolution of animal mating systems (R. L. Smith, ed.). Academic Press, New York.

Groves, C. 1998. Systematics of tarsiers and lorises. Primates 39:13–27.

Groves, C. P. 1989. A theory of human and primate evolution. Oxford Univ. Press, New York.

Groves, C. P. 1993. Order Diprotodontia. Pp. 45–62 *in* Mammal species of the world: a taxonomic and geographic reference, 2nd editioin (D. E. Wilson and D. M. Reeder, eds.). Smithsonian Institution Press, Washington, DC.

Groves, C. P. 2001. Primate taxonomy. Smithsonian Institution Press, Washington. DC.

Groves, C. P. 2005. Order Diprotodontia. Pp. 43–70 *in* D. E. Wilson and D. M. Reeder (eds.). Mammal species of the world: a taxonomic and geographic reference. 3rd edition, 2 vols. Johns Hopkins Univ. Press, Baltimore.

Groves, C. P. 2005. Order Primates. Pp. 111–184 *in* D. E. Wilson and D. M. Reeder (eds.). Mammal species of the world: a taxonomic and geographic reference. 3rd edition. John Hopkins Univ. Press, Baltimore

Groves, C. P., and T. F. Flannery. 1990. Revision of the families and genera of bandicoots. Pp. 1–11 *in* Bandicoots and bilbies (J. H. Seebeck, P. R. Brown, R. W. Wallis, and C. M. Kemper, eds.). Surrey Beatty and Sons, Sydney.

Groves, C. P., Y. X. Wang, and P. Grubb. 1995. Taxonomy of musk deer, genus *Moschus* (Moschidae, Mammalia). Acta Theriol. Sinica 15:181–197.

Grubb, P. 1993a. Order Perissodactyla. Pp. 369–372 *in* Mammal species of the world: a taxonomic and geographic reference, 2nd edition (D. E. Wilson and D. M. Reeder, eds.). Smithsonian Institution Press, Washington, DC.

Grubb, P. 1993b. Order Artiodactyla. Pp. 377–414 in Mammal species of the world: a taxonomic and geographic reference, 2d ed. (D. E. Wilson and D. M. Reeder, eds.). Smithsonian Institution Press, Washington, DC.

Grubb, P. 2001. Review of family-group names of living bovids. J. Mammal. 82:374–388.

Grubb, P. 2005. Order Artiodactyla. Pp. 637–722 in Mammal species of the world: a taxonomic and geographic reference. 3rd edition, 2 vols. (D. E. Wilson and D. M. Reeder, eds.). Johns Hopkins Univ. Press, Baltimore.

Grubb, P. 2005a. Perissodactyla. Pp. 629–636 in Mammal species of the world: a taxonomic and geographic reference, 3rd edition, 2 vols. (D. E. Wilson and D. M. Reeder, eds.). Johns Hopkins Univ. Press, Baltimore.

Grubb, P. 2005b. Artiodactyla. Pp. 637–722 in Mammal species of the world: a taxonomic and geographic reference, 3rd edition, 2 vols. (D. E. Wilson and D. M. Reeder, eds.). Johns Hopkins Univ. Press, Baltimore.

Grubb, P., C. P. Groves, J. P. Dudley, and J. Shoshani. 2000. Living African elephants belong to two species: *Loxodonta africana* (Blumenbach, 1797) and *Loxodonta cyclotis* (Matschie, 1900). Elephant 2:1–4.

Grützner, F., J. Deakin, W. Rens, N. El-Mogharbel, and J. A. M. Graves. 2003. The monotreme genome: a patchwork of reptile, mammal and unique features? Comp. Biochem. Physiol. Part A 136:867–881.

Grützner, F., W. Rens, E. Tsend-Ayush, N. El-Mogharbel, P. C. M. O'Brien, R. C. Jones, M. A. Ferguson-Smith, and J. A. M. Graves. 2004. In the platypus a meiotic chain of ten sex chromosomes shares genes with the bird Z and mammal X chromosomes. Nature 432:913–917.

Guarch-Delmonte, J. M. 1984. Evidencias de la existencia de *Geocapromys* y *Heteropsomys* (Mammalia: Rodentia) en Cuba. Misc. Zool. (Havana). 18:1.

Gubbins, C. 2002. Use of home ranges by resident bottlenose dolphins (*Tursiops truncatus*) in a South Carolina estuary. J. Mammal. 83:178–187.

Guimaraes, P. R., Jr., U. Kubota, B. Z. Gomes, R. L. Fonseca, C. Bottcher, and M. Galetti. 2006. Testing the quick meal hypothesis: the effect of pulp on hoarding and seed predation of *Hymenaea courbaril* by red-rumped agoutis (*Dasyprocta leporine*). Aust. Ecol. 31:95–98.

Gunderson, H. L. 1976. Mammalogy. McGraw-Hill, New York.

Gunderson, H. L., and J. R. Beer. 1953. The mammals of Minnesota. Minn. Mus. Nat. Hist., Occ. Pap. 6.

Gunn, A. 1982. Muskox (*Ovibos moschatus*). Pp. 1021–1035 in Wild mammals of North America: biology, management, and economics (J. A. Chapman and G. A. Feldhamer, eds.). Johns Hopkins Univ. Press, Baltimore.

Gunnell, G. F., and N. B. Simmons. 2005. Fossil evidence and the origin of bats. J. Mammal. Evol. 12:209–246.

Gunson, J. R., and R. R. Bjorge. 1979. Winter denning of the striped skunk in Alberta. Canadian Field-Naturalist 93:252–258.

Gurnell, J. 1987. The natural history of squirrels. Christopher Helm, London.

Gustafson, A. W., and D. A. Damassa. 1985. Annual variation in plasma sex steroid-binding protein and testosterone concentrations in the adult male little brown bat: relation to the asynchronous recrudescence of the testis and accessory reproductive organs. Biol. Reprod. 33:1126–1137.

Guthrie, R. D. 1971. A new theory of mammalian rump patch evolution. Behaviour 38:132–145.

Guyton, A. C. 1986. Textbook of medical physiology, 7th edition. W. B. Saunders, Philadelphia.

Gwinner, E. 1986. Circannual rhythms. Springer-Verlag, Berlin.

Gwynne, M. D., and R. H. V. Bell. 1968. Selection of vegetation components by grazing ungulates in the Serengeti National Park. Nature 220:390–393.

Haarberg, O., and F. Rosell. 2006. Selective foraging on woody plant species by the Eurasian beaver (*Castor fiber*) in Telemark, Norway. J. Zool. Lond. 270:201–208.

Habersetzer, J., G. Richter, and G. Storch. 1994. Paleoecology of early middle Miocene bats from Messel, FRG: aspects of flight, feeding, and echolocation. Hist. Biol. 8:235–260.

Hadley, E. A., M. van Tuinen, Y. Chan, and K. Heiman. 2003. Ancient DNA evidence of prolonged ppulation persistence with negligible genetic diversity in an endemic tuco-tuco (*Ctenomys sociabilis*). J. Mammal. 84:403–417.

Hadley, M. E. 1988. Endocrinology. Prentice-Hall, Englewood Cliffs, NJ.

Haffer, J. 1969. Speciation in Amazonian forest birds. Science 165:131–137.

Haffer, J. 1997. Alternative modles of vertebrate speciation in Amazonia: an overview. Biodivers. Conserv. 6:451–417.

Hafner, D. J., and C. J. Shuster. 1996. Historical biogeography of western peripheral isolates of the least shrew, *Cryptotis parva*. J. Mammal. 77:536–545.

Hafner, D. J., and R. M. Sullivan. 1995. Historical and ecological biogeography of Nearctic pikas (Lagomorpha: Ochotonidae). J. Mammal. 76:302–321.

Hafner, M. S., W. L. Gannon, J. Salazar-Bravo, and S. T. Alvarez-Casteñeda. 1997. Mammal collections in the western hemisphere. American Society of Mammalogists, Provo, UT.

Haigh, G. R. 1979. Sun-compass orientation in the thirteen-lined ground squirrel, *Spermophilus tridecemlineatus*. J. Mammal. 60:629–632.

Haigh, J. C., and R. J. Hudson. 1993. Farming wapiti and red deer. Mosby Publ., St. Louis.

Hain, J. H. W., G. R. Carter, S. D. Kraus, C. A. Mayo, and H. E. Winn. 1982. Feeding behavior of the humpback whale, *Megaptera novaeangliae*, in the western North Atlantic. Fish. Bull. 80:99–108.

Hainsworth, F. R. 1981. Animal physiology, adaptations in function. Addison-Wesley, Reading, MA.

Hainsworth, F. R. 1995. Optimal body temperatures with shuttling: desert antelope ground squirrels. Anim. Behav. 49:107–116.

Haley, M. P., C. J. Deutsch, and B. J. Le Boeuf. 1994. Size, dominance and copulatory success in male northern elephant seals, *Mirounga angustirostris*. Anim. Behav. 48:1249–1260.

Hall, E. R. 1948. Mammals of Nevada. Univ. California Press, Berkeley.

Hall, E. R. 1951. American weasels. Univ. Kansas Pub., Mus. Nat. Hist. 4:1–466.

Hall, E. R. 1962. Collecting and preparing study specimens of vertebrates. Misc. Publ. Univ. Kansas Mus. Nat. Hist. 30:1–46.

Hall, E. R. 1981, 2001. The mammals of North America, 2nd edition, 2 vols. Wiley-Interscience, New York.

Hall, E. R., and K. R. Kelson. 1959. The mammals of North America, 2 vols. Ronald Press, New York.

Hall, R. 1998. The plate tectonics of Cenozoic SE Asia and the distribution of land and sea. Pp. 99–131 in Biological and geological evolution of SE Asia (R. Hall and J.D. Holloway, eds.). Backbuys, Leiden, Netherlands.

Hall-Aspland, S. A., and T. L. Rogers. 2004. Summer diet of leopard seals (*Hydrurga leptonyx*) in Prydz Bay, eastern Antarctica. Polar Biology 27:729–734.

Halle, S. 1995. Diel pattern of locomotor activity in populations of root voles, *Microtus oeconomus*. J. Biol. Rhythms 10:211–224.

Halls, L. K. (ed.). 1984. White-tailed deer ecology and management. Stackpole Books, Harrisburg, PA.

Hamilton, H., S. Caballero, A. G. Collins, and R. L. Brownell, Jr. 2001. Evolution of river dolphins. Proc. Roy. Soc. Lond. 268:549–556.

Hamilton, J. L., R. M. Dillaman, W.A. McLellan, and D. A. Pabst. 2004. Structural fiber reinforcement of keel blubber in harbor porpoise (*Phocoena phocoena*). J. Morphol. 261:105–117.

Hamilton, W. D. 1963. The evolution of altruistic behavior. Am. Nat. 97:354–356.

Hamilton, W. D. 1964. The genetical evolution of social behavior I, II. J. Theoret. Biol. 7:1–52.

Hamilton, W. D. 1971. Geometry for the selfish herd. J. Theoret. Biol. 31:295–311.

Hamilton, W. D., and R. M. May. 1977. Dispersal in stable habitats. Nature 269:578–581.

Hamilton, W. D., and M. Zuk. 1984. Heritable true fitness and bright birds: a role for parasites? Science 218:384–387.

Hamilton, W. J., Jr. 1955. Mammalogy in North America. Pp. 661–688 in A century of progress in the natural sciences. California Academy of Sciences, San Francisco.

Hammill, M. O., and G. B. Stenson. 2000. Estimated prey consumption by harp seals (*Phoca groenlandica*), hooded seals (*Cystophora cristata*), grey seals (*Halichoerus grypus*) and harbour seals (*Phoca vitulina*) in Atlantic Canada. J. NW Atlantic Fish. Sci. 26:1–23.

Hammock, E. A. D., and L. J. Young. 2005. Microsatellite instability generates diversity in brain and sociobehavioral traits. Science 308:1630–1634.

Hammond, E. L., and R.G. Anthony. 2006. Mark-recapture estimates of population parameters for selected species of small mammals. J. Mammal. 87:618–627.

Hammond, K. A., and B. A. Wunder. 1991. Effects of food quality and energy needs: changes in gut morphology and capacity of *Microtus ochrogaster*. J. Mammal. 64:541–567.

Hammond, P. S., S. A. Mizroch, and G. P. Donovan (eds.). 1990. Individual recognition of cetaceans: use of photo-identification and other techniques to estimate population and pod characteristics. Sci. Rep. Whales Res. Inst. 29:59–85.

Han, Z., F. Wei, Z. Zhang, M. Li, B. Zhang, and J.Hu. 2004. Habitat selection by red pandas in Fengtongzhai Natural Reserve. Acta Theriol. Sinica 24:185–192.

Hand, S. J., P. F. Murray, D. Megirian, M. Archer, and H. Godthelp. 1998. Mystacinid bats (Microchiroptera) from the Australian Tertiary. J. Vert. Paleo. 14:375–381.

Handley, C. O., Jr., D. E. Wilson, and A. L. Gardner (eds.). 1991. Demography and natural history of the common fruit bat, Artibeus jamaicensis, on Barro Colorado Island, Panama. Smithsonian Contr. Zool. 511:1–173.

Hanken, J., and P. W. Sherman. 1981. Multiple paternity in Belding's ground squirrel litters. Science 212:351–353.

Hansen, E. W. 1966. The development of maternal and infant behavior in the rhesus monkey. Behaviour. 27:107–149.

Hanski, I. 1996. Metapopulation ecology. Pp. 13–43 in Population dynamics in ecological space and time (O. E. Rhodes, Jr., R. K. Chesser, and M. H. Smith, eds.). Univ. Chicago Press, Chicago.

Hanski, I., L. Hansson, and H. Henttonen. 1991. Specialist predators, generalist predators, and the microtine rodent cycle. J. Anim. Ecol. 60:353–367.

Hanson, A. M., M. B. Hall, L. M. Porter, and B. Lintzenich. 2006. Composition and nutritional characteristics of fungi consumed by *Callimico goeldii* in Pando, Bolivia. Int. J. Primatol. 27:323–346.

Happold, D. C. D., and M. Happold. 1989. Biogeography of small montane mammals in Malawi, Central Africa. J. Biogeogr. 16:353–367.

Harcourt, A. H. 1989. Deformed sperm are probably not adaptive. Anim. Behav. 37:863–864.

Harcourt, A. H., P. H. Harvey, S. G. Larson, and R. V. Short. 1981. Testis weight, body weight, and breeding system in primates. Nature 293:55–57.

Harder, J. D., and R. L. Kirkpatrick. 1994. Physiological methods in wildlife research. Pp. 275–306 in Research and management techniques for wildlife and habitats, 5th edition (T.A. Bookhout, ed.). The Wildlife Society, Bethesda, MD.

Harder, J. D., and A. Woolf. 1976. Changes in plasma levels of oestrone and oestradiol during pregnancy and parturition in white-tailed deer. J. Reprod. Fert. 47:161–163.

Harder, J. D., M. J. Stonerook, and J. Pondy. 1993. Gestation and placentation in two New World opossums: *Didelphis virginiana* and *Monodelphis domestica*. J. Exp. Zool. 266:463–479.

Harlow, H. J. 1981. Torpor and other physiological adaptations of the badger (*Taxidea taxus*) to cold environment. Physiol. Zool. 54:267–275.

Harlow, H. J., T. Lohuis, R. C. Anderson-Sprecher, and T. D. I. Beck. 2004. Body surface temperature of hibernating black bears may be related to periodic muscle activity. J. Mammal. 85:414–419.

Harney, B. A., and R. D. Dueser. 1987. Vertical stratification of activity of two *Peromyscus* species: an experimental analysis. Ecology 68:1084–1091.

Harper, S. J., and G. O. Batzli. 1996. Monitoring use

of runways by voles with passive integrated transponders. J. Mammal. 77:364–369.

Harrington, L. A., D. E. Biggins, and A. W. Alldredge. 2003. Basal metabolism of the black-footed ferret (*Mustela nigripes*) and the Siberian polecat (*M. eversmannii*). J. Mammal. 84:497–504.

Harrington, M. E., D. M. Nance, and B. Rusak. 1985. Neuropeptide Y immunoreactivity in the hamster geniculo-suprachiasmatic tract. Brain Res. Bull. 15:465–472.

Harris, A. H. 1998. Fossil history of shrews in North America. Pp. 133–156 in Evolution of shrews (J. M. Wójcik and M. Wolsan, eds.). Mammal Research Institute, Polish Academy of Sciences, Bialowieza.

Harris, H. 1966. Enzyme polymorphisms in man. Proc. Roy. Soc. Lond. Ser. B 164:298–310.

Harris, H., and D. A. Hopkinson. 1976. Handbook of enzyme electrophoresis in human genetics. North-Holland, Amsterdam.

Harris, R. H. 1959. Small vertebrate skeletons. Museums J. 58:223–224.

Harris, V. T. 1952. An experimental study of habitat selection by prairie and forest races of the deer mouse, *Peromyscus maniculatus*. Contrib. Lab. Vert. Biol., Univ. Mich. 56:1–53.

Harrop, C. J. F., and I. D. Hume. 1980. Digestive tract and digestive function in monotremes and nonmacropod marsupials. Pp. 63–77 in Comparative physiology: primitive mammals (K. Schmidt-Nielsen, L. Bolis, and C. R. Taylor, eds.). Cambridge Univ. Press, Cambridge, UK.

Hart, J. A., and T. B. Hart. 1988. A summary report on the behaviour, ecology, and conservation of the okapi (*Okapia johnstoni*) in Zaire. Acta Zool. Pathol. Antverpiensia. 80:19–28.

Hart, J. S. 1971. Rodents. Pp. 1–149 in Comparative physiology of thermoregulation (G. C. Whittow, ed.). Academic Press, New York.

Harthoorn, A. M. 1976. The chemical capture of animals. Balliére Tindall, London.

Hartman, D. S. 1979. Ecology and behavior of the manatee (*Trichechus manatus*) in Florida. Spec. Pub. No. 5, American Society of Mammalogists.

Hartman, G. and S. Tornlov. 2006. Influence of watercourse depth and width on dam-building behaviour by Eurasian beaver (*Castor fiber*). J. Zool. Lond. 268:127–131.

Hartman, G. D., and T. L. Yates. 2003. Moles. Pp. 30–55 in Wild mammals of North America: biology, management, and conservation (G. A. Feldhamer, B. C. Thompson, and J. A. Chapman, eds.). 2d edition, Johns Hopkins Univ. Press, Baltimore.

Hartwell, L. H., L. Hood, M. L. Goldberg, A. E. Reynolds, L. M. Silver, and R. C. Veres. 2008. Genetics: from genes to genomes, 3rd edition. McGraw-Hill, New York.

Harvey, P. H., and M. D. Pagel. 1991. The comparative method in evolutionary biology. Oxford Univ. Press, New York.

Harvey, P. H., and K. Ralls. 1985. Homage to the null weasel. Pp. 155–171 in Evolution: essays in honour of John Maynard Smith (P. J. Greenwood, P. H. Harvey, and M. Slatkin, eds.). Cambridge Univ. Press, Cambridge, UK.

Hassanin, A., and E. J. P. Douzery. 2003. Molecular and morphological phylogenies of Ruminantia and the alternative position of the Moschidae. Syst. Biol. 52: 206–228.

Hasson, O. 1991. Pursuit-deterrent signals: communication between prey and predator. Trends Ecol. Evol. 6:325–329.

Hausdorf, B. 1998. Weighted ancestral area analysis and a solution of the redundant distribution problem. Syst. Biol. 47:445–456.

Hausdorf, B. 2002. Units in biogeography. Syst. Biol. 51:648–652.

Hausfater, G., and S. B. Hrdy. 1984. Infanticide: comparative and evolutionary perspectives. Aldine, New York.

Hausfater, G., and D. F. Watson. 1976. Social and reproductive correlates of parasite ova emissions by baboons. Nature 262:688–689.

Hawkey, C. M. 1966. Plasminogen activator in the saliva of the vampire bat, *Desmodus rotundus*. Nature 211:434–435.

Hawkins, B.A. 2001. Ecology's oldest pattern? Trends Ecol. Evol. 16:470.

Hawkins, R. E., L. D. Martoglio, and G. G.

Montgomery. 1968. Cannon-netting deer. J. Wildl. Mgmt. 32:191–195.

Hay, K. A., and A. W. Mansfield. 1989. Narwhal Monodon monoceros Linnaeus, 1758. Pp. 145–176 in Handbook of marine mammals, vol. 4: river dolphins and the larger toothed whales (S. H. Ridgway and R. Harrison, eds.). Academic Press, New York.

Hayes, A. R., and N. J. Huntly. 2005. Effects of wind on the behavior and call transmission of pikas (*Ochotona princeps*). J. Mammal. 86:974–981.

Hayes, J. P., and T. Garland, Jr. 1995. The evolution of endothermy: testing the aerobic capacity model. Evolution 49:836–847.

Hayes L. D. 2000. To nest communally or not to nest communally: a review of rodent communal nesting and nursing. Anim. Behav. 59: 677–688.

Hays, J. P. 2001. Mass-specific and whole-animal metabolism are not the same concept. Physiol. Biochem. Zool. 74:147–150.

Hays, J. P., and J. S. Shonkwiler. 1996. Analyzing mass-independent data. Physiol. Zool. 69:974–980.

Hays, W. S. T., and W. Z. Lidicker, Jr. 2000. Winter aggregations, Dehnel effect, and habitat relations in the Suisan shrew *Sorex ornatus sinuosus*. Acta Theriol. 45:433–442.

Haysen, V., and R. C. Lacy. 1985. Basal metabolic rates in mammals: taxonomic differences in the allometry of BMR and body mass. Comp. Biochem. Physiol. 81A:741–754.

Hayssen, V. 1993. Empirical and theoretical constraints on the evolution of lactation. J. Dairy Sci. 76:3213–3233.

Hayssen, V., and T. H. Kunz. 1996. Allometry of litter mass in bats: maternal size, wing morphology, and phylogeny. J. Mammal. 77:476–490.

Hayssen, V., A. van Tienhoven, and A. van Tienhoven. 1993. Asdell's patterns of mammalian reproduction. Comstock Publ. Assoc., Ithaca, NY.

Haythornthwaite, A. S., and C. R. Dickman. 2006. Distribution, abundance, and individual strategies: a multi-scale analysis of dasyurid marsupials in arid central Australia. Ecography 29:285–300.

Hayward, J. S., and P. A. Lisson. 1992. Evolution of brown fat: its absence in marsupials and monotremes. Can. J. Zool. 70:171–179.

Hayward, J. S., and C. P. Lyman. 1967. Nonshivering heat production during arousal from hibernation and evidence for the contribution of brown fat. Pp. 346–355 in Mammalian hibernation III (K. C. Fisher, A. R. Dawe, C. P. Lyman, E. Schönbaum, and F. E. South, eds.). Oliver and Boyd, Edinburgh.

Heaney, L. R. 1978. Island area and body size of insular mammals: evidence from the tri-colored squirrel (*Callosciurus prevosti*) of Southeast Asia. Evolution 32:29–44.

Heaney, L. R. 1991. A synopsis of climatic and vegetational change in Southeast Asia. Climatic Change 19:53–61.

Heath, M. E. 1992. *Manis temminckii*. Mammal. Species No. 415:1–5.

Heath, M. E. 1995. *Manis crassicaudata*. Mammal. Species No. 513:1–4.

Heatwole, H. 1965. Some aspects of the association of cattle egrets with cattle. Anim. Behav. 13:79–83.

Hedrick, P. W. 2000. Genetics of populations, 2nd edition. Jones and Bartlett Publishers, Sudbury, MA.

Hedricks, C., and M. K. McClintock. 1990. Timing of insemination is correlated with the secondary sex ratio of Norway rats. Physiol. Behav. 48:625–632.

Heideman, P. D. 1988. The timing of reproduction in the fruit bat *Haplonycteris fischeri* (Pteropodidae): geographic variation and delayed development. J. Zool. 215:577–595.

Heideman, P. D. 1989. Delayed development in Fischer's pygmy fruit bat, *Haplonycteris fischeri*, in the Philippines. J. Reprod. Fert. 85:363–382.

Heideman, P. D , J. A. Cummings, and L. R. Heaney. 1993. Reproductive timing and early embryonic development in an Old World fruit bat, *Otopteropus cartilagonodus* (Megachiroptera). J. Mammal. 74:621–630.

Heideman, P. D., and K. S. Powell. 1998. Age-specific reproductive strategies and delayed embryonic development in an Old World fruit bat, *Ptenochirus jagori*. J. Mammal. 79:295–311.

Heinsohn, R., and C. Packer. 1995. Complex cooperative strategies in group-territorial African lions. Science 269:1260–1262.

Heithaus, M. R. 2001a. Shark attacks on bottlenose dolphins (*Tursiops truncatus*) in Shark Bay, Western Australia: attack rate, bite scar frequencies, and attack seasonality. Mar. Mammal Sci. 17:526–539.

Heithaus, M. R. 2001b. Predator-prey and competitive interactions between sharks (order Selachii) and dolphins (suborder Odontoceti): a review. J. Zool. 253:53–68.

Heithaus, M. R., and L. M. Dill. 2002. Feeding strategies and tactics. Pp. 412–421 in Encyclopaedia of marine mammals (W. F. Perrin, B. Wursig, and J. G. M. Thewissen, eds.). Academic Press, San Diego.

Heldmaier, G. 1971. Relationship between non-shivering thermogenesis and body size. Pp. 73–80 in Non-shivering thermogenesis (L. Jansky, ed.). Swets and Zeitlinger, Amsterdam.

Helgen, K. M. 2005. Order Scandentia. Pp. 104–109 in Mammal species of the world: a taxonomic and geographic reference, 3rd edition, vol. 1 (D.E. Wilson and D. M. Reeder, eds.). Johns Hopkins Univ. Press, Baltimore.

Heller, H. C., and T. L. Poulson. 1970. Circannian rhythms, II: Endogenous and exogenous factors controlling reproduction and hibernation in chipmunks (*Eutamias*) and ground squirrels (*Spermophilus*). Comp. Biochem. Physiol. 33:357–383.

Hellgren, E. C., and J. A. Bissonette. 2003. Collared peccary, Tayassu tajuca. Pp. 867-576 in Wild mammals of North America: biology, management, and conservation (G. A. Feldhamer, B. C. Thompson, and J. A. Chapman, eds.). Johns Hopkins Univ. Press, Baltimore.

Hellgren, E. C., M. R. Vaughan, R. L. Kirkpatrick, and P. F. Scanlon. 1990. Serial changes in metabolic correlates of hibernation in female black bears. J. Mammal. 71:291–300.

Hellgren, E. C., D. R. Synatzske, P. W. Oldenburg, and F. S. Guthery. 1995. Demography of a collared peccary population in south Texas. J. Wildl. Mgmt. 59:153–163.

Hemmer, H. 1990. Domestication: the decline of environmental appreciation. Cambridge Univ. Press, Cambridge, UK.

Hennig, W. 1966. Phylogenetic systematics. Univ. Illinois Press, Urbana.

Hensel, R. J., W. A. Troyer, and A. W. Erickson. 1969. Reproduction in the female brown bear. J. Wildl. Mgmt. 33:357–365.

Henshaw, R. E., L. S. Underwood, and T. M. Casey. 1972. Peripheral thermoregulation: foot temperature in two arctic canines. Science 175:988–990.

Hensley, A. P., and K. T. Wilkins. 1988. *Leptonycteris nivalis*. Mamm. Species 307:1–4.

Herbert, J. 1972. Initial observations on pinealectomized ferrets kept for long periods in either daylight or artificial illumination. J. Endocrinol. 55:591–597.

Herbst, M., and N. C. Bennett. 2006. Burrow architecture and burrowing dynamics of the endangered Namaqua dune mole rat (*Bathyergus janetta*) (Rodentia: Bathyergidae). J. Zool. Lond. 270:420–428.

Herman, L. M., and W. N. Tavolga. 1980. The communication systems of cetaceans. Pp. 149–209 in Cetacean behavior (L. M. Herman, ed.). Wiley Interscience, New York.

Hermanson, J. W., M. A. Cobb, W. A. Schutt, F. Muiradali, and J. M. Ryan. 1993. Histochemical and myosin composition of vampire bat (*Desmodus rotundus*) pectoralis muscle targets a unique locomotory niche. J. Morphol. 217:347–356.

Hershkovitz, P. 1962. Evolution of Neotropical cricitine rodents (Muridae) with special reference to the phyllotine group. Fieldiana 46:1–524.

Hershkovitz, P. 1971. Basic crown patterns and cusp homologies of mammalian teeth. Pp. 95–150 in Dental morphology and evolution (A. A. Dahlberg, ed.). Univ. Chicago Press, Chicago.

Hershkovitz, P. 1999. *Dromiciops gliroides* Thomas, 1894, last of the Microbiotheria (Marsupialia), with a review of the family Microbiotheriidae. Fieldiana Zool. 93:1–60.

Hess, H. H. 1962. History of ocean basins. Pp. 599–620 in Petrological studies: a volume in honor of A. F. Buddington (A. E. J. Engel, H. L. James, and B .F. Leonard, eds.). Geological Society of America, New York.

Hesse, R. 1937. Ecological animal geography. John Wiley and Sons, New York.

Hestbeck, J. B. 1982. Population regulation of cyclic mammals: the social fence hypothesis. Oikos 39:157–163.

Hestbeck, J. B. 1988. Population regulation of cyclic mammals: a model of the social fence hypothesis. Oikos 52:156–168.

Hewison A. J. M., J. M. Gaillard, P. Kjellander, C. Toigo, O. Liberg, and D. Delorme. 2005. Big mothers invest more in daughters—reversed sex allocation in a weakly polygynous mammal. Ecol. Lett. 8: 430–437.

Heyning, J. E. 1989. Comparative facial anatomy of beaked whales (Ziiphidae) and a systematic revision among the families of extant Odontoceti. Nat. Hist. Mus. Los Angeles County No. 405:1–64.

Heyning, J. E., and J. G. Mead. 1996. Suction feeding in beaked whales: morphological and observational evidence. Nat. Hist. Mus. Los Angeles County No. 464:1–12.

Hickman, C. P., Jr., L. S. Roberts, and A. Larson. 1997. Integrated principles of zoology, 9th edition. Wm. C. Brown, Dubuque, IA.

Hickman, C. P., Jr., L. S. Roberts, and A. Larson. 2000. Animal diversity, 2nd edition. McGraw-Hill, Boston.

Hickman, C. P., Jr., L. S. Roberts, A. Larson, and H. I. Anson. 2004. Integrated principles of zoology. 12th edition. McGraw-Hill, New York.

Hiendleder, S., B. Kaupe, R. Wassmuth, and A. Janke. 2002. Molecular analysis of wild and domestic sheep questions current nomenclature and provides evidence for domestication from two different subspecies. Proc. Roy. Soc. Lond. Ser. B 269:893–904.

Hildebrand, M. 1968. Anatomical preparations. Univ. Calif. Press, Berkeley.

Hildebrand, M. 1980. The adaptive significance of tetrapod gait selection. Am. Zool. 20:255–267.

Hildebrand, M. 1985a. Digging of quadrupeds. Pp. 38–57 in Functional vertebrate morphology (M. Hildebrand, D. M. Bramble, K. F. Liem, and D. B. Wake, eds.). Harvard Univ. Press, Cambridge, MA.

Hildebrand, M. 1985b. Walking and running. Pp. 89–109 in Functional vertebrate morphology (M. Hildebrand, D. M. Bramble, K. F. Liem, and D. B. Wake, eds.). Harvard Univ. Press, Cambridge, MA.

Hildebrand, M. 1995. Analysis of vertebrate structure, 4th edition. John Wiley and Sons, New York.

Hildebrand, M., D. M. Bramble, K. F. Liem, and D. B. Wake (eds.). 1985. Functional vertebrate morphology. Harvard Univ. Press, Cambridge, MA.

Hildebrand, M., G. E. Goslow, Jr., and V. Hildebrand. 2001. Analysis of vertebrate structure, 5th edition. John Wiley, New York.

Hill, J. E. 1974. A new family, genus, and species of bat (Mammalia: Chiroptera) from Thailand. Bull. Br. Mus. Nat. Hist. Zool. 27:301–336.

Hill, J. E., and S. D. Smith. 1984. Bats: A natural history. British Museum, London.

Hill, J. E., and S. E. Smith. 1981. Craseonycteris thonglongyai. Mammal. Species No. 160:1–4.

Hill, R. W. 1983. Thermal physiology and energetics of Peromyscus; ontogeny, body temperature, metabolism, insulation, and microclimatology. J. Mammal. 64:19–37.

Hill, R. W., and G. A. Wyse. 1989. Animal physiology, 2nd edition. Harper and Row, New York.

Hill, R. W., D. P. Christian, and J. H. Veghte. 1980. Pinna temperature in exercising jackrabbits. J. Mammal. 61:30–38.

Hillis, D. M., B. K. Mable, A. Larson, S. K. Davis, and E. A. Zimmer. 1996a. Nucleic acids IV: sequencing and cloning. Pp. 321–381 in Molecular systematics, 2d edition (D.M. Hillis, C. Moritz, and B.K. Mable, eds.). Sinauer Associates, Sunderland, MA.

Hillis, D. M., B. K. Mable, and C. Moritz. 1996b. Applications of molecular systematics: the state of the field and a look to the future. Pp. 515–543 in Molecular systematics, 2nd edition (D. M. Hillis, B. K. Mable, and C. Moritz, eds.). Sinauer Associates, Sunderland, MA.

Hilton-Taylor, C. 2000. IUCN red list of threatened species. IUCN, Gland, Switzerland.

Himms-Hagen, J. 1985. Brown adipose tissue metabolism and thermogenesis. Annu. Rev. Nutrition 5:69–94.

Hinds, D. S., and R. E. MacMillen. 1985. Scaling of energy metabolism and evaporative water loss in heteromyid rodents. Physiol. Zool. 58:282–298.

Hines, E., K. Adulyanukosol, D. Duffus, and P. Dearden. 2005. Community perspectives and conservation needs for dugongs (Dugong dugon) along the Andaman coast of Thailand. Env. Mgmt. 36:654–664.

Hirakawa, H. 2001. Coprophagy in leporids and other mammalian herbivores. Mammal Rev. 31:61–80.

Hiruki, L. M., M. K. Schwartz, and P. L. Boveng. 1999. Hunting and social behaviour of leopard seals (Hydrurga leptonyx) at Seal Island, South Shetland Island, Antarctica. J. Zool. Lond. 249:97–109.

Hiryu, S., K. Katsura, L-K. Lin, H. Riquimaroux, and Y. Watanabe. 2005. Doppler-shift compensation in the Taiwanese leaf-nosed bat (Hipposideros terasensis) recorded with a telemetry microphone system during flight. J. Acoustical Soc. Am. 118:3927–3933.

Hissa, R. 1997. Physiology of the European brown bear (Ursus arctos arctos). Ann. Zool. Fennici 34:267–287.

Hissa, R., J. Siekkinen, E. Hohtola, S. Saarela, A. Hakala, and J. Pudas. 1994. Seasonal patterns in the physiology of the European brown bear (Ursus arctos arctos) in Finland. Comp. Biochem. Physiol. Part A. Sensory, Neural, and Behav. Physiol. 109:781–791.

Ho, Y. Y., and L. L. Lee. 2003. Roost selection by Formosan leaf-nosed bats (Hipposideros armiger terasensis). Zool. Sci. (Tokyo) 20:1017–1024.

Hoare, R. E., and J. Du Toit. 1999. Coexistence between people and elephants in African savannas. Conserv. Biol. 13:633–639.

Hobbs, R. J. 2001. Synergisms among habitat fragmentation, livestock grazing, and biotic invasions in southwestern Australia. Conserv. Biol. 15:1522–1528.

Hoekstra, H. E., J. G. Krenz, and M. W. Nachman. 2005. Local adaptation in the rock pocket mouse (Chaetodipus intermedius): natural selection and phylogenetic history of populations. Heredity 94:217–228.

Hoelzel A. R., R. C. Fleischer., C. Campagna, B. J. Le Boeuf, and G. Alvord. 2002. Impact of a population bottleneck on symmetry and genetic diversity in the northern elephant seal. J. Evol. Biol. 15: 567–575.

Hoffmeister, D. F. 1969. The first fifty years of the American Society of Mammalogists. J. Mammal. 50:794–802.

Hoffmeister, D. F. 1989. Mammals of Illinois. Univ. Illinois Press, Urbana.

Hogg, J. T. 1988. Copulatory tactics in relation to sperm competition in Rocky Mountain bighorn sheep. Behav. Ecol. Sociobiol. 22:49–59.

Hohtola, E. 2004. Shivering thermogenesis in birds and mammals. Pp. 241–252 in Life in the cold: evolution, mechanisms, adaptation, and application. Twelfth International Hibernation Symposium. Biological Papers of the Univ. Alaska, 7. Institute of Arctic Biology, Univ. Alaska, Fairbanks.

Holden, M. E. 2005. Family Gliridae. Pp. 819–841 in Mammal species of the world: a taxonomic and geographic reference. 3rd edition (D. E. Wilson and D. M. Reeder, eds.). Johns Hopkins Univ. Press, Baltimore.

Holden, M. E., and G. G. Musser. 2005. Family Dipodidae. Pp. 871–893 in Mammal species of the world: a taxonomic and geographic reference. 3rd edition (D. E. Wilson and D. M. Reeder, eds.). Johns Hopkins Univ. Press, Baltimore.

Holder, M., and P. O. Lewis. 2003. Phylogeny estimation: traditional and Bayesian approaches. Nat. Rev. Genet. 4:275–284.

Holekamp, K. E., and P. W. Sherman. 1989. Why male ground squirrels disperse. Am. Sci. 77:232–239.

Holekamp, K. E., L. Smale, R. Berg, and S. M. Cooper. 1997. Hunting rates and hunting success in the spotted hyena (Crocuta crocuta). J. Zool. Lond. 242:1–15.

Holekamp, K. E., M. Szykman, E. E. Boydston, and L. Smale. 1999. Association of seasonal reproductive patterns with changing food availability in an equatorial carnivore, the spotted hyaena (Crocuta crocuta). J. Repro. Fertil. 116:87–93.

Holland, R. A., D. A. Waters, and J. M. V. Rayner. 2004. Echolocation signal structure in the megachiropteran bat Rousettus aegyptiacus Geoffroy 1810. J.

Exp. Biol. 207:4361–4369.

Holling, C. S. 1959. The components of predation as revealed by a study of small mammal predation of the European pine sawfly. Can. Entomol. 91:293–320.

Holmala, K. and K. Kauhala. 2006. Ecology of wildlife rabies in Europe. Mammal Rev. 36:17–36.

Holmes, W. G. 1988. Kinship and development of social preferences. Pp. 389–413 in Developmental psychobiology and behavioral ecology (E. M. Blass, ed.). Plenum Press, New York.

Holmes, W. G. 1995. The ontogeny of littermate preferences in juvenile golden-mantled ground squirrels: effects of rearing and relatedness. Anim. Behav. 50:309–322.

Holmes, W. G., and P. W. Sherman. 1983. Kin recognition in animals. Am. Sci. 71:46–55.

Holroyd, P. A., and J. C. Mussell. 2005. Macroscelidea and Tubulidentata. Pp. 71–83 in The rise of placental mammals: origins and relationships of the major extant clades (K. D. Rose and J. D. Archibald, eds.). Johns Hopkins Univ. Press, Baltimore, MD.

Holst, M., I. Stirling, and K. A. Hobson. 2001. Diet of ringed seals (Phoca hispida) on the east and west sides of the North Water Polynya, northern Baffin Bay. Mar. Mammal Sci. 17:888–908.

Honeycutt, R. L. 1992. Naked mole-rats. Am. Sci. 80:43–53.

Honeycutt, R. L., K. Nelson, D. A. Schlitter, and P. W. Sherman. 2000. Genetic variation within and among populations of the naked mole rat: evidence from nuclear and mitochondrial genomes. Pp. 195–208 in The biology of the naked mole rat (P. W. Sherman, J. U. M. Jarvis, and R. D. Alexander, eds.). Princeton Univ. Press, Princeton, NJ.

Hoogland, J. L. 1979a. Aggression, ectoparasitism, and other possible costs of prairie dog (Sciuridae: Cynomys spp.) coloniality. Behaviour 69:1–35.

Hoogland, J. L. 1979b. The effect of colony size on individual alertness of prairie dogs (Sciuridae: Cynomys spp.). Anim. Behav. 27:394–407.

Hoogland, J. L. 1982. Prairie dogs avoid extreme inbreeding. Science 215:1639–1641.

Hoogland, J. L. 2003. Prairie dogs, Cynomys ludovicianus and allies. Pp. 232–247 in Wild mammals of North America: biology, management, and conservation. (G. A. Feldhamer, B. C. Thompson, and J. A. Chapman, eds.). Johns Hopkins Univ. Press, Baltimore.

Hooker, J. J. 2005. Perissodactyla. Pp. 199–214 in The rise of placental mammals: origins and relationships of the major extant clades (K. D. Rose and J. D. Archibald, eds.). Johns Hopkins Univ. Press, Baltimore.

Hope, P. J., D. Pyle, C. B. Daniels, I. Chapman, M. Horowitz, J. E. Morley, P. Trayhurn, J. Kumaratilake, and G. Wittert. 1997. Identification of brown fat and mechanisms for energy balance in the marsupial, Sminthopsis crassicaudata. Am. J. Physiol. 273:R161–R167.

Hopkins, D. R. 1992. Homing in on helminths. Am. J. Tropical Med. Hyg. 46:626–634.

Hopla, C. E., and A. K. Hopla. 1994. Tularemia. Pp. 113–126 in Handbook of zoonoses, Section A: bacterial, rickettsial, chlamydial, and mycotic diseases, 2nd edition (C. W. Beran and J. H. Steele, eds.). CRC Press, Boca Raton, FL.

Hopson, J. A. 1991. Systematics of the nonmammalian Synapsida and implications for patterns of evolution in synapsids. Pp. 635–693 in Origins of the higher groups of tetrapods: controversy and consensus (H.-P. Schultze and L. Trueb, eds.). Comstock Pub. Assoc., Ithaca, NY.

Hopson, J. A. 1994. Synapsid evolution and the radiation of non-eutherian mammals. Pp. 190–219 in Major features in vertebrate evolution: short courses in paleontology (R. S. Spencer, ed.). No. 7, Paleontological Society, Univ. Tennessee, Knoxville.

Hopson, J. A. 1995. Patterns of evolution in the manus and pes of non-mammalian therapsids. J. Vert. Paleontol. 15:615–639.

Horner, B. E., J. M. Taylor, A. V. Linzey, and G. R. Michener. 1996. Women in mammalogy (1940–1994): personal perspectives. J. Mammal. 77:655–674.

Hoscak, D. A., K. V. Miller, R. L. Marchinton, C. M. Wemmer, and S. L. Montfort. 1998. Stag exposure augments urinary progestagen excretion in Eld's deer hinds (*Cervus eldi thamin*). Mammalia 62:341–350.

Hoss, M., A. Dilling, A. Current, and S. Paabo. 1996. Molecular phylogeny of the extinct ground sloth Mylodon darwinii. Proc. Nat. Acad. Sci. USA 93:181–185.

Hotton, N., III, P. D. MacLean, J. J. Roth, and E. C. Roth (eds.). 1986. The ecology and biology of mammal-like reptiles. Smithsonian Institution Press, Washington, DC.

Houck, M. L., S. C. Kingswood, and A. T. Kumamoto. 2000. Comparative cytogenetics of tapirs, genus *Tapirus* (Perissodactyla, Tapiridae). Cytogenet. Cell Genet. 89:110–115.

Houck, M. L., A. T. Kumamoto, D. S. Gallagher, and K. Benirschke. 2001. Comparative cytogenetics of the African elephant (*Loxodonta africana*) and Asian elephant (*Elephus maximus*). Cytogenet. Cell Genet. 93:249–252.

Hoyle, S. D., A. R. Pople, and G. J. Toop. 2001. Mark-recapture may reveal more about ecology than about population trends: demography of a threatened ghost bat (*Macroderma gigas*) population. Austral. Ecol. 26:80–92.

Hovland, N., and H. P. Andreassen. 1995. Fluorescent powder as dye in bait for studying foraging areas in small mammals. Acta Theriol. 40:315–320.

Hoyt, J. A. 1994. Animals in peril. Avery, Garden City Park, NY.

Hoyt, R. A., and R. J. Baker. 1980. Natalus major. Mammal. Species No. 130:1–3.

Hrdina, F., and G. Gordon. 2004. The koala and possum trade in Queensland, 1906–1936. Aust. Zool. 32:543–585.

Hrdy, S. B. 1977a. Infanticide as a primate reproductive strategy. Am. Sci. 65:40–49.

Hrdy, S. B. 1977b. Langurs of Abu: female and male strategies of reproduction. Harvard Univ. Press, Cambridge, MA.

Hsu, M. J., D. W. Garton, and J. D. Harder. 1999. Energetics of offspring production: a comparison of a marsupial (*Monodelphis domestica*) and a eutherian (*Mesocricetus auratus*). J. Comp. Physiol. B 169:67–76.

Hsu, T. C. 1979. Human and mammalian cytogenetics. Springer-Verlag, Berlin.

Hu, Y., Y. Wang, Z. Luo, and C. Li. 1997. A new symmetrodon mammal from China and its implications for mammalian evolution. Nature 390:137–142.

Hubbard, A. L., S. McOrist, T. W. Jones, R. Boid, R. Scott, and N. Easterbee. 1992. Is survival of European wild cats *Felis silvestris* in Britain threatened by interbreeding with domestic cats? Biol. Conserv. 61:203–208.

Huchon, D., F. Delsuc, F. M. Catzeflis, and E. J. P. Douzery. 1999. Armadillos exhibit less genetic polymorphism in North America than in South America: nuclear and mitochondrial data confirm a founder effect in *Dasypus novemcinctus* (Xenarthra). Mol. Ecol. 8:1743–1748.

Huck, U. W., J. Seger, and R. D. Lisk. 1990. Litter sex ratios in the golden hamster vary with time of mating and litter size and are not binomially distributed. Behav. Ecol. Sociobiol. 26:99–109.

Hudson, J. W. 1973. Torpidity in mammals. Pp. 97–165 *in* Comparative physiology of thermoregulation (G. C.Whittow, ed.). Academic Press, New York.

Hudson, J. W. 1978. Shallow, daily torpor: a thermoregulatory adaptation. Pp. 67–108 *in* Strategies in cold: natural torpidity and thermogenesis (L. C. H. Wang and J. W. Hudson, eds.). Academic Press, New York.

Hudson, R. J., and R. G. White. 1985. Bioenergetics of wild herbivores. CRC Press, Boca Raton, FL.

Hudson, R. J., K. R. Drew, and L. M. Baskin. 1989. Wildlife production systems: economic utilization of wild ungulates. Cambridge Univ. Press, Cambridge, UK.

Hughes, J. J., D. Ward, and M. R. Perrin. 1994. Predation risk and competition affect habitat selection and activity of Namib desert gerbils. Ecology 75:1397–1405.

Hughes, R. L. 1984. Structural adaptations of the eggs and the fetal membranes of monotremes and marsupials for respiration and metabolic exchange. Pp. 389–421 *in* Respiration and metabolism of embry-

onic vertebrates (R. S. Seymour, ed.). Junk, Doordrecht.

Hughes, R. L., F. N. Carrick, and C. D. Shorey. 1975. Reproduction in the platypus, *Ornithorhynchus anatinus*, with particular reference to the evolution of viviparity. J. Reprod. Fert. 43:374–375.

Hui, C. A. 1994. Lack of association between magnetic patterns and the distribution of free-ranging dolphins. J. Mammal. 75:399–405.

Huihua, Z., Z. Shuyi, Z. Mingxue, and Z. Jiang. 2003. Correlations between call frequency and ear length in bats belonging to the families Rhinolophidae and Hipposideridae. J. Zool. Lond. 259:189–195.

Huelsenbeck, J.P., B. Rannala, and B. Larget. 2000. A Bayesian framework for the analysis of cospeciation. Evolution 54:352–364.

Hume, I. D. 1989. Optimal digestive strategies in mammalian herbivores. Physiol. Zool. 62:1145–1163.

Hume, I .D. 1994. Gut morphology, body size and digestive performance in rodents. Pp. 315–323 *in* The digestive system in mammals: food, form, and function (D. J. Chivers and P. Langer, eds.). Cambridge Univ. Press, Cambridge, UK.

Hume, I. D. 2003. Nutrition of carnivorous marsupials. Pp. 221–228 *in* Predators with pouches: the biology of carnivorous marsupials (M. Jones, C. Dickman, and M. Archer, eds.). CSIRO Publishing, Collingwood, Australia

Humphries, C. J., and L. R. Parenti. 1999. Cladistic biogeography, 2nd edition. Oxford Univ. Press, New York.

Huntly, N., and R. S. Inouye. 1987. Small mammal populations of an old-field chronosequence: successional patterns and associations with vegetation. J. Mammal. 68:739–745.

Hurst, R. N., and J. E. Wiebers. 1967. Minimum body temperature extremes in the little brown bat, *Myotis lucifugus*. J. Mammal 48:465.

Hutcheon, J. M., and T. Garland, Jr. 2004. Are megabats big? J. Mammal. Evol. 11:257–277.

Hutcheon, J. M., and J. A. W. Kirsch. 2004. Camping in a different tree: results of molecular systematic studies of bats using DNA-DNA hybridization. J. Mammal. Evol. 11:17–47.

Hutchinson, G. E. 1957. Concluding remarks. Cold Spring Harbor Symp. Quant. Biol. 22:415–427.

Hutchinson, G. E. 1959. Homage to Santa Rosalia, or why are there so many kinds of animals? Am. Nat. 93:145–159.

Hutterer, R. 2005a. Order Erinaceomorpha. Pp. 212–219 *in* Mammal species of the world: a taxonomic and geographic reference, 3rd edition, vol. 1 (D. E. Wilson and D. M. Reeder, eds.). Johns Hopkins Univ. Press, Baltimore.

Hutterer, R. 2005b. Order Soricomorpha. Pp. 220–311 *in* Mammal species of the world: a taxonomic and geographic reference, 3rd edition, vol. 1 (D.E. Wilson and D. M. Reeder, eds.). Johns Hopkins Univ. Press, Baltimore.

Hutterer, R. 2005c. Homology of unicuspids and tooth nomenclature in shrews. Pp. 397–404 *in* Advances in the biology of shrews II (J. F. Merritt, S. Churchfield, R. Hutterer, and B. I. Sheftel, eds.). Special Publication of the International Society of Shrew Biologists 01, New York.

Hwang, Y. T., S. Larivière, and F. Messier. 2007. Energetic consequences and ecological significance of heterothermy and social thermoregulation in striped skunks (*Mephitis mephitis*). Physiol. Biochem. Zool. 80:138–145.

Hyvärinen, H. 1969. On the seasonal changes in the skeleton of the common shrew (*Sorex araneus* L.) and their physiological background. Aquilo Ser. Zool. 7:1–32.

Hyvärinen, H. 1984. Winter strategy of voles and shrews *in* Finland. Pp. 139–148 in Winter ecology of small mammals (J. F. Merritt, ed.). Spec. Pub., Carnegie Mus. Nat. Hist. 10:1–380.

Hyvärinen, H. 1994. Brown fat and the wintering of shrews. Pp. 259–266 *in* Advances in the biology of shrews (J. F. Merritt, G. L. Kikland, and R. K. Rose, eds.). Spec. Pub. No. 18, Carnegie Museum of Natural History, Pittsburgh.

Ibanez, C., J. Juste, J. L. Garcia-Mudarra, and P. T. Agirre-Mendi. 2001. Bat predation on nocturnally migrating birds. Proc. Natl. Acad. Sci. USA 98:9700–702.

Illius, A.W., P. Duncan, C. Richard, and P. Mesochina. 2002. Mechanisms of functional response and resource exploitation in browsing roe deer. J. Anim. Ecol. 71:723–734.

Ilse, L. M., and E. C. Hellgren. 1995. Spatial use and group dynamics of sympatric collared peccaries and feral hogs in southern Texas. J. Mammal. 76:993–1002.

Inns, R. W. 1982. Seasonal changes in the accessory reproductive system and plasma testosterone levels of the male tamar wallaby, *Macropus eugenii*, in the wild. J. Reprod. Fert. 66:675–680.

International Commission on Zoological Nomenclature. 1999. International code of zoological nomenclature, 4th ed. Natural History Museum, London.

International Union for the Conservation of Nature. 1988. Red list of threatened animals. IUCN, Cambridge, UK.

International Union for the Conservation of Nature. 2000. 2000 IUCN red list of threatened species. IUCN, Cambridge, UK.

Irving, L. 1972. Arctic life of birds and mammals, including man. Springer-Verlag, Berlin.

Isaac, E. 1970. Geography of domestication. Prentice-Hall, Englewood Cliffs, NJ.

Ivanoff, D. V. 2001. Partitions in the carnivoran auditory bulla: their formation and significance for systematics. Mammal Rev. 31:1–16.

Iverson, S. J. 1993. Milk secretion in marine mammals in relation to foraging: can milk fatty acids predict diet? Symp. Zool. Soc. Lond. 66:263–291.

Iverson, S. J., J. P. Y. Arnould, and I. L. Boyd. 1997. Milk fatty acid signatures indicate both major and minor shifts in the diet of lactating Antarctic fur seals. Can. J. Zool. 75:188–197.

Izzo, A. D., M. Meyer, J. M. Trappe, M. North, and T. D. Bruns. 2005. Hypogeous ectomycorrhizal fungal species on roots and in small mammal diets in a mixed-conifer forest. Forest Sci. 51:243–254.

Jackson, S. M. 1999. Glide angle in the genus *Petaurus* and a review of gliding in mammals. Mammal Rev. 30:9–30.

Jacobs, D. S., and S. Kuiper. 2000. Individual recognition in the Damaraland mole-rat, *Cryptomys damarensis* (Rodentia: Bathyergidae). J. Zool. Lond. 251:411–415.

Jacobs, L., S. J. C. Gaulin, D. Sherry, and G. E. Hoffman. 1990. Evolution of spatial cognition: Sex-specific patterns of spatial behavior predict hippocampal size. Proc. Natl. Acad. Sci. USA 87:6349–6352.

Jacobs, L. F. 1992. Memory for cache locations in Merriam's kangaroo rats. J. Mammal. 43:585–593.

Jacobs, L. L. 1980. Siwalik fossil tree shrews. Pp. 205–216 *in* Comparative biology and evolutionary relationships of tree shrews (W. P. Luckett, ed.). Plenum Press, New York.

Jacquot, J. J., and N.G. Solomon. 2004. Experimental manipulation of territory occupancy: effects on immigration of female prairie voles. J. Mammal. 85:1009–1014.

Jacquot, J. J., and S. H. Vessey. 1994. Non-offspring nursing in the white-footed mouse, *Peromyscus leucopus*. Anim. Behav. 48:1238–1240.

Jaeger, M. M., R. K. Pandit, and E. Haque. 1996. Seasonal differences in territorial behavior by golden jackals in Bangladesh: howling versus confrontation. J. Mammal. 77:768–775.

James, F. C. 1970. Geographic size variation in birds and its relationship to climate. Ecology 51:365–390.

James, W. H. 1996. Evidence that mammalian sex ratios at birth are partially controlled by parental hormone levels at the time of conception. J. Theor. Biol. 180:271–286.

Jameson, E. W., Jr. 1999. Host-ectoparasite relationships among North American chipmunks. Acta Theriol. 44:225 231.

Janke, A., O. Magnell, G. Wieczorek, M. Westerman, and U. Arnason. 2002. Phylogenetic analysis of 18s, rRNA and mitochondrial genomes of the wombat, *Vombatus ursinus*, and the spiny anteater, *Tachyglossus aculeatus*: increased support for the Marsupionta hypothesis. J. Mol. Evol. 54:71–80.

Jansen, T., P. Forster, M. A. Levine, H. Oelke, M. Hurles, C. Renfrew, J. Weber, and K. Olek. 2002. Mitochondrial DNA and the origins of the domestic horse. Proc. Natl. Acad. Sci. USA 99:10905–10910.

Jansky, L. 1973. Non-shivering thermogenesis and its thermoregulatory significance. Biol. Rev. 48:85–132.

Janson, C. H., and M. L. Goldsmith. 1995. Predicting group-size in primates—foraging costs and predation risks. Behav. Ecol. 6:326–336.

Janzen, D. H. 1985. Coevolution as a process: what parasites of animals and plants do not have in common. Pp. 83–99 in Coevolution of parasitic arthropods and mammals (K. C. Kim, ed.). John Wiley and Sons, New York.

Jaquet, N., S. Dawson, and L. Douglas. 2001. Vocal behavior of male sperm whales: why do they click? J. Acoust. Soc. Am. 109:2254–2259.

Jarman, P. J. 1974. The social organization of antelope in relation to their ecology. Behaviour 48:215–267.

Jarvis, J. U. M. 1981. Eusociality in a mammal: cooperative breeding in naked mole-rat colonies. Science 212:571–573.

Jarvis, J. U. M., M. J. O'Riain, N. C. Bennett, and P. W. Sherman. 1994. Mammalian eusociality: a family affair. Trends Ecol. Evol. 9:47–51.

Jay, C. V., S. D. Farley, and G. W. Garner. 2001. Summer diving behavior of male walruses in Bristol Bay, Alaska. Mar. Mammal Sci. 17:617–631.

Jefferson, T. A., S. Leatherwood, and M. A. Webber. 1993. Marine mammals of the world. United Nations Environment Programme, FAO, Rome.

Jeffreys, A. J. 1987. Highly variable minisatellites and DNA fingerprints. Biochem. Soc. Trans. 15:309–317.

Jenkins, F. A., Jr., S. M. Gatesy, N. H. Shubin, and W. W. Amaral. 1997. Haramiyids and Triassic mammalian evolution. Nature 385:715–718.

Jenkins, P. D. 1987. Catalogue of primates in the British Museum (Natural History) and elsewhere in the British Isles, Part 4: Suborder Strepsirhini, including the subfossil Madagascan lemurs and family Tarsiidae. British Museum of Natural History, London.

Jenkins, S. H., and S. W. Breck. 1998. Differences in food hoarding among six species of heteromyid rodents. J. Mammal. 79:1221–1233.

Jenness, R., and E. H. Studier. 1976. Lactation and milk. Pp. 201–218 in Biology of bats of the New World family Phyllostomatidae, part I. (R. J. Baker, J. K. Jones, Jr., and D. C. Carter, eds.). Spec. Pub. No. 10, Museum Texas Tech Univ., Lubbock

Jensen, R. G. (ed.). 1995. Handbook of milk composition. Academic Press, San Diego.

Jepson, G. L. 1966. Early Eocene bat from Wyoming. Science 154:1333–1339.

Jepson, G. L. 1970. Bat origins and evolution. Pp. 1–64 in Biology of bats, vol. I (W. A. Wimsatt, ed.). Academic Press, New York.

Jerison, H. J. 1973. Evolution of the brain and intelligence. Academic Press, New York.

Jike, L., G. O. Batzli, and L. L. Getz. 1988. Home ranges of prairie voles as determined by radiotracking and by powdertracking. J. Mammal. 69:183–186.

Jimenez, I. 2005. Development of predictive models to explain the distribution of the West Indian manatee Trichechus manatus in tropical watercourses. Biol. Con. 125:491–503.

Jiménez, J. A., K. A. Hughes, G. Alaks, L. Graham, and R. C. Lacy. 1994. An experimental study of inbreeding depression in a natural habitat. Science 266:271–273.

Jinling, L., B. S. Rubidge, and C. Zhengwu. 1996. A primitive anteosaurid dinocephalian from China: implications for the distribution of earliest therapsid faunas. S. Afr. J. Sci. 92:252–253.

Joffe, T. H., and R. I. M. Dunbar. 1998. Tarsier brain component composition and its implications for systematics. Primates 39:211–216.

Johnson, C. N. 1986. Philopatry, reproductive success of females, and maternal investment in the red-necked wallaby. Behav. Ecol. Sociobiol. 19:143–150.

Johnson, C. N. 1994. Mycophagy and spore dispersal by a rat-kangaroo: consumption of ectomycorrhizal taxa in relation to their abundance. Funct. Ecol. 8:464–468.

Johnson, C. N., and S. Wroe. 2003. Causes of extinction of vertebrates during the Holocene of mainland Australia: arrival of the dingo, or human impact? Holocene 13:941–948.

Johnson, E. 1984. Seasonal adaptive coat changes in mammals. Acta Zool. Fennica 171:7–12.

Johnson, E., and J. Hornby. 1980. Age and seasonal

coat changes in long haired and normal fallow deer (Dama dama). J. Zool. 192:501–509.

Johnson, K. A. 1995. Marsupial mole, Notoryctes typhlops. Pp. 409–411 in The mammals of Australia (R. Strahan, ed.). Reed Books, Chatswood, Australia.

Johnson, P. M., and S. Delean. 2001. Reproduction in the northern bettong, Bettongia tropica Wakefield (Marsupialia: Potoroidae), in captivity, with age estimation and development of the pouch young. Wildl. Res. 28:79–85.

Johnson, W. E., and J. M. Coffin. 1999. Constructing primate phylogenies from ancient retrovirus sequences. Proc. Nat. Acad. Sci. USA 96:10254–10260.

Johnson, W. E., E. Eizirik, J. Pecon-Slattery, W. J. Murphy, A. Antunes, E. Teeling, and S. J. O'Brien. 2006. The late Miocene radiation of modern Felidae: a genetic assessment. Science 311:73–77.

Jones, F. W. 1968. The mammals of South Australia, parts I–III. Government Printer, Adelaide.

Jones, G. 1990. Prey selection by the greater horseshoe bat (Rhinolophus ferrumequinum): optimal foraging by echolocation. J. Anim. Ecol. 59:587–602.

Jones, G., and E. C. Teeling. 2006. The evolution of echolocation in bats. Trend. Ecol. Evol. 21:149–156.

Jones, G., P. I. Webb, J. A. Sedgeley, and C. F. J. O'Donnell. 2003. Mysterious Mystacina: how the New Zealand short-tailed bat (Mystacina tuberculata) locates insect prey. J. Exp. Biol. 206:4209–4216.

Jones, J. K., and D. C. Carter. 1979. Systematic and distributional notes. Pp. 7–11 in Biology of bats of the New World family Phyllostomatidae, part III (R. J. Baker, J. K. Jones, Jr., and D. C. Carter, eds.). Spec. Pub. No. 16, Museum Texas Tech Univ., Lubbock.

Jones, J. K., Jr., D. M. Armstrong, and J. R. Choate. 1985. Guide to mammals of the Plains States. Univ. Nebraska Press, Lincoln.

Jones, J. S., and K. E. Wynne-Edwards. 2001. Paternal behaviour in biparental hamsters, Phodopus campbelli, does not require contact with the pregnant female. Anim. Behav. 62:453–464.

Jones, M. E., and L. A. Barmuta. 2000. Niche differentiation among sympatric Australian dasyurid carnivores. J. Mammal. 81:434–447.

Jones, W. T. 1987. Dispersal patterns in kangaroo rats (Dipodomys spectabilis). Pp. 119–127 in Mammalian dispersal patterns. The effects of social structure on population genetics (B. D. Chepko-Sade and Z. T. Halpin, eds.). Univ. Chicago Press, Chicago.

Jorde, L. B., M. Bamshad, and A. R. Rogers. 1998. Using mitochondrial and nuclear DNA markers to reconstruct human evolution. Bioessays 20:126–136.

Jorge, M. S. P., and C. A. Peres. 2005. Population density and home range size of red-rumped agoutis (Dasyprocta leporina) within and outside a natural Brazil nut stand in southeastern Amazonia. Biotropica 37:317–321.

Joslin, J. K. 1977a. Rodent long distance orientation ("homing"). Adv. Ecol. Res. 10:63–89.

Joslin, J. K. 1977b. Visual cues used in orientation by white-footed mice, Peromyscus leucopus: a laboratory study. Am. Mid. Nat. 98:308–318.

Jung, K. Y., S. Crovella, and Y. Rumpler. 1992. Phylogenetic relationships among lemuriform species determined from restriction genomic DNA banding patterns. Folia Primat. 58:224–229.

Juskaitis, R. 2006. Interactions between dormice (Gliridae) and hole-nesting birds in nestboxes. Folia Zool. 55:225–236.

Kadwell, M., M. Fernandez, H. F. Stanley, R. Baldi, J. C. Wheeler, R. Rosadio, and M. W. Bruford. 2001. Genetic analysis reveals the wild ancestors of the llama and the alpaca. Proc. Roy. Soc. Lond. 268:2575–2584.

Kalcounis, M. C., and R. M. Brigham. 1995. Intraspecific variation in wing loading affects habitat use by little brown bats (Myotis lucifugus). Can. J. Zool. 73:89–95.

Kalinowski, S. T., P. W. Hedrick, and P. S. Miller. 2000. Inbreeding depression in the Speke's gazelle captive breeding program. Conserv. Biol. 14:1375–1384.

Kalko, E. K. 1998. Organization and diversity of tropical bat communities through space and time. Zoology 101:281–297.

Kalko, E. K. V., E. A. Herre, and C. O. Handley, Jr. 1996. Relation of fig fruit characteristics to fruit-eat-

ing bats in the New and Old World tropics. J. Biogeogr. 23:565–576.

Kam, M., and A. A. Degen. 1992. Effect of air temperature on energy and water balance of Psammomys obesus. J. Mammal. 73:207–214.

Kangas, A. T., A. R. Evans, I. Thesleff, and J. Jernvaal. 2004. Nonindependence of mammalian dental characters. Nature 432:211–214.

Kantongol, C. B., F. Naftolin, and R. V. Short. 1971. Relationship between blood levels of luteinizing hormone, testosterone in bulls, and the effects of sexual stimulation. J. Endocrinol. 50:457–465.

Kanwisher, J., and G. Sundnes. 1966. Thermal regulation in cetaceans. Pp. 397–409 in Whales, dolphins, and porpoises (K. S. Norris, ed.). Univ. California Press, Berkeley.

Kappelar, P. M., R. M. Rasoloarison, L. Razafirmanantsoa, L. Walter, and C. Roos. 2005. Morphology, behavior and molecular evolution of giant mouse lemurs (Mirza spp.) Gray 1870, with description of a new species. Primate Report 71:3–7.

Kardong, K.V. 2006. Vertebrates: comparative anatomy, function, evolution, 4th edition. McGraw-Hill, New York.

Kasangaki, A., Kityo, R., and J. Kerbis. 2003. Diversity of rodents and shrews along an elevational gradient in Bwindi Impenetrable National Park, south-western Uganda. Afr. J. Ecol. 41:115–123.

Kaufman, D. M., D. W. Kaufman, and G. A. Kaufman. 1996. Women in the early years of the American Society of Mammalogists (1919–1949). J. Mammal. 77:642–654.

Kaufmann, D. M. 1995. Diversity of New World mammals: universality of the latitudinal gradients of species and bauplans. J. Mammal. 76:322–334.

Kavanagh, J. R., A. Burk-Herrick, M. Westerman, and M. S. Springer. 2004. Relationships among families of Diprotodontia (Marsupialia) and the phylogenetic position of the autapomorphic honey possum (Tarsipes rostratus). J. Mammal. Evol. 11:207–222.

Kawamichi, M. 1996. Ecological factors affecting annual variation in commencement of hibernation in wild chipmunks (Tamias sibiricus). J. Mammal. 77:731–744.

Kay, R. F., C. Ross, and B. A. Williams. 1997. Anthropoid origins. Science 275:797–804.

Kays, R. W., and J. L. Gittleman. 2001. The social organization of the kinkajou, Potos flavus (Procyonidae). J. Zool. 253:491–504.

Kayser, C. 1965. Hibernation. Pp. 180–278 in Physiological mammalogy, vol. 2. (W. V. Mayer and R. G. van Gelder, eds.). Academic Press, New York.

Keane, B. 1990. The effect of relatedness on reproductive success and mate choice in the white-footed mice Peromyscus leucopus. Anim. Behav. 39:264–273.

Keeley, A. T. H., and B. W. Keeley. 2004. The mating system of Tadarida brasiliensis (Chiroptera: Molossidae) in a large highway bridge colony. J. Mammal. 85:113–119.

Keith, L. B. 1963. Wildlife's ten-year cycle. Univ. Wisconsin Press, Madison.

Keith, L. B. 1983. Role of food in hare populations. Oikos 40:385–395.

Keith, L. B. 1987. Dynamics of snowshoe hare populations. Curr. Mammal. 2:119–195.

Keller, C. A., L. I. Ward-Geiger, W. B. Brooks, C. K. Slay, C. R. Taylor, and B. J. Zoodsma. 2006. North Atlantic right whale distribution in relation to sea-surface temperature in the southeastern United States calving grounds. Mar. Mammal Sci. 22:426–445.

Kelli, M. J. 2001. Computer-aided photograph matching in studies using individual identification: an example from Serengeti cheetahs. J. Mammal. 82:440–449.

Kemp, S., A. C. Hardy, and N. A. Mackintosh. 1929. Discovery investigations: objects, equipment, and methods. Discovery Rep. 1:141–232.

Kemp, T. S. 1982. Mammal-like reptiles and the origin of mammals. Academic Press, London.

Kemp, T. S. 2005. The origin and evolution of mammals. Oxford Univ. Press, New York.

Kemp, T. S. 2006. The origin of mammalian endothermy: a paradigm for the evolution of complex biological structure. Zool. J. Linnean Soc. 147:473–488.

Kenagy, G. J. 1972. Saltbush leaves: excision of hypersaline tissue by a kangaroo rat. Science

178:1094–1096.

Kenagy, G. J., and G. A. Bartholomew. 1985. Seasonal reproductive patterns in five coexisting California desert rodent species. Ecol. Monogr. 55:371–397.

Kenagy, G. J., and D. F. Hoyt. 1980. Reingestion of feces in rodents and its daily rhythmicity. Oecologia 44:403–409.

Kenagy, G. J., and D. F. Hoyt. 1989. Speed and time-energy budget for locomotion in golden-mantled ground squirrels. Ecology 70:1834–1839.

Kenagy, G. J., and S. C. Trombulak. 1986. Size and function of mammalian testes in relation to body size. J. Mammal. 67:1–22.

Kenagy, G. J., S. M. Sharbaugh, and K. A. Nagy. 1989a. Annual cycle of energy and time expenditure in a golden-mantled ground squirrel population. Oecologia 78:269–282.

Kenagy, G. J., R. D. Stevenson, and D. Masman. 1989b. Energy requirements for lactation and post-natal growth in captive golden-mantled ground squirrels. Physiol. Zool. 62:470–487.

Kenagy, G. J., D. Masman, S. M. Sharbaugh, and K. A. Nagy. 1990. Energy expenditure during lactation in relation to litter size in free-living golden-mantled ground squirrels. J. Anim. Ecol. 59:73–88.

Kennelly, J. J., and B. E. Johns. 1976. The estrous cycle of coyotes. J. Wildl. Mgmt. 40:272–277.

Kent, G. C., and R. K. Carr. 2001. Comparative anatomy of the vertebrates, 9th edition. McGraw-Hill, New York.

Kenward, R. 1987. Wildlife radio tagging. Academic Press, San Diego.

Kenward, R. E. 2000. A manual for wildlife radio tagging, 2d edition. Academic Press, San Diego.

Kermack, D. M., and K. A. Kermack. 1984. The evolution of mammalian characters. Kapitan Szabo Publishing, Washington, DC.

Keuroghlian, A., D. P. Eaton, and W. S. Longland. 2004. Area use by white-lipped and collared peccaries (Tayassu pecari and Tayassu tajacu) in a tropical forest fragment. Biol. Conserv. 120:411–425.

Keys, P. L., and M. C. Wiltbank. 1988. Endocrine regulation of the corpus luteum. Annu. Rev. Physiol. 50:465–482.

Khateeb, A., and E. Johnson. 1971. Seasonal changes of pelage in the vole (Microtus agrestis), I: Correlation with changes in the endocrine glands. Gen. Comp. Endocrinol. 16:217–228.

Kido, H., and D. Uemura. 2004. Blarina toxin, a mammalian lethal venom from the short-tailed shrew Blarina brevicauda: isolation and characterization. Proc. Natl. Acad. Sci. USA 101:7542–7547.

Kielan-Jaworowska, Z. 1997. Characters of multituberculates neglected in phylogenetic analyses of early mammals. Lethaia 29:249–266.

Kielan-Jaworowska, Z., R. L. Cifelli, and Z.-X. Luo. 2004. Mammals from the age of dinosaurs: orgins, evolution, and structure. Columbia Univ. Press, New York.

Kierdorf, U., and H. Kierdorf. 2002. Pedicle and first antler formation in deer: anatomical, histological, and developmental aspects. Z. Jagdwiss. 48:22–34.

Killian, J. K., T. R. Buckley, N. Stewardt, B. L. Munday, and R. L. Jirtle. 2001. Marsupials and eutherians reunited: genetic evidence for the Theria hypothesis of mammalian evolution. Mammal. Genome 12:513–517.

Kim, K. C. 1985a. Evolutionary relationships of parasitic arthropods and mammals. Pp. 3–82 in Coevolution of parasitic arthropods and mammals (K. C. Kim, ed.). John Wiley and Sons, New York.

Kim, K. C. 1985b. Parasitism and coevolution: epilogue. Pp. 661–682 in Coevolution of parasitic arthropods and mammals (K. C. Kim, ed.). John Wiley and Sons, New York.

Kimchi, T., A. S. Etienne, and J. Terkel. 2004. A subterranean mammal uses the magnetic compass for path integration. Proc. Natl. Acad. Sci. USA 101:1105–1109.

Kimura, K. A., and T. A. Uchida. 1983.Ultrastructural observations of delayed implantation in the Japanese long-fingered bat, Miniopterus schreibersii fuliginosis. J. Reprod. Fertil. 69:187–193.

Kimura, M. 1983. The neutral theory of molecular evolution. Pp. 208–233 in Evolution of genes and proteins (M. Nei and R. K. Koehn, eds.). Sinauer Associates, Sunderland, MA.

King, C. 1983. Mustela erminea. Mammal. Spec. 195:1–8.

King, C. 1984. The origin and adaptive advantages of delayed implantation in Mustela erminea. Oikos. 42:126–128.

King, C. M. 1989. The advantages and disadvantages of small size to weasels, Mustela species. Pp. 302–334 in Carnivore behavior, ecology, and evolution (J. L. Gittleman, ed.). Comstock Publ. Assoc., Ithaca, NY.

King, C. M. 1990. The natural history of weasels and stoats. Comstock Publ. Assoc., Ithaca, NY.

King, J. A., D. Maas, and R. G. Weisman. 1964. Geographic variation in nest size among species of Peromyscus. Evolution. 18:230–234.

King, J. E. 1983. Seals of the world, 2d edition. British Museum of Natural History, London.

Kingdon, J. 1971. East African mammals, vol. I. Academic Press, New York.

Kingdon, J. 1972. East African mammals: carnivores. vol. IIIA. Univ. Chicago Press, Chicago.

Kingdon, J. 1974a. East African mammals, vol. I. Univ. Chicago Press, Chicago.

Kingdon, J. 1974b. East African mammals: An atlas of evolution in Africa, vol. IIA. Univ. Chicago Press, Chicago.

Kingdon, J. 1977. East African mammals: carnivores, vol. IIIA. Univ. Chicago Press, Chicago.

Kingdon, J. 1979. East African mammals: large mammals, vol. IIIB. Univ. Chicago Press, Chicago.

Kinnaird, M. F., and T. G. O'Brien. 1998. Ecological effects of wildfire on lowland rainforest in Sumatra. Conserv. Biol. 12:954–956.

Kirby, L. T. 1990. DNA fingerprinting. Stockton Press, New York.

Kirkpatrick, R. L. 1980. Physiological indices in wildlife management. Pp. 99–127 in Wildlife management techniques manual, 4th edition. (S. D. Schemnitz, ed.). Wildlife Society, Washington, DC.

Kirsch, J. A. W. 1977. The comparative serology of Marsupialia, and a classification of marsupials. Aust. J. Zool. Suppl. Ser. 52:1–152.

Kirsch, J. A. W. 1984. Vicariance biogeography. Pp. 109–112 in Vertebrate zoogeography and evolution in Australasia. Hesperian Press, Carlisle, Australia.

Kirsch, J. A. W., A. W. Dickerman, O. A. Reig, and M. S. Springer. 1991. DNA hybridization evidence for the Australasian affinity of the American marsupial Dromiciops australis. Proc. Natl. Acad. Sci. USA 88:10465–10469.

Kirsch, J. A. W., T. F. Flannery, M. S. Springer, and F.-J. Lapointe. 1995. Phylogeny of the Pteropodidae (Mammalia: Chiroptera) based on DNA hybridisation, with evidence for bat monophyly. Austral. J. Zool. 43:395–428.

Kirsch, J. A. W., F.-J. Lapointe, and M. S. Springer. 1997. DNA-hybridisation studies of marsupials and their implications for metatherian classification. Aust. J. Zool. 45:211–280.

Kirsch, J. A. W., J. M. Hutcheon, D. G. P. Byrnes, and B. D. Lloyd. 1998. Affinities and historical zoogeography of the New Zealand short-tailed bat, Mystacina tuberculata Gray 1843, inferred from DNA-hybridization comparison. J. Mammal. Evol. 5:33–64.

Kirsch, J. A. W., M. S. Springer, C. Krajewski, M. Archer, K. Aplin, and A. W. Dickerman. 1990. DNA/DNA hybridization studies of the carnivorous marsupials, I: The intergeneric relationships of bandicoots (Marsupialia: Perameloidea). J. Mol. Evol. 30:434–448.

Kishino, H., J. L. Thorne, and W. J. Bruno. 2001. Performance of a divergence time estimation method under a probabilistic model of rate evolution. Mol. Biol. Evol. 18:352–361.

Kistler, H. B., Jr., N. J. Baker, D. Anderson, M. B. B. McNabb, M. M. Cooke, and N. Schafer. 1975. Vertebrates: a laboratory text. William Kaufmann, Los Altos, CA.

Kita, M., Y. Nakamura, Y. Okumura, S.D. Ohdachi, Y. Oba, M. Yoshikuni, H. Kido, and D. Uemura. 2004. Blarina toxin, a mammalian lethal venom from the short-tailed shrew Blarina brevicauda: isolation and characterization. Proc. Natl. Acad. Sci. USA 101:7542–7547.

Kita, M., Y. Okumura, S. D. Ohdachi, Y. Oba, M. Yoshikuni, Y. Nakamura, H. Kido, and D. Uemura. 2005. Purification and characterization of blarinasin, a new tissue kallikrein-like protease from the short-

tailed shrew Blarina brevicauda: comparative studies with blarina toxin. Biol. Chem. 386:177–182.

Kita, M., Y. Nakamura, Y. Okumura, S.D. Ohdachi, Y. Oba, M. Yoshikuni, H. Lewis, E. R., P. M. Narins, J. U. M. Jarvis, G. Bronner, and M. J. Mason. 2006. Preliminary evidence for the use of microseismic cues for navigation by the Namib golden mole. J. Acoust. Soc. Am. 119:1260–1268.

Kitchener, A. C. 2000. Fighting and the mechanical design of horns and antlers. Pp. 291–314 in Biomechanics in animal behaviour (P. Domenici and R.W. Blake, eds.). BIOS Sci. Publ. Ltd., Oxford.

Kitching, R. P. 2000. The role of the World Reference Laboratories for foot-and-mouth disease and for rinderpest. Trop. Vet. Dis. 916:139–146.

Kleiber, M. 1932. Body size and metabolism. Hilgardia 6:315–353.

Kleiber, M. 1961. The fire of life. John Wiley and Sons, New York.

Kleiman, D. G. 1976. International conference on the biology and conservation of the Callitrichidae. Primates 17:119–123.

Kleiman, D. G. 1977. Monogamy in mammals. Q. Rev. Biol. 52:39–69.

Kleiman, D. G., and J. J. C. Mallinson. 1998. Recovery and management committees for lion tamarins: partnerships in conservation planning and implementation. Conserv. Biol. 12:27–38.

Kleiman, D.G. and A.B. Rylands (eds.). 2002. Lion tamarins: Biology and conservation. Smithsonian Institution Press, Washington DC.

Kleiman, D. G., M. E. Allen, K. V. Thompson, and S. Lumpkin (eds.). 1996. Wild animals in captivity. Univ. Chicago Press, Chicago.

Kleiman, D. G., B. B. Beck, A. J. Baker, J. D. Ballou, L. Dietz, and J. M. Dietz. 1990. The conservation program for the golden-lion tamarin, Leontopithecus rosalia. Endangered Species Update 8:82–84.

Kleinebeckel, D., and F. W. Klussmann. 1990. Shivering. Pp. 235–253 in Thermoregulation: physiology and biochemistry (E. Schanbaum and P. Lomax, eds.). Pergamon Press, New York.

Kliman, R. M., and G. R. Lynch. 1992. Evidence for genetic variation in the occurrence of the photore sponse of the Djungarian hamster, Phodopus sungorus. J. Biol. Rhythms 7:161–173.

Klinowska, M. 1985. Cetacean stranding sites relate to geomagnetic topography. Aquat. Mammals 1:27–32.

Klir, J. J., and J. E. Heath. 1992. An infrared thermographic study of surface temperature in relation to external thermal stress in three species of foxes: the red fox (Vulpes vulpes), arctic fox (Alopex lagopus), and kit fox (Vulpes macrotis). Physiol. Zool. 65:1011–1021.

Knott, K. K., P. S. Barboza, and R. T. Bowyer. 2005. Growth in arctic ungulates: postnatal growth and organ maturation in Rangifer tarandus and Ovibos moschatus. J. Mammal. 86:121–130.

Kodric-Brown, A., and J. H. Brown. 1984. Truth in advertising: the kinds of traits favored by sexual selection. Am. Nat. 124:309–323.

Koehler, C. E., and P. R. K. Richardson. 1990. Proteles cristatus. Mamm. Species 363:1–6.

Koeln, G. T., L. M. Cowardin, and L. L. Strong. 1996. Geographic information systems. Pp. 540–566 in Research and management techniques for wildlife and habitats, 5th edition, rev. (T. A. Bookhout, ed.). Wildlife Society, Bethesda, MD.

Koelsch, J. K. 2001. Reproduction in female manatees observed in Sarasota Bay, Florida. Mar. Mammal. Sci. 17:331–342.

Koepfli, K. P., S. M. Jenks, E. Eizirik, T. Zahirpour, B. Van Valkenburgh, and R. K. Wayne. 2006. Molecular systematics of the Hyaenidae: relationships of a relictual lineage resolved by a molecular supermatrix. Mol. Phylo. Evol. 38:603–620.

Kohler, N., M. H. Gallardo, L. C. Contreras, and J. C. Torres-Mura. 2000. Allozymic variation and systematic relationships of the Octodontidae and allied taxa (Mammalia, Rodentia). J. Zool. Lond. 252:243–250.

Kohler-Rollefson, I. U. 1991. Camelus dromedarius. Mamm. Species 375: 1–8.

Kohno, N. 2006. A new Miocene odobenid (Mammalia: Carnivora) from Hokkaido, Japan, and its implications for odobenid phylogeny. J. Vert. Paleo. 26:411–421.

Koop, B. F., M. Goodman, P. Zu, J. L. Chan, and J. L. Slighton. 1986. Primate gamma-globulin DNA sequences and man's place among the great apes. Nature 319:234–238.

Koop, B. F., D. A. Tagle, M. Goodman, and J. L. Slighton. 1989. A molecular view of primate phylogeny and important systematic and evolutionary questions. Mol. Biol. Evol. 6:580–612.

Koopman, H. N. 1998. Topographic distribution of the blubber of harbor porpoises (*Phocoena phocoena*). J. Mammal. 79:260–270.

Koopman, K. F. 1984. Bats. Pp. 145–186 *in* Orders and families of recent mammals of the world (S. Anderson and J. K. Jones, Jr., eds.). John Wiley and Sons, New York.

Koopman, K. F. 1993. Order Chiroptera. Pp. 137–241 *in* Mammal species of the world: a taxonomic and geographic reference, 2d edition (D. E. Wilson and D. M. Reeder, eds.). Smithsonian Institution Press, Washington, DC.

Kooyman, G. K., and P. J. Ponganis. 1997. The challenges of diving to depth. Am. Sci. 85:530–539.

Kooyman, G. L. 1963. Milk analysis of the kangaroo rat, Dipodomys merriami. Science 147:1467–1468.

Kooyman, G. L. 1981. Weddell seal: consummate diver. Cambridge Univ. Press, London.

Kooyman, G. L., R. L. Gentry, and D. L. Urquhart. 1976. Northern fur seal diving behavior: a new approach to its study. Science 193:411–413.

Koren, L., O. Mokady, and E. Geffen. 2006. Elevated testosterone levels and social ranks in female rock hyrax. Hormones and Behavior 49:470–477.

Korhonen, H., and M. Harri. 1984. Seasonal changes in thermoregulation of the raccoon dog (*Nyctereutes procyonoides* Gray 1834). Comp. Biochem. Physiol. 77A: 213–219.

Korhonen, H., M. Harri, and E. Hohtola. 1985. Response to cold in the blue fox and raccoon dog as evaluated by metabolism, heart rate and muscular shivering: a re-evaluation. Comp. Biochem. Physiol. 82A:959–964.

Koteja, P. 1996. The usefulness of a new TOBEC instrument (ACAN) for investigating body composition in small mammals. Acta Theriol. 41:107–112.

Kotler, B. P. 1989. Temporal variation in the structure of a desert rodent community. Pp. 127–139 *in* Patterns in the structure of mammalian communities (D. W. Morris, Z. Abramsky, B. J. Fox, and M. R. Willig, eds.). Texas Tech Univ. Press, Lubbock.

Koukkari, W. L., and R. B. Southern. 2006. Introducing biological rhythms. Springer, New York.

Kowalski, K. 1976. Mammals: an outline of theriology. Panstwowe Wydawnictwo Naukowe, Warsaw (National Technical Information Service translation for the Smithsonian Institution and the National Science Foundation).

Koyasu, K., K. Kawahito, H. Hanamura, and S-I. Oda. 2005. Dental anomalies in *Suncus murinus*. Pp. 405–411 *in* Advances in the biology of shrews II (J. F. Merritt, S. Churchfield, R. Hutterer, and B. I. Sheftel, eds.). Spec. Pub. No. 01 International Society of Shrew Biologists, New York.

Krajewski, C., A. C. Driskell, P. R. Baverstock, and M. J. Braun. 1992. Phylogenetic relationships of the thylacine (Mammalia: Thylacinidae) among dasyuroid marsupials: evidence from cytochrome b DNA sequences. Proc. Roy. Soc. Lond. Ser. B 250:19–27.

Krajewski, C., L. Buckley, and M. Westerman. 1997. DNA phylogeny of the marsupial wolf resolved. Proc. Roy. Soc. Lond. Ser. B 264:911–917.

Krajewski, C., M. J. Blacket, and M. Westerman. 2000a. DNA sequence analysis of familial relationships among dasyuromorphian marsupials. J. Mammal. Evol. 7:95–108.

Krajewski, C., S. Wroe, and M. Westerman. 2000b. Molecular evidence for the timing of cladogenesis in dasyurid marsupials. Zool. J. Linnean Soc. 130:375–404.

Krajewski, C., G. R. Moyer, J. T. Sipiorski, M. G. Fain, and M. Westerman. 2004. Molecular systematics of the enigmatic 'phascolosoricine' marsupials of New Guinea. Austr. J. Zool. 52:389–415.

Krajick, K. 2005. Ecology: Winning the war against invaders. Science 310:1410–1413.

Krakauer, E., P. Lemelin, and D. Schmitt. 2002. Hand and body position during locomotor behavior in the aye-aye (*Daubentonia madagascariensis*). Am. J. Primatol. 57:105–118.

Kramer, A., F.-C. Yang, P. Snodgrass, X. Li, T. E.

Scammell, F. C. Davis, and C. J. Weitz. Regulation of daily locomotor activity and sleep by hypothalamic EGF receptor signaling. Science 294:2511–2515.

Krane, D. E., and M. L. Raymer. 2003. Fundamental concepts of bioinformatics. Benjamin Cummings, San Francisco.

Krapp, F. 1965. Schädel under Kaumuskulatur von Spalax leucodon (Nordmann, 1840). Z. Wissensch. Zool. 173:1–71.

Krasnov, B. R., G. I. Shenbrot, D. Mouillot, I. S. Khokhlova, and R. Poulin. 2006. Ecological characteristics of flea species relate to their suitability as plague vectors. Oecologia Berlin. 149:474–481.

Krebs, C. J. 1963. Lemming cycle at Baker Lake, Canada during 1959–62. Science 140:674–676.

Krebs, C. J. 1989. Ecological methodology. Harper and Row, New York.

Krebs, C. J. 1996. Population cycles revisited. J. Mammal. 77:8–24.

Krebs, C. J. 1999. Ecological methodology, 2d edition. Benjamin Cummings, Menlo Park, CA.

Krebs, C. J. 2001. Ecology: the experimental analysis of distribution and abundance, 5th edition. Benjamin Cummings, San Francisco.

Krebs, C. J., M. S. Gaines, B. L. Keller, J. H. Myers, and R. H. Tamarin. 1973. Population cycles in small rodents. Science 179:35–41.

Krebs, C. J., S. Boutin, R. Boonstra, A. R. E. Sinclair, J. N. M. Smith, M. R. T. Dale, and R. Turkington. 1995. Impact of food and predation on the snow shoe hare cycle. Science 269:1112–1115.

Krebs, C. R. 1999. Ecological methodology, 2d edition. Benjamin Cummings, Menlo Park, CA.

Krebs, J. R., and R. Dawkins. 1984. Animal signals: mind-reading and manipulation. Pp. 380–402 *in* Behavioural ecology: an evolutionary approach (J. R. Krebs and N. B. Davies, eds.). Sinauer Associates, Sunderland, MA.

Krebs, J. W., J. S. Smith, C. E. Rupprecht, and J. E. Childs. 2000. Mammalian reservoirs and epidemiology of rabies diagnosed in human beings in the United States, 1981–1998. Trop. Vet. Dis. 916:345–353.

Krebs, J. W., M. L. Wilson, and J. E. Childs. 1995. Rabies-epidemiology, prevention, and future research. J. Mammal. 76:681–694.

Kripke, D. F. 1974. Ultradian rhythms in sleep and wakefulness. Adv. Sleep Res. 1:305–325.

Krupa, J. J., and K.N. Geluso. 2000. Matching the color of excavated soil: cryptic coloration in the plains pocket gopher (*Geomys bursarius*). J. Mammal. 81:86–96.

Krutzsch, P. H., T. H. Fleming, and E. G. Crichton. 2002. Reproductive biology of male Mexican freetailed bats (*Tadarida brasiliensis mexicana*). J. Mammal. 83:489–500.

Kruuk, H. 1972. The spotted hyena: a study of predation and social behavior. Univ. Chicago Press, Chicago.

Kruuk, H. 1986. Interactions between Felidae and their prey species: a review. Pp. 353–374 *in* Cats of the world: biology, conservation, and management (S. D. Miller and D. D. Everett, eds.). National Wildlife Federation, Washington, DC.

Kruuk, H., and W. A. Sands. 1972. The aardwolf (*Proteles cristatus* Sparrman) 1783 as a predator of termites. E. Afr. Wildl. J. 10:211–227.

Kruuk L. E. B., T. H. Clutton-Brock, S. D. Albon, J. M. Pemberton, and F. E. Guinness. 1999. Population density affects sex ratio variation in red deer. Nature 399: 459–461.

Krzanowski, A. 1967. The magnitude of islands and the size of bats (Chiroptera). Acta Zool. Cracoviensia 15(XI):281–348.

Kudo, H., and R. I. M. Dunbar. 2001. Neocortex size and social network size in primates. Anim. Behav. 62:711–722.

Kuhn, C. E., D. Aurioles-Gamboa, M. J. Weise, and D. P. Costa. 2006. Oxygen stores of California sea lion pups: implications for diving ability. Pp. 31–44 *in* Sea lions of the world (A. W. Trites, S. K. Atkinson, D. P. DeMaster, L. W. Fritz, T. S. Gelatt, L. D. Rea, and K. M. Wynne, eds.). Alaska Sea Grant College Program, Fairbanks.

Kulzer, E. 1969. Das Verhalten von Eidolon helvum (Kerr) in Gefangenschaft. Z. Säugetierk. 34:129–148.

Kumar, S., and S. B. Hedges. 1998. A molecular

timescale for vertebrate evolution. Nature 392:917–920.

Kummer, H. 1968. Social organization of hamadryas baboons: A field study. Univ. Chicago Press, Chicago.

Kunz, T. H. 1982. Roosting ecology of bats. Pp. 1–55 *in* Ecology of bats (T. H. Kunz, ed.). Plenum Press, New York.

Kunz, T. H. 1988. Ecological and behavioral methods for the study of bats. Smithsonian Institution Press, Washington, DC.

Kunz, T. H. 1996. Obligate and opportunistic interactions of Old-World tropical bats and plants. Pp. 37–65 *in* Conservation and faunal biodiversity in Malaysia (Z. A. A. Hasan and Z. Akbar, eds.). Penerbit Universiti Kebangsaan Malaysia, Bangi.

Kunz, T. H., and C. A. Diaz. 1995. Folivory in fruit-eating bats, with new evidence from *Artibeus jamaicensis* (Chiroptera: Phyllostomidae). Biotropica 27:106–120.

Kunz, T. H., and M. B. Fenton. 2003. Bat ecology. Univ. Chicago Press, Chicago.

Kunz, T. H., and K. A. Ingalls. 1994. Folivory in bats: an adaptation derived from frugivory. Funct. Ecol. 8:665–668.

Kunz, T. H., and G. F. McCracken. 1996. Tents and harems: apparent defence of foliage roosts by tent-making bats. J. Tropical Ecol. 12:121–137.

Kunz, T. H., and E. D. Pierson. 1994. Bats of the world: an introduction. Pp.1–46 in Walker's bats of the world (R. Nowak, ed.). Johns Hopkins Univ. Press, Baltimore.

Kunz, T. H., and S. K. Robson. 1996. Postnatal growth and development in the Mexican free-tailed bat (*Tadarida brasiliensis mexicana*): birth size, growth rates, and age estimation. J. Mammal. 76:769–783.

Kunz, T. H., and A. A. Stern. 1995. Maternal investment and post-natal growth in bats. Pp. 123–138 *in* Ecology, evolution, and behaviour of bats (P. A. Racey and S. M. Swift, eds.). Proc. 67th Symp. Zool. Soc. Lond., November 26–27, 1993.

Kunz, T. H., M. S. Fujita, A. P. Brooke, and G. F. McCracken. 1994. Convergence in tent architecture and tent-making behavior among neotropical and paleotropical bats. J. Mammal. Evol. 2:57–78.

Kunz, T. H., J. O. Whitaker, Jr., and M. D. Wadanoli. 1995. Dietary energetics of the insectivorous Mexican free-tailed bat (*Tadarida brasiliensis*) during pregnancy and lactation. Oecologia 101:407–415.

Kurta, A. 1995. Mammals of the Great Lakes region. Univ. Michigan Press, Ann Arbor.

Kurta, A., and T. H. Kunz. 1987. Size of bats at birth and maternal investment during pregnancy. Symp. Zool. Soc. Lond. 57:79–106.

Kurta, A., G. P. Bell, K. A. Nagy, and T. H. Kunz. 1990. Energetics and water flux of free-ranging big brown bats (*Eptesicus fuscus*) during pregnancy and lactation. J. Mammal. 71:59–65.

Kurten, B. 1969. Continental drift and evolution. Sci. Am. 220:54–63.

Kurten, L., and U. Schmidt. 1982. Thermoreception in the common vampire bat (*Desmodus rotundus*). J. Comp. Physiol. 146:223–228.

Lacey, E. A., and J. R. Wieczorek. 2001. Territoriality and male reproductive success in arctic ground squirrels. Behav. Ecol. 12:626–632.

Lacey, E. A., J. R. Wieczorek, and P. K. Tucker. 1997. Male mating behaviour and patterns of sperm precedence in Arctic ground squirrels. Anim. Behav. 53:767–779.

Lacey, E. A., J. L. Patton, and G. N. Cameron (eds.). 2000. Life underground—the biology of subterranean rodents. Univ. Chicago Press, Chicago.

Lahann, P., J. Scmid, and J. U. Ganzhorn. 2006. Geographic variation in populations of *Microcebus murinus* in Madagascar: resource seasonality or Bergmann's rule? Int. J. Primat. 27:983–999.

Laidre, K. L., and M. P. Heide-Jorgensen. 2005. Arctic sea ice trends and narwhal vulnerability. Biol. Conserv. 121:509–517.

Laidre, K. L., M. P. Heide-Jorgensen, M. L. Logsdon, R. C. Hobbs, R. Dietz, and G. R. VanBlaricom. 2004. Fractal analysis of narwhal space use patterns. Zool. Jena 107:3–11.

Lambert, T. D., J. R. Malcolm, and B. L. Zimmerman. 2005. Variation in small mammal species richness by trap height and trap type in southeastern Amazonia. J. Mammal. 86:982–990.

588 References

Lambertsen, R., N. Ulrich, and J. Straley. 1995. Frontomandibular stay of Balaenopteridae: a mechanism for momentum recapture during feeding. J. Mammal. 76:877–899.

Lancia, R. A., J. D. Nichols, and K. H. Pollock. 1996. Estimating the number of animals in wildlife populations. Pp. 215–253 in Research and management techniques for wildlife and habitats, 5th edition, rev. (T. A. Bookhout, ed.). Wildlife Society, Bethesda, MD.

Lander, E., et al. 2001. Initial sequencing and analysis of the human genome. Nature 409:860–921.

Lane, R. S., J. Piesman, and W. Burgdorfer. 1991. Lyme borreliosis: relation of its causative agent to its vectors and hosts in North America and Europe. Annu. Rev. Entomol. 36:587–609.

Lane-Petter, W. 1978. Identification of laboratory animals. Pp. 35–40 in Animal marking (B. Stonehouse, ed.). Univ. Park Press, Baltimore, MD.

Lang, S. L. C., S. J. Iverson, and W. D. Bowen. 2005. Individual variation in milk composition over lactation in harbor seals (Phoca vitulina) and the potential consequences of intermittent attendance. Can. J. Zool. 83:1525–1531.

Langbauer, W. R., Jr. 2000. Elephant communication. Zoo Biol. 19:425–445.

Langbauer, W. R., Jr., K. B. Payne, R. A. Charif, L. Rappaport, and F. Osborn. 1991. African elephants respond to distant playbacks of low-frequency con specific calls. J. Exp. Biol. 157:35–46.

Langer, P. 1996. Comparative anatomy of the stomach of the Cetacea: ontogenetic changes involving gastric proportions-mesentaries-arteries. Z. Säugetierk. 61:140–154.

Langer, P. 2002. The digestive tract and life history of small mammals. Mammal Rev. 32:107–131.

Langevin, P., and R. M. R. Barclay. 1991. Hypsignathus monstrosus. Mammal. Species No. 357:1–4.

Lanyon, L. E. 1981. Locomotor loading and functional adaptations in limb bones. Symp. Zool. Soc. Lond. 48:305–330.

Lanyon, J. M., and G. D. Sanson. 2006. Degenerate dentition of the dugong (Dugong dugon), or why a grazer does not need teeth: morphology, occlusion and wear of mouthparts. J. Zool. 268:133–152.

Lara-Ruiz, P., and A. G. Chiarello. 2005. Life history traits and sexual dimorphism of the Atlantic forest maned sloth Bradypus torquatus (Xenarthra: Bradypodidae). J. Zool. 267:63–73.

Larson, G., K. Dobney, U. Albarella, M. Fang, E. Matisoo-Smith, J. Robins, S. Lowden, H. Finlayson, T. Brand, E. Willerslev, P. Rowley-Conwy, L. Andersson, and A. Cooper. 2005. Worldwide phylo geography of wild boar reveals multiple centers of pig domestication. Science 307:1618–1621.

Lau, S. K. P., P. C. Y. Woo, K. S. M. Li, Y. Huang, H. W. Tsoi, B. H. L. Wong, S. S. Y. Wong, S. Y. Leung, K. H. Chan, and K. Y. Yuen. 2005. Severe acute respiratory syndrome coronavirus-like virus in Chinese horseshoe bats. Proc. Natl. Acad. Sci. USA 102:14040–14045.

Laurin, M., and R. R. Reisz. 1995. A reevaluation of early amniote phylogeny. Zool. J. Linnean Soc. 113:165–223.

Lavigne, D. M., C. D. Bernholz, and K. Ronald. 1977. Functional aspects of pinniped vision. Pp. 135–173 in Functional anatomy of marine mammals, vol. III (R. J. Harrison, ed.). Academic Press, London.

Lavigne, D. M., and K. M. Kovacs. 1988. Harps and hoods: ice-breeding seals of the Northwest Atlantic. Univ. Waterloo Press, Ontario.

Law, B. S. 1992. Physiological factors affecting pollen use by the Queensland blossom bat (Syconycteris australis). Funct. Ecol. 6:257–264.

Law, B. S. 1993. Roosting and foraging ecology of the Queensland blossom-bat (Syconycteris australis) in northeastern New South Wales: flexibility in response to seasonal variation. Wildl. Res. 20:419–431.

Lawlor, T. E. 1973. Aerodynamic characteristics of some Neotropical bats. J. Mammal. 54:71–78.

Lawlor, T. E. 1983. The peninsular effect on mammalian species diversity in Baja California. Am. Nat. 121:432–439.

Lawlor, T. E. 1986. Comparative biogeography of mammals on islands. Biol. J. Linnean Soc. 28:99–125.

Lawrence, B. 1945. Brief comparison of short-tailed shrew and reptile poison. J. Mammal. 26:393–396.

Laws, R. M. 1967. Occurrence of placental scars in the uterus of the African elephant (Loxodonta africanus). J. Reprod. Fert. 14:445–449.

Laws, R. M. 1970. Elephants as agents of habitat and landscape change in East Africa. Oikos. 21:1–15.

Laws, R. M. 1974. Behaviour, dynamics, and management of elephant populations. Pp. 513–529 in The behaviour of ungulates and its relation to management (V. Geist and F. Walther, eds.). IUCN Publ. new series No. 24, Morges, Switzerland.

Laws, R. M., A. Baird, and M. M. Bryden. 2003. Breeding season and embryonic diapause in crabeater seals (Lobodon carcinophagus). Reproduction 126: 365–370.

Lavigne, D. M., and K. M. Kovacs. 1988. Harps and hoods: ice-breeding seals of the Northwest Atlantic. Univ. Waterloo Press, Ontario.

Layne, J. N. 1969. Nest-building behavior in three species of deer mice, Peromyscus. Behaviour 35:288–303.

Layne, J. N. 2003. Armadillo, Dasypus novemcinctus. Pp. 75–97 in Wild mammals of North America: biology, management, and conservation (G. A. Feldhamer, B. C. Thompson, and J. A. Chapman, eds.). Johns Hopkins Univ. Press, Baltimore.

Leach, W. J. 1961. Functional anatomy: mammalian and comparative, 3rd edition. McGraw-Hill, New York.

Leakey, M. G., F. Spoor, F. H. Brown, P. N. Gathogo, C. Karle, L. M. Leakey, and I. McDougall. 2001. New homonid genus from eastern Africa shows diverse middle Pliocene lineages. Nature 410:433–440.

Lean, G., D. Hinrichsen, and A. Markham. 1990. Atlas of the environment. Prentice-Hall, New York.

Leatherwood, S., R. Reeves, and L. Foster. 1983. Sierra Club handbook of whales and dolphins. Sierra Club Books, San Francisco.

Le Boeuf, B. J. 1974. Male–male competition and reproductive success in elephant seals. Am. Zool. 14:163–176.

Le Boeuf, B. J., and J. Reiter. 1988. Lifetime reproductive success in northern elephant seals. Pp. 344–362 in Reproductive success (T. H. Clutton-Brock, ed.). Univ. Chicago Press, Chicago.

Lee, A. K., and A. Cockburn. 1985. Evolutionary ecology of marsupials. Cambridge Univ. Press, New York.

Lee, A. K., and R. W. Martin. 1988. The koala: a natural history. Univ. New South Wales Press, Sydney.

Lee, Y-F., and G. F. McCracken. 2005. Dietary variation of Brazilian free-tailed bats links to migratory populations of pest insects. J. Mammal. 86:67–76.

Lefcourt, A. M., and W. R. Adams. 1996. Radiotelemetry measurement of body temperatures of feedlot steers during summer. J. Anim. Sci. 74:2633–2640.

Lehmann, T., P. Vignaud, A. Likius, and M. Brunet. 2005. A new species of Orycteropodidae (Mammalia, Tubulidentata) in the Mio-Pliocene of Chad. Zool. J. Linnean Soc. 143:109–131.

Lehmkuhl, J. F., L. E. Gould, E. Cazares, and D. R. Hosford. 2004. Truffle abundance and mycophagy by northern flying squirrels in eastern Washington forests. Forest Ecol. Mgmt. 200:49–65.

Lehner, P. N. 1996. Handbook of ethological methods, 2d edition. Cambridge Univ. Press, New York.

Lemen, C. A., and P. W. Freeman. 1985. Tracking mammals with fluorescent pigments: a new technique. J. Mammal. 66:134–136.

Leopold, A. 1966. A Sand County almanac. Oxford Univ. Press, New York.

LePendu, Y., I. Ciofolo, and A. Gosser. 2000. The social organization of giraffes in Niger. Afr. J. Ecol. 38:78–85.

Leslie, P. H., and R. M. Ranson. 1940. The mortality, fertility, and rate of natural increase of the vole (Microtus agrestis) as observed in the laboratory. J. Anim. Ecol. 9:27–52.

Lestrel, P. E. 2000. Morphometrics for the life sciences. World Scientific, Singapore.

Leutenegger, W. 1973. Maternal-fetal weight relationships in primates. Folia Primat. 20:280–293.

Leutenegger, W. 1976. Metric variability in the anterior dentition of African colobines. Am. J. Phys. Anthropol. 45:45–52.

Lev, B., H. Ward, G. T. Keusch, and M. E. A. Perreira. 1986. Lectin activation in Giardia lamblia by host protease: a novel host–parasite interaction. Science 232:71–73.

Lewin, B. 2006. Essential genes. Pearson Prentice Hall, Upper Saddle River, NJ.

Lewis, D. M., and P. Alpert. 1997. Trophy hunting and wildlife conservation in Zambia. Conserv. Biol. 11:59–68.

Lewis, M., and W. Clark (E. Coues, ed.). 1979. The history of the Lewis and Clark expedition, 3 vols. Dover Publications, New York.

Lewis, P. O., and A. A. Snow. 1992. Deterministic paternity exclusion using RAPD markers. Mol. Ecol. 1:155–160.

Lewis, R. E. 1993. Fleas (Siphonaptera). Pp. 529–575 in Medical insects and arachnids (R. P. Lane and R. W. Crosskey, eds.). Chapman and Hall, London.

Lewis-Oritt, N., C. A. Porter, and R. J. Baker. 2001. Molecular systematics of the family Mormoopidae (Chiroptera) based on cytochrome b and recombination activating gene 2 sequences. Mol. Phylogenet. Evol. 20:426–436 .

Lewontin, R. C. 1974. The genetic basis of evolutionary change. Columbia Univ. Press, New York.

Lewontin, R. C., and L. C. Dunn. 1960. The evolutionary dynamics of a polymorphism in the house mouse. Genetics 45:705–722.

Ley, W. 1968. Dawn of zoology. Prentice-Hall, Englewood Cliffs, NJ.

Li, W., Z. Shi, M. Yu, W. Ren, C. Smith, J. H. Epstein, H. Wang, G. Crameri, Z. Hu, H. Zhang, J. Zhang, J. McEachern, H. Field, P. Daszak, B. T. Eaton, S. Zhang, and L. F. Wang. 2005. Bats are natural reservoirs of SARS-like coronaviruses. Science 310:676–679.

Lidicker, W. Z., Jr. 1973. Regulation of numbers in an island population of the California vole, a problem in community dynamics. Ecol. Monogr. 43:271–302.

Lidicker, W. Z., Jr. 1988. Solving the enigma of microtine "cycles." J. Mammal. 69:225–235.

Lidicker, W. Z., Jr. 1995. The landscape concept: something old, something new. Pp. 3–19 in Landscape approaches in mammalian ecology and conservation (W. Z. Lidicker, Jr., ed.). Univ. Minnesota Press, Minneapolis.

Liebig, J. 1847. Chemistry application to agriculture and physiology, 4th edition. Taylor and Walton, London.

Liem, K. F., W. E. Bemis, W. F. Walker, Jr., and L. Grande. 2001. Functional anatomy of the vertebrates: an evolutionary perspective. Harcourt Collge Publ., Fort Worth.

Lillegraven, J. A. 1975. Biological considerations of the marsupial-placental dichotomy. Evolution. 29:707–722.

Lillegraven, J. A. 1979. Introduction. Pp. 1–6 in Mesozoic mammals: the first two-thirds of mammalian history (J. A. Lillegraven, Z. Kielan-Jaworowska, and W. A. Clemens, eds.). Univ. California Press, Berkeley.

Lillegraven, J. A. 1987. The origin of eutherian mammals. Biol. J. Linnean Soc. 32:281–336.

Lillywhite, H. B., and B. R. Stein. 1987. Surface sculpturing and water retention of elephant skin. J. Zool. Lond. 211:727–734.

Lim, M. M., Z. X. Wang, D. E. Olazabal, X. H. Ren, E. F. Terwilliger, and L. J. Young. 2004. Enhanced partner preference in a promiscuous species by manipulating the expression of a single gene. Nature 429: 754–757.

Lin, M., and R. C. Jones. 2000. Spermiogenesis and spermiation in a monotreme mammal, the platypus, Ornithorhynchus anatinus. J. Anat. 196:217–232.

Lindemayer, D. B., M. A. McCarthy, K. M. Parris, and M. L. Pope. 2000. Habitat fragmentation, landscape context, and mammalian assemblages in southeastern Australia. J. Mammal. 81:787–797.

Lindemayer, D. B., A. D. Manning, P. L. Smith, H. P. Possingham, J. Fischer, I. Oliver, and M. A. McCarthy. 2002. The focal-species approach and landscape restoration: a critique. Conserv. Biol. 16:338–345.

Lindsay, S. L. 1987. Geographic size and non-size variation in Rocky Mountain Tamiscurus hudsonicus: significance in relation to Allen's rule and vicariant biogeography. J. Mammal. 68:39–48.

Lindstedt, S. L. 1980. Energetics and water economy of the smallest desert mammal. Physiol. Zool.

53:82–97.

Lindstedt, S. L., and M. C. Boyce. 1985. Seasonality, fasting endurance, and body size in mammals. Am. Nat. 125:873–878.

Linn, I. J. 1978. Radioactive techniques for small mammal marking. Pp. 177–191 in Animal marking (B. Stonehouse, ed.). Univ. Park Press, Baltimore.

Linzey, D. W., and A.V. Linzey. 1967. Maturational and seasonal molts of the golden mouse, Ochrotomys nuttalli. J. Mammal. 48:236–241.

Lister, A. M. 1989. Rapid dwarfing of red deer on Jersey in the last interglacial. Nature 342:539–542.

Lister, A., and P. Bahn. 2000. Mammoths: giants of the ice age. Marshall Editions, London.

Lloyd, S. 2001. Oestrous cycle and gestation length in the musky rat-kangaroo, Hypsiprymnodon moschatus (Potoroidae: Marsupialia). Aust. J. Zool. 49:37–44.

Loeb, S. C., F. H. Tainter, and E. Cazares. 2000. Habitat associations of hypogeous fungi in the southern Appalachians: implications for the endangered northern flying squirrel (Glaucomys sabrinus coloratus). Am. Midl. Nat. 144:286–296.

Lohmann, K. J., and S. Johnsen. 2000. The neurobiology of magnetoreception in vertebrate animals. Trends Neurosci. 23:153–159.

Lombardi, J. R., and J. G. Vandenbergh. 1977. Pheromonally induced sexual maturation in females: regulation by the social environment of the male. Science 196:545–546.

Lomolino, M. V. 1985. Body size of mammals on islands: the island rule reexamined. Am. Nat. 125:310–316.

Lomolino, M. V. 1986. Mammalian community structure on islands: the importance of immigration, extinction and interactive effects. Biol. J. Linnean Soc. 28:1–21.

Lomolino, M. V. 2001. Elevation gradients of species density: historical and prospective views. Global Ecol. Biogeogr. 10:3–13.

Lomolino, M. V. 2005. Body size evolution in insular vertebrates: generality of the island rule. J. Biogeogr. 32:1683–1699.

Lomolino, M. V., B. R. Riddle, and J. H. Brown. 2006. Biogeography, 3rd edition. Sinauer Associates, Sunderland, MA.

Long, J., M. Archer, T. Flannery, and S. Hand. 2002. Prehistoric mammals of Australia and New Guinea: one hundred million years of evolution. Johns Hopkins Univ. Press, Baltimore.

Longland, W. S., and C. Clements. 1995. Use of fluorescent pigments in studies of seed caching by rodents. J. Mammal. 76:1260–1266.

Lopez, A., P. Miranda, E. Tejada, and D. B. Fishbein. 1992. Outbreak of human rabies in the Peruvian jungle. Lancet 339:408–411.

Lord, R. D. 2007. Mammals of South America. Johns Hopkins Univ. Press, Baltimore.

Loughry, W. J., P. A. Prodohl, and C. M. McDonough. 2005. The inadequacy of observation: understanding armadillo biology with molecular markers. Recent Res. Dev. Ecol. 3:55–73.

Louwman, J. W. W. 1973. Breeding the tailless tenrec Tenrec ecaudatus at Wassenaar Zoo. Intern. Zoo Yearb. 13:125–126.

Lovegrove, B. G., J. Raman, and M. R. Perrin. 2001. Heterothermy in elephant shrews, Elephantulus spp. (Macroscelidea): daily topror or hibernation? J. Comp. Physiol. B 171:1–10.

Low, B. S. 1990. Marriage systems and pathogen stress in human societies. Am. Zool. 30:325–339.

Lowe, V. P. W. 1969. Population dynamics of the red deer (Cervus elaphus L.) on Rhum. J. Anim. Ecol. 38:425–457.

Lowery, G. H., Jr. 1974. The mammals of Louisiana and its adjacent waters. Louisiana State Univ. Press, Baton Rouge.

Lu, X. 2003. The annual cycle of molt in the Cape hare Lepus capensis in northern China: pattern, timing, and duration. Acta. Theriol. 48:373–384.

Lucio, V., M. M. Lahr, and J. Cheverud. 2003. Morphometric heterochrony and the evolution of growth. Evolution 57:2459–2468.

Luckett, W. P. 1975. Ontogeny of the fetal membranes and placenta: their bearing on primate phylogeny. Pp. 157–182 in Phylogeny of the primates (W. P. Luckett and F. S. Szalay, eds.). Plenum Press, New York.

Luckett, W. P. (ed.). 1980a. Comparative biology and

evolutionary relationships of tree shrews. Plenum Press, New York.

Luckett, W. P. 1980b. Monophyletic or diphyletic origins of Anthropoidea and Hystriscognathi: evidence of the fetal membranes. Pp. 347–368 in Evolutionary biology of the New World monkeys and continental drift (R. L. Ciochon and A. B. Chiarelli, eds.). Plenum Press, New York.

Luckett, W. P. 1994. Suprafamilial relationships within Marsupialia: resolution and discordance from multi disciplinary data. J. Mammal. Evol. 2:225–283.

Luckett, W. P., and N. Hong. 2000. Ontogenetic evidence for dental homologies and premolar replacement in fossil and extant caenolestids (Marsupialia). J. Mammal. Evol. 7:109–127.

Luckett, W. P., and F. S. Szalay. 1975. Phylogeny of the primates. Plenum Press, New York.

Luckett, W. P. and P. A. Woolley. 1996. Ontogeny and homology of the dentition in dasyurid marsupials: development in Sminthopsis virginiae. J. Mammal. Evol. 3:327–364.

Luikart, G., L. Gielly, L. Excoffier, J. D. Vigne, J. Bouvet, and P. Taberlet. 2001. Multiple maternal origins and weak phylogeographic structure in domestic goats. Proc. Natl. Acad. Sci. USA 98:5927–5932.

Luna, F. and C. D. Antinuchi. 2006. Cost of foraging in the subterranean rodent Ctenomys talarum: effect of soil hardness. Can. J. Zool. 84:661–667.

Luo, Z., and A. W. Crompton. 1994. Transformation of the quadrate (incus) through the transition from non-mammalian cynodonts to mammals. J. Vert. Paleontol. 14:341–374.

Luo, Z.-X., R. L. Cifelli, and Z. Kielan-Jaworowska. 2001. Dual origin of tribosphenic mammals. Nature 409:53–57.

Luo, Z.-X., R. L. Cifelli, and Z. Kielan-Jaworowska. 2002. In quest for a phylogeny of Mesozoic mammals. Acta Palaeontol. Polonica 47:1–78.

Luo, Z.-X., Q. Ji, J. R. Wible, and C.-X. Yuan. 2003. An early Cretaceous tribosphenic mammal and metatherian evolution. Science 302:1934–1940.

Lurz, P. W. W., and A. B. South. 1998. Cached fungi in non-native conifer forest and their importance for red squirrels (Sciurus vulgaris L.). J. Zool. Lond. 246:468–471.

Lydekker, R. A. 1896. A geographical history of mammals. Cambridge Univ. Press, Cambridge, UK.

Lyman, C. P., J. S. Willis, A. Malan, and L. C. H. Wang. 1982. Hibernation and torpor in mammals and birds. Academic Press, New York.

Lynch, G. R. 1973. Seasonal change in the thermogenesis, organ weights, and body composition in the white-footed mouse, Peromyscus leucopus. Oecologia 13:363–367.

Lynch, J. D. 1988. Refugia. Pp. 301–342 in Analytical biogeography (A. A. Myers and P. S. Giller, eds.). Chapman and Hall, New York.

Lynch, M., and T. J. Crease. 1990. The analysis of population survey data on DNA sequence variation. Mol. Biol. Evol. 7:377–394.

Lynch, M., and B. Milligan. 1994. Analysis of population genetic structure with RAPD markers. Mol. Ecol. 3:91–100.

Lyons, S. K., and M. R. Willig. 1997. Latitudinal patterns of range size: methodological concerns and empirical evaluations for New World bats and marsupials. Oikos 79:568–580.

MacArthur, R. A., and M. Aleksiuk. 1979. Seasonal microenvironments of the muskrat (Ondatra zibethicus) in a northern marsh. J. Mammal. 60:146–154.

MacArthur, R. H., and E. R. Pianka. 1966. On the optimal use of a patchy environment. Am. Nat. 100:603–609.

MacArthur, R. H., and E. O. Wilson. 1967. The theory of island biogeography. Princeton Univ. Press, Princeton, NJ.

Macdonald, D. 1976. Food caching by red foxes and some other carnivores. Z. Tierphysiol. 42:170–185.

Macdonald, D. (ed.) 1984. The encyclopedia of mammals. Facts on File, New York.

Macholán, M. 2006. A geometric morphometric analysis of the shape of the first upper molar in mice of the genus Mus (Muridae, Rodentia). J. Zool. 270:672–681.

Machado, T., M. J. deJ. Silva, M. Leal, R. Emygdia, A. P. Carmignotto, and Y. Yonenaga-Yassuda. 2005. Nine karyomorphs for spiny rats of the genus

Proechimys (Echimyidae, Rodentia) from north and central Brazil. Genet. Mol. Biol. 28:682–692.

Macholán, M., M. G. Filippucci, B. Benda, D. Frynta, and J. Sádlová. 2001. Allozyme variation and systematics of the genus Apodemus (Rodentia: Muridae) in Asia Minor and Iran. J. Mammal. 82:799–813.

MacLeod, C. D. 2000. Species recognition as a possible function for variations in position and shape of the sexually dimorphic tusks of Mesoplodon whales. Evolution 54:2171–2173.

MacLeod, C. D. 2000a. Review of the distribution of Mesoplodon species (order Cetacea, family Ziphiidae) in the North Atlantic. Mammal Rev. 30:1–8.

MacLeod, C. D. 2006. How big is a beaked whale? A review of body length and sexual dimorphism in the family Ziphiidae. J. Cetacean Res. Mgmt. 7:301–308.

MacLeod, C. D., W. F. Perrin, R. Pitman, J. Barlow, L. Balance, A. D'Amico, T. Gerrodette, G. Joyce, K. D. Mullin, D. L. Palka, and G. T. Waring. 2006. Known and inferred distributions of beaked whale species (Cetacea: Ziphiidae). J. Cetacean Res. Mgmt. 7:271–286.

MacLulich, D. A. 1957. The place of chance in population processes. J. Wildl. Mgmt. 21:293–299.

MacMillen, R. E. 1965. Aestivation in the cactus mouse, Peromyscus eremicus. Comp. Biochem. Physiol. 16:227–248.

MacMillen, R. E., and T. Garland, Jr. 1989. Adaptive physiology. Pp. 143–168 in Advances in the study of Peromyscus (G. L. Kirkland, Jr., and J. N. Layne, eds.). Texas Tech Univ. Press, Lubbock.

MacMillen, R. E., and D. S. Hinds. 1983. Water regulatory efficiency in heteromyid rodents: a model and its application. Ecology 64:152–164.

MacMillen, R. E., and A. K. Lee. 1969. Water metabolism of Australian hopping mice. Comp. Biochem. Physiol. 28:493–514.

MacMillen, R. E., and A. K. Lee. 1970. Energy metabolism and pulmocutaneous water loss of Australian hopping mice. Comp. Biochem. Physiol., 35:355–369.

MacPhee, R. D. E., and M. J. Novacek. 1993. Definition and relationships of Lipotyphla. Pp. 13–31 in Mammal Phylogeny, vol. 2: Placentals (F. S. Szalay, M. J. Novacek, and M. C. McKenna, eds.). Springer-Verlag, New York.

MacPhee, R. D. E., C. Flemming, and D .P. Lunde. 1999. Last occurrence of the Antillean insectivoran Nesophontes: new radiometric dates and their inter pretation. Am. Mus. Novitates 3261:1–20.

Maddison, W. P., M. J. Donoghue, and D. R. Maddison. 1984. Outgroup analysis and parsimony. Syst. Zool. 33:83–103.

Maderson, P. F. A. (symposium organizer). 1972. The vertebrate integument. Am. Zool. 12:12–17.

Madison, D. M. 1985. Activity rhythms and spacing. Pp. 373–419 in Biology of New World Microtus (R. H. Tamarin, ed.). Spec. Pub. No. 8, American Society of Mammalogists.

Madison, D. M., R. W. FitzGerald, and W. J. McShea. 1984a. Dynamics of social nesting in overwintering meadow voles (Microtus pennsylvanicus): possible con sequences for population cycling. Behav. Ecol. Sociobiol. 15:9–17.

Madison, D. M., J. P. Hill, and P. E. Gleason. 1984b. Seasonality in the nesting behavior of Peromyscus leucopus. Am. Midl. Nat. 112:201–204.

Madsen, O., M. Scally, C. J. Douady, D. J. Kao, R. W. DeBry, R. Adkins, H. M. Amrine, M. J. Stanhope, W. W. de Jong, and M. S. Springer. 2001. Parallel adaptive radiations in two major clades of placental mammals. Nature 409:610–614.

Magnanini, A., A. F. Coimbra-Filho, R. A. Mittermeier, and A. Aldright. 1975. The Tijuca Bank of lion marmosets, Leontopithecus rosalia: A progress report. Int. Zoo Yearb. 15:284–287.

Maguire, J. H., and R. Hoff. 1992. Trypanosoma. Pp. 1984–1991 in Infectious diseases (S. L. Gorbach, J. G. Bartlett, and N. R. Blacklow, eds.). W. B. Saunders, Philadelphia.

Mahady, S. J., and J. O. Wolff. 2002. A field test of the Bruce effect in the monogamous prairie vole (Microtus ochrogaster). Behav. Ecol. Sociobiol. 52: 31–37.

Mahboubi, M., R. Ameur, J. Y. Crochet, and J. J. Jaeger. 1984. Earliest known proboscidean from early Eocene of northwest Africa. Nature 308:543–544.

Maher, C. R. 1991. Activity budgets and mating system of male pronghorn antelope at Sheldon National Wildlife Refuge, Nevada. J. Mammal. 72:739–744.

Maher, C. R., and D. F. Lott. 2000. A review of ecological determinants of territoriality within vertebrate species. Am. Mid. Nat. 143:1–29.

Mahoney, R. 1966. Techniques for the preparation of vertebrate skeletons. Pp. 327–351 in Laboratory techniques in zoology. Butterworth, Washington, DC.

Maier, W., J. Van Den Heever, and F. Durand. 1996. New therapsid specimens and the origin of the secondary hard and soft palate of mammals. J. Zool. Syst. Evol. Res. 34:9–19.

Malcolm, J. R., and J. C. Ray. 2000. Influence of timber extraction routes on Central African small mammal communities, forest structure, and tree diversity. Conserv. Biol. 14:1623–1638.

Maldini, D., L. Mazzuca, and S. Atkinson. 2005. Odontocete stranding patterns in the main Hawaiian Islands (1937–2002): how do they compare with live animal surveys? Pacific Sci. 59:55–67.

Malik, S., P. J. Wilson, R. J. Smith, D. M. Lavigne, and B. N. White. 1997. Pinniped penises in trade: a molecular-genetic investigation. Conserv. Biol. 11:1365–1374.

Malmquist, M. G. 1985. Character displacement and biogeography of the pygmy shrew in Northern Europe. Ecology 66:372–377.

Malo, A. F., E. R. S. Roldan, J. Garde, A. J. Soler, and M. Gomendio. 2005. Antlers honestly advertise sperm production and quality. Proc. Roy. Soc. Lond. Ser. B 272:149–157.

Malthus, R. T. 1798. An essay on the principle of population as it affects the future improvement of society. Johnson, London.

Mangan, S. A., and G. H. Adler. 1999. Consumption of arbuscular mycorrhizal fungi by spiny rats (Prochimys semispinosus) in eight isolated populations. J. Tropical Ecol. 15:779–790.

Manger, P. R., and J. D. Pettigrew. 1995. Electroreception and the feeding behaviour of platypus (Ornithorhynchus anatinus: Monotremata: Mammalia). Phil. Trans. Roy. Soc. Lond. Ser. B 347:359–381.

Manger, P. R., and J. D. Pettigrew. 1996. Ultrastructure, number, distribution, and innervation of electroreceptors and mechanoreceptors in the bill skin of the platypus, Ornithorhynchus anatinus. Brain Behav. Evol. 48:27–54.

Mannen, H., Y. Nagata, and S. Tsuji. 2001. Mitochondrial DNA reveal that domestic goat (Capra hircus) are genetically affected by two sub species of bezoar (Capra aegagurus). Biochem. Genet. 39:145–154.

Manning, J. T., and A. T. Chamberlain. 1993. Fluctuating asymmetry, sexual selection, and canine teeth in primates. Proc. Roy. Soc. Lond. Ser. B 251:83–87.

Manning, J. T., and A. T. Chamberlain. 1994. Fluctuating asymmetry in gorilla canines: a sensitive indicator of environmental stress. Proc. Roy. Soc. Lond. Ser. B 255:189–193.

Manser, M. B. 2001. The acoustic structure of suricates' alarm calls varies with predator type and the level of response urgency. Proc. Roy. Soc. Lond. Ser. B 268:2315–2324.

Mansergh, I. M., and L. S. Broome. 1994. The mountain pygmy-possum of the Australian Alps. New South Wales Univ. Press, Sydney.

Marchand, P. J. 1996. Life in the cold: an introduction to winter ecology, 3rd edition. Univ. Press of New England, Hanover, NH.

Marcus, L. C. 1992. Infections acquired from animals. Pp. 1267–1269 in Infectious diseases (S. L. Gorbach, J. G. Bartlett, and N. R. Blacklow, eds.). W. B. Saunders, Philadelphia.

Mares, M. A. 1992. Neotropical mammals and the myth of Amazonian biodiversity. Science 255:976–979.

Mares, M. A., and G. N. Cameron. 1994. Communities and ecosystems. Pp. 348–376 in Seventy-five years of mammalogy (E. C. Birney and J. R. Choate, eds.). Spec. Pub. No. 11, American Society of Mammalogists.

Mares, M. A., J. K. Braun, R. M. Barquez, and M. M. Diaz. 2000. Two new genera and species of halophytic desert mammals from isolated salt flats in Argentina. Occ. Pap. Mus. Texas Tech Univ. 203:1–27.

Mares, M. A., and T. E. Lacher, Jr. 1987. Ecological, morphological, and behavioral convergence in rock-dwelling mammals. Curr. Mammal. 2:307–348.

Marimuthu, G., J. Habersetzer, and D. Leippert. 1995. Active acoustic gleaning from the water surface by the Indian false vampire bat, Megaderma lyra. Ethology 99:61–74.

Marivaux, L., J.-L. Welcomme, P.-O. Antoine, G. Metais, I. M. Baloch, M. Benammi, Y. Chaimanee, S. Ducrocq, and J.-J. Jaeger. 2001. A fossil lemur from the Oligocene of Pakistan. Science 294:587–591.

Marivaux, L., L. Bocat, Y. Chaimanee, J.-J. Jaeger, B. Marandat, P. Srisuk, P. Tafforeau, C. Yamee, and J.-L. Welcomme. 2006. Cynocephalid dermoptrans from the Palaeogene of south Asia (Thailand, Myanmar and Pakistan): systematic, evolutionary and palaeobiogeographic implications. Zoologica Scripta 35:395–420.

Marler, P. 1968. Aggregation and dispersal: Two functions in primate communication. Pp. 420–438 in Primates: studies in adaptation and variability (P. C. Jay, ed.). Holt, Rinehart, and Winston, New York.

Marler, P. 1973. A comparison of vocalizations of red-tailed monkeys and blue monkeys, Cercopithecus ascanius and C. mitis, in Uganda. Z. Tierpsychol. 33:223–247.

Marmontel, M., S. R. Humphrey, and T. J. O'Shea. 1997. Population viability analysis of the Florida manatee (Trichechus manatus latirostris), 1976–1991. Conserv. Biol. 11:467–481.

Marquardt, W. C., R. S. Demaree, and R. B. Grieve. 2000. Parasitology and vector biology, 2d edition. Harcourt/Academic Press, New York.

Marshall, A .G. 1985. Old World phytophagous bats (Megachiroptera) and their food plants: a survey. Zool. J. Linnean Soc. 83:351–369.

Marshall, C. D., P. S. Kubilis, G. D. Huth, V. M. Edmonds, D. L. Halin, and R. L. Reep. 2000. Food-handling ability and feeding-cycle length of manatees feeding on several species of aquatic plants. J. Mammal. 81:649–658.

Marshall, C. R. 1990. Confidence intervals on stratigraphic ranges. Paleobiology 16:1–10.

Marshall, L. G. 1980. Marsupial paleobiogeography. Pp. 345–386 in Aspects of vertebrate history (L. L. Jacobs, ed.). Museum of Northern Arizona Press, Flagstaff.

Marshall, L. G. 1984. Monotremes and marsupials. Pp. 59–115 in Orders and families of recent mammals of the world. (S. Anderson and J. K. Jones, Jr., eds.). John Wiley and Sons, New York.

Marshall, L. G. 1988. Extinction. Pp. 219–254 in Analytical biogeography (A. A. Myers and P. S. Giller, eds.). Chapman and Hall, New York.

Marshall, L.G. 1988a. Land mammals and the Great American Interchange. Am. Sci. 76:380–388.

Marshall, L. G., J. A. Case, and M. O. Woodburne. 1990. Phylogenetic relationships of the families of marsupials. Pp. 433–506 in Current mammalogy (H. H. Genoways, ed.). Plenum Press, New York.

Marshall, L. G., T. Sempere, and R. F. Butler. 1997. Chronostratigraphy of the mammal-bearing Paleocene of South America. J. South Am. Earth Sci. 10:49–70.

Marshall, L. G., S. D. Webb, J. J. Sepkoski, and D. M. Raup. 1982. Mammalian evolution and the Great American Interchange. Science 215:1351–1357.

Martan, J., and B. A. Shepherd. 1976. The role of the copulatory plug in reproduction of the guinea pig. J. Exp. Zool. 196:79–84.

Martenson, J. 1982. The pregnant rabbit, guinea pig, sheep, and rhesus monkey as models in reproductive physiology. Europ. J. Obstet. Reprod. Biol. 18:169–182.

Martin, A. R., and V. M. F. daSilva. 2006. Sexual dimorphism and body scarring in the boto (Amazon River dolphin) Inia geoffrensis. Mar. Mammal. Sci. 22:25–33.

Martin, A. R., P. Hall, and P. R. Richard. 2001. Dive behaviour of belugas (Delphinapterus leucas) in the shallow waters of western Hudson Bay. Arctic 54:276–283.

Martín H. G., and Goldenfeld N. 2006. On the origin and robustness of power-law species-area relationships in ecology. Proc. Natl. Acad. Sci. USA 103: 10310–10315.

Martin, I. G. 1981. Venom of the short-tailed shrew (Blarina brevicauda) as an insect immobilizing agent. J. Mammal. 62:189–192.

Martin, I. G. 1983. Daily activity of short-tailed shrew (Blarina brevicauda) in simulated natural conditions. Am. Midl. Nat. 109:136–144.

Martin, J. E., J. A. Case, J. W. M. Jagt, A. S. Schulp, and Eric W. A. Mulder. 2005. A new European marsupial indicates a late Cretaceous high-latitude transatlantic dispersal route. J. Mammal. Evol. 12:495–511.

Martin, L. D. 1989. Fossil history of the terrestrial Carnivora. Pp. 536–568 in Carnivore behavior, ecology, and evolution (J. L. Gittleman, ed.). Cornell Univ. Press, Ithaca, NY.

Martin, R. A., M. Florentini, and F. Connors. 1980. Social facilitation of reduced oxygen consumption in Mus musculus and Meriones ungiculatus. Comp. Biochem. Physiol. 65A:519–522.

Martin, R. D. 1979. Phylogenetic aspects of prosimian behavior. Pp. 45–77 in The study of Prosimian behavior (G. A. Doyle and D. Martin, eds.). Academic Press, New York.

Martin, R. D. 1990. Primate origins and evolution. Princeton Univ. Press, Princeton, NJ.

Martin, R. D., A. F. Dixson, and E. J. Wickings (eds.). 1992. Paternity in primates: genetics tests and theories. S. Karger, Basel, Switzerland.

Martin, R. E., R. Pine, and A. F. DeBlase. 2001. A manual of mammalogy with keys to the families of the world, 3rd edition. McGraw-Hill, New York.

Martins, E. G., V. Bonato, C. Q. da-Silva, and S. R. F. dos Reis, 2006. Partial semelparity in the neotropical didelphid marsupial Gracilinanus microtarsus. J. Mammal. 87:915–920.

Maser, C., and Z. Maser. 1987. Notes on mycophagy in four species of mice in the genus Peromyscus. Great Basin Nat. 47:308–313.

Maser, C., J. M. Trappe, and R. A. Nussbaum. 1978. Fungal-small mammals interrelationships with emphasis on Oregon coniferous forests. Ecology 59:799–809.

Maser, Z., C. Maser, and J. M. Trappe. 1985. Food habits of the northern flying squirrel (Glaucomys sabrinus) in Oregon. Can. J. Zool. 63:1084–1088.

Maser, C., Z. Maser, J. W. Witt, and G. Hunt. 1986. The northern flying squirrel: a mycophagist in southwestern Oregon. Can. J. Zool. 64:2086–2089.

Masini, R. J., P. K. Anderson, and A. J. McComb. 2001. A Halodule-dominated community in a sub tropical embayment: physical environment, productivity, biomass, and impact of dugong grazing. Aquat. Bot. 71:179–197.

Mason, M. J. 2003. Morphology of the middle ear of golden moles (Chrysolchloridae). J. Zool. Lond. 260:391–403.

Mason, M. J., and P. M. Narins. 2002. Seismic sensitivity in the desert gold mole (Eremitalpa granti): A review. J. Comp. Psychol. 116:158–163.

Massei, G., and P. V. Genov. 2004. The environmental impact of wild boar. Galemys 16:135–145.

Massey, A. 1977. Agonistic aids and kinship in a group of pigtail macaques. Behav. Ecol. Sociobiol. 2:31–40.

Massey, A., and J. G. Vandenbergh. 1980. Puberty delay by a urinary cue from female house mice in feral populations. Science 209:821–822.

Mate, B., P. Duley, B. Lagerquist, F. Wenzel, A. Stimpert, and P. Clapham. 2005. Observations of a female North Atlantic right whale (Eubalaena glacialis) in simultaneous copulation with two males: evidence for sperm competition. Aquat. Mammals 31:157–160.

Mateo, J. M. 2003. Kin recognition in ground squirrels and other rodents. J. Mammal. 84: 1163–1181.

Mateo, J. M. 2006. The nature and representation of individual recognition odours in Belding's ground squirrels. Anim. Behav. 71:141–154.

Matthee, C. A., J. D. Burzlaff, J. F. Taylor, and S. K. Davis. 2001. Mining the mammalian genome for Artiodactyl systematics. Syst. Biol. 50:367–390.

Matthews, L. H. 1978. The natural history of the whale. Weidenfeld and Nicolson, London.

Matthew, W. D. 1915. Climate and evolution. Ann. N.Y. Acad. Sci. 24:171–318.

Mattson, D. J., B. M. Blanchard, and R. R. Knight. 1991. Food habits of Yellowstone grizzly bears, 1977–1987. Can. J. Zool. 69:1619–1629.

Mauck, B., K. Bilgmann, D.D. Jones, U. Eysel, and G. Dehnhardt. 2003. Thermal windows on the trunk of hauled-out seals: hot spots for thermoregulatory evaporation? J. Exp. Biol. 206:1727–1738.

Maxim, P. E., D. M. Bowden, and G. P. Sackett. 1976. Ultradian rhythms of solitary and social behavior in Rhesus monkeys. Physiol. Behav. 17:337–344.

Maxson, L. R., and R. D. Maxson. 1990. Proteins II: immunological techniques. Pp. 127–155 in Molecular systematics (D.M. Hillis and C. Moritz, eds.). Sinauer Associates, Sunderland, MA.

May, E. L. 2003. Effects of cold acclimation on shivering Intensity In the kowari (Dasyuroides byrnei), a dasyurid marsupial. J. Therm. Biol. 28:477–487.

May, M. 1991. Aerial defense tactics of flying insects. Am. Sci. 79:316–328.

Mayen, F. 2003. Haematophagous bats in Brazil, their role in rabies transmission, impact on public health, livestock industry and alternatives to an indiscriminate reduction of bat population. J. Vet. Med. B. Infectious Diseases and Veterinary Public Health 50:469–472.

Mayle, B. A., and B. W. Staines. 1998. An overview of methods used for estimating the size of deer populations in Great Britain. Pp. 19–31 in Population ecology, management, and welfare of deer (C. R. Goldspink, S. J. King, and R. J. Putnam, eds.). Manchester Metropolitan Univ., Manchester.

Mayle, B.A., R. J. Putnam, and I. Wyllie. 2000. The use of trackway counts to establish an index of deer presence. Mammal Rev. 30:223–237.

Maylon, C., and S. Healy. 1994. Fluctuating asymmetry in antlers of fallow deer, Dama dama, indicates dominance. Anim. Behav. 48:248–250.

Maynard Smith, J., and R. Savage. 1956. Some loco motor adaptations in mammals. J. Linnean Soc. Zool. 42:603–622.

Mayr, E. 1942. Systematics and the origin of species. Columbia Univ. Press, New York.

Mayr, E. 1956. Geographical character gradients and climatic adaptation. Evolution 10:105–108.

Mayr, E. 1963. Animal species and evolution. Harvard Univ. Press, Cambridge, MA.

Mayr, E. 1969. Principles of systematic zoology. McGraw-Hill, New York.

Mayr, E. 1970. Populations, species, and evolution. Belknap Press of Harvard Univ. Press, Cambridge, MA.

Mayr, E., and P. D. Ashlock. 1991. Principles of systematic zoology. McGraw-Hill, New York.

McAleer, K., and L-A Giraldeau. 2006. Testing central place foraging in eastern chipmunks, Tamias striatus, by altering loading functions. Anim. Behav. 71:1447–1453.

McAllan, B. 2003. Timing of reproduction in carnivorous marsupials. Pp. 147–168 in Predators with pouches: the biology of carnivorous marsupials (M. Jones, C. Dickman, and M. Archer, eds.). CSIRO Publishing, Collingwood, Australia.

McBee, R. H. 1971. Significance of intestinal microflora in herbivory. Annu. Rev. Ecol. Syst. 2:165–176.

McCarty, R. 1975. Onychomys torridus. Mamm. Species 59:1–5.

McCarty, R. 1978. Onychomys leucogaster. Mamm. Species 87:1–6.

McConkey, K. M., and D. R. Drake. 2006. Flying foxes cease to function as seed dispersers long before they become rare. Ecology 87:271–276.

McCormick, J. B., and S. P. Fisher-Hoch. 1994. Zoonoses caused by Filoviridae. Pp. 375–383 in Handbook of zoonoses, section B: viral, 2d edition (C. W. Beran and J. H. Steele, eds.). CRC Press, Boca Raton, FL.

McCracken, G. F. 1984. Communal nursing in Mexican free-tailed bat maternity colonies. Science 223:1090–1091.

McCracken, G. F., and J. W. Bradbury. 1977. Paternity and genetic heterogeneity in the polygynous bat, Phyllostomus hastatus. Science 198:303–306.

McCracken, G. F., and G. S. Wilkinson. 2000. Bat mating systems. Pp. 321–362 in Reproductive biology of bats (E. G. Crichton and P. H. Krutzsch, eds.). Academic Press, New York.

McCullough, D. R. 1985. Long range movements of large terrestrial mammals. Pp. 444–465 in Migration: mechanisms and adaptive significance (M. A. Rankin, ed.). Marine Science Institute, Port Aransas, TX.

McDade, J. E., and V. F. Newhouse. 1986. Natural history of Rickettsia rickettsii. Annu. Rev. Microbiol. 40:287–309.

McFarland, W. N., and W. A. Wimsatt. 1969. Renal function and its relation to the ecology of the vampire bat, Desmodus rotundus. Comp. Biochem. Physiol. 28:985–1006.

McIlwee, A. P., and C. N. Johnson. 1998. The contribution of fungus to the diets of three mycophagous marsupials in Eucalyptus forests, revealed by stable isotope analysis. Func. Ecol. 12:223–231.

McKenna, M. C., and S. K. Bell. 1997. Classification of mammals above the species level. Columbia Univ. Press, New York.

McLaren, S. B., and J. K. Braun (eds.). 1993. GIS applications in mammalogy. Univ. Oklahoma Press, Norman.

M'Closkey, R. T. 1975. Habitat dimensions of white-footed mice, Peromyscus leucopus. Am. Midl. Nat. 93:158–167.

M'Closkey, R. T. 1978. Niche separation and assembly in four species of Sonoran Desert rodents. Am. Nat. 112:683–694.

McKenna, M. C. and S. K. Bell. 1997. Classification of mammals above the species level. Columbia Univ. Press, New York.

McManus, J. J. 1974. Didelphis virginiana. Mamm. Species 40:1–6.

McNab, B. K. 1963. Bioenergetics and the determination of home range size. Am. Nat. 97:133–140.

McNab, B. K. 1971. On the ecological significance of Bergmann's rule. Ecology 52:845–854.

McNab, B. K. 1974. The energetics of endotherms. Ohio J. Sci. 74:370–380.

McNab, B. K. 1979a. Climatic adaptation in the energetics of heteromyid rodents. Comp. Biochem. Physiol. 62A:813–820.

McNab, B. K. 1979b. The influence of body size on the energetics and distribution of fossorial and burrowing mammals. Ecology 60:1010–1021.

McNab, B. K. 1980a. Energetics and the limits to a temperate distribution in armadillos. J. Mammal. 61:606–627.

McNab, B. K. 1980b. On estimating thermal conductance in endotherms. Physiol. Zool. 53:145–156.

McNab, B. K. 1984. Physiological convergence amongst ant-eating and termite-eating mammals. J. Zool. Lond. 203:485–510.

McNab, B. K. 1985. Energetics, population biology, and distribution of Xenarthrans, living and extinct. Pp. 219–232 in The evolution and ecology of armadillos, sloths, and vermilinguas (G. G. Montgomery, ed.). Smithsonian Institution Press, Washington, DC.

McNab, B. K. 1988. Complications inherent in scaling the basal rate of metabolism in mammals. Q. Rev. Biol. 63:25–54.

McNab, B. K. 1989. Basal rate of metabolism, body size, and food habits in the Order Carnivora. Pp. 335–354 in Carnivore behavior, ecology, and evolution (J. L. Gittleman, ed.). Comstock Publ. Assoc., Ithaca, NY; Cornell Univ. Press, Ithaca, NY.

McNab, B. K. 1991. The energy expenditure of shrews. Pp. 35–45 in The biology of the Soricidae (J. S. Findley and T. L. Yates, eds.). The Museum of Southwestern Biology, Univ. New Mexico, Albuquerque.

McNab, B. K. 1995. Energy expenditure and conservation in frugivorous and mixed-diet carnivorans. J. Mammal. 76:206–222.

McNab, B. K. 2006. The evolution of energetics in eutherian "insectivorans": An alternate approach. Acta Theriol. 51:113–128.

McNab, B. K., and F. J. Bonaccorso. 1995. The energetics of pteropodid bats. Pp. 111–122 in Ecology, evolution, and behaviour of bats (P. A. Racey and S. M. Swift, eds.). Zool. Soc. Lond., Clarendon Press, London.

McNab, B. K., and P. Morrison. 1963. Body temperature and metabolism in subspecies of Peromyscus from arid and mesic environments. Ecol. Monogr. 33:63–82.

McNair, J. N. 1982. Optimal giving-up times and the marginal value theorem. Am. Nat. 119:511–529.

McNaughton, S. J. 1976. Serengeti migratory wildebeest: facilitation of energy flow by grazing. Science 171:92–94.

McNaughton, S. J. 1985. Ecology of a grazing ecosystem: the Serengeti. Ecol. Monogr. 55:259–294.

McNay, M. E., T. R. Stephenson, and B. W. Dale. 2006. Diagnosing pregnancy, in utero litter size, and fetal growth with ultrasound in wild, free-ranging wolves. J. Mammal. 87:85–92.

McQuade, D. B., E. H. Williams, and H. B. Eichenbaum. 1986. Cues used for localizing food by the grey squirrel (Sciurus carolinensis). Ethology 72:22–30.

Mead, J. G., and R. L. Brownell, Jr. 1993. Order Cetacea. Pp. 349–364 in Mammal species of the world: a taxonomic and geographic reference, 2d edition (D. E. Wilson and D. M. Reeder, eds.). Smithsonian Institution Press, Washington, DC.

McShea, W. J., M. Aung, D. Poszig, C. Wemmer, and S. Monfort. 2001. Forage, habitat use, and sexual segregation by a tropical deer (Cervus eldi thamin) in a dipterocarp forest. J. Mammal. 82:848–857.

McWilliams, L. A. 2005 Variation in the diet of the Mexican free-tailed bat (Tadarida brasiliensis mexicana). J. Mammal. 86:599–605.

Mead, J. G., and R. L. Brownell, Jr. 2005. Order Cetacea. Pp. 723–743 in Mammal species of the world: a taxonomic and geographic reference. 3rd edition, 2 vols. (D. E. Wilson and D. M. Reeder, eds.). Johns Hopkins Univ. Press, Baltimore.

Mead, R. A. 1968. Reproduction in western forms of the spotted skunk (Genus Spilogale). J. Mammal. 49:373–390.

Mead, R. A. 1989. The physiology and evolution of delayed implantation in carnivores. Pp. 437–464 in Carnivore behavior, ecology, and evolution (G. L. Gittleman, ed.). Comstock Publ. Assoc., Ithaca, NY.

Mearns, E. A. 1907. Mammals of the Mexican boundary of the United States. Bull. U.S. Nat. Mus. 56:1–530.

Mech, L. D. 1970. The wolf: the ecology and behavior of an endangered species. Natural History Press, Garden City, NY.

Mech, L. D. 1995. The challenge and opportunity of recovering wolf populations. Conserv. Biol. 9:270–278.

Mech, L. D., and F. J. Turkowski, 1966. Twenty-three raccoons in one winter den. J. Mammal. 47:529–530.

Mech, L. D., U. S. Seal, and G. D. DelGiudice. 1987. Use of urine in snow to indicate condition of wolves. J. Wildl. Mgmt. 51:10–13.

Medellin, R. A., and J. Soberon. 1998. Predictions of mammal diversity on four land masses. Conserv. Biol. 13:143–149.

Medellin, R. A., M. Equihua, and M. A. Amin. 2000. Bat diversity and abundance as indicators of disturbance in neotropical rainforests. Conserv. Biol. 14:1666–1675.

Meier, B., R. Albignac, A. Peyrieras, Y. Rumpler, and P. Wright. 1987. A new species of Hapalemur (Primates) from south east Madagascar. Folia Primatol. 48:211–215.

Meijaard, E., and C. P. Groves. 2004. A taxonomic revision of the Tragulus mouse-deer (Artiodactyla). Zool. J. Linnean Soc. 140:63–102.

Meikle, D. B., and S. H. Vessey. 1981. Nepotism among rhesus monkey brothers. Nature 294:160–161.

Meikle, D. B., and S. H. Vessey. 1988. Maternal dominance rank and lifetime survivorship of male and female rhesus monkeys. Behav. Ecol. Sociobiol. 22:379–383.

Meiri, S., and T. Dayan. 2003. On the validity of Bergmann's rule. J. Biogeogr. 30:331–351.

Meiri, S., T. Dayan, and D. Simberloff. 2004. Carnivores, biases, and Bergmann's rule. Biol. J. Linnean Soc. 81:579–588.

Melis, A. P., B. Hare, and M. Tomasello. 2006. Chimpanzees recruit the best collaborators. Science 311:1297–1300.

Melis, C., P. A. Szafranska, B. Jedrzejewska, and K. Barton. 2006. Biogeographical variation in the population density of wild boar (Sus scrofa) in western Eurasia. J. Biogeogr. 33:803–811.

Mellish, J. E., S. J. Iverson, and W. D. Bowen. 1999. Variation in milk production and lactation performance in grey seals and consequences for pup growth and weaning characteristics. Physiol. Biochem. Zool. 72:677–690.

Meng, J. and A. R. Wyss. 2005. Glires (Lagomorpha, Rodentia). Pp. 145–158 *in* The rise of placental mammals: origins and relationships of the major extant clades (K. D. Rose and J. D. Archibold, eds.). Johns Hopkins Univ. Press, Baltimore.

Mengak, M. T., and D. C. Guynn, Jr. 1987. Pitfalls and snap traps for sampling small mammals and herptofauna. Am. Midl. Nat. 118:284–288.

Menkhorst, P., and F. Knight. 2005. A Field Guide to the Mammals of Australia. Oxford Univ. Press, Oxford.

Mercador, J., M. Panger, and C. Boesch. 2002. Excavation of a chimpanzee stone tool site in the African rain forest. Science 296:1452–1455.

Mercer, J. M., and V. L. Roth. 2003. The effects of Cenozoic global change on squirrel phylogeny. Science 299:1568–1572.

Merriam, C. H. 1890. Results of a biological survey of the San Francisco mountain region and desert of the Little Colorado, Arizona. N. Am. Fauna. 3:1–136.

Merriam, C. H. 1894. Laws of temperature control of the geographical distribution of terrestrial animals and plants. Natl. Geog. 6:229–238.

Merriam, G. 1995. Movement in spatially divided populations: responses to landscape structure. Pp. 64–77 *in* Landscape approaches in mammalian ecology and conservation (W. Z. Lidicker, Jr., ed.). Univ. Minnesota Press, Minneapolis.

Merriam, G., and A. Lanoue. 1990. Corridor use by small mammals: field measurement for three experimental types of *Peromyscus leucopus*. Landscape Ecol. 4:123–132.

Merrick, J. R., M. Archer, G. M. Hickey, and M. S. Y. Lee (eds.). 2006. Evolution and biogeography of Australasian vertebrates. Auscipub, Oatlands, NSW, Australia.

Merritt, J. F. (ed.). 1984. Winter ecology of small mammals. Spec. Pub. No 10, Carnegie Museum of Natural History, Pittsburgh.

Merritt, J. F. 1986. Winter survival adaptations of the short-tailed shrew (*Blarina brevicauda*) in an Appalachian montane forest. J. Mammal. 67:450–464.

Merritt, J. F. 1987. Guide to the mammals of Pennsylvania. Univ. Pittsburgh Press, Pittsburgh.

Merritt, J. F. 1995. Seasonal thermogenesis and changes in body mass of masked shrews, *Sorex cinereus*. J. Mammal. 76: 1020–1035.

Merritt, J. F., and J. M. Merritt. 1978. Population ecology and energy relationships of *Clethrionomys gapperi* in a Colorado subalpine forest. J. Mammal. 59: 576–598.

Merritt, J. F., and S. H. Vessey. 2000. Shrews—small insectivores with polyphasic patterns. Pp. 235–251 *in* Activity patterns in small mammals (S. Halle and N. C. Stenseth, eds.). Ecological Studies 141, Springer-Verlag, Heidelberg, Germany

Merritt, J. F., and D. A. Zegers. 2002. Maximizing survivorship in cold: thermogenic profiles of non-hibernating mammals. Acta Theriol. 47:221–234.

Merritt, J. F., G. L. Kirkland, Jr., and R. K. Rose. 1994. Advances in the biology of shrews. Spec. Pub. No. 18, Carnegie Museum of Natural History, Pittsburgh.

Merritt, J. F., D. A. Zegers, and L. R. Rose. 2001. Seasonal thermogenesis of southern flying squirrels (Glaucomys volans). J. Mammal. 82:51–64.

Merritt, J. F., S. Churchfield, R. Hutterer, and B. I. Sheftel (eds.). 2005. Advances in the biology of shrews II. Special Publication of the International Society of Shrew Biologists 01, New York.

Meserve, P. L. 1976. Food relationships of a rodent fauna in a California coastal scrub community. J. Mammal. 57:300–319.

Meserve, P. L., J. R. Gutiérrez, J. A. Yunger, L. C. Contreras, and F. M. Jaksic. 1996. Role of biotic interactions in a small mammal assemblage in semi-arid Chile. Ecology 77:133–148.

Messenger, S. L., C. E. Rupprecht, and J. S. Smith. 2003. Bats, emerging viral infections, and the rabies paradigm. Pp. 622–679 *in* Bat Ecology (T. H. Kunz and M. B. Fenton, eds.). Univ. Chicago Press, Chicago.

Messer, M., A. S. Weiss, D. C. Shaw, and M. Westerman. 1998. Evolution of the monotremes: phylogenetic relationship to marsupials and eutheri-

ans, and estimation of divergence dates based on alpha-lactalbumin amino acid sequences. J. Mammal. Evol. 1998:95–105.

Messier, F. 1985. Solitary living and extraterritorial movements of wolves in relation to social status and prey abundance. Can. J. Zool. 63:239–245.

Metais, G., Y. Chaimanee, J. J. Jaeger, and S. Ducrocq. 2001. New remains of primitive ruminants from Thailand: evidence for the early evolution of the Ruminantia in Asia. Zool. Scr. 30:231–248.

Michaux, J., A. Reyes, and F. Catzeflis. 2001. Evolutionary history of the most speciose mammals: molecular phylogeny of muroid rodents. Mol. Biol. Evol. 18:2017–2031.

Michaux, J. R., J. G. De Bellocq, M. Sarà, and S. Morand. 2002. Body size increase in insular rodent populations: a role for predators? Glob. Ecol. Biogeogr.11:427–436.

Mikesic, D. G., and L. C. Drickamer. 1992a. Effects of radiotransmitters and fluorescent powders on activity of wild house mice (*Mus musculus*). J. Mammal. 73:663–667.

Mikesic, D. G. and L. C. Drickamer. 1992b. Factors affecting home-range size in house mice (*Mus musculus domesticus*) living in outdoor enclosures. Am. Midl. Nat. 127:31–40.

Miles, A. E. W., and C. Grigson. 1990. Colyer's variations and diseases of the teeth of animals. Cambridge Univ. Press, New York.

Milinkovitch, M. C. 1992. DNA-DNA hybridization support for ungulate ancestry of Cetacea. J. Evol. Biol. 5:149–160.

Milinkovitch, M. C., M. Berube, and P. J. Palsboll. 1998. Cetaceans are highly derived artiodactyls. Pp. 113–131 *in* The emergence of whales: evolutionary patterns in the origin of Cetacea (J. G. M. Thewissen, ed.). Plenum Press, New York.

Milinski, M., and C. Wedekind. 2001. Evidence for MHC-correleated perfume preferences in humans. Behav. Ecol. 12:140–149.

Milius, S. 2002. Wild hair: the suddenly famous science of fur snagging. Sci. News 161:250–252.

Millar, J. S. 1971. Breeding of the pika in relationship to the environment. Ph.D. diss., Univ. Alberta, Edmonton.

Millar, J. S. 1978. Energetics of reproduction in *Peromyscus leucopus*: the cost of lactation. Ecology 59:1055–1061.

Millar, J. S., and F. C. Zwickel. 1972. Characteristics and ecological significance of hay piles of pikas. Mammalia 36:657–667.

Miller, B., D. Biggins, L. Hanebury, and A. Vargas. 1994a. Reintroduction of the black-footed ferret (*Mustela nigripes*). Pp. 455–464 *in* Creative conservation: interactive management of wild and captive animals (P. J. S. Olney, G. M. Mace, and A. T. C. Feistner, eds.). Chapman and Hall, London.

Miller, B., G. Ceballos, and R. Reading. 1994b. The prairie dog and biotic diversity. Conserv. Biol. 8:677–681.

Miller, G. S., Jr., and J. W. Gidley. 1934. Mammals and how they are studied, part II, in warm-blooded vertebrates. Smithsonian Scientific Series, vol. 9. Smithsonian Institution, Washington, DC.

Miller, P. J. O., N. Biassoni, A. Samuel, and P. L. Tyack. 2000. Whale songs lengthen in response to sonar. Nature 405:903.

Miller, P. J. O., M. P. Johnson, and P. L. Tyack. 2004. Sperm whale behaviour indicates the use of echolocation click buzzes "creaks" in prey capture. Proc. Roy. Soc. Biol. Sci. Ser. B 271:2239–2247.

Millien, V. 2004. Relative effects of climate change, isolation, and competiton on body-size evolution in the Japanese field mouse, *Apodemus argenteus*. J. Biogeogr. 31:1267–1276.

Mills, M. G. L. 1985. Related spotted hyaenas forage together but do not cooperate in rearing young. Nature 316:61–64.

Mills, M. G. L. 1996. Methodological advances in capture, census, and food-habits studies of large African carnivores. Pp. 223–266 *in* Carnivore behavior, ecology, and evolution, vol. 2 (J. Gittleman, ed.). Cornell Univ. Press, Ithaca, NY.

Milner, J. M., and S. Harris. 1999. Activity patterns and feeding behaviour of the tree hyrax, *Dendrohyrax arboreus*, in the Parc National des Volcans, Rwanda. Afr. J. Ecol. 37:267–280.

Milner-Gulland, E. J., and R. Mace. 1991. The impact

of the ivory trade on the African elephant *Loxodonta africana* population as assessed by data from the trade. Biol. Conserv. 55:215–229.

Milton, K. 1985. Mating patterns of woolly spider monkeys, *Brachyteles arachnoides*: implications for female choice. Behav. Ecol. Sociobiol. 17:53–59.

Mindell, D. P. 1991. Aligning DNA sequences: homology and phylogenetic weighting. Pp. 73–89 *in* Phylogenetic analysis of DNA sequences (M. M. Miyamoto and J. Cracraft, eds.). Oxford Univ. Press, Oxford.

Mitchell, D., S. K. Maloney, H. P. Laburn, M. H. Knight, G. Kuhnen, and C. Jessen. 1997. Activity, blood temperature, and brain temperature of free-ranging springbok. J. Comp. Physiol. B 167:335–343.

Mitchener, G. R. 2004. Hunting techniques and tool use by North American badgers preying on Richardson's ground squirrels. J. Mammal. 85:1019–1027.

Miththapala, S., J. Seidensticker, and S. J. O'Brien. 1996. Phylogenetic subspecies recognition in leopards (*Panthera pardus*): molecular genetic variation. Conserv. Biol. 10:1115–1132.

Mittermeier, R. A., and D. L. Cheney. 1986. Conservation of primates and their habitats. Pp. 477–490 *in* Primate societies (B. B. Smuts, D. L. Cheney, R. M. Seyfarth, R. W. Wrangham, and T. T. Struhsaker, eds.). Univ. Chicago Press, Chicago.

Mittermeier, R. A., I. Tattersall, W. R. Konstant, D. M. Meyers, and R. B. Mast. 1994. Lemurs of Madagascar. Conservation International, Washington, DC.

Mivart, St. G. J. 1873. On Lepilemur and Cheirogaleus and on the zoological rank of Lemuroidea. Proc. Zool. Soc. Lond. 1873:484–510.

Mizroch, S. A., L. M. Herman, J. M. Straley, D. A. Glockner-Ferrari, C. Jurasz, J. Darling, S. Cerchio, C. M. Gabriele, D. R. Salden, and O. von Ziegesar. 2004. Estimating the adult survival rate of central North Pacific humpback whales (*Megaptera novaeangliae*). J. Mammal. 85:963–972.

Mladenoff, D. J., T. A. Sickley, R. G. Haight, and A. P. Wydeven. 1995. A regional landscape analysis and prediction of favorable gray wolf habitat in the northern Great Lakes Region. Cons. Biol. 9:279–294.

Moehlman, P. D. 1979. Jackal helpers and pup survival. Nature 277:382–383.

Moehlman, P. D. 1983. Socioecology of silverbacked and golden jackals, *Canis mesomelas* and *C. aureus*. Pp. 423–453 *in* Recent advances in the study of mammalian behavior (J. F. Eisenberg and D. G. Kleiman, eds.). Spec. Pub. No. 7, American Society of Mammalogists.

Moehlman, P. D. 1986. Ecology of cooperation in canids. Pp. 64–86 *in* Ecological aspects of social evolution: birds and mammals (D. I. Rubenstein and R. W. Wrangham, eds.). Princeton Univ. Press, Princeton, NJ.

Mohl, B. 2001. Sound transmission in the nose of the sperm whale Physeter catodon: a post mortem study. J. Comp. Physiol. A 187:335–340.

Møller, A. P. 1988. Ejaculate quality, testes size, and sperm competition in primates. J. Hum. Evol. 17:479–488.

Møller, A. P., J. J. Cuervo, J. J. Soler, and C. Zamora-Muñoz. 1996. Horn asymmetry and fitness in gemsbok, *Oryx g. gazella*. Behav. Ecol. 7:247–253.

Moller, H. 1983. Foods and foraging behaviour of red (*Sciurus vulgaris*) and grey (*S. carolinensis*) squirrels. Mammal Rev. 13:81–98.

Montgelard, C., S. Ducrocq, and E. Douzery. 1998. What is a Suiforme (Artiodactyla)? Contribution of cranioskeletal and mitochondrial DNA data. Mol. Phylogenet. Evol. 9:528–532.

Montgomery, G. G. 1978. The ecology of arboreal folivores. Smithsonian Institution Press, Washington, DC.

Moore, J., and R. Ali. 1984. Are dispersal and inbreeding avoidance related? Anim. Behav. 32:94–112.

Moore, R. Y. 1973. Retinohypothalamic projection in mammals: a comparative study. Brain Res. 49:403–409.

Moore, R. Y. 1982. Organization and function of a central nervous system circadian oscillator: the suprachiasmatic nucleus. Fed. Proc. 42:2783–2789.

Moore, S. E., J. M. Grebmeier, and J. R. Davies. 2003.

Gray whale distribution relative to forage habitat in the northern Bering Sea: current conditions and retrospective summary. Can. J. Zool. 81:734–742.

Morales, J. C., and J. W. Bickham. 1995. Molecular systematics of the genus *Lasiurus* (Chiroptera: Vespertilionidae) based on restriction-site maps of the mitochondrial ribosomal genes. J. Mammal. 76:730–749.

Morand, S., M. S. Hafner, R. D. M. Page, and D. L. Reed. 2000. Comparative body size relationships in pocket gophers and their chewing lice. Biol. J. Linnean Soc. 70:239–249.

Morell, V. 1996. New mammals discovered by biology's new explorers. Science 273:1491.

Morey, D. F. 1994. The early evolution of the domestic dog. Am. Sci. 82:336–347.

Morgan, G. S., and N. J. Czaplewski. 2003. A new bat (Chiroptera: Natalidae) from the early Miocene of Florida, with comments on natalid phylogeny. J. Mammal. 84:729–752.

Morgan, L. H. 1868. The American beaver and his works. J. B. Lippincott, Philadelphia.

Mori, J., T. Kubodera, and N. Baba. 2001. Squid in the diet of northern fur seals, *Callorhinus ursinus*, caught in the western and central North Pacific Ocean. Fish. Res. 52:91–97.

Mori, Y., and I. L. Boyd. 2004. The behavioral basis for nonlinear functional responses and optimal foraging in Antarctic fur seals. Ecology 85:398–410.

Moritz, C., T. E. Dowling, and W. M. Brown. 1987. Evolution of animal mitochondrial DNA: relevance for population biology and systematics. Annu. Rev. Ecol. Syst. 18:269–292.

Morone, J. J., and J. V. Crisci. 1995. Historical biogeography: introduction to methods. Annu. Rev. Ecol. Syst. 26:373–401.

Morris, D. W. 1992. Optimum brood size: tests of alternative hypotheses. Evolution 46:1848–1861.

Morris, R. J. 1986. The acoustic faculty of dolphins. Pp. 369–399 *in* Research on dolphins (M. M. Bryden and R. Harrison, eds.). Clarendon Press, Oxford.

Morrison, M. L., W. M. Block, M. D. Strickland, and W. L. Kendall. 2001. Wildlife study design. Springer-Verlag, New York.

Morrison, P. 1960. Some interrelations between weight and hibernation function. Bull. Mus. Zool. 124:75–91.

Morrison, P. 1966. Insulative flexibility in the guanaco. J. Mammal. 47:18–23.

Morrison, P., and B. K. McNab. 1967. Temperature regulation in some Brazilian phyllostomid bats. Comp. Biochem. Physiol. 21:207–221.

Morrison, P., F. A. Ryser, and A. R. Dawe. 1959. Studies on the physiology of the masked shrew *Sorex cinereus*. Physiol. Zool. 32:256–271.

Morrison, P. R. 1948. Oxygen consumption in several small wild mammals. J. Cell. Comp. Physiol. 31:69–96.

Morrison, P. R. 1965. Body temperatures in some Australian mammals. Aust. J. Zool. 13:173–187.

Morton, S. R., and R. E. MacMillen. 1982. Seeds as sources of preformed water for desert-dwelling granivores. J. Arid Environ. 5:61–67.

Morwood, M. J., P. Brown, W. Jatmiko, T. Sutikna, E. Saptomo, K. E. Westway, A. D. Rokus, R. G. Roberts, T. Maeda, S. Wasisto, and T. Djubiantono. 2005. Further evidence for small-bodied hominins from the Late Pleistocene of Flores, Indonesia. Nature 437:1012–1017.

Moss, C. J. 1983. Oestrous behaviour and female choice in the African elephant. Behaviour 86:167–196.

Mossing, T. 1975. Measuring small mammal locomotory activity with passage counters. Oikos 26:237–239.

Mossman, H. W. 1987. Vertebrate fetal membranes. Rutgers Univ. Press, New Brunswick, NJ

Motokawa, M., L. Liang-Kong, M. Harada, and S. Hattori. 2003. Morphometric geographic variation in the Asian lesser white-toothed shrew *Crocidura shantungensis* (Mammalia, Insectivora) in East Africa. Zool. Sci. 20:789–795.

Motulsky, H. 1995. Intuitive biostatistics. Oxford Univ. Press, New York.

Mouchaty, S., J. A. Cook, and G. F. Shields. 1995. Phylogenetic analysis of northern hair seals based on nucleotide sequences of the mitochondrial

cytochrome b gene. J. Mammal. 76:1178–1185.

Mouchaty, S. K., A. Gullberg, A. Janke, and U. Arnason. 2000a. Phylogenetic position of the tenrecs (Mammalia: Tenrecidae) of Madagascar based on analysis of the complete mitochondrial genome sequence of *Echinops telfairi*. Zool. Scripta 29:307–317.

Mouchaty, S. K., A. Gullberg, A. Janke, and U. Arnason. 2000b. The phylogenetic position of the Talpidae within eutheria based on analysis of complete mitochondrial sequences. Mol. Biol. Evol. 17:60–67.

Mouse Genome Sequencing Consortium. 2002. Initial sequencing and comparative analysis of the mouse genome. Nature 420:520–562.

Moyer, K. E. 1976. Psychobiology of aggression. Harper and Row, New York.

Muchala, N. 2006. Nectar bat stows huge tongue in its rib cage. Nature 444:701.

Muchala, N., and P. Jarrin-V. 2002. Flower visitation by bats in cloud forests of western Ecuador. Biotropica 34:387–395.

Muchala, N., P. Mena, and L. Albuja V. 2005. A new species of *Anoura* (Chiroptera: Phyllostomidae) from the Ecuadorian Andes. J. Mammal. 86:457–461.

Muchlinski, A. E. 1980. Duration of hibernation bouts in *Zapus hudsonius*. Comp. Biochem. Physiol. 67A:287–289.

Mueller, H. C. 1966. Homing and distance-orientation in bats. Z. Tierpsychol. 23:403–421.

Muizon, C. de, and B. Lange-Badre. 1997. Carnivorous dental adaptations in tribosphenic mammals and phylogenetic reconstruction. Lethaia 30:353–355.

Mulcahy, N. J., and J. Call. 2006. Apes save tools for future use. Science 312:1038–1040.

Müller, P. 1973. Dispersal centres of terrestrial vertebrates in the Neotropical realm. Junk, The Hague.

Müller, P. 1974. Aspects of zoogeography. Junk, The Hague.

Muller, S., R. Stanyon, P. C. M. O'Brien, M. A. Ferguson-Smith, R. Plesker, and J. Wienberg. 2000. Defining the ancestral karyotype of all primates by multidirectional chromosome painting between tree shrews, lemurs, and humans. Chromosoma 108:393–400.

Muller-Schwarze, D. 1971. Pheromones in black-tailed deer. Anim. Behav. 19:141–152.

Mullican, T. R. 1988. Radiotelemetry and fluorescent pigments: a comparison of techniques. J. Wildl. Mgmt. 52:627–631.

Mullis, K. B., and F. A. Faloona. 1987. Specific synthesis of DNA in vitro via a polymerase catalyzed chain reaction. Methods Enzymol. 155:335–350.

Mullis, K., F. Faloona, S. Scharf, R. Saiki, G. Horn, and H. Erlich. 1986. Specific enzymatic amplification of DNA in vitro: the polymerase chain reaction. Cold Spring Harb. Symp. Quant. Biol. 51:263–273.

Mumford, R. E., and J. O. Whitaker, Jr. 1982. Mammals of Indiana. Indiana Univ. Press, Bloomington.

Munger, J. C., and W. H. Karasov. 1994. Costs of bot fly infection in white-footed mice: energy and mass flow. Can. J. Zool. 72:166–173.

Munn, A. J., and T. J. Dawson. 2001. Thermoregulation in juvenile red kangaroos (*Macropus rufus*) after pouch exit: higher metabolism and evaporative water requirements. Physiol. Biochem. Zool. 74:917–927.

Murie, O. 1944. The wolves of Mount McKinley. U.S. Dept. Interior Nat. Park Serv., Fauna Ser. No. 5, Washington, DC.

Murie, O. 1954. A field guide to animal tracks. Macmillan, New York.

Murie, O. J., and A. Murie. 1931. Travels of *Peromyscus*. J. Mammal. 12:200–209.

Murray, D. L. 2003. Snowshoe hares and other hares, *Lepus americanus* and allies. Pp. 147–175 *in* Wild mammals of North America: biology, management, and conservation (G. A. Feldhamer, B. C. Thompson, and J. A. Chapman, eds.). Johns Hopkins Univ. Press, Baltimore.

Murphy, B. P., K. V. Miller, and R. L. Marchinton. 1994. Sources of reproductive chemosignals in female white-tailed deer. J. Mammal. 75:781–786.

Murphy, R. W., and G. Aguirre-Léon. 2002. Nonavian reptiles: origins and evolution. Pp. 181–220 *in* A new island biogeography of the Sea of

Cortés. Oxford Univ. Press, New York.

Murphy, R.W., J. W. Sites, Jr., D. G. Buth, and G. H. Haufler. 1996. Proteins: Isozyme electrophoresis. Pp. 51–120 *in* Molecular systematics, 2d edition (D. M. Hillis, C. Moritz, and B. K. Mable, eds.). Sinauer Associates, Sunderland, MA.

Murphy, S., A. Collet, and E. Rogan. 2005. Mating strategy in the male common dolphin (*Delphinus delphis*): what gonadal analysis tells us. J. Mammal. 86:1247–1258.

Murphy, W. J., E. Eizirik, W. E. Johnson, Y. P. Zhang, O. A. Ryder, and S. J. O'Brien. 2001a. Molecular phylogenetics and the origins of placental mammals. Nature 409:614–618.

Murphy, W. J., E. Eizirik, S. J. O'Brien, O. Madsen, M. Scally, C. J. Douady, E. Teeling, O. A. Ryder, M. J. Stanhope, W. W. de Jong, and M. S. Springer. 2001b. Resolution of the early placental mammal radiation using Bayesian phylogenetics. Science 294:2348–2351.

Murray, A. L., A. M. Barber, S. H. Jenkins, and W. S. Longland. 2006. Competitive environment affects food-hoarding behavior of Merriam's kangaroo rats (*Dipodomys merriami*). J. Mammal. 87:571–578.

Murray, D. L. 2003. Snowshoe hare and other hares, *Lepus americanus* and allies. Pp. 147–175 *in* Wild mammals of North America: biology, management, and conservation. (G. A. Feldhamer, B. C. Thompson, and J. A. Chapman, eds.). Johns Hopkins Univ. Press, Baltimore.

Murtagh, C. E. 1977. A unique cytogenetic system in monotremes. Chromosoma 65:37–57.

Musser, A. M. 2003. Review of the monotreme fossil record and comparison of palaentological and molecular data. Comp. Biochem. Physiol. Part A 136:927–942.

Musser, A. M., and M. Archer. 1998. New information about the skull and dentary of the Miocene platypus *Obdurodon dicksoni*, and a discussion of ornithorhyncid relationships. Phil. Trans. Roy. Soc. Lond. Ser. B. Biol. Sci. 353:1063–1079.

Musser, G. G., and M. D. Carleton. 2005. Superfamily Muroidea. Pp. 894–1531 *in* Mammal species of the world: a taxonomic and geographic reference. 3rd edition (D. E. Wilson and D. M. Reeder, eds.). Johns Hopkins Univ. Press, Baltimore.

Musser, G. G., and M. Dagosto. 1987. The identity of *Tarsius pumilus*, a pygmy species endemic to the montane mossy forests of central Sulawesi. Am. Mus. Novitates 2867:1–53.

Mutch, G. R. P., and M. Aleksiuk. 1977. Ecological aspects of winter dormancy in the striped skunk (*Mephitis mephitis*). Can. J. Zool. 55:607–615.

Muteka, S. P., C. T. Chimimba, and N. C. Bennett. 2006. Reproductive seasonality in *Aethomys namaquensis* (Rodentia: Muridae) from southern Africa. J. Mammal. 87:67–74.

Mutere, F. A. 1965. Delayed implantation in an equitorial fruit bat. Nature 207:780.

Muul, I. 1968. Behavioral and physiological influences on the distribution of the flying squirrel, *Glaucomys volans*. Misc. Pub., Mus. Zool., Univ. Mich. 124:1–66.

Myers, A. A., and P. S. Giller. 1988. Analytical biogeography. Chapman and Hall, London.

Myers, J., and C. Krebs. 1971. Genetic, behavioral, and reproductive attributes of dispersing field voles *Microtus pennsylvanicus* and *Microtus ochrogaster*. Ecol. Monogr. 41:53–78.

Nadler, R. D. 1975. Sexual cyclicity in captive lowland gorillas. Science 189:813–814.

Nagel, A. 1977. Torpor in the European white-toothed shrews. Experientia 33:1455–1458.

Nagy, K. A. 1987. Field metabolic rate and food requirement scaling in mammals and birds. Ecol. Monogr. 57:111–128.

Nagy, K. A. 1989. Field bioenergetics: accuracy of models and methods. Physiol. Zool. 62:237–252.

Nagy, K. A., and M. J. Gruchacz. 1994. Seasonal water and energy metabolism of the desert-dwelling kangaroo rat (*Dipodomys merriami*). Physiol. Zool. 67:1461–1478.

Nagy, K. A., and M. H. Knight. 1994. Energy, water, and food use by springbok antelope (*Antidorcas marsupialis*) in the Kalahari Desert. J. Mammal. 75:860–872.

Nagy, K. A., and C. C. Peterson. 1980. Scaling of water flux rates in animals. Univ. Calif. Publ. Zool.

120:1–172.

Nagy, K. E., C. Meienberger, S. D. Bradshaw, and R. D. Woller. 1995. Field metabolic rate of a small Australian mammal, the honey possum (*Tarsipes rostratus*). J. Mammal. 76:862–866.

Nagy, T. R. 1993. Effects of photoperiod history and temperature on male collared lemmings, *Dicrostonyx groenlandicus*. J. Mammal. 74: 990–998.

Nams, V. O., and E. A. Gillis. 2003. Changes in tracking tube use by small mammals over time. J. Mammal. 84:1374–1380.

Napier, J. R., and P. H. Napier. 1967. A handbook of living primates. Academic Press, New York.

Naples, V. L. 1999. Morphology, evolution, and function of feeding in the giant anteater (*Myrmecophaga tridactyla*). J. Zool. Lond. 249:19–41.

Narins, P. M., E. R. Lewis, J. U. M. Jarvis, and J. O'Riain. 1997. The use of seismic signals by fossorial southern African mammals: A neuroethological gold mine. Brain Res. Bull. 44:641–646.

Nash, L. T. 1986. Dietary, behavioral, and morphological aspects of gummivory in primates. Yearb. Physical Anthropol. 29:113–137.

Nash, L. T., S. K. Bearder, and T. R. Olson. 1989. Synopsis of Galago species characteristics. Int. J. Primat. 10:57–80.

Nash, R. 1967. Wilderness and the American mind. Yale Univ. Press, New Haven, CT.

National Public Radio (NPR). 2001. Radio Expeditions report by Alex Chadwick concerning coltan mining and eastern Congo's gorillas.

Neal, E. 1986. The natural history of badgers. Christopher Helm, London.

Neff, D. J. 1968. The pellet-group count technique for big game trend, census, and distribution: A review. J. Wildl. Mgmt. 32:597–614.

Negus, N. C., and P. J. Berger. 1977. Experimental triggering of reproduction in a natural population of *Microtus montanus*. Science 196:1230–1231.

Nei, M., and S. Kumar. 2000. Molecular evolution and phylogenetics. Oxford Univ. Press, New York.

Nelson, G., and P.Y. Ladiges. 1991. Three-area statements: standard assumptions for biogeographic analyses. Syst. Zool. 40:470–485.

Nelson, G., and P.Y. Ladiges. 1996. Paralogy in cladistic biogeography and analysis of paralogy–free subtrees. Am. Mus. Nov. 3167:1–58.

Nelson, G., and N. I. Platnick. 1981. Systematics and biogeography: cladistics and vicariance. Columbia Univ. Press, New York.

Nelson, J. F., and R. M. Chew. 1977. Factors affecting seed reserves in the soil of a Mojave Desert ecosystem, Rock Valley, Nye County, Nevada. Am. Midl. Nat. 97:300–320.

Nelson, R.A. 1980. Protein and fat metabolism in hibernating bears. Fed. Proc. 39:2955–2958.

Nelson, R. A., H. W. Wahner, J. D. Jones, R. D. Ellefson, and P. E. Zollman. 1973. Metabolism of bears before, during, and after winter sleep. Am. J. Physiol. 224:491–496.

Nelson, R. J. 1995. An introduction to behavioral endocrinology. Sinauer Associates, Sunderland, MA.

Nelson, S. L., T. H. Kunz, and S. R. Humphrey. 2005. Folivory in fruit bats: leaves provide a natural source of calcium. J. Chem. Ecol. 31:1683–1691.

Nepstad, D., A. Verissimo, P. Lefebvre, P. Schlesinger, C. Potter, C. Nobre, A. Setzer, T. Krug, A. C. Barros, A. Alencar, and J. R. Pereira. 1998. Forest fire prediction and prevention in the Brazilian Amazon. Conserv. Biol. 12:951–953.

Nestler, J. R., G. P. Dieter, and B. G. Klokeid. 1996. Changes in total body fat during daily torpor in deer mice (*Peromyscus maniculatus*). J. Mammal. 77:147–154.

Neubaum, D. J., M. A. Neubaum, L. E. Ellison, and T. J. O'Shea. 2005. Survival and condition of big brown bats (*Eptesicus fuscus*) after radiotagging. J. Mammal. 86:95–98.

Neuhaus, P. 2000. Timing of hibernation and molt in female Columbian ground squirrels. J. Mammal. 81:571–577.

Neuweiler, G. 1990. Auditory adaptations for prey capture in echolocating bats. Physiol. Rev. 70:615–641.

Neuweiler, G. 2000. The biology of bats. Oxford Univ. Press, Oxford.

Newmark, W. D. 1995. Extinction of mammal populations in western North American national parks.

Conserv. Biol. 9:512–526.

Nguyen, V. K., C. Su, S. Muyldermans, and W. van der Loo. 2002. Heavy-chain antibodies in Camelidae: a case of evolutionary innovation. Immunogenetics 54:39–47.

Nguyen, V. P., A. D. Needham, and J. A. Friend. 2005. A quantitative dietary study of the 'critically endangered' Gilbert's potoroo *Potorous gilbertii*. Aust. Mammal. 27:1–6.

Ninomiya, H. 2000. The vascular bed in the rabbit ear: microangiography and scanning electron microscopy of vascular corrosion casts. Anat. Histol. Embryol. 29:301–305.

Nichols, J. D., and J. E. Hines. 1984. Effects of permanent trap response in capture probability of Jolly-Seber capture-recapture model estimates. J. Wildl. Mgmt. 48:289–294.

Nicholson, W. L., C. D. Paddock, L. Demma, M. Traeger, B. Johnson, J. Dickson, J. McQuiston, and D. Swerdlow. 2006. Rocky Mountain spotted fever in Arizona: documentation of heavy environmental infestations of *Rhipicephalus sanguineus* at an endemic site. Ann. N.Y. Acad. Sci. 1078:338–341.

Nicol, S. C., and N. A. Andersen. 1993. The physiology of hibernation in an egg-laying mammal, the echidna. Pp. 56–64 *in* Life in the cold: ecological, physiological, and molecular mechanisms (C. Carey, G. L. Florant, B. A. Wunder, and B. Horwitz, eds.). Westview Press, Boulder, CO.

Nicol, S. C., and N. A. Andersen. 2000. Patterns of hibernation of echidnas in Tasmania. Pp. 21–28 *in* Life in the cold: Eleventh International Hibernation Symposium (G. Heldmaier and M. Klingenspor, eds.). Springer-Verlag, Berlin.

Nicoll, M. E. 1983. Mechanisms and consequences of large litter production in *Tenrec ecaudatus* (Insectivora: Tenrecidae). Ann. Mus. R. Afr. Centr. 237:219–226.

Nicoll, M. E., and P. A. Racey. 1985. Follicular development, ovulation, fertilization, and fetal development in tenrecs (*Tenrec ecaudatus*). J. Reprod. Fertil. 74:47–55.

Nikaido, M., A. P. Rooney, and N. Okada. 1999. Phylogenetic relationships among cetartiodactyls based on insertions of short and long interspersed elements: hippopotamuses are the closest extant relatives of whales. Proc. Natl. Acad. Sci. 96:10261–10266.

Nikaido, M., F. Matsuno, H. Hamilton, R. L. Brownell, Jr., Y. Cao, W. Ding, Z. Zuoyan, A. M. Shedlock, R. E. Fordyce, M. Hasegawa, and N. Okada. 2001. Retroposon analysis of major cetacean lineages: the monophyly of toothed whales and the paraphyly of river dolphins. Proc. Natl. Acad. Sci. 98:7384–7389.

Nisa, C., N. Kitamura, M. Sasaki, S. Agungpriyono, C. Choliq, T. Budipitojo, J. Yamada, and K. Sigit. 2005. Immunohistochemical study on the distribution and relative frequency of endocrine cells in the stomach of the Malayan pangolin, *Manis javanica*. Anat. Histol. Embryol. 34:373–378.

Noll-Banholzer, U. 1979. Body temperature, oxygen consumption, evaporative water loss, and heart rate in the fennec. Comp. Biochem. Physiol. 62A:585–592.

Nonacs, P. 2001. State dependent behavior and the marginal value theorem. Behav. Ecol. 12:71–83.

Noonan, J. P., M. Hofreiter, D. Smith, J. R. Priest, N. Rohland, G. Rabeder, J. Krause, J. C. Detter, S. Pääbo, and E. M. Rubin. 2005. Genomic sequencing of Pleistocene cave bears. Science 309:597–599.

Norberg, U. M. 1985. Flying, gliding, and soaring. Pp. 129–158 *in* Functional vertebrate morphology (M. Hildebrand, D. M. Bramble, K. F. Liem, and D. B. Wake, eds.). Harvard Univ. Press, Cambridge, MA.

Norberg, U. M. 1990. Vertebrate flight: Mechanisms, physiology, morphology, ecology, and evolution. Springer-Verlag, New York.

Norberg, U. M., and M. B. Fenton. 1988. Carnivorous bats? Biol. J. Linnean Soc. 33:383–394.

Nordenskiöld, E. 1928. The history of biology. Tudor Publishing, New York.

Norell, M. A. 1992. Ghost taxa, ancestors, and assumptions: a comment on Wagner. Paleobiology 22:453–455.

Norris, K. S. 1968. The evolution of acoustic mechanisms in odontocete cetaceans. Pp. 297–324 *in* Evolution and environment (E. T. Drake, ed.). Yale

Univ. Press, New Haven, CT.

Norris, K. S., and G. W. Harvey. 1972. A theory for the function of the spermaceti organ of the sperm whale (*Physeter catodon* L.). NASA Spec. Publ. No. 262:397–417.

Norris, K. S., and G. W. Harvey. 1974. Sound transmission in the porpoise head. J. Acoustic. Soc. Am. 56:659–664.

Novacek, M. J. 1985. Evidence for echolocation in the oldest known bats. Nature 315:140–141.

Novacek, M. J. 1992. Mammalian phylogeny: shaking the tree. Nature 356:121–125.

Novarino, W., S. N. Karimah, M. Silma, and S. M. Jarulis. 2004. Habitat use by Malay tapir (*Tapirus indicus*) in west Sumatra, Indonesia. Tapir Conserv. 13:14–18.

Novick, A. 1977. Acoustic orientation. Pp. 77–289 *in* Biology of bats, vol. III (W. A. Wimsatt, ed.). Academic Press, New York.

Novick, A., and D. R. Griffin. 1961. Laryngeal mechanisms in bats for the production of sounds. J. Exp. Zool. 148:125–145.

Novoa, C., and J. C. Wheeler. 1984. Llama and alpaca. Pp. 116–128 *in* Evolution of domesticated animals (I. L. Mason, ed.). Longman, New York.

Novotny, M., B. Jemiolo, S. Harvey, D. Wiesler, and A. Marchlewska-Koj. 1986. Adrenal-mediated endogenous metabolites inhibit puberty in female mice. Science 231:722–725.

Nowak, R. M. 1991. Walker's mammals of the world, 5th edition, 2 vols. Johns Hopkins Univ. Press, Baltimore.

Nowak, R. M. 1994. Walker's bats of the world. Johns Hopkins Univ. Press, Baltimore.

Nowak, R. M. 1999. Walker's mammals of the world, 6th edition, vols. 1 and 2. Johns Hopkins Univ. Press, Baltimore.

Nowak, R. M. 2002. The original status of wolves in eastern North America. Southeastern Naturalist 1:95–130.

Nutt, K. J. 2005. Philopatry of both sexes leads to the formation of multimale, multifemale groups in *Ctenodactylus gundi* (Rodentia: Ctenodactylidae). J. Mammal. 86:961–968.

Oakwood, M., A. J. Bradley, and A. Cockburn. 2001. Semelparity in a large marsupial. Proc. Roy. Soc. Lond. Ser. B 268:407–411.

Oates, J. F. 1984. The niche of the potto, Perodicticus potto. Int. J. Primatol. 5: 51–61.

Oates, J. F., M. Abedi-Lartey, W. S. McGraw, T. T. Struhsaker, and G. H. Whitesides. 2000. Extinction of a West African red colobus monkey. Conserv. Biol. 14:1526–1532.

Obrist, M. K., M. B. Fenton, J. L. Eger, and P. A. Schlegel. 1993. What ears do for bats: a comparative study of pinna sound pressure transformation in chiroptera. J. Exp. Biol. 180:119–152.

Ochocinska, D., and J. R. E. Taylor. 2003. Bergmann's rule in shrews: geographical variation of body size in Palearctic *Sorex* species. Biol. J. Linnean Soc. 78:365–381.

O'Connell-Rodwell, C. E., J. D. Wood, T. C. Rodwell, S. Puria, S. R. Partan, R. Keefe, D. Shriver, B. T. Arnason, and L. A. Hart. 2006. Wild elephant (*Loxodonta africana*) breeding herds respond to artificially transmitted seismic stimuli. Behav. Ecol. Sociobiol. 59:842–850.

O'Connor, B. M. 1988. Host associations and coevolutionary relationships of astigmatid mite parasites of New World primates, I: Families Psoroptidae and Audycoptidae. Fieldiana 39:245–260.

Odell, D. K. 2003. West Indian manatee, *Trichechus manatus*. Pp. 855–864 *in* Wild mammals of North America: biology, management, and conservation (G. A. Feldhamer, B. C. Thompson, and J. A. Chapman, eds.). Johns Hopkins Univ. Press, Baltimore.

O'Farrell, M. J., and E. H. Studier. 1970. Fall metabolism in relation to ambient temperatures in three species of Myotis. Comp. Biochem. Physiol. 35:697–703.

Oftedal, O. T. 1984. Milk composition, milk yield, and energy output at peak lactation: a comparative review. Symp. Zool. Soc. Lond. 52:33–85.

Oftedal, O. T., D. J. Boness, and R. A. Tedman. 1987. The behavior, physiology, and anatomy of lactation in the Pinnipedia. Pp. 175–234 *in* Current mammalogy (H. H. Genoways, ed.). Plenum Press, New

York.

O'Gara, B. W., and G. Matson. 1975. Growth and casting of horns by pronghorns and exfoliation of horns by bovids. J. Mammal. 56:829–846.

O'Hara, R. J. 1993. Systematic generalization, historical fate, and the species problem. Syst. Biol. 42:231–246.

O'Keefe, J., and L. Nadel. 1978. The hippocampus as a cognitive map. Clarendon Press, Oxford.

Olcott, S.P., and R. E. Barry. 2000. Environmental correlates of geographic variation in body size of the east cottontail (*Sylvilagus floridanus*). J. Mammal 81:986–998.

Oleyar, C. M., and B. S. McGinnes. 1974. Field evaluation of diethylstilbestrol for suppressing reproduction in foxes. J. Wildl. Mgmt. 38:101–106.

Oli, M. K. 2004. The fast-slow continuum and mammalian life-history patterns: an empirical evaluation. Basic Appl. Ecol. 5:449–463.

Oliphant. L. W. 1983. First observations of brown fat in birds. Condor 85:350–354.

Olsen, S. J. 1985. Origins of the domestic dog: the fossil record. Univ. Arizona Press, Tucson.

Olson, J. M., W. R. Dawson, and J. J. Camilliere. 1988. Fat from black-capped chickadees: avian adipose tissue? Condor. 90:529–537.

Orcutt, E. E. 1940. Studies on the muscles of the head, neck, and pectoral appendages of *Geomys bursarius*. J. Mammal. 21:37–52.

O'Regan, H. J., and A. C. Kitchener. 2005. The effects of captivity on the morphology of captive, domesticated and feral animals. Mammal Rev. 35:215–230.

O'Riain, M. J., N. C. Bennett, P. N. M. Brotherton, G. McIlrath, and T. H. Clutton-Brock. 2000. Reproductive suppression and inbreeding avoidance in wild populations of cooperatively breeding meerkats (*Suricata suricatta*). Behav. Ecol. Sociobiol. 48:471–477.

Orians, G. H. 1969. On the evolution of mating systems in birds and mammals. Am. Nat. 103:589–603.

Oritsland, N. A., J. W. Lentfer, and K. Ronald. 1974. Radiative surface temperature of the polar bear. J. Mammal 55:459–461.

Øritsland, T. 1977. Food consumption of seals in the Antarctic pack ice. Pp. 749–768 *in* Adaptations with in Antarctic ecosystems (G. A. Llano, ed.). Smithsonian Institution Press, Washington, DC.

Orr, R. T. 1970. Animals in migration. Macmillan, New York.

Orrock, J. L., D. Farley, and J. F. Pagels. 2003. Does fungus consumption by the woodland jumping mouse vary with habitat type or the abundance of other small mammals? Can. J. Zool. 81:753–756.

Ortmann, S., J. Schmid, J. U. Ganzhorn, and G. Heldmaier. 1996. Body temperature and torpor in a Malagasy small primate, the mouse lemur. Pp. 55–61 *in* Adaptations to the cold: Tenth International Hibernation Symposium (F. Geiser, A. J. Hulbert, and S. C. Nicol, eds.). Univ. New England Press, Armidale, NSW, Australia.

Osborne, M. J., and L. Christidis. 2001. Molecular phylogenetics of Australo-Papuan possums and gliders (family Petauridae). Mol. Phylogenet. Evol. 20:211–224.

Osborne, M. J., and L. Christidis. 2002. Systematics and biogeography of pygmy possums (Burramyidae: *Cercartetus*). Aust. J. Zool. 50:25–37.

Osgood, D. W. 1980. Temperature sensitive telemetry applied to studies of small mammal activity patterns. Pp. 525–528 *in* A handbook of radiotelemetry and radio tracking (C. J. Amlaner and D. W. Macdonald, eds.). Pergamon Press, Oxford.

Osgood, W. H. 1909. Revision of mice of the American genus *Peromyscus*. North Am. Fauna 28:1–285.

Oste, C. 1988. Polymerase chain reaction. Biotechniques 6:162–167.

Ostfeld, R. S. 1997. The ecology of Lyme-disease risk. Am. Sci. 85:338–346.

Ostfeld, R. S., C. G. Jones, and J. O. Wolff. 1996a. Of mice and mast. BioScience 46:323–330.

Ostfeld, R. S., M. C. Miller, and K. R. Hazler. 1996b. Causes and consequences of tick (*Ixodes scapularis*) burdens on white-footed mice (*Peromyscus leucopus*). J. Mammal. 77:266–273.

Ostfeld, R. S., C. D. Canham, K. Oggenfuss, R. J. Winchcombe, and F. Keesing. 2006. Climate, deer,

rodents, and acorns as determinants of variation in Lyme-disease risk. PLoS Biol. 4:1058–1068.

Ostro, L. E. T., T. P. Young, S. C. Silver, and F. W. Koonz. 1999. A geographic information system method for estimating home range size. J. Wildl. Mgmt. 63:748–755.

Ostrowski, S., J. B. Williams, E. Bedin, and K. Ismail. 2002. Water influx and food consumption of free-living oryxes (*Oryx leucoryx*) in the Arabian Desert in summer. J. Mammal. 83:665–673.

Ostrowski, S., P. Misochina, and J. B. Williams. 2006. Physiological adjustments of and gazelles (*Gazella subgutturosa*) to a boom-or-bust economy: standard fasting metabolic rate, total evaporative water loss, and changes in the sizes of organs during food and water restriction. Physiol. Biochem. Zool. 79:810–819.

Oswald, C., and P. A. McClure. 1990. Energetics of concurrent pregnancy and lactation in cotton rats and woodrats. J. Mammal. 71:500–509.

Otis, D. L., K. P. Burnham, G. C. White, and D. R. Anderson. 1978. Statistical inference from capture data on closed animal populations. Wildl. Monogr. 62:1–135.

Ovadia, O., and H.zu Dohna. 2003. The effect of intra-and interspecific aggression on patch residence time in Negev Desert rodents: a competing risk analysis. Behav. Ecol. 14:583–591.

Owen, J. G., and R. J. Baker. 2001. The *Uroderma bilobatum* (Chiroptera: Phyllostomidae) cline revisit ed. J. Mammal. 82:1102–1113.

Owen, O. S. 1971. Natural resource conservation: an ecological approach. Macmillan, New York.

Owen-Smith, N. 1975. The social ethology of the white rhinoceros *Ceratotherium simum* (Burchell 1817). Z. Tierpsychol. 38:337–384.

Owen-Smith, N. 1993. Assessing the constraints for optimal diet models. Evol. Ecol. 7:530–531.

Owen-Smith, R. N. 1988. Megaherbivores: the influence of very large body size on ecology. Cambridge Univ. Press, New York.

Oxberry, B. A. 1979. Female reproductive patterns in hibernating bats. J. Reprod. Fert. 56:359–367.

Oxley, D. J., M. B. Fenton, and G. R. Carmody. 1974. The effects of roads on populations of small mammals. J. Appl. Ecol. 11:51–59.

Oxnard, E. 1981. The uniqueness of Daubentonia. Am. J. Physical Anthropol. 54: 1–21.

Pacey, T. L., P. R. Baverstock, and D. R. Jerry. 2001. The phylogenetic relationships of the bilby, *Macrotis lagotis* (Peramelimorphia: Thylacomyidae), to bandicoots—DNA sequence evidence. Mol. Phylogenet. Evol. 21:26–31.

Packard, G. C., and T. J. Boardman. 1988. The misuse of ratios, indices, and percentages in ecophysiological research. Physiol. Zool. 61:1–9.

Packer, C. 1977. Reciprocal altruism in Papio anubis. Nature 265:441–443.

Packer, C. 1986. The ecology of sociality in felids. Pp. 429–451 *in* Ecological aspects of social evolution: birds and mammals (D. I. Rubenstein and R. W. Wrangham, eds.). Princeton Univ. Press, Princeton, NJ.

Packer, C., and A. E. Pusey. 1982. Cooperation and competition within coalitions of male lions: kin selection or game theory? Nature 296:740–742.

Packer, C., D. A. Collins, A. Sindimwo, and J. Goodall. 1995. Reproductive constraints on aggressive competition in female baboons. Nature 373:60–63.

Packer, C., L. Herbst, A. E. Pusey, D. J. Bygott, J. P. Hanby, S. J. Cairns, and M. Borgerhoff-Mulder. 1988. Reproductive success in lions. Pp. 363–383 *in* Reproductive success (T. H. Clutton-Brock, ed.). Univ. Chicago Press, Chicago.

Packer, C., S. Lewis, and A. Pusey. 1992. A comparative analysis of non-offspring nursing. Anim. Behav. 43:265–282.

Packer, C., D. Scheel, and A. E. Pusey. 1990. Why lions form groups: food is not enough. Am. Nat. 136:1–19.

Padilla, M., and R. C. Dowler. 1994. *Tapirus terrestris*. Mammal. Species No. 481:1–8.

Padykula, H. A., and J. M. Taylor. 1982. Marsupial placentation and its evolutionary significance. J. Reprod. Fertil., Suppl. 31:95–104.

Page, R. D. M. 1987. Graphs and generalized tracks: quantifying Croizat's panbiogeography. Syst. Zool.

36:1–17.

Page, R. D. M. 1990. Temporal congruence and cladistic analysis of biogeography and cospeciation. Syst. Zool.39:205–226.

Page, R. D. M. 1993. Parasites, phylogeny, and cospeciation. Int. J. Parasitol. 23:499–506.

Page, R. D. M. 1994. Maps between trees and cladistic analysis of historical associations among genes, organisms, and areas. Syst. Biol. 43:58–77.

Pagel, M. D., R. M. May, and A. R. Collie. 1991. Ecological aspects of the geographical distribution and diversity of mammalian species. Am. Nat. 137:791–815.

Pala, C. 2006. Sea animals get tagged for double-duty research. Science 313: 1383–1384.

Palo, R. T., and C. T. Robins (eds.). 1991. Plant defenses against mammalian herbivory. CRC Press, Boca Raton, FL

Pälike, H., R. D. Norris, J. O. Herrle, P. A. Wilson, H. K. Coxall, C. H. Lear, N. J. Shackleton, A. K. Tripati, and B. S. Wade. 2006. The heartbeat of the Oligocene climate system. Science 314:1894–1898.

Palumbi, S. R. 1996. Nucleic acids II: the polymerase chain reaction. Pp. 205–247 *in* Molecular systematics, 2d edition (D. M. Hillis, C. Moritz, and B. K. Mable, eds.). Sinauer Associates, Sunderland, MA.

Pan American Health Organization. 1978. The armadillo as an experimental model in biomedical research. World Health Organization Scientific Publ. No. 366, Washington, DC.

Paquet, P. C., and L. N. Carbyn. 2003. Gray wolf, *Canis lupus* and allies. Pp. 482–510 *in* Wild mammals of North America: biology, management, and conservation. 2d edition (G. A. Feldhamer, B. C. Thompson, and J. A. Chapman, eds). Johns Hopkins Univ. Press, Baltimore, MD.

Paradis, E. 1995. Survival, immigration and habitat quality in the Mediterranean pine vole. J. Anim. Ecol. 64:579–591.

Paradiso, J. L., and R. M. Nowak. 1982. Wolves, Canis lupus and allies. Pp. 460–474 *in* Wild mammals of North America: biology, management, and economics (J. A. Chapman and G. A. Feldhamer, eds.). Johns Hopkins Univ. Press, Baltimore.

Parag, A., N. C. Bennett, C. G. Faulkes, and P. W. Bateman. 2006. Penile morphology of African mole rats (Bathyergidae): structural modification in relation to mode of ovulation and degree of sociality. J. Zool. Lond. 270:323–329.

Parker, G. A. 1970. Sperm competition and its evolutionary consequences in the insects. Biol. Rev. 45:525–568.

Parmagiano, S., and F. S. vom Saal. 1994. Infanticide and parental care. Harwood Academic, Langhorne, PA.

Parsons, L. M., and C. R. Terman. 1978. Influence of vision and olfaction on the homing ability of the white-footed mouse (*Peromyscus leucopus noveboracensis*). J. Mammal. 59:761–771.

Pascual, R., and F. J. Goin. 2002. Non-tribosphenic Gondwanan mammals and the alternative development of molars with reversed triangle cusp pattern. Asociacion Paleontologica Argentina Publicacion Especial 7:157–162.

Pascual, R., M. Archer, E. O. Jaureguizar, J. L. Prado, H. Godthelp, and S. J. Hand. 1992. The first non-Australian monotreme: an early Paleocene South American platypus (Monotremata: Ornithorhynchidae). Pp. 1–14 *in* Platypus and echidnas (M. L. Augee, ed.). Royal Zoological Society, NSW, Sydney.

Pascual, R., M. Archer, E. Ortiz Jaureguizar, J.L. Prado, H. Godthelp, and S.J. Hand. 1992a. First discovery of monotremes in South America. Nature 356:704–706.

Pastor, J., B. Dewey, and D. P. Christian. 1996. Carbon and nutrient mineralization and fungal spore composition of fecal pellets from voles in Minnesota. Ecography 19:52–61.

Pastorini, J., R. D. Martin, P. Ehresmann, E. Zimmermann, and M. R. J. Forstner. 2001. Molecular phylogeny of the Lemur family Cheirogaleidae (Primates) based on mictochondrial DNA sequences. Mol. Phylogenet. Evol. 19:45–56.

Pastorini, J., M. R. Forstner, and R. D. Martin. 2002. Phylogenetic relationships among Lemuridae (Primates): evidence from mtDNA. J. Hum. Evol. 43:463–478.

Patterson, B. 1965. The fossil elephant shrews (Family Macroscelidae). Bull. Mus. Comp. Zool. Harvard. 133:295–335.

Patterson, B. 1975. The fossil aardvarks (Mammalia: Tubulidentata). Bull. Mus. Comp. Zool. 147:185–237.

Patterson, B. 1978. Pholidota and Tubulidentata. Pp. 268–278 in Evolution of African mammals (V. J. Maglio and H. B. S. Cooke, eds.). Harvard Univ. Press, Cambridge, MA.

Patterson, B. D. 1999. Contingency and determinism in mammalian biogeography: the role of history. J. Mammal. 80:345–360.

Patterson, B.D. and R. Pascual. 1968. The fossil mammal gauna of South America. Pp. 247–309 in Evoluion, mammals, and the southern continents (A. Keast, F.C. Erk, and B. Glass, eds.). State Univ. New York Press, Albany.

Patterson, B. D., P. L. Meserve, and B. K. Lang. 1989. Distribution and abundance of small mammals along an elevational transect in temperate rainforests of Chile. J. Mammal. 70:67–78.

Patterson, B.D, V. Pacheco, and S. Solari. 1996. Distribution of bats along an elevational gradient in the Andes of south-east Peru. J. Zool. Soc. Lond. 240:637–658.

Patterson, B. D., M. R. Willig, and R. D. Stevens. 2003. Trophic strategies, niche partitioning, and patterns of ecological organization. Pp. 536–579 in Bat Ecology (T.H. Kunz and M. B. Fenton, eds.). Univ. Chicago Press, Chicago.

Patterson, C. 1981. Methods of paleobiogeography. Pp. 446–489 in Vicariance biogeography: a critique (G. Nelson and D.E. Rosen, eds.). Columbia Univ. Press, New York.

Patterson, M. A., and S. H. Vessey. 1973. Tapeworm (Hymenolepis nana) infection in male albino mice: effect of fighting among the hosts. J. Mammal. 54:784–786.

Patton, J. L. 2005. Family Heteromyidae. Pp. 844–858 in Mammal species of the world: a taxonomic and geographic reference. 3rd edition (D. E. Wilson and D. M. Reeder, eds.). Johns Hopkins Univ. Press, Baltimore.

Patton, J. L., and J. H. Feder. 1981. Microspatial heterotgeneity in pocket gophers: non-random breeding and drift. Evolution 35:912–920.

Pavey, C. R., and C. J. Burwell. 1997. The diet of the diadem leaf-nosed bat Hipposideros diadema: confirmation of a morphologically-based prediction of carnivory. J. Zool. Lond. 243:295–303.

Pavey, C. R., and C. J. Burwell. 2004. Foraging ecology of the horseshoe bat, Rhinolophus megaphyluss (Rhinolophidae), in eastern Australia. Wildl. Res. 31:403–413.

Pavey, C. R., C. J. Burwell, J. E. Grunwald, C. J. Marshall, and G. Neuweiler. 2001a. Dietary benefits of twilight foraging by the insectivorous bat Hipposideros speoris. Biotropica 33:670–681.

Pavey, C. R., J. E. Grunwald, and G. Neuweiler. 2001b. Foraging habitat and echolocation behaviour of Schneider's leafnosed bat, Hipposideros speoris, in a vegetation mosaic in Sri Lanka. Behav. Ecol. Sociobiol. 50:209–218.

Payne, K. B., and R. S. Payne. 1985. Large scale changes over 19 years in songs of humpback whales in Bermuda. Z. Tierpsychol. 68:89–114.

Payne, K. B., W. R. Langbauer, Jr., and E. M. Thomas. 1986. Infrasonic calls of the Asian elephant (Elephas maximus). Behav. Ecol. Sociobiol. 18:297–301.

Payne, K. B., M. Thompson, and L. Kramer. 2003. Elephant calling patterns as indicators of group size and composition: the basis for an acoustic monitoring system. Afr. J. Ecol. 41:99–107.

Payne, R. S., and S. McVay. 1971. Songs of humpback whales. Science 173:585–597.

Pech, R. P., A. R. E. Sinclair, A. E. Newsome, and P. C. Catling. 1992. Limits to predator regulation of rabbits in Australia: evidence from predator-removal experiments. Oecologia 89:102–112.

Pedler, C., and R. Tilley. 1969. The retina of a fruit bat (Pteropus giganteus Brünnich). Vision Res. 9:909–922.

Peigne, S., L. deBonis, A. Likius, H. T. Mackaye, P. Vignaud, and M. Brunet. 2005. The earliest modern mongoose (Carnivora, Herpestidae) from Africa (late Miocene of Chad). Naturwissenschaften 92:287–292.

Peinke, D. M. and C. R. Brown. 2005. Burrow utilization by springhares (Pedetes capensis) in the Eastern Cape, South Africa. Afr. Zool. 40:37–44.

Pelton, M. R. 1982. Black bear. Pp. 504–514 in Wild mammals of North America: biology, management, and economics (J. A. Chapman and G. A. Feldhamer, eds.). Johns Hopkins Univ. Press, Baltimore.

Penn, D. J., and W. K. Potts. 1999. The evolution of mating preferences and major histocompatibility complex genes. Am. Nat. 153:145–164.

Pepin, D., C. Adrados, C. Mann, and G. Janeau. 2004. Assessing real daily distance traveled by ungulates using different GPS locations. J. Mammal. 85:774–780.

Pereira, H. R. J., Jr., W. Jorge, and M. E. L. Teixeira da Costa. 2004. Chromosome study of anteaters (Myrmecophagidae, Xenarthra) — a preliminary report. Genetics Mol. Biol. 27:391–394.

Peres, C. A. 2001. Synergistic effects of subsistence hunting and habitat fragmentation on Amazonian forest vertebrates. Conserv. Biol. 15:1490–1505.

Perez, M., B. Li, A. Tillier, A. Cruaud, and G. Veron. 2006. Systematic relationships of the bushy-tailed and black-footed mongooses (genus Bdeogale, Herpestidae, Carnivora) based on molecular, chromosomal and morphological evidence. J. Zool. Syst. Evol. Res. 44:251–259.

Perrin, W., B. Wursig, and J. Therwissen. 2002. Encyclopedia of Marine Mammals. Academic Press, Amsterdam.

Perry, J. W. 1972. The ovarian cycle of mammals. Hafner, New York.

Petter, J.-J. 1962a. Ecological and behavioural studies of Madagascan lemurs in the field. Ann. N.Y. Acad. Sci. 102:267–281.

Petter, J.-J. 1962b. Remarques sur l'ecologie et l'ethologie comparées des Lémuriens Malagaches. Mem. Mus. Nat. Hist. 27:1–146.

Petter, J.-J. 1972. Order of primates: Suborder of lemurs. Pp. 683–702 in Biogeography and ecology of Madagascar (R. Battistini and G. Richard-Vindard, eds.). W. Junk, The Hague.

Petter, J.-J. 1977. The aye-aye. Pp. 38–57 in Primate conservation (H. S. H. Prince Ranier III and G .H. Bourne, eds.). Academic Press, New York.

Petter, J.-J., and A. Petter. 1967. The aye-aye of Madagascar. Pp. 195–205 in Social communication among primates (S. A. Altmann, ed.). Univ. Chicago Press, Chicago.

Petter, J.-J., and A. Petter-Rousseaux. 1979. Classification of the prosimians. Pp. 1–44 in The study of prosimian behavior (G. A. Doyle and R. D. Martins, eds.). Academic Press, London.

Pettigrew, J. D. 1986. Flying primates? Megabats have the advanced pathway from eye to midbrain. Science 231:1304–1306.

Pettigrew, J. D. 1995. Flying primates: crashed or crashed through. Pp. 3–26 in Ecology, evolution, and behaviour of bats (P. A. Racey and S. M. Swift, eds.). Oxford Univ. Press, Oxford.

Pettigrew, J. D., B. G. M. Jamieson, S. K. Robson, L. S. Hall, I. I. McAnally, and J. M. Cooper. 1989. Phylogenetic relations between microbats (Mammalia: Chiroptera and Primates). Phil. Trans. Roy. Soc. Lond. Ser. B Biol. Sci. 325:489–559.

Phelan, J., and R. H. Baker. 1992. Optimal foraging in Peromyscus polionotus: the influence of item-size and predation risk. Behaviour 121:95–109.

Phillips, A. V., and I. Stirling. 2001. Vocal repertoire of South American fur seals, Arctocephalus australis: structure, function, and context. Can. J. Zool. 79:420–437.

Phillips, M. J., P. A. McLenachan, C. Down, G. C. Gibb, and D. Penny. 2006. Combined mitochondrial and nuclear DNA sequences resolve the interrelations of the major Australasian marsupial radiations. Syst. Biol. 55:122–137.

Phillips, M. K. 1990. Measures of the value and success of a reintroduction project: red wolf reintroduction in Alligator River National Wildlife Refuge. Endangered Sp. Update. 8:24–26.

Phillips, P. K., and J. E. Heath. 1992. Heat exchange by the pinna of the African elephant (Loxodonta africana). Comp. Biochem. Physiol. 101A:693–699.

Phillips, P. K., and J. E. Heath. 2001. An infrared thermographic study of surface temperature in the euthermic woodchuck (Mamota monax). Comp. Biochem. Physiol. 129A:557–562.

Pianka, E.R. 1966. Latitudinal gradients in species diversity: a review of concepts. Am. Nat. 100:33–46.

Pianka, E. R. 1970. On r- and K-selection. Am. Nat. 104:592–597.

Pianka, E. R. 1994. Evolutionary ecology, 5th edition. HarperCollins, New York.

Pianka, E. R. 2000. Evolutionary ecology, 6th edition. Benjamin Cummings, San Francisco.

Pidancier, N., S. Jordan, G. Luikart, and P. Taberlet. 2006. Evolutionary history of the genus Capra (Mammalia, Artiodactyla): discordance between mitochondrial DNA and Y-chromosome phylogenies. Mol. Phylogenet. Evol. 40:739–749.

Pielou, E. C. 1991. After the ice age. Univ. Chicago Press, Chicago.

Pierce, B. M., V. C. Bleich, and R. T. Bowyer. 2000. Selection of mule deer by mountain lions and coyotes: effects of hunting style, body size, and reproductive status. J. Mammal. 81:462–472

Pierce, S. S., and F. D. Vogt. 1993. Winter acclimatization in Peromyscus maniculatus gracilis, P. leucopus noveboracensis, and P. l. leucopus. J. Mammal. 74:665–677.

Pieres-Ferreira, J. W., E. Pieres-Ferreira, and P. Kaulicke. 1976. Preceramic animal utilization in the Central Peruvian Andes. Science 194:483–490.

Piesman, J., G. O. Maupin, E. G. Campos, and C. M. Happ. 1991. Duration of adult female Ixodes dammini attachment and transmission of Borrelia burgdorferi, with description of a needle aspiration isolation method. J. Infect. Dis. 163:95–97.

Pilleri, G. 1990. Adaptation to water and the evolution of echolocation in the Cetacea. Ethology Ecol. Evol. 2:135–163.

Pitman, R. L., L. T. Balance, S. I. Mesnick, and S. J. Chivers. 2001. Killer whale predation on sperm whales: observations and implications. Mar. Mammal Sci. 17:494–507.

Pitt, J. A., S. Larivière, and F. Messier. 2006. Condition indices and bioelectrical impedance analysis to predict body condition of small carnivores. J. Mammal. 87:717–7222.

Pivorunas, A. 1979. The feeding mechanisms of baleen whales. Am. Sci. 67:432–440.

Platnick, N. I., and G. Nelson. 1988. Spanning tree biogeography: shortcut, detour or dead-end? Syst. Zool. 37:410–419.

Pleasants, J. M. 1989. Optimal foraging by nectarivores: a test of the marginal value theorem. Am. Nat. 134:51–71.

Pluháček, J., L. Bartoš, and J. Víchová. 2006. Variation in incidence of male infanticide within subspecies of plains zebra (Equus burchelli). J. Mammal. 87:35–40.

Poinar, H. N., C. Schwartz, J. Qi, B. Shapiro, R. D. E. MacPhee, B. Buigues, A. Tikhonov, D. H. Huson, L. P. Tomsho, A. Auch, M. Rampp, W. Miller, and S. C. Schuster. 2006. Metagenomics to paleogenomics: large-scale sequencing of mammoth DNA. Science 311:392–394.

Poland, J. D., T. J. Quan, and A. M. Barnes. 1994. Plague. Pp. 93–112 in Handbook of zoonoses, section A: bacterial, rickettsial, chlamydial, and mycotic diseases, 2d edition (C. W. Beran and J. H. Steele, eds.). CRC Press, Boca Raton, FL.

Pond, C.M. 1977. The significance of lactation in the evolution of mammals. Evolution 31:187–199.

Pond, C. M. 1978. Morphological aspects and the ecological and mechanical consequences of fat deposition in wild vertebrates. Pp. 519–570 in Annual review of ecology and systematics, vol. 9 (R. F. Johnston, P. W. Frank, and C. D. Michener, eds.). Annual Reviews, Palo Alto, CA.

Poole, J. H., K. Payne, W. R. Langbauer, Jr., and C. J. Moss. 1988. The social contexts of some very low frequency calls of African elephants. Behav. Ecol. Sociobiol. 22:385–392.

Popper, A. N. 1980. Sound emission and detection by delphinids. Pp. 1–52 in Cetacean behavior (L. M. Herman, ed.). John Wiley and Sons, New York.

Popper, A. N., and R. R. Fay (eds.). 1995. Hearing in bats. Springer-Verlag, New York.

Porter, R. H., V. J. Tepper, and D. M. White. 1981. Experimental influences on the development of huddling preferences and "sibling" recognition in spiny mice. Devel. Psychobiol. 14:375–382.

Porter, W. P., and D. M. Gates. 1969. Thermodynamic equilibria of animals with environment. Ecol. Monogr. 39:227–244.

Posada, D., and K.A. Crandall. 1998. Modeltest: testing the model of DNA substitution. Bioinformatics 14:817–818.

Posadas, P., J.V. Crisci, and L. Katinas. 2006. Historical biogeography: a review of its basic concepts and critical issues. J. Arid Environ. 66:389–403.

Post, D. M., and O. J. Reichman. 1991. Effects of food perishability, distance, and competitors on caching behavior by eastern woodrats. J. Mammal. 72:513–517.

Post, D. M., O. J. Reichman, and D. E. Wooster. 1993. Characteristics and significance of the caches of eastern woodrats (Neotoma floridana). J. Mammal. 74:688–692.

Post, D. M., M. V. Snyder, E. J. Finck, and D. K. Saunders. 2006. Caching as a strategy for surviving periods of resource scarcity; a comparative study of two species of Neotoma. Func. Ecol. 20: 717–722.

Potts, W. K., C. J. Manning, and E. K. Wakeland. 1991. Mating patterns in seminatural populations of mice influenced by MHC genotype. Nature 352:619–621.

Pough, F. H., J. B. Heiser, and W. N. McFarland. 1989. Vertebrate life, 3d edition. Macmillan, New York.

Pough, F. H., J. B. Heiser, and W. N. McFarland. 1996. Vertebrate life, 4th edition. Macmillan, New York.

Pough, F. H., C. M. Janis, and J. B. Heiser. 2005. Vertebrate life, 7th edition. Pearson Prentice-Hall, Upper Saddle River, NJ.

Powell, R. A. 1993. The fisher: life history, ecology, and behavior, 2d edition. Univ. Minnesota Press, Minneapolis.

Prabhakar, S., J. P. Noonan, S. Pääbo, and E. M. Rubin. 2006. Accelerated evolution of noncoding sequences in humans. Science 314:786.

Prager, E. M., and A. C. Wilson. 1978. Construction of phylogenetic tress for proteins and nucleic acids: empirical evaluation of alternative matrix methods. J. Mol. Evol. 11:129–142.

Prager, E. M., A. C. Wilson, J. M. Lowenstein, and V. M. Sarich. 1980. Mammoth albumin. Science 209:287–289.

Prentice, E. F., T.A. Flagg, C.S. McCutcheon, and D.F. Brastow. 1990. PIT-tag monitoring systems for hydroelectric dams and fish hatcheries. Am. Fish. Soc. Symp. 7:232–334.

Prentice, E. F., T. A. Flagg, C. S. McCutcheon, D. Brastow, and D. C. Cross. 1990a. Equipment, methods, and an automated data-entry station for PIT tagging. Am. Fish. Soc. Symp. 7:335–340.

Price, E. O. 1984. Behavioral aspects of animal domestication. Q. Rev. Biol. 59:1–32.

Price, J. S., S. Allen, C. Faucheux, T. Althnaian, and J. G. Mount. 2005. Deer antlers: a zoological curiosity or the key to understanding organ regeneration in mammals? J. Anat. 207:603–618.

Price, M. V., and R. A. Correll. 2001. Depletion of seed patches by Merriam's kangaroo rats: are GUD assumptions met? Ecol. Lett. 4:334–343.

Price, M. V., and P. R. Endo. 1989. Estimating the distribution and abundance of a cryptic species, Dipodomys stephensi (Rodentia: Heteromyidae), and implications for management. Conserv. Biol. 3:293–301.

Price, M. V., and P. A. Kelly. 1994. An age-structured demographic model for the endangered Stephens' kangaroo rat. Conserv. Biol. 8:810–821.

Price, M. V., and J. E. Mittler. 2006. Cachers, scavengers, and thieves: a novel mechanism for desert rodent coexistence. Am. Nat. 168:194–206.

Price, M. V., N. M. Waser, and S. McDonald. 2000. Seed caching by heteromyid rodents from two communities: implications for coexistence. J. Mammal. 81:97–106.

Price, P. W. 1980. Evolutionary biology of parasites. Princeton Univ. Press, Princeton, NJ.

Prins, R. A., and D. A. Kreulen. 1990. Comparative of plant cell wall digestion in mammals. Pp. 109–120 in The rumen ecosystem: the microbial metabolism and its regulation (S. Hoshino et al., eds). Scientific Societies Press, Berlin.

Procheș. 2005. The world's biogeographical regions:

cluster analysis based on bat distributions. J. Biogeog. 32:607–614.

Proctor-Grey, E. 1984. Dietary ecology of the coppery brush-tail possum, green ringtail possum and Lumholt's tree kangaroo in North Queensland. Pp. 129–135 in Possums and gliders (A. P. Smith and I. D. Hume, eds.). Australian Mammal Society and Surrey Beatty and Sons, Sydney.

Promislow, D. E. L., and Harvey, P. H. 1990. Living fast and dying young: a comparative analysis of life history variation among mammals. J. Zool. Lond. 220:417–437.

Proske, U., and E. Gregory. 2003. Electrolocation in the platypus—Some speculations. Comp. Biochem. Physiol. A 136:821–825.

Proske, U., A. Iggo, A. K. McIntyre, and J. E. Gregory. 1993. Electroreception in the platypus: a new mammalian sense. J. Comp. Physiol. A 173:708–710.

Proske, U., J. E. Gregory, and A. Iggo. 1998. Sensory receptors in monotremes. Phil. Trans. Roy. Soc. Lond. Ser. B Biol. Sci. 353:1187–1198.

Prothero, D. R. 1998. Bringing fossils to life: an introduction to paleobiology. McGraw-Hill, Boston.

Pruitt, W. O., Jr. 1957. Observations of the bioclimate of some taiga mammals. Arctic 10:131–138.

Prusiner, S., J. Collinge, J. Powell, and B. Anderton (eds.). 1992. Prion diseases of humans and animals. Ellis Horwood, New York.

Pucek, Z. 1965. Seasonal and age changes in the weight of internal organs of shrews. Acta Theriol. 10:369–438.

Puig, S., M. I. Rosi, M. I. Cona, V. G. Roig, and S. A. Monge. 1999. Diet of a Piedmont population of Ctenomys mendocinus (Rodentia, Ctenomyidae): seasonal patterns and variations according to sex and relative age. Acta Theriol. 44:15–27.

Pulliam, H. R., and B. J. Danielson. 1991. Sources, sinks, and habitat selection: A landscape perspective on population dynamics. Am. Nat. 137:S50–S66.

Purves, P. E. 1967. Anatomical and experimental observations on the cetacean sonar system. Pp. 197–270 in Animal sonar systems: biology and bionics, vol. 1 (R. G. Busnel, ed.). Laboratoire de physiologie acoustique, Jouy-en-Josas, France.

Purves, P. E., and G. Pilleri. 1983. Echolocation in whales and dolphins. Academic Press, London.

Pusey, A. E., and C. Packer. 1987. The evolution of sex-biased dispersal in lions. Behaviour 101:275–310.

Putman, R. 1988. The natural history of deer. Comstock Publ. Assoc., Ithaca, NY.

Puttick, G. M., and J. U. M. Jarvis. 1977. The functional anatomy of the neck and forelimbs of the cape golden mole, Chrysochloris asiatica (Liptophyla, Chrysochloridae). Zool. Afr. 12:445–458.

Pyare, S., and W. S. Longland. 2001. Mechanisms of truffle detection by northern flying squirrels. Can. J. Zool. 79:1007–1015.

Pyare, S., and W. S. Longland. 2002. Interrelationships among northern flying squirrels, truffles, and microhabitat structure in Sierra Nevada old-growth habitat. Can. J. Forest Res. 32:1016–1024.

Pyke, G. H., H. R. Pulliam, and E. L. Charnov. 1977. Optimal foraging: a selective review of theory and tests. Q. Rev. Biol. 52:137–154.

Querouil, S., R. Hutterer, P. Barriere, M. Colyn, J. C. K. Peterhans, and E. Verheyen. 2001. Phylogeny and evolution of African shrews (Mammalia: Soricidae) inferred from 16s rRNA sequences. Mol. Phylogenet. Evol. 20:185–195.

Querouil, S., P. Barriere, M. Colyn, R. Hutterer, A. Dudu, M. Dillen, and E. Verheyen. 2005. A molecular insight into the systematics of African Crocidura (Crocidurinae, Soricidae) using 16s rRNA sequences. Pp. 99–113 in Advances in the biology of shrews II (J. F. Merritt, S. Churchfield, R. Hutterer, and B. I. Sheftel, eds.). Special Publication of their International Society of Shrew Biologists 01, New York.

Quilliam, T. A. 1966. The mole's sensory apparatus. J. Zool. 149:76–78.

Quinn, G. P., and M. J. Keough. 2002. Experimental design and data analysis for biologists. Cambridge Univ. Press, Cambridge, UK.

Quinn, T. H., and J. J. Baumel. 1993. Chiropteran tendon locking mechanism. J. Morph. 216:197–208.

Qumsiyeh, M. B. 1994. Evolution of number and morphology of mammalian chromosomes. J. Hered. 85:455–465.

Qumsiyeh, M. B., and R. J. Baker. 1985. G-and C-banded karyotypes of the Rhinopomatidae (Microchiroptera). J. Mammal. 66:541–544.

Qumsiyeh, M. B., and J. K. Jones, Jr. 1986. Rhinopoma hardwickii and Rhinopoma muscatellum. Mammal. Species 263:1–5.

Rabb, G. B. 1959. Toxic salivary glands in the primitive insectivore Solenodon. Chicago Acad. Sci. Nat. Hist. Mus. 170:171–173.

Racey, P. A. 1982. Ecology of bat reproduction. Pp. 57–104 in Ecology of bats (T. H. Kunz, ed.). Plenum, New York.

Racey, P. A., and S. M. Swift (eds.). 1995. Ecology, evolution, and behaviour of bats. Symp. Zool. Soc. Lond. 67:1–421.

Racey, T. L., P. R. Baverstock, and D. R. Jerry. 2001. The phylogenetic relationships of the bilby, Macrotis lagotis (Peramelimorphia: Thylacomyidae), to bandicoots—DNA sequence evidence. Mol. Phylogenet. Evol. 21:26–31.

Rachlow, J. I., and J. Berger. 1997. Conservation implications of patterns of horn regeneration in dehorned white rhinos. Conserv. Biol. 11:84–91.

Radeloff, V. C., A. M. Pidgeon, and P. Hostert. 1999. Habitat and population modelling of roe deer using an interactive geographic information system. Ecol. Model. 114:287–304.

Rademaker, V., and R. Cerqueira. 2006. Variation in the latitudinal reproductive patterns of the genus Didelphis (Didelphimorphia: Didelphidae). Aust. Ecol. 31:337–342.

Rafael, M., F. Trillmich, and A. Honer. 1996. Energy allocation in reproducing and nonreproducing guinea pigs (Cavia porcellus) females and young under ad libitum conditions. J. Zool. 239:437–452.

Rahn, D. W., and J. Evans. 1998. Lyme disease. American College of Physicians, Philadelphia.

Rainey, W. E., E. D. Pierson, T. Elmqvist, and P. A. Cox. 1995. The role of flying foxes (Pteropodidae) in oceanic island ecosystems of the Pacific. Pp. 47–62 in Ecology, evolution, and behaviour of bats (P. A. Racey and S. M. Swift, eds.). Oxford Univ. Press, Oxford.

Ralls, K. 1971. Mammalian scent marking. Science 171:443–449.

Ralls, K. 1976. Mammals in which females are larger than males. Q. Rev. Biol. 51:245–276.

Ralls, K., K. Brugger, and J. Ballou. 1979. Inbreeding and juvenile mortality in small populations of ungulates. Science 206:1101–1103.

Ralls, K., J. D. Ballou, and A. R. Templeton. 1988. Estimates of lethal equivalents and the cost of inbreeding in mammals. Conserv. Biol. 2:185–193.

Ramsey, E. M. 1982. The placenta: human and animal. Praeger, New York.

Rancourt, S. J., M. I. Rule, and M. A. O'Connell. 2005. Maternity roost site selection of long-eared myotis, Myotis evotis. J. Mammal. 86:77–84.

Randall, J. A. 1984. Territorial defense and advertisement by footdrumming in bannertail kangaroo rats (Dipodomys spectabilis) at high and low population densities. Behav. Ecol. Sociobiol. 16:11–20.

Randall, J. A. 1993. Behavioural adaptations of desert rodents (Heteromyidae). Anim. Behav. 45:263–287.

Randall, D., W. Burggren, and K. French. 2002. Animal physiology, 5th edition. W. H. Freeman and Co., New York.

Randolph, J. C. 1973. Ecological energetics of a homeothermic predator, the short-tailed shrew. Ecology 54:1166–1187.

Ransome, R. 1990. The natural history of hibernating bats. Christopher Helm, London.

Ranta, E., J. Lindström, V. Kaitala, H. Kokko, H. Lindén, and E. Helle. 1997. Solar activity and hare dynamics: a cross-continental comparison. Am. Nat. 149:765–775.

Rapoport, E. 1982. Aerography: geographical strategies of species. Pergammon, New York.

Rasmussen, D. T., G. C. Conroy, and E. L. Simons. 1998. Tarsier-like locomotor specializations in the Oligocene primate Afrotarsius. Proc. Nat. Acad. Sci. USA 95:14848–14850.

Rathbun, G. B., and K. Redford. 1981. Pedal scent-marking in the rufous elephant-shrew Elephantulus rufescens. J. Mammal 62:635–637.

Rathbun, G. B., and C. D. Rathbun. 2006. Social structure of the bushveld sengi (*Elephantulus intufi*) in Namibia and the evolution of monogamy in the Macroscelidea. J. Zool. 269:391–399.

Ratner, S. C., and R. Boice. 1975. Effects of domestication on behaviour. Pp. 3–19 *in* The behaviour of domestic animals, 3d edition (E. S. E. Hafez, ed.). Baillière Tindall, London.

Raum-Suryan, K. L., K. W. Pitcher, D. G. Calkins, J. L. Sease, and T. R. Loughlin. 2002. Dispersal, rookery fidelity, and metapopulation structure of Stellar sea lions (*Eumetopias jubatus*) in an increasing and a decreasing population in Alaska. Mar. Mammal. Sci. 28:746–764.

Raup, D. M. 1991. Extinction: bad genes or bad luck? W. W. Norton, New York.

Rawlins, R. G., M. J. Kessler, and J. E. Turnquist. 1984. Reproductive performance, population dynamics and anthropometrics of the free-ranging Cayo Santiago rhesus macaques. J. Med. Primatol. 13:247–259.

Ray, C. E. 1976. Geography of phocid evolution. Syst. Zool. 25:391–406.

Ray, G., J. McCormick-Ray, P. Berg, and H. E. Epstein. 2006. Pacific walrus: benthic bioturbator of Beringia. J. Exp. Mar. Biol. Ecol. 330:403–419.

Réale, D., A. G. McAdam, S. Boutin, and D. Berteaux. 2003. Genetic and plastic responses of a northern mammal to climate change. Proc. Roy. Soc. Lond. Ser. B Biol. Sci. 270:591–596.

Rebar, C., and O. J. Reichman. 1983. Ingestion of moldy seeds by heteromyid rodents. J. Mammal. 64:713–715.

Rebar, C. E. 1995. Activity of *Dipodomys merriami* and *Chaetodipus intermedius* to locate resource distributions. J. Mammal. 76:437–447.

Redman, P., C. Selman, and J. R. Speakman. 1999. Male short-tailed field voles (*Microtus agrestis*) build better Insulated nests than females. J. Comp. Physiol. B 169:581–587.

Reed, A. W., G. A. Kaufman, and D. W. Kaufman. 2006. Species richness-productivity relationship for small mammals along a desert-grassland continuum: differential responses of functional groups. J. Mammal. 87: 777–783.

Reed, C. A. 1969. Animal domestication in the Near East. Pp. 361–380 *in* The domestication and exploitation of plants and animals (P. J. Ucko and G. W. Dimbleby, eds.). Gerald Duckworth, London.

Reed, C. A. (ed.). 1977. Origins of agriculture. Mouton, Paris.

Reed, C. A. 1984. The beginnings of animal domestication. Pp. 1–6 *in* Evolution of domesticated animals (I. L. Mason, ed.). Longman, New York.

Reed, D. L., and M. S. Hafner. 1997. Host specificity of chewing lice on pocket gophers: a potential mechanism for cospeciation. J. Mammal. 78:655–660.

Reed, D. L., M. S. Hafner, and S. K. Allen. 2000. Mammalian hair diameter as a possible mechanism for host specialization in chewing lice. J. Mammal. 81:999–1007.

Reed, J. M., L. S. Mills, J. B. Dunning, Jr., E. S. Menges, K. S. McKelvey, R. Frye, S. R. Bessinger, M.-C. Anstett, and P. Miller. 2002. Emerging issues in population viability analyses. Conserv. Biol. 16:7–19.

Reep, R. L., M. L. Stoll, C. D. Marshall, B. L. Homer, and D. A. Samuelson. 2001. Microanatomy of facial vibrissae in the Florida manatee: the basis for specialized sensory function and oripulation. Brain Behav. Evol. 58:1–14.

Reeve, H. K., D. F. Westneat, W. A. Noon, P. W. Sherman, and C. F. Aquadro. 1990. DNA "finger printing" reveals high levels of inbreeding in colonies of the eusocial naked mole-rat. Proc. Natl. Acad. Sci. 87:2496–2500.

Reeves, R. R., and N. J. Gales. 2006. Realities of baiji conservation. Conserv. Biol. 20:626–628.

Reeves, R. R., and R. D. Kenney. 2003. Baleen whales, the right whales, *Eubalaena* spp and allies. Pp. 425–463 *in* Wild mammals of North America: biology, management, and conservation (G. A. Feldhamer, B. C. Thompson, and J. A. Chapman, eds.). Johns Hopkins Univ. Press, Baltimore.

Reeves, R. R., and A. J. Read. 2003. Bottlenose dolphin, harbor porpoise, sperm whale and other toothed cetaceans (*Tursiops truncates, Phocoena pho-*

coena, and *Physeter macrocephalus*). Pp. 397–424 *in* Wild mammals of North America: biology, management, and conservation. (G. A. Feldhamer, B. C. Thompson, and J. A. Chapman, eds.). Johns Hopkins Univ. Press, Baltimore.

Rehmeier, R. L., G. A. Kaufman, and D. W. Kaufman. 2006. An automatic activity-monitoring system for small mammals under natural conditions. J. Mammal. 87:628–634.

Reichman, O. J. 1981. Factors influencing foraging in desert rodents. Pp. 195–213 *in* Foraging behavior: ecological, ethological, and psychological approaches (A. C. Kamil and T. D. Sargent, eds.). Garland STPM Press, New York.

Reichman, O. J. 1991. Desert mammal communities. Pp. 311–347 *in* The ecology of desert communities (G. Polis, ed.). Univ. Arizona Press, Tucson.

Reichman, O. J., and D. Oberstein. 1977. Selection of seed distribution types by *Dipodomys merriami* and *Perognathus amplus*. Ecology 58:636–643.

Reichman, O. J., and M. V. Price. 1993. Ecological aspects of heteromyid foraging. Pp. 539–574 *in* Biology of the Heteromyidae (H. H. Genoways and J. H. Brown, eds.). Spec. Pub. No. 10, American Society of Mammalogists.

Reichman, O. J., and C. Rebar. 1985. Seed preferences by desert rodents based on levels of mouldiness. Anim. Behav. 33:726–729.

Reichman, O. J., and S. C. Smith. 1990. Burrows and burrowing behavior of mammals. Pp. 197–244 *in* Current mammalogy, vol. 2 (H. H. Genoways, ed.). Plenum, New York.

Reichman, O. J., and K. M. Van De Graaff. 1975. Association between ingestion of green vegetation and desert rodent reproduction. J. Mammal. 56:503–506.

Reichman, O. J., D. T. Wicklow, and C. Rebar. 1985. Ecological and mycological characteristics of caches in the mounds of *Dipodomys spectabilis*. J. Mammal. 66:643–651.

Reid, W. V., and K. R. Miller. 1989. Keep options alive: the scientific basis for conserving biodiversity. World Resources Institute, Washington, DC.

Reijnders, P., S. Brasseur, J. van der Toorn, P. van der Wolf, I. Boyd, J. Harwood, D. Lavigne, and L. Lowry. 1993. Seals, fur seals, sea lions, and walrus: status survey and conservation action plan. International Union for Conservation of Nature and Natural Resources, Gland, Switzerland.

Reiss, K. Z. 1997. Myology of the feeding apparatus of myrmecophagid anteaters (Xenarthra: Myrmecophagidae). J. Mammal. Evol. 4:87–117.

Reiss, K. Z. 2001. Using phylogenies to study convergence: the case of the ant-eating mammals. Am. Zool. 41:507 525.

Reiter, R. J. 1980. The pineal and its hormones in the control of reproduction in mammals. Endocrinol. Rev. 1:109–131.

Rendell, L., and H. Whitehead. 2005. Spatial and temporal variation in sperm whale coda vocalizations: stable usage and local dialects. Anim. Behav. 70:191–198.

Renfree, M. B. 1993. Diapause, pregnancy, and parturition in Australian marsupials. J. Exp. Zool. 266:450–462.

Renfree, M. B., E. M. Russell, and R. D. Wooller. 1984. Reproduction and life history of the honey possum, Tarsipes rostratus. Pp. 427–437 *in* Possums and gliders (A. P. Smith and I. D. Hume, eds.). Surrey Beatty and Sons, Sydney.

Rens, W., F. Grützner, P. C. M. O'Brien, H. Fairclough, J. A. M. Graves, and M. A. Ferguson-Smith. 2004. Resolution and evolution of the duck-billed platypus karyotype with an X1Y1X2Y2X3Y3X4Y4X5Y5 male sex chromosome constitution. Proc. Natl. Acad. Sci. 101:16257–16261.

Rensch, B. 1936. Studien über klimatische paarallelitat der merkmalsauspragung in vogeln und saugern. Arch. Naturgeschichte (N.F.) 5:17–363.

Rensch. B. 1938. Some problems of geographical variation and species-formation. Proc. Linnean Soc. Lond. 150:275–285.

Repenning, C. A., 1972. Underwater hearing in seals: functional morphology. Pp. 307–331 *in* Functional anatomy of marine mammals (R. J. Harrison, ed.). Academic Press, London.

Revkin, A.C. 2006. China's river dolphin declared

extinct. New York Times, December 17.

Rewcastle, S. C. 1981. Stance and gait in tetrapods: an evolutionary scenario. Symp. Zool. Soc. Lond. 48:239–268.

Rexstad, E. A., and K. P. Burnham. 1991. User's guide for interactive program CAPTURE. Colorado Cooperative Fish & Wildlife Research Unit, Colorado State Univ., Fort Collins.

Reyes, E., G. A. Bubenik, D. Schams, A. Lobos, and R. Enriquez. 1997. Seasonal changes of testicular parameters in southern pudu (*Pudu puda*) in relationship to circannual variation of its reproductive hormones. Acta Theriol. 42:25–35.

Rhodes, M., and G. Wardell-Johnson. 2006. Roost tree characteristics determine use by the white-striped freetail bat (*Tadarida australis*, Chiroptera: Molossidae) in suburban subtropical Brisbane, Australia. Aust. Ecol. 31:228–239.

Rhodes, O. E., Jr., R. K. Chesser, and M. H. Smith (eds.). 1996. Population dynamics in ecological space and time. Univ. Chicago Press, Chicago.

Rhymer, J. M., and D. Simberloff. 1996. Extinction by hybridization and introgression. Annu. Rev. Ecol. Syst. 27:83–109.

Ribble, D. O. 1991. The monogamous mating system of *Peromyscus californicus* as revealed by DNA finger printing. Behav. Ecol. Sociobiol. 29:161–166.

Ribble, D. O., A. E. Wurtz, E. K. McConnell, J. J. Buegge, and K. C. Welch, Jr. 2002. A comparison of home ranges of two species of *Peromyscus* using trapping and radiotelemetry data. J. Mammal. 83:260–266.

Rice, C. G., and P. Kalk. 1996. Identification and marking techniques. Pp. 56–66 *in* Wild mammals in captivity: principles and techniques (D. G. Kleiman, M. E. Allen, K. V. Thompson, and S. Lumpkin, eds.). Univ. Chicago Press, Chicago.

Rice, D. W. 1984. Cetaceans. Pp. 447–490 *in* Orders and families of recent mammals of the world (S. Anderson and J. K. Jones, Jr., eds.). John Wiley and Sons, New York.

Rice, W. R. 1987. Speciation via habitat specialization: the evolution of reproductive isolation as a correlated character. Evol. Ecol. 1:301–314.

Rich, T. H., J. A. Hopson, A. M. Musser, T. F. Flannery, and P. Vickers-Rich. 2005. Independent origins of middle ear bones in monotremes and therians. Science 307:910–914.

Rich, T. H., P. Vickers-Rich, P. Trosler, T. F. Flannery, R. L. Cifelli, and A. Constantine. 2001. Monotreme nature of the Australian early Cretaceous mammal *Teinolophos trusleri*. Acta Palaeontol. Polonica 46:113–118.

Richard, P. B. 1973. Capture, transport, and husbandry of the Pyrenian desman *Galemys pyrenaicus*. Int. Zoo Yearb. 13:175–177.

Richards, G. C. 1990. Rainforest bat conservation: unique problems in a unique environment. Aust. J. Zool. 26:44–46.

Richards, G. C. 1995. A review of ecological interactions of fruit bats in Australian ecosystems. Pp. 79–96 *in* Ecology, evolution, and behaviour of bats (P. A. Racey and S. M. Swift, eds.). Oxford Univ. Press, Oxford.

Richardson, E. G. 1977. The biology and evolution of the reproductive cycle of *Miniopterus schreibersii* and *M. australis* (Chiroptera: Vespertilionidae). J. Zool. 183:353–375.

Richardson, J. 1829. Fauna boreali Americana, 4 vols. J. Murray, London.

Richter, H. V., and G. S. Cumming. 2006. Food availability and annual migration of the straw-colored fruit bat (*Eidolon helvum*). J. Zool. 268:35–44.

Rickart, E. A. 2001. Elevational diversity gradients, biogeography and the structure of montane mammal communities in the inter-mountain region of North America. Global Ecol. Biogeogr. 10:77–100.

Ricklefs, R. E. 1990. Ecology, 3d edition. W. H. Freeman, New York.

Ricklefs, R. E., and G. L. Miller. 1999. Ecology, 4th edition. W. H. Freeman, New York.

Ricklefs, R. E., and D. Schluter. 1993. Species diversity: regional and historical influences. Pp. 350–363 *in* Species diversity in ecological communities (R.E. Ricklefs and D. Schluter, eds.). Univ. Chicago Press, Chicago.

Riddle, B. R. 1995. Molecular biogeography in the pocket mice (*Perognathus* and *Chaetodipus*) and grasshopper mice (*Onychomys*): the late Cenozoic development of a North American aridlands rodent guild. J. Mammal. 76:283–301.

Riddle, B. R. 1996. The molecular phylogeographic bridge between deep and shallow history in continental biotas. Trends Ecol. Evol. 11:207–211.

Riddle, B. R., and D. J. Hafner. 2006. A step-wise approach to integrating phylogeographic and phylogenetic biogeographic perspectives on the history of a core North American warm deserts biota. J. Arid Environ. 66:435–461.

Riddle, B. R., D. J. Hafner, L. F. Alexander, and J. R. Jaeger. 2000. Cryptic vicariance in the historical assembly of a Baja California Peninsular Desert biota. Proc. Natl. Acad. Sci. USA 97:14438–14443.

Ride, W. D. L. 1970. A guide to the native mammals of Australia. Oxford Univ. Press, Melbourne.

Rinn, J. L., and M. Snyder. 2005. Sexual dimorphism in mammalian gene expression. Trends Genet. 21:298–305.

Rismiller, P. D., and M. W. McKelvey. 2000. Frequency of breeding and recruitment in the short-beaked echidna, *Tachyglossus aculeatus*. J. Mammal. 81:1–17.

Rivier, C., J. Rivier, and W. Vale. 1986. Stess-induced inhibition of reproductive functions: role of endogenous corticotropin-releasing factor. Science 231:607–609.

Robbins, C. T. 1993. Wildlife feeding and nutrition, 2d edition. Academic Press, New York.

Roberts, L. S., and J. Janovy. 1996. Foundations of parasitology, 5th edition. Wm. C. Brown, Dubuque, IA.

Robinson, J. G., and K. H. Redford (eds.). 1991. Neotropical wildlife use and conservation. Univ. Chicago Press, Chicago.

Robinson, M. F. 1996. A relationship between echolocation calls and noseleaf widths in bats of the genera *Rhinolophus* and *Hipposideros*. J. Zool. 239:389–393.

Robinson, P. 1992. Rabies. Pp. 1269–1277 in Infectious diseases (S. L. Gorbach, J. G. Bartlett, and N. R. Blacklow, eds.). W. B. Saunders, Philadelphia.

Robinson, S. K., F. R. Thompson III, T. M. Donovan, D. R. Whitehead, and J. Faarborg. 1995. Regional forest fragmentation and the nesting success of migratory birds. Science 267:1987–1990.

Robinson, T. J., and C. A. Matthee. 2005. Phylogeny and evolutionary origins of the Leporidae: a review of cytogenetics, molecular analyses and a supermatrix analysis. Mammal Rev. 35:231–247.

Roby, D. D. 1991. A comparison of two noninvasive techniques to measure total body lipid in live birds. Auk 108:509–518.

Roca, A. L., N. Georgiadis, J. Pecon-Slattery, and S. J. O'Brien. 2001. Genetic evidence for two species of elephant in Africa. Science 293:1473–1477.

Roca, A. L., G. K. Bar-Gal, E. Eizirik, K. M. Helgen, R. Maria, M. S. Springer, S. J. O'Brien, and W. J. Murphy. 2004. Mesozoic origin for West Indian insectivores. Nature 429:649–651.

Rodgers, A. R., R. S. Rempel, and K. F. Abraham. 1996. A GPS-based telemetry system. Wildl. Soc. Bull. 24:559–566.

Rodriguez, M. A., L. López-Sañudo, and B. A. Hawkins. 2006. The geographic distribution of mammal body size in Europe. Global Ecol. Biogeogr. 15:173–181.

Roeder, K. D., and A. E. Treat. 1961. The detection and evasion of bats by moths. Am. Sci. 49:135–148.

Rogers, A. A., and H. Harpending. 1992. Population growth makes waves in the distribution of pairwise genetic differences. Mol. Biol. Evol. 9:552–569.

Rogers, L. 1981. A bear in its lair. Nat. Hist. 90:64–70.

Rogowitz, G. L. 1998. Limits to milk flow and energy allocation during lactation of the hispid cotton rat (*Sigmodon hispidus*). Physiol. Zool. 71:312–320.

Rohde, K. 1992. Latitudinal gradients in species diversity: the search for the primary cause. Oikos 65:514–527.

Rohr, J. J., and F. E. Fish. 2004. Strouhal numbers and optimization of swimming by odontocete cetaceans. J. Exp. Biol. 207:1633–1642.

Roman, J., and S. R. Palumbi. 2003. Whales before whaling in the North Atlantic. Science 301:508–510.

Romer, A. S. 1966. Vertebrate paleontology, 3d edition. Univ. Chicago Press, Chicago.

Romer, A. S., and T. S. Parsons. 1977. The vertebrate body. W. B. Saunders, Philadelphia.

Romer, A. S., and T. S. Parsons. 1986. The vertebrate body, 6th ed. Saunders College Publ., Philadelphia.

Ronquist, F. 1994. Ancestral areas and parsimony. Syst. Biol. 43:267–274.

Ronquist, F. 1997. Dispersal-vicariance analysis: a new approach to quantification of historical biogeography. Syst. Biol. 46:195–203.

Ronquist, F., and J. P. Huelsenbeck. 2003. MRBAYES 3: Bayesian phylogenetic inference under mixed models. Bioinformatics 19:1572–1574.

Ronquist, F., and S. Nylin. 1990. Process and pattern in the evolution of species associations. Syst. Zool. 39:323–344.

Rood, J. P. 1975. Population dynamics and food habits of the banded mongoose. East Afr. Wildl. J. 13:89–111.

Rood, J. P. 1980. Mating relationships and breeding suppression in the dwarf mongoose. Anim. Behav. 28:143–150.

Rood, J. P. 1983. The social system of the dwarf mongoose. Pp. 454–488 in Advances in the study of mammalian behavior (J. F. Eisenberg and D. G. Kleiman, eds.). Spec. Pub. No. 7, American Society of Mammalogists.

Rood, J. P. 1986. Ecology and social evolution in the mongoose. Pp. 131–152 in Ecological aspects of social evolution: birds and mammals (D. I. Rubenstein and R. W. Wrangham, eds.). Princeton Univ. Press, Princeton, NJ.

Rosatte, R., and S. Lariviere. 2003. Skunks, Genera *Mephitis*, *Spilogale*, and *Conepatus*. Pp. 692–707 in Wild mammals of North America: biology, management, and conservation. 2d edition (G. A. Feldhamer, B. C. Thompson, and J. A. Chapman, eds). Johns Hopkins Univ. Press, Baltimore.

Rose, K. D. 2006. The beginning of the age of mammals. Johns Hopkins Univ. Press, Baltimore.

Rose, K. D., and J. D. Archibald (eds.). 2005. The rise of placental mammals: origins and relationships of the major extant clades. Johns Hopkins Univ. Press, Baltimore.

Rose, K. D., and R. J. Emry. 1993. Relationships of Xenarthra, Pholidota, and fossil "edentates": the morphological evidence. Pp. 81–102 in Mammal phylogeny (F. S. Szalay, M. J. Novacek, and M. C. McKenna, eds.). Springer-Verlag, New York.

Rose, K. D., R. J. Emry, T. J. Gaudin, and G. Storch. 2005. Xenarthra and Pholidota. Pp. 106–126 in The rise of placental mammals: origins and relationships of the major extant clades (K. D. Rose and J. D. Archibald eds.). Johns Hopkins Univ. Press, Baltimore.

Rose, R.W., and M. P. Ikonomopoulou. 2005. Shivering and non-shivering thermogenesis in a marsupial, the eastern barred bandicoot (*Perameles gunnii*). J. Therm. Biol. 30:85–92.

Rose, R. W., A. K. West, J.-M. Ye, G. H. McCormack, and E. Q. Colquhoun. 1999. Nonshivering thermogenesis in a marsupial (the Tasmanian bettong *Bettongia gaimardi*) is not attributable to brown adipose tissue. Physiol. Biochem. Zool. 72:699–704.

Rosen, B. R. 1988. From fossils to earth history: applied historical biogeography. Pp. 437–481 in Analytical biogeography (A.A. Myers and P.S. Giller, eds.). Chapman & Hall, New York, NY.

Rosen, D. E. 1975. A vicariance model of Caribbean biogeography. Syst. Zool. 24:431–464.

Rosen, D. E. 1978. Vicariant patterns and historical explanation in biogeography. Syst. Zool. 27:159–188.

Rosenzweig, M. L. 1968. The strategy of body size in mammalian carnivores. Am. Midl. Nat. 80:299–315.

Rosenzweig, M. L. 1985. Some theoretical aspects of habitat selection. Pp. 517–540 in Habitat selection in birds (M. L. Cody, ed.). Academic Press, New York.

Rosenzweig, M. L. 1992. Species diversity gradients: we know more and less than we thought. J. Mammal. 73:715–730.

Rottman, S. J., and C. T. Snowden. 1972. Demonstration and analysis of an alarm pheromone in mice. J. Comp. Physiol. Psychol. 81:483–490.

Rougier, G. W., J. R. Wible, and M. J. Novacek. 1998. Implications of *Deltatheridium* specimens for early marsupial history. Nature 369:459–463.

Rounds, R.C. 1987. Distribution and analysis of colourmorphs of the black bear (*Ursus americanus*). J. Biogeogr. 14:521–538.

Rowe, D. L., and R. L. Honeycutt. 2002. Phylogenetic relationships, ecological correlates, and molecular evolution within the Cavioidea (Mammalia, Rodentia). Mol. Biol. Evol. 19:263–277.

Rowe, R. J. 2005. Elevational gradient analyses and the use of historical museum specimens: a cautionary tale. J. Biogeogr. 32:1883–1897.

Rowe, T. 1993. Phylogenetic systematics and the early history of mammals. Pp. 129–145 in Mammal phylogeny: Mesozoic differentiation, Multituberculates, Monotremes, early Therians, and Marsupials (F. S. Szalay, M. J. Novacek, and M. C. McKenna, eds.). Springer-Verlag, New York.

Rowe, T. 1996. Coevolution and the mammalian middle ear and neocortex. Science 273:651–654.

Rowe, T., and J. Gauthier. 1992. Ancestry, paleontology, and definition of the name Mammalia. Syst. Biol. 41:372–378.

Rowell, T. E. 1974. The concept of social dominance. Behav. Biol. 11:131–154.

Roze, U., and L. M. Ilse. 2003. Porcupine, *Erethizon dorsatum*. Pp. 371–380 in Wild mammals of North America: Biology, management, and conservation. (G. A. Feldhamer, B. C. Thompson, and J. A. Chapman, eds.). Johns Hopkins Univ. Press, Baltimore.

Rubidge, B. S. 1994. Australosyodon, the first primitive anteosaurid dinocephalian from the Upper Permian of Gondwana. Palaeontology 37:579–594.

Rudnai, J. 1973. The social life of the lion. Washington Square East, Wallingford, PA.

Ruedas, L. A., and J. C. Morales. 2005. Evolutionary relationships among genera of Phalangeridae (Metatheria: Diprotodontia) inferred from mitochondrial DNA. J. Mammal. 86:353–365.

Ruf, T., J. Fietz, W. Schlund, and C. Bieber. 2006. High survival in poor years: life history tactics adapted to mast seeding in the edible dormouse. Ecology 87:372–381.

Ruggiero, A. 1994. Latitudinal correlates of the sizes of mammalian geographical ranges in South America. J. Biogeogr. 21:545–559.

Rumpler, Y. S., and B. Dutrillaux. 1986. Evolution chromosoinque des prosimiens. Mammalia. 50:82–107.

Rumpler, Y., S. Warter, J.-J. Petter, R. Albignac, and B. Dutrillaux. 1988. Chromosomal evolution of Malagasy lemurs, XI: Phylogenetic position of Daubentonia madagascariensis. Folia Primat. 50:124–129.

Rusak, B., and I. Zucker. 1979. Neural regulation of circadian rhythms. Physiol. Rev. 59:449–526.

Russell, A. P., and H. N. Bryant. 2001. Claw retraction and protraction in the Carnivora: the cheetah (*Acinonyx jubatus*) as an atypical felid. J. Zool. Lond. 254:67–76.

Russell, E. M. 1982. Patterns of parental care and parental investment in marsupials. Biol. Rev. Cambridge Phil. Soc. 57:423–486.

Ryan, M. J., and M. D. Tuttle. 1983. The ability of the frog-eating bat to discriminate among novel and potentially poisonous frog species using acoustic cues. Anim. Behav. 31:827–833.

Ryder, M. L. 1973. Hair. Edward Arnold, London.

Ryder, O. A. (ed.). 1993. Rhinoceros biology and conservation. San Diego Zool. Soc., San Diego.

Saarela, S., R. Hissa, A. Pyörnilä, R. Harjula, M. Ojanen, and M. Orell. 1989. Do birds possess brown adipose tissue? Comp. Biochem. Physiol. 92A:219–228.

Saarela, S., J. S. Keith, E. Hohtola, and P. Trayhurn. 1991. Is the "mammalian" brown fat-specific mitochondrial uncoupling protein present in adipose tissue of birds? Comp. Biochem. Physiol. 100B:45–49.

Sabol, B. M., and M. K. Hudson. 1995. Technique using thermal infrared-imaging for estimating populations of gray bats. J. Mammal. 76:1242–1248.

Sacks, B. N. 2005. Reproduction and body condition of California coyotes (*Canis latrans*). J. Mammal. 86:1036–1041.

Sade, D. S. 1965. Some aspects of parent-offspring and sibling relations in a group of rhesus monkeys, with a discussion of grooming. Am. J. Phys. Anthro. 23:1–17.

Sadleir, R. M. F. S. 1973. The reproduction of vertebrates. Academic Press, New York.

Safar-Hermann, N., M. N. Ismail, H. S. Choi, E. Möstl, and E. Mamberg. 1987. Pregnancy determination in zoo animals by estrogen determination in feces. Zoo Biol. 6:189–193.

Samuel, M. D., and M. R. Fuller. 1996. Wildlife radiotelemetry. Pp. 370–418 in Research and management techniques for wildlife and habitats, 5th edition, rev. (T. A. Bookhout, ed.). Wildlife Society, Bethesda, MD.

Samuel, W. M., M. J. Pybus, D. A. Welch, and C. J. Wilke. 1992. Elk as a potential host for meningeal worm: implications for translocation. J. Wildl. Manage. 56:629–639.

Samuel, W. M., M. J. Pybus, and A. A. Kocan (eds.). 2001. Parasitic diseases of wild mammals. Iowa State Univ. Press, Ames.

Sanchez-Cordero, V., and T. H. Fleming. 1993. Ecology of tropical heteromyids. Pp. 596–617 in Biology of the Heteromyidae (H. H. Genoways and J. H. Brown, eds.). Spec. Pub. No. 10, American Society of Mammalogists.

Sánchez-Prieto, C. B., J. Carranza, and F. J. Pulido. 2004. Reproductive behavior in female Iberian red deer: effects of aggregation and dispersion of food. J. Mammal. 85:761–767.

Sand, H., B. Zimmerman, P. Wabakken, H. Andren, and H. C. Pederson. 2005. Using GPS and GIS cluster analyses to estimate kill rates in wolf-ungulate ecosystems. Wildl. Soc. Bull. 33:914–925.

Sandell, M. 1984. To have or not to have delayed implantation: the example of the weasel and the stoat. Oikos. 42:123–126.

Sanderson, M. J. 1997. A nonparametric approach too estimating divergence times in the absence of rate constancy. Mol. Biol. Evol. 14:1218–1232.

Sanger, F., S. Nicklen, and A. R. Coulson. 1977. DNA sequencing with chain-terminating inhibitors. Proc. Natl. Acad. Sci. USA 74:5463–5467.

Sanger, V. L. 1971. Toxoplasmosis. Pp. 326–334 in Parasitic diseases of wild mammals (J. W. Davis and R. C. Anderson, eds.). Iowa State Univ. Press, Ames.

Santini-Palka, M. E. 1994. Feeding behavior and activity patterns of two Malagasy bamboo lemurs, *Hapalemur simus* and *Hapalemur griseus* in captivity. Folia Primatol. 63: 44–49.

Santos, M. B., G. J. Pierce, J. Herman, A. Lopez, A. Guerra, E. Mente, and M. R. Clarke. 2001a. Feeding ecology of Cuvier's beaked whale (*Ziphius cavirostris*): a review with new information on the diet of this species. J. Mar. Biol. Assn. UK 81:687–694.

Santos, M. B., G. J. Pierce, C. Smeenk, M. J. Addink, C. C. Kinze, S. Tougaard, and J. Herman. 2001b. Stomach contents of northern bottlenose whales Hyperoodon ampullatus stranded in the North Sea. J. Mar. Biol. Assn. UK 81:143–150.

Sapolsky, R. 1997. Testosterone rules. Discovery 18:44–50.

Sapolsky, R. M. 1990. Adrenocortical function, social rank, and personality among wild baboons. Biol. Psychiatry 28:862–885.

Sapolsky, R. M. 1991. Testicular function, social rank, and personality among wild baboons. Psychoneuroendocrinology 16:281–293.

Sargis, E. 2004. New views on tree shrews: the role of tupaiids in primate suprordinal relationships. Evol. Anthropol. 13:56–66.

Sargis, E. J. 2001. A preliminary qualitative analysis of the axial skeleton of tupaiids (Mammalia: Scandentia): functional morphology and phylogenetic implications. J. Zool. 253:473–483.

Sarich, V., J. M. Lowenstein, and B. J. Richardson. 1982. Phylogenetic relationships of Thylacinus cynocephalus, Marsupialia, as reflected in comparative serology. Pp. 707–709 in Carnivorous marsupials (M. Archer, ed.). Royal Zoological Society, New South Wales, Sydney.

Sarich, V. M., and J. E. Cronin. 1976. Molecular systematics of the primates. Pp. 141–170 in Molecular anthropology (M. Goodman and R. E. Tashian, eds.). Plenum Press, New York.

Sarich, V. M., and A. C. Wilson. 1967a. Immunological time scale for hominid evolution. Science 158:1200–1203.

Sarich, V. M., and A. C. Wilson. 1967b. Rates of albumin evolution in primates. Proc. Natl. Acad. Sci. USA 58:142–148.

Sasaki, T., M. Nikaido, H. Hamilton, M. Goto, H.

Kato, N. Kanda, L. A. Pastene, Y. Cao, R. E. Fordyce, M. Hasegawa, and N. Okada. 2005. Mitochondrial phylogenetics and evolution of mysticete whales. Syst. Biol. 54:77–90.

Sauer, C. O. 1969. Agricultural origins and dispersals: the domestication of animals and foodstuffs, 2d edition. MIT Press, Cambridge, MA.

Savage, R. J. G., and M. R. Long. 1986. Mammal evolution: an illustrated guide. Facts on File, New York.

Savage-Rumbaugh, E. S. 1986. Ape language: from conditioned response to symbol. Columbia Univ. Press, New York.

Savage-Rumbaugh, E. S., and K. E. Brakke. 1990. Animal language: methodological and interpretive issues. Pp. 313–343 in Interpretation and explanation in the study of animal behavior (M. Bekoff and D. Jamieson, eds.). Westview Press, Boulder, CO.

Savolainen, P., Y. Zhang, J. Luo, J. Lundeberg, and T. Leitner. 2002. Genetic evidence for an East Asian origin of domestic dogs. Science 298:1610–1613.

Sawyer, T. G., K. V. Miller, and R. L. Marchinton. 1994. Patterns of urination and rub-urination in female white-tailed deer. J. Mammal. 74:477–479.

Scantlebury, M., M. K. Oosthuizen, J. R. Speakman, C. R. Jackson, and N. C. Bennett. 2005. Seasonal energetics of the Hottentot golden mole at 1500 altitude. Physiol. Behav. 84:739–745.

Schaller, G. B. 1963. The mountain gorilla: Ecology and behavior. Univ. Chicago Press, Chicago.

Schaller, G. B. 1972. The Serengeti lion: a study of predator-prey relations. Univ. Chicago Press, Chicago.

Schaller, G. B., T. Qitao, K. G. Johnson, W. Xiaoming, S. Heming, and H. Jinchu. 1989. The feeding ecology of giant pandas and Asiatic black bears in the Tangjiahe Reserve, China. Pp. 212–241 in Carnivore behavior, ecology, and evolution (J. L. Gittleman, ed.). Cornell Univ. Press, Ithaca, NY.

Schantz, V. S. 1943. Mrs. M. A. Maxwell, a pioneer mammalogist. J. Mammal. 24:464–466.

Schauber, E. M., R. S. Ostfield, and A. S. Evans, Jr. 2005. What is the best predictor of annual Lyme disease incidence: weather, mice, or acorns? Ecological Appl. 15:575–586.

Scheel, D., and C. Packer. 1991. Group hunting behaviour of lions: a search for cooperation. Anim. Behav. 41:697–709.

Scheffer, V. B. 1951. The rise and fall of a reindeer herd. Sci. Mon. 73:356–362.

Scheffer, V. B. 1958. Seals, sea lions, and walruses. Stanford Univ. Press, Stanford, CA.

Scheibe, K. M., M. Dehnhard, H. H. D. Meyer, and A. Scheibe. 1999. Noninvasive monitoring of reproductive function by determination of faecal progestgagens and sexual behaviour in a herd of Przewalski mares in a semireserve. Acta Theriol. 44:451–463.

Scheich, H., G. Langner, C. Tidemann, R. B. Coles, and A. Guppy. 1986. Electroreception and electrolocation in platypus. Nature 319:401–402.

Schemnitz, S. D. 1996. Capturing and handling wild animals. Pp. 106–124 in Research and management techniques for wildlife and habitats, 5th edition, rev. (T. A. Bookhout, ed.). Wildlife Society, Bethesda, MD.

Schenk, A., and K. M. Kovacs. 1996. Genetic variation in a population of black bears as revealed by DNA fingerprinting. J. Mammal. 77:942–950.

Schino, G. 2001. Grooming, competition, and social rank among female primates: a meta-analysis. Anim. Behav. 62:265–271.

Schliemann, H., and B. Maas. 1978. Myzopoda aurita. Mammal. Species No. 116:1–2.

Schlitter, D. A. 2005. Order Pholidota. Pp. 530–531 in Mammal species of the world: a taxonomic and geographic reference. (D. E. Wilson and D. M. Reeder, eds.) 3rd edition, 2 vols. Johns Hopkins Univ. Press, Baltimore.

Schlitter, D. A. 2005a. Order Macroscelidea. Pp. 82–85 in Mammal species of the world: a taxonomic and geographic reference (D.E. Wilson and D. M. Reeder, eds.). Third Edition, volume 1, The Johns Hopkins Univ. Press, Baltimore, MD.

Schmidly, D. J. 2004. The Mammals of Texas (rev. ed.). Univ. Texas Press, Austin.

Schmidly, D. J., K. T. Wilkins, and J. N. Derr. 1993. Biogeography. Pp. 319–356 in Biology of the Heteromyidae (H. H. Genoways and J. H. Brown, eds.). Spec. Pub. No. 10, American Society of

Mammalogists.

Schmidt, G. D. 1971. Acanthocephalan infections of man, with two new records. J. Parasitol. 57:582–584.

Schmidt-Nielsen, K. 1964. Desert animals: physiological problems of heat and water. Oxford Univ. Press, New York.

Schmidt-Nielsen, K. 1979. Desert animals: physiological problems of heat and water. Dover Publications, Inc., New York.

Schmidt-Nielsen, K. 1997. Animal physiology: adaptation and environment, 4th edition. Cambridge Univ. Press, New York.

Schmidt-Nielsen, K., and R. O'Dell. 1961. Structure and concentrating mechanism in the mammalian kidney. Am. J. Physiol. 200:1119–1124.

Schmidt-Nielsen, K., B. Schmidt-Nielsen, S. A. Jarnum, and T. R. Houpt. 1957. Body temperature of the camel and its relation to water economy. Am. J. Physiol. 188:103–112.

Schmidt-Nielsen, K., W. L. Bretz, and C. R. Taylor. 1970a. Panting in dogs: unidirectional air flow over evaporative surfaces. Science, 169:1102–1104.

Schmidt-Nielsen, K., F. R. Hainsworth, and D. E. Murrish. 1970b. Counter-current heat exchange in the respiratory passages: effect on water and heat balance. Resp. Physiol. 9:263–276.

Schnabel, Z. E. 1938. The estimation of the total fish population in a lake. Am. Math. Mon. 45:348–352.

Schneider, H., M. P. C. Schneider, I. Sampaio, M. L. Harada, M. Stanhope, J. Czelusniak, and M. Goodman. 1993. Molecular phylogeny of the New World monkeys (Platyrrhini, Primates). Mol. Phylogenet. Evol. 2:225–242.

Schnitzler, H. U., E. K. V. Kalko, I. Kaipf, and A. D. Grinnell. 1994. Fishing and echolocation behavior of the greater bulldog bat, Noctilio leporinus, in the field. Behav. Ecol. Sociobiol. 35:327–345.

Schoener, T. W. 1974. Resource partitioning in ecological communities. Science 185:27–39.

Scholander, P. F. 1955. Evolution of climatic adaptation in homeotherms. Evolution. 9:15–26.

Scholander, P. F. 1957. The wonderful net. Sci. Am. 196:97–107.

Scholander, P. F., V. Waters, R. Hock, and L. Irving. 1950. Body insulation of some arctic and tropical mammals and birds. Biol. Bull. 99:225–235.

Schonewald-Cox, C. M., S. M. Chambers, B. MacBryde, and L. Thomas (eds.). 1983. Genetics and conservation. Benjamin Cummings, Menlo Park, CA.

Schooley, R. L., B. T. Bestelmeyer, and J. F. Kelly. 2000. Influence of small-scale disturbances by kangaroo rats on Chihuahuan desert ants. Oecologia 125:142–149.

Schradin, C., and G. Anzenberger. 2001. Costs of infant carrying in common marmosets, *Callithrix jacchus*: an experimental analysis. Anim. Behav. 62:289–295.

Schug, M. D., S. H. Vessey, and A. I. Korytko. 1991. Longevity and survival in a population of whitefooted mice (*Peromyscus leucopus*). J. Mammal. 72:360–366.

Schulz, M. 2000. Diet and foraging behavior of the golden-tipped bat, *Kerivoula papuensis*: a spider specialist. J. Mammal. 81:948–957.

Schulze, M. D. 1998. Forest fires in the Brazilian Amazon. Conserv. Biol. 12:948–950.

Schum, M. 1984. Phenetic structure and species richness in North and Central American bat faunas. Ecology 65:1315–1324.

Schutt, W. A., and N. B. Simmons. 2001. Morphological specializations of *Cheiromeles* (naked bulldog bats; Molossidae) and their possible role in quadrupedal locomotion. Acta Chiropt. 3:225–235.

Schutt, W. A., Jr., and J. S. Altenbach. 1997. A sixth digit in *Diphylla ecaudata*, the hairy legged vampire bat (Chiroptera, Phyllostomidae). Mammalia 61: 280–285.

Schutt, W. A., Jr., F. Muradali, N. Mondol, K. Joseph, and K. Brockmann. 1999. Behavior and maintenance of captive white-winged vampire bats, *Diaemus youngi*. J. Mammal. 80:71–81.

Schwagmeyer, P. L. 1988. Scramble-competition polygyny in an asocial mammal: male mobility and mating success. Am. Nat. 131:885–892.

Schwartz, C. C., S. D. Miller, and M. A. Haroldson. 2003. Grizzly bear, *Ursus arctos*. Pp. 556–586 in Wild mammals of North America: Biology, management, and conservation. 2d edition (G. A. Feldhamer, B. C. Thompson, and J. A. Chapman, eds). Johns Hopkins Univ. Press, Baltimore.

Schwartz, J. H. 1986. Primate systematics and a classification of the order. Pp. 1–41 in Comparative primate biology: vol. 1, Systematics, evolution, and anatomy (D. R. Swindler and J. Erwin, eds.). Alan R. Liss, New York.

Schwartz, J. H., I. Tatersall, and N. Eldridge. 1978. Phylogeny and classification of the primates revisit ed. Yearb. Phys. Anthrop. 21:95–133.

Scott, J. P. 1966. Agonistic behavior of mice and rats: a review. Am. Zool. 6:683–701.

Scott, J. P. 1972. Animal behavior, 2d edition. Univ. Chicago Press, Chicago.

Scullard, H. H. 1974. The elephant in the Greek and Roman world. Thames and Hudson, Cambridge, UK.

Seagle, S. W. 1985. Competition and coexistence of small mammals in an east Tennessee pine plantation. Am. Mid. Nat. 114:272–282.

Sealander, J. A. 1952. The relationship of nest protection and huddling to survival of Peromyscus at low temperature. Ecology 33:63–71.

Searle, A.G. 1968. Comparative genetics of coat color in mammals. Academic Press, New York.

Searle, K. R., N. T. Hobbs, and L. A. Shipley. 2005. Should I stay or should I go? Patch departure decisions by herbivores at multiple states. Oikos 111: 417–424.

Sears, K. E., R. R. Behringer, J. J. Rasweiler, and L. A. Niswander. 2006. Development of bat flight: morphologic and molecular evolution of bat wing digits. Proc. Natl. Acad. Sci. USA 103:6581–6586.

Seber, G. A. F. 1965. A note on the multiple recapture census. Biometrika 52:249–259.

Seber, G. A. F. 1982. The estimation of animal abundance and related parameters, 2d edition. Griffin, London.

Sedgeley, J. A. 2001. Winter activity in the tree-roosting lesser short-tailed bat, Mystacina tuberculata, in a cold-temperature climate in New Zealand. Acta Chiropt. 3:179–195.

Sedgeley, J. A. 2003. Roost site selection and roosting behaviour in lesser short-tailed bats (Mystacina tuberculata) in comparison with long-tailed bats (Chalinolobus tuberculatus) in Nothofagus forest, Fiordland. New Zealand J. Zool. 30:227–241.

Seebeck, J. H., and P. G. Johnston. 1980. Potorous longipes (Marsupialia: Macropodidae): a new species from eastern Victoria. Aust. J. Zool. 28:119–134.

Selander, R. K. 1970. Behavior and genetic variation in natural populations. Am. Zool. 10:53–66.

Selander, R. K. 1982. Phylogeny. Pp. 32–59 in Perspectives on evolution (R. Milkman, ed.). Sinauer Associates, Sunderland, MA.

Selander, R. K., M. H. Smith, S. Y. Yang, W. E. Johnson, and J. B. Gentry. 1971. Biochemical polymorphism and systematics in the genus Peromyscus, I: Variation in the old-field mouse (Peromyscus polionotus). Studies in genetics VI, Univ. Texas Pub. 7103:49–90.

Selye, H. 1950. The physiology and pathology of exposure to stress. Acta Montreal.

Sempéré, A. J., V. E. Sokolov, and A. A. Danilkin. 1996. Capreolus capreolus. Mamm. Spec. 538:1–9.

Serpell, J. A. 2000. Domestication and history of the cat. Pp. 179–192 in The domestic cat: the biology of its behaviour (D. C. Turner and P. Bateson, eds.). Cambridge Univ. Press, Cambridge, UK.

Service, R. F. 2006. The race for the $1000 genome. Science 311:1544–1546.

Sessions, S. K. 1996. Chromosomes: molecular cytogenetics. Pp. 121–168 in Molecular systematics, 2d edition (D.M. Hillis, C. Moritz, and B.K. Mable, eds.). Sinauer Associates, Sunderland, MA.

Seyfarth, R. M., D. L. Cheney, and P. Marler. 1980a. Monkey responses to three different alarm calls: evidence for predator classification and semantic communication. Science 210:801–803.

Seyfarth, R. M., D. L. Cheney, and P. Marler. 1980b. Vervet monkey alarm calls: semantic communication in a free-ranging primate. Anim. Behav. 28:1070–1094.

Seymour, R. S., and M. K. Seely. 1996. The respiratory environment of the Namib Desert golden mole. J. Arid Environ. 32:453–461.

Seymour, R. S., P. C. Withers, and W. W. Weathers.

1998. Energetics of burrowing, running, and free-living in the Namib Desert golden mole (Eremitalpa namibensis). J. Zool. Lond. 244:107–117.

Shanahan, M., and S. G. Compton. 2000. Fig-eating by Bornean tree shrews (Tupaia spp.): evidence for a role as seed dispersers. Biotropica 32:759–764.

Shapiro, L. J., C. V. M. Seiffert, L. R. Godfrey, W. L. Jungers, E. L. Simons, and G. F. N. Radria. 2005. Morphometric analysis of lumbar vertebrae in extinct Malagasy primates. Am. J. Phys. Anthropol. 128:823–839.

Sharifi, M., and Z. Hemmati. 2002. Variation in the diet of the greater mouse-tailed bat, Rhinopoma microphyllum (Chiroptera: Rhinopomatidae) in southwestern Iran. Zool. Middle East 26:65–70.

Sharman, G. B. 1963. Delayed implantation in marsupials. Pp. 3–14 in Delayed implantation (A. C. Enders, ed.). Univ. Chicago Press, Chicago.

Sharman, G. G. 1965. The effects of suckling on normal and delayed cycles of reproduction in the red kangaroo. Z. Saugetierk. 30:10–20.

Sharman, G. B. 1976. Evolution of viviparity in mammals. Pp. 32–70 in Reproduction in mammals, vol. 6 (C. R. Austin and R. V. Short, eds.). Cambridge Univ. Press, Cambridge, UK.

Sharman, G. B., and P. J. Berger. 1969. Embryonic diapause in marsupials. Adv. Reprod. Physiol. 4:211–240.

Sharman, G. B., and M. J. Clark. 1967. Inhibition of ovulation by the corpus luteum in the red kangaroo, Megaleia rufa. J. Reprod. Fert. 14:129–137.

Sharp, J. A., K. N. Kane, C. Lefevre, J. P. Y. Arnould, and K. R. Nicholas. 2005. Fur seal adaptations to lactation: insights into mammary gland function. Curr. Top. Devel. Biol. 72:275–308.

Sheffield, S. R., R. P. Morgan II, G. A. Feldhamer, and D. M. Harman. 1985. Genetic variation in white-tailed deer (Odocoileus virginianus) populations in western Maryland. J. Mammal. 66:243–255.

Sheridan, M., and R. H. Tamarin. 1988. Space use, longevity, and reproductive success in meadow voles. Behav. Ecol. Sociobiol. 22:85–90.

Sherman, P. W. 1977. Nepotism and the evolution of alarm calls. Science 197:1246–1253.

Sherman, P. W. 1981. Kinship, demography, and Belding's ground squirrel nepotism. Behav. Ecol. Sociobiol. 8:251–259.

Sherman, P. W., J. U. M. Jarvis, and R. D. Alexander (eds.). 1991. The biology of the naked mole-rat. Princeton Univ. Press, Princeton, NJ.

Sherman, P. W., S. Braude, and J. U. M. Jarvis. 1999. Litter sizes and mammary numbers of naked mole-rats: breaking the one-half rule. J. Mammal. 80:720–733.

Sherrod, S. K., T. R. Seastedt, and M. D. Walker. 2005. Northern pocket gopher (Thomomys talpoides) control of alpine plant community structure. Arctic Antarctic Alpine Res. 37:585–590.

Shields, W. M. 1982. Philopatry, inbreeding, and the evolution of sex. State Univ. New York Press, Albany.

Shilton, L. A., J. D. Altringham, S. G. Compton, and R. J. Whittaker. 1999. Old World fruit bats can be long-distance seed dispersers through extended retention of viable seeds in the gut. Proc. Roy. Soc. Lond. Ser. B 266:219–233.

Shimmin, G. A., J. Skinner, and R. V. Baudinette. 2002. The warren architecture and environment of the southern hairy-nosed wombat (Lasiorhinus latifrons). J. Zool. Lond. 258:469–477.

Shkolnik, A., and A. Borut. 1969. Temperature and water relations in two species of spiny mice (Acomys). J. Mammal. 50:245–255.

Short, J., and B. Turner. 2000. Reintroduction of the burrowing bettong Bettongia lesueur (Marsupialia: Potoroidae) to mainland Australia. Biol. Conserv. 96:185–196.

Short, R. V. 1983. The biological bases for the contraceptive effects of breast feeding. Pp. 27–39 in Advances in international maternal and child health (D. B. Jellife and E. F. B. Jellife, eds.). Oxford Univ. Press, Oxford.

Shoshani, J. 2005a. Order Proboscidea. Pp. 90–91 in Mammal species of the world: a taxonomic and geographic reference. 3rd edition, 2 vols. Johns Hopkins Univ. Press, Baltimore.

Shoshani, J. 2005b. Order Hyracoidea. Pp. 87–89 in Mammal species of the world: a taxonomic and geo-

graphic reference. 3rd edition, 2 vols. Johns Hopkins Univ. Press, Baltimore.

Shoshani, J., and M. C. McKenna. 1998. Higher taxonomic relationships among extant mammals based on morphology, with selected comparisons of results from molecular data. Mol. Phylogenet. Evol. 9:572–584.

Sibley, C. G., and J. E. Ahlquist. 1984. The phylogeny of homonoid primates as indicated by DNA-DNA hybridization. J. Mol. Evol. 20:2–15.

Sibly, R. M., D. Barker, M. C. Denham, J. Hone, and M. Pagel. 2005. On the regulation of populations of mammals, birds, fish, and insects. Science 309:607–610.

Sicuro, F. L., and L. F. B. Oliveira. 2002. Coexistence of peccaries and feral hogs in the Brazilian pantanal wetland: an ecomorphological view. J. Mammal. 83:207–217.

Siegel, S., and N. J. Castellan, Jr. 1988. Nonparametric statistics for the behavioral sciences, 2d edition. McGraw-Hill, New York.

Sikes, S. K. 1971. The natural history of the African elephant. Weidenfeld and Nicolson, London.

Sikorski, M. D., and A. D. Bernshtein. 1984. Geographical and intrapopulation divergence in Clethrionomys glareolus. Acta Theriol. 29:219–230.

Silk, J. B., D. L. Cheney, and R. M. Seyfarth. 1996. The form and function of post-conflict interactions between female baboons. Anim. Behav. 52:259–268.

Sillén-Tullberg, B., and A. P. Møller. 1993. The relationship between concealed ovulation and mating systems in anthropoid primates: a phylogenetic analysis. Am. Nat. 141:1–25.

Simmons, A.H. 1988. Extinct pygmy hippopotamus and early man in Cyprus. Nature 333:554–557.

Simmons, I. G. 1981. The ecology of natural resources. John Wiley and Sons, New York.

Simmons, J. A., D. J. Howell, and N. Suga. 1975. Information content of bat sonar echoes. Am. Sci. 63:204–215.

Simmons, N. B. 1995. Bat relationships and the origin of flight. Pp. 27–43 in Ecology, evolution, and behaviour of bats (P. A. Racey and S. M. Swift, eds.). Oxford Univ. Press, Oxford.

Simmons, N. B. 1998. A reappraisal of interfamilial relationships of bats. Pp. 3–26 in Bat biology and conservation (T. H. Kunz and P. A. Racey, eds.). Smithsonian Institution Press, Washington, DC.

Simmons, N. B., and T. M. Conway. 2001. Phylogenetic relationships of mormoopid bats (Chiroptera: Mormoopidae) based on morphological data. Bull. Am. Mus. Nat. Hist. 258:1–97.

Simmons, N. B., and J. H. Geisler. 1998. Phylogenetic relationships of Icaronycteris, Archaeonycteris, Hassianycteris, and Palaeochiropteryx to extant lineages, with comments on the evolution of echolocation and foraging strategies in microchiroptera. Bull. Am. Mus. Nat. Hist. 235:1–182.

Simmons, N. B., and T. H. Quinn. 1994. Evolution of the digital tendon locking mechanism in bats and dermopterans: a phylogenetic perspective. J. Mammal. Evol. 2:231–254.

Simons, E. L. 1972. Primate evolution. Macmillan, New York.

Simpson, C. D. 1984. Artiodactyls. Pp. 563–587 in Orders and families of recent mammals of the world (S. Anderson and J. K. Jones, Jr., eds.). John Wiley and Sons, New York.

Simpson, G. G. 1940. Mammals and land bridges. J. Wash. Acad. Sci. 30:137–163.

Simpson, G. G. 1945. The principles of classification and a classification of mammals. Bull. Am. Mus. Nat. Hist. 85:1–350.

Simpson, G. G. 1951. The species concept. Evolution 5:285–298.

Simpson, G. G. 1961. Principles of animal taxonomy. Columbia Univ. Press, New York.

Simpson, G. G. 1964. Species density of North American recent mammals. Syst. Zool. 13:57–73.

Simpson, G. G. 1965. The geography of evolution. Chilton Book Co., Philadelphia.

Sinclair, A. R. E., and J. M. Gosline. 1997. Solar activity and mammal cycles in the northern hemisphere. Am. Nat. 149:776–784.

Sinclair, A. R. E., J. M. Gosline, G. Holdsworth, C. J. Krebs, S. Boutin, J. N. M. Smith, R. Boonstra, and M. Dale. 1993. Can the solar cycle and climate synchronize the snowshoe hare cycle in Canada? Am.

Nat. 141:173–198.

Sinclair A. R. E., D. Chitty, C. I. Stefan, and C. J. Krebs 2003. Mammal population cycles: evidence for intrinsic differences during snowshoe hare cycles. Can. J. Zool.–Rev. Can. Zool. 81: 216–220.

Sinclair, E. A., B. Costello, J. M. Courtnay, and K. A. Crandall. 2002. Detecting a genetic bottleneck in Gilbert's potoroo (*Potorous gilbertii*) (Marsupialia: Potoroidae), inferred from microsatellite and mitochondrial DNA sequence data. Cons. Genet. 3:191–196.

Sinclair, J. A., R. L. Lochmiller, C. W. Qualls, Jr., and C. C. Cummings. 1998. Alopecia associated with postjuvenile molt in a wild population of *Peromyscus maniculatus*. Southwest. Nat. 43:405–407.

Singaravelan, N., and G. Marimuthu. 2004. Nectar feeding and pollen carrying from *Ceiba pentandra* by pteropodid bats. J. Mammal. 85:1–7.

Singer, S. 1959. A history of biology to about the year 1900. Abelard-Schuman, London.

Singleton, G. R., L. K. Chambers, and D. M. Spratt. 1995. An experimental field study to determine whether *Capillaria hepatica* (Nematoda) can limit house mouse populations in Eastern Australia. Wildl. Res. 22:31–53.

Siniff, D. B., and S. Stone. 1985. The role of the leopard seal in the tropho-dynamics of the Antarctic marine ecosystem. Pp. 555–560 *in* Antarctic nutrient cycles and food webs (W. R. Siegfried, et al., eds.). Springer-Verlag, Berlin.

Sipe, T. W., and R. A. Browne. 2004. Phylogeography of the masked (*Sorex cinereus*) and smoky shrews (*Sorex fumeus*) in the southern Appalachians. J. Mammal. 85:875–885.

Sjöstedt, A. B. 2005. *Fransicella*. Pp. 200–210 *in* The proteobacteria, part B. Bergey's manual of systematic bacteriology (D. J. Brenner, N. R. Krieg, J. T. Stanley, G. M. Garrity, eds.). 2d edition. Springer, New York.

Skelpkovych, B. O., and W. A. Montevecchi. 1996. Food availability and food hoarding behavior by red and arctic foxes. Arctic 49:228–234.

Skjenneberg, S. 1984. Reindeer. Pp. 128–138 *in* Evolution of domesticated animals (I. L. Mason, ed.). Longman, New York.

Slade, N. A., and L. A. Russell. 1998. Distances as indices to movements and home-range size from trapping records of small mammals. J. Mammal. 79:346–351.

Slater, P. J. B. 1983. The study of communication. Pp. 9–42 *in* Animal behaviour (T. R. Halliday and P. J. B. Slater, eds.). W. H. Freeman, New York.

Slijper, E. J. 1979. Whales, 2d edition. Cornell Univ. Press, Ithaca, NY.

Smale, L., K. E. Holekamp, M. Weldele, L. G. Frank, and S. E. Glickman. 1995. Competition and cooperation between litter-mates in the spotted hyaena, *Crocuta crocuta*. Anim. Behav. 50:671–682.

Smallwood, P. D., M. A. Steele and S. H. Faeth. 2001. The ultimate basis of the caching preferences of rodents, and the oak dispersal syndrome: tannins, insects, and seed germination. Am. Zoologist 41:840–851.

Smil, V. 2000. Laying down the law. Nature 403:597.

Smiseth, P., and S. H. Lorentsen. 2001. Begging and parent-offspring conflict in grey seals. Anim. Behav. 62:273–279.

Smith, A. B. 1994. Rooting molecular trees: problems and strategies. Biol. J. Linnean Soc. 31:279–292.

Smith, A. L., G. R. Singleton, G. M. Hansen, and G. Shellam. 1993. A serologic survey for viruses and *Mycoplasma pulmonis* among wild house mice (*Mus domesticus*) in southeastern Australia. J. Wildl. Dis. 29:219–229.

Smith, A. T. 1987. Population structure of pikas: dispersal versus philopatry. Pp. 128–142 *in* Mammalian dispersal patterns: the effects of social structure on population genetics (B. D. Chepko-Sade and Z. T. Halpin, eds.). Univ. Chicago Press, Chicago.

Smith, A. T., N. A. Formozov, R. S. Hoffmann, Z. Changlin, and M. A. Erbajeva. 1990. The pikas. Pp. 14–60 *in* Rabbits, hares, and pikas: status survey and conservation action plan (J. A. Chapman and J. E. C. Flux, eds.). IUCN/SSC Lagomorph Specialist Group, Gland, Switzerland.

Smith, B. D., B. Ahmed, M. E. Ali, and G. Braulik. 2001. Status of the Ganges river dolphin or shushuk *Platanista gangetica* in Kaptai Lake and the southern

rivers of Bangladesh. Oryx 35:61–72.

Smith, C. C., and O. J. Reichman. 1984. The evolution of food caching by birds and mammals. Annu. Rev. Ecol. Syst. 15:329–351.

Smith, F. A., and D. M. Kaufman. 1996. A quantitative analysis of the contributions of female mammalogists from 1919 to 1994. J. Mammal. 77:613–628.

Smith, F. D. M., R. M. May, and P. H. Harvey. 1994. Geographical ranges of Australian mammals. J. Anim. Ecol. 63:441–450.

Smith, J. D. 1972. Systematics of the chiropteran family Mormoopidae. Univ. Kansas Publ. Mus. Nat. Hist. 56:1–32.

Smith, J. D. 1980. Chiropteran phylogenetics: Introduction. Pp. 233–244 *in* Proc. 5th International Bat Research Conference (D. E. Wilson and A. L. Gardner, eds.). Texas Tech Univ. Press, Lubbock.

Smith, M. M. Manlove, and J. Joule. 1978. Spatial and temporal dynamics of the genetic organization of small mammal populations. Pp. 99–113 *in* Populations of small mammals under natural conditions (D. Snyder, ed.)., Spec. Pub., Ser. 5, Pymatuning Lab Ecology, Univ. Pittsburgh.

Smith, M. H., R. K. Selander, and W. E. Johnson. 1973. Biochemical polymorphism and systematics in the genus *Peromyscus*, III: variation in the Florida deer mouse (*Peromyscus floridanus*), a Pleistocene relict. J. Mammal. 54:1–13.

Smith, P., and E. Tchernov (eds.). 1992. Structure, function, and evolution of teeth. Freund Publishing House, London.

Smith, P. A., J. A. Schaefer, and B. R. Patterson. 2002. Variation at high latitudes: the geography of body size and cranial morphology of the muskox, *Ovibos moschatus*. J. Biogeogr. 29: 1089–1094.

Smith, R. E., and A. B. Horwitz. 1969. Brown fat and thermogenesis. Physiol. Rev. 49:330–425.

Smith, W. J. 1984. Behavior of communicating, 2d edition. Harvard Univ. Press, Cambridge, MA. Smithsonian Institution. 2004. *http://nationalzoo.si.edu/ConservationAndScience/ EndangeredSpecies*

Smuts, B. B. 1985. Sex and friendship in baboons. Aldine, New York.

Smythe, N. 1977. The function of mammalian alarm advertising: social signals or pursuit invitation? Am. Nat. 111:191–194.

Sober, E. 1984. The nature of selection. MIT Press, Cambridge, MA.

Sober, E. 1988. Reconstructing the past: parsimony, evolution, and inference. MIT Press, Cambridge, MA.

Soholt, L. F. 1973. Consumption of primary production by a population of kangaroo rats (*Dipodomys merriami*) in the Mojave desert. Ecol. Monogr. 43:357–376.

Sokal, R. R., and F. J. Rohlf. 1981. Biometry, 2d edition. W. H. Freeman, San Francisco.

Sokal, R. R., and F. J. Rohlf. 1995. Biometry: the principles and practice of statistics in biological research. W.H. Freeman, New York.

Solari, S., R. A. Van den Bussche, S. R. Hoofer, and B. D. Patterson. 2004. Geographic distribution, ecology, and phylogenetic affinities of *Thyroptera lavali* Pine 1993. Acta Chiropterol. 6:293–302.

Soltis, J., R. Thomsen, and O. Takenaka. 2001. The interaction of male and female reproductive strategies and paternity in wild Japanese macaques, *Macaca fuscata*. Anim. Behav. 62:485–494.

Soltis, J., K. Leong, and A. Savage. 2005. African elephant vocal communication I: antiphonal calling behaviour among affiliated females. Anim. Behav. 70:579–587.

Sondaar, P. Y. 1977. Insularity and its effect on mammal evolution. Pp. 671–707 *in* Major patterns of vertebrate evolution (M. K. Hecht, P. C. Goody, and B. M. Hecht, eds.). Plenum Press, New York.

Sosa, R. A., and J. H. Sarasola. 2005. Habitat use and social structure of an isolated population of guanacos (*Lama guanicoe*) in the Monte Desert, Argentina. Europ. J. Wildl. Res. 51:207–209.

Soulé, M. E. 1987. Viable populations for conservation. Cambridge Univ. Press, Cambridge, UK.

Soulé, M. E., and B. A. Wilcox (eds.). 1980. Conservation biology: an evolutionary-ecological perspective. Sinauer Associates, Sunderland, MA.

Sousa-Lima, R. S., A. P. Paglia, and G. A. B. DaFonseca. 2002. Signature information and indi-

vidual recognition in the isolation calls of Amazonian manatees, *Trichechus inunguis* (Mammalia: Sirenia). Anim. Behav. 63:301–310.

Southgate, R. I., P. Christie, and K. Bellchambers. 2000. Breeding biology of captive, reintroduced, and wild greater bilbies, *Macrotis lagotis* (Marsupialia: Peramelidae). Wildl. Res. 27:621–628.

Sparti, A. 1992. Thermogenic capacity of shrews (Mammalia, Soricidae) and its relationship with basal rate of metabolism. Physiol. Zool. 65:77–96.

Speakman, J. R. 1993. The evolution of echolocation for predation. Pp. 39–63 in Mammals as predators (N. Dunstone and M. L. Gorman, eds.). Proc. Zoological Society of London, Clarendon Press, Oxford.

Speakman, J. R. 1995. Chiropteran nocturnality. Pp. 187–201 *in* Ecology, evolution, and behaviour of bats (P. A. Racey and S. M. Swift, eds.). Oxford Univ. Press, Oxford.

Speakman, J. R. 2001. The evolution of flight and echolocation in bats: another leap in the dark. Mammal Rev. 31:111–130.

Speakman, J. R., and D. W. Thomas. 2003. Physiological ecology and energetics of bats. Pp. 430–490 *in* Bat ecology (T. H. Kunz and M. B. Fenton, eds.). Univ. Chicago Press, Chicago.

Sperber, I. 1944. Studies on the mammalian kidney. Zool. Bidrag Fran Uppsala 22:249–430.

Spinage, C. 1994. Elephants. Poyser Natural History, London.

Spratt, D. M. 1990. The role of helminths in the biological control of mammals. Int. J. Parasitol. 20:543–550.

Springer, M. S., and C. Krajewski. 1989. DNA hybridization in animal taxonomy: a critique from first principles. Q. Rev. Biol. 64:291–318.

Springer, M. S., J. A. W. Kirsch, and J. A. Case. 1997. The chronicle of marsupial evolution. Pp. 129–161 *in* Molecular evolution and adaptive radiation (T. J. Givnish and K. J. Sytsma, eds.). Cambridge Univ. Press, New York.

Springer, M. S., G. C. Cleven, O. Madsen, W. W. de Jong, V. G. Waddell, H. M. Amrine, and M. J. Stanhope. 1997a. Endemic African mammals shake the phylogenetic tree. Nature 388:61–64.

Springer, M. S., M. Westerman, J. R. Kavanagh, A. Burk, M. O. Woodburne, D. J. Kao, and C. Krajewski. 1998. The origin of the Australasian marsupial fauna and the phylogenetic affinities of the enigmatic monito del monte and marsupial mole. Proc. Roy. Soc. Lond. 265:2381–2386.

Springer, M. S., R. W. DeBry, C. Douady, H. M. Amrine, O. Madsen, W. W. de Jong, and M. J. Stanhope. 2001a. Mitochondrial versus nuclear gene sequences in deep-level mammalian phylogeny reconstruction. Mol. Biol. Evol. 18:132–143.

Springer, M. S., E. C. Teeling, O. Madsen, M. J. Stanhope, and W. W. de Jong. 2001b. Integrated fossil and molecular data reconstruct bat echolocation. Proc. Natl. Acad. Sci. 98:6241–6246.

Springer, M. S., W. J. Murphy, E. Eizirik, and S. J. O'Brien. 2003. Placental mammal diversification and the Cretaceous-Tertiary boundary. Proc. Natl. Acad. Sci. USA 100:1056–1061.

Springer, M. S., M. J. Stanhope, O. Madsen, and W. W. DeJong. 2004. Molecules consolidate the placental mammal tree. Trends Ecol. Evol. 19:430–438.

Spritzer, M. D., Solomon, N. G., and Meikle, D. B. 2005a. Influence of scramble competition for mates upon the spatial ability of male meadow voles. Anim. Behav. 69: 375–386.

Spritzer, M. D., Meikle, D. B., and Solomon, N. G. 2005b. Female choice based on male spatial ability and aggressiveness among meadow voles. Anim. Behav. 69: 1121–1130.

Stafford, B. J. 2005. Order Dermoptera. P. 110 *in* Mammal species of the world: a taxonomic and geographic reference, 3rd edition, vol. 1 (D.E. Wilson and D. M. Reeder, eds.). Johns Hopkins Univ. Press, Baltimore.

Stafford, B. J. 2005. Order Dermoptera. P. 110 *in* Mammal species of the world: a taxonomic and geographic reference, 3rd edition, vol. 1 (D.E. Wilson and D. M. Reeder, eds.). Johns Hopkins Univ. Press, Baltimore.

Stafford, B. J., and F. S. Szalay. 2000. Craniodental functional morphology and taxonomy of dermopterans. J. Mammal. 81:360–385.

Stafford, K. M., S. E. Moore, and C. G. Fox. 2005. Diel variation in blue whale calls recorded in the eastern tropical Pacific. Anim. Behav. 69:951–958.

Stains, H. J. 1984. Carnivores. Pp. 491–521 in Orders and families of recent mammals of the world (S. Anderson and J. K. Jones, Jr., eds.). John Wiley and Sons, New York.

Stallings, R. L., A. F. Ford, D. Nelson, D. C. Torney, C. E. Hildebrand, and R. K. Moyzis. 1991. Evolution and distribution of (GT)n repetitive sequences in mammalian genomes. Genomics 10:807–815.

Stancampiano, A. J., and G. D. Schnell. 2004. Microhabitat affinities of small mammals in south western Oklahoma. J. Mammal. 85:948–958.

Stanhope, M. J., V. G. Waddell, O. Madsen, W. de Jong, S. B. Hedges, G. C. Cleven, D. Kao, and M. S. Springer. 1998. Molecular evidence for multiple origins of Insectivora and for a new order of endemic African insectivore mammals. Proc. Natl. Acad. Sci. 95:9967–9972.

Stanley, H. F., M. Kadwell, and J. C. Wheeler. 1994. Molecular evolution of the family Camelidae: a mitochondrial DNA study. Proc. Roy. Soc. Lond. 256:1–6.

Stapp, P. 1992. Energetic influences on the life history of *Glaucomys volans*. J. Mammal. 73:914–920.

Stapp, P., P. J. Pekins, and W. W. Mautz. 1991. Winter energy expenditure and the distribution of southern flying squirrels. Can. J. Zool. 69:2548–2555.

Start, A. N. 1972. Pollination of the baobab (*Adansonia digitata* L.) by the fruit bat *Rousettus aegyptiacus* Geoffroy. E. Afr. Wildl. J. 10:71–72.

Stearns, S. C. 1976. Life-history tactics: a review of the ideas. Q. Rev. Biol. 51:3–47.

Stearns, S. C. 1992. The evolution of life histories. Oxford Univ. Press, New York.

Stebbings, R. E. 1978. Marking bats. Pp. 81–94 in Animal marking (B. Stonehouse, ed.). Univ. Park Press, Baltimore.

Steele, M. A., and J. L. Koprowski. 2002. North American tree squirrels. Smithsonian Institution Press, Washington, DC.

Steele, M. A., G. Turner, P. D. Smallwood, J. O. Wolff, and J. Radillo. 2001. Cache management by small mammals: experimental evidence for the significance of acorn-embryo excision. J. Mammal. 82:35–42.

Steele, M. A., S. Manierre, T. Genna, T. A. Contreras, P. D. Smallwood, and M. E. Pereira. 2006. The innate basis of food-hoarding decisions in grey squirrels: evidence for behavioural adaptations to the oaks. Anim.l Behav. 71:155–160.

Stein, B. R. 1996. Women in mammalogy: the early years. J. Mammal. 77:629–641.

Steiner, C., M. Tilak, E. J. P. Douzery, and F. M. Catzeflis. 2005. Mol. Phylo. Evol. 35:363–379.

Stephen, C. L., J. C. Devos, Jr., T. E. Lee, Jr., J. W. Bickham, J. R. Heffelfinger, and O. E. Rhodes, Jr. 2005. Population genetic analysis of Sonoran pronghorn (*Antilocapra Americana sonoriensis*). J. Mammal. 86:782–792.

Stephens, D. W. and J. R. Krebs. 1986. Foraging theory. Princeton Univ. Press, Princeton, NJ.

Stephenson, P. J., and P. A. Racey. 1994. Seasonal variation in resting metabolic rate and body temperature of streaked tenrecs, *Hemicentetes nigriceps* and *H. semispinosus* (Insectivora, Tenrecidae). J. Zool. 232:285–294.

Steudel, K., W. P. Porter, and D. Sher. 1994. The biophysics of Bergmann's rule—a comparison of the effects of pelage and body-size variation on metabolic rate. Can. J. Zool. 72:70–77.

Stevens, C. E., and I. D. Hume. 1996. Comparative physiology of the vertebrate digestive system, 2d edition. Cambridge Univ. Press, New York.

Stevens, G. C. 1989. The latitudinal gradient in geographical range: how so many species coexist in the tropics. Am. Nat. 133:240–256.

Stevenson, R. D. 1986. Allen's rule in North American rabbits (*Sylvilagus*) and hares (*Lepus*) is an exception, not a rule. J. Mammal. 67:312–316.

Stevens, R. D., and M. R. Willig. 2002. Geographical ecology at the community level: perspectives on the diversity of New World bats. Ecology 83:545–560.

Stewart, B. S., and R. L. DeLong. 1995. Double migrations of the northern elephant seal, *Mirounga*

angustirostris. J. Mammal. 76:196–205.

Stewart, C.-B., and T. R. Disotell. 1998. Primate evolution—in and out of Africa. Curr. Biol. 8:R582–R588.

Stiles, D. 2004. The ivory trade and elephant conservation. Env. Conserv. 31:309–321.

Stockwell, E. F. 2001. Morphology and flight maneuverability in New World leaf-nosed bats (Chiroptera: Phyllostomidae). J. Zool. 254:505–514.

Stonehouse, B. (ed.) 1978. Animal marking. Univ. Park Press, Baltimore.

Stoner, C. J., O. R. P. Bininda-Emonds, and T. Caro. 2003a. The adaptive significance of coloration in lagomorphs. Biol. J. Linnean Soc. 79:309–328.

Stoner, C. J., T. M. Caro, and C. M. Graham. 2003b. Ecological and behavioral correlates of coloration in artiodactyls: systematic analyses of conventional hypotheses. Behav. Ecol. 14:823–840.

Storer, T. I. 1969. Mammalogy and the American Society of Mammalogists, 1919–1969. J. Mammal. 50:785–793.

Storer, T. I., F. C. Evans, and F. G. Palmer. 1944. Some rodent populations in the Sierra Nevada of California. Ecol. Monogr. 14:165–192.

Storey, K. B., and J. M. Storey. 1988. Freeze tolerance in animals. Physiol. Rev. 68:27–84.

Storrs, E. E. 1971. The nine-banded armadillo: a model for leprosy and other biomedical research. Int. J. Leprosy 39:703–714.

Storz, J. F., H. R. Bhat, and T. H. Kunz. 2000. Social structure of a polygynous tent-making bat, *Cynopterus sphinx* (Megachiroptera). J. Zool. Lond. 251:151–165.

Stouffer, R. L. (ed.). 1987. The primate ovary. Plenum Press, New York.

Strahan, R. (ed.). 1995. The mammals of Australia. Reed Books, Chatsworth, Australia.

Strauss, J. F., III, F. Martinez, and M. Kiriakidou. 1996. Placental steroid hormone synthesis: unique features and unanswered questions. Biol. Reprod. 54:303–311.

Strickler, T. L. 1978. Functional osteology and myology of the shoulder in the Chiroptera. S. Karger, New York.

Strier, K. B. 1992. Faces in the forest. Oxford Univ. Press, New York.

Struhsaker, T. T. 1967. Auditory communication among vervet monkeys (*Cercopithecus aethiops*). Pp. 281–324 in Social communication among primates (S. A. Altmann, ed.). Univ. Chicago Press, Chicago.

Stuart, C., and T. Stuart. 1991. The feral cat problem in southern Africa. Afr. Wildl. 45:13–15.

Stuart, M. D., K. B. Strier, and S. M. Pierberg. 1993. A coprological survey of parasites of wild muriquis, *Brachyteles arachnoides* and brown howling monkeys, *Alouatta fusca*. J. Helminth. Soc. Wash. 60:111–115.

Studier, E. H., S. H. Sevick, D. M. Ridley, and D. E. Wilson. 1994. Mineral and nitrogen concentrations in feces of some neotropical bats. J. Mammal. 75:674–680.

Stuenes, S. 1989. Taxonomy, habits, and relationships of the subfossil Madagascan hippopotami *Hippopotamus lemerlei* and *H. madagascariensis*. J. Vertebr. Paleontol. 9:241–268.

Su, B., Y. Wang, H. Lan, W. Wang, and Y. Zhang. 1999. Phylogenetic study of complete cytochrome b genes in musk deer (Genus *Moschus*) using museum samples. Mol. Phylogenet. Evol. 12:241–249.

Su, B., Y. Fu, Y. Wang, L. Jin, and R. Chakraborty. 2001. Genetic diversity and population history of the red panda (*Ailurus fulgens*) as inferred from mitochondrial DNA sequence variations. Mol. Biol. Evol. 18:1070–1076.

Su, C., V. K. Nguyen, and M. Nei. 2002. Adaptive evolution of variable region genes encoding an unusual type of immunoglobulin in camelids. Mol. Biol. Evol. 19:205–219.

Suga, N. 1990. Biosonar and neural computation in bats. Sci. Am. 262:34–41.

Sulkin, S. E., and R. Allen. 1974. Virus infections in bats. S. Karger, New York.

Sullivan, B. J., G. S. Baxter, A. J. Lisle, L. Pahl, and W. M. Norris. 2004. Low-density koala (*Phascolarctos cinereus*) populations in the mulgalands of southern Queensland. IV. abundance and conservation status. Wildl. Res. 31:19–29.

Sullivan, J., E. Arellano, and D.S. Rogers. 2000. Comparative phylogeography of Mesoamerican

highland rodents: concerted versus independent response to past climatic fluctuations. Am. Nat. 155:755–768.

Sunquist, M., and F. Sunquist. 2002. Wild Cats of the World. Univ. Chicago Press, Chicago.

Surlykke, A., L. A. Miller, B. Mohl, B. B. Andersen, J. Christiansen-Dalsgaard, and M. B. Jorgensen. 1993. Echolocation in two very small bats from Thailand: *Craseonycteris thonglongyai* and *Myotis siligorensis*. Behav. Ecol. Sociobiol. 33:1–12.

Sutherland, W. J. 1996. Ecological census techniques. Cambridge Univ. Press, New York.

Suthers, R. A. 1970. Vision, olfaction, taste. Pp. 265–309 in Biology of bats, vol. II (W. A. Wimsatt, ed.). Academic Press, New York.

Sutton, D. A., and B. D. Patterson. 2000. Geographic variation of the western chipmunks *Tamias senex* and *T. siskiyou*, with two new subspecies from California. J. Mammal. 81:299–316.

Suzuki, R., J. R. Buck, and P. L. Tyack. 2006. Information entropy of humpback whale songs. J. Acoust. Soc. Am. 119:1849–1866.

Svare, B. B., and M. A. Mann. 1983. Hormonal influences on maternal aggression. Pp. 91–104 in Hormones and aggressive behavior (B. Svare, ed.). Plenum, New York.

Svendsen, G. E. 2003. Weasels and Black-footed ferret, *Mustela* species. Pp. 650–661 in Wild mammals of North America: biology, management, and conservation. 2d edition (G. A. Feldhamer, B. C. Thompson, and J. A. Chapman, eds). Johns Hopkins Univ. Press, Baltimore.

Swart, J. M., P. R. K. Richardson, and J. W. H. Ferguson. 1999. Ecological factors affecting the feeding behaviour of pangolins (*Manis temminckii*). J. Zool. Lond. 247:281–292.

Swartz, S. M., M. B. Bennett, and D. R. Carrier. 1992. Wing bone stresses in free flying bats and the evolution of skeletal design for flight. Nature 359:726–729.

Sweeney, J. R., J. M. Sweeney, and S. W. Sweeney. 2003. Feral hog, *Sus scrofa*. Pp. 1164–1179 in Wild mammals of North America: biology, management, and conservation (G. A. Feldhamer, B. C. Thompson, and J. A. Chapman, eds.). Johns Hopkins Univ. Press, Baltimore.

Swihart, R. K., and J. P. Bryant. 2001. Importance of biogeography and ontogeny of woody plants in winter herbivory by mammals. J. Mammal. 82:1–21.

Swihart, R. K., and N. A. Slade. 1984. Road crossing in *Sigmodon hispidus* and *Microtus ochrogaster*. J. Mammal. 65:357–360.

Swofford, D. L. 2002. PAUP*: phylogenetic analysis using parsimony (*and other methods), version 4.0b10. Sinauer Associates, Sunderland, MA.

Swofford, D. L., and G. J. Olsen. 1990. Phylogeny reconstruction. Pp. 411–501 in Molecular systematics (D. M. Hillis and C. Moritz, eds.). Sinauer Associates, Sunderland, MA.

Swofford, D. L., G. J. Olsen, P. J. Waddell, and D. M. Hillis. 1996. Phylogenetic inference. Pp 407–514 in Molecular systematics (D. M. Hillis, C. Moritz, and B. K. Mable, eds.). Sinauer Associates, Sunderland, MA.

Symonds, M. R. E. 2005. Phylogeny and life histories of the 'Insectivora': controversies and consequences. Biol. Rev. 80:93–128.

Szalay, F. S. 1994. Evolutionary history of the marsupials and an analysis of osteological characters. Cambridge Univ. Press, Cambridge, UK.

Szalay, F. S., and E. Delson. 1979. Evolutionary history of the primates. Academic Press, New York.

Taber, A. B., C. P. Doncaster, N. N. Neris, and F. H. Colman. 1993. Ranging behavior and population dynamics of the Chacoan peccary, *Catagonus wagneri*. J. Mammal. 74:443–454.

Taber, R. D., and I. McT. Cowan. 1969. Capturing and marking wild animals. Pp. 277–317 in Wildlife management techniques, 3rd edition (R. H. Giles, ed.). Wildlife Society, Washington, DC.

Tabuce, R., B. Coiffait, P. E. Coiffait, M. Mahboubi, and J. J. Jaeger. 2000. A new species of *Bunohyrax* (Hyracoidea, Mammalia) from the Eocene of Bir El Ater (Algeria). C. R. Acad. Sci. Ser. II A 331:61–66.

Tabuce, R., M. Mahboubi, and J. Sudre. 2001. Reassessment of the Algerian Eocene hyracoid *Microhyrax*. Consequences on the early diversity and basal phylogeny of the order Hyracoidea

(Mammalia). Eclogae Geol. Helv. 94:537–545.

Tabuce, R., B. Coiffait, P-E. Coiffait, M. Mahboubi, and J-J. Jaeger. 2001a. A new genus of Macroscelidea (Mammalia) from the Eocene of Algeria: a possible origin for elephant-shrews. J. Vert. Paleontol. 21:535–546.

Taggert, D. A., G. A. Shimmin, J. R. Ratcliff, V. R. Steele, R. Dibben, J. Dibben, C. White, and P. D. Temple-Smith. 2005. Seasonal changes in the testis, accessory glands and ejaculate characteristics of the southern hairy-nosed wombat, *Lasiorhinus latifrons* (Marsupialia: Vombatidae). J. Zool. Lond. 266:95–104.

Talmage, R. V., and G. D. Buchanan. 1954. The armadillo (*Dasypus novemcintus*): a review of its natural history, ecology, anatomy and reproductive physiology. Rice Inst. Pamphlet 41:1–135.

Tamarin, R.H. 1977. Dispersal in island and mainland voles. Ecology 58:1044–1054.

Tamarin, R. H. 1980. Dispersal and population regulation in rodents. Pp. 117–133 *in* Biosocial mechanisms of population regulation (M. N. Cohen, R. S. Malpass, and H. G. Klein, eds.). Yale Univ. Press, New Haven, CT.

Tattersall, I. 1972. Of lemurs and men. Nat. Hist. 81:32–43.

Tattersall, I. 1982. The primates of Madagascar. Columbia Univ. Press, New York.

Taube, E., J. Keravec, J-C. Vie, and J-M. Duplantier. 2001. Reproductive biology and postnatal development in sloths, *Bradypus* and *Choloepus*: review with original data from the field (French Guiana) and from captivity. Mammal Rev. 31:173–188.

Taulman, J. F., and L. W. Robbins. 1996. Recent range expansion and distributional limits of the nine-banded armadillo (*Dasypus novemcinctus*) in the United States. J. Biogeogr. 23:635–648.

Tautz, D. 1989. Hypervariability of simple sequences as a general source for polymorphic DNA markers. Nucleic Acid Res. 17:6463–6471.

Tautz, D., P. Arctander, A. Minelli, R. H. Thomas, and A. P. Vogler. 2003. A plea for DNA taxonomy. Trends Ecol. Evol. 18:70–74.

Tavare, S., C. R. Marshall, O. Will, C. Soligo, and R. D. Martin. 2002. Using the fossil record to estimate the age of the last common ancestor of extant primates. Nature 416:726–729.

Taylor, C. R. 1970. Strategies of temperature regulation: effect of evaporation in East African ungulates. Am. J. Physiol. 219:1131–1135.

Taylor, C. R. 1972. The desert gazelle: a paradox resolved. Pp. 215–227 *in* Comparative physiology of desert animals (G. M. O. Maloiy, ed.). Symp. Zool. Soc. Lond., 31.

Taylor, C. R., and C. P. Lyman. 1972. Heat storage in running antelopes: independence of brain and body temperatures. Am. J. Physiol. 222:114–117.

Taylor, J. M. 1984. The Oxford guide to mammals of Australia. Oxford Univ. Press, New York.

Taylor, J. R. E. 1998. Evolution of energetic strategies of shrews. Pp. 309–346 *in* Evolution of shrews (J. M. Wójcik and M. Wolsan, eds.). Mammal Research Institute, Polish Academy of Sciences, Bialowieza, Poland.

Taylor, R. H., and R. M. Williams. 1956. The use of pellet counts for estimating the density of populations of the wild rabbit, *Oryctolagus cuniculus* (L.). N. Z. J. Sci. Tech. 38:236–256.

Taylor, R. J. 1992. Seasonal changes in the diet of the Tasmanian bettong (*Bettongia gaimardi*), a mycophagous marsupial. J. Mammal. 73:408–414.

Taylor, R. J., and L. A. Pfannmuller. 1981. A test of the peninsular effect on species diversity. Can. Field Nat. 95:144–148.

Taylor, R. J., and P. J. Regal. 1978. The peninsular effect on species diversity and the biogeography of Baja California. Am. Nat. 112:583–593.

Taylor, W. A., and J. D. Skinner. 2000. Associative feeding between aardwolves (*Proteles cristatus*) and aardvarks (*Orycteropus afer*). Mammal Rev. 30:141–143.

Taylor, W. A., P. A. Lindsey, and J. D. Skinner. 2002. The feeding ecology of the aardvark *Orycteropus afer*. J. Arid Environ. 50:135–152.

Tedford, R. H., M. R. Banks, N. Kemp, I. McDougal, and F. L. Sutherland. 1975. Recognition of the oldest known fossil marsupials from Australia. Nature 255:141–142.

Teeling, E. C., M. Scally, D. J. Kao, M. L. Romagnoll, M. S. Springer, and M. J. Stanhope. 2000. Molecular evidence regarding the origin of echolocation and flight in bats. Nature 403:188–192.

Teeling, E. C., O. Madsen, W. J. Murphy, M. S. Springer, and S. J. O'Brien. 2003. Nuclear gene sequences confirm an ancient link between New Zealand's short-tailed bat and South American noctilionoid bats. Mol. Phylogenet. Evol. 28:308–319.

Teeling, E. C., M. S. Springer, O. Madsen, P. Bates, S. J. O'Brien, and W. J. Murphy. 2005. A molecular phylogeny for bats illuminates biogeography and the fossil record. Science 307:580–584.

Tejedor, A. 2005. A new species of funnel-eared bat (Natalidae: *Natalus*) from Mexico. J. Mammal. 86:1109–1120.

Tekiala, S. 2005. Mammals of Michigan. Adventure Publications, Cambridge, MN.

Templeton, A.R. 1998. Nested clade analysis of phylogeographic data: testing hypotheses about gene flow and population history. Mol. Ecol. 7:381–397.

Templeton, A. R. 2002. Out of Africa again and again. Nature 416:45–51.

Terborgh, J., L. Lopez, P. Nunez, M. Rao, G. Shahabuddin, G. Orihuela, M. Riveros, R. Ascanio, G. Adler, T. Lambert, and L. Balbas. 2001. Ecological meltdown in predator-free forest fragments. Science 294:1923–1926.

Terman, C. R. 1973. Reproductive inhibition in asymptotic populations of prairie deermice. J. Reprod. Fertil., Suppl. 19:457–463.

Tevis, L. 1958. Interrelations between the harvester ant *Veromessor pergandei* (Mayr) and some desert ephemerals. Ecology 39:695–704.

Thabah, A., G. Li, Y. Wang, B. Liang, K. Hu, S. Zhang, and G. Jones. 2007. Diet, echolocation calls and phylogenetic affinities of the great evening bat *Ia io* (Vespertilionidae): another carnivorous bat. J. Mammal. 88:

Thackway, R., and I. D. Cresswell (eds.). 1995. An interim biogeographic regionalisation for Australia: a framework for establishing the national system of reserves, version 4.0. Australian Nature Conservation Agency, Canberra.

Theodor, J. M., K. D. Rose, and J. Erfurt. 2005. Artiodactyla. Pp. 215–233 *in* The rise of placental mammals: origins and relationships of the major extant clades (K. D. Rose and J. D. Archibald, eds.). Johns Hopkins Univ. Press, Baltimore.

Thewissen, J. G. M., and E. L. Simons. 2001. Skull of *Megalohyrax eocaenus* (Hyracoidea, Mammalia) from the Oligocene of Egypt. J. Vert. Paleontol. 21:98–106.

Thom, M. D., D. D. P. Johnson, and D. W. Macdonald. 2004. The evolution and maintenance of delayed implantation in the Mustelidae (Mammalia: Carnivora). Evolution 58:175–183.

Thomas, D. W. 1983. The annual migrations of three species of West African fruit bats (Chiroptera: Pteropodidae). Can. J. Zool. 61:2266–2272.

Thomas, D. W. 1995. The physiological ecology of hibernation in vespertilionid bats. Pp. 233–244 *in* Ecology, evolution, and behaviour of bats (P. A. Racey and S. M. Swift, eds.). Zool. Soc. Lond., Clarendon Press, London.

Thomas, R. H., W. Schaffner, A.C. Wilson, and S. Pääbo. 1989. DNA phylogeny of the extinct marsupial wolf. Nature 340:465–467.

Thomas, T. R., and L. R. Irby. 1990. Habitat use and movement patterns by migrating mule deer in southeastern Idaho. Northw. Sci. 64:19–27.

Thompson, D. 1942. On growth and form: a new edition. Cambridge Univ. Press, Cambridge, UK.

Thorington, R. W., Jr. and R. S. Hoffmann. 2005. Family Sciuridae. Pp. 754–818 *in* Mammal species of the world: a taxonomic and geographic reference. 3rd edition (D. E. Wilson and D. M. Reeder, eds.). Johns Hopkins Univ. Press, Baltimore.

Thornton, D. H., M. E. Sunquist, and M. B. Main. 2004. Ecological separation within newly sympatric populations of coyotes and bobcats in south-central Florida. J. Mammal. 85:973–982.

Thornton, S. J., D. M. Spielman, N. J. Pelc, W. F. Block, D. E. Crocker, D. P. Costa, B. J. LeBoeuf, and P. W. Hochachka. 2001. Effects of forced diving on the spleen and hepatic sinus in northern elephant seal pups. Proc. Natl. Acad. Sci. 98:9413–9418.

Thorpe, W. H. 1945. The evolutionary significance of

habitat selection. J. Anim. Ecol. 14:67–70.

Thurber, J. M., and R. O. Peterson. 1993. Effects of population density and pack size on the foraging ecology of gray wolves. J. Mammal. 74:879–889.

Tiedemann, R., S. Hammer, F. Suchentrunk, and G. B. Hartl. 1996. Allozyme variability in medium-size and large mammals: determinants, estimators, and significance for conservation. Biodiv. Lett. 3:81–91.

To, L. P., and R. H. Tamarin. 1977. The relation of population density and adrenal gland weight in cycling and non-cycling voles (*Microtus*). Ecology 58:928–934.

Toldt, K. 1935. Aufbau under natürliche Färbung des Haarkledes der Wildsäugetiere. Deuts. Gesch. Kleintier Pelztierzucht, Leipzig.

Tomasi, T. E. 1978. Function of venom in the short-tailed shrew, *Blarina brevicauda*. J. Mammal. 59:852–854.

Tomasi, T. E. 1979. Echolocation by the short-tailed shrew, *Blarina brevicauda*. J. Mammal. 60:751–759.

Tomlinson, R. F., H. W. Calkins, and D. F. Marble. 1976. Computer handling of geographical data. UNESCO Press, Paris.

Tong, H. W., J. Y. Liu, and L. G. Han. 2002. On fossil remains of early Pleistocene tapir (Perissodactyla, Mammalia) from Fanchang, Anhui. Chin. Sci. Bull. 47:586–590.

Tonnessen, J. N., and A. O. Johnsen. 1982. The history of modern whaling. Univ. California Press, Berkeley.

Tougard, C., T. Delefosse, C. Hanni, and C. Montgelard. 2001. Phylogenetic relationships of the five extant rhinoceros species (Rhinocerotidae, Perissodactyla) based on mitochondrial cytochrome b and 12S rRNA genes. Mol. Phylogenet. Evol. 19:34–44.

Tracy, R. L., and G. E. Walsberg. 2001. Intraspecific variation in water loss in a desert rodent, *Dipodomys merriami*. Ecology 82:1130–1137.

Trappe, J. M., and C. Maser. 1976. Germination of spores of *Glomus macrocarpus* (Endogonaceae) after passage through a rodent digestive tract. Mycologia 68:433–436.

Travis, S. E., C. N. Slobodchikoff, and P. Keim. 1997. DNA fingerprinting reveals low genetic diversity in Gunnison's prairie dog (*Cynomys gunnisoni*). J. Mammal. 78:725–732.

Trayhurn, P., and D. G. Nicholls (eds.). 1986. Brown adipose tissue. Edward Arnold, London.

Trayhurn, P., J. H. Beattie, and D. V. Rayner. 2000. Leptin-signals and secretions from white adipose tissue. Pp. 459–469 *in* Life in the cold: Eleventh International Hibernation Symposium, August 13–18, 2000, Jungholz, Austria. Springer-Verlag, Berlin.

Tributsch, H., H. Goslowsky, U. Küppers, and H. Wetzel. 1990. Light collection and solar sensing through the polar bear pelt. Solar Energy Material 21:219–236.

Trivers, R. L. 1971. The evolution of reciprocal altruism. Q. Rev. Biol. 46:35–57.

Trivers, R. L. 1972. Parental investment and sexual selection. Pp. 136–179 *in* Sexual selection and the descent of man, 1871–1971 (B. Campbell, ed.). Aldine, Chicago.

Trivers, R. L. 1974. Parent-offspring conflict. Am. Zool. 14:249–264.

Trivers, R. L., and D. E. Willard. 1973. Natural selection of parental ability to vary the sex ratio of off spring. Science 179:90–92.

Trombulak, S. C., and C. A. Frissell. 1999. Review of ecological effects of roads on terrestrial and aquatic communities. Conserv. Biol. 14:18–30.

Troy, C. S., D. E. McHugh, J. F. Bailey, D. A. Magee, R. T. Loftus, P. Cunningham, A. T. Chamberlain, B. C. Sykes, and D. G. Bradley. 2001. Genetic evidence for Near-Eastern origins of European cattle. Nature 410:1088–1091.

Tschapka, M., A. P. Brooke, and L. T. Wasserthal. 2000. Thyroptera discifera (Chiraptera: Thyropteridae): A new record for Costa Rica and observations on echolocation. Z. Saugetierkd. 65: 193–198.

Tudge, C. 1992. Last animals at the zoo. Oxford Univ. Press, Oxford.

Turner, C. D., and J. T. Bagnara. 1976. General endocrinology. W. B. Saunders, Philadelphia.

Turner, D. C. 1975. The vampire bat: a field study in

behavior and ecology. Johns Hopkins Univ. Press, Baltimore.

Turner, T. R., M. L. Weiss, and M. E. Pereira. 1992. DNA fingerprinting and paternity assessment in Old World monkeys and ringtailed lemurs. Pp. 96–112 *in* Paternity in primates: genetic tests and theories (R. D. Martin, A. F. Dixson, and E. J. Wickings, eds.). S. Karger, Basel, Switzerland.

Tuttle, M. D., and M. J. Ryan. 1981. Bat predation and the evolution of frog vocalizations in the Neotropics. Science 214:677–678.

Twigg, G. I. 1978. Marking animals by tissue removal. Pp. 109–118 *in* Animal marking (B. Stonehouse, ed.). Univ. Park Press, Baltimore.

Twigg, G. I. 1978a. The role of rodents in plague dissemination: a worldwide review. Mammal Rev. 8:7–110.

Tyndale-Biscoe, C. H. 2005. Life of marsupials. CSIRO Publishing, Collingwood, Australia.

Tyndale-Biscoe, H., and M. Renfree. 1987. Reproductive physiology of marsupials. Cambridge Univ. Press, New York.

Umetsu, F., L. Naxara, and R. Pardini. 2006. Evaluating the efficiency of pitfall traps for sampling small mammals in the Neotropics. J. Mammal. 87:757–765.

Ure, D. C., and C. Maser. 1982. Mycophagy of red-backed voles in Oregon and Washington. Can. J. Zool. 60:3307–3315.

Utzurrum, R. C. B. 1995. Feeding ecology of Philippine fruit bats: patterns of resource use and seed dispersal. Pp. 63–77 *in* Ecology, evolution and behaviour of bats (P. A. Racey and S. M. Swift, eds.). Oxford Univ. Press, Oxford.

Uzzell, T., and D. R. Pilbeam. 1971. Phyletic divergence dates of homonoid primates: a comparison of fossil and molecular data. Evolution 25:615–635.

Valdés, A. M., M. Slatkin, and N. B. Freimer. 1993. Allele frequencies at microsatellite loci: the stepwise mutation model revisited. Genetics 133:737–749.

Valdespino, C., C. S. Asa, and J. E. Bauman. 2002. Estrous cycles, copulation, and pregnancy in the fennec fox (*Vulpes zerda*). J. Mammal. 83:99–109.

Valenzuela, D., and G. Ceballos. 2000. Habitat selection, home range, and activity of the white-nosed coati (*Nasua narica*) in a Mexican tropical dry forest. J. Mammal. 81:810–819.

Vandenbergh, J. G. 1967. Effect of the presence of a male on the sexual maturation of female mice. Endocrinology 81:345–348.

Vandenbergh, J. G. 1969a. Endocrine coordination in monkeys: male sexual responses to the female. Physiol. Behav. 4:261–264.

Vandenbergh, J. G. 1969b. Male odor accelerates female sexual maturation in mice. Endocrinology 84:658–660.

Vandenbergh, J. G. 1983. Pheromonal regulation of puberty. Pp. 95–112 *in* Pheromones and mammalian reproduction (J. G. Vandenbergh, ed.). Academic Press, New York.

Vandenbergh, J. G., and D. M. Coppola. 1986. The physiology and ecology of puberty modulation by primer pheromones. Adv. Stud. Behav. 16:71–108.

Vandenbergh, J. G., and L. C. Drickamer. 1974. Reproductive coordination among free-ranging rhesus monkeys. Physiol. Behav. 13:373–376.

Van Den Bussche, R. A., and S. R. Hoofer. 2000. Further evidence for inclusion of the New Zealand short-tailed bat (*Mystacina tuberculata*) within Noctilionoidea. J. Mammal. 81:865–874.

Van Der Merwe, M., and A. Van Zyl. 2001. Postnatal growth of the greater cane rat *Thryonomys swinderianus* (Thryonomyidae: Rodentia) in Gauteng, South Africa. Mammalia 65:495–507.

Vander Wall, S. B. 1990. Food hoarding in animals. Univ. Chicago Press, Chicago.

Vander Wall, S. B. 1995. Influence of substrate water on the ability of rodents to find buried seeds. J. Mammal. 76:851–856.

Vander Wall, S. B., M. I. Borchert, and J. R. Gworek. 2006. Secondary dispersal of bigcone Douglas-fir (*Pseudotsuga macrocarpa*) seeds. Acta Oecol. 30:100–106.

van Rheede, T., T. Bastiaans, D. N. Boone, S. B. Hedges, W. W. de Jong, and O. Madsen. 2006. The platypus in its place: nuclear genes and indels confirm the sister group relation of monotremes and therians. Mol. Biol. Evol. 23:587–597.

van Schaik, C. P., M. Ancrenaz, G. Borgen, B. Galdikas, C. D. Knott, I. Singleton, A. Suzuki, S. S. Utami, and M. Merrill. 2003. Orangutan cultures and the evolution of material culture. Science 299:102–105.

Vanselow K. H., and K. Ricklefs 2005. Are solar activity and sperm whale *Physeter macrocephalus* strandings around the North Sea related? J. Sea Res. 53: 319–327.

van Tienhoven, A. 1983. Reproductive physiology of vertebrates, 2d edition. Cornell Univ. Press, Ithaca, NY.

van Tuinen, P., T. J. Robinson, and G. A. Feldhamer. 1983. Chromosome banding and NOR location in sika deer. J. Heredity 74:473–474.

Van Valen, L. 1973. A new evolutionary law. Evol. Theory 1:1–33.

Van Valen, L., and R. E. Sloan. 1966. The extinction of the multituberculates. Syst. Zool. 15:261–278.

Van Valkenburgh, B. 1989. Carnivore dental adaptations and diet: a study of trophic diversity within guilds. Pp. 410–436 *in* Carnivore behavior, ecology, and evolution (J. L. Gittleman, ed.). Cornell Univ. Press, Ithaca, NY.

van Zyl, A., R. V. Rambau, and M. van der Merwe. 2005. Aspects of the anatomy and histology of the alimentary canal of the greater cane rat, *Thryonomys swinderianus*, with reference to its feeding physiology. Afr. Zool. 40:25–36.

Vartanyan, S. L., V. E. Garutt, and A. V. Sher. 1993. Holocene dwarf mammoths from Wrangel Island in the Siberian Arctic. Nature 362:337–340.

Vaughan, T. A. 1959. Functional morphology of three bats: *Eumops*, *Myotis*, and *Macrotus*. Univ. Kansas Publ. Mus. Nat. Hist. 12:1–153

Vaughan, T. A. 1966. Morphology and flight characteristics of molossid bats. J. Mammal. 47:249–260.

Vaughan, T. A. 1967. Food habits of the northern pocket gopher on shortgrass prairie. Am. Midl. Nat. 77:176–189.

Vaughan, T. A. 1970a. The muscular system. Pp. 139–194 *in* Biology of bats, vol. I (W. A. Wimsatt, ed.). Academic Press, New York.

Vaughan, T. A. 1970b. The skeletal system. Pp. 97–138 *in* Biology of bats, vol. I (W. A. Wimsatt, ed.). Academic Press, New York.

Vaughan, T. A. 1970c. Flight patterns and aerodynamics. Pp. 195–216 *in* Biology of bats, vol. I (W. A. Wimsatt, ed.). Academic Press, New York.

Vaughan, T. A. 1978. Mammalogy, 2d edition. W. B. Saunders, Philadelphia.

Vaughan, T. A. 1986. Mammalogy, 3rd edition. W. B. Saunders, Philadelphia.

Vaughan, T. A., J. M. Ryan, and N. J. Czaplewski. 2000. Mammalogy, 4th edition. Saunders College Publishing, Fort Worth, TX.

Veech, J. A. 2001. The foraging behavior of granivorous rodents and short-term apparent competition among seeds. Behav. Ecol. 12:467–474.

Veitch, C. E., J. Nelson, and R. T. Gemmell. 2000. Birth in the brushtail possum, *Trichosurus vulpecula* (Marsupialia: Phalangeridae). Aust. J. Zool. 48:691–700.

Venter, J. C., et al. 2001. The sequence of the human genome. Science 291:1304–1351.

Verme, L. J., and J. J. Ozoga. 1981. Sex ratio of white-tailed deer and the estrus cycle. J. Wildl. Mgmt. 45:710–715.

Veron, G., and S. Heard. 2000. Molecular systematics of the Asiatic Viverridae (Carnivora) inferred from mitochondrial cytochrome b sequence analysis. J. Zool. Syst. Evol. Res. 38:209–217.

Veron, G., P. Gaubert, N. Franklin, A. P. Jennings, and L. I. Grassman, Jr. 2006. A reassessment of the distribution and taxonomy of the endangered otter civet *Cynogale bennettii* (Carnivora: Viverridae) of south-east Asia. Oryx 40:42–49.

Vessey, S. H. 1971. Free-ranging rhesus monkeys: Behavioural effects of removal, separation, and reintroduction of group members. Behaviour. 40:216–227.

Vessey, S. H. 1973. Night observations of free-ranging rhesus monkeys. Am. J. Phys. Anthropol. 38:613–620.

Vickery, W. L., and J. S. Millar. 1984. The energetics of huddling by endotherms. Oikos 43:88–93.

Vidya, T. N. C., P. Fernando, D. J. Melnick, and R. Sukumar. 2005. Population differentiation within and among Asian elephant (*Elephas maximus*) populations in southern India. Heredity 94:71–80.

Vidya, T. N. C., and R. Sukumar. 2005. Social organization of the Asian elephant (*Elephas maximus*) in southern India inferred from microsatellite DNA. J. Ethol. 23:205–210.

Viguier, B. 2004. Functional adaptation in the craniofacial morphology of Malagasy primates: shape variations associated with gummivory in the family Cheirogaleidae. Ann. Anat. 186:495–501.

Vilà, C., J. A. Leonard, A. Götherström, S. Markland, K. Sandberg, K. Lidén, R. K. Wayne, and H. Ellegren. 2001. Widespread origins of domestic horse lineages. Science 291:474–477.

Vispo, C. R., and G. S. Bakken. 1993. The influence of thermal conditions on the surface activity of thirteen-lined ground squirrels. Ecology 74:377–389.

Voelker, W. 1986. The natural history of living mammals. Plexus, Medford, OR.

Vogel, G. 2006. Tracking Ebola's deadly march among wild apes. Science 314:1522–1523.

Vogel, P. 1976. Energy consumption of European and African shrews. Acta Theriol. 21:195–206.

Vogel, S. 1994. Life in moving fluids: The physical biology of flow. Princeton Univ. Press, Princeton, NJ.

Vogt, F. D., and G. R. Lynch. 1982. Influence of ambient temperature, nest availability, huddling, and daily torpor on energy expenditure in the white-footed mouse *Peromyscus leucopus*. Physiol. Zool. 55:56–63.

Vogt, F. D., G. R. Lynch, and S. Smith. 1983. Radiotelemetric assessment of diel cycles in euthermic body temperature and torpor in a free-ranging small mammal inhabiting man-made nest sites. Oecologia 60:313–315.

Vogt, J. L. 1984. Interactions between adult males and infants in prosimians and New World monkeys. Pp. 346–376 *in* Primate paternalism (D. M. Taub, ed.). Van Nostrand Reinhold, New York.

Voigt, C. C., and D. H. Kelm. 2006. Host preference of the common vampire bat (*Desmodus rotundus*: Chiroptera) assessed by stable isotopes. J. Mammal. 87:1–6.

Voigt, C. C., G. Heckel, and O. von Helversen. 2005. Conflicts and strategies in the harem—polygynous mating system of the sac-winged bat, *Saccopteryx bilineata*. Pp. 269–289 *in* Functional and evolutionary ecology of bats (A. Zubaid, G. F. McCracken, and T. H. Kunz, eds.). Oxford Univ. Press, New York.

vom Saal, F. 1979. Prenatal exposure to androgen influences morphology and aggressive behavior of male and female mice. Horm. Behav. 12:1–11.

vom Saal, F. 1989. Sexual differentiation in litter bearing mammals: influence of sex of adjacent fetuses in utero. J. Anim. Sci. 67:1824–1840.

von Engelhardt, W., P. Haarmeyer, and M. Lechner-Doll. 2006. Feed intake, forestomach fluid volume, dilution rate and mean retention of fluid in the forestomach during water deprivation and rehydration in camels (*Camelus* sp.). Comp. Biochem. Physiol. Part A Mol. Integr. Physiol. 143:504–507.

Vonhof, M. J., and M. B. Fenton. 2004. Roost availability and population size of *Thyroptera tricolor*, a leaf-roosting bat, in northeastern Costa Rica. J. Trop. Ecol. 20:291–305.

Vorbach, C., M. R. Capecchi, and J. M. Penninger. 2006. Evolution of the mammary gland from the innate immune system. BioEssays 28:606–616.

Waage, J. K. 1979. The evolution of insect/vertebrate associations. Biol. J. Linnean Soc. 12:187–224.

Wade-Smith, J., and B. J. Verts. 1982. *Mephitis mephitis*. Mamm. Species 173:1–7.

Wahlberg, M., A. Frantzis, P. Alexiadou, P. T. Madsen, and B. Mohl. 2005. Click production during breathing in a sperm whale (*Physeter macrocephalus*) (L). J. Acoust. Soc. Am. 118:3404–3407.

Wahrman, J., and A. Zahavi. 1955. Cytological contributions to the phylogeny and classification of the rodent genus *Gerbillus*. Nature 175:600–602.

Waite, J. N., and V. N. Burkanov. 2006. Steller sea lion feeding habits in the Russian Far East, 2000–2003. Pp. 223–234 *in* Sea lions of the world (A. W. Trites, S. K. Atkinson, D. P. DeMaster, L. W. Fritz, T. S. Gelatt, L. D. Rea, and K. M. Wynne, eds.). Alaska Sea Grant College program, Fairbanks.

Walker, M. M., J. L. Kirschvink, G. Ahmed, and A. E.

Dizon. 1992. Evidence that fin whales respond to the geomagnetic field during migration. J. Exp. Biol. 171:67–78.

Wall, C. C., and E. W. Krause. 1992. A biomechanical analysis of the masticatory apparatus of *Ptilodus* (Multituberculata). J. Vert. Paleontol. 12:172–187.

Wallace, A.R. 1876. The geographical distribution of animals. Macmillan, London.

Wallis, R. L. 1979. Responses to low temperature in small marsupial mammals. J. Thermal Biol. 4:105–111.

Walsberg, G. E. 1983. Coat color and solar heat gain in animals. BioScience 33:88–91.

Walsberg, G. E. 1988. Evaluation of a nondestructive method for determining fat stores in small birds and mammals. Physiol. Zool. 61:153–159.

Walsberg, G. E. 1991. Thermal effects of seasonal coat change in three subarctic mammals. J. Thermal Biol. 16:291–296.

Walther, F. R. 1984. Communication and expression in hoofed mammals. Indiana Univ. Press, Bloomington.

Wan, Q-H., H. Wu, and S-G. Fang. 2005. A new subspecies of giant panda (*Ailuropoda melanoleuca*) from Shaanxi, China. J. Mammal. 86:397–402.

Wang, D., T. Oakley, J. Mower, L. C. Shimmin, S. Yim, K. L. Honeycutt, H. Tsao, and W-H. Li. 2004. Molecular evolution of bat color vision genes. Mol. Biol. Evol. 21:295–302.

Wang, L. C. H. 1978. Energetic and field aspects of mammalian torpor: the Richardson's ground squirrel. Pp. 109–145 *in* Strategies in cold: natural torpidity and thermogenesis (L. C. H. Wang and J. W. Hudson, eds.). Academic Press, New York.

Wang, L. C. H. 1979. Time patterns and metabolic rates of natural torpor in the Richardson's ground squirrel. Can. J. Zool. 57:149–155.

Wang, L. C. H., and M. W. Wolowyk. 1988. Torpor in mammals and birds. Can. J. Zool. 66:133–137.

Ward, S. J. 1990. Reproduction in the western pygmy possum, *Cercartetus concinnus* (Marsupialia: Burramyidae), with notes on reproduction of some other small possum species. Aust. J. Zool. 38:423–438.

Warneken, F., and M. Tomasello. 2006. Altruistic helping in human infants and young chimpanzees. Science 311:1301–1303.

Warren, W. S., and V. M. Cassone. 1995. The pineal gland: photoreception and coupling of behavioral, metabolic, and cardiovascular circadian outputs. J. Biol. Rhythms 10:64–79.

Waser, P. M. 1975. Diurnal and nocturnal strategies in the bushbuck (*Tragelaphus scriptus*) (Pallas). East Afr. Wildl. J. 13:49–63.

Wasser, S. K., L. Risler, and R. A. Steiner. 1988. Excreted steroids in primate feces over the menstrual cycle and pregnancy. Biol. Reprod. 39:862–872.

Wasser, S. K., S. L. Montfort, and D. E. Wildt. 1991. Rapid extraction of faecel steroids for measuring reproductive cyclicity and early pregnancy in free-ranging yellow baboons (*Papio cynocephalus cynocephalus*). J. Reprod. Fert. 92:415–423.

Wasser, S. K., A. M. Shedlock, K. Comstock, E. A. Ostrander, B. Mutayoba, and M. Stevens. 2004. Assigning African elephant DNA to geographic region of origin: Applications to the ivory trade. Proc. Natl. Acad. Sci. USA 101:14847–14852.

Waters, J. R., and C. J. Zabel. 1995. Northern flying squirrel densities in fir forests of northeastern California. J. Wildl. Mgmt. 59:858–866.

Watt, P. D., N. A. Oritsland, C. Jonkel, and K. Ronald. 1981. Mammalian hibernation and the oxygen consumption of a denning black bear (*Ursus americanus*). Comp. Biochem. Physiol. Part A. Sens. Neur. Behav. Physiol. 69:121–123.

Wayne, R. K., and S. M. Jenks. 1991. Mitochondrial DNA analysis implying extensive hybridization of the endangered red wolf *Canis rufus*. Nature 351:565–568.

Weary, D. M., and D. L. Kramer. 1995. Response of eastern chipmunks to conspecific alarm calls. Anim. Behav. 49:81–93.

Webb, P. W., and R. W. Blake. 1985. Swimming. Pp. 110–128 *in* Functional vertebrate morphology (M. Hildebrand, D. M. Bramble, K. F. Liem, and D. B. Wake, eds.). Harvard Univ. Press, Cambridge, MA.

Webb, S. D. 1985a. The interrelationships of tree sloths and ground sloths. Pp. 105–112 *in* The

evolution and ecology of armadillos, sloths, and vermilinguas (G. G. Montgomery, ed.). Smithsonian Institution Press, Washington, DC.

Webb, S. D. 1985b. Late Cenozoic mammal dispersals between the Americas. Pp. 357–386 *in* The great American biotic interchange (F. G. Stehli and S. D. Webb, eds.). Plenum Press, New York.

Webb, P. I., and J.Ellison. 1998. Normothermy, torpor, and arousal in hedgehogs (*Erinaceus europaeus*) from Dunedin. New Zealand J. Zoology 25:85–90.

Webb, P. W., and R. W. Blake. 1985. Swimming. Pp. 110–128 *in* Functional vertebrate morphology (M. Hildebrand, K.F. Liem, and D.B. Wake, eds.). Harvard Univ. Press, Cambridge, MA.

Webster, A. B., and R. J. Brooks. 1981. Daily movements and short activity periods of free-ranging meadow voles, *Microtus pennsylvanicus*. Oikos 37:80–87.

Wecker, S. C. 1963. The role of early experience in habitat selection by the prairie deer mouse, *Peromyscus maniculatus bairdi*. Ecol. Monogr. 33:307–325.

Weckerly, F. W., M. A. Ricca, and K. P. Meyer. 2001. Sexual segregation in Roosevelt elk: cropping rates and aggression in mixed-sex groups. J. Mammal. 82:825–835.

Wegener, A. L. 1912. Die Entstehung der Kontinente. Geolog. Rundsch. 3:276–292.

Wegener, A. L. 1915. Die Entstehung der Kontinente und Ozeane. Vieweg, Braunschweig, Germany.

Wegner, J., and G. Merriam. 1990. Use of spatial elements in a farmland mosaic by a woodland rodent. Biol. Conserv. 54:263–276.

Wegner R. E., Begall S., and Burda H. 2006. Magnetic compass in the cornea: local anaesthesia impairs orientation in a mammal. J. Exp. Biol. 209:4747–4750.

Wehausen, J. D., and R. R. Ramey II. 2000. Cranial morphometric and evolutionary relationships in the northern range of *Ovis canadensis*. J. Mamm. 81:145–161.

Wei, F., Z. Feng, Z. Wang, and M. Li. 1999. Feeding strategy and resource partitioning between giant and red pandas. Mammalia 63:417–430.

Wei-Dong, L., and A. T. Smih. 2005. Dramatic decline of the threatened Ili pika *Ochotona iliensis* (Lagomorpha: Ochotonidae) in Xinjiang, China. Oryx 39:30–34.

Weinreich, D. M. 2001. The rates of molecular evolution in rodent and primate mitochondrial DNA. J. Mol. Evol. 52:40–50.

Weir, B. J., and I. W. Rowlands. 1973. Reproductive strategies of mammals. Pp. 139–163 *in* Annual review of ecological systems ,4 (R. F. Johnston, P. W. Frank, and C. D. Michener, eds.). Annual Reviews, Palo Alto, CA.

Weir, B. S. 1996. Genetic data analysis II. Sinauer Associates, Sunderland, MA.

Weissengruber, G. E., G. Forstenpointner, A. Kubber-Heiss, K. Riedelberger, H. Schwammer, and K. Ganzberger. 2001. Occurrence and structure of epipharyngeal pouches in bears (Ursidae). J. Anat. 198:309–314.

Weldon, P. J. 2004. Defensive anointing: extended chemical phenotype and unorthodox ecology. Chemoecology 14:1–4.

Weller, D. W., A. M. Burdin, B. Wuersig, B. L. Taylor, and R. L. Brownell, Jr. 2002. The western gray whale: a review of past exploitation, current status and potential threats. J. Cetacean Res. Mgmt. 4:7–12.

Wells, R. T. 1978. Field observations of the hairynosed wombat (*Lasiorhinus latifrons* [Owen]). Aust. Wildl. Res. 5:299–303.

Wenstrup, J. J., and R. A. Suthers. 1984. Echolocation of moving targets by the fish-catching bat, *Noctilio leporinus*. J. Comp. Physiol. Ser. A 155:75–89.

Werman, S. D., M. S. Springer, and R. J. Britten. 1996. Nucleic acids I: DNA-DNA hybridization. Pp. 169–203 *in* Molecular systematics, 2d edition (D. M. Hillis, C. Moritz, and B. K. Mable, eds.). Sinauer Associates, Sunderland, MA.

West, G. B., and J. H. Brown. 2005. The origin of allometric scaling laws in biology from genomes to ecosystems: towards a quantitative unifying theory of biological structure and organization. J. Exp. Biol. 208:1575–1592.

West, P. M. 2005. The lion's mane. *Am. Sci.*

93:226–235.

West, S. D., and H. T. Dublin. 1984. Behavioral strategies of small mammals under winter conditions: solitary or social? Pp. 293–299 *in* Winter ecology of small mammals (J. F. Merritt, ed.). Spec. Pul. No. 10, Carnegie Museum of Natural History, Pittsburgh.

Westerman, M., M. S. Springer, J. Dixon, and C. Krajewski. 1999. Molecular relationships of the extinct pig-footed bandicoot *Chaeropus ecaudatus* (Marsupialia: Perameloidea) using 12S rRNA sequences. J. Mammal. Evol. 6:271–288.

Westerman, M., M. S. Springer, and C. Krajewski. 2001. Molecular relationships of the New Guinean bandicoot genera *Microperoryctes* and *Echymipera* (Marsupialia: Peramelina). J. Mammal. Evol. 8:93–105.

Westlake, R. L., and G. M. O'Corry-Crowe. 2002. Macrogeographic structure and patterns of genetic diversity in harbor seals (*Phoca vitulina*) from Alaska to Japan. J. Mammal. 83:1111–1126.

Weston, E. M. 2000. A new species of hippopotamus *Hexaprotodon lothagamensis* (Mammalia: Hippopotamidae) from the late Miocene of Kenya. J. Vert. Paleontol. 20:177–185.

Wetzel, R. M. 1985. The identification and distribution of recent Xenarthra (Edentata). Pp. 5–21 *in* The evolution and ecology of armadillos, sloths, and vermilinguas (G. G. Montgomery, ed.). Smithsonian Institution Press, Washington, DC.

Wetzel, R. M., R. E. Dubos, R. L. Martin, and P. Myers. 1975. Catagonus, an "extinct" peccary, alive in Paraguay. Science 189:379–381.

Wheaton, C., M. Pybus, and K. Blakely. 1993. Agency perspectives on private ownership of wildlife in the United States and Canada. Trans. North Am. Wildl. Nat. Resourc. Conf. 58:487–494.

Wheeler, W. C., and R. L. Honeycutt. 1988. Paired sequence differences in ribosomal RNAsL evolutionary and phylogenetic implications. Mol. Biol. Evol. 5:90–96.

Whitaker, J. O., Jr. 1963. Food of 120 Peromyscus leucopus from Ithaca, New York. J. Mammal. 44:418–419.

Whitaker, J. O., Jr. 1966. Food of *Mus musculus*, *Peromyscus maniculatus bairdi* and *Peromyscus leucopus* in Vigo County, Indiana. J. Mammal. 47:473–486.

Whitaker, J. O., Jr. 1968. Parasites. Pp. 254–311 *in* Biology of *Peromyscus* (Rodentia) (J. A. King, ed.). Spec. Pub. No. 2, American Society of Mammalogists.

Whitaker, J. O., Jr. 1988. Food habit of insectivorous bats. Pp. 171–189 *in* Ecological and behavioral methods for the study of bats (T.H. Kunz, ed.). Smithsonian Institution Press, Washington, DC.

Whitaker, J. O., Jr. 1994a. Academic propinquity. Pp. 121–138 *in* Seventy-five years of mammalogy, 1919–1994 (E. C. Birney and J. R. Choate, eds.). Spec. Pub. No. 11, American Society of Mammalogists.

Whitaker, J. O., Jr. 1994b. Food availability and opportunistic versus selective feeding in insectivorous bats. Bat Res. News 35:75–77.

Whitaker, J. O., Jr. 1995. Food of the big brown bat *Eptesicus fuscus* from maternity colonies in Indiana and Illinois. Am. Midl. Nat. 134:346–360.

Whitaker, J. O., Jr. and L. J. Rissler. 1993. Do bats feed in winter? Am. Midl. Nat. 129:200–203.

Whitaker, J. O. Jr., C. Neefus, and T. H. Kunz. 1996. Dietary variation in the Mexican free-tailed bat (*Tadarida brasiliensis mexicana*). J. Mammal. 77:716–724.

White, D. J. 1993. Lyme disease surveillance and personal protection against ticks. Pp. 99–125 in Ecology and environmental management of Lyme disease (H. S. Ginsberg, ed.). Rutgers Univ. Press, New Brunswick, NJ.

White, G. C., and R. A. Garrott. 1990. Analysis of wildlife radio-tracking data. Academic Press, San Diego.

Whittaker, R.H. 1975. Communites and ecosystems, 2d ed. Macmillan, New York, NY.

Wible, J. R. 1991. Origin of Mammalia: the craniodental evidence re-examined. J. Vert. Paleontol. 11:1–28.

Wible, J. R., G. W. Rougier, M. J. Novacek, M. C. McKenna, and D. Dashzeveg. 1995. A mammalian petrosal from the Early Cretaceous of Mongolia:

implications for the evolution of the ear region and mammaliamorph interrelationships. Am. Mus. Novitates 3149:1–19.

Wickler, S. J. 1980. Maximal thermogenic capacity and body temperatures of white-footed mice (*Peromyscus*) in summer and winter. Physiol. Zool. 53:338–346.

Wiens, F., A. Zitzmann, and N. A. Hussein. 2006. Fast food for slow lorises: Is low metabolism related to secondary compounds in high-energy plant diet? J. Mammal. 87:790–798.

Wiens, J. A. 1976. Population responses to patchy environments. Annu. Rev. Ecol. Syst. 7:81–120.

Wiens, J. A. 1977. On competition and variable environments. Am. Sci. 65:590–597.

Wiens, J. J., and M.J. Donoghue. 2004. Historical biogeography, ecology, and species richness. Trends Ecol. Evol. 19:639–644.

Wiggins, S. M., E. M. Oleson, M. A. McDonald, and J. A. Hildebrand. 2005. Blue whale (*Balaenoptera musculus*) diel call patterns offshore off southern California. Aquat. Mammals 31:161–168.

Wildt, D. E., M. Bush, K. L. Goodrowe, C. Packer, A. E. Pusey, J. L. Brown, P. Joslin, and S. J. O'Brien. 1987. Reproductive and genetic consequences of founding isolated lion populations. Nature 329:328–331.

Wiley, E. O. 1987. Methods in vicariance biogeography. Pp. 283–306 *in* Systematics and evolution: a matter of diversity (P. Hovenkamp, ed.). Utrecht Univ., Utrecht, Netherlands.

Wiley, R. W. 1980. *Neotoma floridana*. Mamm. Species 139:1–7.

Wilkinson, G. S. 1984. Reciprocal food sharing in the vampire bat. Nature 308:181–184.

Wilkinson, G. S. 1985. The social organization of the common vampire bat, I: Pattern and cause of association. Behav. Ecol. Sociobiol. 17:111–121.

Wilkinson, G. S. 1987. Altruism and cooperation in bats. Pp. 299–323 *in* Recent advances in the study of bats (M. B. Fenton, P. A. Racey, and J. M. V. Rayner, eds.). Cambridge Univ. Press, New York.

Wilkinson, G. S. 1990. Food sharing in vampire bats. Sci. Am. 262:64–70.

Willi, U. B., G. N. Bronner, and P. M. Narins. 2006. Middle ear dynamics in response to seismic stimuli in the Cape golden mole (*Chrysochloris asiatica*). J. Exp. Biol. 209:302–313.

Williams, E. M. 1998. Synopsis of the earliest Cetaceans. Pp. 1–28 *in* The emergence of whales: evolutionary patterns in the origin of Cetacea (J. G. M. Thewissen, ed.). Plenum Press, New York.

Williams, E. S., and I. K. Barker, eds. 2001. Infectious diseases of wild mammals. Iowa State Univ. Press, Ames.

Williams, E. S., E. T. Thorne, D. R. Kwiatkowski, and B. Oakleaf. 1992. Overcoming disease problems in the black-footed ferret recovery program. Trans. North Am. Wildl. Nat. Resourc. Conf. 57:474–485.

Williams, E. S., M. W. Miller, T. J. Kreeger, R. H. Khan, and E. T. Thorne. 2002. Chronic wasting disease of deer and elk: a review with recommendations for management. J. Wildl. Mgmt. 66:551–563.

Williams, G. C. 1966. Adaptation and natural selection: a critique of some current evolutionary thought. Princeton Univ. Press, Princeton, NJ.

Williams, G. C. 1975. Sex and evolution. Princeton Univ. Press, Princeton, NJ.

Williams, J. B., M. D. Anderson, and P. R. K. Richardon. 1997. Seasonal differences in field metabolism, water requirements, and foraging behaviour of free-living aardwolves. Ecology 78:2588–2602.

Williams, T. C., and J. M. Williams. 1967. Radio tracking of homing bats. Science 155:1435–1436.

Williams, T. C., and J. M. Williams. 1970. Radio tracking of homing and feeding flights of a neotropical bat, *Phyllostomus hastatus*. Anim. Behav. 18:302–309.

Williams, T. C., J. M. Williams, and D. R. Griffin. 1966. The homing ability of the neotropical bats, *Phyllostomus hastatus*, with evidence for visual orientation. Anim. Behav. 14:468–473.

Williams, T. M. 1990. Heat transfer in elephants: thermal partitioning based on skin temperature profiles. J. Zool. Lond. 222:235–245.

Willig, M. R., and S. K. Lyons. 1998. An analytical model of latitudinal gradients of species richness

with an empirical test for marsupials and bats in the New World. Oikos 81:93–98.

Willig, M. R., and M. A. Mares. 1989. A comparison of bat assemblages from phytogeographic zones of Venezuela. Pp. 59–67 *in* Patterns in the structure of mammalian communities (D. W. Morris, Z. Abramsky, B. J. Fox, and M. R. Willig, eds.). Texas Tech. Univ. Press, Lubbock.

Willig, M. R., and M. P. Moulton. 1989. The role of stochastic and deterministic processes in structuring Neotropical bat communities. J. Mammal. 70:323–329.

Willig, M. R., and K. W. Selcer. 1989. Bat species density gradients in the New World: a statistical assessment. J. Biogeogr. 16:189–195.

Willig, M.R., D. M. Kaufman, and R. D. Stevens. 2003. Latitudinal gradients of biodiversity: pattern, process, scale and synthesis. Annu. Rev. Ecol. Evol. Syst. 34:272–309.

Willis, K., M. Horning, D.A. S. Rosen, and A.W. Trites. 2005. Spatial variation of heat flux in Steller sea lions: evidence for consistent avenues of heat exchange along the body trunk. J. Exp. Mar. Biol. Ecol. 315:163–175.

Wilson, D. E., and F. R. Cole. 2000. Common names of mammals of the world. Smithsonian Institution Press, Washington, DC.

Wilson, D. E., and J. F. Eisenberg. 1990. Origin and applications of mammalogy in North America. Pp. 1–35 *in* Current mammalogy, vol. 2 (H. H. Genoways, ed.). Plenum, New York.

Wilson, D. E., and G. L. Graham. 1992. Pacific island flying foxes: proceedings of an international conservation conference. USDI, USFWS, Biol. Rep. 90(23).

Wilson, D. E., and D. M. Reeder (eds.). 1993. Mammal species of the world: a taxonomic and geographic reference, 2d edition. Smithsonian Institution. Press, Washington, DC.

Wilson, D. E., and D. M. Reeder (eds.). 2005. Mammal species of the world: a taxonomic and geographic reference, 3rd edition. 2 vols. Johns Hopkins Univ. Press, Baltimore.

Wilson, D. E., F. R. Cole, J. D. Nichols, R. Rudran, and M. S. Foster. 1996. Measuring and monitoring biological diversity: standard methods for mammals. Smithsonian Institution Press, Washington, DC.

Wilson, D. S. 1980. The natural selection of populations and communities. Benjamin/Cummings, Menlo Park, CA.

Wilson D. S., and Dugatkin L. A. 1997. Group selection and assortative interactions. Am. Nat. 149: 336–351.

Wilson, E. O. 1975. Sociobiology: the new synthesis. Harvard Univ. Press, Cambridge, MA.

Wilson, J. W., III. 1974. Analytical zoogeography of North American mammals. Evolution 28:124–140.

Wilz, M., and G. Heldhaier. 2000. Comparison of hibernation, estivation and daily torpor in the edible dormouse, *Glis glis*. J. Comp. Physiol. Part B 170:511–521.

Wimsatt, W. A. 1945. Notes on breeding behavior, pregnancy, and parturition in some vespertilionid bats of the eastern United States. J. Mammal. 26:23–33.

Wimsatt, W. A., and A. L. Guerriere. 1962. Observations on the feeding capacities and excretory functions of captive vampire bats. J. Mammal. 43:17–27.

Wimsatt, W. A., and B. Villa-R. 1970. Locomotor adaptations in the disc-winged bat, *Thyroptera tricolor*. Am. J. Anat. 129:89–119.

Winn, H. E., T. J. Thompson, W. C. Cummings, J. Hain, J. Hudnall, H. Hays, and W. W. Steiner. 1981. Song of the humpback whale-population comparisons. Behav. Ecol. Sociobiol. 8:41–46.

Winslow, J. T., N. Hastings, C. S. Carter, C. R. Harbaugh, and T. R. Insel. 1993. A role for central vasopressin in pair bonding in monogamous prairie voles. Nature 365:545–548.

Winter, Y., and O. von Helversen. 2003. Operational tongue length in phyllostomid nectar-feeding bats. J. Mammal. 84:886–896.

Wirsing, A., T. Steury, and D. Murray. 2002. Noninvasive estimation of body composition in small mammals: a comparison of conductive and morphometric techniques. Physiol. Biochem. Zool. 75:489–497.

Wischusen, E. W., and M. E. Richmond. 1998. Foraging ecology of the Philippine flying lemur (*Cynocephalus volans*). J. Mammal. 79:1288–1295.

Wishmeyer, D. L., G. D. Snowder, D. H. Clark, and N. E. Cockett. 1996. Prediction of live lamb chemical composition utilizing electromagnetic scanning (TOBEC). J. Anim. Sci. 74:1864–1872.

Withers, P.C. 1992. Comparative animal physiology. Saunders College Publishing, Philadelphia.

Witmer, L. M., S. D. Sampson, and N. Solounias. 1999. The proboscis of tapirs (Mammalia: Perissodactyla): a case study in novel narial anatomy. J. Zool. Lond. 249:249–267.

Wojcik, J. M., and M. Wolson (eds.). 1998. Evolution of shrews. Mammal Res. Inst., Polish Acad. Sci., Bialowieza.

Wolfe, J. L. 1970. Experiments on nest building behavior in *Peromyscus* (Rodentia: Cricetinae). Animal Behav. 18:613–615.

Wolfe, J. L., and S. A. Barnett. 1977. Effects of cold on nest-building by wild and domestic mice, *Mus musculus*. L. Biol. J. Linnean Soc. 9:73–85.

Wolff, J. O. 1980. Social organization of the taiga vole (Microtus xanthognathus). Biologist 62:34–45.

Wolff, J. O. 1985. Maternal aggression as a deterrent to infanticide in *Peromyscus leucopus* and *P. maniculatus*. Anim. Behav. 33:117–123.

Wolff, J. O. 1989. Social behavior. Pp. 271–291 *in* Advances in the study of *Peromyscus* (G. L. Kirkland, Jr., and J. N. Layne, eds.). Texas Tech. Univ. Press, Lubbock.

Wolff, J. O. 1993. Why are female small mammals territorial. Oikos 68: 364–370.

Wolff, J. O. 1996. Population fluctuations of mast-eating rodents are correlated with production of acorns. J. Mammal. 77:850–856.

Wolff, J. O. 1997. Population regulation in mammals: An evolutionary perspective. J. Anim. Ecol. 66:1–13.

Wolff, J. O. 2003. Laboratory studies with rodents: Facts or artifacts? Bioscience 53: 421–427.

Wolff, J. O., M. H. Freeberg, and R. D. Dueser. 1983. Interspecific territoriality in two sympatric species of *Peromyscus* (Rodentia: Cricetidae). Behav. Ecol. Sociobiol. 12:237–242.

Wolff, J. O., and D. S. Durr. 1986. Winter nesting behavior of *Peromyscus leucopus* and *Peromyscus maniculatus*. J. Mammal. 67:409–412.

Wolff, J. O., and W. Z. Lidicker, Jr. 1981. Communal winter nesting and food sharing in taiga voles. Behav. Ecol. Sociobiol. 9:237–240.

Wolff, J. O., R. D. Dueser, and K. S. Berry. 1985. Food habits of sympatric *Peromyscus leucopus* and *Peromyscus maniculatus*. J. Mammal. 66:795–798.

Wolff, J. O., K. I. Lundy, and R. Baccus. 1988. Dispersal, inbreeding avoidance, and reproductive success in white-footed mice. Anim. Behav. 36:456–465.

Womble, J. N., and M. F. Sigler. 2006. Temporal variation in Steller sea lion diet at a seasonal haul-out in Southeast Alaska. Pp. 141–154 *in* Sea lions of the world (A. W. Trites, S. K. Atkinson, D. P. DeMaster, L. W. Fritz, T. S. Gelatt, L. D. Rea, and K. M. Wynne, eds.). Alaska sea Grant College program, Fairbanks.

Wong, K. 2002. The mammals that conquered the seas. Sci. Am. (May):70–79.

Wood, A. E. 1965. Grades and clades among rodents. Evolution. 19:115–130.

Wood, B., and B. G. Richmond. 2000. Human evolution: taxonomy and paleobiology. J. Anat. 196:19–60.

Wood, J. D., B. McCowan, W. R. Langbauer, Jr., J. J. Viljoen, and L. A. Hart. 2005. Classification of African elephant *Loxodonta africana* rumbles using acoustic parameters and cluster analysis. Bioacoustics 15:143–161.

Wood, W. F. 2001. Antibacterial compounds in the interdigital glands of pronghorn, Antilocapra americana. Biochem. Syst. Ecol. 29:417–419.

Wood, W. F. 2002. 2-pyrrolidinone, a putative alerting pheromone from rump glands of pronghorn, Antilocapra americana. Biochem. Syst. Ecol. 30:361–363.

Woodburne, M. O., and J. A. Case. 1996. Dispersal, vicariance, and the late Cretaceous to early Tertiary land mammal biogeography from South America to Australia. J. Mammal. Evol. 3:121–161.

Woodburne, M. O., and R. H. Tedford. 1975. The

first Tertiary monotreme from Australia. Am. Mus. Novitates 2588:1–11.

Woodburne, M. O., B. J. MacFadden, J. A. Case, M. Springer, N. S. Pledge, J. D. Power, J. M.

Woodburne, and K. Johnson. 1993. Land mammal biostratigraphy and magnetostratigraphy of the Etadunna Formation (late Oligocene) of South Australia. J. Vert. Paleontol. 13:132–164.

Woolaver, L., R. Nichols, W. F. Rakotombololona, A. T. Volahy, and J. Durbin. 2006. Population status, distribution and conservation needs of the narrow-striped mongoose *Mungotictis decemlineata* of Madagascar. Oryx 40:67–75.

Woolley, P. 1974. The pouch of Planigale subtilissima and other dasyurid marsupials. J. Roy. Soc. West. Aust. 57:11–15.

Woolley, P. A. 2005. Revision of the three-striped dasyures, genus *Myoictis* (Marsupialia: Dasyuridae), of New Guinea, with description of a new species. Rec. Austral. Mus. 57:321–340.

Woolley, P.A., and A. Valente. 1992. Hair structure of the dasyurid marsupials of New Guinea. Science in New Guinea 18:29–49.

World Health Organization. 1984. Expert committee on rabies, seventh report. WHO Tech. Report No. 709, Geneva.

Worthington, W. J., C. Moritz, L. Hall, and J. Toop. 1994. Extreme population structuring in the threatened ghost bat, Macroderma gigas: evidence from mitochondrial DNA. Proc. Roy. Soc. Lond. 257:193–198.

Worthington, W. J., L. Hall, E. Barratt, and C. Moritz. 1999. Genetic structure and male-mediated gene flow in the ghost bat (*Macroderma gigas*). Evolution 53:1582–1591.

Wozencraft, W. C. 1989a. The phylogeny of the Recent Carnivora. Pp. 495–535 *in* Carnivore behavior, ecology, and evolution (J. L. Gittleman, ed.). Cornell Univ. Press, Ithaca, NY.

Wozencraft, W. C. 1989b. Classification of the Recent Carnivora. Pp. 569–593 *in* Carnivore behavior, ecology, and evolution (J. L. Gittleman, ed.). Cornell Univ. Press, Ithaca, NY.

Wozencraft, W. C. 1993. Order Carnivora. Pp. 279–348 *in* Mammal species of the world, 2d edition (D. E. Wilson and D. M. Reeder, eds.). Smithsonian Institution Press, Washington, DC.

Wozencraft, W. C. 2005. Order Carnivora. Pp. 532–628 *in* Mammal species of the world: a taxonomic and geographic reference. 3rd edition, 2 vols. (D. E. Wilson and D. M. Reeder, eds.). Johns Hopkins Univ. Press, Baltimore.

Wrabetz, M. J. 1980. Nest insulation: a method of evaluation. Can. J. Zool. 58:938–940.

Wright, P. L., and M. W. Coulter. 1967. Reproduction and growth in Maine fishers. J. Wildl. Mgmt. 31:70–87.

Wright, S. 1940. Breeding structure of populations in relation to speciation. Am. Nat. 74:232–248.

Wright, S. 1951. The genetical structure of populations. Ann. Eugen. 15:114–138.

Wroe, S. 2001. Maximucinus muirheadae, gen. et sp. nov (Thylacinidae: Marsupialia), from the Miocene of Riversleigh, north-western Queensland, with estimates of body weights for fossil thylacinids. Aust. J. Zool. 49:603–614.

Wroe, S., and M. Archer. 2006. Origins and early radiations of marsupials. Pp. 551–574 *in* Evolution and biogeography of Australasian vertebrates (J. R. Merrick, M. Archer, G. M. Hickey, and M. S. Y. Lee, eds.). Auscipub, Oatlands, NSW, Australia.

Wroe, S., M. Ebach, S. Ahyong, C. de Muizon, and J. Muirhead. 2000. Cladistic analysis of dasyuromorphian (Marsupialia) phylogeny using cranial and dental characters. J. Mammal. 81:1008–1024.

Wroe, S., C. McHenry, and J. Thomason. 2005. Bite club: comparative bite force in big biting mammals and the prediction of predatory behaviour in fossil taxa. Proc. Roy. Soc. Biol. Sci. 272:619–625.

Wroot, A. 1984. Hedgehogs. Pp. 751–757 *in* The encyclopedia of mammals (D. Macdonald, ed.). Facts on File, New York.

Wu, C., J. Wu, T. D. Bunch, Q. Li, Y. Wang, and Y. Zhang. 2005. Molecular phylogenetics and biogeography of *Lepus* in eastern Asia based on mitochondrial DNA sequences. Mol. Phylogenet. Evol. 37:45–61.

Wu, J., L. Zhou, and L. Mu. 2006. Summer habitat

selection by Siberian musk deer (*Moschus moschiferus*) in Tonghe forest area in the Lesser Khingan Mountains. Acta Theriol. Sinica 26:44–48.

Wu, R. K., and Y. R. Pan. 1985. Preliminary observations on the cranium of *Laccopithecus robustus* from Lufeng, Yunnan with reference to its phylogenetic relationship. Acta Anthropol. Sinica 4:7–12.

Wu, S., N. Liu, Y. Zhang, and G. Ma. 2004. Assessment of threatened status of Chinese pangolin (*Manis pentadactyla*). Chin. J. Appl. Environ. Biol. 10:456–461.

Wunder, B. A. 1978. Implications of a conceptual model for the allocation of energy resources by small mammals. Pp. 68–75 *in* Populations of small mammals under natural conditions (D. P. Snyder, ed.). Spec. Pub., Ser. 5, Pymatuning Lab. Ecol., Univ. Pittsburgh.

Wunder, B. A. 1984. Strategies for, and environmental cueing mechanisms of, seasonal changes in thermoregulatory parameters of small mammals. Pp. 165–172 *in* Winter ecology of small mammals (J. F. Merritt, ed.). Spec. Pub. No. 10, Carnegie Museum of Natural History, Pittsburgh.

Wunder, B. A. 1985. Energetics and thermoregulation. Pp. 812–844 *in* Biology of New World Microtus (R. H. Tamarin, ed.). Spec. Pub. No. 8, American Society of Mammalogists.

Wunder, B. A., and R. D. Gettinger. 1996. Effects of body mass and temperature acclimation on the non-shivering thermogenic response of small mammals. Pp. 131–139 *in* Adaptation to the cold: Tenth International Hibernation Symposium (F. Geiser, A. J. Hulbert and S. C. Nicol, eds.). Univ. New England Press, Armidale, NSW, Australia.

Wunder, B. A., D. S. Dobkin, and R. D. Gettinger. 1977. Shifts of thermogenesis in the prairie vole (Microtus ochrogaster): strategies for survival in a seasonal environment. Oecologica 29:11–26.

Würsig, B. 1988. The behavior of baleen whales. Sci. Am. 258:102–107.

Würsig, B., and M. Würsig. 1977. The photographic determination of group size, composition, and stability of coastal porpoises (*Tursiops truncatus*). Science 198:755–756.

Wynen, L. P., S. D. Goldsworthy, S. J. Insley, M. Adams, J. W. Bickham, J. Francis, J. P. Gallo, A. R. Hoelzel, P. Majluf, R. W. G. White, and R. Slade. 2001. Phylogenetic relationships within the eared seals (Otariidae: Carnivora): implications for the historical biogeography of the family. Mol. Phylogenet. Evol. 21:270–284.

Wyner, Y., R. DeSalle, and R. Absher. 2000. Phylogeny and character behavior in the Family Lemuridae. Mol. Phylogenet. Evol. 15:124–134.

Wynne-Edwards, V. C. 1962. Animal dispersion in relation to social behavior. Oliver and Boyd, Edinburgh.

Wynne-Edwards, V. C. 1986. Evolution through group selection. Blackwell Scientific Publ., Boston.

Wyss, A. R., and J. J. Flynn. 1993. A phylogenetic analysis and definition of the Carnivora. Pp. 32–52 *in* Mammal phylogeny: placentals (F. S. Szalay, M. J. Novacek, and M. C. McKenna, eds.). Springer-Verlag, New York.

Xia, X., and J. S. Millar. 1991. Genetic evidence of promiscuity in Peromyscus leucopus. Behav. Ecol. Sociobiol. 28:171–178.

Xiao, Z., and Z. Zhang. 2006. Nut predation and dispersal of Harland tanoak Lithocarpus harlandii by scatter-hoarding rodents. Acta Oecol. 29:205–213.

Yager, D. D., and M. L. May. 1990. Ultrasound-triggered, flight-gated evasive manoeuvres in the praying mantis, *Parasphendale agrionina*, II: Tethered flight. J. Exp. Biol. 152:41–58.

Yager, D. D., M. L. May, and M. B. Fenton. 1990. Ultrasound-triggered, flight-gated evasive escape manoeuvres in the praying mantis, *Parasphendale agrionina*, I: Free flight. J. Exp. Biol. 152:17–39.

Yagil, R. 1985. The desert camel. Karger, Basel.

Yamazaki, K., E. A. Boyse, V. Mike, H. T. Thaler, B. J. Mathieson, J. Abbott, J. Boyse, and Z. A. Zayas. 1976. Control of mating preferences in mice by genes in the major histocompatibility complex. J. Exp. Med. 144:1324–1335.

Yamazaki, K., G. K. Beauchamp, C. J. Wysocki, J. Bard, L. Thomas, and E. A. Boyse. 1983.

Recognition of H-2 types in relation to the blocking of pregnancy in mice. Science 221:186–188.

Yang, F., E. Z. Alkalaeva, P. L. Perlman, A. T. Pardini, W. R. Harrison, P. C. M. O'Brien, A. S. Graphodatsky, M. A. Ferguson-Smith, and T. J. Robinson. 2003. Reciprocal chromosome painting among human, aardvark, and elephant (superorder Afrotheria) reveals the likely eutherian ancestral karyotype. Proc. Natl. Acad. Sci. USA 100:1062–1065.

Ya-Ping, Z., and S. Li-Ming. 1993. Phylogenetic relationships of macaques inferred from restriction endonuclease analysis of mitochondrial DNA. Folia Primat. 60:7–17.

Yaskin, V. A. 1994. Variation in brain morphology of the common shrew. Pp. 155–161 *in* Advances in the biology of shrews (J. F. Merritt, G. L. Kirkland, Jr., and R. K. Rose, eds.). Spec. Pub., Carnegie Mus. Nat. Hist. 18:1–458.

Yates, T. L. 1984. Insectivores, elephant shrews, tree shrews, and Dermopterans. Pp. 117–144 *in* Orders and families of recent mammals of the world (S. Anderson and J. K. Jones, Jr., eds.). John Wiley and Sons, New York.

Yates, T. L., J. N. Mills, C. A. Parmenter, T. G. Ksiazek, R. R. Parmenter, J. R. Vande Castle, C. H. Calisher, S. T. Nichol, K. D. Abbott, J. C. Young, M. L. Morrison, B. J. Beaty, J. L. Dunnum, R. J. Baker, J. Salazar-Bravo, and C. J. Peters. 2002. The ecology and evolutionary history of an emergent disease: hantavirus pulmonary syndrome. BioScience 52:989–998.

Yeboah, S., and K. B. Dakwa. 2002. Colony and social structure of the Ghana mole-rat (*Cryptomys zechi*, Matchie) (Rodentia: Bathyergidae). J. Zool. 256:85–91.

Yoder, A. D., M. M. Burns, S. Zehr, T. Delefosse, G. Veron, S. M. Goodman, and J. J. Flynn. 2003. Single origin of Malagasy Carnivora from an African ancestor. Nature 421:734–737.

Yom-Tov, Y., and E. Geffen. 2006. Geographic variation in body size: the effects of ambient temperature and precipitation. Oecologia 148:213–218.

Yom-Tov, Y., and J. Yom-Tov. 2005. Global warming, Bergmann's rule and body size in the masked shrew Sorex cinereus Kerr in Alaska. J. Anim. Ecol. 74:803–808.

Young, J.Z., and M. J. Hobbs. 1975. The life of mammals. Oxford Univ. Press, Oxford, U.K.

Young, L. J., M. M. Lim, B. Gingrich, and T. R. Insel. 2001. Cellular mechanisms of social attachment. Hormones Behav. 40:133–138.

Young, L. J., Z. X. Wang, and T. R. Insel. 1998. Neuroendocrine bases of monogamy. Trends Neurosci. 21:71–75.

Yousef, M. K., S. M. Horvath, and R. W. Bullard (eds.). 1972. Physiological adaptations, desert and mountain. Academic Press, New York.

Yuill, T. M. 1987. Diseases as components of mammalian ecosystems: mayhem and subtlety. Can. J. Zool. 65:1061–1066.

Zahavi, A. 1975. Mate selection—a selection for a handicap. J. Theor. Biol. 53:205–214.

Zalmout, I. S., M. Ul-Haq, and P. D. Gingerich. 2003. New species of *Protosiren* (Mammalia, Sirenia) from the early Middle Eocene of Balochistan (Pakistan). Contrib. Mus. Paleo. Univ. Michigan 31:79–87.

Zar, J. H. 1996. Biostatistical analysis, 3d edition. Prentice-Hall, Upper Saddle River, NJ.

Zar, J. H. 2006. Biostatistical analysis, 5th edition. Prentice Hall, Upper Saddle River, NJ.

Zedrosser, A., B. Dahle, and J. E. Swenson. 2006. Population density and food conditions determine female body size in brown bears. J. Mammal. 87:510–518.

Zenuto, R. R., A. D. Vitullo, and C. Busch. 2003. Sperm characteristics in two populations of the subterranean rodent Ctenomys talarum (Rodentia: Octodontidae). J. Mammal. 84:877–885.

Zeuner, F. E. 1963. History of domesticated animals. Hutchinson, London.

Zeveloff, S. I. 1988. Mammals of the inter mountain West. Univ. Utah Press, Salt Lake City.

Zeveloff, S. I., and M. S. Boyce. 1980. Parental investment and mating systems in mammals. Evolution 34:973–982.

Zhang, Y.-P., and O. A. Ryder. 1998. Mitochondrial cytochrome b gene sequence of Old World monkeys: with special reference on evolution of Asian colobines. Primates 39:39–49.

Zhang, Y. Y., R. Proenca, M. Maffei, M. Barone, L. Leopold, and J. M. Friedman. 1994. Positional cloning of the mouse obese gene and its human homolog. Nature 372:425–432.

Zhang, Z., F. Wei, M. Li, and J. Hu. 2006. Winter microhabitat separation between giant and red pandas in *Bashania faberi* bamboo forest in Fengtongzhai Reserve. J. Wildl. Mgmt. 70:231–235.

Zhu, X. J., D. G. Lindburg, W. S. Pan, K. A. Forney, and D. J. Wang. 2001. The reproductive strategy of giant pandas (*Ailuropoda melanoleuca*): infant growth

and development and mother-infant relationships. J. Zool. 253:141–155.

Zima, J., and M. Macholán. 1995. B chromosomes in the wood mice (genus *Apodemus*). Acta Theriol. Suppl. 3:75–86.

Zimmer, W. M. X., P. T. Madsen, V. Teloni, M. P. Johnson, and P. L. Tyack. 2005. Off-axis effects on the multipulse structure of sperm whale usual clicks with implications for sound production. J. Acoust. Soc. Am. 118:3337–3345.

Zink, R.M. 1996. Comparative phylogeography in North American birds. Evolution 50:308–317.

Zucker, I. 1985. Pineal gland influences period of

circannual rhythms of ground squirrels. Am. J. Physiol. 249:R111–R115.

Zucker, I., M. Boshes, and J. Dark. 1983. Suprachiasmatic nuclei influence circannual and circadian rhythms of ground squirrels. Am. J. Physiol. 244:R472–R480.

Zuckerkandl, E., and L. Pauling. 1965. Evolutionary divergence and convergence in proteins. Pp. 97–116 *in* Evolving genes and proteins (V. Bryson and H.J. Vogel, eds.). Academic Press, New York.

Credits

Photographs

Part Openers

Part 1: Photo by J.D. Harweeli/American Society of Mammalogists; **Part 2:** Ken Lucas/Visuals Unlimited; **Part 3:** Photo courtesy of Romina Vidal-Russell; **Part 4:** Dr. Marc Bekoff; **Part 5:** © Len Rue, Jr./Visuals Unlimited.

Chapter 1

Chapter Opener: Photo courtesy of Rex Lord; **Figure 1.1a:** Lee C. Drickamer; **1.1b:** Photo courtesy of Justin O'Rain; **1.1c:** Lee C. Drickamer; **1.2:** Photo courtesy of Rex Lord; **1.3:** © Smithsonian Institution, NMNH Chip Clark, Photographer.

Chapter 2

Chapter Opener: Illustration by Mark Catesby; **Figure 2.1:** © Hubertus Kanus/Photo Researchers, Inc.; **2.2:** Art Resource/Field Museum of Natural History, Chicago, USA; **2.3:** Special Collections, University of South Florida Library; **2.4:** © Bettmann/CORBIS; **2.5:** © Bettmann/CORBIS; **2.6(1) (2):** Photo by J.D. Harweeli/American Society of Mammalogists; **2.7:** © Francois Gohier/Photo Researchers, Inc.; **2.8:** © Des Bartlett/Photo Researchers, Inc.; **2.9:** American Society Mammalogists; **2.19c:** Lee C. Drickamer; **2.10d,e:** American Society Mammalogists; **2.11:** © Hal H. Harrison/Photo Researchers, Inc.; **2.12:** © Leonard Lee Rue III/Visuals Unlimited.

Chapter 3

Chapter Opener: Ó Image 100/Royalty-Free/CORBIS; **Figure 3.1:** George Feldhamer; **3.2:** © Nicholas De Bore III/Bruce Coleman, Inc.; **3.5:** © Image 100/Royalty-Free/CORBIS; **3.9:** Photos by M. Westerman and M. Schwartzman; **3.10:** R.P. Canham, "Serum Protein Variations and Selections in Fluctuating Populations of Cricetid Rodents," Ph.D. thesis, University of Alberta, 1969.

Chapter 4

Chapter Opener: Courtesy A. S. Romer, 1966.

Chapter 5

Chapter Opener: Leonard Lee Rue III/Visuals Unlimited; **Figure 5.14:** Leonard Lee Rue III/Visuals Unlimited.

Chapter 6

Chapter Opener: © Nigel J. Dennis/Photo Researchers, Inc.; **Figure 6.1:** © Mark Newman/Bruce Coleman, Inc.; **6.4:** Photo courtesy of P.A. Wooley; **6.6:** Lee Drickamer; **6.9a:** © Ken Lucas/Visuals Unlimited; **6.9b:** © Charlie Ott/Photo Researchers, Inc.; **6.9c:** © Verna R. Johnston, Photo Researchers. Inc.; **6.9d:** Science VU Visuals Unlimited; **6.11a:** © Ken Lucas/Visuals Unlimited; **6.15:** © Bill Bachman 1986/Photo Researchers, Inc.; **6.16a:** © Jen and Des Bartlett/Bruce Coleman, Inc.; **6.16b** © Nigel J. Dennis/Photo Researchers, Inc.; **6.17:** © Leonard Lee Rue III/Visuals Unlimited; **6.18:** © Science VU/Visuals Unlimited; **6.20:** © Paul A. Souders/CORBIS.

Chapter 7

Chapter Opener: © R. Van Nostrand from National Audubon Society/Photo Researchers, Inc. **Figure 7.4a:** © Dwight Kuhn; **7.4b:** © Dr. Edwin Gould (National Zoological Park)/Dwight Kuhn Photography; **7.5a:** © Studio Carlo Dani/Animals Animals/Earth Scenes; **7.5b:** © Nancy Adams/Tom Stack & Associates; **7.12:** Dr. Joseph F. Merritt; **7.13:** © R. Van Nostrand from National Audubon Society/Photo Researchers, Inc.; **7.16:** © John Alcock/Visuals Unlimited.

Chapter 8

Chapter Opener: © J.O.Wolff Mammal Images Library of the American Society of Mammalogists; **Figure 8.8:** © Des Bartlett/Photo Researchers, Inc.; **8.9a:** © J. Kirkley/Mammal Images Library of the American Society of Mammalogists; **8.9b:** © J.O.Wolff Mammal Images Library of the American Society of Mammalogists.

Chapter 9

Chapter Opener: Leonard Lee Rue III/Visuals Unlimited; **Figure 9.7:** Leonard Lee Rue III, from National Audubon Society/Photo Researchers, Inc.; **9.8:** Courtesy of P.F. Scholander, University of California, San Diego; **9.11:** © Leonard Lee Rue III/Visuals Unlimited; **9.12:** © Carnegie Museum of Natural History; **9.13a,b:** Courtesy Robert MacArthur.

Chapter 10

Chapter Opener: John D. Cunningham/Visuals Unlimited; **Figure 10.17:** © John D. Cunningham/Visuals Unlimited; **10.19a:** © Jen and Des Bartlett/Photo Researchers, Inc.

Chapter 11

Figure 11.6: Photo courtesy of Mike Westerman; **11.17A:** Photo by Peter L. Meserve; **11.18:** Photo courtesy of Romina Vidal-Russell; **11.21:** Photo courtesy of Mike Westerman; **11.22:** Photo courtesy of Carey Krajewski; **11.24:** Photo by H. J. Aslin with permission pf Mammal Images Library of the American Society of Mammalogists; **11.25:** Photo by H.J. Aslin with permission of Mammal Images Library of the American Society of Mammalogists; **11.27:** Photo courtesy of Carey Krajewski; **11.28:** Photo courtesy of Mike Westerman; **11.32:** Photo courtesy of Mike Westerman; **11:34:** Photo by Graham J.B. Ross; **11.35:** Photo courtesy of Mike Westerman.

Chapter 12

Figure 12.3a,b: American Society Mammalogists; **12.5a:** © A.W. Ambler/Photo Researchers, Inc.; **12.5b:** © Irene Vandermolen/Visuals Unlimited; **12.6:** © Kim Taylor Bruce Coleman, Inc.; **12.12a:** © William J. Weber/Visuals Unlimited; **12.14a:** Galen Rathbun; **12.14b:** Gene Maliniak, National Zoological Park; **12.17:** Photo by C.J. Phillips/American Society Mammalogists.

Chapter 13

Chapter Opener: © Meril D. Tuttle, Bat Conservation International; **Figure 13.12:** © Meril D. Tuttle, Bat Conservation International; **13.14:** © David L. Pearson/Visuals Unlimited; **13.15:** George Feldhamer; **13.16:** Photo courtesy of Rexford D. Lord Jr.; **13.19:** Photo courtesy of Rexford D. Lord, Jr.; **13.20a:** Photo by PV August/Mammal Images Library of the American Society of Mammalogists; **13.20b:** © Stephen Dalton. Animals Animals/Earth Scenes; **13.22:** Photo courtesy of Rexford D. Lord, Jr.; **13.26:** Photo by Timothy Carter; **13.27:** Photo courtesy of Rexford D. Lord, Jr.

Chapter 14

Chapter Opener: © Pat Crowe/Animals Animals/Earth Scenes; **Figure 14.6:** © James Burke, Time Life Nature Library/Getty Images; **14.9:** © David Haring/ Oxford Scientific Films; **14.10:** © Rob Williams/Bruce Coleman, Inc.; **14.11:** © A.W. Ambler/Photo Researchers, Inc.; **14.13:** Sidney Bahrt/Photo Researchers, Inc.; **14.16:** © Nigel J. Dennis/Photo Researchers, Inc.; **14.17:** © Doug Wechsler/Animals Animals/Earth Scenes; **14.20:** © Norman Owen Tomalin/Bruce Coleman, Inc.; **14.22:** Bates Littlehales/Animals Animals /Earth

Science; **14.23a:** Rod Williams/ Animals Animals/Earth Sciences; **14.24b:** © Stewart D. Halperin/Animals Animals/Earth Scenes; **14.24c:** Pat Crowe/Animals Animals/Earth Sciences.

Chapter 15

Chapter Opener: Gary Milburn/Tom Stack and Associates; **Figure 15.7:** Photo courtesy of Rexford D. Lord, Jr.; **15.8a:** Photo courtesy of Rexford D. Lord, Jr.; **15.8.b:** Photo by P. Myers/Mammal Images Library of the American Society of Mammalogists; **15.9a:** Photo courtesy of Rexford D. Lord, Jr. **15.9b:** Photo courtesy of Rexford D. Lord, Jr.; **15.9c:** Photo courtesy of Rexford D. Lord, Jr.; **15.12:** Lee C. Drickamer; **15.13a:** © Roy Pinney, Photo Researchers, Inc. **15.14a:** © Gary Milburn/Tom Stack and Associates.

Chapter 16

Chapter Opener: Courtesy Kay Holekamp; **Figure 16.2:** Photo courtesy of Dan Roby; **16.3a:** Photo courtesy of Rexford D. Lord, Jr.; **16.3b:** Photo courtesy of Rexford D. Lord, Jr.; **16.12:** Photo courtesy of Kay Holekamp; **16.13:** Photo courtesy of Rexford D. Lord, Jr.; **16.15:** Photo courtesy of Rexford D. Lord, Jr.; **16.16a,b:** Photos courtesy of Rexford D. Lord Jr. **16.17:** Photo courtesy of Rexford D. Lord, Jr. **16.18:** Photo courtesy of Rexford D. Lord, Jr.; **16.21:** Photo by K R Gordon/Mammal Images Library of the American Society of Mammalogists; **16.22:** Photo courtesy of Rexford D. Lord, Jr.;

Chapter 17

Chapter Opener: © M. DeMocker/Visuals Unlimited; **Figure 17.13:** © M. DeMocker/Visuals Unlimited; **17.20:** Reprinted with permission from JASA, vol. 118, p. 3338. Fig. 1 Zimmer et al., "Bent-horn model of sperm whale sound production." Copyright 2005, Acoustical Society of America.

Chapter 18

Figure 18.8: © Karl H. and Stephen Maslowski/Visuals Unlimited; **18.9:** © Woodrow Goodpaster/Photo Researchers, Inc.; **18.10a:** © Woodrow Goodpaster/Photo Researchers, Inc.; **18.11:** DG Huckaby/Mammal Images Library of the American Society of Mammalogists; **18.14a:** Society of Mammalogists. Photo by Timothy Carter; **18.14b:** ©Joe Whittaker; **18.15:** © Society of Mammalogists, Photo by M Andera/Mammal Images Library Of the American Society of Mammalogists; **18.16:** Photo by M Anders/Mammal Images Library of the American Society of Mammalogists; **18.17:** Photo by Justin Oirian; **18.18a:** Photo courtesy of Rexford D. Lord, Jr.; **18.19a:** Photo by Rexford D. Lord, Jr.; **18.19b:** Photo courtesy of

Rexford D. Lord, Jr.; **18.20:** Courtesy of Rexford D. Lord, Jr.; **18.21A:** ©Tom McHugh/Photo Researchers, Inc.; **18.21b:** Photo courtesy of Rexford D, Lord. Jr.; **18.22:** Courtesy of Rexford D. Lord, Jr.; **18:25:** Leonard Lee Rue III, Visuals Unlimited; **18.28:** © James des Lauriers/Mammal Images Library of the American Society of Mammalogists; **18.29:** George Feldhamer.

Chapter 19

Chapter Opener: © Len Rue, Jr./Visuals Unlimited; **Figure 19.4:** Photo by African Conservation Experience; **19.9:** © Leonard Lee Rue III/Visuals Unlimited; **19.11:** Len Lee La Rue Visuals; **19.11:** Leonard Lee LaRue II, Visuals Unlimited; **19.12:** © Dr. Hendrik Hoeck; **19.18:** © Daniel K. Odell.

Chapter 20

Chapter Opener: Photo courtesy of Rexford D. Lord, Jr; **Figure 20.6a:** Photo courtesy of Rexford D. Lord, Jr.; **20.6b:** Photo by J D Haweeli/Mammal Images Library of the Society of Mammalogists; **20.6c:** George Feldhamer; **20.8:** Photo by G.C. Hickman with permission of Mammal Images Library of the American Society of Mammalogists; **20.12:** Photo by P Waser/Mammal Images Library of the American Society of Mammalogists; **20.13:** Photo courtesy of Rexford D. Lord, Jr.; **20.15:** George Feldhamer; **20.16a:** Photo courtesy of Rexford D. Lord, Jr.; **20.16b:** Photo courtesy of Rexford D. Lord, Jr. **20.16c:** Courtesy of Rexford D. Lord, Jr.; **20.18:** Photo by H.G. Haweeli/Mammal Images Library of the American Society of Mammalogists; **20.20:** Photo by B.E. Joseph/Mammal Images Library of the American Society of Mammalogists; **20.21a:** George Feldhamer; **20.21b:** George Feldhamer; **20.23:** Image Provided by John Fischer/Photo by H. L. Gunderson/Mammal Images Library of the American Society of Mammalogists.

Chapter 21

Chapter Opener: Stephen H. Vessey; **Figure 21.3:** Courtesy Anne L. Enge **21.5:** Stephen H. Vessey; **21.6:** Photo by D. Meikle; **21.8:** © Fiona Guinness; **21.9:** Photo by Douglas B. Meikle; **21.10:** © Scott and Nancy Creel; **21.11:** Dr. Marc Bekoff; **21.13:** Photo by Stephen H. Vessey.

Chapter 22

Chapter Opener: Stephen H. Vessey; **Figure 22.4:** Photo by Stephen H. Vessey; **22.5:** Photo by Stephen H. Vessey; **22.8:** Burney Le Boeuf; **22.10:** Stephen H. Vessey; **22.11:** Dr. Laurence Frank/University of California, Berkeley; **21.13:** © Meril D. Tuttle, Bat Conservation International.

Chapter 23

Chapter Opener: © Patricia D. Moehlman; **Figure 23.1:** © Craig Packer; **23.2:** © Scott and Nancy Creel; **23.3:** Kelly C. Fletcher; **23.4:** Leanne T. Nash; **23.5:** Photo by Jennifer U. M. Jarvis; **23.6:** Photo by G.D. Lepp **23.8:** © Patricia D. Moehlman.

Chapter 24

Chapter Opener: © Len Rue, Jr. Visuals Unlimited; **Figure 24.7:** © Len Rue, Jr./Visuals Unlimited.

Chapter 25

Chapter Opener: © Fiona Guinness.

Chapter 26

Chapter Opener: Photo by Stephen H. Vessey; **Figure 26.8:** Leonard Lee Rue III/Visuals Limited; **26.9:** Photo by Stephen H. Vessey.

Chapter 27

Chapter Opener: Photo from Centers from Disease Control and Prevention; **Figure 27.1:** Courtesy of Roy C. Anderson; **27.4:** Photo from Centers from Disease Control and Prevention; **27.5a:** Photo from Centers of Disease Control and Prevention; **27.5b:** © Dr. Stanley Flegler/Visuals Unlimited; **27.7:** © Bruce Iverson/Science Photo Library/Photo Researchers, Inc.; **27.8:** © Bruce Iverson Science Photo Library/Photo Researchers, Inc.; **27.11:** Photo from Centers for Disease Control and Prevention; **27.13:** Science VU-FIP/Visuals Unlimited.

Chapter 28

Chapter Opener: Photo from American Society of Mammalogists; **Figure 28.6:** Judy Rains; **28.7:** Judy Rains; **28.8,a,b:** Courtesy of George Feldhamer; **28.9:** George Feldhamer.

Chapter 29

Chapter Opener: Photo by D. G. Huckaby. Mammal Slide Library/American Society of Mammalogists; **Figure 29.1a:** © Len Rue, Jr./Visuals Unlimited; **29.2:** © Frank Lambrecht/Visuals Unlimited; **29.5a:** © Asa C. Thoresen/Photo Researchers, Inc.; **29.5b:** © Michael Freeman/Bruce Coleman, Inc.; **29.6:** Photo by D.G. Huckaby, Mammal Slide Library American Society Mammalogists; **29.7:** © Rod Williams/Bruce Coleman, Inc.; **29.10:** Lee C. Drickamer; **29.12:** Photo by A.H. Shoemaker/American Society Mammalogists.

Line Art

Line art in this book were adapted and inspired from a number of scholarly sources, some new to the third edition, some redrawn from earlier editions. Although we have endeavored to give credit whenever possible, we are certain that some sources of inspiration have been missed. We apologize for any oversights.

Chapter Opener 2: Illustration from Mark Catesby

Figure 4.12: From Cleveland P. Hickman, Jr., et al., Integrated Principles of Zoology, 10th edition. Copyright © 1996 McGraw-Hill Companies, Inc., Dubuque, Iowa. All Rights Reserved. Reprinted by permission.

Figure 5.3: After W. Hamilton, U.S. Geological Survey as appeared in Carla W. Montgomery and David Dathe, Earth Then and Now, 3rd edition. Copyright © 1997 McGraw-Hill Company, Inc., Dubuque, Iowa. All Rights Reserved. Reprinted by permission.

Figure 8.10: From D. M. Madison, "Activity Rhythm and Spacing" in Biology of New World *Microtus*, R. Tamarin (ed.), 1985, 8:373–419, 1985 American Society of Mammalogists. Reprinted by permission of the author.

Figure 10.18: After C.R. Austin and R.V. Short, eds., Reproduction in Mammals, vol. 4, Reproductive Patterns, 1972 in Cleveland P. Hickman, Jr., et al., Integrated Principles of Zoology, 10th edition. Copyright © 1996 McGraw-Hill Company, Inc., Dubuque, Iowa. All Rights Reserved. Reprinted by permission.

Figure 13.18: From J.E. Hill and J.D. Smith, Bats: A Natural History. Natural History Museum, London. Reprinted by permission.

Figure 14.24: From Cleveland P. Hickman, Jr., et al., Integrated Principles of Zoology, 10th edition. Copyright © 1996 McGraw-Hill Companies, Inc., Dubuque, Iowa. All Rights Reserved. Reprinted by permission.

Figure 19.8: From H.F. Osborn, Proboscidea Volume II, American Museum Press. Reprinted by permission of American Museum of Natural History.

Figure 20.4: From Storer and Usinger, Elements of Zoology, 2nd edition. Copyright © 1961 California Academy of Sciences. Reprinted by permission.

Figure 22.1: From D. Muller-Schwarze, "Pheromones in Black-tailed Deer" in Animal Behavior, 19:141–152, 1971. Academic Press, London. Reprinted by permission.

Indexes

Subject Index

f indicates a figure (an image, its caption, or a table)

Aardvark, 311–312, 311f
  and disease, 504
  myrmecophagy of, 121
  taxonomy of, 243, 244
  teeth of, 63
Aardwolf, 320
  insectivory of, 121
  metabolic rate of, 30
Abert's squirrel, 24f
Abiotic processes
  in biogeography, 72–74
  in habitat selection, 452
Ablation, 151
Abomasum, 120f, 128f, 129
  and disease, 507
Abrocomidae, 365
Acanthocephalans, 508, 508f
Accidental hosts, 501
Acetabulum, 107
Achilles' tendon, 107, 108f
Acid rain, 533
Acouchi, 363
Acrobatidae, 237–238, 238f
ACTH. See Adrenocorticotrophic hormone (ACTH)
Active dispersal, 75
Activity level
  and cold adaptations, 170
Adaptations. See Environmental adaptations
Adaptive communication, 407
Adaptive hypothermia, 171
Adaptive radiations, 75, 283, 388f
Addax, 401f
Adenosine triphosphate (ATP), 175
Adipose tissue, 97, 98f
  brown, 174, 175, 175f
  white, 163
Adirondack Forest Preserve, 530
Adrenocortical hormones, 207, 417
Adrenocorticotrophic hormone (ACTH), 471
Aerial carnivores, 124–126
Aerial insectivores, 119–121
Aerodynamics, 259–260, 260f, 261f
African antelope
  sperm competition in, 424
African ass, 388
  domestication of, 523
African blue monkeys
  alarm calls of, 413
African buffalo
  and plant-herbivore interactions, 489, 489f
  as prey, 437
African elephant, 24f, 371, 371f, 373f
  as exploited captive, 525
  gestation period of, 204
  and infrasound, 410
  as keystone species, 487, 488f
  as seed disperser, 136
  thermal window in, 183
African grass rat, 356
African green monkey
  dominance in, 417
African grivet
  and disease, 514
African hedgehog, 246f
African honey badger
  and mutualism, 488–489
African mole-rat, 360
  digging by, 113
  and social behavior, 444
  taxonomy of, 357
African palm civet, 314, 322
African sleeping sickness, 504
African spiny mouse, 356
African wild cat
  introgression in, 538
African wild dog, 161
  dominance in, 417

hunting by, 315, 413–414, 414f
  natal dispersal of, 451
  polyandry in, 431
  social behavior of, 437, 438
Afrosoricida, 215, 242, 243–245
Afrotheria, 244, 311, 370
Age estimation
  from horn, 104, 463
  from teeth, 63, 64
Age structures, 465, 465f
Agenesis, 65
Aggression, 414–415
  and hormones, 149
  and social status, 413
Aging. See Senescence
Agonistic behavior, 414–415
  in groups, 438
Agouti, 363
  Coiban, 363
  red-rumped, 363
  Ruatan Island, 363
Agriculture
  and conservation, 284, 531, 532f, 539, 539f
AIDS-HIV, 19
Ailuridae, 324–325
Alarm calls, 413, 435, 437, 440, 440f, 441f
Alaskan red-backed vole
  nonshivering thermogenesis in, 176
Albinism, 99
Alexander, Annie M., 15
Alexander the Great, 15, 523, 525
Alignment of sequences, 42
Alimentary tracts. See Digestive systems
Allee effect, 437
Allelopathic agents, 452
Allen's rule, 87–88, 164–165
Allen's swamp monkey, 294
Alliances, 417
Allografts, 200
Allometry, 160–161
Allopatric species, 483
Alpaca, 395
  domestication of, 517f, 523–524
Alpine marmot
  sex ratio of, 422
Altricial young, 192, 207
  in lagomorphs, 367
  in marsupials, 225, 225f
Altruistic behavior, 435. See also Social behavior
  and group selection, 438–439
  reciprocal, 442–443, 443f
Alveolus, 63
  remnant, 380
Amami rabbit, 368
Amazonian manatee, 381
  conservation of, 380
  digits of, 370
Ambergris, 127, 345, 536
Ambulatory locomotion, 109
American Beaver and His Works, The [Morgan], 16
American bison
  and agriculture, 533
  and hybridization, 538
American Museum of Natural History, 14
American pika, 367, 367f
American Society of Mammalogists, 4, 17, 23
American trypanosomiasis, 504
Amniota, 48, 49f
  evolution of, 59–61
Amphibious species, 114–115
Amphilestids, 55–56
Ampullary gland, 194, 194f
Anal sacs, 317, 320, 321, 322
Analytical biogeography, 77–79
Anapsida, 48, 49f
Andean caenolestid, 229
Andean hairy armadillo, 307f
Anderson's gerbil
  in ecological communities, 484–485
Andrew's beaked whale, 346f
Androgens, 195

Anestrus species, 197
Angle of attack, 260, 260f
Angora hair, 98
Angwantibo, 286
Anisogamy, 420–421
Ankle, 106f, 107, 385
Annual molting, 100
Anomaluridae, 355
Anomodontia, 50f, 51
Anosomism, 457
Antagonistic actions
  in the endocrine system, 149
Antarctic minke whale, 342
Anteater, 306–307
  banded, 534
  dentition of, 63, 119f
  gestation period of, 204
  giant, 119f, 121, 307, 307f
  myrmecophagy of, 121
  scaly, 63, 136, 309–311, 310f
  silky, 121, 303, 308
  taxonomy of, 303
  tongue of, 121f
Antechinus
  marsupium of, 225f
Antelope. See also Pronghorn antelope
  African, 424
  evolution of social behavior in, 445
  four-horned, 400
  hunting of, 392–393
  locomotion of, 110f
  royal, 400
  sable, 536
  thermal window in, 183
  water conservation by, 179, 188–189, 189f
Antelope ground squirrel
  comparative phylogeography of, 84, 84f
  and heat avoidance, 186, 187f
Anterior pituitary glands, 208
Anthropoid primates, 281
Antigens, 150
Antilocapridae, 400
Antitragus, 259, 259f
Antlers, 103–104, 390f, 398–399, 399f
  and runaway selection, 423, 423f, 424
Aotidae, 292–293
Apes, 297–299
Aplodontidae, 352
Apocrine sweat glands, 101, 180
Apophyses, 106
Appendages
  and Allen's rule, 87–88
  as cold adaptations, 163–165
  as heat adaptations, 183–184
Appendicular muscles, 108, 108f
Appendicular skeleton, 104, 106f, 107
Aquatic carnivores, 126–127
Aquatic mole
  kidney of, 178f
Arabian oryx
  conservation of, 541–542, 542f
  trapping, 23
  water conservation by, 188
Arboreal locomotion, 111
Arboreal theory, 283
Arctic fox
  appendages of, 165f
  early descriptions of, 14
  fur of, 161, 166
  as introduced species, 488
  pelage dimorphism in, 165
  as predator, 475
  seasonal molting of, 100
Arctic ground squirrel
  early descriptions of, 14
  hibernation by, 171, 173–174, 174f
  territory of, 416
Arctic hare, 367
  fur of, 165, 166f
  teeth of, 366f
Area cladograms, 78–79, 78f

Scientific Names Index

f indicates a figure (an image, its caption, or a table)

About the Authors

George A. Feldhamer is Professor of Zoology as well as Coordinator of the Environmental Studies Program at Southern Illinois University at Carbondale. His research has focused exclusively on mammalian populations, ecology, and management; biology of introduced deer; and threatened and endangered species of rodents and bats. He is a former associate editor for Forest Biology and Ecology for the *Journal of Forest Research* and a former associate editor of the *Wildlife Society Bulletin*. He is the senior editor of *Wild Mammals of North America: Biology, Management, and Conservation* (2003), coauthor of *Mammals of the National Parks* (2005), both published by Johns Hopkins University Press, and coeditor of *Ecology, Behavior, and Conservation of the Golden Mouse: A Model Species for Research* (2007). He is curator of the mammal collection at SIUC and has 30 years of experience teaching an upper-division mammalogy course. In 2000, he was named Outstanding Teacher in the College of Science at SIUC.

Lee C. Drickamer was Professor of Biology and Chair of the Department of Biological Sciences at Northern Arizona University, Professor of Biology at Williams College for 15 years, Professor of Zoology at Southern Illinois University at Carbondale for 11 years, and is currently Regents' Professor of Biological Science and Vice President for Research at Northern Arizona University. He is a past president of the Animal Behavior Society, past secretary-general of the International Council of Ethnologists, past chair of the Division of Animal Behavior of what is now the Society for Integrative and Comparative Biology, and former editor of *Animal Behaviour*. His research emphases have included social factors affecting development and reproduction in house mice and swine, behavioral ecology of house mice and deer mice, social biology of primates, intrauterine position effects on behavior and reproduction of mice and swine, the consequences of mate selection for offspring viability in house mice, and prairie dog ecology and population biology.

Stephen H. Vessey is Professor Emeritus of Biological Sciences at Bowling Green State University. His research interests include the behavioral ecology of mammals, especially primates and rodents. He has been studying a population of white-footed mice in northwestern Ohio for more than 30 years. He is a former associate editor of the *Journal of Mammalogy* and is a Fellow of the Animal Behavior Society. He taught mammalogy and animal behavior at Bowling Green for 30 years and coauthored a textbook in animal behavior with Lee Drickamer. After retiring from Bowling Green in 2000, he served as Program Director and Deputy Division Director, Division of Integrative Biology and Neurosciences, National Science Foundation.

Joseph F. Merritt is senior mammalogist with the Illinois Natural History Survey. He is the former Director of Powdermill Biological Station of the Carnegie Museum of Natural History and served as Distinguished Visiting Professor at the US Air Force Academy in Colorado Springs, Colorado, during academic year 2004 to 2005. Dr. Merritt is a physiological ecologist and functional morphologist specializing in adaptations of mammals to cold. He is the author of *Guide to the Mammals of Pennsylvania*, published by the University of Pittsburgh Press. Dr. Merritt is also editor of several technical monographs on specific taxa of mammals and the upcoming book *Biology of Small Mammals*, to be published by the Johns Hopkins University Press in 2008. He has served on the Publications Committee of the American Society of Mammalogists since 1990 and is currently the Editor for Special Publications and Assistant to the Journal Editor for the ASM. He is Editor for the Western Hemisphere for the journal *Acta Theriologica*, published by the Polish Academy of Sciences. Dr. Merritt instructs mammalogy at the University of Colorado Mountain Research Station and courses in mammalian ecology and winter ecology at Antioch New England Graduate School, Indiana State University, and at the Adirondack Ecological Center, SUNY College of Environmental Science and Forestry.

Carey Krajewski is Professor of Zoology and Director of Graduate Studies in Zoology at Southern Illinois University at Carbondale. His research in mammalogy involves molecular studies of marsupial phylogeny, particularly the systematics of dasyuromorphians, but he also maintains an active research program on the evolution of gruoid birds. He is an associate editor for the *Journal of Mammalogy* and *Journal of Mammalian Evolution*, and a member of the editorial board of Molecular Phylogenetics and Evolution. In 2004, he was named Outstanding Scholar in the College of Science at SIUC and, since 1993, has been an honorary research fellow in the Genetics Department at La Trobe University, Australia. During his 17 years at SIUC he has taught courses in molecular evolution, population genetics, vertebrate zoology, and vertebrate anatomy.

Equivalent Weights and Measures

Length

Unit	Abbreviation	Equivalent
meter	m	approximately 39 in
centimeter	cm	10^{-2} m
millimeter	mm	10^{-3} m
micrometer	μm	10^{-6} m
nanometer	nm	10^{-9} m

Volume

Unit	Abbreviation	Equivalent
liter	L	approximately 1.06 qt
milliliter	mL	10^{-3} L (1 mL = 1 cm^3 = 1 cc)
microliter	μL	10^{-6} L

Mass

Unit	Abbreviation	Equivalent
kilogram	kg	10^3 g (approximately 2.2 lb)
gram	g	approximately 0.035 oz
milligram	mg	10^{-3} g
microgram	μg	10^{-6} g
nanogram	ng	10^{-9} g
picogram	pg	10^{-12} g

Common Prefixes

Prefix	Equivalent	Examples
kilo	1000	a kilogram is 1000 grams
centi	0.01	a centimeter is 0.01 meter
milli	0.001	a milliliter is 0.001 liter
micro (μ)	one-millionth	a micrometer is 0.000001 (one-millionth) of a meter
nano (n)	one-billionth	a nanogram is 10^{-9} (one-billionth) of a gram
pico (p)	one-trillionth	a picogram is 10^{-12} (one-trillionth) of a gram

Conversion Chart

Length Conversions

1 in = 2.5 cm	1 mm = 0.039 in
1 ft = 30 cm	1 cm = 0.39 in
1 yd = 0.9 m	1 m = 39 in
1 mi = 1.6 km	1 m = 1.094 yd
1 km = 0.6 mi	

To convert	*Multiply by*	*To obtain*
inches	2.54	centimeters
feet	30	centimeters
centimeters	0.39	inches
millimeters	0.039	inches

Volume Conversions

1 tsp = 5 mL	1 mL = 0.03 fl oz
1 tbsp = 15 mL	1 L = 2.1 pt
1 fl oz = 30 mL	1 L = 1.06 qt
1 cup = 0.24 L	1 L = 0.26 gal
1 pt = 0.47 L	
1 qt = 0.95 L	
1 gal = 3.79 L	

To convert	*Multiply by*	*To obtain*
fluid ounces	30	milliliters
quart	0.95	liters
milliliters	0.03	fluid ounces
liters	1.06	quarts

Mass Conversions

1 oz = 28.3 g	1 g = 0.035 oz
1 lb = 453.6 g	1 kg = 2.2 lb
1 lb = 0.45 kg	

To convert	*Multiply by*	*To obtain*
ounces	28.3	grams
pounds	453.6	grams
pounds	0.45	kilograms
grams	0.035	ounces
kilograms	2.2	grams

Energy Conversions

calorie (cal) = energy required to raise the temperature of 1 g of water (at 16°C) by 1°C
1 calorie = 4.184 joules
1 kilocalorie (kcal) = 1000 cal
1 joule = 0.24 cal
1 kilocalorie = 4.184 kJ

Temperature Conversions

Celsius (Centigrade) $°C = \dfrac{(°F-32) \times 5}{9}$

Fahrenheit $°F = \dfrac{°C \times 9}{5} + 32$

Kelvin $K = °C + 273$